SECOND EDITION
HANDBOOK OF HYDRAULIC FLUID TECHNOLOGY

SECOND EDITION

HANDBOOK OF HYDRAULIC FLUID TECHNOLOGY

Edited by

George E. Totten
Victor J. De Negri

CRC Press
Taylor & Francis Group
Boca Raton   London   New York

CRC Press is an imprint of the
Taylor & Francis Group, an **informa** business

CRC Press
Taylor & Francis Group
6000 Broken Sound Parkway NW, Suite 300
Boca Raton, FL 33487-2742

First issued in paperback 2017

© 2012 by Taylor & Francis Group, LLC
CRC Press is an imprint of Taylor & Francis Group, an Informa business

No claim to original U.S. Government works
Version Date: 20110715

ISBN 13: 978-1-138-07734-8 (pbk)
ISBN 13: 978-1-4200-8526-6 (hbk)

Library of Congress Cataloging-in-Publication Data

Handbook of hydraulic fluid technology, second edition / editors, George E. Totten, Victor J. De Negri. -- 2nd ed.
    p. cm.
  Includes bibliographical references and index.
  ISBN 978-1-4200-8526-6 (hardback)
  1. Fluid power technology--Handbooks, manuals, etc. 2. Hydraulic fluids--Handbooks, manuals, etc. I. Totten, George E. II. Negri, Victor J. de. III. Title.

TJ843.H36 2011
621.2'0424--dc23
                    2011019738

Visit the Taylor & Francis Web site at
http://www.taylorandfrancis.com

and the CRC Press Web site at
http://www.crcpress.com

This book is dedicated to our families, without whose continued support the completion of this work would not have been possible:

For my wife Alice.

G.E.T.

For my wife Rosely and my daughter Fernanda.

V.J.D.N.

Contents

Preface to the Second Edition

This book is a significant revision of the first edition of the *Handbook of Hydraulic Fluid Technology*, which was edited by Dr. George E. Totten and published 10 years ago. Since the original publication of this text, no other similar book has been published that treats hydraulic fluids as a component of a hydraulic system and addresses all the major aspects of hydraulic fluid technology. In view of the unique position of the *Handbook of Hydraulic Fluid Technology*, a decision was made to significantly update this invaluable text.

The *Handbook of Hydraulic Fluid Technology—Second Edition* contains 21 chapters. Chapter 1: Fundamentals of Hydraulic Systems and Components, Chapter 5: Control and Management of Particle Contamination in Hydraulic Fluids, Chapter 11: Noise and Vibration of Fluid Power Systems, and Chapter 18: Biobased and Biodegradable Hydraulic Oils have been completely rewritten to more effectively address and expand coverage of critical new technology developments. Chapter 21: Food-Grade Hydraulic Fluids, is a newly added chapter to the book. The remaining chapters of the book have been revised and updated, and in many cases substantially. The updated and expanded coverage necessitated the elimination of three chapters from the first edition: Lubricant Additives for Mineral Oil–Based Hydraulic Fluids, Bearing Selection, and Lubrication and Electro-Rheological Fluids. With the exception of the chapter on electro-rheological fluids, the necessary content has been integrated into the remaining chapters of the book as appropriate. In general, the *Handbook of Hydraulic Fluid Technology—Second Edition* is a substantially new text on this very important critical hydraulics technology.

The editors of the *Handbook of Hydraulic Fluid Technology—Second Edition* are George E. Totten, PhD and Victor De Negri, D.Eng. Both editors are deeply indebted to the contributing authors for their vital assistance in completing this project. The editors also express appreciation to the staff of CRC Press for the opportunity to undertake this task and for their ongoing encouragement and vital support during all aspects of the book, from concept to production. Most importantly, the encouragement of our families is particularly appreciated.

George E. Totten
Texas A&M University
College Station, TX, USA

Victor J. De Negri
Federal University of Santa Catarina
Florianópolis, SC, Brazil

Preface to the First Edition

One of the most frustrating practices of my career has been the search for information on hydraulic fluids, which includes information on fluid chemistry; physical properties; maintenance practices; and fluid, system, and component design. Although some information on petroleum oil hydraulic fluids can be found, there is much less information on fire resistant, biodegradable, and other types of fluids. Unfortunately, with few exceptions, fluid coverage in hydraulic texts is typically limited to a single-chapter overview intended to cover all fluids. Therefore, it is often necessary to perform a literature search or a time-consuming manual search of my files. Some time ago, it occurred to me that others must be encountering the same problem. There seemed to be a vital need for an extensive reference text on hydraulic fluids that would provide information in sufficient depth and breadth to be of use to the fluid formulator, hydraulic system designer, plant maintenance engineer, and others who serve the industry.

Currently, there are no books dedicated to hydraulic fluid chemistry. Most hydraulic fluid treatment is found in handbooks, which primarily focus on hydraulic system hardware, installation, and troubleshooting. Most of these books fit into one of two categories. One type of book deals with hydraulic equipment, with a single, simplified overview chapter covering all hydraulic fluids, but with a focus on petroleum-derived fluids. The second type of book provides fluid coverage with minimal, if any, discussion of engineering properties of importance in a hydraulic system.

The purpose of the *Handbook of Hydraulic Fluid Technology* is to provide a comprehensive and rigorous overview of hydraulic fluid technology. The objective is not only to discuss fluid chemistry and physical properties in detail, but also to integrate both classic and current fundamental lubrication concepts with respect to various classes of hydraulic fluids. A further objective is to integrate fluid dynamics with respect to their operation in a hydraulic system in order to enable the reader to obtain a broader understanding of the total system. Hydraulic fluids are an important and vital component of the hydraulic system.

The 21 chapters of this book are grouped into three main parts: hardware, fluid properties and testing, and fluids.

HARDWARE

Chapter 1 provides the reader with an overview of basic hydraulic concepts, a description of the components, and an introduction to hydraulic system operation. In Chapter 2, the rolling element bearings and their lubrication are discussed. An extremely important facet of any well-designed hydraulic system is fluid filtration. Chapter 3 not only provides a detailed discussion of fluid filtration and particle contamination and quantification, but also discusses fluid filterability.

An understanding of the physical properties of a fluid is necessary to understand the performance of a hydraulic fluid as a fluid power medium. Chapter 4 features a thorough overview of the physical properties, and their evaluation and impact on hydraulic system operation, which includes: viscosity, viscosity-temperature and viscosity-pressure behavior, gas solubility, foaming, air entrainment, air release, and fluid compressibility and modulus.

FLUID PROPERTIES AND TESTING

Viscosity is the most important physical property exhibited by a hydraulic fluid. Chapter 5 presents an in-depth discussion of hydraulic fluid viscosity and classification. The hydraulic fluid must not only perform as a power transmission medium, but also lubricate the system. Chapter 6 provides a thorough review of the fundamental concepts involved in lubricating a hydraulic system. In many applications, fluid fire resistance is one of the primary selection criteria. An overview of historically important fire-resistance testing procedures is provided in Chapter 7, with a discussion of currently changing testing protocol required for industry, national, and insurance company approvals. Ecological compatibility properties exhibited by a hydraulic fluid is currently one of the most intensive research areas of hydraulic fluid technology. An overview of the current testing requirements and strategies is given in Chapter 8.

One of the most inexpensive but least understood components of the hydraulic system is hydraulic seals. Chapter 9 provides a review of mechanical and elastomeric seal technology and seal compatibility testing. An often overlooked but vitally important area is adequate testing and evaluation of hydraulic fluid performance in a hydraulic system. Currently, there is no consensus on the best tests to perform and what they reveal. Chapter 10 reviews the state-of-the-art of bench and pump testing of hydraulic fluids. Vibrational analysis is not only an important plant maintenance tool, but it is also one of the most important diagnostic techniques for evaluating and troubleshooting the operational characteristics of a hydraulic system. An introductory overview of the use of vibrational analysis in fluid maintenance is given in Chapter 11. No hydraulic system operates trouble-free forever. When problems occur, it is important to be able to identify both the problem and its cause. Chapter 12 provides a thorough discussion of hydraulic system failure analysis strategies.

FLUIDS

Although water hydraulics do not constitute a major fluid power application, they are coming under increasing scrutiny as ecocompatible alternatives to conventional hydraulic fluids. Chapter 13 offers an overview of this increasingly important technology.

The largest volume fluid power medium is petroleum oil. In Chapter 14, the reader is provided with a thorough overview of oil chemistry, properties, fluid maintenance, and change-out procedures. Chapter 15 reviews additive technology for petroleum oil hydraulic fluids. There are various types of synthetic hydraulic fluids. A description of the more important synthetic fluids, with a focus on aerospace applications, is given in Chapter 16.

Chapters 17 to 20 describe fire-resistant hydraulic fluids. Emulsions, water glycols, polyol esters, and phosphate esters are discussed individually and in depth in Chapters 17, 18, 19, and 20, respectively. This discussion includes fluid chemistry, physical properties, additive technology, maintenance, and hydraulic system conversion.

Vegetable oils are well-known lubricants that have been examined repeatedly over the years. Currently, there is an intensive effort to increase the utilization of various types of vegetable oils as an ecologically sound alternative to mineral oil hydraulic fluids. Chapter 21 provides a review of vegetable oil chemistry, recovery, and properties. The applicability of these fluids as hydraulic fluid basestocks is examined in detail.

Chapter 22 discusses electrorheological fluids, which are becoming increasingly interesting for use in specialized hydraulic applications. In Chapter 23, various standardized fluid maintenance procedures are discussed and a summary of equivalent international testing standards is provided.

The preparation of a text of this scope was a tremendous task. I am deeply indebted to many colleagues for their assistance, without whom this text would not have been possible. Special thanks go to Dr. Stephen Lainer (University of Aachen), Professor Atsushi Yamaguchi (Yokohama National University), Professor Toshi Kazama (Muroran Institute of Technology), K. Mizuno (Kayaba Industrial Ltd.), and Jürgen Reichel (formerly with DMT, Essen, Germany).

Special thanks also goes to my wife, Alice, for her unending patience, and to Susan Meeker, who assisted in organizing and editing much of this material; to Glenn Webster, Roland J. Bishop, Jr., and Yinghua Sun, without whose help this text would never have been completed; and to Union Carbide Corporation for its support.

George E. Totten

Editors

George E. Totten received his BS and MS degrees from Fairleigh Dickinson University in New Jersey and his PhD from New York University. Dr. Totten is past president of the International Federation for Heat Treating and Surface Engineering (IFHTSE) and a fellow of ASM International, SAE International, IFHTSE and ASTM International. Dr. Totten is an adjunct professor at Texas A&M University in College Station, TX and he is also president of G.E. Totten & Associates LLC, a research and consulting firm specializing in thermal processing and industrial lubrication problems.

Dr. Totten is the author or coauthor (editor) of over 500 publications including patents, technical papers, book chapters, and books, which include *Handbook of Hydraulic Fluid Technology; Handbook of Aluminum* Vol. 1 and Vol. 2; *Handbook of Lubrication and Tribology – Volume 1: Application and Maintenance; Handbook of Quenchants and Quenching Technology, Quenching Theory and Technology,* 2nd edition; *Steel Heat Treatment Handbook; Handbook of Residual Stress and Deformation of Steel; Handbook of Metallurgical Process Design;* and the *ASTM Fuels and Lubricants Handbook: Technology, Properties, Performance, and Testing (MNL 37).*

Victor Juliano De Negri, D.Eng. received his mechanical engineering degree in 1983, from UNISINOS, Brazil, a M.Eng. degree in 1987 and a D.Eng. degree in 1996, both from UFSC, Brazil. Since 1995, he has been associate professor in the mechanical engineering department at the Federal University of Santa Catarina (UFSC). He is currently the head of the Laboratory of Hydraulic and Pneumatic Systems (LASHIP). He is a member of the Brazilian Society of Mechanical Sciences and Engineering (ABCM) and LASHIP official company representative to the National Fluid Power Association (NFPA). His research areas include analysis and design of hydraulic and pneumatic systems and components and design methodologies for automation and control of equipment and processes. He has coordinated several research projects with industry and governmental agencies in the areas of hydraulic components, power-generating plants, mobile hydraulics, pneumatic systems, and positioning systems. He supervised 40 academic works including master's and doctorate theses and final term projects. He has 2 patents and written more than 90 journal and technical papers, conference papers, and magazine articles.

Contributors

Christa M.A. Chilson
PolyMod® Technologies Inc.
Fort Wayne, Indiana

Donald L. Clason
The Lubrizol Corporation
Mentor, Ohio

Victor J. De Negri
Department of Mechanical Engineering
Federal University of Santa Catarina
Florianópolis, Brazil

Acires Dias
Department of Mechanical Engineering
Federal University of Santa Catarina
Florianópolis, Brazil

Brian B. Filippini
The Lubrizol Corporation
Wickliffe, Ohio

Jim C. Fitch
Noria Corporation
Tulsa, Oklahoma

Samir N.Y. Gerges
Department of Mechanical Engineering
Federal University of Santa Catarina
Florianópolis, Brazil

Lois J. Gschwender
Air Force Research Laboratory
Wright-Patterson Air Force Base, Ohio

Sibtain Hamid
Lubriplate Lubricants Co.
Toledo, Ohio

Lou A.T. Honary
Department of Marketing
University of Northern Iowa
Cedar Falls, Iowa

D. Nigel Johnston
Department of Mechanical Engineering
University of Bath
Bath, U.K.

Toshi Kazama
Department of Mechanical Systems
 Engineering
Muroran Institute of Technology
Muroran, Japan

Bernard G. Kinker
Consultant to Evonik RohMax USA, Inc.
Kintnersville, Pennsylvania

Kari T. Koskinen
Department of Intelligent Hydraulics and
 Automation
Tampere University of Technology
Tampere, Finland

Irlan von Linsingen
Mechanical Engineering Department
Federal University of Santa Catarina
Florianópolis, Brazil

Kenneth C. Ludema
Department of Mechanical Engineering
University of Michigan
Ann Arbor, Michigan

Paul W. Michael
Fluid Power Institute
Milwaukee School of Engineering
Milwaukee, Wisconsin

John J. Mullay
The Lubrizol Corporation
Wickliffe, Ohio

W.D. Phillips
W David Phillips & Associates
Stockport, U.K.

In-Sik Rhee
Development and Engineering Center
U.S. Army TARDEC
Warren, Michigan

Leonardo Zanetti Rocha
Department of Mechanical Engineering
Federal University of Santa Catarina
Florianópolis, Brazil

John R. Sander
Lubrication Engineers, Inc.
Wichita, Kansas

John V. Sherman
BASF Corporation
Wyandotte, Michigan

Ronald L. Shubkin
Albemarle Corporation
Baton Rouge, Louisiana

D. Smrdel
The Lubrizol Corporation
Wickliffe, Ohio

Carl E. Snyder, Jr.
University of Dayton Research Institute
Dayton, Ohio

Yinghua Sun
Union Carbide Corporation
Tarrytown, New York

George E. Totten
Department of Mechanical Engineering
Texas A&M University
College Station, TX

Paula R. Vettel
Primagy
Downers Grove, Illinois

Matti J. Vilenius
Department of Intelligent Hydraulics and
    Automation
Tampere University of Technology
Tampere, Finland

Thomas S. Wanke
Milwaukee School of Engineering
Fluid Power Institute
Milwaukee, Wisconsin

Lavern D. Wedeven
Wedeven Associates, Inc.
Edgmont, Pennsylvania

Ronald E. Zielinski
PolyMod® Technologies Inc.
Fort Wayne, Indiana

1 Fundamentals of Hydraulic Systems and Components

Irlan von Linsingen and Victor J. De Negri[*]

CONTENTS

---

[*] Some parts of this chapter are based on the chapter titled "Basic Hydraulic Pump and Circuit Design" by Richard K. Tessmann, Hans M. Melief, and Roland J. Bishop, Jr. from the *Handbook of Hydraulic Fluid Technology*, 1st Edition of this book.

1.1   INTRODUCTION

A hydraulic system, from a general perspective, is an arrangement of interconnected components that uses a liquid under pressure to provide energy transmission and control. It has an extremely broad range of applications covering basically all fields of production, manufacturing and service. Consequently, the energy transmission and control requirements are very diverse and thus the structure of each hydraulic system has its specificities.

However, on analyzing the current hydraulic systems, one can identify four main functions [1], as presented in Figure 1.1, which are: primary energy conversion, energy limitation and control, secondary energy conversion, and fluid storage and conditioning.

Furthermore, this figure shows the main resources that flow through a hydraulic system and which can be grouped into the classes: information, material, and energy [2].

The input of mechanical energy (M), which is a result of the external conversion of primary electrical or chemical (combustion) energy, is converted into hydraulic energy (H). Using signals or data (S, D) from an operator or from other equipment, the hydraulic energy (H) is limited and controlled such that it becomes appropriate for conversion into mechanical energy (M). This mechanical energy is the desired output of the hydraulic system and will be used to drive or move external devices.

The hydraulic energy is carried by the hydraulic fluid (F) and thus its storage and conditioning, including contamination and temperature control, are also essential functions.

As a consequence of the physical phenomena, construction characteristics, and circuit arrangement, part of the useful energy is dissipated in a hydraulic system. Therefore, all functions transfer thermal energy (T) to the fluid and to the environment.

Since this *Handbook* is concerned with fluid technology, the objective of this chapter is to characterize hydraulic systems, that is, applications in which hydraulic fluids are used.

The construction characteristics and the functioning principles of the main hydraulic components are presented, with the aim of providing an overview of the interaction between the fluid and the mechanical parts.

Moreover, the main equations that govern the component and circuit behavior are presented, where one can identify the influence of the fluid parameters, which, in turn, are a consequence of the physical-chemistry proprieties.

An important aspect of this chapter is the symbol notation that is used in the diagrams and equations. Both the hydraulic circuit diagrams and the component identification codes are in accordance

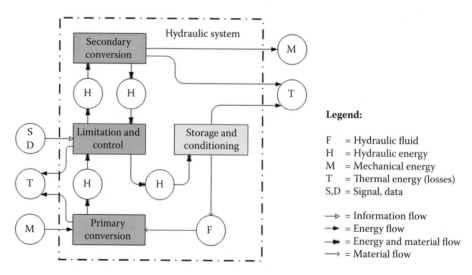

Legend:

F   = Hydraulic fluid
H   = Hydraulic energy
M   = Mechanical energy
T   = Thermal energy (losses)
S,D = Signal, data

⇢   = Information flow
⟶ = Energy flow
⟶ = Energy and material flow
⟶   = Material flow

FIGURE 1.1   Generic hydraulic system: Functions and resource flows.

with ISO 1219-1 [3] and ISO 1219-2 [4]. The quantities (variables and parameters) used in the circuit diagrams, component illustrations, and equations are represented by letter symbols, including subscripts and superscripts, in compliance with ISO 4391 [5], IEC 27-1 [6], and ISO 1219-2 [4] standards.

1.2　HYDROMECHANICAL PRINCIPLES

Essentially, a hydraulic system consists of mechanical parts operating together with a hydraulic fluid. Hence, its behavior is described by the classic laws of both mechanics and fluid mechanics. Although it is not the focus of this text, it is important to remember that several hydraulic components comprise electromechanical converters, such as solenoids, linear motors and torque motors and/or electro-electronic systems like sensors, power amplifiers and controllers. Therefore, the principles of electricity, electronics and electro-magnetism are also required for their modeling.

1.2.1　Hydrostatics: Pascal's Principle

Fluids (gases or liquids) are compressible, which means that their mass density varies with the pressure to which they are submitted. Consequently, an abrupt local pressure variation will be propagated through the fluid with a velocity equal to the fluid sound velocity until the equilibrium has been re-established. This means that the fluid will have a dynamic behavior alternating between the two equilibrium states.

When a fluid is treated as incompressible it is assumed that a local pressure perturbation is instantaneously transmitted throughout the fluid. This means that considering a fluid as being compressible or incompressible is dependent on the observer's viewpoint and its validation depends on the use of the system and the particular design or analysis that is being carried out.

Pascal's principle states that "a change in the pressure of an enclosed incompressible fluid is conveyed undiminished to every part of the fluid and to the surfaces of its container" [1,7]. Hence, when a fluid is in a state of equilibrium, that is, in a steady state, the whole system is under the same internal pressure.

The practical use of Pascal's principle can be exemplified by the hydrostatic press principle whose objective is to amplify the force. As shown in Figure 1.2a [1], it consists of two cylinders (actuators) (A1 and A2) that are connected by a pipe.

In this press, the resistive force (F^{A2}* [N]) offered by the material to be pressed must be compensated by the input force (F^{A1} [N]) such that the equilibrium occurs. Since in a steady state the pressure (p [N/m^2] or [Pa]) is equal throughout the volume, one has

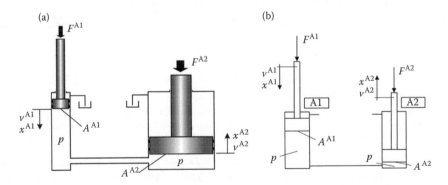

FIGURE 1.2　Hydrostatic press principle: (a) Illustration of the hydraulic circuit; (b) Hydraulic circuit diagram.

---

* The kernel (central part of the letter symbol) represents the generic quantity. The subscript indicates the quantity application and the superscript is used to indicate to which component or system the quantity is associated (ISO 4391, *ISO 1219-2 - Fluid Power Systems and Components – Graphic symbols and circuit diagrams – Part 2: Circuit diagrams,* Switzerland, 1991.

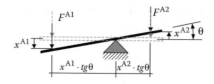

FIGURE 1.3   Mechanical system of force amplification.

$$p = \frac{F^{A1}}{A^{A1}} = \frac{F^{A2}}{A^{A2}} \Rightarrow \frac{F^{A2}}{F^{A1}} = \frac{A^{A2}}{A^{A1}} \text{ or } F^{A2} = \left(\frac{A^{A2}}{A^{A1}}\right) \cdot F^{A1}, \tag{1.1}$$

where A^{A1} [m²] and A^{A2} [m²] are the piston areas.

Equation 1.1 shows that, for $A^{A2}/A^{A1} \gg 1$, a low force F^{A1} is sufficient to overcome a higher force like F^{A2}, which is the objective of most hydraulic systems.

Moreover, considering the incompressible fluid, the volume variations in the two cylinders (ΔV^{A1} [m³] and ΔV^{A2} [m³]) are equal. According to Equation 1.2, in this case the displacements x^{A1} [m] and x^{A2} [m] are different, their relationship being determined by the area ratio:

$$\Delta V^{A1} = \Delta V^{A2} \Rightarrow x^{A1} \cdot A^{A1} = x^{A2} \cdot A^{A2} \text{ or } x^{A2} = \left(\frac{A^{A1}}{A^{A2}}\right) \cdot x^{A1}. \tag{1.2}$$

Considering an efficiency of 100%, the work required of cylinder A1, determined by the product of the force and displacement, is equal to the work applied to cylinder A2. Hence, according to Equation 1.3, the correlation between F^{A1} and F^{A2} is given by the displacement ratio:

$$W = F^{A1} \cdot x^{A1} = F^{A2} \cdot x^{A2} \Rightarrow \frac{F^{A2}}{F^{A1}} = \frac{x^{A1}}{x^{A2}} \text{ or } F^{A2} = \left(\frac{x^{A1}}{x^{A2}}\right) \cdot F^{A1}. \tag{1.3}$$

Equation 1.3 is designed as the hydraulic lever equation [1], since the same force amplification could be obtained through a mechanical system—such as that shown in Figure 1.3.

These hydrostatic relationships allow the static behavior of a system to be determined—that is, the relationships between the forces and displacements in the equilibrium condition. The behavioral description with temporal variation is carried out using the laws of hydrodynamics [1].

1.2.2 HYDRODYNAMICS: CONSERVATION OF MASS

The steady-state and transient behavior of hydraulic components and systems is described by the basic principles of hydrodynamics and thermodynamics [1,8]. In this chapter, two of these principles are studied; namely, the conservation of mass (continuity equation) and the conservation of energy (Bernoulli's equation), which are essential to the comprehension of the hydraulic component behavior.

From the conservation of mass principle, an important expression is obtained which describes the behavior of pressure in volumes. Consider the hydraulic device shown in Figure 1.4 [1], which has an inlet port (1), an outlet port (2) and a movable piston.

The mass density (ρ [kg/m³]), the pressure (p [Pa]), and the temperature (T [K] or [°C]) of the fluid are considered constant in the space defined by the chamber, but they vary over time. The flow rate in the inlet port is considered positive when entering the chamber and the flow rate in the outlet port is positive when leaving the chamber. The chamber volume changes with the piston movement.

The result of the continuity equation [1,8,9] applied to this case is [5].

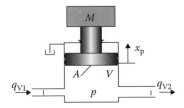

FIGURE 1.4 Chamber with variable volume.

$$q_{V1} - q_{V2} = \frac{dV}{dt} + \frac{V}{\beta} \cdot \frac{dp}{dt}, \tag{1.4}$$

where V [m³] is the chamber volume and q_{V1} [m³/s] and q_{V2} [m³/s] are the volumetric flow rates (commonly referred to as the "flow rate") at the inlet and outlet ports, respectively. β [Pa] is the bulk modulus (inverse of compressibility), which characterizes the mass density variation with the fluid pressure.

In this equation, the terms on the right are related to the mass accumulation in the volume, where dV/dt represents the variation in the chamber volume over time and $(V/\beta)(dp/dt)$ the variation in the pressure over time associated with the fluid compressibility.

Therefore, Equation 1.4 describes the dynamic behavior of the pressure in the chamber as a consequence of the change in flow rate at port 1 and/or port 2. The pressure change will take the piston out of equilibrium, causing its movement. As a consequence, the first term on the right will be different from zero, in turn changing the pressure.

It is important to note that the continuity equation, as presented in Equation 1.4, is the basic form used in the hydraulics area to model the dynamic behavior of a fluid in cylinders, accumulators, motors, pipes and so forth.

Studying again the hydrostatic press (Figure 1.2), one can observe that the volume variation in cylinders A1 and A2 is dependent on the displacement direction of the pistons, which means that volume V^{A1} will be decreasing and volume V^{A2} increasing toward the positive directions indicated in this figure, that is

$$\frac{dV^{A1}}{dt} = -A^{A1} \cdot v^{A1} \text{ and } \frac{dV^{A2}}{dt} = A^{A2} \cdot v^{A2}, \tag{1.5}$$

where v^{A1} [m] and v^{A2} [m²] are the piston velocities.

Appling Equations 1.4 and 1.5 to cylinders A1 and A2 for a constant pressure condition and taking into account that the flow rate that leaves cylinder A1 is the same as that entering cylinder A2, one can obtain

$$q_V = A^{A1} \cdot v^{A1} = A^{A2} \cdot v^{A2} \Rightarrow \frac{v^{A2}}{v^{A1}} = \frac{A^{A1}}{A^{A2}} \text{ or } v^{A2} = \left(\frac{A^{A1}}{A^{A2}}\right) \cdot v^{A1}. \tag{1.6}$$

Equation 1.6 describes the velocity relationship for the hydraulic press, completing the set of equations together with Equations 1.1 and 1.2.

1.2.3 HYDROSTATIC PRESS: LINEAR MOTION

By means of the circuit in Figure 1.2, it is possible to have an upward moving cylinder A2 when cylinder A1 is moving downward. The displacement relationship (Equation 1.2) and velocity

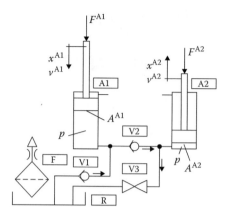

FIGURE 1.5   Hydraulic circuit diagram of a real hydrostatic press.

relationship (Equation 1.6) imply that a movement of cylinder A1 with displacement and velocity according to human capacity results in a press operation with both small displacement and velocity. Cylinder A1, having reached the required displacement, will reach its stroke end much earlier than cylinder A2.

Therefore, this basic circuit is not valuable for real uses. A typical circuit found in hydrostatic presses and hydraulic jacks is presented in Figure 1.5, where some components are added to the original circuit (Figure 1.2).

In this circuit an external reservoir (R), which compensates for the difference between the cylinder volumes, and two non-return valves (V1 and V2) are included. These valves allow fluid suction from the reservoir on the upward movement of cylinder A1 and fluid pumping to cylinder A2 on the downward movement. Valve (register) V3, when opened, allows the fluid in cylinder A2 to return to the reservoir as a consequence of the external force (F^{A2}) applied to the piston.

Correlating Figure 1.5 and 1.1, the arrangement constituted by A1, V1, and V2 performs the primary energy conversion function, V3 the energy control, and A2 the secondary energy conversion. The fluid storage and conditioning is performed by both the reservoir (R) and the air filter (F). The filter establishes the connection between the fluid and the external environment in order to keep the reservoir cleaned and at atmospheric pressure.

1.2.4 HYDROSTATIC TRANSMISSION: ROTARY MOTION

The principles presented previously for linear motion are now applied to rotary motion transmission using a pump and a motor (hydrostatic machines) as presented in Figure 1.6. According to ISO 1219-2 [4], the pump has its own symbol, P, while the hydraulic motor is an actuator and, for this reason, it is designed as A.

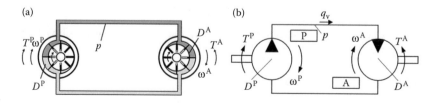

FIGURE 1.6   Hydrostatic transmission: (a) Illustration of the hydraulic circuit; (b) Hydraulic circuit diagram.

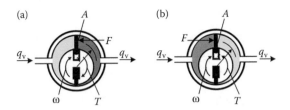

FIGURE 1.7 Principles of a hydrostatic machine: (a) Functioning as a pump; (b) Functioning as a motor.

The hydrostatic pump driven by an electric motor, for example, runs at an angular speed $(\omega^P \text{ [rad/s]})^*$ supplying a flow rate $(q_V \text{ [m}^3\text{/s]})$ to the hydraulic motor that causes an angular speed $(\omega^A \text{ [rad/s]})$ at the motor axis. At the same time, a loading applied to the axis causes a torque (T^A) in the opposite direction to the movement, inducing a pressure (p) increase. This pressure, which is transmitted to the whole system, acts on the pump increasing the mechanical torque T^P.

In fact, the pressure in the motor inlet is not the same as that in the pump outlet, as a consequence of the flow energy losses. However, as an ideal system is being considered, the load losses, leakages, and mechanical friction are neglected. In the same way as the hydrostatic press (Figure 1.2), both the pump suction port and motor discharge port are at atmospheric pressure, which means that the gauge pressure is equal to zero.

At each complete revolution of a hydrostatic machine rotor (Figure 1.7) (1 revolution = 2π rad) a certain fluid volume displacement $(V \text{ [m}^3\text{]})$ occurs. From this effect, the volumetric displacement $(D \text{ [m}^3\text{/rad]})$ is defined as

$$D = \frac{V}{2\pi}. \tag{1.7}$$

The volume displaced in one complete revolution is a function of the rotor geometry. For a rotor with vanes, as shown in Figure 1.7, this volume is the product of the vane area and the mean perimeter—that is, $V = A \cdot 2\pi \cdot r$. Hence, the volumetric displacement is $D = A \cdot r$.

Moreover, the torque on the pump or motor axis can be calculated by the product of the resulting force on the vanes and the mean radius, that is, $T = F \cdot r$. Thus, the pressure in a pump or motor chamber can be written as

$$p = \frac{F}{A} = \frac{T/r}{D/r} = \frac{T}{D}. \tag{1.8}$$

Equivalently to the hydrostatic press (Equation 1.1), the pump and motor torques can be related by

$$p = \frac{T^M}{D^M} = \frac{T^A}{D^A} \Rightarrow \frac{T^A}{T^M} = \frac{D^A}{D^M} \text{ or } T^A = \left(\frac{D^A}{D^M}\right) \cdot T^M. \tag{1.9}$$

Since the tangential velocity $(v \text{ [m/s]})$ at a distance r [m] from the rotor axis is related to the angular velocity $(\omega \text{ [rad/s]})$ and to the rotational frequency $(n \text{ [rps]})$ by $v = r \cdot \omega$ and $v = r \cdot 2\pi \cdot n$, respectively, Equation 1.6 can be modified to describe the relationship between the pump and motor velocities as

$$q_V = D^M \cdot \omega^M = D^A \cdot \omega^A \Rightarrow \omega^A = \left(\frac{D^M}{D^A}\right) \cdot \omega^M \Rightarrow n^A = \left(\frac{D^M}{D^A}\right) \cdot n^M. \tag{1.10}$$

---

* Observe that the quantity rotational frequency (or just rotation) $(n \text{ [rps]})$ is commonly used instead of angular velocity $(\omega \text{ [rad]})$ and these are correlated by $\omega = 2\pi \cdot n$.

1.2.5   HYDRODYNAMICS: CONSERVATION OF ENERGY

To understand the energy transmission and control in hydraulic systems it is fundamental to apply Bernoulli's equation [8,9]. According to this equation the sum of all forms of mechanical energy in a steady and unidimensional flow of an ideal and incompressible fluid is the same at all points in the stream line.

One fundamental use of Bernoulli's equation is to describe the flow behavior through a sharp-edge orifice in a pipe, which causes an abrupt reduction in the flow cross section, as shown in Figure 1.8 [1].

In this case, the stream lines converge to a point where the diameter of the stream is the smallest. This point is called *vena contracta* and corresponds to cross section 2 in the figure. By applying Bernoulli's equation to cross section 1 (orifice upstream) and cross section 2 (orifice downstream), one obtains

$$p_1 + \frac{1}{2} \cdot \rho \cdot v_1^2 + \rho \cdot g \cdot z_1 = p_2 + \frac{1}{2} \cdot \rho \cdot v_2^2 + \rho \cdot g \cdot z_2, \tag{1.11}$$

p [Pa] being the static pressure, $1/2 \cdot \rho \cdot v^2$ [Pa] the dynamic pressure and $\rho \cdot g \cdot z$ [Pa] the gravitational pressure.

Since Bernoulli's equation is valid for steady flow, the use of Equation 1.4 implies that the inlet and outlet flow rates are the same, that is, $q_V = A_1 \cdot v_1 = A_2 \cdot v_2$. Furthermore, since the orifice area (A_0) and, consequently, the *vena contracta* area (A_2), are much smaller than the inlet area (A_1), the velocity in the inlet cross section (v_1) is neglected.

Therefore, since the change in the $\rho \cdot g \cdot z$ term along the stream line is very small compared with the other terms it can be ignored and Equation 1.11 can be written as

$$q_V = A_2 \cdot \sqrt{\frac{2 \cdot (p_1 - p_2)}{\rho}}. \tag{1.12}$$

Aiming at its practical use, this equation must be corrected to include viscosity losses. Additionally, experimental data from the literature [9,10] correlate the *vena contracta* area (A_2) with the real orifice area (A_0) such that Equation 1.10 can be rewritten as

$$q_V = cd \cdot A_0 \cdot \sqrt{\frac{2 \cdot \Delta p}{\rho}}, \tag{1.13}$$

where cd is the discharge coefficient whose value is dependent on the orifice geometry and flow type.

Another important aspect is that the turbulence downstream of the orifice causes a significant energy loss such that the velocity reduction in cross section 3 (Figure 1.8), as a consequence of the cross-sectional area increase, does not cause a static pressure increase. Hence p_3 is very close to p_2.

Therefore, Equation 1.13, known as the orifice flow equation, is appropriate to calculate the flow rate through an orifice as a function of the cross-sectional area and the pressure drop between the cross sections of the inlet (1) and outlet (3).

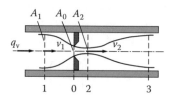

FIGURE 1.8   Flow through an orifice. (From Linsingen, I. von, *Fundamentos de Sistemas Hidráulicos*, 3rd ed., Florianópolis, Brazil: UFSC Ed., 2008. With permission.)

Finally, since the hydraulic power is defined as

$$P_h = p \cdot q_V, \tag{1.14}$$

the fact that the input pressure (p_1) is greater than the output pressure (p_3) implies that the hydraulic power is reduced with the fluid passing through an orifice. This hydraulic power difference is transformed into thermal energy, heating the fluid and the environment.

1.3 HYDRAULIC CIRCUITS

Hydraulic circuits are comprised of interconnected components so as to perform the four functions as identified in Figure 1.1. Typically, these circuits are represented by diagrams composed of graphical symbols that represent fluid power components and devices.

ISO 1219-1 [3] establishes basic elements for symbols and rules for devising fluid power symbols for use in components and circuit diagrams. ISO 1219-2 [4] establishes the rules for drawing fluid power diagrams using symbols from ISO 1219-1 [3], including rules for identification of equipment.

Table 1.1 presents the symbols according to ISO 1219-1 [3] for the hydraulic components used in this chapter. Furthermore, an identification code will be associated with these symbols following the rules shown in Figure 1.9.

TABLE 1.1
Some Symbols of Hydraulic Components

Primary Energy Conversion

Hydraulic pumps	Fixed-displacement	
	Variable-displacement	
	Variable-displacement, with pressure compensation, external drain line, one direction of rotation	
	Variable-displacement, two directions of flow, external drain line, one direction of rotation	
Electric motor		

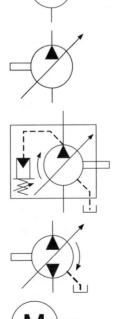

(*continued*)

TABLE 1.1   (Continued)
Some Symbols of Hydraulic Components

Energy Limitation and Control

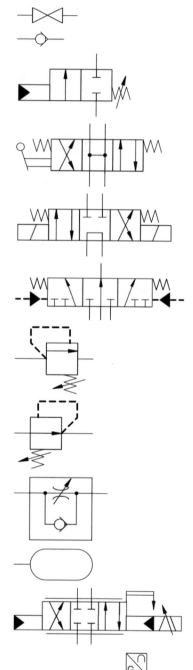

Directional control valves	Manual shut-off	
	Non-return (check)	
	2-port, 2-position, controlled by hydraulic pilot control, opening pressure adjusted by spring	
	4-way, 3-position, controlled by lever, with spring-centered central position	
	4-way, 3-position, directly controlled by two solenoids with spring-centered central position	
	5-way, 3-position, hydraulically controlled, with spring-centered central position	
Pressure control valves	Pressure relief, directly controlled, opening pressure adjusted by a spring (See Figure 1.45)	
	Pressure reducing valve, directly operated, closing pressure adjusted by a spring	
Flow control valves	Flow control adjustable, with reverse free flow	
Accumulators	(See Figure 1.55)	
Directional continuous control valves	Servo-valve, pilot-operated, pilot stage with electrical control mechanism with two coils, continuously controlled in both directions, with mechanical feedback of the main stage to the pilot stage	
	Proportional directional control valve, directly operated, with closed-loop position control of the main stage	

TABLE 1.1 (Continued)
Some Symbols of Hydraulic Components

Secondary Energy Conversion

Hydraulic cylinders	Single-acting (See Figure 1.32)	
	Double-acting (See Figure 1.33)	
Hydraulic motors	Fixed-displacement	
	Fixed-displacement, two directions of flow, two directions of rotation, with external drain	
	Variable-displacement	
Hydraulic filters	Filter	
	Filter with bypass valve	
	Filter with air exhausting	
Reservoir	Reservoir with return line / Reservoir with drain line	
Heat exchanger	Cooler	

Figure 1.10 shows a typical hydraulic circuit where the fixed-displacement pump (P) runs at a constant rotational frequency driven by the electric motor (M). Since the pump theoretically supplies a constant flow rate, it is necessary to direct part of the flow through the relief valve (V1) aiming to obtain velocity control in the cylinder (A). Therefore, the effect of the flow control valve (V3) is to cause a pressure loss such that the supply pressure (p_P) is above the setting pressure (p_{Pset}) at the relief valve (V1), and it opens. The directional control valve (V2) directs the fluid from the supply

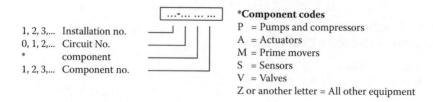

1, 2, 3,... Installation no.
0, 1, 2,... Circuit No.
* component
1, 2, 3,... Component no.

*Component codes
P = Pumps and compressors
A = Actuators
M = Prime movers
S = Sensors
V = Valves
Z or another letter = All other equipment

FIGURE 1.9 Identification code according to ISO 1219-2.

line (P) to cylinder chamber A or B and from cylinder chamber B or A to the reservoir line (This type of circuit is considered an open circuit since the fluid does not return directly to the pump suction port but to the reservoir, where it is stored before undergoing suction by the pump. The motion control of the actuators is fundamentally dissipative since it is carried out by directional, pressure, and flow control valves. The functioning principle of these valves is described by the orifice flow equation (Equation 1.13).

By comparing Figures 1.10 and 1.1, one can observe that the pump (P), together with the electric motor (M), performs the primary conversion function; the actuator (A) performs the secondary conversion; and the pressure relief valve (V1), directional control valve (V2) and flow control valve (V3) perform the energy limitation and control. The fluid storage and conditioning is performed by the reservoir (R) and filter (F1).

The open-loop circuit is by far the most popular design. The advantage of an open-loop design is that, if necessary, a single pump can be used to operate several different actuators simultaneously. The main disadvantage is its large reservoir size.

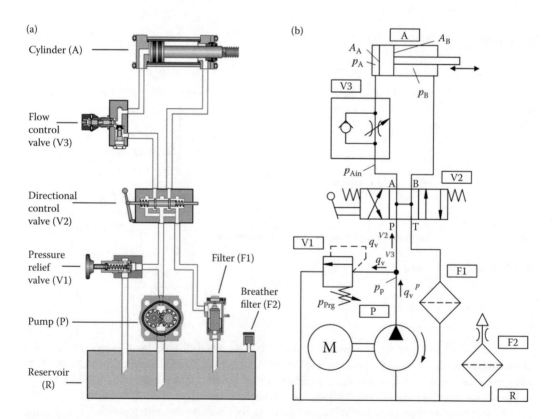

FIGURE 1.10 Open-loop hydraulic circuit: (a) Illustration; (b) Circuit diagram.

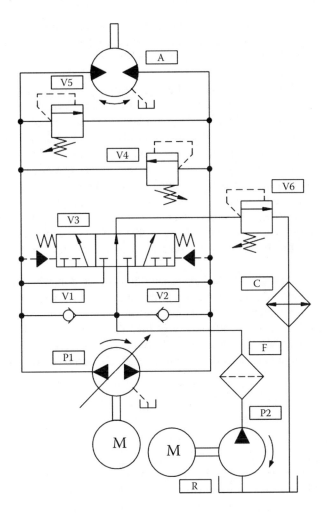

FIGURE 1.11   Closed-loop circuit diagram.

A second general type of hydraulic circuit is the closed-loop circuit [7], whose main operational difference relates to the means of hydraulic energy control. As can be observed in the example in Figure 1.11, it is not only the pump discharge but also the pump suction that is directly connected to the motor ports. Therefore, the motor rotational frequency will be modified if the volumetric displacement of either the motor or the pump is varied or the pump rotational frequency is changed. The relationship between the flow rate, volumetric displacement, and rotational frequency of a pump or motor is described by Equation 1.10.

In the circuit shown in Figure 1.11 [7], a variable-displacement pump (P1) is used to drive a fixed-displacement hydraulic motor (A). A closed-loop circuit is always used in conjunction with a smaller replenishing circuit. The replenishing circuit consists of a small fixed-displacement pump (P2) (usually about 15% of the displacement of the main pump), a small fluid reservoir (R), filter (F), and a heat exchanger (cooler) (C).

The replenishing circuit always works on the low-pressure side of the main loop. Its function is to pump freshly filtered fluid into the closed loop through non-return valves (V1 and V2) while bleeding-off a percentage of the hot fluid through a directional control valve (V3). This hot fluid is then cooled by a cooler (H) and stored in a small reservoir (R) before returning to the main system. The pressure in the replenishing circuit is limited to 10–20 bar (1–2 MPa) by the supercharge relief valve (V6). The pressure setting of this valve is determined by the requirements of the pump/motor combination and the operating conditions of the system. The cross-port relief valves (V4 and V5) on the motor are there only to protect the actuator from load-induced pressure spikes. They are not intended to function like those found in open-loop circuits, which would cause severe overheating of the circuit due to the diverting of the unnecessary flow through the relief valve.

The advantages of a closed-loop circuit are that high-power systems are compact and efficient and require less hydraulic fluid storage. The high efficiency of this circuit is the result of the pump control being designed to supply only the fluid flow required by the actuator to operate at the load-induced pressure. The pump is the heart of the system and controls the direction, acceleration, speed, and torque of the hydraulic motor, thus eliminating the need for pressure and flow control components.

In this type of circuit the energy control is transformative, instead of dissipative as in open-loop circuits, since it is the energy transformed in the pump or motor that is controlled. However, the secondary valves (pressure, directional and flow-control valves) impose energy losses—besides the internal mechanical and fluid flow losses—in pumps and motors, thereby reducing the overall efficiency.

A major disadvantage of a closed-loop circuit is that a single pump can only operate a single output function or actuator. In addition, this type of hydraulic circuit is generally used only with motor actuators.

The third general configuration is the half-closed-loop circuit as shown in Figure 1.12 [7]. This circuit is similar to the closed-loop circuit except that it can be used with cylinder actuators with different areas. As can be seen from the figure, during cylinder extension, the pump (P) must generate a higher flow rate from its left-hand port than that being returned to its right-hand port from the cylinder (A). The extra fluid needed by the pump (P) is supplied by its left-hand inlet non-return valve, which is an integral part of the pump. When the pump control moves the pump over

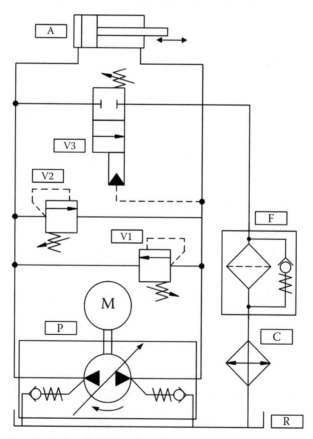

FIGURE 1.12   Half-closed-loop circuit diagram.

the center, the flow from the pump (P) is reversed and the cylinder (A) begins to retract. During retraction, the larger area of the cylinder piston causes a higher flow rate than needed at the inlet of the pump (P). This excess flow is directed to the reservoir (R) through the unloading valve (V3). The unloaded fluid is filtered and cooled prior to its return to the reservoir. In this way, a portion of the closed-loop fluid is filtered (by F) and cooled (by C) in an open-loop circuit each time the cylinder (A) is cycled.

In this case, the fluid volume and reservoir size reductions are not as significant as in the closed-loop scenario.

As can be seen in the above examples, each hydraulic component has a basic function, but it is the circuit itself that determines the hydraulic system behavior. Hence, for a designer to conceive a hydraulic system he/she needs to have an understanding of the functional and behavioral characteristics of the components which, in turn, are dependent on the fluid-mechanical interaction inside the component.

1.4 HYDRAULIC COMPONENTS

1.4.1 HYDROSTATIC MACHINES: PUMPS AND MOTORS

The energy conversion functions in a hydraulic system are performed by pumps and actuators (basically motors and cylinders). The pumps perform the primary conversion, transforming mechanical energy into hydraulic energy. The actuators retransform the hydraulic energy into mechanical energy to be used by the machine or the equipment.

There are two classes of hydraulic machines: hydrodynamic and hydrostatic machines. They differ in the way the internal energy is transformed and, consequently, in their form of construction [1].

In hydrodynamic machines (such as centrifugal pumps, turbines, and fans), the fluid energy involved on the transformation process is fundamentally kinetic, due to the variation in the fluid velocities of the impeller blades. In these machines there is a gap between the pump housing and the impeller (or rotor) leading to a high internal leakage even with low differential pressure.

In the centrifugal pumps, as shown in Figure 1.13a, when the output fluid flow resistance is increased (e.g., as a consequence of the load loss in the discharge line) the output flow rate is reduced until it drops to zero, as shown in the characteristic curve in Figure 1.14a.

(a)   (b)

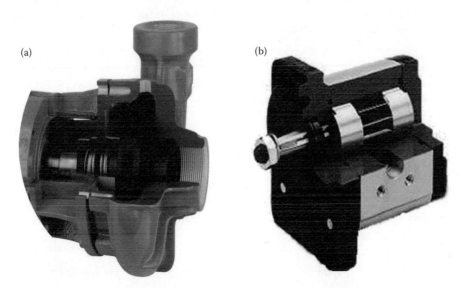

FIGURE 1.13 Classes of pumps: (a) Hydrodynamic pump (centrifugal pump) (Courtesy of Franklin Electric–Joinville–SC-Brazil); (b) Hydrostatic pump (gear pump). (Courtesy of Bosch Rexroth–Pomerode–SC-Brazil).

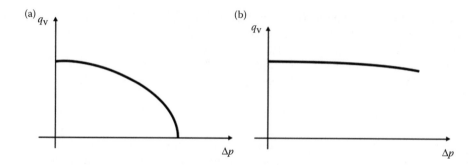

FIGURE 1.14 Characteristic curves of pumps: (a) Hydrodynamic pump; (b) Hydrostatic pump.

In hydrostatic machines, also referred to as "positive displacement machines," the fluid energy involved in the transformation process is mainly related to the variation between the inlet and outlet pressures through the rotor. Since the pressure in a system is caused by the fluid flow resistance, the effective pump outlet pressure increase is dependent on the valves and actuators downstream of the pump. In turn, the pressure in an actuator inlet is dependent on the rotor movement resistance caused by an external mechanical loading.

In hydrostatic pumps the clearance between the housing and the rotor is very small and thus the suction and discharge chambers are basically isolated. As a consequence, the pump flow rate is slightly influenced by the downstream pressure, as illustrated by the characteristic curve shown in Figure 1.14b.

Since the construction principle of hydrostatic (rotary) motors is the same as that of pumps, an increase in the mechanical axis loading leads to a small leakage increase. Hence, the motor rotational frequency can be considered constant in several applications [1].

The fact that the hydrostatic pumps are an almost ideal flow rate source and operate under high pressures makes this class of hydraulic machine basically the only one used in fluid power systems [1]. At same time, to attain the requirements of the several application fields, different construction principles of hydrostatic machines have been developed, as shown in Table 1.2.

In the right column of this table, an important feature of hydrostatic machines is indicated. According to Equation 1.10, the volumetric displacement establishes the proportionality between the flow rate and the rotational frequency. Machines whose construction characteristics do not allow changes in the volumetric displacement are named *fixed-displacement machines*, and those where is possible to obtain different flow rates at the same rotational frequency are named *variable-displacement machines*.

TABLE 1.2

Classification of Hydrostatic Machines According to Construction Principle and Volumetric Displacement

Constructive Principle			Volumetric Displacement
Gear	External		Fixed
	Internal	Crescent	Fixed
		Gerotor	Fixed
Screw			Fixed
Vane	Balanced		Fixed
	Unbalanced		Fixed or Variable
Piston	Radial		Fixed or Variable
	Axial	Swash Plate	Fixed or Variable
		Bent-Axis	Fixed or Variable

TABLE 1.3
Typical Pump Performance Parameters

Pump Type	Max. Working Pressure [MPa (bar)]	Flow Rate [dm³/s (Lpm)]	Rotational Frequency [rps (rpm)]	Global Efficiency [%]
External Gear	15–25 (150–250)	0.08–9.5 (5–570)	8.3–83.3 (500–5,000)	80–90
Internal gear	3.5–20 (35–200)	0.08–12.7 (5–760)	15–41.7 (900–2500)	70–90
Screw	0.4–40 (4–400)	0.017–350 (1–21,000)	16.7–58.3 (1,000–3,500)	80–85
Vane	7–21 (70–210)	0.08–10 (5–600)	10–45 (600–2,700)	80–95
Radial piston	7–815 (70–815)	0.08–12.7 (5–760)	16.7–56.7 (1,000–3,400)	85–95
Axial piston	14–81.5 (140–815)	0.08–12.7 (5–760)	8.33–71.7 (500–4,300)	90–95

In Table 1.3 some typical values of the operational characteristics of pumps are presented. Similar values are applicable to hydraulic motors.

In the next sections, the functional and construction principles of these hydrostatic machines are presented. Although pumps and motors are very similar, some specific construction aspects—such as internal channels for lubrication, external leakage drain, seals, and so forth—differ since motors do not have a port under low pressure all the time, as in the case of pumps.

Therefore, a pump cannot be used as a motor and vice-versa, unless the component has been designed to carry out both functions.

1.4.1.1 Gear Pump and Motors

External gear pumps and motors. This type of hydrostatic machine consists of a pair of equal gears assembled in housing with one inlet and one outlet, enclosed by two side plates. The drive gear is responsible for the external motion transmission and the driven gear runs free in its shaft.

According to Figure 1.15 (pump), fluid transport cells are formed between two consecutive teeth of each gear and the housing through the rotational movement. At the same time, the ungearing

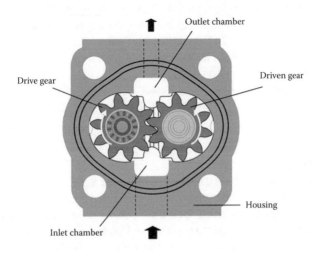

FIGURE 1.15   External gear pump (and motor).

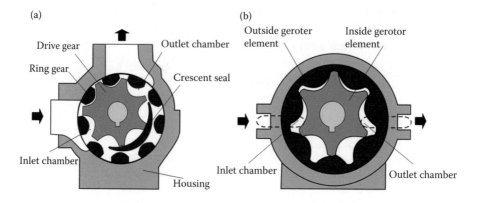

FIGURE 1.16   Internal gear pumps: (a) Crescent-seal type; (b) Gerotor type.

produces new cells to which the fluid is suctioned. In the outlet chamber the continuous gearing pushes the fluid out to the outlet port.

It is generally agreed that the gear pump is the most robust and rugged type of fluid power pump and thus its use is predominant in hydraulic services and also very intensive in industrial machines.

Gear pumps and motors are not very sensitive to fluid viscosity variations and to fluid contamination. However, since the outlet and inlet ports are opposite to one other, the forces over the gear axis are unbalanced. This limits the maximum values of pressure and flow rate.

As a consequence of the friction between the gears and the side plates, and the fluid leakage between the tips of the gears and across the side plates, the overall efficiency is lower than that of solutions based on the other construction principles.

Internal gear pumps. Given the possibility of operating under high pressures with low ripple pressures and low noise, these pumps are used in several systems such as injection machines, hydraulic presses, machine tools, and so forth. The operational principle is the same as that of external gear machines—that is, the continuous tooth ungearing and gearing of a gear pair.

The crescent seal internal gear pump consists of a small internal gear and a larger ring gear (see Figure 1.16a). The small internal gear is driven by the prime mover. The internal gear meshes with the ring gear and turns it in the same direction. The sealing of the high-pressure chamber from the pump inlet is achieved by a crescent seal between the upper teeth of the internal small gear and the upper teeth of the ring gear. In the gerotor gear pump, the inner gerotor has one less tooth than the outer element (Figure 1.16b). The internal gear is driven by the prime mover and, in turn, drives the outer element in the same direction [7].

In the same way as in external gear pumps, internal gear pumps are unbalanced, limiting the maximum pressure and efficiency. Furthermore, the gear pump design does not allow the displacement to be varied.

1.4.1.2 Screw Pumps

Screw pumps for fluid power systems are composed of two or more helical screws assembled inside housing. The relative movement of the screws can be obtained driving one shaft where the movement is transmitted to the others by either their own gearing or by external gears mounted on the shafts. An illustration of this type of pump is shown in Figure 1.17.

Each screw thread is matched to carry a specific volume of fluid. Fluid is transferred through successive contact between the housing and the screw flights from one thread to the next. Its

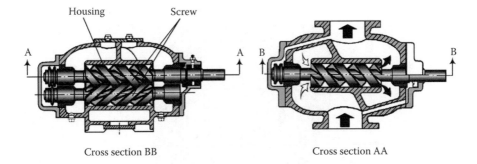

Housing	Screw

Cross section BB	Cross section AA

FIGURE 1.17	Screw pump. (From Linsingen, I. von, *Fundamentos de Sistemas Hidráulicos*, 3rd ed., Florianópolis, Brazil: UFSC Ed., 2008. With permission.)

operational characteristics imply that the flow does not present pulsation and the unbalanced forces are axial, being compensated for easily.

Screw pumps are generally used for hydraulic systems where high flow rates are necessary and they are also suitable for high pressures. The disadvantages are their low efficiency and high cost.

1.4.1.3 Vane Pumps and Motors

Fixed-displacement vane pumps and motors. Vane machines are comprised of a cylindrical rotor with vanes sliding in its grooves. This set runs inside a cam ring and the sides of the rotor and vanes are sealed by side bushings (port plates). Figure 1.18 presents an illustration of this type of machine.

The vanes are forced against the internal surface of the cam ring due to centrifugal force and either high pressure applied on the vane bottom or the force of the spring mounted on the vane bottom.

Between two consecutive vanes, rotor, cam ring and port plates, fluid transport cells are formed that increase in the inlet chamber and decrease in the outlet chamber. The port plates include apertures connecting these chambers to the external ports of the machine.

In the construction principle shown in Figure 1.18 the low and high pressures act appositively over the axis, causing unbalanced forces and limiting the maximum work pressure of the pump or motor. An alternative is the balanced vane pump shown in Figure 1.19 where there are two low-pressure chambers and two high-pressure chambers and thus the resultant radial forces tend to be null.

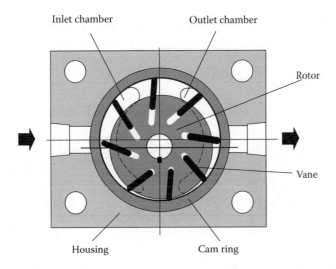

Inlet chamber	Outlet chamber

Rotor

Vane

Housing	Cam ring

FIGURE 1.18	Vane pump. (From Linsingen, I. von, *Fundamentos de Sistemas Hidráulicos*, 3rd ed., Florianópolis, Brazil: UFSC Ed., 2008. With permission.)

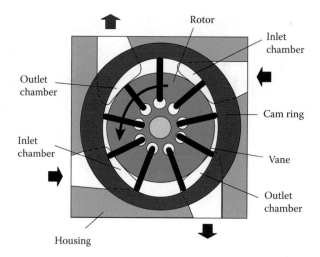

FIGURE 1.19 Balanced vane pump.

The total pumping flow results from the superposition of the flow rate from the two outlet chambers. The resulting amplitude and frequency at the outlet port is dependent on the number of vanes, where an odd number of vanes is advantageous, since the volumes discharged from each discharge chamber are not in phase.

Variable-displacement vane pumps. The variation in the volumetric displacement in vane pumps is achieved by moving the cam ring and, therefore, changing the eccentricity between it and the rotor. This can be seen in Figure 1.20, where the flow direction can also be inverted without changing the rotational frequency direction. The hydraulic circuit shown in Figure 1.11 is an example of the use of this type of pump.

Variable-displacement pumps can also include internal pressure compensation as shown in Figure 1.21. In this case, the maximum eccentricity is obtained while the internal pressure in the discharge chamber produces a force lower than the spring force. When the outlet pressure increases over the pre-load force of the spring, the cam ring moves against the spring, changing the flow rate delivered.

In general, fluid leakage in vane pumps occurs between the high- and low-pressure sides of the vanes and across the side bushings, which results in decreased volumetric efficiency and, hence, reduced flow output. The unbalanced design suffers from shortened bearing life because of the unbalanced thrust force within the pump.

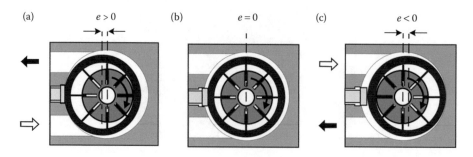

FIGURE 1.20 Illustration of the volumetric displacement variation: (a) Regular flow; (b) Null flow; (c) Reverse flow. (From Linsingen, I. von, *Fundamentos de Sistemas Hidráulicos*, 3rd ed., Florianópolis, Brazil: UFSC Ed., 2008. With permission.)

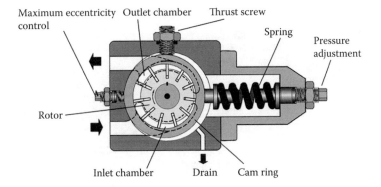

FIGURE 1.21   Variable-displacement vane pump with pressure compensation. (From Linsingen, I. von, *Fundamentos de Sistemas Hidráulicos*, 3rd ed., Florianópolis, Brazil: UFSC Ed., 2008. With permission.)

1.4.1.4   Piston Pumps and Motors

Piston machines have radial clearances in their main movable parts of between 2 and 5 mm. Consequently, they can operate under higher pressures and lower volumetric losses when compared with other hydrostatic machines.

According to the position of the pistons in relation to the shaft, these machines are classified as axial piston pumps (swash plate and bent-axis) and radial piston pumps.

Fixed-displacement axial piston machines. In this type of machine the pistons run in cylindrical holes machined in a cylinder block. The alternative movement of each piston is obtained by the rotary movement of the cylinder block or the swash plate.

a. Swash plate design

As shown in Figure 1.22, axial machines can be constructed with either rotary or stationary swash plates. In the motor shown in Figure 1.22a, the cylinder block is stationary and the swash plate is rigid with the shaft. In Figure 1.22b, the swash plate is stationary and the cylinder block rotates with the shaft. The swash plate angle defines the piston stroke and, hence, the volumetric displacement [1,11].

The valve plate identified in this figure consists of a plate with circumferential apertures and its function is to connect the inlet and outlet ports to the bottom of each piston.

In this type of machine there is a continuous leakage that is necessary for the lubrication of parts with relative mechanical movement such as that between the valve plate and the cylinder block, and that between the cylinder block and the swash plate. Therefore, a port for external drainage is required.

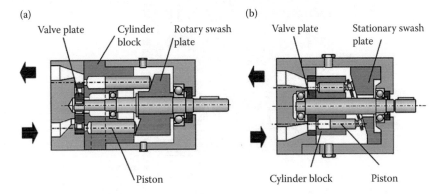

FIGURE 1.22   Swash plate design: (a) Motor with rotary swash plate; (b) Pump with stationary swash plate. (From Linsingen, I. von, *Fundamentos de Sistemas Hidráulicos*, 3rd ed., Florianópolis, Brazil: UFSC Ed., 2008. With permission.)

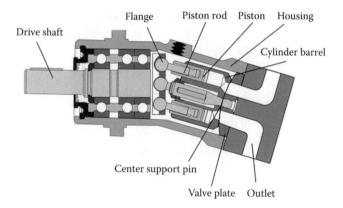

FIGURE 1.23 Illustration of axial piston machine bent-axis design. (From Frankenfield, T.C. *Using Industrial Hydraulics*, 2nd ed., Penton Publishing, 1985, ISBN-13: 9780932905017. With permission.)

b. Bent-axis design

In this design the cylinder block is mounted obliquely in relation to the driven shaft (Figure 1.23). The piston rods are coupled to the driven shaft by spherical articulations such that the rotary movement of the cylinder block produces the alternating piston movement. The connection between the pistons and the inlet and outlet ports is through the valve plate, as shown in this figure.

Since pistons have no lateral forces, angles of around 25°, and even 40°, are allowable. In relation to the swash plate, the bent-axis type has as disadvantages a greater occupied volume and higher moment of inertia. On the other hand, it has higher efficiency and less sensitivity to contaminants.

Variable-displacement axial piston machines. The swash plate machines can also have variable volumetric displacement by changing the swash plate angle. An angle equal to zero corresponds to null flow rate and the maximum positive angle produces the maximum volumetric displacement and, consequently, the maximum flow rate supplied by a pump or consumed by a motor. When a negative angle is allowed, the machine has two flow directions. In the same way, in the bent-axis type the angle between the cylinder block/valve plate axis and the shaft can also be controlled.

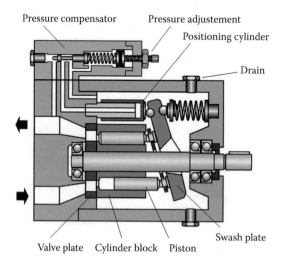

FIGURE 1.24 Variable-displacement axial piston pump, swash plate design, with pressure compensation. (From Linsingen, I. von, *Fundamentos de Sistemas Hidráulicos*, 3rd ed., Florianópolis, Brazil: UFSC Ed., 2008. With permission.)

Variable-displacement piston pumps lend themselves to the incorporation of various mechanisms that will alter their performance. One typical example is the pressure-compensated pump where the hydraulic mechanism will alter the pump displacement to limit the outlet pressure to some pre-adjusted value. Figure 1.24 presents a pressure-compensated axial piston pump (swash plate type).

Other commercial solutions allow the control of the hydraulically supplied power according to the system demand. Electro-hydraulic pumps using proportional valves are also available to design circuits with transformation control, which means directly through the primary conversion function. The circuits presented in Figure 1.11 and 1.12 are examples of the use of variable-displacement pumps.

Radial piston machines (fixed- and variable-displacement). In these machines, the piston axes are perpendicular to the driven shaft. Depending on the construction principle, the pistons can be mounted in a star format around the shaft or in line on a crankshaft.

Figure 1.25 shows the basic configuration of a three-piston pump. Each hollow piston consists of an inlet non-return valve, a spring, a piston barrel, a pumping chamber, an outlet non-return ball, and a support bearing. As the driven shaft is rotated, the spring holds the base of the piston in contact with the eccentric cam shaft. The downward motion of the piston causes the volume to increase in the pumping chamber. This creates a reduced pressure that enables the inlet check valve to open, thereby allowing oil to enter the pump chamber. The oil enters the chamber by way of a groove machined into the cam-shaft circumference. Further rotation of the cam shaft causes the piston to move back into the cylinder barrel. The rapid rise in chamber pressure closes the inlet check valve. When the rising pressure equals the system pressure, the outlet check valve opens, allowing flow to exit the piston and pass to the pressure port of the pump. The resulting flow is the sum of all the piston displacements. The number of pistons that a radial pump can have is only limited by the spatial restrictions imposed by the size of the pistons, housing, and cam shaft.

Figure 1.26 illustrates a variable-displacement pump with pressure compensation composed of a cam ring eccentrically mounted relative to a cylinder block. The alternative movement of the pistons is obtained by the rotary movement of the cylinder block reaming the pistons in contact with the cam ring through shoes. The shoes slide on a trail fixed on the cam ring. The fluid suction and discharge occurs via semicircular ports and pipes machined on a stationary piece inside the driven shaft.

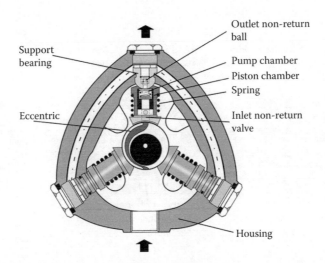

FIGURE 1.25 Radial piston pump. (From Frankenfield, T.C. *Using Industrial Hydraulics*, 2nd ed., Penton Publishing, 1985, ISBN-13: 9780932905017. With permission.)

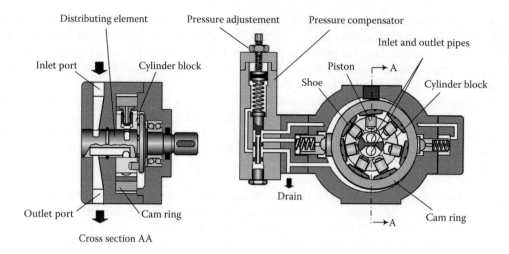

FIGURE 1.26 Radial piston pump with pressure compensation. (From Linsingen, I. von, *Fundamentos de Sistemas Hidráulicos*, 3rd ed., Florianópolis, Brazil: UFSC Ed., 2008. With permission.)

In general, the radial piston pump has a higher continuous-pressure capability than any other type of pump (Table 1.3). However, it should be noted that for extremely high-pressure applications, the volumetric displacements of radial pumps are usually not larger than 2.4×10^{-6} m³/rad (0.015 dm³/rev).

1.4.1.5   Pump and Motor Performance Characteristics

In Section 1.2.4 the flow rate and torque equations of pumps and motors was presented, where they were considered as ideal machines, without internal or external leakage and friction. However, these losses are present in real machines and they are identified in a general way by the volumetric, mechanical and overall efficiencies.

Consider Figure 1.27, where the main variables associated with pumps and motors are presented. Based on this figure, the equations given below describe the steady-state behavior and efficiency expressions valid for pumps and motors.

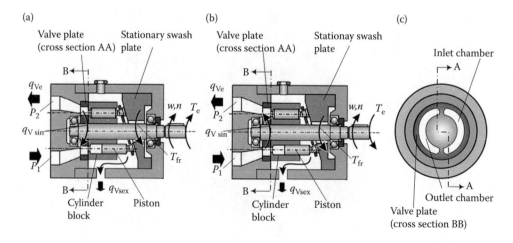

FIGURE 1.27   Main variables associated with: (a) Pumps; and (b) Motors (c) Cross section BB.

Flow rate and volumetric efficiency. The volumetric losses in hydrostatic machines occur as a consequence of the mechanical clearances, pressure drops and relative velocity between movable parts. Cavitation and fluid aeration also induce flow losses. However, since these phenomena should not occur under normal operational conditions, they are not considered in the mathematical description of volumetric efficiency [12].

The theoretical flow rate given by Equation 1.10 and rewritten in Equation 1.15 is dependent on the volumetric displacement (D). This parameter is calculated according to the geometric dimensions or by measuring the absorbed or discharged volume for a complete revolution with differential pressure close to zero.

$$q_{Vtc} = D \cdot \omega = D \cdot 2\pi \cdot n. \tag{1.15}$$

The effective flow rate (discharged) (q_{Ve} [m³/s] or [L/min]) in pumps is lower than the theoretical flow rate (q_{Vtc} [m³/s] or [L/min]) and can be determined by

$$q_{Ve}^{P} = q_{Vtc}^{P} - q_{Vs}^{P}, \tag{1.16}$$

where q_{Vs} [m³/s] or [L/min] is the flow rate loss that can be due to internal leakage ($q_{V\,sin}$) (between the pump chambers), or external leakage ($q_{V\,sex}$), as in vane and piston pumps that have a drain port.

In motors, the effective flow rate (inlet) (q_{Ve} [m³/s] or [L/min]) is higher than the theoretical flow rate (q_{Vtc} [m³/s] or [L/min]), since part of the fluid is lost through leakage (q_{Vs}). Therefore:

$$q_{Ve}^{M} = q_{Vtc}^{M} + q_{Vs}^{M}. \tag{1.17}$$

The volumetric efficiency is then calculated through the following expressions:
For pumps:

$$\eta_V^P = \frac{q_{Ve}^P}{q_{Vtc}^P}. \tag{1.18}$$

For motors:

$$\eta_V^M = \frac{q_{Vtc}^M}{q_{Ve}^M}. \tag{1.19}$$

The leakage in pumps and motors is approximately laminar and thus under operational conditions, with approximately constant temperature, the leakage is proportional to the pressure difference ($q_{V\,sin} \propto \Delta p$). Hence, the volumetric efficiency also changes proportionally to the pressure difference.

Torque and mechanical efficiency. Based on Equation 1.8, the theoretical torque (T_{tc} [N·m]) can be expressed as

$$T_{tc} = D \cdot \Delta p, \tag{1.20}$$

where Δp [Pa] is the pressure difference between the inlet and outlet ports of the pump or motor.

However, this is not the real torque in the machine shaft, since there are losses associated with mechanical friction and fluid viscous friction [12].

For pumps, the effective torque required in the driven shaft (T_e [N · m]) is higher than the theoretical torque (T_{tc} [N · m]), that is

$$T_e^P = T_{tc}^P + T_{fr}^P, \tag{1.21}$$

where T_{fr} [N · m] is the friction torque.

In the case of motors, the effective torque available in the shaft (T_e) is lower than the theoretical torque (T_{tc}), such that

$$T_e^A = T_{tc}^A - T_{fr}^A. \tag{1.22}$$

Consequently, the mechanical efficiencies are defined by the following expressions:

For pumps:

$$\eta_m^P = \frac{T_{tc}^P}{T_e^P}. \tag{1.23}$$

For motors:

$$\eta_m^A = \frac{T_e^A}{T_{tc}^A}. \tag{1.24}$$

Power and overall efficiency. The useful power of a pump is the hydraulic power at the outlet port and for a motor it is the mechanical power at the driven shaft. The useful power can be described by:

For pumps:

$$P_h = q_{Ve} \cdot \Delta p \cong q_{Ve} \cdot p_2 = q_{Vtc} \cdot p_2 \cdot \eta_V, \tag{1.25}$$

where $\Delta p = p_2 - p_1$, p_2 p_2 is the pressure in the outlet port (discharge) and p_1 is the pressure in the inlet port (suction). Since the pressure p_1 is close to atmospheric pressure ($p_1 \approx 0$ Pa [gauge pressure]), this expression can be written considering only the output pressure (p_2).

For motors:

$$P_m = T_e \cdot \omega = T_e \cdot 2\pi \cdot n = T_{tc} \cdot 2\pi \cdot n \cdot \eta_m. \tag{1.26}$$

Or applying Equation 1.20:

$$P_m = D \cdot (p_1 - p_2) \cdot 2\pi \cdot n \cdot \eta_m = q_{Vtc} \cdot (p_1 - p_2) \cdot \eta_m. \tag{1.27}$$

The drive power is the mechanical power at the shaft for a pump and the hydraulic power at the inlet port for a motor. Hence:

For pumps:

$$P_m = T_e \cdot \omega = T_e \cdot 2\pi \cdot n = \frac{T_{tc} \cdot 2\pi \cdot n}{\eta_m}. \tag{1.28}$$

Or applying Equation 1.20:

$$P_m = \frac{D \cdot (p_2 - p_1) \cdot 2\pi \cdot n}{\eta_m} = \frac{q_{Vtc} \cdot (p_2 - p_1)}{\eta_m}. \tag{1.29}$$

For motors:

$$P_h = q_{Ve} \cdot \Delta p = q_{Ve} \cdot (p_1 - p_2) = \frac{q_{Vtc} \cdot (p_1 - p_2)}{\eta_V}. \tag{1.30}$$

Consequently, the overall efficiency is defined as
For pumps:

$$\eta_t^P = \frac{P_h^P}{P_m^P} = \eta_V^P \cdot \eta_m^P. \tag{1.31}$$

For motors:

$$\eta_t^M = \frac{P_m^M}{P_h^M} = \eta_V^M \cdot \eta_m^M. \tag{1.32}$$

1.4.1.6 Characteristic Curves

The variables presented in the section above are frequently presented in graphs as a function of the pressure difference to which the hydrostatic machine will be submitted. Moreover, operational conditions like temperature and rotational frequency, and fluid specification, need to be pre-fixed when these operating curves are obtained experimentally.

Fixed-displacement pumps. A typical characteristic curve is shown in Figure 1.28 where the curve of the effective flow rate (q_{Ve}) represents the basic characteristic of a pump, where its scope shows the operating pressure influence on the leakage. From this curve the volumetric efficiency curve (η_V) is obtained using Equation 1.18.

The mechanical efficiency (η_m) increases with the fluid leakage, improving the lubrication and reducing the friction torque (Equations 1.22 and 1.23). The useful power (P_h) is a linear function of the effective flow rate and the output pressure (Equation 1.25) and, in turn, the drive power (P_m) is dependent on the mechanical losses (Equation 1.28). According to Equation 1.31, the curve of the overall efficiency (η_t) is determined by either the useful power to drive power ratio, or the volumetric and mechanical efficiency product.

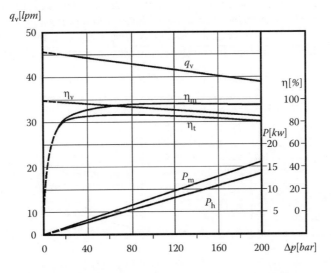

FIGURE 1.28 Operating curves of a fixed-displacement pump. (From Linsingen, I. von, *Fundamentos de Sistemas Hidráulicos*, 3rd ed., Florianópolis, Brazil: UFSC Ed., 2008. With permission.)

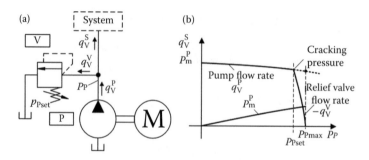

FIGURE 1.29 Fixed-displacement pump with relief valve: (a) Hydraulic circuit; (b) Characteristic curve.

The strong reduction in the overall efficiency at low pressures is a consequence of poor lubrication and high friction in this operational range. For this reason, the manufacturers recommend a minimal operation pressure, with the aim of not reducing the useful life of the pump. In the case of Figure 1.28, the pump must operate above 2 MPa (20 bar) [1].

Fixed-displacement pumps with relief valve. The fixed-displacement pump is frequently used together with a relief valve since the flow from the pump needs to be diverted to a reservoir when it is not being used by the system (Figure 1.29a).

The effective flow rate supplied to the system (q_V^S) is obtained by combining the characteristic curves of the two components, as can be seen in Figure 1.29b. The cracking pressure is the pressure adjusted at the relief valve ($p_{P_{set}}$) at which it opens. From this operational point onward, any increase in the system pressure (p_P) causes a significant decrease in the flow rate to the system (q_V^S).

For pressures lower than the cracking pressure the system flow rate is equal to the pump flow rate ($q_V^S = q_V^P$), and for higher pressures part of the flow is diverted to the relief valve ($q_V^S = q_V^P - q_V^V$). At the maximum supply pressure ($p_{P_{max}}$) the relief valve flow rate is equal to the pump flow rate ($q_V^V = q_V^P$), which means that all the hydraulic power (P_h^P) is being dissipated at the relief valve, increasing the temperature of the fluid that returns to the reservoir. Consequently, the pump drive power (P_m^P) continues to increase after the cracking pressure has been reached.

Variable-displacement pumps with pressure compensation. In the case of variable-displacement pumps with pressure compensation, as shown in Figure 1.21, 1.24, and 1.26, the use of a relief valve is not required, although it can be installed in the hydraulic circuit for safety reasons.

As shown in Figure 1.30a, the system flow rate is always equal to the pump flow rate. When there is no demand from the system, the pressure increases above the set pump pressure, changing

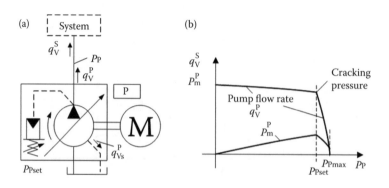

FIGURE 1.30 Variable-displacement pump with pressure compensation: (a) Hydraulic circuit; (b) Characteristic curve.

its volumetric displacement (D). Therefore, the power consumption is reduced when the cracking pressure is surpassed, as illustrated in Figure 1.30b. One can observe that this power is not null when $q_V^S = q_V^P = 0$ since there is always a small lubrication flow rate (q_{Vs}^P), which is drained to the reservoir [13].

1.4.2 Hydraulic Cylinders

Hydraulic systems are designed to provide controlled mechanical energy through linear or angular movement. The action over the external environment occurs on the last block of the functional chain shown in Figure 1.1, the secondary energy conversion, and it is performed by the hydraulic actuators, which in this case are the motors, oscillators and cylinders.

The basics of motors were described in the section above, since their construction principles are the same as those of pumps.

The hydraulic oscillators also produce angular movement but they do not provide continuous rotation and the angle is limited to a value below 360°. Their construction is derived from the hydraulic motor design (from the vane motor, for example) or from double-acting hydraulic cylinders with mechanical transmission converting linear into angular displacement.

In turn, cylinders are the hydraulic actuators most used in hydraulic systems. They are typically comprised of (1) a barrel, (2) piston assembly, (3) piston rod, (4) end caps, (5) ports, and (6) seals, as shown in Figure 1.31. The piston provides the effective area against which the fluid pressure is applied and supports the piston assembly and rod. The opposite end of the rod is attached to the load. The cylinder bore, end caps, ports, and seals maintain a fluid-tight chamber in which the fluid energy is contained. Whether the rod will extend or retract is dependent on the port to which the fluid is directed.

Hydraulic cylinders are classified according to different premises, with two of them being particularly important in terms of understanding the use and behavior of cylinders. Hence, in relation to the operating principle they are sub-divided into single- and double-acting (single- and double-effect) and, considering the area ratio, they are classified as either symmetrical or asymmetrical (non-differential or differential) cylinders.

In Figure 1.32, the several types of single-acting cylinders are symbolically represented. In this construction principle, the hydraulic power is available in only one direction of movement—that is, on either extension or retraction. In the opposite direction the movement results from an external force (including gravitational force) as shown in Figure 1.32a, b and e, or from an internal spring force as in Figure 1.32c and d.

Unlike the other types, telescopic cylinders have two or more stages which, when fully extended, can produce a stroke that exceeds the length of the cylinder when fully retracted. The symbol shown in Figure 1.32e presents a two-stage model.

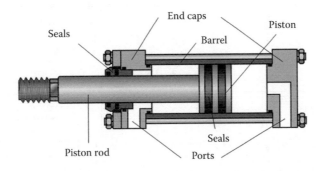

FIGURE 1.31 Main parts of a hydraulic cylinder. (From Linsingen, I. von, *Fundamentos de Sistemas Hidráulicos*, 3rd ed., Florianópolis, Brazil: UFSC Ed., 2008. With permission.)

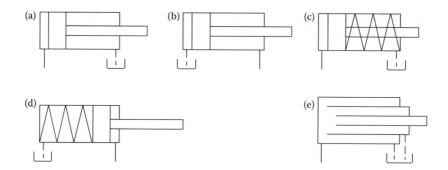

FIGURE 1.32 Single-acting cylinders: (a) Retraction by external force; (b) Extension by external force; (c) Retraction by spring; (d) Extension by spring; (e) Telescopic cylinder with retraction by external force.

As a consequence of the inevitable leakage between the piston and barrel, the non-active chambers must have an external drain avoiding counter-pressure and cylinder blocking.

Some examples of double-acting cylinders are shown in Figure 1.33. In this type of cylinder, the effective work is carried out in both directions of movement (extension and retraction).

The most common double-acting cylinder is the single-rod cylinder (Figure 1.33a), which is classified as an asymmetric (differential) cylinder since the piston areas on the bottom-side and the rod-side are different. As a consequence, the velocity and hydraulic force are generally different during the extension and retraction movements.

The double-rod cylinders (Figure 1.33b) can be designed with rods of the same diameter (symmetric [non-differential] cylinder) and with different diameters (asymmetric [differential] cylinder). In the case of symmetric cylinders the hydraulic force and velocity are the same, considering the same loading and supplied flow rate, during extension and retraction.

Tandem actuating cylinders (Figure 1.33c) consist of two or more cylinders arranged one behind the other but designed as a single unit. The main operational characteristic is the greater force when compared with a regular cylinder of the same diameter.

In the same way as in telescopic single-acting cylinders, the double-acting cylinders (Figure 1.33d) have the advantage of being compact. However, since their construction costs are higher than those of other designs, their use is somewhat limited.

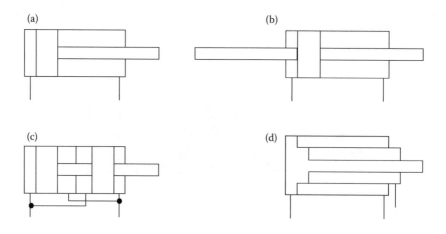

FIGURE 1.33 Double-acting cylinders: (a) Single rod; (b) Double rod; (c) Tandem; (d) Telescopic.

1.4.2.1 Hydraulic Cylinder Behavior

The hydraulic cylinders are intended for use under several operational conditions, including motion with a constant velocity, positioning control, force control, or just to provide a force to fix something.

In all these situations, the motion achieved is influenced by factors such as inertia, fluid compressibility and friction, and must be considered in the analysis and design of the hydraulic system [14,15].

By observing Figure 1.34 one can identify two main parts to be modeled: the movable piston and the fluid in the cylinder chambers.

The linear motion of the piston is described by Newton's second law, which establishes that the sum of the forces must be equal to the product of the mass and acceleration ($M_t \cdot a = M_t \cdot d_2x/dt_2$). Therefore, for an asymmetric double-acting cylinder, as shown in Figure 1.34, the motion equation is

$$(A_A \cdot p_A) - (A_B \cdot p_B) = M_t \cdot \frac{d^2 x_p}{dt^2} + F_{fr} + F_e, \tag{1.33}$$

$A_A \cdot p_A$ being the force in area A_A caused by the pressure in chamber A (p_A), $A_B \cdot p_B$ is the force in area A_B caused by the pressure in chamber B (p_B), and x_p is the piston displacement. F_{fr} is the friction force associated with the cylinder and external load and F_e is the effective force available at the rod piston to move the load. The total mass (M_t) includes the piston mass (M_p) and external mass (load) (M_{ex}).

Equation 1.33 demonstrates that a hydraulic force $(A_A \cdot p_A) - (A_B \cdot p_B)$ is necessary in order to overcome the external forces, friction force and inertia. Therefore, for the piston to achieve a new position or velocity the chamber pressures must change.

The dynamic behavior of the pressure in the chambers is determined by the conservation of mass principle as presented in Section 1.2.2. Hence, applying Equation 1.4 to chamber A (Figure 1.34) the following expression is obtained:

$$q_{VA} = A_A \cdot \frac{dx_p}{dt} + q_{V\sin} + \frac{V_A}{\beta} \cdot \frac{dp_A}{dt}. \tag{1.34}$$

For the cylinder extension, the input flow rate at port A (q_{VA}) leads to a pressure increase (dp_A/dt) caused by the fluid compression. With the pressure increase internal leakage ($q_{V\sin}$) can occur and the cylinder will start to move. The product of the area and velocity ($A_A \cdot dx_p/dt = A_A \cdot v_p$) establishes the chamber volume variation with the piston movement and this volume is occupied by the fluid.

For chamber B the fluid behavior is expressed by Equation 1.35, such that, on the cylinder extension q_{VB} is the flow rate induced by the piston motion at which the fluid exits the cylinder in the direction of a directional valve.

$$q_{VB} = A_B \cdot \frac{dx_p}{dt} + q_{V\sin} - \frac{V_B}{\beta} \cdot \frac{dp_B}{dt}. \tag{1.35}$$

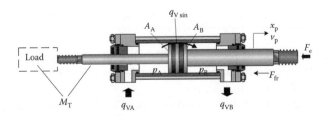

FIGURE 1.34 Parameters and variables associated with a hydraulic cylinder.

One can observe that Equations 1.33 through 1.35 are suitable for any type of cylinder, symmetrical or asymmetrical, single- or double-acting. In the case of symmetrical cylinders, the piston areas are equal ($A_A = A_B$). In the case of single-acting cylinders, the continuity equation is applied only to the controlled chamber. In the other chamber, the pressure is considered to be constant or the spring force is included in Equation 1.33.

1.4.2.2 Cylinder Performance Characteristics

Mechanical efficiency. The mechanical efficiency of the cylinder is the ratio between the theoretical force (hydraulic force) and the effective force available for the external system motion. Since the efficiency characterizes the steady-state performance of the cylinder, the cylinder is considered to have a constant velocity, equal to or differing from zero, and null acceleration. Therefore, the mechanical efficiency can be expressed by

$$\eta_m = \frac{F_e}{F_{tc}} = \frac{F_e}{F_h} = \frac{F_e}{\left(A_A \cdot p_A\right) - \left(A_B \cdot p_B\right)}. \tag{1.36}$$

Volumetric efficiency. Similarly to hydraulic motors, the volumetric efficiency is the ratio between the geometric (theoretical) flow rate and the effective flow rate through the cylinder ports, that is

$$\eta_V = \frac{q_{Vtc}}{q_{Ve}} = \frac{A_A v}{q_{VA}} = \frac{A_B v}{q_{VB}}. \tag{1.37}$$

However, cylinders remain stopped for some periods of time, as they can stay at the stroke end or in a controlled position when enclosed in a closed-loop system. Therefore, aiming to obtain representative values, this efficiency must be calculated beyond these specific operational conditions.

Power and overall efficiency. Considering the cylinder at constant velocity, the useful power (mechanical power) present at the piston rod is

$$P_m = F_e \cdot v. \tag{1.38}$$

Or applying Equation 1.36:

$$P_m = F_{tc} \cdot v \cdot \eta_m = (A_A p_A - A_B p_B) \cdot v \cdot \eta_m. \tag{1.39}$$

The drive power of a cylinder is the net hydraulic power at the cylinder ports, such that

$$P_h = q_{VA} \cdot p_A - q_{VB} \cdot p_B = \frac{A_A \cdot v}{\eta_V} \cdot p_A - \frac{A_B \cdot v}{\eta_V} \cdot p_B. \tag{1.40}$$

The overall efficiency of the cylinder is expressed by

$$\eta_t = \frac{P_m}{P_h} = \eta_V^M \cdot \eta_m^M. \tag{1.41}$$

Natural frequency and dynamic performance. The concept of efficiency is a direct way to evaluate the steady-state performance of a system. Thus, the dynamic performance can also be characterized through a simplified analysis as follows.

As seen above, the hydraulic cylinder behavior is described by differential equations. Dynamic systems like this do not respond instantaneously to an input and a behavior analysis must be carried out according to system control theory.

Most mathematical models of systems can be reduced to a second-order equation, such as that presented in Equation 1.42.

$$\frac{1}{\omega_n^2} \cdot \frac{d^2 y}{dt^2} + \frac{2 \cdot \zeta}{\omega_n} \cdot \frac{dy}{dt} + y = K_{ST} \cdot u, \qquad (1.42)$$

where u is the input, y is the output, ω_n [rad/s] is the natural frequency, ζ [1 (non dimensional)] is the damping ratio and K_{ST} [output unit/input unit] is the steady-state gain of the system [16].

The response time of a second-order system to a step input is shown in Figure 1.35a. Since the abscissa is $\omega_n \cdot t$, these curves show how both the natural frequency and the damping ratio influence the dynamic response.

In Figure 1.35b the time-domain specifications used in hydraulic system design are shown. According to ISO 10770-1 [17] and ISO 10770-2 [18], the response time (t_{re}) is defined as the time required for the response to reach 90% of the final value. The settling time (t_s) is defined as the time required for the response to decrease to and remain at a specified percentage of its final value. The settling time definition is well known from control theory [16] and 5% is the percentage recommended by the standards mentioned above.

The natural frequency can be correlated with the settling time by [16]:

$$t_s = \frac{3}{\zeta \cdot \omega_n} \text{ for 5\% error.} \qquad (1.43)$$

Since there is no algebraic correlation with the time response as defined by ISO 10770-1 [17], one can use the rise time, defined as the time required to change from 0% to 100% of the final value [16]. The expression associated with the rise time is presented in Equation 1.44 [16] and can be used to approximately calculate the natural frequency when the time response is known.

$$t_r = \frac{1}{\omega_n \cdot \sqrt{1 - \zeta^2}} \cdot \arctan\left(\frac{\sqrt{1 - \zeta^2}}{\zeta}\right). \qquad (1.44)$$

On continuing the study of the hydraulic cylinder and its loading, Equations 1.33 through 1.35 can be combined such that, for ports A and B closed ($q_{VA} = q_{VB} = 0$), the system model is

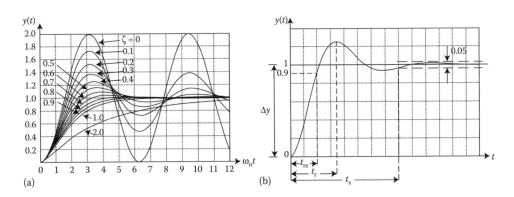

FIGURE 1.35 Response of a second-order system to a unit step input: (a) Influence of the natural frequency and damping ratio; (b) Time-domain specifications.

$$\frac{M_t}{\beta e \cdot \left(\dfrac{A_A^{\,2}}{V_A} + \dfrac{A_B^{\,2}}{V_B} \right)} \cdot \frac{d^2 x}{dt^2} + x = 0. \tag{1.45}$$

Through comparing this equation with Equation 1.42, it can be concluded that the natural frequency of the cylinder with loading is expressed by

$$\omega_n = \left[\frac{\beta_e}{M_t} \cdot \left(\frac{A_A^{\,2}}{V_A} + \frac{A_B^{\,2}}{V_B} \right) \right]^{1/2}. \tag{1.46}$$

Besides Equation 1.46 being valid for asymmetrical double-acting cylinders, it can also be applied to symmetrical double-acting cylinders considering $A_A = A_B$. For application to single-acting asymmetrical cylinders the term related to the non-controlled chamber needs to be excluded (A_A^2/V_A or A_B^2/V_B).

1.4.3 DIRECTIONAL CONTROL VALVES

One of the main functions of the directional control valves is the connection or isolation of one or more flow paths. These valves are identified according to their specific function, as will be presented below, but some characteristics are common to all of them, such as the number of ports, number of positions, and the type of control mechanism [19].

The port means the terminus of a flow path in a component, to which connections can be made. The number of ports refers only to those related to the power flow paths, thus excluding drain and pilot ports. For example, a valve with four ports [19] is commercially identified as a four-way valve.

The number of valve positions refers to the number of pre-defined states in which the valve can operate and it is related to the feasible stable positions of a movable valve element. Designations such as two-position valve or three-position valve are used in valve identifications.

Finally, control mechanisms are devices that provide an input signal to a component. Levers, solenoids, plungers, and pilots are examples of control mechanisms that are used in directional valves.

1.4.3.1 Non-return Valves (Check Valves)

The simplest type of directional control valve is a non-return valve or check valve. Its function is to permit free flow in one direction and prevent flow in the opposite direction. Figure 1.36a shows a simple non-return valve for line mounting which consists of a seat, a poppet, and a spring.

The valve remains closed to the flow until the pressure at its inlet port (A) creates sufficient force to overcome the spring force. Once the poppet leaves its seat, hydraulic fluid is permitted to flow

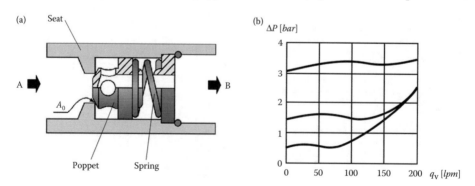

FIGURE 1.36 Single non-return valve: (a) Illustration; (b) Characteristic curve. (From Linsingen, I. von, *Fundamentos de Sistemas Hidráulicos*, 3rd ed., Florianópolis, Brazil: UFSC Ed., 2008. With permission.)

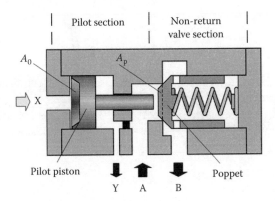

| Pilot section | Non-return valve section |

A_0 A_p

X

Pilot piston   Poppet

Y   A   B

FIGURE 1.37 Pilot-operated non-return valve. (From Linsingen, I. von, *Fundamentos de Sistemas Hidráulicos*, 3rd ed., Florianópolis, Brazil: UFSC Ed., 2008. With permission.)

around and through the poppet to the valve outlet port (B). For this reason, a simple non-return valve can only allow flow in one direction. By changing the spring, cracking pressures between 0.05 MPa (0.5 bar) and 0.5 MPa (5 bar) can be obtained. For special applications, a no-spring version is also available.

In Figure 1.36b, characteristic curves for three different springs are presented. The cracking pressures are 0.05, 0.15, and 0.3 MPa (0.5, 1.5, and 3 bar). In each curve the pressure drop remains basically constant until a specified flow rate. Above this value the load loss in the valve increases and the valve behaves like a fixed orifice, as described by Equation 1.13.

Examples of circuits using non-return valves are shown in Figure 1.5 and 1.11. In Figure 1.12 the pump is designed with two internal non-return valves allowing the fluid suction through one port without fluid return through the other. This type of valve is also enclosed in filters, as shown in Figure 1.12, to prevent line blocking in the case of filter obstruction.

For load holding and in decompression-type hydraulic press circuits, a pilot-operated non-return valve is used. This performs the same function as the simple non-return valve described above. However, in contrast, a pilot-operated non-return valve can be piloted to remain opened when a reverse flow is required. Figure 1.37 illustrates the components of a pilot-operated non-return valve. The valve has two distinct sections—the non-return valve section and the pilot section. The non-return valve section allows free fluid flow from port A to port B while preventing reverse flow from B to A without leakage. However, if a pilot pressure signal is supplied to port X, then a force is applied to the pilot piston, which forces the piston rod against the non-return valve poppet. This force then unseats the poppet, allowing free flow of fluid from port B to port A.

1.4.3.2 Spool-type Directional Control Valves

As presented in Sections 1.4.1 and 1.4.2, the actuators normally have two ports. If hydraulic fluid is pumped into one of the ports while the other is connected to the reservoir, the actuator will move in one direction. In order to reverse its direction of motion, the pump and reservoir connections must be reversed. The sliding spool-type directional control valve has been found to be the best way to achieve this change.

These valves have a cylindrical shaft called a "spool," which slides into a machined bore in the valve housing. The housing has ports to connect the valve to the hydraulic circuit.

The sliding spool-type directional control valves can be designed with different combinations of spool and housing. Therefore, two-way and two-position, either normally closed or normally open valves (2/2 NC or 2/2 NO), are available as well as three-way and four-way, or with more ports with three or more positions and different configurations of valve center positions.

Because of their construction characteristics, these valves present internal leakage, which can be a serious restriction in some applications. The use with pilot-operated non-return valves or counterbalanced valves is a common solution.

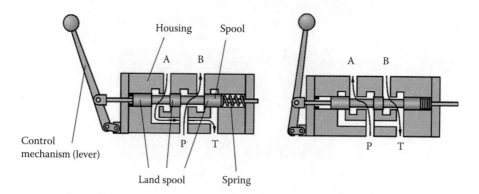

FIGURE 1.38 4/2 sliding spool-type directional control valve. (From Linsingen, I. von, *Fundamentos de Sistemas Hidráulicos*, 3rd ed., Florianópolis, Brazil: UFSC Ed., 2008. With permission.)

Two-position directional control valves. Figure 1.38 shows an illustration of a four-way, two-position, lever-controlled, spring return sliding spool-type directional control valve.

In the solution shown in this figure, the normal position (non-actuated position) establishes the flow paths *P-B* and *A-T*. While actuated by the lever the flow paths *P-A* and *B-T* are maintained.

A common use for such a valve is in a cylinder application which only requires the cylinder to extend or retract to its fullest positions. Another application would be in hydraulic motors, which only run in forward or reverse directions.

Three-position directional control valves. A three-position valve is similar in operation to a two-position valve except that it can be stopped in a third or centered position. While in the centered

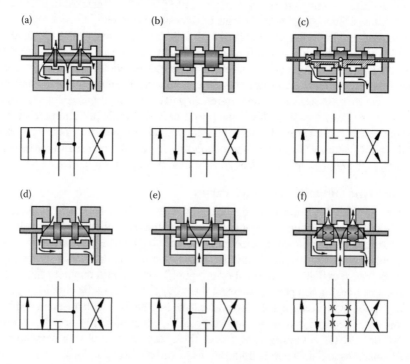

FIGURE 1.39 Typical center flow paths for four-way, three-position valves: (a) Open center; (b) Closed center; (c) Tandem center; (d) Pressure closed center; (e) Reservoir closed center; (f) Restricted open center. (From Linsingen, I. von, *Fundamentos de Sistemas Hidráulicos*, 3rd ed., Florianópolis, Brazil: UFSC Ed., 2008. With permission.)

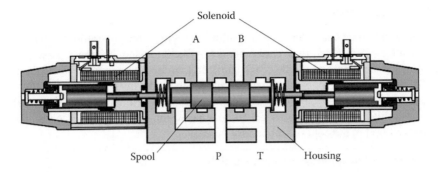

FIGURE 1.40 4/3 directional control valve, directly controlled by two solenoids with spring-centered central position. (From Linsingen, I. von, *Fundamentos de Sistemas Hidráulicos*, 3rd ed., Florianópolis, Brazil: UFSC Ed., 2008. With permission.)

or neutral position, flow may or may not be possible, depending on the spool design of the center position. Figure 1.39 shows some common three-position spool designs.

The open center valve (Figure 1.39a) and the tandem center valve (Figure 1.39c) divert the pump flow to the reservoir keeping the supply pressure low. In the closed center valve (Figure 1.39b), all ports are blocked in the centered position, preventing the actuator movement. At the same time, the pump flow can be used for other parts of the circuit. The restricted open center valve shown in Figure 1.39f avoids both the complete actuator relaxation and peak pressures during the valve commutation.

The pressure closed center design (Figure 1.39d) allows low pressure at ports A and B to be maintained while the reservoir closed center design (Figure 1.39e) means that the supply pressure is applied to both working ports (Figure 1.33a). This has a regenerative effect when an asymmetrical cylinder is used, causing the cylinder to extend rapidly due to the difference in the effective areas at opposite sides of the piston. The cylinder extension velocity is determined by the sum of the pump flow rate (q_V^P) and the flow rate at the rod end of the cylinder (q_{VB}^A), that is, $q_{VA}^A = q_V^P + q_{VB}^A$. When the cylinder chambers are interconnected, the pressure has the tendency to be the same but, as the areas are different, the hydraulic force (Equation 1.33) differs from zero, causing movement.

Control mechanisms, flow and pressure in directional valves. Besides the mechanical control mechanisms, as exemplified in Figure 1.38, hydraulically-controlled and solenoid-controlled valves are common. An example of a solenoid-controlled directional control valve is shown in Figure 1.40.

A typical characteristic curve of directional control valves is the graph of the pressure drop (Δp) versus the flow rate (q_V) through each flow path, as shown in Figure 1.41. This steady-state behavior is described by Equation 1.13 presented above, and shows that the load loss can be different for each valve position (P–A, P–B, A–T, B–T).

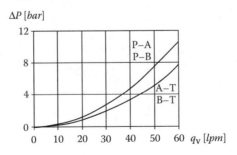

FIGURE 1.41 Characteristic curve of the steady-state behavior of directional control valves. (From Linsingen, I. von, *Fundamentos de Sistemas Hidráulicos*, 3rd ed., Florianópolis, Brazil: UFSC Ed., 2008. With permission.)

1.4.4 Pressure Control Valves

One of the most important characteristics of hydraulic systems is the possibility for pressure control. Besides providing security against overloading, the hydraulic system has the capability of limiting and/or controlling the force and torque of the actuator, thereby avoiding mechanical damage.

Basically, there are two groups of pressure control valves: the *normally closed* (NC) valves and the *normally open* (NO) valves. In the first group the pressure at the inlet port is controlled and in the second the outlet pressure is controlled. In both cases, the valve begins to control the pressure when the pressure set in the control mechanism is reached.

1.4.4.1 Normally Closed Pressure Control Valves

This group includes valves that have the same operational principle but with a few construction differences and which thus can perform different functions in the hydraulic circuit. These are the pressure relief valve, counterbalance valve, unloading valve and sequence valve [20].

The pressure relief valve is usually installed in parallel with the hydrostatic pump and remains closed until the system pressure surpasses the set pressure, when pump flow is partially or completely diverted to the reservoir. Figure 1.10 shows this situation where the fixed-displacement pump (P) runs at a constant rotational frequency, driven by the electrical motor (M), supplying a basically constant flow rate to the circuit. As discussed in Section 1.3, for the effective velocity control of the cylinder (A) the cracking pressure of the pressure relief valve (V1) must be reached and, in this way, the flow rate to the cylinder is reduced.

A typical design of a pressure relief valve is shown in Figure 1.42, which is composed of a poppet held in the valve seat by a spring force. In operation, the flow enters from the bottom of the valve (port A). When the inlet pressure (p_A) reaches the value such that the pressure times the exposed area of the poppet is greater than the spring setting ($F_{k0} = Kx_0$), the valve will begin to pass hydraulic fluid. Note that the spring must be compressed in order for the poppet to move and provide a greater flow area.

Characteristic curves. The steady-state characteristic curve of a pressure relief valve is given in Figure 1.42b, which shows that the inlet pressure increases as the flow rate through the valve increases. The pressure at which the valve first begins to open is called the "cracking pressure" and

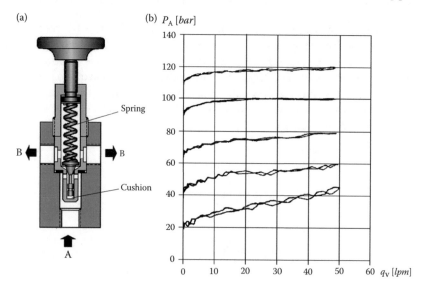

FIGURE 1.42 Directly-operated pressure relief valve: (a) Illustration; (b) Steady-state characteristic curve. (From Linsingen, I. von, *Fundamentos de Sistemas Hidráulicos*, 3rd ed., Florianópolis, Brazil: UFSC Ed., 2008. With permission.)

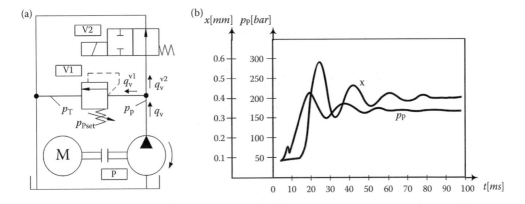

FIGURE 1.43 Dynamic behavior of a directly-operated pressure relief valve: (a) Test circuit; (b) Dynamic response.

it corresponds to the set pressure through the control mechanism (screw) ($p_A = P_{Aset}$). The override pressure is essentially a result of the spring force and flow force in the valve.

The dynamic behavior of a pressure relief valve has a strong influence on the system pressure behavior, as shown in Figure 1.43. Observing the circuit in Figure 1.43a, when the directional control valve (V2) is closed rapidly, the displacement of the valve element (poppet) and the system pressure oscillate as shown in Figure 1.43b. The cushion in the valve shown in Figure 1.42a must be designed to reduce the pressure spikes while at same time reaching the steady state as quickly as possible.

Since the pressure in a hydraulic system is described by the mass conservation principle (Equation 1.4) the pressure behavior is dependent on the circuit fluid volume and the fluid compressibility (bulk modulus) and not only on the valve behavior.

Pilot-operated valve. The pilot-operated pressure relief valve, as shown in Figure 1.44, increases pressure sensitivity and reduces the pressure override normally found in relief valves using only the direct-acting force of the system pressure against a spring element. In operation, the fluid pressure acts on both sides of the piston because of the small orifice through the piston, and the piston is held in the closed position by the light-bias piston spring. When the pressure increases sufficiently

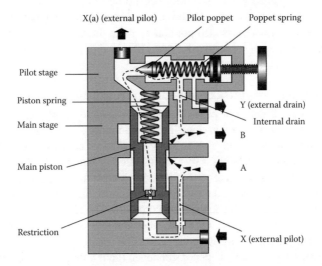

FIGURE 1.44 Pilot-operated pressure relief valve. (From Linsingen, I. von, *Fundamentos de Sistemas Hidráulicos*, 3rd ed., Florianópolis, Brazil: UFSC Ed., 2008. With permission.)

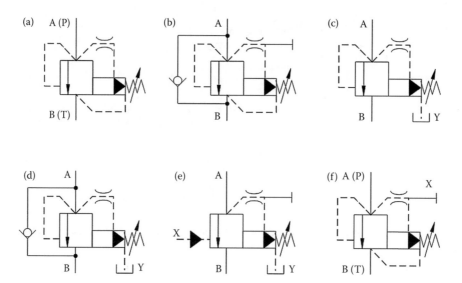

FIGURE 1.45 Pressure control valves according to ISO 5781. (a) Pressure relief valve; (b) Counterbalance valve; (c) Sequence valve; (d) Sequence valve with bypass non-return valve; (e) Unloading valve; (f) Remote-controlled pressure relief valve. (From ISO, ISO 5781 - *Hydraulic fluid power – Pressure-reducing valves, sequence valves, unloading valves, throttle valves and check valves – Mounting surfaces*, Switzerland, 2000. 20p. With permission.)

to move the pilot poppet from its seat, the fluid behind the piston will be directed to a low-pressure area, such as the return line. The resulting pressure imbalance in the piston will cause it to move in the direction of the lower-pressure area, compressing the piston spring and opening the discharge port. This action will effectively prevent any additional increase in pressure. The setting of the pilot-operated pressure relief valve is adjusted by the preload of the poppet spring.

The valve design shown in Figure 1.44 allows different operational configurations. In the configuration presented, the valve can be used as a pressure relief valve and when a non-return valve is incorporated it becomes a counterbalance valve. Closing the internal drain and using the external drain (Y) results in a sequence valve, with the incorporation of a bypass non-return valve being optional. When the internal pilot line is closed and an external pilot signal (X) is used, the valve is utilized as an unloading valve. It is also possible to open the valve at low pressure or promote a remote control using the another external pilot port (X(a) in Figure 1.44). The symbolic representation of these valves is shown in Figure 1.45.

1.4.4.2 Normally Open Pressure Control Valves (Pressure-Reducing Valves)

Pressure-reducing valves (directly- or pilot-operated) are used to supply fluid to branch circuits at a pressure lower than that of the main system. Their main purpose is to bring the pressure down to the requirements of the branch circuit by restricting the flow when the branch reaches some preset limit. One example of pressure-reducing valve is illustrated in Figure 1.46. In operation, a pressure-reducing valve permits fluid to pass freely from port A to port B until the pressure at port B becomes high enough to overcome the force of the spring. At this point, the spool will move, obstructing the flow to port B and thus regulating the downstream pressure. The direction of flow is irrelevant with a pressure-reducing valve, as the spool will close when the pressure at port B reaches the set value. If free reverse flow is required, a non-return valve must be used.

The reduced pressure (p_B) must be kept constant even though there is no flow downstream. Since the valve operational principle is based on the pressure drop control, an internal leakage (port Y) is required so that there is a continuous flow through the control orifice.

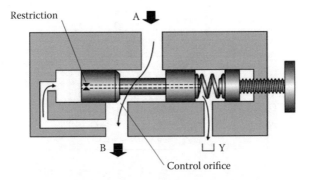

FIGURE 1.46 Directly-operated pressure-reducing valve. (From Linsingen, I. von, *Fundamentos de Sistemas Hidráulicos*, 3rd ed., Florianópolis, Brazil: UFSC Ed., 2008. With permission.)

1.4.5 FLOW CONTROL VALVES

Flow rate control in a hydraulic system is commonly used to control the rod velocity of linear actuators or the shaft rotational frequency of hydraulic motors. There are three ways to carry out flow rate control. One is to vary the speed of a fixed-displacement pump; another is to regulate the volumetric displacement of a variable-displacement pump. The third way is with the use of flow control valves.

Flow control valves may vary from a simple orifice to restrict the flow to a complex pressure-compensated flow control valve or flow divider. In all designs the flow rate control is carried out according to Equation 1.13, which means that the hydraulic energy is dissipated through the valve.

Uncompensated flow control valves. The simplest uncompensated flow control is the fixed-area orifice. Normally, these orifices are used in conjunction with a non-return valve so that the fluid passes through the orifice in one direction, but in the reverse direction the fluid may pass through the non-return valve, thus bypassing the orifice. Another design incorporates a variable-area orifice so that the effective area of the orifice can be increased or decreased (usually manually). One example of a variable-area orifice with a reverse-flow non-return valve is shown in Figure 1.47. These uncompensated flow control valves are used where exact flow control is not critical.

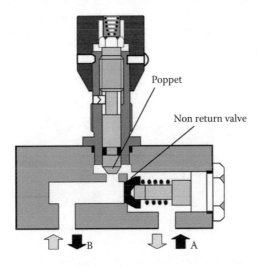

FIGURE 1.47 Uncompensated flow control valve.

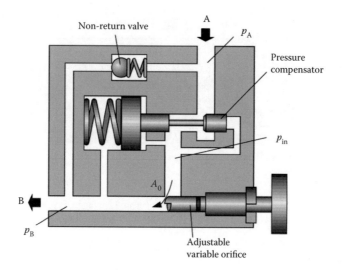

FIGURE 1.48 Example of pressure-compensated flow control valve. (From Linsingen, I. von, *Fundamentos de Sistemas Hidráulicos*, 3rd ed., Florianópolis, Brazil: UFSC Ed., 2008. With permission.)

Recalling Equation 1.13, the flow rate through an orifice is dependent on the pressure drop across the orifice. Therefore, if the pressure differential increases or decreases, the flow will also increase or decrease. To avoid this, a compensated flow control valve must be used.

Pressure-compensated flow control valves. A pressure-compensated flow control valve is shown in Figure 1.48. In this valve, as the pressure differential across the valve from the inlet to the outlet increases, the flow would also increase. However, any increase in flow will be accompanied by a resulting increase in the pressure drop across the control orifice (A_0) ($\Delta p = p_{in} - p_B$). When this pressure differential begins to produce a force larger than the spring preload, the valve spool will shift and the secondary orifice (A_1) will be restricted. These valves normally incorporate a non-return valve for a free inverse flow.

Flow dividers. Flow dividers are also a form of flow control valve. There are at least two types of flow dividers: One is called a "priority flow divider"; the other is a "proportional flow divider." The priority type of flow rate control provides flow to a critical circuit at the expense of other circuits in the system. Figure 1.49 [11] illustrates a priority flow divider. In operation, the flow will enter the priority flow divider from port *B*. When the flow reaches a value and the

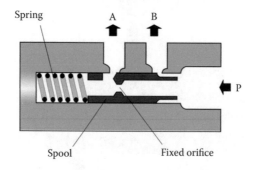

FIGURE 1.49 Priority flow divider. (From Sullivan, J.A. *Fluid Power: Theory and Application*, 2nd ed., USA: Prentice–Hall International, 1982, ISBN 013907668-9. With permission.)

pressure drop across the fixed orifice produces a force larger than that provided by the spring, the spool will move to the left. This action will begin to close the priority outlet port (*A*) and open the secondary outlet (*B*). When the flow rate is below the designed priority flow rate, the spool will be all the way to the right, the secondary outlet will be closed, and the priority outlet will be wide open. The proportional-type flow divider follows the same principle as the priority flow divider, except that two orifices are used and the spool is normally spring-loaded to a particular flow split ratio.

1.4.6 DIRECTIONAL CONTINUOUS CONTROL VALVES

As established by ISO 5598 [19], continuous control valves are valves "that control the flow of energy of a system in a continuous way in response to a continuous input signal.'

Moreover, according to the function performed by the valve in the system, these valves can be classified as directional continuous control valves, pressure continuous control valves and flow continuous control valves.

Observing the directional control valves described in Section 1.4.3, it can be seen that there is an intrinsic possibility for continuous movement of the valve element (typically the spool). However, several of the control mechanisms used for directional control valves, like a solenoid, detent lever, hydraulic pilot, and so forth, only allow the valve to move to specific positions.

With directional continuous control valves, continuous position changing is possible; for example, from the *P-A/B-T* position to the blocked port center position and then to the *P-B/A-T* position.

Directional continuous control valves with mechanical control are well known in mobile hydraulics where the position of the command lever is defined by a human operator based on his or her own observation of the position or velocity of the cylinder or motor.

Valve technology with continuous electrical input started with the servo-valves in the early 1940s [21]. Another notable event was the development of the proportional directional control valves in the late 1970s [22]. Encompassing technological principles from both these valve types, new products are being offered on the market, such as servo-proportional valves [23]. Regardless of their commercial identification or construction principle, according to ISO 10770-1 [17] and ISO 10770-2 [18] these are electrically-modulated hydraulic flow control valves, since they provide a degree of proportional flow control in response to a continuously variable electrical input signal.

1.4.6.1 Servo-valves

Since their beginning in the 1940s, different conceptions have been developed and the two-stage valve is a representative servo-valve concept. The first stage (pilot stage) is composed of either a jet pipe valve or flapper-nozzle valve driven by a torque motor (a permanent magnet, variable reluctance actuator). The second stage is a spool valve, its position being fed back in order to place the torque motor armature at the null position.

Figure 1.50 shows a typical servo-valve with mechanical feedback or force feedback. Other methods of position feedback are the spring-centered spool, direct position feedback or hydraulic follower, and electric feedback using a position transducer [24].

Frequently, the spool slides into a sleeve where the ports were machined. The relative position between the spool lands and sleeve ports then determines the flow control orifices. The same solution is adopted for directly operated valves with electrical feedback, driven by a linear force motor. This valve design is referred to as the "servo-proportional valve" [23,25].

Advances in the manufacturing process and changes in the user requirements have led to changes in the construction details. For example, pilot-operated servo-valves like that shown in Figure 1.50 but without a sleeve are also available.

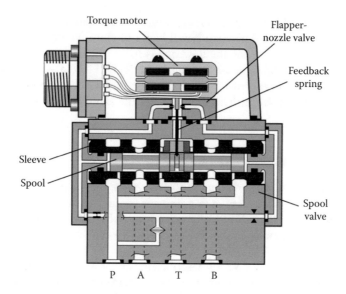

FIGURE 1.50   Pilot-operated servo-valve with mechanical feedback.

1.4.6.2   Proportional Directional Control Valves

The conception of proportional directional control valves comes from two distinct fields: mobile hydraulics and industrial hydraulics. In both cases, the objective was to obtain the same functional characteristics as servo-valves—that is, the continuous control of flow direction and rate, but with a distinct mechanical design.

The proportional valves are controlled by proportional solenoids, which, unlike the torque motor and linear force motor, do not comprise a permanent magnet and the force is provided in only one direction for any current polarity.

Figure 1.51 shows a proportional directional valve, directly controlled by two solenoids, with a spring-centered central position and a spool position transducer. The operation of this type of valve requires an electronic controller/amplifier that receives both the external reference signal and the feedback signal from the position transducer, processes them and sends electrical signals to the solenoids.

There is a significant diversity of proportional directional valves on the market, including valves without feedback position, valves with only one solenoid acting against a spring and valves with controller/amplifier assembled together in the valve (on-board electronics). The metering notches on the spool, as shown in Figure 1.51, can be of different types and are used to define the curve of the flow rate against the spool displacement. However, they are not machined on all valve designs.

Valve designs with spool-sleeve mounting are also available with both smaller machining tolerances and radial clearances. Usually these valves include position feedback optimizing their static

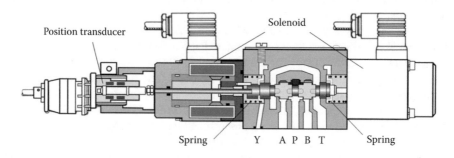

FIGURE 1.51   Proportional directional control valve.

and dynamic behavior. The servo-proportional valve designation has also been used by valve manu-factures for these construction solutions [23,26,27].

1.4.6.3 Fundamental Model and Characteristic Curves

Considering a directional continuous control valve as being the valve itself with the controller/amplifier, on-board or not, its main function is to control the flow rate (output) in response to a input voltage (reference signal).

The valve behavior can be described through the composition of two parts—with feedback or without feedback. The first block corresponds to the transformation of the input voltage into spool displacement. The second one refers to the output flow rate as a consequence of the spool displacement and the pressures in the supply (P), return (T), and working (A and B) ports of the valve (Figure 1.52).

In essence, the valve amplifier controls the current applied to each proportional solenoid or to the pair of coils of a torque motor or linear motor. According to electromechanical principles, this current produces a force (or torque) that is transmitted to a valve element.

In the case of a pilot-operated servo-valve, as shown in Figure 1.50, the torque produces the pipe motion (on jet-pipe valves) or the flapper motion (on flapper-nozzle valves) which, in turn, changes the pressure on the spool sides. The pressure difference makes the resting spool change its position, which is fed back to the pilot valve. In directly-operated valves, as shown in Figure 1.51, the force produced by the electromagnetic actuator is applied directly on the spool.

Based on these principles, a dynamic relationship between the control voltage (U_c) and the spool displacement (x_s) can be expressed by

$$K_{RP} \cdot U_c = \frac{1}{\omega_n^2} \cdot \frac{d^2 x_s}{dt^2} + \frac{2 \cdot \zeta}{\omega_n} \cdot \frac{dx_s}{dt} + x_s, \tag{1.47}$$

where K_{RP} [m/V] is the steady-state gain (ratio between the spool displacement and control voltage in a steady state), ω_n [rad/s] is the natural frequency and ζ [1 (non-dimensional)] is the damping ratio.

The parameter values of Equation 1.47 can be obtained from valve data sheets; for example, from the response time curves shown in Figure 1.53 [28]. Comparing these curves with the general response time of a second-order system (Figure 1.35b), it can be concluded that this valve has a damping ratio (ζ) close to 0.8 and a settling time (t_s) of approximately 50 ms for an input of 50% of the maximum amplitude. Using Equation 1.43, the natural frequency is determined as 53.6 rad/s (8.5 Hz).

The valve catalogs also inform the response time defined according to ISO 10770-1 [17] and shown in Figure 1.35b. The approximate calculation of the natural frequency based on the response time is carried out using Equation 1.44, where $\zeta = 0.7$ can be used when the value is not given in the catalog.

Another way to present the valve dynamic response is through a frequency response diagram (Bode diagram), where it is possible to extract directly the values of the natural frequency and damping ratio [16].

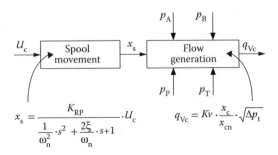

FIGURE 1.52   Block diagram of the directional continuous control valve.

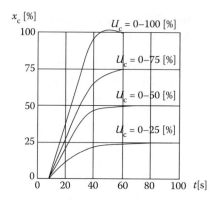

FIGURE 1.53 Response time of a directional proportional valve.

The second block in Figure 1.52 refers to the flow rate control as a function of the orifice opening and the pressures at the valve ports. By applying the concepts related to Equation 1.13, the following expression is valid for directional continuous control valves [14,29,30]:

$$q_{\mathrm{Vc}} = Kv \cdot \frac{x_c}{x_{\mathrm{cn}}} \cdot \sqrt{\Delta p_t}, \tag{1.48}$$

where q_{Vc} [m³/s] is the control flow rate, Kv [(m³/s)/(Pa)$^{1/2}$] is the flow coefficient, x_{cn} [m] is the nominal spool displacement and Δp_t [Pa] is the total pressure drop at the valve.

By combining Equations 1.47 and 1.48, one obtains the general expression for a directional continuous control valve—that is

$$q_{\mathrm{Vc}} = Kv \cdot \left(\frac{1}{\dfrac{1}{\omega_n^2} \cdot D^2 + \dfrac{2\xi}{\omega_n} \cdot D + 1} \right) \cdot \frac{U}{U_n} \cdot \sqrt{\Delta p_t}, \tag{1.49}$$

where $D = d/t$ is the differential operator.

When the valve is under a steady-state condition, this equation takes the following form:

$$q_{\mathrm{Vc}} = Kv \cdot \frac{U}{U_n} \cdot \sqrt{\Delta p_t}. \tag{1.50}$$

The total pressure drop at the valve (Δp_t) corresponds to the pressure drop between the supply port (P) and the return port (T), which, for the flow paths P–A/B–T, is expressed by

$$\Delta p_t = \Delta p_{P-A} + \Delta p_{B-T} = (p_P - p_A) + (p_B - p_T), \tag{1.51}$$

where $\Delta p_{P-A} = p_P - p_A$ is the pressure drop between ports P and A and $\Delta p_{B-T} = p_B - p_T$ is the pressure drop between ports B and T.

For the flow paths P–B/A–T, the total pressure drop is:

$$\Delta p_t = \Delta p_{P-B} + \Delta p_{A-T} = (p_S - p_B) + (p_A - p_T), \tag{1.52}$$

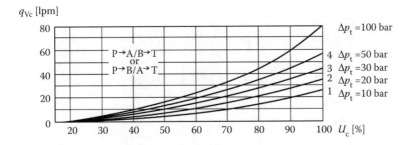

FIGURE 1.54 Flow rate versus input voltage of a directional proportional valve.

where $\Delta p_{P-B} = p_S - p_B$ is the pressure drop between ports P and B and $Dp_{A-T} = p_A - p_T$ is the pressure drop between ports A and T.

The valve catalogs inform the nominal flow rate (q_{Vcn}) at a determined pressure drop that can be either 1 MPa (10 bar), 7 MPa (70 bar), or 1/3 of the nominal supply pressure [17,18]. The nominal flow occurs when the valve is operating with nominal voltage, that is, with the nominal opening. The flow coefficient (Kv [$(m^3/s)/(Pa)^{1/2}$] or [$(lpm/(bar)^{1/2}$]) can be calculated as:

$$Kv = \frac{q_{Vcn}}{\sqrt{\Delta p_{tn}}}. \tag{1.53}$$

The data for the Kv calculation can also be obtained from curves, as shown in Figure 1.54 [28], at 100% of the input signal. In this case, the nominal flow rate presented on the data sheet is 25 lpm@10 bar (41×10^{-3} m^3/s@1 MPa) (which corresponds to curve 1).

It is important to observe that for some valves the nominal flow is specified at a partial pressure drop (Dp_{P-A}) and this must be multiplied by two to allow the flow coefficient calculation.

Constructive aspects of the directional control valves, like different center position arrangements (Figure 1.39) and the existence of symmetrical and asymmetrical designs, are also applicable to directional continuous control valves.

1.4.7 Hydraulic Accumulators

The purpose of a hydraulic accumulator is to store fluid or provide fluid at a certain pressure in order to minimize short-duration pressure spikes or to reach a short-duration high-flow demand. The accumulators used in hydraulic systems can be grouped into three categories: weight-loaded or gravity type, spring-loaded type, and gas-loaded type [31] (Figure 1.55). The weight-loaded type consists of a cylinder with a piston where a mass is attached to its top. The gravitational action on the mass creates a constant fluid pressure, irrespective of the flow rate and fluid volume in the cylinder chamber.

The spring-loaded accumulator simply uses the spring force to load the piston. When the fluid pressure increases to a point above the preload force of the spring, fluid will enter the accumulator to be stored until the pressure reduces. In this type of accumulator, the fluid pressure varies with the piston position and, consequently, with the fluid volume in the accumulator.

The gas-loaded accumulator can be either without separation between liquid and gas, a piston type or a bladder and diaphragm type, as shown in Figure 1.55. In the gas-loaded accumulator, an inert gas, such as dry nitrogen, is used as a pre-charge medium. In operation, this type of accumulator contains the relatively incompressible hydraulic fluid and the more readily compressible gas. When the hydraulic pressure exceeds the pre-charge pressure exerted by the gas, the gas will compress, allowing hydraulic fluid to enter the accumulator. The hydraulic pressure changes with the volume occupied by fluid as a consequence of the pressure gas variation caused by its compression/decompression.

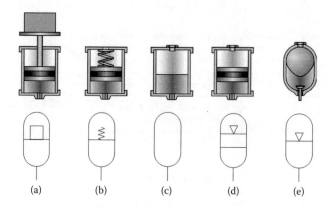

(a)   (b)   (c)   (d)   (e)

FIGURE 1.55 Basic types of accumulators. (From Linsingen, I. von, *Fundamentos de Sistemas Hidráulicos*, 3rd ed., Florianópolis, Brazil: UFSC Ed., 2008. With permission.)

1.4.8 RESERVOIR AND ITS ACCESSORIES

A typical design for an industrial reservoir is shown in Figure 1.56 where the main parts can be identified. The reservoir should be sized to both afford adequate fluid cooling and to enclose a sufficient volume of oil to permit air bubbles and foam to escape during the residence time of the fluid in the reservoir. Commonly, the reservoir is sized to hold at least three times the volume of fluid that can be supplied by the pump in one minute. Baffles are also provided to prevent channeling of the fluid from the return line to the inlet line and the bottom of the return line is usually cut at a 45° angle to assist in the redirection of the fluid away from the inlet.

The reservoir depth must be adequate in order to assure that during peak pump demands, the oil level will not drop below the pump inlet level. Moreover, the pump should be mounted below the reservoir so that a positive head pressure is available at all times. This is critical when water-based hydraulic fluids are used, as these fluids can have a higher mass density as well as a much higher vapor pressure than mineral-oil-based fluids.

Sight gauges are normally used to monitor the fluid level and a cleanout plate is provided to promote cleaning and inspection. A breather system with a filter is also provided to admit clean air and to maintain atmospheric pressure as fluid is pumped into and out of the reservoir. With water-based hydraulic fluids, a pressurized reservoir is recommended. Special breather caps can be installed to vent between 0.005 MPa (0.05 bar) and 0.1 MPa (1 bar). If one of these is used, it

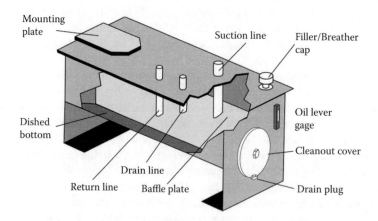

FIGURE 1.56 A typical design for an industrial reservoir. (From Norvelle, F.D. *Fluid Power Technology*, New York, NY, West Publishing Company, 1995. With permission.)

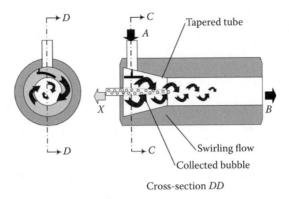

FIGURE 1.57 Bubble eliminator.

must have a vacuum brake to vent at approximately −0.003 MPa (−0.03 bar). This is an important feature to have so that when the reservoir is cooling down, no appreciable vacuum develops in the reservoir. This feature will minimize pump cavitation upon start-up and also prevent a possible reservoir implosion.

Recent trends in industrial manufacturing are to compact machines and equipment in order to economize materials, energy consumption, and required space. A reduction in the size of fluid power systems is encouraged in order to conserve energy and reserve oil. It is somewhat inevitable in designing these systems to minimize the size of the oil reservoir, meaning that the bubbles entrained in the oil may not be removed effectively during the fluid sojourn time in the reservoir. As mentioned above, in order to remove bubbles big vessels are generally used, but it takes a long time to eliminate minute bubbles from fluids by flotation alone.

Another solution is the device shown in Figure 1.57, which has the capacity to eliminate bubbles and decrease dissolved gases using a swirl flow [33,34]. This device, called a "bubble eliminator," consists of a tapered tube where the fluid containing bubbles flows tangentially from the inlet port (port A) and generates a swirling flow. Due to the difference in centrifugal forces created in the swirl flow, the bubbles tend to move toward the central axis (port B) where they are collected and ejected through the vent port (port X).

1.4.9 Filters

As discussed throughout this chapter, hydraulic components are composed of mechanical elements with relative movement and small clearances between them. The hydraulic fluid is expected to create a lubricating film, thereby keeping precision parts separated. Particulate contaminants can break this film, cause erosion on the surfaces or even block the relative movement. Consequently, the hydraulic component life expectancy is reduced, impairing its performance or even causing its complete failure.

The contaminants in hydraulic systems come from several sources, such as the degradation of the circuit components, the external environment, the circuit assembly, and from the new hydraulic fluid which can have a standard contamination level below the system requirements.

The removal of particulate matter and silt from a hydraulic fluid is performed by filters that can be installed at different locations in the hydraulic circuit, characterizing the following types of filtration: suction, pressure, return and off-line filtration [35,36].

Suction line filtration: Suction filters are located before the suction port of the pump and provide pump protection against fluid contamination (Figure 1.58a). Some may be inlet strainers, submersed in the fluid. Others may be externally mounted. In either case, they utilize relatively coarse elements

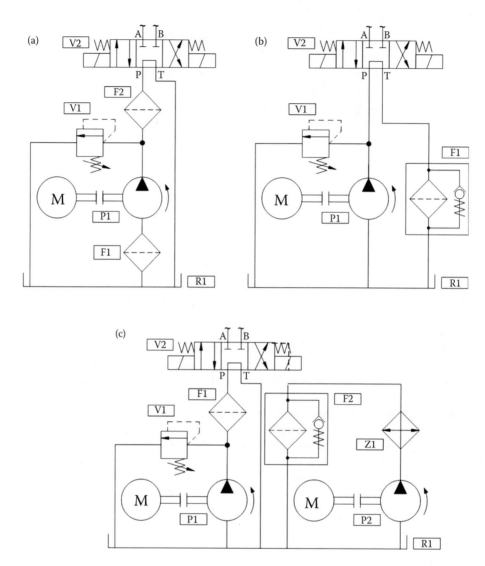

FIGURE 1.58 Types of filtration: (a) Suction filter (F1) and Pressure filter (F2); (b) Return filter (F1); (c) Pressure.

to avoid high pressure drops that can cause cavitation on the pump. Some pump manufacturers do not recommend the use of a suction filter.

Pressure line filtration: Pressure filters are located downstream of the pump (Figure 1.58a and c). They usually produce the lowest system contamination levels to assure clean fluid for sensitive high-pressure components and provide protection of downstream components from pump-generated contamination.

Return line filtration: In most systems, the return filter is the last component through which fluid passes before entering the reservoir (Figure 1.58b). Therefore, it captures wear debris from system working components and particles entering through worn cylinder rod seals before such contaminants can enter the reservoir. A special concern in applying return filters is sizing for a potential flow rate greater than the pump output, since large rod cylinders and other components can cause induced return line flows. Return lines can have substantial pressure surges, which need to be taken into consideration when selecting filters and their locations. The relatively low cost and the cleanliness of the fluid suctioned by the pump are factors that make the use of these filters attractive.

Re-circulating or off-line filtration: Off-line filtration consists of a hydraulic circuit with at least a pump and its prime mover and a filter. These components are installed off-line as a small subsystem separate from the working lines or can be included in a fluid-cooling loop (Figure 1.58c). As with a return line filter, this type of system is best suited to the maintenance of overall cleanliness, but does not provide specific component protection. An off-line filtration loop has the added advantage of being relatively easy to retrofit on an existing system that has inadequate filtration. Also, the filter can be serviced without shutting down the main system.

The circuits shown in Figure 1.10 through 1.12 also present some examples of filter installations. In general, the systems can incorporate multiple filtration techniques, using a combination of suction, pressure, return, and off-line filters.

1.4.10 Hydraulic Fluid

The main characteristic of hydraulic systems, as well as of pneumatic systems, is their requirement that matter flow in such a way as to promote the flow of energy. As discussed in Section 1.1, the hydraulic system must perform three fundamental functions in terms of the energy: primary conversion, limitation and control, and secondary conversion. A fourth function is related to fluid storage and conditioning. This function is required because the fluid must be available for the energy transmission, and since the fluid is continuously in contact with the hydraulic components its proprieties must be controlled.

Fluid proprieties such as viscosity, mass density, vapor pressure, contamination, gas solubility, and bulk modulus change the physical relations modeled by the continuity equation, and conservation of energy, among others. Therefore, besides causing component degradation, the modifying of physical proprieties also changes the hydraulic system behavior.

Throughout the chapters of this *Handbook* the proprieties of different fluids that are used in hydraulic systems are analyzed as well as their effect on the life and behavior of the components.

ACKNOWLEDGMENTS

We are thankfully to Mr. Luis Alberto Galaz Mamani for his assistance with the preparation of the figures shown herein.

REFERENCES

1. Linsingen, I. von, *Fundamentos de Sistemas Hidráulicos,* 3rd ed., Florianópolis, Brazil: UFSC Ed., 2008.
2. Belan, H.C., Szpak, R., Cury, J.E.R. and De Negri, V.J., "Channel/Instance Petri net for structural and functional modeling of industrial equipment," In: *Proceedings of the 20th International Congress of Mechanical Engineering—COBEM 09*, 2009, Gramado Brazil, Brazil: ABCM, 2009.
3. International Organization for Standardization, *ISO 1219-1 - Fluid Power Systems and Components – Graphic symbols and circuit diagrams – Part 1: Graphic symbols for conventional use and data-processing applications,* Switzerland, 2nd ed., 2006.
4. ISO, *ISO 1219-2 - Fluid Power Systems and Components – Graphic symbols and circuit diagrams – Part 2: Circuit diagrams,* Switzerland, 1991.
5. ISO, *ISO 4391 - Hydraulic Fluid Power – Pumps, motors and integral transmissions – Parameter definitions and letter symbols*, Switzerland, 2nd ed., 1983.
6. International Electrotechnical Commission, *IEC 27-1 – Letter Symbols to be Used in Electrical Technology – Part 1: General*, Switzerland, 1971.

7. Frankenfield, T.C. *Using Industrial Hydraulics*, 2nd ed., Penton Publishing, 1985, ISBN-13: 9780932905017.
8. Fox, R.W. and McDonald, A.T., *Introduction to Fluid Mechanics*, 5th ed., John Wiley & Sons, 1998, ISBN: 0471124648.
9. Merritt, H.E. *Hydraulic Control Systems*, John, Wiley & Sons, 1967, New York.
10. Blackburn, J.F., Reethof, G. and Shearer, J.L., *Fluid Power Control*, Cambridge, MA: The M.I.T. Press, 1960.
11. Sullivan, J.A. *Fluid Power: Theory and Application*, 2nd ed., USA: Prentice–Hall International, 1982, ISBN 013907668-9.
12. Dalla Lana, E. and De Negri, V.J., "A New Evaluation Method for Hydraulic Gear Pump Efficiency through Temperature Measurements," In: *SAE Commercial Vehicle Engineering Congress and Exhibition*, 2006, Chicago, *SP-2054 - Fluid Power for Mobile, In-Plant, Field and Manufacturing*. USA: SAE International, 2006, pp. 53-60.
13. Retzlaff, L. and De Negri, V.J., "Performance Analysis of a Load-Sensing Hydraulic System," In: *51st National Conference on Fluid Power (NCFP)*, in conjunction with IFPE 2008, 2008, Las Vegas, *Proceedings* USA: NFPA, 2008.
14. Schwartz, C., De Negri, V.J. and Climaco, J.V., "Modeling and Analysis of an Auto-Adjustable Stroke End Cushioning Device for Hydraulic Cylinders," *Journal of the Brazilian Society of Mechanical Sciences and Engineering*, 2005, V. XXVII, N. 4, pp. 415–425.
15. Valdiero, A.C., Guenther; R., De Pieri, E.R. and De Negri, V.J., "Cascade Control of Hydraulically Driven Manipulators with Friction Compensation," *International Journal of Fluid Power*, 2007, V. 8, pp. 7–16.
16. Ogata, K., *Modern Control Engineering*, 5th ed., Prentice Hall, 2008, ISBN 0136156738.
17. ISO, *ISO 10770-1 – Hydraulic Fluid Power – Electrically modulated hydraulic control valves — Part 1: Test methods for four-way directional flow control valves*, Switzerland, 1998.
18. ISO, *ISO 10770-2 - Hydraulic Fluid Power – Electrically modulated hydraulic control valves — Part 2: Test methods for three-way directional flow control valves*, Switzerland, 1998.
19. ISO, *ISO 5598 – Fluid Power and Components – Vocabulary*, Switzerland, 2nd ed., 2008.
20. ISO, *ISO 5781 - Hydraulic Fluid Power – Pressure-reducing valves, sequence valves, unloading valves, throttle valves and check valves – Mounting surfaces*, Switzerland, 2000. p. 20.
21. Maskrey, R.H. and Thayer, W.J., "A Brief History of Electrohydraulic Servomechanisms" *(Technical Bulletin 141)*, 1978, USA: MOOG ING, p. 7.
22. Henke, R.W., "Proportional Hydraulic Valves Offer Power, Flexibility," *Control Engineering*, April. 1981, pp. 68–71.
23. Penton. *Electrohydraulic valves - Part 1, 2 and 3*, http://www.hydraulicspneumatics.com/200/GlobalSearch/Article/True/6413/.
24. DeRose, D., "Proportional and Servo Valve Technology," *Fluid Power Journal*, March/April 2003, pp. 8–12.
25. Moog Inc. *Electrohydraulic Valves...A Technical Look*, Technical brief, www.moog.com.
26. Atos. Servoproportional valves type DLHZO and DLKZOR. Data sheet, www.atos.com.
27. Hannifin, P., *Servo Proportional Valve DFplus® Pilot Operated*, Data sheet, www.parker.com.
28. Rexroth, B., *RE 29 115/02.02 - 4/2, 4/3 and 5/2, 5/3 proportional directional valves, pilot operated without electrical position feedback Types .WRZ, .WRZE and WRH*, Germany, 2002.
29. Johnson, J.L., *Design of Electrohydraulic Systems for Industrial Motion Control*, Parker Hannifin, USA, 1995.
30. De Negri, V.J., Ramos Filho, J.R.B. and de Souza, A.D.C., "A Design Method for Hydraulic Positioning Systems," In: *Proceedings of the 51st National Conference on Fluid Power (NCFP)*, in conjunction with IFPE 2008, 2008, Las Vegas.
31. Doddannavar, R., Barnard, A. and Ganesh, J., 2005, Elsevier. *Practical Hydraulic Systems: Operation and Troubleshooting for Engineers and Technicians*, 2005, Elsevier, ISBN-13: 978-0-7506-6276-5, p. 240
32. Norvelle, F.D. *Fluid Power Technology*, New York, NY, West Publishing Company, 1995.
33. Suzuki, R., Tanaka, Y., Arai, K. and Yokota, S., "Bubble Elimination in Oil for Fluid Power Systems." In: *International Off-Highway and Power plant Congress and Exposition*, Milwaukee, Wisconsin September 14-16, 1998. *SAE Technical Paper Series 982037*. ISSN 0148-7191.
34. Suzuki, R., Tanaka, Y., Totten, G.E. and Bishop Jr., R.J., "Removing Entrained Air in Hydraulic Fluids and Lubrication Oils," *Machinery Lubrication* Magazine, July 2002.
35. Parker. *The Handbook of Hydraulic Filtration*, 2005, http://www.parker.com/filtration.
36. Schroeder. *Contamination Control Fundaments*, 1999, http://www.schroeder-ind.com.

2 Seals and Seal Compatibility

Ronald E. Zielinski and Christa M.A. Chilson

CONTENTS

2.1 INTRODUCTION

The majority of seals used in industrial applications are either "elastomer contact" or "elastomer energized." Elastomer contact seals are those that rely solely on elastomer contact for sealing. Elastomer energized seals are those that use the elastomer to force a plastic seal element against one of the faces that must be sealed. Because both types of seals utilize an elastomer as an essential

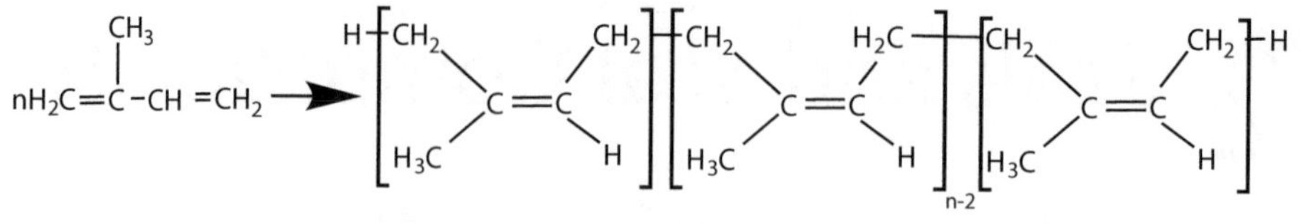

Isoprene (monomer)   Repeat unit structure

FIGURE 2.1 *cis*-Polysioprene (Naultural Rubber).

part of the seal, the choice of a proper elastomeric material and verification of its compatibility with the system's fluids is essential for sealing performance. Therefore, the first topic to be discussed is elastomer materials and fluid compatibility.

The discussion on elastomer contact seals will focus on the O-ring as the principal elastomer seal. The O-ring was chosen because it is still the most widely-used elastomer seal configuration. Specialty seal configurations are derived from the O-ring. These configurations were driven by attempts to improve the seal geometry by addressing O-ring geometry deficiencies. Elastomer energized seals will be discussed because the O-ring is the principal energizer.

Additional subjects to be discussed in this chapter include lip or pressure-energized seals, oil seals, and mechanical seals. Basic principles of seal technology will be provided to facilitate optimum selection seal.

2.2 SEAL CHEMISTRY

In this section, a brief primer on seal chemistry and terminology will be provided in order to facilitate an understanding of the following discussion for the non-chemist.

Elastomeric seal materials are derived from either natural or synthetic polymers. A *polymer*, or *macromolecule*, is a large molecule that is built up by the repetition of smaller molecules or *repeat units*. For example, natural rubber, *cis*-polyisoprene, is synthesized in nature from the *monomer* isoprene (1, 3-butadiene) (See Figure 2.1). Most, but not all, polymers used for seal manufacture have greater than 10,000 repeat units.

One method of characterizing polymers is by the *sequence* that the monomer units are connected together. If the polymer is composed by only one *monomer* (A), it is classified as a *homopolymer* (See Figure 2.2). If the polymer is composed of two or more monomers, it is called a *copolymer*. Copolymers can be further classified according to the sequence of the constituent monomers. For a copolymer composed of monomers A and B, these classifications include the following:

- *Random*: The placement of monomers A and B occur randomly within the polymer chain; for example,
  —A—B—A—A—B—A—B—B—B—A—B—B—A— and so forth.
- *Block*: The monomer units A and B occur together in the polymer, as shown in Figure 2.3.

Polymer properties are dependent on polymer structure. Polymers may exist in a number of *configurations*. The most common are *linear, branched,* and *cross-linked*. Cross-linked polymers are synthesized by introducing chemical linkages between linear or branched polymers to form a three-dimensional *network* structure (See Figure 2.4).

FIGURE 2.2 Homopolymer.

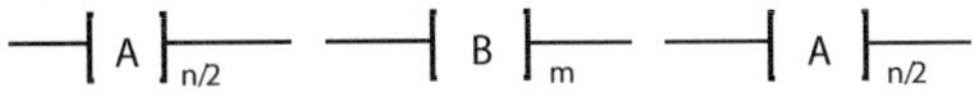

$$\dashv A \vdash_{n/2} \quad \dashv B \vdash_{m} \quad \dashv A \vdash_{n/2}$$

FIGURE 2.3   Copolymer block.

Rubbers or elastomers are high-molecular-weight linear polymers (M_w >1,000,000) which have been minimally cross-linked to eliminate flow. Elastomers typically exhibit long-range reversible extensibility, often ≥ 600% extension, under relatively small applied stress.

Polymers used for seal materials include *natural* polymers such as *cis*-polyisoprene (natural rubber) or *synthetic* polymers, which are produced industrially. With the exception of natural rubber, most seal materials are derived from synthetic polymers.

Some of the monomers used to synthesize polymers used for seals are shown in Figure 2.5.

2.3   ELASTOMER COMPOUNDS

Elastomer materials used for hydraulic applications include natural rubber, poly (*cis*-1, 4-isoprene), or synthetic polymers compounded (mixed) with several different ingredients to facilitate ease of molding and to impart desirable properties in the molded seal [1].

An elastomer compound typically consists of the following types of ingredients:

- Elastomer
- Processing aid
- Vulcanizing agent
- Accelerator
- Activator
- Antidegradant
- Filler
- Plasticizer

Elastomer polymers possess some common characteristics: They are elastic, flexible, impermeable to air and water, and are also tough. However, each elastomer exhibits properties unique to its polymer structure and it is these properties that influence the selection of an elastomer for a specific application.

Processing aids are added to the elastomer compound to aid in mixing and molding the compound. Low-molecular-weight polyethylene can be added to serve as a release agent to facilitate the removal of molded parts from the mold. It can also serve as a lubricant to reduce the inherently high

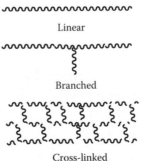

Linear

Branched

Cross-linked

FIGURE 2.4   Polymer configurations.

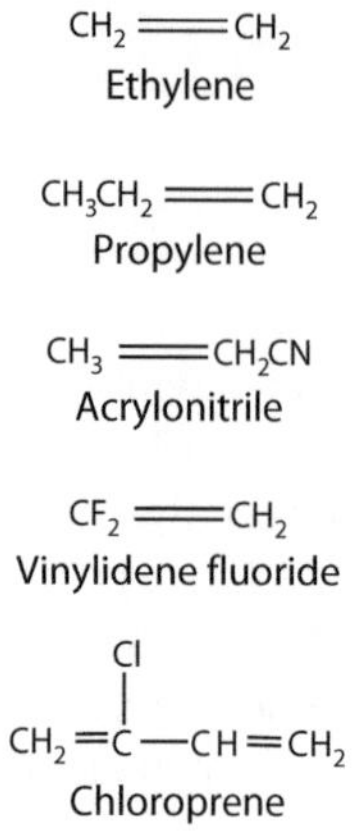

$$CH_2 = CH_2$$
Ethylene

$$CH_3CH_2 = CH_2$$
Propylene

$$CH_3 = CH_2CN$$
Acrylonitrile

$$CF_2 = CH_2$$
Vinylidene fluoride

$$CH_2 = \overset{\overset{\displaystyle Cl}{|}}{C} - CH = CH_2$$
Chloroprene

FIGURE 2.5 Examples of monomers.

friction of the elastomer part. In the molded part, the polyethylene would migrate to the surface and form a lubricating layer between the elastomer and the surface against which it is sealing. Paraffin wax and petroleum hydrocarbons serve the same purpose. Aliphatic-aphthenic-aromatic resins can be added to aid in homogenizing the elastomer compound during mixing.

Vulcanizing agents are necessary additives to thermoset elastomers. They cause the chemical reactions which result in cross-linking of the elastomer. The chemical cross-linking changes the elastomer compound from a soft, tacky material to one that is thermally stable and has a defined set of physical properties: tensile strength, resilience, and elongation.

Accelerators are added to reduce the curing cycle. Most accelerators are organic substances that contain nitrogen and sulfur. The most widely-used ones are thiazoles such as benzothiazyldisulfides.

Activators are used to activate the accelerator and improve its efficiency. Zinc oxide and stearic acid are added with sulfur and organic accelerators to ensure good cross-linking efficiency.

Antidegradants are added to slow down the aging process in molded parts. The antidegradants act as sacrificial ingredients to slow down the attack on the molded elastomer which can result from exposure to oxygen, ozone, heat, light, and metal catalysts. One of the best antidegradants is *para*-phenylene-diamines. Waxes are often used in conjunction with antidegradants. The waxes migrate to the surface of the molded part and form a coating, which protects the part from ozone attack.

Fillers are normally used to reinforce or modify the physical properties of the molded elastomer. Reinforcing filler, usually carbon black or a fine-particle mineral pigment such as fumed silica, is used to enhance hardness, tensile strength, and abrasion resistance. Carbon blacks are the best reinforcing fillers. Non-carbon black fillers are typically used to impart heat resistance. Many oil seals are mineral filled to resist heat build-up at the sealing lip. Fillers may also be used as extenders. Calcium carbonate and talc may be added to the compound to reduce cost and improve processing.

Plasticizers are added to aid in mixing the compound or to provide flexibility in the molded part at low temperature. Ester plasticizers, such as dioctyl phthalate, provide for good low-temperature flexibility. The wrong choice of plasticizer may adversely affect performance. The plasticizer may leach out of the part when it is exposed to the hydraulic fluid which substitutes for the plasticizer, causing high elastomeric volume swell. When plasticizers are removed from the elastomer, a significant change in the elastomer's physical properties may occur and the elastomer may become tacky. This can adversely affect the elastomer's performance.

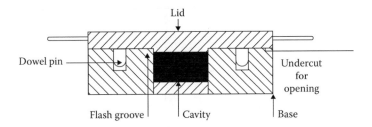

FIGURE 2.6 Illustration of a compression mold.

2.4 MOLDING

Once the ingredients are mixed into the elastomer compound, the fully formulated elastomer is molded into a finished part. There are three primary types of molding processes [1]: compression molding, transfer molding, and injection molding.

Compression molding consists of placing a precut blank, a "preform," into a two-piece mold that is then closed. Pressure is applied to allow the preform to fill the mold cavity. The excess material flows out of the mold cavity into flash grooves. The mold remains under pressure at an elevated temperature for a specified time to allow vulcanization to occur. The pressure is then released and the mold is opened to remove the finished elastomer part. A typical compression mold is shown in Figure 2.4.

Transfer molding, illustrated in Figure 2.6, forces the uncured elastomer from the pot into the mold cavity as pressure is applied to the mold. Although transfer molds are more costly than compression molds, the transfer process exhibits shorter cure times because higher molding pressures result in better heat transfer.

In injection molding (See Figure 2.7), the elastomer stock is temperature controlled as it is forced into the mold. This results in lower vulcanization times. Injection molds are expensive, but for high-volume parts, reduction in cycle time is an economic advantage.

In molding operations, curing or vulcanization is the most important process. The curing process converts the elastomer material from a shapeless, gummy material into a strong elastic product with a definite geometric shape with desired engineering properties. In the curing process, randomly-oriented polymer chains are cross-linked to produce the desired physical properties. Characteristically, a cross-linked elastomer will recover quickly and forcibly [2]. During the vulcanization process, the elastomeric polymer develops improved resistance to degradation by heat, light, and chemical aging.

The most common curing system is sulfur [3]. When sulfur is used as a curative, sulfur atoms form a bond between carbon atoms on adjacent polymer chains. This bond effectively cross-links the molecules as shown in Figure 2.8. It is possible to have multiple bonds, such as —C—S—S—S—C—,

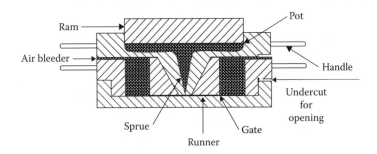

FIGURE 2.7 Illustration of a transfer mold.

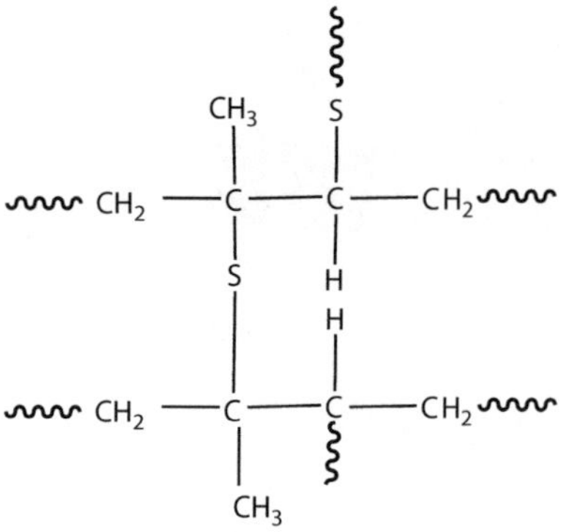

FIGURE 2.8   Sulfur cross-linked natural rubber.

when sulfur is used as the vulcanization agent. Sulfur donors, such as thiuram disulfide and dithiodimorphaline, are also used. These materials make sulfur atoms available for the curing process but limit the amount of sulfur atoms that are available, thus bonds are shorter. Sulfur bonds achieved with sulfur donor cures have only one or two sulfur atoms and are more stable than longer-chain sulfur bonds (Table 2.1).

A non-sulfur cure is required for saturated rubber because it contains no carbon double bonds to accommodate sulfur-bonding [3]. Peroxide curatives promote the formation of radicals to form the carbon-to-carbon bonds on adjacent polymer chains. The peroxide undergoes thermal decomposition to produce oxyradicals:

$$ROOR \xrightarrow{\Delta} 2RO \bullet$$

Peroxide   oxy radicals

The radical interaction abstracts peroxide "oxy radicals" hydrogen atoms from the polymer chain. This results in a free-radical site on the polymer chain as shown in Figure 2.9. Two free-radical sites then couple to form a C—C cross-link as shown in Figure 2.10. Other cure systems also work in this manner. Carbon-to-carbon vulcanization is relatively stable as shown in Table 2.1.

The thermal stability of the peroxide —C—C bond is equivalent to any of the carbon–carbon bonds in the polymer backbone. Peroxide cross-links are stable to oxidation, whereas sulfur cross-links may oxidize and result in cross-link rupture. Peroxide-cured elastomers generally have better compression set resistance and better low-temperature flexibility. On the other hand, sulfur-cured elastomers generally have the better abrasion resistance and tear strength.

TABLE 2.1
Vulcanization Bond Strength

Bond	Disassociation Energy (kcal/mol)
—C—C	80
—C—S—C—	74
—C—S—S—C—	74
—C—S—S—S—C—	54
—C—S—S—S—S—C—	34

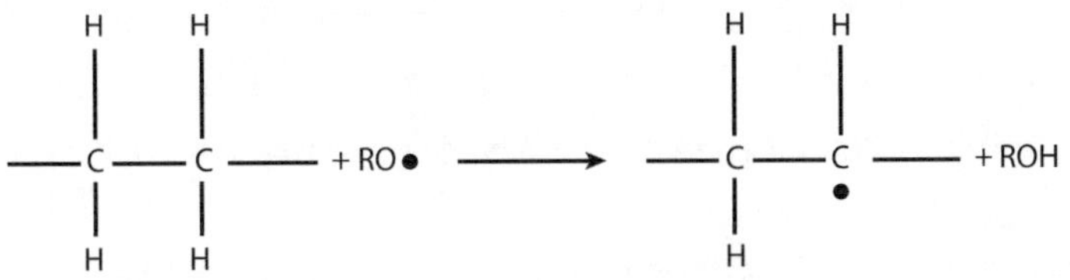

FIGURE 2.9 Free radical formation.

Some curatives may remain unreacted in the elastomer after processing [4]. This can cause further curing when the elastomer is in service, which results in significant property changes.

The choice of vulcanization or curing agent is influenced by elastomer selection. Sulfur or sulfur-donor cure systems are used with natural rubber, isoprene, butyl, ethylene propylene, and nitrile elastomers. Peroxide cure systems are used with urethane, silicone, nitrile, ethylene propylene, and fluorocarbon elastomers. Organic amines are used with fluorocarbon, epichlorohydrin, and ethylene–acrylic elastomers. Metallic oxides are used with chloroprene and chlorosulfonated polyethylene elastomers.

2.5 ELASTOMER SEAL MATERIALS

There are numerous elastomer seal materials [5]. However, nitrile (NBR), fluorocarbon (FKM), ethylene propylene (EPR, EPDM), chloroprene (CR), and urethane (AU, EU) are the most commonly-used materials for industrial applications. The designations for the various elastomers, such as NBR, EPDM, and so forth, were developed in 1955 and are detailed in Reference 5.

2.5.1 NITRILE

Nitrile (NBR) seals are the most widely-used industrial elastomeric seals because of their use temperature of −65°F (−54°C) to 250°F (121°C) and their oil and fuel resistance [6]. The actual temperature range and degree of chemical resistance for particular seals are dependent on the seal's polymer composition and curing agent.

NBR elastomers are emulsion copolymers of butadiene and acrylonitrile (see Figure 2.11). The ratio of these two monomers defines seal performance. NBR is polar because of the presence of the "polar" monomer acrylonitrile (ACN). Acrylonitrile functionality imparts fuel and oil resistance. ACN content may vary from 20% to 50%. High levels of ACN content provide increased strength, decreased gas permeability, and lower volume swell when exposed to fluids. These benefits are obtained at the expense of low-temperature resilience. Low ACN content results in better low-temperature characteristics and higher volume swell [6], as shown in Table 2.2. As there is unsaturation in the butadiene, NBR is also susceptible to oxygen and ozone attack, and to consequent degradation.

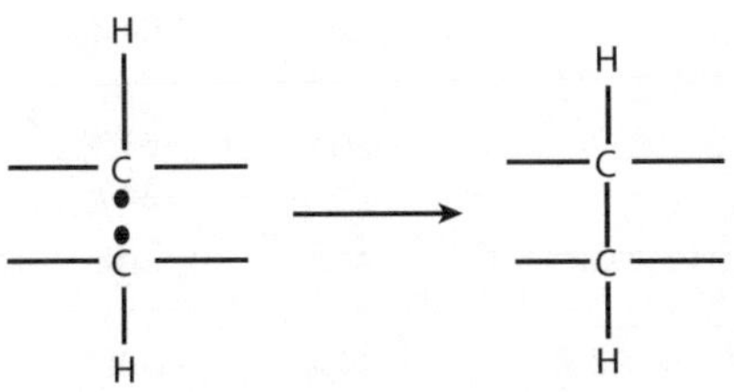

FIGURE 2.10 C—C cross link.

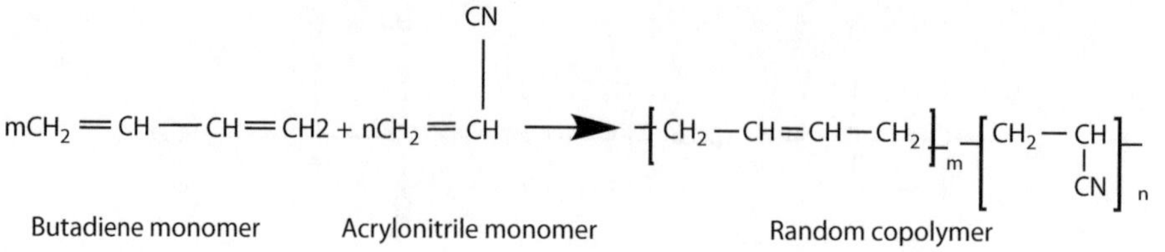

FIGURE 2.11   Nitrile.

In addition to conventional nitrile elastomers, carboxylated nitrile (XNBR) and highly satu-rated nitrile (HNBR) may also be used. Carboxylated nitrile contains carboxyl groups (COOH) distributed in the rubber molecule. This structure change results in an elastomer with superior abra-sion resistance than conventional nitrile elastomers [6] at the expense of compression set resis-tance. Compatibility problems with gasoline and high-swell, petroleum-based hydraulic fluids are observed. However, reducing abrasive wear and prolonging seal life is sometimes a very important consideration in reducing costly downtime between seal replacements.

Nitrile rubber contains unsaturated double bonds, as shown previously. Highly saturated nitrile has pendant ethyl groups. Highly saturated nitrile offers improved aging resistance and high-temperature resistance (300°F). HNBR is prepared by grinding the base NBR polymer, dissolving it in a suitable solvent, and selectively hydrogenating the NBR polymer under pressure with a suit-able catalyst. The solvent and catalyst are then recovered after the formation of HNBR [7]. HNBR may be partially (Figure 2.12a) or fully saturated (Figure 2.12b).

Hydrogenation of medium- and low-ACN NBR reduces low-temperature resilience. HNBR exhib-its equivalent oil, fuel, and solvent resistance and better water, ozone, and oil-additive resistance than conventional NBR. HNBR may be used as an alternative to fluorocarbon elastomers because of its high-temperature capability; however, as with any material fluid, compatibility must be verified throughout the operating temperature range for sufficiently long periods of time using appropriately-sized test samples [8]. Lower temperature and short-term test results are poor models for comparison of elastomer compatibility because they do not adequately reflect the operational environment [8]. Consequently, lower-priced alternatives may not be the most cost-effective solution for long-term use.

2.5.2   Ethylene Propylene

Two types of ethylene propylene (EP) polymers are used: ethylene propylene rubber (EPR)—a random copolymer of ethylene and propylene monomers (See Figure 2.13)—and ethylene propylene

TABLE 2.2
Effect of ACN Content on NBR Elastomers

	ACN Content				
	20%	**28%**	**33%**	**40%**	**50%**
Tensile strength (psi)	2500	2800	2800	3100	2200
Compression set (%), 70 h at 212°F (100°C)	25	27	27	32	40
Low-temp. brittleness, ASTM	−71°F	−57°F	−43°F	−15°F	+5°F
D 746	−57°C	−49°C	−42°C	−26°C	−15°C
Volume change—Fuel B, 4 weeks at room temperature [approx. 70°F (21°C)]	85%	52%	41%	27%	23%
Volume change—ASTM No. 3 oil, 70 h at 212°F (100°C)	61%	32%	21%	10%	7%

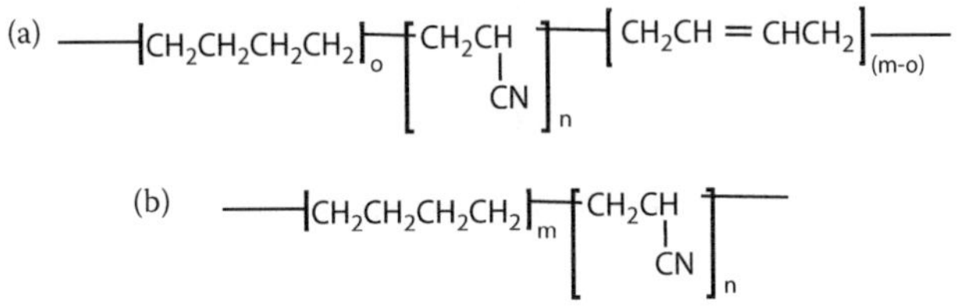

(a) $-\left[CH_2CH_2CH_2CH_2\right]_o\left[\begin{array}{c}CH_2CH\\|\\CN\end{array}\right]_n\left[CH_2CH=CHCH_2\right]_{(m-o)}-$

(b) $-\left[CH_2CH_2CH_2CH_2\right]_m\left[\begin{array}{c}CH_2CH\\|\\CN\end{array}\right]_n-$

FIGURE 2.12 Partially (a) and fully (b) Hydrogenated HNBR.

diene (EPDM) (See Figure 2.14). The diene is the principal variant in this polymer. The ethylene propylene diene usually combines ethylene and propylene monomers with a comonomer such as 1,2 butadiene or ethyldiene norbornene (ENB) as shown in Figure 2.15. A postcure converts the diene into a phenol complex, which is a stabilizer. Other dienes that are used include 1,4 hexadiene and dicyclopentadiene. EP and EPDM elastomers have a useful temperature range of −67°F to 302°F. They possess poor solvent and petroleum oil resistance.

2.5.3 URETHANE

Polyurethanes are either ether-based (EU) or ester-based (AU) [9] (See Figure 2.16). A common polyester is polyethylene adipate as shown in Figure 2.17. A common polyether is polytetramethylene glycol as shown in Figure 2.18. The most common diisocyanates are 2,4- and 2,6-toluene diisocyanate (TDI) and 4,4′-diphenylmethane diisocyanate (MDI).

These elastomers are becoming more popular due to their excellent abrasion resistance in dynamic seal applications. Their principal limitation has been their high-temperature capability. Conventional urethanes exhibit fluid compatibility characteristics equivalent to that of nitrile up to temperatures of 158°F, although they are sometimes rated for service up to 212°F. As operating temperatures increase the frictional heating of the seal causes softening, a loss of physical properties, and accelerated compatibility problems because the higher temperatures accelerate chemical reactions. Water and high humidity may damage urethane elastomers by hydrolyzing the ester, such as —COOC, in the polymer. Polyether-based materials are, by their nature, more hydrolytically stable than polyester-based materials [9]. New grades of polyurethanes are being introduced that provide dry heat resistance capability up to 250°F. However, it is essential that fluid compatibility and physical property retention also be confirmed by testing at the higher temperatures before these materials be considered for use as seals above 158°F.

2.5.4 FLUOROCARBON

Fluorocarbon elastomers (FKM) are categorized as a class of polymers comprising copolymers of vinylidene and hexafluoropropylene, as shown in Figure 2.19, or terpolymers of vinylidene, hexafluoropropylene and tetrafluoroethylene, as shown in Figure 2.20. An exception to this is tetrafluoroethylene/hexafluoropropylene (See Figure 2.21).

These materials can be effective seals over the temperature range of −40°F to 437°F. The actual temperature range for a specific compound, however, is dependent on polymer type and fluorine content of the polymer. By varying the ratio of the three monomers, fluorocarbon elastomers with

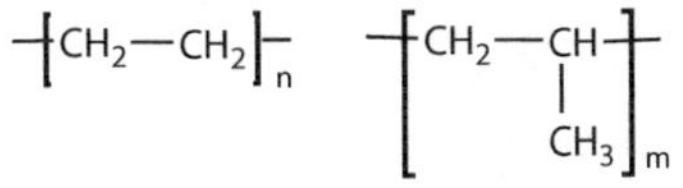

$-\left[CH_2-CH_2\right]_n\quad\left[\begin{array}{c}CH_2-CH\\|\\CH_3\end{array}\right]_m$

FIGURE 2.13 Ethylene propylene rubber (EPR).

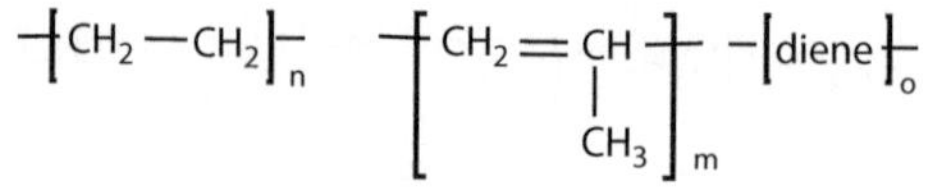

$$\left[CH_2 - CH_2 \right]_n \qquad \left[\begin{matrix} CH_2 = CH \\ | \\ CH_3 \end{matrix} \right]_m \quad -\left[diene \right]_o$$

FIGURE 2.14   Ethylene propylene diene (EPDM).

65% to 70% fluorine content can be produced. The fluorine content of the polymer backbone affects chemical resistance and low-temperature capability, as shown in Table 2.3. In this table [10], T_g is the glass transition temperature. This is the temperature at which an elastomer loses its elasticity and becomes glassy and brittle [11].

Fluorocarbon elastomers can be used with petroleum fluids, diester lubricants, silicone fluids, halogenated hydrocarbons, and some acids. They cannot be used with phosphate esters, amines, ketones, hot water, steam, and brake fluids. However, tetrafluoroethylene/hexafluoropropylene elastomers are being used successfully with newer brake fluids. Fluorocarbon elastomers are relatively high cost and are normally used where high temperature and longer-term compatibility requirements are essential.

2.6 SEAL COMPATIBILITY

2.6.1 EMPIRICAL PREDICTION OF ELASTOMER COMPATIBILITY

The compatibility of an elastomer with hydraulic fluid is a critical factor in system performance and reliability. Although considerable effort has been expended in the attempt to empirically predict seal compatibility, these efforts have not been successful to date. One reason is the inability to accurately model the broad range of environmental conditions and chemical interactions encountered in an actual hydraulic system. Another reason is the inability to adequately define the term "compatibility."

A compatible elastomer is one that will function as an effective seal in an actual hydraulic system for an effective period of time. The property changes experienced by the elastomer exposed to the system's fluid and environment should not prematurely affect its ability to function as a seal.

The empirical approach principally addresses the chemical interaction between the elastomer and the fluid base stock only. This will identify elastomers that are obviously incompatible with the fluid base stock, but it does not address subtle chemical reactions due to the presence of additives, nor does it address the thermal effects that occur in actual systems. Several of the empirical approaches to identifying large variation in seal incompatibility will be provided here.

Fluid interaction with elastomeric polymers has been addressed by Hertz [12]. Most elastomers approximate super condensed gases because they are long-chain polymers: C_2 (alkene and vinyl) and C_4 (diene) structures. Hertz stated: "The C_2 monomers range from ethene ($CH_2 = CH_2$) to ethenyl ($CH_2 = CH$-) or vinyl ethylene (-$CH = CH$-) or vinylene, or ethenylidiene ($CH_2 = C$) or vinylidene.

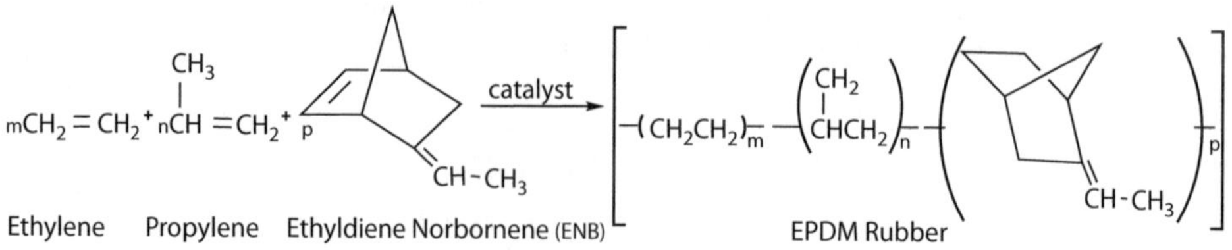

FIGURE 2.15   EDPM formed from ethylene and propylene monomers with ENB co-monomer.

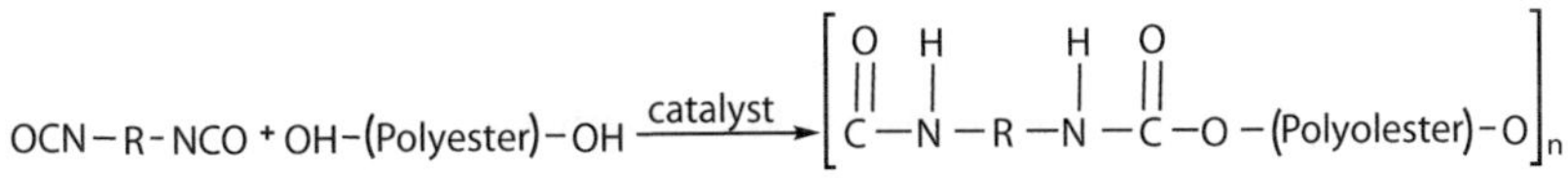

FIGURE 2.16   Polyurethane.

C_4 examples are conjugated dienes, specifically butadiene, 2-methyl-1,3-butadiene (chloroprene)." Because elastomer densities approach those of comparable liquids, the solubility concept of "like dissolves like" is the initial basis for discussing elastomer–liquid compatibility.

Although this simplistic approach initially seems to have merit, Hildebrand and Scott [13] explained that the problem is more complex. They concluded that the heat of vaporization (ΔH_V) minus the volume work (RT) is the estimate of energy necessary to maintain the liquid state. Dividing this value by molar volume (V) corrects for density and the square root of this result is the solubility parameter (δ):

$$\delta = \left(\frac{\Delta H_v - RT}{V} \right)^{1/2}. \tag{2.1}$$

This approach is inadequate except for vapors that obey the ideal-gas law or non-polar fluids (non-electrolytes).

Barton [14] attempted to address polar (aqueous and non-aqueous electrolytes) fluids by assuming that they have three intermolecular forces: dispersion, hydrogen-bonding, and dipole moment. Hansen and Beerbower [15] tried to clarify Barton's concept but were unable to adequately address solubility problems with solvent mixtures. The modification that was proposed, namely:

$$\delta = \delta_d^2 + \delta_p^2 + \delta_H^2, \tag{2.2}$$

would result in only positive or endothermic values for the solubility parameter. A mixture that is exothermic upon mixing is not addressed by this approach because the solubility parameter would be negative. Therefore, this approach was incapable of accurately predicting non-ideal fluid–elastomer interactions.

Jensen [16] noted that most solubility parameter concepts use the historic "similarity matching" of properties rather than the more appropriate "complementary matching" of properties and, consequently, do not address real situations in which elastomers and fluids have complementary properties.

Although considerable effort has been expended in the attempt to adequately predict seal compatibility, these have ignored the broad range of conditions encountered in a hydraulic system. The empirical approach falls short of adequately predicting seal–fluid compatibility in a hydraulic system. The solubility parameter concept is not valid for solvent–solute conditions which are not in thermodynamic equilibrium.

Beerbower and Dickey [17] developed the "swell criterion," which states that if an elastomer swells by 25% or less in a fluid, it is compatible based on these solubility premises. However, this does not address true chemical compatibility, which involves chemical reactions, such as oxidation, not just the physical absorption of fluid. For example, if an elastomer swelled by

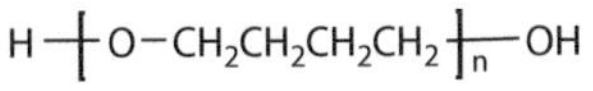

$$H \!-\!\!\left[O-CH_2CH_2CH_2CH_2 \right]_n\!\!-OH$$

FIGURE 2.17   Polyurethane adipate.

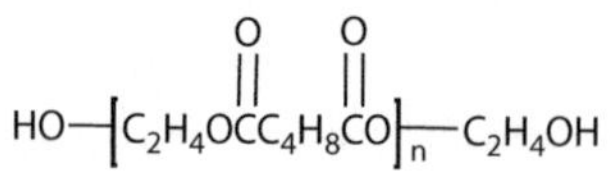

$$HO-\left[C_2H_4O\overset{\overset{O}{\|}}{C}C_4H_8\overset{\overset{O}{\|}}{C}O\right]_n-C_2H_4OH$$

FIGURE 2.18 Polytetramethylene glycol.

25% and had no significant physical property degradation, it would be considered incompatible, even though it would perform as well as a seal. Thus, the fundamental difficulty with solubility theories is the assumption that no chemical reactions occur between the elastomer and fluid [18]. This is a greater deficiency when considering their failure to account for the potential effects that can occur because of high pressures and temperatures encountered in a hydraulic system and how these may affect the actual seals in the system differently than test samples. Many compatibility listings [19] use room-temperature, volume-swell data to define compatibility. Since hydraulic systems may need to function reliably under demanding operational parameters, including temperatures other than room temperature, the user of these charts must determine how the data were generated and carefully assess the relevance of that data to the user's situation.

Unfortunately, there is no universally accepted empirical method for assessing compatibility because elastomers and fluids are not simple two-component systems. Elastomeric polymers are normally multi-component polymers containing other organic materials. Fluids are also complex containing not only base stocks, but performance additives. Additives may interact with the polymers to affect compatibility. The chemical reactions that may occur between all these can also be affected by temperature, pressure, and time. Various thicknesses and configurations of elastomer test specimens may react differently, and thus differ in compatibility, under the same test parameters [8]. This is why compatibility tests should be performed with elastomer test specimens in the seal configuration and size used in the system under test parameters which accurately reflect service conditions; actual fluids, maximum operating temperature and pressure, and long test periods.

2.6.2 ELASTOMER COMPATIBILITY TESTS

Standard elastomer compatibility tests normally involve measuring the physical properties of an elastomer specimen prior to testing in order to obtain tensile strength, elongation, and hardness. Another specimen is soaked for a short period of time in a standard test fluid then removed and tested for volume swell and physical properties. If the volume swell and physical properties fall within a specified range, then the elastomer is considered compatible. Although the test procedure is straightforward, the problem is to properly select test fluids, test times, and property change limits to adequately define compatibility for the system being modeled. Standard test fluids do not contain performance additives of a fully formulated fluid, and these additives may exhibit a deleterious effect on the elastomer.

There is normally no rigorously determined technical basis for the property change limits that have been established. For example, some specifications permit the volume swell to range from a negative number (−8%) to a positive number (20%), whereas others allow a range of 0% to 40% and still others a range of 0.5% to 15% [20–23]. Unfortunately, there is no adequate guidance in the literature that defines meaningful volume-swell ranges. These specifications were typically created using test results from the first elastomer to work in an application and not based upon an

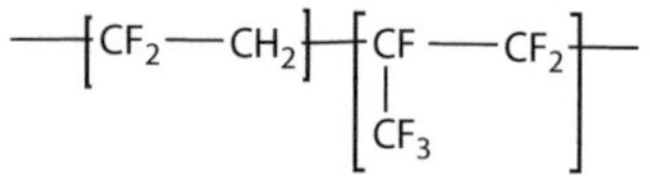

$$-\left[CF_2-CH_2\right]\left[\underset{\underset{CF_3}{|}}{CF}-CF_2\right]-$$

FIGURE 2.19 FKM copolymer.

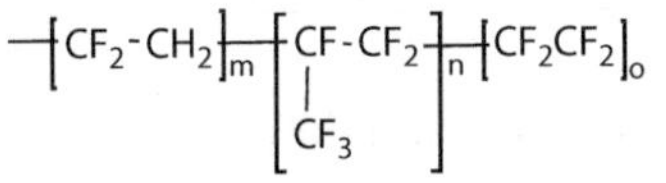

$$-[CF_2\text{-}CH_2]_m \left[\begin{matrix} CF\text{-}CF_2 \\ | \\ CF_3 \end{matrix} \right]_n [CF_2CF_2]_o$$

FIGURE 2.20 FKM terpolymer.

established or rigorously determined technical basis. They ignore such questions as: Was this the best elastomer for the application? What properties really affected performance? If some of the limits were changed, would the result be a better seal? Are the correct properties to accurately assess compatibility being monitored? The importance of using technically established property changes to determine elastomer compatibility is slowly being recognized as new specifications are created. The International Organization for Standardization (ISO) 6072:2002(E) [24] and American Society for Testing and Materials (ASTM) D 6546-00 [25] are two specifications that set property change limits for different soak times in fluids that are the same regardless of the material or fluid being tested. These limits were technically determined.

Elastomer compatibility tests are normally short-term 24 h, 48 h, and 72 h tests. Fluid tests such as ASTM D 943 [26] are normally 70 h or longer tests. If the elastomer degrades in the short-term test, then it is obviously incompatible. If it does not, then compatibility cannot be assumed unless the service life is 72 h or less. Elastomers are principally organic materials and, therefore, undergo aging. An example of misleading information from short-term tests is shown in Figure 2.22. At 24 h, the change in Young's modulus for both elastomer specimens was approximately the same. As the tests continued, the difference in chemical aging of the elastomers is dramatic. It took much longer for one elastomer to become saturated before chemical aging began. In both of these cases, a short-term test would have been very misleading.

Many companies rely on ASTM D 2000 line callouts [27] to specify their elastomers. This standard was developed to classify elastomers into general classes using standard tests and a documented range of values for the properties that are being monitored. Each of the line callouts represents the fact that this elastomer was tested to a particular set of test conditions and the test results were within the limits established for that test. An example of a line callout would be M2BG614B14EO16. Referencing ASTM D 2000, this callout can be translated as follows:

M   Metric units; 2 = grade (medium to low ACN content)
BG   NBR heat-aged for 70 h at 100°C (212°F): in air, to result in a maximum change in elongation of −50%, a change in tensile strength of no more than ±30%, a change in hardness of no more than ±15 points; in ASTM Oil No. 3, to result in a 40% maximum volume swell
6   Hardness (60 ± 5 Shore A); 14 = 14 MPa (2031 psi) tensile strength, minimum value
B14   Compression set test required on sheet sample for 22 h at 100°C (212°F)
EO14 Fluid resistance test required in ASTM Reference Fuel A for 70 h at 100°C (212°F).

The above information categorizes a nitrile elastomer based on a series of standard tests performed in standard test fluids. The property values listed are for sheet material, not seals. The standard test fluids are reference fluids, such as ASTM Reference Fuel A, not fully formulated

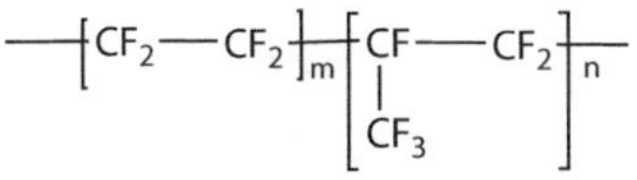

$$-[CF_2 \text{---} CF_2]_m \left[\begin{matrix} CF \text{---} CF_2 \\ | \\ CF_3 \end{matrix} \right]_n -$$

FIGURE 2.21 Tetrafluoroethylene/hexafluoropropylene.

TABLE 2.3
Effect of Fluorine Concentration in the Polymer on its Low Temperature and Chemical-Resistance Characteristics

Property	65% Fluorine	67% Fluorine	69% Fluorine
T_g [°F (°C)]	−8 (−22)	−2 (−19)	+16 (−9)
Volume swell in Methanol (%) [7 days at room temperature, approx. 70°F (21°C)]	120	25	2

hydraulic fluids. The tests are short term, no longer than 70 h. The specified limits are based on the maximum limits at which an elastomer could be considered as nonfunctional. ASTM stresses that these line callouts are not to be used in order to establish compatibility or performance acceptability.

Line callouts are general classification methods and not specification requirements. Another example would be the line callout for the ASTM No. 3 Oil Test Requirement. This test is a 70 h test at 212°F. The acceptable limits are as follows: hardness change of −10 points to 5 points; tensile strength change of −45%, maximum; elongation change of −45%, maximum; volume change of

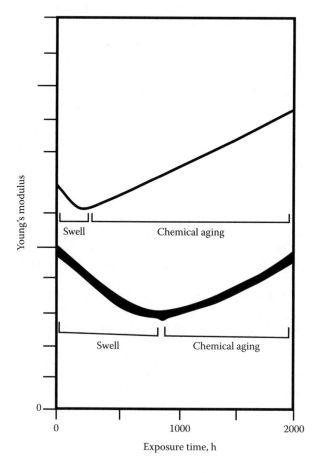

FIGURE 2.22 A comparison of the change in Young's modulus for two different elastomer compounds as a function of time.

0% to 25%. One elastomer has a hardness change of −1 point, a tensile strength change of −6%, an elongation change of −3%, and a volume change of 4%. Another elastomer has a hardness change of −9 points, a tensile strength change of −42%, an elongation change of −37%, and a volume change of 24%. Both of these elastomers qualify for the same ASTM line callout even though they are significantly different and, almost certainly, would not perform the same in service. Table 2.4 presents an ASTM line callout and test data for two elastomers that meet the callouts. Elastomer A performed well in actual dynamic service in a Naphthenic Mid VI fluid, whereas Elastomer B failed shortly after being installed. When the failed elastomer was removed from the seal gland the O-ring

TABLE 2.4
Evaluation of Compounds A and B to ASTM D 2000
M2BG714B14EA14EF11EF21EO14EO34

Properties	Specification	Compound A	Compound B
Original			
Hardness (A)	65 to 75	70	69
Tensile (psi)	2031 (min.)	2220	2050
Elongation (%)	250 (min.)	300	375
Air Aging, 70 h at 100°C			
Hardness change (pts.)	+15 to −15	+5	+11
Tensile change (%)	+30 to −30	+8	+21
Elongation change (%)	−50 (max.)	−10	−24
Compression set, 22 h at 100°C (% of original)	25 (max.)	12	19
Water Immersion, 70 h			
Hardness change (pts.)	+10 to −10	−2	−5
Volume change (%)	+15 to −15	+3	+6
Fuel A 70 h at 23°C			
Hardness change (pts.)	−10 to +10	0	+5
Tensile change (%)	−25 (max.)	−16	−22
Elongation change (%)	−25 (max.)	−10	−17
Volume change (%)	−5 to +10	0	−4
Fuel B, 70 h at 23°C			
Hardness change (pts.)	0 to −30	−6	−12
Tensile change (%)	−60 (max.)	−25	−40
Elongation change (%)	−60 (max.)	−18	−32
Volume change (%)	0 to +40	+17	+27
ASTM No. 1, 70 h at 100°C			
Hardness change (pts.)	−5 to +10	+2	−3
Tensile change (%)	−25 (max.)	−10	−19
Elongation change (%)	−45 (max.)	−8	+5
Volume change (%)	−10 to +5	−4	+4
ASTM No. 3, 70 h at 23°C			
Hardness change (pts.)	−10 to 5	−6	−9
Tensile change (%)	−45 (max.)	−10	−23
Elongation change (%)	−45 (max.)	−5	−18
Volume change (%)	0 to +25	+15	+22

was swollen into a rectangular shape which fully occupied the gland volume. The O-ring was very soft and abraded. This example illustrates that ASTM line callouts do not accurately assess seal compatibility.

2.6.3 GUIDELINES

Basic guidelines that may be used to facilitate the search for suitable elastomers involve short-term testing of the actual elastomer in standard test fluids. From these test results, it is possible to categorize a general compatibility table. General compatibility guidelines for initial selection of the elastomer polymer for the sealing application are presented in Table 2.5 [28,29].

After the elastomer polymer has been selected, specific application problems are then addressed. This is done by reviewing the system requirements and the properties of elastomers that affect sealability. Elastomers are both engineered materials and organic polymers; therefore, their engineering and age-resistance properties are important. The engineering moduli—shear and Young's, elastic recovery, and the chemical degradation of the elastomer are important in assessing compatibility. Engineering moduli address seal stiffness and effectiveness in shearing the fluid film and keeping the fluid in the system.

TABLE 2.5
General Elastomer Compatibility

Polymer	Continuous Service Temp. (°F)	Service Temp. Limits (°F)	Compatible Fluids	Incompatible Fluids
NBR-1[a]	−20 to +210	−20 to +240	Aliphatic hydrocarbons, mineral oils and greases, oil-in-water and water-in-oil emulsions, water–glycol, ethylene glycol, soybean oil, sunflower oil, corn oil	Phosphate esters, aromatic hydrocarbons, halogenated solvents, esters
NBR-2	−40 to +195	−40 to +230	Aliphatic hydrocarbons, mineral oils and greases, oil-in-water and water-in-oil emulsions, water–glycol, ethylene glycol, soybean oil, sunflower oil, corn oil	Phosphate esters, aromatic hydrocarbons, halogenated solvents, esters
NBR-3	−60 to +195	−40 to +230	Aliphatic hydrocarbons, mineral oils and greases, oil-in-water and water-in-oil emulsions, water–glycol, ethylene glycol, soybean oil, sunflower oil, corn oil	Phosphate esters, aromatic hydrocarbons, halogenated solvents, esters
EPDM	−60 to +240	−60 to +300	Phosphate esters, air, acids, bases, water–glycol, steam, water, rapeseed oil, silicon fluids and greases	Hydrocarbons, halogenated hydrocarbons
FKM	−40[b] to +390	−40[b] to +480	Aliphatic and aromatic hydrocarbons, mineral oils and greases, halogenated hydrocarbons, polyglycol, diesters, silicone fluids	Phosphate esters, hot water, steam
HNBR	−20 to +265	−29 to +300	Mineral oils and greases, aliphatic hydrocarbons, air, steam water-in-oil and oil-in-water emulsions, glycol	Aromatic hydrocarbons, acids, halogenated solvents

[a] NBR compounds are listed in order of aliphatic hydrocarbon compatibility, with NBR-1 being the best.
[b] This is the low-temperature limit for the low-temperature grade, −20°F is the acceptable limit for the other grades.

Peacock [30] presents a good explanation of the value of shear modulus in elastomer assessment. Unless there is immediate deterioration of the elastomer, these properties must be evaluated in long-term static soak tests followed by functional tests. They should also be performed in the actual hydraulic fluid at the operating temperature of interest. Testing time should be sufficient for fluid saturation of the elastomer and chemical reactions to occur; this can be as long as 500 h to 1000 h. Elastomer seals of the actual size that will be used in the hydraulic system should be the test specimens.

To illustrate this point, soak tests were performed on two different size EPDM O-rings. The first O-ring (-214) had a cross-section of 0.139 in. and an inside diameter of 0.984 in. The second O-ring (-316) had a cross-section of 0.210 in. and an inside diameter of 0.850 in. The first test involved soaking the O-rings at 225°F in phosphate ester fluid and measuring the time to saturation. The -214 O-ring reached 99% saturation in 19 h, whereas it took 42 h for the -316 O-ring to reach 99% saturation. The second test was performed in the same fluid at 185°F and it took 36 h for the -214 O-ring to reach 99% saturation. Testing times of 79 h for the -316 O-ring to reach 99% saturation were required. If a compatibility test was performed at 185°F for 70 h, the -316 O-ring would not be saturated and chemical interaction would not be accurately evaluated.

Elastomers are organic materials. Compatibility involving chemical processes cannot be fully evaluated until the specific elastomer being evaluated is saturated with the actual fluid of use in an equilibrium condition. This can take several days (or longer) depending on the polymer, the specific compound formulation, the test temperature, pressure, and the service fluid. The test temperature needs to be the actual service temperature or maximum operating temperature in order to accurately fix the chemical system. The test time needs to be a certain period of time beyond saturation in order to allow chemical reactions to occur in the elastomer matrix. This should be 1000 h to ensure that saturation has occurred and chemical reactions have been allowed to continue for a sufficient period of time to represent service time periods. Actual service fluid–elastomer interaction needs to be assessed because of subtleties in compositions.

2.7 PHYSICAL PROPERTIES

Elastomer seals must resist chemical degradation and maintain their engineering properties over their service lifetime. Typically, a stress–strain curve is obtained each time a tensile test is performed. Three parameters [31] are reported from the data: tensile strength, which is the stress at the strain point that the elastomer breaks; it is recorded as the ultimate elongation—how far it stretched before it broke. Another value that is often recorded is the "modulus at 100% elongation." This is the stress at the particular strain point. Stress is the force acting across a unit area to resist separation (tensile), compression (compressive), or shearing (shear). Strain is the change in length of a material in the direction of force per unit undistorted length. Thus, in tensile, as a load is applied to pull the sample apart, strain results as the sample becomes stretched. A typical stress–strain curve is shown in Figure 2.23. Young's modulus is the ratio of the stress applied to the elastic strain which results. Because a force is moving through a distance, integrating under the stress–strain curve would yield the value of work that is being done. In any engineered material, Young's modulus is a key parameter. Young's modulus (E) and shear modulus (G) are engineering values, and for an isotropic solid such as rubber,

$$E = 3G. \tag{2.3}$$

The shear modulus for an unfilled elastomer is defined as [32]

$$G = \frac{\rho RT}{M_c}, \tag{2.4}$$

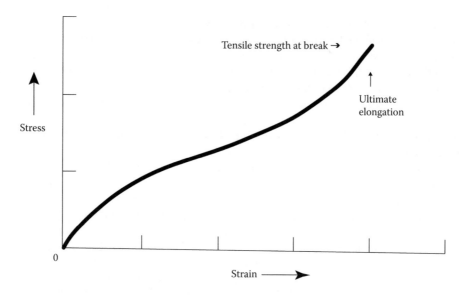

FIGURE 2.23 A typical stress–strain curve for an elastomeric O-ring.

where ρ is the density, R is the universal gas constant, T is the temperature (in K), and M_c is the molecular weight between cross-links. M_c is the variable that reflects the chemical changes that would occur during the aging process and, to some degree, the changes can be monitored in real time by monitoring the value of G before, after, and during the aging process. Work is a product of shear modulus times and the extension ratio, δ. The extension ratio is a nonlinear function:

$$\delta = 1 + \text{elongation.} \tag{2.5}$$

Work can be determined by integrating the area under a stress–strain curve. A finite value of work will give an accurate representation of the changes that are occurring during the aging process. This can be called the work function [30] and can be determined by integrating the area under the stress–strain curve from 0% to 20% elongation. Thus, the work function can be used to assess elastomer aging and the loss of engineering performance.

The elements that should be in a test program include the following: tests should be conducted at the expected maximum operating temperature in the actual operational fluid for a sufficient amount of time to allow elastomer saturation and chemical reactions to occur. Only after saturation can chemical compatibility be assessed. The actual operational fluid is required because assumptions based on test results using standard test fluids without performance additives being present can be misleading. In the studies conducted to date, it appears that 500 h is a minimal test time and tests should be conducted to 1000 h for a full assessment. This will also indicate the elastomer's compatibility over a meaningful service period. It also appears that a change of greater than 20% in the starting value of the work function has a deleterious effect on elastomer engineering performance as a seal [33]. The elastomer test specimens should be O-rings of the appropriate cross-sections to be used in the system in order to adequately assess actual operational situations.

In addition to the chemical or aging changes of elastomers, volume changes must also be considered. Elastomers can swell with the uptake of fluid or they can shrink with the depletion or withdrawal of components in the elastomer compound. Most seal glands are designed to accommodate 15% volume swell of the elastomer [34–36]. If elastomer swelling of beyond 15% occurs, the seal not only fully occupies the gland volume but also flows out of the gland and can become damaged. Excessive swelling leads to softening and abrasion, loss of resilience, extrusion and nibbling of the seal as it flows out of the gland, and, consequently, seal failure. Elastomer seal shrinkage leads to hardening and a reduction in seal force, which results in seal failure. Shrinkage may occur when a system is idle and the seal is allowed to "dry out." If the seal has taken a set during operation and then dries out, it may not function when the system is restarted. If a seal shrinks away from a mating surface and has taken a set or hardened on the surface, when it swells again on exposure to fluid, it may not mate correctly against the surface it is supposed to seal. This results in leakage.

Volume swell and volume shrinkage should also be assessed. Actual volume swell in a gland is normally considered to be approximately 50% of the value obtained in a free volume swell test, where the elastomer is soaked in the free state in fluid. Thus, values of free volume swell of less than 25% are considered acceptable. Allowing for variability and margin of error, a material that exhibits a maximum of 20% free volume swell would be required. Some volume swell is beneficial because it compensates for compression set. Compression set is the tendency of seal materials to return only part of the way to their free state after being compressed; this reduces the sealing force. If an elastomer shrinks in service, the sealing force is reduced and shrinkage, when combined with compression set, always leads to leakage; therefore, no shrinkage should be allowed.

Also, as the elastomer reswells, its surface conformance changes, resulting in wear and leakage. After a soak test, the seal should be allowed to dry out for a 24 h period, and if shrinkage is ≥4% from the swelled value, then there is a high potential for seal leakage.

In summary, the factors for elastomer seal compatibility are as follows:

- Fluid dissolves into the sealing surface and then diffuses into the interior.
- An equilibrium fluid concentration occurs.
- Fluid in the elastomer interacts physically and chemically.
- The elastomer ages through local fracturing of bonds, oxidation, and other chemical reactions.

The compatibility program should address seal volume change, chemical reactions that occur in the seal after fluid saturation, and changes in Young's modulus, and should be conducted in the actual fluid for a sufficient amount of time to assess changes. The test time should be no less than 500 h and the test temperature should be the actual maximum temperature the seal will see in service.

2.8 SEAL COMPATIBILITY TEST PROGRAM

A seal compatibility test program that can be easily performed will be provided here. Mechanical test equipment and laboratory ovens are required. The results will assist in evaluating optimal elastomers for a specific application. The following tests should be performed on test specimens as part of the seal compatibility test program:

- Hardness change
- Volume swell
- Tensile strength change
- Elongation change
- Work function change

During the course of testing, the results from each test should be plotted as a function of time to detect any radical property changes or definitive trends. For a critical seal application, property changes limits as detailed below should be monitored. These limits have been established in the ASTM [33] and National Fluid Power Association (NFPA) [37] compatibility programs and will indicate good chemical compatibility. The compatibility test limits are as follows:

- Hardness change: ± 8 points
- Volume swell: 0 to +15%
- Shrinkage after swell test: −4%, maximum
- Tensile strength change: ± 20%
- Elongation change: ± 20%
- Work function change: ± 12%

If all of the changes are within these limits, the elastomer should be considered compatible.

Once a seal material is found to be compatible, all seals for that system should be ordered by specific compound and not by ASTM D 2000 Line Callout or generic polymer designation. Carefully selecting the initial seal helps to avoid problems later on.

To illustrate the property changes that can be experienced in a compatibility test, AS568-120 [38] NBR O-ring specimens were submitted to the compatibility test program. The O-rings were tested in a naphthenic fluid at 100°C. The average test results for the specimens are shown in Table 2.6. An examination of the complete test results shows several interesting points. The work function value exceeded 12% after just 24 h, even though saturation did not occur until approximately 500 h, as indicated by the volume swell reaching a maximum. The lower the test temperature, the longer it takes to saturation. This is why short-term tests at low temperatures are meaningless. Work function data also showed a reaction trend with the value starting off as a significant negative change, gradually becoming positive until the significant positive change at 1000 h. This indicated that the chemical reaction, once it had begun, continued to develop. If the work function was not monitored, the tensile and elongation data would have not indicated incompatibility until the 500 h test; thus, any short-term test would have indicated full compatibility. Had only the hardness change and volume change been monitored, there would have been no indication of incompatibility. This test sequence illustrates the problem of relying solely on volume-swell and hardness changes as compatibility indicators.

2.8.1 Test Conditions and Specimens

The test fluid should be the actual system fluid and the test temperature should be the maximum seal exposure temperature. Test specimens should be analyzed at 24 h, 70 h, 100 h, 250 h, 500 h, and 1000 h to assess the compatibility of the seal over the minimum service life of the fluid. The test is concluded when the property changes exceed the established limits. The test samples should approximate the cross-section of the actual seal used so that the saturation effect is properly considered. The test samples should be either dash/size code -021, -120, -214, or -320 O-rings per AS568 [38] or ISO 3601-1 [39], which are specifications that identify the actual sizes of O-rings. These have an approximate inside diameter of 1 inch (25 mm) and represent the most popular cross-sections of seals used in industrial systems. The test specimens need to be molded from the same elastomer

TABLE 2.6
Compatibility Test Results for AS568-120 NBR O-Rings

Test Time (h)	Tensile Change (%)	Elongation Change (%)	Work Function Change (%)	Hardness Change (pts.)	Volume Change %
0	0	0	0	0	0
24	+7.5	+2.9	−24.1	−1	+3.61
100	+4.3	−1.3	−12.4	−1	+4.16
250	−13.3	−17.0	−7.0	+1	+4.21
500	−63.9	−54.1	+0.5	+3	+4.7
1000	−73.0	−79.3	+126.0	+8	+2.78

compound as the actual system's seals. There should be at least three test replicates for each sample point and test.

2.8.2 DIMENSIONAL MEASUREMENTS OF SPECIMENS

Prior to performing any tests, dimensional measurements (cross-section, axial thickness, and inside diameter) should be taken at ambient (room) temperature and recorded for each specimen. Care should be taken to ensure specimens are not dimensional-distorted while taking these measurements.

The cross-section may be determined by: micrometer (ball-type anvil); vernier calipers; optical comparator; rotating-type fixture with dial indicator; or visual or laser dimensioning equipment [39]. Measurements should be made at four points 90° apart around the circumference of the specimen; the average for each specimen should be used for calculations.

Axial thickness may be measured by either: micrometer (ball-type anvil); dial indicator with a maximum contact force of 0.29 N (29 gf); or electronic micrometer with a spring force of no more than 0.1 N (10 gf) [40]. Measurements should be made at four points of equal distance apart around the circumference of the specimen; the average for each specimen should be used for calculations.

The inside diameter may be measured by either a stepped cone with diametric intervals not exceeding 2% of the diameter to be measured [40], or calibrated visual or laser dimensioning equipment [39]. Three measurements should be made at starting points 120° apart around the circumference of the specimen; the average for each specimen should be used for calculations.

2.8.3 HARDNESS CHANGE

Measure the hardness per ASTM D 1414 [40] Section 16 using a microhardness tester and record the mean value. Measurements are to be taken before and after exposure to fluid. The hardness change is calculated as:

$$\Delta H = H_1 - H_2. \tag{2.6}$$

where ΔH is the hardness change, H_1 is the hardness before fluid exposure, and H_2 is the hardness after fluid exposure.

The units are given as Shore A points. The Shore hardness scale was developed for elastomers and has no correlation to other hardness scales, although it does overlap a portion of the IRHD scale [41], which is used in ISO specifications and does have a relationship to Young's modulus.

2.8.4 VOLUME SWELL

Apparatus: A quart glass canning jar with a standard two-piece lid to prevent liquid and vapor from escaping.
Heating Device: A forced-air oven with the temperature maintained within at ±1°C (1.8°F).
Test Specimen: The test specimen shall consist of an entire O-ring. The same specimen may be used for all tests with hardness and volume determinations made prior to stress–strain tests.

Procedure

1. Weigh each test specimen in air, M_1, to the nearest 1 mg and then weigh each specimen immersed in water, M_2, at room temperature. It is important that all air bubbles clinging to the test specimen be removed before reading the weight in water. Blot the specimen dry on a lint-free paper towel.

2. Suspend the specimens in the test apparatus (glass jar) by the use of corrosion-resistant wire. Separate the specimens by the use of glass beads or corrosion-resistant washers or by bending small loops in the wire.

3. Suspend the specimen vertically so that 1 inch of test fluid is between the lower extremity of the specimen and the bottom of the apparatus. Add enough test fluid to cover the specimen to at least a depth of 1 inch over the upper extremity of the specimen. Carefully tighten the apparatus's lid.

4. Place the sealed test apparatus in the oven at the test temperature for the appropriate test time. At the end of the immersion period, remove the specimen from the apparatus. Cool the specimen to room temperature by immersing it in a fresh amount of the test fluid for 45 minutes.

5. At the end of the cooling period, remove the specimen from the fluid and blot dry. Weigh each test specimen in air, M_3 and then weigh each specimen immersed in water, M_4, as in step 1.

6. The change in volume is calculated as

$$\Delta V(\%) = \frac{(M_3 - M_4) - (M_1 - M_2)}{M_1 - M_2} \times 100, \tag{2.7}$$

where

M_1 = initial mass of specimen in air (g)
M_2 = initial mass of specimen in water (g)
M_3 = mass of specimen in air after immersion (g)
M_4 = mass of specimen in water after immersion (g)

2.8.5 SHRINKAGE

Apparatus: A screen or oven rack that allows air circulation around the test specimen.
Heating Device: A forced-air oven capable of maintaining the temperature at $23 \pm 1°C$ ($73.4 \pm 1.8°F$).
Test Specimen: The test specimen shall consist on an entire O-ring. The same specimen that was used for the volume-swell test may be used provided that this specimen was not used for the stress–strain tests.

Procedure

1. Weigh each test specimen in air, M_1, to the nearest 1 mg and then weigh each specimen immersed in water, M_2, at room temperature. It is important that all air bubbles clinging to the test specimen be removed before reading the weight in water. Blot the specimen dry on a lint-free paper towel.

2. Place each test specimen on the oven rack and maintain the oven at the test temperature for 22 ± 0.25 h. At the end of the required period, remove the specimen(s) from the oven and allow to air cool.

3. Weigh each test specimen in air, M_5, and then weigh each specimen immersed in water, M_6.

4. The change in volume or shrinkage is calculated as:

$$\Delta V(\%) = \frac{(M_3 - M_4) - (M_1 - M_2)}{M_1 - M_2} \times 100, \tag{2.8}$$

where

M_3 = initial mass of volume-swell specimen in air (g)
M_4 = initial mass of volume-swell specimen in water (g)

M_5 = mass of specimen in air after drying out (g)
M_6 = mass of specimen in water after drying out (g)

2.8.6 TENSILE STRENGTH CHANGE

Testing Machine: The testing machine shall conform to the requirements as specified in Section 3 of ASTM D 412 [42], with the exception of grips. Grips for testing O-rings shall consist of ball-bearing spools at least 0.35 in. in diameter and be capable of being brought within 0.75 in. center-to-center distance at closest approach. *(Note: These grips should be readily available from the manufacturer of the testing machine.)* Stresses within the specimen shall be minimized by rotating one spool of the grips or by lubricating the contact surface of the spools with castor oil.

Test Specimen: The test specimen shall consist of an entire O-ring.

Procedure

1. Measure each test specimen to determine its axial thickness (See Section 2.8.2 above).
2. Bring the grips close enough together so the specimen can be installed without stretching. Separate the grips to remove any slack in the specimen. Care should be taken that no load is placed on the specimen.
3. Pull the specimen at a rate of 20 inches/min. Record the breaking force value, *F*, at the time of rupture.

Calculations

Tensile strength

$$T = \frac{F}{A}, \tag{2.9}$$

where
  T = tensile strength (psi)
  F = breaking force (lb.)
  A = twice the cross-sectional area calculated from axial thickness, W (in.^{2}), as follows:

$$A = \frac{\pi W^2}{2} = 1.57W^2, \tag{2.10}$$

Tensile strength change

$$\Delta T = \frac{T_2 - T_1}{T_1} \times 100, \tag{2.11}$$

where
  ΔT = tensile strength change (%)
  T_2 = tensile strength after immersion
  T_1 = tensile strength prior to immersion

2.8.7 ELONGATION CHANGE

Testing Machine: Same as that used for tensile strength change (see Section 2.8.6).
Test Specimen: Same as for tensile strength change (see Section 2.8.6).

Procedure: Same as for tensile strength change (see Section 2.8.6), except record the center-to-center distance between the spools at rupture to the nearest 0.1 in. and record the value for D.

Calculations

Ultimate elongation

$$E(\%) = \frac{2D + G - C}{C} \times 100, \tag{2.12}$$

where

D = distance between centers of the spool grips at the time of rupture of the specimen
G = circumference of one spool (spool diameter × 3.14)
C = inside circumference of the specimen (inside diameter × 3.14)

Change in elongation

$$\Delta E(\%) = \frac{E_2 + E_1}{E_1} \times 100, \tag{2.13}$$

where

E_2 = elongation after immersion
E_1 = elongation prior to immersion

2.8.8 WORK FUNCTION MODULUS CHANGE

Testing Machine: Same as for tensile strength change (see Section 2.8.6).
Procedure: Same as for tensile strength change (see Section 2.8.6).
Calculations: Calculate the work function (WF) as the energy per unit volume at 20% elongation. This value is determined as the area under the stress–strain curve from 0–20% strain.

Change in work function

$$\Delta WF = \frac{WF_2 + WF_1}{WF_1} \times 100, \tag{2.14}$$

where

ΔWF = change in work function (%)
WF_2 = work function after immersion (psi)
WF_1 = work function prior to immersion (psi)

2.9 SEALS

2.9.1 INTRODUCTION

A seal is used between two surfaces to maintain a pressure drop from a positive fluid (liquid or gas) pressure area to a relatively lower fluid pressure area for a specific amount of time [43]. A seal's effectiveness is usually thought of in terms of leakage. For example, a zero leak seal is one that theoretically allows no leakage across its interface or sealing contact surface. However, leakage is only one aspect to a seal's effectiveness. Seal effectiveness should be measured using several requirements. Good seals should have the following characteristics:

- Integrity: Must not leak;
- Reliability: Must ensure system integrity for a desired period of time;
- Long life: Must minimize downtime for change-outs;
- Ease of installation: Must be easy to install and replace; and
- Compatibility: Must be compatible with the fluid it seals throughout the operational range of pressures and temperatures.

The earliest seals were nothing more than stuffing boxes packed with waxed string (or cord), leather strips, pieces of rubber, cotton, wool, or even paper (see Figure 2.24). If leakage was observed, more material was stuffed in until the leakage stopped or became acceptable. These sealing methods were acceptable because the inefficiency of the seal was a small contributor to the overall inefficiency of the equipment being used.

The first advance in sealing technology was the development of V-Rings (Chevron Rings) and U-Cups (See Figure 2.25). These seals are reasonably effective but require hardware modifications, which may be complex. Another disadvantage is that these seals are unidirectional because they are pressure-activated. Thus, two seals are needed to seal in both directions. As pressure is increased, more V-seals must be added. Thus, three or more V-seals can be required to seal system pressures in excess of 2000 psi. U-Cups typically have a pressure limitation of 1500 psi.

The next advance in sealing technology was the O-ring (See Figure 2.26). The O-ring was patented in the nineteenth century. The original application called for use of the O-ring in a very long groove. The O-ring was intended to roll during the motion of the parts and act like a wiper and seal barrier. Around 1940, an elastomeric O-ring was developed. The sizes of the O-rings were then standardized for universal use [20,38,44–46]. The simple design allowed the O-ring to be used both as a static and dynamic seal. Because it was a compression seal, it was also a bi-directional seal in that it was elastic and distorted with pressure and acted as an effective barrier to prevent leakage. During the Second World War, the O-ring became the seal of choice because the war effort forced the simplification and standardization of hydraulic components. During this time, many new applications were developed for O-rings because of their simplicity and effectiveness in sealing.

Currently, the majority of seals contain elastomer components. This is true because elastomers are still very effective in maintaining system integrity. Whether the seal is an elastomer, such as an O-ring, square ring, U-cup, and so forth, or a cap seal which has a plastic component energized

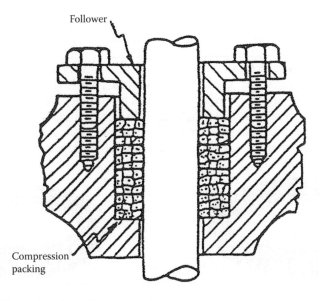

FIGURE 2.24   An illustration of a stuffing-box sealing arrangement.

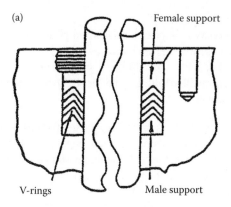

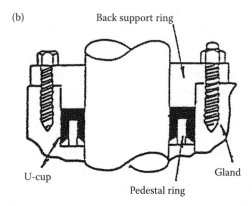

FIGURE 2.25   (a) Illustration of chevron V-rings used as a static seal. (b) Illustration of a U-cup used as static seal.

(squeezed against the surface to be sealed) by an elastomer, the elastomer is an essential component of the majority of seals that are used in today's systems.

2.9.2   Seal Types

2.9.2.1   Static Seals

A static seal is one that maintains a pressure drop across two stationary surfaces. Although the surfaces are considered stationary in theory, in fact none are stationary due to vibration, seal set, and so forth. Static seals are sometimes referred to as joints or gaskets because they are predominantly face seals in that they seal axially in a plane located at right angles to the centerline of the seal (See Figure 2.27). Static seals can also be radial seals where they seal on their inner diameter (ID) as rod type seals or outer diameter (OD) as piston-type seals. The requirements for static seals generally are not as rigorous

FIGURE 2.26   An O-ring profile.

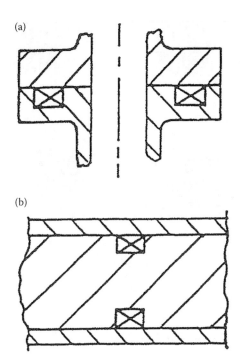

FIGURE 2.27 (a) Illustration of the seal gland location for a static face seal application. (b) Illustration of the seal gland location for a static rotary seal application.

as for dynamic seals: Groove finish can be rougher; higher squeeze is used, as friction is not a factor; the ID stretch for radial seals can also be slightly greater, by up to 8% [36]. Static seal geometries include O-rings, square rings (lathe cuts), D-rings, X-rings, T-seals, and cap seals (See Figure 2.28). Many other geometries are also used. For large static seals, there is a tendency to use extruded cord stock. This stock is cut to size and bonded together to form a ring. The profile can vary. This lowers cost because a static seal is typically not a high-performance seal. While extruded material can be used, it does not have the same physical properties as molded seals. Extruded material has lower physical properties than molded material because it is processed differently. The resulting lower density, strength, resistance to compression set, and different compatibility behavior can adversely affect sealability. If extruded stock is to be used, it should be tested in the fluid of use at the temperature of operation to verify that it will perform satisfactory service. Compression set and low temperature elasticity should also be verified.

The most common causes of static seal leakage are installation damage, extrusion, sealing finish, overfill, splice, poor seal fit, and use of the wrong seal material [43]. Installation damage can occur to any seal and care has to be used during installation in order to ensure that it is installed correctly and is not cut, twisted, or pinched during the process. After the seal has been installed, correct installation should be verified before the hardware is assembled. This caution saves time and money.

Extrusion, or the tendency of the seal material to flow out of the seal gland, is normally associated with dynamic seals. A static seal is also dynamic because relative movement occurs between hardware components, and pressure pulsations can force the static elastomeric seal into the clearance between the sealing faces. To avoid this, the gap between the sealing faces is minimized. A harder elastomer should be used as the static seal; caps and backup rings can also be used. As pressure is increased, backup rings should be considered. Again, the initial incremental cost is low when measured against downtime costs.

Surface roughness of the seal gland and mating surfaces has a significant affect on seal life and performance [36]. While finer (smoother) finishes are usually more of a concern with dynamic seals, it is still important to ensure a proper mating finish for static seals. Sealing finishes should not

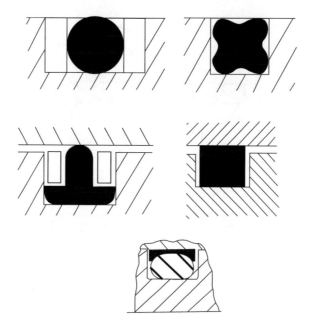

FIGURE 2.28 Illustration of the various static seal types.

exceed 63 micro inches Ra (arithmetic average roughness), which is the arithmetic average height of roughness-component irregularities from the mean line measured within a sampling length, and preferably should be 32 micro inches Ra or better. An elastomer seal must conform to the surfaces it is sealing, and if the sealing surface is very rough, conformance will not occur and leakage will result. If cap seals are used, harder cap material cannot conform to real rough finishes with radical peak-to-valley ratios (the ratio of the highest point on a surface to the lowest point).

Overfill is also a problem. The elastomer fills only a portion of the seal groove. As a result of swell upon exposure to fluid, the elastomer expands into the void area of the groove. If the groove becomes over occupied, the elastomer is forced into the gap between the sealing surfaces where cutting and nibbling can occur. Fluid exposure results in elastomer swell and softening and this can accelerate seal failure. Initial occupancy of the gland should not exceed 85%. Low occupancy (less than 60%) could also be a problem, especially for an O-ring, allowing the seal too much room to move in the gland.

For extremely large-sized seals used in static applications, it is economical to use extruded rubber stock cut to size and bonded together at the splices to form a seal ring. However, splice failures frequently occur [47]. Often, no attention is paid to the splice. There are many different ways to splice elastomeric materials, and the splice geometry, bonding technique, and bond strength must be addressed so that the seal withstands operational conditions. Leakage and operational safety require that care be exercised in choosing spliced seals. The specific application and system requirements can be discussed with the seal supplier.

A compatibility test such as that described above is a wise investment if optimum performance and zero leakage are requirements. By verifying the bond strength of the splice on a test sample before and after exposure to the operational temperature and fluid, it can be verified that the correct seal has been chosen for the application. This test also serves to characterize the compatibility of the elastomer material.

Although the O-ring has been the most popular shape for static seals, some engineers suggest that square rings are more effective static seals because they permit a wider seal contact area and more insurance for zero leakage [48]. The square ring was initially proposed as a dynamic seal to overcome the tendency of an O-ring to roll and twist. Because of its geometry, the square ring creates a wider footprint, and in dynamic situations with identical squeeze, higher friction results.

This is not the case in static sealing conditions where friction is a not an important factor. In static systems, the O-ring is still the most popular seal shape, especially because it is easier to extrude and bond a circular shape. The square ring, however, can offer some advantages if it is prepared from a good elastomeric material and it is proven to be compatible in long-term tests. The square ring should also be more extrusion-resistant because of its geometry.

For square rings or O-rings in face seal applications, the groove should be designed to assure seal contact with the low-pressure side of the groove assembly. For internal pressure applications, the maximum seal groove should be the same as the nominal outside diameter of the seal. When pressure is external, the minimal seal groove should be designed to be the same as the nominal inside diameter of the seal. A design chart for O-ring face (axial) seals is presented in Table 2.7 [36]. In this table, the term "squeeze" appears. This is the amount of distortion put on a seal when it is installed in the system. Squeeze is necessary in order to allow the seal to maintain a positive seal force. The seal is actually distorted that amount when it is installed between the sealing members. Other elastomer seals generally use the same guidelines.

O-rings in static piston and rod applications have the same "rules of thumb" and guidelines for groove design and determining seal fit as those for dynamic piston and rod seals. These will be discussed in the next section. Guidelines to help choose the correct seal size and tables containing hardware dimensions for standard size (AS 568/ISO 3601-1) O-rings in static applications can be found in ISO 3601-2 [36].

2.9.2.2 Dynamic Seals

A dynamic seal is one that is used to maintain the pressure drop across two surfaces moving relative to one another. Dynamic seals can be further categorized as ID or rod seals, and OD or piston seals (See Figure 2.29). These are traditionally reciprocating seals. Dynamic seals can also be rotary seals when they prevent leakage between one surface which has an angular movement relative to the other surface (See Figure 2.30) [43].

TABLE 2.7
Static O-Ring Guidelines; Design Chart for O-Ring Face Seal Gland

Dimensions in Inches

O-ring Dash Size/Size Code	Cross-Section		Gland Height	Squeeze		Groove Width		Groove Radius
	Nominal	Actual		Actual	%	Liquids	Vacuum or Gases	
			0.051	0.013	19 to 30	0.126	0.114	0.005
004–050	1/16	0.070 ± 0.003	to	to		to	to	to
			0.054	0.022		0.134	0.122	0.015
			0.079	0.017	17 to 25	0.157	0.142	0.005
102–178	3/32	0.103 ± 0.003	to	to		to	to	to
			0.083	0.027		0.165	0.150	0.015
			0.106	0.025	19 to 26	0.209	0.189	0.010
201–284	1/8	0.139 ± 0.004	to	to		to	to	to
			0.110	0.037		0.217	0.197	0.025
			0.165	0.036	18 to 23	0.299	0.256	0.020
309–395	3/16	0.210 ± 0.005	to	to		to	to	to
			0.169	0.050		0.307	0.264	0.035
			0.224	0.041	15 to 20	0.354	0.335	0.020
425–475	1/4	0.275 ± 0.006	to	to		to	to	to
			0.228	0.057		0.362	0.343	0.035

Note: These dimensions are intended primarily for face-type seals and low-temperature applications.

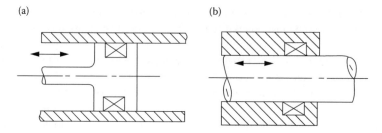

FIGURE 2.29 Illustration of the seal gland locations for (a) dynamic piston and (b) rod seal applications.

Seals can be further categorized as "squeeze-type" or "pressure-activated" (lip-type) seals (See Figure 2.31). The squeeze-type seals are principally elastomeric seals of various geometries. The most commonly used squeeze-type seal is the O-ring. This is because it is readily available in a wide range of sizes and materials; it is a low-cost seal easily installed; and, most importantly, it is a very effective seal. In general, O-rings can be used to effectively seal up to 1500 psi by imparting 15% to 30% squeeze on the ring. Sealing above 1500 psi is possible with the use of 80 to 90 Shore A elastomeric materials and reduced clearance gaps and/or the use of backup rings [43]. The clearance gap, also called the extrusion or e-gap, is the maximum difference between the dimensions of the hardware components creating a gap in which the seal could extrude. For a piston application, this is the maximum difference between the piston diameter, P, and the bore diameter, B:

$$\text{e-gap}_{\text{min}} = \frac{B_{\text{min}} - p_{\text{max}}}{2} \qquad \text{e-gap}_{\text{max}} = \frac{B_{\text{max}} - p_{\text{min}}}{2} \qquad (2.15)$$

For a rod application, this is the maximum difference between the rod diameter, R, and the bore diameter, B:

$$\text{e-gap}_{\text{min}} = \frac{B_{\text{min}} - R_{\text{max}}}{2} \qquad \text{e-gap}_{\text{max}} = \frac{B_{\text{max}} - R_{\text{min}}}{2}. \qquad (2.16)$$

Table 2.8 lists the recommended clearance gaps for elastomer seals with different levels of hardness. These gaps can be widened when backup rings are used with the elastomer. The backup rings are anti-extrusion devices that function as elastomer protection, blocking the flow of the elastomer into the gap where the elastomer can be torn and nibbled. Backup rings should be used when system pressure exceeds 1500 psi.

An elastomer can act as if it remembers the original shape it was molded into and can return to this shape for relatively long periods of time while undergoing deformation due to pressure (memory effect). In low-pressure situations, an elastomer squeezed into a gland has a tendency to maintain

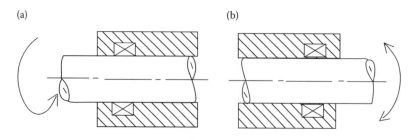

FIGURE 2.30 Illustration of the seal gland locations for (a) rotary and (b) oscillatory seal applications.

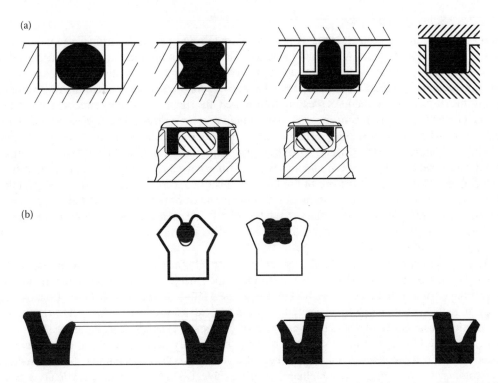

FIGURE 2.31 Illustrations of (a) the various squeeze - (compression) type dynamic seals and (b) typical lip (pressure-activated) seals.

its original shape and this creates the seal. In high-pressure situations, the elastomer is deformed against one side of the gland, but maintains a sealing force against the surface it is sealing. This force is equal to the force deforming it [43].

If an elastomer is treated as an incompressible fluid, its viscosity is directly correlated to its hardness. Thus, an 80 Shore A elastomer would have a higher viscosity than a 70 Shore A material and would exhibit a higher compressive force. Compressive force relates directly to the sealing force that an elastomer exhibits. Compression modulus is the ratio of compressive stress to compressive strain [49]. This is simply compressive force versus deflection.

TABLE 2.8
Recommended e-Gap Between Hardware Components for Different Hardness Elastomer Seals at Different Pressures

Dimensions in Inches

| Pressure (psi) | Elastomer Shore Hardness | | | |
	60A	70A	80A	90A
5000	n/a	n/a	0.002	0.002
3000	n/a	0.002	0.004	0.008
2000	n/a	0.0015	0.0055	0.010
1000	0.002	0.007	0.012	0.018
500	0.010	0.015	0.020	0.025
100	0.030	0.033	0.035	0.036

2.9.2.2.1 Squeeze-Type Seals

The O-ring is a toroidal shaped ring (Figure 2.26). Most O-rings are molded from elastomeric materials to always be consistent in size and tolerance. The O-ring is installed in a groove designed to accommodate it and to allow firm contact with both the inner and outer walls of the groove. Thus, it is squeezed out of round when confined in the groove. As pressure is applied, the ring deforms against the surface to be sealed. As pressure is increased, the tighter the seal becomes (Figure 2.32).

The O-ring is pressure-responsive and seals at high pressures despite minor wear or cuts in the seal or somewhat irregular sealing surfaces. The O-ring is easily installed in one piece glands, thus simplifying engineering designs, and it makes an excellent static seal though it has some limitations as a dynamic seal. In high-pressure applications, O-rings have a tendency to roll because of their high friction. This can limit the life of the O-ring. This effect can be minimized by lowering the frictional characteristics of the elastomer using internal lubes or polymer modification techniques that reduce the friction of the elastomer to the range of PTFE. Friction reduction of the O-ring by minimizing squeeze is not recommended. The sealing force is changed and it is a short-term solution that will produce premature leakage.

At a high pressure, O-rings exhibit a tendency to flow into the space between the hardware elements. This extrusion can lead to nibbling or cutting of the O-ring, resulting in premature failure. To protect the O-ring from this effect, a backup ring is used [50–54]. The sole purpose of the backup ring is to maintain zero diametrical clearance, preventing extrusion of the O-ring. An O-ring with a low-friction elastomer and backup rings can be a very effective dynamic seal (Figure 2.33).

Because the O-ring was developed and became the primary seal in hydraulic and pneumatic systems, other compression-type seals have been designed primarily to accommodate the perceived shortcomings of the O-ring. These include T-seals, X-rings, D-rings, and many proprietary shapes. However, the O-ring has been a key element in the advancement of highly efficient fluid power systems.

The "rules of thumb" or general guidelines for dynamic elastomer seals are the same regardless of geometric shape. These include the following:

- Choose a seal material that provides reliable, functional service at the system's maximum temperature in the system fluid for at least 1000 h.
- Determine the low-temperature seal capability from its T_g and TR_{10} values. Under normal circumstances, the seal should be effective up to 15°F below the TR_{10} value.

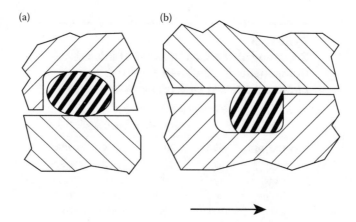

FIGURE 2.32 Illustration of an O-ring's location in the seal gland at (a) zero pressure and when (b) high pressure is applied, forcing the O-ring against the gland wall to seal the gap between the hardware components.

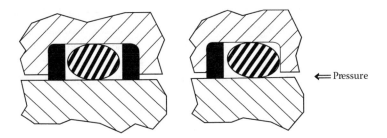

FIGURE 2.33 Illustration of the use of backup rings with elastomer seals. Two backup rings are used when the pressure is reversible and one backup ring is used when the pressure is always unindirectional; this backup ring is positioned at the downstream side of the pressure direction.

- During installation, the elastomer seal should not be stretched beyond 100%. If stretch exceeds 20% during installation, the seal should be allowed to fully recover before final assembly of the hardware.
- Installed stretch of a dynamic seal should not exceed 5%; ideally stretch should be between 2% and 5%.

For piston applications, stretch is the difference between the groove, G, and the seal I.D., ID, and is calculated as follows:

$$\%\text{Stretch}_{min} = \left[\frac{(G_{min} - ID_{Max})}{ID_{Max}}\right] \times 100, \tag{2.17a}$$

$$\%\text{Stretch}_{Max} = \left[\frac{(G_{Max} - ID_{min})}{ID_{min}}\right] \times 100. \tag{2.17b}$$

For rod applications, stretch is the difference between the rod, R, and seal I.D., ID, and is calculated as follows:

$$\%\text{Stretch}_{min} = \left[\frac{(R_{min} - ID_{Max})}{ID_{Max}}\right] \times 100, \tag{2.18a}$$

$$\%\text{Stretch}_{Max} = \left[\frac{(R_{Max} - ID_{min})}{ID_{min}}\right] \times 100. \tag{2.18b}$$

Seals deteriorate under excessive stretching. EPDM, FKM, CR and urethane elastomers are best for high stretch; NBR elastomers are the worst. Excessive stretch leads to the Joule effect [31] and may result in seal leakage. The Joule or Gow-Joule effect can be demonstrated with a thick rubber band and a match. Stretch the rubber band between your fingers and bring the match close to the rubber band. The rubber band might be expected to stretch further. However, the opposite occurs. When a piston seal is stretched excessively, it heats up due to friction and system temperature, pulls away from the bore, and tightens its grip on the piston. The sealing force on the bore is reduced and leakage occurs. In a rod situation, the seal would tighten on the rod and pull away from the base of the groove which would also create more friction.

The cross-section of a seal is reduced by stretch. This is sometimes ignored because significant stretching is thought to produce a better seal based on the belief that stretching exhibits more sealing force on the two sealing elements. However, this should not be ignored, not only due to the

Gow-Joule effect, but due to the fact that the seal cross-section is reduced approximately 50% of the diametral stretch. Thus, a seal that is stretched 16% suffers an approximate 8% reduction in the cross-section. This reduction in the cross-section also affects the installed squeeze of the seal and should be considered when designing sealing systems. For seals stretched by 0% to 3%, the percent of reduction in cross-section, RCS, is determined using the following equation [55]:

$$RCS = 0.01 + 1.06(Stretch) - 0.1(Stretch)^2. \tag{2.19}$$

For example, if a seal is stretched 2%: RCS = 0.01 + 1.06(2) − 0.1(2)² = 1.73%
For seals stretched from 3% to 25%, RCS is determined using the following equation [55]:

$$RCS = 0.56 + 0.59(Stretch) - 0.0046(Stretch)^2. \tag{2.20}$$

For RCS_{min}, use the minimum stretch and for RCS_{Max}, use the maximum stretch.
The installed cross-section of the seal due to stretch, MCS, then can be determined using the following equations:

$$MCS_{min} = ID_{min} - \left(\frac{RCS_{Max}}{100} \right) \times ID_{min}, \tag{2.21a}$$

$$MCS_{Max} = ID_{Max} - \left(\frac{RCS_{min}}{100} \right) \times ID_{Max}, \tag{2.21b}$$

Installed Squeeze, Sq, can then be determined:

$$\%Sq_{min} = \left(\frac{MCS_{min} - t_{Max}}{MSC_{min}} \right) \times 100, \tag{2.22a}$$

$$\%Sq_{Max} = \left(\frac{MCS_{Max} - t_{min}}{MSC_{Max}} \right) \times 100, \tag{2.22b}$$

where t is the gland depth.
For piston applications the gland depth is the difference between the bore, B, and groove diameter, G, and is calculated:

$$t_{min} = \frac{B_{min} - G_{Max}}{2}, \tag{2.23a}$$

$$t_{min} = \frac{B_{min} - G_{Max}}{2}. \tag{2.23b}$$

For Rod applications the gland depth is the difference between the groove diameter, G, and Rod, R, and is calculated:

$$t_{min} = \frac{G_{min} - R_{Max}}{2}, \tag{2.24a}$$

$$t_{Max} = \frac{G_{Max} - R_{min}}{2}.$$ (2.24b)

Radial displacement and eccentricity can also be factored in to determine squeeze to establish worse case conditions.

- Squeeze less than 5% and greater than 25% can result in enhanced compression set and leakage.
- Differential pressure must be greater than 50 psi to distort a seal beyond installation squeeze. Squeeze needs to be at least 5% for a seal to have sufficient seal force to seal under no or low-pressure conditions. Recommended ranges for squeeze [36] for O-rings in hydraulic and pneumatic dynamic and static systems are presented in Table 2.9. While these squeeze recommendations are for O-rings, they can be used as general guidelines for different seal geometries; some geometries may work equally well with slightly less squeeze.
- Surface finish of mating parts is more important in dynamic applications than static ones. A rough surface with motion can abrade the seal and decrease performance and life. A rough surface can also increase seal friction. The roughness for the mating surface that seal needs to seal against should be a maximum of 16 micro inch Ra and 63 micro inch Rz. The material ratio, R_{mr}—also sometimes referred to as the bearing ratio—for mating surfaces is also important in dynamic seal performance. This should be 50% to 80% [36], determined at a cut depth of C = .25 Rz, relative to a reference profile line of C0 = .05R_{mr}.

Dynamic seals are normally designed for a maximum of 85% gland occupancy [56–60]; thus, seal swell is an important factor. A seal that swells excessively can soften and abrade. It can leave the gland, and tearing and extrusion can result. An over occupied gland can change seal force and system frictional characteristics. A gland with occupancy of less than 60% can provide the seal with too much room to move in the gland. This can actually increase the friction; in addition to the friction between the seal and its mating surface, the seal moving in the gland creates friction. O-rings can be more prone to roll and twist in the gland in low-occupancy conditions. For reference purposes, the approximate gland fill can be calculated by dividing the cross-sectional area of the installed O-ring, CSA_1, by the minimal cross-sectional area of the gland, CSA_2:

$$CA_{1min} = \pi \times \left[\frac{MCS_{min}}{2} \right]^2,$$ (2.25a)

TABLE 2.9
Recommended Squeeze (Compression) for O-rings by Type of System

Dimension in Inches

| O-ring Dash/Size Code Series | O-ring Cross-Section, (Nominal) | Squeeze % | | Static Radial |
| | | Dynamic | | Hydraulic & Pneumatic |
		Hydraulic	Pneumatic	
0xx	0.070	13 to 27	10 to 24	14 to 35
1xx	0.103	12 to 24	8 to 22	13 to 30
2xx	0.139	11 to 22	7 to 20	13 to 30
3xx	0.201	11 to 20	7 to 18	12 to 28
4xx	0.275	9 to 19	6 to 17	10 to 25

$$CA_{1Max} = \pi \times \left[\frac{MCS_{Max}}{2} \right]^2, \tag{2.25b}$$

$$CA_{2min} = t_{min} \times Gw_{min}, \tag{2.26a}$$

$$CA_{2min} = t_{min} \times Gw_{min}, \tag{2.26b}$$

$$\% GlandFill_{min} = \left[\frac{CA_{1min}}{CA_{2Max}} \right] \times 100, \tag{2.27a}$$

$$\% GlandfFill_{Max} = \left[\frac{CA_{1Max}}{CA_{2min}} \right] \times 100, \tag{2.27b}$$

where MCS is the installed cross-section of the seal due to stretch, t is the gland depth, and Gw is the groove width.

Seals can shrink due to incompatibility and can dry out. This can also lead to leakage on start-up and during operation. Seal shrinkage of 4% or more can lead to leakage.

Seal friction is not normally considered in system design because in a standard hydraulic operating system, the seal eventually rides on an oil film which lubricates the seal and reduces friction. However, short strokes, long extent strokes after long downtimes, eccentricity, and wear prevent a uniform oil layer from fully developing. In these cases, low-friction elastomers offer performance benefits. The additional costs are minimal when considered against the downtime costs due to seal wear and leakage. In a zero leak system, low-friction elastomeric seals and scheduled preventive maintenance are necessities.

The following guidelines should be adopted in order to reduce dynamic friction [43]:

* Surface finish of the bore or rod should be better than 16 micro inch Ra
* Increase the speed of moving parts
* Decrease seal cross-section and hardness
* Decrease the environmental temperatures for pistons and increase it for rods
* Use reduced-friction elastomers

When considering reduced-friction elastomers, make sure the following are considered: Additives that are used to reduce friction affect the seal's properties and compatibility; make sure the new seal is compatible. These additives can also contaminate systems. Surface treatments such as chlorination and ion etching can quickly wear away and friction returns. Greases used for installation are washed away, increasing friction. Greases can also sometimes harden the elastomer when the system heats up. Polymer-modified elastomers have a polymer modification inherent in the seal that does not leach out nor is it a surface coating. Good hardware design and the proper choice of seal material and design are important in order to reduce system friction.

Cap seals are also used to reduce friction [61,62]. A cap seal is typically an elastomer-energized PTFE shape (See Figure 2.34). Although this seal design reduces breakaway or static friction, it results in higher running friction. It has been felt that the cap seal was the only way to reduce seal friction without compromising hardware design. The cap seal, however, is not a zero leak seal if it is functioning correctly. The cap seal would have a limited lifespan in a system if it were not for the oil layer on which the cap eventually rides. As the cap wears, leakage can occur and this is why a cap seal is sometimes used in tandem with an elastomer contact seal. The cap seal is a buffer seal that reduces pressure and the elastomer seal is the primary seal which sees pressures below the system pressure. In this way, the elastomer seal becomes an effective zero leak seal because it only seals what gets by the buffer seal. Pressure build-up between the seals is addressed in the hardware and/or cap design. The primary seal in this system must have low friction in order to minimize wear, heat build-up, and hysteresis. A polymer-modified elastomer with its inherent low-friction properties is often used in this position for optimum performance.

Specifications for hardware dimensions using standard size seals (AS 568/ISO 3601-1:2008) exist for both general/industrial applications [36] and military/aerospace applications [34,35,63]. In the past the only widely used specifications for designing sealing systems were military/aerospace specifications, in particular MIL-G-5514 and its replacement SAE AS4716. There was not a standard specification for industrial or general use applications until ISO 3601-2 was released in 2008. Until then, guidelines and hardware dimensions for industrial applications were mainly listed in seal manufacturers' catalogs or literature, and these sometimes differed significantly. Historically, industrial guidelines resulted in more stretch and squeeze of the seal, whereas the aerospace guidelines utilized less stretch and squeeze. For static applications the industrial guidelines were often preferred; for dynamic applications the military/aerospace guidelines were preferred.

ISO 3601-2 contains technical guidelines for seal fit, including squeeze and stretch, and gland design for dynamic and static O-rings. ISO 3601-2 further divides dynamic applications into pneumatic and hydraulic, and static into pneumatic/hydraulic and axial (face seal) applications with applicable guidelines. The hardware tolerances listed in this specification are often larger than the tighter tolerances typically used in military/aerospace specifications, so the aerospace standards, SAE AS 4716 for dynamic applications, and SAE AS 5857 for static applications, are still used for industrial applications depending on the designers' preference and tolerance situation. The guidelines in ISO 3601-2 are similar to those used in the aerospace specifications with the main

FIGURE 2.34 Illustration of typical cap seal geometries. These are called cap seals because the cap, which is a plastic element, usually a PTFE material, is energized (or forced against the sealing surface) by an elastomer which, because of its elastomeric nature, can be called an energizer because it exerts a sealing force on the cap.

difference being hardware tolerances and groove widths. The groove widths for both ISO 3601-2 and AS 4716 are listed in Table 2.10.

When designing a sealing system or trying to determine the best size seal, gland design specifications are good places to start; however, they do not always provide the optimal design for every application or seal size, nor designs for non-standard size seals. These specifications also have been based on ambient room temperature (70°F) and do not take into account the significant differences in coefficients of thermal expansion and contraction of the seal material and the hardware materials. Coefficients of expansion for elastomers and plastics are generally several times higher than those for metals such as steel. It is best to always check the actual seal fit in the gland throughout the temperature range. When using cap seals, remember that the squeeze on the elastomer is increased by the thickness of the cap, so make sure that for the particular application the squeeze does not exceed 25%.

Although the above information addresses O-rings specifically, it is also true for all elastomer contact seals. Whether an O-ring or other geometric design is being used, these basic guidelines still apply and it is important to review them with the seal supplier during the design phase. Quite often during system design, the sealing system is considered as an afterthought and this is when problems arise. Seals might be the least expensive component in the hardware; however, they are engineered components of the system and require design attention. Seal material section and design should be supported by valid technical data. Do not accept seal substitutes until it can be verified that their performance is equivalent or better.

2.9.2.2.2 Lip Seals

Lip seals are also used in hydraulic systems. These seals are appropriate for systems that are low speed and operate at low pressure. At higher pressures, they produce increasing friction and accelerated wear [43]. They are pressure-activated in the sense that they have low squeeze to begin with, and when exposed to pressure, their lips spread to affect a wider sealing face [43]. Lip seals have to be aligned with their lips facing towards the pressure direction because they are pressure-activated.

TABLE 2.10
Groove Widths for Radial O-rings

Dimensions in Inches

O-ring Dash/Size Code Series	O-ring Cross-Section, (Nominal)	ISO 3601-2 Groove Width Number of Backup Rings			SAE AS4716 Groove Width Number of Backup Rings		
		0	1	2	0	1	2
0xx	0.070	0.110 to 0.120	0.165 to 0.175	0.220 to 0.230	0.094 to 0.103	0.150 to 0.164	0.207 to 0.220
1xx	0.103	0.150 to 0.160	0.205 to 0.215	0.260 to 0.270	0.141 to 0.151	0.183 to 0.193	0.245 to 0.255
2xx	0.139	0.197 to 0.207	0.252 to 0.262	0.307 to 0.317	0.188 to 0.198	0.235 to 0.245	0.304 to 0.314
3xx	0.201	0.283 to 0.293	0.354 to 0.364	0.429 to 0.439	0.281 to 0.291	0.334 to 0.344	0.424 to 0.434
4xx	0.275	0.374 to 0.384	0.484 to 0.494	0.594 to 0.604	0.375 to 0.385	0.475 to 0.485	0.579 to 0.589

In the case of systems that operate with pressure reversal, two seals must be used. Lip seals also require special glands [64]. This means that lip seals cannot be retrofitted with a standard squeeze-type seal. This is a definite disadvantage when compared to squeeze-type seals, as most squeeze-type seals can be interchanged in the same gland.

Lip seals can be used up to 800 psi without backup rings. Above 800 psi, backup rings should be used. Some designers consider 1,500 psi as the maximum service pressure for lip seals; however, the seals with backup rings have operated at 2,500 psi without problems [43]. Because of their light initial squeeze and the way they function, lips seals require a surface finish of 8 to 16 Ra.

Lip seals vary in lip design and configuration (See Figure 2.35): Lips can be rounded or flat; the seals can be symmetrical or asymmetrical. Depending on the manufacturer, a case can be made for each of the configurations and geometry. Lip seals are available in the common polymers, as are squeeze-type seals.

2.9.2.3 Radial Lip Seals

Radial lip seals [65,66] are primarily designed for rotary applications [47,67]. Figure 2.36 illustrates how they operate. Essentially, they act like bearings [68]. An oil film is generated between the lip and shaft. If the oil film thickness is minimal, the meniscus formed at the air side of the interface will not break and leakage will be avoided [47].

An important rule regarding radial lip seals is that they are essentially low-pressure seals [47,69]. Conventional radial lip seals operate between 5 and 10 psi [47,70]. Special seals are available that will operate up to 100 psi. These special seals have broader lips and must operate at slower speeds. Figure 2.37 illustrates how the lip is deformed as pressure is increased. Figure 2.38 shows the influence of pressure on seal life [47,71].

Normally, a manufacturer will provide peripheral speed limits for his seals, and any rubbing speed up to these limits is acceptable. This speed is typically on the order of 3500 ft/min. From this limitation, it is obvious that shaft size needs to be considered. For shaft sizes under 1.25 inches in diameter, standard seals will work up to 7000 to 8000 revolutions per minute (rpm). For shaft sizes between 2.5 inches and 3 inches in diameter, the rpm are reduced to 3000 to 4000. Figure 2.39 shows the effect of shaft speed on seal life [47].

Seal life can be affected by other factors: shaft size (the smaller the shaft, the longer the seal life), oil viscosity (as oil ages and viscosity changes, seal life is reduced), and fluid additives (seal lips run hot, facilitating fluid additive reaction with the elastomer, causing hardening or softening, and, consequently, affecting seal balance which results in leakage). Friction and heat can result in

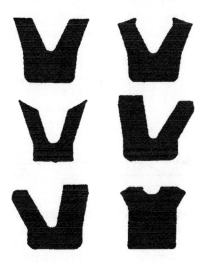

FIGURE 2.35 Illustration of typical profiles of lip seals.

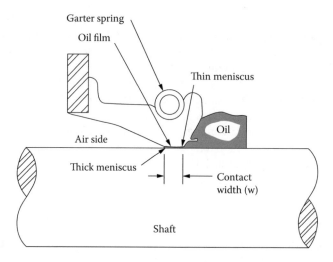

FIGURE 2.36 Illustration of the principle of operation of a radial lip seal.

compatibility problems, resulting in hardening and cracking of the seal lip, which leads to seal leakage [47].

Corrosion of the shaft affects seal performance. Nicking, cutting, and pitting of the shaft can affect the nature of the oil film or cause mechanical leak paths, which affect the ability of the seal to work effectively [72]. Case leakage and excessive eccentricity can also affect sealability.

The frictional nature of the elastomer seal lip can radically affect the seal's performance. High friction results in stick-slip, frictional heat build-up, excessive lip wear, hardening of the lip, and elastomer degradation—all of which adversely affect the seal's performance.

Wave seals, helix seals, and other design modifications allow more oil at the lip–shaft interface and, therefore, effect lower friction. This is true during operation; however, the friction and wear problems begin at start-up. It takes time for the oil film to develop, and during this time, the seal lip is running on a dry shaft, generating frictional heat. Some of the newer seal designs minimize the time for the oil film to develop, but they do not eliminate dry start-up. To minimize dry start-up effects, low-friction elastomers should be used. This is an area where polymer-modified materials have an advantage. They do not adversely affect the properties of the elastomer lip and provide low friction until the oil film develops. They also serve to minimize or eliminate stick-slip.

Some seals also include wiper lips (See Figure 2.40). These lips are designed to keep the shaft free from dust and dirt, which can work its way under the seal and cause leakage.

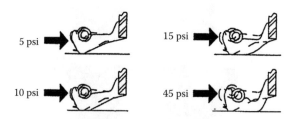

FIGURE 2.37 Illustration of how the radial seal lip deforms as a function of pressure: The greater the deformation, the greater the seal wear and, consequently, the shorter the seal life.

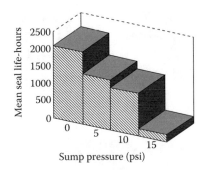

FIGURE 2.38   Illustration of how sump pressure affects the mean life of radial lip seals.

When using radial lip seals, the following cautions should be taken heed of [47,73,74]:

- Verify the seal quality prior to installation.
- Verify that there is no shaft damage that will result in mechanical leakage or will change the properties of the oil film.
- Maintain clean oil in the system—dirty oil affects the viscosity and adversely changes the properties of the oil film.
- Abrasive debris may result in seal lip abrasion and cutting.
- Choose the correct elastomer for the application. Incompatibility can affect the thin seal lip more dramatically in this type of seal. It is extremely important to verify the compatibility of the elastomer in the fluid of use at the maximum lip temperature.
- Remember that the thermal expansion of an elastomer is an order of magnitude greater than that for most metals, and as the elastomer expands, the nature of the lip changes.
- Properly handle and install the seal.

If seal failure occurs, it is important that a detailed failure analysis is performed so that there is sufficient information available to address the failure and correct it [47]. A failure analysis checklist follows. It should be used to accumulate sufficient information to review and analyze so that future failures can be avoided. By following the checklist, information is accumulated at the time of failure

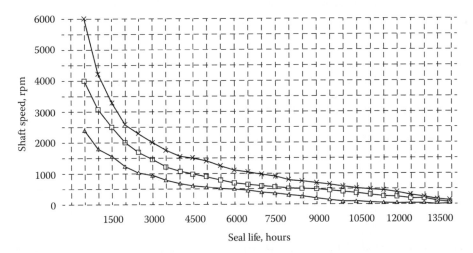

FIGURE 2.39   Illustration of how shaft speed affects the life of nitrile radial lip seals of various sizes. Higher speed means shorter life; radial lip seals for larger shafts inherently have a shorter life than radial lip seals for smaller shafts. △: 3-in.-diameter shaft; ×: 0.5-in.-diameter shaft; □: 1.75-in.-diameter shaft.

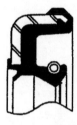

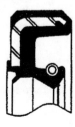

FIGURE 2.40   Illustration of typical seal lip geometries of radial lip seals with a wiper lip designed to keep external contamination out of the system.

and not trusted to memory to be recalled at a later time. Similar checklists can be used for all seal failures. This would aid in failure analysis.

Radial Lip Seal Failure Analysis Checklist

1. Note the condition of the system prior to removal

Amount of leakage:	☐Slight	☐Heavy	☐Damp
Condition of the area:	☐Dirty	☐Clean	☐Dusty
Leakage observed at:	☐Area between lip & shaft		☐Between seal elements
	☐Between O.D & bore		☐At retainer gasket
	☐Between wear sleeve & shaft		☐At retainer bolt holes

2. After wiping the area clean, inspect for the following:

☐Nicks on bore chamfer   ☐Seal installed improperly

☐Shaft to bore misalignment   ☐Seal cocked in bore

☐Seal loose in bore   ☐Seal case deformed

☐Paint spray or other deposits on seal

☐Other

3. If shaft can be rotated, observe for:

☐Excessive end play   ☐Excessive run out

4. If you could not determine the seal leak, dust the area with talcum powder and operate the system for approximately 15–20 min and see if the location of the leak can be determined.

☐Location

5. Mark the seal so you know how it was positioned and retain an oil sample for cleanliness and viscosity determination.

6. Inspect the hardware upon seal removal for:

Shaft:	☐Clean	☐Discolored	☐Corroded	☐Damaged
	☐Deposits or coke on shaft			
	☐Rough bore surface		☐Flaws or pits in bore	

7. Seal observations:

Primary Lip:	☐Soft	☐Normal	☐Hardened	☐Damaged
Lip wear:	☐Excessive	☐Eccentric	☐Normal	☐None
Seal O.D:	☐Damaged	☐Axial scratches or scores		☐Normal
Spring:	☐In place	☐Missing	☐Separated	☐Corroded

The effects of leakage are accumulative. When a few drops of leakage appear, sometimes they are ignored because after all: "It's just a few drops." Table 2.11 shows what a few drops can really mean. Considering that there are 45 gal per barrel, a few drops of leakage affects the overall operating cost. Once a leak starts, it never gets smaller. Making sure that operations start with the correct seal and a leak-free system is important, as downtime costs to change out the seal can be high.

Radial lip seals may also be used in reciprocating systems. In this case, selected seals must be designed for this use. Normally, these seals have thicker lips and will work in systems where the surface speed is limited to 40 ft/min.

O-rings can also be used as a rotating shaft seals [43] and will give satisfactory performance up to speeds of 1500 ft/min. To use an O-ring in a rotary application, the following guidelines need to be employed:

- The O-ring should be located in a groove in the mating member.
- The O-ring should be located as close to the lubricating fluid as possible.
- There should be little to no stretch on the O-ring.
- The O-ring should be used only as a seal and not as a load-bearing device.
- The concentricity should be better than 0.005 in.
- The smallest possible cross-section O-ring should be used.

Design guidelines for O-rings in rotary applications are listed in Table 2.12.

2.9.2.4   Mechanical Face Seals

Mechanical face seals are used to prevent leakage in rotating shaft applications that exceed the capabilities of radial lip shaft seals and packings [47]. The seals have no universal standards because designs and applications vary significantly. Four typical seal configurations are shown in Figure 2.41. The seals tend to be customized for the specific application and differ widely based on specific application. They have significant advantages over radial shaft seals; they can handle all types of fluids—acids, salts, some abrasive media, and so on. They function in slightly misaligned and non-concentric situations, and work with bi-directional shaft rotation, high pressure, temperature, and speed excursions. The condition of the shaft is not as critical for their performance and they normally have long operating lives. However, mechanical face seals require more space and unique configurations. The seal must be optimized for the specific applications, and because they replace an existing seal with a new design, hardware modifications are normally required. Mechanical face seals cannot handle axial end play. The sealing faces must be finished smooth (0.08–0.4 μm) and can easily be damaged [47]. They cost significantly more initially because they are highly specialized and require hardware modifications.

The performance of mechanical face seals is dependent on geometric, hydrodynamic, and thermal effects [43]. All of these factors interact and make mechanical face seal designs complex. The geometric factors include face deformation, shaft eccentricity, misalignment, and vibration.

TABLE 2.11
Oil Volume Lost by Leakage

Leakage Rate	Loss per Day (gal)	Loss per Week (gal)	Loss per Month (gal)
1 drop/10 s	0.09	0.66	2.7
1 drop/5 s	0.19	1.32	5.6
1 drop/s	1.05	7.32	31.5
3 drops/s	3.15	22	94
5 drops/s	5.25	36.6	157.5
Drips	20.0	140	420

TABLE 2.12
O-Ring Design Data for Rotary Applications—Groove Dimensions

Dimensions in Inches

Cross-Section		Max. Rubbing Speed (ft/min)	Gland Depth	Diametral Clearance (E)[a]	Groove Width	Groove Radius (R)	Eccentricity Max.[b]	Bearing Length Min.
Nominal	Actual							
1/16	0.070 ± 0.003	1500	0.065 to 0.067	0.012 to 0.016	0.075 to 0.079	0.005 to 0.015	0.002	0.700
3/32	0.103 ± 0.003	600	0.097 to 0.099	0.012 to 0.016	0.108 to 0.112	0.005 to 0.015	0.002	1.030
1/8	0.139 ± 0.004	400	0.133 to 0.135	0.016 to 0.020	0.144 to 0.148	0.010 to 0.025	0.003	1.390

[a] For maximum working pressure of 500–750 psi; for higher pressures, clearances would be reduced.
[b] Total indicator reading between the groove O.D., and shaft and adjacent bearing surface.

Hydrodynamic effects concern the fluid to be sealed and include viscosity, lubricity, and surface tension. Thermal effects include vapor pressure and the sealing face heat build-up due to friction and thermal conductivity.

Mechanical face seals are not zero leak seals. For a 2 inch seal operating at 3000 rpm, typical leakage is 0.1–10 cm³/h. When volatile fluids are sealed, liquid-phase leakage may appear to be

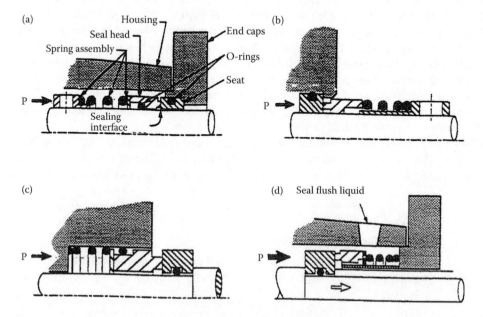

FIGURE 2.41 Illustrations of typical mechanical seal configurations. (a) Rotating seal head pressure on the outside diameter of the faces; (b) rotating externally mounted seal head pressure on the inside diameter of the faces; (c) stationary internally mounted seal head pressure on the inside diameter of the faces; (d) stationary externally mounted seal head pressure on the outside diameter of the faces.

zero, but vapor-phase leakage on the order of 0.1–10 g/h is possible. Vapor-phase leakage may also occur when the seal is stationary. A first approximation of leakage can be estimated by:

$$L\left(cm^3/h\right) = 2\times10^{-4}\eta^{0.5}\left(\frac{N}{P}\right)^{1.5} D_0pb^2 \qquad (2.28)$$

where
  η = absolute viscosity of the liquid being sealed (N · s/m^2)
  N = rotational speed (rpm)
  P = closing force per unit area of interface (MPa)
  D_0 = outside diameter of seal interface (mm)
  p = sealed pressure differential (MPa)
  b = interface radial width (mm)

Mechanical face seals are best specified by working with the seal manufacturer and it is critical that during the specification process, all system parameters and requirements are reviewed. Key requirements that the mechanical face seal should meet are specific to the application [43,47]. These are listed below and it is important that the seal manufacturer demonstrate how these requirements are to be met for the seal being recommended for the application.

For *water and fuel applications*, low shaft erosion, high-temperature operational capability, and abrasion resistance are key seal requirements. For *slurries*, the seal should be abrasion-resistant. For *food and medical applications*, the seal must include all approved materials and needs to be abrasion-resistant. It should have a profile that eliminates trapping of material and it should be capable of withstanding multiple system sterilizations.

For mildly-corrosive fluids and refrigerants, the seal needs to have low shaft erosion; it must have corrosion-resistant materials and be abrasion-resistant. It must have demonstrated low leakage. For highly-corrosive fluids, it should meet the same requirements for seals used with mildly-corrosive fluids and, in addition, have predictable life so that preventive maintenance can be performed prior to seal failure.

For hydrocarbons, seal requirements are dictated by the operation temperature of the system. For hot systems, the seal should have low shaft erosion, abrasion resistance, and low leakage, and be capable of high-temperature operation. For lower-temperature systems, the seal must have low leakage and be capable of sealing high pressures at low temperature.

For gas compressors, the seal must be capable of operating at high temperature and pressure with low leakage. It must also accommodate high shaft speeds.

Although most of the discussion on seal requirements, design, and performance are specific to the application, there are some basic properties of the seal face:

Hardness of seal face materials is expressed in values on the Moh scale [47]. Hardness for a seal face material is a measure of the material's resistance to scratching. The Moh scale ranges from 1 for talc to 10 for diamond. For a mechanical face seal to function properly, the wear plates should have a higher hardness than the mating seal face. The wear plate should also have a higher hardness than any particulate matter suspended or found in the fluid.

Stiffness in the seal face resists distortion caused by pressure gradients, thermal gradients, and speed distortion. In some applications, low stiffness is desirable in order to allow the seal to conform to the mating surface. Stiffness can be a function of both material selection and design. Stiffness is reflected by the elastic modulus of the material.

Tensile strength is important in calculating hoop stress, which is the result of internal hydraulic pressure. A seal face must have sufficient tensile strength to withstand the hoop stresses present in the system. Seal design can accommodate a lower-tensile-strength material, but it is important that this is demonstrated prior to seal selection.

A *low coefficient of thermal expansion* is required to resist the risk of failure by thermal shock. This is very important when the elastic modulus is high and the tensile strength and thermal conductivity are low.

Thermal conductivity controls the temperature gradient from localized hot spots. If the thermal conductivity is low, spalling, heat checking, and seal failure can result.

A *high-density sealing face* increases the mass of the rotating parts and can make it difficult to obtain dynamic balance of the seal assembly.

Temperature limitation and chemical compatibility are obvious requirements. The sealing face material must be capable of operating in the required temperature range without a drastic change of properties. If there is system incompatibility, seal failure will result.

Finally, the *seal face must exhibit low wear.* Unfortunately, seal face wear data are not always true indicators of seal wear because of the influence of the actual operating conditions and environment. Laboratory-controlled wear tests on the seal face materials often underestimate the actual seal face wear in an actual service.

As with all other seals, mechanical face seals do have finite life and it is important to analyze the seal failure to make changes or to at least assess predicted life to avoid unanticipated system shutdowns. If premature leakage occurs, this could be the result of improper seal selection, use, or installation. It could also be the result of incorrect material usage. If a seal fails due to leakage, it is important to note the conditions prior to failure and the condition of the hardware, fluid, and seal on disassembly. This information is important for determining the cause of failure and implementing corrective action. The items that need to be observed and noted include the following: Does the seal leak steadily whether the shaft is rotating or stationary? Is there steady leakage at low pressure and no leakage at high pressure? Is there steady leakage when the shaft is rotating but little to no leakage when the shaft is stationary? Is there visible seal damage? If there are elastomers present, what are their conditions? Is there hardware damage? Did the seal appear to be installed correctly?

2.10   SUMMARY

The subject of seals is often ignored when a system is being designed. This is true more often than not because the seals represent the lowest cost items of the system. They are the afterthought: as long as it is black and round, it should work. There was an advert that showed a picture of a $600,000 O-ring. It had failed prematurely and the resulting cost in system damage, downtime, lost revenue, and lost customers totaled this value. Although this might have been an exaggeration, the message should be clear: As much care and effort should be put into finding the optimum seal for a system as goes into the rest of the system design. A system is only as good as the seals used to make it work. *Cost-effective* is a very important term that seems to be ignored in today's hectic environment. Care should be taken in selecting and maintaining the best seals for the system to avoid the $600,000 O-ring.

REFERENCES

1. Morton, M., ed., *Rubber Technology*, 3rd ed. Van Nostrand Reinhold: New York, 1987.
2. ASTM D 1566, "Standard Terminology Relating to Rubber," *ASTM,* 100 Barr Harbor Drive, West Conshohocken, PA 1428-2959.
3. Fath, M.A., "Vulcanization of Elastomers," *Rubber World*, October 1993, pp. 22–25.
4. Boyum, B.M. and Rhoads, J.E., "Elastomer Shelf Life: Aged Junk or Jewels," *IEEE Transactions on Energy Conversion*, 1989, 4(2), pp. 197–203.
5. ASTM D 1418, "Standard Practices for Rubber and Rubber Lattices—Nomenclature," *ASTM*, 100 Barr Harbor Drive, West Conshohocken, PA 1428-2959.
6. Babbit, R.O., ed., *Vanderbilt Handbook*, R. T. Vanderbilt Co.: Norwalk, CT, 1978.
7. Thoermer, J., Mirzo, J., Szentivanyi, Z., Obrecht, W. and Rohde, E., *Effect of Cross Linking System on the Processing Behavior and Performance Profile of Hydrogenated Rubber (HNBR*, Miles Inc.: Akron, OH, 1988.

8. Chilson, C.M.A., Zielinski, R.E. and Crandell, K., "Importance of Long Term Test Times in Determining Seal Material Compatibility with Hydraulic Fluid at Service Temperature," *17th International Conference on Fluid Sealing;* J. Hoyes, ed., BHR Group Ltd., Cranfield, UK, 2003.

9. Hepburn, C., *Polyurethane Elastomers,* Applied Science Publishers: London, 1982.

10. Worm, A.T. and Brullo, R.A., "A High Fluorine-Containing Tetrapolymer for Harsh Chemical Environments," *Rubber Division, ACS Meeting,* 1983.

11. Schaefer, R.J., "Dynamic Properties of Rubber," *Rubber World,* September 1994, pp. 17–18.

12. Hertz, D.L., *Chemtech,* September 1990, pp. 574–576.

13. Hildebrand, J.H. and Scott, R.L., *Solubility of Non-electrolytes,* 3rd ed., Rheinhold Publishing: New York, 1949.

14. Barton, A.F.M., *Handbook of Solubility Parameters and Other Cohesion Parameters,* CRC press Inc.: Boca Raton, FL, 1983.

15. Hansen, C. and Beerbower, A., *Kirk Othmer Encyclopedia in Chemical Technology,* Standey, A., ed., *Supplementary Vol.,* 2nd ed., Wiley–Interscience: New York, 1971.

16. Jensen, W.B., *Surface and Colloid Science in Computer Technology,* Mittal, K.L., ed., Plenum Press: New York, 1987, pp. 27–59.

17. Beerbower, A. and Dickey, J.R., *American Society of Lubrication Engineers Transactions,* January 1969, p. 12.

18. Eleftherakis, J., "A New Method of Determining Hydraulic Fluid/Elastomer Compatibility," *40th Annual Earthmoving Industry Conference,* 1989.

19. Pruett, K.M., *Chemical Resistance Guide for Elastomers,* Compass Publications: La Mesa, CA, 1988.

20. J-120 "Rubber Rings for Automotive Applications," *SAE International,* 400 Commonwealth Dr. Warrendale, PA 15096-0001.

21. SAE AMS7270, "Rings, Sealing, Butadiene-Acrylonitrile (NBR) Rubber Fuel Resistant 65-75," *SAE International,* 400 Commonwealth Dr. Warrendale, PA 15096-0001.

22. MIL-P-25732, "Packing, Preformed, Petroleum Hydraulic Fluid Resistant, Limited Service at 275°F (132°C)," *Document Automation and Production Service (DAPS),* Building 4/D, 700 Robbins Ave., Philadelphia, PA 19111-5094

23. SAE AMS-R-7362, "Rubber, Synthetic, Solid, Sheet and Fabricated Parts, Synthetic Oil Resistant," *SAE International,* 400 Commonwealth Dr. Warrendale, PA 15096-0001.

24. ISO 6072:2002 (E) "Hydraulic Fluid Power – Compatibility Between Fluids and Standard Elastomeric Materials," *International Organization for Standardization,* Case postale 56, CH-1211 Geneva 20, Switzerland.

25. ASTM D 6546-00 "Standard Test Methods and Suggested Limits for Determining the Compatibility of Elastomer Seals for Industrial Hydraulic Fluid Applications," *ASTM,* 100 Barr Harbor Drive, West Conshohocken, PA 1428-2959.

26. ASTM D 943, "Standard Test Method for Oxidation Characteristics of Inhibited Mineral Oils," *ASTM,* 100 Barr Harbor Drive, West Conshohocken, PA 1428-2959.

27. ASTM D 2000, "Classification System for Rubber Products in Automotive Applications," *ASTM,* 100 Barr Harbor Drive, West Conshohocken, PA 1428-2959.

28. ISO 3601-5:2002, "Fluid power systems—O-rings—Part 5: Suitability of elastomeric materials for industrial applications," *International Organization for Standardization,* Case postale 56, CH-1211 Geneva 20, Switzerland.

29. NF T 47-503, "Rubber O-Rings—Material Requirements for the Common O-Ring Types," *French Commission of Normalization,* 1996.

30. Peacock, C.R., "Quality Control Testing of Rubber Shear Modulus," *Elastomerics,* 1992, 42, pp. 42–45.

31. *The Language of Rubber,* E. I. du Pont de Nemours & Co. (Inc.): Wilmington, DE, 1957.

32. Gent, A.N., ed., *Engineering with Rubber—How to Design Rubber Components,* Oxford University Press: New York, 1992.

33. Seabury, M., "Compatibility of Elastomer Seals and Industrial Hydraulic Fluids," in *ASTM D02 Meeting,* 1995.

34. MIL-G-5514, Rev G. "Gland Design; Packing; Hydraulics, General Requirements For," *Document Automation and Production Service (DAPS),* Building 4/D, 700 Robbins Ave., Philadelphia, PA 19111-5094.

35. SAE AS4716, "Aerospace Standard, Gland Design, O-Ring and Other Elastomeric Seals," *SAE International,* 400 Commonwealth Dr. Warrendale, PA 15096-0001.

36. ISO 3601-2:2008, "Fluid power systems—O-Rings—Part 2: Housing dimensions for general applications," *International Organization for Standardization,* Case postale 56, CH-1211 Geneva 20, Switzerland.

37. Vander Laan, D., "Compatibility of Elastomer Seals and Industrial Hydraulic Fluids," in *National Fluid Power Association Meeting,* 1996.

38. SAE AS568, "Aerospace Size Standards for O-rings," *SAE International,* 400 Commonwealth Dr. Warrendale, PA 15096-0001.

39. ISO 3601-1:2008, "Fluid power systems—O-rings—Part 1: Inside diameters, cross-sections, tolerances and designation codes," *International Organization for Standardization,* Case postale 56, CH-1211 Geneva 20, Switzerland.

40. ASTM D 1414, "Test Methods for Rubber O-rings," *ASTM,* 100 Barr Harbor Drive, West Conshohocken, PA 1428-2959.

41. ASTM D 1415, "Test Methods for Rubber Property—International Hardness," *ASTM,* 100 Barr Harbor Drive, West Conshohocken, PA 1428-2959.

42. ASTM D 412, "Test Methods for Vulcanized Rubber and Thermoplastic Rubbers and Thermoplastic Elastoemers—Tension," *ASTM,* 100 Barr Harbor Drive, West Conshohocken, PA 1428-2959.

43. *Seals and Sealing Handbook,* 1986, DuPont de Nemours International S. A. Geneva.

44. SAE J-515, "Specification for Hydraulic O-Ring Materials, Properties and Sizes for Metric and Inch Stud Ends, Face Seal Fitting and Four-Screw Flange Tube Connections," *SAE International,* 400 Commonwealth Dr. Warrendale, PA 15096-0001.

45. SAE MA2010, "Packing, Preformed—O-Ring Seal Standard Sizes & Size Codes, Metric," *SAE International,* 400 Commonwealth Dr. Warrendale, PA 15096-0001.

46. ANSI/B93.35M, "Cavity Dimensions for Fluid Power Exclusion Devices (Inch Series)," *American National Standards Institute,* 25 West 43rd Street, New York, NY 10036.

47. Brink, R.V., ed., *Handbook of Fluid Sealing,* 1993, McGraw-Hill: New York.

48. *O P Round Table: Leakages - Are They Inevitable?,* Germany, 1996.

49. Smith, L.P., *The Language of Rubber,* Butterworth–Heinemann Ltd.: Oxford, 1993.

50. SAE AS879, "Retainer, Packing, Hydraulic and Pneumatic Polytetrafluoroethylene Resin (Single Turn)," *SAE International,* 400 Commonwealth Dr. Warrendale, PA 15096-0001.

51. MS 27595, "Retainer, Packing Back-up, Continuous Ring, Polytetrafluoroethylene," *Document Automation and Production Service (DAPS),* Building 4/D, 700 Robbins Ave., Philadelphia, PA 19111-5094.

52. MS 28782, "Retainer, Packing Back-up, Teflon," *Document Automation and Production Service (DAPS),* Building 4/D, 700 Robbins Ave., Philadelphia, PA 19111-5094.

53. MS 28783, "Ring, Gasket, Back-up Teflon," *Document Automation and Production Service (DAPS),* Building 4/D, 700 Robbins Ave., Philadelphia, PA 19111-5094.

54. SAE ARP1802, "Selection and Application of Polytetrafluoroethylene (PTFE or TFE) Backup Rings for Hydraulic and Pneumatic Fluid Power Applications," *SAE International,* 400 Commonwealth Dr. Warrendale, PA 15096-0001.

55. SAE MAP3440, "O-ring groove design for packing preformed, elastomeric O-ring seals, static, radial squeeze, metric," *SAE International,* 400 Commonwealth Dr. Warrendale, PA 15096-0001.

56. ANSI/B93.76M, "Hydraulic Fluid Power—Cylinder Rod and Piston Seals for Reciprocating Applications—Dimensions and Tolerances of Housing," *American National Standards Institute,* 25 West 43rd Street, New York, NY 10036.

57. ANSI/B93.93M, "Hydraulic Fluid Power—Cylinders—Piston Seal Housings Incorporating Bearing Rings—Dimensions and Tolerances," *American National Standards Institute,* 25 West 43rd Street, New York, NY 10036.

58. SAE ARP1232, "Gland Design, Elastomeric O-Ring Seals, Static Radial," *SAE International,* 400 Commonwealth Dr. Warrendale, PA 15096-0001.

59. SAE ARP1233, "Gland Design, Elastomeric O-Ring Seals, Dynamic Radial, 1500 psi Max," *SAE International,* 400 Commonwealth Dr. Warrendale, PA 15096-0001.

60. SAE ARP1234, "Gland Design, Elastomeric O-Ring Seals, Static Axial, Without Back-up Rings," *SAE International,* 400 Commonwealth Dr. Warrendale, PA 15096-0001.

61. SAE AIR1244, "Selecting Slipper Seals for Hydraulic–Pneumatic Fluid Power Applications," *SAE International,* 400 Commonwealth Dr. Warrendale, PA 15096-0001.

62. SAE AIR1243, "Anti-Blow-By Design Practice for Cap Strip Seals," *SAE International,* 400 Commonwealth Dr. Warrendale, PA 15096-0001.

63. SAE AS5857, "Gland design, O-ring and Other Elastomeric Seals, Static Applications," *SAE International,* 400 Commonwealth Dr. Warrendale, PA 15096-0001.

64. *Hydraulic Seals, Engineering Manual and Catalog,* Disogrin Industries, Manchester, NH, 1988.

65. SAE J-111, "Seals—Terminology of Radial Lip," *SAE International,* 400 Commonwealth Dr. Warrendale, PA 15096-0001.

66. ISO 6194-2:2009, "Rotary shaft lip-type seals incorporating elastomeric sealing elements - Part 2: Vocabulary," *International Organization for Standardization*, Case postale 56, CH-1211 Geneva 20, Switzerland.

67. OS-4, "Technical Bulletin: Application Guide for Radial Lip Type Shaft Seals," *Rubber Manufacturers Association*, 1400 K Street, NW, Suite 900, Washington, DC 20005.

68. Jagger, E. "Rotary Shaft Seals—The Sealing Mechanism of Synthetic Rubber Seals Running at Atmospheric Pressure," in *Proceedings of the Institute of Mechanical Engineers*, 1957.

69. OS-6, "Radial Lip Seales, Shaft Seals, Radial Force Measurement," *Rubber Manufacturers Association*, 1400 K Street, NW, Suite 900, Washington, DC 20005.

70. SAE J-110, "Sales—Testing of Radial Lip," *Society of Automotive Engineers*, Warrendale, PA.

71. OSU-HS-1, "Method for Determining the Pressure Sealing Capabilities of a Reciprocating Hydraulic Seal," *Fluid Power Research Center*, Oklahoma State University, Stillwater, OK, 1972.

72. OS-1, "Handbook: Shaft Finishing Techniques for Rotating Shaft Seals," *Rubber Manufacturers Association*, 1400 K Street, NW, Suite 900, Washington, DC 20005.

73. OS-7, "Technical Bulletin: Storage and Handling Guide for Radial Lip Type Shaft Seal," *Rubber Manufacturers Association*, 1400 K Street, NW, Suite 900, Washington, DC 20005.

74. OS-16, "Recommended Methods for Assuring Quality of Radial Lip Seal Characteristics," *Rubber Manufacturers Association*, Washington, DC.

BIBLIOGRAPHY

ANSI/B93.98M, "Rotary Shaft Lip Seals—Nominal Dimensions and Tolerances," American National Standards Institute, 25 West 43rd Street, New York, NY 10036.

ANSI/B93.58M, "Fluid Systems—O-Rings—Inside Diameters, Cross-Sections, Tolerances and Size Identification," American National Standards Institute, 25 West 43rd Street, New York, NY 10036.

ANSI/B93.62M, "Method of Testing, Measuring and Reporting Test Results for Reciprocating Dynamic Hydraulic Fluid Power Sealing Devices," American National Standards Institute, 25 West 43rd Street, New York, NY 10036.

ANSI/B93.111M, "Fluid Power Systems and Components—Cylinders—Housings for Rod Wiper Rings in Reciprocating Applications—Dimensions and Tolerances," American National Standards Institute, 25 West 43rd Street, New York, NY 10036.

Bhowmick, A.K., ed., *Rubber Products Manufacturing Technology,* 1994, Marcel Dekker, Inc.: New York.

Fein, R.S., "Boundary Lubrication," Lubrication, 1971, 57(1), pp. 3–12.

Johnston, P.E., "Using the Frictional Torque of Rotary Shaft Seals to Estimate the Film Parameters and the Elastomer Surface Characteristics," in *8th International Conference of Fluid Sealing* (BHRA), 1978.

O-Ring Handbook ORD5700A, 1999, Parker Seal Company, Lexington, KY.

T3.19.25-1995, "Information Report—Fluid Power Systems—Sealing Devices—Storage, Handling and Installation of Elastomeric Seals and Exclusion Devices," National Fluid Power Association, Milwaukee, WI.

ISO 1302:2002, "Geometrical Product Specifications (GPS)—Indication of Surface Texture in Technical Product Documentation," International Organization for Standardization, Case postale 56, CH-1211 Geneva 20.

ISO 3601-3:2005, "Fluid power systems—O-rings—Part 3: Quality Acceptance Criteria," International Organization for Standardization, Case postale 56, CH-1211 Geneva 20.

ISO 3601-4:2008, "Fluid Power Systems—O-rings—Part 4: Anti-Extrusion Rings (back-up rings)," International Organization for Standardization, Case postale 56, CH-1211 Geneva 20.

ISO 4287:1997, "Geometrical Product Specifications (GPS) - Surface Texture: Profile Method - Terms, Definitions and Surface Texture Parameters," International Organization for Standardization, Case postale 56, CH-1211 Geneva 20.

ISO 6194-4:2009, "Rotary Shaft Lip-Type Seals Incorporating Elastomeric Sealing Elements—Part 4: Performance Test Procedures," International Organization for Standardization, Case postale 56, CH-1211 Geneva 20.

ISO 6194-5:2008, "Rotary-Shaft Lip-Type Seals Incorporating Elastomeric Sealing Elements—Part 5: Identification of Visual Imperfections," International Organization for Standardization, Case postale 56, CH-1211 Geneva 20.

MIL-P-19918 (1), "Packing V Ring," Document Automation and Production Service (DAPS), Building 4/D, 700 Robbins Ave., Philadelphia, PA 19111-5094.

OS-5, "Technical Bulletin: Garter Springs for Radial Lip Seals," Rubber Manufacturers Association, 1400 K Street, NW, Suite 900, Washington, DC 20005.

OS-8, "Handbook: Visual Variations Guide for Rotating Shaft Seals," Rubber Manufacturers Association, 1400 K Street, NW, Suite 900, Washington, DC 20005.

OR-1, "Handbook: O-Ring Inspection Guide: Surface Imperfections Control," Rubber Manufacturers Association, 1400 K Street, NW, Suite 900, Washington, DC 20005.

OR-2, "Technical Bulletin: Compression Set and Its Relationship to O-Ring Performance," Rubber Manufacturers Association, 1400 K Street, NW, Suite 900, Washington, DC 20005.

OR-6, "Technical Bulletin: O-Ring Standard Dimensional Measurement Practices," Rubber Manufacturers Association, 1400 K Street, NW, Suite 900, Washington, DC 20005.

OR-7, "Technical Bulletin: Test Methods for O-Ring Compression Set in Fluids," Rubber Manufacturers Association, 1400 K Street, NW, Suite 900, Washington, DC 20005.

SAE AIR1707, "Patterns of O-Ring Failures," SAE International 400 Commonwealth Dr. Warrendale, PA 15096-0001.

SAE AS708, "Top Visual Quality (TVQ) O-Ring Packings and Gaskets, Surface Inspection Guide and Acceptance," SAE International 400 Commonwealth Dr. Warrendale, PA 15096-0001.

3 Physical Properties and Their Determination

Toshi Kazama and George E. Totten[*]

CONTENTS

---

[*] This chapter is a revision of the chapter titled "Physical Properties and Their Determination" by George E. Totten, Glenn M. Webster, and F.D. Yeaple from the *Handbook of Hydraulic Fluid Technology*, 1st Edition.

3.1 INTRODUCTION

Hydraulic fluid is a component of the hydraulic system. This component is used for power transmission. The most common fluids used in a hydraulic application are shown in Table 3.1.

Hydraulic system performance is affected by the physical properties of the fluid component, which include: density, viscosity, specific heat, thermal conductivity, thermal expansivity, air entrainment and air release, foaming, cavitation, and various electrical properties such as conductivity, resistivity and permittivity. In this chapter, a review of these and other important physical properties will be provided. The role of these physical properties on fluid performance in hydraulic systems will also be discussed.

3.2 DISCUSSION

3.2.1 Viscosity

One of the most important fundamental physical properties of a hydraulic fluid is viscosity. Fluid viscosity, which is a measure of the "thickness" of a fluid, affects both the operational characteristics of the hydraulic system and the lubrication properties exhibited by the fluid.

The minimum viscosity limits for hydraulic pumps are dependent on pump design. The maximum fluid viscosity that can be used, also known as "cold-start" viscosity, is dependent on the [1]:

- Maximum viscosity that the pump will tolerate.
- Maximum viscosity creating a pressure drop in the suction line, which is dependent on pipe resistance, pipe bends, line strainers, and so forth, and should equal the pump suction capacity.
- Maximum viscosity creating maximum pressure drops acceptable in long high-pressure lines.

3.2.1.1 Hydraulic Pump Viscosity Limits

Typical hydraulic fluid viscosity limits for vane, gear and piston pumps are provided in Table 3.2 [1]. Similar recommendations have been provided for various vane, gear, radial and axial piston pumps

---

TABLE 3.1
Illustrative Examples of Hydraulic Fluids Used in Industrial Applications

Petroleum oil
Polyol ester
Phosphate ester
Vegatable oil
Water-glycol
Water-in-oil-emulsion
Chorinated hydrocarbon
Systhetic hydrocarbon
Silicate ester
Silicone
Water

TABLE 3.2
Typical Recommended Hydraulic Fluid Viscosity Ranges

Typical Data	Vane	High Pressure Gear Pump		Un-boosted Piston	
		Plain Bearings	Bearing	Sliding Valves	Seated Valves
Minimum viscosity (mm^2/s)	20	25	16	16	12
Optimum viscosity (mm^2/s)	25	25	20	30	20
Maximum viscosity (mm^2/s)	850	850	850	500	200
Suction capacity (in. Hg)	10	17	17	5	1

by a number of specific manufactures [2]. Although these data illustrate the importance of viscosity on the operation of a hydraulic system, specific viscosity limits for the pump and hydraulic circuit are dependent on the pump and hydraulic circuit in use.

3.2.1.2 Newtonian Viscosity

The "viscosity" (η) for the fluid is linearly proportional to shear stress (τ) and shear rate (dv/dz):

$$\eta = \frac{\tau}{\left(dv/dz\right)}.$$

The value (dv/dz) represents the change in fluid velocity with respect to the change in position in the z direction [3,4].

3.2.1.3 Newtonian/Non-Newtonian Behavior

Fluid viscosity is classified as Newtonian or non-Newtonian. As shown above, Newtonian fluid viscosity does not vary with shear rate. Viscosities of non-Newtonian fluids do vary with shear rate, as illustrated in Figure 3.1 [5]. Hydraulic fluids typically exhibit varying degrees of non-Newtonian viscosity behavior. At high shear rates, which are estimated to approach 10^6 s^{-1} in the inlet zone of the wear contact [6], even hydraulic fluids which would normally exhibit Newtonian behavior at low shear rates may be non-Newtonian [7]. The fluid shown in Figure 3.2 exhibits two Newtonian regions referred to as the "first" and "second" Newtonian regions, respectively [7]. The two Newtonian regions are connected by a non-Newtonian transition.

Non-Newtonian viscosity behavior is particularly important when considering the lubrication properties of a fluid. Aderin et al. have reported that non-Newtonian viscosity behavior at the Hertzian contact inlet region may account for significantly lower-pressure viscosity coefficients than predicted from low shear viscosity measurements [8].

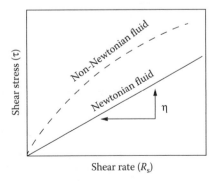

Shear rate (R_s)

FIGURE 3.1 Illustration of shear-stress variation with shear rate for Newtonian and non-Newtonian fluids [5].

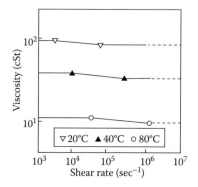

FIGURE 3.2   High shear viscosity profile of a water-glycol hydraulic fluid [7].

A general equation that has been used for non-Newtonian, "power law" fluids is the Ostwald-de Waele equation [3]:

$$\tau = \eta_0 \left[\frac{dv}{dz} \right]^n,$$

where η_0 is the viscosity of the fluid at a reference low shear rate and n is the power law exponent, which is unity for Newtonian fluids and less than unity for non-Newtonian fluids [9].

The power law relationship may be modified to describe the two Newtonian and the non-Newtonian regions depicted in Figure 3.3 [9]. One modification is the truncated power law, whereby the three regions are simply connected (as shown in Figure 3.3) [9]. In this model, η_0 is the low shear viscosity in the first Newtonian region, η_∞ is the viscosity in the second Newtonian region and the power law is used to describe the viscosity behavior during the transition from the first to the second Newtonian regions.

One of the disadvantages of the truncated power law is the discontinuous first derivative. There are various alternative models that have been used to provide a continuous transition between the first and second Newtonian regions [10]. An alternative model to the truncated power law is the Carreau model [11]:

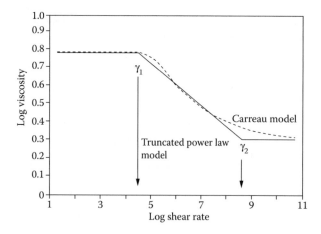

FIGURE 3.3   Illustration of the truncated power law model and the Carreau model for non-Newtonian fluids [9].

$$\left[\frac{\eta-\eta_{\infty}}{\eta_0-\eta_{\infty}}\right]=\left[1+\lambda\left(\frac{dv}{dz}\right)\right]^{(n-1)/2},$$

where η is a time constant and n is the power law coefficient.

Under a very high-pressure condition, such as elastohydrodynamic lubrication (EHL), the viscosity η is expressed using the relationship proposed by Roelands et al. [15-2]:

$$\eta=\eta_{0T}\exp\left\{\left[\left(\eta_{0T}\right)+9.67\right]\times\left[-1+\left(1+5.1\times10^{-9}p\right)^{z_t}\right]\right\}, \tag{3.1}$$

whereas η_{0T} is the viscosity at atmospheric pressure, defined as follows:

$$\log\left[\log\left(\eta_{0T}\right)+4.2\right]=-S_0\log\left(1+\frac{T}{135}\right)+\log G_0. \tag{3.2}$$

Additionally, z_t, the viscosity index parameter, is expressed as:

$$z_t=D_0+C_z\log\left(1+\frac{T}{135}\right). \tag{3.3}$$

Therein, C_z, D_z, G_0 and S_0 are constants and depend on the lubricant considered [12].

3.2.1.4 Viscosity-Temperature Relationship

Fluid viscosity varies with both the temperature and chemical composition of the fluid, as illustrated in Figure 3.4 [4]. The temperature dependence of viscosity is described by Eyring and Ewell's equation [13,14]:

$$\eta=\frac{hN}{V}e^{\Delta E/RT},$$

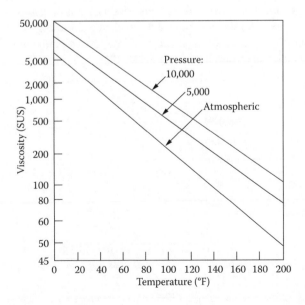

FIGURE 3.4 Viscosity versus pressure for a hydraulic oil [4].

where h is Planck's constant, N is Avogadro's number, V is molecular volume (molecular weight/density), ΔE is the activation energy for viscous flow, R is the gas constant, and T is the absolute temperature (K).

Andrade's equation has been used as an alternative to Eyring and Ewell's equation:

$$\eta = Ae^{b/T},$$

where A and b are constants.

Currently, one of the most widely used equations to predict viscosity temperature relationships of mineral oils is the Walther equation [15,16]:

$$\log \log (\eta + c) = a - b \, \log T,$$

where a and b are constants for a particular fluid and C varies with viscosity. This use of the equation is described in ASTM D 341–87 [17].

A more general treatment of the viscosity–temperature relationships is to curve-fit experimental viscosity–temperature data to the Vogel equation [8] and solve for a temperature–viscosity coefficient:

$$\eta = Ae^{B/(T-C)},$$

where A, B, and C are constants by least squares curve fitting constants, and T is the absolute temperature (K). The temperature–viscosity coefficient can then be determined at any temperature from:

$$\frac{d\ln \eta}{dT} = \frac{-B}{(T-C)^2}.$$

3.2.1.5   Viscosity–Pressure Relationships

In addition to shear rate and temperature, the viscosity of hydraulic fluids may also be significantly affected by pressure, as shown in Figure 3.4 [4]. This relationship is dependent on the chemical composition of the fluid, as shown for the paraffinic and naphthenic oils in Figures 3.5 [18] and 3.6 [19].

The pressure–viscosity variation of naphthenic and paraffinic petroleum oils has been studied and a number of general observations have been made [18]:

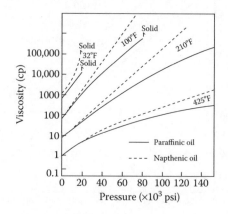

FIGURE 3.5   Viscosity–pressure relationship at different temperatures for paraffinic and naphthenic hydraulic oils [18].

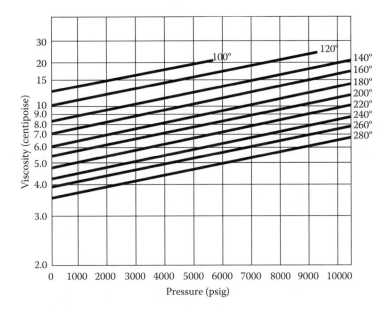

FIGURE 3.6 Viscosity–pressure relationship for MIL 5606 aerospace hydraulic fluid [19].

- Viscosity increases with increasing pressure.
- The increase of viscosity with pressure decreases with increasing temperature.
- At room temperature, the viscosity of most petroleum oils increases by approximately 2.3 times when pressure is increased to 35 MPa (5000 psi).
- The same increases in viscosity will exhibit a greater effect on viscosity at higher pressure.
- Increasing the pressure will increase the viscosity index of an oil, with naphthenic oils exhibiting a greater effect than paraffinic oils.

The effect of pressure on viscosity is modeled by the Barus equation [20]:

$$\eta_P = \eta_0 e^{\alpha P},$$

where η_P is the centipoise viscosity at pressure (P); η_0 is the viscosity at atmosphere pressure; P is the pressure; and α is the pressure–viscosity coefficient.

Unfortunately, as Jones et al. have shown, a linear graphical solution to the Barus equation by plotting $\ln \eta$ as a function of P is not obtained, except at low pressures (see Figure 3.7) [21]. The slope of the tangent to $\log \eta$ as a function of P isotherm (α_{0T}) at atmospheric pressure is often used for the calculation of the pressure–viscosity coefficient used for EHD lubrication analysis:

$$\alpha_{0T} \equiv \frac{d \ln \eta}{dP} \Big|_{T, P = 1 \text{ atm}} = \frac{1}{\eta} \frac{d\eta}{dP} \Big|_{T, P = 1 \text{ atm}}.$$

However, α_{0T}, which is obtained by graphical differentiation, is dependent on relatively few low-pressure data points which contribute substantially to overall error. An alternative solution that provides a more reliable viscosity–pressure response is to solve for α^* as shown in Figure 3.7 by graphically integrating [22]:

$$\alpha^* \equiv \left[\int_0^{P \to \infty} \frac{\eta \, (T, P = 1 \text{ atm})}{\eta \, (P, T)} \, dP \right]^{-1} \Big|_T.$$

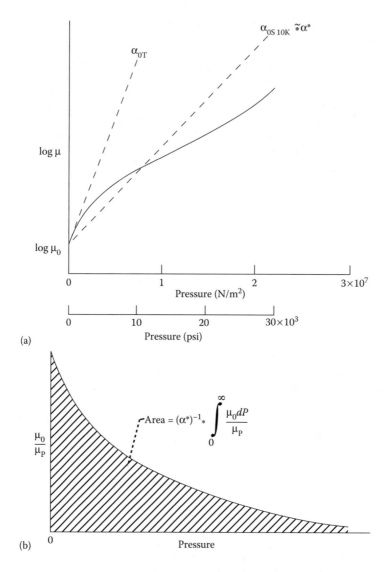

FIGURE 3.7   (a) Modeling viscosity–pressure behavior with α_{0T} and α^*. (b) Typical viscosity–pressure isotherm [21].

The advantage of α^* is that all of the variations of viscosity with pressure over the entire pressure range are included in the calculation, as illustrated in Figure 3.7 [21].

The ASTM slope is the slope of the viscosity temperature line of the Walther equation, which is plotted as described in ASTM D 341 (This is not equivalent to the value B in the Walther equation). In addition to determining the ASTM slope graphically, it can be calculated from [16]:

$$\text{ASTM slope} = \frac{\log \log(\eta_1 / \eta_2)}{\log(T_2 / T_1)}.$$

So and Klaus subsequently developed a single correlation that can be used for a broad range of fluids and fluid blends [23]. This correlation is based on the viscosity and density of the fluid at the temperature of interest. The viscosity–temperature behavior of the fluid utilizes the term m_0, which is calculated from:

$$m_0 = \frac{\text{ASTM slope}}{0.2}.$$

The So–Klaus correlation is [23,24,25]

$$\alpha = 1.030 + 3.509\,(\log\eta_0)^{3.0627} + 2.412\times10^{-4}\,m_0^{5.1903}(\log\eta_0)^{1.5976} - 3.387\,(\log\eta_0)^{3.0975}\rho^{0.1162},$$

where α is the pressure–viscosity coefficient (kPa^{-1} × 10^5) = $\partial(\log\eta)/\partial P$ at $P = 0$, $\log\eta_0$ = base 10 logarithm of the atmospheric viscosity in centistokes at the temperature of interest, ρ is the atmospheric density in gram per milliliters at the temperature of interest, and m_0 is the viscosity–temperature property based on atmospheric kinematic viscosities at 37.8°C (100°F) and 98.9°C (210°F) of the fluid of interest.

Wu, Klaus and Duda have developed an even more simplified equation for predicting α based on free-volume theory [24]:

$$\alpha = \left(0.1657 + 0.2332\log\eta_0\right)\times m_0$$

where α is the pressure–viscosity coefficient, kPa^{-1} × 10^5, m_0 is the viscosity–temperature property determined from the ASTM-Walther equation and is equal to (ASTM slope/0.2), and η_0 is the atmospheric kinematic viscosity at the temperature of interest (millimeter squared per second). Other methods that have been used to predict the value of α are summarized in Table 3.3.

The Appledoorn equation has been proposed for the calculation of the simultaneous effects of both temperature and pressure on viscosity [25]:

$$\log\left(\eta/\eta_{\mathrm{A}}\right) = a\log\left(F/F_0\right) + \alpha P.$$

where η is the viscosity at temperature F, expressed in °F, and pressure P, η_{A} is a reference viscosity at atmospheric pressure and temperature F_0, a is the viscosity–temperature (Appledoorn) coefficient, and α is the pressure–viscosity coefficient.

TABLE 3.3
Various Methods Used in the Comparisons of the Effectiveness in Predicting the Pressure Effects on Viscosity

Developer (Reference)	Correlation	Reference Number
Kouzel	$\log\dfrac{\eta_p}{\eta_0} = \dfrac{P}{1000}\left(0.0239 + 0.01638\eta_0^{0.278}\right).$	[26]
Roelands, Vlugter and Wateman	For non-polymer fluids $\log\dfrac{\eta_p}{\eta_0} = \left(\dfrac{P}{5000}\right)^{y}\left[\left(0.002C_{\mathrm{A}} + 0.003C_{\mathrm{N}} + 0.055\right)\log\eta_o + 0.228\right],$ $\log\left(y - 0.890\right) = 0.00955\left(C_{\mathrm{A}} + 1.5C_{\mathrm{N}}\right) - 1.930$ For polymer fluids $\log\left(\log\eta_p + 1.200\right) = Z_{\mathrm{Bi}}\log\left(1 + \dfrac{P}{28400}\right) + \log\left(\log\eta_o + 1.200\right),$ $Z_{\mathrm{BI}} = 0.6Z_{\mathrm{qr}} + 0.4Z_{\mathrm{BI}}.$	[27]

(continued)

TABLE 3.3 (Continued)
Various Methods Used in the Comparisons of the Effectiveness in Predicting the Pressure Effects on Viscosity

Fresco
$$\log \frac{\upsilon}{\upsilon_0} = P\alpha'^{\alpha}\left(10^{-4}\right). \alpha = \frac{560}{^\circ R}, \ a = A + B\log(V_0) + C(\log V_0)^2.$$
[28]

A, B, and C are functions of ASTM slope

Kim
$$\log \frac{\upsilon}{\upsilon_0} = P\left(\alpha' + \beta\right)\left(10^{-4}\right)', \ \alpha' \text{ is calculated from Fresco's Method,}$$
[29]

$$\beta = A'\log Vr + B'\left(\log V_r\right)^2 \ A' \text{ and } B' \text{ are function of temperature,}$$

graphical method is also available

So and Klaus
$$\log\left(\frac{\eta_p}{\eta_0}\right) = \alpha\left(98.1 \times P\right).$$
[30]

$$\alpha = \frac{10^{-5}}{2.3025}\left\{1.216 + 4.143\left[\log\left(V_0\right)\right]^{3.0627} + s - t\right\},$$

$$s = 2.848 \times 10^{-4} m^{5.1903}\left[\log\left(V_0\right)\right]^{1.5976},$$

$$t = 3.999\left[\log\left(V_0\right)\right]^{3.0975} p^{0.01162},$$

Chu and Cameron
$$\log\left(\frac{\eta_p}{\eta_0}\right) = 16\log\left[1 + \frac{0.062\left(10^{+a}\right)P10^{-3}}{\eta_0^{0.062}}\right] + a$$
[31]

$[0.183 + 0.2951\ \log(\eta_0)]$ for naphtenics.

Worster
$$\log\left(\frac{\eta_p}{\eta_0}\right) = aP\left(10^{-4}\right) + a = \left[0.183 + 0.2951\ \log\left(\eta_0\right)\right] \text{ for naphtenics}$$
[32]

Johnston
$$\alpha = \frac{\beta E_r}{\alpha_r 2CRT^2}.$$
[33]

Both temperature and pressure exhibit potentially large effects on viscosity and the simultaneous inclusion of both in viscosity calculations is desirable.

3.2.1.6 Shear Degradation

Many hydraulic fluids contain higher molecular weight viscosifier or thickener additives which will degrade under the high shear rate conditions present in hydraulic pump applications. Molecular degradation due to high shear rates is referred to as "mechanodegration" [34–36].

High molecular weight additives such as viscosifiers and thickeners do not exist in solution. Instead, the hydrocarbon or polymeric portions of the molecules coil around each other, providing a relatively long-range order in solution that increases viscosity (thickening effect) with increasing concentration. The degree of intermolecular hydrocarbon or polymer chain entanglement increases with increasing molecular weight (chain length).

When mechanical energy (shear) is applied to thickened solution—for example, by circulation in a hydraulic pump circuit—a shear-induced thinning of the solution occurs. If this is a reversible process, it is called non-Newtonian behavior, as discussed previously. However, if sufficient energy is applied, it may take less energy to break covalent C—C bond in the entangled hydrocarbon or polymer chains (mechanodegradation) than it would take to pull them apart (non-Newtonian shear thinning).

When C—C bond scission occurs, which is a non-reversible, permanent effect, the average molecular weight of the hydrocarbon or polymer is reduced. If there is a reduction in molecular weight, there will be a corresponding reduction in solution viscosity. It is important to note that although other processes may produce molecular degradaton, such as thermal and oxidative degradation, only shear degradation is a mechanodegradation process.

Hydraulic fluid shear stability has been studied using an Eaton hydraulic pump shear stability test conducted according to SAE 71R "Recommended Practices for Petroleum Fluid" [37]. The pump circuit used for this test is shown in Figure 3.8 [37].

Fluid viscosity stability decreases with increasing pump rotational speed beyond a threshold value, as shown in Figure 3.9 [38]. Figure 3.10 shows that fluid viscosity stability decreases with increasing pump pressure [38].

Another viscosity shear stability test involves the use of a multiple-pass full injection or diesel injection nozzle system [39]. At present, a Fuel Injection Shear Stability Test (FISST), described in ASTM D 3945–93, is used [40].

Hydraulic fluid shear stability may also be modeled by a sonic shear stability test such as that discussed in ASTM D 2603 [37,41]. This test is conducted using a Raytheon Model DF-101 magnetostrictive oscillator rated at 200–250 W and 10 kHz. Potentially useful data have been obtained when the test is conducted at full power. An alternative test is the ultrasonic shear stability test [37].

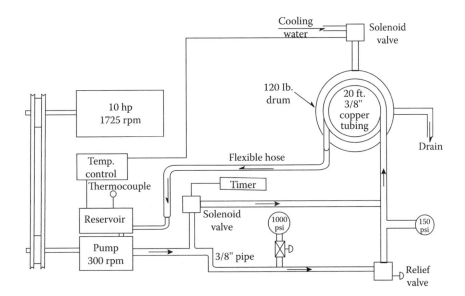

FIGURE 3.8 Schematic of hydraulic pump circuit for prediction of hydraulic fluid shear stability [37].

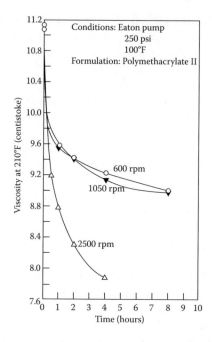

FIGURE 3.9   Effect of hydraulic pump motor rotational speed on shear stability [38].

3.2.2 VAPOR PRESSURE

If the vapor pressure of a hydraulic fluid is too high, gas pockets may form in areas of low pressure, such as the inlet of the hydraulic pump. This will aggravate cavitation. Fluid volatility also affects lubrication since hydrocarbon lubricants with higher volatility may result in higher wear [13].

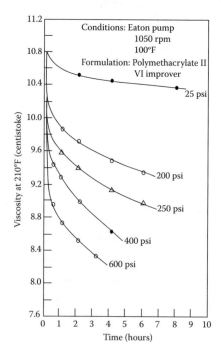

FIGURE 3.10   Effect of hydraulic pump pressure on shear stability [38].

The vapor pressure of hydraulic fluids is dictated by the most volatile components. For aqueous fluids, this would be water. For mineral oils, volatility is determined by the more volatile component of the oil. The chart shown in Figure 3.11 may be used for determining the vapor pressure from the boiling point of the fluid [42].

For mineral oils exhibiting vapor pressures approaching 10^{-6} mm Hg, the Maxwell-Bonnell chart (shown in Figure 3.12) may be used [51]. Vapor pressure for low-volatility synthetic oils should be appropriately corrected [43]. The variation of vapor pressure of various fluids with temperature is shown in Figure 3.13.

3.2.3 GAS SOLUBILITY

When a gas, such as air, comes into contact with a hydraulic fluid, a finite amount will become soluble in the fluid. Gas solubility is primarily dependent on the chemical composition of the fluid, concentration of the gas, and system pressure. Examples are provided in Figure 3.14 [4].

Estimation of gas solubility is important when considering the effect of air entrainment on lubrication, cavitation and potential foaming, especially upon rapid depressurization [44]. In this section, the prediction of gas solubilization in various fluids will be discussed.

Solubility of a dissolved gas in a fluid is quantitatively described by Henry's Law [45]:

$$K_1 = \frac{P_g}{X_1},$$

where K_1 is Henry's Law constant, P_g is the pressure of this gas, and X_1 is the mole fraction of the dissolved gas in solution.

This relationship indicates that the mole fraction of the dissolved gas will increase as the pressure of the gas increases. This is illustrated by the family of curves hown in Figure 3.15 [46].

Gas solubility is often described by the Ostwald coefficient (L), where the solubility of a gas is expressed as volume of a gas (V_g) dissolved in a given volume of liquid (V_1) at equilibrium [45]:

$$L = \frac{V_g}{V_1}.$$

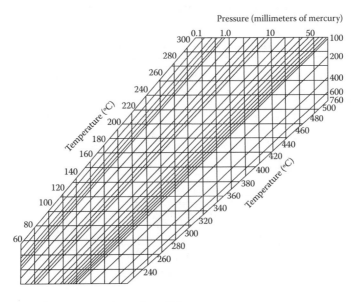

FIGURE 3.11 Hydrocarbon vapor pressure chart [42].

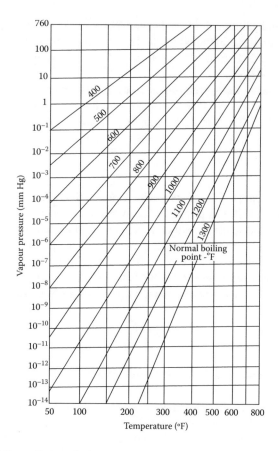

FIGURE 3.12   Maxwell-Bonnell chart for low vapor pressure.

One procedure for calculating the Ostwald coefficient (L_R) of a reference fluid (R) is provided in ASTM D 2779:

$$L_R = 0.300 \exp\left[\left(0.6399\left(700-T\right)/T\right)\ln 3.333\, L_0\right].$$

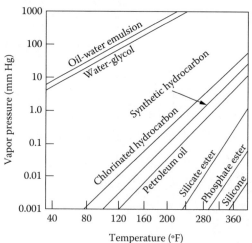

FIGURE 3.13   Typical vapor pressure values for different classes of hydraulic fluids [4].

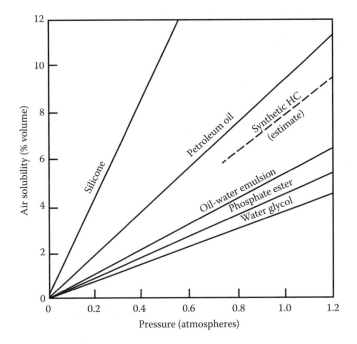

FIGURE 3.14 Air solubility in different hydraulic fluid types [4].

where L_o is the Ostwald coefficient at 273K and may be selected from Table 3.4 (see ASTM D 2779 for details). The Ostwald coefficient for the described fluid (L) is calculated from:

$$L = 7.70\,L_R\,(980 - \rho),$$

where ρ is the density of the fluid (kg/L) at 288K (59°F).

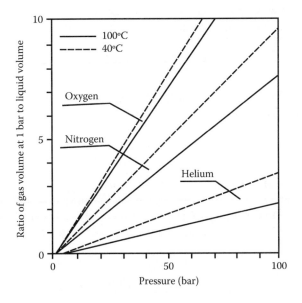

FIGURE 3.15 Gas dissolution in oil as a function of pressure and temperature [46].

TABLE 3.4
Gas Solubility Parameters

Gas	Ostwald Coefficient (L) at 0°C	Beerbower Solubility Parameter (δ_2) (MPa)$^{1/2}$
Helium	0.012	3.35
Neon	0.018	3.87
Hydrogen	0.040	5.52
Nitrogen	0.069	6.04
Air	0.098	6.69
Carbon monoxide	0.12	7.47
Oxygen	0.16	7.75
Argon	0.18	7.77
Krypton	0.60	10.34
Carbon dioxide	1.45	14.81
Ammonia	1.7	–
Xenon	3.3	–
Hydrogen sulfide	5.0	–

Beerbower has proposed an alternative model to calculate the Ostwald coefficient for a broader range of fluids [47]:

$$\ln L = [0.0395(\delta_1 - \delta_2)^2 - 2.66](1 - 273/T) - 0.303\delta_1 - 0.0241(17.60 - \delta_2)^2 + 5.731,$$

where L is the Ostwald coefficient, T is the absolute temperature in K, δ_1 is the solubility parameter of the fluid in $(MP_a)^{1/2}$ and is defined as

$$\delta_1 = \left[\frac{\Delta H_v - RT}{V_1} \right]^{1/2},$$

where ΔH_v is the heat of vaporization, and V_1 is the volume of the fluid.

The solubility parameter (δ_1) may also be calculated from [47]:

$$\delta_1 = 7.36 + 0.01203\rho,$$

for densities of up to 0.886 kg/m³ at 288K (60°F). For higher density fluids, the solubility parameter is calculated from the refractive index (n_D) at 298K [47]:

$$\delta_1 = 8.63 n_D^2 + 0.96.$$

The value of δ_2 for the gas may be taken from Table 3.4.

Another common parameter used to describe the gas solubility in a liquid is the Bunsen coefficient (α), which is the volume of a gas at 273K (32°F) and 101.3 kPa (1 atm) dissolved by one volume of the liquid at a given temperature at 101.3 kPa (1 atm). The Bunsen coefficient may be calculated from ideal gas law [45]:

$$\alpha = \left[V_g \frac{273.15}{T} \frac{P}{760} \frac{1}{V_s} \right] \left[\frac{760}{P_g} \right],$$

where P_g is the partial pressure of the gas; V_g is the volume of the gas absorbed at standard temperature and pressure; V_s is the volume of the absorbing solvent; T is the absolute temperature (K); and P is the total pressure.

The Bunsen coefficient may also be calculated according to ASTM D 3827:

$$\alpha = 2697\,\frac{P - P_v\,L}{T},$$

where P is the partial pressure of the gas (MP_a); P_v is the vapor pressure of the fluid (MP_a); T is the temperature in K; and L is the Ostwald coefficient. Bunsen coefficients for a number of hydraulic fluids are provided in Table 3.5 [67].

The variation of the Bunsen coefficients with temperature for various gases in hydraulic oil is illustrated in Figure 3.16 [48]. Interestingly, there is only minimal variation of the Bunsen coefficient for air in hydraulic oil with variation in temperature.

An example of the use of the Bunsen coefficient to predict air solubility in a mineral oil is provided by Figure 3.17 [49].

$$V_A = \alpha_{oil}V_{oil}\,\frac{P_2}{P_1},$$

where V_A is the volume of air in cm^3 at 760 mm Hg and 20°C; V_{oil} is the volume of oil in cm^3; α_{oil} is the Bunsen coefficient for mineral oil (~0.09); P_1 is the initial pressure in kilopascal per cm^2; and P_2 is the final pressure in kilopascal per cm^2. This example illustrates that the partial pressure of air of 1 kPa/cm^2 will result in 9% volume of air in the oil.

Another approach (and one of the most commonly used) to quantify gas solubility is the Hildebrand and Scott equation [50]:

$$\ln X_2 = -\ln P^{\circ}_2 - V_2\phi^2(\alpha_1 - \alpha_2)^2/RT\,,$$

where X_2 is the mole fraction of gas in the fluid; P°_2 is the vapor pressure of the gas at temperature T (absolute); and ϕ is the volume fraction of the gas.

X_2 is the partial molal volume of the gas and α_1 and α_2 all the solubility parameters of the fluid and gas, respectively. The values for α may be taken from Table 3.6, Hildebrands' reference [50], or an extensive listing provided by Hoy [51].

TABLE 3.5
Bunsen Coefficients for Dissolved Air in Lubricating Oils

Fluids	Bunsen Coefficient	Ref
Water	0.187	[48]
Mineral oil	0.07–0.09	[48]
Phosphate ester (HFD-R)	~0.09	[48,79]
Dicarboxylic ester	~0.09	[48]
Silicone oil	0.15–0.25	[48]
Polychlorinated biphenyls,(HFD-S)	~0.04	[48]
Oil in water emulsion (HFA)	~0.05	[79]
Water - glycol (HFC)	~0.04	[79]

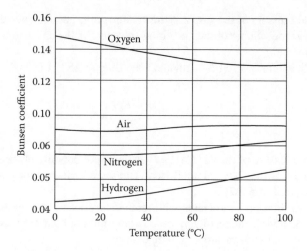

FIGURE 3.16 Variation of the Bunsen coefficient as a function of temperature for air, oxygen, nitrogen, and hydrogen [48].

3.2.3.1 Sources of Foaming and Air Entrainment

Excessive foaming and air entrainment may lead to numerous problems in a hydraulic system, including [52–56]:

- Spongy controls
- Loss of horsepower
- Reduction in bulk modulus
- Loss of system fluid
- Temperature effects
- Erosion and cavitation
- Noise
- Reduction in the load-bearing capacity of the lubricating film [57]

Increasing pump noise is one of the best indicators of fluid aeration problems. This is illustrated in Figure 3.18 [58]. There are various sources of air entrainment, some of which are discussed below [59].

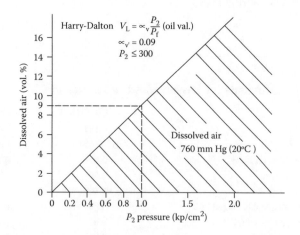

FIGURE 3.17 Illustration of the dependence of air dissolution on pressure [49].

TABLE 3.6
Viscosity Dependence of Foaming Tendency of Mineral Base Oils

Viscosity Grade	Foaming Tendency (mL)[a]			Air Release 50° min[b]
	24°C	93.5°C	After	
VG 7	40/0	20/0	30/0	0
VG 22	100/0	20/0	110/0	0.5
VG 46	420/0	20/0	380/0	1.5
VG 460	200/0-	230/0	120/0	160.0

[a] Foaming behavior was determined according to JIS K 2518.
[b] Air release was determined according to ASTM D 3427.

3.2.3.2 Improperly-Designed Reservoirs

Hydraulic fluid air entrainment may be due to improperly designed fluid reservoirs [53,60,61]. Two commonly encountered errors in reservoir design are fluid return lines above the liquid level and insufficient immersion of the suction line as shown in Figure 3.19 [53]. Insufficient immersion of the return line causes splashing, increased foaming, and air entrainment. Insufficient immersion of the suction line causes surface vortexing, which will draw air into the fluid facilitating air entrainment.

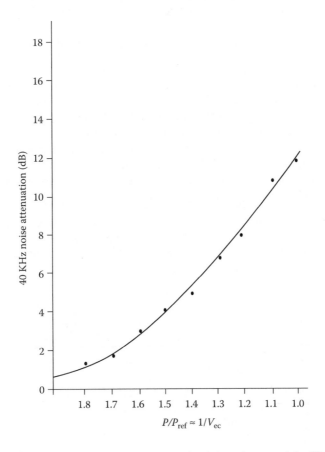

FIGURE 3.18   Hydraulic pump noise dependence on entrained air at the pump inlet [58].

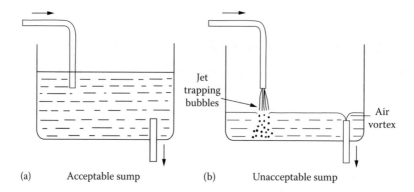

(a)   Acceptable sump   (b)   Unacceptable sump

FIGURE 3.19 Illustration of the effect of reservoir design on air entrainment [53].

In some cases, where it is difficult to remove air, a fine wire gauze screen can be placed in the sump as shown in Figure 3.20, which will "filter" all but the finest air bubbles in the system, thereby preventing recirculation through the hydraulic system [53].

Air solubility decreases with decreasing pressure. One low-pressure point in the hydraulic system is at the inlet of the pump. Figure 3.21 illustrates how air bubbles form with decreasing pressure and then re-dissolve on the pressure side of the pump. This is also illustrated for a gear pump in Figure 3.22 [62]. In order to avoid pressure drops at the pump inlet, the pipe diameter from the reservoir should be maximized and the length of the line should be as short as possible. Similar pressure drops may be encountered at sharp bends, elbows, pinched or kinked flexible hoses, and banjo joints.

A recommended reservoir design is illustrated in Figure 3.23 [60]. Hallet recommended the following:

1. The main return line should be approximately 8 inches. below the fluid surface.
2. The suction line should be located 6 inches. from the bottom of the reservoir in order to minimize settled solid contaminants from entering the system.
3. A 3 inches. air filter should be located at the top of the reservoir.
4. A float switch should be installed at the top of the tank to shut the system down if the oil level drops below the recommended working range.
5. A 4 inches. suction fan should be installed at the top of the reservoir to prevent water condensation in cold weather by removing heated air.

If the reservoir is pressurized with compressed air or nitrogen, direct contact of the gas with the fluid should be minimized with the use of a perforated plate as shown in Figure 3.24. This will help prevent saturation of the hydraulic oil by the compressed gas. Another way of minimizing the

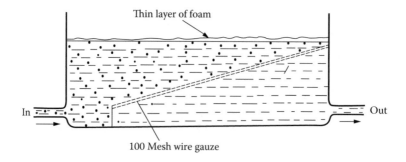

FIGURE 3.20 Illustration of the use of wire gauze to facilitate air-bubble removal from the reservoir [53].

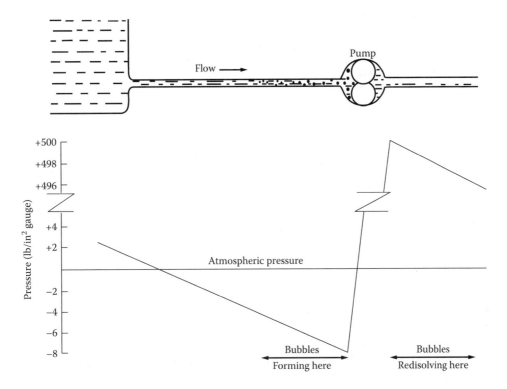

FIGURE 3.21 Suction-line pressure drop facilitation of air-bubble release from entrained air [53].

potential saturation of the hydraulic fluid with a gas is to pressurize the reservoir hydraulically, as shown in Figure 3.25 [53].

It may not be possible to design the pump reservoir and suction line as shown in Figure 3.26 due to space limitation. In these cases, a possible alternative is to put the suction line inside of the reservoir, as shown in Figure 3.26 [61]. A plug is placed at position A to break the siphon effect by the entry of air. An alternative design is provided by Figure 3.27, which includes a suction screen chamber at the top of the reservoir.

Ingvast reported the use of a specially designed reservoir (see Figure 3.28) that provides continuous degassing of a hydraulic fluid [63]. In addition to continuous degassing, it is also possible to continuously feed pressurized fluid to the pump inlet, thus minimizing the potential for both fluid air entrainment and cavitation.

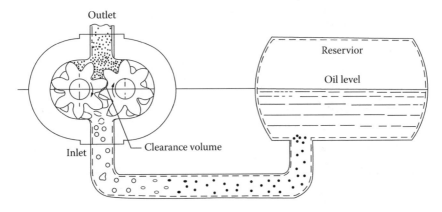

FIGURE 3.22 Illustration of air release at the inlet of a gear pump.

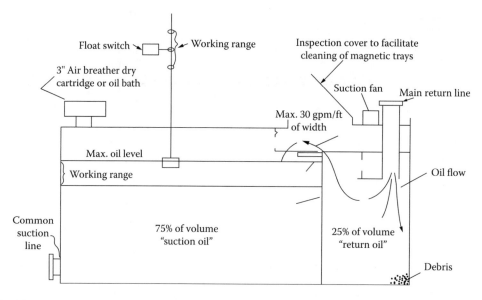

FIGURE 3.23   Dofasco 6000-gal reservoir design [60].

Improper installation of air vents is another potentially significant source of air entrainment in a hydraulic system. Figure 3.29 illustrates that air pockets of entrapped air can be avoided by the use of properly designed air vents [64].

3.2.3.3   Foaming Versus Air Entrainment

When fluids are agitated or mixed in reservoirs and pumping through the hydraulic system they are in contact with a gas, usually air. Gases in a hydraulic fluid may be present in one or more of these possible forms: free air, entrained air, and dissolved air [55].

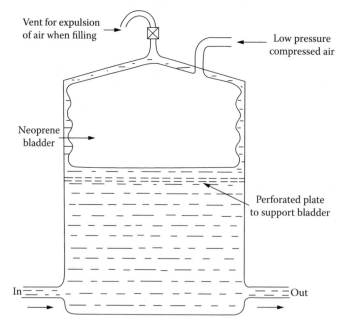

FIGURE 3.24   Procedure for hydraulic reservoir pressurization [53].

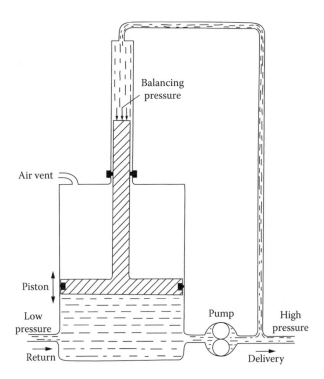

FIGURE 3.25	Reservoir pressurization using hydraulic pressure [53].

Free air refers to air trapped in a system, such as an air pocket in a hydraulic line, but not totally in contact with the fluid. Entrained air is air that is suspended in the fluid, usually in the form of small bubbles. Dissolved air is in solution and is not free or entrained. It is estimated that petroleum oil hydraulic fluids may contain as much as 10% volume of dissolved air [54]. Although dissolved air is thought to have no significant effect while dissolved, processes that reduce air solubility such as decreasing pressure and increasing temperature will result in foaming or the release of entrained air, both of which are deleterious processes [55].

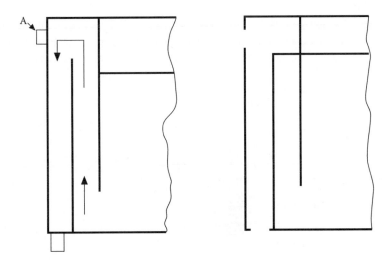

FIGURE 3.26	Illustration of pump suction-line placement inside of the reservoir [61].

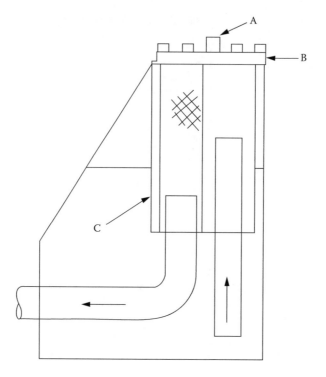

FIGURE 3.27   Illustration of a suction screen chamber [61].

The solubility of a gas in a fluid was treated quantitatively in the previous section. Problems of "free air" will discussed in a subsequent section. In this section, the fundamental properties of the formation and elimination of air bubbles, either as foam or entrained air, will be discussed. Although there is often an identity drawn between foam and air entrainment, these are actually different processes and are associated with different problems.

It should be noted that it is possible to have a high amount of unstable foam with relatively little air entrainment. Similarly, it is possible for air entrainment and high foam to coexist. The objective in fluid formulation is to produce foaming with relatively little air entrainment.

3.2.3.4   Foaming

Foam is a substance formed through the dispersion of an entrapped gas in a liquid. To produce stable foam, a surface-active component must be adsorbed into the gas–liquid interface. This surface film prevents the coalescence of the air bubbles, thus stabilizing the film. Foam stability is dependent on [65]:

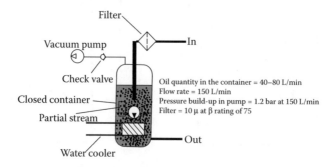

FIGURE 3.28   Facilitation of air release using a vacuum pump [63].

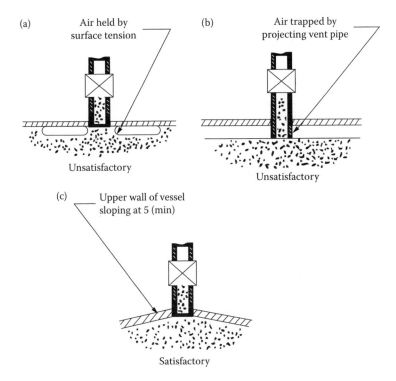

FIGURE 3.29 Air-vent designs that prevent air entrainment [64].

1. Surface area: Foam stability increases with increasing surface area. Foam destabilization is dependent on bubble coalescence. There is an energy balance of the Gibbs Free Energy (ΔG) of the system, which is quantitatively dependent on:

$$\Delta G = \sigma \Delta A,$$

where: σ is the surface energy and ΔA is the change in surface area. In foam, there is an increase in energy due to the surface area increase and a decrease in energy is due to the transfer of the profoamant (surfactant) molecules from the fluid to the gas–liquid interface [66].

2. Surface tension: Foam stability increases as the surface tension decreases relative to the fluid film. More importantly, is the way the surface tension changes with the surface area of the film which follows the Gibbs relationship of:

$$E = 2A \frac{d\gamma}{dA},$$

where: E is the surface elasticity, A is the surface area of the fluid film, and γ is the surface tension of liquid film. Stable foams have variable (elastic) surface films.

3. Viscosity: It is more difficult to form foams in higher viscosity fluids. However, once formed, these foams are more difficult to break. Table 3.6, Figure 3.30 show that this is true up to approximately 100 mm²/s (cS). Above 100 mm²/s (cS), profoaming behavior decreases [67].

4. The temperature dependence of foam stability for various hydraulic fluids is shown in Figure 3.31 [68]. Foam stability generally decreases with increasing temperature (decreasing viscosity).

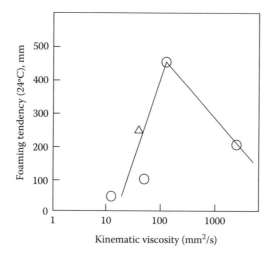

FIGURE 3.30   Effect of viscosity on foaming tendency [67].

5. Contamination concentration: The magnitude and stability of foams that have formed in a fluid increase to a plateau point with the increasing concentration of the profoaming contaminant.

The antifoam mechanism is shown in Figure 3.32. The first step is the introduction of the defoamant into the surface film that is stabilizing the foam. The defoamant then spreads throughout the surface film surrounding the air bubbles. As the defoamant spreads, a shearing force results, which causes a flow of the stabilizing film away from the gas bubble interface, resulting in a thinning of the interfacial film which then continues until the gas bubble ruptures, releasing the gas of bubble [69].

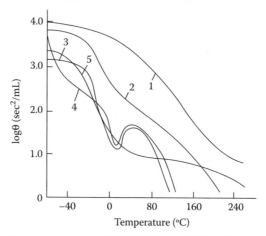

Composition	Designation	Surface tension (dyne/cm)
1. Polyalkylchlorarylsiloxanes	PAKhS	21.7
2. Polyalkylsiloxanes 60% dioctylsebacate 50%	PAS	28.4
3. Petroleum fraction 200–300 °C.+8% Butyl ester of polyvinyl alcohol	NF	27.7
4. Alkylarylphosphates	AAF	34.1
5. Polyalkyl-gamma-Trifluoropropylsiloxanes	PAFS	18.8

FIGURE 3.31   Temperature dependence of foam stability of different fluids designated [68].

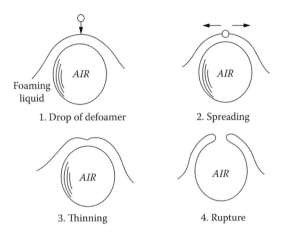

1. Drop of defoamer

2. Spreading

3. Thinning

4. Rupture

FIGURE 3.32 Mechanism of deforming by foam-control agents [66].

Antifoam compositions vary from single to multiple-component systems. Single component systems are typically water-insoluble and surface active since they must displace the profoaming surfactant in the interfacial film stabilizing the gas bubbles. Examples of single-component systems include: fatty acids and their glycerides or ethoxylates, polypropylene glycols, and higher alkylalcohols. Typically, this class of antifoams is used at concentrations of 0.1%–0.4% [66].

A multi-component antifoam system typically contains a uniform dispersion of a mineral oil-based material, hydrophobic silica, and a surfactant such as a fatty acid or alcohol ethoxylate composition in the fluid throughout the hydraulic system [66,70–73].

The rising velocity (U) and drag coefficients (C_d) of gas bubbles in a fluid can be calculated from Hadamard–Rybczynski equations [74]:

$$C_d = \frac{8}{Re} \frac{2\eta + 3\eta'}{\eta + \eta'},$$

where ρ and η are the respective density and viscosity of a fluid surrounding a spherical gas bubble; ρ' and η' are the respective density and viscosity of the gas; d is the diameter of the bubble; g is the acceleration due to gravity; and Re is the Reynolds number for the bubble, which may be calculated from:

$$Re = \frac{ud}{\eta\rho}.$$

For a gas bubble, $\rho \gg \rho'$ and $\eta \gg \eta'$, the well-known Stokes equations are derived, which state that the velocity of bubble movement is dependent on the density, viscosity, and diameter of the bubbles.

$$U = \frac{\rho g d^2}{12\eta},$$

$$C_D = \frac{16}{Re}.$$

This model is represented by Figure 3.33a.

An alternative form of this equation is [75]:

TABLE 3.7
Ascent Rates of Naked Bubbles [67]

Bubble Diameter (mm)	Terminal Rates of Ascent (MM/S)
10^0	11.00
10^{-1}	0.11
10^{-2}	0.0011
10^{-3}	0.000011

$$U = \frac{2r^2 g}{9\eta},$$

where r is the radius of the gas bubble. The rate of ascent of the "naked" bubble is a function of diameter as shown in Table 3.7 [74]. Typically, air bubbles range from 1^{-6} to 10^{-3} m for a gas lyosol, 10^{-6} to 10^{-3} m for an aero emulsion such as entrained air, and $>10^{-3}$ for coagulation for dispersed gas [76].

If a surface active agent such as an antifoam composition is present, the stream lines around the bubble follow the Hadamard–Rybcznski model as shown in Figure 3.33b [74]. Slightly different forms of the Stokes equations should be used for this model:

$$U = \frac{\rho g d^2}{18\eta},$$

$$C_d = \frac{24}{Re}.$$

The drag coefficients of three different antiwear hydraulic oils as a function of varying Reynolds number are shown in Figure 3.34a. Similar data for two fire-resistant hydraulic fluids, phosphate ester (SFR-C), and the water-glycol fluid Irus 504 are provided in Figure 3.34b. In both cases, the data correlates within the Stokes equation, which means that gas bubbles move as solid spheres in these fluids by the Stokes model [74].

Ida performed similar experiments with a silica-supported defoaming agent and a single-component higher-alcohol-type defoaming agent. His results showed that the silica-supported defoaming agent followed the Stokes model and the higher alcohol defoaming agent (and other additives) followed the Hadamard–Rybczynski model [74].

Fowle has studied the affect of a silicone antifoam on the air release properties of a turbine oil [75]. Although these fluids initially exhibited both antifoaming and air release properties, the air

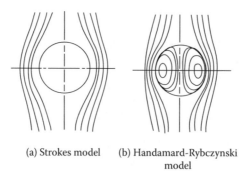

(a) Strokes model   (b) Handamard-Rybczynski
                              model

FIGURE 3.33 Comparison of the (a) Stokes and (b) Hadmard-Rybczynski models of fluid movement around an air bubble in solution [74].

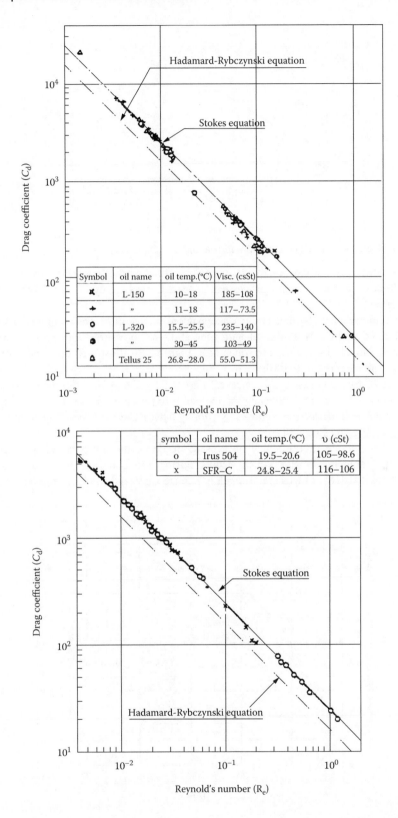

The following data is shown in the figure tables:

Symbol	oil name	oil temp.(°C)	Visc. (csSt)
x	L-150	10–18	185–108
+	"	11–18	117–.73.5
o	L-320	15.5–25.5	235–140
●	"	30–45	103–49
△	Tellus 25	26.8–28.0	55.0–51.3

symbol	oil name	oil temp.(°C)	υ (cSt)
o	Irus 504	19.5–20.6	105–98.6
x	SFR–C	24.8–25.4	116–106

FIGURE 3.34 Comparison of drag coefficients for (a) different petroleum oils and (b) fire-resistant fluids [74].

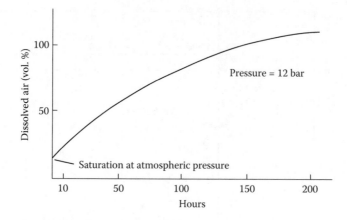

FIGURE 3.35 Dissolution rate of air in hydraulic oil under pressure [79].

release property was lost with increasing use over time. Other work [77–78] has shown that although silicones exhibit excellent antifoam properties, at least initially, they do not exhibit superior air release properties.

The dissolution of air in a fluid is a time-dependent diffusion process. Figure 3.35 shows the dissolution rate of air in a hydraulic oil with time at 12 bar pressures with no flow, which helps to explain the relatively long diffusion times [79]. Figure 3.36 illustrates the solubilization of air bubbles with increasing pressure [80].

The reverse of this process is "air release." Figure 3.37 shows that the release of air from a hydraulic fluid is enhanced by increasing the temperature (lower viscosity) and reducing the pressure [4]. Figures 3.38 illustrate the volume reduction of a hydraulic oil containing entrained air with respect to increasing pressure [79].

When an air-ignitable mixture may be present inside of the air bubbles, ignition can take place from the temperature rise which occurs during the compression in a pressure cycle (see Figure 3.39) [81]. This process requires only nanoseconds and the localized temperature may

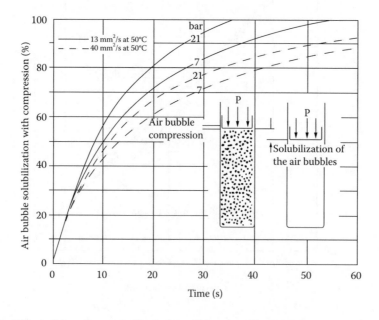

FIGURE 3.36 Effect of viscosity on air dissolution with compression [80].

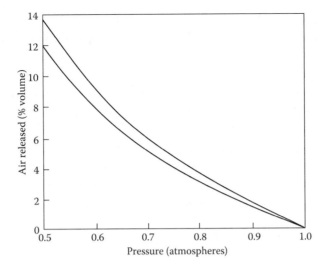

FIGURE 3.37 Air released from petroleum oil [4].

be a high as 1100°C. These high temperatures may lead to subsequent oil degradation [82]. This is called the "micro-diesel effect" and may cause localized hot spots and pressure spikes, which may subsequently lead to structural damage of the hydraulic pump [79].

3.2.3.5 Foam Tests

In this section, an overview of various laboratory tests that have been used to measure the foaming tendency of hydraulic and other industrial fluids will be discussed. The Tamura classification of these tests will be followed [83]. These test classifications include: static tests such as pouring, shaking, beating, rotational, and stirring; and dynamic tests which include: air injection and circulation.

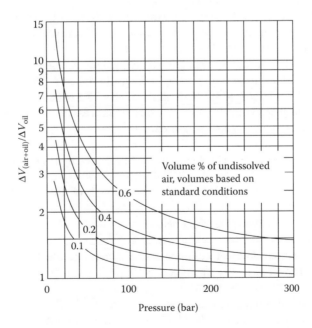

FIGURE 3.38 Volume reduction of H-L hydraulic oil (H-LP36) permeated with air bubbles relative to bubble-free oil with increasing pressure [99].

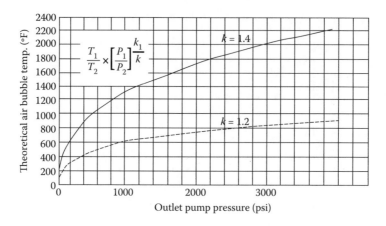

FIGURE 3.39 .Theoretical bubble temperature increase with compression at atmospheric pressure [81].

This discussion will be followed by a description of pump tests which have been reported for the measurement of the foaming tendency of hydraulic fluids.

A few methods of static tests are provided here:

1. Pouring Tests: One common test used to determine the foaming tendency of a fluid is the Ross–Miles method. [84,100]. The initial foam head and the foam head after 5 min is measured. These values provide a measure of foaming tendency and foam stability.
2. Shake Test: Perhaps the simplest foam test, although not very reproducible, is the bottle shake test [85] where 200 mL of the fluid is placed in a 16-ounce wide-mouth, screw-top glass bottle and then shaken with an up and down motion. The initial foam height is recorded, and again after a period of 5 min. The time it takes for the foam head to drop to 1 cm above the initial fluid level is recorded.
3. Beating Method: A standard volume (200 mL) of the fluid is placed in a 1-L cylinder and beaten 30 times in an up and down motion with a porous dish using the apparatus [86]. The foam volume immediately after beating is recorded.
4. Stirring Method: The ASTM D 3519 [87] test is used for the measurement of foaming of low-viscosity solutions under high shear conditions. In this test, 200 mL of fluid is placed in a Waring blender and stirred at a high speed for 30 s. The initial foam height is measured along with the foam height after 5 min. The time it takes for the foam head to drop to 1 cm above the initial fluid level is also measured.

A variation of this test is the "Stirring Plate" method described by JIS K 2241 [88]. In this test, 60 mL of solution is placed in a 100-mL cylinder and stirred at 1500 rpm for 5 min. The foam volume is measured 15 min after the stirring has stopped. An example of this test apparatus was reported by Bhat and Harper [89].

Similary, dynamic test methods are given below:

1. Air Injection Method: Another common test to determine the foaming tendency of a hydraulic fluid is ASTM D 892 (see Figure 3.40) [90]. A simplified version of this test is used to measure the foaming properties of engine coolants [83]. A similar version of this test, using requested compressed gas flow, was patented [91,92].
2. Circulation Method: A French Method [93] has been reported which involves continuously dropping the test fluid from a 750 mm height through a nozzle placed in a 40 mm-diameter 1080 mm-height glass cylinder [83]. A pump is used to re-circulate the fluid.

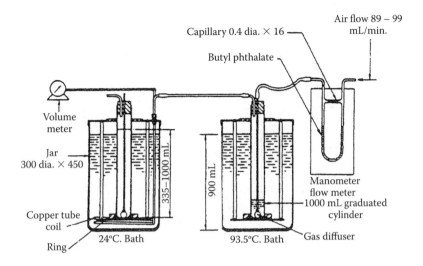

FIGURE 3.40 ASTM D 892 foam testing apparatus [90].

3.2.3.6 Experimental Procedures of Air Entrainment

Numerous experimental procedures have been reported which determine the propensity for a fluid to entrain air [94]. Some procedures are based upon relatively simple laboratory glassware apparatus. More often, proprietary experimental apparatus, which have been custom-built for this specific purpose, are used. An overview of experimental procedures that have been reported to study hydraulic fluid air entrainment will be provided here.

1. "Bubbly" oil viscometer: In order to study the effect of air–bubble entrainment on the viscosity of mineral-oil based hydraulic fluids, Hayward developed the "bubbly" oil viscometer [94].

   Bubble content was determined by measuring the volume of the bubbles in the bubbly oil. Hayward found that bubbles imparted a relatively small increase in viscosity over the range of 30–170 cS at 30°C according to

$$\frac{\eta_b}{\eta_0} = 1 + 0.015\beta,$$

   where η_b is the viscosity of bubbly oil, η_0 is the viscosity of bubble-free oil, and β is the percentage of bubble content.

2. Deutsche Shell Air Release Test: An air/oil emulsion is prepared using an aspirator, usually fitted to water line, to provide a vacuum source for filtration [95]. Oil is fed through the entry, which is usually attached to the water line. The rate of air release from a 5% air/oil emulsion is determined graphically by plotting the change in density with time. Poor temperature control, which is vital, is one of the disadvantages of this procedure.

3. Allgemeine Elek Tricitats—Gesellschaft (AEG) Method: This is basically the Waring blender test, where 700 mL of the fluid at the test temperature (usually 50°C) is stirred at a "high" agitation rate (approaching 20,000 rpm) for 25 sec. The fluid is then poured into a 1000 mL graduated cylinder and the rate of air bubble loss is measured [95].

4. Technischer Überwachungs/Verein (TÜV) method: This method DIN51 381, involves the preparation of an air–water emulsion by introducing compressed air through a 2-mm capillary tube for 7 min using the apparatus illustrated in Figure 3.41.

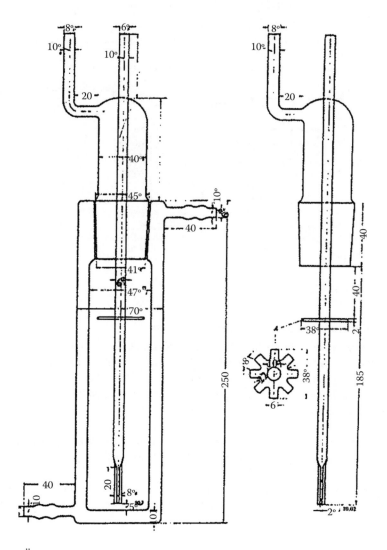

FIGURE 3.41   TÜV air-in-oil dispersion method.

The test temperature is usually 25°C, 50°C, or 75°C. Air release is determined by monitoring fluid density with time. The volume of dispersed air is calculated from:

$$L = \frac{(\rho_0 - \rho_x)100}{\rho_0 - \rho_L},$$

where L is the volume % of dispersed air, ρ_p is the density (gram per milliliter) of the bubble-free fluid, ρ_x is the density after time (x) in minutes, and ρ_L is the density of air in gram per milliliter at the test temperature.

5. Rowland's Air Entrainment Method: In this test, which is a variation of the AEG method discussed previously, 600 mL of the fluid at 100°F is mixed in a 1200 mL reservoir of a Waring blender at 9200 rpm for 10 sec and then poured into a rectangular reservoir. The filled reservoir is immediately placed in a 100°F constant temperature bath. The test cell is withdrawn periodically to measure bubble content by light absorption using the photocell and is then replaced in the constant temperature bath.

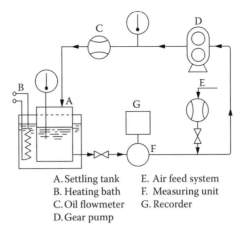

A. Settling tank   E. Air feed system
B. Heating bath   F. Measuring unit
C. Oil flowmeter   G. Recorder
D. Gear pump

FIGURE 3.42 Schematic diagram for apparatus used for dynamic study of air entrainment [97].

6. Dynamic L.T.G. Method: The L.T.G. apparatus is illustrated in Figure 3.42 [97]. This method is based on a circuit utilizing a gear pump. A "rest time" (TR) value is calculated from:

$$TR_i = V_i Q,$$

where TR_i is the time of the oil in the settling tank in seconds, Q is the volumetric capacity [L/s], and V_i is the volume of oil (L) in the tank being used.

The advantage of this method is that it not only measures air release, but also the stability of the aerated fluid under operating conditions.

7. Dead Weight Apparatus: The principle of measurement of bubble content for the dead weight apparatus, which is shown in Figure 3.43, is based on Henry's Law, where the solubility of a gas is proportional to the absolute pressure [98]. With sufficient compression, all of the gas bubbles will dissolve in the oil. The reduction in volume is a function of the bubble content.

This is one variation of the general class of piston-and-plunger methods where a cylinder is filled to overflowing. A cap is then screwed on, and a 15 lb plunger is inserted which creates a pressure of 300 psi [94]. The reduction in volume is proportional to the bubble content of the oil. Other variations of the general method have been reported by Magorien [55] and Liddell et al. [99]. An interesting variation of the system was reported where gas content of a water-glycol fire-resistant hydraulic fluid was measured while the fluid was in operation in an actual hydraulic circuit using ram pressure and displacement curves [100].

8. Zander Method: Zander et al. have recently reported an apparatus procedure which is capable of measuring relatively short air release times [101]. This procedure is based on the continuous measurement of the differential hydrostatic pressure between the points shown in Figure 3.44 [101]. The volume fraction of air bubbles is calculated from the differential pressure divided by the fluid density and is then plotted as a function of time. This procedure is sensitive for air release times and of less than 10 sec.

9. Hydrostatic Balance: A relatively simple procedure for the reproducible measurement of air bubble volume is to apply Archimedes' principle [98]. In this procedure a heavy object, of known volume (V) and weight (W) is immersed into the bubbly oil and the displaced fluid weight is then measured (the volume will be equal to the volume of the object). The bubble content is calculated from:

$$\text{Bubble content} = \frac{W_b - W_0}{W - W_0},$$

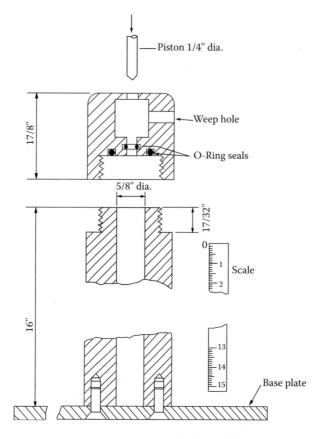

FIGURE 3.43 Dead-weight apparatus for measurement of air entrainment [98].

where W_b is the apparent weight of an object when immersed in bubbly oil, W is the apparent weight of an object when immersed in bubble-free oil, and W_0 is the weight of the object.

10. Sample Bottle Method: The steel bottle (shown in Figure 3.45) is filled with bubble-free oil at atmospheric pressure and then the cap is immediately screwed on with the attached valve in the open position [102]. For either bubbly oil or bubble-free oil under pressure (or

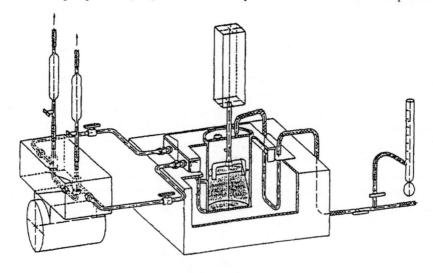

FIGURE 3.44 Measurement of entrained air by differential pressure [101].

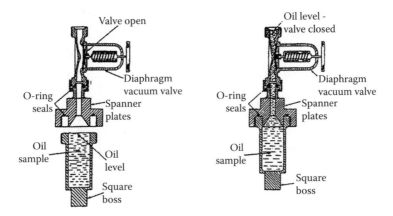

FIGURE 3.45   Open-type sample bottle [102].

a vacuum) in a hydraulic system, the pressure-type sample bottle shown in Figure 3.46 is used. The "bottles" are essentially 1/2 inch bare pipes with a 1/4 inch vacuum diaphragm valve at each end that are capable of withstanding pressures of up to 200 psi. After de-aerating at 0.01–0.02 mm Hg, the gas content at standard temperature (273K, 760 mm Hg), and the volume of air content (1/1 by volume) are determined from:

$$\text{Air content} = 35.92 \; \frac{V_c}{V_b} \; \frac{\left(P_2 - P_1\right)}{T},$$

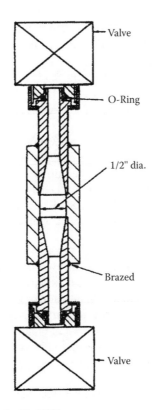

FIGURE 3.46   Pressure-type sample bottle [102].

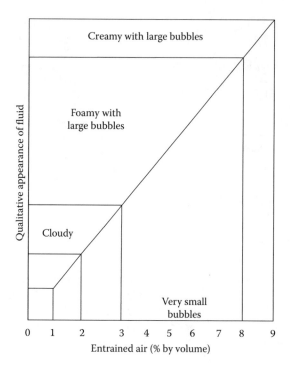

FIGURE 3.47   Quantification of entrained air by appearance [103].

where: V_c is the volume of the de-aeration chamber (mL); V_b is the volume of the sample
bottle (mL); P_2 is the final pressure of the de-aeration chamber (mm Hg); and P_1 is the
initial pressure of the de-aeration chamber (mm Hg).

11. **Turbidity Measurement:** Turbidity of hydraulic fluids varies with both the quantity and size
    of entrained air bubbles as shown in Figure 3.47 [103]. However, this figure also shows that
    turbidity measurements may possibly yield erroneous results due to instrument sensitivity
    for very low levels of entrained air.

12. **In-Line Analysis of Fluid Aeration:** Tsuji and Katakura photographically examined and
    measured the effect of dissolved air on the changes of bubble size in a hydraulic system
    [104]. This study was performed using a measuring apparatus which was inserted in the
    hydraulic circuit as shown in Figure 3.48a. A schematic of the measuring apparatus is pro-
    vided in Figure 3.48b. In this work, de-aerated oil was used and bubbles were injected into
    the system through a "bubble injection port." The characteristics of the injected bubbles
    were then studied photographically.

In another study, Tsuji and Matsui quantitatively measured the bubble content of a hydraulic fluid
during operation [105]. The experimental test circuit is shown in Figure 3.49. Air is drawn into the
hydraulic fluid with the flow by the suction with pressure of the pump and air content is controlled
through with valve. Aerated fluid is sent to the measurement system. Bubbles in the fluid are photographed through a sight glass in the system shown in Figure 3.49. The volume distribution of the
bubbles in the fluid in the hydraulic circuit is shown in Figure 3.50.

3.2.3.7 Air Release Specifications
Barber and Perez have summarized the current national specifications for air release. These are
summarized in Table 3.8 [77].

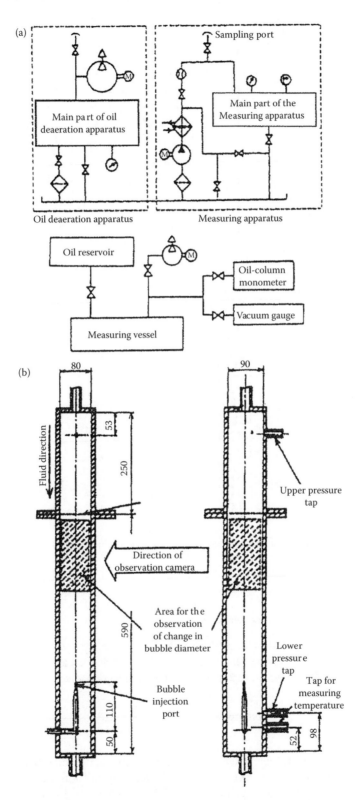

FIGURE 3.48 (a) Apparatus for measurement of air content of hydraulic oil (b) observation cell of air-content measurement apparatus [104].

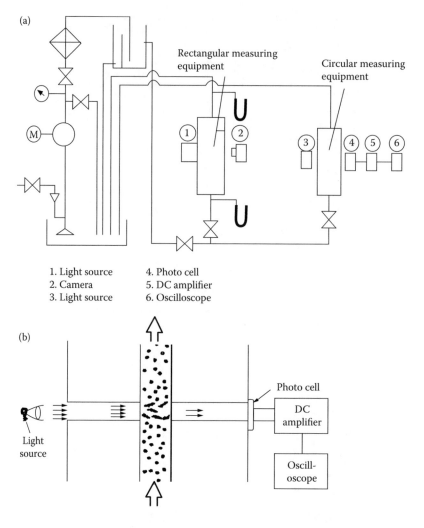

1. Light source   4. Photo cell
2. Camera   5. DC amplifier
3. Light source   6. Oscilloscope

FIGURE 3.49   (a) Hydraulic circuit for measurement of air content (b) optical cell for quantification of air-bubble content [105].

3.2.4   COMPRESSIBILITY AND BULK MODULUS

When a hydraulic system is activated, the hydraulic fluid is compressed as shown in Figure 3.51. Compression is defined as: "volume percent decrease under adiabatic conditions" [106]. (Note that in adiabatic processes, it is assumed that there is no loss of heat due to compression.)

$$\text{Compressibility } (X) = -\frac{\Delta V}{V_0} \cdot \Delta P.$$

In the absence of specific technical data, it is often assumed that the compressibility of petroleum oil is approximately 0.5% for each 100 psi pressure increase up to 4000 psi [107]. However, more specific data for representative fluids is provided in Table 3.9 [108] and Figure 3.52 [109].

Bulk modulus is defined as "the resistance to a decrease in volume when subjected to pressure" and is the inverse of compressibility [106].

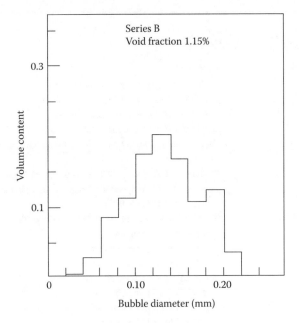

FIGURE 3.50   Bubble distribution in hydraulic oil [105].

$$\text{Bulk modulus } (K) = -V_0 \frac{\Delta P}{\Delta V}.$$

Although the bulk modulus of a fluid affects hydraulic system performance, it is especially important in servo-valve operation. If the bulk modulus is too high, fluid is not sufficiently compressible and excessive fatigue failures of the valve housing may result.

The change in volume at a constant pressure is dependent on the type of compression process; isothermal compression with no change in temperature or adiabatic (isoentropic). The differences between isentropic and isothermal compressibility are shown in Figure 3.53.

Hayward has reported the isoentropic compressibilities for the selected fire-resistant fluids shown in Table 3.10. For mineral oils, the following expression for isoentropic mineral oil was given [110–111]:

$$\beta @ \ 20°C \text{ and } 10,000 \ \text{lb/in}^{-2} = 3.5 - 0.2 \log\eta,$$

where β is in 10^{-6} in $^{-2}$/lb and η is the kinematic viscosity at 22°C. Bulk modulus does not follow Hooke's Law, since the relationship of pressure to specific volume ($\Delta V / V_0$) is not linear, as shown in Figure 3.53.

There are two general graphical solutions that can be used to determine the bulk modulus. In the secant bulk modulus (K_{sec}), a straight line is drawn between atmospheric pressure and the pressure of interest on the pressure-specific volume curve as shown in Figure 3.54 [109].

$$K_{sec} = -\left[\frac{V_0}{\Delta V}\right] \Delta P \text{ at constant temperature.}$$

The isothermal secant bulk modulus can be calculated for any fluid from density or viscosity as follows [5]:

TABLE 3.8
National Specifications for Air Release [77]

Specification Number	Comment

DIN 51524 Parts 1, 2, and 3	For ISO VG grade, air release limit up to
	    ISO VG 10, 22, and 32, 5 min maximum
	    ISO VG 46 and 68, 10 min maximum
	    ISO VG 100, 14 min maximum

Dension Hydraulic ATP-30283-A	De-aeration by AFNOR NFT 60–149 shall be 7 min maximum for any viscosity grade

Case Poclain SA	De-aeration by AFNOR NFT 60–149 shall be 7 min maximum for all
	    grades. The test temperature varies according to the grade (PPF @ 30°C,
	    PPTC 50°C, PPC C 70°C)

Svensk Standard ss 15 54 34	The air release requirement varies with the grade:
	    SH15, 5 min maximum.
	    SH32 & 46, 10 min maximum
	    SH68, 12 min maximum

AFNOR NF E 48–603	For the HM fluids, the requirement varies with ISO grade: up to
	    ISO VG 32, 5 min maximum
	    ISO VG 46, 7 min maximum
	    ISO VG 68, 10 min maximum

ISO CD 11158	For up to
	    ISO VG 32, 5 min maximum
	    ISO VG 46 and 68 grades, 10 min maximum and for ISO VG 100 and
	    above, 14 min maximum

From fluid kinematic viscosity:

$$K_{\text{sec}} = [1.30 + 0.15 \log v] \times [\text{antilog } 0.0023(20 - T)] \times 10^4 + 5.6\, P\,(\text{bar}).$$

From fluid density:

$$K_{\text{sec}} = [1.51 + 7(\rho - 0.86)] \times [\text{antilog } 0.0023(20 - T)] \times 10^4 + 5.6\, P\,(\text{bar}),$$

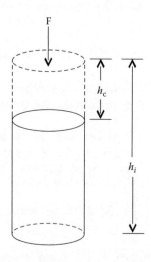

FIGURE 3.51   Illustration of fluid compressibility.

TABLE 3.9
Volume Reduction with Increasing Pressure for Common Hydraulic Fluids [108]

Fluid	% Reduction in Volume at 1000 psi	% Reduction in Volume at 10,000 psi
Water	0.34	3.3
Water-in-Oil	0.35	3.5
Water-Glycol	0.26	2.6
Mineral Oil	0.35	3.4
Phosphate Ester	0.25	2.5
Chlorinated Aromatic Fluids	0.24	2.4
Silicone Fluids	0.66	6.8

where P is the pressure in bar; T is the temperature in °C; v is the kinematic viscosity at atmospheric pressure, in cS at 20°C; ρ is the density at 20°C and atmospheric pressure in kg/L.

If sudden variations in pressure are encountered, as is typical in hydraulic systems, the use of the isoentropic (dynamic) secant bulk modulus is preferred:

$$K_{\sec} = -\Delta P\left[\frac{V_0}{\Delta V}\right] \text{ at constant entropy,}$$

where V_0 is the specific volume at atmospheric pressure at constant entropy, and ΔV is the difference is specific volume ($V_0 - V$) at pressure P.

Isoentropic second bulk modulus may also be calculated from either fluid viscosity or density [5]. From fluid viscosity:

$$K_{\sec} = [1.57 + 0.15\log v]\times[\text{antilog}\,0.0024(20-T)]\times 10^4 + 5.6\,P(\text{bar}),$$

$$K_{\sec} = [1.78 + 7(\rho-0.86)]\times[\text{antilog}\,0.0024(20-T)]\times 10^4 + 5.6\,P(\text{bar}).$$

The secant bulk modulus is dependent on both temperature, as shown in Figure 3.55 and Table 3.11 [112], and on pressure, as shown in Figure 3.56 [113].

The isothermal tangent bulk modulus is determined by drawing a line tangent to the pressure of interest on the pressure-specific volume curve as shown in Figure 3.57.

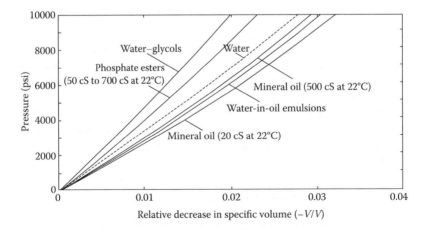

FIGURE 3.52 Relative decrease in specific volume for different hydraulic fluids [109].

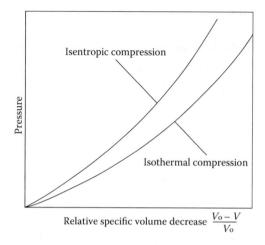

FIGURE 3.53   Illustration of the differences in specific volume for isentropic and isothermal compression [109].

$$K_{tan} = -V \left[\frac{dp}{dv} \right] \text{ at constant temperature,}$$

where V is the specific volume at pressure P.

The isoentropic tangent bulk modulus may be calculated by multiplying the isothermal bulk modulus by the heat capacity ratio C_p/C_r. Typical values for mineral oils are provided in Table 3.12 [94]

All of the illustrative compressibility and bulk modulus data provided so far has been for non-aerated fluids. However, as discussed previously, it is common for hydraulic fluids to undergo some aeration during use. Since air is more than 10,000 times as compressible as oil, aeration will significantly affect bulk modulus as shown in Figure 3.58 [114]. The Hayward equation that describes the effect of air bubbles on secant bulk modulus is:

$$\frac{K_{fluid/air}}{K_{fluid}} = \frac{\left[\dfrac{V_f}{V_a} + 1 \right]}{\left[\dfrac{V_f}{V_a} + \dfrac{KP_0}{P^2} \right]},$$

where V_f is the total volume at atmospheric pressure, P_0; V_a is the air volume at atmospheric pressure, P_0; P_0 is atmospheric pressure; P is the system pressure; K is the secant bulk modulus of the un-aerated oil [94,115,116]. This equation should be used with caution due to the solubility of air in the fluid at higher pressures.

Hodges has provided monograms to determine the secant and tangent bulk modulus of aerated fluids when the relative volume of the entrained air is known. The secant and tangent bulk modulus values are calculated from the value of ϕ taken from Figures 3.59 and 3.60 respectively, and by applying the appropriate equation [5]:

$$\text{Secant Bulk Modulus (aerated)} = K_{sec}\phi_{sec},$$

$$\text{Tangent Bulk Modulus (aerated)} = K_{tan}\phi_{tan}.$$

TABLE 3.10

Compressibility of Fire-Resistant Hydraulic Fluid [111]

Supplier Fluid Type	Description Additives	Water Content (%)	NEL Measurements Viscosity cS at 22°C	Isentropic secant compressibility at 10,000 psi and 20°C × 10^6
Water-glycol	Liquid phase and vapor phase corrosion inhibitors, metal deactivator and lubricity improver	35–50	61.9	2.01
Water-glycol	Liquid phase and vapor phase corrosion inhibitor, metal deactivator and lubricity improver	35–50	100	1.98
Diethylene glycol polyalkylene glycol mixture in water	Corrosion inhibitor and anti-form	55	43.4	2.07
Water-glycol	Liquid phase corrosion inhibitor	50	91.7	2.05
Water-glycol	Liquid and vapor phase corrosion inhibitor	40	81.6	2.06
Water-glycol	Corrosion inhibitor anti-wear and anti-oxidant	–	22.7	2.01
Water-Glycol	Corrosion inhibitor, anti-wear and anti-oxidant	–	52.4	2.02
Water-glycol	Corrosion inhibitor, anti-wear and anti-oxidant	–	139	1.99
Water-glycol solution	Anti-rust	40	117	1.97
Water-in-oil emulsion	Emulsifier, corrosion inhibitor, anti-wear and oxidation inhibitor	40	117	3.15
Water-in-oil emulsion	Emulsifiers, anti-wear and anti-rust	43	77.7	3.00
Water-in-oil emulsion	Emulsifiers	37.5	1200	3.05
Water-in-oil emulsion	Emulsifier, gum inhibitor, anti-rust and anti-wear	40 (by vol.)	222	3.09
Water-in-oil emulsion	Corrosion inhibitor and anti-wear	–	145	3.08
Water-in-oil emulsion	Corrosion inhibitor and anti-wear	–	182	3.07
Phosphate ester	Viscosity index improver, corrosion inhibitor, anti-oxidant, dye and other multi-purpose additives	–	18.2	2.93
Phosphate ester (mixture of triaryl phosphates)	None	–	148	2.93
Phosphate ester	None	–	135	2.27
Phosphate ester	Anti-oxidant, copper deactivator, anti-rust, demulsifier and anti-form	–	73.5	2.29

(continued)

TABLE 3.10 (Continued)
Compressibility of Fire-Resistant Hydraulic Fluid [111]

Supplier	Description	NEL Measurements		
Phosphate ester	Viscosity index improver, anti-oxidant, copper deactivator, anti-rust, demulsifier and anti-former	85.5	—	2.27
Phosphate ester	Anti-oxidant and anti-form	142	—	2.34
Phosphate ester	Anti-form	342	—	2.33
Aryl phosphate ester	None	723	—	2.32
Aryl phosphate ester	None	149	—	2.38
Aryl phosphate ester	None	91.9	—	2.28
Aryl phosphate ester	None	32.5	—	2.28
Aryl phosphate ester	None	245	—	2.37
Mixed aryl/alkyl phosphate ester	Viscosity index improver and traces of other additives	15.9	—	3.20
Mixed chlorinated diphenyl	Rust inhibitor, form control additive, and lubricity additive	182	—	2.27
Chlorinated poyphenyl	Viscosity index improver, anti-wear, anti-form and anti-rust	347	—	2.28
Sebacate diester	None	21.7	—	3.14
Adipate diester	None	22.6	—	3.28
Silicate ester	None	11.5	—	3.61
Chlorophenylmethyl polysiloxa	None	74.4	—	4.27

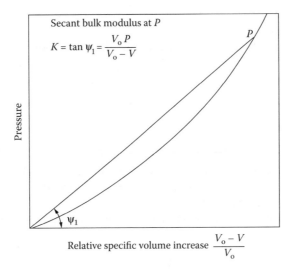

Secant bulk modulus at P

$$K = \tan \psi_1 = \frac{V_o P}{V_o - V}$$

Pressure

Relative specific volume increase $\dfrac{V_o - V}{V_o}$

FIGURE 3.54   Illustration of the calculation of secant bulk modulus [109].

3.2.4.1   Prediction of Bulk Modulus

Since modulus varies with temperature and pressure, it is of interest to be able to calculate the value at a temperature or pressure other than the one available. Klaus and O'Brien have shown that the use of the following equations with Figure 3.61 permits the calculation of secant bulk modulus at any temperature between 32°F and 425°F and pressure between 0 and 150,000 psig. [117].

$$\left[K_{\text{sec}} = \left(K_{\text{sec}} \right)_0 + 5.30 P \right]_{\text{T}},$$

where K_{sec} is the isothermal secant bulk modulus at pressure, P (psi) and temperature T (°F); $(K_{\text{sec}})_0$ is the isothermal secant bulk modulus at 0 psig and temperature (T); and P is the pressure (psig).

$$\left[\log \left[\frac{(K_{\text{tan}})_{T1}}{(K_{\text{tan}})_{T2}} \right] = \beta (T_2 - T_1) \right]_p,$$

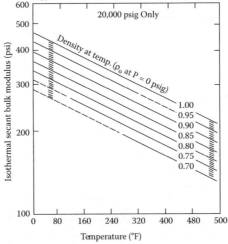

FIGURE 3.55   Secant bulk modulus versus temperature for petroleum oils [113].

TABLE 3.11
Variation of Bulk Modulus with Pressure and Temperature [112]

	Secant Bulk Modulus Increase with Pressure (0–5000 psi) %	
Fluid type	100°F	400°F
Petroleum base	30	100
Phosphate ester	20	47

where $(K_{tan})_{T1}$ is the isothermal secant bulk modulus at pressure P (psig) and temperature T_1 (°F); $(K_{tan})_{T2}$ is the secant isothermal bulk modulus at pressure P (psig) and temperature T_2 (°F). The value of β is obtained from Figure 3.62 [117].

Klaus also showed that the tangent bulk modulus and secant bulk modulus are interrelated by through the equation [117]:

$$\left[K_{tan}\right]_P - \left[K_{sec}\right]K_{tan}.$$

Wright has provided a monogram for the prediction of the secant bulk modulus of a petroleum oil at different temperatures (as shown in Figure 3.55) and pressures (as shown in Figure 3.56) [118].

Bulk modulus varies with oil chemistry; paraffinic versus naphthenic oil, as shown in Figure 3.63 [118].

3.2.4.2 Methods of Measurement

Various methods of determining the bulk modulus of fluids have been reported. These include: mercury piezometer, single-liquid piezometer, metal bellows, an acoustic method, and a thermodynamic conversion method. [110]. Goldman et al. have reported the use of a piston method employing the apparatus shown in Figure 3.64 [119].

A similar system has been developed by Hayward, where a load is applied through a rod entering the fluid being tested and the penetration of the rod is proportional to the compressibility of the fluid. A schematic of the instrument is shown in Figure 3.65 [110].

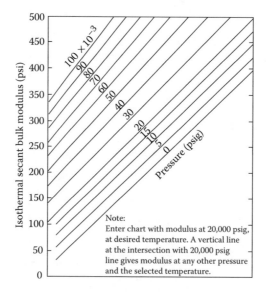

FIGURE 3.56 Secant bulk modulus versus pressure for petroleum oils [113].

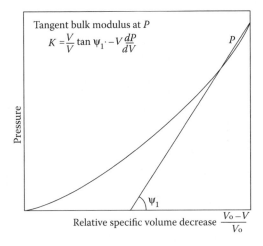

The figure shows a graph with "Pressure" on the vertical axis and "Relative specific volume decrease $\dfrac{V_0 - V}{V_0}$" on the horizontal axis.

Tangent bulk modulus at P

$$K = \frac{V}{V} \tan \psi_1 \cdot - V \frac{dP}{dV}$$

P

ψ_1

FIGURE 3.57 Illustration of the calculation of tangent bulk modulus [109].

Ultrasonic measurement of compressibility has also been applied using an apparatus similar to that shown in Figure 3.66 [120–121].

3.2.5 CAVITATION

Cavitation is defined as "the dynamic process of gas cavity growth and collapse in a liquid" [122]. These cavities are due to the presence of dissolved gases or vaporizable liquids and they are formed when the pressure is less than the saturation pressure of the gas (gaseous cavitation) or the vapor pressure of liquid (vaporous cavitation) [123].

3.2.5.1 Gaseous Cavitation

According to Henry's Law, the solubility of a dissolved gas, such as dissolved (not entrained) air in a hydraulic fluid, is directly proportional to the pressure on the fluid. A decrease in pressure will decrease the solubility of the gas, thereby releasing it from solution in the form of bubbles. When the system is pressurized again the gas will re-dissolve with the generation of heat. This process is called *gaseous cavitation*.

3.2.5.2 Vaporous Cavitation

If the pressure of a system is reduced below the vapor pressure of the fluid, vapor bubbles (cavities) will form. This is called *vaporous cavitation*. The vapor pressure of water, which is present in many fire-resistant fluids or as a contaminant in mineral oil, is shown in Figure 3.67 [124]. The greater propensity for cavitation of water and of water-containing fluids when compared to mineral oil is due to the higher density and vapor pressure of water.

TABLE 3.12
Typical c_p/c_v Ratios for Mineral Oils [94]

Temperature (°C)	c_p/c_v (Atmospheric Pressure)	c_p/c_v (70 MPa)
10	1.175	1.15
60	1.166	1.14
120	1.155	1.13

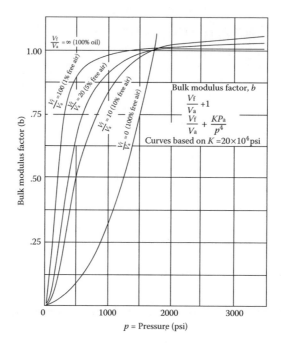

FIGURE 3.58   Bulk modulus factor-pressure curves for air-oil mixture [114].

Of the two forms of cavitation, gaseous and vaporous, gaseous cavitation is most often encountered in hydraulic systems. However, even with relatively large amounts of entrained air, gaseous cavitation may still not occur, although hydraulic response times will be affected [125].

3.2.5.3   Cavitation Numbers

A common indicator of the cavitation potential of a fluid at the cavitating surface is the cavitation number (K) [126,127]:

$$K = \frac{P_0 - P_v}{\frac{1}{2}\rho V_0^2},$$

where P_0 is the static pressure at the inlet; P_v is the vapor pressure of the fluid; V_0 is the velocity at the inlet; and ρ is the fluid density. The smaller the number, the greater the cavitation potential will be.

Hara et al. evaluated the effect of water content of a water-glycol and a W/O emulsion on the "critical" cavitation numbers (K_c) using an orifice in a hydraulic circuit test as shown in Figure 3.68 [128]. For this work, (K_c) was defined as:

$$K_c = \frac{P_2}{P_1 - P_2},$$

where P_1 is the orifice inlet pressure and P_2 is the orifice outlet pressure.

The results of this study showed that increasing the water content produced a general increase in critical cavitation numbers for both the water-glycol and W/O emulsion fluids as shown in Figures 3.69 and 3.70, respectively.

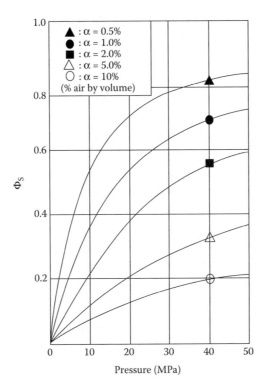

FIGURE 3.59 Correction coefficient for secant bulk modulus of any fluid containing undissolved air [5].

3.2.5.4 Cavitation Equations

A dimensionless equation has been developed which provides an interrelationship between the type of pump and the required fluid velocity inside of the pump in order to ensure that there is no cavitation. This equation is [4]:

$$\left[\frac{Q_L}{Q_T}\right] = 1 - \left[\frac{P_A}{\rho \, CV_P^2}\right]\left[\frac{P_i - P_V}{P_A}\right],$$

where Q_L is the volume flow loss due to cavitation (ft³/s); Q_T is the total theoretical pump discharge flow rate (ft³/s) which is calculated from $A_P V_P$ (A_P = area in ft, velocity of piston, vane to gear tooth in ft/s); P_A is atmospheric pressure in lb/ft² (1 in Hg = 70.7 lb/ft²); P_i is the pump inlet pressure (lb/ft²); P_V is the vapor pressure of the fluid (lb/ft²); ρ is the fluid density (slugs/ft³); and C is the cavitation number which is calculated from $(V_2 - V_1)/V_P$ where V_1 and V_2 is the velocity of the fluid in the pumping cavity between points 1 and 2 (see Figure 3.71).

The dimensionless term Q_L/Q_T indicates the loss of flow caused by cavitation, $(P_i - P_V)/P_A$ is a measure of the static pressure available at in pump inlet to increase the fluid velocity, and $P_A/(\rho CV_P^2)$ is a coefficient that includes the effect of fluid density and flow rate. This equation was used to derive Figure 3.72 which show the probability and nature of cavitation, respectively.

The dimensionless cavitation equation may be rewritten as

$$Q_d = \frac{(P_i - P_V)\, A_P}{\rho C V_P},$$

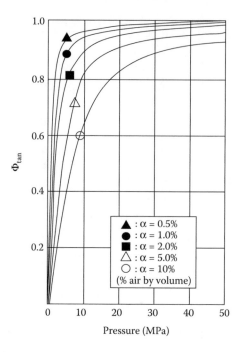

FIGURE 3.60 Correction coefficient for tangent bulk modulus of any fluid containing undissolved air [5].

where Q_d is the actual pump discharge flow volume (cubic feet per second), which is ($Q_T - Q_2$). The limitations to the use of these equations are [4].

1. A fluid that is cavitating may not completely flash to vapor. In this case the pump will perform less than predicted because the back pressure is less. Conversely, fluid with vapor pressure (P_V) that is higher than the inlet pressure (P_i) can prevent pumping entirely, and pressurization or cooling of the inlet may be required.
2. Air or gas entrainment may increase cavitation, particularly with large inlet pressure drops, because trapped air expands and occupies the pumping space. These equations assume that there is no entrained or dissolved air and that the theoretical vapor pressure is achieved.
3. Although the temperature variation will affect the vapor pressure, the slope of the curve Q_V/Q_T vs ($P_i - P_V$)/P_A), shown in Figure 3.73, will remain the same. Any fluid at any temperature may be analyzed as long as there is a position suction head.
4. Viscosities up to 2,000 (SUS) do not affect performance, but above that, cavitation increases. For fluid viscosities greater than 20,000 SUS, it may be necessary to pressurize the fluid.
5. Any or all of these separate effects may produce a separate family of curves. For example, if the speed is increased but other factors remain the same, cavitation may become "excessive" instead of "moderate." However, since inlet restriction or entrained air will give similar results, careful analysis is necessary.

3.2.5.5 Gaseous and Vaporous Cavitation in Pumps and Orifices

Cavitation problems often occur when pressure losses in the suction line to the pump inlet are sufficient to reduce air solubility, thus producing an entrained air mixture. This mixture is then fed

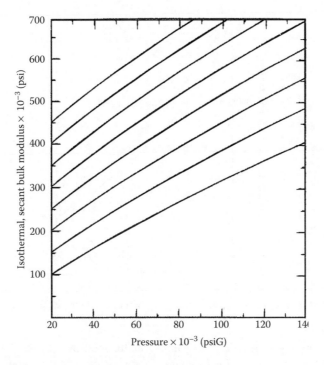

FIGURE 3.61   Generalized bulk modulus-pressure relationship [117].

into the pump resulting in gaseous cavitation. If the pressure drop is sufficient, vaporous cavitation may result.

One example of a potential vaporous cavitation problem is in a gear pump when gear teeth come out of mesh on the suction side, as shown in Figure 3.74. An increase in void volume occurs due to the rotation of the gears. This volume must be filled through the orifice between the tip of the driving tooth and the face of the driven tooth. If there is insufficient fluid flow, vacuum cavities which have formed within the fluid may implode on the discharge side [123].

In cases where a fluid flows through an orifice, there must be a pressure drop at the constriction according to Bernoulli's theorem, which states:

$$P_1 + \frac{1}{2}\rho V_1^2 = P_2 + \frac{1}{2}\rho V_2^2,$$

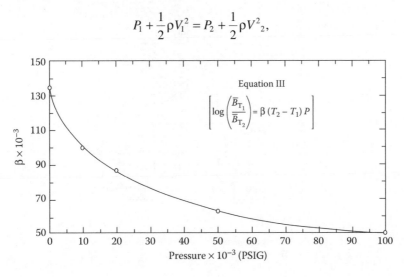

FIGURE 3.62   Effect of temperature and pressure on bulk modulus [117].

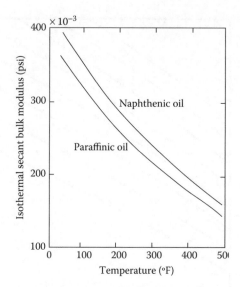

FIGURE 3.63 Effect of oil type and temperature on secant bulk modulus at 20,000 psig [118].

where ρ is the fluid density, P_1 and P_2; and V_1 and V_2 are the pressures (P) and flow rates (V), respectively, at points 1 (upstream) and 2 (downstream). To provide the necessary equality, the pressure at position 2 must be less than the pressure at position 1 since the fluid velocity is greater at position 2 than position 1. This pressure decrease may be sufficient to cause vaporous cavitation.

Hara also studies the effect of orifice shape on cavitation using the hydraulic circuit test illustrated in Figure 3.68. The results of this work, summarized in Figure 3.75, showed that shape edges produce a higher critical cavitation number (K_c) than rounded edges [128].

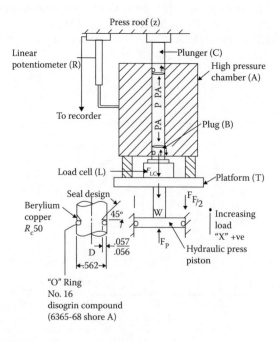

FIGURE 3.64 Apparatus for measurement of fluid compressibility [119].

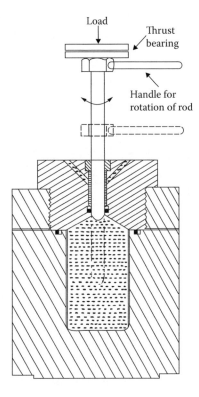

FIGURE 3.65 NEL bulk modulus tester [110].

Ying used flow numbers (f) to illustrate the effect of orifice shape on cavitation numbers (K) [129]. The equation for the cavitation number of a jet pumped through a long orifice where there is a constriction with a pressure difference across the constriction is [127]:

$$K = \frac{P_d - P_v}{P_u - P_d},$$

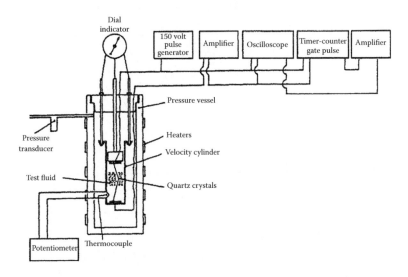

FIGURE 3.66 Ultrasonic velocity apparatus [121].

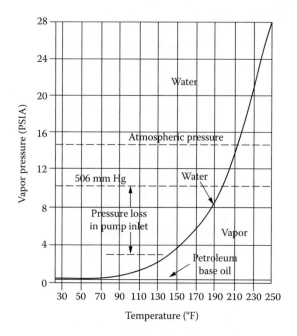

FIGURE 3.67   Vapor pressure of hydraulic fluids [124].

where P_d is the downstream pressure, P_v is the vapor pressure of the fluid, and P_u is the upstream pressure.

The equations for flow numbers are:

$$f = \frac{2x}{\eta} \left[\frac{2(P_u - P_d)}{\rho} \right]^{0.5} = \frac{Re}{C_d},$$

$$f = \frac{2H \sin a}{\eta} \left[\frac{2(P_u - P_d)}{\rho} \right],$$

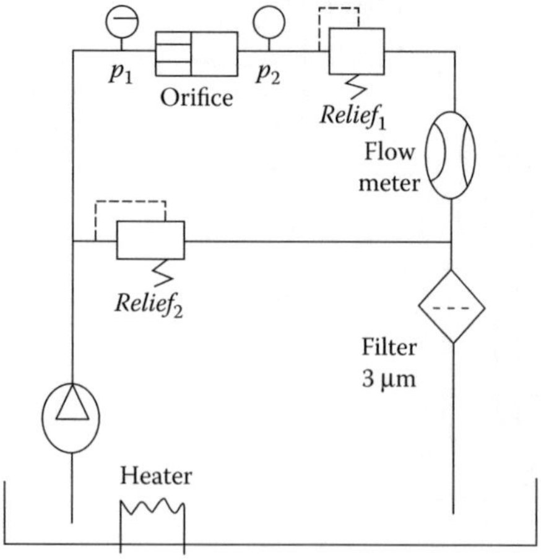

FIGURE 3.68   Hydraulic circuit for cavitation test [128].

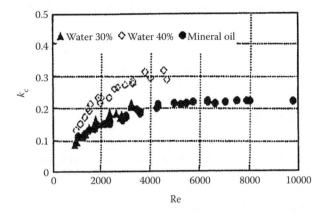

FIGURE 3.69   Effect of water content on cavitation of a water-glycol and a mineral-oil hydraulic fluid [128].

where a is the half cone apex angle of the poppet valve, x is the valve displace relative to the sleeve, H is the lift of a poppet valve, C_d is the discharge coefficient, P_ρ is the mass density, η is the kinematic viscosity, Re is the Reynolds number.

Cavitation should not occur since atmospheric pressure is greater than the vapor pressure. However, cavitation does occur due to presence of "flash" pressure drops (jet cavitation), as shown in Figure 3.76, which are caused by localized pressure drops in vortices that arise from turbulent mixing [123,130].

3.2.5.6   Bubble Collapse Mechanism

Figure 3.77 shows that bubbles begin to collapse with increasing pressure. Bubbles collapse non-symmetrically, forming a liquid micro-jet, which subsequently ruptures, thus producing a "water hammer" [131].

Kleinbreuer modeled the bubble collapse as a symmetrical process. The results are shown in Figure 3.78 [132]. The wide range of bubble collapse pressures have been reported up to >70 MPa, and are summarized in Table 3.13 [133–134].

The impact force of the imploding bubble during cavitation is dependent on a number of factors, such as the vacuum within the bubble at the time of collapse. For example, the bubble may be partially filled with either a gas (gasesous cavitation) or a liquid (vaporous cavitation), thus causing

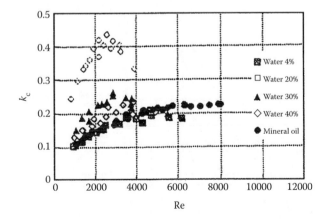

FIGURE 3.70   Effect of water content on a W/O emulsion and a mineral-oil hydraulic fluid [128].

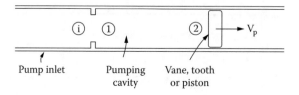

FIGURE 3.71 Schematic for derivation of the cavitation equation [4].

the implosion to be less intense [135]. Although cavitation damage is often associated with vaporous cavitation, which is expected to occur when the hydraulic system pressure is equal or less than the vapor pressure of the fluid, this actually may not occur [136]. This is because cavitation failure is related to the true tensile strength of the fluid. This is affected by contamination of the surface of the metal, which will serve as sites to facilitate the bubble nucleation process, thus also facilitating cavitation [123,137].

Deshimaru studied the effect of Reynolds number (Re) on the cavitation of fluids with or without additives [138]. The results of this study showed that:

1. The cavitation number increases with turbulence at Reynolds number (Re $= d\,[2\Delta P/\rho]^{1/2}/\eta$) values less than 8000. Higher Re numbers produce slightly decreasing cavitation numbers.
2. Cavitation decreased with the addition of viscosity-modifying additives, which was attributed to the "Thoms Effect"—that is, the inhibition of cavitation by the addition of viscosity-enhancing additives which promote more laminar flow, thus reducing Re numbers.

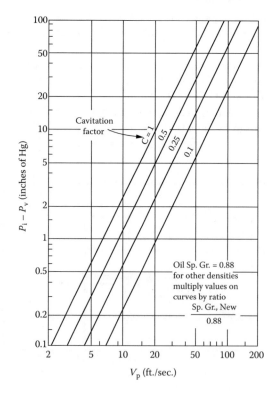

FIGURE 3.72 Probability of cavitation [4].

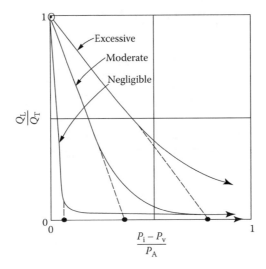

FIGURE 3.73 Nature of cavitation [4].

3.2.5.7 Material Effects

Cavitation erosion from bubble collapse occurs primarily through fatigue fracture due to repeated bubble implosions on the cavitating surface, if the implosions have sufficient impact force [139]. Therefore, the potential for cavitation is dependent on the material properties— especially tensile strength—and the initial condition of the material surface [139–140] and, in some cases, material structure [141]. Some illustrative examples are provided in Table 3.14 [139,165].

Okada et al. have determined that impact loads of 9.1 N (aluminum) 9.7 N (copper) and 13.7 N (mild steel) are required to form a 4 μm pit from the impact of a single bubble [142]. However, the bubble collapse pressures will vary with the bubble size, shape and location.

Talks and Moreton evaluated the volumetric cavitation erosion rates of different hydraulic fluids toward different steels using an ultrasonic cavitation erosion test (see Table 3.15) [143]. Their results, summarized in Table 3.16, illustrate that the general order of the cavitation of different fire-resistant hydraulic fluids toward the different steels was: water > w/o emulsion > water-glycol> invert emulsion > mineral oil. Although cavitation erosion rates were dependent on fluid properties, the erosion damage was only a function of material properties [143].

Tsujino et al., in another ultrasonic laboratory study with a different group of hydraulic fluids, found the order of erosion rates shown in Figure 3.79 to be [144]: Water > HWBF > water-glycol > phosphate ester > mineral oil.

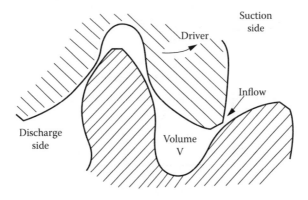

FIGURE 3.74 Suction mechanism of a gear pump [123].

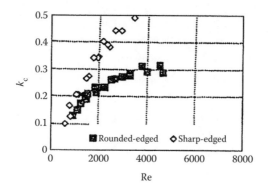

FIGURE 3.75   Influence of orifice shape on cavitation [128].

These data are consistent with the results predicted by Hara et al. from the cavitation number calculations shown in Figure 3.75 [128].

Typically, cavitation bubbles exist in a large "clouds." Figure 3.80 illustrates a cavity formed by cloud of bubbles from a high water base fluid (HWBF). The actual eroded test specimen is shown in Figure 3.81 [145]. The varying distribution and shape of the eroded surfaces formed by different fluids were examined by scanning electron microscopy (SEM), as illustrated in Figure 3.82 [145].

Different studies have shown that every bubble in the cloud does not cause damage upon collapse. The damage frequency has been reported to vary between 1 in 16,000 [146] and 1 in 30,000 [147].

Cavitation bubbles may undergo multiple collapse and reformation processes. Knapp and Hollander have reported that a single bubble may undergo as many as seven collapse and reformation processes [148]. The breakage and reformation process will cease as the viscosity of the surrounding fluid dissipates the energy generated from bubble breakage.

Backè has reported that cavitation wear rate (mm³/h) can be estimated from material properties [149].

$$V_{max} = 27252 \left[\frac{E^{0.562} \, R_{p\,0.2}^{0.618}}{\rho_w \, R_m^{1.071} \, W^{0.125} \, H^{1.971}} \right],$$

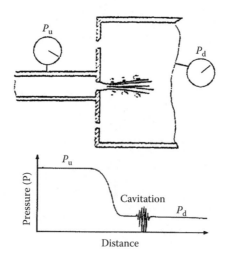

FIGURE 3.76   Cavitation in the mixing zone of a submerged jet [123].

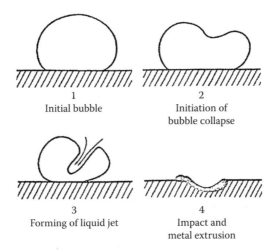

1
Initial bubble

2
Initiation of
bubble collapse

3
Forming of liquid jet

4
Impact and
metal extrusion

FIGURE 3.77   Mechanism of cavitation bubble collapse [131].

where E (N/mm^2) is the elastic modulus of the material; $R_{p0.2}$ is the maximum yield strength at 0.2% elongation; R_m(N/mm^2) is tensile strength; W (N/mm^2) is the work to deform the test specimen for the tensile test ($A_0 \cdot L_0$); H is Vickers hardness and ρ_w (g/cm^3) is the material density.

3.2.5.8 Test Methods

Although there have been a number of test procedures reported, including: spark-induced bubble formation and cavitation collapse [150], hydraulic cylinders, [151–152], hydraulic circuits [128,153], and other devices. Although these tests, particularly the use of hydraulic circuitry, continue to be of interest, there are primarily two types of laboratory screening tests used to determine the cavitation potential of fluids: submerged jets and ultrasonic vibratory apparatus.

Two common submerged jet test configurations are in use: One is the configuration reported by Lichtarowicz [154–160].

Another one is the ultrasonic vibratory test [161–164]. The ultrasonic vibrational cavitation test may be conducted in one of two possible configurations; the specimen or the fluid may be vibrated

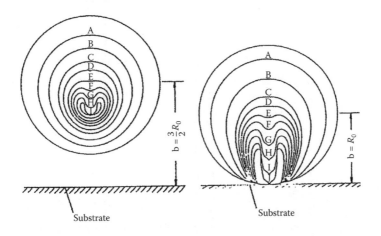

FIGURE 3.78   Theoretical bubble-rupture pattern during cavitation as bubble approaches a solid surface [132].

TABLE 3.13
Bubble Collapse Pressures

Investigators	Bubbles	Methods for Calculation or Measurement of Collapse Pressure	Results	References
		Theory		
Rayleigh	Empty bubble	Spherical; incompressible	1260 atm at the stage of 1/20 of initial radius	[190]
Hickling and Plesset	Gas bubble; initial gas pressure $Pg = 10^{-3}$ atm	Spherical; compressible	$< 2 \times 10^4$ atm	[191]
Ivany and Hammitt	Gas bubble; initial gas pressure $Pg = 10^{-3}$ atm $Pg = 10^{-4}$ atm	Spherical; compressible	6.77×10^4 atm 5.82×10^5 atm	[192]
Plesset and Chapman	Vapor bubble	Based on micro jet velocity	2×10^3 atm	[193]
		Experiment (Single Bubble)		
Jones and Edwards	Spark-induced hemispherical bubble	Piezoelectric pressure- bar gauge	10^4 atm	[194]
Akamatsu and Fujikawa	H_2 gas in water shock tube	Pressure gauge; holographic interferometry	Time duration: 2–3 μ s 10^4–10^5 atm	[195]
Tomita and Shima	Spark-induced bubble	Pressure transducer; photoelasticity	Several 10 MPa	[196]
		Experiment (Cavitation Cloud)		

Investigators	Apparatus to Generate Cavitation	Methods for Measurement of Collapse Pressure	Results	References
Sutton	Acoustic	Photoelasticity	Collapse time 2 μ s 1.36×10^4 atm	[197]
Endo and Nishimura	Vibration	Observation of pit on steel surface	1.2–1.4 Gpa	[198]
Sanada et al.	Vibration	Holographic interferometry	>1 GPa	[199]
Kato et al.	Hydrofoil Model propeller	Pressure-detecting film	Max. 50 MPa Max. 10 MPa	[200]
Okabe et al.	Vibration	Same as above	Apparent impact pressure Max. 15 MPa	[201]
Oba et al.	Jet-flow gate-valve	Same as above	Erosive impact pressure > 70 MPa	[202]

TABLE 3.14
Mechanical Properties of Various Materials [139]

Steel JIS AISI	Tensile Strength σ (MPa)	Modulus of Elasticity (GPa)	Vickes Hardness, Hv
S10C	420	204	130
S35C	615	204	180
S55C	706	204	240
HT80	853	204	300
Cu	196	122	98
Al	145	71	41

[163], the test configuration of the ASTM G 32 test procedure. A classification procedure has been proposed using a vibrational cavitation testing system based on the following criteria [165]:

- Cavitation threshold
- Cavitation power input
- Maximum cavitation erosion; and
- Slope of the cavitation erosion line

3.2.5.9 Bubble Eliminator

Entrained air may cause major problems, such as a reduction in bulk modulus, cavitation and cavitation erosion, as well as rise in oil temperature. When air bubbles in oil are compressed quickly at high pressure, the temperature of the bubble rises drastically. The process is often adiabatic since the period is too short for an exchange of heat to occur. The ignition takes place because of the rising temperature on the boundary surface between the bubbles and the oil, which is known as the "diesel effect." The high temperature accelerates the oils to be aged. The temperature can be estimated as:

$$\left(\frac{T_2}{T_1}\right) = \left(\frac{p_2}{p_1}\right)^{(\kappa-1)/\kappa},$$

where p is the absolute pressure (Pa), T is absolute temperature (K), and κ is the ratio of the specific heat ($\kappa = 1.4$ for air). The subscript 1 is the initial and 2 is the compressed state [166].

One of the devices used to eliminate bubbles from hydraulic fluids. The device consists of a tapered tube which is connected with a cylindrical-shaped chamber. The fluids containing bubbles flow tangentially into the tapered tube from the inlet port, generating a swirling flow. The fluid pressure along the central axis of the flow decreases and the bubbles move to the central axis based on the centrifugal force [167–169].

TABLE 3.15
Volumetric Cavitation Erosion Rates for Different Hydraulic Fluids toward Different Steels' Volumetric Erosion Rates (mm³/h) [143]

Steel	Distilled Water	W/O Emulsion	Water-Glycol	Invert Emulsion	Mineral Oil
Mild steel	5.54	5.5	0.73	0.21	0.27
Stainless steel	1.47	1.2	0.34	0.05	0.02
Bearing steel	0.27	0.22	0.02	0.02	0.01

TABLE 3.16
Cavitation Rating of Different Fluids [165]

		1	2	3	4
Cavitation Rating	Ce_2 (mg)	Straight Mineral	Mineral Oil with Additives	Biodegradable Hydraulic Oils	Synthetic-base Oil
Very low	4			BD 32	
Low	3	Tellus-68			Cassida-HF68
Medium low	2	Turbo-T68			
		Tonna-T220			
Medium high	1	Vitrea-150	Naturelle-HFE46		Cassida-GL150
			Naturelle-HFR32		
High		Ondina-	Omala-320	Cassida-GL460	Cassida-HF46
		Vitrea-9	Tellus-32	Cassida-HF32	Cassida-GL460
			Tellus-T32		Cassida-HF32
					Cassida-HF15

3.2.6 Density and Thermal Expansivity

The density, ρ, of the lubricant under high pressure like in the EHL conjunction is often defined by the conventional Dowson and Higginson's formula [170]. A thermal expansion term is, however, added to this expression to account for the change in volume of the lubricant with temperature [171]. The three coefficients—ρ'_0, D_1, and D_2—required for the definition of the density can be adjusted using the measurement data given as, for example, $D_1 = 0.6$ GPa^{-1} and $D_2 = 1.7$ GPa^{-1}.

$$\rho = \rho'_0 \left(1 + \frac{D_1 p}{1 + D_2 p} \right) \left[1 - \beta \left(T - T'_0 \right) \right].$$

As proposed by Ghosh and Hamrock [172], the thermal expansion coefficient β is represented as an exponential function of pressure:

$$\beta = \beta_0 e^{-cp},$$

where β_0 is the thermal expansion coefficient at the reference temperature and under atmospheric pressure. For example, c can be a constant equal to 1.5 GPa^{-1}.

Following the equation derived by Kuss and Taslimi [173] and Kjølle [174], the density, ρ, is given by

$$\rho = \left[E_0 \left(1 - e^{-p/E_1} - E_2 p \right) + D \right] \left(T - T^0 \right) + \frac{\rho^0}{1 - C \log \dfrac{B + p}{B}},$$

whereas, for the Mobil oil DTE-M, $\rho^0 = 886.86$ kg/m^3, $B = 1.71801 \times 10^8$ Pa, $C = 0.215003$, $D = -0.618199$ kg/(m$^3 \cdot$ K), $E_0 = 0.269774$ kg/(m$^3 \cdot$ K), $E_1 = 1.15299 \times 10^8$ Pa, $E_2 = -0.134291 \times 10^{-9}$ Pa^{-1}.

Following the above equation, the thermal expansion coefficient, β, is given by:

$$\beta = -\rho \frac{\partial}{\partial T} \left(\frac{1}{\rho} \right).$$

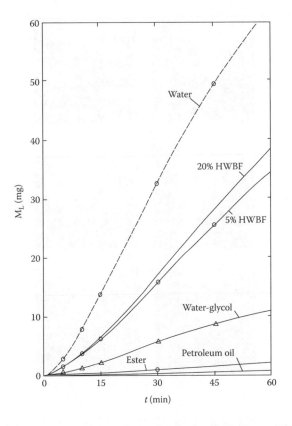

FIGURE 3.79 Propensity for cavitation damage for various hydraulic fluid types [144].

Thus,

$$\beta = \left[E_0 \left(1 - e^{-p/E_1} + E_2 p \right) + D \right] / \rho.$$

3.2.7 SPECIFIC HEAT

The specific heat, c_p, at constant pressure is calculated using the relation proposed in [171]:

$$c_p = \left(\rho c_p \right)_0 \left[1 + \beta_0 \left(1 + b_1 p + b_2 p^2 \right) \left(T - T_0' \right) \right] \left(1 + \frac{k_1 p}{1 + k_2 p} \right),$$

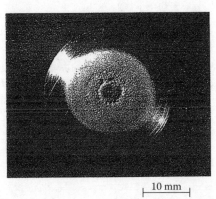

FIGURE 3.80 Illustration of a cavity formed by a cloud of bubbles from a high-water-based fluid [145].

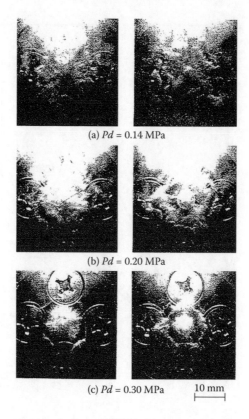

(a) $Pd = 0.14$ MPa

(b) $Pd = 0.20$ MPa

(c) $Pd = 0.30$ MPa　　　　$\underset{\longmapsto\quad\longmapsto}{10\text{ mm}}$

FIGURE 3.81 Impinging cavity clouds for a HWBF. (Courtesy of A. Yamaguchi, Yokohama National University, Yokohama, Japan.)

where $(\rho c_p)_0$ is the specific heat per unit volume at T_0' under atmospheric pressure, while b_1, b_2, k_1, and k_2 are constants.

Another one is [181]

$$c_p = A_0 + A_1 T - T \cdot p \left\{ \frac{\left[E_0 \left(1 - e^{-p/E_1} - E_2 p \right) + D \right]^2}{\rho^3} + \frac{D^2}{\rho_T^3} \right\},$$

where $A_0 = 640.157$ J/(kg · K), $A_1 = 4.4324$ J/(kg · K^2).

3.2.8 THERMAL CONDUCTIVITY

The experimental data [175] reveal that temperature has little effect on the thermal conductivity of the lubricant since this quantity decreases by less than 5% when the temperature increases from 298K to 380K. The thermal conductivity, λ, of a lubricant is therefore regarded as a function of pressure only:

$$\lambda = \lambda_0 \left(1 + \frac{c_1 p}{1 + c_2 p} \right),$$

where c_1 and c_2 are constants.

The thermal conductivity of hydraulic fluids may be determined experimentally by various methods [176–180].

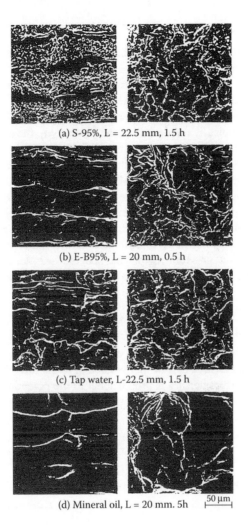

(a) S-95%, L = 22.5 mm, 1.5 h

(b) E-B95%, L = 20 mm, 0.5 h

(c) Tap water, L-22.5 mm, 1.5 h

(d) Mineral oil, L = 20 mm. 5h   50 μm

FIGURE 3.82 SEM photographs of eroded surfaces. (Courtesy of A. Yamaguchi, Yokohama National University, Yokohama, Japan.)

Thermal conductivity (k_L) is calculated from:

$$k_L = \frac{A}{\left[m - \dfrac{B}{kg} \right]} \, J \, cm^{-1} \, sec^{-1} \, {}^\circ C^{-1},$$

where A and B are cell constants; kg is the mass of the fluid in the cell; and m is the rate of change of cell resistance with power input, which is determined using regression analysis of the experimentally-determined cell resistance (R) and power input (P).

$$R = R_0 + mP,$$

where R_0 is the resistance of the cell with zero current.

A typical range for thermal conductivity for mineral oil and synthetic hydrocarbon base oils is: 0.14 W/m · K at 0°C (273K), and 0.11 W/m · K at 400°C (673K) [13].

3.2.9 Lubricant Characteristics

Five lubricants were examined which corresponded respectively to a set of blended mineral oils, a polyalphaolefin (PAO) lubricant, and a polyglycol (PG) lubricant [181]. The viscosity grades of the mineral oils were 32, 68 and 150, which were designated as VG32, VG68 and VG150, respectively. The grades of both PAO and PG were 150. The characteristics of these lubricants are given in Table 3.17 [181] and are considered to be valid for pressures up to nearly 1.2 GPa and for temperatures from 313K to 393K. Typically the density ρ, the viscosity η, the specific heat c_p, and the conductivity λ, increase with pressure while the coefficient β decreases. As temperature increases both ρ and η decrease. The measurements also show that increasing pressure from atmospheric value to 1 GPa leads λ to increase by a factor higher than 2.

The specific heat of liquids at constant pressure increases as the temperature increases. However, the specific heat rather depends on an increase in pressure [182–183]. Under the relatively low pressure, the specific heat decreases as pressure increases; in a certain pressure condition, as pressure increases, the specific heat reaches a minimum, and then increases [184–185]. Therefore, for measuring the data of the low pressure region, the specific heat decreases as pressure increases [186–187], while for the data of the high pressure region, the specific heat increases [188].

3.2.10 Electrical Properties

3.2.10.1 Electrical Conductivity

The amount of electrical charge that passes any point in a conductor per unit time is the current. The current passing through an area of 1 m² perpendicular to the direction of the current is current density (j, amp/m²). The current density is proportional to the potential gradient (dU/dx) and they are related by Ohm's Law [189]

$$j = -\kappa \frac{\partial U}{\partial x}.$$

Conductivity (κ) is the proportionality constant.

TABLE 3.17
Reference Physical Properties of Lubricant at 313K Under Atmospheric Pressure [181]

	Density ρ_0 [kg/m³]	Viscosity μ_0 [kg/m³]	Specific Heat c_{p0} [kJ/(kg . K)]	Thermal Conductivity λ_0 [W/(m . K)]	Thermal Expansivity β_0 [K⁻¹]	Pressure-Viscosity Index α_0 [GPa⁻¹]
Blended mineral oil (VG32)	866	27.6	1.97	0.123	0.72	23.3
Blended mineral oil (VG68)	872	50.5	1.96	0.123	0.69	25
Blended mineral oil (VG150)	886	145	1.92	0.129	0.67	27
Polyalphaolefin	839	127	2.14	0.154	0.68	19.7
Polyglycol	981	136	1.94	0.148	0.76	21

Source: From Kazama, T. "A Comparative Newtonian and Thermal EHL Analysis Using Physical Lubricant Properties, Boundary and Mixed Lubrication: Science and Application," *Proceedings of the 28th Leeds-Lyon Symposium on Tribology*, 2002, pp. 435–446.© 2002, with permission from Elsevier.

Since the electrical field (E) is defined as

$$E = -\frac{\partial U}{\partial x},$$

another expression for Ohm's Law is

$$j = k.$$

If the conductor has a length l' and a cross-sectional area of A, the EMF (E) across the ends is

$$\varepsilon = V_2 - V_1,$$

$$E = \frac{V_2 - V_1}{l'} = \frac{\varepsilon}{l'}.$$

The current carried by the conductor is related to the current density by

$$I = \frac{\kappa A \varepsilon}{l'}.$$

Conductance (L) is defined as

$$L = \frac{\kappa A}{l'}.$$

The current is related to conductance by

$$I = L\varepsilon.$$

The resistance (R) of the conductor is

$$R = \frac{1}{L} = \frac{l'}{\kappa A} = \frac{\rho l'}{A},$$

where ρ is the resistivity which is the inverse of conductivity ($\rho = 1/\kappa$). Ohm's Law becomes

$$\varepsilon = IR.$$

Fluid conductivity is measured using a conductivity cell. Two platinum electrodes, which are typically coated with platinum black, are sealed into the ends of the cell. The cell is filled with the fluid and the resistance is measured by placing the cell in one arm of a Wheatstone bridge. Typically, the frequency is 1000 cycles/s. The resistance of the cell (R) is

$$R = \rho \frac{l'}{A} = \left[\frac{1}{\kappa}\right]\left[\frac{l'}{A}\right].$$

The cell constant ($K = l'/A$) is determined by measuring the resistance (R_s) using a fluid of known conductivity (κ_s).

$$K = \frac{l'}{A} = \kappa_s R_s.$$

The conductivity (κ) of an unknown fluid is calculated from:

$$\kappa = \kappa_s \left(\frac{R_s}{R} \right).$$

It should be noted that fluid conductivity is dependent on temperature and composition.

3.2.10.2   DC Resistivity

DC resistivity of an insulating material is the quotient of a DC electrical field strength and steady-state current within the material [189]. A DC test voltage is applied to the test fluid to provide an electrical stress of 250 V.

3.2.10.3   Volume Resistivity

Volume resistivity of an insulating material is the quotient of a DC electrical field strength and steady-state current within the material [189]. A DC test voltage is applied to a test fluid to provide an electrical stress od 250 V $\cdot$ mm over 60 sec. After the test, the current (I) and test voltage (E) are recorded. The resistivity (ρ) is calculated from:

$$\rho = K \frac{I}{E},$$

where E = test voltage in volts, I = test current in amps, K = cell constant in meters.

3.2.10.4   Relative Permittivity

The relative permittivity (εf_n) of a fluid is the ratio of capacitance (C_x) of a capacitor in which the space between and around the electrodes is entirely filled with the insulating fluid of interest to the capacitance of the some cell in vacuum [189]. The relative permittivity (ε_x) of a fluid is obtained from capacitance measurements of an unknown fluid and a known fluid, and is calculated from:

$$\varepsilon_x = \frac{C_x}{C_a},$$

where C_x is the capacitance of the cell filled the test fluid, and C_a is the capacitance of the cell filled with air as the dielectric.

REFERENCES

1. Jackson, T.L. "Viscosity Requirements of Mineral Oil in Hydraulic Systems," *Hydraulic Pneumatic Power and Controls*, 1963, February, pp. 122–129.
2. Zino, A.J. "What to look for in Hydraulic Oils - II: Viscosity," *American Machinist*, 1947, November, pp. 112–116.
3. Briant, J., Denis, J. and Parc, G. *Rheological Properties of Lubricants*, 1989, Editions Technip, Paris, France, pp. 23–63.
4. Yeaple, F. *Fluid Power Design Handbook - 2nd Edition*, Marcel Dekker Inc., New York, 1990; Chapter 1, pp. 1–22.
5. Hodges, P.K.B. *Hydraulic Fluids*, John Wiley and Sons, Inc., New York, 1996, p. 43.
6. Van Oene, H. "Discussion of Papers 73046 and 730487," *SAE Transaction*, 1973, 1982, p. 1580.

7. Totten, G.E. and Webster, G.M. "High Performance Thickened Water-Glycol Hydraulic Fluids," *Proceedings of the 46th National Conference on Fluid Power*, March 23–24,1994, National Fluid Power Association, Milwaukee,WI, pp. 185–194.
8. Aderin, M., Johnston, G.J., Spikes, H.A. and Caporiccion, G. "The Elastohydrodynamic Properties of Some Advanced Non-Hydrocarbon-Based Lubricants," *Lubrication Engineering*, 1992, 48, pp. 633–638.
9. Gershick, F. "Non-Newtonian Fluid Dynamics in High Temperature High Shear Capillary Viscometers," *Rheology and Tribology of Engine Oils SP-936*, SAE International, Warrendale, PA, 1992, pp. 75–86.
10. Bair, S., Winer, W.O. and Qureshi, F. "Lubricant Rheological Properties at High Pressure," *Lubrication Science*, 1993, pp. 189–203.
11. Carreau, J.P. "Rheological Equations from Molecular Metwork Theories," *Ph.D. Thesis, University of Wisconsin*, pp. 196–198.
12. Roelands, C.J.A., Vlugter, J.C. and Waterman, H.I., "The Viscosity-Temperature-Pressure Relationship of Lubricating Oils and Its Correlation with Chemical Constitution," *Journal of Basic Engineering, Transactions of ASME*, 1963, 101, pp. 601–610.
13. Klaus, E.E. and Tewksbury, E.J. "Liquid Lubricants in CRC Handbook of Lubrication, Theory and Practice of Tribology - Vol. II: Theory and Design" ed. Booser, E.R., *CRC Press Inc.* 1986, Boca Raton, Fl, pp. 229–254.
14. Ewell, R.H. and Eyring, H.J., *Journal of Chemical Physics*, 1937, 5, p. 726.
15. Walther, C. "The Viscosity–Temperature Diagram," *Petroleum Zeitschrift*, 1930, 26, p. 755.
16. Klaus, E. and Fenske, M.R. "The Use of ASTM SIope for Predicting Viscosities," *ASTM Bulletin*, No. 215, July, 1956, pp. TP143–TP150.
17. ASTM D 341–87 – "Standard Viscosity–Temperature Charts for Liquid Petroleum Products," *American Society for Testing and Materials*, Conshohocken, PA, 1987.
18. Anon. "Viscosity – II," *Lubrication*, 1961, 47, pp. 13–27.
19. Dransfield, P. and James, C. "Measuring the Absolute Viscosity of Hydraulic Oils at High Pressure," 1969, February, pp. 83–84.
20. Barus, C. "Note on the Dependence of Viscosity on Pressure and Temperature," *Proceedings of the American Academy of Arts and Sciences*, 1981–1982, 27, pp. 13–18.
21. Jones, W.R., Johnson, R.L., Winer, W.O. and Sanburn, D.M. "Pressure Viscosity Measurements for Several Lubricants to 5.5×10^8 Newtons Per Square Meter (8×10^4PSI) and 149°C (300°F)," *ASLE Transactions*, 1974, 18(4), pp. 249–262.
22. Roelends, C.H.A. "Correlation Aspects of the Viscosity–Temperature–Pressure Relationship of Lubricating Oils"; *O.P. Books Program*, University Microfilm, Ann Arbor, Michigan, 1966.
23. So, B.Y.C. and Klaus, E.F. "Viscosity–Pressure Correlation of Liquids," *ASLE Transactions*, 1979, 23(4), pp. 409–421.
24. Wu, C.S., Klaus, E.E. and Duda, J.L. "Development of a Method for the Prediction of Pressure–Viscosity Coefficients of Lubricating Oils Based on Free-Volume Theory," *Journal of Tribology*, 1989, Vol. III, pp. 121–128.
25. Appledoorn, J.K., *SAE Journal*, 1963, 71, p. 108.
26. Kouzel, B., Hydrocarbon Processing and Pet. Refiner, 1965, 443, 120.
27. Roelands, C.J.A. and Druk, V.R.B. Kleine der A3–4 Groningen, Holland, 1966.
28. Fresco, G.P. *M.S. Thesis, The Pennsylvania State University*, University Park, Pennsylvania, 1962.
29. Kim, H.W. *Ph.D. Thesis, The Pennsylvania State University*, University Park, Pennsylvania, 1970.
30. So, B.Y. and Klaus, E.E. "Viscosity-Pressure Correlation of Liquids," *ASLE Transactions*, 23(4), pp. 409–421.
31. Chu, P.S.Y. and Cameron, A. J. "Pressure Viscosity Characteristics of Lubricating Oils," *Journal of the Institute of Petroleum*, 48, 461, 147 (1962).
32. Worster, R.C. Discussion to Paper by A. E. Bringham: *Proceedings of the Institution of Mechanical Engineers*, 1951, 165, 269.
33. Johnson. W.G. *ASLE Transactions*, 1981, 24(2), p. 232.
34. Zharkin, L.S., Sheberstov, S.V., Panfilovich, N.V. and Manevich, L.I., *Russian Chemical Reviews*, 1989, 58(4), pp. 381–392.
35. Knight, J. "Mechanical Shear Degradation of Polymers in Solution, A Review," *Royal Aircraft Establishment, Technical Report 76073*, June 1976.
36. Foster, T.D. and Mueller, E.R. "Effect of Polymer Structure and Shear Stability of Polymer-Thickened Power Transfer Fluids," *ASTM Special Technical Publication*, 1965, 382, pp. 14–32.
37. Nejak, R.P. and Dzuna, E.R. "Mechanical, Sonic, Ultrasonic and Radiation Studies of Polymer-Thickened Oil Shear Characteristics," *1061 SAE International Congress and Exposition of Automotive Engineering*, Cobo Hall, Detroit, MI, January 9–13, 1961.

38. Stambaugh, R.L. and Preuss, A.F. "Laboratory Methods for Predicting the Viscosity Loss of Polymer-Thickened Hydraulic Fluids," *SAE Technical Paper Series*, Paper Number 680438, 1968.

39. Crail, I.R.H. and Neville, A.L. "The Mechanical Shear Stability of Polymeric VI Improvers," *Journal of the Institute of Petroleum*, 1969, 55(542), pp. 100–108.

40. ASTM D 3945–93, "Standard Test Method for Shear Stability of Polymer-Containing Fluids Using a Diesel Injector Nozzle," *American Society for Testing Materials*, Conshohocken, PA, 1993.

41. ASTM D 2603–91, "Test Methods for Sonic Shear Stability of Polymer-Containing Oils," *American Society for Testing and Materials*, 1991.

42. Myers, H.S. Jr., "Volatility Characteristics of High-Boiling Hydrocarbons," *Ph.D. Thesis, Pennsylvania State University*, University Park, PA, 1952.

43. Maxwell, J.B. and Bonnell, L.S. *Industrial and Engineering Chemistry*, 1957, 49, p. 1187.

44. ASTM D 3827, "Standard Test Method for Estimation of Solubility of Gases in Petroleum and Other Organic Liquids," *American Society for Testing and Materials*, Conshohocken, PA.

45. Wilkinson, E.L. "Measurement and Prediction of Gas Solubilities in Liquids," *Ph.D. Thesis, Pennsylvania State University*, 1971.

46. Blok, P. "The Management of Oil Contamination," *Koppen and Lethem Anadrizjftechniek B.V.*, The Netherlands, 1995, p. 44.

47. Beerbower, A. "Estimating the Solubility of Gases in Petroleum and Synthetic Lubricants," *ASLE Transactions*, 1980, 23(4), pp. 335–342.

48. Hamann, W., Menzel, O. and Schrooder, H. "Gase in ölen—Grundlagen für Hydrauliken," *Fluid*, September, 1978, pp. 24–28.

49. GUlker, E. "Schaumbildung und deren Ursachen," *VDI-Berichte*, 1964, 85, pp. 47–52.

50. Hildebrand, J.H. and Scott, R.L. *The Solubility of Nonelectrolytes*, Dover Publications, New York, 1964.

51. Hoy, H.L. "New Values of the Solubility Parameters from Vapor Pressure Data," *Journal of Paint Technology*, 1970, 42(541), pp. 76–118.

52. Magorien, V.G. "Effects of Air on Hydraulic Systems," *Hydraulics & Pneumatics*, October, 1967, pp. 128–131.

53. Hayword, A.T.J. "How to Keep Air Out of Hydraulic Circuits," *The National Engineering Laboratory*, East Kilbride, Glasgow, 1963.

54. Anon. "Air Entrainment and Foaming," *Fluid Power International*, January/February, 1974, pp. 43–49.

55. Magorien, V.G. "How Hydraulic Fluids Generate Air," *Hydraulics & Pneumatics*, June, 1968, pp. 104–108.

56. Wood, W.D. and Lindsay, W.C. "Dynamic Foam and Aeration Test Apparatus," *Midwest Research Institute, Technical Report AFAPL-TR71–83*, Report prepared for the Air Force, June 1972.

57. Tonder, K. "Effect on Bearing Performance of a Bubbly Lubricant," *Proceedings JSLE/ASLE Lubrication Conference* Tokyo, 1975, pp. 213–221.

58. Honeyman, M.A. and Maroney, G.E. "Air in Oil-Available Measurement Methods," *The BFPR Journal*, 1978, 11(3), pp. 275–281.

59. Anon. "Air Entrainment and Foaming," *Fluid Power International*, January/February, 1974, pp. 43–49.

60. Hallet, B. "Hydraulic Systems at a New High Speed Roll Mill," *Lubrication Engineering*, 1968, 24(4), pp. 173–181.

61. Rood, A.A. "Hydraulic Reservoir Design and Filtration," *SAE Technical Paper Series*, Paper No. 902A, September, 1964.

62. Banks, R.B. "A Comparison Open-Center and Close-Center Hydraulic Systems," in *Proceedings of National Conference on Industrial Hydraulics*, 1956, pp. 35–49.

63. Ingvast, H. "De-Aeration of Hydraulic Oil Offers Many Effects." *The Third Scandinavian International Conference on Fluid Power*, 1993, 2, pp. 535–546.

64. Hayward, A.T.J. "How to Avoid Aeration in Hydraulic Circuits," *Hydraulic & Pneumatics*, November, 1963, pp. 79–83.

65. Bowman, L.J. "Foam Control Agents - Technology and Application," *Speciality Chemicals*, 1982, 2(4), pp. 4–10.

66. Willig, D.N. "Foam Control in Textile Systems," *American Dyestuff Reporter*, June, 1980, pp. 42–50.

67. Okada, M. "Anti-Foaming Properties of Lubricating Oils," (in Japanese), *Journal of Oleao Science*, 1993, pp. 807–810.

68. Prigodorov, V.N. and Gornets, L.V. "Foaming Properties of Hydraulic Fluids," *Chemistry and Technology of Fuels and Oils*, 1970, 9–10, pp. 688–691.

69. Bowman, L.J. "Foam Control Agents - Technology and Application," *Speciality Chemicals*, 1982, 2(4), pp. 5–9.

70. Beerbower, A. and Barnum, R.E. "Studies on the Dispersion of Silicone Defoamant in Non-Aqueous Fluids" *Lubrication Engineering*, June, 1961, pp. 282–285.

71. Salb, F.E. and Lea, F.K. "Foam and Aeration Characteristics of Commercial Aircraft Lubricants," *Lubrication Engineering*, 1975, 31(3), pp. 123–131.

72. Kakstra, R.D. and Sosis, P. "Controlled Foam Laundry Formulation," *Tenside Surfactant Detergents*, 1972, 9(2), pp. 69–72.

73. Grower, G.K. "Current Status of Fuels and Lubricants from Construction Use's Viewpoint," *SAE Technical Paper Series*, Paper No. 775A, October, 1963.

74. Ida, T. "Drag Coefficients of a Single Gas Bubble Rising in Hydraulic Fluids," *Journal of Japan Hydraulics and Pneumatics Society*, 1978, 9(4), pp. 6l–269.

75. Fowle, T.I. "Aeration in Lubrication Oils," *Tribology International*, 198l, 14, pp. 151–157.

76. Möller, U.J. and Boor, U. *Lubricants in Action*, VDI Verlag, London, UK, 1996.

77. Barber, A.R. and Perez, R.J. "Air Release Properties of Hydraulic Fluids," *NFPA Technical Paper Series*, 196, April, 1996.

78. Mang, T. and Junemann, H. "Evaluation of the Performance Characteristics of Mineral Oil-Based Hydraulic Fluids," *Erdol and Kohie-ErUgas - Petrochemic Vereinight mit Brennstoff-chemie*, 1972, 25, pp. 459–464.

79. Staeck, D. "Gases in Hydraulic Oils," *Tribologie & Schmierungstechnik*, 1983, 34(4), pp. 201–207.

80. Hamann, W. "Gase in Olen," *Fluid*, 1978, September, 1978, pp. 24–28.

81. Schanzlin, E.H. "Higher Speeds and Pressures for the Hydraulic Pump," *Proceedings of National Conference on Industrial Hydraulics*, 1956, pp. 35–48.

82. Blok, P. "The Management of Oil Contamination," *Koppen & Lethem Aandriftechnek B.V.*, The Netherlands, 1994, pp. 43–45.

83. Tamura, T. "Test Methods for Measuring Foaming and Antifoaming Properties of Liquids," *Yukagaku*, 1993, 42(10), pp. 737–745.

84. Ross, J. and Miles, G.D. *Oil & Soap*, 1941, 18, p. 19.

85. ASTM D 3601, "Standard Test Method for Foam in Aqueous Media (Bottle Test)," *American Society for Testing Materials*.

86. DIN Standard 53902 Part I, "Prufung von Tensiden und Textilhi-fsmittein Bestimming des Schaum-ermogens LocI,scheibenSchlagverfahren," *Normenausschuss Materialprufung (NMP) Lm Deutches Institut fur Normung e.V.*

87. ASTM D 3519, "Standard Test Method for Foam in Aqueous Media (Blender Test)," *American Society for Materials Testing*.

88. JIS K 2241–86, "Cutting Fluid," 1986, *Japan Industrial Standard*.

89. Bhat, G.R. and Harper, D.L. "Measurement of Foaming Properties of Surfactants and Surfactant Products," *Surfactants in Solution*, ed. by Mihal, K.L., Vol. 10, Plenum Press, New York, 1989, pp. 381–399.

90. ASTM D 892–92, "Standard Test Method for Foaming Characteristics of Lubricating Oils," *American Society for Testing of Materials*, Conshohocken, PA.

91. Saito, I. "Simple Method for Testing Foaming Tendency of Lubricating Oils," Japan Patent, JP 8–62206, August 18, 1994.

92. ASTM D 1881–96, "Standard Test Method for Foaming Tendencies of Engine Coolants in Glassware," *American Society for Testing and Materials*, Conshocken, PA.

93. AFNOR Draft T73–412.

94. Hayward, A.T.J. "Aeration in Hydraulic Systems - Its Assessment and Control," *Proceedings of Institute of Mechanical Engineers Conf. Oil Hydraulic Power and Control*, 1961, pp. 216–224.

95. Claxton, P.D. "Aeration of Petroleum-Based Steam Turbine Oils," *Tribology*, February, 1992, pp. 8–13.

96. ISO/DIS 9120, Petroleum-Type Steam Turbine and Other Oils—Determination of Air Release Properties—Impinger Method, *Draft International Standard*, 1987.

97. Volpato, G.A., Manzi, A.G. and Del Ross, S. "A New Method to Study Air Entrainment in Lubricating Oils," *Proceedings of the First European Tribology Conference*, 1973, pp. 335–342.

98. Hayward, A.T.J. "Methods of Measuring the Bubble Content of Bubbly Oil," *NEL Fluids Note 92*, August, 1960.

99. Liddell, R.E., Rimmer, R.F. and Orr, R.E.H. "Paper No. 5: Design of Lubricating Oil System," *Proceedings of Institute of Mechanical Engineers*, 1969–1970, 184, pp. 41–52.

100. Tat 'kov, V.V. and Proizvodstvo, L. "Hydraulic Drive Performance in the Injection Mechanism of Pressure Diecasting Machines," *Soviet Coating Technology*, 1986, 6, pp. 43–46.

101. Zander, R., Ruppprath, P., Schneider, G.M. and Rohne, E. "New Laboratory Measurement Method for Evaluating Air Release Properties of Fluids - Part I," *Tribologie und Schmierungstechnik*, 1995, 42–5, pp. 263–268.

102. Hayward, A.T.J. "Two New Instruments for Measuring the Air Content of Oil," *Journal of Institute of Petroleum*, 1961, 47(447), pp. 99–106.
103. Honeyman, M.A. and Maroney, G.E. "Air in Oil - Available Measurement Methods," *The BFPR Journal*, 1978, 11(3), pp. 275–281.
104. Tsuji, S. and Katakura, H. "A Fundamental Study of Aeration in Oil 2nd Report: The Effects of the Diffusion of Air on the Diameter Change of a Small Bubble Rising in a Hydraulic Oil," *Bulletin of Japan Society of Mechanical Engineers*, 1978, 21, pp. 1015–1021.
105. Tsuji, S. and Matsui, K. "On the Measurement of Void Fraction in Hydraulic Fluid with Entrained Bubbles," *Bulletin of Japan Society of Mechanical Engineers*, 1978, 21(152), pp. 239–245.
106. Denherder, M.J. "Important Properties of Hydrostatic Transmission Fluids," *SAE Technical Paper Series*, Paper Number 650593, September, 1966.
107. Anon. "Hydraulic Fluids," *Machine Design*, 1995, June, p. 125.
108. Anon. "Hydraulic Fluids: Their Application and Selection," *Hydraulic Pneumatic Power & Control*, August, 1963, pp. 572–584.
109. Hayward, A.T.J. "Compressibility Measurement on Hydraulic Fluids," *Hydraulic Pneumatic Power*, November, 1965, pp. 642–646.
110. Hayward, A.T.J. "The Compressibility of Hydraulic Fluids," *Institute of Petroleum*, 1965, 51, pp. 35–52.
111. Hayward, A.T.J., Martins, R.R. and Robertson, J. "Compressibility Measurements on Hydraulic Fluids Part I: Isoentropic Measurements on Thirty-Four Fire-Resistant Fluids at 20°C and Pressures up to 10,000 lb/in-2," Report issued by *the Department of Scientific and Industrial Research, National Engineering Laboratory*, Glasgow, Scotland.
112. Noonan, J.W. "Ultrasonic Determination of the Bulk Modulus of Hydraulic Fluids," *Material and Standards*, December, 1965, pp. 615–621.
113. Wright, W.A. "Prediction of Bulk Moduli and Pressure Volume - Temperature Data for Petroleum Oils," *ASLE Transaction*, 1967, 10, pp. 349–356.
114. Rendel, D. and Allen, G.R. "Air in Hydraulic Transmission Systems," *Aircraft Engineering*, 1951, 23, pp. 337–346.
115. Hayward, A.J.T. "Air Entrainment and Compressibility of Hydraulic Fluids," *Mechanical World*, October, 1961, p. 332.
116. Hayward, A.T.J. "How Air Bubbles Affect the Compressibility of Hydraulic Oil," *Hydraulic Power Transmission*, June, 1962, pp. 384–388, p. 419.
117. Klaus, E.E. and O'Brien, J.A. "Precise Measurement and Prediction of Bulk - Modulus Values for Fluids and Lubricants," *Journal of Basic Engineering, ASME Transactions*, 1964, 86(D-3), pp. 469–474.
118. Wright, W.A. "Prediction of Bulk Modulus and Pressure Volume-Temperature Data for Petroleum Oils," *ASLE Transactions*, 1967, 10, pp. 349–356. These data are the basis for an ANSI specification ANSI/B 93.63m - 1964, "Hydraulic Fluid Power Petroleum Fluids - Prediction of Bulk Moduli," printed by the National Fluid Power Association, Inc, Milwaukee, WI.
119. Goldman, I.B., Ahmect, N., Nlenkatesan, P.S. and Cartwright, J.S. "The Compressibility of Selected Fluids at Pressures up to 230,000 psi," *Lubrication Engineering*, October, 1971, pp. 334–341.
120. Peeler, R.L. and T. Green, "Measurement of Bulk Modulus of Hydraulic Fluids," *ASTM Bulletin*, January, 1959, pp. 51–57.
121. Noonan, J.W. "Ultrasonic Determination of the Bull: Modulus of Hydraulic Fluids," *Materials Research and Standards*, December, 1965, pp. 615–621.
122. Silva, G. "A Study of the Synergistic Effects of Pump Wear," *Ph.D. Thesis, Oklahoma State University*, Stillwater, OK, 1987.
123. Hobbs, J.M. and McCloy, D. "Cavitation Erosion in Oil Hydraulic Equipment," *Methods and Materials*, January, 1972, pp. 27–35.
124. Li, Z.Y. "Cavitation in Fluid Power Equipment Operating with Fire-Resistant Fluids," *The FRH Journal*, 1984, 4(2), pp.191–199.
125. Ramakrishnan, P. and Sundaram, S. "Effect of Entrained Air on Peak Pressures and Cavitation in a Linear Hydraulic System," *Journal of Institution of Engineers, India - ME*, 1983, 63, pp. 213–219.
126. Hutton, S.P. and Lobo Guerrero, J. "The Damage Capacity of Some Cavitating Flows," *Proceedings of the 5th Conference on Fluid Machinery*, Budapest, 1975.
127. Colemonl, S.L., Scott, V.D., McEnaney, B., Angell, B. and Stokes, K.R. "Comparison of Tunnel and Jet Methods for Cavitation Erosion Testing," *Wear*, 1995, 184, pp. 73–81.
128. Hara, S., Deshimaru, J. and Kasai, M. "Study on Cavitation of Water-Soluble Hydraulic Fluid"; *Proceedings of the International Tribology Conference Yokohama*, 1995, 2, pp. 909–914.

129. Ying, G.L. "Cavitation and Cavitation Erosion of Spool and Popet Valves in Fire Resident Fluids," *FRHJ*. 1984, 5(1), pp. 61–68.

130. Rouse, H. "Cavitation in the Mixing Zone of a Submerged Jet," *La Houille Blanche*, 1953, 8(1), pp. 9–19.

131. Bose, R.E. "The Effect of Cavitation on Particulate Contamination Generation," *Ph.D. Thesis, Oklahoma State University*, Stillwater, OK, 1966.

132. Kleinbreuer, W. "Untersuchung der WerkstaffzerstOrung durch Kavitation in Olhydraulischen Systemen," *Ph.D. Thesis, Rheinisch Westfalischen Technischen Hochschule Aachen*, Aachen, Germany, 1997.

133. Okada, T. and Iwai, Y. "Cavitation Erosion," *Japan Society of Mechanical Engineers International Journal*, 1990, 33(2), pp. 128–135.

134. Okada, T., Iwai, Y. and Awazu, K. "Study of Cavitation Bubble Collapse Pressures and Erosion Part I: A Method for Measurement of Collapse Pressure," *Wear*, 1989, 133, pp. 219–232.

135. Riddel, V., Pacor, P. and Appeldoorn, J.K. "Cavitation Erosion and Rolling Contact Fatigue," *Wear*, 1974, 27, pp. 99–108.

136. Leshchenko, V.A. and Yu. I. Gudilkin. "Cavitation in Self-Induced Vibration Conditions in Hydraulic Servo Systems," *Machines & Tooling*, 1967, 38(6), pp. 19–22.

137. Schweitzer, P.H. and Szebehely, V.G. "Gas Evaluation in Liquids and Cavitation," *Journal of Applied Physics*, 1950, 21, pp. 1218–1224.

138. Deshimaru, J. "A Study of Cavitation on Oils: Effects of Base Oils and Polymers," *Japanese Journal of Tribology*, 1994, 36(4), pp. 531–542.

139. Iwai, Y., Okada, T. and Tanako, S. "A Study of Cavitation Bubble Collapse Pressures and Erosion Part 2: Estimation of Erosion from the Distribution of Bubble Collapse Pressures," *Wear*, 1989, 133, pp. 233–243.

140. Tichler, J.W., van den Elsen, J.B. and de Gee, A.W.J. "Resistance Against Cavitation Erosion of 14 Chromium Steels," *Transactions of ASME, Journal of Lubrication Technology*, April, 1970, pp. 220–227.

141. Okada, T., Iwai, Y., Hattori, S. and Tanimura, N. "Relation between Impact Load and the Damage Produced by Cavitation Bubble Collapse," *Wear*, 1995, 184, pp. 231–239.

142. Okada, T., Iwai, Y. and Awazu, K. "Study of Cavitation Bubble Collapse Pressures and Erosion Part I: A Method for Measurement of Collapse Pressures," *Wear*, 1989, 133, pp. 219–232.

143. Talks, M.G. and Moreton, G. "Cavitation Erosion of Fire-Resistant Hydraulic Fluids," *Proceedings of ASME Symposium on Cavitation Erosion in Fluid Systems*, 1981, pp. 139–152.

144. Tsujino, T., Shima, A. and Oikawa, Y. "Cavitation Damage and Generated Noise in High Water Base Fluids," (in Japanese), *Transactions of Japan Society of Mechanical Engineers, Ser. B*, 1990, 56(532), pp. 3592–3596.

145. Yamaguchi, A. and Shimizu, S. "Erosion Due to Impergement of Cavitating Jet," *Transactions of ASME, Journal of Fluids Engineering*, 1987, 109(4), pp. 442–447.

146. Knapp, R.T. "Recent Investigations of the Mechanics of Cavitation and Cavitation Damage," *Transactions of ASME*, October, 1955, 77, p. 1045–1054.

147. Robinson, J. and Hammitt, F.G. "Detailed Damage Characteristics in Cavitation Venturi," *Transactions of ASME, Journal of Basic Engineering*, 1967, 89, pp. 511–517.

148. Knapp, R.T. and Hollander, A. "Laboratory Investigations of the Mechanism of Cavitation," *ASME*, 1948, 40, pp. 419–475.

149. Backe, W. and Berger, J. "Kavitation Erosion bei I-FA Flussigkeiten," *Olhdraulich and Pneumatik*, 1984, 28(5), pp. 288–296.

150. Kling, C.L. and Hammit, F.G. "A photographic Study of Spark-Induced Cavitation Bubble Collapse," *University of Michigan, Report No. UMICH 03371-4-T*, July 1970.

151. Yamaguchi, A. "Cavitation in Hydraulic Fluids: Part I - Inception in Shear Flow," *Fluids Quarterly*, 1980, 12(3), p. 115.

152. Yamaguchi, A. "Cavitation in Hydraulic Fluids: Part 2 - Delay Time for Stepwise Reduction in Pressure," *Fluids Quarterly*, 12(3), pp. 16–28.

153. Deshimaru, J. "A Study on Cavitation in Fluid Oils," (in Japanese), *Journal of Japanese Society of Tribologists*, 1994, 36(4), pp. 531–542.

154. Lichtarowicz, A. "Use of a Simple Cavitating Nozzle for Cavitation Erosion Testing and Cutting," *Nature*, Physical Science, 1972, 239(91), pp. 63–64.

155. Lichtarowicz, A. "Cavitating Jet Apparatus for Cavitation Erosion Testing, Erosion: Prevention and Useful Application," *ASTM STP 664*, 1977, Adlex, W.F., ed, American Society for Testing and Materials, 1979, pp. 530–549.

156. Lichtarowicz, A. "Erosion Testing with a Cavitating Jet," *Cavitation Erosion in Fluid System, ASME Fluids Engineering Conference*, Boulder, Colorado, USA, 1981.

157. Momma, T. and Lichtarowicz, A. "Some Experiences on Cavitation Damage Produced by a Submerged Jet," *ASME/JSME Nuclear Engineering Conference,* 1993, 2, pp. 877–884.

158. Momma, T. and Lichtarowicz, A. "A Study of Pressures and Erosion Produced by Collapsing Cavitation," *Wear,* 1995, 186–187, pp. 425–436.

159. Kleinbreuer, W. "Werkstoffzerstarung durch Kavitation in Ölhydraudichen Systemen," *Industrie – Anzeriger,* 1976, 98(61), pp. 1096–1100.

160. Yamaguchi, A. and Shimizu, S. "Erosion Due to Impingement of Cavitation Jet," *Transactions of ASME, Journal of Fluid Engineering.,* 1987, 109(4), pp. 442–447.

161. Rheingans, W.J. "Accelerated - Cavitation Research," *Transactions of ASME,* 1950, 72, pp. 705–724.

162. Jones, L.R. and Edwards, D.H. "An Experimental Study of the Forces Generated by the Collapse of Transient Cavities in Water," *Journal of Fluid Mechanics,* 1960, 7, pp. 596–609.

163. Sakamoto, A. Funaki, H. and Matsumura, M. "Influence of Galvanic Macro-Cell Corrosion on the Cavitation Erosion Durability Assessment of Metallic Materials - International Cavitation Erosion Test of Gdansk," *Wear,* 1995, 186–187, pp. 542–547.

164. ASTM G32–92, "Standard Test Method for Cavitation Erosion Using Vibratory Apparatus," *American Society for Testing and Materials,* 1992.

165. Meged, Y., Venner, C.H. and ten Napel, W.E. "Classification of Lubricants According to Cavitation Criteria," *Wear,* 1995, 186–187, pp. 444–453.

166. Tanaka Y., Suzuki, R., Arai, K., Iwamoto, K. and Kawazura, K. "Visualization of Flow Fields in a Bubble Eliminator," *Journal of Visualization,* 2001, 4(1), pp. 81–90.

167. Nagaishi K., Tanaka, Y. and Suzuki, R. "Bubble Elimination for Hydraulic Systems – New Design of Hydraulic System for Environmental Compatibility," *Proceedings of 5th FPNI Ph.D. Symposium Krakow* (2008).

168. Suzuki, R. and Tanaka, Y. "Bubble Elimination in Hydraulic Fluids: Part I – Basic Principle and Technology Overview," *Proceedings of, IFPE2005 Technical Conference* (2005), *NCFP105–17.2.*

169. Suzuki, R. and Tanaka, Y. "Bubble Elimination in Hydraulic Fluids: Part II – A New Technology for Downsizing of Reservoirs," *Proceedings of, IFPE2005 Technical Conference* (2005), *NCFP105–17.4.*

170. Dowson, D. and Higginson, G. R. "Elasto-Hydrodynamic Lubrication," *The Fundamentals of Roller Gear Lubrication, Pergamon Press Ltd,* 1966.

171. Larsson, R., Larsson, P.O., Eriksson, E., Sjoberg, M. and Hoglund, E. "Lubricant Properties for Input to Hydrodynamic and Elastohydrodynamic Lubrication Analyses," *Journal of Engineering Tribology, Proceedings of Institution of Mechanical Engineers, Part J,* 2000, 214, pp. 17–27.

172. Ghosh, M.K. and Hamrock, B.J. "Thermal Elastohydrodynamic Lubrication of Line Contacts," *ASLE Trans.,* 1985, 28, pp. 159–171.

173. Kuss, E. and Taslimi, M. "pV,T Measurements on Twenty Organic Liquids," *Chemie-Ingenieur-Technik,* 1970, 42, pp. 1073–1081.

174. Kjølle, A. "Relations of the Properties of Fluid Power Oils for Thermodynamic Measurements," *VL Rapport,* Nr.10.145 (1976).

175. Larsson, R. and Andersson, O. "Lubricant Thermal Conductivity and Heat Capacity Under High Pressure," *Journal of Engineering Tribology, Proceedings of Institution of Mechanical Engineers, Part J,* 2000, 214, pp. 337–342.

176. Jamieson, D.T. and Tudhope, J.S. "A Simple Device for Measuring the Thermal Conductivity of Liquids with Moderate Accuracy," *Journal of the Institute of Petroleum,* 1964, 50(486), pp. 1-Q-153.

177. Cecil, O.B. and Munch, R.H. "Thermal Conductivity of Some Organic Liquids," *Industrial and Engineering Chemistry,* 1956, 48(3), pp. 437–440.

178. Mason, H.L. "Thermal Conductivity of Some Industrial Liquids from 0–100°C," *Transactions of ASME,* 1954, pp. 817–821.

179. Powell, R.W. and Challoner, A.R. "Thermal Conductivity Measurement on Oils," *Journal of the Institute of Petroleum,* 1960, 446–440, pp. 267–271.

180. Schmidt, A.F. and Spurlock, B.H. "The Thermal Conductivity of Fluids," *Transactions ASME,* 1954, pp. 823–830.

181. Kazama, T. "A Comparative Newtonian and Thermal EHL Analysis Using Physical Lubricant Properties, Boundary and Mixed Lubrication: Science and Application," *Proceedings of the 28th Leeds-Lyon Symposium on Tribology,* 2002, pp. 435–446.

182. Safarov, M.M., Usupov, S.T. and Tagoev, S.A. "Thermophysical Properties of Vegetable Oils in a Wide Range of Temperatures and Pressures," *High Temperatures – High Pressures,* 1999, 31, pp. 43–48.

183. Bessieres, D., Saint-Guirons, H. and Daridon, J.L. "Thermophysical Properties of n-Tridecane from 313.15 to 373.15°K and up to 100 MPa from Heat Capacity and Density Data," *Journal of Thermal Analysis and Calorimetry*, 2000, 62, pp. 621–632.

184. Randzio, S.L. "An Attempt to Explain Thermal Properties of Liquids at High Pressures," *Physics Letters A*, 1986, 117(9), pp. 473–476.

185. Czarnota, I. "Heat Capacity of 2-methylpentane at High Pressures," *Journal of Chemical Thermodynamics*, 1998, 30, pp. 291–298.

186. Dzida, M. and Prusakiewicz, P. "The Effect of Temperature and Pressure on the Physicochemical Properties of Petroleum Diesel Oil and Biodiesel Fuel," *Fuel*, 2007, doi:10.1016/j.fuel.2007.10.010.

187. Wiryana, S., Slutsky, L.J. and Brown, J.M. "The Equation of State of Water to 200 °C and 3.5 GPa: Model Potentials and the Experimental Pressure Scale," *Earth and Planetary Science Letters*, 1998, 163, pp. 123–130.

188. Czarnota, I. "Heat Capacity of Water at High Pressure," *High Temeratures – High Pressures*, 1984, 16, pp. 295–302.

189. "Measurement of Relative Permeativity, Dielectric Dissipation Factor and Resisting of Insulating Fluids," *IEC Standard 247, International Electrotechnical Commission*, Geneva, Switzerland.

190. Rayleigh, L. "On the Pressure Developed in a Liquid during the Collapse of a Spherical Cavity," *Philosophical Magazine*, 1914, 34, p. 94.

191. Hickling, R. and Plesset, M.S. "Collapse and Rebound of a Spherical Bubble in Water," *Physics of Fluids*, 1964, 7(1), p. 7.

192. Ivany, R.D. and Hammitt, F.G. "Cavitation Bubble Collapse in Viscous Compressible Liquids-Numerical Analysis," *Transactions of ASME, Ser. D*, 1965, 87, p. 977.

193. Plesset, M.S. and Chapman, R.B. "Collapse of an Initially Spherical Vapor Cavity in the Neighborhood of a Solid Boundary," *Journal of Fluid Mechanics*, 1971, 2, p. 283.

194. Jones, I.R. and Edwards, D.H. "An Experimental Study of the Forces Generated by the Collapse of Transient Cavities in Water," *Journal of Fluid Mechanics*, 1960, 7, p. 596.

195. Fujikawa, S. and Akamatsu, T. "Experimental Investigations of Cavitation Bubble Collapse by a Water Shock Tube," *Bulletin of Japan Society of Mechanical Engineers*, 1978, 21(152), p. 233.

196. Tomita, Y. and Shima, A. "Mechanisms of Impulsive Pressure Generation and Damage Pit Formation by Bubble Collapse," *Journal of Fluid Mechanics*, 1986, 169, p. 535.

197. Sutton, G.W. "A Photoelastic Study of Strain Waves Caused by Cavitation," *Transactions of ASME, Journal of Applied Mechanics*, 1957, 24(3), p. 340.

198. Endo, K. and Nishimura, Y. "Fundamental Studies of Cavitation Erosion (In the Case of Low Cavitation Intensity)," *Bull. Japan Society of Mechanical Engineers*, 1973, 16(91), p. 22.

199. Sanada, N., Takayama, K., Onodera, O. and Ikeuchi, J. "Observation of Cavitation Induced Shock Waves in an Ultrasonic Vibratory Test," *Transactions of Japan Society of Mechanical Engineers* (in Japanese), 1984, 50(458), p. 2275.

200. Kato, H., Maeda, M. and Nakashima, Y. "A Comparison and Evaluation of Various Cavitation Erosion Test Methods," *Cavitation Erosion in Fluid Systems*, ASME Fluids Engineering Conference, Boulder, Colorado, USA, 1981, p. 83–94.

201. Okabe, Y., Kitajima, A., Koishikawa, A. and Takeuchi, Y. "Experimental Studies on Relationship Between Erosion Rate and Apparent Impact Pressure and Cavitation Monitoring System by Acoustic Detector," *Proceedings of International Symposium on Cavitation*, Sendai, Japan, 1986–4, p. 351.

202. Oba, R., Takayama, K., Ito, Y., Miyakura, H., Nozaki, S., Ishige, T., Sonoda, S. and Sakamoto, K. "Spatial Distribution of Cavitation Shock Pressure Around a Jet-flow Gate Valve," *Transactions of Japan Society of Mechanical Engineers* (in Japanese), 1987, 53(487), p. 671.

4 Fluid Viscosity and Viscosity Classification

Bernard G. Kinker

CONTENTS

4.1 VISCOSITY

Viscosity is the most important property of a hydraulic fluid because it relates directly to both hydrodynamic lubrication and power transmission. For adequate hydrodynamic lubrication, it is necessary to maintain a fluid film between moving surfaces under load. A lubricant of insufficient viscosity will fail to do this, which will lead to wear problems; a fluid of excessive viscosity may also lead to wear as it may not flow adequately to lubricated contacts. For power transmission, fluids of low, inadequate viscosity will be volumetrically inefficient, while a fluid with excessive viscosity will make startup difficult and then result in sluggish, hydro-mechanically inefficient operation. The importance of viscosity is far more complex than described here, but these simple descriptions are a starting point for a discussion of viscosity and its relevance to hydraulic fluids.

4.1.1 ABSOLUTE VISCOSITY AND NEWTON'S LAW

Viscosity is a fluid's resistance to flow. The viscosity of a fluid, either liquid or gas, describes the opposition to a change in shape or to movement. A mathematical description was first developed by Sir Isaac Newton as a special case of his second law of motion:

$$\text{Shear stress} = (\text{Coefficient of viscosity}) \times (\text{Shear rate}), \tag{4.1}$$

where the shear stress is the force per area and the shear rate is the velocity gradient. This leads to a description of viscosity as the force that must be overcome to cause a given fluid motion. Newton's viscosity law applies when laminar, or non-turbulent, flow occurs. Laminar flow can be imagined as numerous, discrete layers (lamina) of the fluid moving in the same direction, but with the velocity of each layer varying, depending on the distance from a system boundary. The concept of velocity as a function of distance is known as the "velocity gradient."

Fluid flow through a tube caused by an applied pressure, known as "Poiseuelle flow," is represented in Figure 4.1. The arrows represent layers moving in streamlines between stationary, parallel walls. The layer closest to a wall is assumed to adhere to the wall and, in turn, will cause a frictional drag on the next closest layer, which is moving because of the applied force. Successive layers experience gradually reduced frictional drag and thus flow at successively higher velocities. The central layer encounters the least friction from adjacent layers and, consequently, has the highest velocity. This change in velocity—from zero at the boundaries to the highest at the center—describes the term velocity gradient.

Another way to induce flow is to provide a tangential shearing force to one surface that moves with respect to a second surface (Couette flow). In Figure 4.2, the upper, movable plate is being forced at a uniform velocity over a fluid film so as to exert a shearing force on the fluid. The bottom

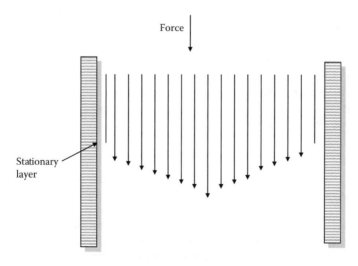

FIGURE 4.1 Schematic of streamline flow between two stationary boundaries.

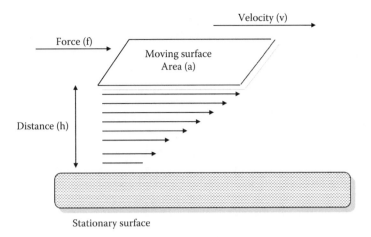

FIGURE 4.2   Schematic of streamline flow between moving and stationary surfaces.

surface is stationary. Assuming adhesion, the fluid layer next to the moving surface experiences the same velocity as the moving plate, while the layer immediately adjacent to the bottom surface is stationary. Moving up from the stationary layer, intervening layers experience gradually reduced drag and gradually higher velocities until arriving at the topmost layer, which has a velocity in concert with the moving surface.

If the dimensions of the system in Figure 4.2 are force = 1 Newton; velocity = 1 m/s; area of the moving plate = 1 m^2; and the distance between the two plates is 1 m, then rearranging Equation 4.1 and substituting

$$\text{Absolute viscosity} = \frac{\text{Force/Area}}{\text{Velocity/Distance}} = \frac{\text{N/m}^2}{\text{m/s/m}} = \frac{\text{Pa}}{\text{s}^{-1}} = \text{Pa} \cdot \text{s}. \tag{4.2}$$

Newton's law describes absolute, or dynamic, viscosity which has the units of Pascal·seconds. Other, commonly used units are Poise (P), which is the centimeter-gram-second (cgs) unit named after Jean Louis Marie Poiseuelle, investigator of streamlined flow, and the centipoise (cP); 1000 cP equals 10 P, which equals 1 Pa · s. The usual symbol for viscosity is η.

Newtonian fluids—that is, those conforming to Newton's law—have a linear shear stress–shear rate relationship and the value of the slope is the coefficient of viscosity as shown in Figure 4.3. By examining the shear stress–shear rate properties of the two fluids in Figure 4.3, one can begin to gather a sense of the importance of viscosity in maintaining a fluid film between moving surfaces. A higher-viscosity fluid provides more resistance to flow out of a potential contact zone and generates a larger lifting force. By definition, Newtonian fluids have a constant viscosity over all values of shear rate (see Figure 4.4). However, under certain conditions, not all fluids are Newtonian; important deviations relevant to lubricant behavior will be discussed in Section 5.

Fluidity is the reciprocal of viscosity and conveys the meaning of ease of flow. Rearrangement of Equation 4.1 yields:

$$\begin{aligned} \text{Shear rate} &= (\text{Shear stress})\,(1/\text{coefficient of viscosity}) \\ &= (\text{Shear stress})\,(\text{Fluidity}). \end{aligned} \tag{4.3}$$

When fluidity is high, the velocity gradient is large at constant shear stress.

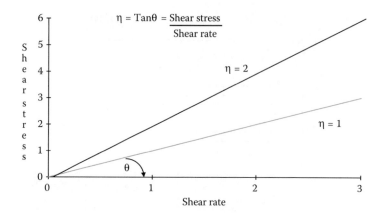

FIGURE 4.3 Newtonian liquids: relationship of shear rate versus shear stress.

4.1.2 RHEOLOGY AND KINEMATIC VISCOSITY

Rheology, the study of fluid flow, can be accomplished by a variety of techniques. The most common include measuring the time-to-flow through a capillary, measuring the force necessary to rotate a cylinder at a given angular velocity through a fluid, or measuring the time for a falling sphere to move through a fluid. Absolute viscosity is usually measured by the rotating cylinder technique. A more common technique, often used for lubricants, is to determine kinematic viscosity by passing a liquid through a capillary tube. Kinematic viscosity is calculated from Poiseuelle's equation:

$$\eta = \frac{P\pi r 4t}{8Vl},$$ (4.4)

where t is the time for a volume V to flow through a capillary of radius r and length l under a pressure of P. However, the method is difficult to apply experimentally, so indirect measurements are usually made.

A very common, indirect procedure used to measure kinematic viscosity is ASTM D 445 (Kinematic Viscosity of Transparent and Opaque Liquids) [1]. The procedure utilizes a modified Ostwald viscometer, which is a U-shaped tube with a capillary near the bottom of one side of the U. When a liquid is higher on one side of the tube, the pressure that causes flow is proportional to the difference in height between the two levels, the density of the liquid, and the acceleration due to gravity. If the viscosity of one liquid is known, then the time-to-flow between two reference points on the tube can be compared to the efflux time of an unknown liquid. Viscosity can be calculated by the simple equation:

FIGURE 4.4 Newtonian liquids: constancy of viscosity with increasing shear rate.

$$\frac{\eta_1}{\eta_2} = \frac{t_1}{t_2}, \tag{4.5}$$

where subscripts 1 and 2 represent values for the two liquids. The CGS units of kinematic viscosity are reported as millimeters squared per second, commonly known as centistokes (cSt) (named for George Gabriel Stokes, who investigated the motion of falling spheres in liquids). Because kinematic viscosity is measured only under gravity and at atmospheric pressure, the shear rate is relatively low at about 100 s^{-1}. Kinematic viscosity should only be measured at temperatures above the cloud point, the temperature at which small, nascent wax crystals begin to cause haze. Insoluble wax crystals can lead to serious errors in the measurement as the particles may interfere with flow through the capillary.

Kinematic viscosity is related to absolute viscosity by the density of the liquid:

$$\text{Absolute viscosity} = (\text{Kinematic viscosity}) \times (\text{Density}). \tag{4.6}$$

4.2 VISCOSITY–TEMPERATURE RELATIONSHIP

An ideal lubricant would be equiviscous at all temperatures in order to provide a constant degree of lubrication and/or power transmission. However, viscosity is highly dependent on temperature as well as pressure. In the extreme, at hot temperatures, a fluid may no longer have sufficient viscosity to form a film of adequate strength between moving surfaces, whereas at cold temperatures, it may become too viscous to flow at all, thereby unable to flow into a contact zone to provide lubrication.

4.2.1 EFFECT OF TEMPERATURE

It is a well-known phenomena that the viscosity of a liquid decreases with increasing temperature; gases have the opposite relationship. An empirical relationship of viscosity to temperature for petroleum oils and hydrocarbons in general is given by the McCoull–Walther–Wright equation first described in 1921 [2], which is the relationship that is the basis for ASTM D 341 (Viscosity–Temperature Charts for Liquid Petroleum Products):

$$\log \log (\eta + 0.7) = A - B \log T, \tag{4.7}$$

where η is the kinematic viscosity, T is the absolute temperature, and both A and B are constants for a given liquid. The negative slope B indicates that higher temperatures result in lower viscosities, as shown for some selected liquids in Figure 4.5. Taking the derivative of Equation 4.7 yields the slope $B = (1/\eta) \, (\delta\eta/\delta T)$, usually referred to as the "viscosity–temperature coefficient."

4.2.2 VISCOSITY INDEX

It is apparent from Figure 4.5 that different liquids may have distinctly different viscosity–temperature relationships, as some lose far more viscosity than others when the temperature increases. The extent of decreasing viscosity with increasing temperature is described by the viscosity index (VI), a dimensionless number, derived from ASTM D 2270 (Calculating Viscosity Index from Kinematic Viscosity at 40°C and 100°C). This arbitrary VI scale was originally devised by Dean and Davis [3]. The calculation of a VI number requires measurement of 100°C and 40°C kinematic viscosities which are then compared to a reference scale. In the original reference scale, a Pennsylvania paraffinic oil, which for its era had excellent viscosity–temperature properties, was assigned a VI of 100. A very different oil, a Texas Gulf naphthenic oil with poor viscosity–temperature properties, was assigned a VI of 0. VI is calculated by

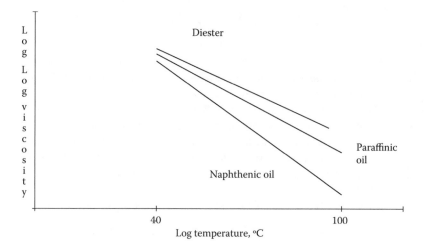

FIGURE 4.5   Viscosity–temperature properties of various liquids.

$$\text{Viscosity index } = \left(\frac{L - U}{L - H} \right) \times 100, \qquad (4.8)$$

where L is the kinematic viscosity at 40°C of a reference oil with VI = 0 and kinematic viscosity at 100°C equivalent to the oil being evaluated; H is the kinematic viscosity at 40°C of a reference oil with VI = 100 and kinematic viscosity at 100°C equivalent to the oil being evaluated; and U is kinematic viscosity at 40°C of the oil being evaluated. In practice, one rarely calculates VI since extensive tabulations are provided in ASTM DS 39B and numerous computer-based programs are also available. For any given liquid, one can calculate its viscosity (between the reference temperatures) with a knowledge of its VI and one reference viscosity.

High-VI lubricants tend to be preferred for applications that experience wide temperature variations, notably outdoor applications, because viscosity changes less with temperature. As shown in

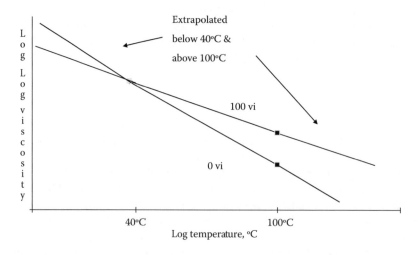

FIGURE 4.6   Viscosity–temperature properties of high- and low-viscosity-index liquids.

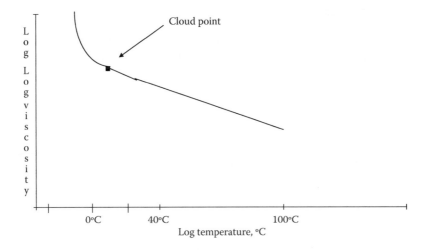

FIGURE 4.7   Effect of wax crystal formation on viscosity–temperature properties below the cloud point.

Figure 4.6, for two oils with equivalent 40°C kinematic viscosities but with different VIs, the higher-VI oil is more viscous as temperature increases and is less viscous at lower temperatures. The VI of a base oil can be enhanced by the use of polymeric additives known as "viscosity index improvers" (VI improvers), which are further described in Section 4.2. Oils containing VI improvers are often known as "multigrade oils" or "HV hydraulic fluids."

The VI is often used as an approximate quality indicator of lubricants. However, some caution is necessary when interpreting the importance of VI when considering temperatures outside the reference temperature range of 40°C to 100°C; this applies in particular for multigrade oils. At sufficiently cold temperatures, below the cloud point of paraffinic oils, viscosity may be much higher than implied by VI because of the onset of wax crystallization, as shown in Figure 4.7. Additionally, multigrade oils may be less viscous than implied by VI because of polymer shear thinning or shear degradation in equipment operating at shear rates that are far higher than those associated with the determination of kinematic viscosity. Both cold-temperature and shear-thinning effects will be discussed further in Section 4.5 on non-Newtonian behavior.

In addition to describing the viscosity–temperature behavior, VI may also be used to infer the chemical nature and quality of base oils. This should be apparent from the definition of VI, as high-naphthenic and aromatic-content oils have low VI values, whereas substantially paraffinic oils have higher VIs. In comparing the VIs of two oils, one might assume that the higher-VI oil has a greater paraffin content and is, therefore, more oxidatively stable. However, some caution must be used because VI does not impart any detailed knowledge of the distribution of naphthenic to aromatic molecules, nor to unstable structures.

4.3   VISCOSITY–PRESSURE RELATIONSHIP

Just as viscosity varies with temperature, it also varies with pressure, but in this case, viscosity increases with increasing pressure. This behavior is an important aspect of lubrication because a liquid is significantly more viscous, or may even become an amorphous solid, under sufficiently high loads. Peak Hertzian contact pressures are often in the 6.9×10^4 kPa (100,000 psi) range, well above the typical solidification points of naphthenic [6.9×10^4 kPa (10,000 psi)] or paraffinic oils [1.4×10^4 kPa (20,000 psi)]. It is under such high-pressure conditions that elastohydrodynamic lubrication can occur, thus providing the film thickness necessary to maintain sufficient lubrication.

4.3.1 Effect of Pressure

High-pressure viscosity can be measured directly by various instruments: a falling-body high-pressure viscometer, a rolling-ball high-pressure viscometer, or a high-pressure capillary viscometer. In 1893, Barus [4] described an empirical isothermal viscosity–pressure relationship in which viscosity increases exponentially with pressure:

$$\eta_p = \eta_0 e^{\alpha p}, \tag{4.9}$$

where η_p is the viscosity at pressure p, η_0 is the viscosity at atmospheric pressure, and α is the pressure–viscosity coefficient at pressure p. It must be noted that viscosity–pressure properties do not vary uniformly; rather, α depends on the magnitude of pressure, the chemical nature of the liquid, and on the temperature.

4.3.2 Mathematical Relationships of Viscosity and Pressure

Viscosity–pressure relationships for liquids are normally expressed by the viscosity–pressure coefficient α in Equation 4.9. A better sense of the nature of the coefficient comes from rearranging Equation 4.9 to:

$$\mathrm{Ln}\left(\frac{\eta}{\eta_0}\right) = \alpha p. \tag{4.10}$$

This indicates that the viscosity–pressure coefficient is the change in viscosity relative to the change in pressure. Higher values of α predict a more viscous liquid as pressure increases. For a change of pressure equal to $1/\alpha$, viscosity will change by a factor of 2.71.

Further defining α, the derivative of Equation 4.10, $(1/\eta)\,(\delta\eta/\delta p)$, expresses the viscosity–pressure differential relative to viscosity at atmospheric pressure. Although Equation 4.10 implies a linear relationship of ln (viscosity) with pressure, this is seldom the case for most liquids except at low pressures. This is illustrated in Figure 4.8, where a plot of isothermal viscosity–pressure properties indicates a decreasing slope (decreasing viscosity–pressure coefficient) with increasing pressure [5]. For most liquids useful as lubricants or hydraulic fluids, the approximate range of the

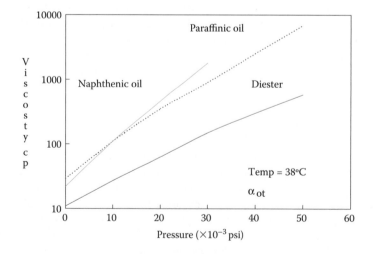

FIGURE 4.8 Viscosity–pressure properties of various liquids.

viscosity–pressure coefficient is $(1.5–5.0) \times 10^{-8}$ Pa^{-1} [$(1–3) \times 10^{-4}$ psi^{-1}]. At pressures typically less than 2.8×10^4 kPa (4000 psi), the difference among viscosity–pressure coefficients of different fluids is relatively small, but at high pressures the differences can be quite marked.

Because α is often a function of pressure, the methodology used to obtain its value must be clearly stated. One well-known technique assumes α is the value of the tangent, at atmospheric pressure, to a log (viscosity) versus pressure plot. This form of the viscosity–pressure coefficient, termed α_{OT}, multiplied by atmospheric viscosity, is often used to predict film thickness tendencies. Another technique is to evaluate the integral of $(\eta_0/\eta_p \delta p)^{-1}$ over the entire pressure range. This value is designated as α^* (reciprocal asymptotic isoviscous pressure) and is the least dependent on measurement procedures and errors.

An empirical relationship has been developed by Roelands [6] which smoothes viscosity–pressure data for a given liquid at a given temperature:

$$\log \eta_p + 1.200 = (\log \eta_0 + 1.200)\left(1 + \frac{P}{2000}\right)^z, \tag{4.11}$$

where η is the absolute viscosity in centipoises, and P is gauge pressure in kilograms cm^2. A simplified form of Equation 4.11 is:

$$H = Zp + H_0, \tag{4.12}$$

where $H = \log(\log \eta_p + 1.200)$, $p = \log(1 + P/2000)$, $H_0 = \log(\log \eta_0 + 1.200)$, and Z, the slope of the relationship, is the pressure–viscosity index. A chart with axes proportional to H and p will linearize pressure–viscosity data for many fluids. The pressure–viscosity index and α_{OT} are related by:

$$\log \alpha_{OT} = \log Z + (H_0 - 2.9388). \tag{4.13}$$

4.3.3 Factors Affecting Viscosity–Pressure Relationships

Viscosity–pressure properties vary greatly with the chemical nature of a liquid. Paraffinic and naphthenic oils, for example, exhibit different degrees of viscosity increase under pressure, as shown in Figure 4.8, where the naphthenic oil, relative to the paraffinic oil, has a greater viscosity increase with increasing pressure. Here one is reminded of the viscosity–temperature properties of different chemical classes and there is indeed a relationship with viscosity-pressure properties. Generally, naphthenic oil viscosity increases more with increasing pressure and decreases less with decreasing temperature than a correspondingly viscous paraffinic oil does. These behaviors lead to a general rule that when comparing two liquids of similar atmospheric viscosity, but in different chemical classes, those with a lower viscosity index will have a higher value of α. Even within the same chemical class, higher-viscosity materials tend to have higher viscosity–pressure coefficients, as indicated by the data for naphthenic oils A and B in Table 4.1 [5].

Temperature also influences viscosity–pressure properties. Lower temperatures tend to give higher values of α and to produce larger differences of α within a chemical class. At sufficiently high temperatures, the differences become quite small, as shown by the 149°C data in Table 4.1. Viscosity–temperature and viscosity–pressure properties have been linked through an empirical equation [7] that relates viscosity–pressure coefficient to atmospheric viscosity and density at a given temperature, and the viscosity–temperature property [the ASTM slope for Equation 4.7 divided by 0.200]. It has also been shown that temperature and pressure effects on viscosity can be combined in a free-volume treatment (see Section 4.1 for a discussion of free volume) of viscosity [8].

The chemical nature, temperature, and pressure effects on viscosity are, indeed, quite complex, but they should become clearer after a more fundamental discussion of viscosity in the following section.

TABLE 4.1
Selected Viscosity–Pressure Coefficients

Fluid	η, 38°C (cP)	α_{OT} (N/m²)⁻¹ 38°C	α_{OT} (N/m²)⁻¹ 99°C
2 Ethylhexyl sebecate	11	1.39	1.19
Paraffinic oil	29	2.18	1.78
Naphthenic oil A	22	2.15	1.44
Naphthenic oil A + VI improver 1.21[a]	60	1.83	1.24
Naphthenic oil B 1.33	68	3.07	1.81
Polybutene (M_n = 409)	90	3.18	2.22

[a] VI Improver = 4% PMA (Mw$_v$ = 560,000)

4.4 MOLECULAR BASIS OF VISCOSITY

Viscosity is also identified with the internal friction of a fluid—that is, the friction generated by molecules as they move by each other during flow. How do molecules move during viscous flow? Why do some flow more slowly? Why do liquids become less viscous with increasing temperature, and more viscous with increasing pressure? Detailed scientific treatments have provided a significant understanding of the physical chemistry and the molecular movements during viscous flow. However, the theory is not so well advanced that accurate predictions can be made for liquids composed of complex molecules. Still, the principles help one to comprehend the nature of viscosity and the behavior of different types of molecules during viscous flow.

4.4.1 SMALL MOLECULES

Base fluids or base oils are made up of relatively small molecules. It should be recognized that most fluids contain mixtures of many different smaller molecules; nevertheless flow properties can still be described by general theories of viscosity.

What is known as the "lattice theory" of viscosity states that liquids consist of a matrix of molecules and that between these molecules there are vacancies or "holes" scattered throughout. During liquid viscous flow, the flowing unit, which may be a group of molecules, a single molecule, or a segment of a molecule, "jumps" into an existing hole and thereby creates a new hole. Some source of energy is required for a molecule to jump over an energy barrier and into a hole. Under a shearing force, holes are filled so as to relieve the stress; thus, flow is in the direction away from the stress source. At higher temperatures, more energy is available and more jumps can be made per unit time, resulting in lower viscosity. At higher pressure, a fluid is compressed, resulting in smaller holes, so fewer jumps can be made, thus causing higher viscosity.

Energy of activation for viscous flow is related to the latent heat of vaporization of the molecule or molecular segment making the jump, because the molecule must be removed from its surroundings. Activation energy for viscous flow tends to be about one-third to one-fourth of the heat of vaporization because the molecule remains in the liquid state and intermolecular forces are replaced. Eyring's equation [9,10] combines viscosity and energy–temperature terms

$$\eta = A e^{\Delta G_v / 2.45 RT},$$

(4.14)

where
   R = gas constant
   T = temperature

ΔG_v = standard free energy of activation for viscous flow (related to standard free energy of vaporization)

A = related to the molar volume of a given material

Several important viscosity relationships can be found in Equation 4.14. The inverse relationship with temperature is clearly present in the exponential term and can be seen more clearly by differentiation of Equation 4.14 at constant pressure:

$$\left(\frac{\delta(\ln \eta)}{\delta T}\right)_p = -\frac{\delta \Delta G_v}{RT^2}. \tag{4.15}$$

The direct relationship of molecular weight, for a homologous chemical series, is in ΔG_v because higher molecular weight translates to higher heat of vaporization and, thus, higher viscosity. The validity of the viscosity–molecular weight relationship for linear paraffins [11], from pentane (C_5H_{12}) through hexadecane ($C_{16}H_{34}$), is shown in Figure 4.9.

The constant A from Equation 4.14 is defined by

$$A = \frac{hN_0}{MW/\rho}, \tag{4.16}$$

where

h = Planck's constant
N_0 = Avagadro's number
MW = molecular weight
ρ = density
MW/ρ = molar volume (volume/mole)

The term A does not contain energy or temperature parameters, but does incorporate molecular size. As molecular weight increases, the value of A decreases, leading to the somewhat surprising implication, at least at first, that viscosity is lower as molecular weight increases. A physical sense of this phenomenon is that higher molecular weight yields a lower molar volume, thus fewer—although larger—molecules are contained in a given volume. Accordingly, there would be more free volume or holes available for jumps. However, the exponential term by far dominates in Equation 4.14, so that molecular weight, by its relationship to energy of activation, has a profound influence on viscosity and in this case it appears intuitively correct as higher MW produces higher viscosity.

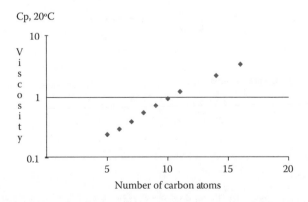

FIGURE 4.9 Effect of increasing molecular weight on viscosity of n-paraffins (cP) at 20°C.

TABLE 4.2
Molecular Structure Effects and Barriers to Internal Rotation on Molecular Flexibility

Compound	Structure Effect	V_0 (kcal/mol of linkages) (barriers)	Flexibility[a]
n-C_9H_{20}	Branching	2.8 (8)	2.1
$(C_2H_5)_4C$		2.8 (4), 4.9 (4)	0.3
2,2,4,4-Tetramethylpentane		4.9 (8)	0.5
n-$C_{14}H_{30}$		2.8 (13)	4.2
2,2,3,3,5,6,6-Heptamethylheptane		4.9 (all)	0.3
n-C_6H_{14}	Double bonds	2.8 (5)	0.84
1-Hexene		2.8 (3), 2.0 (1)	0.85
1,5-Hexadiene		2.8 (2), 2.0 (2)	1.0
1,3,5-Hexatriene, cis		1.2 (2)	>1.7
Benzene, naphthalene	Aromatic rings	Infinite	0
Cyclopentane, cyclohexane, decalins	Naphthene rings	10 (1, 2)[b]	0.5
Diphenyl	Jointed rings	4.5 (1)	0.5
Dicyclohexyl		3.6 (1)	0.3
Tricyclopentylmethane	Crowding of rings	5 (all)	0.95
Tricyclohexylmethane	on alkane	5 (all)	1.4
Tri(2-cyclohexylethyl)methane		3.6 (4), 2.8 (3)	5.1

[a] External degrees of freedom including internal rotation, from density and vapor pressure data.
[b] Pseudorotation for cyclopentane and boat/chair transition for cyclohexane.

Further development of the theory [12] suggests that viscosity on a molecular scale can be related to the energy required for internal rotation of molecular linkages and the barriers to such rotation. If the energy barrier is high, then the molecule tends to be rigid and have a high-energy barrier for translational motion of the molecule or a molecular segment, thus hindering a "jump" into a hole. Molecular structures can range from rigid to flexible. For example, in Table 4.2, the aromatic compound benzene is completely rigid because the individual linkages cannot bend out-of-plane. On the other hand, linear paraffins are quite flexible because rotation around the C—C linkage is possible. However, even with paraffins, these molecules become more rigid as branching is introduced into the molecule creating barriers to free rotation, thus increasing rigidity. Viscosity equations are similar to Equation 4.14 but differ for rigid and flexible molecular structures:

$$\eta_{rigid} = Ae^{kE_0/5CRT}, \tag{4.17}$$

$$\eta_{flexible} = A'e^{k'E_0/5CRT}, \tag{4.18}$$

where
  E_0 = standard energy of vaporization
  k, k´ = constants
  C = flexibility
  R = gas constant
  T = temperature
  A, A´ = $(E_0m)^{1/2}/v$
  m = molecular mass

For rigid molecules, v is related to the molecular surface area because a rigid surface must move in concert as the flow unit, and for flexible molecules, v is related to molecular cross-sectional thickness because a segment of the molecule may be the flow unit. The exponential energy term dominates and depends on the value of C, which is low for rigid molecules; therefore, viscosity is high. A physical sense is that rigid structures must move as a whole unit (in the extreme when there is complete rigidity) or in a highly coordinated state; whereas flexible molecules may experience segmental translations. The viscosity predictions for similar MW compounds are aromatic > naphthenic $\cong$ highly branched paraffins > n-paraffins. Double bonds increase flexibility by allowing free rotation around the bond alpha (adjacent) to the double bond, thereby reducing viscosity.

The viscosity–temperature relationship clearly depends on activation energy, but an additional element—density change associated with temperature change—needs to be considered. If ΔG_v from Equation 4.14 is replaced by the relationship $\Delta H_v - T\Delta S_v$, and density is transferred to the left-hand side of the equation, then differentiation at constant pressure yields:

$$\left(\frac{\delta \ln (\eta / \rho)}{\delta T^{-1}} \right)_p = \frac{\Delta H}{R}, \tag{4.19}$$

which indicates a viscosity–density relationship. As the temperature rises, the packing density of molecules will decrease, creating greater free volume or more "holes." The relationship of viscosity to free volume is expressed as

$$\eta = Ae^{(BV_0/V_f)}, \tag{4.20}$$

where V_0 is the molal volume at 0 K, and V_f is the free (unoccupied) volume that relates to the quantity and size of the holes in lattice theory. As the temperature rises, so does the free volume (as well as energy), with a corresponding reduction in viscosity of both rigid and flexible molecules. However, rigid molecules experience a greater loss of viscosity as translational motion into additional and larger holes becomes more favored. One will recall that aromatic and naphthenic structures—which are, generally speaking, more rigid molecules—tend to exhibit relatively low viscosity indices.

The viscosity–pressure coefficient is also a function of free volume, molecular rigidity, and the size of a flow unit participating in a jump. The viscosity–pressure relationship is approximated by:

$$\ln \left(\frac{\eta}{\eta_0} \right) = \left(\frac{V_{rtz}}{RT} \right) P, \tag{4.21}$$

where P is the pressure and V_{rtz} represents the size of holes per mole and is directly related to the viscosity–pressure coefficient at a constant temperature. Clearly, pressure impacts free volume and the size of holes. Molecular rigidity considerations are the same as for viscosity–temperature, but with decreased hole-size under higher pressure, viscosity will increase.

4.4.2 Polymers in Solution

Polymeric additives, known as VI improvers, are used to increase both the viscosity and VI of base stocks. VI improvers are widely used to formulate high-VI, or multigrade, fluids for use where improved viscosity–temperature properties are desirable. Multigrade fluids are typically used in automotive, off-highway vehicles (such as tractors and aircraft), and industrial equipment exposed to wide temperature ranges. To formulate a multigrade fluid, a lower-viscosity base stock is thickened with a VI improver to a high-temperature viscosity target; and, ideally, the low-temperature viscosity of the formulated oil experiences only a relatively modest increase in viscosity (Figure 4.10).

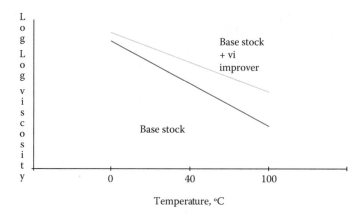

FIGURE 4.10 Thickening effect of viscosity index improver at high and low temperatures.

Because of very large molecular sizes and the possibility of chain entanglement, polymer flow is far more complex than that of small molecules, so that viscosity theories and treatments are far less precise. Although Equation 4.9 still applies to describe viscosity properties of polymers, the heat of vaporization of the polymer is not the correct parameter to describe the activation energy of viscous flow [12]. It has been found that the ΔH_v of paraffins rises fairly uniformly with chain length to an asymptotic limit at about 30 carbons, suggesting that polymer segments of about 20–40 carbons may participate in translational motion in concert. This segment length is roughly similar to the size of base oils used to formulate hydraulic fluids and lubricants. Polymer flow appears to be controlled largely by the negative entropy of activation for viscous flow, perhaps due to the large molecular size. Similar negative entropies have been noted for small rigid molecules with the possible physical sense that insufficient space is available for translational motion, though the exact meaning is obscure.

Polymer thickening of base oil is due to its molecular size, which is immensely greater than that of the solvent or base oil in which it is dissolved. The long polymeric strand, the "backbone" of the polymer, is configured in a random coil shape. Coil size, or hydrodynamic volume, is proportional to polymer molecular weight, as a first approximation, but it is more exactly related to the cube of the root mean square end-to-end distance of the polymer chain [13]. From a simplistic microscopic viewpoint, smaller, less viscous solvent molecules do not readily flow around or through the large polymer coil [14]. Another viewpoint is that larger molecules more completely fill holes, thus limiting the ability of other molecules or segments to participate in movement. The degree of viscosity increase depends on coil size so that higher-molecular-weight polymers provide more thickening. The overall viscosity of a polymer-thickened solution is related to VI improver concentration and molecular weight through the following equation derived by Stambaugh [15]:

$$\ln \eta = K M_v^a c - k''(M_v^a)^2 c^2 + \ln \eta_0, \qquad (4.22)$$

where

M_v = VI improver viscosity average molecular weight
c = VI improver concentration
η_0 = solvent viscosity
a = solubility of the specific polymer chemistry, solvent and temperature

The viscosity increase by temperature displayed in Figure 4.10 may seem anomalous at first glance, since one new to lubricant rheology would normally think that viscosity contribution would be greater at lower rather than higher temperatures. However, this seemingly odd behavior is due to

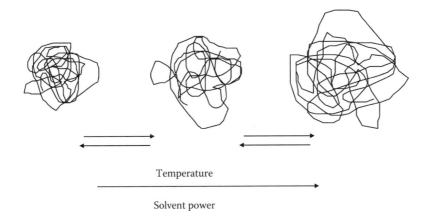

Temperature

Solvent power

FIGURE 4.11   VI Improver coil expansion.

a VI improver's detailed chemical nature resulting in a different relative solubility at different temperatures. A relatively low viscosity contribution at a cold temperature (also known as Higher Viscosity Index) is provided by chemical structures with relatively poor solubility at cold temperatures. With this relatively poor solubility comes a contracted, lower-volume polymer coil contributing relatively little to the viscosity of the solution. However, as temperature increases the coil expands [16] due to improving solubility. With better solvation, coils expand (eventually to a maximum size) and, occupying more volume, now donate more viscosity. The process of coil expansion is entirely reversible, as coil contraction occurs with decreasing temperature (see Figure 4.11). Polymer chemistry and molecular-weight effects have significant influence on coil expansion/contraction and, ultimately, VI lift. Polymers with polar chemical compositions (i.e., polymethacrylates and styrene polyester) undergo larger coil expansion/contraction because solubility in hydrocarbon solvents changes with temperature. By contrast, hydrocarbon polymers (e.g., olefin copolymers, polyisobutene, hydrogenated styrene–diene, and polyisoprene) are well solvated by hydrocarbon solvents at all temperatures so as to experience lesser degrees—sometimes none—of coil expansion/contraction [14]. These same chemical factors also relate to the value of the exponent a in Equation 4.22. For a given chemistry, higher-molecular-weight polymers, being less soluble, thus impart higher VI lifts than chemically equivalent lower-molecular-weight materials.

Commercial VI improvers are available in various chemical compositions and molecular weights ranging from about 20,000 to 800,000 Daltons. The higher-molecular-weight materials are the most efficient thickeners and provide the greatest VI lift, but are also the most susceptible to shearing effects (see Section 5.3). Selection criteria should focus on the impact of the application on stability [17] as well as thickening efficiency and VI lift.

4.5 NON-NEWTONIAN BEHAVIOR

Under certain conditions, some liquids may not always obey Newton's viscosity law. "Non-Newtonian" behavior occurs when the viscosity changes, depending on the value of shear stress or shear rate. Because viscosity may vary, it is often referred to as the apparent viscosity. There are several non-Newtonian behaviors, but those exhibited by lubricants generally fall into only two classes: Bingham fluids and pseudoplastic or shear-thinning fluids.

4.5.1 Bingham Fluids

Bingham fluids do not flow at lower values of shear stress (force). They require the application of sufficient shear stress before flow is initiated; however, once flow begins, the liquid generally

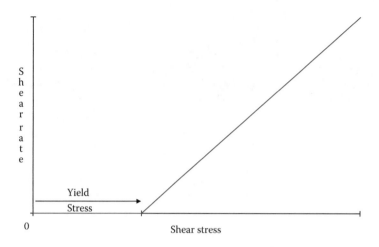

FIGURE 4.12   Bingham fluids: yield stress necessary to initiate flow.

conforms to Newton's viscosity law if Equation 4.1 is modified to include the yield stress (shear stress necessary to initiate flow):

$$\text{(Shear stress} - \text{Yield stress)} = \eta(\text{Shear rate}).\tag{4.23}$$

This behavior is represented in Figure 4.12 when an adequate shearing force initiates fluid flow [18]. Bingham fluids contain a structure extending throughout the material that prevents flow at low values of shear stress because there is insufficient force to perturb the structure. Application of enough shear stress, relaxes or, in some cases, shatters, the structure so as to permit flow. For the case of a shattered structure, smaller elements may persist beyond the yield stress value, and increasing the shear stress may cause further destruction until only the smallest structure elements persist. The apparent viscosity decreases until stability or Newtonian behavior is achieved at a shear stress, referred to as the upper Newtonian, which is sufficiently high to complete the disintegration of the structure (Figure 4.13).

4.5.2   Low-Temperature Rheology of Lubricants

Hydraulic fluids and lubricants can act as Bingham fluids at cold temperatures. This behavior is most often associated with fluids based on paraffinic mineral oils which by nature and by virtue of

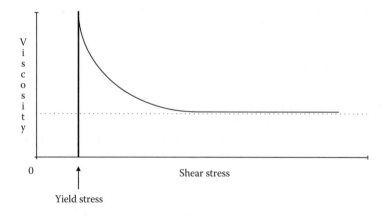

FIGURE 4.13   Bingham fluids: viscosity after structure disintegration.

refinery processing contain small concentrations of naturally occurring waxes. At sufficiently low temperatures, these waxes begin to form crystals. The temperature at which an oil becomes visually hazy is known as the "cloud point" and is determined by ASTM D 2500 (Cloud Point of Petroleum Products). The crystallization process continues as wax solidifies in relatively flat plates that can eventually interlock with each other to form a three-dimensional network. The resulting wax matrix traps the remaining liquid and prevents flow, even though the trapped liquid constitutes a large preponderance of the total mass. This phenomenon is sometimes referred to as gelation of the oil. Two analogies come to mind: the entrapment of honey within a honeycomb and the entrapment of water in a sponge. In both cases, the liquid portion can be quite voluminous but cannot readily flow because of occlusion within the three-dimensional structure of the system.

Sufficient application of force, often simple shaking or stirring, can destroy the relatively weak wax structure, and flow is again possible. In hydraulic system reservoirs (and with other types of lubricants residing in static sumps) at cold ambient temperatures, there may not be sufficient force or suction to destroy wax structures and, in these cases, pump starvation can occur. Wax-impeded flow can severely restrict the low-temperature operating window of paraffinic oils and is of obvious concern for hydraulic fluids that are used during cold conditions.

Paraffinic oils are desirable lubricants because of good VI and oxidative stability properties, but they are susceptible to wax gelation. Gel problems can be alleviated by a small concentration of an additive type, known as pour point depressant (PPD), which can dramatically delay the onset of gelation. These additives, usually low-molecular-weight polymers, contain linear hydrocarbon segments that are wax-like and function by cocrystallizing with waxy paraffins in oil. The additive's attachment to the wax crystal edge and large size hinder further growth in-plane. Additional growth is redirected; leading to crystal sizes and shapes other than the usual plates. Thus, the formation of a wax gel or wax crystal matrix is substantially delayed and fluidity is maintained until the temperature becomes so cold that wax crystal growth is overwhelming. The temperature at which wax-related flow ceases, usually described as the pour point, can be improved substantially with proper PPD additive treatment of a paraffinic fluid; a change of pour point from $-12°C$ for untreated oil to $-40°C$ for an appropriately-treated oil is not uncommon. However, the degree of improvement is highly dependent on the specific oil and its interaction with PPD additive chemistry and concentration.

As previously discussed, wax crystallization phenomena restrict the usefulness of kinematic viscosity measurements and viscosity index applications below the cloud point. A more meaningful test to describe wax-impeded flow is the aforementioned pour point as described in ASTM D 97 (Pour Point of Petroleum Products). The measurement is made by placing a sample in a glass cylinder and cooling rapidly at a rate of approximately $0.6°C/min$. Visual observations of flow are made at $3°C$ intervals after tilting the cylinder 90°. When visually-perceptible flow can no longer be observed, this temperature plus $3°C$ gives the "pour point," the lowest-observed temperature at which flow still occurred. The procedure, by virtue of a relatively large-diameter container and manual tilting, is conducted at low shear rate ($\sim10^{-1}$ sec^{-1}), which is important so as to not destroy the wax structures that are the focus of the test. The ASTM Pour Point is a quick, convenient test well suited for screening wax gelation properties, but it has limited utility in understanding viscosity at cold temperatures as it only answers a flow/no flow question. There are numerous automated methods used to simulate the results of the D 97 Pour Point Test: ASTM D 5949 (Automatic Pressure Pulsing Method); ASTM D 5959 (Automatic Tilt Method); ASTM D 5985 (Rotational Method); and ASTM D 6749 (Automatic Air Pressure Method).

The ASTM Pour Point alone is inadequate to understand the cold-temperature rheology of a lubricant. There are two major limitations of the pour point test: its rapid cooling rate and its qualitative nature. It is fairly obvious that cooling rates can have great influence on wax crystallization; thus, wax gel structure and strength depend on cooling conditions and time. ASTM D 97 employs a rapid cooling rate that may not allow sufficient time for ultimate crystal growth and strength. For instance, vegetable oils used in environmentally friendly hydraulic fluids often have acceptable

ASTM pour points but can solidify at higher temperatures than the "pour point temperature" on extended storage time. Nor does the qualitative nature of ASTM D 97 provide knowledge of fluid viscosity; it is only a determination of the temperature at which flow ceases under rapid cool conditions. Even if wax is absent or under complete control via PPD addition, viscosity is still increasing significantly with decreasing temperature. Any liquid will eventually become so viscous that it can cause sluggish equipment operation or fail to be pumped because viscosity exceeds a pump's upper viscosity limit. This is often noted as the "viscous pour point."

Because the pour point does not provide a high level of information, additional rheological evaluations are often obtained and cited in order to better predict and provide a better understanding of a fluid's low-temperature operating window. The typical test for a hydraulic fluid is ASTM D 2983 (Low-Temperature Viscosity of Automotive Fluid Lubricants Measured by Brookfield Viscometer) usually referred to simply as "Brookfield viscosity." The procedure requires that a sample be placed in a glass tube and then shock-cooled by immersion in a cold air bath where it remains for 16–18 h at the desired test temperature. Still held at test temperature, a measurement is made with a Brookfield viscometer, which rotates a spindle at a low shear rate in the test fluid. The torque required to turn the spindle at constant angular velocity is measured and used to calculate viscosity. Low-shear-rate operation is important in order to measure any wax-gel-contributed viscosity. In fact, the procedure can even be adapted to measure the yield stress as defined in Equation 4.23.

Different cooling rates can exert a profound influence on wax-related low-temperature rheology; as a consequence, numerous other tests have been developed in order to predict lubricant viscosity and, ultimately, a fluid's pumpability properties. These include Federal Standard 791b, Method 203 (Cycle C Stable Pour Point), ASTM D 2532 (Viscosity and Viscosity Change After Standing at Low Temperature of Aircraft Turbine Lubricants), ASTM D 3829 (Predicting the Borderline Pumping Temperature of Engine Oil), ASTM D 4684 (Determination of Yield Stress and Apparent Viscosity of Engine Oils at Low Temperature), ASTM D 5133 (Low Temperature, Low Shear Rate, Viscosity/Temperature Dependence of Lubricating Oils Using a Temperature-Scanning Technique); Federal Standard 791, Method 3456 (Channel Point), JDQ73 Cold Soak, and JDQ74 Slow Cool. These low-temperature tests differ by cooling rate and/or techniques of measurement, but all are conducted at low shear rates to evaluate the potential influence of wax structure as it relates to viscosity. Individual tests are often cited in pertinent lubricant specifications.

Cold-temperature operation is of great importance for hydraulic fluids that are used out-of-doors or under cold ambient conditions. Knowledge of fluid low-temperature viscosity and equipment manufacturer's recommendation of maximum viscosity are important criteria for proper selection of a hydraulic fluid.

4.5.3 Loss of Viscosity in Polymer-Thickened Fluids

Shear thinning, or pseudoplasticity, is a non-Newtonian behavior usually associated with polymer-thickened liquids. At high shear rates, these fluids have lower viscosity than would have been predicted from measurements obtained at low shear rates such as kinematic viscosity determinations. When the shear rate rises above a critical level (Figure 4.14), apparent viscosity decreases as shear rate continues to rise, but then viscosity stabilizes and remains at a constant level [18]. It should be noted that the phenomena is entirely reversible as shear rate is reduced, thus the loss of viscosity under such conditions is only temporary.

Another phenomenon, permanent loss of viscosity, is known to occur with fluids containing VI improvers. In this case polymeric materials undergo mastication via mechanical degradation, leading to lower molecular weight (but more) polymer species, resulting in lower viscosity contribution.

Both shear thinning and mastication effects depend very much on the severity of the operating equipment that defines the ability to shear polymers, and on VI improver molecular weight, which defines a polymer's susceptibility to shearing.

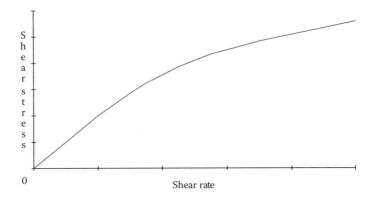

FIGURE 4.14 Shear-thinning behavior of a non-Newtonian fluid.

4.5.4 POLYMER VISCOSITY-LOSS PHENOMENA

4.5.4.1 Temporary Viscosity Loss

Individual polymer molecules useful as oil-soluble VI improvers are quite large on a molecular basis but are also quite flexible. At sufficiently high shear rates, typically greater than 10^4 s^{-1}, a randomly coiled polymer molecule can distort from its at-rest configuration and become oriented along the axis of flow. The polymer's shape changes from a spherical coil to an elongated configuration, which occupies a smaller hydrodynamic volume and thus contributes less viscosity to the solution. With further increases of shear rate, molecules increasingly deform, leading to a corresponding greater loss of viscosity contribution. Eventually, maxima for polymer distortion and loss of viscosity are reached and viscosity contribution, while diminished, does becomes stable (Figure 4.15). Shear thinning of lubricants is often referred to as "temporary loss of viscosity," as the process is entirely reversible upon removal of the high shear rate. When the high shear stress is removed, distorted polymer molecules resume spherical, random coil shapes and reoccupy the original hydrodynamic volumes, and then again, provide the original viscosity contribution as sketched in Figure 4.16. The low-shear-rate, stable-viscosity region of this graph is often referred to as the "First Newtonian," whereas the higher-shear-rate, stable-viscosity region is known as the "Second Newtonian," because viscosity in both regions obey Newton's law.

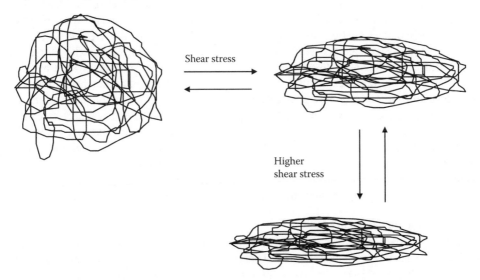

FIGURE 4.15 Schematic of polymer distortion under high shear stress.

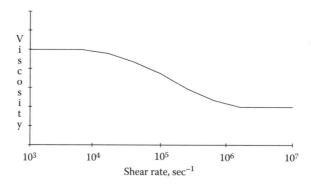

FIGURE 4.16   Shear thinning of apparent viscosity.

Shear thinning can occur in hydraulic system pumps and across pressure relief valves where shear rates can be on the order of 10^6 s^{-1} [19] and produce multigrade fluid viscosities lower than predicted by the low-shear rate measurement of kinematic viscosity. Similarly, in internal combustion engines, shear rates are on the order of 10^5–10^6 s^{-1} in main bearings and at the piston ring cylinder interface, and at 10^6–10^7 s^{-1} in cams and followers. Because the effective, or "in-service," viscosity of multigrade fluids is lowered in these high-shear-rate regions, film thickness is also reduced. Thus, it is necessary to consider high-shear-rate viscosity as an important factor in hydraulic and lubrication systems.

The degree of temporary viscosity loss depends on both the level of shear stress provided by the mechanical equipment and the molecular weight of the polymer. Higher-pressure hydraulic systems generate higher shear stresses so as to cause greater temporary viscosity losses in a given polymer-thickened fluid. Low-molecular-weight polymers, because of their small coil sizes, are less susceptible to elongation and thus to temporary viscosity losses. Depending on system severity, a polymer of a low enough molecular weight may experience relatively little or even no shear thinning. Overall, severe operating conditions can be counterbalanced by multigraded fluids formulated with low-molecular-weight VI improvers.

Equipment shear stress must be sufficiently high to cause shear thinning of polymer-thickened oils. So too must laboratory techniques intended to measure shear thinning be energetic enough to cause the phenomenon. A variety of high-shear-rate rotational and high-pressure capillary techniques, usually operated at 10^6 s^{-1}, are available: ASTM D 4624 (Measuring Apparent Viscosity by Capillary Viscometer at High Temperature and High Shear Rates), ASTM D 4683 (Measuring Viscosity at High Shear Rate and High Temperature by Tapered Bearing Simulator), and ASTM D 5481 (Measuring Apparent Viscosity at High Temperature and High Shear Rate by Multicell Capillary Viscometer). Some care must be taken when comparing results from different procedures because capillary viscometers may give a marginally higher result when compared to those from rotational devices [20]. The above methods are frequently used to evaluate and specify engine oils such as those described in Table 4.3, in which low-shear-rate kinematic and high-shear-rate viscosities are compared for VI improvers of different chemistries and molecular weights (or different shear stabilities). Although hydraulic fluid pumping efficiency has been related to high-shear-rate viscosity [17,21], other evaluation techniques are normally used. A more typical practice is to induce a permanent loss of viscosity by mechanical degradation and then measure kinematic viscosity.

4.5.4.2 Permanent Viscosity Loss

Very high shear stresses, perhaps coupled with turbulent flow, can lead to extreme polymer coil distortion and concentrate enough vibrational energy to cause polymer rupture. Cavitation may also play a role in polymer chain rupture, more commonly referred to as "polymer shearing," by producing intense velocity gradients [22]. Polymer shearing occurs by cleavage of a carbon–carbon bond

TABLE 4.3
High-Shear-Rate Viscosities

Fresh Fluid	Kinematic Viscosity (mm²/sec)		High-Shear-Rate Viscosity (mPa.sec)
	100°C	150°C	150°C
A (base oil plus additive package)	5.4		
A plus VII 1[a]	10.5	4.81	2.99
A plus VII 2[a]	10.5	4.43	2.99
A plus VII 3[a]	10.5	3.72	2.58
A plus VII 4[b]	10.5	4.78	3.19
A plus VII 5[b]	10.5	4.47	3.06
A plus VII 6[b]	10.5	4.08	2.96

[a] Different VI improver chemistries but of equivalent permanent shear stability (i.e., ASTM D 7109 SSI = 45).
[b] Different VI improver chemistries but of equivalent permanent shear stability (i.e., ASTM D 7109 SSI = 25).

located, statistically speaking, near the middle of the polymer chain. The result of shearing a single polymer molecule is two molecules each having approximately half of the molecular weight of the original molecule. The total hydrodynamic volume of the two smaller molecules is less than that of the single starting molecule, resulting in lower viscosity contribution. Since scissioning of polymer backbone carbon bonds is not reversible, the viscosity loss is permanent. Figure 4.17 represents the molecular elongation and rupture concepts leading to polymer shearing

Higher-molecular-weight polymers are more susceptible to distortion and shearing during mechanical degradation processes. Conversely, polymers of sufficiently low-molecular weight may not undergo permanent shearing at all, depending on the severity of the application. Polymer molecules produced by shearing effects are, of course, of a lower molecular weight than the starting material and, depending on the shear stress of the application, may not be susceptible to further degradation which is usually the case. Thus, the shearing process is self-limiting and viscosity

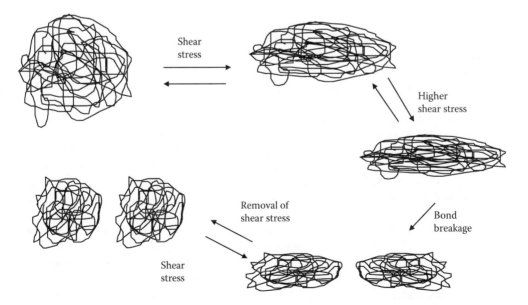

FIGURE 4.17 Schematic representation of polymer temporary and permanent shear thinning.

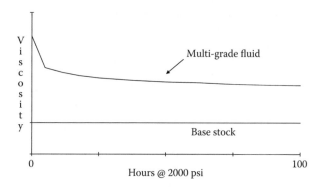

FIGURE 4.18   Permanent viscosity loss with time.

eventually becomes stable, as displayed in Figure 4.18. It is also important to note that a VI improver characteristically contains a wide array of different molecular-weight species which have a Gaussian distribution around an average value. Therefore, not all of the molecular species are necessarily as susceptible to shearing forces, and if the average molecular weight is appropriate for the application severity, only a minor portion may be sheared. The degree of permanent viscosity loss, just as with temporary loss, is a function of equipment severity and polymer molecular weight.

Viscosity loss after mechanical degradation can be expressed mathematically as:

$$\% \, \text{Overall viscosity loss} = \left(\frac{\eta_i - \eta_s}{\eta_i} \right) \times 100, \tag{4.24}$$

where η_i is the viscosity before shear and η_s is the viscosity after shear. If the base-stock viscosity is known, a more meaningful calculation can provide a means to rank the shear stability of various VI improvers:

$$\text{SSI} = \left(\frac{\eta_i - \eta_s}{\eta_i - \eta_0} \right) \times 100, \tag{4.25}$$

where SSI is the shear stability index and η_0 is the viscosity of the base stock and other additives; the remaining terms are the same as in Equation 4.19. The SSI is useful for VI improver stability ranking because it is an expression of the viscosity lost due to polymer shearing, relative to the initial viscosity contributed by polymer. Caution must be exercised when comparing SSIs, as different applications—for example different pressure pumps and/or bench test procedures—can produce widely divergent shearing severity and thus widely different results.

Various devices with sufficient energy to shear polymers—pumps, engines, ball mills, ultrasonic oscillators, high-speed stirring, transmissions, gear sets, fuel injectors, and roller bearings—have been used to study the mechanical degradation of polymer-thickened fluids in the laboratory. Although there is no standard, most hydraulic fluid shear stability work focuses on three procedures. The now obsolete ASTM D 2882 (Indicating the Wear Characteristics of Petroleum and Non-Petroleum Hydraulic Fluids in a Constant Volume Vane Pump) was primarily used to measure wear performance but has often been used to evaluate shear stability. Equipment problems have forced recent workers to modify the test by substituting a more durable pump for 100-hour 13,800-kPa (2000-psi) operation; the newer method is described in ASTM D 7043. ASTM D 5621 (Sonic Shear Stability of Hydraulic Fluid) utilizes an ultrasonic oscillator which irradiates a small sample for 40 minutes. A similar test has long been used to specify military aircraft hydraulic fluids per MIL-H-5606. ASTM D 7109 (Shear Stability of Polymer-Containing Fluids Using a Diesel Injector Nozzle)

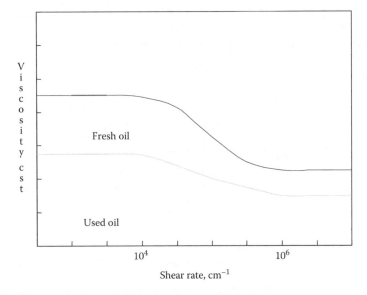

FIGURE 4.19 Combined permanent and temporary viscosity loss.

specifies a double-plunger European injector. The test was originally designed to evaluate engine oils and the procedure requires 30 passes of test fluid through the equipment; hydraulic fluids are often exposed to 250 passes because the VI improvers used in multi-grade hydraulic fluids tend to be more shear stable than those used in engine oils. ASTM D 5275 utilizes Fuel Injector Shear Stability Test (FISST) equipment, where the injector is a single-plunger type and requires 20 passes through the injector. Despite the similarities of ASTM D 7109 and ASTM D 5275, the results obtained on a given fluid in each procedure may differ. CEC L-45-T-93 Method C, a 20-hour test, is based on tapered roller bearing (TRB) equipment and provides very high shearing; incidentally this particular test has many variations where time is varied in order to provide more or less shearing exposure. The severity of these methods can be ranked as TRB > pump > ultrasonic oscillator >> fuel injectors, but there is a reasonable degree of correlation among all three [21]. Care must be taken in interpreting results from the non-pump tests, as not only severity levels can differ but results can also be biased relative to the chemistry of the polymeric VI improver. For instance, polymethacrylates are more severely sheared in the ultrasonic device than are other VI improver chemistries.

4.5.4.3 Permanent and Temporary Viscosity-Loss Effects

Returning to the concept of shear thinning or molecular elongation, polymer-thickened fluids that have undergone permanent loss of viscosity can still undergo shear-thinning effects, albeit to a far lesser extent than an unsheared fluid. A schematic of this molecular concept is also present in Figure 4.17. Permanently sheared, smaller molecules are less susceptible to elongation so that shear-thinning effects are now far less pronounced. The high-shear-rate viscosity of a fluid that is already mechanically degraded in a severe application is only marginally lower than its low-shear-rate viscosity as represented in Figure 4.19, and the data in Table 4.4 [19] illustrate this fact. Thus, the simple kinematic viscosity of appropriately-degraded hydraulic fluids could be considered sufficient to evaluate in-service viscosity and has been shown to correlate well with vane pump flow rates [21].

4.6 VISCOSITY CONTROL AND IMPACT

In order to achieve an appropriate viscosity for a lubricant or hydraulic fluid, it is necessary to carefully select the base fluids and additives. The formulated oil must meet all important viscosity criteria, whether at high or low temperature or at high or low shear rate.

TABLE 4.4
Viscosities of Fluids at Pump Running Temperature

	Viscosity, 65.6°C (150°F)			
	Kinematic (mm²/sec)		High Shear Rate at 106 s⁻¹ (mPa.sec)	
Fluid	Fresh	100 hour	Fresh	100 hour
HSO-01	19.9	17.9	15.9	14.4
HSO-02	16.9	13.3	12.5	11.3
HSO-03	13.6	10.1	9.8	9.0
HSO-04	16.4	9.1	9.0	7.9
HSO-05	27.3	17.8	15.6	14.6
HSO-06	27.4	15.9	13.5	13.2

After 100 h in a 2000-psi vane pump circuit.

4.6.1 VISCOSITY CONTROL

A hydraulic fluid is formulated to a desired viscosity primarily by choice of base stock(s) and, if used, by VI improver and pour point depressant. To a first approximation, other additives do not contribute significantly to viscosity. The simplest case in formulating a fluid is to select a base stock which meets the requirements of the lubricant specification. However, if a base stock meeting the required viscosity is not commercially available, one can simply mix a lower-viscosity base fluid with a higher-viscosity fluid in the proper ratio to achieve the target viscosity.

To prepare polymer-thickened multigrade fluids, a VI improver is added in sufficient quantity to thicken a low-viscosity liquid to the desired viscosity target. Temporary and/or permanent viscosity losses introduced by VI improvers can be easily accommodated by allowing for shearing effects and the concomitant loss of viscosity through choice of VI improver with suitable shear stability. By choosing a VI Improver of appropriate shear stability index for the given application, one may formulate a multigrade fluid to an initial viscosity such that the fully sheared fluid will still retain sufficient viscosity. The initial viscosity target is dictated in large part by the magnitude of the shear stability index of the VI improver.

If paraffinic base oils are involved in the formulation, particularly lubricants intended for use at ambient, out-of-doors temperatures, a pour point depressant is usually added in order to protect against wax gelation.

4.6.2 VISCOSITY IMPACT

Hydrodynamic lubrication and power transmission principles are also discussed in other chapters in this volume, but a brief review here should help to link their relationships to viscosity.

Hydrodynamic lubrication can be simply defined as the separation of two surfaces in relative motion to one another by a fluid film between them. The fluid film is of sufficient thickness to prevent significant contact of the surfaces (e.g., the mating surfaces in a hydraulic system pump). At constant speed and load, film thickness depends on the presence of a fluid having some minimum viscosity sufficient to maintain film strength and prevent contact. Newton's relationship equating shear stress to shear rate explains how a fluid of sufficient viscosity, under a shearing force (motion), can maintain film strength and maintain surface separation. The upshot of all this is that a lubricant must be of some minimum viscosity to provide at least minimum film thickness. On the other hand, excessive viscosity is undesirable because energy must be expended to overcome the internal friction of the lubricant. Within the hydrodynamic lubrication regime, the coefficient of friction, f, depends on:

$$f = \frac{k\eta V}{W},\tag{4.26}$$

where η is the dynamic viscosity, W is the load, and V is the sliding velocity. This equation is known as the "Stribeck relationship." At constant load and velocity, increased friction results from a higher-than-necessary viscosity. Thus, selection of lubricant viscosity is a compromise between providing sufficient film strength at equipment-operating conditions and temperature and then minimizing friction from the lubricant itself by taking care to not use fluids of excessive viscosity. In the extreme, very high viscosity, for instance at cold temperatures, may even limit the flow of the lubricant to critical lubricated areas.

At cold ambient temperatures, consideration must be given to adequate flow. A fluid should not be so viscous as to cause pump starvation, nor should it be prone to wax gelation at temperatures within the normal operating-temperature window. Because of these important factors, most pump OEMs list specific startup maximum viscosity limits for their equipment.

Hydraulic fluids are also power transmission fluids. Assuming the usual incompressibility and non-viscous flow idealizations, the overall efficiency of the hydraulic system then depends on fluid viscosity. Low viscosity provides good system response, but some minimum level of viscosity is required to internally seal pump mating surfaces and prevent excessive levels of internal leakage. This phenomenon occurs as fluid leaks back from the high-pressure to the low-pressure side of a pump mechanism, causing a loss of volumetric efficiency. For a given mechanical system, the higher the operating pressure, then the higher the internal leakage will be; similarly, lower viscosity leads to more internal leakage. These concepts and their relationships are embodied in Equation 4.27:

$$Q_a = Q_n - k(P/\eta),\tag{4.27}$$

where Q_a = actual flow rate, Q_n = nominal flow rate, k = geometric constant for the pump, P = discharge pressure, and η = kinematic viscosity after shearing. A further consequence of high leakage rates is fluid overheating which, in effect, further reduces viscosity and in turn further reduces volumetric efficiency—clearly a self-accelerating process which may have consequences on wear. On the other hand, overly-viscous fluids have low hydromechanical efficiency because of frictional losses and the energy spent just moving the fluid itself. In addition, high viscosity causes sluggish equipment response, poor flow characteristics, and with extremely high viscosities a consequence may be cavitation.

An optimum fluid viscosity is then a balance between good volumetric efficiency and good mechanical efficiency.

This concept is demonstrated in Figure 4.20 where the relative trade-off of lower versus higher viscosity is superimposed on typical optimum and maximum working viscosity ranges.

Ideally, a fluid with an optimum viscosity could be selected for any given piece of equipment operating under set conditions. The problem in the real world is that conditions vary, particularly temperature, which obviously impacts viscosity. In the case of cold ambient temperature startup, usually associated with mobile equipment, fluid viscosity is undoubtedly far higher than the ideal for efficient operation; in an extreme case, a fluid may be so viscous as to prevent startup or provide conditions leading to mechanical breakage. The usual solution for mobile equipment is to employ a high VI hydraulic fluid which, because of its lower viscosity at cold temperature, allows easier startup. Even after startup, as the system is warming but below the normal operating temperature, the reduced viscosity of high VI hydraulic fluids results in lower hydromechanical energy losses

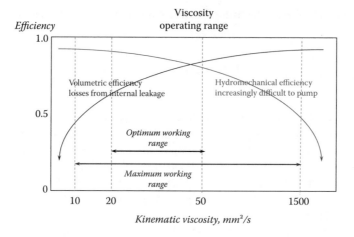

FIGURE 4.20   Efficiency and viscosity working ranges.

because of reduced viscous drag compared to monograde hydraulic fluids. Even during normal operation, high VI hydraulic fluids also expand the temperature zone in which the fluid viscosity is in the optimum operating range compared to corresponding monograde, lower VI fluids. An outcome of these efficiencies is reduced energy demand.

Since overall efficiency is a balance of lower viscosity for better mechanical efficiency versus higher viscosity for better volumetric efficiency, there should be an optimum viscosity that maximizes overall efficiency in a given piece of equipment. An ideal fluid providing this optimum viscosity is, of course, a real-world illusion in that it will not be equiviscous at all temperatures experienced during its use in equipment. A practical way to better approach the ideal is to reduce temperature effects by employing an appropriate high-viscosity index fluid to minimize viscos-

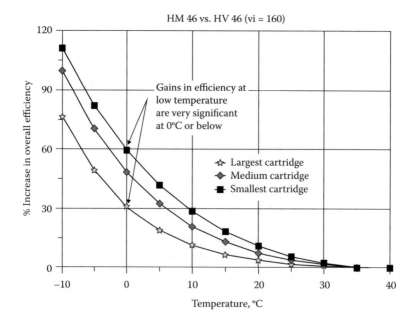

FIGURE 4.21   Increase in overall efficiency in vane pump due to higher VI fluids.

ity change with temperature. Figure 4.21 demonstrates the improvement of overall efficiency with higher VI fluids.

Selection of hydraulic fluid viscosity is a careful balance of lubrication and power transmission requirements related to the specifics of the equipment, particularly the pump, operating pressure and temperatures encountered both at startup and while operating. OEMs provide viscosity recommendations for their equipment, often suggesting a maximum at startup in order to avoid cavitation; then another maximum to commence equipment operation; an optimum range for efficient and economical operation; and finally a minimum at operating conditions to avoid not only inefficient operation but also the consequences of poor lubrication. Various OEM viscosity recommendations are gathered in National Fluid Power Association (NFPA) document T2.13.13 2002 Table 3, which is reproduced in this chapter's appendix.

4.7 VISCOSITY CLASSIFICATION

Viscosity requirements differ greatly depending on the application. Without a systematic approach and well known classification lexicon, the world would be a confusing place to consumers of lubricants and power transmission fluids. In order to aid end-user selection and to provide consistency, various viscosity classification systems have been devised and, of necessity, focus on specific end-uses. These systems range from the simple ISO system, defining only a 40°C kinematic viscosity range, to the more complex and informative hydraulic, gear and engine oil classification systems with viscosity ranges for high and low temperatures and for high and low shear rates. Some encompass permanent or temporary loss of viscosity criteria as well.

Also, be aware that there are many other classification systems dealing with fluid "performance" issues, such as wear or seal compatibility, and so forth. This section is intended to describe only viscometric classification.

4.7.1 ISO VISCOSITY GRADES

Hydraulic fluid viscosity is often classified according to ASTM D 2422 (Industrial Fluid Lubricants by Viscosity System) and the resulting fluids are commonly referred to as ISO viscosity grades. The ISO system stems from a co-operation between ASTM, STLE (formerly ASLE), BSI, and DIN in 1975, which resulted in ISO 3448 (Viscosity Classification for Industrial Liquid Lubricants). The ISO system (Table 4.5) classifies fluids solely on the kinematic viscosity measured at 40°C. The choice of 40°C as the reference temperature is a compromise between maximum operating and ambient temperatures as well as being convenient since it is a reference temperature for the determination of the viscosity index, a commonly-reported property of hydraulic fluids. The system is an orderly mathematical construction of 18 different ISO viscosity grades, usually noted as "ISO VG," starting at ISO VG 2, which has a minimum viscosity of 1.98 mm²/s, and going up to ISO VG 1500, which has a maximum viscosity of 1650 mm²/s. Each grade is named by a whole number which is the rounded, midpoint viscosity of its associated range of viscosity. Each range is ±10% of the midpoint viscosity.

Each ISO VG is approximately 50% more viscous than the next lower grade and, since each grade has a range of 20% (±10%), the grades are not continuous. Rather, there are viscosity "gaps" between the grades. As a result, there are clear viscometric differences between grades, ensuring that moving from one grade to another brings with it a significant difference in viscosity. A disadvantage is that a fluid with a viscosity that does not fall into an ISO VG range cannot be formally classified. For example, some vegetable oils cannot be classified in the ISO system; for instance, rapeseed (canola) oil, with a typical viscosity of 38 mm²/s, falls in the gap between ISO VG 32 and 46 and thus cannot be formally classified.

The ISO viscosity classification is a simple, readily understood, and well-known system. It does not imply any consideration of quality or performance other than to identify kinematic viscosity at

TABLE 4.5
ASTM D 2422 ISO Viscosity Grade

ISO VG	Kinematic Viscosity, 40°C (mm²/s)		
	Midpoint	Minimum	Maximum
2	2.20	1.98	2.42
3	3.20	2.88	3.52
5	4.60	4.14	5.06
7	6.80	6.12	7.48
10	10.0	9.00	11.0
15	15.0	13.5	16.5
22	22.0	19.8	24.2
32	32.0	28.8	35.2
46	46.0	41.4	50.6
68	68.0	61.2	74.8
100	100	90.0	110
150	150	135	165
220	220	198	242
320	320	288	353
460	460	414	506
680	680	612	748
1000	1000	900	1100
1500	1500	1350	1650

40°C. There are newer, more elaborate, and more informative classification systems, but even these are still based on ISO viscosity grades and nomenclature.

4.7.2 ASTM Hydraulic Fluid Viscosity Classification

Hydraulic fluid viscosity classification ASTM D 6080 (Standard Practice for Defining the Viscosity Characteristics of Hydraulic Fluids) was developed in 1997 to build upon ISO VG classification in order to better describe rheological properties. An enhanced classification standard was desirable because many workers viewed the ISO system (ISO 3448 and ASTM D 2422) as incomplete because it does not address the following important rheology issues:

- Cold-temperature requirements
- Multigrade (HV) oils
  - Identification as multigrades
  - Actual viscosity after shearing
- Viscosity at operating temperature (viscosity index)
- Classification of fluids outside of ISO ranges

ASTM D 6080 is a more comprehensive classification system but still utilizes the well-recognized ISO VG system while also building upon the ISO system through the addition of the missing information described above. ASTM D 6080 retains the current ISO VG definitions and nomenclature but then adds a second tier of information indicating the following: cold-temperature grade; 40°C kinematic viscosity after shearing; and viscosity index after shearing. The second classification tier does not require viscosity to fall within the usual ISO ranges. This system is limited to ISO VG 5 through ISO VG 150, as these grades represent the great majority

TABLE 4.6
Hydraulic Fluid Low-Temperature Grades

ISO VG	°C for Brookfield Viscosity of 750 mPa. sec, max.
L10	−33 to −41.9
L15	−23 to −32.9
L22	−15 to −22.9
L32	−8 to −14.9
L46	−2 to −7.9
L68	4 to −1.9
L100	10 to 3.9
L150	16 to 9.9

of hydraulic fluids. An hydraulic fluid producer might utilize this voluntary system to label an ISO VG 32 fluid as:

ISO 32

L22-30 (150)

where L22 is the low-temperature grade; 30 is the 40°C viscosity after shearing; and 150 is the viscosity index after shearing.

The cold-temperature grade is the appropriate ISO grade for the fluid being described and it is preceded by "L" to indicate low-temperature classification. The L grade is determined by measuring Brookfield viscosity (ASTM D 2983) at two temperatures (at least) and interpolating the temperature at which the fluid corresponds to 750 mPa·sec. (or 860 mm^2/sec) which represents a relatively conservative case for the maximum allowable viscosity at pump startup. Each ISO VG has an associated cold-temperature range based on extrapolation of current 40°C kinematic viscosity ranges. Using the higher and lower kinematic viscosity limits of an ISO VG and applying the calculated viscosity changes associated with a 100 Viscosity Index fluid down to the 750 mPa·sec limit, one can then state the temperatures (higher and lower) associated with the limit. For instance, an ISO L22 has a temperature range of −15°C to −22.9°C; additional grade descriptions can be found in Table 4.6.

With this system, all fluids can be classified into both lower- (cold) and higher- (warm) temperature grades.

Additionally, the system allows clear identification of multigrade oils, those containing VI improvers. These are also classified by 40°C kinematic viscosity but after 40 minutes of sonic shearing (ASTM D 5621). This particular bench test was selected for use based on published work indicating good correlation with mechanical shearing of polymer-thickened oils in pump service [19].

A third classification element, viscosity index, has been added in order to provide a basis for better understanding the viscosity of a fluid at the operating temperatures encountered in equipment. Knowledge of a fluid's viscosity index, coupled with its 40°C viscosity, can be used to calculate the viscosity at any temperature from 40°C to 100°C. For oils containing VI improvers, the viscosity index would be reported after exposure to the sonic shearing procedure described above. This after-shear value provides the possibility of more accurately understanding fluid in-service viscosity. After all, real fluids during their entire life cycle operate as sheared oils.

One aspect of ASTM D 6080 is that the 40°C viscosity reported in the second information tier need not match a current ISO VG range. This allows reporting of a sheared viscosity that no longer matches the original ISO grade and serves to identify oils that are not stay-in-grade after shear. Perhaps another positive aspect of reporting an actual viscosity in the sheared viscosity category is

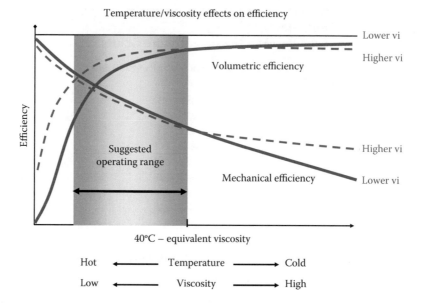

FIGURE 4.22 Temperature and viscosity effects on efficiency [Equivalent ISO VGs one high VI, the other low VI].

that fluids that do not now fit into the current ISO ranges would have a basis for classification; some vegetable oils with viscosities intermediate to ISO VG ranges fall into this category.

The ability to adequately classify multigrade (high VI) fluids is an important step forward to provide rheological information from these fluids, since they are often recommended for mobile equipment (cold ambient temperature) and/or when operating temperatures vary significantly. Compared to corresponding ISO VG fluids with lower VI (monogrades), multigrades are well known for good cold temperature startup properties simply because of reduced cold-temperature viscosity at cold startup. This fact, combined with similar (by definition) but more uniform viscosity over the operating (warm/hot) temperature range, can eliminate the need for seasonal oil changes. Even after startup but still under cold conditions, higher VI fluids reduce the likelihood of cavitation and lubricant starvation. Should equipment be operated under cold conditions, then high VI fluids flow more quickly to provide better mechanical efficiency and avoid sluggish system response. On the other side of the coin, high VI fluids maintain better pumping efficiency at operating temperatures. As temperature rises, pumping efficiency drops as decreasing fluid viscosity leads to increasing internal leakage. Multigrade fluid viscosity decreases less with increasing temperature, thus maintaining better pumping efficiency. A fluid with improved system efficiency should then achieve improved economy through lower power consumption. These efficiency concepts are represented in Figure 4.22.

4.7.3 NFPA Viscosity Grade Selection

The document, National Fluid Power Association T2.13.13-2002 "Recommended Practice – Hydraulic Fluid Power – Fluids – Viscosity Selection Criteria for Hydraulic Motors and Pumps" [23], was assembled to aid users with the selection of appropriate viscosity grades. An important element is a tabulation of viscosity recommendations provided by many OEM pump and motor component manufacturers. As stated previously, these are reproduced in the appendix "Equipment Builder's Viscosity Guidelines for Hydraulic Fluids" where one can find the minimum, maximum and optimal viscosity recommendations sorted by manufacturers' pumps and

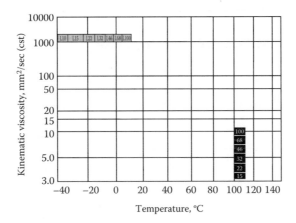

FIGURE 4.23 An ALTOW chart.

motors. After one selects the equipment of interest, NFPA's recommended practice provides the methodology to determine an appropriate fluid viscosity grade for the given pump and its presumed temperature operating range. There are two processes described to use these data to select the proper fluid. The first is the so-called "Temperature Operating Window" (TOW) method and the second is known as "ALTOW" (Alternate TOW)—a modification of the fairly conventional TOW methodology.

TOW method. The TOW method graphically presents the usable temperature ranges for typical hydraulic ISO VG grades 10 through 100. These ranges are constructed by extrapolating mid-range ISO VGs with a 100 VI to viscometric targets of 13 mm²/s (cSt) minimum for operation and 860 mm²/s maximum at startup [approximately 750 mPa×s (cP)]. These targets are taken from the builder's viscometric guidelines as they would reasonably match those of many of the other listed equipment. To use the TOW method, one would first nominate the lowest ambient temperature at startup and the highest fluid temperature when in use. These are compared to the temperature ranges in the TOW chart to select an ISO VG fluid that encompasses the temperature range needed for the application.

ALTOW method. Should an application not fall within the TOW operating temperature ranges, or should the pump's recommended viscosity range be substantially different than the 13 to 860 mm²/s range, then the more complex ALTOW method may be employed. This is a variation of the ASTM D 341 viscosity-temperature chart used to graph viscosity temperature properties according to the McCoull-Walther-Wright relationship described in Equation 4.7. Figure 4.23 presents an ALTOW chart; notice, for the various ISO VGs, the *viscosity* ranges at high temperatures and the *temperature* ranges for cold, startup viscosities. The low-temperature ranges are those of ASTM D 6080; high-temperature ranges are based upon an extrapolation of 40°C viscosities to 100°C, assuming a VI of 100.

To use ALTOW, one would identify recommended viscosities from the Builder's Guidelines for the equipment of interest and then estimate the minimum cold startup temperature and the maximum operating temperature. These combinations of cold-temperature/maximum viscosity and high-temperature/minimum viscosity are plotted on the ALTOW chart. A line is drawn through the points and extended such that it passes through the high- and low-temperature ISO VG "boxes." Where it does pass through the high- and low-temperature ISO VG boxes, these would be the recommended grades. Another important advantage of ALTOW is that it is useful for determining multigrade fluid needs.

TABLE 4.7
Proposed NFPA Viscosity Grades

NFPA Grade	Kinematic Viscosity, 100°C, mm²/sec After shearing by ASTM D 5621	
	Minimum	Maximum
15	3.2	<4.0
22	4.0	<5.0
32	5.0	<6.3
46	6.3	<8.1
68	8.1	<10.5
100	10.5	<14.0
150	14.0	<18.2

Potential changes to the NFPA viscosity selection practice. The 2002 NFPA Recommended Practice did not take into consideration the probable shearing of multigrade fluids based on VI improvers and their consequent loss of viscosity. Thus, fluids susceptible to shearing might be misapplied in applications where the used-oil viscosity becomes too low. Currently, the NFPA Fluids Technical Committee is considering a change to the recommended practice which would create viscosity classifications for sheared fluids [24]. These new NFPA VGs would still utilize the well-known ISO VG nomenclature and would be based on 100°C rather than 40°C kinematic viscosity after shearing via ASTM D 5621. Although not yet implemented as a recommended NFPA practice, the Table 4.7 is included here so as to maintain this chapter's relevance, assuming that the NFPA moves forward with this proposal.

The proposed NFPA grade lower limits are based on equivalent ISO VG lower limits (KV at 40°C) extrapolated to 100°C using a value of 100 VI. The upper limits are set just below the next higher grades' lower limit. One might note this classification system is now continuous, as opposed to the ISO system with its intentional gaps between grades.

There is also consideration being given [24] to adding a VI requirement of 160 minimum after D 5621 shearing. The rational to include a relatively higher VI limit is to provide criteria favoring higher VI, more efficient fluids as discussed in Section 6.2. Again, at the time of writing, NFPA had not yet endorsed this as a recommended practice.

4.7.4   GEAR OIL CLASSIFICATION

Industrial gear oils are often classified according to AGMA (American Gear Manufacturers Association) standards: 250.04 for enclosed gears and 251.02 for open gears (Table 4.8). The AGMA classifications also rely on kinematic viscosity measured at 40°C but with two exceptions where viscosity is classified according to 100°C kinematic viscosity because of these two grades' extremely viscous nature. The AGMA system is referenced to the ISO VG system for all grades that can be so accommodated. The lowest grade is AGMA 1, equivalent to an ISO VG 46, then up through AGMA 9, equivalent to ISO VG 1500. There are six more viscous grades, up to AGMA 15, but these can not be referenced to the ISO VG system as it does not go beyond ISO VG 1500.

Gear oils used in transportation are classified according to SAE J306 JUN 05 [25], shown in Table 4.9. This system differs significantly from the previous two in that it incorporates two viscosity determinations, one at 100°C, rather than 40°C, and one at cold temperature. Kinematic viscosity is the method of choice for the 100°C measurement. This higher reference temperature may be a better approximation of the more severe conditions experienced by an automotive gear

TABLE 4.8
AGMA Viscosity Classifications

AGMA Standard 250.04 Enclosed Gears			AGMA Standard 251.02 Open Gears			Viscosity (mm²/sec), 40°C	
AGMA Number	AGMA Number		AGMA Number	AGMA Number			
R&O oil	EP oils	ISO VG	R&O oils	EP oils	ISO VG	Min.	Max.
1	—	46	4	4 EP	150	135	165
2	2 EP	68	5	5 EP	220	198	242
3	3 EP	100	6	6 EP	320	288	352
4	4 EP	150	7	7 EP	460	414	506
5	5 EP	220	8	8 EP	680	612	748
6	6 EP	320	9	9 EP	1,500	1,350	1,650
7 Comp	7 EP	460	10	10 EP	—	2,880	3,520
8 Comp	8 EP	680	11	11 EP	—	4,140	5,060
8A Comp	8A EP	1,000	12	12 EP	—	6,120	7,480
			13	13 EP	—	25,600	38,400

						Viscosity (mm2/sec), 100°C	
			14R		—	428.5	857.0
			15R		—	857.0	1,714

lubricant. In addition, the high-temperature limit must be met after shearing; this testing is done according to CEC L-45-T-93 Method C (20 hours). On the low temperature side, Brookfield viscosity is measured at a temperature suitable for the intended service of the lubricant. Cold-temperature viscosity is a major feature of SAE J306 JUN 05 and its inclusion as a requirement seeks to ensure that these lubricants, intended for outdoor use, will have an appropriately low viscosity during cold temperature service.

The viscosity grade descriptions of SAE J306 JUN 05 are based on whole number designations, with higher numbers indicating higher viscosities. The lower-numbered grades, those including the

TABLE 4.9
Axle and Manual Transmission Lubricant Viscosity Classification SAE J306 JUN 05

SAE Viscosity Grade	Maximum Temperature for Viscosity of 150,000 mPa · sec (°C)	Kinematic Viscosity at 100°C (mm²/sec)	
		Min.	Max.
70W	−55	4.1	—
75W	−40	4.1	—
80W	−26	7.0	—
85W	−12	11.0	—
80	—	7.0	<11.0
85	—	11.0	<13.5
90	—	13.5	<18.5
110	—	18.5	<24.0
140	—	24.0	<32.5
190	—	32.5	<41.0
250	—	41.0	—

letter W (for winter), require a cold temperature measurement which varies by grade. The lower the SAE W grade, the lower the temperature of measurement and the better the ability to provide adequate lubrication under cold conditions. For example, a SAE 75W will be less viscous than a SAE 85W gear oil in cold-temperature service. An alternative way to look at this low-temperature classification is that if 150,000 mPa · sec is the maximum viscosity to allow sufficient flow and lubrication, then SAE 75W oil should provide operability as low as −40°C, whereas SAE 85W oil will presumably operate only as low as −12°C. In practice, an OEM specifies (or a user selects) a grade suitable for prevailing conditions.

Other important aspects of SAE J306 JUN 05 are: (1) it allows for a description of multigrade oils, and (2) viscosity grades are continuous. At high temperature, W grades have only minimum requirements, so an oil meeting the low-temperature requirements of a "W grade" also meets the high-temperature requirements of the higher SAE grades and so can be labeled as meeting both classifications. For instance, a SAE 75W-90 oil has the low-temperature properties of a SAE 75W and the high-temperature properties of a SAE 90. The continuity of viscosity from grade to grade essentially allows any reasonable viscosity fluid to be classified.

4.7.5 ENGINE OIL CLASSIFICATION

Engine oils are classified according to SAE J300 NOV 07 [25], as displayed in Table 4.10. This classification system shares many of the features of the SAE gear oil classification: continuous viscosity descriptions (no gaps between the grades); W grades; cold-temperature requirements that vary by W grade; high-temperature requirements that approximate actual service conditions; and nomenclature and limits to accommodate multigrading. However, the engine oil classification is more complex, including viscosity measurements under high shear rates at both cold ambient and hot operating temperatures.

SAE J300 NOV 07 requires the well-known kinematic viscosity measurement done at 100°C and, in addition, a high-shear-rate viscosity measured at 150°C and 10^6 s^{-1} (ASTM D 4863 or ASTM D 4741 or ASTM D 5481) to ascertain fluid viscosity under temporary shearing

TABLE 4.10
Engine Oil Viscosity Classification SAE J300 NOV 07

SAE Viscosity Grade	Low-Temperature (°C) Cranking Viscosity (mPa.sec), max.	Low-Temperature (°C) Pumping Viscosity (mPa.sec), max. with no yield stress	Kinematic Viscosity mm²/ sec at 100°C Min.	Kinematic Viscosity mm²/ sec at 100°C Max.	High-Temperature, High-Shear Viscosity (mPa.sec), 150°C, 10^6 s^{-1} min.
0W	6,250 at −35	60,000 at −40	3.8	–	–
5W	6,600 at −30	60,000 at −35	3.8	–	–
10W	7,000 at −25	60,000 at −30	4.1	–	–
15W	7,000 at −20	60,000 at −25	5.6	–	–
20W	9,500 at −15	60,000 at −20	5.6	–	–
25W	13,000 at −10	60,000 at −15	9.3	–	–
20	–	–	5.6	<9.3	2.6
30	–	–	9.3	<12.5	2.9
40	–	–	12.5	<16.3	3.5 - 0W-40, 5W-40 & 10W-40
40	–	–	12.5	<16.3	3.7 -15W-40, 20W-40, 25W-40, 40
50	–	–	16.3	<21.9	3.7
60	–	–	21.9	<26.1	3.7

conditions. Since a preponderance of engine oils are multigrades, imposing a requirement for high-shear-rate viscosity minima is of particular importance in assuring acceptable in-service viscosity. The specific high temperature and high shear rate are meant to simulate typical operating conditions that occur in critical lubrication areas of engines, such as bearings, piston rings, or valve systems.

On the cold temperature side, the low-temperature, low-shear-rate viscosity requirement is intended to describe and thus classify (by "W" grade) "pumpability" properties. This property implies an oil's ability to be quickly and adequately pumped from the engine sump under cold ambient conditions. This viscometric determination is done according to ASTM D 4684 in a Mini Rotary Viscometer which provides two results. The first is yield stress (recall Bingham fluids), which is carried out to check for the absence of wax gel-induced flow retardation and then viscosity, to ensure that the oil is not so viscous that it can not, in a reasonable time, flow to the sump pump. Another cold-temperature viscometric measurement seeks to ensure the "crankability" of the oil. This requirement is to help in assessing whether an oil's viscosity is sufficiently low (specific to its "W" grade) so as to allow an engine to crank and thus start. This, of course, presumes that the engine systems are in good mechanical and electrical condition. The measurement is done according to ASTM D 5293 Cold Cranking Simulator, a high-shear-rate viscometer set up to mimic the critical shear rates found in engine-starting conditions. A "W" grade is assigned by virtue of its viscosity falling into one of the cracking viscosity brackets. Once the "W" designation is established, the oil must meet pumping requirements of that particular "W" grade. Even a casual examination of SAE J300 shows that the lower numbered "W" grades, for example, SAE 5W, carry the implication of better operating windows under cold conditions.

All four requirements are included to describe rheological performance under internal combustion engine conditions present during low-temperature cranking and pumping regimes and then at high-temperature operating conditions.

It might be noted that SAE J300 does not address permanent shear stability and resulting used-oil kinematic viscosity, but bear in mind that high-shear-rate viscosity is, to a fair degree, a reasonable surrogate for this.

REFERENCES

1. *Annual Book of ASTM Standards 2008, Volumes 5.01, 5.02*, and *5.03*, American Society for Testing and Materials, Philadelphia, PA, 2008.
2. McCoull, N. and Walther, C. "Viscosity Temperature Chart," *Lubrication*, June, 1921.
3. Dean, E.W. and Davis, G.H.B. "Viscosity Variations of Oils with Temperature," *Chemical and Metallurgical Engineering*, 1929, 36, pp. 618–19.
4. Barus, C. "Isothermals, Isopiestics and Isometrics Relative to Viscosity," *American Journal of Science*, 1893, 45, pp. 87–99.
5. Winer, W.O. "The Mechanical Properties of Fluids in High Pressure Hydraulic Systems," In *National Conference on Fluid Power*, 1974, pp. 412–21.
6. Roelands, C., Vluger, J. and Waterman, H. "Correlational Aspects of the Viscosity–Temperature–Pressure Relationship of Lubricating Oils and in Correlation with Chemical Constitution," Trans. ASME., *Journal of Basic Engineering*, 1963, 11, pp. 601–19.
7. So, B.Y.C. and Klaus, E.E. "Viscosity–Pressure Correlation of Liquids," *ASLE Trans.*, 1980, 23, pp. 409–21.
8. Yasutomi, S., Bair, S. and Winder, W.O. "An Application of a Free Volume Model to Lubricant Rheology I—Dependence of Viscosity on Temperature and Pressure," *Journal of Tribology*, 1984, 106, pp. 291–303.
9. Eyring, H. J. *Chemical Physics*, 1936, 4, p. 283.
10. Powell, R.H. and Eyring, *H. J. Chem. Phys.*, 1937, 5, p. 729.
11. Weast, R.C. (ed.), *CRC Handbook of Chemistry and Physics,* 50th ed., Chemical Rubber Co., Cleveland, OH, 1969, pp. F37–F42.
12. Bondi, A. "Viscosity and Molecular Structure," in *Rheology Theory and Applications, Vol 4* (Eirich, F.R., ed.), Academic Press, New York, 1967, pp. 1–82.
13. Billmeyer, F.W. *Textbook of Polymer Science*, Wiley–Interscience, New York, 1971, p. 28.

14. Bondi, A. "Rheology of Lubrication and Lubricants," in *Rheology Theory and Applications, Vol 3* (Eirich, F.R., ed.), Academic Press, New York, 1960, pp. 444–78.
15. Stambaugh, R.L. "Viscosity Index Improvers and Thickeners," in *Chemistry and Technology of Lubricants* (Mortimer and R.M., Orszulik, S.T. ed.), Blackie Academic & Professional, London, 1992, pp. 124–59.
16. Selby, T.W. "The Non-Newtonian Characteristics of Lubricating Oils," *Trans. ASLE*, 1958, 1, pp. 68–81.
17. Kopko, R.J. and Stambaugh, R.L. "Effect of VI Improver on the In-Service Viscosity of Hydraulic Fluids," 1975, *SAE Paper* 750683.
18. Brodkey, R.S. *The Phenomena of Fluid Motions*, Addison-Welsley, Reading, MA, 1967, pp. 365–72.
19. Stambaugh, R.L. and Kopko, R.J. "Behavior of Non-Newtonian Lubricants in High-Shear-Rate Applications," 1973, *SAE Paper* 730487.
20. Girshick, F. "Non-Newtonian Fluid Dynamics in High-Temperature High-Shear Capillary Viscometers," 1992, *SAE Paper* 922288.
21. Stambaugh, R.L., Kopko, R.J. and Roland, T.F. "Hydraulic Pump Performance—A Basis for Fluid Viscosity Classification," 1990, *SAE Paper* 901633.
22. Ram, A. "High-Shear Viscometry," in *Rheology Theory and Applications, Vol 4* (Eirich, F.R., ed.), Academic Press, New York, 1967, pp. 281–83.
23. National Fluid Power Association T2.13.13 – 2002 "Recommended Practice – Hydraulic Fluid Power – Fluids – Viscosity Selection Criteria for Hydraulic Motors and Pumps," April, 2002.
24. Herzog, S.N., Marougy, T.E. and Michael, P.W. "Hydraulic Fluid Viscosity Selection to Improve Equipment Fuel Economy and Productivity," *International Fluid Power Association Paper NCFP* I08 – 2008.
25. *SAE Handbook 208, Vol 3, 2008, Society of Automotive Engineers*, Warrendale, PA.

Appendix

Equipment Builder's Viscosity Guidelines for Hydraulic Fluids
NFPA T2.13.13 2002

n	Equipment	Operating Minimum mm²/s (cSt)	Operating Maximum mm²/s (cSt)	Startup (Under Load) Maximum mm²/s (cSt)	Optimum mm²/s (cSt)
Bosch (see Rexroth Corporation)					
Commercial Intertech (see Parker Hannifin)					
Danfoss (see Sauer-Danfoss, USA)					
Denison Hydraulics	**Piston Pumps**	10	162	1618	30
SPO-AM305	**Vane Pumps**	10	107	860 (low speed and pressure)	30
Dynex/Rivett	PF4200 Series	1.5	372	372	20-70
axial piston pumps	PF2006/8, PF/PV4000, and PF/PV6000 Series	2.3	413	413	20-70
	PF 1000, PF2000 and PF3000 Series	3.5	342	342	20-70
Eaton	**Heavy-Duty Piston Pumps and Motors, Medium-Duty Piston Pumps and Motors Charged Systems, Light-Duty Pumps**	6	–	2158	10-39
	Medium-Duty Piston Pumps and Motors – Non-charged Systems	6	–	432	10-39
	Gear Pumps, Motor, and Cylinders	6	–	2158	10-43
Eaton - Vickers	**Mobile Piston Pumps**	10	200	860	16-40
	Industrial Piston Pumps	13	54	220	16-40
	Mobile Vane Pumps	9	54	860	16-40
	Industrial Vane Pumps	13	54	860	16-40
Eaton - Char-Lynn	**J, R, and S Series Motors, and Disc Valve Motors**	13	–	2158	20-43
	A Series and H Series Motors	20	–	2158	20-43
Haldex Barnes	W Series Gear Pumps	11	–	750	21
Kawasaki					
P-969-0026	Staffa Radial Piston Motors	25	150	2000 (no load)	50
P-969-0190	**K3V/G Axial Piston Pumps**	10	200	1000	
Linde	**All**	10	80	1000	15-30
Mannesmann Rexroth (see Rexroth Corporation)					

(continued)

Equipment Builder's Viscosity Guidelines for Hydraulic Fluids
NFPA T2.13.13 2002   (Continued)

n	Equipment	Operating Minimum mm²/s (cSt)	Operating Maximum mm²/s (cSt)	Startup (Under Load) Maximum mm²/s (cSt)	Optimum mm²/s (cSt)
Parker Hannifin	**Roller and Sleeve-Bearing Gear Pumps**	10	–	1600	20
	Gerotor Motors	8	–	–	12-60
	Gear Pumps PGH Series	–	–	**1000**	**17-180**
	Gear Pumps D/H/M Series	–	–	**1000**	**17-180**
	Hydraulic Steering	8	–	–	12-60
	PFVH / PFVI Vane Pumps	–	–	**1000**	**17-180**
	Series T1	**10**	–	**1000**	**10-400**
	VCR2 Series	**13**	–	**1000**	–
	Low-Speed High-Torque Motors	**10**	–	–	–
	Variable Vol Piston Pumps	–	–	**1000**	**17-180**
	PVP and PVAC	–	–	**1000**	**17-180**
	Axial-Fixed Piston Pumps	–	–	**850**	**12-100**
	Variable Vol Vane - PVV	–	–	440	16-110
Poclain Hydraulics	**H and S Series Motors**	**9**	**–**	**1500**	**20-100**
Rexroth Corporation	FA, RA,; K	15	216	864	26-45
Form No S/106 US	Q, Q-6, SV-10, 15, 20, 25, VPV 16, 25, 32	21	216	864	32-54
	SV-40, 80 and 100				
	VPV 45, 63, 80, 100, 130,164	32	216	864	43-64
	Radial Piston (SECO)	10	65	162	21-54
	Axial and RKP Piston	14	450	647	32-65
	V3, V4, V5, V7 Pumps	25	–	800	25-160
	V2 Pumps	16	160	800	25-160
	R4 Radial Piston Pumps	10	200	–	25-160
	G2, G3, G4 Pumps and Motors				
	G8, G9, G10 Pumps	**10**	**300**	**1000**	**25-160**
Rotary Power	"SMA" Radial Piston Motor	15	–	1000	20-200
Sauer-Danfoss, USA	**Steering and Valves**	**10**	**–**	**1000**	**12-60**
	PVG Valves	4	–	460	12-75
	Gear Pumps and Motors	**10**	**–**	**1600**	**20-40**
	Closed-Circuit Axial Piston Pumps and Motors	**7**	**–**	**1600**	**12-60**
	Open-Circuit Axial Piston Pumps				
	Bent Axis Motors	**6**	**–**	**1000**	**9-110**
	LSHT Motors	**7**	**–**	**1600**	**12-60**
		10	**–**	**1000**	**20-75**
Sauer-Danfoss, GmbH	Series 10 and 20, RMF(Hydrostatic Motor)	7	–	1000	12-60
	Series 15 Open Circuit	**12**	**–**	**860**	**12-60**
	Series 40, 42, 51 and 90				
	CW S-8 Hydrostatic Motor	**7**	**–**	**1600**	**12-60**
	Series 45	**9**	**–**	**1000**	**12-60**
	Series 60, LPM (Hydrostatic Motor)	**9**	**–**	**1600**	**12-60**
	Gear Pumps plus Motors	**10**	**–**	**1000**	**12-60**
Sundstrand (see Sauer-Danfoss, USA)					

5 Control and Management of Particle Contamination in Hydraulic Fluids

Jim C. Fitch

CONTENTS

5.1 INTRODUCTION

It has been extensively documented and widely stated that particle contamination is the number one cause of wear and failure of hydraulic components. The problem is generally more pronounced than in other types of machinery incorporating circulating systems that use similar types of oils. This heightened contaminant sensitivity is due to the high pressures and tight tolerances which are characteristic of modern hydraulic machines. Pressure is known to have a disproportionate effect on contaminant sensitivity.

Much has been learned in the past three decades about contamination control at both a laboratory research level as well as the real-world deployment of this knowledge in machinery-intensive industries. Case studies have flourished on the practical and economic benefits of maintaining hydraulic systems and fluids at extreme levels of cleanliness. Hence, the speculation is gone relating to the business case and strategies that produce savings and benefits to user organizations. For many owners of hydraulic systems the opportunities of planned cleanliness are like low-hanging fruit that is ripe for picking. This chapter summarizes this body of knowledge and the value-producing strategies needed to control particle contamination in hydraulic fluids.

5.2 CONTAMINANT SENSITIVITY AND PARTICLE-INDUCED COMPONENT FAILURE

The tribology field is replete with published studies on the damage caused by fluid-borne particles in hydraulic fluids and lubricating oils. Therefore, the focus of this chapter is not on the numerous pathways and modes of particle-induced failure, but rather on establishing and discussing some well-grounded strategies to control the damage and mitigate the risk. For those who are new to the field of tribology and machine reliability, the following is a concise summary of the four ways in which particles can rob a company of precious productivity and profits:

1. **Surface removal.** This is the product of three-body abrasion in sliding contact zones and surface-fatigue in rolling contacts. Hydraulic machines that are exposed to dense terrain dust from ambient air are at the greatest risk. These particles are harder than internal frictional surfaces causing plowing, cutting, and pitting. A hydraulic component can only tolerate so much material loss. For instance, a 20 GPM gear pump will have lost over 30% of its volumetric efficiency when just 10 g of wear metal has been generated [1]. High-pressure systems are far less tolerant to particle-induced wear than low-pressure systems. Figure 5.1 illustrates the influence of key factors that define particle contamination destructive potential in frictional zones of a hydraulic component (e.g., pumps and actuators).
2. **Restriction of oil flow and part movement.** Particles can form deposits, impede part movement and starve systems of oil. While no or limited wear may have occurred, this too can contribute to business interruption and expensive repairs. The most notorious example of this type of failure is the silt-lock of electro-hydraulic valves. These valves can become jammed due to particles lodged between the spool and bore [1].
3. **Increased consumption of lubricants and filters.** The ways in which particles can shorten lubricant service life and impair its performance are numerous. Particles accelerate additive depletion, leading to premature oil oxidation, oil-water emulsion problems, impaired corrosion protection, and poor film strength. The result is higher fluid consumption and distress to the hydraulic system. Likewise, undeterred particle ingression will lead to wastefully high filter consumption.
4. **Higher energy consumption and environmental impact.** There are many ways in which particles increase mechanical friction, impair antifriction additive performance, and decrease volumetric efficiencies in hydraulic components. The more energy and fuel that are consumed due to these losses, the more waste stream will be produced, which pollutes our atmosphere [2].

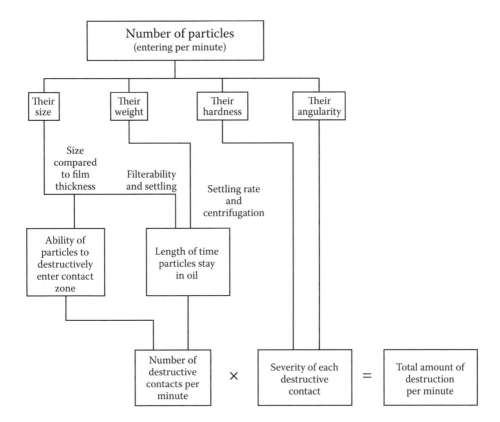

FIGURE 5.1   Particle-induced wear risk factors.

5.3   CHARACTERIZATION OF HARD PARTICLES

There is a lot to know about particles other than their size and count. The following section examines the intricacies of the physical, chemical and electrical properties that make up and characterize solid particle contamination. Knowledge of this information is useful in understanding the source of particles and their destructive potential. Not included in this discussion are soft particles and organic insolubles associated with additive precipitation, base oil oxidation, thermal degradation byproducts and chemical contamination.

The ten particle characteristics described below should be important to tribology analysts and lubrication professionals. Each of these characteristics or traits can influence the health and performance of hydraulic machinery. While the name of the trait may be familiar to many people, the damage it causes may be less so.

Particle size. Particle size is usually defined as a particle's equivalent spherical diameter in microns (micrometers). This relates to the diameter of a sphere as having the equivalent two-dimensional projected area as an irregular-shaped particle in question. Automatic particle counters size particles on this principle, using the projected blockage of light from particles.

Particle size is important because it characterizes the particle's ability to bridge the working clearances of moving machine surfaces. When large particles get crushed into smaller particles, they tend to get closer in size to a machine's working clearances. The closer the particle size is to these working clearances, the more readily it can enter the gap and cause abrasion or surface fatigue to opposing surfaces. For instance, a single 40-μm particle can theoretically be broken into 512 individually-destructive 5-μm particles. Hence, restricting the ingression of this 40-μm particle, or removing it before crushing can occur, should be a maintenance priority [3].

Surface area. When large particles break into many smaller particles, the cumulative surface area in contact with the oil increases many fold. For instance, if you break a particle into 100 equal-size pieces, you have roughly 4.5 times more interfacial surface area. So, in the previous example, a 40-μm particle, when broken down into 5-μm particles, will produce eight times more surface area in contact with the oil. The more surface area relative to particle mass, the slower the particle settles (longer residence time in the oil), the more it attracts and emulsifies water, the more it can incite catalytic chemical reactions with the oil, the more it can tie up the performance of polar additives (like antiwear agents, rust inhibitors and the like), and the more air bubbles it can nucleate, thus inhibiting their efficient detrainment from the oil.

Particle shape/angularity. Particle shape/angularity is a central risk-factor relating to the wear and damage caused by particles. Spherical-shaped particles are like ball bearings: They may cause surface indentations but are much less likely to cut or abrade. On the other hand, particles with high annularity (possess sharp, acute angles between facets) are more prone to impart three-body abrasion, leading to material removal. This is characterized by the study done by Hamblin and Stachowiak as shown in Figure 5.2 [4].

Angular particles are generally caused by the crushing (comminution) of large particles into smaller particles (Figure 5.3). If a spherical particle were broken into 100 smaller particles having the general shape of cubes, this would expose sliding machine surfaces to 800 angular edges.

Hardness. Hardness relates to a particle's compressive strength, that is, its resistance to deformation (plastically or elastically) or fragmentation by crushing. Particle hardness relative to surface hardness largely defines its ability to cause wear and fatigue. As a point of reference, common dirt consists largely of silica and alumina particles, which are harder than all metallic surfaces used in hydraulic components. Only ceramic surfaces found in some bearings would be harder. The relative hardness of common particles is shown in Table 5.1.

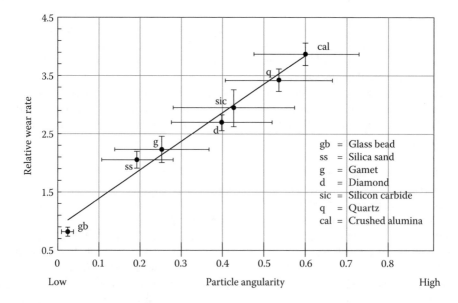

FIGURE 5.2   Wear rate versus particle angularity.

FIGURE 5.3 Angular rock dust from mining, quarry or excavation.

Density. Density, or specific gravity, influences how buoyant particles are in lubricating oils. Heavy particles will settle more rapidly in tanks and sumps. The densities of common particles are shown in Table 5.1 above. It takes only 2.8 minutes for a 20-μm babbitt particle to settle one-half inch in an ISO 22 hydraulic fluid, based on the principles of Stokes Law. Heavy particles are also more prone to cause particle impingement erosion, as oil circulates at high velocity, sending heavy and hard particles on destructive trajectories [1].

Composition. While terrain dust is known for its wear-inducing potential due to its hardness, it is also rather chemically inert. However, the wear particles generated by this dust (plowed up by abrasive and surface fatigue) that become suspended are typically not inert. This is due to the fact that these nascent wear particles are often composed of iron, copper, or tin. Although less hard and abrasive, wear metals aggressively promote oil oxidation, which in turn contributes to the formation of corrosive acids, varnish, and sludge.

Polarity. Many particles have unique polar affinities or possess ionic charges. This can lead to the mass transfer and depletion of polar oil additives such as rust inhibitors, antiwear agents, detergents, dispersants and extreme pressure additives, which are more prone to hitch a ride on these particles. Also, polar particles are apt to cluster and obliterate fine oil passages, oil ways, weep holes, and silt lands. This is compounded if water is present, which has a tendency to cling to polar solid contaminants, thus further promoting obliteration and the formation of emulsions and sludge.

Magnetic susceptibility. Permanent magnets are used in some filters and on line wear particle sensors. Particles of iron or steel that are attracted to a magnetic field are preferentially separated from the oil by these devices. Later, any particles that may have sloughed off these separators and sensors (due to shock or surge-flow conditions) are often left magnetized. They can then magnetically grip onto steel orifices, glands, and oil ways restricting flow or simply interfering with part movement. Additionally, directional control and servo-valves commonly used in hydraulic systems deploy the use of electro-magnets in their solenoids. The actuation of these valves can be adversely affected by the magnetic susceptibility of iron and steel particles that are attracted by the solenoid [5].

TABLE 5.1
Examples of Particle Hardness and Density

Particle Type	Typical Specific Gravity	Mohs Hardness*
Burrs and machining swarf	6–9	3–7
Grindings	6–9	3–7
Abrasives	3–6	7–9
Floor dust	1–5	2–8
Road dust (mostly silica)	2–6	2–8
Mill scale	5	NA
Coal dust	1.3–1.5	NA
Ore dust	Various	Various
Wood pulp	0.1–1.3	1.5–3
RR ballast dust (limestone)	2.68–2.8	5–9
Quarry dust (limestone)	2.68–2.8	5–9
Foundry dust	2.65	7
Fibers	Various	Various
Slag particles (blast furnace)	2.65	7
Aluminum oxides	NA	9
Red iron oxides (rust)	2.4–3.6	5–6
Black iron oxides (magnitite)	4–5.2	5–6
Copper oxides	6.4	3.5–4
Tool steel	7–8	6–7
Forged steel	7–8	4–5
Cast iron	6.7–7.9	3–5
Mild steel	7–8	3
Alloys of copper, bronze	7.4–8.9	1–4
Alloys of aluminum	2.5–3	1–3
Babbitt particles	7.5–10.5	1
Soot	1.7–2.0	NA

* Mohs hardness scale 1–10, diamond = 10, fingernail = 1.

Conductivity. There are some positive characteristics of particle contamination. For instance, in recent years the electrification of hydraulic fluids and lubricating oils has become a greater and more common problem due to the high purity of basestocks which are frequently used by formulators. Base oils in the categories of API Groups II to IV present the highest risk.

Circulating oil can build a static charge in the oil due to molecular friction. This can lead to electrical arcing within the body of the oil, charring the oil in its path. Conductive particles are effective at dissipating charges, preventing damage to the oil from static discharge. According to one study, particle contamination equivalent to an ISO 18/15 was sufficient to dissipate static charge buildup in contrast to low contaminant levels of ISO 13/10 or cleaner, which led to strong spark discharges [6].

Particle count. As previously discussed, a single particle of the right size, shape, and hardness is a potentially destructive contact (Figure 5.1). Two such particles proportionally multiply the risk wear rate, and so forth. In fact, the total amount of surface material removed could be four to ten times the weight of the original offending particle. This risk is greatest for unfiltered or poorly-filtered systems. This is due to the reproductive cycle of particle contamination—particles make more particles. With each successive generation of particles there is increasing risk of wear and lubricant degradation. This is because of the cumulative growth of particles, total interfacial surface area, and nascent metal composition (causing catalytic chemical reactions). Controlling particle population growth is a fundamental and effective strategy in stabilizing machine reliability and fluid health.

5.4 PROACTIVE MAINTENANCE AS THE CONTAMINATION CONTROL STRATEGY

While it is not practical to eradicate all contamination from new and in-service hydraulic fluids, control of contaminant levels within acceptable limits is both accomplishable and important. Controlled systems, by definition, are those which include measurement and feedback loops. This explains why instruments called particle counters may be the most widely-used on site oil analysis tools today. Proactive maintenance, employing the use of particle counters, is the central strategy for success in reducing maintenance costs and increasing machine reliability [7].

While the benefits of detecting abnormal machine wear or an aging lubricant condition are important and frequently achieved with oil analysis programs, they should be regarded as being lower on the scale of importance when compared to the more rewarding objective of failure avoidance. This is achieved by treating the causes of failure (proactive maintenance), and not simply the symptoms (predictive maintenance). In fact, proactive maintenance is the only effective way to achieve simple solutions to complex machine maintenance problems. Restated, it is far easier to prevent a failure through root cause control than to troubleshoot an incipient or impending failure control that is already occurring [8].

Whenever a proactive maintenance strategy is applied, three steps are necessary to ensure that its benefits are achieved (Figure 5.4). Since proactive maintenance, by definition, involves the continuous monitoring and controlling of machine failure root causes, the first step is simply to set a target, or standard, associated with each root cause. In oil analysis, the root causes of greatest importance relate to fluid contamination (particles, moisture, heat, coolant, etc.). This target should be sufficiently rigorous as to reduce wear and increased reliability.

1 Set cleanliness targets
Target cleanliness level should reflect reliability goals

2 Take specific actions to achieve targets
1. Reduce ingression
2. Improve filtration

3 Measure contaminant levels frequently
1. What gets measured gets done
2. Post control charts of measured results

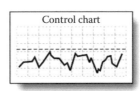

FIGURE 5.4 Three steps to implementation of proactive maintenance.

However, the process of defining precise and challenging targets (e.g., high particle cleanliness) is only the first step (discussed in the following section). Control of the fluid's conditions within these targets must then be achieved and sustained. This is the second step and often includes an audit of how fluids become contaminated and then systematically eliminating these entry points. Often, better filtration and the use of separators may also be required.

The third step is the vital action element of providing feedback to the oil analysis program. When exceptions occur (e.g., over-target results) remedial actions can then be immediately commissioned. Using the proactive maintenance strategy, contamination control becomes a disciplined activity of monitoring and controlling fluid cleanliness, not a reactive activity of responding to high dirt and wear debris levels.

The relationship between proactive and predictive maintenance is perhaps best illustrated in the graph shown in Figure 5.5 below. The Proactive Domain is influenced by the control of root causes such as particle contamination, with the goal of extending this domain indefinitely, if possible. The Predictive Domain starts at failure inception, which is also the end of the Proactive Domain. Its goal is early detection, while there is still considerable Remaining Useful Life (RUL) of the system components. The closer the point of failure detection is to the point of failure inception, the more effective the maintenance response will be.

If an impending failure goes undetected, then catastrophic failure is imminent. During this failure (Protective Domain) the objective is to minimize the failure severity (repair costs) and to prevent collateral damage to other system components. When the life extension benefits of proactive maintenance are flanked by the early warning benefits of predictive maintenance, a comprehensive condition-based maintenance program can result [9].

5.5 SETTING RELIABILITY-BASED CLEANLINESS TARGETS

While there are numerous methods used to arrive at target cleanliness levels for fluids and lubricants in different applications, most consider both the importance of machine reliability and the general contaminant sensitivity of the machine or system. This approach enables customization of the target to: (a) the reliability goals of the machine owner, (b) the risk of contamination from the operating environment, and (c) the contaminant tolerance of the hydraulic system. A common example of this approach is shown in Table 5.2 and is referred to as the Target Cleanliness Grid (TCG) [10].

The TCG utilizes the Reliability Penalty Factor (RPF) and the Contaminant Severity Factor (CSF), which are arrived at through a subjective scoring system (see Figures 5.6 and 5.7). The RPF

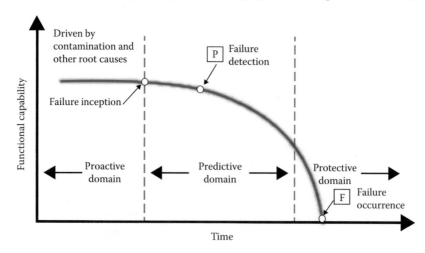

FIGURE 5.5   Condition-monitoring domains in the PF interval curve.

TABLE 5.2
Target Cleanliness Grid (TCG) for Setting Cleanliness Targets

Reliability Penalty Factor (RPF) / Cost, Safety, and Business Interruption Penalty from Failure	CSF 1	CSF 2	CSF 3	CSF 4	CSF 5	CSF 6	CSF 7	CSF 8	CSF 9	CSF 10
10	19/16/13	18/15/12	17/14/12	16/13/11	15/12/10	14/11/19	13/10/8	12/10/8	11/9/7	10/9/7
9	19/16/13	19/16/13	18/15/12	17/14/11	16/13/10	15/12/10	14/11/9	13/10/8	12/9/7	11/9/7
8	20/17/14	20/17/13	19/16/13	18/15/12	16/13/11	15/12/10	14/11/9	13/10/8	12/9/8	12/9/7
7	21/17/14	20/17/14	19/16/13	18/15/12	17/14/11	16/13/10	15/11/9	14/11/9	13/10/8	12/10/8
6	21/18/15	21/18/14	19/16/13	18/15/12	17/14/11	16/13/10	15/12/9	14/11/9	13/10/8	12/10/8
5	22/18/15	21/18/15	20/17/14	19/16/13	18/15/12	17/14/11	16/14/11	15/13/11	14/11/10	13/11/9
4	22/19/16	22/19/16	20/17/14	19/17/14	18/15/13	17/14/11	16/14/11	15/13/10	14/12/9	13/11/9
3	22/19/16	22/19/16	21/18/15	20/17/14	19/16/13	18/15/12	17/14/11	16/14/11	15/13/10	14/12/9
2	23/20/17	23/20/17	22/19/16	21/18/15	20/17/14	19/16/13	18/15/12	17/14/11	16/14/11	15/13/10
1	24/20/17	23/20/17	22/19/16	21/18/15	20/17/14	19/16/13	17/14/11	18/15/11	17/14/11	16/14/11

Contaminants Severity Factor (CSF)
Sensitivity of Machine to Contaminant Failure

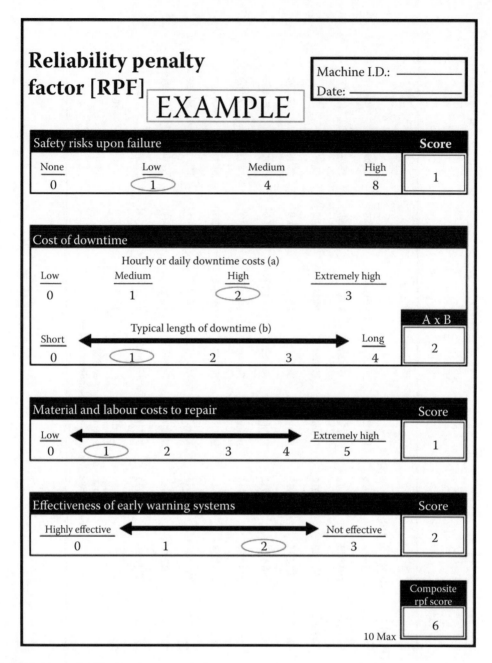

FIGURE 5.6 Reliability Penalty Factor (RPF).

scores system reliability needs based on repair cost, safety, and business interruption risks. For instance, the score for flight control hydraulics on commercial aircraft would be markedly different to that for agricultural hydraulics. The CSF scores the sensitivity of the system and its components to particle contamination and the likelihood of contaminant ingression from the work environment. High-pressure hydraulics with servo-valves operating in a mining environment would score much differently than a low-pressure punch-press in an automotive plant. The RPF and CSF combine on the TCG to select a target cleanliness utilizing the ISO Solid Contaminant Code (ISO 4406:99) [10,11] (see Figure 5.8).

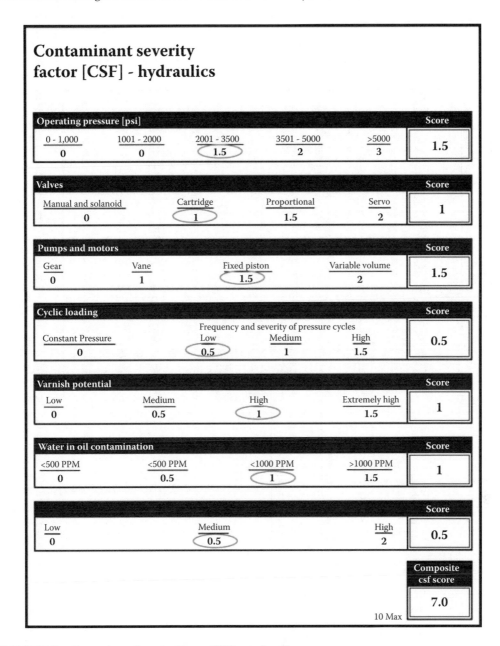

FIGURE 5.7 Contaminant Severity Factor (CSF)—hydraulics.

Target cleanliness can also be viewed in terms of filtration and contaminant exclusion as seen in the conceptual Contamination Control Balance [1] of Figure 5.9. Starting at the left scale, the desired machine service life is defined (say in thousands of hours). With the balance pointed to this reliability goal, the corresponding target cleanliness is defined on the vertical scale to the right. Maintaining the balance at that angle requires adjustments to the ingression rate, filter flow rate, and filter capture efficiency. From the balance, we can see that the reliability objective defines the target cleanliness which, in turn, defines the ingression and filtration needs of the system.

What does not work is using universal cleanliness targets for all machinery. Precision maintenance is about customized choices, not generalized default choices. Once these precision targets are

Example particle count			Number of particles per ml		
Size (microns)	Count larger than size per ml		More than	Up to and including	Range number (r)
			80,000	160,000	24
			40,000	80,000	23
			20,000	40,000	22
			10,000	20,000	21
			5,000	10,000	20
			2,500	5,000	19
4	1752		1,300	2,500	18
6	517		640	1,300	17
10	144		320	640	16
14	55		160	320	15
20	25		80	160	14
50	1.3		40	80	13
75	0.27		20	40	12
100	0.08		10	20	11
			5	10	10
			2.5	5	9
			1.3	2.5	8
			0.64	1.3	7
			0.32	0.64	6
			0.16	0.32	5
			0.08	0.16	4
			0.04	0.08	3
			0.02	0.04	2
			0.01	0.02	1

ISO 4406:99
$R_4/R_6/RI_4$

1752 particles > 4 μm/ml

517 particles > 6 μm/ml

55 particles > 14 μm/ml

ISO 18/16/13

FIGURE 5.8　Under ISO 4406:99, a sample is given a fluidcleanliness rating using the above table. To do this, thenumber of particles greater than three size ranges, 4, 6 and 14 μm are determined in the equivalent of onemilliliter of sample. In the above example, the particlecount distribution shown in the table on the leftcount translates to an ISO 4406:99 rating of 18/16/13. The particlecount distribution shown in the table on the left translates to an ISO 4406:99 rating of 18/16/13.

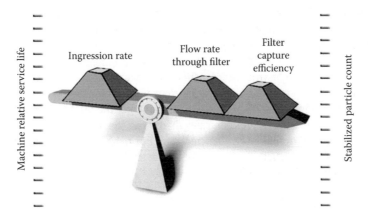

FIGURE 5.9 Contamination control balance.

set they should be communicated to the maintenance staff and each system or machine should be labeled, with the targets made conspicuously visible.

5.6 CONTAMINANT INGRESSION AND EXCLUSION

For many machines, the exclusion of contamination is the only practical way to control contamination. This is because these machines either have no filter or the filter in use is coarse, providing no real protection in the particle size range of critical oil films and surfaces. When particles are not removed by filtration or by settling, a lubricant's contaminant level equals the machine's service hours multiplied by the number of particles ingressed per hour (ingression rate). For machines exposed to high ambient dust, particle counts can exceed target levels in just a few hours. After days of exposure, a fluid can turn into more of a honing compound than a lubricating medium.

5.6.1 INGRESSION AND MASS BALANCE

Even hydraulic systems with good filters are often faced with ingression challenges. To maintain contaminant levels within targets, the filter must remove particles at a rate equal to the ingression rate (mass balance). The lower the target cleanliness level (higher cleanliness), the more difficult this becomes. This is because, in order for a fluid to stay within these high cleanliness targets, particles are not densely packed in the oil, but rather are sparsely distributed—few and far between. This means that for every gallon of fluid that enters the filter, there are few particles from that gallon that are available to be removed. Yet the filter must still remove particles at a rate equal to the ingression rate, otherwise the contaminant level will rise. This places increasing demand on the quality and capture efficiency of the filter (percent particles removed above a certain size).

Also sharply influencing this is the flow rate of the oil entering the filter. The flow provides the necessary conveyance of particles to the filter. If flow rates are low, filters with even 100% capture efficiency (Beta ratio equal to infinity) cannot remove enough particles to keep pace with ingression, causing contaminant levels to exceed targets. The higher the target cleanliness (dirty oil), the higher the minimum required flow rate for a given filter.

5.6.2 COST OF EXCLUDING DIRT

It is often said that the cost of excluding a gram of dirt is only about 10% of what it will cost you once you let it enter the oil. Dirt puts stress on additives, the base oil, and machine surfaces. Also, the cost to filter a gram of dirt from the oil is much higher than the cost of filtering a gram of dirt from the air intake/breather.

The word "ingression" refers to the introduction of particles into lubricants and hydraulic fluids, regardless of the source (external and internal). Figure 5.10 organizes common ingression sources into three subcategories: (1) built-in, (2) ingested, and (3) generated. Depending on the nature of the machine, the ingression rate and sources can vary considerably. For clean-environment indoor equipment, the primary sources can be from process fluids and internal generation (wear, corrosion, etc.).

Hydraulic systems that operate in outdoor work environments can encounter high levels of ambient dust, representing as much as 95% of all particles that enter the oil. Machines that operate close to the ground are prone to higher ingression rates than those that operate high in the air (including aviation) or away from the ground altogether (marine). For outdoor machinery, climate conditions have a marked influence on particle ingestion. For instance, rain and damp soil keep particles from becoming airborne. High winds and dry climates do just the opposite.

5.6.3 CONTROLLING TOP-END INGRESSION

For many machines, reducing ingression means reducing top-end ingression—that is, the particles entering through fill ports, vents, breathers, hatches, inspection ports, and other headspace openings. There are numerous ways to control top-end ingression, such as:

- **Purge methods.** This involves the introduction of a clean gas or aerosol into the headspace of the reservoir. A slight positive pressure is maintained to prevent the entry of ambient air. Examples include instrument air purge, oil mist purge, and nitrogen purge.
- **Isolation methods.** Expansion chambers, piston/cylinder reservoirs and bladders have been used to isolate headspace air from ambient air in order to prevent contamination. One disadvantage is that moisture (humid air) is often unable to escape from the headspace. This also locks moisture into the oil as well. In some cases, users have reported that this has led to heavy corrosion.

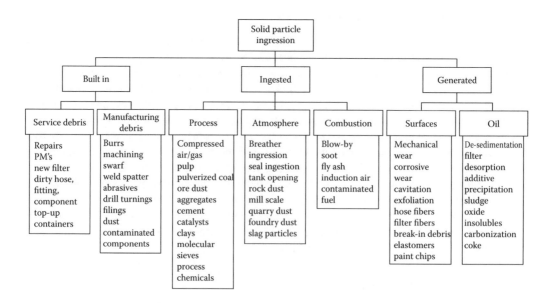

Ingression ➡ All new particles entering a lubricant, regardless of source.

FIGURE 5.10 Categories of particle ingression.

- **Filter breathers.** If reservoirs and sumps can be sealed tightly, such that all air exchanged between the atmosphere and the headspace can be directed through a single port, then high-quality filter breathers can be used to remove dust from incoming air at that port (vent). The quality of the filter (capture efficiency) should be no less than that of the oil filter in use.

Figure 5.11 presents a table of the headspace management options for both particulate and moisture ingression risks. The ingression control strategy needs to correspond to the machine design, operating conditions, and exposures [12].

5.6.4 Map Contaminant Ingression Sources

The first step of a contamination control program is to identify a machine's target cleanliness level as previously mentioned. Next, identify the source and entry points of particles. This generally involves conducting a contaminant ingression study.

Because particles are often internally generated, a contaminant ingression study is not simply a matter of doing a walk-down inspection to look for top-end ingression points. For many machines there is a need to examine particles found in used filters, bottom sediment, oil drains, and live zone oil samples as a means to determine their origin. This can be done using microscopic methods and by element analysis (testing particles for copper, lead, iron, silicon, etc.). Multiple oil sampling points in circulating equipment can help to isolate ingression to certain components like hydraulic cylinders. Additionally, taking particle counts up and downstream of filters while the machine is in normal service can be helpful in identifying the approximate ingression rate (number of particles entering per unit time).

Figure 5.12 shows how this information can be used to map the contaminant sources for a hydraulic system. In the hypothetical example, the figure shows particle and moisture entry relative to six contributing sources. Furthermore, the headspace and ventilation contributing source shows a breakout of six sub-entry points for these contaminants. This same detailed breakout could be charted for the other main contributing sources. Once these sources are understood, plans should be developed and deployed to systematically restrict ingression, starting at the highest ingression points [12].

	Environmental dust	Humidity or steam	Rain or washdown sprays	Ingression past open hatches, inspection ports, etc	Initial costs	Maintenance costs
Spin-on filter or enclosed cartridge filter used as a breather	Y	N	N	N	L	M
Desiccant filter breather	Y	Y	N	N	L	M
Dry instrument air or nitrogen purge	Y	Y	Y	Y	M	M
Oil mist purge	Y	Y	Y	Y	M to H	M
Desiccant headspace dryer	N	Y	M	N	M	M
Mechanical dehumidifier system	N	Y	M	N	H	L
Expansion chambers, bladder, etc.	Y	Y	Y	N	H	L

Controls these contaminants or these sources of contaminants entering headspace · Other factors

Y = Yes   N = No   M = Marginal   L = Low   H = High

FIGURE 5.11 Headspace management options for both particulate and moisture Ingression.

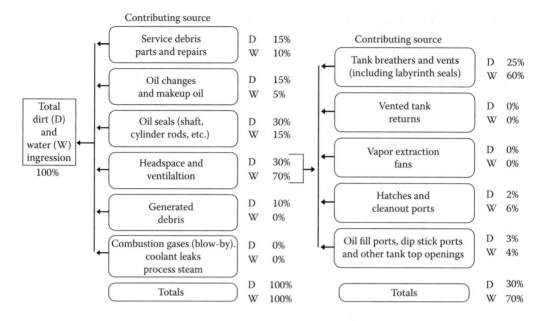

FIGURE 5.12 Sample contaminant ingression mapping chart.

5.6.5 ROLL-OFF CLEANLINESS

Roll-off cleanliness is the level of cleanliness expected to be achieved when assembling any component or system with clean parts in a controlled manufacturing environment. The purpose of maintaining a high level of roll-off cleanliness is to minimize the overall original system contamination and reduce the premature damage caused to various system components upon startup.

Built-in contamination is the inadvertent contamination left in a system or component during initial assembly or system rebuild. The quality and cleanliness of the manufacturing environment is paramount to limiting the amount of built-in contamination. Therefore, the important role of cleanliness begins with the control of contamination in manufacturing environments and the associated work and assembly practices. Much like the in-service control of hydraulic fluid contamination, the work zones of a manufacturing facility should also be viewed as "controlled areas." Some of the areas that must be controlled include:

- Assembly, rebuild, cleaning and repair areas;
- Component, parts and fluid storage areas;
- Shipping and receiving, purchasing and the supplier areas.

Two standards relating to roll-off cleanliness for hydraulic systems are the following:

- ISO/TR 10949 – Hydraulic fluid power – Component cleanliness – Guidelines for achieving and controlling cleanliness of components from manufacture to installation.
- ISO/TS 16431 – Hydraulic fluid power – Assembled systems – methods for achieving roll-off cleanliness.

5.7 FILTRATION AND REMOVAL OF PARTICLE CONTAMINATION

Referring back to the contamination control balance (Figure 5.9), after all efforts have been expended to reduce ingression, the only remaining areas of focus are in the decisions related to the quality, performance, and economy of filtration. There is a price tag for removing dirt from oil. For large plants and fleets operating in dusty environments, the cost can be substantial—hundreds of thousands of dollars per year. With that said, where quality filtration is needed to achieve cleanliness targets, there are still several options to get the most cleanliness for the fewest filtration dollars, referred to as "filter economy."

5.7.1 SELECT FILTERABLE HYDRAULIC FLUID

Consider testing fluid filterability, especially for filters with mean pore sizes of 5 μm or below. Even if new fluids are relatively clean, they may be simply non-filterable (or poorly filterable). Many hydraulic fluids exhibit unique differences when it comes to filterability. Refer to the standard ISO 13357 for filterability testing for both wet and dry conditions [12].

There are several contributing factors that cause impaired fluid filterability. For instance, many new lubricants may have soft, organic impurities or metal soaps that contribute to premature filter plugging. Some of this filterable material may be undissolved additives or perhaps stringy polymeric additives (e.g., VI improvers or pour point depressants) which partially restrict flow through fine pore-size filter media.

Another cause of poor filterability relates to old oils that suspend a high population of very small particles. Often these particles fall below the size detection limit of optical particle counters. While these particulates may be smaller than the mean pore of the filter media (say less than 2 μm), through a mechanism known as secondary and tertiary flow restriction, the filter can become rapidly plugged by these particles. In such cases, an oil change may be a more economical solution than filtration.

Undissolved moisture (above the oil's saturation point) can also shorten a filter's life. Water has a tendency to absorb into the pores of cellulose media or adsorb onto filter media fibers. In either case, the presence of water can shorten filter life and can even impair the structural integrity of the filter media. Water also contributes to oxidation and hydrolysis of the oil, which can produce gums and resins, leading to premature filter plugging.

5.7.2 SEEK ECONOMIC FILTER AND FILTRATION CHOICES

The process of making economic filtration choices can be broken down into two categories: economic filters and economic filtration, which are similar concepts although with certain differences. Economic filters relate to such considerations as filter size, media type, dirt-holding capacity, and so forth. Economic filtration relates to the system and operating conditions such as flow density, pressure, filter location, use of multiple filters, use of centrifuges, and so on.

The following is a list of factors and conditions that, with some exceptions, can improve filter and/or filtration economy [13].

1. **Low filter pressure.** Spin-on filters and even disposable cartridge filters that are designed for high-pressure systems generally cost more for the same dirt-holding capacity than most low-pressure filters.
2. **Low-collapse filters.** High-collapse hydraulic filters generally do not have bypass valves. These filter elements are robustly constructed to resist desorption, media migration, and collapse. These often unneeded attributes are more expensive than common low-collapse filters.
3. **Oversized filters.** The lower the oil's flow rate relative to the maximum allowable element flow rate (catalog flow rate), the better the filter economy (see Figure 5.13). This is also referred to as "flow density." For instance, doubling the size of a filter may triple the dirt-holding capacity (and triple the filter's service life) but may cost less than twice the price (per filter element). Additionally, low-filter flow density also reduces energy consumption costs [14].
4. **High dirt-holding capacity elements.** The technologies used in filter media and filter element construction vary considerably. For instance, mean fiber diameter, fiber composition, pore density, pore depth, tapered pore structure and cladded media are design factors that

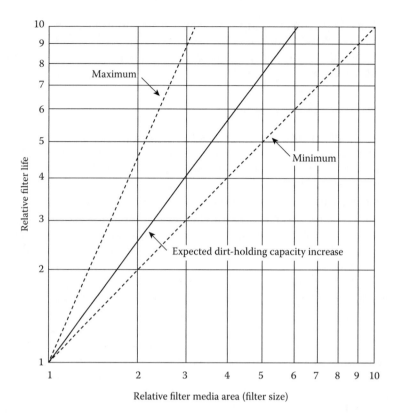

FIGURE 5.13   Filter size effect (media area) on dirt-holding capacity.

influence the dirt-holding capacity of a filter. There are also differences in pleat geometry, flow direction and element construction that influence the total media area per filter element unit volume, which in turn influences oil flow density. The element construction also influences the risk of pleat movement, flow channeling, fatigue and structural integrity. As mentioned above, the lower the effective flow density (flow rate per unit media area), the higher the dirt-holding capacity and the longer the filter's service life. Most hydraulic filters are tested to ISO 16889, which reports information on dirt-holding capacity and capture efficiency (Beta ratio).

5. **Series filtration.** Two or three filters arranged in series have been found, in certain cases, to improve filtration economy. The oil passes through coarse, lower-cost filters before reaching the final polishing filter. Most of the dirt is removed by these lower-cost filters first, allowing the more expensive polishing filter to have extended service life.

6. **Warm oil filtration.** Configuring filters upstream of heat exchangers can extend service life as well. The lower viscosity of the warm oil enables the oil to flow with less restriction through the filter media, delaying the time it takes to reach the terminal pressure drop (filter change alarm). As mentioned, warm oil also improves fluid filterability. However, exceedingly high oil temperatures present many other challenges such as premature oil oxidation and thermal fatigue of the filter media.

5.7.3   Key Filter Selection and Location Considerations

Selecting and locating hydraulic filters to meet cleanliness and reliability objectives is an engineering process and the details of which are beyond the scope of this chapter. However, an overview

TABLE 5.3
Filter Selection Decision Table

Filter Selection Decision	Why It Matters	Inputs Needed to Make the Decision
Beta (time-weighted average)	Defines the life cycle particle capture efficiency at certain particle sizes. Helps to determine what target cleanliness can be achieved.	Target cleanliness level.
Beta (life cycle minimum)	Defines worst-case capture efficiency. Helps define reliability risk in protecting system.	Contaminant sensitivity of high-risk components, for example, servo valves.
Duty cycle tolerance (surge flow, flow cycling, vibration, shock, etc.)	Some machines produce high duty cycle and therefore are at a high risk of losing filtration performance.	Machine design and application information. Other related factors are pressure/flow cycles (frequency and amplitude), temperature extremes, etc.
Dirt-holding capacity	The dirt-holding capacity helps define the service interval (how often filter must get changed) and the average cost to remove a gram of dirt (filter economy).	Ingression rate, maximum permissible size of the filter, importance of filter economy (budget constraints, etc.).
Filter media type	Some filter media types are incompatible with the expected service life of the filter, temperature extremes, fluid chemistry, duty cycle and pressure differential.	Information on the fluid, duty cycle, required service interval, temperature extremes, and minimum required collapse pressure.
Housing and element size, metallurgy, surface coatings/treatments, maximum allowable operating pressure, ports, canister configuration, mounting brackets/configuration, collapse strength, etc.	Determines whether a filter selection will satisfy filter selection objectives.	Filter location, flow rate at that location, maximum new element differential pressure, maximum required terminal differential pressure, duty cycle, pipe/hose connectors/size/threads, viscosity, fluid type, coldest startup temperature, normal operating temperature.
Need for drainback blocking valves. Need for duplex filtration with switching valve (change-over value)	Relates to whether the filter can be changed on the run and with minimal fluid loss.	Machine operating conditions, frequency of downtime, risk of fluid loss during filter changeouts, etc.
Bypass valve type and setting	Determines what differential pressure permits oil to bypass to avoid starvation, yet risking downstream contamination. Aids in determining the lowest cold-start temperature.	Information on temperature extremes, reliability/safety risks of lubricant starvation, downstream contaminant sensitivity, availability of other filters (serial or parallel).
Filter change indicator and type	Determines the criteria for changing the filter and how the need will be communicated to maintenance staff.	Reliability/safety risks of lubricant starvation, downstream contaminant sensitivity, availability of other filters (serial or parallel), and availability of maintenance staff.

table of the decision steps needed to select filters for hydraulic systems is shown in Table 5.3. Most filter suppliers can provide the engineering support to facilitate the design and selection process. However, it needs to be emphasized that good filter selection at the outset can save considerable costs down the road in terms of filter replacement and machine reliability [11].

There are three primary options for locating a filter on a conventional hydraulic system: (1) pressure-line, (2) return-line, and (3) off-line (see Figure 5.14). In rare cases all three locations have been fitted with filters. If just two locations are selected, it could be any duplex combination of the three options. If just one filter location is fitted with a filter it is usually either the pressure-line or the return-line. The following is a discussion of these three locations:

Pressure-line filtration. Filters located on the pressure line receive the full flow and pressure delivered by the pump. As such, both the filter element and the housing must be designed to handle these often extreme operating conditions. Other unique field and duty cycle conditions can put stress on the pressure-line filter as well including vibration, pressure ripple, shock loading, and temperature cycling.

Pressure-line filters are often selected to mitigate the risk of tank contamination from being dispersed into sensitive work-end system components. Additionally, pumps in failure mode are protected from shelling out a debris field into downstream components when pressure-line filters are used. Because these filters are more expensive than return-line and off-line filters the resultant cost per gram of dirt removed is consequently higher. Many designers put slightly coarser filter elements (say 15 μm) in pressure-line filter housings to gain their protective attributes but rely on other finer filters (say 3 μm) elsewhere for dirt removal.

Return-line filtration. Like pressure-line filters, return-line filters are subjected to extreme operating conditions. However, these conditions have unique differences. Instead of high pump pressure

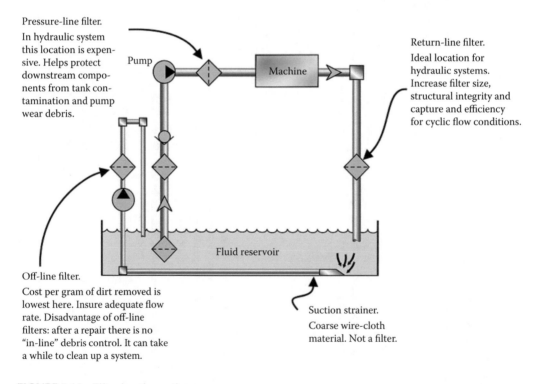

FIGURE 5.14   Filter location options.

and flow, return-line filters generally see only mild line pressures but widely varying flows. The flow rate on the return-line is often not defined by the system pump but rather the load conditions from actuators. Large hydraulic cylinders can induce flow surges that often exceed maximum pump volume by a factor of three or more. Surge-flow conditions can wreak havoc on return-line filter elements causing structural fatigue and impaired capture-efficiency.

Filters specifically designed to resist surge-flow stresses are sometimes specified. In other cases the solution may simply be the use of oversized filters. The benefit of return-line filters is their downstream proximity to the largest particle ingression sites on most hydraulic systems (cylinder rods ingression past wiper seals). Using return-line filters, these particles can be stripped from the oil before reaching the reservoir. Once particles reach the reservoir they present a high risk to the pump since, for a number of reasons, suction-line filters are not a practical reality.

Off-line filtration. Off-line filters are a relatively modern alternative or addition to conventional full-flow filters. These filters sit off the main operating system as a side-loop from the reservoir. A necessary supplemental component to the off-line filter is a pump and motor. Because it does not depend on the hydraulic system, this can run independently, even when the main system is off. The following are some of the additional benefits and attributes of the off-line filtration option:

- Higher initial cost (pump, motor, valves, piping);
- Constant flow optimizes dirt-holding capacity and capture efficiency for a given type of filter;
- Easy to service on the run (filter changes, repairs, etc.);
- Heat exchanges can be built in the loop;
- Sample ports can be installed for sampling on the run;
- Lowest cost to remove a gram of dirt (expensive pressure-line and surge-resistant filters are required);
- Can double for an oil transfer system for adding makeup oil.

5.7.4   FILTER PERFORMANCE TESTING AND RATINGS

Gone are the days when filter manufacturers described the performance of their filters in terms of nominal and absolute micron ratings. Modern hydraulic filters by reputable manufacturers have been tested to assess performance attributes across a range of criteria. These include collapse strength, burst pressures, and structural integrity. However, the contaminant removal characteristics of a filter come from testing that is done in accordance with ISO 16889 (formerly ISO 4572).

Information from this standard includes pressure versus flow characteristics, dirt-holding capacity and filtration ratio (also known as the "Beta ratio"). The filtration ratio is a measure of the particle capture performance of a filter at standardized test conditions (maximum rated flow, constant flow, constant temperature, constant contaminant injection rate, standardized test dust, and fluid). The simplified schematic of a multipass test stand that runs a filter to ISO 16889 is shown in Figure 5.15. The Beta ratio is calculated as the number of particles above a specific micron size (per unit volume of fluid) upstream of the test filter, divided by the number of particles above that same micron size downstream of the filter (see Figure 5.16). The standard calls for the micron size to be reported for filtration ratios of 2, 20, 75, 100, 200 and 1000.

5.7.5   OPTIMIZING FILTER CONSUMPTION

Changing a filter too late puts the oil and machine in jeopardy. Changing a filter too soon wastes valuable dirt-holding capacity. It has been reported that in many cases the real cost of a common oil change can exceed 10 times the apparent cost of the oil and associated labor. This multiplier may hold equally true for the cost of a filter change. In addition to the cost of the filter, there are

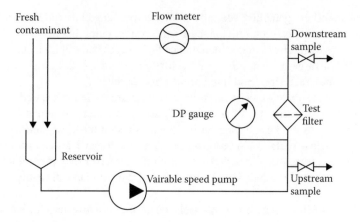

FIGURE 5.15   Multipass filter test stand.

additional costs for labor, inventory, scheduling, used-filter disposal, waste oil disposal, and oil top-off costs (some oil is lost when changing filters) [15].

There are many available technologies to help improve the timing of a filter change. These include pressure-rise profile monitoring, delta-P indicators, bypass indicators, on-line particle counting, and time-out alerts. Multiple methods used together may be the best choice in certain cases. Nonetheless, changing filters "on condition" should be a primary objective towards achieving filter economy.

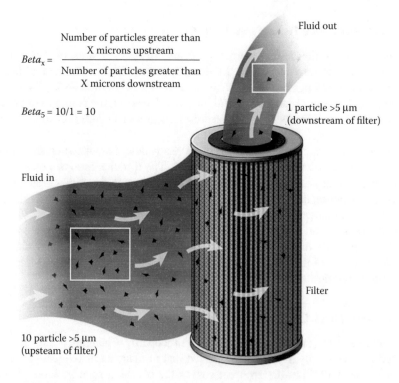

FIGURE 5.16   Filtration ratio (Beta).

5.8 OBTAINING REPRESENTATIVE FLUID SAMPLES FOR CONTAMINANT MONITORING

Oil sampling is one of the most critical factors of successful fluid analysis. Errors in obtaining a representative sample can impair all further analytical efforts, especially particle counting. There are two primary goals in obtaining a representative oil sample. The first is to sample at a location and in a manner that maximizes data density. For instance, the data could be dust particles, moisture, additive levels, or wear debris [10,16–18].

The second goal is to minimize data disturbance. Samples should be extracted in such a way that the concentration of information is uniform, consistent and unaltered by the sampling process. It is important to make sure that the sample does not become contaminated during the sampling. This can distort the data, making it difficult to distinguish what was originally in the oil from what has come into the oil during the sampling process.

To ensure good data density and minimum data disturbance in oil sampling, one should consider the following factors:

- **Sampling location:** Not all locations in a hydraulic machine will produce the same concentration of data. Complex hydraulic systems require multiple sampling locations in order to answer specific questions related to system condition, usually on an exception basis (troubleshooting). Primary sampling points are used for routine sampling and analysis. Secondary sampling ports are used only for troubleshooting to isolate the contaminant-generating/ingressing source.
- **Sampling method:** The procedure by which a sample is drawn is critical to the success of lubricant analysis. Sampling procedures can vary substantially and therefore should be documented and followed uniformly. Technicians should be trained to follow the documented standardized method without variation.
- **Sampling hardware:** The hardware used to extract the sample should not disturb sample quality. It should be easy to use, clean, rugged, and cost-effective.
- **Sample container:** The type and size of bottle and cleanliness help ensure that a representative sample is achieved and fluid volume is sufficient to perform the intended analyses.

It is always advised that one expend the necessary resources for critical systems in order to install proper sampling hardware (valves, access ports, etc.) and ensure that the above goals in oil sampling are achieved. Experience has shown that sampling hardware is not a place to economize; oil analysis is too expensive and unrepresentative samples lead to costly false positives and false negatives.

5.8.1 STRATEGY FOR OPTIMUM SELECTION OF SAMPLING LOCATION(S)

There are several rules for properly locating oil sampling ports on hydraulic systems. These rules cannot always be precisely followed because of various constraints in the machine's design, application, and work environment. However, the rules outlined below should be followed as closely as is reasonably possible [10,16–18]:

- **Turbulence:** The best sampling locations are highly turbulent areas where the oil is not flowing in a straight line but is turning and rolling in the line. Sampling valves located at right angles to the flow path in long straight sections of pipe can result in particle fly-by, especially where there is high fluid velocity and low fluid viscosity. Such conditions can lead to a marked reduction of the particle concentration entering the sample bottle. This can generally be avoided by instead locating sampling valves at elbows and sharp bends in the flow line (Figure 5.17).

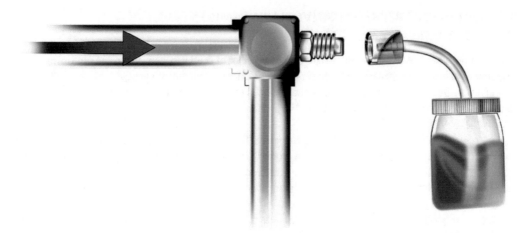

FIGURE 5.17   Sample valve located in a highly turbulent fluid zone.

- **Ingression Points:** Where possible, sampling ports should be located downstream of the components that produce wear particles and potentially ingress particles and moisture. Return-line sample port locations usually offer the most representative levels of wear debris and contaminants in hydraulic systems. Once the fluid reaches the reservoir, wear debris and contamination can become sharply diluted (and potentially undetected).
- **Filtration:** Filters and separators are contaminant removers and, as such, they can remove valuable data from the oil prior to sampling. Sampling valves should be located upstream of filters, separators, dehydrators, and settling tanks unless the performance of the filter is being specifically evaluated.

Just as there are factors that can improve the quality of a sample, there are also other factors which can diminish a sample's quality and should thus be avoided. For example, it is important not to sample from dead pipe legs, hose ends, and stand pipes where the fluid is not moving or circulating. Samples should not be collected after filters or separators, or after an oil change, filter change, or at some time when the fluid would not represent typical conditions. Samples should not be taken when the machine is cold and has not been operating or has been idle. In addition, samples should not be taken from laminar flow zones where a lack of fluid turbulence occurs.

5.8.2   SAMPLING FROM PRESSURIZED LINES

When samples need to be taken from pressurized lines the sampling method is often simplified. Figure 5.18 shows four different configurations for sampling from pressurized lines.

- **Portable high-pressure tap sampling:** The uppermost configuration on Figure 5.18 is a high-pressure zone where a ball valve or needle valve is installed and the outlet is fitted with a piece of stainless steel helical tubing. The purpose of the tubing is to reduce the pressure of the fluid to a safe level before it enters the sampling bottle. A similar effect can be achieved using a small, hand-held pressure-reduction valve.
- **Minimess tap sampling:** This option requires the installation of a minimess valve or similar sampling valve, preferably on an elbow. Minimess valves are probe-style valves commonly used for oil analysis and pressure diagnostics. The sampling bottle has a tube fitted with a probe protruding from its cap. The probe attaches to the minimess valve, allowing the oil to flow into the bottle. There is a vent hole on the cap of the sampling bottle so that when the fluid enters the bottle the air can exhaust. This particular sampling method requires lower pressures (less than 500 psi) for safety.

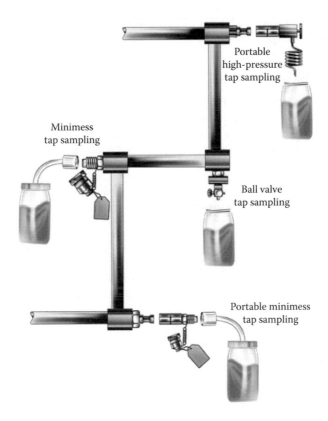

FIGURE 5.18   Options for sampling pressurized lines.

- **Ball valve tap sampling:** This configuration requires the installation of a ball valve or one of many closely-related valves specific for oil sampling. They are typically actuated by push-button or knob rotation to extract the fluid sample. When sampling, the valve should be opened and adequately flushed. Extra flushing is required if the exit extension from the valve is uncapped. Once flushed, the sample bottle's cap is removed and a sample is collected from the flow stream before closing the valve. Care should be taken when removing the bottle cap to prevent the entry of contamination. This technique is generally not suitable for high-pressure applications.
- **Portable minimess tap sampling:** This option requires installing a minimess valve onto the female half of a standard quick-connect coupling. This assembly is portable. The male half of a quick-connect is permanently fitted to a pressurized line of the system at the desired sampling location. To sample, the portable female half of the quick-connect is screwed or snapped (depending on adapter type) onto the male piece affixed to the machine. As the adapter is threaded onto the minimess valve, a small spring-loaded ball is depressed within the minimess valve, thereby allowing oil to flow through the valve and into the sample bottle. In many cases, these male quick-connect couplings are pre-existing on the equipment. A helical coil or pressure reduction valve, previously described, should be used on high-pressure lines for safety reasons.

5.8.3   Sampling from Low-Pressure Circulating Lines

Occasionally a return line is not sufficiently pressurized to take a sample. In such cases, sampling requires assistance from a vacuum pump equipped with a special adapter allowing it to attach momentarily to a sampling port, such as a minimess valve. With the adapter threaded onto the minimess valve, fluid can then be drawn by vacuum into the bottle (Figure 5.19) [10,16–18].

5.8.4 DROP-TUBE VACUUM SAMPLING

One of the most common methods for sampling sumps and reservoirs is to use the drop-tube vacuum sample method. A tube is inserted through a fill port, hatch or dipstick port and lowered into the fluid, usually about midway into the oil level. This sampling method has a number of drawbacks and should be avoided if the sampling methods previously described can be applied instead.

5.8.5 SAMPLING BOTTLES AND HARDWARE

An important factor in obtaining a representative sample is to make sure that the sampling hardware is completely flushed prior to obtaining the sample. This is usually accomplished using a spare bottle to catch the purged fluid. It is important to flush five to ten times the dead space volume before obtaining the sample. All hardware in which the oil comes into contact is considered dead space and must be flushed, including:

- System dead-legs,
- Sampling ports, valves and adapters,
- Probe on sampling devices,

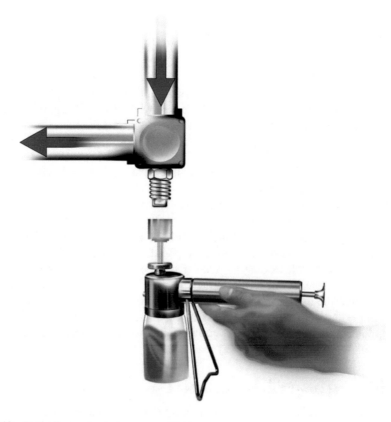

FIGURE 5.19   Drain line vacuum-pump sampling.

- Adapters for using vacuum sample extraction pumps,
- Plastic tubing used for vacuum pumps (this tubing should not be reused to avoid cross-contamination between oils).

There is an assortment of sampling bottle types that are widely used in oil analysis. An appropriate bottle needs to be selected for the application and the test slate that is planned. Several features, including size, material and cleanliness must be considered when selecting a sample bottle [10,16–18].

Bottle material. Modern sample bottles are made of PET plastic (polyethylene terephthalate) due to their chemical compatibility with most base oils and additives, along with the fact that they are clear, strong (fracture resistance), inexpensive, and widely availability. The primary disadvantage of using PET bottles is the risk that they will melt or become soft when sampling high temperature fluids, say above 200°F.

Bottle size. Bottle size should correspond directly to the minimum amount of fluid specified by the laboratory or onsite instrument(s). This volume should be sufficient to:

- Perform the routine test slate;
- Repeat one or two tests in the event of aberrant data;
- Perform exception tests (e.g., ferrography) triggered by a routine test such as a particle count;
- Leave a sufficient residual amount for retesting (confirmation testing) in the future, if required (these are often called "retains").

Bottle cleanliness. The sample bottle cleanliness requirement is dependent on the target cleanliness of the hydraulic fluid. Modern oil analysis programs typically specify that bottles be ten times cleaner than the fluid target cleanliness for the same volume. The nomograph in Figure 5.20 is helpful in defining bottle cleanliness requirements [1].

5.8.6 SAMPLING FREQUENCY

The use of scheduled sampling intervals is common in lubricant and hydraulic fluid analysis. The sampling frequency is generally tied to oil drain intervals, operating hours or usage events. Standard recommended intervals reported by OEMs, laboratories, and in technical literature are often used initially as a default frequency. These intervals can be later adjusted based on experience, or customized, by taking into account the following machine and application-specific conditions [10]:

- Consequence of failure: Safety, downtime costs, repair costs, and general business interruption costs should be considered.
- Operating environment: Operation and fluid environment conditions influence the frequency and rate of machine and fluid failure. Among other factors, these include pressures, loads, temperature, speed, contaminant ingression rate, and duty cycle severity.
- Lubricant age: In many cases problems occur right after lubricants are serviced (drains and refills). A problem can be associated with the accidental entry of the wrong oil and an incompatible oil. Hydraulic fluids approaching the end of useful life are also high risk. Aged oils will often have depleted additives, incipient oxidation, and high levels of various types of contaminants.
- Machine age and maintenance factors: For most machines the chances of failure are greatest during break-in and after major repairs, rebuilds, and extended downtime. The risk may also increase as a machine approaches the end of its expected life.

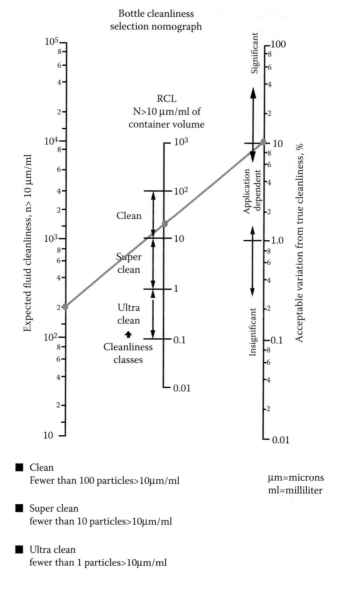

FIGURE 5.20   Nomograph for specifying required bottle cleanliness.

5.9   QUANTIFICATION AND CHARACTERIZATION OF PARTICLES

When particle contamination is monitored and reported routinely, not only is proactive maintenance generally achieved but also many of the goals of predictive maintenance. As such, particle counting (and other similar methods) is an important first line of defense in machinery reliability. Because of its value, it is not uncommon to find organizations testing the cleanliness of their oils as frequently as once a week, especially for high-criticality machines.

The following are common proactive and predictive uses for particle counting and analysis in condition monitoring [19]:

Proactive Maintenance:

1. Routinely verify that in-service oils are within targeted cleanliness levels.
2. Check the cleanliness of new oil deliveries.

3. Quickly identify failed or defective filters.
4. Confirm that seals and breathers are effectively excluding contaminants.
5. Confirm that systems are properly cleaned and flushed after repair.
6. Confirm that new hydraulic systems are cleaned and flushed before use (roll-off cleanliness).
7. Identify the improper use of dirty top-up containers and poor maintenance practices.
8. Identify the need and timing for portable filtration systems.

Predictive Maintenance:

1. Identify early-stage abnormal machine wear.
2. Identify the location/source of abnormal wear by multi-point isolating methods.
3. Verify the effectiveness of corrective maintenance and botched repair jobs.
4. Monitor machine break-in progress by wear particle generation.
5. Identify abnormal rust and corrosion debris generation.
6. Serve as an effective screen for wear debris analysis (e.g., analytical ferrography).

5.9.1 Particle Monitoring and Analysis Methods

The following are the most significant and effective particle monitoring and analysis methods commonly used in the oil analysis field:

Particle counting. Particle counting is considered to be one of the most valuable test methods in fluid analysis and its use dates back to the 1960s. The particle count test reports the number of particles above specified size ranges (in microns) per fluid volume (usually per ml or 100 mL). Also, particle concentration and distribution data may be expressed in terms of ISO 4406:99 Cleanliness Codes (Figure 5.8) or by other less frequently-used codification systems, such as the revised SAE AS 4059E (formerly NAS 1638). Particles can be counted manually using optical microscopy (ISO 4407 and ASTM F312-97). In this method an aliquot of fluid is passed through a membrane. Afterwards, particles on the membrane are manually counted under a microscope. The method is similar to the patch test procedure discussed below. There are commercial methods available which enable membranes to be optically scanned and digitally analyzed for particle size, count and shape [20].

Most laboratories use automatic particle counters, which can report a particle count or ISO Code in just a couple of minutes. The two methods are laser optical (ISO 11500) particle counters and pore-blockage (BS 3406 & ISO/DIS 21018). Optical particle counters direct a laser light source at passing particles in the sensor cell [21], see Figure 5.21. The amount and frequency of light blockage is measured by a photodiode. This signal is converted to particle size and count by the use of standardized calibration methods. Pore-blockage particle counters use calibrated screens through which the sample flows during a test (Figure 5.22). The profile of the pressure rise or flow decay, caused by particle blockage of the screen's pores, is measured [22]. This profile is mathematically converted to an estimated particle count or ISO Code [23]. Some modern particle counting technologies also have the ability to characterize particle shape. With this added information, interpretation of the source, type, and severity of the particles can be estimated [24].

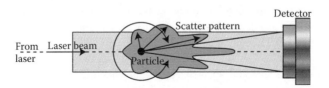

FIGURE 5.21   Light-blockage particle counter.

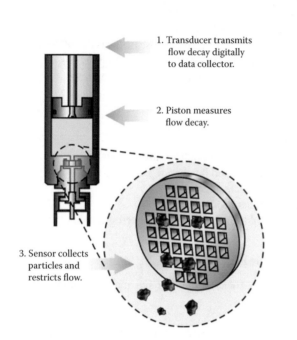
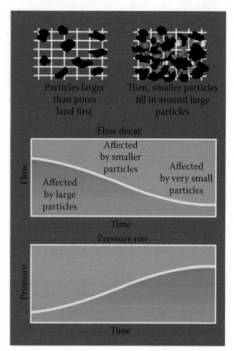

1. Transducer transmits flow decay digitally to data collector.

2. Piston measures flow decay.

3. Sensor collects particles and restricts flow.

Particles larger than pores land first

Then, smaller particles fill in around large particles

Flow decay

Affected by smaller particles

Affected by large particles

Affected by very small particles

Flow

Time

Pressure rise

Pressure

Time

FIGURE 5.22 Pore-blockage particle counter.

Because of the differences between particle counting methods (manual, laser and pore-blockage), as expected, there will be differences in results on the same samples. The reasons for this are many and should not be a point of major concern for users if sample preparation and instrument procedures are used correctly. These methods are not absolute measurements, but instead estimate size and concentration using standard practice and assumptions. Regardless of the method, it is recommended that the same one is always used in order to ensure that results are consistent and repeatable. In the right application with the right procedure, most methods are very suitable in the context of maintenance and machine reliability [1,25,26].

Ferrous density. A sudden and significant increase in the population of large (greater than 5 μm) ferrous particles can signal the presence of an abnormal wear condition and perhaps of impending component failure. Contamination, poor lubrication and adverse mechanical conditions are the usual causes of high ferrous particles. Typically, at least one surface in a frictional pair is ferrous (iron or steel) and it is usually the surface most critical to machine reliability. For this reason the monitoring of ferrous particle density in used lubricants can provide valuable machine-health information. The need is further magnified by the fact that elemental analysis becomes less accurate with larger size particles (larger than 5 μm), which is usually the critical size range in monitoring and detecting impending failure.

Several instruments and methods are used by onsite and full-service laboratories for determining the concentration of ferrous debris. These methods are typically only able to detect concentrations of ferro-magnetic particles, but others employing the magnetic-induction principle can quantify non-ferrous metal particles as well. The ferrous density measurement units reported by laboratories vary by instrument type [10,25,26].

Elemental spectroscopy. Elemental spectroscopy quantifies the presence of dissolved and some undissolved inorganic materials by element in the lubricant (both oil and grease). Most elemental spectrometers used today for lubricants and hydraulic fluid analysis are the atomic emission type, either Inductive Coupled Plasma (ICP), or Rotating Disc Electrode (RDE). These instruments work by exposing the sample to extreme temperatures generated by an arcing electrode (ASTM D6595) or by an argon plasma torch (ASTM D5185). The extreme heat vaporizes the atoms, causing them to emit energy in the form of light. Each atomic element emits light at specific and characteristic frequencies. The spectrometer quantifies the amount of light generated at each frequency (spectral line) and calculates the concentration of each element (iron, lead, tin, etc.) in parts per million (ppm) based on calibration curves.

Most elemental spectrometers report the concentration of 15 or more elements. The elements reported can provide an indication of increased generation of wear debris, ingression of various types of contamination or depletion of certain additive elements (see Table 5.4). Dissolved metals and suspended particles up to approximately 2 µm are detected with high accuracy. The accuracy diminishes as particle size increases to more than 2 µm. Elemental concentrations can be greatly understated for particles larger than 5 µm.

It is important that critical machines have metallurgical maps which show where elemental families (unique groups of elements) typically emerge during wear and corrosion. Additionally, elemental data from close-proximity contaminants should also be characterized in terms of their major, minor, and trace elements that can be used as markers for identification purposes [10,25,26].

Microscopic contaminant and wear particle identification. When abnormal wear metals have been identified by other methods, including particle counting, elemental spectroscopy, and/or ferrous density analysis, a common and important exception test to perform next is the microscopic particle examination and identification. The most common version of the procedure is referred to as "analytical ferrography". Analytical ferrography involves the analysis of debris deposited onto a ferrogram slide or alternatively a filtergram membrane. Analysis of particle morphology (shape), color, size, reflectivity, surface appearance, edge detail, angularity, elemental content, and relative concentration provides the analyst with clues about the nature, severity and root cause of the contaminant ingression or wear problem. Scanning Electron Microscopy (SEM) can also used to examine particles as well as their elemental composition using an Energy Dispersive Spectroscopy (EDS) feature [27].

TABLE 5.4
Common Elements Found in Lubricants and Hydraulic Fluids

Element	Wear	Contamination	Additive
Iron (Fe)	X	X	
Copper (Cu)	X	X	X
Chromium (Cr)	X		
Tin (Sn)	X		
Aluminum (Al)	X	X	
Lead (Pb)	X		
Silicon (Si)		X	X
Sodium (Na)		X	X
Boron (B)		X	X
Calcium (Ca)		X	X
Magnesium (Mg)		X	X
Zinc (Zn)	X		X
Phosphorous (P)		X	X
Molybdenum (Mo)			X
Potassium (K)		X	

1. Introduce sample   2. Pull vacuum to   3. Rinse with solve   4. Remove patch for
   with solvent into   transfer fluid   inspection
   patch-making   through patch
   funnel

FIGURE 5.23 Patch preparation sequence.

Although largely a qualitative technique, the analyst typically reports the presence and concentration of wear particles, friction polymers, dirt and sand, fibers, and other solid contaminants on either a 1 to 10 or a 1 to 100 scale in order to illustrate severity. Descriptive text and photomicrographs usually accompany the enumerated values to clarify conclusions and recommend corrective actions.

It is important to determine the root cause of machine failure and abnormal wear problems so they can be eliminated, thus avoiding recurrence. By combining information from analytical ferrography with other lubricant analysis and maintenance technology evaluations, the analyst attempts to answer the following questions [10,25,26,28]:

- Where in the machine does the contaminant or wear debris originate?
- What is causing it (forcing function)?
- How severe or threatening is it (residual life)?
- Can the condition be mitigated or arrested without downtime or loss production?

Patch Test. This method is similar to microscopic contaminant and wear particle identification. A small amount of sample is pulled by vacuum through a porous membrane (typically around 5 μm) to enable suspended particles to become deposited on the membrane's surface. A solvent is used to rinse any residual oil from the surface of the membrane (Figure 5.23). Afterwards the

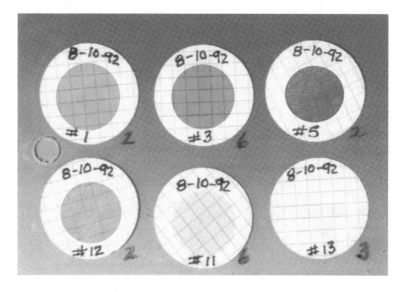

FIGURE 5.24 Patches of varying colors and densities.

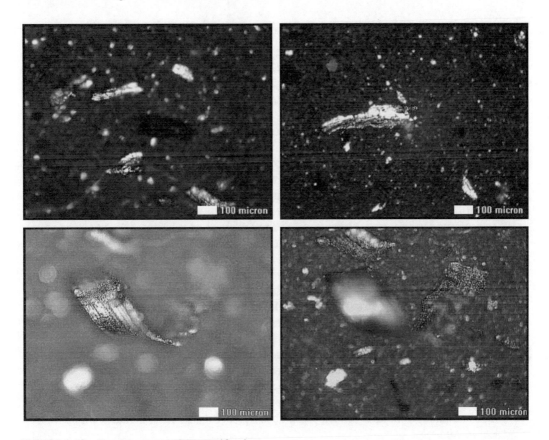

FIGURE 5.25   Patch images under magnification.

membrane can be visually inspected for overall particle density and color (Figure 5.24) [28]. If an abnormal debris field is encountered then the membrane can be placed under a top-lit microscope for a detailed analysis and characterization of the particles (Figure 5.25) [29]. Many analysts will estimate the fluid's ISO Code (ISO 4406:99) based on the overall appearance of particles, sometimes using comparator standards. One such comparator standard used in patch analysis of aviation fuels is discussed in Appendix A3 of ASTM D2276 and is easily applied to hydraulic fluids as well. However, unlike optical particle counters, patch testing allows particle shape, color, edge detail, and organic particles to be inspected. In addition, unlike analytical ferrography, patch testing is relatively inexpensive and can easily be performed in the field [29].

Blotter spot test. This simple test, also known as "paper chromatography" or "radial planar chromatography," is used to examine soft insoluble suspensions in oil using blotter paper to which a small aliquot of sample is applied. Varnish-producing impurities will form distinction deposits and rings on the blotter paper as the oil wicks outward in a radial direction by capillary action. These impurities include carbon insolubles, oxide insolubles, additive degradation products, and glycol contamination. This is a good quality field and laboratory test [10,30].

Ultracentrifuge and other varnish-potential tests. These methods use centrifugation or coagulation to separate and estimate the concentration of varnish-producing oxide-insolubles and other soft impurities in hydraulic fluids and other oils. Various methods are currently used by oil analysis labs [31].

5.10 DATA INTERPRETATION AND TROUBLESHOOTING

The following are common examples of different data-alarming strategies used in lubricant analysis. To one extent or another all of these strategies have an application in the analysis and interpretation of solid particle contamination in hydraulic fluids. When used together, across several testing schemes, early detection and diagnosis of data anomalies can be easily achieved [32].

- **Goal-based limits:** These are targets applied to fluid parameters like contamination to achieve machine life extension. Target cleanliness levels, as previously discussed, are goal-based alarms. For example, a hydraulic machine running at ISO 18/15 (per ISO 4406:99) cleanliness may experience a triple life extension if the fluid is cleaned to an improved ISO 15/12. Setting the limit at ISO 15/12 is a goal-based strategy; the goal being increased component service life. This type of limit is usually applied to particle count, moisture level (e.g., ASTM D6304), glycol level (e.g., ASTM D4291), fuel dilution, Acid Number (AN) (e.g., ASTM D664), and other common root cause conditions. The setting of such limits is highly dependent on the reliability goals of the equipment owner.

- **Aging limits**: Another type of limit or alarm relates to the progressive aging of a lubricant or hydraulic fluid. From the moment that a fluid is placed in service, its chemical and physical properties transition away from the ideal (i.e., those of the newly-formulated oil). Some properties transition very slowly, while others transition more dynamically. Limits keyed to the symptoms of lubricant deterioration are referred to as "aging limits". Aging limits can be effectively applied to such parameters as acid number, viscosity, oxidation stability tests (e.g., ASTM D2272), elemental spectroscopy for additives, infrared spectroscopy (oxidation, nitration, sulfation, and additives) and dielectric constant. The collective use of aging limits helps to estimate the Remaining Useful Life (RUL) of the lubricant. The level of contaminants (root cause) in a lubricant has a great influence on the rate at which RUL decays (effect).

- **Rate-of-change alarms**: Rate-of-change alarms are typically set to measure properties that are being progressively introduced into the oil, such as wear debris or contamination. The add rate (change) can be calculated per unit of time, hours, cycles, and so forth. For example, a 100 ppm increase in iron over a period of 100 operating hours could be stated as one ppm per hour of operation. When the parameter is plotted against time, the rate-of-change (add rate) equals the current slope of the curve. Unlike level limits, rate-of-change limits ignore the absolute value of the data parameter, emphasizing instead the speed at which the level is changing. Rate-of-change limits are effectively applied to particle counting (unfiltered systems), elemental wear metals, ferrous density, AN and oxidation stability. It can also be effectively applied to monitor abnormal degradation of additives with elemental and infrared spectroscopy.

- **Statistical alarms**. For many years, statistical alarms have been used effectively in lubricant analysis. The practice requires the availability of a sufficient quantity of machine and application-specific historical data from which to draw meaningful statistical benchmarks. The statistical alarming approach is simple. A population mean and associated standard deviation are generated from the available data. The data from a sample is compared to the mean of the population. If the value falls within one standard deviation of the mean, it is considered normal. If it falls outside of one standard deviation from the mean, but within two standard deviations, it is considered a caution, or simply reportable. If the result exceeds two standard deviations, the value is considered in critical alarm as it is higher, or lower as the case may be, than 95% of the population. Should the value exceed three standard deviations, it is a critical alarm, as the value exceeds the ninty-ninth percentile of the historic population. Statistical alarming methods are commonly applied to ferrous density, elemental metals and other predictive lubricant analysis measurements.

5.10.1   Analytical Strategy for Detecting and Troubleshooting Common Problems

The following is a discussion of how oil analysis can be used to detect and troubleshoot particle-related problems that are commonly found in fluid samples. This methodology combines sampling strategy, onsite oil analysis tools, routine laboratory analysis, exception testing, inspections, and companion technologies.

Wear debris detection: When hydraulic systems and components are operating abnormally due to misalignment, fluid degradation, contamination, corrosive conditions, and so on, microscopic pieces of the system's components become suspended in the oil in the form of wear debris. Fluid analysis provides very early warning of this occurrence and increases the planning time and the number of options with which to troubleshoot and correct the problem.

Sampling Strategy	In close proximity to the wearing component or directly downstream of it.
Machine Inspections	Wear debris on filters, metallic sediment on tank bottoms, abnormal noise, high running temperatures.
Onsite Tests	Patch tests, particle counting, used-filter inspections, ferrous density tests.
Primary Lab Tests	Particle counting, ferrous density, elemental analysis.
Alarming Strategy	Statistical limits, rate-of-change, trend plots.
Exception or Confirming Tests	Analytical ferrography, patch testing, machine internal inspections.
Confirming Companion Technologies	Vibration, thermography.

Wear debris analysis. When an abnormal wear condition is encountered, it should be analyzed to provide an indication of the nature, severity, and root cause of the problem. This requires an investigation of the wear particles themselves along with a review of collateral information such as vibration analysis, operational information, lubricant analysis, system inspection, used-filter inspection, and so on.

Sampling Strategy	In close proximity to the wearing component or directly downstream of it.
Machine Inspections	Wear debris on filters, metallic sediment on tank bottoms, abnormal noise, high running temperatures.
Onsite Tests	Patch tests, particle counting, used-filter inspections, ferrous density tests.
Primary Tests	Analytical ferrography, elemental analysis.
Alarming Strategy	None, qualitative.
Exception, Supporting or Confirming Tests	Ferrous density, particle counting, elemental analysis of filter debris.
Confirming Companion Technologies	Vibration, thermography.

Solid particle contamination. An alarm on particle contamination, using ISO Codes—for instance, signals an increase in suspended particles due to such occurrences as the failure of a filter, ingestion of contaminants from the environment through seals, vents, new oil or an increase in the generation of wear debris.

Sampling Strategy	In close proximity to high-risk ingression points or directly downstream of them.
Machine Inspections	Sediment on tank bottoms, unsealed reservoirs/sumps, filters in bypass, defective breathers, etc.
Onsite Tests	Patch tests, onsite particle counter, used-filter inspections, ferrous density tests.
Primary Tests	Particle count, elemental analysis.
Alarming Strategy	Cleanliness targets (ISO Code, NAS, etc.) based on reliability goals.
Exception, Supporting or Confirming Tests	Patch tests, analytical ferrography.
Confirming Companion Technologies	None.

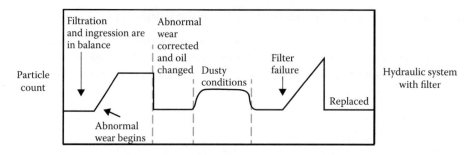

FIGURE 5.26   Typical particle count trends.

Figure 5.26 shows how particle count trends can vary over time. Because particle counters monitor particles in the general size range controlled by filters, particle concentration equilibrium (steady-state condition) is usually achieved—that is, particles entering the oil from ingression minus particles exiting from filtration will leave behind a steady-state concentration. When filters are properly specified and ingression is under control this steady-state concentration will typically be within the cleanliness target. For systems with no continuous filtration, or poor filtration, the equilibrium is usually not effectively established (there is no continuous or reliable particle removal). This can cause the particle concentration to be continuously rising, or moving erratically.

5.10.2   TROUBLESHOOTING A HIGH PARTICLE COUNT RESULT

High particle counts generally have one of four possible explanations and outcomes when investigated:

1. The system is not in any immediate danger; however, either a filter has failed or there is a new source of particle ingression. The problem is solved by correcting the offending filter or ingression source.
2. There is a new ingression source or filter failure and the machine is in immediate danger due to the resulting high particle count. This problem is solved by a rapid clean-up of the oil, followed by correcting the failed filter or ingression source.
3. The high particle count is due to abnormal wear particle generation constituting a potential threat to machine reliability. This can be solved by performing a root cause failure analysis followed by appropriate remediation and clean-up as required.
4. The high reading was due to sampling error (including dirty sample bottle), analytical error (particle counter calibration, sample preparation, etc.) or soft particles (dead additives, oxide insolubles, etc.) that were read as hard particles.

5.11   SUMMARY

Contamination control requires a critical amount of planning, preparation and deployment. It is sometimes referred to as "planned cleanliness". Success depends heavily on behavior-based strategies and execution. This fact is widely validated by the many case studies that have been published on this subject in recent years.

Some of the main elements for achieving planned cleanliness as discussed in this chapter are summarized below:

1. Educate organizational players and stakeholders on the virtues of cleanliness and the tactics for achieving it. This puts everyone on the same page, aligned with a single objective.
2. Keep target cleanliness for critical machines front and center—the more conspicuous, the better. Make ISO Codes a part of the company's reliability vocabulary. Put highly visible

cleanliness targets on, or near, machinery to which they relate. Communicate clearly the ways in which contamination control plays a strategic role in achieving business objectives.

3. Invest in onsite particle counting or patch testing. Install live-zone sampling ports to ensure representative samples. Monitor machine cleanliness vigorously. People work the metric, so make particle counting an important one. Talk it up and celebrate cleanliness at every opportunity.

4. Post green, yellow and red tags on all program machines to enunciate cleanliness status. Any fluid that is noncompliant gets a yellow or red flag (depends on severity) tagged to the machine until the aberrant condition is remedied. Take immediate action to correct non-compliant machines.

5. Pursue every reasonable opportunity to exclude contaminant ingression. Upgrade filtration prudently.

6. Put oil suppliers, workshop technicians, parts suppliers, and rebuild contractors on notice regarding roll-off cleanliness. Develop rigorous inspection procedures and follow through for all nonconforming oils or equipment.

7. Keep track of program costs and savings.

REFERENCES

1. Fitch, J.C. *Fluid Contamination Control*. Stillwater: FES, Inc., 1988.
2. Fitch, J.C. "New Lubrication Commandments - Conserve Energy, Protect the Environment," *Machinery Lubrication Magazine*, July 2002.
3. Fitch, J.C. "A Much Closer Look at Particle Contamination," *Practicing Oil Analysis Magazine*, September, 2005.
4. Hamblin, M. and Stachowiak, G. "Description of Abrasive Particle Shape and Its Relation to Two-Body Abrasive Wear," *Tribology Transactions*, Vol. 39, 4, pp. 803–10.
5. Fitch, J.C. "Applications and Benefits of Magnetic Filtration," *Machinery Lubrication Magazine*, September, 2005.
6. White, L. "Snap, Crackle & Pop," *Internal Bulletin*, The Hilliard Corporation.
7. Borden, H. and Fitch, J.C. "Use of Lubricant Proactive Maintenance and Contamination Control to Achieve Predictable Machine-Life Extension," *47th Annual Meeting of the Society of Tribologists and Lubrication Engineers (STLE)*, May, 1992.
8. Fitch, J.C. and Borden, H. "Interpreting Contaminant Analysis Trends Into a Proactive and Predictive Maintenance Strategy," *The 4th International Conference on Profitable Conditioning Monitoring*, Stratford-Upon-Avon, UK, December, 1992.
9. Moubray, J. *Reliability-Centered Maintenance – RCM II*, Industrial Press Inc., 1997.
10. Fitch, J.C. *Oil Analysis for Maintenance Professionals (Coursebook)*. Tulsa, OK: Noria Corporation, 1998.
11. Fitch, J.C. *Contamination Control Seminar (Coursebook)*. Tulsa, OK: Noria Corporation, 2004.
12. Day, M. "Filterability Testing of Paper Machine Oils," *Machinery Lubrication Magazine*, November-December, 2001.
13. Fitch, J.C. "Filter Economy - Insider Tips on Managing the Costs of Lubrication Filtration," *Machinery Lubrication Magazine*, May, 2005.
14. Bensch, L.E. "Dirt Capacity: The Overrated Filter Rating Factor," *Machine Design*, 23 June, 1983.
15. Brown, K., Utility Service Associates, "The Hidden Cost of Oil Changes," *Practicing Oil Analysis Magazine*, May, 1999.
16. Fitch, J.C. "Elements of a Successful Oil Analysis Program - Part I Oil Sampling," *Lubrication Engineering*, August, 1998.
17. Fitch, J.C. "Sampling Methods for Used Oil Analysis," *Lubrication Engineering*, March, 2000.
18. Fitch, J.C. and Troyer, D., "The Basics of Used Oil Sampling," *Practicing Oil Analysis Magazine*. Tulsa, OK: Noria Corporation, September, 2004.
19. Troyer, D. and Fitch, J.C. "An Introduction to Fluid Contamination Analysis," *P/PM Technology*, June, 1995.
20. Leavers, V.F. "Observing Precipitated Wear Debris Particles Technological Advances for Particle Counting," *Practicing Oil Analysis Magazine*, July, 2007.
21. Sommer, H.T. "Advancements in Oil Condition Monitoring," *Oil Analysis*, 99, 26–27 October, 1999.

22. Williamson, M. "Mesh Blockage Particle Monitoring," *Machinery Lubrication Magazine*, March, 2009.

23. Barnes, M. "Particle Counting - Oil Analysis 101," *Practicing Oil Analysis Magazine*, July, 2002.

24. Lukas, M., Anderson, D.P., Sebok, T. and Filicky, D. "LaserNet Fines – A New Tool for the Oil Analysis Toolbox," *Practicing Oil Analysis Magazine*, September, 2002.

25. Fitch, J.C. "Elements of an Oil Analysis Program," *STLE/CRC Tribology Data Handbook*, ed. Booser, E.R., CRC Press, 1996, Chapter 75.

26. Fitch, J.C. "Elements of a Successful Oil Analysis Program-Part II Selection of Tests," *Lubrication Engineering*, September, 1998.

27. Luckhurst, T. "Scanning Electron Microscopy for Wear Particle Identification," *Practicing Oil Analysis Magazine*, September-October, 1999.

28. Fitch, J.C. "Best Practices in Maximizing Fault Detection in Rotating Equipment Using Wear Debris Analysis," *Proceedings of the International Conference on Condition Monitoring*, Swansea, Wales, April, 1999.

29. Jones, M.H. "Effective Use of the Patch Test for Simple On-site Analysis," *Practicing Oil Analysis Magazine*, September, 2004.

30. Herguth, W.R. "Spotting Oil Changes Using Radial Planar Chromatography," *Practicing Oil Analysis Magazine*, July-August 2000.

31. Fitch, J.C. "Using Oil Analysis to Control Varnish and Sludge," *Practicing Oil Analysis Magazine*, May-June, 1999.

32. Fitch, J.C. "Proactive and Predictive Strategies for Setting Oil Analysis Alarms and Limits," *Joint Oil Analysis Program Conference Proceedings*, U.S. Dept. of Defense, May, 1998.

6 Lubrication Fundamentals

Lavern D. Wedeven and Kenneth C. Ludema

CONTENTS

6.1   INTRODUCTION

Hydraulics is associated with the transmission and control of forces and movement by means of a functional fluid. Hydraulic power is derived from a pump and transmitted to terminal elements such as rotary motors or linear cylinders. Although the primary demands on hydraulic fluid are associated with the power transmission, there are many secondary demands, including corrosion protection, heat dissipation, environmental compatibility, and lubrication.

Advanced hydraulic systems place severe demands on hydraulic fluid. For the most part, the lubricating attributes of a hydraulic fluid are not the primary considerations during the selection of a hydraulic fluid. Yet, the performance and life of hydraulic components, especially the pump, are critically dependent on the lubricating quality of the fluid. Compared with other mechanical hardware that must be lubricated, hydraulic equipment is unique in that hydraulic fluids cover an enormous range of fluid types. These include mineral oils, oil-in-water emulsions, various synthetic fluids, as well as non-typical fluids which are highly environmentally acceptable. Additives, which have contributed significantly to wear protection, further expand the performance range of hydraulic fluids. Added to the broad scope of fluids is the complex nature of lubrication and wear mechanisms and their interactions with an enormous range of contacting materials in hydraulic equipment. Lubricants, and especially hydraulic fluids, have multidimensional lubricating characteristics. These characteristics are derived from both the physical properties and chemical attributes of the fluid. Although physical properties and chemical attributes of fluids can be described, their connection to lubrication performance and hydraulic equipment life is somewhat elusive.

The understanding and prediction of lubricating performance are subjects in the field of tribology. Tribology is the science and technology of contacting surfaces in relative motion. Because of the complex nature of tribology within hydraulic systems, component design and fluid selection are accomplished with some degree of risk. Consequently, research and development rely heavily on testing. The use of pump tests for the selection and development of hydraulic fluids provides limited reliability. Although bench tests may be more efficient to operate, and perhaps easier to understand, their results can sometimes be even more remote from predicting performance in service than pump tests.

Because of the complex state of affairs with respect to the lubricating characteristics of hydraulic fluids and how they play themselves out in hardware, this chapter presents a new approach by which lubricated contacts can be viewed. The new approach identifies key structural elements that control performance. It focuses on lubrication and failure mechanisms, and on their connection to various types of lubricated contacts within the hydraulic systems. What is important—and, unfortunately, what is generally lacking—is an understanding of how these mechanisms play themselves out in service. The approach taken in this chapter supports the concept of systematic tribology. Systematic tribology is the process of understanding and developing the structural elements of a lubricated contact system and its mechanisms of interaction in a systematic way.

Armed with an understanding of the mechanistic processes for lubrication and wear, a rationale can be developed for the selection of fluids and their expected performance behavior. In addition, a fundamental understanding of lubrication and failure mechanisms is essential for failure diagnostics and the development of new technologies to enhance performance.

6.2 LUBRICATED CONTACTS IN HYDRAULIC SYSTEMS

Hydraulic fluids supply the lubricant to an enormous variety of load-bearing contacts which are constructed from a wide range of materials. Some of the most highly stressed and difficult-to-lubricate surfaces are found in hydraulic pumps, as shown in Figure 6.1. Under the rather harsh conditions of high pump pressure, high temperature, and contamination, the lubricating attributes of the hydraulic fluid must maintain the integrity of the contacting surfaces. Surface integrity is essential for pump life with respect to loss of clearances caused by wear, leakage, component alignment or position, friction, and torque. Because it is desirable to have hydraulic equipment that is small in size and light in weight, there is generally little margin for lubrication in many of the critical load-bearing contacts.

Some of the critical contacts in a piston pump shown in Figure 6.1 are the sliding contacts between the valve plate and cylinder block, piston and cylinder wall, piston and slipper socket, and slipper and swash plate. The critical areas in the gear pump shown in Figure 6.1 are near the tip and root of the gear teeth, where high-contact stresses accompany high sliding velocities. Other critical areas are associated with the highly-stressed bearings supporting the rotating components. Pump life and durability is intimately associated with each lubricated contact system that consists of hydraulic fluid, materials or surface treatments, and design. The modes of deterioration and failure analysis of these contacts are discussed in Chapter 8. The general mechanical features of the various contact systems are discussed in the next section.

6.3 BASIC MECHANICAL FEATURES OF LUBRICATED CONTACTS

To understand the tribology of hydraulic fluid-wetted parts, it is helpful to identify the types of contacts, their associated stress fields, and kinematics.

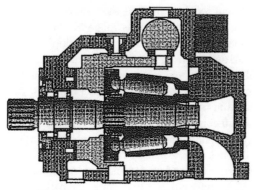

Variable displacement pump for open loop circuit

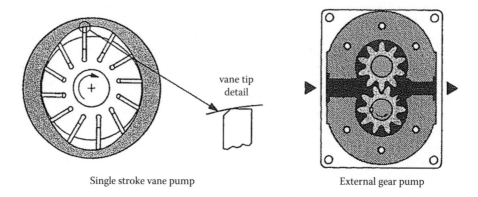

Single stroke vane pump   External gear pump

FIGURE 6.1   Types of hydraulic pumps and critical areas for lubrication.

6.3.1   TYPES OF CONTACTS

The geometry of contacting bodies creates the character of the contact with respect to stress and the mechanisms of lubrication. Conformal contacts, where contacting bodies conform to one another, distribute the load over relatively large areas. Conformal contacts, like piston or cylinder contacts, have clearance spaces that must be preserved. Concentrated contacts are non-conformal. They concentrate the load over small areas. The gear mesh in the gear pump of Figure 6.1 and the rolling element/raceway contacts in bearings are examples of concentrated contacts.

It is helpful to grasp the perspective of tribological contacts for their scale. The lubricated length of a conformal contact is generally less than a centimeter (10^{-2} m). The lubricated length of a concentrated contact is generally less than a millimeter (10^{-3} m). These load-bearing areas are on a "macroscale" where they can be viewed with the unaided eye.

Within the world of macroscale contacts are "microscale" features which control tribology behavior. Surface roughness features (asperities), and the grain size of metals, are on the order of microns (10^{-6} m). These microscopic features are critically linked to a yet lower scale of things which control lubrication and failure phenomena. Adsorbed films of molecular size that prevent adhesion between surfaces are on the order of nanometers (10^{-9} m). Here lies a fundamental difficulty within the tribology of hydraulic system hardware. Lubrication and failure mechanisms are controlled by microscale phenomena, but the parameters that control loads and motions of hydraulic system contacts are on the macroscale engineering level. To connect lubrication and failure mechanisms, which take place on a microscale, surface stresses and kinematics on the engineering level should be considered as input to a tribological contact system.

6.3.2 KINEMATICS AND DYNAMICS OF LUBRICATED CONTACTS

The relative motion of surfaces in contact invokes lubrication mechanisms as well as failure mechanisms. Kinematics is a description of these motions, without reference to the forces or mass which cause these motions. The hydraulic pump hardware in Figure 6.1 encounters linear sliding, rotational sliding, reciprocating sliding, rolling with sliding, and near-pure rolling motions. Some of these motions are mechanically controlled and easily calculated, like the gear mesh in the gear pump. Other motions are indeterminate, such as the rotational motions of the piston or slipper in the piston pump. These motions are controlled by friction forces acting on each body.

Dynamics treats the actions of forces on bodies in motion or at rest. Pressure forces and the rotating mass of pump components create dynamic forces that must be restrained by the contacting bodies. The durability of these contacts relies on other dynamic forces. However, these forces are generated between surfaces in contact and are created by hydrodynamic or elastohydrodynamic (EHD) pressures generated by the kinematics of the contacts themselves. Since the kinematics of contacts is controlled by hardware operation, duty cycle—in terms of motions and loads—becomes an important consideration in tribological design and operating life.

6.3.3 STRESS FIELDS

The dynamics of a tribological contact are composed of two major stress fields, illustrated in Figure 6.2. The load applied is distributed across the contact area as a normal stress. The strain within the contact gives rise to a tangential stress. The concepts of normal stress and tangential stress are valid for the entire macroscale contact as well as the microscale contacts within, which are created by interacting asperities between surfaces.

The normal stress of conformal contacts is typically in the range of megaPascals (MPa), 1,000,000 N/m². The normal stress of concentrated contacts is typically in the range of gigaPascals (GPa), 1,000,000,000 N/m². Concentrated contacts encounter significant elastic deformation. The elastic contact area and stress distribution can be calculated by using Hertzian theory [1].

Tangential stress is directly linked to the coefficient of friction, which is defined as the ratio of the tangential force to the normal force on the contact. Tangential stress may vary between 0.01 and 0.5 of the normal stress, depending on the degree of lubrication. The tangential stress, or coefficient of friction, of a contact can be the result of shearing an interposed material throughout the extent of the contact, or it can be the summation of individual frictional events associated with interacting asperities between the surfaces. The friction of many of the contacts within hydraulic systems is mostly a combination of the two.

Because lubrication and failure mechanisms are frequently controlled by asperity-scale phenomena, it is important to remember that normal and tangential stresses at asperities may not be that different for conformal contacts as they are for highly-stressed concentrated contacts. Asperity stresses depend on how the load is momentarily distributed among the surface features. Conformal contacts,

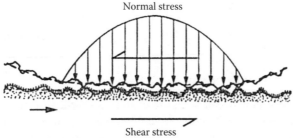

FIGURE 6.2 Two stress fields control the dynamics of a tribological contact.

with rough surfaces and a few contacting asperities, can have stresses similar to the asperities found in concentrated contacts with smoother surfaces. Because concentrated contacts tend to have finer surface finishes than conformal contacts, this situation may be quite common. This means that asperity stresses for both conformal and concentrated contacts may be on the same order of magnitude, even if their "global" or overall pressure across the contact area is substantially different. The major difference between conformal and concentrated contacts is that the stress field of the latter penetrates deeper into the material. In addition, high global pressure for concentrated contacts can dramatically change hydraulic fluid viscosity, even before it is consumed by the contact.

Whether normal and tangential stresses are on a global scale or asperity scale, the objective of lubrication is to create or preserve interposing material that bears the normal stress and accommodates the tangential shear. This appears to be somewhat of a paradox since the interfacial film must then possess both high compressive strength and low shear strength. The lubrication mechanisms described in Sections 6.6 and 6.8 illustrate the dynamic power by which this can occur. Because it is difficult to link engineering parameters with microscale lubrication and failure mechanisms, it would seem that problem solving and technical developments of hydraulic systems should focus on capturing lubricating mechanisms and suppressing failure mechanisms.

6.4   APPROACH

Because engineering parameters are difficult to link to lubrication phenomena, this chapter deals with the mechanisms of lubrication and failure. The use of mechanisms allows for the development of theory or models by which tribology performance of hydraulic fluids and hardware can be explained through the laws of mechanics, physics, and chemistry.

Lubrication and failure mechanisms are described within the context of a tribological system. A tribological system can be composed of four major structural elements: (1) hydrodynamic film, (2) surface film, (3) near-surface material, and (4) subsurface material. These structural elements create the operating environment for lubrication and failure mechanisms. They also establish the interconnections between the commercial suppliers of materials, fluids, and designs for hydraulic systems. Because one or two of the structural elements are dynamically generated, the mechanisms of hydrodynamic lubrication, EHD lubrication, and boundary lubrication are described, along with typical material and hydraulic fluid properties. The interactions of lubrication mechanisms with failure mechanisms are described within the framework of a dynamic tribological system. The mechanism leading to failure processes of wear, scuffing, and surface fatigue are also described and are illustrated with performance maps, based on engineering parameters. The material in this chapter introduces the concept of systematic tribology. Systematic tribology is the process of understanding and developing the structural elements of practical tribological contacts and their mechanisms of interaction in a systematic way.

6.5   STRUCTURAL ELEMENTS AND THE TRIBOSYSTEM

The performance of lubricated contacts is derived from the integrity of four general elements or regions, as shown in Figure 6.3. Each region performs certain functions with respect to lubrication and failure mechanisms. The success of lubrication depends on how well lubrication mechanisms handle the normal stress and the accommodation of tangential shear within these regions.

6.5.1   Hydrodynamic Film Region

The formation of an oil film between bodies in contact is a structural element which is dynamically generated. Its creation is a function of motion, which generates a pressure within a viscous fluid. Hydrodynamic films generated within conformal contacts may be tens of micrometers (μm) thick. Hydrodynamic or elastohydrodynamic (EHD) films generated in non-conforming

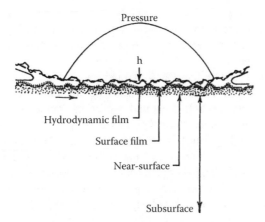

FIGURE 6.3 Structural elements of a tribosystem controlling lubrication and failure mechanisms.

contacts may be on the order of 1 μm thick. On a global scale, the EHD film is derived from the hydrodynamic pressure generated in the inlet region of the contact. On a local or asperity scale, hydrodynamic pressure is derived from the micro-EHD lubrication action associated with the local topographic features of the surfaces. These micro-EHD films are typically much less than 1 μm thick.

6.5.2 SURFACE FILM REGION

The surface film region contains the thin outer layers of the surface. They may consist of surface oxides, adsorbed films, and chemical reaction films derived from the lubricant and its additives. These surface films are almost always less than 1 μm thick. Although the surface film region is extremely thin, its formation by mechanisms of surface forces and chemical reactions has a profound effect on surface lubrication.

6.5.3 NEAR-SURFACE REGION

The near-surface region contains the inner layers of the surface. This region may include a finely structured and highly worked or mechanically mixed layer. It may also include compacted wear debris or transferred material from a mating surface. The deformed layers, which are of a different microstructure than the material below them, may arise from surface preparation techniques such as grinding and honing. They may also be induced during operation—for example, during run-in. Hardness and residual stress may vary significantly in this region. They may also be substantially different from the bulk material below. The near-surface region may be on the order of 50 μm below the surface.

6.5.4 SUBSURFACE REGION

The subsurface region is particularly important for highly-stressed concentrated contacts. The subsurface region can be on the order of 50–1000 microns below the surface. This region is not significantly affected by the mechanical processes that produce the surface or the asperity-induced changes that occur during operation. Its microstructure and hardness may still be different from the bulk material below it, and significant residual stresses may still be present. These stresses and microstructures, however, are the result of macro-processes such as heat treatment, surface hardening, and forging. For typical Hertzian contact pressures, and neglecting asperity pressures, the maximum shear stress is located within the subsurface region (see Figure 6.3). In other words, the detrimental global contact stresses are communicated to the subsurface region where subsurface-initiated fatigue commences.

These structural elements define the operating environment for lubrication and failure mechanisms. They also form the ingredients by which surface life and durability are determined. Because some of the structural elements are self-generated within the contact, it is clear that the structural elements constitute a dynamic system. When we consider the practical construction of this dynamic system, we find the wide-ranging businesses of hydraulic technology intimately linked. The lubricated contact is a tribosystem in which everything is joined together and operated in a dynamic fashion. This tribosystem is a melting pot of many commercial contributions which involve materials, surface treatments, hydraulic fluids, additives, equipment designs, and manufacturing.

The dynamics of this interdisciplinary tribosystem are described in Section 6.10. However, it can be said from the outset that the dynamics of practical systems are not adequately known. This state of affairs makes it difficult to solve problems and has resulted in disjointed efforts in developing hydraulic system technology. This is a common problem among non-integrated businesses and disciplines that supply materials to the interdisciplinary field of tribology. It is clear that the assembly of new tribology technology has not taken advantage of the potential synergism that can be obtained by constructing load-carrying materials, fluids, and designs as a system that operates on mechanisms.

6.6 HYDRODYNAMIC AND ELASTOHYDRODYNAMIC MECHANISMS

This section describes the fundamental mechanisms associated with the self-generation of a fluid film between contacting surfaces in motion. Elastohydrodynamic (EHD) lubrication is a "miracle" mechanism that is able to separate surfaces under enormous stress on both a macroscale and microscale. The mechanism of EHD lubrication is essentially an extension of ordinary hydrodynamic lubrication, which was described by Osborne Reynolds in 1886 [1]. The concepts involved in hydrodynamic lubrication are fundamental to the understanding of EHD lubrication. These fundamental concepts are basic to hydraulic fluid properties and how they lubricate many of the load-bearing contacts in hydraulic systems.

6.6.1 Viscous Flow – Parallel Surface

First, consider the flow of fluid between two parallel surfaces as shown in Figure 6.4. The fluid molecules adjacent to the solid surfaces adsorb or stick to it. The adsorption of liquids onto solid surfaces is caused by a fundamental atomic attraction. The formation of adsorbed material on solid surfaces is fundamental to the mechanism of boundary lubrication. The mechanism of adsorption is described in Section 6.8.2.

Because fluid molecules are attached to the bottom surface, they are forced to move with that surface at a velocity, u. The molecules on the top surface remain stationary. The molecules of fluid in between are dragged along by the bottom surface. The velocity of the fluid at various positions within the film is shown by the arrows in Figure 6.4. Note that the velocity of the fluid increases linearly from the top surface to the bottom.

The engagement of molecules in a fluid and their relaxation under strain creates an inherent resistance to flow, which is expressed in terms of its viscosity. Therefore, it requires a force to move the bottom surface relative to the top surface. The amount of shear force required is a function of the viscosity of the fluid. The force also depends on the rate at which the fluid molecules are being sheared. The rate of shear for the linear velocity distribution shown in Figure 6.4 is proportional to the surface velocity u and inversely proportional to the thickness of the film, h. Therefore, an increase in u or a decrease in h will require a greater force to move the bottom surfaces. The shear

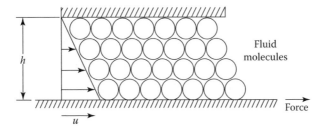

FIGURE 6.4 The flow of fluid between parallel surfaces in motion. Flow in cannot be larger than flow out.

force is parallel to the surfaces; the relationship between shear force and shear rate is the definition of viscosity.

The situation of flow between parallel surfaces provides no lifting force, which is required for lubrication. On the other hand, if the surfaces were allowed to collapse, it would take a finite time for the fluid to squeeze out. This momentary lifting force of the fluid while the surfaces are normally approaching is called a "hydrodynamic squeeze film." The hydrodynamic squeeze film is a useful concept to remember in connection with the mechanism of elastohydrodynamic lubrication.

6.6.2 CLOSING THE GAP

Consider now the flow of fluid between two converging surfaces, as shown in Figure 6.5. The top surface, which is of finite length, is at an angle to the bottom one. The gap between the surfaces is thicker at the inlet than it is at the outlet. The fluid adjacent to the lower surface moves with a velocity u, and the fluid adjacent to the upper surface is stationary. The inlet region would like to uniformly drag in fluid with a velocity distribution that is linear. At the same time, the outlet region would like to drag out fluid with a linear velocity distribution. The amount of fluid coming in is represented by the area of the triangle at the inlet. The amount of fluid going out is represented by the area of the triangle at the outlet. Because the triangle in the inlet is bigger, we find that the inlet would like to take more fluid in than the outlet will allow out. This pumping of the fluid into a converging region creates a pressure between the surfaces, as shown in Figure 6.6. The high pressure in the center forces the fluid to slow down in the inlet and makes the fluid go faster at the outlet so that the flow coming in will equal the flow going out.

The pressure generated by this pumping action creates a lifting force which separates the surfaces with a fluid film. The dynamically-generated pressure is a hydrodynamic pressure. The mechanism by which this pressure separates surfaces under load is called "hydrodynamic lubrication." The load that a bearing of the configuration represented in Figure 6.6 is able to carry is a function of the hydrodynamic pressure that is generated. If the viscosity of the fluid is increased, then more work is required in order to slow the fluid down in the converging inlet space and, as a consequence, a greater pressure is generated. Increasing the velocity u will also increase the hydrodynamic pressure.

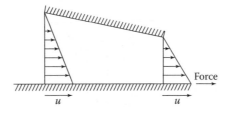

FIGURE 6.5 The flow of the fluid between two converging surfaces in motion (hypothetical).

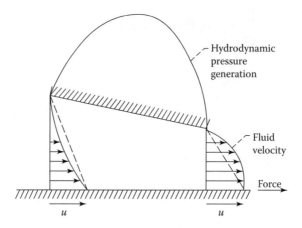

FIGURE 6.6 Hydrodynamic pressure generated between converging surfaces in motion.

If the surfaces pump fluid into the convergent space at a greater rate, a greater pressure is needed to slow the fluid down so that continuity of flow between the inlet and outlet is maintained.

There are three basic requirements for the generation of hydrodynamic pressure of the type shown in Figure 6.6: (1) the surfaces must be moving; (2) the surfaces must be converging; and (3) a viscous fluid must be between them.

The bearing shown in Figure 6.6 is a simple bearing. It is called a "slider bearing" or "pad bearing." Several of these pads arranged in a circle make a thrust bearing of the type found in many machines.

Another common bearing that uses hydrodynamic film generation is the journal bearing shown in Figure 6.7. Here, the converging surfaces are formed by virtue of the fact that the journal is not concentric with the bearing housing. When the journal rotates, it drags fluid into the converging region, thus generating a pressure that separates the surfaces.

6.6.3 Properties in Hydrodynamic Lubrication

When the motion of the surfaces draws the fluid into a converging space, the viscous nature of the fluid resists this motion and generates an internal pressure which is able to resist the normal stress applied to the contact. The pressure generated is governed by the Reynolds equation:

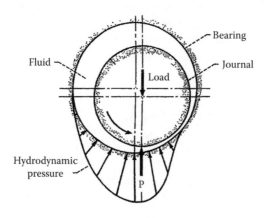

FIGURE 6.7 Hydrodynamic film generation in a journal bearing.

$$\frac{dp}{dx} = 12\mu U \left[\frac{(h - h_0)}{h^3} \right] \tag{6.1}$$

The pressure gradient, dp/dx, is proportional to the viscosity μ, the surface speed U, and a geometry term $(h - h_0)/h^3$.

The pressure generated within a convergent space creates a hydrodynamic lubricant film. The thickness of lubricating films can be calculated by using equations derived from simplifying assumptions for certain types of contact geometry. All film thickness equations illustrate the important role of viscosity. The role of viscosity is easily seen in an equation for film thickness, assuming a contact of rigid cylinders and a viscosity independent of temperature or pressure [2].

$$\frac{h_0}{R} = 4.896 \frac{U_{\mu_0}}{W} \tag{6.2}$$

where:
h_0 = film thickness on the line of centers
R = equivalent radius of curvature
U = one-half the sum of the surface speeds
μ_0 = viscosity
W = load

Because film thickness is directly related to viscosity, the level of viscosity can be selected for a given application to provide sufficient film thickness for complete surface separation. As hydraulic system contacts seldom operate at constant speed, constant load, or constant temperature, fluid viscosity decisions must be made to select an appropriate level of viscosity for the entire range of conditions. In many cases, design decisions associated with geometry, surface finish, material, and normal stress are made to accommodate the use of typical hydraulic fluid properties.

The process of generating a hydrodynamic film between contacting surfaces consumes power and results in a temperature rise in the oil film. Both internal and external thermal effects can decrease the viscosity substantially. The rate of change of viscosity with temperature can be significant. The thermal effect on viscosity also varies with lubricant type (see fluid property data in other chapters).

6.6.4 THE NON-CONFORMISTS

Hydrodynamic lubrication is characterized by conformal contacting surfaces where the normal stress is distributed over a relatively large area. The hydrodynamic mechanism is actively involved in the lubrication of piston/cylinder, slipper/swash plate, and many other sliding contacts. There are many hydraulic system contacts that are non-conforming where the load is concentrated on a small area. With concentrated contacts, the full burden of the load on a small area can cause considerable elastic deformation in and around the contact. Some examples of these non-conforming surfaces in hydraulic systems are rolling element bearings and gear teeth in gear pumps.

Figure 6.8 illustrates the non-conforming surfaces that can be found in a ball bearing and gear mesh. The ball and race conform to some degree in one direction, but the side view of the bearing shows that curvatures of the ball and race have little degree of conformity. Up until the early 1950s there was little belief that these non-conforming surfaces, with their extremely small area of contact, could be separated by an oil film. The lubrication of these non-conforming surfaces leads us to a remarkable and powerful film-generating mechanism called "elastohydrodynamic lubrication."

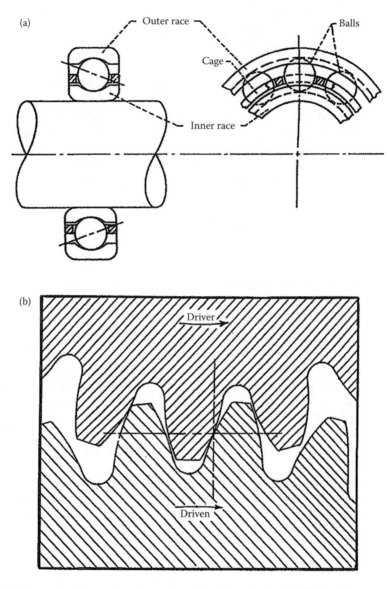

FIGURE 6.8 Non-conforming surfaces found in a: (a) ball bearing, and (b) gear mesh.

6.6.5 HERTZ: THE OPPOSITION

Let us first consider the contact condition between two non-conforming surfaces to examine what the mechanism of elastohydrodynamic lubrication is up against. The contacting surfaces of bearings and gears may be represented by a sphere or cylinder loaded against a flat surface as shown in Figure 6.9. The load, which presses the two bodies together, causes the surfaces to elastically deform and contact each other over a small but finite region of contact. This condition of elastic contact is named after Heinrich Hertz [1] who, in 1881, analyzed the contact between elastic bodies under conditions where the region of contact is much smaller than the radius of curvature. A close-up view of the intimate region of contact is shown in Figure 6.9. The load gives rise to a pressure, called "Hertzian pressure," which is distributed over a small region of contact called the "Hertzian region." The pressure has a parabolic distribution which is high in the middle and diminishes to zero at the edges of contact. Typical maximum Hertzian pressures found in bearing and gear contacts

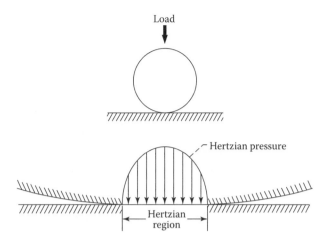

FIGURE 6.9 Hertzian pressure and deformation for non-conforming bodies in elastic contact.

are very high, on the order of 1.5×10^9 N/m^2. Note that this Hertzian pressure is an elastic pressure caused by the elastic deformation of the surfaces.

The Hertzian condition of contact is a dominating feature of EHD lubrication. It establishes the overall shape of the contacting surfaces. Thus, if we were to follow the journey of a fluid particle passing between the surfaces, it would first encounter a converging region, followed by a flat or parallel region, and finally it would be exposed to a diverging region. As stated above, one of the requirements for the generation of hydrodynamic pressure is that the surfaces must be converging. Therefore, we should expect all dynamic pressure-generating action to be in the converging region. The hydrodynamic pressure generated in this region has the task of separating the surfaces, which are forced together by the enormous pressure in the Hertzian region. When we consider that typical maximum Hertzian pressure may be on the order of 1.5 GPa, and that the usual hydrodynamic pressures generated in journal bearings are only on the order of 7 MPa, there does not seem to be much hope for the establishment of an oil film under these conditions.

6.6.6 ANKLE-DEEP IN CHEESE

The power of elastohydrodynamics is derived from two important considerations, which together provide the teamwork necessary to separate the surfaces. The first is that typical EHD films are a great deal thinner (about one thousand times thinner) than they are long. If, for example, the Hertzian region was the length of a football field and you were standing on the field, a typical EHD oil film would only be about ankle deep.

The second important consideration is the effect of pressure on the viscosity of the fluid. Figure 6.10 shows how the application of pressure influences the viscosity of a synthetic fluid at 38°C. The viscosity, which is given in centipoise (cp), is plotted on a logarithmic scale because it increases rapidly with pressure. At atmospheric pressure, the viscosity is about 350 cp (approximately the same consistency as a typical motor oil). At 0.14×10^9 N/m^2 the viscosity has increased by one order of magnitude to 3500 cp. The fluid at this pressure has a consistency approaching that of molasses. At typical Hertzian pressures of 1.5×10^9 N/m^2, the fluid viscosity can increase by several orders of magnitude so that its consistency may be more like butter or cheese.

The influence of pressure on viscosity is usually represented by a straight line, as shown in Figure 6.10, which has the equation:

$$\mu = \mu_0 e^{\alpha p}, \tag{6.3}$$

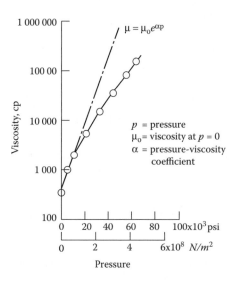

FIGURE 6.10 The effect of pressure on viscosity for a synthetic fluid at 38°C.

where:

μ is the viscosity at pressure p

μ_0 is the viscosity at atmospheric pressure

The exponent α is a property of the fluid called the pressure–viscosity coefficient. When α is large, the viscosity rises rapidly with pressure. When α is small, the viscosity rises more slowly with pressure.

6.6.7 HYDRODYNAMICS VERSUS HERTZ

Figure 6.11 shows the flow of fluid in the convergent inlet region where hydrodynamic pressure is generated. It should be noted that this figure was drawn by using two conventions which are frequently used to graphically illustrate an EHD-lubricated concentrated contact. First, because typical EHD oil films are much thinner than they are long, the vertical dimensions are usually expanded about one thousand times more than the horizontal dimensions. If this was not done, the thickness of a typical EHD oil film drawn to scale would be less than the thickness of the horizontal line shown in the Hertzian region of Figure 6.9. The second convention illustrates the contacting surfaces of machine elements with an equivalent sphere or cylinder on a flat surface with all the elastic deformation represented in the curved body.

The shape of the convergent region as shown in Figure 6.11 must be viewed with this in mind. The real shape of the convergent inlet region, where hydrodynamic pressure is generated, is actually long and narrow.

The surfaces shown in Figure 6.11 are in pure rolling, where each surface is moving with the same velocity, u. Each surface carries with it a certain quantity of fluid, which joins together at some location to fill the gap between the surfaces. Fluid adjacent to each surface is attached to it and travels with it. Because the surfaces are converging, the fluid in the interior of the film is forced to slow down and may even flow backward. As previously explained, this slowing down of the fluid or pumping action of the surfaces to draw fluid into a converging region generates a hydrodynamic pressure. As the pressure rises in the inlet region, the viscosity rises with it. The higher viscosity produces even higher pressure. When the fluid reaches the leading edge of the Hertzian region, the viscosity of the fluid may have increased by an order of magnitude and the hydrodynamic pressure may have reached typical values of 0.15×10^9 N/m^2. This hydrodynamic pressure must compete with the Hertzian pressure. While the hydrodynamic pressure is trying to separate the surfaces, the Hertzian pressure is trying to force them together.

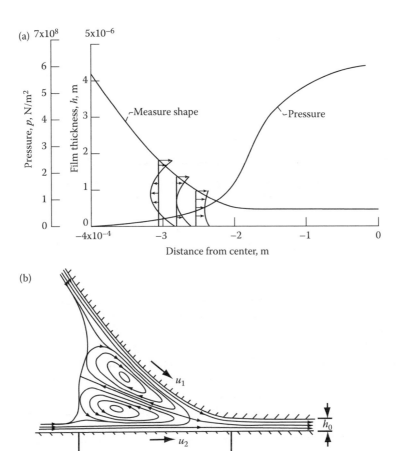

FIGURE 6.11   (a) Pressure and velocity distribution, and (b) flow distribution in convergent inlet region.

The hydrodynamic pressure generated in the convergent inlet region is much lower than the maximum Hertzian pressure. Nevertheless, the hydrodynamic pressure is capable of separating the surfaces, and it does so in a very subtle way. It cannot compete with the Hertzian pressure in the center of the contact where the pressure is very high, but it can overcome the Hertzian pressure at the leading edge of the Hertzian region where the pressure is much lower. If it does this, and if it separates the surfaces at the leading edge of the Hertzian region, the hydrodynamic pressure will have achieved total surface separation. This is achieved as a result of time (i.e., the dynamics of the moment). Once the fluid gets into the leading edge of the Hertzian region, it cannot escape because the viscosity becomes too high and the film is too thin. There will not be enough time for the Hertzian pressure to squeeze the fluid out because the motion of the surfaces passes the fluid through the Hertzian region very quickly (typically on the order of milliseconds).

The final pressure achieved, and the overall shape, are shown in Figure 6.12. The final pressure and shape are very similar to the Hertzian pressure and shape, except at the leading and trailing edges of the Hertzian region. The EHD pressure generated in the inlet region elastically deforms the surfaces to a greater extent than the original Hertzian shape. The EHD pressure near the exit region goes into a sudden perturbation with enormous pressure gradients. The sudden drop in pressure at the exit is associated with a drop in film thickness, which creates a fluid flow constriction. The local constriction at the trailing edge of the contact is the result of a requirement to maintain the continuity of flow. If there was no constriction, the sudden drop in pressure would force more fluid out than is coming in. Therefore, the surfaces deform in such a manner that they restrict the flow going out.

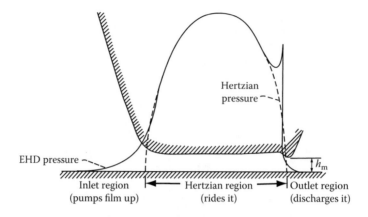

FIGURE 6.12   Typical pressure distribution and elastic deformation during EHD lubrication.

6.6.8   THE TRINITY

The conjunction zone of a typical EHD contact can be conveniently divided into three separate regions, as shown in Figure 6.12. Each region performs a unique function. The inlet region pumps the film up; the Hertzian region rides the film; and the outlet region discharges the film. The viscous character of the fluid changes drastically while passing through these regions, going from an easy-flowing liquid to a pseudosolid and back again within a matter of milliseconds.

The viscous properties of the fluid in each region are determined by the temperature, pressure, and shear conditions the fluid encounters or creates in each region. Because viscosity plays an important role in how these regions function, it is important to understand how the viscous character of the fluid is influenced by the environmental conditions in these regions as it passes through them.

For example, the film-forming capability of the hydrodynamic pressure generated in the inlet region is governed by the local viscosity throughout this region. Because the inlet region is extremely narrow, the viscosity of the fluid is controlled by the temperature of the solid surfaces. The variation of viscosity with pressure is usually accounted for by the α-parameter in Equation 6.3. While the pressure–viscosity coefficient, α, as defined in Figure 6.10, departs from experiment at high pressures, its characterization of fluid behavior is sufficiently accurate over the pressure range that is generally encountered in the inlet region. The viscous properties of the fluid, as governed by the pressure and temperature conditions in the inlet region, influence the thickness of the fluid film, which is observed in the Hertzian region.

6.6.9   STRESSED TO THE LIMIT

By the time the fluid has entered the Hertzian region, it has completed its task of pumping the film up. The viscous (or shear strength) character of the fluid in the Hertzian region now becomes important with respect to sliding friction (or traction). Because the fluid film in the Hertzian region has high viscosity, relative motion between the surfaces is resisted by tangential stress which has built up within the fluid in the Hertzian region. Because enormous pressures in the Hertzian region transform the fluid into a pseudosolid, the internal friction or traction of the fluid becomes governed by the shear strength of the solidified fluid instead of its viscosity. The tremendous enhancement of viscosity with pressure is offset, to some degree, by frictional heating of the fluid in the Hertzian region as well as possible shear and time-dependent effects on fluid behavior. If the fluid in the Hertzian region is liquid-like, frictional heating reduces its viscosity. If the fluid in the Hertzian region is solid-like, frictional heating reduces the limiting shear stress of the fluid. In either case, sliding friction is reduced with temperature. Friction or traction coefficients of hydraulic fluids may

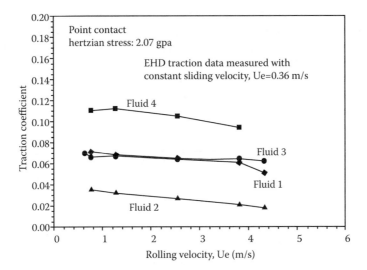

FIGURE 6.13 Traction coefficients of four fluids measured under high-pressure EHD conditions. Fluids 1 and 2 are model hydraulic fluids. Fluid 3 is a polyolester, and Fluid 4 is a perfluoropolyalkyl ether.

range from 0.1 to 0.001. The level of traction is a function of the pressure and temperature in the Hertzian region. Because traction under high-pressure EHD conditions reflects the limiting shear strength of the fluid, traction coefficient is a function of the molecular structure of the fluid. Figure 6.13 shows EHD traction coefficients of four fluid types. The EHD traction measurements were conducted at a maximum Hertzian stress of 2.07 GPa and ambient temperature [3]. The traction data were taken over a range of rolling velocities, but with a constant sliding velocity (0.36 m/s). With a constant sliding velocity and Hertzian stress, the frictional heating in the contact is the same for all rolling velocities. It is interesting that the traction coefficient is almost independent of the rolling velocity. From the discussion above, EHD film thickness should increase with surface entraining velocity. This means that the traction coefficient under high-pressure EHD conditions is essentially independent of film thickness (or shear rate). The independence of traction with shear rate confirms that the fluid is shearing like a solid rather than a viscous liquid.

The fact that fluids under Hertzian pressure shear like solids has profound technological implications. If fluids remained viscous liquids (i.e., took on Newtonian behavior), the high shear rates typically found in EHD contacts would result in enormous friction coefficients. EHD-lubricated contacts would then have to be operated at low shear rates or they would suffer meltdown by frictional heating. The transformation of the fluid to a pseudosolid limits frictional resistance caused by the limiting shear strength of the interposed material. Furthermore, as the shear rate and frictional heating in the contact increases, traction decreases. This is because the limiting shear strength of fluids, like ordinary solids, decreases with temperature.

6.6.10 PREDICTING FILM THICKNESS

The practical importance of the mechanism of EHD lubrication lies in the thickness of the oil film between the surfaces. Its thickness is controlled by the operating conditions expressed in terms of various operating parameters such as surface velocity, load, and fluid viscosity.

The influence of these parameters on film thickness should be obvious if the basic concepts of EHD have been understood. For example, a change in any parameter, which causes a greater hydro-dynamic pressure to be generated in the inlet region, will result in a larger film thickness. Thus, an increase in surface velocity or fluid viscosity will result in a larger film thickness.

An important feature of EHD lubrication is that the influence of load on film thickness is very small. This is not surprising when one considers that an increase in load merely increases the maximum Hertzian pressure and makes the Hertzian region larger. It does very little to the inlet region where the hydrodynamic pressure is generated.

The influence of various operating parameters on film thickness can be shown with an equation. Equation 6.4 has been derived from theory [4] for a line contact geometry presented by a cylinder on a plane as shown in Figure 6.9.

$$h_{m} = 2.65 \frac{(\mu_0 u)^{0.7} \alpha^{0.54} R^{0.43}}{E^{0.03} w^{0.13}} \tag{6.4}$$

where:

h_m = film thickness at the rear constriction

μ_0 = viscosity at atmospheric pressure

α = pressure–viscosity coefficient as defined in Equation 6.3

u = velocity, defined as $u = 1/2 (u_1 + u_2)$ where u_1 and u_2 are the individual velocities of the moving surfaces

R = radius of equivalent cylinder

w = load per unit width

E = elastic modulus of equivalent cylinder (flat surface assumed completely rigid)

This equation shows that film thickness is most sensitive to the velocity (u), the lubricant properties (μ_0 and α) and the radius of curvature (R). An increase in any one of these parameters, which are in the numerator of Equation 6.4, will result in a larger film thickness. The relationship between viscosity, μ_0, and temperature is well known for hydraulic fluids and is covered in other chapters.

While the pressure–viscosity coefficient, α, is an essential fluid parameter for EHD film generation, its value is not frequently given for hydraulic fluids. Pressure–viscosity coefficients can be measured with high-pressure capillary viscometers. They can also be obtained from actual EHD film thickness measurements [3]. Pressure–viscosity coefficient is a function of molecular structure. Because hydraulic fluids cover a wide variety of fluid types, their pressure–viscosity coefficients cover a wide range. The pressure–viscosity coefficients of four fluid types are plotted as a function of temperature in Figure 6.14. Fluids 1 and 2 are model hydraulic fluids. Fluid 3 is a polyolester and Fluid 4 is a perfluoropolyalkyl ether. Most mineral oils, and some synthetic fluids, have pressure–viscosity coefficients similar to Fluids 1 and 3. The data in Figure 6.14 show that pressure–viscosity coefficient decreases with temperature. This is particularly true for high-viscosity fluids. As a general rule, pressure–viscosity coefficient tends to increase with viscosity for a given fluid type.

The denominator in Equation 6.4 has two parameters (E and w) which tend to decrease film thickness. However, neither the load, w, nor the elastic modulus, E, have much influence on film thickness because their exponents are very small (0.13 and 0.03). It is somewhat ironic that the lubrication mechanism that bears the name elastohydrodynamic shows film thickness to have little dependence on the elasticity of the materials. As long as the applied load causes elastic deformation similar to the Hertzian deformation, it does not matter if the elastic modulus is high—like that of tungsten carbide, or low—like that of aluminum or brass. As previously discussed, the region of EHD pressure generation is the inlet region, the shape of which is only slightly affected by elastic modulus. However, if the elastic modulus is extremely low, like that of elastomeric seal materials, then the pressure will not be sufficient to enhance the viscosity of the fluid. The mechanism of lubrication, while still elastohydrodynamic, cannot be described by Equation 6.4, since it includes the pressure–viscosity coefficient α. Other equations that account for low elastic modulus materials can be used to calculate film thickness.

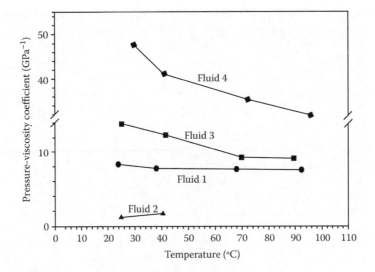

FIGURE 6.14 Pressure–viscosity coefficients of four fluid types. Fluids 1 and 2 are model hydraulic fluids. Fluid 3 is a polyolester and Fluid 4 is a perfluoropolyalkyl ether.

6.6.11 EXPERIMENTAL VERIFICATION OF EHD THEORY

For moderately hard materials, the theoretical prediction of film thickness as represented by Equation 6.2 has been verified by experimental measurements. Because typical EHD films are extremely thin (generally less than 1 μm), film thickness measurements require special techniques.

One measurement technique that has been used is optical interferometry, which uses the wavelength of light as a unit of measure. Figure 6.15 shows a photomicrograph of light interference fringes that have formed between a steel ball rolling against a glass surface. The circular area is the Hertzian region. The fluid passes through the Hertzian region from left to right. A fluid cavitation pattern is clearly visible in the outlet region. The interference colors in the Hertzian region provide a contour map of the thickness of the oil film. The uniformity of color in the center of the contact indicates that the film is extremely parallel in the high-pressure region—that is, the elastic shape is Hertzian. Some exceptions to the Hertzian shape are seen near the rear and sides of the contact region where the thinner films are observed. The thinner films at the sides of the EHD contact are the result of side leakage of the fluid immediately upstream in the inlet region. This side leakage inhibits the generation of hydrodynamic pressure and causes a thinner film to form in the portion of the Hertzian region immediately downstream. The thinner film at the rear of the EHD contact causes a restriction in flow immediately downstream of the high pressure in the center of the Hertzian region. This restriction is a necessary requirement to maintain continuity of flow, as previously discussed.

6.6.12 EXIT REGION

After the fluid leaves the Hertzian region, it enters a diverging region. As the surfaces drag the fluid into the diverging space of the outlet region, the situation of fluid dynamics is opposite to that found in the inlet region. The outlet region attempts to generate a negative pressure. When sub-ambient pressure is reached, dissolved gases in the fluid come out of solution and fill the space between the surfaces. This action ruptures the fluid film and terminates the hydrodynamic pressure. It is fortunate that fluids *out-gas* rather than respond by way of true cavitation. Although fluid cavitation can result in considerable damage to hydraulic system valves, the gentle cavitation that occurs at the termination of hydrodynamic pressure causes no trauma to the surfaces. The fluid merely "passes gas."

FIGURE 6.15 Photomicrograph of light interference fringes showing EHD film formed between a steel ball rolling against a transparent surface. Fluid passes through the Hertzian region from left to right.

6.6.13 STARVATION

The oil film rupture in the exit region leaves thin residual ribs of lubricant downstream of the contact. Because most lubricated contacts repeatedly travel over the same track, there is a concern that the surfaces drawing fluid into the inlet region might not provide an adequate supply. Without some degree of external replenishment into the track, the inlet region can become *starved*. Experimental work on starvation [5] shows that a less than fully flooded inlet region decreases the EHD film thickness according to the degree of inlet starvation. Many lubricated contacts, like grease-lubricated bearings, run starved. So long as the EHD film thickness provides sufficient surface separation (or entrainment of chemistry), some degree of starvation can be tolerated. Fortunately, most highly stressed contacts in hydraulic equipment tend to be adequately supplied with fluid.

With regard to starvation, one should appreciate a little-known fact that allows high-speed contacts to survive without starvation. The inlet flow field illustrated in Figure 6.11 shows that fluid entering the inlet region is re-circulated. It "hangs around" awhile before it is consumed by the Hertzian region. The inlet region accumulates fluid and actually moves with the contact as it travels along the surface. This "in-flight refueling" is one of several phenomena that make EHD lubrication a "miracle mechanism."

6.6.14 Just a Tickle?

The thickness of EHD lubricating films that are found in lubricated contacts is frequently not much greater than the height of individual asperity features on the surfaces. If total surface separation can be achieved, the life of hydraulic pump contacts can be long, limited ultimately by fatigue of the metal surfaces. However, when total separation of the surface is not achieved, the load is partially supported by the EHD film, and partially by local areas of asperity contact. These local areas are vulnerable sites for the initiation of surface wear and failure. The calculated thickness of the EHD film, relative to the individual asperity height, is an important criterion used in design. This is a first-order estimate of whether the intermittent contact between asperities is just a "tickle" or something more traumatic.

6.6.15 Micro-EHD

The power of the EHD mechanism, which has been described above with smooth surfaces, also carries its influence to the micro-world of surface roughness features. The realities of typical lubricated contacts are that EHD-generated films are on the same order of magnitude as the roughness heights of machine-finished surfaces. It is also interesting to note that typical "slopes" of surface roughness features are frequently on the same order of magnitude as the angle of convergence in the inlet region. As discussed above, the generation of a hydrodynamic pressure requires relative motion where a fluid is drawn into a convergent space. Therefore, undulations on the surface which are of a scale less than the length of the inlet region or contact region, are subject to the same principles of EHD film generation as the global or bulk shape of contacting bodies. When these EHD principles become active on topographical features, it is called "*micro-EHD* lubrication." Micro-EHD action can be superimposed on a high-pressure EHD-lubricated contact as well as a low-pressure hydrodynamic contact. In the latter case, elastic deformation is confined to roughness features only, and the effect of pressure on viscosity is confined to the neighborhood of the roughness features.

An example of micro-EHD lubrication at work is shown in Figure 6.16. Artificially produced undulations are seen traveling through an EHD contact region with a nominal EHD film thickness of 0.16 μm. The data points, which were obtained with light interference [6], give the actual shape

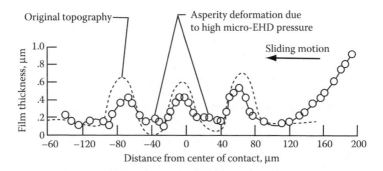

FIGURE 6.16 Micro-EHD pressures at roughness features with sliding motion "flatten" surface topography in Hertzian contact region.

of the undulations as these features go through the contact region under sliding motion. The dashed line shows the original geometry of the undulations, as measured with a stylus instrument. As these undulations move through the high-pressure and highly viscous Hertzian region, they create significant local EHD pressures at the converging sides of the undulations. The micro-EHD pressures are sufficient to cause a noticeable compression or flattening of the "peaks" to avoid local contact. At the same time, the micro-EHD action on the downstream or diverging side of the undulations tends to create a negative pressure. If cavitation does not occur, the "valleys" are relieved of pressure and they deform in a direction toward the mean plane of the surface. So, if topographical features are of an appropriate scale, micro-EHD action can momentarily flatten the roughness features. A severe encounter between surface roughness features may turn out to be only a tickle rather than a scratch.

6.6.16 SUMMARY OF EHD

Elastohydrodynamic lubrication deals with highly stressed surfaces which are elastically deformed by loads carried over small areas. The overall pressures and deformations are similar to the Hertzian conditions for dry contact. The dynamics of fluid entrained into the converging space upstream of the Hertzian contact generates pressure between the surfaces. The "macho" Hertzian pressure opposes this hydrodynamic pressure and, while the pressure of the battle builds up, the viscosity of the fluid increases. The enhancement of viscosity with pressure gives a tremendous boost to the generation of hydrodynamic pressure. Although the hydrodynamic pressure may still be much smaller than the maximum Hertzian pressure, it is sufficient to elastically deform and separate the surfaces at the leading edge of the Hertzian region. Once this is achieved, the fluid becomes captured by the contact: It cannot escape because its viscosity is too high, the gap between the surfaces in the Hertzian region is too thin, and the time is too short for it to "leak" out. Because Hertzian pressures are high, the momentarily-captured fluid is transformed into a solid-like material. Friction is almost never more than 10% of the normal load. Because EHD films are generated upstream of the Hertzian contact, their thickness is almost independent of load. Their *load capacity* can easily exceed the plastic flow of metals.

The formation of hydrodynamic and EHD films has a profound effect on the life and durability of highly-stressed hydraulic system components. Although the power of hydrodynamic and EHD mechanisms is somewhat amazing, it alone is not sufficient to preserve the integrity of surfaces under all conditions of operation. These dynamic mechanisms create a structural element of the lubricated contact, which comes and goes with surface motion. The *attachment* of films, which do not require surface motion for their creation, is an equally important mechanism in the whole scheme of lubrication. The orchestration of surface films and dynamically generated films is the heart of lubrication practice.

Before we describe the mechanisms that create the structural element of *surface films*, the general features of surfaces and near-surface material should be understood. Surface films are part of the surface. In many cases, they are also part of the fluid. At the interface between surfaces, fluids can become "solid" and the solid surface can "flow" to accommodate stress or strain. Without some sort of modeling or understanding of the processes that go on between surface films and the surface, it is difficult to manage fluid chemistry, engineer surfaces, design contacting hardware, or solve problems.

6.7 THE SURFACE AND NEAR-SURFACE REGION

The near-surface region is a structural element that lubrication mechanisms attempt to preserve. The near-surface region, which includes surface topography, becomes the crime scene of almost all failure processes. These failure processes fall under general categories with descriptive terms, such as wear, scuffing, and surface fatigue. The near-surface region is a "constructed" element, as opposed to a dynamically generated element like a hydrodynamic film. However, the near-surface region has many dynamic features. The manufacture of a surface is not a benign process. Surfaces can acquire residual stresses, and they can possess "energy." The material composition and microstructural features, which define the near-surface region of practical materials, cover an enormous scope and provide great opportunity. These opportunities will continue to grow with advancements in surface treatments and with developments in engineered surfaces. Whether the initial creation of contacting surfaces is high or low tech, an understanding of the characteristic features of the near-surface region is important for understanding its neighboring structural elements: the surface films and subsurface region.

6.7.1 ROUGHNESS, ASPERITIES, AND SUBSTRATES

All technological surfaces are rough or non-smooth on several arbitrarily-defined dimensional scales. A nominally flat surface has "undulations" on the surface itself. This scale of roughness is often referred to as "*waviness*." The next scale of roughness is the microscale, and this scale is referred to as "*surface roughness*." Surface roughness with distinctive three-dimensional features may have *texture*. A third scale of roughness is the "*nanoscale*," a scale not necessarily confirmed to be as important as the others in the lubrication of most technological surfaces. The microscale entities on surfaces that make up roughness are called "*asperities*." The microscale is important because differences in surface roughness in this range strongly influence the functioning of heavily-loaded, lubricated surfaces in relative motion.

Electronic surface-tracer instruments yield a variety of surface parameters that describe the height and spatial distribution of surface topography. Their numerical values in the microscale range, such as Ra (arithmetic average), are the most commonly used. Computer and optical technology have fostered the development of optical topographic sensors, which enhance surface characterization in three dimensions.

Graphical traces from the old electronic instruments popularized the notion that asperities are jagged mountain peaks that are readily "broken off" when lateral forces are applied to them. This came about because the vertical scale on the trace was usually amplified at least one hundred times more than the horizontal scale so that less paper was used to portray the important features of surfaces. Actually, asperities resemble rolling hills more than jagged mountain peaks, having slopes that average in the range of 1° to 5° or so. Asperity contact is better represented as occurring on very low slopes rather than the commonly drawn slopes of 30° or 45°.

6.7.2 MODELING SURFACES

The modeling of surfaces in contact requires some sort of numerical characterization of surface topography. The simplest notion of surface roughness is the sinusoidal ridge pattern, or perhaps the close-packed array of spheres. Actual roughness is much more complicated and quite difficult to express satisfactorily in numerical form. Roughness is thought of mostly in terms of asperity height, but distribution and shape of asperities are also important, and all of these attributes appear to be more random than regular across a surface. From a tribological perspective, peak height and slope of roughness features are important. Surfaces with a *plateau* texture are almost always desired over equally-distributed *peaks* and *valleys*. Available instruments yield a host of ways to express surface roughness, most often statistical in nature. However, few people, if any, can surely connect any single value or group of values with surface performance. As with so many other topics in lubrication, the better course of action in using surface roughness-measuring instruments is to gain experience with only one or two surface roughness expressions (parameters) and to connect these with the functioning of the machine component of concern.

6.7.3 THE CREATION OF ASPERITIES

Asperities are usually taken as entities that simply "exist" on surfaces, but it is useful to consider how they are formed. Surfaces in mechanics are taken simply to be the boundary of a *semi-infinite* solid of no particular internal structure. Actually, metallic and ceramic solids are made up of atoms, and polymeric solids are made of identifiable arrays of atoms (i.e., molecules). Solids that are solidified (cast) from the molten state take on a surface topography (e.g., roughness, waviness) that is some combination of the surface features of the container (mold) plus the shape of surface grains, as they attempt to grow in their natural crystallographic directions. Surfaces cut by tools or abrasive substances take shapes that result from the material being torn or fractured by the passing cutting edges. This leaves torn edges and "ironed" folds on the surfaces of ductile materials, which are severely cold-worked in the substrate. It leaves grooves and cracks on and in the surfaces of brittle materials. Surfaces influenced by corrosion assume shapes that result from having different grains corroding at different rates along different crystallographic directions.

6.7.4 Correcting Surface Deficiencies

Manufactured surfaces are certainly not perfect for their future encounters with mating partners in lubricated contacts. Because many processing steps are required to improve surface topography and near-surface integrity for their ultimate use, economic considerations play a large part in the final condition of the surface. As a practical matter, the contact just has to live with whatever surface quality it is given. To survive the initial operation with less-than-perfect surfaces, chemical treatments such as phosphating are used.

Economic considerations also bring into play many low-cost materials which are surface-hardened for improved durability. The diffusion of carbon (carburizing) or nitrogen (nitriding) into steel improves hardness and wear resistance. However, hard iron nitrides, which appear as thin white layers at the surface when etched, can occur in nitrided steels. These *white layers* have low ductility and are easily fractured under stress.

If economic considerations permit, surfaces may be coated with hard or soft coatings. Silver coatings are used to provide "compliance" to the surface and improve heat transfer. Hard coatings, such as titanium nitride (TiN) or thin dense chrome (TDC), are used for wear resistance and corrosion resistance. Diamond-like coatings (DLC) have the potential to make their way into use for wear resistance and low friction.

The near-surface region of lubricated contacts is enormously complex in terms of its topography, composition, and mechanical properties. To make matters worse, this structural element does not stand alone. Its neighboring structural element, the surface film region, is intimately connected to the near-surface region. This intimate connection is described in the following section.

6.8 SURFACE FILM REGION AND BOUNDARY LUBRICATION MECHANISMS

The structural element that we identify as the "surface film region" is a whisper in thickness, but mighty in influence. It is well known that surface films are important to the mechanism of boundary lubrication. These elusive films are the last defense mechanism to prevent adhesion and accommodate shear. These films may be in the form of oxides, adsorbed films from surfactants, and chemical reaction films from other additives. They can also be derived from bulk fluid or the surrounding atmosphere. The thin region of a surface film is illustrated in Figure 6.17.

6.8.1 Incompatible Assumptions about Surfaces: Mechanics Versus Chemistry

In the mechanics of lubrication, a solid body ends at a plane and the next substance is a lubricant; the solid has its (bulk) properties, the lubricant has its (bulk) properties, and there is no interaction between them. Peculiarly, although successive layers of fluid between moving surfaces are sheared past one another, a common assumption in hydrodynamics is that the layer of fluid touching the solid surface does not "slip." This is a most remarkable phenomenon. It is an essential feature of the mechanism of hydrodynamic lubrication, and it is the "engine" of boundary lubrication.

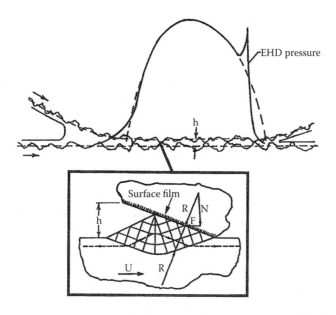

FIGURE 6.17   The surface film region.

There is, in fact, an interaction between a lubricant *in contact* with a solid. Solid surfaces are atoms with their outer "neighbors" removed from them. These outer atoms exist in a higher state of energy than when they were submerged in the solid. The increased energy is what was required to fracture the bonds that formed this new surface. In a tensile test, for example, this is most readily seen in the case of a perfectly brittle solid because no energy is expended in plastic deformation in the fracture region.

Simply stated, a surface atom, in an elevated state of energy, can be restored to its original energy state by returning neighbors of its own kind to its vicinity. However, if a neighbor of its own kind is not available, an atom of another kind will restore a fraction of the energy rise that occurred during fracture. The new neighbor becomes attached or adsorbed to the solid surface, with a strength of attachment roughly relative to the amount of energy released on adsorption. On introduction of shear, the strength of attachment may be sufficient to resist or impede a large-scale slip between a fluid and solid surface.

6.8.2   TWO LEVELS OF SURFACE ENERGIES IN ADSORPTION

Surface interactions are actually surface *regions* of interactions. The outer layer of atoms in a surface is displaced from its equilibrium crystallographic lattice location. This displacement distorts the lattice array in the next layer, and so on, decreasingly into the next layers. The *surface energy* of a solid is the sum of the energy changes that occur in all affected layers of atoms into the solid substrate. In exactly the same way, all fluid molecules near the solid are altered in form and properties by the bonding energies in the region of interaction. The following paragraphs discuss the lubricating qualities of adsorbed substances. The discussion applies to all liquid and gaseous substances that approach a solid surface.

Some molecules adsorb to a solid surface with low energy, on the order of 0.2 kJ/mol of the adsorbate. There is no change in the chemical state of the attached molecules. Liquid molecules at the solid surface adjust their natural structure to the atomic structure of the solid. The range of adjustment extends with diminishing effect into the volume of attached molecules. The result is a significant increase in flow resistance or viscosity of the adsorbate. A large adjustment of physical properties occurs when gases adsorb to a solid. These gases build in several layers and pack so tightly on a solid surface that the adsorbed layer has properties similar to those of a liquid. These layers exist on all surfaces that are exposed to atmospheric gases, including water vapor. Their presence significantly affects sliding friction.

Some substances are chemically-altered on adsorption to a solid. These chemical alterations are found to lower the surface energy by an amount in the range of 2.5–25 kJ/mol of the free adsorbate. This type of reaction is referred to as *"chemical adsorption."* The adsorption of oxygen to all solids (except gold) results in the formation of oxides. Other reactive gases also form new compounds.

6.8.3 ADHESION

When two solid surfaces are brought close to one another within their normal atomic spacing of about 0.5 nm, they will either adhere or bond together. No other action, rubbing, or heating is required. The strength of the bond varies widely. The ultimate bond strength is that in the substrate itself, but the relative lattice orientation of the two surfaces can influence the strength of the bond to a great degree. In the case of metal, the atomic array or lattice pattern in the two surfaces need not align perfectly. They will create strong bonds—that is, the adhesion strength is high. In the case of ionic and covalent structures, as in most ceramic materials, adhesion strength is reduced to 2%–3% of the substrate strength—provided that the lattice patterns in the surfaces are out of alignment by only a degree or two.

However, if the solids cannot be brought to within 0.5 or even 2 nm of one another, they will not bond. Adsorbed films are usually thicker than these dimensions, and they prevent high adhesion. Physically adsorbed films can be displaced over time, but chemically adsorbed films are not likely to be displaced unless there is some sliding. Adhesion between solids is thus influenced by two factors: lattice alignment and *screening* by adsorbed substances. Two metals in close proximity can potentially produce the highest adhesion. However, compared with ceramic materials, they are more likely to have chemically-adsorbed films which prevent adhesion. The bonding between polymeric surfaces is attributable to the weak van der Waals forces, which assure that only very weak adsorption occurs on these surfaces. Adhesion is important because the factors that influence adhesion also influence friction to a great extent.

6.8.4 THE LUBRICATING (OR SEPARATING) QUALITIES OF ADSORBED SUBSTANCES

The scale of physical adsorption extends a distance into the adsorbed layer on the order of 5–10 nm in the case of simple fluid molecules. For long molecular chain fluids, the scale of influence extends to a micrometer. These long-chain molecules are often depicted as being attached to the solid surface by active end groups with long tails extending away from the solid. Whereas it is easy to refer to these scales of influence, it is much more difficult to develop sound models for the behavior of the altered fluid layers in practical systems. The properties of these fluid layers have not yet been measured. Quite simply, however, these films can be modeled as liquids with spatially-increasing viscosity, to a high value approaching a thousand times the bulk viscosity in the first liquid "layer." In this sense, these surface films cannot be defined in terms of thickness.

The effective scale of chemical adsorption is measured by the thickness of the new substance formed on the surface. This can range from nanometers (nm) to tens of micrometers (μm). The mechanical properties of the new film depend on what forms on the solid. In some instances, a long-chain molecular substance is formed (sometimes referred to as a "polymer"), and in other cases an ordered structure or solid is formed. Often a molecular substance forms on top of an ordered structure. The long-chain molecular substance could probably be characterized as a viscous substance, with the usual (but not confirmed) property of firm attachment to the solid surface. The solid would behave as a solid, but the strength of its attachment to the substrate also influences its lubricating function with imposed sliding. Many of the common solid substances that are formed, such as oxides and others materials, are brittle in bulk form. However, when they are present in the form of thin films, and particularly when they are under high three-dimensional compressive stresses, they become more likely to flow plastically.

6.8.5 SECRET COMPOSITIONS AND MYSTERIOUS PROPERTIES

Despite the tremendous importance of surface films to boundary lubrication, the exact properties of adsorbed films are not well known. There is some value in speculating on these properties and incorporating reasonable properties into models and hypotheses for their lubricating qualities. There is far more work devoted to the chemical composition of surface films than to their physical properties. The act of formulating fluids for the creation of surface films is an art more than it is a science. The process involves selecting chemistry rather than physical properties or mechanisms. Therefore, composition and concentration reign in the storehouse of secret formulations.

Because surface film properties are mostly a mystery, technical progress is derived from an experience base gathered from the notions of field results or laboratory tests. Screening tests in the laboratory, as well as major qualification tests for the introduction of new fluid formulations into service, are more historical than rational. The difficulties of testing for performance attributes of new surface film formulations are partially attributable to the mysteries of surface films themselves. There is also a surprising lack of understanding as to how the lubrication mechanisms actually play themselves out in service. A systematic approach to this is summarized in later sections. In the meantime, the fourth structural element of a lubricated contact, the *subsurface region*, is highlighted in Section 6.9.

6.9 SUBSURFACE REGION

The subsurface region, which may be on the order of 50–1000 µm below the surface, does not appear at first to be connected to the lubrication mechanisms which are associated with hydrodynamic film generation or surface film formation. Although the subsurface region is remote from the tribological actions taking place at the surfaces, it still "feels" these actions by way of transmitted stress. The subsurface region is important with respect to failure mechanisms that result from plastic flow and fatigue. These failure mechanisms can have their initiating sites "subsurface." The subsurface region is the foundation or "roadbed" of the tribosystem. Material properties in the subsurface region must be able to tolerate the transmitted stresses at the stress levels and stress cycles demanded by the duty cycle of the equipment.

6.9.1 DEMANDS ON SUBSURFACE REGION

The demands on the subsurface region are derived from the normal load, which gives rise to a pressure distributed over an area of "contact." The subsurface region becomes an important element of the tribosystem when the applied pressure is high, relative to the strength of the subsurface material. When the applied pressure is low compared with the strength of the subsurface region, the demands on the subsurface region are diminished. The stress field may then be confined to the near-surface region only.

Friction, which creates a tangential stress "at the surface," can be thought of as a modifier to the stress field created by the normal load. The degree of modification on the normal stress field is directly related to the coefficient of friction.

If the load on a lubricated contact is being shared by asperities, or if surface motions generate micro-EHD pressures, the normal stress field is modified to yet another degree. The degree of asperity modification is influenced by asperity shape, height distribution, the number of asperities sharing the load, and the elusive *asperity friction*. Asperity pressure and friction put a "ripple" on the pressure distribution across the area of contact. If the contact is finely divided among asperities, the subsurface region may not feel this ripple. If a few isolated asperities are at work, the subsurface region will be annoyed by the hammering upstairs.

This discussion shows that the depth of tribology technology in the surface to be concerned with is a function of the depth of the stress field below the surface. Although friction coefficient and asperity phenomena may complicate the picture, at least the size of the subsurface playing field can be identified so that the approach to problem solving and technical developments can be better focused.

6.9.2 DIMENSIONS OF PRESSURE AND STRESS FIELDS

Because the purpose of lubrication is to bear the stress and accommodate the strain in the contact, it is important to quantify the *dimensions* of the pressure and stress fields. The engineering dimension of the pressure of a conformal contact is simply the load divided by the area of contact. Design calculations for conformal contact are frequently based on the *projected* area of contact, although the actual area of contact may be significantly smaller. In addition, the *real* area of contact may be even smaller. Care must always be given to *edge effects* where the termination of one body does not coincide with its mating surface.

The pressure and stress fields of non-conforming bodies with elastic deformations are calculated from Hertzian theory [1]. The theory of Heinrich Hertz was motivated by his study of Newton's optical interference fringes (such as those shown in Figure 6.15) and the problem of elastic deformation between two glass lenses in contact. His theory was worked out in 1880 during Christmas vacation when he was 23 years old. From his theory, a combined elastic modulus E' can be defined as:

$$\frac{1}{E'} = \left[\frac{(1-\sigma_1^2)}{E_1} + \frac{(1-\sigma_2^2)}{E_2} \right] \tag{6.5}$$

where:

σ_1 and E_1 are Poison's ratio elastic modulus of body 1
σ_2 and E_2 are Poison's ratio and elastic modulus of body 2

6.9.3 POINT CONTACT DIMENSIONS

For a *point contact* where two spherical bodies are in elastic contact, a combined radius of curvature, R, is defined as:

$$\frac{1}{R} = \frac{1}{R_1} + \frac{1}{R_2} \tag{6.6}$$

where R_1 and R_2 are the radii of curvatures of bodies 1 and 2, respectively.

In engineering practice, the normal load, W, is usually available, and it is used in Equation 6.7 to calculate the radius, a, of contact and in Equation 6.8 to calculate the maximum pressure, P_{max}.

$$a = \left(\frac{3WR}{4E'} \right)^{1/3} \tag{6.7}$$

$$P_{max} = \left(\frac{6WE'^2}{\pi^3 R^2} \right)^{1/3} \tag{6.8}$$

A characteristic of point contact elastic behavior is that the maximum pressure and Hertzian radius of contact are proportional to the cube root of the load.

Sometimes the maximum pressure is replaced with a *mean* pressure, which is 2/3 P_{max}. The pressure distribution over a point contact area is given by:

$$P = P_{max} \left[1 - \left(\frac{r}{a} \right)^2 \right]^{-1/2} \tag{6.9}$$

where P is the pressure at any radial point r within the contact. From Equation 6.9, the pressure is maximum (P_{max}) at the center of the contact, where $r = 0$. The pressure is zero at the edge of the contact, where $r = a$. The distribution of pressure produces a uniform normal displacement within the circular contact; that is, the contact is flat (the surfaces are parallel).

It is also of interest to calculate the total elastic deformation, δ, in the normal direction as a result of a normal load, W.

$$\delta = \frac{a^2}{R} = \left(\frac{W^2}{16RE'^2} \right)^{1/3} \tag{6.10}$$

When the elastic deformation δ is calculated for typical EHD contacts, we find that the amount of elastic deformation is much greater than the thickness of the EHD film itself. This result is compatible with the notion that typical EHD-lubricated contacts are Hertzian with respect to pressure and elastic shape.

6.9.4   LINE CONTACT DIMENSIONS

Line contacts are generated when one or more of the contacting bodies are cylindrical and the radii of curvature are generated around parallel axes. A line contact has a length L, a half-width of contact b, and a combined radius of curvature R. The characteristic line contacts can be calculated by using Equations 6.11 and 6.12.

$$b = \left(\frac{4WR}{\pi LE'} \right)^{1/2} \tag{6.11}$$

$$P_{max} = \left(\frac{WE'}{\pi LR} \right)^{1/2} \tag{6.12}$$

A characteristic of line contact elastic behavior is that the maximum pressure and Hertzian half-width of contact are proportional to the square root of the load. The pressure distribution for a line contact is given by:

$$P = P_{max} \left[1 - \left(\frac{x^2}{b^2} \right) \right]^{-1/2} \tag{6.13}$$

where x is a distance from the centerline of the contact.

6.9.5 Stresses Below the Surfaces

From the Hertzian pressures described above, it is possible to calculate the stresses in the contacting bodies. At the surface, the stress components are all compressive, except for the very edge of the contact where the stress is tensile for a point contact. This is the greatest tensile stress anywhere. This localized tensile stress is responsible for what are called "*Hertzian stress cracks*" in brittle materials.

A stress component of importance to fatigue is the *principal shear stress*, which has a value of approximately $0.3P_{max}$. For point contact, the maximum shear stress is at a depth below the surface of about $0.48a$. For line contact, the maximum shear stress is located at about $0.78b$. The maximum shear stress of about $0.3P_{max}$ is the greatest in the entire stress field, exceeding the shear stress at the edge of the contact. Therefore, plastic yielding is expected to initiate below the surface, rather than at the surface.

Surface friction modifies the stress field. As the coefficient of friction increases, the stress field becomes skewed, and the maximum shear stress moves closer to the surface. Asperities with *over-pressure* (i.e., above Hertzian pressure) create additional near-surface stresses. If surface friction is high enough, local asperity stress fields may join the subsurface stresses created by the Hertzian pressure. This condition may be critical. Surface-initiated cracks caused by asperity and other defects can then be driven deeper into the surface by the subsurface Hertzian stresses.

For Hertzian contacts, it is convenient to think of two major stress fields. The global stress field is created by the Hertzian pressure distribution, and the near-surface stress field is created by surface asperities, local depressions, debris, and other "defects" at the surfaces. Between these two stress fields one can also define a *quiescent zone*, which is located between the near-surface region and the subsurface region. The quiescent zone resides at a depth below the surface in which the local asperity and surface defect stresses are not significant and in which the stress field from the macroscopic Hertzian contact stress is not yet appreciable. This zone is quiescent from the point of view of stress and the accumulation of plastic flow and fatigue damage. The existence of the quiescent zone is important with regard to rolling contact fatigue. It inhibits the propagation of cracks between the stress field in the near-surface region and the stress field in the subsurface region.

6.9.6 The Core Region

Below the subsurface region is sometimes a *core material*, which is not really a structural element of the tribosystem. The core region is highlighted because materials, which must be selected for purposes other than tribological, are frequently surface-modified to achieve desired qualities. There are many examples where it is desired to have properties of the core material which are different form the *case material*. To impart structural toughness, hydraulic pump gear teeth are usually case-hardened to achieve wear and scuff resistance. The case may be hardened to a depth of approximately 1 mm, which is usually well below the subsurface region. The hardness of the core is controlled to achieve ductility for tooth strength. Although case-hardened materials generally provide good wear protection, they can lead to a catastrophic wear mode when the softer core material is exposed.

6.10 DYNAMICS OF A TRIBOSYSTEM

The preceding sections have described the structural elements of a tribosystem. The construction of the near-surface and subsurface regions must bear the applied stress. The generation of hydrodynamic films and the formation of surface films function to accommodate shear and control tangential friction forces. When the tribosystem is set into operation, the mechanistic processes within the structural elements begin. There are many lubrication and failure pathways that can be taken. From the very start, some of the structural elements will never be the same again. If the structural elements are viewed as a dynamic system, it is possible to anticipate or explain some of these mechanistic pathways. The following sections describe some of the dynamic interactions which are likely to take place and their consequences in terms of failure processes.

6.10.1 DYNAMICS OF ASPERITY STRESSES DURING INITIAL SLIDING

The most important interactions for the initiation of failure are those between the structural elements of the near-surface region and surface films. The controlling interactions are on a microscale, and are associated with the dynamic interaction of asperities.

At the very start, many asperities on one surface encounter asperities on the opposite surface and transfer load from one to the other. New surfaces, which require some sort of "run-in," transmit high pressure to the asperities approaching three times the yield strength of the softest of the two bodies. The surfaces are likely to have 90% of the normal load transferred through plastically-deforming asperities. This makes it reasonably certain that most asperities are straining progressively during initial sliding.

In addition to the normal load, sliding introduces shear forces to the contacting asperities. This produces a result that cannot be predicted from the principles of elasticity. Shear forces, in addition to the normal load, influence the amount of plastic flow. When a shear force is applied, the contacting asperities tend to "collapse" inward, and the micro-area of contact increases. This behavior is associated with a *yield criterion* that is expressed in several equations, one of which is that of von Mises:

$$(\sigma_x - \sigma_y)^2 + (\sigma_y - \sigma_z)^2 + (\sigma_z - \sigma_x)^2 + 6(\tau_{xy}^2 + \tau_{zx}^2) = 2Y^2 \qquad (6.14)$$

where:
σ is a normal stress
τ is a shear stress
Y is the yield strength of a material in uniaxial tension.

Taking only a single normal stress and a single shear stress, the equation could be simplified to $\sigma_x^2 + 3\tau_{xy}^2 = Y^2$. This result shows that when a shear stress is introduced, the normal stress on asperities is diminished, not by reducing the applied normal load but by increasing the area over which the load is carried.

6.10.2 DYNAMICS OF ASPERITIES DURING CONTINUED SLIDING

During initial sliding, the highest asperities experience severe deformation. After a few cycles these asperities will fail and produce loose particles in the interface. These entrapped particles can, in turn, inflict plastic deformation in other regions. If the "new" and exposed material created by plastic flow at the surface is not protected against adhesion, the surfaces will soon become unacceptable for further service unless the load is extremely light.

The protection of existing and newly created asperities is important with respect to adhesion and friction. This point is dramatically illustrated by experiments conducted in a vacuum. If the interaction of asperities is operating in the empty space of a vacuum, many asperities and loose particles may weld or bond together, thus affecting a pure state of adhesion. In air, which often contains water vapor in addition to oxygen, adhesion is inhibited considerably. Many soft metals slide together with no lubricant deliberately applied, but are nonetheless lubricated by oxides and physically adsorbed gases on the surface. Some ceramic materials also acquire a surface layer by reaction with ambient gases, which act to inhibit adhesion.

The formation of these surface films is an essential event in the control of friction or shear force at the asperity site. The discussion in the previous section explains the role of shear force on the yielding of asperities to accommodate the normal load over a larger asperity area. However, there is a further—even more important—effect of the shear force.

Values of the ratio of shear force to normal force which are below about 0.3 tend to "flatten" or reduce colliding asperities in height, whereas values of the ratio above 0.3 tend to "push" metal along and increase asperity height. This ratio of shear force to normal force is parallel to the notion of coefficient of friction, but in this case it is applied to asperity encounters. When the ratio of shear force to normal force is below 0.3, the sliding surface becomes smoother and, over time, fewer and fewer asperities will continue to plastically deform. However, although most asperities continue to deform elastically under load, some are likely to have high loads and high friction, and thus will plastically deform.

It is important to understand that the coefficient of friction at an asperity site may be quite different from the *global* friction coefficient measured across the entire contact. When the coefficient of friction on many asperities exceeds 0.3 by a good margin, the asperities will build up. The heights of asperities are built up by "pushing" metal in the sliding direction. The highest asperities will begin to carry a larger fraction of load, building up still more and remaining in this state over some distance of sliding. Temperatures in these asperities will increase, the metal will soften, the adsorbed films become thinner or softer, and the coefficient of friction on such asperities will rise still more. This is a progression leading to gross loss of surface integrity, as a growing number of asperities build up material in the same direction and at the same time.

Most manufactured surfaces for lubricated contacts are designed, perhaps unintentionally, to narrowly skirt disaster. They will be loaded to the point just short of failure. A small fraction of asperities on such surfaces is always progressing toward failure, but the loads upon them and the surface friction are momentarily relieved. If a sufficient supply of the *proper* chemical species is available to the surface during that short interval, a low-friction chemically adsorbed film will form to keep friction below 0.3 or so. The asperities in this sequence may be said to have "healed," and there is evidence to suggest that this takes place continuously on surfaces of many long-life lubricated contacts in service.

The availability of the *proper* chemical species is, incidentally, an uncommon occurrence across the wide range of industrial materials and lubricants. Low-cost iron-based materials with low total alloys, operating in slightly acidic lubricants, seem to form films of the right durability and at the right rate. While low-alloy steels may be good candidate combinations for surface film formation, they are generally not the materials of choice for reasons associated with the requirement to accommodate high normal stresses in the near-surface and subsurface regions. In the whole scheme of things, hardness and strength are important trade-off attributes with surface film formation. Some of the worst candidates for surface film formation are stainless steels, where chrome oxides limit the access of surface film-forming chemical species.

6.10.3 Run-in and the Bifurcation

If the initial running of a lubricated contact is moderately well lubricated, the interacting asperities undergo plastic flow. As previously discussed, a friction coefficient at asperity encounters below about 0.3 is likely to "iron down" the most protruding asperities. Over time, the number of asperities that are subjected to plastic flow diminishes, and asperity friction is reduced. It is likely that most of these asperities will now be cycling in the elastic range of stress, although some will continue toward plastic failure.

Run-in is the conditioning of surface features which initially stand proud above the average terrain of the original surface topography. Run-in of the type that seems to produce remarkably durable surfaces involves two interactive processes: *Topographical* run-in reduces asperity height and tends to create a "plateau" surface texture; *chemical* run-in creates durable and perhaps low-friction surface films. Surfaces that are properly run-in are less vulnerable to overload, start/stop events, and momentary liquid lubricant interruption.

If the initial running of a contact is not properly lubricated, high asperity pressures and shear forces can lead the contact down a pathway of surface roughening. In many cases, only minor changes in operating conditions will determine whether the asperity events lead to run-in or complete loss of surface integrity. This elusive bifurcation is associated with many puzzling early component failures.

6.10.4 THE THERMAL EFFECT

Thermal conditions on a micro- and global scale drive the fires of interactions among the lubricant films, surface films, and the near-surface region. Surface temperature is an important ingredient in these interactions and a key link between lubrication and failure. Temperature significantly influences the viscous properties of the fluid that control the thickness of a hydrodynamic or EHD-generated lubricant film. It is also a major driving force in the formation of chemical reaction films. Temperature influences the rate of fluid degradation, the strength of surface films, as well as the flow properties of the material in the near-surface region. Consequently, it is not surprising that the "total" temperature in a contact is frequently used as a failure criterion for a catastrophic failure event, such as scuffing.

From a simplistic point of view, the total temperature (T) is the sum of the bulk temperature (T_b) of the component and the "flash" temperature (T_f) associated with the instantaneous temperature rise derived from the friction in the lubricated contact. Flash temperature may arise from the traction of the lubricant film as well as from the energy dissipated at asperity encounters. The temperature rise at an asperity encounter may be derived from micro-EHD film generation, shear of surface films, adhesion between asperities, and deformation of the material in the near-surface region. The magnitude of T_f can be predicted if simplifying assumptions about the coefficient of friction and convection heat transfer are made.

6.10.5 FEEDING THE FRENZY

The survival of sliding surfaces in severe service is clearly dependent on the availability of a *protective film* in the vicinity of the asperities of sliding surfaces. This protective substance can be inserted into the tribosystem in its final form by at least two methods: One method is to coat a surface before use with a substance; if it "lubricates," it will be called a "solid lubricant." The solid lubricant is bonded to the surface and accommodates interfacial shear generally as a sacrificial film, which eventually wears away. Another method is to insert useful substances into a stream of liquid carrier and hope that enough of it gets to the right place to "lubricate."

A more common method of providing a protective film is to add a physically or chemically-active substance into a liquid carrier. The active substance, or *additive*, adsorbs to the surface or reacts with the solid surface to create a new substance. The liquid carrier or fluid is itself a lubricant and serves to generate hydrodynamic or EHD films. The formulated fluid has inherent physical properties as represented by its temperature–viscosity, pressure–viscosity, and traction behavior. The formulated fluid also has chemical attributes which are derived from small concentrations of active ingredients as well as from the base fluid itself. It is important to understand that physical adsorption and chemical attributes for lubrication are realized in the lubricated contact. The chemically derived lubricating properties are not inherent in the fluid itself; they are developed as part of the tribosystem.

Additives or constituents of the base fluid having physical adsorption to the surface no doubt provide some effectiveness over a limited range of pressure. Fundamental studies have been conducted by surface scientists in which a very thin film of some simple molecule, a solvent for example, is placed between flat mica sheets. Sliding resistance and film thickness are usually measured. There is clear evidence that the molecules of the first layer near the solid assume an ordered lattice arrangement of the solid, with decreasing order in succeeding layers of the liquid. This order enhances the near-surface viscosity for a limited range into the fluid. On sliding, the order becomes momentarily disrupted. It is likely that the same order could be seen in commercial fluids in practical systems, but the practical effectiveness of adsorbed layers is not clear.

It seems quite clear that if the proper chemical reaction occurs between some constituent in the lubricant and a metal (or metal oxide) surface, the surface will then endure much greater severity than without that constituent in the lubricant. Some of the products of the chemical reaction have been identified, but the physical properties of these products in commercial systems are not understood.

The actions of the chemical attributes of fluids in the lubricated contact are sometimes referred to as *"chemical boundary lubrication."* The chemical pathway by which the effective boundary lubricating films are formed is mostly a mystery. It appears that effective surface films require high "confinement" pressures and/or high shear rates. Although surface films are observed in areas where high pressures or temperatures do not exist, they are much less effective. A common notion is that high flash temperatures generated in the contact are the driving force behind the formation of effective surface films. The action of confinement pressures, which are generated at local sites by entrapped fluid under high shear, may play a more significant role than has been conventionally thought. This *tribochemical* action is no doubt accelerated by temperature rise at the surface. Along with the assistance of temperature, the catalytic actions of the near-surface or contaminant materials help to control chemical film formation.

Asperity encounters under severe operation are producers of chemical products which react and "combine" with the surface. Some chemical products may be lost to the system by dissolving into the fluid and being transported away. Some of the chemical products generated during operation can have ample time to dissolve into the fluid after the equipment is shut down.

6.10.6 ALWAYS VULNERABLE

A major point to understand is that the formation of an adequate chemical boundary film requires some sliding, during which time the surfaces are unprotected and vulnerable to failure by scuffing or galling. The rate of chemical reaction is thus important. Hydraulic fluid chemistry is carefully formulated to simultaneously avoid severe corrosive reactions at the surfaces and achieve chemical reactivity responsive enough for protection. If original surfaces put into service are adequately prepared, the formulated reactivity of the fluid should be sufficient during the initial run-in as well as in repairing damage during operation. Operational damage may be caused by overload, by the entrapment and abrasive action of debris, and by moments of inadequate supply of reactive chemical species to the sites that need to be repaired. Although there is incomplete understanding of the formation of protective films, there is evidence that some plastic flow of asperities and flaking off of native metal oxide are necessary events in the formation of protective substances.

The dynamic interactions and "competition" between chemically active materials in the fluid and the surrounding environment of lubricated contacts present a significant difficulty for fluid formulation, testing, and performance prediction. The formation of surface films and their composition and "properties" are part of the tribosystem itself. There is little understanding of how these films play out in lubricating the near-surface region. There are also no engineering parameters that connect to the dynamic actions in the tribosystem. The adequacy of film regeneration depends on an adequate supply of the necessary chemical species in the liquid carrier, adequate flow of the carrier through the contact region, and adequate time in microscopic contact regions to allow the completion of chemical reaction. The adequacy of lubrication depends on the thickness and mechanical properties of these films.

6.10.7 FUNCTIONAL ROLE OF LUBRICATION: STRESS RELIEF

Having taken an inside view of asperity encounters, the functional role of lubrication becomes clear: Lubrication action must relieve normal stress and tangential stress (friction) at asperity sites.

The easiest means of lubrication for the normal stress relief of asperities is the generation of hydrodynamic or EHD films. Although adequate hydrodynamic or EHD films can reduce asperity stresses to minimal values, the pressure of commerce usually results in a final manufactured design that is barely adequate. There will always be asperity encounters in essentially all practical contacts. The dynamic generation of films diminishes to nothing during starts and stops. It is hoped that at least partial surface separation is achieved during operation in order to ease the pain. This action moderates the pressure in the contact from high local pressures at asperity sites to minor pressure ripples along the expanse of the contact. The dynamic imposition of a fluid also creates an alternative means to accommodate shear. The degree of normal stress relief is, to a first approximation, a function of the hydrodynamic film thickness relative to the roughness heights of asperities. Remember also that when surfaces are in relative motion, micro-EHD-generated pressure at asperity sites can offset what is anticipated from the average film thickness generated. Hydrodynamic and EHD films transform load support from the microscale to the global scale and, in the process, reduce the maximum pressure everywhere. The lubricated contact becomes relieved of its "headache" and is less vulnerable to failure.

It is important to understand the scale of dynamic pressure generation and surface stress fields. Pressure generation in conformal contacts is low (MPa range) and distributed over a relatively large area. Concentrated contacts form high-pressure (GPa range) EHD films which are concentrated in small Hertzian deformation regions. The stressed region for concentrated contacts is transmitted much deeper into the supporting materials than are conformal contacts. Even with fully separated surfaces, high subsurface stress for concentrated contacts is not relieved; however, in both cases, the dynamic generation of a film relieves asperity stress.

The functional role of lubrication to relieve tangential stress (friction) at asperity sites is a more complex situation than normal stress relief. One can rationalize the notion that the dynamic imposition of a fluid creates an alternative substance to help accommodate shear in a field of asperity sites. Without a "full film," the shear is now distributed in the contact between a fluid and surface film. This situation is commonly known as "*mixed-film lubrication*"—that is, the lubrication is mixed between hydrodynamic, or EHD, mechanisms and the so-called boundary lubrication mechanisms.

We must now address the most common dynamic interaction between the two major structural elements of a lubricated contact found in practice: hydrodynamic, or EHD, films and surface films. The dynamics of this interaction have enormous implications for the success of lubrication and the prevention of failure. Hydrodynamic films provide the chemistry to enter the workplace. The chemistry that is carried by the fluid then "conditions" the surfaces to allow the hydrodynamic, or EHD, mechanisms to flourish. If the dynamic condition for hydrodynamic film generation is limited, the transported chemistry steps in to help carry the load and bear the shear. The dynamics of this interaction are reflected in friction. Although the sources of this friction in the contact are important for survival, they can only be inferred from gross friction measurements and asperity damage that appear on the surfaces.

6.10.8 Mechanisms of Global Friction

Although friction is rarely measured in lubricated contacts of operating equipment, it is an important internal parameter that governs the generation of heat and the onset of failure. Friction can be measured on a global scale, encompassing everything in the contact. Friction reflects the state of the material that is being sheared in the contact. From experiments, we know that characteristic friction behavior also reflects *what* is being sheared. Monitoring friction coefficient is like monitoring the heartbeat of the contact.

Starting from the simplest form of friction, the shear force of a full hydrodynamic film is directly connected to its viscosity and shear rate. Shear rate varies with film thickness and the relative velocity between the surfaces. The friction coefficient of hydrodynamic films may be on the order of 0.01. One must be cognizant of the fact that the total viscous friction of high-speed or high-viscosity hydrodynamic films is greatly influenced by the viscous churning of the fluid upstream of the point of closest approach. The churning component hides the component of viscous friction in the area where additional friction forces may be coming from asperity encounters. With high-viscous churning friction, asperity friction may be lost in the noise until catastrophic asperity friction is on its way.

As contact pressure increases, the viscosity of the fluid may also increase, as a result of the generated pressure. Viscous friction, then, is a function of pressure as well as of temperature. As discussed above, fluids under Hertzian-like pressures in the range of GPa shear-like pseudosolids with a shear limit. For most hydraulic fluids, the shear limit gives a maximum friction or traction coefficient of about 0.08 for mineral oils and most synthetic fluids. The shear limit of some types of hydraulic fluids gives a friction coefficient as low as 0.02. The shear limit decreases linearly with temperature, so that the EHD friction coefficient for high-temperature contacts may be on the order of 0.01. Note that high-pressure EHD friction coefficients on the low end (0.01) may actually be about the same as hydrodynamic friction coefficients. In addition, the EHD friction coefficient on the high end (0.1) is approximately equal to the friction coefficient of surface films. Although friction coefficient is the heartbeat of the contact, it may only reflect the general health of the contact without saying much about exactly what is wrong or right. This is unfortunate, because surface failure is primarily attributable to events on the asperity scale, and the asperity scale friction can sometimes be hiding from detection. Nevertheless, friction measurements can be made which separate EHD friction, micro-EHD-generated tangential forces, and some sort of surface film or plastic flow friction. Notions of friction should be centered on the thickness of hydrodynamic films relative to asperity height and the pressure in these films. With respect to friction coefficients over the entire contact that can be tolerated without gross failure, the sliding velocity plays an important role. Low-speed sliding can generally tolerate high friction coefficients better than high-speed sliding.

6.10.9 MECHANISMS OF ASPERITY FRICTION

The friction or resistance to sliding of contacts is primarily attributable to events on a global scale, whereas surface failure is primarily attributable to events on the asperity scale. Asperity friction refers to resistance to sliding of one asperity "over" another. The summation of friction on all asperities in a contact system that survive for even a short time is not likely to exceed two or three percent of the overall friction of that contact. This is because the sum of asperity contact areas is usually only a small fraction of the total contact area between two mating surfaces. However, asperity friction profoundly affects the progression of surface failure by its combined influence on the formation of chemically-adsorbed films and on the rate of progression toward failure of material in and around asperities.

The relatively small contribution of asperity friction to overall friction seems counterintuitive because the friction in failing systems usually increases considerably just before catastrophic failure. However, surface failure has usually progressed irreversibly toward system failure by the time a friction increase is observed. Small *failing events* occur continuously, but in systems that survive, these failing sites do not expand or generate new failures downstream. Rather, they recover in time to continue to function during subsequent contact events.

Surface failure occurs when both normal and shear (frictional) stresses are applied that are of sufficient magnitude and number of cycles to cause movement of substrate material or solid films, either along or from the surface. The action begins on, and in, asperities. Removal of material (wear) may occur, along with the displacement of material. Both may alter surface roughness and, depending on the direction that the roughness takes, the surfaces may run-in or lose surface integrity.

When fluid films in lubricated contact become progressively thin, asperity encounters may start off as benign events, resulting in only small perturbations in pressure. With thinner overall films, one asperity would be expected to sweep out an area occupied by a passing asperity and, as they approach one another, the fluid between them becomes highly compressed and shears at a high rate. Asperity encounters with severe interference can progress in a variety of ways depending on the "energy" of the encounter. At this point, it is convenient to consider the outcome of events, depending on whether the encounter is a low-frictional energy event or a high-frictional energy event. Remember that frictional energy is the product of frictional force and shear rate.

6.10.9.1 Low-Frictional Energy Events

Asperity encounters with low frictional energy include low plastic deformation rates at asperity sites. As hydrodynamic or micro-EHD films become thinner, their thickness approaches that of adsorbed films. Because of the influence of the surface forces from the substrate, any physically-adsorbed films that are present may have effective viscosity greater than the bulk fluid. Chemically-adsorbed films are likely to have an oxide attached to the solid and other products of chemical conversion adsorbed to the oxide. One can envision a process involving the behavior of a squeezed-down fluid film merging with adsorbed films. At some point, it may be possible that the adsorbed film is controlling asperity friction, and thus an interposed fluid film may not be necessary. The lack of interposed fluids is usually referred to as "dry sliding," but so-called dry surfaces in air are always covered with oxides (except gold) plus adsorbed liquids and gases. This explains why solids in long-term static contact do not weld or bond together.

Low asperity pressure and shear stress characterize low-frictional energy events. They survive on weak physically-adsorbed films, or on chemically-adsorbed films from mild reactions.

6.10.9.2 High-Frictional Energy Events

Asperity encounters with high frictional energy include high rates of plastic flow in asperities. The events described in the previous section are altered by high-energy input caused by high sliding velocities or high asperity pressure. The high shear rate in both the fluid and asperities heats the surrounding fluid, thereby reducing its viscosity, which results in thinner fluid films. It should be noted that for EHD contacts, heating of the fluid in a load-carrying region does not affect the EHD film thickness, which is determined in the inlet region. The local heating of the fluid affects the micro-EHD film thickness at the inlet of the asperity encounter. At the same time, adsorption strength of physically-adsorbed substances diminishes when temperature rises, thereby reducing the effective viscosity in the adsorbed film.

Chemical adsorption is also influenced by temperature, depending on the availability of proper chemical species in the vicinity of the asperities. Higher temperature increases the rate of oxidation and the reaction rates leading to the formation of the polymeric substances on top of oxides. Although the latter reaction rates increase with temperature rise, the physical properties are diminished with temperature rise. Thus, there is an optimum combination of the thickness of surface films and physical properties of these films for achieving minimum asperity friction.

High asperity pressure and shear stress characterize high-frictional energy events. They survive only by chemical adsorption and reaction films, which are quick to respond to the severe demands of the moment. Although the temperature rise and plastic flow of high-energy encounters increase the rate of chemical reaction, the shear strength of the reaction films may decrease and chemical composition may be altered.

6.10.10 Failure Without Touching

Material failure in surfaces that are in sliding or even rolling can, and likely often do, take place without actual contact. Contact is in the sense of the atoms in opposing surfaces being subject to one another's force fields. This seems like a trivial distinction, but it is important when it comes to describing the sequence of events that leads to either successful lubrication or failure of systems on the scale of asperities. A common notion in engineering is that surfaces are either adequately lubricated or fail by adhesion. If this were completely true, debris would never fall from a sliding pair! Failed surfaces often show evidence of some severe sliding, which may look like adhesion has taken place, but these events usually occur in the last seconds of failure. These are not the initiating events.

Surface failure without contact may be seen to occur in cavitation erosion, which is common in hydraulic systems. Surface-initiated rolling contact fatigue with micro-EHD films can be easily initiated without contact. This is a common occurrence on rolling contact surfaces which have been damaged by debris dents. Cracks initiate near the shoulders of the dents, where high micro-EHD pressures are generated. Sufficient stress can be imposed on a solid by a liquid to plastically deform ductile materials and fracture brittle materials.

6.10.11   Failure Processes and Descriptive Terms

There are numerous failure processes emanating directly from asperity encounters and many other failure processes not directly related to asperity events. In essentially all cases, the source of failure initiation is confined to localized and microscale events. Failure processes are controlled by four basic mechanisms: (1) adhesion, (2) plastic flow, (3) fatigue, and (4) chemical reaction. The user ultimately defines the success of lubrication and the failure of surfaces. Failed surfaces are characterized with descriptive terms, not always reflecting the failure process. Most wear and fatigue failure terms reflect only one of several failure processes along the pathway to the final surface condition, not to mention the initiating failure event. Although a host of surface deterioration and failure terms are in common use, they all seem to fall into three major categories: wear, scuffing, and fatigue.

6.10.11.1   Wear

Continued plastic flow of asperities can lead to loss of surface material by wear. Actually, the removal of material may occur by low cycle fatigue. It may not be necessary to completely eliminate all low cycle fatigue of asperity material to maintain satisfactory performance. However, it is necessary to keep the size and flow rate of loosened particles down. Low-rate wear particle generation, also be called "*mild wear*," tends to polish the surfaces and is therefore also called "*polishing wear*." Polishing wear can be associated with a continual process of removal and reformation of chemically-adsorbed species, like oxides. If chemical reactions occur, the reaction products may be removed by a sacrificial wear process to prevent adhesion. If the sacrificial wear process occurs at a high rate, the result may be called "corrosive wear." At the other extreme, local adhesion, or adhesive wear, can occur if chemical reaction products are not present to accommodate shear at the interface.

Failure by wear is generally attributable to a loss of engineering tolerance. The consequence for hydraulic systems is excessive leakage. The rate of wear on a surface can easily be accelerated by its own wear particles. Once bits of material loosen, they can migrate in the contact regions. Some may re-adhere, although not very firmly in most instances as the "active" surfaces on newly-formed particles can become protected from adhesion by adsorbed films. Wear particles are usually hard and will form grooves when entrapped between softer surfaces. The hard particles will also remove chemically-adsorbed films or chemical-reaction films (see Figure 6.18). The availability of fluid to flush out wear debris particles, along with the access of channels through which loosened particles may escape from the system, are essential to the prevention of accelerated wear.

6.10.11.2   Scuffing

Scuffing is a catastrophic form of wear derived from a high-frictional energy event. Scuffing is driven by high sliding velocities. Scuffing initiated from high asperity pressures can result in a sudden and total loss of surface integrity and gross failure of the near-surface region (see Figure 6.19). When scuffing is initiated from low asperity pressures, the damage may be confined to the surface roughness features only. This is sometimes called "*microscuffing*." Scuffing is also sometimes called "*galling*," but galling seems to be used more frequently with unlubricated sliding.

6.10.11.3   Fatigue

Concentrated contacts with high Hertzian pressures or weak subsurface material can lead to crack initiation in the region of maximum subsurface shear stress. If cracks propagate under cyclic stresses

(a)

(b)

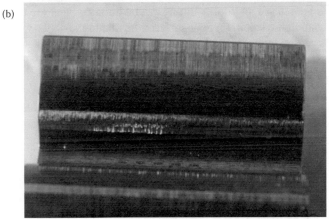

FIGURE 6.18 Entrapped micro-pitting debris removes surface films, causes wear grooves, and makes the surface vulnerable to scuffing.

to the surface, sections of subsurface and near-surface material are removed, which causes *spalling*. Spalling is a subsurface-initiated failure caused by the Hertzian stress field, which can be influenced by frictional stress at the surface.

When the fatigue process initiates at or near the surface, the loss of small fatigue particles is called "*pitting*." Pitting may initiate at asperity sites where local pressures are high. When pitting occurs on roughness features during or shortly after run-in, it is called "*initial pitting*." After run-in, and with marginal hydrodynamic or EHD film thickness, microcracks can form on the surface because of high surface friction. Microcracks are frequently formed in the presence of one-way

FIGURE 6.19 Scuffing is a high-energy event associated with a sudden and catastrophic loss of surface integrity.

sliding friction, where the near-surface material is strained or "ratcheted" in one-direction (see Figure 6.20). This progression to failure is equivalent to adding loads to a tensile specimen progressively until it fails. The details may not be important, but the result is that cracks grow and join others. The rate of progression in crack propagation increases logarithmically with applied stress, which is to say that asperity cracking rates will increase logarithmically with applied load. Furthermore, acidic environments have been found to increase the rate of crack propagation in fatigue cycling.

Surface-initiated fatigue is complex and is almost always influenced by wear. It can be both an interactive and competitive process. The wearing away of surface roughness features during initial running reduces the propensity for pitting. Pitting can be initiated by at least three factors: (1) high asperity stress without high friction, (2) poor surface film and high friction at asperity sites, and (3) microcorrosion pits caused by reactive additives in the fluid.

6.11 SYSTEMATIC TRIBOLOGY

The previous sections show that most of the fundamental mechanisms of lubrication and surface failure deal with events on a microscale. The mechanisms involve four major structural elements

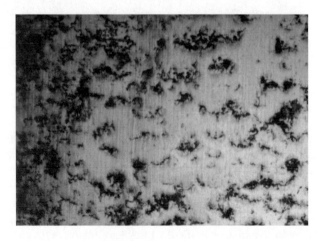

FIGURE 6.20 Microcracks on gear surface following many cycles of tangential and normal stress.

which are dynamically linked in a tribosystem. The development of hydraulic systems requires the management of these mechanisms on engineering terms. Because hydraulic systems encompass a multitude of conformal and concentrated contacts composed of structural elements with numerous materials, the engineering task is unwieldy. When one considers the complexity of each lubricated contact in hydraulic systems, it is not surprising that fluid formulation, material development, and design sometimes turn out to be incompatible. The lack of correlation among fluid bench tests, along with their disconnect from field service, is a testimony of the tribology nightmare.

The complexities of the situation can be overcome, to a large degree, by introducing the concept of *systematic tribology*. Systematic tribology is the process of understanding and developing the structural elements of practical tribological contacts and mechanisms of interaction in a systematic way. The challenge of fluid formulation, material development, and design is to "capture" mechanisms that control performance. The focus is on the creation of structural elements whose attributes are able to carry high normal stress and bear the severe tangential shear in the setting of a tribosystem. With this approach, external engineering parameters take on a new meaning, which gives a more rational connection to what is happening on a microscale.

6.11.1 Tribosystem Input and Output

The framework of thinking for systematic tribology is illustrated in Figure 6.21. The tribosystem is identified with its major structural elements:

HF = hydrodynamic film
SF = surface film
NS = near-surface region
SS = subsurface region

The tribosystem is driven by "input" parameters, which can be appropriately characterized as stress, kinematics, and environment. Stress is the normal stress applied to the contacting bodies. If available, it can also be the normal asperity stress. Kinematics defines the motions within the tribosystem, in particular, the sliding velocity and entraining velocity. The environment is defined

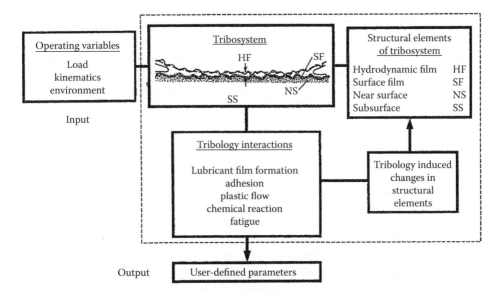

FIGURE 6.21 Dynamic interactions among the structural elements of a tribosystem in operation.

by thermal and chemical conditions in and around the tribosystem. The chemical environment includes the surrounding gases and fluid-supplied chemistry that are introduced into the system.

Once the tribosystem is put into operation, mechanistic processes like hydrodynamic film generation, adsorption, plastic flow, and fatigue are set into action. These mechanistic processes induce changes in the structural elements which, in turn, influence the mechanistic processes. This cycle of activity defines the dynamic interactions among the structural elements.

From an engineering perspective, the dynamic interactions among the structural elements can be considered a so-called "black box." The black box is driven by input parameters defined in engineering parameters. The output parameters to the black box can also be defined in engineering terms. The output parameters are user-defined. Depending on the functional demands of the lubricated contact required for service, they may vary considerably. Examples of output parameters are scuffing resistance for high load, wear resistance for low-speed reciprocating motion, abrasive wear resistance for debris-contaminated hydraulic fluids, or long-life pitting resistance.

The purpose of the tribosystem, as illustrated in Figure 6.21, is to connect engineering parameters with mechanistic processes. The internal workings of the black box need to be exposed in understandable mechanistic processes, which relate directly to the construction of structural elements and how they perform in service. Without this understanding, fluid formulation and material development will continue to be conducted on a trial-and-error basis with bench tests having little correlation with service.

6.11.2 CONNECTING ENGINEERING PARAMETERS WITH MECHANISTIC PROCESSES

Engineering parameters can be connected to both mechanistic processes and their interactions among the structural elements by considering the basic functions of lubrication. The structural elements must carry the load and bear the shear without losing surface integrity and ability to function. The task is to select the engineering parameters which can be most appropriately linked to normal stress and tangential strain at asperity sites.

6.11.2.1 Entraining Velocity

Among all the tribological mechanisms, the generation of a hydrodynamic, or EHD, film is by far the most mature with respect to theory and predictability. Film thickness controls the degree of asperity encounter. The separation of surfaces, particularly with EHD-generated films, is highly precise and predictable. The film is remarkably "stiff" with respect to the elasticity of the surfaces and the applied load.

Film thickness is directly related to the fluid entraining motion of the surfaces. Because the kinematic motions of load-carrying components are generally known, the entraining velocity provides a useful engineering connection to fluid film-forming mechanisms. The entraining velocity U_e is defined as:

$$U_e = \frac{1}{2}(U_1 + U_2) \tag{6.15}$$

where U_1 and U_2 are the surface velocities of the contacting bodies at the point of contact. The entraining velocity is directly linked to the flow of fluid through the contact. For a "pure" sliding contact, the entraining velocity is equal to half the velocity of the moving surface. A contact with a combination of rolling and sliding has an entraining velocity that is half the sum of the surface velocities. Surface velocities are vector quantities, meaning that an entraining velocity for surfaces moving at an angle relative to each another can be calculated as a vector sum.

The engineering value of the entraining velocity results from how it controls the dynamic generation of a structural element (HF) of the tribosystem and how its thickness varies with the duty cycle of the equipment. Because the thickness of a hydrodynamic film determines the degree of engagement between asperities, the entraining velocity is directly linked to asperity-controlled mechanisms. Another engineering value of the entraining velocity and its connection to hydrodynamic, or EHD, film generation is that it is directly linked to the suppliers of base fluids. Fluid properties representing temperature–viscosity and pressure–viscosity characteristics are key ingredients for the construction of the dynamically-generated HF element.

6.11.2.2 Sliding Velocity

The sliding velocity across the contact controls the shear between the contacting surfaces. The sliding velocity U_s is defined as:

$$U_s = (U_1 - U_2) \qquad (6.16)$$

Sliding velocity is also a vector quantity. The magnitude of sliding is a frequently-known engineering parameter that can be connected to the duty cycle of the equipment. The sliding velocity, in conjunction with the entraining velocity, provides an engineering link to the degree of strain at asperity encounters. Another engineering value of sliding velocity is that for EHD contacts; the sliding velocity invokes traction behavior, which is a fundamental property of base fluids.

6.11.2.3 Normal Load and Pressure Distribution

The engineering parameters of entraining velocity and sliding velocity control the level of penetration during asperity encounters and their rate of shear. The all-important pressure at asperity sites is controlled by the distribution of pressure among asperities and the fluid film. The contribution of pressure that is distributed directly to asperities is an engineering unknown. The best that can be done is to simply address the normal load or pressure applied on a global scale.

The applied load is a frequently-known or estimated engineering quantity. For concentrated contacts, the applied pressure is calculated from the load through Hertzian theory. For conformal contacts, the pressure may simply be calculated by assuming that the load is distributed over a "projected" area. The pressure calculated is sometimes used to judge the lubrication severity of a contact through the use of a "PV" factor. The PV factor is the product of the pressure (P) and sliding velocity (V). Unfortunately, PV is not directly linked to mechanisms that control performance.

6.11.3 Mapping Dynamic Interactions with Entraining and Sliding Velocity

To demonstrate the engineering control afforded by entraining velocity and sliding velocity, hydraulic fluids have been mapped over a range of U_e and U_s [3]. Mapping is conducted with a single contact, where the entraining velocity and sliding velocity are independently controlled. The mapping results for two hydraulic fluids are shown in Figures 6.22 and 6.23. The contacting surfaces are a steel ball running against a steel disc with surface roughness of 0.25 µm, Ra and 0.075 µm, Ra, respectively. The applied load gives a Hertzian contact stress of 2.07 GPa. A series of tests is conducted for ten minutes, each over a range of entraining velocities and sliding velocities. New specimens are introduced for each test. Friction coefficient and surface temperatures are monitored to detect failure transitions.

High entraining velocities and low sliding velocities define a region of operation where shear is accommodated entirely by an EHD film (no asperity encounters). Lower entraining velocities and thinner EHD films relative to surface roughness, reduce asperity height and create surface films (run-in). As sliding velocity increases, the surfaces eventually fail by scuffing. These experiments illustrate the dynamic generation of the structural elements for lubrication (HF and SF). The tests also evaluate the durability of the entire tribosystem. Failure is initiated at asperity sites, and the scuffing failure limit is clearly approached as the degree of asperity encounter increases.

The results in Figures 6.22 and 6.23 show that hydraulic fluids can have widely different lubrication performance because of the quality of the structural elements. Lubrication performance is

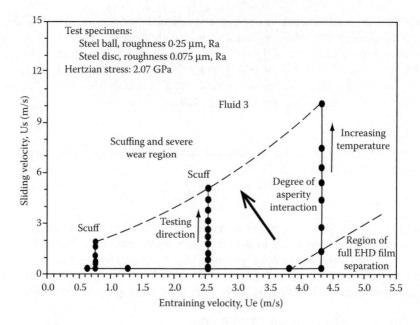

FIGURE 6.22 Performance mapping of a fluid using entraining velocity and sliding velocity to determine failure tolerance to degree of asperity encounter.

derived from EHD films and surface films (elements of HF and SF). Fluid 2 shows much less EHD film-forming capability than fluid 3. However, fluid 2 is able to tolerate a much greater degree of asperity encounter than fluid 3.

These tests use engineering parameters to probe the individual structural elements of a tribo-system. The results have linkage to hydraulic gear pump lubrication and failure mechanisms. This linkage is made through the entraining velocities and sliding velocities that vary across a tooth

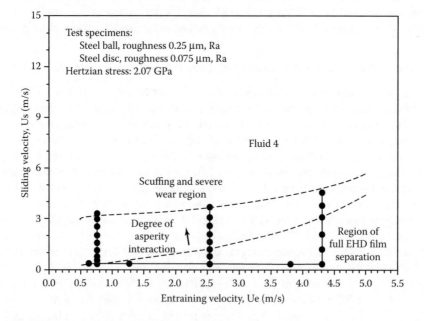

FIGURE 6.23 Performance mapping of a fluid using entraining velocity and sliding velocity to determine failure tolerance to degree of asperity encounter.

face from its root to tip. The wide range of asperity shear above and below the pitch line gives rise to scuffing failure at the root or tip of the tooth along with pitting damage near the pitch line. Simulation of entraining and sliding velocities, along with other environmental conditions, allows for the replication of the same failure mechanisms found in service. If the appropriate failure mechanisms are simulated, fluid formulation and material developments can be orchestrated and efficiently tested in order to achieve predictable performance enhancement.

6.11.4 MAPPING FAILURE MECHANISMS IN ENGINEERING TERMS

By considering dimensionless parameters with good engineering connection to real systems, the mapping of dynamic interactions of the structural elements, as described in the previous section, can be carried a little further. Because we need to have engineering connection to the asperity encounters, we can substitute the entraining velocity for the ratio of hydrodynamic of EHD film thickness (h) to surface roughness (σ). This is a more direct way to make an engineering link to asperity encounters. The roughness parameter, σ, reflects the combined roughness of the contacting pair. It is defined as:

$$\sigma = (\sigma_1^2 + \sigma_2^2)^{1/2} \tag{6.17}$$

where σ_1 and σ_2 are the centerline average heights of body 1 and body 2. The ratio h/σ is known as the lambda (σ) ratio. The lambda ratio is commonly used to judge the quality of lubrication. Because surface roughness (σ) is characterized by average height departure from a mean plane and does not include three-dimensional texture, the lambda ratio has many difficulties. In addition, the truncation of asperities caused by run-in is frequently ignored. Nevertheless, the ratio h/σ is a first-order parameter that serves a good purpose for now.

The sliding velocity used above reflects the shear between the surfaces. Another way to handle sliding, which is particularly appropriate when there is combined rolling and sliding, is to use the ratio of sliding velocity to entraining velocity (U_s/U_e). The ratio (U_s/U_e) is called the "*slide-to-roll ratio.*" With the definitions of U_s and U_e given above, pure sliding, which has one surface stationary, gives a slide-to-roll ratio of 2. In pure rolling, where both surfaces are moving with equal (and collinear) velocities, the slide-to-roll ratio is 0.

From testing experience gathered over time, a number of life-limiting surface failure mechanisms can be mapped with the parameters for h/s and the slide-to-roll ratio (U_s/U_e). Such a map is illustrated in Figure 6.24. When a large portion of the duty cycle allows operation at high h/s, asperity pressures are low and surface life is long. With high Hertzian stress or low fatigue-resistant materials, fatigue initiation is subsurface, and the surface ultimately fails by spalling or pitting.

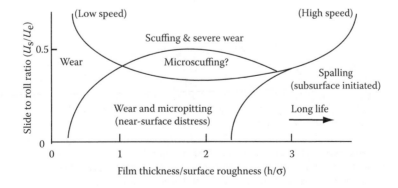

FIGURE 6.24 Postulated life-limiting failure modes controlled by engineering parameters.

Relatively low h/s, along with high sliding, can initiate asperity scuffing, which can transition into catastrophic scuffing. This phenomenon is only observed at high surface speeds.

Low values of h/s create high asperity pressure. When U_s/U_e is low, tangential stress on asperities is low, and asperity failure may teeter between asperity wear and fatigue (micropitting). Near-surface damage under low h/s and low sliding, which is common to rolling element bearings, is frequently called "surface distress" or "frosting." If higher sliding is present, asperities will tend to *wear* rather than fail by fatigue. Still higher sliding velocities can cause asperities to transition into scuffing or severe wear. Sometimes an intermediate region of microscuffing is present, when failure is confined to asperities only and the surface still has a chance to recover.

Operation at the extreme low end of h/s only seems possible at low surface speeds without severe wear or scuffing. Surface durability is limited by wear, but wear may also be influenced by low cycle fatigue. The wear process tends to increase surface roughness.

6.11.5 THE CONSEQUENCE OF SYSTEMATIC TRIBOLOGY

If asperity scale mechanisms can be linked to engineering parameters as illustrated above, the dynamic processes among the structural elements can be put on a more rational basis. The tribosystem becomes more understandable and is less of a black box. With better understanding, surface inspection of service hardware and failure analysis can now be much more revealing.

By using a systematic approach, actions taken to improve viscosity or pressure–viscosity properties of base fluids can be immediately evaluated with respect to failure modes mapped out. Additive enhancements can be evaluated for their impact on wear, scuffing, or fatigue processes. The performance gains with new materials or surface-engineering technologies can also be put into perspective. With a better correlation between testing failure modes and service failure modes, there is greater assurance that corrective actions taken will actually be realized in service.

If the structural elements of hydraulic system contacts are addressed as a tribological system, the introduction of new materials and fluid formulations can be accomplished with greater effectiveness. Systematic tribology helps to integrate the diverse technologies that are introduced into a tribosystem. Consideration of the structural elements helps identify the connection between technical businesses which are normally disconnected. The hydrodynamic film element is associated with base stock suppliers. The surface-film element is associated with additive suppliers. The near-surface element is associated with heat treatment, coating, surface treatment, and surface finishing businesses. The subsurface element is associated with material suppliers. If design is focused more on capturing lubrication mechanisms and avoiding failure mechanisms, the potential outcome will be a synergistic construction of the technologies available. The consequence of systematic tribology is superior output from the tribosystem, even with greater input requirements. To do this requires practical engineering parameters, which reach down into asperity-scale events. This is clearly a work in progress.

REFERENCES

1. Hertz, H.R. and Angew, J.R. Math (Crelle's j.) 1881, 92, pp. 156–171.
2. Martin, H.M. "The Lubrication of Gear Teeth," *Engineering*, 1916, 102, pp. 119–121.
3. Wedeven, L.D., Totten, G.E. and Bishop, R.J. "Testing Within the Continuum of Multiple Lubrication and Failure Mechanisms," ASTM STP 1310, *Tribology of Hydraulic Pump Testing* (Totten, G.E., Kling, G.H., and Smolenski, D.J. eds.), ASTM, West Conshohocken, PA, 1997, pp. 3–20.
4. Dowson, D. "Elastohydrodynamics," *Proc. Inst. Mech. Engr.*, 1967–1968, 182, part 3A, paper no. 10.
5. Wedeven, L.D., Evans, D. and Cameron, A. "Optical Analysis of Ball Bearing Starvation," ASME Trans., *Journal of Lubrication Technology*, Series F, 93(3), pp. 349–363.
6. Cusano, C. and Wedeven, L.D. "Elastohydrodynamic Film Thickness Measurements of Artificially Produced Non-Smooth Surfaces," *ASLE Trans.*, 1981, 24(1), pp. 1–14.

7 Hydraulic Fluid and System Standards

Paul W. Michael and Thomas S. Wanke

CONTENTS

7.1 INTRODUCTION

Standard specifications for hydraulic fluids and components serve an important function in the fluid power industry. Fluid standards validate the safety, durability, compatibility, cleanliness, and functionality of hydraulic media [1–5]. Dimensional standards for hoses, tubes, seals, and fittings ensure interchangeability of components and simplify the assembly process [6–9]. Standard mounting geometries for pumps, motors, valves, and cylinders provide flexibility in design and reduce costs [10–12]. Standard test methods for evaluating hydraulic component efficiency, power output, dynamic response, fatigue life, and reliability provide benchmarks for product performance [13–17]. Standard terminology and symbols in fluid power create a common language for fluid power system designers and users [18,19]. These standards are especially important in the fluid power industry because hydraulic systems incorporate components produced by a variety of manufacturers from across the globe.

The International Organization for Standardization (ISO) governs global standards for safety, reliability, efficiency, performance, and interchangeability in wide-ranging fields including aviation, building materials, communication, medical devices, photography, and fluid power. ISO is based in Geneva, Switzerland, and is composed of nationally-chartered standards institutes. These member bodies represent their nation's interests and retain a single vote, regardless of the size of the population or economy. Table 17.1 lists the member bodies for some of the nations that participate in the ISO. These nations integrate government and industry groups within its member body to represent national interests within the ISO. For instance, in the United States, the American Society for Testing and Materials (ASTM) represents American National Standards Institute (ANSI) with respect to fuel and lubricant standards, while the National Fluid Power Association (NFPA) manages fluid power systems and component standards under the ANSI umbrella. These organizations

TABLE 7.1
Member Body Representatives in Fluid Power Technical Advisory Groups

Nation	Acronym	Member Body
Algeria	IANOR	Institut Algérien de Normalization
Argentina	IRAM	Instituto Argentino de Normalización y Certificación
Belgium	NBN	Bureau de Normalisation
Brazil	ABNT	Associação Brasileira de Normas Técnicas
Canada	SCC	Standards Council of Canada
China (PRC)	SAC	Standardization Administration of China
France	AFNOR	Association Française de Normalisation
Germany	DIN	Deutsches Institut für Normung
India	BIS	Bureau of Indian Standards
Italy	UNI	Ente Nazionale Italiano di Unificazione
Japan	JICS	Japanese Industrial Standards Committee
Mexico	DGN	Dirección General de Normas
Russian Federation	GOST R	Federal Agency on Technical Regulating and Metrology
Spain	AENOR	Asociación Española de Normalización y Certificación
South Korea	KATS	Korean Agency for Technology and Standards
United Kingdom	BSI	British Standards Institution
United States	ANSI	American National Standards Institute

hold secretariat positions within the ISO and therefore are responsible for setting agendas, coordinating meetings, distributing ballots and administrating standards.

ISO standards are established through the collaboration of national delegations of experts within a Technical Committee (TC). Technical Committee 28 (TC-28) is responsible for hydraulic fluid standards, while fluid power equipment standards are developed within TC-131. As can be seen in Tables 7.2 and 7.3, different Subcommittees (SC) within the TC are responsible for specific areas. For instance, within TC 28, SC 1 is responsible for terminology.

Within the technical committee, standard development takes place in subcommittees and Working Groups (WG). Usually the working groups prepare a draft of the standard for the working groups' parent committee. These technical committees meet, discuss, and debate until they reach consensus on a draft agreement. This first consensus document is circulated as a Draft International Standard (DIS) to ISO's membership for ballot and comment. If the voting is affirmative, the document, typically with modifications, is circulated to the ISO members as a Final Draft International Standard (FDIS). If that vote is positive, the document is then published as an International Standard. The process of creating a consensus standard requires a good deal of time and effort, as well as many compromises. Nonetheless, these standards play an important role in the safety, efficiency, and reliability of fluid power systems.

TABLE 7.2
ISO Technical Committee 28 – Subcommittees

TC 28 - Petroleum Products and Lubricants

TC 28/SC 1	Terminology
TC 28/SC 2	Measurement of petroleum and related products
TC 28/SC 4	Classifications and specifications
TC 28/SC 5	Measurement of liquefied gaseous fuels
TC 28/SC 7	Liquid biofuels

TABLE 7.3
ISO Technical Committee 131 – Subcommittees

TC 131 - Fluid Power Systems

TC 131/SC 1	Terminology, classification and symbols
TC 131/SC 2	Pumps, motors and integral transmissions
TC 131/SC 3	Cylinders
TC 131/SC 4	Connectors and similar products and components
TC 131/SC 5	Control products and components
TC 131/SC 6	Contamination control
TC 131/SC 7	Sealing devices
TC 131/SC 8	Product testing
TC 131/SC 9	Installations and systems

7.2 HYDRAULIC FLUIDS

The primary purpose of a hydraulic fluid is to transfer power. The concept of fluid power is based on a principle articulated by Blaise Pascal, which is given as follows: "Pressure applied to an enclosed fluid is transmitted undiminished to every portion of that fluid and the walls of the containing vessel." [20] Within the context of fluid power, pressure is related to the force acting on a confined fluid. This principle has given rise to modern hydraulics, which entails highly-engineered systems for efficiently controlling fluid flow to transfer energy and accomplish work.

The heart of any hydraulic system is the pump, which draws fluid through its inlet and forces the fluid through its outlet, usually against pressure created by valves, plumbing, and actuators downstream of the pump. Pumps, actuators, and other system components have surfaces that move relative to one another, often at high speeds, pressures, and temperatures. These components require cooling and lubrication for efficient performance and durability. Consequently, hydraulic fluids not only transmit power, they also serve a critical function as a lubricant and heat transfer medium.

Most hydraulic fluids consist of a base fluid and a combination of additives that have been optimized to impart chemical characteristics and functionality to the finished product. Operating conditions and equipment builder specifications generally dictate the type of fluid that is needed and thus, the kind of base stocks and additives that must be employed. ISO 6743 establishes a classification system for lubricants, industrial oils, and related items [21]. As shown in Table 7.4, there are 15 categories of lubricants within this classification system.

ISO 6743-4 describes categories of lubricants within the hydraulic fluid class of lubricants [22]. The categories of hydraulic fluids within this classification include mineral oil, fire-resistant fluids, and environmentally-acceptable fluids [23–25]. The specification for these types of fluid are defined within the following ISO standards:

ISO 11158 Lubricants, industrial oils, and related products (Class L)—Family H (Hydraulic systems)—Specifications for categories HH, HL, HM, HR, HV, and HG.

ISO 12922 Lubricants, industrial oils, and related products (Class L)—Family H (Hydraulic systems)—Specifications for categories HFAE, HFAS, HFB, HFC, HFDR, and HFDU.

ISO 15380 Lubricants, industrial oils, and related products (Class L)—Family H (Hydraulic systems)—Specifications for categories HETG, HEES, HEPG, and HEPR.

7.2.1 Mineral Oil-Based Fluids (ISO 11158 Lubricants)

The majority of hydraulic fluids in service are mineral oil-based because they generally provide satisfactory performance at a relatively low cost. Within the mineral hydraulic oil category there is a wide range of fluid types. The ISO 11158 defines the physical properties and performance

TABLE 7.4
ISO Industrial Fluid and Lubricant Classes

Specification	Family	Type
ISO 6743-1	A	Total loss systems
ISO 6743-2	F	Spindle bearings
ISO 6743-3	D	Compressors
ISO 6743-4	H	Hydraulic systems
ISO 6743-5	T	Turbines
ISO 6743-6	C	Gears
ISO 6743-7	M	Metalworking
ISO 6743-8	R	Rust protection
ISO 6743-9	X	Greases
ISO 6743-10	Y	Miscellaneous
ISO 6743-11	P	Pneumatic tools
ISO 6743-12	Q	Heat transfer
ISO 6743-13	G	Slideways
ISO 6743-14	U	Heat treatment
ISO 6743-15	E	IC Engine

requirements of five types of mineral oil-based hydraulic fluids listed in Table 7.5. These fluids are described below.

HH   Type HH fluids are mineral oils without additives. HH fluids may be used in air-over-oil hydraulic systems, such as those found in car lifts at automotive service centers. They are also used in manual hydraulic pumps, jacks, and other low-pressure hydraulic systems. While type HH fluids are able to perform the primary function of a hydraulic fluid, which is to transmit power, they are unable to withstand high temperatures and have limited lubricating capabilities. Thus, these HH fluids find limited application in industry.

HL   Type HL fluids are also mineral oil-based, but they contain rust and oxidation inhibitors to protect equipment from the detrimental effects of water contamination and chemical deterioration due to heat. These fluids are also known as R&O oils because they contain rust and oxidation inhibitors. Type HL fluids are often recommended for use in machine tool applications where system pressures are limited to 2000 psi or less. They are also recommended for some piston pump applications. This is because some antiwear oils can be aggressive to yellow metal (brass and bronze) and silver alloyed components in piston pumps.

TABLE 7.5
ISO 11158 Designations for Mineral Oil-Based Hydraulic Fluids

Symbol	Classification	Commercial Designations
HH	Non-inhibited refined mineral oils	Straight base oils
HL	Refined mineral oils with improved rust protection and oxidation stability	R&O oils
HM	Oils of the HL type with improved antiwear properties	Antiwear oils
HG	Oils of the HM type with improved slip/stick properties	Hydraulic/way lubes
HV	Oils of the HM type with improved viscosity index properties	Multigrade oils

Type HL fluids often are formulated using a rust inhibitor chemistry that contains succinic acid derivatives [26]. These additives are incompatible with the calcium-based rust inhibitors routinely used in antiwear hydraulic fluids. Comingling HL and HM fluids may result in the formation of precipitates that can cause valve sticking and filter plugging [27].

HM Type HM fluids contain antiwear additives in addition to the rust and oxidation inhibitors found in HL fluids. They are the most widely-used mineral oil-based hydraulic fluids because antiwear additives provide enhanced performance in high-pressure hydraulic applications. While early versions of HM oils lacked the thermal stability that is necessary for satisfactory piston pump performance, modern fluids are able to perform quite well in piston pump applications.

Zinc dialkyldithiophosphate (ZDTP) is the most widely-used antiwear additive for hydraulic applications. Global concerns about the environmental effects of ZDTP have led to the development of zinc-free or ashless antiwear hydraulic fluids. These products utilize sulfur and phosphorus compounds to achieve satisfactory antiwear performance. Thus, a type HM fluid may contain zinc or some other type of antiwear additive chemistry. It is important to note that these fluids may not be compatible with each other [28,29].

HG Type HG fluids contain similar antiwear additives to HM fluids plus additives for improved slip/stick performance. These fluids are used in machine tools that utilize a common reservoir for the hydraulic system and slideway lubrication. Friction modifier additives are incorporated into type HG fluids. These additives absorb on the surfaces of slideways, reducing static friction. This prevents the carriage from surging forward as the machine tool feeds into the work piece, and then hesitating as the drive catches up; thus ensuring smooth machine operation.

HV Type HV fluids contain the same basic chemistry as HM fluids plus a viscosity index (VI) improver. Viscosity index improvers impart multigrade functionality to type HV fluids. While a wide range of polymers may be used for VI enhancement, these additives all function in the same basic manner. At low temperatures, VI improvers have a minimal effect upon fluid viscosity, while at high temperatures they have a thickening effect. This enables the fluid to provide satisfactory performance at a wider operating temperature range [28]. HV fluids have also been shown to improve energy efficiency [29].

7.2.2 FIRE-RESISTANT FLUIDS (ISO 12922 LUBRICANTS)

Fire-resistant hydraulic fluids are used in power generation, mining, die casting, and in military and foundry applications. They may be found in any application where a ruptured hydraulic line presents a potential fire hazard. Fire-resistant hydraulic fluids are formulated with materials that have a lower energy of combustion than mineral oils, such as polyol esters, phosphate esters, and water-glycol solutions. As a result, when ignited, they burn with less heat generation than mineral hydraulic oils. As with mineral hydraulic fluids, the ISO has established a classification system for fire-resistant fluids based upon composition. Table 7.6 provides a list of the ISO 12922 designations for fire-resistant hydraulic fluids.

While power transmission, heat transfer, and lubrication are essential requirements for all types of hydraulic fluids, it may be necessary to compromise these properties in order to accommodate a critical fluid characteristic. This is especially true of fire-resistant hydraulic fluids. Fire-resistant fluids differ from mineral hydraulic fluids in density, compatibility, and lubricating properties. As a result, hydraulic systems are often modified when utilizing a fire-resistant fluid. To optimize the performance of fire-resistant fluids, ISO has published guides for their use [30]. This ISO document details the operational characteristics of fire-resistant fluids and provides suggestions for storage, use, and handling. The four main categories of fire-resistant hydraulic fluids are described below:

TABLE 7.6
ISO 12922 Designations for Fire-Resistant Hydraulic Fluids

Symbol	Classification	Commercial Designations
HFAE	Oil-in-water emulsions containing typically > 80% water	Soluble oils
HFAS	Chemical solutions in water containing typically > 80% water	High water-based fluids
HFB	Water-in-oil emulsions containing approximately 45% water	Invert emulsions
HFC	Water-polymer solutions containing approximately 45% water	Water-glycols
HFDR	Synthetic fluids containing no water and consisting of phosphate esters	Phosphate esters
HFDU	Synthetic fluids containing no water and of other compositions	Polyol esters

HFA HFA fluids contain more than 80% water. These products are sometimes referred to as 95:5 fluids because 5% concentrations are commonly employed. The ISO 6743-4 classification divides HFA into two subcategories: HFAE and HFAS. HFAE fluids are oil-in-water emulsions. HFAS fluids are chemical solutions or blends of selected additives in water. Typically, these products are sold as concentrates and diluted prior to use in service. Because of the high vapor pressure of water, the maximum recommended bulk fluid temperature for HFA fluids is 50°C. At higher tempertures, pump cavitaion may occur. Some HFA fluids have a kinematic viscosity that is comparable to water; approximately 1 centistoke at 20°C. Performance is satisfactory with HFA fluids when suitable components are used but is apt to be poor if used in conventional hydraulic systems. Special precautions are also required in the selection of filter construction materials and plumbing of pump inlets. As a result, it is advisable to work closely with fluid and component suppliers when utilizing HFA fluids. NFPA T2.13.5 provides recommendations for use of this class of fluid [31].

HFB HFB fluids are water-in-oil emulsions consisting of petroleum oil, emulsifiers, selected additives, and water. They are commonly referred to as "invert emulsions". In an invert emulsion, oil is the continuous phase and water is the dispersed phase. The oil phase incorporates rust inhibitors and antiwear additives, while the water phase provides fire resistance. The water content of an HFB fluid is normally in the 43% to 45% range (w/w). When the water content of these fluids decreases due to evaporation, the fire resistance of the invert emulsion deteriorates. In addition, the viscosity is reduced. Consequently, maintaining the water content of these fluids is critical to system safety and performance. Generally, a high shear mixing device is necessary for proper dispersion of water into the invert emulsion. Therefore, replacing a portion of the system volume with fresh HFB is preferable to adding neat water. As with HFA fluids, special precautions are required in the selection of filter construction materials and plumbing of pump inlets. In addition, manufacturers must generally derate pump speeds and operating pressures in systems that use HFB fluids.

HFC HFC fluids are solutions of water, glycols, additives, and thickening agents. They are commonly referred to as "water-glycol" hydraulic fluids. Typically, water-glycol fluids are formulated with diethylene glycol or propylene glycol and a polyalkylene glycol-based thickening agent. The low-molecular weight glycol reduces the vapor pressure of the fluid (relative to water) while high-molecular weight polyalkylene glycol acts as a thickening agent, much like a viscosity index (VI) improver. This combination of thickeners and glycols enhances the lubricating properties of a water–glycol and reduces the likelihood of cavitation erosion. Nonetheless, operating temperatures for water-glycols are limited to a maximum of 50°C because of the effect of temperature on vapor pressure.

Water–glycol fluids are highly alkaline due to the presence of amine-based corrosion inhibitors. As a result, these fluids can attack zinc, cadmium, magnesium, and non-anodized aluminum, forming sticky or gummy residues. Fluorocarbon-based elastomers, such as viton, are also incompatible with amines. Consequently, these materials should be avoided when selecting system components. Special precautions are also required in the selection of filter construction materials and plumbing of pump inlets. Thus, it is advisable to work closely with fluid and component suppliers when utilizing HFC fluids [32].

HFD HFD fluids are non-water-containing fire-resistant fluids. Two types of HFD fluids are described in ISO 6743: HFDR and HFDU. HFDR fluids are composed of phosphate esters; HFDU fluids are typically composed of polyol esters or polyalkylene glycols. Polyol ester-based HFDU fluids are commonly used in industrial hydraulic systems. Phosphate ester and polyalkylene glycols are less common. This is primarily due to the cost and compatibility advantages of polyol esters. Phosphate esters and polyalkylene glycols may provide safety and performance advantages in certain applications. In general, HFD fluids have excellent lubricating properties but are prone to hydrolysis in the presence of water [33,34]. HFD fluids are denser than mineral oil-based fluids, and the chemical compatibility differs widely within this class of fluids [35].

7.2.3 ENVIRONMENTALLY-ACCEPTABLE HYDRAULIC FLUIDS (ISO 15380 LUBRICANTS)

Environmentally-acceptable hydraulic fluids are increasingly employed in hydraulic systems where there is the risk of fluid entering the environment (especially waterways) affecting aquatic and terrestrial life. Some industries that employ environmentally-acceptable hydraulic fluids include agriculture, forestry, construction, hydroelectric power, turf-care, mining and off-shore oil exploration. Most environmentally-acceptable hydraulic fluids exhibit two key environmental characteristics: virtual non-toxicity to aquatic life, and aerobic biodegradability [25]. Organizations such as the Organization for Economic Co-operation and Development (OECD), the Co-ordinating European Council (CEC), and the U.S. Environmental Protection Agency (EPA) have developed standard test methods to determine the toxicity and biodegradability of substances. ASTM has also developed a *Guide for Assessing Biodegradability of Hydraulic Fluids* [38] (ASTM D 6006), and a *Classification of Hydraulic Fluids for Environmental Impact* [39] (ASTM D 6046), based on the above organizations' methods. Utilizing the methodology from these organizations, standard classifications, and performance requirements for environmental fluids have also been established by the ISO. ISO environmental hydraulic fluid classifications are described in Table 7.7.

HETG Type HETG fluids are based on naturally-occurring vegetable oils or triglyceride esters. Without the addition of a thickener, vegetable oils are limited to a narrow viscosity range of between ISO 32 and 46. While HETG fluids biodegrade rapidly, have excellent natural lubricity, and have a natural VI in excess of 200, they are unsuitable for use at high- and low-temperature extremes. This is because they tend to gel at low temperatures and oxidize at high temperatures. The practical temperature limits for HETG fluids ranges from –25°F to 165°F [36].

HEES Type HEES fluids are based on unsaturated to fully-saturated synthetic esters. Common ester chemistries utilized for hydraulic fluids consist of trimethylol propane (TMP) oleates,

---

TABLE 7.7
ISO 15380 Designations for Environmentally-Acceptable Hydraulic Fluids

Symbol	Classification	Commercial Designations
HETG	Vegetable oil types	Vegetable oils, triglycerides, and natural esters
HEES	Synthetic ester types	Polyol esters, neopentylglycols, synthetic adipate esters
HEPG	Polyglycol types	Polyglycols
HEPR	Polyalphaolefin types	Polyalphaolefins (PAO) or synthetic hydrocarbons (SHC)

neopentylglycols, pentaerythritol esters, adipate esters, and complex esters. The synthetic esters provide better performance than HETG type hydraulic fluids, with wider operating temperature ranges, a broader range of ISO viscosity grades, and better oxidation stability, while still maintaining biodegradability [37].

HEPG   Type HEPG fluids are polyethyleneglycols (PEG), which possess good oxidation stability and low-temperature flow characteristics. At molecular weights of 600 – 800, HEPG-type fluids are ecotoxicologically harmless and readily biodegradable (>90% in 28 days) [38]. Some disadvantages of this class of fluids include miscibility with water, incompatibility with mineral oils, and aggressiveness toward some common types of elastomer seal materials.

HEPR   Type HEPR fluids are polyalphaolefins (PAO) or synthesized hydrocarbon (SHC) base fluids, which have significantly better viscometric properties over a wider range of temperatures than mineral base fluids with the same standard viscosity classification. Some low-viscosity PAOs have shown acceptable primary biodegradability, though not as rapid as vegetable or synthetic ester base fluids.

7.3   HYDRAULIC COMPONENT AND SYSTEM STANDARDS

A hydraulic system consists of components such as pumps, valves, conductors, cylinders, motors, filters, heat exchangers, accumulators, and reservoirs arranged to perform a specific task. In order to select the proper components, it is necessary to define machine load conditions and control objectives. These objectives must be implemented while adhering to government safety and environmental regulations. Based upon these requirements, designers will create a circuit diagram using the standard fluid power graphic symbols in ISO 1219-1 [19]. The circuit diagram provides a template from which to create a bill of materials and system models. These models are the fundamental tool for understanding the behavior of the hydraulic system [39]. Safety, functionality, reliability, efficiency, and cost are the key considerations in the design of a hydraulic circuit [40].

7.3.1   SAFETY

Safety is an important consideration in equipment design. Since fluid power systems are capable of generating and storing a very high level of energy, the importance of safety is magnified. Fluid power systems should incorporate an independent and reliable means to safely and verifiably de-energize all components. A breach of a hydraulic system under pressure can not only lead to oil spills, but can result in an explosive release of energy causing injury or death. The Material Safety Data Sheet (MSDS) provides instructions for the safe storage and handling of fluids. As with most chemicals, minimizing exposure through the use of oil-impervious gloves and other personal protection equipment is recommended. One unique risk from hydraulic fluids is high-pressure injection due to a pin-hole leak in a pressurized system. *Report to an emergency room if hydraulic fluid injection occurs—regardless of how insignificant the wound appears. Such injuries must be surgically treated within six hours to avoid permanent tissue damage* [41]. ISO 4413 provides guidelines for incorporating essential health and safety requirements within a hydraulic fluid power system design [42]. The principles of ISO 4413 apply to the design, construction, and modification of systems, taking into account assembly, installation, adjustment, ease of maintenance, and reliability.

7.3.2   RELIABILITY TESTING

Equipment manufacturers and machine users need to know the reliability of fluid power components. Armed with this knowledge, informed decisions can be made regarding capital acquisitions, service intervals, spare part inventories, and design modifications [43]. ISO has been slow to adopt reliability standards for hydraulic equipment. This is largely due to the costs involved in long-term testing. Consequently, reliability standards for hydraulic equipment tend to be based upon Original Equipment Manufacturer (OEM) protocols or national standards. For instance, NFPA T2.12.11 [17] provides guidance for the reliability testing of accumulators, hydraulic cylinders, filters, pumps, motors, and valves. This standard defines reliability as "the probability that a component can perform continuously, without failure, for a specified interval of time or number of cycles when operating under a stated condition." Failure is defined as "the point at which the component reaches a threshold level." The threshold level is not necessarily a failure of the component, but rather a standardized point where failure is defined for statistical purposes. Examples of test parameters for which threshold levels have been established include:

- Dynamic leakage, both internal and external
- Static leakage, both internal and external
- Fatigue failure of the pressure-containing envelope
- Changes in performance characteristics, such as deterioration of flow rate, increase in response time, or change in electrical characteristics.

In addition to defining threshold levels, this document provides detailed procedures for reliability testing. The test recommendations given in the annexes of NFPA T2.12.11 have been established such that the failure mode or failure mechanism is the same as might typically occur in fielded equipment. For example, it has been determined that excessive rod seal leakage is the most common failure mode for a hydraulic cylinder. The test circuit and operating conditions for the reliability testing of cylinders provided in Annex A of NFPA T2.12.11 are designed to specifically evaluate rod seal leakage.

7.3.3 Component Specifications

Dimensional interchange standards exist for nearly all hydraulic system components. Dimensional standardization is required because without standard fluid port sizes, fittings, threads, O-rings, and sealing methods it would be difficult to assemble hydraulic systems that do not leak. Metric port dimension and design standards for hydraulic components are defined in ISO 6149 [6]. O-ring diameters, cross sections, tolerances, and designation codes are defined in ISO 3601-1 [9]. Metallic tube connections and elastomeric hose fittings are described in ISO 8434 and ISO 12151 respectively [7,8]. Many fluid power components are also manufactured to Society of Automotive Engineers (SAE) dimensional and performance specifications. Standards for hydraulic pumps, valve mounting surfaces, hoses, cylinders, and filter assemblies not only simplify the assembly process, but also allow the system designer to interchange components from competing manufacturers. From a product cost standpoint, the interchangeability of components is obviously advantageous to equipment builders [44].

In addition to dimensional interchange standards, NFPA, SAE, and ISO have established standardized component performance requirements. These standards provide detailed test procedures and the format for reporting results. This helps the system designer compare and select components from competing manufacturers. One of the most fundamental requirements is to determine the rated fatigue pressure of a hydraulic component. As with hydraulic reliability testing, achieving international consensus on the method for determining the rated fatigue pressure for hydraulic components has also proven difficult. Consequently, standards in this area tend to be based upon OEM and individual country requirements.

7.3.4 Accelerated Endurance Tests

Traditionally, the fluid power industry has used accelerated testing to evaluate the endurance of components. Accelerated testing exposes components to conditions that are more severe (higher

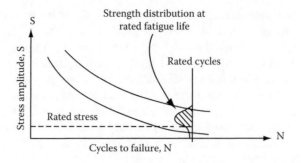

FIGURE 7.1   S-N curve showing relation between fatigue stress and component life.

pressure) than those for which the component is rated. This is often necessary to keep the test time to a reasonable length. The failure mode in an accelerated test can differ from the mode of failure under normal operating conditions. Nonetheless, accelerated endurance tests provide a useful means for comparing hydraulic components and establishing their pressure-fatigue ratings.

The endurance life of fluid power components is often limited by metal fatigue. Metal fatigue is localized structural damage that occurs as the result of cyclic loading below the yield strength of the material. The mechanism of fatigue failure is a progressive cracking that leads to fracture [45]. Since cyclic internal pressurization is inherent in fluid power applications, pressure fatigue tests are required to ensure safe and reliable performance.

Several standard methods are used to establish the rated fatigue pressure of fluid power components. The rated fatigue pressure is the maximum pressure that a component has been verified to sustain for the rated life without failure, with a known probability. The procedure provided in NFPA T2.6.1 [46] is used to establish the fatigue and burst pressure ratings of metal fluid power components. This procedure employs a statistical methodology to evaluate metal fatigue. Since it is not possible to determine the fatigue life from single point, a static pressure burst test is combined with a cyclical pressure test to determine the rated fatigue pressure. As shown in Figure 7.1, at each level of cyclic life (N) in a fatigue diagram there is a distribution of the fatigue strength (S). This value—the coefficient of variation of the fatigue strength (k_o) for a specific metal—is used to determine the number of specimens that must be tested, as well as the cyclic test pressure. As shown in Table 7.8, the coefficient of variation for plain carbon steel is lower than that of low alloy steel. As a result, low alloy steel must be subjected to more severe fatigue test conditions than plain carbon steel in order to obtain an equal fatigue pressure rating.

A different procedure is used to validate the rated working pressure of hydraulic hoses. Many of the hoses used in hydraulic systems are manufactured to dimensional and performance requirements

TABLE 7.8
Coefficient of Variation for Metal Specimens in Fatigue Testing

Metal	k_o
Plain carbon steel	0.08
Stainless steel	0.09
Nickel steel	0.10
Tool steel	0.10
Titanium	0.12
Low alloy steel	0.14

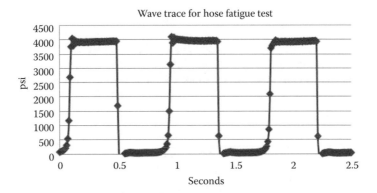

FIGURE 7.2 Pressure trace for fatigue testing of hoses.

of 100R series hoses specified in SAE J517 [51]. There are 19 different classifications of 100R series hoses. For instance, SAE 100R1 hoses are steel wire-reinforced rubber-covered hydraulic hoses that are designed for use with petroleum-based hydraulic fluids in a range of −40 to +100C. SAE 100R2 hoses have the same general description but they have higher working pressure ratings. The procedures for testing and evaluating SAE 100R series hoses are defined in SAE Standard J343, which is equivalent to the ISO 6605 standard [47,48]. SAE J343 contains 13 test procedures including change in length, proof, burst, and impulse tests. Proof, burst, and pressure impulse fatigue test methods are used to validate the pressure rating of 100R series hoses. In a proof test, hoses are subject to a static pressure that is double the working pressure of the hose for a period of one minute. In a burst test, the internal pressure is gradually increased until the hose fails and the maximum pressure value is recorded. In the pressure fatigue test, hoses are subjected to cyclic internal pressurization. A typical pressure wave trace is shown in Figure 7.2. Four hose specimens are tested for a predetermined number of pressure cycles or until they fail. For most 100R series hoses, the rated pressure is half that of the proof test pressure and one-quarter that of the burst test pressure in order to provide a safety factor.

7.3.5 Pump and Motor Testing

The fundamental precept in fluid power is that pumps generate flow while resistive loads from cylinders, rotary actuators, and components within the system generate pressure. In most hydraulic applications, positive displacement pumps are used. This is because, relative to centrifugal pumps, positive displacement pump flow is independent of system pressure. Thus, the task of the hydraulic pump is to convert rotating mechanical shaft power into fluid power that can be used downstream of the pump. Due to friction, fluid shear, and internal pump flow losses (leakage), this energy conversion process is less than 100% efficient. Overall pump efficiency η_o, is defined as the useful output power divided by the supplied input power [49]. This definition may be expressed mathematically, as shown in Equation 7.1:

$$\eta_o = \frac{P_d Q_d}{T\omega} \tag{7.1}$$

where P_d is the discharge pressure of the pump; Q_d is the volumetric flow rate measured at the discharge port of the pump; T is the input torque on the pump shaft; and ω is the angular velocity of the pump shaft. The overall efficiency of a pump consists of two components—volumetric efficiency η_v and mechanical or torque efficiency η_t. This relationship is given by Equation 7.2:

$$\eta_o = \eta_v \eta_t \tag{7.2}$$

The efficiency equations for hydraulic motors are similar to those of hydraulic pumps because motors convert fluid power to rotary motion. Again, overall motor efficiency η_{om}, is defined as the useful output power divided by the supplied input power. This relationship is given by Equation 7.3:

$$\eta_{om} = \frac{T\omega}{P_i Q_i} \tag{7.3}$$

where T is the output torque on the motor shaft; ω is the angular velocity of the shaft; P_i is the differential pressure across the motor; and Q_i is the volumetric flow rate into the motor. As in the pump, overall efficiency is related to mechanical and volumetric efficiency via Equation 7.2. Pump and motor efficiency are important in fluid power systems because efficiency must be taken into consideration when sizing the prime mover. In addition, efficiency has an obvious impact upon energy costs.

Several ISO methods have been developed for the purpose of determining hydraulic pump and motor efficiency. ISO 4409 describes methods for testing and presenting the steady-state performance of positive displacement pumps and motors [13]. In this test method, fluid viscosity, density, and temperature are carefully monitored as the efficiency of power transformation is measured. For pumps tested at one constant rotational frequency, the effective outlet flow rate and efficiency curves are plotted over a range of measured outlet pressures. For pumps tested at several different constant rotational frequencies, a family of curves is plotted depicting effective outlet flow rate and efficiency as a function of rotational frequency and measured outlet pressure. Motor output torque and efficiency values are plotted in opposition to rotational frequency. This enables the performance of different pumps and motors to be compared for system design and selection purposes.

In order to measure volumetric efficiency using the ISO 4409 method, it is necessary to determine the volume of fluid displaced by the pump or motor per shaft revolution. Due to inaccuracies in the manufacturing process and deflection under load, volumetric displacement may vary from published specifications. ISO 8426 provides a method for deriving the volumetric capacity of pumps and motors [50]. In this standard test method, pump and motor displacement are derived from the measured pump outlet or motor inlet flow rates at a number of pressures. Using a least squares approximation, the derived capacity is determined from the zero pressure intercept. This value thus becomes the basis for determining volumetric efficiency under steady-state conditions.

Hydraulic motors, such as those found in off-highway, forestry, and construction equipment, must produce high levels of torque under startup conditions as well as when engaging heavy payloads. ISO 4392-1 and 4392-2 were developed to evaluate the low-speed performance characteristics of motors. In the ISO 4392-1 method, the torque output of the hydraulic motor is measured as it rotates at 1 rpm while maintaining a constant differential pressure across the motor [14]. The results of this test determine the variation in torque output at constant speed and pressure. In the ISO 4392-2 method, the pressure required to start a hydraulic motor against a fixed torque load as a function of shaft position is determined [51]. The results of this testing determines the motor's true starting torque efficiency and the corresponding differential pressure required to start the load. These performance characteristics are important because they determine the minimum displacement (size) and operating pressure of the hydraulic system. Typical results from startability testing of axial and Geroler motors are shown in Figure 7.3.

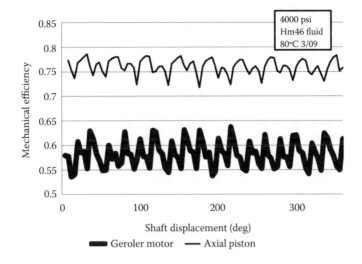

FIGURE 7.3 Mechanical efficiency of hydraulic motors during startability test.

7.4 CONCLUSIONS

Standard specifications for hydraulic fluids and components serve an important function in the fluid power industry. These standards validate the safety, durability, compatibility, cleanliness, and functionality of hydraulic fluid media, components, and systems. Standardized bench tests serve as a cost-effective indicator of fluid response to varying conditions of temperature, pressure, contamination, and so forth. A significantly higher number of variables influence component and system performance. Standardized test methods will continue to evolve as more sophisticated models and techniques are developed to predict performance. The challenge for the standardization community is to develop test protocols that are truly representative of fluid power system conditions.

REFERENCES

1. ISO 15029-1, Petroleum and related products – Determination of spray ignition characteristics of fire-resistant fluids – Part 1: Spray flame persistence – Hollow cone nozzle method.
2. ISO 2719, Petroleum products and lubricants – Determination of flash point – Pensky – Martens closed cup method.
3. ISO 6072, Hydraulic fluid power – Compatibility between elastomeric materials and fluids.
4. ISO 4406, Hydraulic fluid power – Fluids – Method for coding the level of contamination by solid particles.
5. ASTM D6973, Standard Test Method for Indicating Wear Characteristics of Petroleum Hydraulic Fluids in a High Pressure Constant Volume Vane Pump.
6. ISO 6149, Connections for hydraulic fluid power and general use.
7. ISO 8434-1, Metallic tube connections for fluid power and general use – Part 1: 24 degree cone connectors.
8. ISO 12151-2, Connections for hydraulic fluid power and general use – Hose fittings Part 2: Hose fittings with ISO 8434-1 and ISO 8434-4 24 degree cone – connector ends with O-rings.
9. ISO 3601-1, Fluid power systems – O-rings – inside diameters, cross-sections, tolerances and designation codes.
10. SAE J744, Hydraulic Pump and Motor Mounting and Drive Dimensions Standard.
11. NFPA T3.5.1, Hydraulic fluid power – Valves – Mounting surfaces.
12. ISO 6099, Fluid power systems and components – Cylinders – Identification code for mounting dimensions and mounting types.

13. ISO 4409, Hydraulic fluid power – Positive displacement pumps, motors and integral transmissions – Methods of testing and presenting basic steady state performance.
14. ISO 4392-1, Hydraulic fluid power – Determination of characteristics of motors – Part 1: At constant low speed and constant pressure.
15. NFPA T3.5.16, Hydraulic fluid power – Pressure compensated flow control valves – Method for measuring and reporting regulating characteristics.
16. NFPA T2.6.1, Fluid power components – Method for verifying the fatigue and establishing the burst pressure ratings of the pressure containing envelope of a metal fluid power component.
17. NFPA T2.12.11, Hydraulic fluid power components – Assessment of reliability by testing.
18. ISO 5598, Fluid power systems and components – Vocabulary.
19. ISO 1219-1 Fluid power systems and components – Graphic symbols and circuit diagrams – Part 1: Graphic symbols.
20. Bishop, M. *Makers of Modern Thought*, American Heritage Publishing, NY, 1972, p. 140.
21. ISO 6743-99, Lubricants, industrial oils and related products (Class L) — Classification – Part 99: General.
22. ISO 6743-4, Lubricants, industrial oils and related products (Class L) — Classification – Part 4: Family H (Hydraulic systems).
23. ISO 11158, Lubricants, industrial oils and related products (Class L) — Family H (Hydraulic systems) — Specifications for categories HH, HL, HM, HR, HV and HG.
24. ISO 12922, Lubricants, industrial oils and related products (Class L) — Family H (Hydraulic systems) — Specifications for categories HFAE, HFAS, HFB, HFC, HFDR, and HFDU.
25. ISO 15380 Lubricants, industrial oils and related products (Class L) — Family H (Hydraulic systems) – Specifications for categories HETG, HEES, HEPG, and HEPR.
26. Rizvi, S.Q.A., Lubricant Additives and Their Functions, *ASM Handbook*, 10th ed. *Vol. 18, Friction Lubrication, and Wear Technology*, ASM International; Materials Park, OH, 1992, pp. 98–112.
27. Sharma, S.K., Snyder Jr., C.E., Gschwender, L.J., Liang, J.C. and Schreiber, B.F., Stuck Servo Valves in Aircraft Hydraulic Systems, *Lubrication Engineering*, vol. 55, no. 7, July, 1999, p. 27-32.
28. NFPA T2.13.13 Recommended practice – Hydraulic fluid power – Fluids – Viscosity selection criteria for hydraulic motors and pumps.
29. Michael, P.W. and Herzog, S.N. *Hydraulic Fluid Selection for Improved Fuel Economy*, Proceedings of the STLE/ASME International Joint Tribology Conference, IJTC2008-71305, October 20-22, 2008, Miami, Florida.
30. ISO 7745, Hydraulic fluid power – Fire resistant fluids – Guidelines for use.
31. NFPA T2.13.5 Hydraulic Fluid Power – Hydraulic Systems – Practice for the use of high water content fluids.
32. NFPA T2.13.1, Recommended Practice – Hydraulic Fluid Power – Use of Fire Resistant Fluids in Industrial Systems.
33. Wachter, D.A., Bishop, R.J., McDaniels, R.L. and Totten, G.E. Water–Glycol Hydraulic Fluid Performance Monitoring: Fluid Performance and Analysis Strategy, *SAE Tech. Paper Series*, Paper 952155, 1995.
34. Murrrenhoff, H., Gohler O-C. and Meindorf, T. Hydraulic Fluids in *Handbook of Lubrication and Tribology Volume 1*, CRC Taylor & Francis, Boca Raton, 2006, p. 11-6.
35. Givens, W.A. and Michael, P.W. Hydraulic Fluids in *Fuels and Lubricants Handbook: Technology, Properties, Performance and Testing*, ASTM, Conshohocken, 2003, p. 377.
36. NFPA T2.13.14, Recommended Practice – Hydraulic Fluid Power – Use of environmentally Acceptable Fluids.
37. Eastwood, J., Swallow, A. and Colmery, S., *Selection Criteria of Esters in Environmentally Acceptable Hydraulic Fluids*, Proceedings of the 50th National Conference on Fluid Power, NCFP I05-4.2, Las Vegas, 2005.
38. Matlock, P., Brown, W. and Clinton, N. Polyalkylene Glycols, *Synthetic Lubricants and High Performance Functional Fluids*, 2nd ed., ed. by Rudnick and Shubkin, Marcel Dekker, NY, 1999, p. 190.
39. Manring, N.D. *Hydraulic Control Systems*, John Wiley & Sons, Hoboken, NJ, 2005, p.103.
40. Esposito, A. *Fluid Power with Applications*, 7th ed. Prentice Hall, Englewood Cliffs, NJ, 2008.
41. *The Lethal Strike*, DVD, Fluid Power Safety Institute.
42. ISO 4413, Hydraulic fluid power – General rules and safety requirements for systems and their components.
43. Berninger, J.F. *The Initiative to Measure Reliability of Fluid Power Products*, Proceedings of the 50th National Conference on Fluid Power, NCFP 105 – 3.2, Las Vegas, 2005.
44. Rubczynski, W. *One Good Turn: A Natural History of the Screwdriver and Screw*, NY: Scribner, 2000.

45. Morrow, H.W. and Kokernak, K.P. *Statics and Strength of Materials*, 4th ed., Prentice Hall, 2001, p. 383.
46. NFPA T2.6.1, Fluid power components – Method for verifying the fatigue and establishing the burst pressure ratings of the pressure containing envelope of a metal fluid power component.
47. SAE J343, Test and Test Procedures for SAE 100R Series Hydraulic Hose and Hose Assemblies.
48. ISO 6605, Hydraulic fluid power – Hoses and hose assemblies – Test methods.
49. Manring, N.D. *Hydraulic Control Systems*, John Wiley & Sons, Hoboken, NJ, 2005, p. 268.
50. ISO 8426, Hydraulic fluid power – Positive displacement pumps and motors –Determination of derived capacity.
51. ISO 4392-2, Hydraulic fluid power – Determination of characteristics of motors – Part 2: Startability.

8 Biodegradable Hydraulic Fluids

In-Sik Rhee*

CONTENTS

8.1 INTRODUCTION

Although mineral oils have traditionally been the most commonly-used hydraulic fluids in the fluid power industry, they are being subjected to ever-increasing controls, particularly because of the increasingly stringent governmental regulations regarding the impact of hydraulic fluid spill and fluid leakage on the environment [1]. Improper disposal, even if it is incidental, may be the source of large penalties or even litigation [2].

European studies have identified hydraulic fluid leakage as one of the primary sources of groundwater contamination [3]. This has been followed by a worldwide effort to identify hydraulic fluids that will exhibit reduced environmental and toxicological impact upon incidental contact with the environment [4,5]. (*Note*: The term "environmental impact" includes biodegradation and persistence, as well as toxicity.) Additional impetus for the use of biodegradable lubricants has been provided by the use of national environmental labeling criteria. Figure 8.1 illustrates various examples of ecolabels [6].

There are a number of national environmental standards and labeling procedures being developed. The most well known are summarized in Table 8.1 [7]. One of the first and most stringent environmental labeling procedures is the German "Blue Angel" label [8]. The philosophy of this labeling procedure is to avoid consumer products, including hydraulic fluids that are hazardous to the environment. Although the Blue Angel labeling requirements have not been finalized, an approved hydraulic fluid will require greater than 80% biodegradability in twenty-one days by the CEC-L-33-A-93 test, or greater than 70% biodegradation by the Modified Sturm Test.

Details of these testing procedures will be provided later in this chapter. All components must be non-polluting and conform to Water Hazard Class 0 or 1 [9]. The components of this coding procedure are as follows [3]:

---

* This chapter is a revision of the chapter titled "Ecological Compatibility" by Gary H. Kling, Dwayne E. Tharp, George E. Totten and Glenn M. Webster from the *Handbook of Hydraulic Fluid Technology*, 1st Edition.

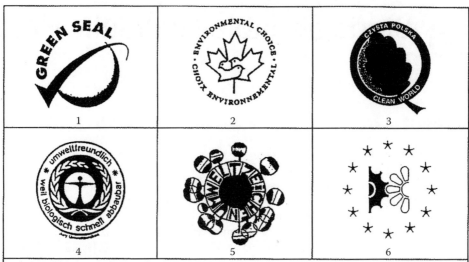

1	2	3
4	5	6

1. The United States privately-licensed Green Seal
2. Canada's Bilingual Maple Leaf
3. The "Clean Poland" logo
4. The German government's Blue Angel Symbol
5. Austria's logo, over-designed by the noted artist Hunderwasser
6. The European Union's Gear-and-Bloom Symbol.

FIGURE 8.1　Illustration of different classes of molecular structures present in mineral oil. (From L. Tocci, "Mother Nature on Lubes: No Simple Choices," *Lubes Greases,* 1995, 1(5), pp. 16–19. With permission.)

CEC　Coordinating European Council.
L　Lubricant testing.
33　These two digits refer to the order of the procedure.
T or A　T indicates "tentative" procedures. Correlation programs have been completed. "A" indicates "accepted."
93　Indicates the year in which the procedure was agreed upon and accorded status by the Council.

Figure 8.2 shows the structure of the CEC European Industry Organization, which is composed of the following:

TABLE 8.1
Environmental Criteria for Various Ecolabeling Schemes

Country	Scheme	Environmental Criteria
Germany	Blue Angel	Biod-Ecotox
Canada	Environmental Choice	Recycling
Japan	Eco-Mark	Biod-Ecotox
France	N-F Environment	Biod-Ecotox
Scandinavia	White Swan	Biod-Ecotox
United States	Green Seal and Green Cross (privately-run schemes)	Recycling
India	Ecomark	Biod-Ecotox

Source: Mang, T. "Environmentally Friendly Biodegradable Lube Base Oils—Technical and Environmental Trends in the European Market," in *Adv. Prod. Appl. Lube Base Stocks, Proc. Int. Symp.*, 1994, pp. 66–80. With permission.

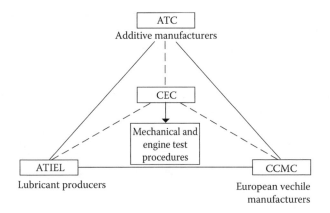

FIGURE 8.2   Structure of CEC european industry organization.

ATIEL   Concerned with technical matter related to the lubricating oil industry in
               Europe.
ATC   Concerned with lubricant testing.
CCMC   Operates on a wide front of matters and is particularly interested in the unifor-
               mity in technical regulations.

Another program currently in development is the Environmental Choice Program of Canada [9]. This program was initially developed to provide guidance for the disposal or reclamation of used industrial oil, as indicated in Table 8.1 [10]. Interestingly, the term "EcoLogo" is used instead of the term "environmentally friendly" because it was not intended to suggest that products with the EcoLogo were perfectly harmless to the environment. Very few products have *no* impact on the environment [10]. However, the Canadian Environmental Choice program is being used to provide criteria for the "environmental friendliness" of industrial oils such as chainsaw lubricants, hydraulic fluids, and others. *Note*: The term "environmental impact."

In the United States, work is underway by the ASTM D.02 N.03 Subcommittee to establish a standard for the classification of environmentally friendly hydraulic fluids. However, instead of "environmentally friendly," these fluids will be classified as "eco-evaluated" (EE) [9,11]. Also, the ASTM D.02 N.03 Subcommittee is working to establish standards for the classification and testing of eco-evaluated fluids.

In addition to base stock, a hydraulic fluid typically contains additives. The additives are used to enhance the fluid properties, which include antioxidant, anti-wear, reduced foaming, and improved air-release properties. The presence of these additives affect hydraulic fluid bio-degradation and toxicological properties [12–15]; therefore, additive selection is of paramount importance [16].

Thus far, the most commonly-cited base stocks used for the formulation of "environmentally friendly" hydraulic fluids are either vegetable oils or synthetic esters [7,9,17]. The most common vegetable oils that have been identified for hydraulic fluid formulations are based on soybean oil [18–20], rapeseed oil [21], and high oleic sunflower oil [22]. Canola oil is also used. The amount of unsaturation in a vegetable oil is climate-dependent [22]. For example, rapeseed grown in colder northern climates contains less unsaturation (canola oil) than when grown in warmer southern cli-mates. This is important because the level of unsaturation affects physical properties, especially oxidative stability. For this reason, canola oil is one of the preferred vegetable oils for hydraulic fluid formulation [22].

Work by Rhee et al. has shown that synthetic esters exhibit a number of advantages over veg-etable oils, such as a broader pour point temperature range, typically −40°C to 150°C versus veg-etable-oil-derived fluids, which typically exhibit a temperature range of −10°C to 90°C [23]. This limited range of operational temperatures for vegetable oils may render them unsuitable for some cold-weather applications. Other workers have reported that, in addition to a broader operational temperature range, synthetic esters provide substantial improvement in thermal/oxidative stability relative to vegetable-oil-based fluids [24,25].

The use of various vegetable-oil-derived hydraulic fluids has been published [21,26–29]. Typically, either vane or piston pumps were used for performance testing. Although the antiwear results were generally good, these tests showed that vegetable oils exhibit poorer oxidative stability than mineral-oil-based hydraulic fluids. In view of the significant market pressures for the availability of "environmentally friendly" hydraulic fluids, most hydraulic pump manufacturers, and many equipment suppliers, have developed guidelines for the selection and use of these fluids [30–34]. Because of the variability of the performance of vegetable-oil-based hydraulic fluids that are currently available, some manufacturers appear to have developed their own privately-labeled fluids to provide greater quality assurance [33].

Because of their favorable low-temperature properties (pour point of −63°C), water solubility, and fire resistance, water-glycol hydraulic fluid compositions provide a potential alternative to vegetable oils and synthetic esters as biodegradable hydraulic fluids [35,36]. Unfortunately, although there are numerous references describing the various biodegradability properties of vegetable oils and synthetic esters, there are considerably fewer references for the biodegradability properties of water-glycols using OECD GLP Principles (OECD, 1992a) [37]. In some cases, it is not clear how the reported data was obtained or if it was previously published in other sources.

This chapter will provide a selected overview of biodegradation and toxicity data published to date for various classes of hydraulic fluids. The general subject of biodegradation and testing procedure will be discussed. Selected examples of published comparative test data will be provided.

8.2 DISCUSION

8.2.1 BIODEGRADABLE FLUIDS

The bio-based hydraulic fluids (BHFs) are currently formulated with renewable products such as rapeseed, sunflower, corn, soybean, and canola or synthetic ester. All bio-based fluids are biodegradable products. These types of fluids are currently considered less toxic and more biodegradable than conventional hydraulic fluids. The U.S. Department of Agriculture's (USDA) bio-based product guideline defines exactly what products—and how much concentration of the renewable product associated with the final product—would be considered as a bio-based product [62]. Some petroleum-based fluids such as Glycol-based fluids and 2 cSt Polyalpaolefin (PAO) are also considered as biodegradable products. To evaluate biodegradable fluids, many studies have been conducted in comparison with the petroleum-based fluids [22,63].

Vegetable oils have excellent lubrication qualities, are nontoxic, and are biodegradable. They are made from renewable resources such as rapeseed, sunflower, corn, canola and soybean, and are much less expensive than synthetic fluids. Their chemical structures are triglycerides in which a variety of saturated, monounsaturated or polyunsaturated fatty acids are esterified to a glycerol backbone. The physical properties of a vegetable oil depend on the nature of its fatty acid composition. These oils tend to oxidize at temperatures above 90°C and have a shorter life when compared with conventional petroleum-based fluids. Also, they have limited low-temperature operability (−15°C). This significantly affects the outdoor mobility applications where hydraulic systems may sit for an extended period at sub-zero temperatures.

Synthetic esters, mainly based on trimethylopropane, polyol ester, and pentaerythritol, are regarded as the best among the biodegradable base fluids. The biodegradability of these oils is comparable to vegetable oils and their lubrication properties are very similar to mineral oils. The advantages of these oils include excellent fluidity, more than adequate low-temperature operability, and excellent stability. Because of these, they provide wide operational operability under varying ambient temperatures (−54 to 150°C), and have long shelf and service lives. On the other hand, the cost of synthetic esters is much higher than those of mineral oils. Their differences are summarized in Table 8.2 [63].

Polyglycol oils are the oldest biodegradable fluids and have been used in construction machinery, the vicinity of ground water, and the food-processing industry. Currently, the usage of these fluids is apt to decline due to their tendency to be aquatically toxic when admixed with lubricating additives and their incompatibility with mineral oils and other materials such as elastomer.

BHFs also require additive ingredients to enhance their performance. Antioxidants, corrosion inhibitors, and pour point stabilizers can improve the lubricating properties of some vegetable oils, such as rapeseed oil and synthetic esters. However, the use of conventional additives in BHFs may pose potential problems for the fluid's biodegradability and ecotoxicological properties due to the toxicity of the chemicals used in the additive compounds. Thus, domestic additive manufacturers are also investing in the development of BHF additives which are compatible with BHFs. Some environmentally acceptable additives, such as sulfur-carriers, have been developed and are currently available in the domestic market.

A tabulation of various vegetable-based and synthetic ester-based hydraulic fluids that are available in the United States is provided in Table 8.3 [23,64]. An international listing is provided in Table 8.4 [38]. In the biodegradable hydraulic fluid marketplace, rapeseed-oil-based hydraulic fluid compositions may be referred to as "European fluids", and canola-oil-based hydraulic fluids may be referred to as "Canadian fluids" [19,20,26,39,40]. There is a movement underway to make soybean-oil-based fluids the preferred American biodegradable vegetable-oil-based fluid composition [40].

Another class of hydraulic fluids to be discussed here is water–glycol fluids, also known as hydrolubes [41–43]. Water–glycol hydraulic fluids are part of a larger class of fire-resistant fluids. Additional fluids classified as "fire-resistant" include invert emulsions, polyol esters, and phosphate esters. Whereas polyol esters and phosphate esters derive their fire-resistance properties by either chemical structure or additives, or both, water-glycols and other water-containing fluids derive their fire-resistance properties from the presence of water, also known as a "snuffer," in the fluid formulation.

A summary of the functions of the various components of a "typical" water–glycol hydraulic fluid is provided in Table 8.5. In addition to the components shown, other additives may be used as needed, such as dyes for leak detection, antifoam, air release agents, and so forth.

TABLE 8.2
Comparison of Base Fluids

	Mineral Oils	Vegetable Oils	Biodegradable Synthetic Esters	Poly–Glycol (HEPG)
Biodegradability ASTM D 5864, %	10–40	40–80	30–80	40–80
Viscosity, cSt	15–150	32–68	20–300	10–100
Viscosity Index	90–100	100–250	120–220	100–200
Pour points,°C	−54 to −15	−20 to 10	−60 to −20	−40 to 20
Compatibility with Mineral oils	−	Good	Good	Not miscible
Oxidation stability	Good	Poor to Good	Poor to Good	Poor
Service life	2 years	6 months to 1 year	3 years	2 years
Relative cost	1	2 to 3	4 to 5	2 to 4

TABLE 8.3
Domestic "Environmentally Acceptable" Hydraulic Fluids

Product Name	Base Stock	Description	Company Name	Address
EL 146	Seed oil and mineral oil	Readily biodegradable and nontoxic ISO 46 hydraulic oil for use in heavy-duty hydraulic systems	EarthRight Technologies	33307 Curtis Blvd. Eastlake, OH 44095
Plantohyd 40	Rapeseed & sulfurized fatty vegetable oil	Universal, multigrade hydraulic oil for agricultural, forestry, and construction machinery	Fuchs Metal Lubricants Company	Metal Lubricants Co. 17050 Lathrop Ave. Harvey, IL 60426
Functional 9403071	Rapeseed	Biodegradable, corrosion-inhibited, antiwear hydraulic oils for mobile, industrial, and marine hydraulic systems	Greenoco Functional Products, Inc.	Functional Products, Inc. 24000 Mercantile Rd, Unit 5 Cleveland, OH 44122
Greenwood Hydraulic Fluids	Vegetable; rapeseed	Biodegradable, corrosion-inhibited, antiwear hydraulic oils for mobile, industrial, and marine hydraulic systems	Greenoco Functional Products, Inc.	The Green Oil Company P.O. Box 577 Blue Bell, PA 19422
HYD271, 272	Vegetable; canola	Biodegradable, antiwear, corrosion-inhibited, and oxidation-inhibited hydraulic oils for industrial hydraulic systems	International Lubricants, Inc.	7930 Occidental South Seattle, WA 98108
Mobile EAL 244H	Vegetable; rapeseed	Biodegradable, antiwear, corrosion-productive hydraulic oil for moderate or severe operating conditions	Mobil Oil	3225 Gallows Rd., RM 5W806 Fairfax, VA 22037-0001
Mobil Xrl 1711-78	Synthetic ester	Antiwear, corrosion-protective hydraulic oil	Mobil Oil	3225 Gallows Rd., RM 5W806 Fairfax, VA 22037-0001
Quintolubric 822-220, 330, 450	Polyol ester	Fire-resistant hydraulic fluid to replace phosphate esters	Quaker Chemical Corp.	Conshohocken, PA 19428-0809
HVO-46 Hydraulic Vegetable Oil	Vegetable	For hydraulic systems that require both antiwear and rust and oxidation properties	Renewable Lubricants, Inc.	476 Griggy Rd., P.O. Box 474 Hartville, OH 4432
Royco 3100, 3046, RTJ27, RTJ41	Polyol ester	Biodegradable, antiwear hydraulic fluids	Royal Lubricants	P.O. Box 518, Merry Lane East Hanover, NJ 07936
OS 107086	Sunflower oil	Based on Sunyl PF 311; sunflower oil with alkylated phenol	SVO Specialty Products, Inc.	35585-B Curtis Blvd. Eastlake, OH 44095
OS 106575	Sunflower oil	Based on Sunyl PF 311; sunflower oil with alkylated phenol	SVO specialty Products, Inc.	35585-B Curtis Blvd. Eastlake, OH 44095
Biostar Hydraulic 46, Code 1616	Rapeseed; ethyl hexyl oleate, C18 fatty acids	Environmentally friendly, zinc-free, antiwear oil for high-pressure hydraulic equipment	Texaco	Texaco Lubricants Co. N.A. P.O. Box 4427 Houston, TX 77210-4427
Synstar Hydraulic 46, Code 2073	Trimethyl-propane (TMP) and Penta-erythritol ester blend, BHT	Environmentally friendly, zinc-free, antiwear oil for high-pressure hydraulic equipment	Texaco	Texaco Lubricants Co. N.A. P.O. Box 4427 Houston, TX 77210-4427

TABLE 8.3   (Continued)
Domestic "Environmentally Acceptable" Hydraulic Fluids

Product Name	Base Stock	Description	Company Name	Address
Terra-Lube ECO 2000	Natural esters, triglycerides	Biodegradable and nontoxic natural-ester-based universal hydraulic fluid designed for use in all equipment	CoChem Inc.	7555 Bessemer Ave. Cleveland, OH 44127
Calgene Q1093, 1094, 1095	Rapeseed, canola oil	Vegetable-oil-based VI Index, wear and oxidation additives	Calgene Chemical, Inc.	7247 N. Central Park Ave. Skokie, IL 60076-4093
Clark Cone oil	Cone oil	Environmentally friendly product	CoChem, Inc.	7555 Bessemer Ave. Cleveland, OH 44127
Cognis Proeco Eaf 422 LL	Polyol ester	Bio-based, MIL-PRF-32073	Cognis	5150 Estecreek Drive Cincinnati, OH 45232 P.O. Box 5700. MS 66
Novus 100 ISO 46	Canola	Bio-based, MIL-PRF-32073	Cargill	Minneapolis, MN 55440 13626 Pear Wood Drive
Hydro Safe ISO VG 46M4	Canola	Bio-based, MIL-PRF-32073	Hydro Safe Oil	Dewitt, MI 48820 P.O. Box 5700. MS 66
Novus 100 ISO 68	Canola	Bio-based, MIL-PRF-32073	Cargill	Minneapolis, MN 55440 8282 Bavaria Road
Biodegradable Hydraulic Fluid Grade 4	Canola	Bio-based, MIL-PRF-32073	Functional Product	Macedonia, OH 44056 13626 Pear Wood Drive
Hydro Safe ISO VG 22M2	Polyol ester	Bio-based, MIL-PRF-32073	Hydro Safe Oil	Dewitt, MI 48820 13626 Pear Wood Drive
Hydro Safe ISO VG68M5	Canola	Bio-based, MIL-PRF-32073	Hydro Safe Oil	Dewitt, MI 48820 35585 Curits Blvd.
Envirologic 146	Canola	Bio-based, MIL-PRF-32073	Terresolve	East Lake, OH 44095 13626 Pear Wood Drive
Hydro Safe ISO VG32M3	Canola	Bio-based. MIL-PRF-32073	Hydro Safe Oil	Dewitt, MI 48820 13626 Pear Wood Drive
Hydro Safe ISO 15M1	Synthetic ester	Bio-based. MIL-PRF-32073	Hydro Safe Oil	Dewitt, MI 48820

Source: In-Sik Rhee, Velez, C. and Von Bernewitz, K. "Evaluation of Environmentally Acceptable Hydraulic Fluids," TARDEC Technical Report No. 13640, U.S. Army Tank-Automotive Command Research, Development and Engineering Center, Warren, MI, March 1995.

Note: The term "environmentally acceptable" (or "environmentally friendly") is not a well-defined term. There is no universally-accepted criterion for "environmental acceptability" at the present time, although various national standards are in the process of being written. Therefore, the term "environmentally acceptable" is essentially a marketing term with little or no technical support. In fact, "environmentally acceptable" is often used synonymously with "vegetable oil."

8.2.2   BIODEGRADATION

In this section, the meaning of biodegradability will be discussed. An overview of the test methods used to determine biodegradability will be provided. This discussion will be followed by an overview of published data on biodegradable hydraulic fluids.

TABLE 8.4
Environmentally Acceptable Hydraulic Fluid HETG, HEPG, and HEE for Axial Piston Units

Type of Fluid ISO Viscosity Class Maker	Vegetable-Oil-Based Hydraulic Fluids, HETG				Synthetic Hydraulic Fluids Based on Polyglycol, HEPG				Synthetic Hydraulic Fluids Based on Esters, HEE		
	VG 22	VG 32	VG46	VG 68	VG22	VG 32	VG 46	VG 22	VG 32	VG 46	VG68
ARAL							BAF-46 Vitam			EHF-46 Vitam	
ASEOL AGIP						Aqua VG 32	Aqua VG 46	Terra 15		Terra 46 Agip Amica S 46	Terra 68 Agip Amica S 68
AUTOL			Autol Bio HVI 46								
AVIA		Avilub Hydraulic Bio 32	Avilub[a] Hydraulic Bio 46		Avia Hydrosynth 22	Avia[a] Hydrosynth 32	Avia Hydrosynth 46			Avia Syntofluid 46[a]	Avia Syntofluid 68
BECHEM		Bio-Hydraulik-öl 32			Hydrostar UWF 22	Hydrostar UWF 32	Hydrostar UWF 46		Hydrostar TMP 32	Hydrostar TMP 46	Hydrostar TMP 68
BINOL Biosttar Kellersberg			Hydrap 46 Biostar								
BLASER										Blasol LP8905	
BP			Bartran Biohyd 46				Biohyd PEG 46			Biohyd 46 SE[a]	Biohyd 68 SE
BRENNTAG			Hydraulic V 32		Hydraulic TR 22	Hydraulic TR 32[a]	Hydraulic TR 46				
BUCHER & Cie MOTOREX			Oekohydro 3268						Oekosynt 2246		
CASTROL	Biotec Alpin 22	Biotec HVX									

Company	Products
DEA	Econa R 32; Econa PG 32; Econa PG 46; Econa E 46
DELTIN	Deltinol Bio-hydraulic-öL HVI 46
ELF	Hydroelf Bio 46
ESSO	Hydraulik-öl PFL; Hydraulik-öl PGK 32[a]; Hydraulik-öl PGK 46; EGL 45947
FINA	Biohydran RS 38; Hydrauliköl D3031.46; Biohydran TMP 32; Biohydran TMP 46; Biohydran TMP 68
FINKE	Aviaticon HV-BD 36
FUCHS	Plantohyd N 32; Plantohyd N 46; Plantohyd N 68; Renodiol PGE 46; Plantohyd S 32; Plantohyd S 46; Plantohyd S 68
GLOBOIL	BHF 32; BHF 46; BHF 68
HOUGHTON	Trigolubric 32; Syntolubric 32; Syntolubric 46
KENDAL (Demmler & Co. Schweiz)	Synth. Natura 46HV
KOMPRESSOL	UW 500/32; UW 500/46; Biovis; Holbein 32; Holbein 46
Kuwait Petroleum Q8	
MOBIL	EAL 224H; Biofluid HLP 32[a]; Biofluid HLP 46
MOLYDUVAL	Chemlube 5126
OMV	Biohyd M 15; Biohyd M 32; Biohyd M 46; Biohyd M 68; Biohyd MS 15; Biohyd MT 32, MS 32; Biohyd MT 46 MS 46; Biohyd MS 68
OEST	Bio-Hydraulik-öl HVI 34; Bio Synth. HYD 46

(continued)

TABLE 8.4 (Continued)
Environmentally Acceptable Hydraulic Fluid HETG, HEPG, and HEE for Axial Piston Units

Type of Fluid	Vegetable-oil-based hydraulic fluids, HETG				Synthetic hydraulic fluids based on polyglycol, HEPG				Synthetic hydraulic fluids based on esters, HEE		
ISO Viscosity Class Maker	VG 22	VG 32	VG46	VG 68	VG22	VG 32	VG 46	VG 22	VG 32	VG 46	VG68
PANOLIN Schweitz								HLP Synth. 15	HLP Synth. 32	HLP Synth 46	HLP Synth. 68
QUAKER CHEMICAL		Quintolubric Greensave N 30								Quintolubric Greensave 46	
RAISON TEHTAAT, Finland		Florahyd RT-HVI 32								Esterhyd HE 46	
TEBIOL, BRD New Process Schweiz											
SHELL		Naturelle HF-R-32[a]			Fluid BD 22	Fluid BD 32[a]	Fluid BD 46			Naturelle HF-E 46	Naturelle HF-E 68
SOLLNER		Connexol HD 32-68									
STRUB & CO Schmiertechnik CH-Reiden									Hydrosint HLP ISO 32	Hydrosint HLP ISO 46	Hydrosint HLP ISO 68
TOTAL										Equivis Bio 46	Equivis Bio 68
VALVOLINE			Ultraplant 40								
WENZEL & WEIDMANN		Ukabiol HY 32	Ukabiol HY 46			Ukadol 32 NG	Ukadol 46 NG[a]			Ukabiol HE 46	
WESTFALLEN AG		Bio Forbex R 32	Bio Forbex R 46							Bio-Forbex E-46	

WINTERSHALL	Wiolgan HR 32	
WISURA	Hydroma UWF 46	
YORK Ginouves		LT 777 Bio
ZELLER & GMELIN	Biovinol HTG 46	

Source: OECD standards and methods are available from the Organization for Economic Cooperation and Development, Paris. With permission.

[a] Tested by Hydromatik (this does not imply that such tests were exhaustive, or that a recommendation of such fluids is implied). All rights reserved, subject to revision.

TABLE 8.5
Components of a "Typical" Water–Glycol Hydraulic Fluid

Component	Purpose
Water	Fire protection
Glycol	Freeze-point reduction and some thickening
Thickener	Thicken the formulation and to provide adequate film viscosity at the wear contact
Antiwear additives	Provide mixed film and some boundary lubrication
Corrosion inhibitors	Vapor and liquid corrosion protection

8.2.2.1 Biological Degradation

Biodegradation is a natural process caused by the action of micro-organisms. In the presence of oxygen, nitrogen, phosphorous, and trace minerals, organic pollutants can support microbial growth and are converted into a series of oxidation products that generally conclude with carbon dioxide and water [65].

There are various national standards currently being developed to define "biodegradability" and how it should be experimentally determined [44–47]. There are two general classifications that are often used to define biodegradability. One classification is based on the ultimate fate of the material after biodegradation. The other classification is based on the biodegradation rate. Recently, Wilkinson offered the following definitions and excerpts will be directly quoted from [47].

1. **Primary biodegradation**. This is a measure of the conversion by biological systems of the original organic material into different products from the starting materials. This is not a measure of the biological impact of the substance, but merely of the efficacy of the first step in what may be a long series of steps. In some cases, nontoxic substances that exhibit facile primary biodegradability can biodegrade into environmentally toxic materials.
2. **Ultimate biodegradation**. This is often referred to as "mineralization" and is the complete conversion by biological systems of the original organic material into carbon dioxide, water, and microbial biomass (and any inorganic ions, if present). (*Note*: Unfortunately, the current biodegradation test methods do not recognize biomass incorporation.)
3. **Ready biodegradation**. Substances which are classified as "readily biodegradable" exhibit greater than a certain fixed-percentage conversion in a standard test (see Table 8.5) [48]. This is usually measuring a parameter related to the degree of "ultimate biodegradation."
4. **Inherent biodegradation**. This is the potential to eventually biodegrade, although at a rate lower than that required for classification as "readily biodegradable."

A number of standard biodegradation tests have been developed to evaluate the biodegradability of lubricants. These biodegradation tests were designed to determine the degree of aerobic aquatic biodegradation of fully-formulated lubricants, or their components, on exposure to inoculum under controlled laboratory conditions. All of these tests can measure the biodegradability of lubricants based on oxygen uptake or carbon dioxide production of micro-organisms (inoculum). Recently, ASTM D-2 Subcommittee 12 on Environmental Standard of Lubricants has developed several biodegradation test methods (ASTM D 5864, ASTM D 6731, ASTM D 6139) for evaluating the biodegradability of petroleum-based and renewable-based lubricants [60,66,67]. The ASTM D 5864 and ASTM D 6731 are very similar to OECD procedures and ASTM D 6139 is the EPA Shake Flask Test. Their procedures are very similar to one another and simulate the biodegradation process used in the waste treatment center [65].

The ASTM D 5864 Biodegradation Test is a version of the Organization for OECD 301 B, Modified Sturm Test, which closely simulates the waste water biodegradation conditions. This test was designed to determine the degree of aerobic aquatic biodegradation of lubricants on exposure to inoculum under laboratory conditions. In this test, the biodegradability of a lubricant is expressed as the percentage of maximum carbon conversion under well-controlled conditions for a period of 28 days. The test apparatus consists of four separate units: the free carbon dioxide air system, bio-degradation batch reactor, a carbon dioxide collector, and a titrator. This test apparatus can be easily constructed by the testing laboratory as most of these components involve regular laboratory glass-wares (i.e., flasks, etc.). This test apparatus is considerably less expensive than the other types of bio-degradation tests (i.e., respirometer). For the biodegradation test, the five test stock solutions for the test medium were prepared in order to provide nutrition for micro-organisms. These stock solutions were ammonium sulfate solution, calcium chloride solution, ferric chloride solution, and magnesium. It should be noted that these solutions do not contain any carbon material in order to avoid an extra source of carbon dioxide production. Canola cooking oil was used as a positive controller to verify the micro-organism's performance during the test. In addition, the initial carbon content of the test lubricants was measured to establish the maximum theoretical biodegradability of that lubricant. The sewage micro-organisms from a local wastewater treatment plant were used as inocula in this test. The test was performed for twenty-eight days in a dark environment, and the titration for carbon dioxide accumulation was performed every day for the first 10 days and then every other day for the remaining 18 days or until a plateau of carbon dioxide evaluation was reached. This test has been widely used for many decades by industry and is considered a reliable biodegradation test. The disad-vantages of this test include its poor precision due to the various sources of inocula, unknown micro-organisms' behavior, and long test times. In addition to these, the test requires skilled manpower and sufficient laboratory space. Currently, ASTM D 6046 (Hydraulic Fluids for Environmental Impact) specifies that the persistence designation of readily biodegradable is Pw 1 and must be greater than, or equal to, 60% of CO_2 evolution within 28 days.

The ATSM D 6731 Biodegradation Test is another version of OECD 301F, the Manometric Respirometry Test, and is known as the modified "Biochemical Oxygen Demand (BOD) Test". This closed-respirometer test was also designed to determine the degree of biodegradability of lubricants or their components in an aerobic aqueous medium on exposure to an inoculum under laboratory conditions. Unlike the ASTM D 5864 test, the biodegradation of a lubricant is determined by mea-suring the oxygen consumption of micro-organisms instead of the carbon conversion of the test sample. This approach was developed based on the assumption that a large amount of oxygen uptake of micro-organisms indicates the growth or generation of more micro-organisms and takes more carbon conversion of the test sample, leading to carbon dioxide production by an enzyme process. For this reason, the respirometer test is currently considered as an indirect biodegradation test of lubricants, and its biodegradability is expressed as the percentage of maximum oxygen consumption under well-controlled conditions for a period of twenty-eight days. The test apparatus of this test method is a well-designed automatic system and requires less manpower. In this test, the oxygen consumption is measured based on the pressure drop of the manometer, which produces a signal that results in the electrolytic generation of oxygen. The sample and medium preparation is almost identical to the ASTM D 5864 test sample. Some advantages of this method lie in the fact that less manpower is required, and also in its closed system, which is suitable for evaluating the biodegrada-tion of volatile lubricants. The disadvantages are its indirect measurement of the biodegradation of lubricants and its poor test precision. Because of its measuring technique and the cost of test appa-ratus, this method is not widely utilized within industry. Currently, the ASTM D 6046 (Hydraulic Fluids for Environmental Impact) specifies that the readily biodegradability of fluid (classification: Pw1) in this test must be greater than or equal to 67% of O_2 consumption within twenty-eight days. This value is a little higher than that of the ASTM D 5864 test because of its total oxygen consump-tion of micro-organisms, including biomass production. Many test apparatuses are available today. Among them is the CES respirometer apparatus, which has an option to measure carbon dioxide evolution directly during a respirometer test. In this option, the micro-organisms consume oxy-gen from the respirometer and produce carbon dioxide. This carbon dioxide is then trapped in an absorbent solution (i.e., sodium hydroxide) and measured by a CO_2 sensor instead of a titration. This technique is very similar to that of the ASTM D 5864 test, while the test results are much closer to those of ASTM D 6731 because of its closed system, which can minimize the loss of carbon dioxide during the biodegradation process. This optional test is not yet developed as a standard test method, but can be considered as a closed automatic ASTM D 5864 test. The advantage of this test over the ASTM D 5864 test is in its automatic system, but it still has a test precision problem. Generally, the closed biodegradation system produces a higher biodegradability of the tested fluids than those of a semi-open system.

The ASTM D 6139 Biodegradation Test is a version of the EPA (Gledhill) Shake Flask Test and is very similar to the ASTM D 5864 biodegradation test except for agitating solution. The test was designed to determine the degree of aerobic aquatic biodegradation of lubricants upon exposure to an inoculum under laboratory conditions. The biodegradability of lubricants express as percentage of maximum (theoretical) carbon conversion (or carbon dioxide generation) under well-controlled conditions for twenty-eight days. The test apparatus is specially designed to agitate the test solution and the carbon dioxide is collected using a Gledhill Shake Flask System. In this test, 60%, or above, of biodegradability is considered as readily biodegradable. However, this test also requires a long testing time (twenty-eight days), knowledge of micro-organisms, and skilled manpower in some cases. In addition, it has very poor test precision due to the various and multiple sources of inocula. For these reasons, it is very difficult to use in petroleum laboratories for assessing the biodegradability of lubricants.

The ASTM D 7373 test method is a bio-kinetic model which is employed to predict the biodegradability of lubricants using an ASTM compositional analysis technique and the fundamental microbiological theory [68]. This bio-kinetic model requires compositional analysis data of lubricants and some formulation information related to the types of base oils used in the lubricants. The bio-kinetic model does not require any biodegradation test apparatus and inocula. The advantages of this model are: (1) its capability to predict the biodegradability of lubricants within a day, and (2) its excellent correlation with results obtained from ASTM D 5864 (Modified Sturm Test) and the ASTM D 6731 test. Table 8.6 shows that their correlation coefficients were found to be from 0.8 to 0.95 [65].

In the European test, the OECD procedure also incorporates a "10-day window", which requires "that the percent theoretical CO_2 must reach 60% within 10 days after reaching the 10% level" [11] if the fluid is to be classified as biodegradable. This requirement applies for all of the OECD 301 methods (A–E) shown in Table 8.7.

The OECD is currently re-evaluating this "10 day window" criterion for the following reasons:

1. This requirement is of questionable applicability to substances that are poorly water soluble [11].
2. Inherently biodegradable substrates may produce a time lag in the degradation process but still undergo ultimate biodegradation [49]. Therefore, this requirement appears to be unduly restrictive.

TABLE 8.6
Correlation Coefficients (r^2) between Biodegradation Tests

	ASTM D 5864	ASTM D 6731	CES CO_2 Evolution Test	Bio-Kinetic Model
ASTM D 5864	1	0.89	0.8	0.95
ASTM D 6731	0.89	1	0.82	0.92
CES CO_2 evolution test	0.8	0.82	1	0.87

TABLE 8.7
Standard Biodegradability Tests

Method	Time (days)	Factor Measured	Criterion
Ready Biodegradability			
ASTM D 5864	28	Production of carbon	>60%[a]
ASTM D 6731	28	dioxide	>60%[a]
ASTM D 6139	28	Oxygen demand	>60%[a]
		Production of carbon dioxide	>70%[a]
Modified AFNOR (OECD 301 A)	28	Loss of dissolved organic carbon	
Modified Sturm Test (OECD 301 B)	28	Production of carbon dioxide	>60%[a]
Modified MITI(1) (OECD 301 C)	28	Oxygen demand	>60%[a]
Closed Bottle (OECD 301 D)	28	Oxygen demand	>60%[a]
Modified OECD Screening Test (OECD 301 E)	28	Loss of dissolved organic carbon	>70%[a]
Inherent Biodegradability			
Modified Semi-Continuous	>28	Loss of dissolved organic carbon	>20%[b]
Activated Sludge (SCAS) (OECD 302 A)			>20%[b]
Zahn–Wellens (OECD 302 B)	28	Loss of dissolved organic carbon	>67%[c]
Primary Biodegradability			
CEC-L-33-T-82	21	Loss of hydrocarbon	>80%[d]
Bio-kinetic Model	1	infrared bands	>60%[a]
ASTM D 7373		Calculation based on composition analysis	

Source: Baggott, J. "Biodegradable Lubricants," in *Institute of Petroleum Symposium: Life Cycle Analysis and Eco-Assessment in the Oil Industry*, 1992. With permission.

[a] For classification as readily biodegradable.
[b] For classification as inherently biodegradable.
[c] ICOMIA standard.
[d] German Blue Angel requirement for biodegradable hydraulic fluids.

The experimental strategies for the various tests listed in Table 8.5 involved the measurement of: (a) the amount of the original carbon consumed, (b) the consumption of oxygen used in the degradation test, or (c) the evolution of carbon dioxide at the end of the biological degradation process. This is illustrated in Figure 8.3 [50].

Of the tests for "ready biodegradability" shown in Table 8.5, those most frequently used include OECD 301A, 301B, and 301E [49]. The use criteria test for these tests are as follows [49]:

1. OECD 301A—AFNOR Test. This test is used for water-soluble compounds that are relatively non-volatile and non-inhibiting to bacteria at 40 mg/L dissolved organic carbon (DOC). The DOC of the test material is the sole source of COC for the aerobic microorganisms. The test criteria are based on the loss of DOC after twenty-eight days.
2. OECD 301B—Modified Sturm Test. This test is applicable for both water-soluble and water-insoluble compounds. Biodegradation is based on the yield of carbon dioxide produced over the 28-day duration of the test.

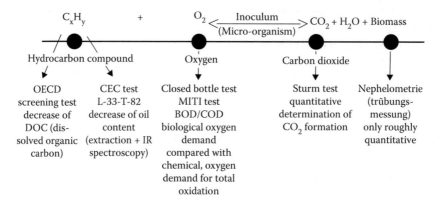

FIGURE 8.3 Test strategies for determining and quantifying biodegradation. (From Völtz, M., Yates, N.C. and Gegner, E. "Biodegradability of Lubricant Base Stocks and Fully Formulated Products," *J. Synth Lubric.*, 1995, 12(3), pp. 215–30. With permission.)

   3. OECD 301E—OECD Screening Test. This test is designed for water-soluble, non-volatile, and non-inhibitory compounds (5–40 mg/L DOC). This test also monitors the change in DOC with respect to time and is similar to OECD 301A.

OECD 301B is often used to determine the biodegradability of hydraulic fluids for the following reasons [49]:

- The method allows for the determination of biodegradability at the lowest concentration possible.
- The method is useful for both water-soluble and water-insoluble compounds.
- Because titrations are used, the method is relatively simple and inexpensive to perform.

One of the difficulties in conducting these aqueous tests is the insolubility of hydraulic fluids, with the exception of the water-containing fluids such as a water-glycol hydraulic fluid. Girling has reviewed the following procedures that are used to enhance oil solubility: vigorous shaking, blending (Waring blender), homogenizing, ultra-sonification, and chemical dispersion [51]. The factors which are known to influence the composition of the aqueous test media include the following [51]:

- The bulk properties: viscosity, interfacial tension, and so on, of the OBP (oil-based products);
- The mixing environment;
- Properties of the aqueous phase, such as pH, water temperature, hardness, and salinity;
- Physical and chemical properties of the OBP constituents;
- The OBP/water ratio.

These difficulties skew BOD data in favor of water-insoluble fluids because the percentage of biodegraded substrate will increase as the solubility of the substrate decreases, assuming the degradation rate remains unchanged. One way that this can be envisioned is by examining the effect of oil-film thickness on biodegradability rate, which would presumably increase with increasing concentration and which should then result in lower degradation half-times. This was observed by Völtz et al., as illustrated in Figure 8.4 [50]. Völtz et al. also showed that the type and concentration of microorganisms which may be present in the aquatic environment will also significantly affect biodegradation rates, as illustrated in Figure 8.5 [50].

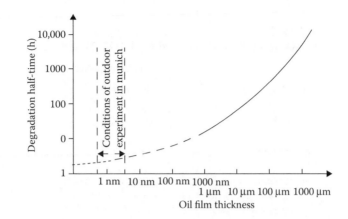

FIGURE 8.4 Dependence of oil-film thickness on biodegradation rate. (From Völtz, M., Yates, N.C. and Gegner, E. "Biodegradability of Lubricant Base Stocks and Fully Formulated Products," *J. Synth Lubric.*, 1995, 12(3), pp. 215–30. With permission.)

Biodegradation rates are often conducted by a number of possible tests, as indicated in Table 8.5. Two of the most commonly-encountered tests are the CEC-L-33-A-94 test and the Modified Sturm Shake Flask Test [11]. Table 8.6 provides a correlation of the two tests with respect to the vegetable oil and synthetic ester that are shown [29]. These data are interesting because they illustrate the similarity and differences between the two commonly-encountered testing methods. The following comments are offered with respect to these results:

- The CEC-L-33-A-94 method is used to determine primary biodegradation (see Table 8.5.). Primary biodegradation refers to the first stage of degradation and corresponds to the initial changes in the molecular composition of the lubricant upon degradation. This is a useful screening test to assess the potential for degradation.
- The Modified Sturm Test is used to assess ultimate biodegradability. It is assumed that the lubricant components have completely "mineralized."
- The data in Table 8.8 are interesting in that they show that primary degradation is always greater than the values for ultimate degradation [29]. Thus, it is possible for a substrate that only partially degrades to exhibit a high percentage of degradation in the primary biodegradability test. This helps to explain the relatively poor correlation between the two tests shown in Figure 8.6 [47].

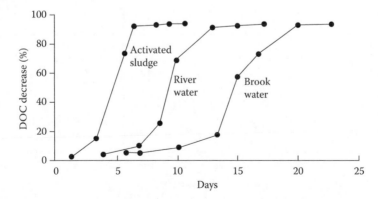

FIGURE 8.5 The effect of microflora on biodegradation rates. (From Völtz, M., Yates, N.C. and Gegner, E. "Biodegradability of Lubricant Base Stocks and Fully Formulated Products," *J. Synth Lubric.*, 1995, 12(3), pp. 215–30. With permission.)

TABLE 8.8
Biodegradability of Selected Hydraulic Fluids

| | Biodegradability | |
Product Base Stock	Shake Flask[a]	CEC Test[b]
Mineral oil	42–48	N.T.[c]
Vegetable oil	72–80	>90
Polyglycol	6–38	N.T.[c]
Synthetic ester	55–84	>90

Source: Cheng, V.M., Wessol, A.A., Baudouin, P., BenKinney, M.T. and Novick, M.J. "Biodegradable and Non-Toxic Hydraulic Oils," *SAE Technical Paper Series*, Paper 910964, 1991. With permission.

[a] CO_2 evolution—EPA Method 560/6-82-003.
[b] CEC Method CEC-L-T-82
[c] Not tested.

- These data explain the occasional confusion that occurs in the marketplace if data from dissimilar test procedures are compared.

8.2.2.2 Published Biodegradation and Toxicity Data

Aquatic biodegradability. Using the CEC-L-33-A-94 test procedure, the biodegradability of various mineral-oil base stocks was reported by Singh and selected examples are shown in Table 8.9. These data show that mineral-oil biodegradability is composition-dependent. These data indicate that if aromatic, aliphatic, and heterocyclic derivatives are minimized and the saturation level is increased, mineral-oil-derived compositions may be produced with biodegradability properties that rival vegetable oil and polyol esters—the closest functional competitors. As expected, vegetable oil and polyol ester-based fluids exhibited considerably higher biodegradation rates [52].

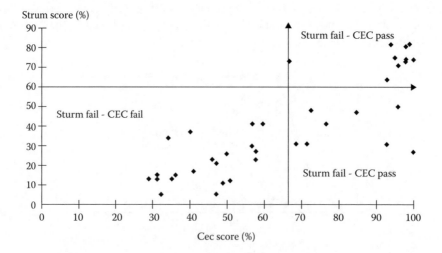

FIGURE 8.6 Comparison of CEC and Sturm Test results. (From Wilkinson, J. "Biodegradable Oils—Design, Performance, Environmental Benefits and Applicability," *SAE Technical Paper Series*, Paper 941077, 1994. With permission.)

TABLE 8.9
Biodegradability of Selected Fluids using the CEC-L-33-A-94 Test Procedure

Fluid	% Biodegradability
LVI VG 100	35
Parafinic VG 100	60
Hydrotreated HVI VG 100	30
Naphthenic VG 100	47
Formulated predominantly mineral oil (R&O)	77
Hydrocracked very high VI	80
Formulated vegetable oil	95
Synthetic polyol ester	95

Source: Singh, M.P., Chhatwal, V.K., Rawat, B.S., Sastry, M.I.S., Srivastava, S.P. and Bhatnagar, A.K. "Environmentally Friendly Base Fluids for Lubricants," in *Adv. Prod. Appl. Lube Base Stocks, Proc. Int. Symp.*, Singh, H., Rao, P. and Tata, T.S.R. eds., McGraw-Hill, New Delhi, 1994 pp. 362–70. With permission.

In a different study, Kiovsky, using the updated procedure CEC-L-33-A-94, also examined the effect of mineral-oil base stock on biodegradability [53]. The results in Figure 8.7 show that hydrocracking produced relatively good biodegradation rates of approximately 80%.

Naegely has provided a comparison of the biodegradability of white oils with typical ranges for vegetable oils, polyol esters, and diesters, which is summarized in Figure 8.8a [22]. The values for polyethers (which are anhydrous) are significantly less than those published by Völtz (as shown in Figure 8.8b) [50], which are less than those reported by Henke (shown in Figure 8.8c) [54]. All of the data were obtained by the CEC-L-33-A-94 biodegradability test. Figure 8.9 illustrates the variation in the limits required by different European specifications using the same CEC biodegradability test procedure to define "rapidly" biodegradable fluids [7].

The rate of biodegradation of a mineral oil and a rapeseed oil, with respect to time, using the CEC test procedure is compared in Figure 8.10 [3]. Clearly, the vegetable oil undergoes most of its bio-oxidation within ten days. These are interesting data and indicate the potential impact of the so-called "ten-day window" that was discussed earlier on the classification of hydraulic fluid biodegradability. Vegetable oils, as a class, undergo a relatively rapid initial biodegradation.

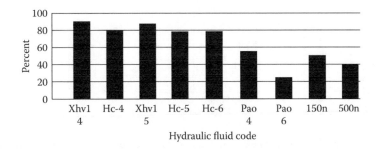

FIGURE 8.7 Effect of mineral-oil composition on biodegradability. Numbers following the designations are viscosities (cSt at 100°C) and those of solvent neutrals are SUS at 100°F. VHVI = high-VI oils; XHVI = shell wax isomerized-based oils; HC = hydrocracked synthetics; PAO = polyalphaolefin; N = solvent-refined neutral oils. (Kiovsky, T.E., Murr, T. and Voeltz, M. "Biodegradable Hydraulic Fluids and Related Lubricants," *SAE Technical Paper Series*, Paper 942287, 1994.)

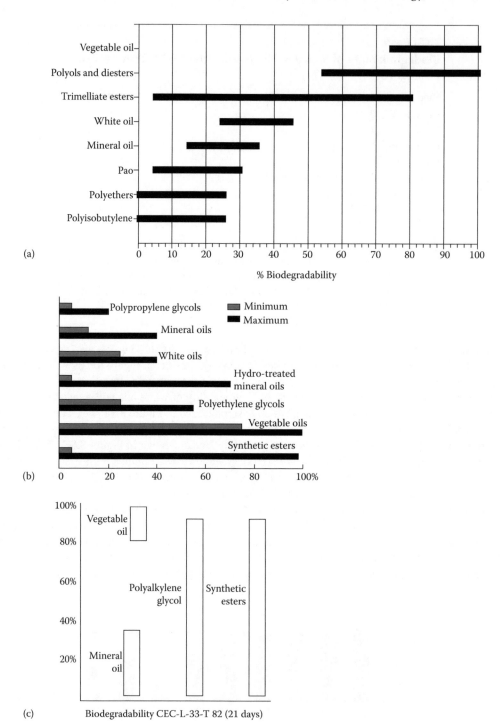

FIGURE 8.8 (a) Naegly biodegradation test data. (From Naegly, P.C. "Environmentally Acceptable Lubricants," in *Seed Oils for the Future*, MacKenzie, S.L. and Taylor, D.C. eds., 1992, pp. 14–25. With permission.), (b) Völtz test data. (From Völtz, M., Yates, N.C. and Gegner, E. "Biodegradability of Lubricant Base Stocks and Fully Formulated Products," *J. Synth Lubric.*, 1995, 12(3), pp. 215–230. With permission.), (c) Henke test data. (From Henke, R. "Increased Use of 'ECOFLUIDS' May Put a Veggie in Your Hydraulic Reservoir," *Diesel Progr., Engines Drives*, September, 1994, pp. 7–9. With permission.)

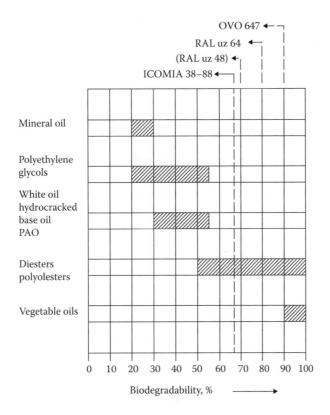

FIGURE 8.9 Illustration of differing biodegradation European regulations. OVO = Austrian law, 1.5. 1992; RAL = German authority for labeling; ICOMIA = International Council of Marine Industry Associates. (From Mang, T. "Environmentally Friendly Biodegradable Lube Base Oils—Technical and Environmental Trends in the European Market," in *Adv. Prod. Appl. Lube Base Stocks, Proc. Int. Symp.*, 1994, pp. 66–80. With permission.)

Aquatic toxicity. Toxicity (ecotoxicity) is defined as the "propensity" of a substance to produce adverse biochemical and physiological effects in a living organism. "It incorporates both acute and chronic effect but does not include physical effects. . . . An acute effect is one which occurs after a short exposure to the test substance, short being some time period which does not constitute a large function of the life span of the organism. A chronic effect is one which occurs after a long contact with the test substance" [46]. An example of a "physical effect" is the deprivation of oxygen to a living organism.

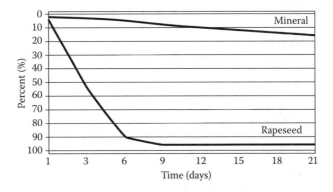

FIGURE 8.10 Comparison of the biodegradation rates of a mineral-oil and rapeseed-oil hydraulic fluid. (From Eichenberger, H.F. "Biodegradable Hydraulic Lubricant—An Overview of Current Developments in Central Europe," *SAE Technical Paper Series*, Paper 910962, 1991. With permission.)

Toxicity is a complex subject to address adequately because different organisms react differently to the same substance. Toxicity is usually expressed in the following terms [46]:

LCXX: The "lethal concentration" to cause the death of *XX%* of the test organisms.

ICXX: The "inhibition concentration" for some inhibitory effect to occur in *XX%* of the test organisms.

ECXX: The "effect concentration" for some environmental effect, such as growth or deformity, to occur in *XX%* of the test organisms.

Comparative toxicity data on various biodegradable hydraulic fluids have been reported by Cheng et al. [29]. (Only base stocks were evaluated.) These data are summarized in Table 8.10 [29]. Rainbow trout were used for the study under mechanical dispersion conditions to facilitate an oil–water dispersion (as the oils evaluated were water-insoluble). In general, none of the base stocks were determined to be toxic in this test.

Soil biodegradability and toxicity. Another area in which there is relatively little biodegradation data for all classes of hydraulic fluids is the fate and effects of biological degradation processes in soil [55]. Recently-determined results of the biodegradation of the hydrolube in soil will be reported here.

Currently, there are no international standards describing in-soil biodegradation testing procedures for hydraulic fluids. Also, there are relatively few oil biodegradation test results compared to aquatic biodegradation for the same series of hydraulic fluids. The objective here is to provide an overview of the different soil biodegradation results that have been published to date.

Völtz et al. described the Hamburg University Test, which measures CO_2 production (mmole CO_2/100 g soil) after seven days with 4% oil. The data in Figure 8.11 compares the biological rate of a conventional 15W-40 mineral oil and an "eco-technology" 5W-40 fresh oil, which is based on a mixture of an ester and hydro-cracked base stock [50]. As expected, the eco-technology base-stock-derived oil degraded fastest.

Figure 8.12 illustrates the biodegradation rate of a series of mineral oils, synthetic esters, and a vegetable oil (sunflower cooking oil). Biodegradation rates were ascertained by determining the loss of the original fluids in soil [56]. The vegetable oil was the most rapidly degraded fluid in the series. In general, natural fats and oils were not persistent in the environment.

To enhance the soil degradation rates, a nutrient (KNO_3) was added to the soil in one set of experiments (see Figure 8.13). These data show the following:

- The vegetable oil exhibited the fastest degradation rate.
- The TTE mix, a formulated synthetic ester containing mineral spirits (see Figure 8.13 caption), exhibited the second fastest biodegradation rate. The commercial mineral oil (Com A) was the most persistent.

TABLE 8.10
Comparative Toxicity of Selected Hydraulic Fluids

Product Base Stock	Trout LC_{50}
Mineral oil	389–>5000
Vegetable oil	633–>5000
Polyglycol	80–>5000
Synthetic ester	>5000

Source: Cheng, V.M., Wessol, A.A., Baudouin, P., BenKinney, M.T. and Novick, M.J. "Biodegradable and Non-Toxic Hydraulic Oils," *SAE Technical Paper Series*, Paper 910964, 1991. With permission.

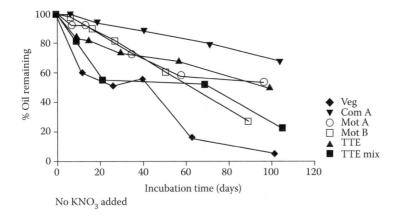

No KNO$_3$ added

FIGURE 8.11 Biodegradation rates of various oils in soil. See Figure 8.12 for an explanation of oils. (From Völtz, M., Yates, N.C. and Gegner, E. "Biodegradability of Lubricant Base Stocks and Fully Formulated Products," *J. Synth Lubric.*, 1995, 12(3), pp. 215–30. With permission.)

In summary, these results showed that although synthetic esters biodegrade faster in soil studies than mineral oil, they are still more persistent than a vegetable oil.

Haigh described the use of a L/m^2 section of an agricultural field [55] to which oil was added at concentrations varying from 0.25 to 5.0 L/m^2 of soil through an inverted test tube to a depth of 6 cm. Spring wheat, Broom variety, was planted. Seed germination (%) and growth rate (mm/day) was determined.

In another Haigh study [56], which was conducted on the lubricants shown in Table 8.11, the following observations were made:

- The synthetic ester did not biodegrade as extensively as the vegetable oil. However, although the vegetable oil biodegraded rapidly, it took more than one year for the extractable residues to decrease to levels comparable to the naturally-occurring lipids present in the soil.
- The mineral-oil-based fluids biodegraded less rapidly than synthetic esters. As with the other fluids, the biodegradation rate increased with soil temperature.

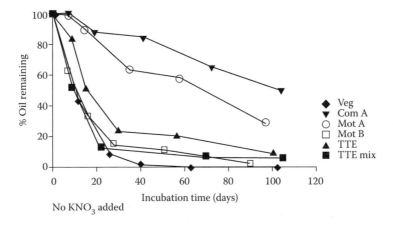

No KNO$_3$ added

FIGURE 8.12 Illustration of the biodegradation rate of 1% of the following: TTE—a commercial mixture by weight, isostearic acid ester of trimethylolpropane; TTE mix—a mixture of 60% TTE, 20% white spirit, and 20% of a substituted imidazoline additive; Mot A—dibasic acid esters of dodecanoic acid; Mot B—analogous to Mot A except using a longer-chain acid; Com A—a highly-refined formulated motorcycle mineral-oil lubricant; Veg—"pure" sunflower cooking oil. To the soil samples were added: (a) 2 mL of distilled water and (b) 2 mL of KNO$_3$ (14 mg NO$_3$–N). The samples were incubated at 25°C and moisture levels were maintained by distilled water addition.

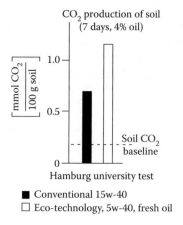

FIGURE 8.13   Comparative biodegradation test results.

- Crop damage was not necessarily related to the level of the oil residues. For example, the dibasic esters, although biodegrading rapidly, caused a complete inhibition of seed germination.
- The white spirit and vegetable oil permitted seed germination but significantly inhibited both growth rate and yield.

There are a number of mechanisms by which an oil can retard plant growth. These include the following [57]:

1. Oils may coat the seeds, thus creating a physical barrier inhibiting germination.
2. Growth inhibition may occur by root suffocation or by oxygen depletion of the soil caused by the oil biodegradation process.

TABLE 8.11
Plant Yields of Spring Wheat Treated with Different Lubricating Oils

Fluid	Ave. Germination rate[a]	Ave. Yield (g/plant)	Ave. Growth Rate (mm/day)
Trimethylolpropane triisostearate (TTE)	83.5	95.0	99.9
TTE + substituted imidazoline (TTE + additive)	74.0	95.0	99.9
TTE + white spirit (20%) + 20% substituted imidazoline (TTE mix)	63.0	NS[b]	99.9
"Pure" sunflower cooking lubricant (Veg)	81.5	99.9	99.9
Highly-refined mineral-oil motorcycle lubricant (Com A)	80.0	90.0	NS[b]
Highly-refined mineral-oil automotive lubricant (Com B)	70.0	95.0	
White spirits	63.5	99.9	99.9
Dibasic acid C-12 ester (Mot A)	0.0	Complete inhibition of germination	
Dibasic acid C-12 with a longer-chain ester (Mot B)	0.0	Complete inhibition of germination	

Source: Haigh, S.D. "Determination of Synthetic Lubricant Concentrations in Soil During Laboratory-Based Biodegradation," *J. Synth. Lubric.*, 1994, 11(2), pp. 83–93. With permission.

[a] The control for untreated soil was 88.5 germination.

[b] NS = not significant.

3. Poor growth may also be due to reduced water uptake by the plants and/or immobilization of nitrogen and phosphorus in the soil.
4. In general, the smaller the molecule, the more toxic the behavior due to increased ease of penetration into the plant tissue.

The results of Haigh's work [55,56] showed that, as expected, synthetic esters are degraded more rapidly than mineral oil, but not as rapidly as vegetable-oil-derived lubricant. In general, even vegetable oil requires one year for extractable residues to decrease to levels comparable to the natural lipids present in the soil. The effects of these oils on growth rates varied from small reductions to complete inhibition. The degree of inhibition was not related to the persistence of the oil residues. This means that it is important to both biodegradation and toxicity.

8.2.2.3 Biodegradability and Toxicity Data for a Water–Glycol Hydraulic Fluid

Extensive biodegradability and toxicity data for one water–glycol hydraulic fluid formulation has been reported [36]. This fluid exhibited 80% biodegradation after twenty-eight days, according to the OECD Modified Sturm Test protocol 301B shown in Table 8.12 [36].

Mammalian acute toxicity and irritancy of this fluid was also determined. The acute oral LD_{50} for rats was determined to be >20.0 g/kg. Dermal irritation determined on rabbits exhibited a Draize score of 0, which means that, according to this test, it is not a skin irritant.

Toxicity toward aquatic and terrestrial organisms was likewise determined. The aquatic organisms tested included fish, invertebrates, and algae. The terrestrial organisms included plants (radish) and invertebrates (earthworms). As the results in Tables 8.13 and 8.14 indicate, this water–glycol fluid was practically nontoxic to the aquatic organisms evaluated. For reference, the aquatic toxicity reference scale is shown in Table 8.15.

A hazard-evaluation model was developed to determine the toxicity and exposure levels if an accidental spill were to occur. The assumptions used for this model are summarized in Table 8.16. The results of this evaluation showed the following [58]:

- Terrestrial environment—small boreal mammals. The model provided an estimated dose for red squirrels of 2690–11,920 mg/kg/day; the acute oral LD_{50} for rats was determined to be >20 g/kg. These data indicate that there is no significant hazard for a single event under the conditions modeled.
- Terrestrial environment—invertebrates. The model estimated the concentration of the high-performance hydrolube in the soil to be 3400 mg/kg soil.

TABLE 8.12
Biodegradability and Mammalian Toxicity and Irritancy Data for a High-Performance Water–Glycol Hydraulic Fluid[a]

Test	Result
OECD 301B—Modified Sturm Test	80% in 28 days
Acute oral LD_{50}—rats	>20.0 g/kg
Dermal irritation—rabbits	Draize score of 0

Source: Totten, G.E., Cerf, J., Bishop, R.J. and Webster, G.M. "Recent Results of Biodegradability and Toxicology of Water–Glycol Hydraulic Fluids," *SAE Technical Paper Series*, Paper 972789, 1997. With permission.

[a] UCON Hydrolube HP-5046 (Union Carbide Corporation, Danbury, CT.).

TABLE 8.13
Ecotoxicity toward Aquatic Organisms for a High-Performance Water–Glycol Hydraulic Fluid[a]

Test	Test Type	Endpoints LC50/IC50
Rainbow trout	Survival acute/96 h	LC50—10,607 μL/L
Daphnia magna	Survival acute/96 h	LC50—10,607 μL/L
Selenastrum capricornutum	Growth chronic/96 h	IC50—467 μL/L

Source: Totten, G.E., Cerf, J., Bishop, R.J. and Webster, G.M. "Recent Results of Biodegradability and Toxicology of Water–Glycol Hydraulic Fluids," *SAE Technical Paper Series*, Paper 972789, 1997. With permission.
[a] UCON Hydrolube HP-5046 (Union Carbide Corporation, Danbury, CT.).

The LC_{50} for earthworm survival is 26,574 mg/kg soil; the LC_{50} for seedling emergence is 6388 mg/kg soil; and the LC_{50} for seedling growth is 493 mg/kg soil. These results indicate that there is no significant hazard for invertebrates and plants.

- Aquatic environment. The estimated concentration of the high-performance hydrolube in water is 0.16 μL/L. The LC_{50} for rainbow trout survival is 10,607 μL/L; the LC_{50} for *Daphnia magna* survival is 10,607 μL/L; and the IC_{50} for *Selenastrum capricornutum* growth is 467 μL/L. On the basis of this model, there is no significant hazard for fish, invertebrates, or algae.

8.2.3 Current Specification Status

Cheng et al. first proposed the following biodegradability criteria to determine if a lubricant would be acceptable for use in environmentally sensitive areas at the 1991 SAE Earth Moving Conference [29]:

- >60% Conversion to CO_2 in twenty-eight days;
- Aquatic toxicity for rainbow trout of >1000 ppm.

There are a number of national standards under development that currently have various levels of international acceptance. These will be reviewed briefly in the next subsection.

8.2.3.1 Ecolabel Testing Requirements

The testing requirements for three ecolabels will be discussed here. They include: German Blue Angel, Canadian EcoLogo, and the method currently being developed by ASTM in the United

TABLE 8.14
Ecotoxicity toward Terrestrial Organisms for a High-Performance Water-Glycol Hydraulic Fluid[a]

Test	Test Type	Endpoints LC50
Radish seedling	Emergence chronic/21 days	6,388 mg/kg soil
	Growth chronic/21 days	493 mg/kg soil
Earthworm	Survival chronic/14 days	26,574 mg/kg soil

Source: Totten, G.E., Cerf, J., Bishop, R.J. and Webster, G.M. "Recent Results of Biodegradability and Toxicology of Water–Glycol Hydraulic Fluids," *SAE Technical Paper Series*, Paper 972789, 1997. With permission.
[a] UCON Hydroluble HP-5046 (Union Carbide Corporation, Danbury, CT.).

TABLE 8.15
Aquatic Toxicity Classification Scale

Classification	LC50 (mg/L or ppm)
Super toxic	>0.01
Extremely toxic	0.01–0.1
Highly toxic	>0.1–1.0
Moderately toxic	>1.0–10.0
Slightly toxic	>10–100
Practically nontoxic	>100–1000
Relatively harmless	>1000

Source: U.S. Fish and Wildlife Service, *Research Information Bulletin* 84-78, 1984.

States to define eco-evaluated (EE) fluids. In addition, the Caterpillar BF-1 specification [34] will be discussed. These procedures will be discussed here in order.

German Blue Angel. The current German Blue Angel testing requirements are summarized in Table 8.17. The objective is that the claim for reduced environmental impact should be supported by test data. The Blue Angel procedure developed in Germany incorporates a water hazard classification system which requires measurement of mammalian, fish, and bacterial toxicity [48]. The Blue Angel requirement is dependent on the water hazard classification (WGK) number, which is summarized in Table 8.18 [48].

Canadian EcoLogo testing requirements. The Canadian EcoLogo is the most commonly-encountered national environmental standard for hydraulic fluid use in North America [59]. The acceptance criteria to receive this label are summarized in Table 8.19.

ASTM "eco-evaluated" testing criteria. The ASTM D.02N.03 Subcommittee is developing a testing criterion to evaluate the potential impact of the use of unused hydraulic fluids on the environment. Table 8.20 summarizes the current major classification designations [59]. Because the standard is under development, the complete selection of required test methods is not yet firmly established. Nevertheless, the most current test methods under consideration are reported here. Fluids are designated as Pw_1, Pw_2, and so forth, according to their performance in specified tests.

TABLE 8.16
Hazard Evaluation Assumptions

Total area of forest impacted by spill	200 m²
Depth of soil impacted	0.1 m
Total volume of HP-5046 released during spill	100 L
Loss of HP-5046 in environment due to biodegradation	0%
Weathering rate of HP-5046 on vegetation (s⁻¹)	5.73×10^{-7}
Vegetation yield (kg plant/m²)	2.8 (value for grass, wet weight)
Density of HP-5046	1.09 mg/µL
Solid loads in water of Canadian forested regions	50 mg/L
Soil ingestion (small animals)	13% of food ingestion food ingestion = 0.0124 kg/day
Red squirrel body weight	145–260 g

TABLE 8.17
Summary of German Blue Angel Criteria

- The product shall meet all relevant technical specifications.
- Every component, including additives, must be Water Hazard Class 0 or 1. (See Table 8.21.)
- The product must be free of organic chlorine, nitrite-containing compounds, and metals (with the exception of calcium up to 100 ppm). No definition of "free" (in terms of an upper concentration limit) is given.
- Readily biodegradable components can be incorporated in the product in any concentration but must collectively account for at least 93% of the product composition. Such components are defined in terms of one of the following tests and pass levels:

Test	Pass Level
Modified AFNOR	70%
Modified OECD Screening Test	70%
Modified Sturm	70%
Modified MITI (1)	70%
Closed Bottle	70%
CEC-L-33-A-93 (poorly soluble components only)	80%

- Total concentration of "inherently" biodegradable components cannot exceed 5%. Such components are defined in terms of any one of the following tests and pass levels:

Test	Pass Level
Modified SCAS	20%
Modified MITI (2)	20%
Zahn–Wellens	20%

- Total concentration of non-biodegradable components (i.e., components that fail to achieve the pass levels in the tests for "inherent" biodegradability) cannot exceed 2%.
- All "inherently" biodegradable and non-biodegradable components must satisfy the following eco-toxicological criteria.
- They must be nontoxic to aquatic organisms.

Potential for Bioaccumulation	No	Yes
OECD 202 Part 1 (Daphnia) EC_{50}/LC_{50}	≥ 1 mg/L	≥ 100 mg/L
OECD 203 (fish) EC_{50}/LC_{50}	≥ 1 mg/L	≥ 100 mg/L
OECD 203 (fish) NEC	≥ 0.01 mg/L	≥ 1 mg/L

- They must be nontoxic to higher plants:

OECD 208 (growing test) WEL	≥ 1 mg/L	≥ 1 mg/L

- In addition to the above, all non-biodegradable components must be nontoxic to bacteria:

OECD 209 or OECD 301D	EC_{50}	≥ 100 mg/L

- All non-biodegradable polymer components must be insoluble in water:

(solubility ≥ 1 mg/L)

TABLE 8.18
German Water Hazard Classification System

Acute Oral Mammalian Toxicity (AOMT)

Measured as an LD_{50} for laboratory animals. The LD_{50} is used to determine a "Water Endangering Number" (WEN) using the following scale:

LD_{50} (mg/kg)	WEN (AOMT)
<25	7
25–200	5
200–2000	3
>2000	1

Acute Bacterial Toxicity (ABT)

Measure as a "no-effect" concentration (NEC). The WEN (ABT) is determined from
$$\text{WEN (ABT)} = -\log(\text{NEC in ppm}/10^6\text{ppm})$$

Acute Fish Toxicity (AFT)

Measured as a WEC. The WEN (AFT) is determined from
$$\text{WEN (AFT)} = -\log(\text{NEC in ppm}/10^6\text{ppm})$$

Water Endangering Number (WEN)

The overall WEN is calculated from
$$\text{Overall WEN} = 1/3[\text{WEN(AOMT)} + \text{WEN(ABT)} + \text{WEN(AFT)}]$$

WGK Number

The Water Hazard Classification is then determined from

WEN	WGK Number	Classification
0–1.9	0	Not hazardous to water
2–3.9	1	Slightly hazardous to water
4–5.9	2	Moderately hazardous to water
>6	3	Highly hazardous to water

"Readily" biodegradable substances are assigned to the next lower Water Hazard Class. (Rapeseed oils and polyethylene glycols are WGK 0.)

Source: Baggott, J. "Biodegradable Lubricants," in *Institute of Petroleum Symposium: Life Cycle Analysis and Eco-Assessment in the Oil Industry*, 1992. With permission.

The current test method under development will only focus on environmental persistence (Category P) and acute ecotoxicity (Category T) of hydraulic fluids. As only test methods for the determination of aerobic freshwater have been developed to date [60], this new standard will be limited to the evaluation of unused hydraulic fluids in aerobic fresh water (Pw). The proposed classification system is provided in Table 8.21. The test methods are summarized in Table 8.22.

The proposed classification of "ecotoxicity" properties of hydraulic fluids is summarized in Table 8.23 and the proposed uses for ecotoxicity are provided in Table 8.24.

8.2.3.2 Biodegradable/Bio-based Hydraulic Fluid Specification

Numerous biodegradable/bio-based hydraulic fluid specifications have been developed, or are being developed, to replace the conventional petroleum-based hydraulic fluids that are not compatible with environments. Among them, Military Bio-based Hydraulic Fluid Specification, MIL-PRF-32073 and Caterpillar BF-1 specification are presented in this Section.

TABLE 8.19
Environmental Choice EcoLogo-Product-Specific Requirements

1. Demonstrate (a) or (b) below:
   (a) Have a C_{50} or an EC_{50} not lower than 1000 mg/L when the OWD (fish and the WAFs (*Dapnhia* and algae) prepared from the whole formulation are tested according to:

   - Biological Test Method: Acute Lethality Test Using Rainbow Trout, Report EPS1/RM/9, July 1990, Environment Canada; and Biological Test Method: Acute Lethality Test Using *Daphia* Spp., Report EPS 1/RM/11 July 1990; Environment Canada and Biological Test Method: Growth Inhibition Test Using the Freshwater Algae *Selenastrum Capricornumtum*, Report EPS 1/RM/25 November 1992, Environment Canada; or
   - Test acceptable to the ECP (Environmental Choice Program).
   (b) Have a EC_{50} of LC_{50} not lower than 2500 ppm when the whole formulation is tested according to Microtox™ test.
2. Be biodegradable according to CEC-L-33-T82.
3. Not contain more than 5% (w/w) additives.
4. Not contain more than 3% (w/w) of an additive that is not proven to be biodegradable.
5. Not contain more than 0.1% petroleum oil or additives containing petroleum oil.
6. Not contain:
   (a) organic chlorine or nitrite compounds;
   (b) lead, zinc, chromium, magnesium, or vanadium.
7. Not have to be labeled according to Class D, "Poisonous and Infectious Material," or set out in the "Controlled Products Regulation of the Hazardous Products Set."
8. Be packaged in a container which bears a label indicating both the pour point as determined by ASTM D97 "Standard Test for Pour Point of Petroleum Oils" and the low-temperature fluidity performance.
9. Yield negative results when tested against ASTM D665 "Standard Test Method for Rust Preventing Characteristics of Inhibited Mineral Oil in the Presence of Water."
10. Not have a flash point lower than 200°C if ISO grade VG 32 and higher, and not lower than 190°C if ISO grade VG 15-22 when measured according to ASTM Test Method D92 "Standard Test Method for Flash and Fire Points by Cleveland Open Cup"; D 93 "Standard Test Method for Flash Point by Pensky-Martens Closed Tester"; or D 56 "Standard Test Method for Flash Points by TAG Closed Tester."
11. Be proven to have good oxidation stability when tested according to ASTM D 525 "Standard Test Method for Oxidation Stability of Gasoline (Induction Period Method)," or modified according to U.S. Patent 4,783,274 (1983) Hydraulic Fluids, K.V. Jokinon et al., the pressure drag is not greater than 35 psi.
12. Demonstrate a low tendency for foaming according to ASTM Test D-892.

---

1. *Military Bio-based Hydraulic Fluid, MIL-PRF-32073 Specification*

In 2001, the U.S. Army issued a new military performance specification, MIL-PRF-32073, Hydraulic fluid, Biodegradable. This specification was developed to cover the hydraulic fluid requirements for military construction and tactical equipment. In 2007, the title of specification was changed from "biodegradable "to "bio-based" because some biodegradable fluids are derived from petroleum, which is considered to be a somewhat toxic material [69]. This specification was originally designed for use in environmentally sensitive areas such as construction sites, forestry, rivers, and mining. Its hydraulic fluids were formulated to be used in construction equipment, bridging, tactical vehicles, shipboard hydraulic systems, and hydraulic systems for metal tool applications. The specification consists of five different ISO grades to cover a wide array of applications and contains a Qualified Product List (QPL) requirement.

In 2005, a fluid compatibility study between QPL products of MIL-PRF-32073 specification was performed because the bio-based fluids were formulated from different renewable oil resources. The study clearly indicated that the BHFs qualified under the MIL-PRF-32073 specification and were compatible with each other. It is anticipated that these BHFs will not create compatibility problems in the field [70].

TABLE 8.20
Classification Designation of ASTM 'Eco-evaluated' Fluids

| Environmental Component | Categories of Environmental Impact | | |
	Environmental Persistence	Ecotoxicity	Bioaccumulation
Freshwater	Pw	Tw	Bw
Marine	Pm	Tm	Bm
Soil	Ps	Tx	Bs
Anaerobic	Pa	Ta	Ba

Source: ASTM D 6046-97, "Standard Classification of Hydraulic Fluids for Environmental Impact," *American Society for Testing and Materials*, West Conshohocken, PA. With permission.

To verify the performance of Bio-based Hydraulic Fluids (BHFs) which qualified under the MIL-PRF-32073 specification in the existing military construction equipment, a second phase of the field demonstration was initiated in 2005 using the five qualified BHFs and ten pieces of construction equipment (i.e., Bulldozers, Scrapers, Graders, Loader, Excavator, Cranes, etc.) utilized at the Engineering School at Fort Leonard Wood, MO (Midwest environments) [71]. The duration of this field test was one year. If the MIL-PRF-32073 fluids performed acceptably during the initial test period, the field demonstration was planned to extend by two more years in order to observe their storage and service life. In 2007, the first portion of these field demonstrations were successfully completed and the MIL-PRF-32073 fluids did not show any abnormal behavior with existing hydraulic systems and provided excellent performance equivalent to the existing petroleum-based oils (i.e., MIL-PRF-2104 oil and 10W oil). No incompatibility issues were observed for the tested construction equipment with the MIL-PRF-32073 fluids. In addition, no signs of incompatibility between BHFs and the existing petroleum-based hydraulic oils were observed and all laboratory data supported these results. The results implied that the existing conventional petroleum-based oils could easily be changed to BHFs without any major cleaning effort to the system. To determine the BHFs' service and storage life, the field demonstration was extended by a further two years at Fort Leonard Wood, MO. No failures have been observed in any of the tested hydraulic systems for three years [72]. In addition, the BHFs have not caused any abnormal behavior compared to the conventional petroleum-based hydraulic oils. Table 8.25 lists the physical and chemical property requirements of MIL-PRF-32073 specification [69].

2. Caterpillar BF-1 Specification

Need for comprehensive performance requirements. In the third quarter of 1991, Caterpillar released its first recommendations for the use of biodegradable hydraulic oils specifically for Caterpillar hydraulic systems. At that time, vegetable oils, particularly those extracted from rapeseed, were widely used and generally available. Over time, customers required a fluid that exceeded the capabilities of conventional rapeseed-based oils. Increased demands on the hydraulic system and on the fluid require biodegradable hydraulic fluids which demonstrate better oxidation resistance. In applications where better oxidation stability of the oil was required, it became necessary to use synthetic ester-based hydraulic fluids.

A survey of the available industry standards revealed no existing industry performance category for high-performance biodegradable hydraulic fluids meeting the Caterpillar system needs. With the advent of higher-performing components in hydraulic systems, hydraulic fluids capable of better lubricity, higher-temperature stability, and better oxidation stability became necessary. Although such products are now commercially available, there is a wide range of performance characteristics even among these new products. For this reason, Caterpillar published a comprehensive set of

TABLE 8.21
Environmental Persistence Classification—Aerobic Freshwater

A. **For Hydraulic Fluids Containing Less than 10 wt% O_2**

Persistence Designation	Ultimate Biodegradation Test Results	
	% Theoretical CO_2	% Theoretical O_2
Pw 1	≥60% in 28 days	≥67% in 28 days
Pw 2	≥60% in 84 days	≥67% in 84 days
Pw 3	≥40% in 84 days	≥45% in 84 days
Pw 4	<40% in 84 days	<45% in 84 days

B. **For Hydraulic Fluids Containing 10 wt% or More O_2**

Persistence Designation	Ultimate Biodegradation Test Results % Theoretical Co_2 or Theoretical O_2
Pw 1	≥60% in 28 days
Pw 2	≥60% in 84 days
Pw 3	≥40% in 84 days
Pw 4	≥40% in 84 days

C. **For All Hydraulic Fluids**

Persistence Designation	Primary Biodegradation Test Results % Loss of Starting Material
Pw-C	≥80% in 21 days
Pw 4	<80% in 21 days

hydraulic fluid requirements for a high-performance biodegradable hydraulic fluid, known as "BF-1, Biodegradable Hydraulic Fluid Requirements", which are shown in Table 8.26.

BF-1 requiremants. The performance properties of VDMA 24,568 were used as a starting point for developing the high-performance biodegradable hydraulic fluid requirement [61]. Because the scope of the VDMA 24,568 document covered the minimum requirements for all hydraulic fluids and not specifically for a high-performance product, Caterpillar selected test procedures and performance levels that differed from the VDMA requirement. Reaction from the survey participants indicated that several test procedures in the VMDA requirement proved to be adequate for evaluating vegetable-based biodegradable hydraulic oils but did not provide a good yardstick for comparing the properties of all available biodegradable hydraulic fluids. Better test methods were required. For example, a new test procedure covering fluid performance at low temperature was needed because the existing procedures were not adequate.

The performance parameters listed in VDMA 24,568 are: viscosity, pour point, flash point, insolubles, water content, a steel corrosion test, a copper corrosion test, oxidation stability, elastomer compatibility, air release, foaming, demulsification, FZG load test, vane pump test, density, ash content, and neutralization value (TAN). Although these tests gave a good first look at the performance of a hydraulic fluid, additional requirements, including some Caterpillar procedures, were added for a more complete representation of fluid performance. These included test procedures for homogeneity, fluid compatibility, foaming characteristics, humidity corrosion, elastomer compatibility, and frictional properties. These test methods will be described below.

In addition, this document also contains requirements for the cleanliness of new fluid, compatibility between fluids, the foaming characteristics of the oil when a small quantity of water is

TABLE 8.22
Tests of Biodegradability in Aerobic Aquatic Environments

Test Title	Measurement	Sponsoring Organization
D 5864, Modified Sturm Test Method for Determining the Aerobic Biodegradation of Lubricants	% Theoretical CO_2	ASTM
D 6731, Monometric Respirometry Test Method for Determining the Aerobic Biodegradation of Lubricants	% Theoretical O_2	ASTM
D 6139, EPA (Gledhill) Shake Flask Test Method for Determining the Aerobic Biodegradation of Lubricants		
	% Theoretical CO_2	ASTM
9429:1990, Technical Corrigendum 1, Water Quality	% Theoretical CO_2	ISO
—Evaluation in an aqueous medium of the "ultimate biodegradability of organic compounds"		
—Method by analysis of released carbon dioxide		
301 B, CO_2 Evolution Test (Modified Sturm Test)	% Theoretical CO_2	OECD
301 C, Modified MITI Test (I)	% Theoretical O_2	OECD
301 F, The Monometric Respirometry Test	% Theoretical O_2	OECD
Aerobic Aquatic Biodegradation Test	% Theoretical CO_2	US EPA
C:4-C: Carbon Dioxide (CO_2) Evolution	% Theoretical CO_2	EU
C:4-D: Monometric Respirometry[a]	% Theoretical O_2	EU
Primary Biodegradation Tests L-33-A-93, Biodegradability of Two Stroke Cycle Outboard Engine Oils in Water (Formerly L-33-T-82)	% Loss of extractable CH_2 groups	CEC

[a] Indicated OECD equivalent.

present, low-temperature storage requirements, the iodine number, demulsibility of water, viscosity index, four-ball wear, static and dynamic friction properties, and, most importantly, requirements for biodegradability and toxicity. With the exception of the iodine number, biodegradability, and toxicity, knowledge of these additional requirements is important in mobile hydraulic systems. Together, these tests are intended to ensure that a fluid described by BF-1 would provide high system performance and be truly biodegradable with low toxicity.

TABLE 8.23
Acute Ecotoxicity Classification

Ecotoxicity in Soil Designation	Ecotoxicity in Water Designation	Loading Rate (wppm), LL_{50}, IL_{50}, or EL_{50}
Ts 1	Tw 1	>1000
Ts 2	Tw 2	1000–100
Ts 3	Tw 3	100–10
Ts 4	Tw 4	<10

TABLE 8.24
Test for Ecotoxicity

Test Title	Sponsoring Organization
Soil Invertebrates	
A.8.5 Earthworm Survival (*Eisenea foetida*)	US EPA
207 Earthworm Acute Toxicity Test	OECD
Soil Plants	
A.8.7 Lettuce Root Elongation (*Lactuca sativa*)	US EPA
A.8.6 Lettuce Seed Germination (*Lactuca sativa*)	US EPA
208, Terrestrial Plants Growth Test	OECD
797.2750 of TSCA, Seed Germination/Root Elongation Toxicity Test	US EPA
797.2800 of TSCA, Early Seedling Growth Toxicity Test	US EPA
Water Plants	
Section 797.1050 of TSCA, Algae Acute Toxicity	US EPA
Section 797.1060 of TSCA, Freshwater Algae Acute Toxicity	US EPA
Section 797.1075 of TSCA, Freshwater & Marine Acute Toxicity Test	US EPA
C.3 Algae Inhibition Test	EU
Biological Test Method: Growth Inhibition Test using Freshwater Algae (*Selenastrum capricornutum*)	
201, Algae Growth Inhibition Test	OECD
A.8.4 Algae growth (*Selenasrum Capricornutum*)	US EPA
Water: Invertebrates	
20.2 *Daphnia* sp. Acute Immobilization Test and Reproduction Test	DECD
Biological Test Method: Acute Lethality Test using *Daphnia* spp.	Canada
C.2 Acute Toxicity for *Daphnia*	EU
A.8.2 *Daphnia pubex* and *Daphnia magna* Survival	US EPA
797.1300 of TSCA, *Daphnia* Acute Toxicity Test	ASTM
E 1440, Acute Toxicity Test Rotifer Brachions	ASTM
MENVIQ.92.03/800-D.Mag. 1.1 Determination de la Toxicite Lethal CL 500-48h (*Daphnia magna*)	Quebec
Water: Vertebrates	
203, Fish, Acute Toxicity Test	OECD
797.1400 of TSCA, Fish, Acute Toxicity Test	US EPA
797.1400 of TSCA, Fish, Acute Toxicity Test[a]	US EPA
C.1. Acute Toxicity for Fish[a]	EU
Biological Test Method: Acute Lethality Test Rainbow Trout	Canada
Biological Test Method: Referenced Method for Determining Acute Lethality of Effluents Trout	Canada
A.8.3 Fathead Minnow Survival	US EPA

[a] Denotes an OECD equivalent test.

TABLE 8.25
Chemical and Physical Requirements for MIL-PRF-32073 Product

Test	Method	Requirement				
		Grade				
		1	2	3	4	5
Viscosity, cSt	ASTM D445					
@ 40°C		13.5–16.5	19.8–24.2	28.8–41.3	41.4–50.6	61.2–74.8
@ −15°C, max.		300	500	1000	1600	2000
Viscosity index, min.	ASTM D2270	135	135	184	184	184
Evaporation loss, %, max. 100 C, 1 hr	ASTM E1131	1.5	1.5	1.5	1.5	1.5
Flash point,°C, min.	ASTM D92	160	160	240	250	265
Pour point,°C, max.	ASTM D97	−54	−42	−30	−25	−23
Water content, %, max.	ASTM D1744	0.05	0.05	0.05	0.05	0.05
Acid number, mg KOH, max	ASTM D664	2	2	2	2	2
Galvanic corrosion	FTM 5322	pass	pass	pass	pass	pass
Oxidation stability (PDSC), min.	ASTM D6186	15 (180°C)	15 (180°C)	15 (155°C)	15 (155°C)	15 (155°C)
Low temperature stability	FTM 3458	pass (−54°C)	pass (−40°C)	pass (−25°C)	pass (−15°C)	pass (−15°C)
Rust prevention	ASTM D665B	pass				same for all grades
Swelling of synthetic rubber, %	FTM 3603	10–30				same for all grades
Solid particle contamination:						
Particle size, micron, max.	ISO 4406	21,19,15				same for all grades
Foaming:						
Sequence 1, ml, max.	ASTM D892	65/10				
Sequence 2, ml, max.	ASTM D892	65/10				same for all grades
Sequence 3, ml, max.	ASTM D892	65/10				
Lubricity, scar diameter, mm, max.	ASTM D4172					
10 kg load		0.30				same for all grades
40 kg load		0.65				
Trace sediment, %, max.	ASTM D2273	0.005				same for all grades
Copper strip corrosion, max.	ASTM D130	1b				same for all grades
Biodegradability, %, min.	ASTM D5864 or ASTMD6731	60				same for all grades
Toxicity, LC50, mg/l, min.	OECD 203	1000				same for all grades
Storage stability, 12 months	FTM 3465 and 4.35.3*	pass				same for all grades
Fluid clarity	3.3.6*	pass				same for all grades

* These are paragraphs in MIL-PRF-32073 where the method is outlined.

TABLE 8.26
BF-1, High-Performance Biodegradable Oil Requirements for New Fluid Chemical and Physical Properties

Measured Property	Standard Test Procedure	Required Value	
Fluid cleanliness	ISO 4406	15/13 maximum	
Homogeneity	Caterpillar Procedure	Max. 0.01 Vol.% sedimentation	
Fluid compatibility	Caterpillar Procedure	No sedimentation	
Foaming characteristics	ASTM D892	Sequence I—25/0	
		Sequence II—50/0	
		Sequence III—25/0	
Foaming characteristics (with 0.1% water added)	ASTM D892 (modified)	Sequence I—25/0	
		Sequence II—50/0	
		Sequence III—25/0	
Humidity corrosion	Caterpillar Procedure	Minimum of 200 h to failure	
Copper strip corrosion	ASTM D1401	1: a minimum rating	
Low-temperature storage	Caterpillar Procedure	No precipitation after 168 h at –25°C	
Demulsibility	ASTM D1401	Minimum of 37 mL of water separated in 20 min.	
Iodine number	AOCS Da 15-48 (WIJS Method)	Report values	
Flash point	ASTM D92	Minimum 200°C	
Pour point	ASTM D97	Maximum –35°C	
Water content	ASTM D1744	Maximum of 0.1 vol.%	
Oxidation stability	ASTM D943 (modified—no water added)	Report values	
Viscosity	ASTM D445	**Temperature (°C)**	**Kinematic Viscosity (cSt)**
		0	780 max.
		40	41.4 min.
		40	50.6 max.
		100	6.1 min.
Viscosity index	ASTM D2270	150 min.	
Van pump	Vickers 35VQ25	Vane weight loss: 15 mg, maximum Ring weight loss: 75 mg, maximum	
FZG Rating	ASTM D5182	Minimum 11 stages	
Four-ball wear	ASTM D4172	0.40 mm, maximum scar	
Biodegradability	EPA 560/6-82-003	Minimum 60% (weight) biodegraded to carbon dioxide in 28 days	
Toxicity, water hazard	German Classification	WGK 0	
Toxicity, fish	OECD 203	LC_{50} >1000 ppm	
Friction properties	Caterpillar Procedure	See BF-1 requirements	

The measured chemical and physical properties, test method, and requirements as published in the first edition of BF-1 are listed in Table 8.26. The elastomer compatibility properties and limits are listed in Table 8.27. The specific values chosen for each of the requirements were obtained either through the recommendations of the reviewers of BF-1 or through testing high-performance fluids at the Caterpillar laboratories. As with all work, lessons were learned through the evaluation process. A further discussion of the values chosen, some of the problems encountered when developing the requirements, and some of the lessons learned on the tests will be given below.

Fluid cleanliness. Caterpillar chose ISO 15/13 cleanliness as the minimum cleanliness requirement for oil that is to be used for factory filling of Caterpillar machines. This requirement is intended to be applied only to new oil, not to oil that has been or is currently being used in a Caterpillar hydraulic system.

Homogeneity. A test method was developed to evaluate the compatibility of additives with the high-performance biodegradable fluid. This method is specified to measure the homogeneity of the formulated oil and to determine how well the additives will remain in solution under variable-temperature conditions. The test method requires that the fluid must be cooled to −32°C for 24 h, warmed to room temperature, and then centrifuged at 100,000 m/s^2. The presence of any sedimentation must be reported and the volume of the sedimentation must not exceed 0.01% of the total fluid volume.

Fluid compatibility. Customers owning and using Caterpillar machines may choose to covert their current equipment in order to use environmentally friendly oils. To do this, they must drain the system of its hydraulic fluid and replace it with new biodegradable oils. Unfortunately, the draining process does not completely eliminate all of the old oil from the system. No matter how carefully the oil is changed, residual oil will remain in the system. Because of this, the mixing of chemically-different types of oil will occur. For this reason, it is important that the oils that become mixed are compatible.

To avoid problems with additive dropout and immiscibility between fluids, Caterpillar has required that any fluid evaluated against the BF-1 requirements should be tested for compatibility with a mineral oil, a vegetable oil, and a synthetic ester. This is done by mixing 50 mL of the test fluid with 50 mL of each of the above-mentioned fluids. The mixture should be heated to 204°C and cooled to room temperature. The cooled fluid should be centrifuged at 100,000 m/s^2 for 30 min. A fluid that successfully passes this requirement will not show any sedimentation or precipitation following centrifuging.

Foaming characteristics. During thermal cycling, hydraulic systems are designed to draw in air in order to replace the lost volume previously occupied by the fluid. If the air being drawn into the system is humid, the hydraulic oil may become contaminated with water, which may act to help stabilize the formation of foam. Obviously, the presence of foam will make the system respond slowly or not at all, causing system malfunction and poor performance of the equipment. For this reason, Caterpillar requires that the fluid be evaluated for its foaming tendency by ASTM D892 dry and with 0.1% water added.

Humidity corrosion. Because water may be introduced into a hydraulic system over time, it is important to evaluate how well the fluid protects the hydraulic components from water corrosion. The test method used to evaluate this ability is a Caterpillar "in-house" procedure. Carbon-steel rods are prepared by cleaning any residual oils and dirt from their surface. Removing the thick oxide layers exposes a clean surface by abrasion. The clean rod is then immersed in the test fluid and subsequently suspended above a water bath which is being held at 32 ± 1°C. Fluids that prevent pitting to a level of six or more spots per 2.54 cm (linear inch) for more than 200 h have passed the test.

Low-temperature storage. Prior experience with rapeseed-based (HETG) hydraulic fluids indicated that the low-temperature properties of the fluids were time dependent. A fluid in the machine exposed to low temperatures for short periods of time (up to 12 h) would likely flow at machine start-up. However, extended cold-temperature exposure times would result in fluid gelling, causing resistance to flow and ultimately resulting in pump damage due to the lack of lubrication at start-up.

TABLE 8.27
BF-1, High-Performance Biodegradable Oil Requirements for New Fluid Elastomer Compatibility

Compound	Test Temp. (°C)	Shore Hardness Change (PTS)	Relative Volume Change (%)	Loss in Tensile Strength (% max.)	Loss in Elongation (% max.)	Residual Elongation (% min.)
HNBR	100	+10/−15	−3/+20	50	50	80
NBR	100	+10/−15	−3/+20	50	50	80
FKM	100	+10/−15	−3/+20	50	50	80
AU	80	+10/−15	−3/+20	50	50	80

Note: Table Values are valid as of 1 January 1998.

The test method for low-temperature storage requires that before testing, the fluid should be heated to remove any trapped air. The fluid is then cooled to −25°C and held at that temperature for a full week. The fluid should be examined every 24 h to determine its fluidity and to ensure that no precipitation has occurred. The fluid has passed this test if it has remained fluid for 168 h and has not had any solid precipitation over the same period. This procedure was developed by an ASTM committee and has been adapted to meet Caterpillar's requirements.

Iodine number. The degree of unsaturation present within a molecule will indicate how well a base fluid is capable of resisting oxidation and how well the fluid will flow at low temperatures. A measure of the iodine number for the fluid will provide information on the degree of unsaturation within the molecules. This requirement was added in order to help evaluate the fluid's ability to resist oxidation and estimate its low-temperature properties.

Water content. It is a well-known fact that the presence of water in ester-based fluids may be detrimental to the chemical stability of the fluid. At high temperature and in the presence of water, esters can revert back to an alcohol and an acid. However, field evaluations on different ester-based fluids have indicated that the critical concentration of water is higher than the water level measured in the machines. Because of this, Caterpillar does not believe it is necessary to include a test for hydrolytic stability in their BF-1 requirements.

Although hydrolysis of the fluid does not appear to be a problem in hydraulic systems, it is important to keep the amount of water to a minimum. The presence of water in a hydraulic system may cause corrosion on the mechanical components as well as foaming. Several hydraulic component manufacturers have determined that fluids having water content of around 0.1% may cause damage to mechanical hydraulic components. Because of this, Caterpillar requires that new fluids meeting BF-1 have less than 0.1% water content.

Elastomer compatibility. To avoid sealing problems, elastomers that are commonly used in hydraulic systems should be compatible with the fluid. BF-1 requires that the compatibility between hydrogenated nitrile (HNBR), nitrile (NBR), fluorocarbon rubber (FKM), urethane (AU), chloroprene, and the hydraulic fluid be evaluated by measuring the hardness, volume change, loss in tensile strength, loss in elongation, and residual elongation in the elastomers. The BF-1 requirements are shown in Table 8.27.

Oxidation stability. DIN 51554-3, better known as the "Baader Oxidation Test", was chosen as the standard test by the VDMA in its requirement. At the time its requirement was being developed, this test was considered to be one of the best oxidation tests available for biodegradable fluids. Although the Baader test had been applied to vegetable-oil-based hydraulic fluids, it could not distinguish performance levels among synthetic esters.

Oxidation tests similar to the ASTM D943 (TOST) test were considered "state-of-the-art" for mineral oils, but they were not capable of discriminating between biodegradable oils. The primary reason was that the TOST test required the addition of water to accelerate the degradation process. Biodegradable hydraulic fluids will oxidize by a reaction similar to mineral oils, but the presence of water at elevated temperatures will cause the ester-based oils to hydrolyze. The TOST test measures the acid content of the fluid. Because both oxidation and hydrolysis contribute to the acid content, this test does not provide a valid measure for oxidation alone.

One solution to the problem was to perform the TOST oxidation test without any added water ("dry" TOST). This would eliminate the hydrolysis problem experienced with ester-based oils. To determine if this approach was sound, a well-known oil company formulated three different synthetic ester-based hydraulic fluids having known differences in their oxidation stability. The laboratory had conducted the Baader test and the TOST test with water on a rapeseed fluid and the two least stable synthetic ester fluids, and performed the "dry" TOST test on all four fluids. The data obtained by that laboratory are listed in Table 8.28.

The fluids were designed to demonstrate a wider performance gap than measured by the Baader test. Note that the viscosity and acid value increases obtained by the Baader test on the rapeseed fluid was expected. The data obtained for the synthetic esters indicated that little oxidation occurred. The only conclusion was that the Baader test was not capable of adequately distinguishing between ester fluids which demonstrate differences in their oxidation stability. The data obtained on the fluids using the regular D943 TOST test are also given. Note that the TOST test does not clearly indicate whether the two synthetic esters exhibited different oxidation stabilities.

Further testing with a modified TOST test provided the results represented in Figure 8.14. The data show the change in the acid value over time. These results indicate that the "dry" TOST test produces data that discriminate and are consistent with the known oxidation stability of the four oils. Caterpillar has chosen to require that the "dry" TOST test be performed in the BF-1 requirement.

The information obtained from the above-described oxidation test evaluation has been shared with many groups that have been working on biodegradable hydraulic fluid specifications. The

TABLE 8.28
Comparison of Oxidative Performance of Biodegradable Oils in Baader, TOST, and Dry TOST Tests

	Fluid A Rapeseed	Fluid B Synthetic Ester	Fluid C Synthetic Ester
Baader Test			
Based on DIN 51554, Part 3			
Acid value change	+1.8[a]	−0.3[b]	0.1[c]
Viscosity increase at 40°C	24%	−0.20%	0.40%
ASTM D942, TOST Test			
Hours to acid value of 2 mg KOH/g	72	79	95
ASTM D942, TOST Test			
Modified, no water			
Hours to acid value of 2 mg KOH/g	<50	168	1260

[a] 72 h at 95°C.
[b] 72 h at 110°C.
[c] 100 h at 120°C.

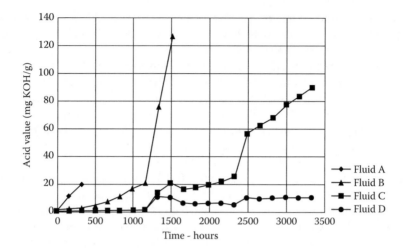

FIGURE 8.14 Oxidative stability of biodegradable hydraulic fluid: ASTM D943—"dry" TOST (no water added).

VDMA and ISO both responded by further evaluation of the "dry" TOST procedure. Also, the ASTM is investigating other candidate test procedures.

Viscosity. The initial release of BF-1 required only ISO viscosity grade 46 because it covers the best set of viscometric properties over a wide temperature range. The addition of requirements for fluids with ISO viscosity grade 32 is being considered, but further testing for its low-temperature storage, oxidation, and elastomer compatibility properties are required.

Friction and wear testing. An understanding of the friction and wear characteristics of a fluid is necessary in order to ensure that the lifetime of the hydraulic components is maximized. Caterpillar has extensive experience in evaluating these properties. Some hydraulic systems have been designed to use the fluids in multiple roles. It is not uncommon to find a machine that uses its hydraulic fluid in power-transmission compartments or in braking applications. To avoid performance problems, it is necessary to know the function properties of the fluid and to design machines to compensate for differences in friction coefficients demonstrated by environmentally friendly fluids. Caterpillar has developed a standard test method and its associated machinery to evaluate the frictional properties of all hydraulic fluids. Further information may be found in BF-1.

The specific wear tests (vane pump, FZG, and four-ball wear) have been specified on the basis of earlier work performed on antiwear hydraulic fluids at Caterpillar. It was important to understand the extreme pressure characteristics of the fluid by performing a FZG gear test, the lower-pressure wear-prevention properties of fluids by the four-ball wear test, and, finally, how the fluid will perform when it has been evaluated in a laboratory-scale component test. Each test provides a piece of the information necessary to adequately describe the wear-prevention characteristics of the fluids.

Environmental testing. BF-1 includes standard requirements for determining the environmental impact that a high-performance biodegradable fluid will have on the environment. The biodegradability and the toxicity of the fluid must be known in order to determine its affect on the environment. The levels of the biodegradability and toxicity requirements were selected to best represent the current attitude of the community.

The test methods required by BF-1 were selected on the basis of their applicability. It was believed that the CEC methods for biodegradability did not define biodegradability well enough and, therefore, the US EPA methods had better application. The German Water Hazard Classification (WGK rating) provides an excellent method for determining the overall toxicity of a substance to water-dwelling creatures. The inclusion of this parameter is very important. Additionally, it was determined that setting a limit for fish toxicity was necessary.

The BF-1 requirement is intended to provide guidelines for the physical and chemical performance of a high-performance biodegradable fluid for use in mobile hydraulic systems. This set of requirements is believed to be comprehensive and includes pertinent information about the system requirements for a biodegradable fluid. This does not mean that the BF-1 document is final. A substantial amount of additional refinement must be carried out to ensure that the requirements outlined in BF-1 adequately represent the needs of mobile-product hydraulic systems.

8.3   CONCLUSIONS

Although mineral-oil-based hydraulic fluids exhibit some of the best overall properties, they have two significant deficiencies: fire resistance and biodegradability, which may preclude their continued use in many applications. While there is little that can be done about the poor fire resistance of mineral oils—except engineering or hardware modifications—biodegradability may be improved. In general, hydrogen reduction, or the removal of aromatic compounds and increase of overall paraffinic composition, will improve aquatic biodegradability, though not to vegetable oil or synthetic ester levels. However, even highly-refined mineral oil compositions are among the most persistent in soil biodegradability tests.

Two biodegradable alternatives to mineral-oil-based hydraulic fluids are vegetable oils and synthetic esters. They are more biodegradable than mineral oil. Of the two base stocks, vegetable oils are somewhat more biodegradable, although that depends somewhat on molecular structure. In soil tests, vegetable oils are clearly more biodegradable. One vegetable oil (sunflower oil)—although biodegradable—exhibited significant toxicity to seedling growth and yield.

Polyol esters, usually those based on trioleate esters of trimethylolpropane, exhibit excellent thermal stability relative to vegetable oils. Formulated fluids, although exhibiting excellent biodegradability, exhibit relatively poor soil biodegradability—a notable disadvantage in addition to their relatively higher cost.

Another class of synthetic fluids is the diesters. Like the polyol esters, all diester-based hydraulic fluids demonstrate excellent thermal stabilities when compared with vegetable oils and exhibit a range of oxidative stabilities. In diester fluids, the degree of oxidation stability is related to the amount of unsaturation on the fatty acid chain. The more saturated the fluid, the greater the oxidative stability. Therefore, a diester hydraulic fluid with good thermal and oxidative stabilities would be one that has few or no double bonds in its fatty acid side chains. Although diesters are not as biodegradable as vegetable oils, they demonstrate good to excellent biodegradability. The rate at which diester-based fluids degrade varies widely, depending on the length and number of branches of the fatty acid side chain.

It was shown that water–glycol hydraulic fluids offer an excellent alternative to vegetable oils, as they provide *both* excellent fire resistance and biodegradability. In addition to exhibiting potentially excellent biodegradability properties, water-glycol hydraulic fluids are also relatively nontoxic.

REFERENCES

1. Jones, N. "Managing Used Oil," *Lubes Greases*, 1996, 2(6), pp. 20–23.
2. Mustokoff, M.M. and Baylinson, J.E. "No Case Is Too Small," *Hydraulics Pneumatics*, February, 1995, pp. 35–37.
3. Eichenberger, H.F. "Biodegradable Hydraulic Lubricant—An Overview of Current Developments in Central Europe," *SAE Technical Paper Series*, Paper 910962, 1991.
4. Meni, J. "Selection of an Environmentally Friendly Hydraulic Fluid for Use in Turf Equipment," *SAE Technical Paper Series*, Paper 941759, 1994.

5. Ohkawa, S. "Rough Road Ahead for Construction Machinery Lubes," *Lubes Greases*, 1995, 1(2), pp. 20–23.

6. Tocci, L. "Mother Nature on Lubes: No Simple Choices," *Lubes Greases*, 1995, 1(5), pp. 16–19.

7. Mang, T. "Environmentally Friendly Biodegradable Lube Base Oils—Technical and Environmental Trends in the European Market," in *Adv. Prod. Appl. Lube Base Stocks, Proc. Int. Symp.*, 1994, pp. 66–80.

8. Fischer, H. "Environmental Labeling in German Award Criteria for Hydraulic Fluids," *SAE Technical Paper Series*, Paper 941078, 1994.

9. Chien, J.Y. "The Dirt on Environmentally Friendly Fluids," *Hydraulic Pneumatics*, May, 1995 pp. 47–48.

10. *Environmental Choice*, Brochure available from Environmental Choice, Environment Canada, Ottawa, Ontario, K1A 0H3.

11. ASTM D 6006-96, "Standard Guide for Assisting Biodegradability of Hydraulic Fluids," *Annual Book of ASTM Standards, Vol. 05.03*, American Society for Testing and Materials, Conshohocken, PA, pp. 1290–94.

12. Clark, D.G. "The Toxicology of Some Typical Lubricating Oil Additives," Erdol Kohle, 1978, 31(12), p. 584.

13. Hewstone, R.K. "Environmental Health Aspects of Additives for the Petroleum Industry," *Regul. Technol. Pharmacol.*, 1985, 5, pp. 284–94.

14. Cisson, C.M., Rausina, G.A. and Stonebraker, P.M. "Human Health and Environmental Hazard Characterization of Lubricating Oil Additives," *Lubric. Sci.*, 1996, 8(2), pp. 145–177.

15. Biggin, R.J.C. "Additives for Lubricants with Improved Environmental Compatibility," in *Adv. Prod. Appl. Lube Base Stocks, Prod. Int. Symp.*, 1994.

16. Busch, C. and Backè, W. "Development and Investigation in Biodegradable Hydraulic Fluids," *SAE Technical Paper Series*, Paper 932450, 1993.

17. Hydrick, H. "Synthetic vs. Vegetable," *Lubric. World*, 1995, May, pp. 25–26.

18. Padavich, R.A. and Honary, L. "A Market Research and Analysis Report on Vegetable-Based Industrial Lubricants," *SAE Technical Paper Series*, Paper 952077, 1995.

19. Honary, L.A.T. "Potential Utilization of Soybean Oil as an Industrial Hydraulic Oil," *SAE Technical Paper Series*, Paper 941760, 1994.

20. Honary, L.A.T. "An Investigation of the Use of Soybean Oil in Hydraulic Systems," *Bioresource Technology*, 1996, 56, pp. 41–47.

21. Scott, S.D. "Biodegradable Fluids for Axial Piston Pumps & Motors—Application Considerations," *SAE Technical Paper Series*, Paper 910963, 1991.

22. Naegly, P.C. "Environmentally Acceptable Lubricants," in *Seed Oils for the Future*, MacKenzie, S.L. and Taylor, D.C. eds., 1992, pp. 14–25.

23. In-Sik Rhee, Velez, C. and Von Bernewitz, K. "Evaluation of Environmentally Acceptable Hydraulic Fluids," TARDEC Technical Report No. 13640, U.S. Army Tank-Automotive Command Research, Development and Engineering Center, Warren, MI, March 1995.

24. Legisa, I., Picek, M. and Nahal, K. "Some Experiences with Biodegradable Lubricants," *J. Synth. Lubric.*, 1997, 13(4), pp. 347–60.

25. Anon, "Hydraulic Fluids Are Getting More Friendly," *Fluid Power*, 1992, No. 7, pp. 68–73.

26. Honary, L. "Performance of Selected Vegetable Oils in ASTM Hydraulic Tests," *SAE Technical Paper Series*, Paper 952075, 1995.

27. Ohkawa, S., Konishi, A., Hatano, H., Ishihama, K., Tanaka, K. and Iwamura, M. "Oxidation and Corrosion Characteristics of Vegetable-Base Biodegradable Hydraulic Oils," *SAE Technical Paper Series*, Paper 951038, 1995.

28. Cheng, V.M., Galiano-Roth, A., Marougy, T. and Berezinski, J. "Vegetable-Based Hydraulic Oil Performance in Piston Pumps," *SAE Technical Paper Series*, Paper 941079, 1994.

29. Cheng, V.M., Wessol, A.A., Baudouin, P., BenKinney, M.T. and Novick, M.J. "Biodegradable and Non-Toxic Hydraulic Oils," *SAE Technical Paper Series*, Paper 910964, 1991.

30. Reichel, J. "Biologically Quickly Degradable Hydraulic Fluids," *Sauer Sundstrand Technical Application Information* ATI 9101 (Status 03/91).

31. "Environmentally Compatible Fluids for Hydraulic Components," *Mannesmann Rexroth*, Technical Bulletin No. 03 145/05.91.

32. "Guide to Alternative Fluids," Vickers Technical Bulletin No. 579, 11/92.

33. "Cat Biodegradable Hydraulic Oil," Caterpillar (BIO HYDO) Product Data Sheet, No. PEHP1021.

34. Erdman, K.D., Kling, G.H. and Tharp, D.E. "High Performance Biodegradable Fluid Requirements for Mobile Hydraulic Systems," *SAE Technical Paper Series*, Paper 981518, 1998.

35. Thiel, D. "Experiences with Sealing Materials, Hydraulic and Lubricating Oils and Biologically Decomposable Hydraulic Fluids," *Gepgyartastechnologia*, 1993, 33(9–10), pp. 433–41.

37. "Environmentally Acceptable Hydraulic Fluids HETG, HEPG, HEE for Axial Piston Units," Brochure No. RE 90221/02.92, available from *Mannesmann Rexroth*, Hydromatik GmbH, Elchingen, Germany.

38. OECD standards and methods are available from the Organization for Economic Cooperation and Development, Paris.

39. Honary, L.A.T. "Soy-Based Hydraulic Oil: A Step Closer," *Off-Highway Eng.*, April, 1995, pp. 15–18.

40. Honary, L.A.T. Director of the ABIL (Ag-Based Industrial Lubricants) program at the University of Northern Iowa, Waverly, personal communication.

41. Roberts, F.H. and Fife, H.R. U.S. Patent 2,425,755 (1947).

42. Lewis, W.E.F., U.S. Patent 4,855,070 (1989).

43. Totten, G.E. and Webster, G.M. "High Performance Thickened Water–Glycol Hydraulic Fluids," in *Proceedings of the 46th National Conference on Fluid Power*, 1994, pp. 185–93.

44. Litt, F.A. "Standards for Environmentally-Friendly Hydraulic Fluids," in *National Fluid Power Conference*, 1996.

45. Hoel, D.I. "Lubricant Development Meets Biology," *ASTM Standardization News*, June, 1994, pp. 42–45.

46. Hooper, D.L. and Hoel, D.I. "Lubricants, the Environment and ASTM D02," *SAE Technical Paper Series*, Paper 961727, 1996.

47. Wilkinson, J. "Biodegradable Oils—Design, Performance, Environmental Benefits and Applicability," *SAE Technical Paper Series*, Paper 941077, 1994.

48. Baggott, J. "Biodegradable Lubricants," in *Institute of Petroleum Symposium: Life Cycle Analysis and Eco-Assessment in the Oil Industry*, 1992.

49. Gilron, G. (Beak Internation, Brompton, Ontario) "Review of Draft ISO Standard Criteria for Toxicity and Biodegradability for Ecologically Acceptable Water Glycol Hydraulic Fluids," Letter to J. Cerf (UCC-Canada), 1996.

50. Völtz, M., Yates, N.C. and Gegner, E. "Biodegradability of Lubricant Base Stocks and Fully Formulated Products," *J. Synth. Lubric.,* 1995, 12(3), pp. 215–30.

51. Girling, A.E. "Preparation of Aqueous Media for Aquatic Toxicity Testing of Oils and Oil-Based Products: A Review of the Published Literature," *Chemosphere*, 1989, 19(10/11), pp. 1635–41.

52. Singh, M.P., Chhatwal, V.K., Rawat, B.S., Sastry, M.I.S., Srivastava, S.P. and Bhatnagar, A.K. "Environmentally Friendly Base Fluids for Lubricants," in *Adv. Prod. Appl. Lube Base Stocks, Proc. Int. Symp.*, Singh, H., Rao, P. and Tata, T.S.R. eds., McGraw-Hill, New Delhi, 1994 pp. 362–70.

53. Kiovsky, T.E., Murr, T. and Voeltz, M. "Biodegradable Hydraulic Fluids and Related Lubricants," *SAE Technical Paper Series*, Paper 942287, 1994.

54. Henke, R. "Increased Use of 'ECOFLUIDS' May Put a Veggie in Your Hydraulic Reservoir," *Diesel Progr., Engines Drives*, September, 1994, pp. 7–9.

55. Haigh, S.D. "Fate and Effects of Synthetic Lubricants in Soil: Biodegradation and Effect on Crops in Field Studies," *Sci. Total Environ.*, 1995, 168, pp. 71–83.

56. Haigh, S.D. "Determination of Synthetic Lubricant Concentrations in Soil During Laboratory-Based Biodegradation," *J. Synth. Lubric.*, 1994, 11(2), pp. 83–93.

57. Zhou, E., Shanahan, A., Mammel, W. and Crawford, R.L. "Biodegradability Study of High-Erucic-Acid-Rapeseed-Oil-Based Lubricant Additives," in *Monitoring and Verification of Bioremediation*, Hinchoe, R.E., Douglas, G.S. and Ong, S.K. eds., Battella Press, 1995, pp. 97–103.

58. Gilron, G. (Beak Internation, Brompton, Ontario), UCON® Hydrolube HP-5046 Ready Biodegradability Criteria," Letter to J. Cerf (UCC-Canada), 1997.

59. ASTM D 6046-97, "Standard Classification of Hydraulic Fluids for Environmental Impact," *American Society for Testing and Materials*, West Conshohocken, PA.

60. ASTM D 5864-95, "Standard Test Method for Determining Aerobic Aquatic Biodegradation of Lubricants or Their Components," *Annual Book of ASTM Standards, Vol. 05.03, American Society for Testing and Materials*, Conshohocken, PA, pp. 1135–41.

61. VDMA 24,568, "VDMA Harmonization Sheet, Fluid Power, Rapidly Biologically Degradable Hydraulic Fluids, Minimum Technical Requirements," March, 1994.

62. "Guidelines for Designating Bio-based Products for Federal Procurement", *Federal Resister*, Vol.71, No. 51, USDA, March 16, 2006.

63. In-Sik Rhee, "Development of Military Biodegradable Hydraulic Fluids", SAE 2002-01-1503, 2002.

64. Qualification Product List (QPL) for MIL-PRF-32073 Specification, August, 2005.

65. In-Sik Rhee, "Assessing the Biodegradability of Hydraulic Fluids Using a Bio-kinetic Model", *STLE Tribology Transactions*, Vol.51, Issue 1, PP-68-73, 2008.

66. ASTM D 6731, "Standard Test Method for Determining the Aerobic Aquatic Biodegradability of Lubricants or Components in a Closed Respirometer", *Annual Book of ASTM Standards, Vol. 05.03*, American Society for Testing and Materials, Conshohocken, PA.

67. ASTM D 6139, "Standard Test Method for Determining the Aerobic Aquatic Biodegradation of Lubricants or Their Components Using the Gledhill Shake Flask", *Annual Book of ASTM Standards, Vol. 05.03*, American Society for Testing and Materials, Conshohocken, PA.

68. ASTM D 7373, "Standard Test Method for Predicting Biodegradability of Lubricants Using a Bio-kinetic Model", *Annual Book of ASTM Standards, Vol. 05.03*, American Society for Testing and Materials, Conshohocken, PA.

69. Military Specification, MIL-PRF-32073, April, 2007.

70. In-Sik Rhee, "A Compatibility Study for Bio-based Hydraulic Fluids under Laboratory Environments", *National Conference on Fluid Power (NCFP)*, 105-4.3. 2005.

71. In-Sik Rhee, "Evaluation of Bio-based Hydraulic Fluids in Military Construction Equipment", National *Conference on Fluid Power (NCFP)*, 2008.

72. In-Sik Rhee, Trip Report for Bio-based Hydraulic Fluids Field Demonstration, 2005-2008.

9 Fire-Resistance Testing Procedures and Standards of Hydraulic Fluids

John V. Sherman[*]

CONTENTS

9.1 INTRODUCTION

Fire is a major hazard in many industrial, transportation, and military applications, requiring the use of approved fire-resistant hydraulic fluids either utilized alone, or in conjunction with other fire-suppression techniques. Trends in the hydraulic industry for compact, higher-pressure hydraulic systems with smaller reservoirs to increase efficiency result in hydraulic fluids operating at higher temperatures, higher pressures, and with less time in the reservoir to cool and de-aerate. These factors may increase the risk of fire in the hydraulic system due to the greater potential for initial

---

[*] This chapter is a revision of the chapter titled "Fire-Resistance Testing Procedures" by Glenn M. Webster and George E. Totten from the *Handbook of Hydraulic Fluid Technology*, 1st Edition.

combustibility of the fluid and an increased rate of fluid degradation which may also affect the fire-resistance properties of the fluid.

The need has perhaps never been greater for test methods and industry standards that are reproducible, can accurately discriminate the level of fire-resistance between hydraulic fluids, and that can directly relate to a specific application fire hazard or hazards. Many test methods have been developed to reproducibly simulate industrial hydraulic fluid ignition hazards. Numerous factors must be modeled to adequately predict hydraulic fluid fire risk, including the type of flame or source of ignition, amount of available energy relative to the amount of fluid, and the physical state of the fluid [1]. The selection of appropriate tests to model fire resistance is usually mandated either by governments, industry associations, insurance companies, or standards organizations.

Hydraulic fluid fires have been disastrous with respect to the loss of buildings, equipment, and, most importantly, human lives. Many of these fire losses have been detailed in the literature [2–6]. Although fire is the major hazard that may be encountered with pressurized hydraulic equipment, injuries and fatalities have also occurred due to the force of uncontrolled high-pressure, high-velocity fluid jets through ruptured hoses [7,8]. Many hydraulic systems must operate in an environment where potential sources of ignition are present and hydraulic fluids may be ignited. Hazards of hydraulic fluid fires include heat release rates, generation of corrosive and toxic combustion compounds, and smoke.

The Bois du Cazier mining disaster of 1956 at Marcinelle, Belgium, in which 262 miners died [9], was the initiator that prompted the European governments to action. As a result of this disaster, ECSO countries, the National Coal Board, and the Ministry of Power in the United Kingdom initiated procedures to investigate the use of mineral oil hydraulic fluids underground. This study revealed the following:

- Improperly-designed or imperfect hydraulic systems may substantially increase the temperature of the equipment and, subsequently, the temperature of the hydraulic fluid.
- The ignition of high-pressure fluid emissions due to equipment failure or line rapture can result in hydraulic fluid fires.
- Hydraulic fluid mists spraying from a high-pressure hose fracture could produce an electrostatic charge, causing fluid ignition.
- Underground fluid storage poses an increased fire risk and may serve as a fuel source during a fire.

A committee was subsequently formed by the National Coal Board and the Ministry of Power to produce a draft of testing requirements for fire-resistant fluids. The following criteria for defining a fire-resistant hydraulic fluid were proposed:

- The fluid shall demonstrate fire-resistance characteristics in compliance with adopted standards.
- The performance and longevity of hydraulic equipment will not be reduced below acceptable limits.
- Fire-resistant fluids shall not exhibit health hazards and will be environmentally acceptable.

The 1st "Luxembourg Report" was published by the European Mines Safety Commission in 1961 [10]. The report set out hydraulic fluid requirements for use in coal mines, including test methods and limits for fire-resistance, physical, chemical, and health properties. In the United States, fire-resistant hydraulic fluid regulations for the coal mining industry were published under Title 30 of the U.S. Code of Federal Regulations in 1959. In 1965, the Factory Mutual Research Corporation (FMRC) published their standard 6930 "Less Flammable Hydraulic Fluids" to provide the insurance industry with a means of fire-resistant hydraulic fluid assessment. The ASTM D.02 N.06 Subcommittee was formed in 1966 to develop "improved" testing methods for hydraulic fluid applications. A 1966

ASTM symposium on hydraulic fluid flammability was held to examine fire-safety testing procedures which were in existence at that time for the potential incorporation into national standards [11]. A second international symposium was held in 1996 to update current international activities [12].

In this chapter, an overview of the test procedures which have been reported to evaluate and quantify the flammability properties of hydraulic fluids with respect to different types of ignition sources will be provided, along with an outline of the standards utilized for the assessment of fire-resistant hydraulic fluids for use in general industrial applications.

9.2 DISCUSSION

9.2.1 Fire Mechanisms

Fluid flammability is dependent on flammability mechanisms. Fire or combustion is an oxidation reaction consisting of three stages:

1. Initiation
2. Development
3. Termination

Many factors influence the duration and severity of these three stages [13]. Fire occurs in the vapor phase at the liquid surface and is dependent on the fuel availability, energy required to sustain the flame, and the presence of oxygen. The absence or removal of any one of these components will cause extinction of the fire.

Ease of ignition is a function of fluid vapor, which is dictated by fluid viscosity and molecular weight, and by the reactivity of the vapor with oxygen, which, in turn, is dependent on chemical structure and thermal stability.

Reaction among the three components—fuel, energy, and oxygen—are required for fire generation and heat release. Heat may be accompanied by a cool (barely visible) flame or a hot (visible) flame.

The combustion reaction will continue after the initiation stage if the heat released is sufficient to provide the energy required to sustain the reaction of the fuel with oxygen. Therefore, heat release is an important variable in predicting fire potential. Continued fluid combustion is dependent on the rate of heat loss due to convection and radiation of heat through the air conduction and through the liquid which must be less than the sum of the rate of heat emitted by the ignition source and that available from the combustion process [13].

9.2.2 Ignition Sources

Hydraulic fluid ignition often occurs when there is a leak or break in the hydraulic system, which is usually operating at high pressures, and results in a spray, stream, or mist of hydraulic fluid. The fluid may then ignite if it encounters an ignition source such as an electrical spark, flame, or hot surface [14].

The potential for combustion of these various ignition sources must also be considered. For example, hot exhaust gases, hot metal surfaces, pre-existing fire, or a spark may all potentially cause hydraulic fluid ignition [15].

A hydraulic leak or spill may wet a hot surface or form a pool, which may then be ignited by a nearby heat source. During ignition, a distillation process occurs which separates the hydraulic fluid into several components. One or more of these volatile components may be easily ignited by a spark or flame.

Porous or wick-like material—such as pipe insulation, cardboard boxes, textiles, or even coal from mining operations—may absorb hydraulic fluid. This fluid may slowly oxidize and form peroxide by-products, which have the potential to undergo subsequent spontaneous combustion [14].

TABLE 9.1
Hydraulic Fluid Flammability Conditions

Source of Ignition	Fluid Condition	Environment
Flame	Pool or puddle	Air temperature
Spark	Spray, stream or atomized	Air current
Hot metal surface	Vaporized	Equipment location
Electrical contact	Wicking	Equipment insulation

Table 9.1 summarizes several fire hazards to consider when selecting appropriate flammability test procedures [1].

Factors to consider when selecting a hydraulic fluid include the following [1]:

- Proximity of the fire hazard to the hydraulic equipment
- Availability of fire-resistant fluids
- Fluid properties
- Required equipment design changes

Various causes have been reported for naval fires and explosions of hydraulic fluid on shipboard systems. Potential causes of ignition include the following [16]:

- Rapid compression or shock-wave ignition
- Formation of carbon-oxygen complexes
- Presence of a porous, oil-soaked medium
- Peroxide formation
- Spontaneous ignition
- Spray ignition
- Electrostatic discharge

Because various sources of ignition may be present in a hydraulic application, it is important that the test method selected will reasonably model the fire risk where the fluid will be used.

9.2.3 FLUID CLASSIFICATION

All fire-resistant hydraulic fluids may be divided into two basic categories: those that derive their fire resistance from the presence of water, and those that demonstrate fire-resistant qualities by their chemical composition or molecular structure [17].

The general designation for hydraulic fluids is HF [18]. Fire-resistant hydraulic fluids are further classified by composition according to ISO 6734/4 [19], as shown in Table 9.2.

TABLE 9.2
ISO Classification of Fire-Resistant Hydraulic Fluids

Fluid Classification	Fluid Description
HFAE	Oil-in-water emulsions, typically more than 80% water content
HFAS	Chemical solutions in water, typically more than 80% water content
HFB	Water–in-oil emulsions
HFC	Water-polymer solutions, typically less than 80% water
HFDR	Water-free synthetic fluids consisting of phosphate esters
HFDU	Water-free synthetic fluids of other compositions than HFDR

9.2.4 Fire-Resistance Testing Strategies

One of the greatest concerns in fire-resistance testing is the selection of a procedure that adequately models the application of interest. Traditionally, most tests conducted for modeling fluid flammability characteristics are the following [20]:

- Flash point
- Fire point
- Spontaneous-ignition temperature

Some of the earliest work on fire-resistance testing and testing strategies for hydraulic fluid took place at the U.S. Naval Research Laboratory during World War II [21]. Combustion of hydraulic oils—long known as a major source of fire hazard in military equipment—increased dramatically with the use of high-velocity and incendiary ammunition during World War II. The first step in the development of fire-resistant hydraulic fluids at the U.S. Naval Research Laboratory was to determine the type and number of tests required to efficiently simulate the fire hazards of greatest concern. The laboratory developed a new method for measuring the flammability of a fluid under controlled conditions. The method determined the volume percentage of oxygen necessary to permit an arc-ignited flame to propagate along a metal tube filled with a dispersed mist of the test fluid. The minimum oxygen concentration required to form an explosive mixture was designated as the "oxygen demand". A second verification test simulating the real combat hazards to military equipment was the use of 50-caliber incendiary bullets to ignite the test fluid. The set of test methods determined to evaluate a wide selection of hydraulic fluids at the laboratory were:

- Flash point
- Fire point
- Spontaneous-ignition temperature
- Oxygen demand flammability
- Incendiary fire

Two fundamental parameters should be considered when selecting a test to model an application [22]: ignition resistance and flame propagation. These two parameters model fluid ignition and the propensity of the fluid to continue to burn once ignited. The experimental strategy is twofold: First, the test should provide information regarding resistance to ignition; second, if ignited, flame propagation must be considered.

Because each application has its own unique exposure (such as pool, spray, or mist) and ignition conditions (such as a hot metal surface or open flame), multiple tests are typically conducted [2]. The complexity of the thermal ignition process has led to the development of a variety of tests designed to model particular combustion mechanisms. These tests attempt to simulate flammability hazards typically encountered in industry with respect to exposed hot pipes, ventilated surfaces, surface material variations, and scale. Although there are many test procedures available, the testing, quantification, and interpretation of thermal ignition data are some of the most difficult problems in fire hazard evaluation [13,23].

9.2.5 Fluid Flammability Tests

9.2.5.1 Fluid Volatility Characterization

This section will summarize the testing procedures normally required to model flammability potential due to fluid volatility.

Open-cup flash point and fire point. The relative fire resistance of non-aqueous hydraulic fluids may be characterized by their flash and fire points. These procedures are described in ASTM D 92-05

(Standard Test Method for Flash and Fire Points by Cleveland Open Cup) [22–24]. Flash and fire points are determined by passing a flame over the surface of the fluid at constant time intervals during constant temperature rise. The flash point is the temperature at which the volatile vapors above the fluid surface ignite. The fire point is the temperature where the fluid itself ignites and burns for at least five seconds. These results are primarily dependent on fluid chemistry and volatility.

Increasing flash points and fire points are indicative of increasing fire resistance, particularly in applications where high-temperature fluid volatility is important, such as in many aircraft applications. These tests are inappropriate for aqueous fire-resistant fluids because it is difficult to assess proper heat-up times. This leads to non-reproducible data because of variable evaporation rates. It is also difficult to directly relate the flash and fire point data to specific end-use fire-risk potential.

Oxygen index. The oxygen index, as defined in ISO 4589, "is the minimum concentration of oxygen, by percentage volume in a mixture of oxygen and nitrogen, introduced at 23°C ± 2°C, that will just support combustion of a material under specified test conditions" [25]. The procedure for determining the oxygen index of a fluid is performed by placing a small quantity of test fluid into a glass cup positioned in a glass chimney, as shown in Figure 9.1 [26]. An upwardly-flowing mixture of oxygen and nitrogen is introduced and then ignited. The oxygen index is determined by the minimum amount of oxygen concentration required to sustain combustion for sixty seconds or more [27].

Oxygen index determination is a simple and relatively inexpensive test to perform and may be used as a quality control procedure, but it is not evident that this procedure offers any significant advantage over flash and fire points [13]. Some illustrative examples of the oxygen indices of various model compounds are provided in Table 9.3 [26].

It has been suggested that the measurement of the oxygen index is actually a measure of the "ease of flame extinction" and does not produce meaningful data with respect to fluid ignitability or flame propagation tendencies.

Auto-ignition temperature. The auto-ignition temperature (AIT) is one of the most widely-known thermal ignition tests [23]. AIT was determined according to ASTM D 2155 (Standard Test Method for Auto-ignition Temperature of Liquid Petroleum Products), where a test fluid is injected via a syringe onto the surface of an Erlenmeyer flask heated in an oven as shown in Figure 9.2 [22–32]. AIT measures the geometry-dependent spontaneous-ignition temperature of a fluid in air at 1 atm of pressure. These measurements are difficult to relate to end-use applications [22]. MacDonald found that AIT varies with the size of the Erlenmeyer flask [33].

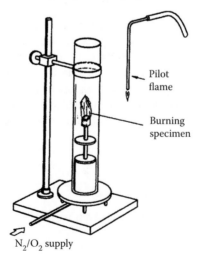

Pilot flame

Burning specimen

N$_2$/O$_2$ supply

FIGURE 9.1 Apparatus for determining oxygen index.

TABLE 9.3
Oxygen Indices for Selected Compounds

Trioctylphosphate	18.6
Ethylene glycol	14.8
Diethylene glycol	13.6
Trimethyl phosphate	23.7
Aroclor 1016	31.3
Aroclor 1221	20.5
Aroclor 5460	61.0
Castor oil	22.9
Dibutylsebacate	15.6
Ethylene glycol	14.8
Fyrquel 150 (Stauffer Chemical)	25.5
Fyrquel 20 (Stauffer Chemical)	25.9
Houghto Safe 1120 (EF Houghton Co.)	22.1
Houghto Safe P C F G 15	24.0
Hydraulic fluid MIL-H-83282	17.7
Hydraulic fluid MIL H 5606B	16.0
MIL-H-5606B hydraulic fluid	16.0
MIL-H-83282 hydraulic fluid fluorocarbon	17.7
Mineral oil USP	16.1
Mineral oil 5314A (Socony Molil Co.)	15.2
Olive oil	16.6
Paraffin oil, mineral oil (white, heavy)	16.4
SF96 (50)—GE Silicone fluid	21.4
SF96 (350)—GE Silicone fluid	27.3
SF1029—GE Silicone fluid	23.0
XF1050—GE Silicone fluid	13.3
Therminol 66	16.1
Trioctyl phosphate	18.6
Triphenyl phosphate	22.5
Tripropylene glycol	17.7
Vegetable oil	19.0

Source: ISO 4589: "Plastics-Determination of Flammability by Oxygen Index," 1984.

Although the ASTM D 2155 method for auto-ignition temperature was the standard method for the aircraft industry and was referenced in many aircraft OEM standards, it is not currently an active standard. Another method for auto-ignition temperature is ASTM E 659, which was developed and has been used as a replacement for ASTM D 2155 in standards including those of the aircraft industry. Aircraft industry scientists cited through their work that auto-ignition temperatures determined for fluids by ASTM D 659 did not correlate with values determined by ASTM D 2155 and, in effect, changed the classifications of fluids currently used in the industry. ASTM members currently participating in ASTM D02.N0.06 Subcommittee "Fire-Resistant Fluids" are working to have the ASTM D 2155 standard reactivated [34]. Table 9.4 shows AIT values for several common hydraulic fluids with corresponding fire and flash point data [13].

Rapid compression. Ignition or explosion may occur when air in contact with a hydraulic fluid is rapidly heated by compression. Hydraulic fluid fires have occurred due to fluid ignition by rapid compression [35]. Because of this particular type of ignition hazard, tests have been developed to model the susceptibility of a hydraulic fluid to undergo compression ignition. A variety of compression-ignition

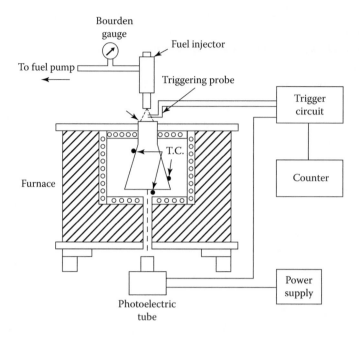

FIGURE 9.2 Apparatus for auto-ignition temperature (AIT) determination.

tests are conducted by introducing air, at pressures comparable to those found in hydraulic systems, through a valve into a small-diameter pipe whose inner surface has been wetted with the test fluid [35].

A rapid-compression-ignition test, developed by Faeth and White, simulates a pipe containing high- and low-pressure air separated by a rapidly-opening valve, as illustrated in Figure 9.3 [36]. When the valve is opened, pressure equalization occurs by pressure waves generated by opening the valve, as illustrated in Figure 9.4 [36]. During this pressure equalization, a contact zone of rapidly-changing temperature and density is formed which separates the expanded and cooled "upstream" gas from the compressed and heated "downstream" gas. This contact zone travels the length of the pipe, eventually compressing the "downstream" gas against the closed end of the pipe. It is in this region where pressure increases could potentially lower the spontaneous-ignition temperature (S.I.T.) of a fluid, as illustrated in Figure 9.5 [36]. Lowering the S.I.T. in conjunction with the air-temperature increase ahead of the contact zone contributes to the fluid's potential for ignition.

This test has been performed using both 10- and 30-ft. sections of pipe. The conclusions of these tests show that combustion is more difficult to obtain in the longer (30-ft.) dead-ended sections,

TABLE 9.4
Relative Auto-Ignition Temperature and Open-Cup Flash/Fire Point Data of Common Hydraulic Fluids

Fluid	Auto-Ignition Temperature (°C)	Open-Cup Flash Point (°C)	Open-Cup Fire Point (°C)
ISO VG 10 mineral oil	320	166	180
ISO VG 46 triaryl phosphate	580	246	365
ISO VG 46 polyol ester	450	275	325
ISO VG 46 silicone fluid	470	300	340
ISO VG 46 polyglycol[a]	400	274	321

[a] Unpublished data, October 13 and 14, 2005, ASTM D 92 and ASTM E 659.

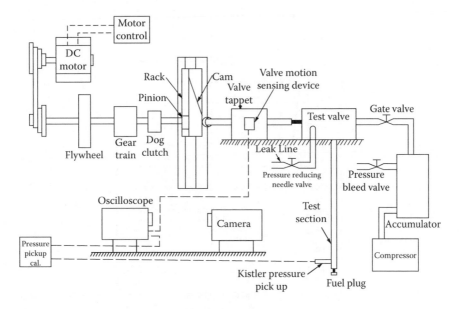

FIGURE 9.3 Illustration of the Faeth and White rapid-compression apparatus.

where the rate of pressure increase is decreased due to the larger volume of the test section and the energy loss by convection and conduction through the pipe walls. This lowering of combustion potential can be attributed to the relatively low heat capacity of air, where a small amount of energy loss represents a relatively large decrease in temperature.

Another combustion test is "the combustion indicator test". This combustion test, which will not be reviewed in detail here, involves the determination of the compression ratio of the fluid in an engine where combustion occurs [37].

Shock-tube ignition test. Although shock-tube tests have been used extensively in the study of gas-phase reactions, a limited amount of testing has been performed with gas-liquid reactions. A shock-tube apparatus [38] was used by Skinner to determine the ignition characteristics of different aerospace hydraulic fluids. Skinner's test results did not correlate well with conventional ignition-time data, indicating that the relative importance of drop breakup, evaporation, and convective and diffusive mixing, with respect to chemical reactivity, is poorly understood [39].

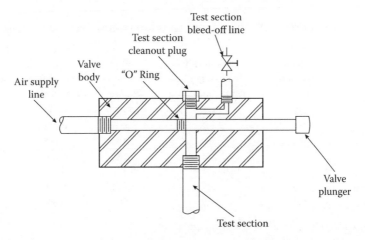

FIGURE 9.4 Illustration of the valve used for the rapid-compression test shown in Figure 9.3.

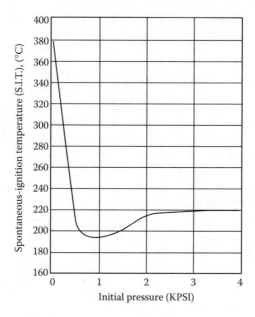

FIGURE 9.5   The effect of pressure on spontaneous-ignition temperature.

Gunfire resistance. The gunfire-resistance test (MIL-H-83282—Fire-Resistant Hydraulic Fluid, Synthetic Hydrocarbon Base, Aircraft, Metric, NATO Code Number H-537) is a specialized ignition test in which the heat of ignition is provided by incendiary ammunition. Testing is performed by firing 50-caliber, armor-piercing incendiary ammunition into aluminum cans partially filled with the test fluid. Five shots are fired and the number of ignitions and severity of the fires are reported [40]. Results of gunfire-resistance tests, as reported by Snyder and Krawetz, are summarized in Table 9.5 [24].

An alternative procedure to the gunfire resistance test is performed to simulate a realistic flammability study of hydraulic fluids selected for use in aircraft weapons system. These fluid applications require full survivability studies against hostile gunfire and are simulated utilizing mock-up hydraulic systems functioning at actual weapon system operating pressures and fluid flow rates [40]. The ammunition used as the ignition source in these tests is chosen based on the threat that the hydraulic system may be subjected to under combat conditions.

TABLE 9.5
Flammability Characteristics of Current and Developmental Aerospace Hydraulic Fluids

Fluid	Spontaneous-Ignition Temperature (°C)	Stream Hot-Manifold Ignition Temperature (°C)	Heat of Combustion (kcal/kg)	Horizontal Flame Propagation Rate (cm/s)	Gunfire Resistance Number of Fires Per 5 Shots
MIL-H-5606	232	388	10,100	0.733	5
MIL-H-83282	354	322	9,800	0.212	1
Phosphate ester	524	760	7,100	0.00[a]	0
Silicate ester	400	371	8,162	0.300	3
Silicone	409	477	5,411	0.218	0
Chlorofluorocarbon	630	927	1,328	0.00[b]	0
Fluoroalkyl ether	669	927	989	0.00[b]	0

[a]   Sample self-extinguished (would not stay lighted) on asbestos wire.
[b]   Sample could not be lighted on asbestos wire.

9.2.5.2 Pool Fires

A pool fire may be defined as "a buoyant diffusion flame in which the fuel is configured horizontally" [41]. This includes not only flammable liquids, but also solids and gases. Although pool fires can form any geometric shape, controlled studies are usually performed on circular geometries characterized by the pool diameter and depth.

One method of determining the ignitability of a pool fire has been developed for the evaluation and comparison of less flammable insulating liquids [42]. This procedure is performed by igniting 4 gal. of heptane in an open circular trough which surrounds 40 gal. of the test fluid contained in an inner concentric pan. Conductive and radiative heat-release rates of the pool fire at a "steady-state" condition are determined during the testing procedure.

Other flammability characteristics may be determined during pool fire testing and may include flash point, resistance to fire suppression, total heat-release rate, flame-spread rate, and heat radiation to surrounding areas. Factors affecting pool fires may also be introduced during testing and can include heated surfaces, the absence or presence of an enclosure, and variable ambient factors such as room temperature, wind, or ventilation.

As with other common bulk fluid flammability tests, it is difficult to compare water-containing fluids to non-water-containing fluids because of variable evaporation rates, which produce non-reproducible data. Fluid ignition may occur after water evaporation; however, such data would not be representative of the original fluid condition.

9.2.5.3 Calorimetric Procedures

Heat of combustion. Heat of combustion is not a measure of fluid flammability but a measure of the heat that is generated during combustion. This property is considered a significant factor in the determination of overall flammability characteristics [43]. Snyder and Krawetz [24], and Parts [28], have illustrated the correlation of heat of combustion with the fire resistance of a fluid. The heat of combustion may be determined according to ASTM D 240 (Standard Test Method for Heat of Combustion of Liquid Hydrocarbon Fuels by Bomb Calorimeter). Some illustrative examples for hydraulic fluids used in aerospace applications are shown in Table 9.5. As expected, fire resistance increases as the heat of combustion decreases. Marzani developed a "static bomb"-type reactor apparatus for evaluating the spontaneous-ignition temperature for fluids at elevated pressures [44].

Another method for measuring heat of combustion had been used by Monsanto to assess flammability characteristics of aircraft hydraulic fluids [45]. This method incorporates the heat of combustion of benzoic acid, which is combined with the sample being tested to attain a more complete combustion for those fluids that exhibit incomplete combustion tendencies. These test results are corrected for the quantity of benzoic acid added to the test fluid. The heat of combustion may be estimated from the summation of the bond strengths of the chemical compositions. The estimated heat of combustion values for some commercially-available hydraulic fluids are shown in Table 9.6 [46].

TABLE 9.6
Heat of Combustion Values for Commercial Hydraulic Fluids

Fluid Type	Heat of Combustion (kcal/g)
Mineral oil	10.7–11.0
Esters	8.5–9.6
Phosphate esters	7.6–7.7
Polyglycols[a]	7.1–7.6
Thickened water-glycol	2.6–3.5

[a] Unpublished data, August 15 2005, ASTM D 240

Heating time. The fire resistance of a fluid is reflected by the time it takes to heat a given volume of fluid to 90°C from ambient temperature and comparing the heating time to a petroleum oil hydraulic fluid and distilled water [30]. The experiment is conducted by heating a given volume of fluid in an insulated, stirred, 2-L stainless-steel beaker equipped with a 56-W immersion heater and a 76-mm immersion thermometer. Polack et al. reported a heating time of 622 s for a hydrocarbon oil, 923 s for a water-in-oil emulsion, and 1815 s for distilled water. These differences in heating times are related to the heat capacity of the fluid [30].

Cone calorimeter. Cone calorimeter tests have evolved from procedures which were initially developed to investigate the heat-release rates resulting from the combustion of solids such as building materials, plastics, and textiles. A standard cone calorimeter device has been described in ASTM E 1354 (Standard Test Method for Heat and Visible Smoke Release Rates for Materials and Products Using an Oxygen Consumption Calorimeter) and in ISO 5660, and is illustrated in Figure 9.6 [47]. The cone calorimeter is based on the principle of "oxygen consumption calorimetry", where the net heat of combustion is directly related to the amount of oxygen consumed during combustion.

9.2.5.4 Evaporation (Wick) Tests

Evaporation, or wick, tests have been developed to identify the flammability characteristics of fluids which have impregnated porous or wick-like materials, such as steam pipe lagging or foam insulation. The flammability hazard arises once the fluid has soaked into these porous materials, slowly oxidizing

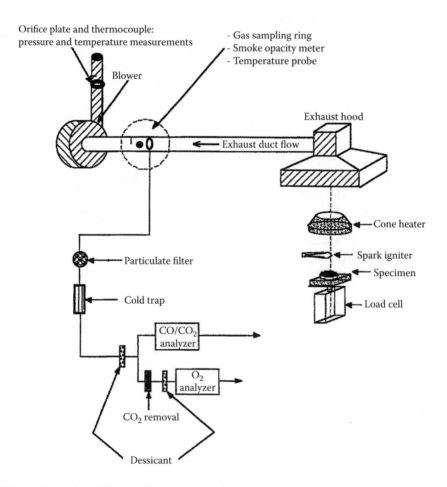

FIGURE 9.6   Illustration of Oxygen Consumption Calorimeter.

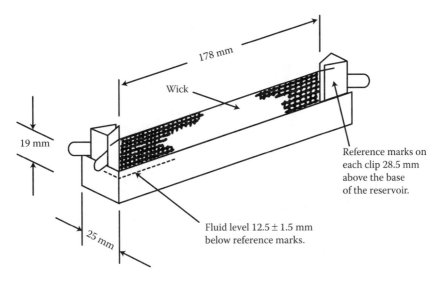

FIGURE 9.7 Illustration of the ceramic wick flammability test.

with increasing time. Fluids which—under normal conditions—are less likely to ignite may be easily ignited when retained by a wick-like material. The insulating properties of a porous material decrease the potential for heat loss and the fluid can sometimes self-heat and spontaneously ignite [14].

In certain industries, such as the mining industry, it is of interest to evaluate the fire-risk potential associated with hydraulic fluid leakage on absorbent material (e.g., coal) which would subsequently be exposed to an ignition source [22]. Evaporation, or wick, tests are used to model this risk.

There are primarily two evaporation tests. One test specified that an asbestos tape be used as the wick. In view of the health hazards associated with the handling of asbestos, a ceramic (aluminosilicate) fiber "tape" replaced the asbestos in more current test procedures, including the 7th Luxembourg Report method lux. 3.2.1, "Determination of Persistence of Flame on a Wick" [48–51]. The fiber tape has been replaced with a length of non-flammable aluminosilicate board in ISO method 14935, "Determination of wick flame persistence of fire-resistant fluids" [52]. Alternatively, a pipe cleaner has been used as the "wick" in some tests [22,53,54].

Ceramic wick test. The ceramic wick test is conducted by immersing one end of a piece of ceramic tape in the hydraulic fluid, as illustrated in Figure 9.7. After equilibration, one end of the wick is ignited and the persistence of the flame after the removal of the ignition source is recorded. This is primarily a measure of the relative ease of hydraulic fluid flammability after the saturation of a porous substratum such as pipe insulation, coal, and so forth.

"Pipe cleaner" wick test. The pipe cleaner wick test is conducted by soaking a pipe cleaner in the hydraulic fluid and then conditioning it at 65°C (150°F) for 2 or 4 h in an oven. The conditioned pipe cleaner is then repeatedly passed through the flame of a Bunsen burner until ignition occurs, as illustrated in Figure 9.8. The number of passes without ignition is recorded [29].

The water content of hydraulic fluids will affect the results of both of the wick tests which have been described. This is illustrated in Figure 9.9, which shows the relationship between the flame duration time and the water content of a water-in-oil emulsion for the ceramic tape test [54].

Soaked-cube flammability test. The soaked-cube flammability test models an exothermic reaction that may occur if a flammable fluid is absorbed into thermal insulation. The fluid is placed into a well that has been drilled into a 1 × 1 × 1-in. (25.4 × 25.4 × 25.4-mm) cube of asbestos-free calcium

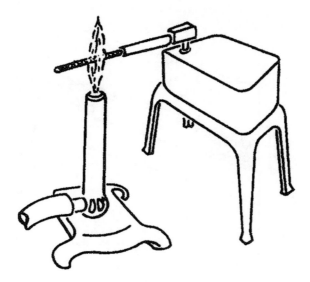

FIGURE 9.8 Illustration of the pipe cleaner flammability test.

silicate insulation material (Johns-Manville-12 molded and block 2-in. [50.8-mm] pipe insulation) with a thermocouple mounted 1/8 in. (3.175 ram) below the bottom of the well, as illustrated in Figure 9.10. The cube is placed into a furnace and the temperature is increased until an exothermic reaction occurs [55].

9.2.5.5 Flame Propagation

One of the greatest risks associated with petroleum-based hydraulic fluids upon ignition is flame propagation back to the ignition source [1]. Fire-resistant fluids are more difficult to ignite, and exhibit a reduced tendency for flame propagation. Therefore, it is not only important to measure the fire resistance but also flame propagation. Both properties are typically determined by the hot surface and the spray flammability tests, which will be discussed subsequently.

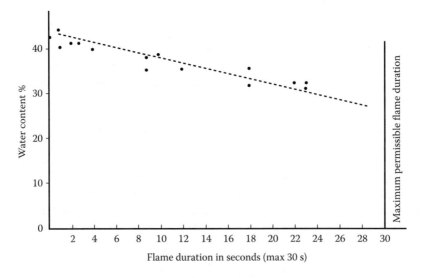

FIGURE 9.9 Illustration of the relationship between the flame duration time and the water content of a water-in-oil emulsion for the ceramic tape test.

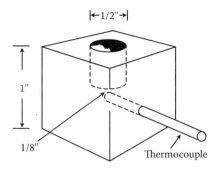

FIGURE 9.10 Schematic of the test specimen used for the soaked-cube flammability test.

Linear flame propagation test. Linear flame propagation measures the rate of flame travel on a ceramic fiber test sample presoaked with the hydraulic fluid (ASTM D 5306-92—re-approved in 2007—Standard Method for Linear Flame Propagation Rate of Lubricating Oils and Hydraulic Fluids). Linear flame propagation rates are determined by placing a 500-mm ceramic string into the fluid to be tested for 60 s. The fluid-saturated ceramic string is then placed on the test apparatus, as shown in Figure 9.11. Two differential thermocouples are attached to the string 15.24 cm apart.

One end of the string is ignited and the thermocouples are used to obtain the time required for the flame to propagate from one thermocouple position to the other. The linear flame propagation rate is reported in meters per second and is calculated as follows:

$$\text{Linear flame propagation rate} = dv/p,$$

where
d = distance between thermocouples (in mm)
v = chart speed (in mm/s)
p = distance measured peak to peak between thermal effects
Linear flame propagation rates are dependent on relative flammability or ignitability of the fluid and do not relate to the flammability properties of materials under actual fire-risk conditions. The

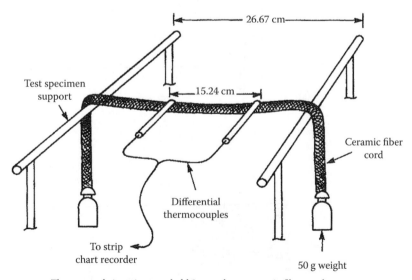

Themocouple junctions are held 2 mm above ceramic fiber cord

FIGURE 9.11 Apparatus to determine linear flame propagation rates.

linear flame propagation test is well suited for polyol esters and phosphate esters. This test is not utilized for volatile water-containing hydraulic fluids such as water-glycols, invert emulsions, and high-water-content hydraulic fluids.

9.2.5.6 Hot-Surface Ignition Tests

The purpose of modeling the ignition of a fluid when contacted with a hot surface is to determine if a hydraulic fluid would ignite and burn if it were sprayed on a hot surface, such as the manifold of earth-moving equipment, or a pool of molten metal.

"Monsanto" molten metal test. Monsanto developed a test to evaluate burning and flame propagation that may occur when a hydraulic fluid is sprayed on molten metal, either zinc at 800°F (427°C), or aluminum at 1200°F (649°C) [22,56]. Fluids may be poured, sprayed, or added dropwise to the molten metal (the mode and rate of addition are not specified). Burning rates and flame propagation to adjacent fluid-wetted areas is determined. This test may be performed with or without an external ignition source. An illustration of the Monsanto molten metal test is presented in Figure 9.12a. The European version of the molten metal test involves pouring the hydraulic fluid onto the molten metal, as illustrated in Figure 9.12b [57]. In both tests, although visual differences between the various fire-resistant hydraulic fluids were observed, reproducibility of these tests is poor [57].

"Houghton" hot-surface ignition test. Houghton International also developed a hot-surface ignition test. This test is performed by introducing a sample onto the hot surface and recording the elapsed time (in seconds) before ignition and the temperature at which ignition occurs [58]. Illustrations of the differentiation of various viscosity classes of hydraulic fluids are shown in Figure 9.13.

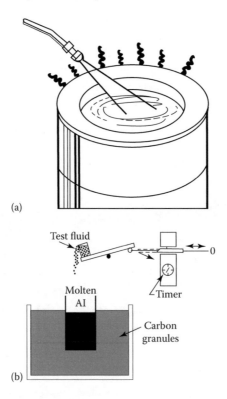

FIGURE 9.12 (a) Monsanto molten metal fire-resistance test; (b) European version of the molten metal fire-resistance test.

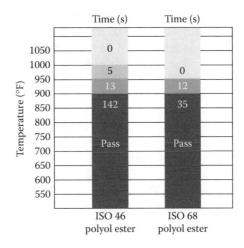

FIGURE 9.13 Illustrations of the differentiation of various classes of hydraulic fluids as determined by the Houghton hot-surface ignition test.

Hot-manifold test. In hydraulic systems, one of the primary concerns is the potential fire risk that may occur if a line ruptures near a heat source. One of the major tests used to model this situation is the hot-manifold test [22,28,59,60]. The hot-manifold test models the situation where hydraulic fluid may leak dropwise, or as a stream, directly onto a hot surface. The test method published in the Aeronautical Material Standard as AMS-3150 C has also been published as ISO 20823. In this test, 10 mL of the hydraulic fluid is applied dropwise onto the surface of a heated stainless-steel manifold (tube) at 1300 °F (704°C), as illustrated in Figure 9.14. The flashing or burning of the fluid upon contact with the hot tube, and the flame propagation to the fluid residues that are collected in the bottom of the enclosure are observed. This test was originally of particular interest to the aviation industry [24] and it is now required as 4.1.1 Exhaust Manifold Test in the aerospace standard SAE AS1241 Revision C [61]

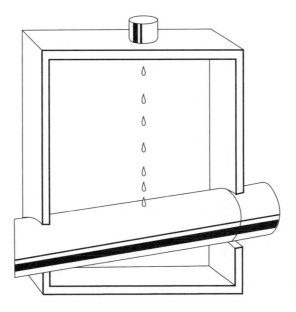

FIGURE 9.14 Hot-manifold test according to AMS-3150C.

Factory Mutual Research Corporation (FMRC) hot-channel ignition test. When a hydraulic hose ruptures, it is more likely to form a spray, which may then contact a hot surface. This condition is modeled by a "hot-channel ignition" test [62,63] shown in Figure 9.15. In this test, the hydraulic fluid is sprayed onto a steel surface 7 in. (7.78 cm) wide and 27 in. (68.58 cm) long inclined at a 30° angle. The hot channel is heated by propane burners from below, to a minimum temperature of 1300°F (704°C). The burners are turned off prior to spray application of the fluid onto the hot surface. The fluid is sprayed from a distance of 6 in. (15.24 cm) from the hot surface for 60 s. To "pass" this test, the flame must not propagate and must not follow the spray source. Although this is an excellent end-use test, it is very difficult to quantify and is reported on a pass/fail basis.

It was shown that although polyol esters, phosphate esters, and water-glycols all pass the FMRC hot-channel ignition test, they all exhibit substantially different flammability properties, which are not indicated by the "pass" notation [64]. This is illustrated in Figure 9.16. Interestingly, the fire-resistant properties of these fluids were similar to those obtained by the molten metal test [56]. The hot-channel ignition test, which is a requirement for the FMRC Standard 6930 (revision 1975) was deleted in the 2002 revision of the FM Approvals (formerly FMRC) Standard 6930, along with the FMRC spray ignition flammability test. These tests were replaced in the standard with the Spray Flammability Parameter determination protocol [65].

One variant of these tests is to conduct them in a wind tunnel as described by Goodall and Ingle [60]. This work was conducted to model aerospace applications. The result showed that the risk of fire was dependent on the temperature of the critical volume of the fluid—the volume at which spontaneous ignition of the fluid will occur—and was not dependent on the hot-surface temperature.

9.2.5.7 Spray Ignition Flammability Tests

Perhaps the most common test to model the potential fire risk of a hydraulic line break is the spray flammability test. A variation of this test is incorporated in the national standards of many countries including the United Kingdom, France, and Canada, and is also incorporated into industry standards including ISO 12922 [18,48,49,62,66]. Although this test has been accepted as a standard, there are some deficiencies, such as the pass/fail ranking and the lack of repeatability when performed at different laboratories using similar test rigs. Although there is no national spray flammability test in the United States, a commonly-encountered insurance industry standard was the Factory Mutual Research Corporation spray flammability test [62,63] until their revised standard 6930 was published in 2002, which did not include the spray flammability test. Yuan studied the ignition of mineral oil and fire-resistant hydraulic fluid sprays by open flames and hot surfaces [67]. Of the four types of fire-resistant hydraulic fluids tested, only high-water-containing fluids and water-glycol fluids demonstrated strong fire-resistant characteristics with open flame and hot surface temperatures.

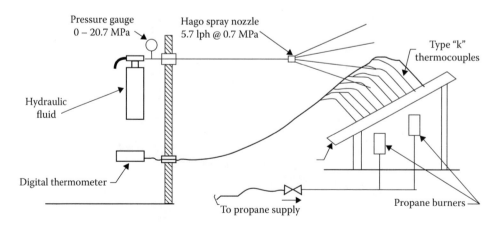

FIGURE 9.15 Illustration of the FMRC hot-channel ignition test apparatus.

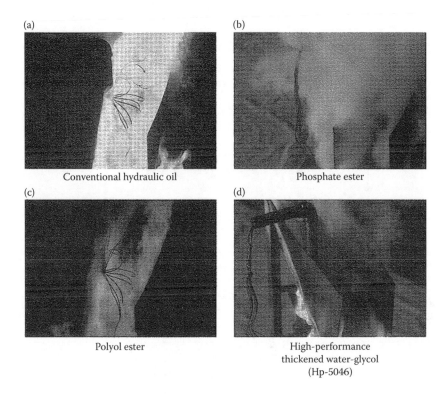

(a) Conventional hydraulic oil

(b) Phosphate ester

(c) Polyol ester

(d) High-performance thickened water-glycol (Hp-5046)

FIGURE 9.16 Hot-channel ignition test fire-resistance properties for (a) mineral oil, (b) phosphate ester, (c) polyol ester, and (d) water-glycol hydraulic fluids.

9.2.5.7.1 Low- and High-Pressure Spray Flammability Tests

The ALCOA low-pressure spray flammability test. ALCOA developed a relatively low-pressure spray flammability test [67]. The low-pressure spray source is a Binks No. 7 paint spray gun (1.8 cm orifice and pressurized to 0.28 MPa with air) [22,68]. The hydraulic fluid is sprayed over an ignition source such as oil-soaked rags, as illustrated in Figure 9.17 [64], and the flame characteristics of the flame are visually recorded [22].

A number of high-pressure spray flammability tests are currently used throughout the world. The most common low-pressure and high-pressure spray flammability and heat release tests are summarized in Table 9.7 and Table 9.8 respectively which includes the source of ignition, fluid

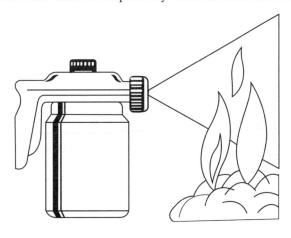

FIGURE 9.17 Low-pressure Binks No. 7 spray gun flammability test.

TABLE 9.7

Summary of Low-Pressure Spray Flammability Test Methods

Test Authority	Ignition Source	Fluid Temp. (°C)	Pressure and Source [MPa (psig)]	Fluid Spray Nozzle	Orifice Diameter	Rated Flow	Distance From Ignition [m (in.)]	Failure Criterion or Results Reported as
MSHA	1. Electric spark 2. Propane torch 3. Flaming trough	65 ± 5	1.034 (150) (N₂ cylinder)	90-m spray	0.635 (0.025)	3.28 gph	0.45 (17.7)	Flame not to exceed 6 s. at 18 in.
Navel Research	Electric spark	Ambient 15.5–26.6	0.17 (25) (N₂ and O₂)	Poasche Model ML-8	–	158 gph	–	Spray flammability limit (min. % O₂)
NBS/CFR	Kerosene-soaked cotton cheesecloth	65	1.034 (150) (N₂ cylinder)	90-m spray Binks Model F-12–15	0.64 mm (0.025)	3.28 gph	0.68 (26.7)	Rated by temperature rise 204–296 mC considered high heat potential

TABLE 9.8
Summary of High-Pressure Spray Flammability and Heat Release Test Methods

Test Authority	Ignition Source	Fluid Temp. (°C)	Pressure and Source [MPa (psig)]	Fluid Spray Nozzle	Orifice Diameter	Rated Flow	Distance From Ignition [m (in.)]	Failure Criterion or Results Reported as
FM Approvals Heat Release Spray Test	12kW, 0.14-m-diameter propane-air ring burner	60	6.9 (1000)	80-m hollow cone	0.38 (0.015)	1.5 gph	Sprayed through ring burner	Spray flammability parameter
AFNOR	A type REX No. 1 Charledave torch size 750 tip	65 ±5	7 ± 0.3 (1015) (N_2 cylinder)		1.6–0.4 with screen (0.063–0.016)	–	1.20 m (47.2)	1 = no ignition 2 = ignites but does not reach screen 3 = ignites and reached screen
ISO 15029-1 Former UK National Coal Board Test	Ocyacetylene torch with No. 10 nozzle	Water-containing –65 all others 85	6.89 (1000) (N_2 cylinder)	80-m hollow cone Monarch Manufacturing Type F80	–	2.5 gph	rVaries	Length of spray and duration of flame
ISO 15029-2	Vertical propane flame	10–25	<2 (N_2 cylinder)	CT Ltd.. twin fluid atomizer nozzle ¼ type JBC-12B	–	90mL/min	0.0425 (1.673)	Ignitability, flame length, smoke emissions

test temperature, spray pressure, nozzle description, pass/ fail criteria, and references for the tests.

The Factory Mutual Research Corporation (FMRC) spray ignition flammability test. The most commonly-used spray ignition flammability test encountered in the United States was the FMRC spray ignition flammability test [62]. The test was performed by rotating a propane torch at 8 rpm in a mist of the hydraulic fluid preheated to 40°C, which is sprayed through a nozzle at 6.9 MPa. If the fluid ignites, the time is recorded, and if the fluid continues to burn 5 s after the removal of the flame, the fluid is considered to "fail" the test. An illustration of the FMRC spray flammability test apparatus is shown in Figure 9.18.

Apart from mineral-oil hydraulic fluids, all of the fire-resistant hydraulic fluids shown in Figure 9.19 exhibit very different flammability properties (although each was awarded a "pass" rating in the Factory Mutual Research Corporation [FMRC] spray flammability ignition test) [62]. Therefore, there has been an ongoing effort to identify a spray test to better differentiate between the various hydraulic fluid flammability properties. One of the most commonly-used fluid flammability testing standards utilized a combination of hot-channel and spray flammability tests as specified by the FMRC [62]. The FMRC spray flammability ignition test—a requirement for the FMRC Standard 6930 (revision 1975)—was deleted in the 2002 revision of the FM Approvals (formerly FMRC) Standard 6930, along with the FMRC hot-channel ignition test. These tests were replaced in the standard with the Spray Flammability Parameter determination protocol [65].

9.2.5.8   Heat Release Spray Tests

Two test procedures quantify the heat release obtained when a test fluid is ignited. FM Approvals (formerly FMRC) developed a protocol to determine the "Spray Flammability Parameter" (or SFP) of a test fluid, which was published in the FM Approvals Standard 6930 (2002) [65]. One component value required to calculate the SFP of a fluid is the chemical heat release rate of the fluid. The FM Approvals Fire Products Collector apparatus is utilized to determine the chemical heat release rate of the fluid. Developed at the University of Manchester in England the "stabilized flame heat release test" (also known as the "Buxton test") also evaluated hydraulic fluids by their heat release. This test was incorporated in the 7th Luxembourg Report and is a requirement in ISO 12922, which specifies the requirements for hydrostatic and hydrodynamic hydraulic systems in general industrial applications [69].

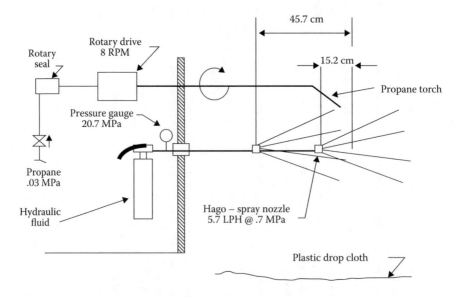

FIGURE 9.18   Illustration of the FMRC spray flammability test.

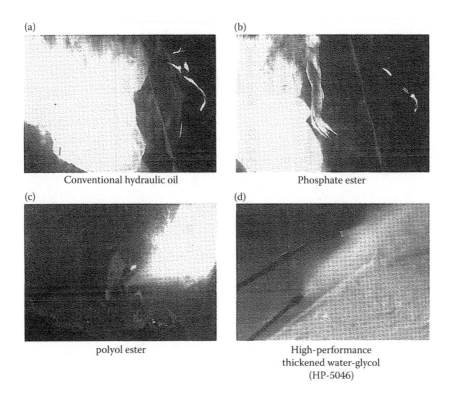

(a)   (b)

Conventional hydraulic oil   Phosphate ester

(c)   (d)

polyol ester   High-performance
thickened water-glycol
(HP-5046)

FIGURE 9.19   Spray flammability properties of: (a) mineral oil, (b) phosphate ester, (c) polyol ester, and (d) water-glycol hydraulic fluids.

The FM Approvals spray test to determine heat release rate. The FM Approvals Standard 6930 (2002) [65] classified the fire resistance of fluids by their Spray Flammability Parameter (SFP) value. SFP [63,70,71] is a calculated value derived from the evaluation of the flammability characteristics of the fluid. The SFP is obtained by determination of the fluid's chemical heat release rate using the FM Approvals 200 kW- Scale Fire Calorimeter and the fluid's critical heat flux for ignition using the methods outlined in the FM Approvals standard. The SFP is calculated from the following equation:

$$SFP_{(normalized)} = 11.02 \times 10^6 \times Q_{ch} / (p_f q_{cr} m_f)$$

where
   Q_{ch} is the chemical heat release rate
   q_{cr} is the critical heat flux for ignition
   p_r is the density of the fluid
   m_f is the fluid mass flow rate during the chemical heat release measurement

The fluid is heated to 60 °C, pressurized to a constant 6.9 MPa and discharged upward through an 80° hollow cone spray nozzle having an exit as shown in Figure 9.20. The nozzle is centered within a propane ring burner, which is adjusted to produce a 12 kW (approximately) flame [65]. Temperature and humidity conditions at the time of testing are recorded. The chemical heat release rate is calculated through the measurement of generated carbon monoxide and carbon dioxide. At a minimum, the first and last 30 s of each test are excluded from the measurements used in the calculation and, prior to testing, the gas analyzers (CO, CO_2) in the apparatus are calibrated using standard gas-nitrogen mixtures in the concentration range expected in the test. In the FM

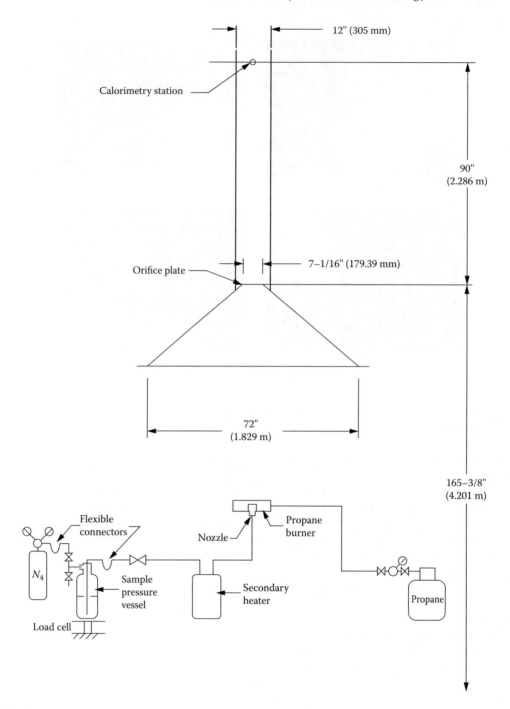

FIGURE 9.20 FM Approvals 200 kW Scale Fire Calorimeter. (Courtesy of FM Approvals: a subsidiary of FM Global.)

Approvals Standard 6930 it is noted that an approximation of the chemical heat release rate may be obtained through the use of an oxygen bomb calorimeter in order to determine the net heat of complete combustion which can be used as a proxy for the chemical heat release rate [65]. The successful use of the SFP to characterize and differentiate the potential fire risk of various hydraulic fluids is shown in Figure 9.21.

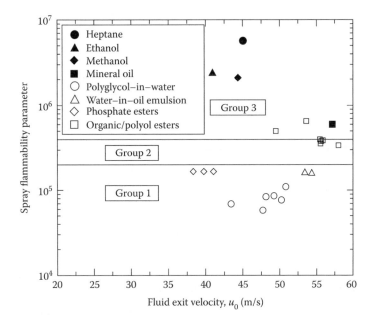

FIGURE 9.21   SFP characterization of various types of hydraulic fluids.

Stabilized heat-release spray test or "Buxton test" (ISO 15029-2). Work by Yule and Moodie was performed using the twin-fluid spray apparatus shown in Figure 9.22 [72]. A horizontal combustion chamber is utilized. Fluids were ranked by exhaust temperature, flame length, and particulate emissions. The fundamental ranking parameters were the exhaust temperature and the "ignitability", which were used to account for both ease of ignition and flame stabilization. Figure 9.23 illustrates Yule and Moodie's correlation between the heat of combustion and ignitability [72]. This test was specified in the 7th Luxembourg Report as 3.1.3 "Stabilised Flame Heat Release" spray test and is the preferred fire-resistance testing method to determine the relative fire resistance of a fluid [51], specified as ISO 15029-2, "Spray test—Stabilized flame heat release method", and is a required test in ISO 12922 [69].

The test apparatus is the same as that used for Yule and Moodie's work, shown schematically in Figure 9.22. In the ISO 15029-2 test method (formerly known as the "Buxton test" and Lux. 3.1.3),

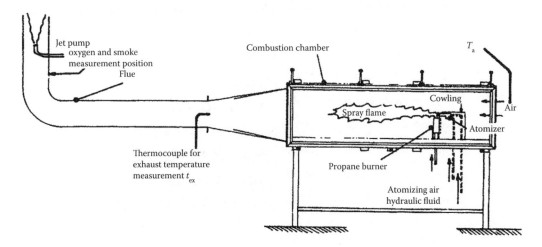

FIGURE 9.22   Yule and Moodie twin-fluid spray flammability apparatus ("Buxton test").

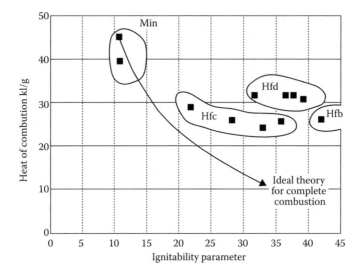

FIGURE 9.23 Correlation between heat of combustion and ignitability.

the relative fire resistance of a fluid is reflected by the RI index (Relative Ignitability index), which is determined from the exhaust temperatures of the hydraulic fluid during combustion. Figure 9.24 illustrates the RI index obtained for various hydraulic fluids and also illustrates interlaboratory reproducibility. The two laboratories shown are DMT (Gesellschaft fur Forschung and Prufung mbH) and HSE (Health Safety Executive in Buxton, UK) [73].

In general, the RI index reflects the expected fire resistance indicated by Figure 9.24. It also illustrates excellent interlaboratory reproducibility. However, it is important to note that interlaboratory reproducibility can only be shown after appropriate calibration procedures have been applied [73]. This is done by developing a linear regression calibration equation, as illustrated in Figure 9.25. The data obtained by the HSE laboratory are taken as the reference.

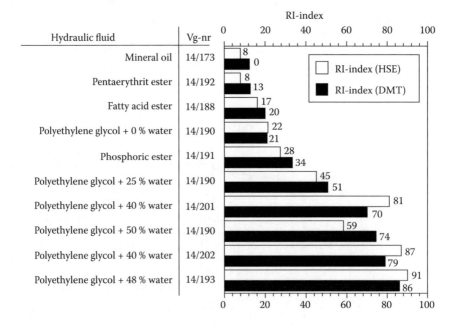

FIGURE 9.24 Comparison of the RI indices of various hydraulic fluids produced by two laboratories.

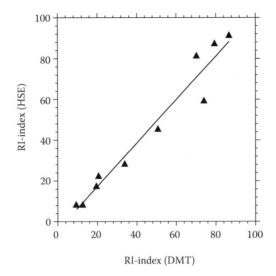

FIGURE 9.25 Plot of RI index values obtained by DMT and HSE.

Holmstedt and Persson reported a procedure which multi-parametrically ranked the relative fire risk of a hydraulic fluid using heat-release rates, heat of combustion, combustion efficiency, generation of carbon monoxide (CO), and smoke [74].

9.2.6 Hydraulic Fluid Flammability Standards for General Industrial Applications

Standards of test methods for the evaluation and approval of fire-resistant hydraulic fluids have been developed for many applications, including coal mining [75], aviation [61], automobile production [76], steel manufacture [65,77], and military equipment [37,78–80]. The two major standards for the evaluation of fire-resistant hydraulic fluids in general industrial applications will be reviewed in this section.

9.2.6.1 Approval Standard for Flammability Classification of Industrial Fluids Class Number 6930 by FM Approvals, a Subsidiary of FM Global

The first 6930 standard, titled "Less Flammable Hydraulic Fluids", was published by FMRC in 1965. This and the 1975 revision of FMRC 6930 standard [63] utilized two methods for the evaluation of fluid fire resistance: a spray ignition test and a hot channel ignition test (detailed earlier in Sections 9.2.5.6 and 9.2.5.7). Standard spray ignition tests, including the FMRC method, in attempting to simulate a specific hazard condition were qualitative pass/fail assessments. A study on repeatability and reproducibility of a spray ignition test conducted by FMRC indicated only a 36% agreement in the pass/fail conclusion among four different test facilities evaluating fourteen hydraulic fluids using identical spray ignition test equipment [64]. Results of the study also indicated that some hydraulic fluid types comprised of polymeric thickeners could achieve a pass on the spray ignition test due to the increase in droplet size under the ignition spray test conditions. This effect was not reproducible however when the same fluid type was tested as per the ASTM 2882-83 vane pump test prior to the spray ignition test and subsequently failed the spray ignition test. The increase in fluid droplet size gained through there incorporation of a polymer was lost due to shearing of the polymer compound during the vane pump test [64].

A study was initiated at FMRC to develop a quantative test method for identifying less flammable fluid [64]. A revised standard, "Flammability Classification of Industrial Fluids", was published in January 2002 [66] by FM Approvals. The standard defined their requirements for the flammability classification rating of industrial fluids intended for use as, but not limited to, lubricants, hydraulic power transmission, turbine governor control, transformer insulation, and cooling. The new standard classified the fire resistance of fluids by their Spray Flammability Parameter (SFP) value. SFP is a calculated value derived by the evaluation of the flammability characteristics of the fluid. The SFP is calculated according to an equation (see Equation 1 in Section 9.2.5.8) obtained by determining the fluid's chemical heat release rate using the FM Global 200 kW—Scale Fire Products Collector, and the fluid's critical heat flux for ignition using the methods outlined in the FM Approvals Standard [66].

The chemical heat release rate (Q_{ch}) of a finely-atomized spray of the test fluid is determined for calculation of the SFP through the use of the 200 kW scale fire calorimeter pictured in Figure 9.26. The fluid is heated to 60 °C, pressurized to a constant 6.9 Mpa and discharged upward through a hollow cone spray nozzle. The nozzle is centered within a propane ring burner adjusted to produce a 12 kW (approximately) flame. Temperature and humidity conditions at the time of testing are recorded. The chemical heat release rate is calculated through the measurement of generated carbon monoxide and carbon dioxide. The critical heat flux for ignition (q_c) of the test fluid was determined by subjecting the test fluid to radiant heat while testing for ignition of the resultant gases by the apparatus described in FM Approvals Standard 6930 (2002). A 100 mL sample of the test fluid was placed in a 100 mm diameter by 25-mm deep cylindrical aluminum dish positioned on a platform surrounded by electrical radiant heaters. The radiant heaters are capable of providing a radiant heat flux at the surface of the fluid of 0 to 60 kW/m². The fire-resistance classifications under the FM Approvals Standard CN 6930 (2002) are in Table 9.9.

Fire-resistance classifications for FM Global Standard CN 6930 were revised in February, 2005 to "Approved" and "Specification Tested Industrial Fluids" categories, as shown in Table 9.10.

(a)   (b)

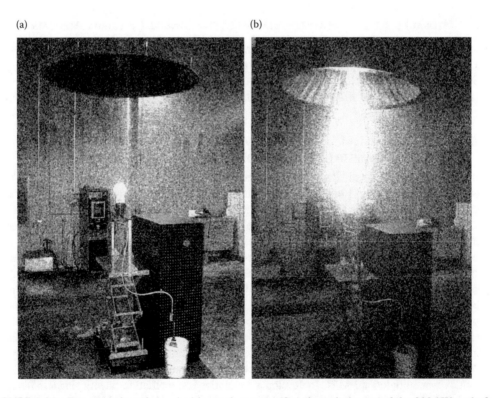

FIGURE 9.26 Determination of chemical heat release rate (Q_{ch}) through the use of the 200 kW scale fire calorimeter: (a) adjustment of propane flame to approximately 12 kW prior to the start of fluid flow, (b) ignition of atomized fluid spray by propane flame. (Courtesy of FM Approvals; a subsidiary of FM Global.)

TABLE 9.9
Spray Flammability Parameter (SFP) Fluid Classifications (FM Approvals Standard 6930 January, 2002)

Group 0	Non-flammable.
Group 1	Having a normalized SFP of 5×10^4, or less, usually unable to stabilize a spray flame.
Group 2	Having a normalized SFP greater than 5×10^4 but no more than 10×10^4
	Less flammable than mineral oil fluids, but may stabilize a spray flame under certain conditions.

FM Approvals determined that the critical heat flux would not be established by the apparatus described in the 2002 publication, but would be calculated from the average of three fire point determinations of the fluid according to the equation:

$$q_{cr} = \alpha \times \sigma \times T^4,$$

where
q_{cr} is the chemical heat release rate (kW)
α is the fluid surface resistivity (assume to be unity)
σ is the Stefan-Boltzman constant (5.67×10^{-11} kW/m$^2 \times$ K^4)
T is the fire point temperature (K)

The evaluation of commercial fire-resistant hydraulic fluids, according to the 2002 standard, indicated that the protocol as stated was not applicable for industrial fluids not having a fire point (typically, water-containing fluids) due to non-reproducibility of the critical heat flux value determinations. The standard was amended in 2006 to state that fluids not having a fire point (water-containing) would now be categorized as shown in Table 9.11.

A new FM Approvals Standard CN 6930 was published in April, 2009 [77].

In summary, the FM Approvals Standard CN 6930 [77] has two protocols for the determination of fluid fire-resistance: one for fluids that have a fire point (non-water-containing) which utilizes the determination of SFP; and one for fluids that do not have a fire point (water-containing fluids) which does not use the determination of SFP, but the determination of the adiabatic stoichiometric flame temperature as the basis for the classification requirements.

Summary of FM Approvals Standard CN 6930 protocol for industrial fluids having a fire point according to Approvals Standard CN 6930 (April, 2009):

The SFP is calculated as:

$$SEP_{(normalized)} = 11.02 \times 10^6 \times Q_{ch} / (p_f q_{cr} m_f),$$

where
Q_{ch} is the chemical heat release rate (kW)
p_f is the density of the fluid (kg/m^3)

TABLE 9.10
Spray Flammability Parameter (SFP) Fluid Classifications (FM Approvals Standard 6930, February, 2005)

FM Approved	Comprised of former Group 0 and 1 fluids.
Industrial Fluids	Having a normalized SFP of $\leq 5 \times 10^4$.
FM Specification Tested Fluids	Comprised of former Group 2 fluids.
	SFP greater than 5×10^4 but no more than 10×10^4.

TABLE 9.11

Fluid Classifications for Fluids Not Having a Fire Point (FM Approvals Standard 6930, 2006)

FM Approved Industrial Fluids	Comprised of former Group 0, 1 fluids. Having an adiabatic, stoichiometric flame temperature of $\leq 2100\text{K}$ and an average chemical heat release rate of $\leq 130 \text{ kW}$.

m_f is the fluid mass flow rate during the chemical heat release measurement (g/s)

q_{cr} is the chemical heat release rate (kW)

Where

$q_{cr} = \alpha \times \sigma \times T^4$

α is the fluid surface resistivity (assume to be unity)

σ is the Stefan-Boltzman constant ($5.67 \times 10^{-11} \text{ kW/m}^2 \times \text{K}^4$)

T is the fire point temperature (K) (average of three fire point tests)

Industrial fluids that have a fire point and a normalized SFP of 5×104 or less are eligible for approval.

Industrial fluids that have a fire point and a normalized SFP greater than 5×10^4, but no more than 10×10^4, are eligible to be listed as a FM Specification Tested Product. FM Specification Tested industrial fluids, which are less flammable than mineral oil fluids but may stabilize a spray flame, can be considered less flammable. Sprinkler protection may still be needed to control a fire involving these fluids.

Summary of FM Approvals Standard CN 6930 protocol for industrial fluids not having a fire point according to Approvals Standard CN 6930 (April, 2009):

The elemental composition, of the industrial fluid (concentrate) is determined. This analysis includes the presence of hydrogen and oxygen for any water in the tested sample, therefore a correction is necessary to account for water in the measured mass percentage of oxygen (O atom). For accuracy, FM Approvals requires that the concentrate be tested. The concentrate shall contain all component parts of the finished fluid and a water percentage not to exceed 18 mass percent.

To correct for the water in the elemental analysis, the following methodology was developed:

(a) Yi is the reported mass fraction of element i, where i can be carbon, hydrogen, oxygen, calcium, phosphorous, potassium, water content, and so on.

(b) The sum of all elements, including the measured water, (YW) is:

$$\Sigma i = YK + YP + YC + ... + YW \tag{9.1}$$

(c) The error in the sum is Error $= \Sigma i - 1$. Note that $error < YW$ because the reported oxygen and hydrogen is also included in the measured water mass fraction.

(d) The actual mass fraction of hydrogen and oxygen is corrected to account for the measured water as follows:

$$Y_{O.a} = \frac{Y_O - \frac{16}{18} E_{error}}{1 - Y_W}; \quad \text{and} \quad Y_{H.a} = \frac{Y_H - \frac{2}{18} E_{error}}{1 - Y_W} \tag{9.2}$$

(e) The corrected actual mass fractions of each element in the absence of any water are calculated as:

$$Y_{K.a} = \frac{I_K}{1 - Y_W}$$

$$Y_{P.a} = \frac{I_P}{1 - Y_W}$$

$$Y_{Ca.a} = \frac{I_{Ca}}{1 - Y_W} \tag{9.3}$$

By using the corrected mass for each element, the adiabatic stoichiometric flame temperature can be determined by employing the NASA Combustion Equilibrium and Application Code. The adiabatic stoichiometric flame temperature is then corrected for the determined water content for the fluid concentrate and the industrial fluid.

Industrial fluids that do not have a fire point are eligible for approval if they meet the following requirements:

- The industrial fluid chemical heat release rate (Q_{ch}) of a highly-atomized spray shall be equal to, or less than, 130 kW.
- The adiabatic stoichiometric flame temperature adjusted for the total mass of water in the finished fluid, shall be equal to, or less than, 2100°K.
- Testing may not be required for a water-based fluid with a water content equal to, or greater than, 60% by mass.

Industrial fluids having a chemical heat release rate (Q_{ch}), or an adiabatic stoichiometric flame temperature exceeding that allowed above are not approved.

9.2.6.2 ISO 12922 Lubricants, Industrial Oils and Related Products (Class L) – Family H (Hydraulic Systems) – Specifications for Categories HFAE, HFAS, HFB, HFC, HFDR and HFDU

This standard provides the technical requirements for all classes of hydraulic fluids used in hydrostatic and hydrodynamic hydraulic systems for general industrial applications. It is not intended for use in power generation or aerospace applications. It is a guide for suppliers and end users of fire-resistant hydraulic fluids and a source of directions and information for equipment manufacturers of hydraulic systems [69]. The standard was based on the technical requirements outlined in the 7th Luxembourg Report [51]. The flammability test methods required in ISO 12922 are outlined in Table 9.12.

TABLE 9.12
ISO 12922 Flammability Test Methods

Manifold Ignition Test	ISO 20823, CETOP RP 65 H, DIN 51373
Spray Ignition	ISO 15029-1 Spray flame persistence—Hollow-cone nozzle method
Characteristics[a]	ISO 15029-2 Spray test—Stabilized flame heat release method
	NT Fire 031 Nordic Regional Test Nordtest Fire 031
Wick Flame Persistence	ISO 14935 Determination of wick flame persistence of
	fire-resistant fluids

[a] Performance under one test condition only would normally be required. Method to be agreed upon by supplier and end user.

The ISO 15029-2 Spray test—Stabilized flame heat release method [81] may be the most important fire-resistant spray ignition method within ISO 12922 because of its ability to reproducibly discriminate levels of fire resistance among types of fire-resistant hydraulic fluids tested. As stated in the method, a pre-conditioned flux of the test fluid is delivered to a test chamber through a twin-fluid atomizer. Compressed air is used to produce an atomized spray which is exposed to a defined flame derived from a gas burner. The gas flame acts to produce a stabilized spray flame, so that the combustion properties, including rate of energy release, are steady and time-averaged values may be measured. Temperatures are measured both at the entry to the combustion chamber and in the exhaust, with the burner operating without, and then with, the release of the test fluid. The flame length and smoke opacity of the exhaust are also measured. Calculations on values such as the ignitability factor, flame length index, and smoke density are made from these measurements. A grading system for the performance of fire-resistant hydraulic fluids is developed from the determined values.

The minimum number of tests is set on the basis of obtaining an ignitability factor, RI, which is the most sensitive property measured in the method. The minimum level of accuracy requires that the average of N individual RI measurements lies within 5% of the true mean, within 67% confidence.

Measurements for the ignitability factor, RI, on a test fluid are determined with an adjusted propane flow rate of 0.130 $Nm^3/h \pm 0.005$ Nm^3/h. If the ignitability factor determined as per the procedure is above 50, or if the flame length is less than 100 mm, then the test must be repeated using a propane flow rate of 0.4 Nm^3/h.

The uncorrected ignitability factor, RI_w, at the low propane flow rate of 0.130 $Nm^3/h \pm 0.005$ Nm^3/h is calculated by:

$$RI_w = 500(T_p - T_{A1})/[7(T_{EX} - T_{A2})],$$

where

T_p is the exhaust temperature without release of the test fluid, expressed in degrees Celsius;

T_{A1} is the air temperature at the chamber inlet without release of the test fluid, expressed in degrees Celsius;

T_{EX} is the exhaust temperature with the release of the test fluid, expressed in degrees Celsius

T_{A2} is the air temperature at the chamber inlet with the release of the test fluid, expressed in degrees Celsius;

RI_w is corrected to RI from the low propane flow rate calibration curve developed as described in Annex C of the ISO 15029-2 method.

The uncorrected ignitability factor, RI_w, at the high propane flow rate; 0.4 $Nm^3/h \pm 0.005$ Nm^3/h is calculated by:

$$RI_w = [100(T_p - T_{A1})/(T_{EX} - T_{A2})] + 30$$

RI_w is corrected to RI from the high propane flow rate calibration curve developed as described in Annex C of the ISO 15029-2 method.

The ranges for Relative Ignitability (RI) Index Grades in the ISO 15029-2 method are outlined in Table 9.13. [81] The typical RI Index Grades for common fire-resistant hydraulic fluid types according to the ISO 15029-2 method, described by Phillips et al. [82], are shown in Table 9.14

9.3 CONCLUSIONS

The fire-resistance properties of a hydraulic fluid cannot be adequately defined by a single test because there are often multiple potential sources for fluid flammability in a single industrial application. For example, spray ignition may occur from a spark or flame source or it may be ignited by spraying on a hot surface. Unfortunately, spontaneous ignition of a fluid-soaked lagging may

TABLE 9.13
Ranges for Relative Ignitability Index Grades according to ISO 15029-2

Relative Ignitability Index Grades	Range of Relative Ignitability Index
A	>100
B	100–80
C	79–65
D	64–50
E	49–36
F	35–25
G	24–14
H	<14

occur in air, or ignition may occur from prolonged heating of a fluid, resulting in a decrease of the auto-ignition temperature or flash point. Results from the different standards for assessing the fire-resistance of fluids may provide confusing or even conflicting conclusions on a particular fluid. A comparison of typical test results on fire-resistant fluid types between the FM Approvals Standard 6930 and ISO 15029-2 (Buxton test) are summarized in Table 9.15 [82,83].

Because it is difficult to model the fire potential of an industrial process by one or two tests, it is prudent to conduct a number of relevant fire-resistance tests to select the hydraulic fluid with the lowest overall risk. One approach is to utilize a number of standardized, application-specific tests, taking into account the potential flammability hazards associated with the particular application of interest. Some of these tests include the following [84]:

- Autogeneous ignition temperature
- High-pressure spray
- Manifold ignition test
- Ignition on CFR motor
- Flame propagation
- Soaked lagging
- Heat release spray
- Wick test

The Health and Safety Laboratory (H&SL) in the United Kingdom undertook a project to develop a test program to provide a more comprehensive understanding of commonly-used fire-resistant hydraulic fluids [85]. The H&SL also investigated the utilization of these test results in a risk-assessment-based fluid selection process. The test program was selected using the following criteria:

TABLE 9.14
Typical Relative Ignitability Index Grades for Fire-Resistant Hydraulic Fluid Types

Fire-Resistant Hydraulic Fluid Types	Typical Relative Ignitability Index Grade
HFA	A
HFB	E
HFC	B-C
HFDR	D-E
HFDU	G-H
Mineral Oil	H

TABLE 9.15

Classification Comparison of Fire-Resistant Hydraulic Fluid Types by FM Approval Standard 6930 and ISO 15029-2

Fire-Resistant Hydraulic Fluid Types	Typical FM Approvals Fire-Resistance Classification by CN 6930 (2009)	Typical RI Grade according to ISO 15029-2
HFA	FM Approved	A
HFB	FM Specification Tested *	E
HFC	FM Approved	B-C
HFDR	FM Approved or Specification Tested	D-E
HFDU	FM Approved or Specification Tested	G-H

* For typical invert emulsions containing 40% water. New formulations containing ≥45% water may be FM Approved.

- Reflect end-use conditions, in particular, the types of releases;
- Use ignition sources of representative sizes and energies;
- Tests required to provide a range of measured data. (e.g., ease of ignition, rate of fire growth, heat release, smoke, and toxic species).

Fluids were ranked according to the results for each test and then were ranked for the entire set of tests by calculation of the mean score for each fluid. Fluids were ranked numerically, in the order in which they performed, with no weightings applied to any test result. The study identified a number of tests which could be applied to the full range of types with good discrimination. Furthermore, it was concluded that to combine fluid fire-resistance performance over a range of tests was difficult and best conducted within the framework of a risk assessment.

Guidance for test selection is available from various insurance underwriters, government agencies, industry associations, and independent laboratories. The information provided in this chapter will assist the user in making informed decisions regarding the selection of hydraulic fluids which exhibit adequate fire-resistance properties for the application of interest.

REFERENCES

1. Hatton, R.E. and Stark, L.R. "Fire-Resistant Hydraulic Fluids—A Guide to Selection and Application," *Chem. Age India*, 1977, 28(9), pp. 765–772.
2. Math, J. "Fire-Resistant Hydraulic Fluids for the Plastics Industry," SPE J., 1967, pp. 17–20.
3. Sullivan, J.M. "Wanted: Cheap, Safe, Enviro-Friendly Fluids," *Lubr. World*, September, 1992 pp. 49–53.
4. White, D.F "The Unintentional Ignition of Hydraulic Fluids Inside High Pressure Pneumatic Systems," ASNE. J., August, 1960, pp. 405–413.
5. Polack, S.P. "Progress in Developing Fire-Resistant Hydraulic Fluids," *Iron Steel Eng.*, August, 1958, pp. 87–92.
6. Davis, R. "Fire-Resistant Hydraulic Fluids," *Quart. J. NFPA*, 1959, July, pp. 44–49.
7. Myers, M.B. "Fire-Resistant Hydraulic Fluids—Their Application in British Mines," *Colliery Guardian*, October, 1977.
8. Anon., *Wall Street Journal*, December 12, 1995, p. B 1.
9. *Tutti Cadaveri, Le procès de la catastrophe du Bois du Cazier à Marcinelle (The Bois du Cazier mine disaster trial*) by Marie Louise De Roeck, Julie Urbain and Paul Lootens, Editions Aden, collection EPO, Brussels, 2006, p. 280.
10. Luxembourg Report, 20 December 1960—Commission of the European Communities—Safety and Health Commission for the Mining and Other Extractive Industries. First Report on the Specifications and Testing Conditions Relating to "Fire-Resistant Hydraulic Fluids Used for Power Transmission in Mines", Luxembourg.

11. Early, C.L. and Hatton, R.E. "Industrial and ASTM Fluid Fire-Test Programs," in *Fire Resistance of Hydraulic Fluids*, American Society for Testing and Materials, Philadelphia, PA, 1966, pp. 105–34.

12. Totten, G.E. and Reichel, J. (eds.), *Fire Resistance of Industrial Fluids*, American Society of Testing and Materials, Philadelphia, PA, 1996.

13. Philips, W.D. "Fire-Resistance Tests for Fluids and Lubricants—Their Limitations and Misapplication," in *Fire Resistance of Industrial Fluids*, 1996, American Society of Testing and Materials, Philadelphia, PA, 1996, pp. 78–101.

14. Bai, J.P. "Fire Protection Gained from Synthetic Fluids," FRH J., 1984, 4(2), pp. 139–45.

15. Anon. "Determination of Ignition Characteristics of Hydraulic Fluids Under Simulated Flight and Crash Conditions," *U.S. Civil Aeronautics Administration*, Tech. Development Report No. 64, 1947.

16. Brown, C.L. and Haillwell, H. "Fire Resistance Fluid Development," *SAE Technical Paper Series*, Paper 656B, 1963.

17. Harrison, A.J. *Fire resistant hydraulic fluids their development and use in the mining industry. (Des fluides hydrauliques resistant au feu et leurs developpements et utilisations dans l'industrie miniere).* Potash '83, Proc 1st International Potash Technology Conf, Oct. 3–5, 1983, Saskatoon, Sask. Edited by R.M. McKercher. p. 459, 8 pp., 3 ref., 1983. (In English).

18. Commission of the European Communities, Safety and Health Commission for the Mining and Extractive Industries, Working Party Rescue Arrangements, Fires and Underground Combustion, Committee of Experts on Fire-resistant Fluids, Sixth Report on Specifications and Testing Conditions Relation to Fire-Resistant Hydraulic Fluids Used for Power Transmission (Hydrostatic and Hydrokinetic) in Mines, *Doc. 2786/8/81 E,* Luxembourg, 1983.

19. ISO 6743-4:2001 "Lubricants, Industrial Oils, and Related Products (Class LJ)—Classification—Part 4: Family H—(Hydraulic Systems)."

20. Sullivan, M.V., Wolfe, J.K. and Zisman, W.A. "Flammability of the Higher Boiling Liquids and their Mists," *Ind. Eng. Chem.*, 1947, 39, pp. 1607–14.

21. Murphy, C.M. and Zisman, W.A. "Non-Flammable Hydraulic Fluids," *Lubr. Eng.*, October, 1949, pp. 231–235.

22. Smith, D.F. "Assessment of Flammability of Aircraft Hydraulic Fluids," *Hydraulics Pneumatics*, June, 1970, p. 77.

23. Snyder, C.E. and Schwenker, H. "Materials on the Move," in *6th National Sampe Tech. Conf*, 1974, p. 428.

24. Snyder C.E. and Krawetz A.E. Review of Testing Methods for Hydraulic Fluid Flammability, *J. Soc. of Lubrication Engineers*, December 1981, 705.

25. ISO 4589:1984 "Plastics-Determination of Flammability by Oxygen Index."

26. Nelson, G.L. "Ease of Extinction—An Alternative Approach to Liquid Flammability," *Fire Resistance of Industrial Fluids*, American Society of Testing and Materials, Philadelphia, PA, 1996, pp. 174–88.

27. IEC Standard No. 1144, "Test Method for the Determination of Oxygen Index of Insulating Fluids," 2001.

28. Parts, L. "Assessment of Flammability of Aircraft Hydraulic Fluids," Technical Report, AFAPL-TR-79-2055, 1978.

29. Loftus, J.J. "An Assessment of Three Different Fire-Resistant Tests For Hydraulic Fluids," Report NBSIR 81-2395, I981.

30. Polack, S.P., Smith, A.F. and Barthe, H.P. "Recent developments in fire-resistant hydraulic fluids for underground use", *Bureau of Mines*, Pittsburgh, Pa. (USA), Technical Report, BM-IC-8043, 1961 Jan 01.

31. DIN 51,794, "Determination of Ignition Temperature," 1978.

32. Pollack, S.P. "Bureau of Mines Evaluates Fire Resistance of Hydraulic Fluids," *Iron Steel Eng.*, 1964, 4(8), pp. 105–10.

33. Macdonald, J.A. "Assessment of the Flammability of Aircraft Fluids," *Fire Resistance of Hydraulic Fluids*, American Society of Testing and Materials, Philadelphia, PA, 1966, pp. 44–52.

34. Sherman, J.V. Minutes of ASTM D02.N0.06 Fire Resistant Fluids, December 10, 2008 Marriot Waterside Hotel, Tampa, Florida.

35. *Fire Resistant Hydraulic Fluids in Enclosed Marine Environments*, (U.S.) National Materials Advisory Board (NRC), Washington, DC, 1979.

36. Faeth, G.M. and White, D.F. "Ignition of Hydraulic Fluids by Rapid Compression," *ASNE J.*, August, 1961, pp. 467–75.

37. MIL-H-19457D, "Military Specification—Hydraulic Fluid, Fire resistant, Non-Neurotoxic,"1981.

38. Skinner, G.B. and Ruehrwein, R.A. "Shock Tube studies on the Pyrolysis and Oxidation of Methane," *J. Phys. Chem.*, 1959, 63, pp. 1736–42,

39. Skinner, G.B. "Shock Tube Evaluation of Hydraulic *Fluids,*" *J. Am. Chem. Soc.*, Div. Fuel Chem., 1964, 8(2), pp. 166–70.
40. Snyder, C.E. and Gschwender, L.J. "Fire Resistant Hydraulic Fluids and Fire Resistance Test Methods Used by the Air Force," *Fire Resistance of Industrial Fluids*, American Society of Testing and Materials, Philadelphia, PA, 1996, pp. 72–77.
41. Hamins, A., Kashiwagi, T. and Buch, R. "Characteristics of Pool Fire Burning," *Fire Resistance of Industrial Fluids*, American Society of Testing and Materials, Philadelphia, PA, 1996, pp. 15–41.
42. Factory Mutual Standard, "Less Flammable Transformer Fluids—Class 6933."
43. "Flammability of Aircraft Hydraulic Fluids—A Bibliography," CRC Report No. 545, Coordinating Research Council, Inc., Atlanta, GA, 1986.
44. Marzani, J.A. "An Apparatus for Studying the Fire Resistance of Hydraulic Fluids at Elevated Temperatures," *Fire Resistance of Hydraulic Fluids*, American Society for Testing and Materials, Philadelphia, PA, 1966, pp. 3–18.
45. Leo Parts, "Assessment of the Flammability of Aircraft Hydraulic Fluids", Air Force Aero Prolusion Laboratory, AFAPL TR 79-2055, 7/1/1979, 85 pages.
46. Clinton, N.A., "Relationship of Chemical Composition to Fire Resistance in Hydraulic Fluids," oral presentation at the 50th STLE Annual Meeting, 1995.
47. Grand, A.F. and Trevino, J.O. "Flammability, Screening and Fire Hazard of Industrial Fluids Using the Cone Calorimeter," *Fire Resistance of Industrial Fluids*, American Society of Testing and Materials, Philadelphia; PA, 1996, pp. 157–73.
48. *Fire Resistant Fluids for Use in Machinery and Hydraulic Equipment,* 1981, National Coal Board, London (NCB Specification No. 570/1981).
49. CAN/CSA- M423-M87 (R2007), "Fire Resistant Hydraulic Fluids," Canadian National Standard.
50. "Wick Test," CETOP Provisional Recommendation RP-66H, 1974-05-31.
51. "Specification and Testing Conditions Relating Fire-Resistant Hydraulic Fluids for Power Transmission (Hydrostatic and Hydrokinetic)," 7th Luxembourg Report, March 3, 1994, *Doc. No. 4746/10/91*, Safety and Health Commission for Mining and Other Extractive Industries. L-2920 Luxembourg, Commission of the European Economic Communities DG VPA/4.
52. ISO 14935:1998(E) "Petroleum and Related Products-Determination of Wick Flame Persistence of Fire-Resistant Fluids".
53. "Effect of Evaporation on Flammability," CETOP Provisional Recommendation RP-64H, 1974-05-31.
54. Meyers, M.B. "Fire Resistant Hydraulic Fluids—Their Use in British Mines," *Colliery Guardian*, October, 1977, pp. 796–808.
55. Mr. R. J. Windgassen (Amoan Oii) to Mr. W. E. E Lewis (Union Carbide Corporation), personal communication, on July 16, I986, concerning a new ASTM D.02N.06 working group for the development of "The Soaked Cube Flammability Test," chaired by J. Anzenberger (Stauffer Chemical Company).
56. Faulkner, H. personal communication, July 14, 1985.
57. Norvelle, D. "A Discussion of Fluid Flammability Tests—Part 2: Molten Metal Tests," FRFI J., I985, Volume 4(2), pp. 133–38.
58. Faulkner, C.H., Totten, G.E. and Webster, G.M. "Fire Resistance Testing: A Technology Overview and Update," in *Proceedings of the 47th National Conference on Fluid Power*, 1996, Vol. 1, pp. 75–81.
59. "Manifold Ignition Test," CETOP Provisional Recommendation RP-65H, 1974-05-31.
60. Goodall, D.G., and Ingle, R. "The Ignition of Flammable Fluids by Hot Surfaces," in *Fire Resistance of Hydraulic Fluids*, American Society of Testing and Materials, Philadelphia, PA, 1966, pp. 66–104.
61. SAE - AS1241 Revision C (1997-09) "Fire Resistant Phosphate Ester Hydraulic Fluid for Aircraft" Aerospace Standard)
62. Factory Mutual Research Corporation, *Approval Standard: Less Hazardous Hydraulic Fluid*, Factory Mutual Research Corporation, Norwood, MA, 1975.
63. Khan, M. "Spray Flammability of Hydraulic Fluids and Development of a Test Method," Technical Report FMRC J.1.0TOW3.RC, Factory Mutual Research Corporation, Norwood, MA, 1991.
64. Totten, G.E. "Thickened Water–Glycol Hydraulic Fluids for Use at High Pressures," *SAE Technical Paper Series*, Paper 921738, 1992.
65. FM Approvals, "Approval Standard: Flammability Classification of Industrial Fluids", (Class 6930), Factory Mutual Global, January 2002.
66. "Hydraulic Transmission Fluids: Determination of the Ignitability of Fire-Resistant Fluids Under High Pressure with a Jet spray on a Screen," Report AFNOR NF E 48-618, 1973.
67. Yuan, L. "Ignition of Hydraulic Fluid Sprays by Open Flames and Hot Surfaces," *Journal of Loss Prevention in the Process Industries 19* (2006), pp. 353–61.

68. Rowand, H.H. and Sargent, L.B. "A Simplified Spray-Flammability Test for Hydraulic Fluids!" in *Fire Resistant Hydraulic Fluids,* 1966, American Society for Testing of Materials, Philadelphia, PA, 1966, pp. 28–43.
69. ISO 12922:1999(E), "Lubricants Industrial Oils and Related Products—Specifications for Categories HFAE, HFAS, HFB, HFC, HFDR and HFDU."
70. Khan, M. and Tewarson, V. "Characterization of hydraulic fluid spray combustion," *Fire Technol.,* 1991, November, p. 321.
71. Kahn, M. and Brandao, A.V. "Method of Testing the Spray Flammability of Hydraulic Fluids," *SAE Technical Paper Series*, Paper 921737, 1992.
72. Yule, A.L. and Moodie, K. *Fire Safety J.,* 1992, 18, p. 273.
73. Holke, K. "Testing and Evaluation of Fire-Resistant Hydraulic Fluids using the Stabilized Heat Release Spray Test," in *Fire Resistance Industrial Fluids*, American Society of Testing and Materials, Philadelphia, PA, 1996, pp. 157–173.
74. Holmstedt, G. and Persson, H. "Spray fire tests with hydraulic fluids," in *Fire Safety Science—Proceedings of the First International Symposium*, 1985, p. 869.
75. CFR Part 35, "Fire Resistant Fluids" (U.S. Mine Safety and Health Administration).
76. GM Lubricant Standard LS2 (2004) Version 5, L5 Individual Lubricants Standards, Document No. GM 1721
77. FM Approvals, "Approval Standard: Flammability Classification of Industrial Fluids (Class 6930)", April 2009, Factory Mutual Global.
78. MIL-H-22072C, "Military Specification-Hydraulic Fluid, Catapult, NATO Code Number H- 579", 1984.
79. MIL-PRF-87257 Revision A, "A Hydraulic Fluid, Fire-Resistant; Low Temperature, Synthetic Hydrocarbon Base, Aircraft and Missile," 1997.
80. MIL-PRF-83282 Revision D, "Hydraulic Fluid, Fire-Resistant, Synthetic Hydrocarbon Base, Metric, NATO CodeNumber H-537," 1997.
81. ISO 15029-2:1999, "Petroleum and Related Products—Determination of Spray Ignition Characteristics of Fire-Resistant Fluids Part 2 – Spray Test-Stabilized Flame Heat Release Method."
82. Phillips, W.D., Goode, M.J. and Winkeljohn, R. "Fire-Resistant Hydraulic Fluids and the Potential Impact of New Standards for General Industrial Applications," *National Fluid Power Association* (100-1.12), 2000.
83. Ferron, "FM Approvals Standard for Industrial Fluids (FM 6930)", Society of Tribologists and Lubrication Engineers (STLE), *62nd Annual Meeting*, Philadelphia, Pennsylvania, USA, May 8, 2007.
84. "Schedule of Fire Resistant Tests for Fire Resistant Fluids," CETOP Provisional Recommendation RP 55 H, 1974-01-04, *British Fluid Power Association*, Oxfordshire, United Kingdom.
85. Jagger, S., Nicol, A., Sawyer, J. and Thyer, A. "Assessing Hydraulic Fluid Fire Resistance", *Machinery Lubrication Magazine*, Health and Safety Laboratory, United Kingdom, September 2007.

10 Bench and Pump Testing Procedures*

D. Smrdel and Donald L. Clason

CONTENTS

10.1 INTRODUCTION

One method of evaluating lubrication properties of a hydraulic fluid is to perform a test in the hydraulic pump (or motor) of interest. However, this is clearly impractical in view of the numerous pump manufacturers and models that are available. This issue may be further complicated because various types of pumps may be configured differently or manufactured with different materials, depending on the pump manufacturer. This problem could be greatly simplified with the use of one or at least a limited number of "standard" hydraulic pump tests.

One standard test, ASTM D-2882 [1], has been developed for this use. This test is conducted for 100 h with a Eaton-Vickers V-104 vane pump with a 5-gal. (19.9-L) reservoir at 65°C, 13.8 MPa (2000 psi), and 1200 rpm, using an 8-gal./min (30.3-L/min) cartridge. The German specification, DIN 51389, is the same except that the test is conducted for 250 h, at 50°C, 10.3 MPa (1500 psi), and 1500 rpm.

---

* This chapter is a revision of the chapter titled "Bench and Pump Testing of Hydraulic Fluids" by L. Xie, R.J. Bishop Jr., and G.E. Totten from the *Handbook of Hydraulic Fluid Technology*, 1st Edition.

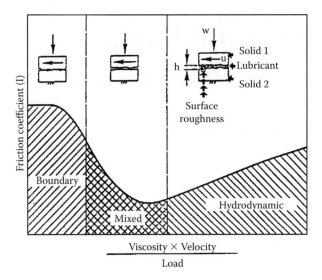

FIGURE 10.1 Illustration of the effect of wear contact loading and speed, and fluid viscosity on wear (Stribeck curve).

The D2882 and DIN 51389 are to be used as reference only due to Eaton-Vickers no longer manufacturing the pump components. The new method for the D2882 is ASTM D-7043. The test procedure and conditions are the same but the 104C components used are manufactured by Conestoga USA*. There is no direct correlation in results to the Eaton-Vickers hardware due to a lack of test data. Ongoing correlation and round robin testing are being conducted in the industry along with a precision and bias statement for the ASTM D7043.

Although the Eaton-Vickers V-104 vane pump has wide acceptance and has served the industry well, it does possess a number of disadvantages, such as the relatively high cost per test and a test duration that is sufficiently long to preclude its use for such applications as quality control testing and general fluid development, as well as used fluid performance troubleshooting. Therefore, it would be desirable to develop a faster, lower-cost "bench test" as a viable alternative to routine pump testing.

Silva published a thorough review of the wear mechanisms in hydraulic pump operation [3]. The role of cavitation, adhesion, corrosion, and abrasion wear was described. Also discussed was the role of fluid viscosity and the speed and loading at the wear contact as modeled by the classic Stribeck curve, illustrated in Figure 10.1. However, discriminating methods of experimental modeling of pump wear were not discussed.

There are numerous references to the use of hydraulic pump tests to evaluate component durability [4,5] or to evaluate some aspect of component design features on either the efficiency or mechanism of energy transfer. However, there are fewer references on the development and use of hydraulic pumps as "tribological tests" to evaluate fluid lubrication wear.

There have been some references describing the impact of fundamental lubrication properties of hydraulic fluids—such as film thickness—on pump wear using the hydraulic pump as a tribological test [6]. However, there are a number of problems with such testing procedures, including pump cost, energy and components, relatively long testing times, relatively poor manufacturing

---

* Conestoga USA manufactures replacement vane pump parts and apparatus necessary for testing according to ASTM D7043 (formerly ASTM D2882), ISO 20763, DIN 51389 and others. Conestoga USA Inc., P.O. Box 3052, Pottstown, PA 19464; Tel. 610-327-2882, FAX: 610-327-1676, www.conestogausa.com.

precision of some of the components for use in reproducible tribological testing, and the volumes of fluid required and their subsequent disposal. Therefore, there has been a long-standing effort to develop "bench test" alternatives to evaluate hydraulic fluid lubricity [7].

Because standardized bench tests have been generally shown to be unreliable [8–15], pump lubrication must often be examined using hydraulic pumps with a range of test conditions and appropriate material pairs [13]. Section 2.3 of this chapter will provide an overview of various pump testing procedures that have been used to study hydraulic fluid lubrication. This discussion will include vane-, piston-, and gear-pump testing procedures. A summary of evaluation criteria will also be provided.

10.2 PART I: DISCUSSION

10.2.1 BENCH TESTS

Various forms of the Shell four-ball test are frequently encountered in fluid lubrication tests. This test is frequently used to set performance criteria such as purchasing specifications for hydraulic fluids. It also used for evaluating the wear characteristics of new and used fluids. There are two main procedures of the four-ball test. First is the ASTM D2783, which determines the load-carrying properties of lubricating fluids. The determinations made are load-wear index and weld point by means of the four-ball extreme pressure tester. The four-ball test is conducted by mounting four balls; three balls are fixed in the bottom assembly and the top ball is rotated at a constant speed of 1760 rpm. See Figure 10.2 for test ball configuration. The fluid is heated to 65–95°C and a series of increasing loads are made at ten-second increments until welding occurs. This test is a good differentiator between fluids having low, medium, and high extreme pressure properties.

The second four-ball test is the ASTM D4172, which covers a procedure for making a preliminary evaluation of anti-wear properties of fluid lubricants in sliding contact. As in the D2783, the test is conducted by mounting four balls; three balls are fixed in the bottom assembly and the top

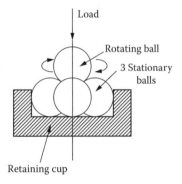

FIGURE 10.2 Illustration of the four-ball test configuration.

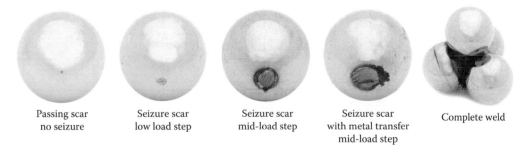

| Passing scar no seizure | Seizure scar low load step | Seizure scar mid-load step | Seizure scar with metal transfer mid-load step | Complete weld |

FIGURE 10.3 Four-ball wear scar examples.

ball is rotated at a constant speed but at 1200 rpm. A load of either 147 or 392 N (15 or 40 kgf) is applied. The temperature is maintained at 75°C (167°F) for a test duration of 1 hour. Lubricants are evaluated by using the average size of the scars worn on the lower balls. See Figure 10.3 for four-ball wear scar examples. This test can be used to determine the relative wear preventative properties of lubricating fluids in sliding contact.

10.2.2 FZG Gear Test

Reichel has reported the use of an FZG gear test to compare the antiwear performance of a hydraulic oil and vegetable oil with and without the addition of antiwear additives [34,35]. The test configuration of the FZG gear apparatus is illustrated in Figure 10.4. It is used to measure wear with

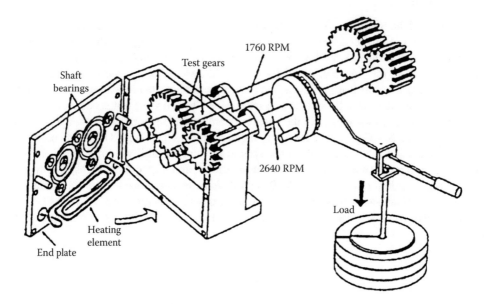

FIGURE 10.4   Schematic illustration of the FZG gear test machine.

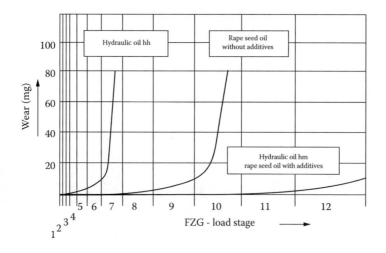

FIGURE 10.5   Evaluation of the antiwear properties of a rapidly biodegradable hydraulic fluid using the FZG gear test.

increasing increments of load. The results of this work, shown in Figure 10.5 [34], exhibited the expected differences in the antiwear properties of these fluids. However, direct correlations with pump wear was not provided, nor was the ability of the FZG gear test to discriminate between the small differences in antiwear properties demonstrated.

10.2.3 Custom Bench Tests

Although it would be desirable to conduct bench tests as alternatives to hydraulic pump testing, with few exceptions these tests have provided poor correlation with pump lubrication. For example, Renard and Dalibert found no correlation between either the Shell four-ball (ASTM D-2596-87 [36]) or "extreme pressure" (ASTM D-2783 [37]) tests and wear results obtained with an Eaton-Vickers V-104 vane-pump test (ASTM D-2882) [1]. Knight reported, "... Regrettably the data from these tests not only failed to give quantitative correlation with data determined from machines, but even the ranking between different fluids failed to agree... ." [9]. More recently, Lapotko et al. reported that four-ball test results "... may deviate considerably from the data obtained in actual service..." [10]. Urata also found no correlation between four-ball wear and the V-104C pump wear results for a series of high-water-base hydraulic fluids [38].

These problems were confirmed in the above discussion in which the inability of "standard" bench tests to model the ASTM D-2882 Eaton-Vickers V-104 hydraulic vane pump wear was described. However, in some cases, bench tests have been either specially modified or even custom designed to model wear of a specific contact within a pump. Selected examples will be provided in the following discussion.

University of Aachen sliding wear bench test. A custom-designed bench test was reported by Jacobs et al. which was used to examine sliding wear occurring in hydraulic pumps with different material pairs [39]. This machine is schematically illustrated in Figure 10.6. The wear contact is a rotating disk.

University of Leeds testing machine. Priest et al. have also reported a bench test that is custom designed, as illustrated in Figure 10.7, to model the sliding wear contact in the V-104 vane pump

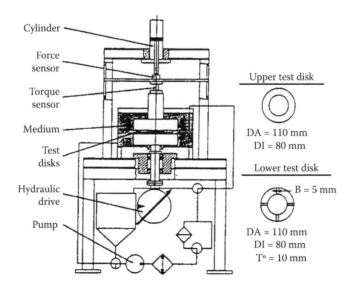

FIGURE 10.6 Aachen mixed-friction sliding wear machine.

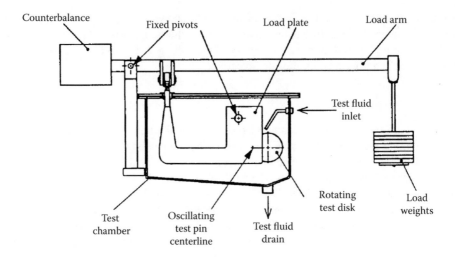

FIGURE 10.7   University of Leeds test machine.

[40]. A schematic comparison of the wear contacts in the bench and V-104 vane-pump tests is illustrated in Figures 10.8 and 10.9, respectively. Excellent correlation between the bench and vane-pump tests has been reported [40].

Cameron–Plint wear test apparatus. The effect of metallurgy on sliding wear using various hydraulic fluids for aerospace applications has been performed using the Cameron–Plint wear testing apparatus shown in Figure 10.10 [41]. In this test, the sliding motion of an upper cylinder on a lower plate surface is controlled by a variable-speed motor.

Unisteel rolling fatigue machine. Another bench test that has been used to model hydraulic fluid wear is the Unisteel rolling fatigue machine shown in Figure 10.11. In this test, a flat ring forms the upper part of a thrust bearing which contains the cage, half of the balls, and one race from a production thrust bearing assembly [42].

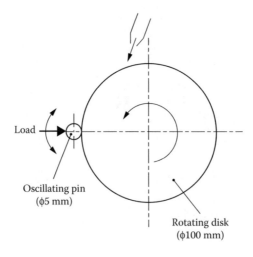

FIGURE 10.8   Wear contact for the University of Leeds test machine.

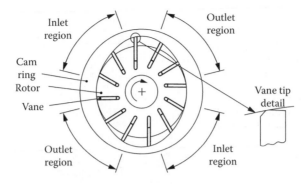

FIGURE 10.9 Wear contact for Vickers V-104 vane pump.

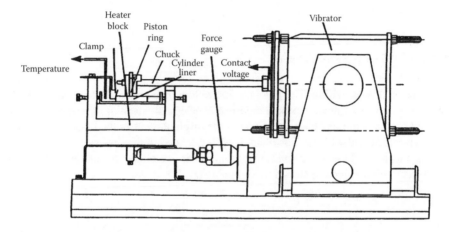

FIGURE 10.10 Cameron–Plint wear test machine.

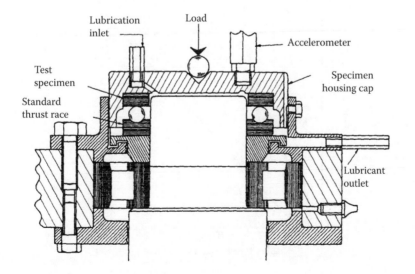

FIGURE 10.11 Unisteel rolling fatigue machine.

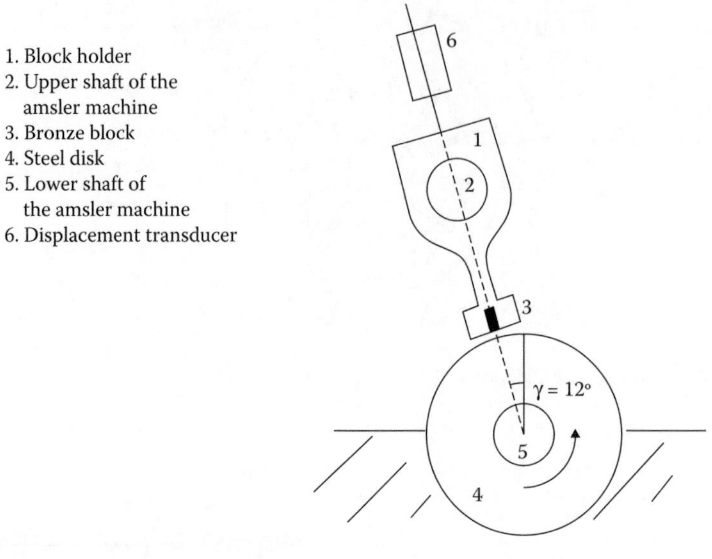

1. Block holder
2. Upper shaft of the
   amsler machine
3. Bronze block
4. Steel disk
5. Lower shaft of
   the amsler machine
6. Displacement transducer

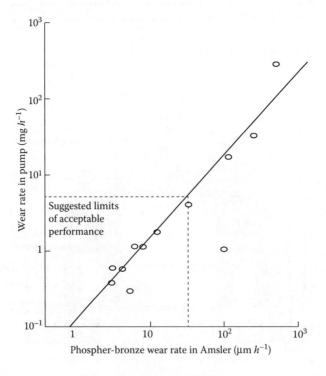

FIGURE 10.12   The setup of the Amsler rig.

Modified Amsler disk machine. The potential for yellow metal wear that may occur with bronze or brass slippers may be evaluated using a modified Amsler disk machine. The modification, shown in Figure 10.12, entails the replacement of one of the original disks in the machine by a phosphor–bronze or high-tensile-brass block. The block is spring-loaded to the disk surface. Wear is continuously recorded using an inductive displacement transducer. Excellent correlation with piston-pump, slipper-pad wear was obtained as shown in Figure 10.13 [43].

FIGURE 10.13   Correlation between the results of pump and Amsler tests.

10.2.4 Hydraulic Fluid Evaluation Using Fundamental Lubrication Parameters

Stribeck–Hersey curve. Lubrication of hydraulic pumps may occur by one of at least four mechanisms: (1) hydrodynamic, (2) elasto-hydrodynamic (EHD), (3) mixed-film, and (4) boundary. The particular mechanism encountered is dependent on the film viscosity, velocity, and applied load as illustrated in Figure 10.1 [3]. Hydrodynamic lubrication is characterized by relatively thick films, typically >300 nm, which are substantially thicker than the asperity contacts of the wear surface. EHD lubrication is characterized by thin-film lubrication. Although the lubrication films are only ~30 nm thick, they are greater than the asperity contacts. Film thicknesses for boundary lubrication, typically approximately 3 nm, are less than the height of the asperity contacts. Mixed-film lubrication occurs at the transition from EHD to boundary lubrication. Ideally, hydraulic pumps operate in the hydrodynamic lubrication regime [3].

Under these conditions, the lubricating capability of the fluid is primarily dependent on the fluid film viscosity. However, it is not possible to assure hydrodynamic lubrication under all operating conditions. For example, start-up conditions, oscillating motion, and so forth, can create contact speeds insufficient to support hydrodynamic lubrication. Thus, a hydraulic fluid must exhibit some antiwear characteristics.

Performance map characterization. It is possible to construct "performance maps" to identify EHD, mixed-film, and boundary lubrication properties of a hydraulic fluid [14,44,45]. A typical performance map is illustrated in Figure 10.14.

As Figure 10.14 shows, performance maps are constructed in terms of rolling speeds (R) and sliding speeds (S). This is important because the generation of an EHD film is primarily a function of the entraining velocity (R) in the inlet region of the Hertzian contact (see Chapter 3). In this region, lubricating film generation is primarily a function of the physical properties (viscosity, pressure–viscosity coefficient, etc.) of the hydraulic fluid. The sliding speed (S) determines the shear strain within the high-pressure Hertzian contact region. This region is important with respect to heat generation, surface film formation, wear, and scuffing within the tribocontact (see Chapter 6). The magnitude of the degree of surface interaction achieved, as a result of thin EHD films,

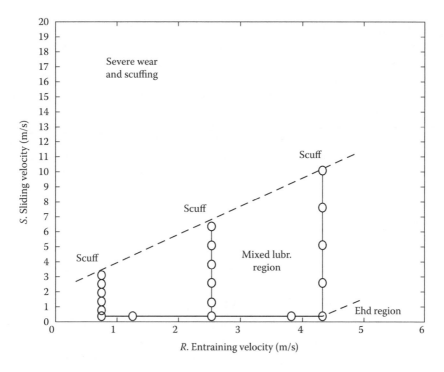

FIGURE 10.14 Illustration of a typical performance map for a hydraulic fluid.

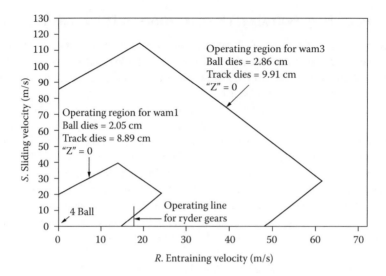

FIGURE 10.15 Map of entraining velocity (R) and sliding velocity (S).

influences the chemical properties of the fluid (e.g., adsorbed films, chemical reaction films, tribochemical reactions, and thermal/oxidative stability).

A correlation between the Ryder gear test, Shell four-ball wear test, and their predictive placement on a performance map has been made [14]. As Figure 10.15 shows, the lubrication results of the test represent just one point on the overall performance map and perhaps this point may be outside of the *range* required for the various lubricated surfaces in a particular hydraulic pump design. In fact, this is a deficiency of conducting bench tests relative to the varying lubrication requirements in a machine, such as a hydraulic pump.

Wear contact geometry. The contact geometry of a tribocontact should also be considered. There are typically three characteristic types: (1) point contact (such as ball-on-disk), (2) line contact (such as a roller-on-disk), and (3) an area contact (such as a flat surface-on-disk). All of these represent uniquely different lubrication problems and, ideally, the contact geometry of the tribocontact should reasonably model the actual system being studied [46]. The common tribocontact surface geometries are shown in Figure 10.16 [47].

It is recommended that the contact geometry of the bench test that has been selected should reasonably model the actual system [47,48]. Recently, Voitik has recommended the use of "Tribological Aspect Numbers (TAN)" to quantitatively and systematically characterize a wear contact [49].

There are four characteristics to be determined in the calculation of a TAN [40]:

1. Contact velocity
2. Contact area
3. Contact pressure
4. Entry angle

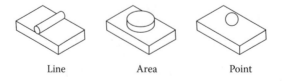

Line   Area   Point

FIGURE 10.16 Tribocontact surface-geometry classification.

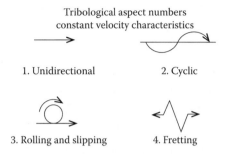

FIGURE 10.17 Contact-velocity characteristics for TAN calculation.

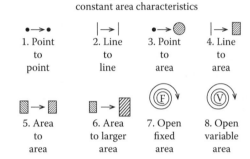

FIGURE 10.18 Contact-area characteristics for TAN calculation.

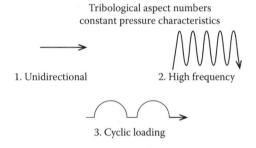

FIGURE 10.19 Contact-pressure characteristics for TAN calculation.

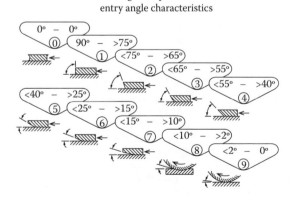

FIGURE 10.20 Interfacial entry angles for TAN calculation.

Contact-velocity characterization. There are four contact-velocity characteristics, which are illustrated in Figure 10.17: (1) unidirectional, (2) cyclic, (3) roll/slip, and (4) fretting. The first digit of the TAN number is selected from 1 to 4.

Contact-area characterization. The second digit of the TAN represents the contact-area characteristic and is selected from one of the eight possible values illustrated in Figure 10.18: (1) point to point, (2) line to line, (3) point to area (circle), (4) line to area (rectangle), (5) area to area, (6) smaller area to larger area, (7) open fixed area, and (8) open variable area.

Contact-pressure characteristic. The third digit of the TAN number is selected from one of the three contact-pressure designations shown in Figure 10.19: (1) unidirectional, (2) high frequency, and (3) cyclic loading.

TABLE 10.1
Comparison of Tribological Aspect Numbers of Various Bench-Test Configurations

TAN	Contact Geometry
1418	Timken
1317	Shell four-ball
1419	Vane-on-disk
1519	Pin-on-disk
1318	Ball-on-disk
3229	FZG test

TABLE 10.2
Wear Aspects of Significant Parts and Subassemblies

Item no.	Description	Material	TAN no. A	B	C	D
01	Cylinder barrel bushing	Brass	3	4	2	9
02	Cylinder control surface	Brass	2	7	2	1
03	Valve plate	Nitrided steel	2	7	2	1
04	Piston	Nitrided steel	3	4	3	8
05	Slipper shoe	Brass	3	6	3	1
06	Swash plate	Nod. cast iron GGG-60	2	5	3	1
07	Sliding plate	Nitrided steel	2	5	3	1
08	Saddle bearings	Brass	2	5	3	1
09	Retainer ball	Nitrided steel	2	4	3	9
10	Retainer plate	Tool steel	2	4	3	9
11	Drive shaft	Steel, induction hardened	2	5	3	0
12	Cylinder roller bearing	Steel, hardened	2	4	3	9
13	Shaft seal	FKM/PTFE	2	4	3	1
14	Seals	BUNA-N/FKM	2	6	3	0
15	Control piston	Tool steel, induction hardened	2	5	3	1
16	Piston housing	Cast iron GGG-40	2	5	3	1
17	Slide stone	Steel, hardened	2	5	3	1
18	Control-valve housing	Cast iron GGG-40	3	5	2	8
19	Control-valve spool	Tool steel	3	5	2	8

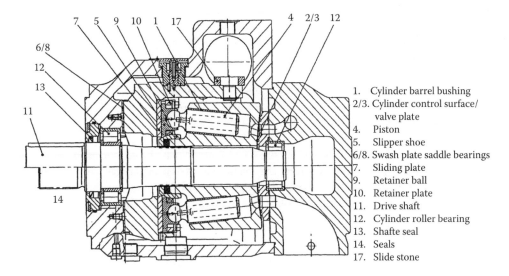

1. Cylinder barrel bushing
2/3. Cylinder control surface/ valve plate
4. Piston
5. Slipper shoe
6/8. Swash plate saddle bearings
7. Sliding plate
9. Retainer ball
10. Retainer plate
11. Drive shaft
12. Cylinder roller bearing
13. Shafte seal
14. Seals
17. Slide stone

FIGURE 10.21 Wear surface TANs for the axial piston pump. See Table 10.2 for part labels.

Interfacial entry angle. Nine characteristic entry angles have been characterized between 0° and 90° and are shown in Figure 10.20. These values reflect hydrodynamics, plowing, starvation, debris entry, thermal conductivity, and other effects on tribological wear.

The values reported by Voitik for some of the bench tests reported here are summarized in Table 10.1 [49]. None of the test configurations exhibit the same TAN as the vane-on-disk, which models the vane-on-ring of the vane pump. Also, different TANs would be expected for bearing lubrication (e.g., cylinder in bore) encountered in other pump designs.

Table 10.2 summarizes various TANs for different wear contacts which may be encountered in hydraulic components. The wear surfaces (1–19) from Table 10.2 are illustrated in Figure 10.21 (1–14), Figure 10.22a (5–17), and Figure 10.22b (18 and 19).

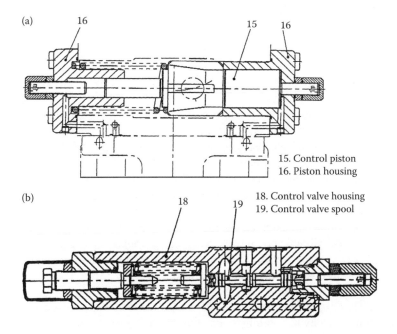

15. Control piston
16. Piston housing

18. Control valve housing
19. Control valve spool

FIGURE 10.22 (a, b) TAN characterization of valve wear surfaces. See Table 10.2 for part labels.

10.2.5 BENCH-TEST CONSIDERATIONS

In addition to modeling the actual wear mechanism in the hydraulic pump, a successful bench test should do the following [47]:

- Reproduce the wear mechanisms in the application of interest
- Reproduce the temperature level of the material during normal wear
- Reproduce the metallurgy and geometry of the contacts

Accelerated tests may rate lubricants in the correct order (which did not occur in the work described earlier in this chapter); however, the magnitude of the difference between the fluids studied may not be proportional with actual field experience [48].

If a bench test is to model field experience, it should do the following [48]:

- Provide reasonable reproducibility.
- The results obtained must have some degree of correlation with the results observed in actual service.

One of the primary reasons for running a bench test is to screen fluids so that expensive pump testing can be used for those fluids that have a reasonable chance of being acceptable in the field. If so, then one criteria that *must* be applied is that a bench test must not fail a fluid that will have acceptable field performance. It may, however, pass a fluid that will ultimately fail in the pump tests. This will be sorted out during the pump test phase of fluid acceptance. Any reasonable bench test should screen out at least 50% of the failing fluids. Hopefully this can be raised to over 70% for a good screening test. If the correlation approximates 100% then there is no need for actual pump testing.

Although the bench tests discussed here were reasonably reproducible, they provided poor, if any, correlation with hydraulic pump results. By this criteria, none of the tests evaluated were suitable models of hydraulic performance.

It is also important to recognize the various forms of wear that may occur in pump operation. Some of these include the following [3]:

- Abrasive wear
- Adhesive wear
- Cavitation wear
- Corrosion wear

With bench testing of hydraulic fluids under atmospheric pressure, it is difficult to model exactly the fluid pressure conditions encountered in hydraulic pumps. Thus, potential for wear by cavitation may not be observed. Similarly, corrosive processes or rolling contact fatigue often take many hours, days, or even months to occur in actual operation. They simply will not have time to occur under the accelerated conditions of the bench test. Furthermore, the promotion of the accelerated wear conditions may actually be accompanied by a change in the wear mechanism; for example, from abrasive to adhesive failure.

Recently, Mizuhara and Tsuya studied the ability of a block-on-ring test (ASTM D 2714–68[50]) to model three different hydraulic pump (vane, gear, and piston) tests. The hydraulic fluids studied were antiwear oil, water–glycol, oil-in-water emulsion, water-in-oil emulsion, phosphate ester, and polyol ester. The conclusions from this study were as follows [13]:

- It was necessary to evaluate the fluids under a wide range of conditions.
- Load-carrying capacity has nothing to do with the antiwear properties.
- Accelerated tests usually provide the wrong results.

- Materials for the actual test pieces must be similar to the actual machine.
- To successfully evaluate the wear characteristics of hydraulic fluids, a wear test in a hydraulic pump must be conducted.

The following test strategy was recommended for the use of laboratory wear-testing machines:

1. Rank hydraulic fluids according to hydraulic pump tests.
2. Evaluate hydraulic fluids in a wear-testing machine under a wide variety of test conditions and material pairs.
3. Determine the testing condition that provides the proper fluid ranking according to hydraulic pump tests.
4. If the proper testing conditions are not found, expand the test variables.
5. Repeat procedure number 3; if inadequate correlations are obtained, change the wear-testing apparatus.

These conclusions, while cumbersome, are reasonable in view of the tribological principles reviewed earlier.

Ludema has stated that "the best approach to wear modeling is to develop an organized way to accumulate empirical results from tests that simulate practical systems and build models from those data" [51]. Clearly, if such a database were available, it would greatly facilitate the correlation of standard wear test results to those obtained with hydraulic pumps under widely varying wear conditions.

10.3 PART II: HYDRAULIC FLUID FILTERABILITY

10.3.1 Filterability Tests

There are a number of tests used in the industry to evaluate the filterability characteristics of hydraulic fluids. It is crucial that the system filter remove contaminants in order to minimize wear which can adversely affect pump performance. The ability of a hydraulic fluid to pass through fine filters, without plugging them, is known as its "filterability".

The ISO 13357 test consists of two parts, with (wet) and without (dry) the presence of water. This procedure only applies to mineral-based oils since other fluid types may not be compatible with the test membranes. This procedure is not suitable for some hydraulic fluids which have specific properties conferred by the use of insoluble/soluble additives, or by particularly large molecular species.

The preparation of the wet samples involves adding 0.2% water to a test fluid and placing it in a 70°C oven for 70 hours. Both parts of the test, wet and dry, pass through a 0.8 μm mean pore diameter filter under specific conditions. As the sample passes through the filter, times are recorded at specified volumes. Two different filterability stages are calculated from ratios of the filtration rate near the start of the test compared to the filtration rate at specified higher filtered volumes. Both wet and dry tests are run in triplicate, with the average of the three determined values being reported. Figure 10.23 shows the configuration for the test.

The results of the tests at each stage will normally be reported as pass or fail against a limiting filterability number of 50. This filterability number is generated by an equation based on the amount of time it takes the fluid to pass through the filter media. Pass results above 50 may be accompanied by the average determined value in parentheses. Results below 50 generally have poor reproducibility, and the average value should not be reported.

The AFNOR Filterability is a standard that defines a laboratory test which evaluates the filterability of mineral hydraulic fluids. The test is in two parts—without water (NF-E-48-690) and with water (NF-E-48-691). The experimental conditions are the same as the ISO 13357 test. However, the AFNOR test calculates a filterability index as opposed to the two filterability stages of the ISO

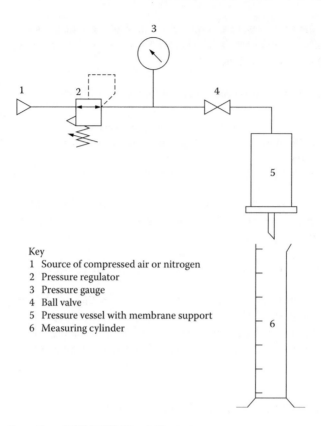

Key
1 Source of compressed air or nitrogen
2 Pressure regulator
3 Pressure gauge
4 Ball valve
5 Pressure vessel with membrane support
6 Measuring cylinder

FIGURE 10.23 Configuration of ISO 13357 filterability test.

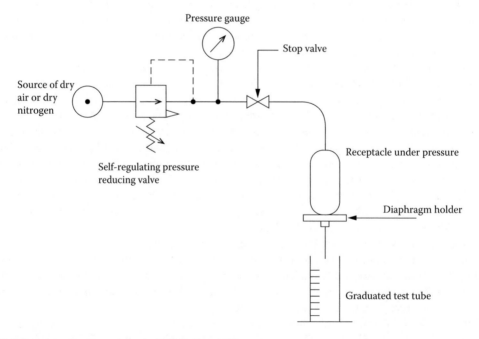

FIGURE 10.24 Configuration of AFNOR filterability test.

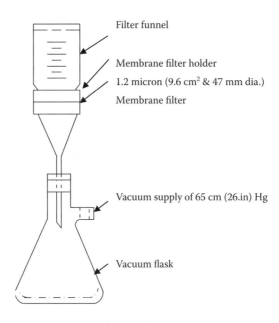

FIGURE 10.25 Configuration of the Parker Denison Filterability Test.

13357. Figure 10.24 shows the configuration for the test. The test is run in triplicate with the average of the three determined values being reported.

The Parker Denison Filterability Test (A-TP-02100) is another filterability test that evaluates the filterability characteristics of mineral oil and synthetic hydraulic fluids. This test also is performed with and without water, using two 100 ml samples of fluid as received and two 100 ml samples with 2% distilled water added. Each sample is tested at room temperature using a 1.2 µm absolute membrane filter. A vacuum of 65 mm Hg is applied to the test apparatus while 75 ml of each fluid is filtered and timed. Figure 10.25 shows the configuration for the test. The reported values are the average of the two dry samples and the average of the two wet samples. If the test time exceeds 600 seconds, the run time is recorded as 600+ seconds.

10.4 PART III: LUBRICATION CHARACTERIZATION BY PUMP TESTING

10.4.1 PUMP-FAILURE MODES

There are numerous reasons for conducting laboratory hydraulic pump tests. These include [15]:

1. To compare the differences in equipment performance
2. To obtain controlled performance experience with new fluids
3. To troubleshoot field performance problems
4. To obtain comparative lubrication data on various hydraulic fluid types and manufacturers

Pump failures are typically accompanied by increased noise and loss of volumetric efficiency. During the course of conducting pump tests, as a minimum, observations must be made to determine the cause of pump performance loss or failure. Preferably, more quantitative wear determinations, coupled with photographic record-keeping and a continuous analysis of pump operation, should be performed. A summary of some common failure modes for gear pumps, axial piston pumps, and vane pumps are provided in Table 10.3 [15,52,53] and Chapter 12.

TABLE 10.3
Common Pump-Failure Modes

Pump Type	Failure Mode
Gear	Roller-bearing fatigue, gear-tooth surface pitting and fatigue, seal-plate scoring and seizure.
Piston (axial)	Roller-bearing fatigue, valve-plate scoring and erosion, slipper-pad failures, piston and cylinder wear, cavitation.
Vane	Severe vane and ring, port plate, bearing, and cavitation wear.

10.4.2 FLUID-SPECIFIC PROPERTIES

In addition to pump testing, and preferably prior to it, a number of fluid-specific properties should be determined. These include metal and nonmetal compatibility. For example, vapor, liquid, and dry film corrosion properties should be performed [53]. Copper and other soft-metal corrosion should also be determined [54,55]. Propensity for sludge formation, viscosity and viscosity stability, pH, and wet ability are other important variables. Feicht has recommended that, because of compressibility differences among hydraulic fluids, valve-housing stability be examined. Nonmetal or seal compatibility is vitally important [55].

10.4.3 VANE-PUMP TESTING

The output pressure of a vane pump is directed to the back of the vanes which hold them against the ring. The leading edge of the vane forms a *line contact* with the ring and the rotation of the vane against the ring generates a *sliding motion*, as shown in Figure 10.26. The challenge is to maintain adequate lubrication between the vane tip and the cam ring with increasing pressures and rotational speeds [6].

Ueno et al. have used the vane pump as a tribological test stand to study the effects of pump delivery pressure, rotational speed, eccentricity, hardness of the cam ring, vane-tip curvature, and the length of action range of vane equilibrium force during rotation on the total wear of the vanes and ring [6,56]. Their results showed that the wear rate simply cannot be estimated from the friction work due to these factors. However, the occurrence of severe wear is proportional to $P_H V$, where P_H is the maximum Hertzian contact stress and V is the sliding velocity.

Hemeon reported the application of a "yardstick formula" to quantitatively analyze and report vane-pump wear. The critical part of the analysis was the weight loss of the ring, as dimensional changes due to wear will significantly affect volumetric efficiency and lead to noise and pulsation [57]. The yardstick formula is:

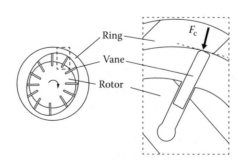

FIGURE 10.26 Illustration of the sliding line contact with vane-on-ring.

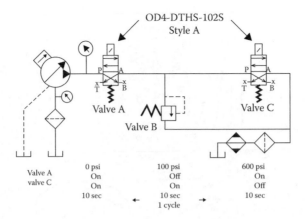

FIGURE 10.27 Test circuit for Bosch Rexroth cycled-pressure, vane-pump test utilizing a 7.5-hp electric motor, SV-10 vane pump, and a 20-gal. reservoir.

$$Y = K(1.482)(\text{mep})(\text{gpm}) \tag{10.1}$$

where:

Y = duty load on the pump in Btu/h.

K = a constant to correct for air entrainment, degraded or contaminated oil, and fluid turbulence

mep = mean effective pressure

gpm = flow in gallons per minute

Although K may be as high as 1.4, a value of 1.03 is typical. The conversion constant 1.482 permits the use of pressures in pounds per square inch and weight loss in grams. Interestingly, in order to obtain reproducible and reliable weight-loss data, it was reported that the ring had to be washed and baked at 200°F for 24 h because the porous metal adsorbed the solvent and gave incorrect weight data.

Bosch Rexroth also utilized a cycled-pressure, vane-pump test. The pressure–time sequence and test circuit is illustrated in Figure 10.27 [58]. This is reported to be a more representative test because it better incorporates pressure spikes that will invariably occur in a hydraulic system during circuit activation and deactivation. At the conclusion of the test, the weight loss of the ring, vanes,

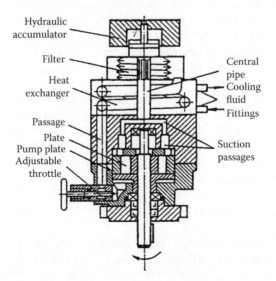

FIGURE 10.28 Schematic illustration of the Russian MP-1 vane pump.

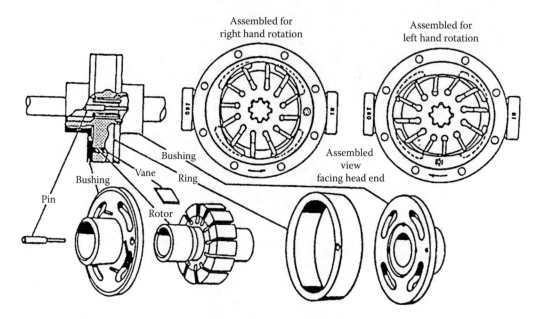

FIGURE 10.29   Illustration of the Eaton-Vickers V-104 vane pump.

port and cover plate, and body and cover bearings were measured. The ring and bearings were inspected for unusual wear patterns and for evidence of corrosion, rusting, and pitting.

Lapotko et al. have reported an alternative vane-pump test designated as the "MP-1 test" [10,59]. A schematic of the MP-1 vane pump is shown in Figure 10.28. Although the MP-1 may be run at pressures up to 10 MPa (1450 psi), the reported test pressure is 7 MPa, with a total fluid volume of 0.7 L. In addition to lower volume, the MP-1 test is conducted for only 50 h (and in some cases, for only 10 h). The wear rate is based on the weight loss of the vanes only after the test is completed. In view of the relatively small size, this test comes as close to a "bench hydraulic pump test" as has been reported to date.

The Eaton-Vickers V-104 vane pump shown in Figure 10.29 and the test circuit in Figure 10.30 continue to be the most commonly utilized hydraulic fluid pump test [60].

There are at least three national standards based on the use of this pump: ASTM D-2882 (obsolete), DIN 51389, and BS 5096 (IP 281/77). A comparison of the test conditions is provided in Table 10.4. The total weight loss of the vanes plus ring at the conclusion of the test is the quantitative value of wear. It is not yet known whether the new Conestoga Pump used in the ASTM D7043 test will be an acceptable alternative for the Eaton-Vickers V-104C pump used in the ASTM D2882, DIN51389, and BS 5096.

Although the Eaton-Vickers V-104C vane pump has, in any of its various forms (ASTM D-2882 (obsolete), DIN 51389, and IP 281/77), served the industry reasonably well in its ability to characterize antiwear properties of hydraulic fluids, it is notorious for providing poor interlaboratory reproducibility. Another common problem is rotor breakage during the test. The ASTM D7043 uses Conestoga hardware.

Recently, Gent conducted a study to identify the greatest sources in testing variation found using the Eaton-Vickers V-104C pump [61].

1. Parts preparation: This was identified as the single most important factor in obtaining a successful test run.
2. Rotor failure: Failure occurred regardless of inspection or testing procedure.
3. ASTM D2882 advises using 10 in.-lb. increments up to 100–140 in.-lb. Feeler gauges and shim stock should be used to ensure that the head is not cocked, resulting in seating against the pump cartridge.

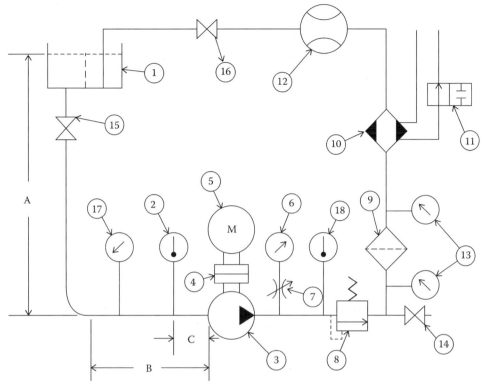

FIGURE 10.30 Test circuit for the Eaton-Vickers V-104 vane-pump test.

1. Reservoir
2. Pump inlet temperature sensor
3. Pump
4. Flexible coupling
5. Motor
6. Pressure gage (high)
7. Snubber valver
8. Relief valve (pressure control)
9. Filter
10. Heat exchanger
11. Temperature control valve
12. Flow meter
13. Pressure gages (low)
14. Fluid sampling port
15. –Pump–inlet valve*

16. Return line valve*
17. Inlet vacuum gage*
18. Outlet temperature sensor*

*= not required

Required dimensions
A. 61–66 cm (24–26 in), vertical
B. 15.2 cm (6 in) minimum, horizontal
C. 10.2 cm (4 in)

4. Bushing failure can be essentially eliminated by proper parts preparation, maintenance of proper clearances in the cartridge, and use of proper torquing techniques.
5. Wear results: Variations in material chemistry and geometry may affect wear results.
6. Rotor reuse: It is common practice to reuse the rotor until breakage.
7. Pump maintenance: According to ASTM D2882, seals must be replaced after each test and the shaft and bearing must be replaced after five tests.
8. Fluid volume, reservoir shape, and baffling vary significantly among laboratories.
9. Filtration: Although ASTM D2882 requires 20-μm filtration, most laboratories use 10-μm filtration.
10. Test conditions: Many laboratories have reduced the pump pressure.
11. Flushing: No laboratories use the flushing required by ASTM D-2882.

TABLE 10.4
Comparison of Vane-Pump Testing Procedures

Test Parameter	ASTM D 2882 (obs.) ASTM D7043	DIN 51389	BS 5096 IP 281/77
Pressure	14 MPa	10 MPa	14 MPa[a]
	2000 psi	1500 psi	2088 psi
			11 Mpa[b]
			1540 psi
Rev./min.	1200	1500	1500
Time (h)	100	250	250
Fluid volume (L)	26.5	56.8	55–70
Fluid temperature	150°F < = ISO 46	c	c
	175°F all other lubes		

 [a] For mineral-oil-type fluids.
 [b] For HFA, HFB, and HFC fluids.
 [c] The temperature is selected to give 46 cSt viscosity.

Totten et al. has provided additional recommendations to obtain improved intralaboratory reproducibility [62]. In addition to those variables identified by Gent, the following were also recommended:

1. The ring and vanes should be inspected for machining irregularities and precision. Machining and scoring marks are unacceptable.
2. The system pressure gauge should be recalibrated before and after every run.
3. Fluid composition such as water and reserve alkalinity for aqueous fluids must be monitored and controlled.
4. To minimize the potential for system contamination using water–glycol, the hydraulic system must be dismantled, scrubbed, and water washed followed by solvent cleaning (use clean solvent) after every run.
5. Water-compatible filters, preferably 3–5-μm fiberglass should be used for water-containing fluids.
6. System temperature variation should be controlled to ±1°C.
7. The complete system, including end plates and bearings, must be inspected and observations recorded, preferably photographically, in addition to the weight loss measurements.

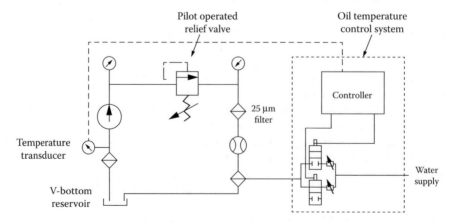

FIGURE 10.31 Low-volume hydraulic and temperature-control system schematic.

TABLE 10.5

Comparison of Test Conditions for Low-Volume Eaton-Vickers V-10 Vane-Pump Test with ASTM D2882 (Eaton-Vickers V-104C Vane Pump)

Operating Parameter	ASTM D 2882 (obs.)	Low-volume System
Pump outlet pressure		
(MPa)	13.8 ± 0.28	13.79 or 17.18 ± 0.03
(psi)	2000 ± 40	2000 or 2500 ± 4
Fluid temp. (°C)	$65.6 \pm 3 < = $ ISO 46	65.5 ± 1.5
	79.4 ± 3 all other fluids	
Pump speed (rpm)	1200 ± 60	1200 ± 40

A 100-h low-volume hydraulic pump test that utilizes only 1.3 L (0.35 gal.) of fluid has been reported by Glancey et al. [63]. This test utilizes a Vickers V10-1P3P1A20 vane pump with a 9.84-cm³/rev. displacement. The hydraulic circuit is shown in Figure 10.31 [63]. A comparison of the test conditions is shown in Table 10.5.

A regression equation was developed to correlate wear obtained when tested according to ASTM D2882 and when tested using the low-volume V-10 vane-pump test [63]:

$$Y = -4.45 + 1.76X$$

where:

Y is the equivalent wear for the ASTM D2882 test (mg)

X is the wear (mg) from the low-volume test system.

Recently, there has been interest in the replacement of the Eaton-Vickers V-104 pump with the newer Eaton-Vickers 20VQ5 vane pump. However, it is important to acknowledge that the tribological conditions are significantly different for both pumps. A comparison of the recommended test conditions are summarized in Table 10.6. In addition, it is also important to recognize that the vane design is fundamentally different for the two pumps. The 20VQ5 pump has an "intravane" configuration, whereas the V-104 pump has a single-vane arrangement. These vane arrangements and the configuration of the vane cartridge in the two pumps are compared in Figures 10.32 through 10.34.

In addition, to the physical differences in these pumps, it should also be noted that under the recommended test conditions, the V-104 pump is being operated at 2000 psi for the ASTM D2882 test protocol, which is outside of the recommended pressure limit of 1000 psi for the pump. The recommended pressure for the test using the 20VQ5 pump is 3000 psi, which is also the recommended pressure limit of the pump [64]. This is particularly important because every pump has its own characteristic wear regions as a function of speed and load [65]. It is almost a certainty that, under these conditions, the V-104 pump would have a significantly greater hydraulic film-load requirement to provide adequate lubrication than the 20VQ5 pump would. This is especially true in

TABLE 10.6

Comparison of the Test Conditions of the Eaton-Vickers V-104 and 20VQ5 Pumps

Parameter	V-104	20VQ5
Vane load (lbs./100 psi)	47	25
Max. pressure (psi)	1000	3000
Max. speed (rpm)	1500	2700
End plates (bronze)	Cast	Sintered

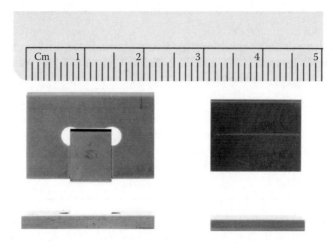

FIGURE 10.32 Comparison of the vanes in the Eaton-Vickers V-104 and 20VQ5 pumps.

V-104   20VQ5

FIGURE 10.33 The internal view of the Eaton-Vickers V-104 and 1 20VQ5 pumps.

FIGURE 10.34 Comparison of the assembled cartridges of the Eaton-Vickers V-104 (left) and 20VQ5 (right) pumps.

TABLE 10.7
Comparison of Pump Operating Parameters

Parameter	Eaton-Vickers V-104 Vane Pump		Eaton-Vickers 20VQ5 Vane Pump		
Flow (gpm)	8	10	5	10	11.3
Pressure (psi)	2000	1500	3000	3000	3000
Speed (rpm)	1200	1500	1200	2400	2700[a]
Power (hp)	9.3	8.8	8.8	17.5	19.7
Torque (in.-lbs.)	490	370	460	460	460

[a] Maximum rated speed at 3000 psi using a 5-gpm cartridge.

view of the different vane designs of the two pumps. Therefore, it is not likely that there is a direct wear correlation between the two pumps.

Experimental water–glycol hydraulic fluids were selected that would provide significant differences in vane pump total wear (vanes plus ring) after completion of the test. Table 10.7 shows the operating parameters used for the V-104 and 20VQ5 pumps in this study.

TABLE 10.8
Results of Pump Tests

Fluid	Flow (gpm)	Pressure (psi)	Speed (rpm)	Ring/vane wt. loss (total wear)[a]	End-plate wt. loss (total wear)[a]	V-104 Pump	20VQ5
A	8.2	2000	1200	21		X	
	10.3	1500	1500	31	23	X	
	5.7	3000	1200	13			X
	12.8	3000	2700	34			X
B	8.2	2000	1200	205	7	X	
	10.3	1500	1500	199	38	X	
	5.7	3000	1200	12	60		X
	11.4	3000	2400	10	482		X
C	8.2	2000	1200	22	20	X	
	10.3	1500	1500	116	14	X	
	5.7	3000	1200	18	40		X
	8.7	3000	1200	37	25		X
	11.4	3000	2400	7.3	130		X
	11.4	3000	2400	16	168		X
D	8.2	2000	1200	37		X	
	5.7	3000	1200	57			X
E	8.2	2000	1200	5110		X	
	5.7	3000	1200	16			X
F	8.2	2000	1200	83		X	
	5.7	3000	1200	93			X
G	8.2	2000	1200	40		X	
	5.7	3000	1200	9.5			X
H	8.2	2000	1200	34		X	
	5.7	3000	1200	20			X
I	8.2	2000	1200	25		X	
	5.7	3000	1200	22			X

[a] These weights were determined after 100 h under the conditions shown.

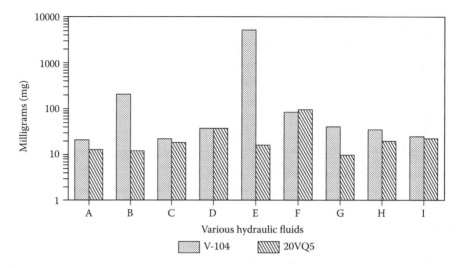

FIGURE 10.35 Chart of ring plus vane wear (in milligrams) from V-104 and 20VQ5 pumps running at 1200 rev/min with nine fluids at pressures of 2000 and 3000 psi for the V-104 and 20VQ5 pumps, respectively.

A comparison of the wear results obtained for these fluids using a variety of test conditions is provided in Table 10.8. The results show the following:

- The worst formulations tested in the V-104 pump, fluids B and E, gave very low wear in the 20VQ5 pump.
- Generally for a given fluid, the 20VQ5 pump gives lower wear than the V-104 pump.

Figure 10.35 provides a graphical comparison of the wear data obtained with the V-104 pump using a "modified" ASTM D-2882 protocol (2000 psi, 1200 rpm, 1-gal. reservoir) with that obtained using the 20VQ5 pump at 3000 psi and 1200 rpm [66]. These data illustrate that the 20VQ5 pump, at least under the recommended test conditions, provides insufficient discrimination of the lubrication properties of fluids.

In another test, a fluid formulation was tested at two different speeds (1200 and 2400 rpm) using the 20VQ5 pump. Although the fluid exhibited very low total wear of the vanes plus ring, fluid B exhibited dramatic cavitation of the bushing at 2400 rpm, as shown in Figure 10.36. This is

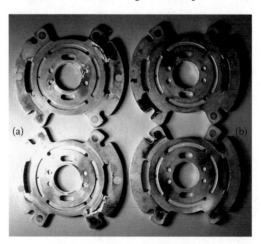

FIGURE 10.36 Comparison of 20VQ5 bushing wear (by cavitation) obtained at two speeds: (a) 1200 rev/min, and (b) 2400 rev/min at 3000 psi, using a fluid that exhibited very low wear on the ring and vanes (12 and 10 mg total wear, respectively after 100 h).

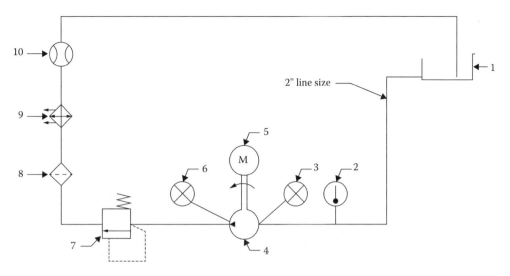

Description of components:

1 Reservoir (50 gal. of oil; elevated above pump centerline to provide gravity feed)
2 Temperature gage or thermocouple
3 Inlet pressure gage
4 Pump: 35VQ25A-11*20 (Cartridge kit P/N 4998040–002)
5 Electric motor (125 hp)
6 Outlet pressure gage
7 Pressure relief valve
8 Filter
9 Cooler
10 Flowmeter

FIGURE 10.37   Schematic of the Eaton-Vickers 35VQ test stand.

also important because neither the ASTM D-2882 nor the proposed 20VQ5-pump-testing protocols requires the reporting of any wear except on the vanes and ring.

A test has been developed using a Chinese-built YB-6 dual-function vane pump operating under ASTM D-2882 testing conditions [67]. In addition to measuring the total wear of the cam ring and vanes, the temperature rise, noise level, and volumetric efficiency were measured.

One of the deficiencies of all of the previously described vane-pump tests is that they are either conducted at relatively low pressure or rotational speed (rpm) when compared to vane pumps in industrial use. The ASTM D6973 Eaton-Vickers 35VQ pump test is often acknowledged to provide a more rigorous and, in some cases, more realistic accelerated industrial-pump wear test (20.7 MPa, 3000 psi, 2400 rpm) [67–73,97]. Test cartridges can be purchased through Eaton-Vickers. A schematic of this test is provided in Figure 10.37 [86]. The maximum weight loss of the combined cam ring and vanes for each of the test cartridges must not exceed 90 mg. If the weight loss of any of the cartridges exceeds these values, two more tests with new cartridges must be run, in which four out of five cartridges will need to pass the weight loss and visual requirements. The cam ring should show no evidence of chatter or excessive transition region wear; the vanes should show no signs of excessive vane tip wear; and the side plates should only show a trace sign of cavitation. Excessive catitation on the side plates could indicate poor air release properties of the lubricant, or an air leak on the inlet side of the hydraulic circuit. A comparison of the test conditions for the V-104C, 20VQ5, and 35VQ25 pump tests is provided in Table 10.9 [68].

Parker Denison (Verizon, France) has developed a combination high-pressure vane and piston-pump test. The pump is a model T6H20C, which houses both a vane and piston pump on one common shaft. This test is widely used in the hydraulics industry but is not a national standard test. The vane pump will be discussed in this section. The piston pump portion of the test will be covered in the next section.

TABLE 10.9
Comparison of Pump Test Conditions

Pump	Rotational Speed (rpm)	Surface (m/s)
V-104	1200	4.2
20VQ5	1200	3.9
35VQ25	2400	11.4

The test circuit for the Parker Denison T6H20C is shown in Figure 10.38. This drawing shows both the vane and piston pump in the circuit. This pump is a special pre-measured pump which can be purchased through Parker Denison in Verizon, France. The test runs for a total of 600 h in two separate phases. Test conditions are as follows:

Outlet Vane Pressure:	250 bar (3600 psi)
Outlet Piston Pressure:	280 bar (4000 psi)
Inlet Temperature:	110°C phase 1, 80°C phase 2
Pump Speed:	1700 rpm
Flow @ 250 bar:	84 lpm minimum vane
	59 lpm minimum piston
	8 lpm maximum piston drain
Fluid Volume:	210 L maximum phase 1,
	160 minimum phase 2
Test Hours:	7 hour break-in
	300 hour phase 1
	300 hour phase 2 with 1% water
ISO Cleanliness:	NAS 8 maximum

The vane pump is evaluated both visually and by weight loss. There must be no evidence of seizing, burnishing, or rippling on the cam ring surface. The side or port plates must show no evident marks of seizing or abnormal wear. The maximum allowable weight loss for the vanes and push-out pins is 15 mg.

10.4.4 PISTON-PUMP TESTING

There are numerous wear surfaces in a piston pump. In addition to sliding wear, such as pistons in cylinders, there is mixed rolling and sliding, which would occur with rolling element bearings, rolling contact fatigue, which also may occur with rolling element bearing, and corrosion and cavitation wear, which might occur on the swash plate and other positions within the hydraulic pump. The relative amount of wear that would occur would also be critically dependent on the material pairs used for construction of the wear contacts. In view of the wide range of materials used for the construction and design of piston pumps, this is one reason why most of the "standard" pump tests developed to date have been vane-pump tests.

The objective of piston-pump design is to minimize energy consumption while optimizing hydrodynamic lubrication in order to minimize wear and internal leakage. The performance parameters are fluid flow, speed, torque, pressure, viscosity, and inlet pressure. To minimize friction and internal leakage, wear contact loading (pressure), speed, and viscosity must be optimized, as shown in Figure 10.1 [3]. In a piston pump, the piston clearances may vary with eccentricity due to load and fluid viscosity. This may produce a change in the lubrication mechanism (e.g., hydrodynamic to boundary), resulting in increased wear and friction.

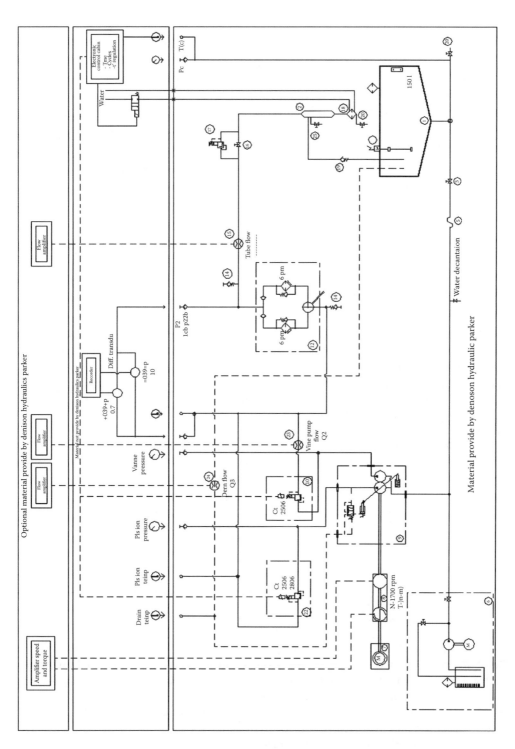

FIGURE 10.38 Parker Denison T6H20C test circuit.

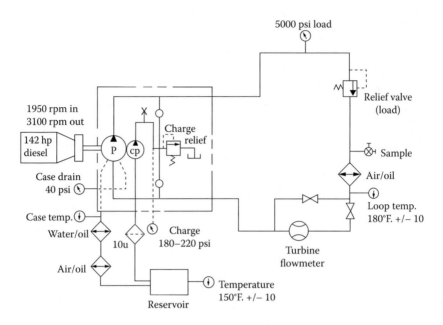

FIGURE 10.39   Schematic of the Sundstrand water stability test circuit.

An ASTM standard which is a valuable reference is the ASTM D6813 – 02a "Standard Guide for Performance Evaluation of Hydraulic Fluids for Piston Pumps". This can be obtained through ASTM, located in Conshohocken, PA.

One piston-pump test commonly used in the hydraulic fluid industry is the Sauer Danfoss Water Stability Test Procedure [75]. The test circuit is shown in Figure 10.39. The test was originally developed using a series 22 pump but currently utilizes a series 90 pump.

The test conditions of the current test are as follows:

Input speed:	2475–2525 rpm
Load pressure:	5000 psi (hours 1 and 2)
	3000 psi (hours 3–25)
	500 psi (hours 25–26)
	5500 psi (hours 26–225)
Charge pressure:	180–220 psi
Case pressure:	40 psi max.
Stroke:	Full
Reservoir temperature:	$150 \pm 10°F$
Loop temperature:	$180 \pm 10°F$
Maximum inlet vacuum:	5 in. Hg

The objective of this test is to determine the effect of water contamination on mineral-oil hydraulic fluid performance. The test is run for the first 25 hours in a dry phase (no water added). At hour 25, 1% water is added to the oil and run on to 225 hours.

In addition to disassembly and inspection for wear, corrosion, and cavitation, the test criterion is "flow degradation". A flow degradation of 10% is considered a failure.

A cyclic loading variation of this test was recently reported to evaluate the performance of a water–glycol hydraulic fluid under these relatively high-loading conditions [76,77]. In this test, a Sauer Danfoss swash-plate, axial piston pump was driven by a Sauer Danfoss Series 20 swash-plate

TABLE 10.10
The 500-h Cycled-Pressure Sauer Danfoss Pump Test

Duration[a] (s)	Vickers (MPa)	Sundstrand (MPa)
130	1.17	21.37
325	0.72	17.24
60	2.07	31.03
85	0.72	17.24

[a] The total time per cycle is 600 s.

piston motor. The load was supplied by a Vickers vane pump. The test sequence is summarized in Table 10.10. The rotating components are visually inspected for wear.

Parker Denison T6H20C piston pump. The Parker Denison T6H20C utilizes a piston pump as described in the vane-pump testing section above (Section 2.3.3). This pump is operated in the same housing as the vane pump as also described above. The piston pump is also evaluated on visual inspection and weight loss criteria. The piston shoes, wear plate, port plate and barrel should show no signs of abnormal wear. The maximum allowable weight loss for all nine pistons is 300 mg.

In addition to the pump components, the fluid must show satisfactory filterability results to be awarded a passing result. (See Section 2.2 on filterability testing.)

Another piston-pump test has recently been proposed by The Bosch Rexroth Corporation [78]. The test circuit is provided in Figure 10.40 and is based on a Breuninghaus A4VSO swash-plate, axial piston pump. The objective of this test is to better discriminate and classify hydraulic and wear performance of a hydraulic fluid. It is proposed that this would be done by prescribing performance levels to be achieved. Establishment of these performance levels has not been developed as of this time.

A less well-known cycled-pressure, piston-pump test is the Vickers AA 65560-ISC-4 piston-pump test [79]. Previous work has shown that if there is no cavitation resulting from inlet starvation and if the hydraulic fluid is clean, then there are four principle modes of failures: (1) spalling of the yoke and spindle-bearing group, (2) fatigue failure of the control piston assembly, (3) rotating group assembly failure due to worn front and tail drive bearings, and (4) static and dynamic O-ring failures. In a sense, piston-pump tests are excellent bearing and cavitation tests.

Janko used high-speed piston-pump tests conducted for 1250 h to supplement successful preliminary vane-pump testing [80]. This test was conducted at constant 140 bars pressure for 1000 h and then completed by cycling the pressure between 70 and 140 bars at 0.1 Hz. The pump was disassembled every 250 h and visually inspected. The stressed components, including the pistons, piston slippers, cylinder barrel, and reversing plate, were inspected and measured for wear. In addition, it was learned that the hydraulic fluids being studied caused such severe bearing wear that they had to be replaced every 250–500 h. The same fluids produced severe wear on the drive shaft with spline ring and thrust pins. These results show the value of piston-pump testing to significantly increase the wear stress of hydraulic fluids.

Edghill and Rubbery studied the correlation of laboratory testing with the type and frequency of field failures [79]. Their analysis showed that the most critical areas for failure are as follows:

For strength:

1. Piston necks
2. Shafts
3. Body kidney ports

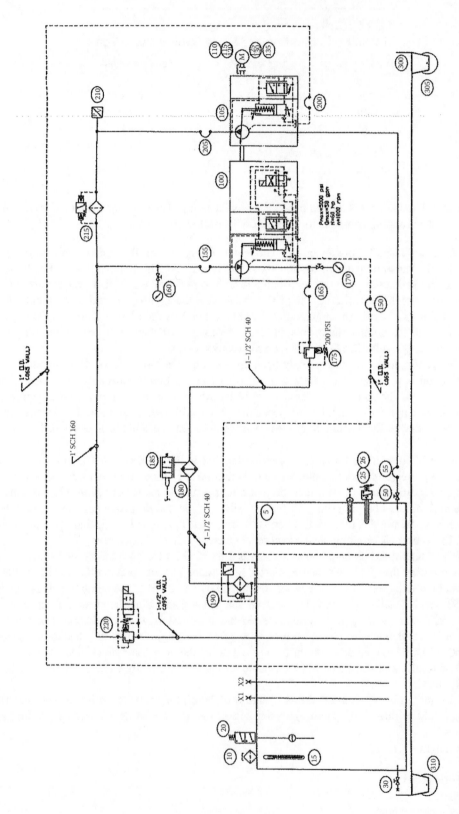

FIGURE 10.40 Schematic of the proposed Rexroth piston-pump test.

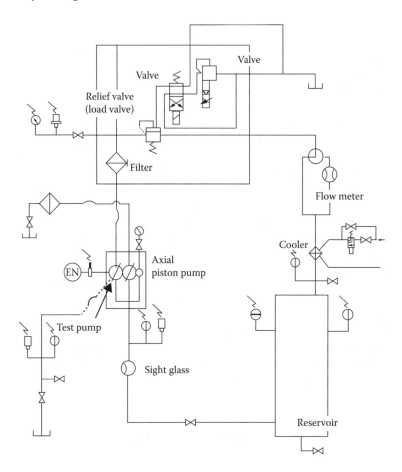

FIGURE 10.41 Komatsu HPV35 + 35 piston-pump test stand circuit.

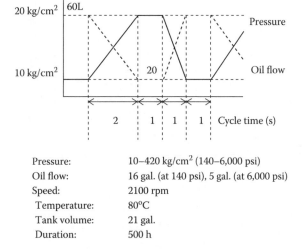

Pressure:	10–420 kg/cm^2 (140–6,000 psi)
Oil flow:	16 gal. (at 140 psi), 5 gal. (at 6,000 psi)
Speed:	2100 rpm
Temperature:	80°C
Tank volume:	21 gal.
Duration:	500 h

FIGURE 10.42 Summary of Komatsu test conditions.

For bearing surfaces:

1. Slipper-cam plate interface
2. Piston skirt-cylinder bore
3. Cam-plate trunion liners

Recently, Ohkawa et al. developed a piston-pump test to evaluate biodegradable vegetable-oil-derived hydraulic fluids [80]. This test is now a standardized test for various fluids. This test stand (see Figure 10.41) utilized a Komatsu HPV35 + 35 twin-piston pump under the following cycled-pressure test conditions (Figure 10.42).

The test criteria included pump efficiency change, wear and surface roughness, formation of lacquer and varnish, and hydraulic-oil deterioration.

Some piston pumps are used in a very broad-range temperature environment (−46°C to 204°C). There are two tests that have been reported for these applications. Hopkins and Benzing utilized the test circuit shown in Figure 10.43, which uses a Manton Gaulin Model 500 HP-KL6-3PA, three-piston pump [81]. The modifications used for this pump did not facilitate the analysis of pump wear surfaces. Instead, the tests, which were conducted at 3000 psi, 550°F for 100 h were designed to evaluate fluid degradation, corrosion, lacquering, and sludging tendencies.

Gschwender et al. used a test circuit utilizing a Vickers model PV3-075-15 pump to evaluate the wear of high-temperature (122°C/250°F) poly (alpha-olefin)-based hydraulic fluids [82]. The test circuit is illustrated in Figure 10.44. The tests were conducted at 20.4 MPa (2960 psi) and 5400 psi with the throttle valve closed, and 5000 psi with the throttle valve open.

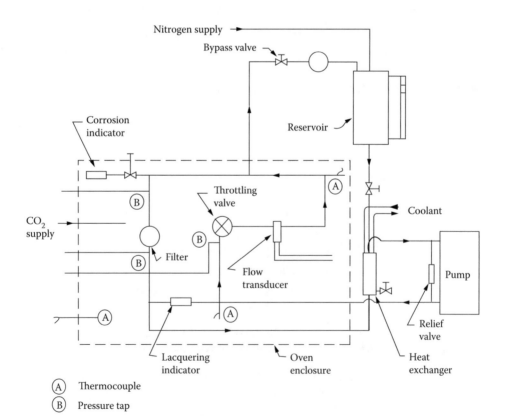

FIGURE 10.43 Hopkins–Benzing high-temperature modified Manton Gaulin piston-pump test circuit.

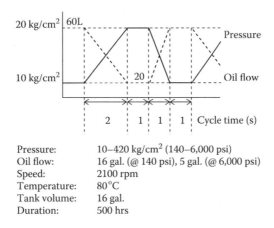

FIGURE 10.44   Gschwender piston-pump test stand.

Pressure: 10–420 kg/cm² (140–6,000 psi)
Oil flow: 16 gal. (@ 140 psi), 5 gal. (@ 6,000 psi)
Speed: 2100 rpm
Temperature: 80°C
Tank volume: 16 gal.
Duration: 500 hrs

10.4.5   GEAR-PUMP TESTING

In the above discussion regarding vane pumps, it was observed that the primary mode of wear, although certainly not the only one, was sliding wear of the vanes on the ring. For piston-pump testing, failure analysis was more complex because there were numerous surfaces to be inspected. Sliding with hydrodynamic wear was still the primary component. However, although sliding wear is still important in gear pumps, the wear mechanisms are even more variable. For example, in open gears, hydrodynamic, EHD, mixed-EHD, and boundary lubrication mechanisms may all occur simultaneously, depending on the position and speed of the gear, as seen in Figure 10.45 [83].

Frith and Scott performed a detailed theoretical analysis of gear-pump wear. They noted that the primary areas of wear are the side plate near the suction port, the gear meshing zone, the gear tips, and casing, especially near the suction port [84]. In addition, the imbalance of pressure across the pump may cause the gear shaft to deflect toward the inlet, creating a reduction in gear-tip clearance at the inlet. Also, a hydrodynamic wedge between the gear ends and the side plate, in combination with the pressure behind the plate, tends to force the side plate against the gears in the inlet region, as shown in Figure 10.46. All of these conditions

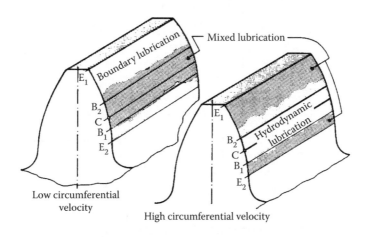

FIGURE 10.45   Lubrication mechanisms on the gear tooth as a function of speed.

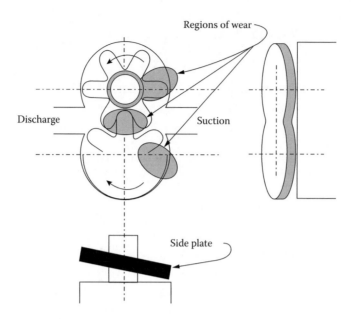

FIGURE 10.46   Side-plate action in a gear pump.

may affect the efficiency of gear-pump operation. Interestingly, there have been relatively few reports of hydraulic fluid lubrication in a gear pump and there are no industry-standard gear-pump lubrication tests.

One study that has been reported was conducted by Knight, who used the multiple-gear-pump test stand shown in Figure 10.47 [85]. Seven Hamworthy Hydraulics Ltd. Type PA 2113 gear pumps

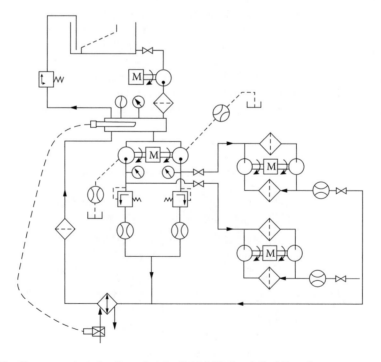

FIGURE 10.47   Gear-pump test circuit used at the British National Coal Board.

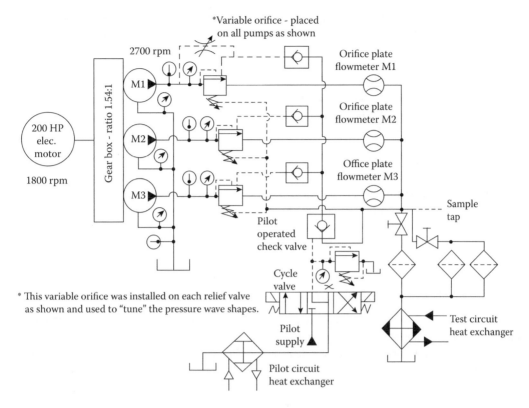

FIGURE 10.48 Wanke gear-pump test stand.

were run at 14.3 MPa (2075 psi). The test is run until pump failure, usually due to needle-bearing fatigue. The condition of the roller bearings was monitored at least once every 24 h.

In another study, a cycled load over 500,000 cycles from zero to the maximum rated pressure, speed, and temperature for the pump and the fluid was recommended [86]. In addition to the cycled loading test, an endurance test (maximum pressure, speed, and temperature for the fluid for 250 h, "proof" test where the pump is operated under extreme conditions at relatively short periods of time (e.g., 5 h) and an initial run-in test. Toogood stated that although degradation in performance would occur under extreme conditions, it was not clear if the damage was greatest under constant-pressure or cycled-pressure conditions [87].

Wanke conducted a study of the effects of fluid cleanliness with a multiple-gear-pump test stand (shown in Figure 10.48) [88]. Two test sequences were used. One was a cycled-pressure test (1 cycle/10 s). The other sequence was an endurance test. As a result of this work, it was recommended that although flow was an adequate measure of the pump's integrity, monitoring the torque throughout the test would provide greater insight into overall pump integrity during the test.

10.4.6 EVALUATION OF PUMP-LUBRICATION RESULTS

The most commonly used analytical methods to monitor hydraulic pump operation have been reviewed in detail [89]. These methods include flow analysis, volumetric and overall efficiency, cavitation pressure, minimum inlet pressure, and others. Johnson has provided a detailed review of pump-testing methods to monitor pump durability and hydraulic operation [90].

Hunt has viewed pump-lubrication analysis from the standpoint of failure correlation. For example, hydraulic pumps are periodically inspected throughout the test for the following:

- Cavitation of port plate
- Bearing wear and breakup
- Piston slipper pad wear and blockage
- Blockage of control devices
- Piston and cylinder wear
- Case seal defects
- Gear teeth wear or fracture

In addition to inspection, temperatures, pressures, and flow throughout the test should be monitored, as they are indicative of friction generation and fluid contamination. Also, wear analysis and vibration analysis should be performed as continuous monitors of wear [91–93]. Common vibrational and acoustical-monitoring procedures and interpretive methods have been reviewed previously [94,95].

10.4.7 A "Universal" Hydraulic-pump Test Stand as a Tribological Test

Feldmann et al. [96] have attempted to develop a "universal" hydraulic-pump test stand that will provide a means of characterizing the expected wear at various wear contacts within the hydraulic system. This test, although still in development, will be based on tribological analysis. Wear contacts for a vane pump and an axial piston pump are illustrated in Figure 10.49. Wear contact geometries and characteristic wear mechanisms have also been classified as shown in Tables 10.11 and 10.12, respectively.

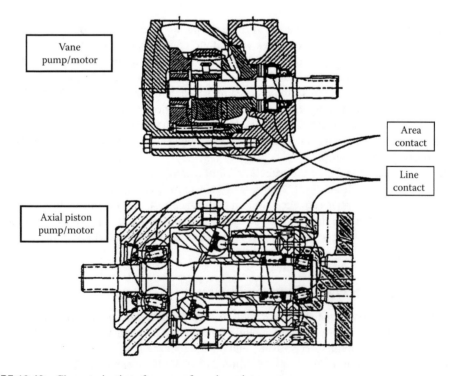

FIGURE 10.49 Characterization of wear surfaces in a piston pump.

TABLE 10.11
Tribological Stresses

	Type of Motion			Contact Geometry			Operational Sequence of Motions		Other Tribologically Relevant Parameters				
	Rolling	Sliding	Impacting	Point Contact	Line Contact	Area Contact	Continuous	Oscillating	Pressure Difference	Dynamic Normal Force	Asymmetric Load	Superimposed Motion	Variability
Vane pumps/motors													
Vane–ring	(E)		(E)		(E)		(E)	(E)	Yes	Yes	No	No	Yes
Vane–slot		(E)			(E)			(E)	Yes	Yes	No	No	No
Vane–sides		(E)				(E)	(E)		Yes	No	No	Yes	No
Gear pumps/motors													
Gear wheel–sides		(E)				(E)	(E)		Yes	No	No	No	No
Gear wheel work	(E)				(E)			(E)	No	Yes	No	No	Yes
Piston pumps/motors													
Piston pin bushing (A/R)		(E)				(E)		(E)	Yes	Yes	Yes	Yes	No
Sliding block–plate (A)		(E)				(E)	(E)		Yes	Yes	No	Yes	Yes
Sliding block–plate (R)		(E)				(E)	(E)		Yes	Yes	No	No	Yes
Sliding block–piston head (A)		(E)				(E)			No	Yes	No		Yes
Sliding block–piston head (R)		(E)				(E)			No	Yes	No		Yes

(continued)

TABLE 10.11 (Continued)

	Type of Motion			Contact Geometry			Operational Sequence of Motions		Other Tribologically Relevant Parameters				
	Rolling	Sliding	Impacting	Point Contact	Line Contact	Area Contact	Continuous	Oscillating	Pressure Difference	Dynamic Normal Force	Asymmetric Load	Superimposed Motion	Variability
Valve plate–piston barrel (A)		(E)				(E)	(E)		Yes	No	Yes	No	Yes
Support for captive C-ring (A)	(E)				(E)			(E)	No	No	No	No	Yes
Axial distributor (A)		(E)				(E)	(E)		No	No	Yes	No	No
Rotary slide valve		(E)				(E)	(E)		No	No	No	No	Yes
Retainer plate–guide ball (A)		(E)			(E)			(E)	No	No	No	No	Yes
Sliding block–retainer plate (A)		(E)				(E)	(E)		No	No	No	No	Yes
Cylinders													
Piston–cylinder tube		(E)				(E)		(E)	Yes	No	No	No	No
Rod seal		(E)				(E)		(E)	Yes	No	No	No	Yes
Others													
Roller bearings	(E)				(E)		(E)		No	No	No	No	No
Shaft–radial shaft sealing ring		(E)			(E)		(E)		Yes	No	No	No	Yes
Serrated shaft				(E)		(E)	(E)		No	No	No	No	Yes
Plain bearing		(E)				(E)	(E)		Yes	No	No	No	Yes

Note: (A): axial piston machine; (R): radial piston machine; (E): extremely important with respect to failure behavior.

TABLE 10.12
Tribological Stresses in Axial Piston Machines (Test Runs in the Working Range with Synthetic Esters)

Pairings	Sliding Wear	Rolling Wear	Grain slip Wear	Erosive Wear	Cavitation	Wear by Impact	Vibration Wear	Additional Observation	Wear Behavior Relative to Mineral Oil
Piston pin bushing	(W)		(W)	(W)	(W)				
Sliding block–plate	(C)		(C)	(W)	(W)			Erosion or cavitation not significant	Comparable
Sliding block–piston head	(W)		(W)	(C)	(C)				
Valve plate–piston barrel	(C)							In some cases, discoloration and transfer of material	Comparable or better
Retainer plate–holding-down clamp for the plate	(W)		(W)						
Sliding block–retainer plate	(C)							Dug in up to 25 μm deep	Worse
Shaft–radial shaft sealing ring	(C)			(W)	(W)			Track worn in (up to 24 μm) leakage	Worse
Plain bearing	(C)		(C)	(W)				In some cases, PTFE layer worn away	Comparable
Roller bearing	(W)	(C)						Tracks, pitting, pressure marks	Comparable or worse
Serrated/polygonal shaft						(W)	(C)	Pitting development, removal of material	Worse

Note: (W): pairings where wear is expected; (C): pairing with conspicuous wear.

10.5 Conclusions

A number of standard and custom bench tests have been discussed in this chapter. It is generally known that the bench tests discussed do not directly correlate to realistic pump performance. Therefore, if bench tests are to be used, they should be custom designed to model the pump-wear contact of interest, and the correlation between the bench test and hydraulic pump wear must be established.

An overview of selected vane-, piston-, and gear-pump testing circuits and procedures from the viewpoint of tribological testing have been provided here. The standard pump tests mainly used in the industry are the ASTM D6973, D7043, the Parker Denison T6H20C, Sauer Danfoss and the Komatsu HPV35 + 35. Most of the tests are pass/fail tests based either on inspection for cavitation and wear, fluid flow leakage, or weight loss of critical rotating components. Some of the non-standard tests discussed do not employ continuous monitoring of hydraulic efficiencies and pressures, torque, vibrational analysis, and so forth. Therefore, although it may be desirable to either use or modify some of the tests reported here in the development of national standards, they should all be updated to reflect current engineering practice and possibilities in order to better model the tribological performance of the pump during use with a particular hydraulic fluid.

REFERENCES

1. ASTM D2882-83, "Standard Method for Indicating the Wear Characteristics of Petroleum and Non-Petroleum Hydraulic Fluids in a Constant Volume Vane Pump," *American Society for Testing and Materials*, Conshohocken, PA.
2. Steiger, J. "AAMA 524 Part 2: Anti-Wear Hydraulic Oils," *American Automobile Manufacturers Association*, 1995.
3. Silva, S. "Wear Generation in Hydraulic Pumps," *SAE Trans.*, 1990, 99, pp. 635–52.
4. Blanchard, R. and Hulls, L.R. "Automated Test Facility for Aircraft Hydraulic Pumps and Motors," *RCA Eng.*, 1975, 20(6), pp. 36–39.
5. Hibi, A., Ichikawa, T. and Yamamura, M. "Experimental Investigation on Torque Performance of Gear Motor in Low Speed Range," *Bull. JSME*, 1976, 19, pp. 179–86.
6. Ueno, H. and Tanaka, K. "Wear in a Vane Pump," Junkatsu (*J. Jpn. Soc. Lubr. Eng.*), 1988, 33, pp. 425–30.
7. Shrey, W.M. "Evaluation of Fluids by Hydraulic Pump Tests," *Lubr. Eng.*, 1959, 15, pp. 64–67.
8. Renard, R. and Dalibert, A. "On the Evaluation of Mechanical Properties of Hydraulic Oils," *J. Inst. Petrol.*, 1969, 55, pp. 110–16.
9. Knight, G.C. "The Assessment of the Suitability of Hydrostatic Pumps and Motors for Use With Fire-Resistant Fluids," in *Rolling Contact Fatigue: Perform. Test. Lubr., Pap. Int. Symp*, Tourret, R. and Wright, E.P., eds., 1977, pp. 193–215.
10. Lapotko, O.P., Shkolnikov, V.M., Bogdanov, Sh. K., Zagorodni, N.G. and Arsenov, V.V., "Evaluation of Antiwear Properties of Hydraulic Fluids in Pump," *Chem. Tech. Fuels and Oils*, 1981, 17, pp. 231–34.
11. Tessmann, R.K. and Hong, I.T. "An Effective Bench Test for Hydraulic Fluid Selection," *SAE Technical Paper Series*, Paper 932438, 1993.
12. Perez, J.M., Hanson, R.C. and Klaus, E.E. "Comparative Evaluation of Several Hydraulic Fluids in Operational Equipment, A Full-Scale Pump Stand Test and the Four-Ball Wear Tester: Part II—Phosphate Esters, Glycols and Mineral Oils," *Lubr. Eng.*, 1990, 46, pp. 249–55.
13. Mizuhara, K. and Tsuya, Y. "Investigation for a Method for Evaluating Fire-Resistant Hydraulic Fluids by Means of an Oil-Testing-Machine," in *Proc. of the JSLE Int. Tribol. Conf.*, 1985, pp. 853–58.
14. Wedeven, L.D., Totten, G.E. and Bishop, R.J. Jr. "Performance Map Characterization of Hydraulic Fluids," SAE Technical Paper Series, Paper 941752, 1994.
15. Platt, A. and Kelley, E.S. "Life Testing of Hydraulic Pumps and Motors on Fire Resistant Fluids," in *Proc. of the 1st Fluid Power Symposium, The British Hydromechanics Research Association*, 1969, Paper SP 982.
16. Totten, G.E., Bishop, R.J. Jr. and Kling, G.H. "Evaluation of Hydraulic Fluid Performance: Correlation of Water–Glycol Fluid Performance by ASTM D2882 Vane Pump and Various Bench Tests," *SAE Technical Paper Series*, Paper 952156, 1995.

17. ASTM D4172-88, "Standard Method for Wear Preventative Characteristics of Lubricating Fluids (Four-Ball Method)," *American Society for Testing and Materials*, Conshohocken, PA.

18. Korycki, J. and Wislicki, B. "Criteria the Lubricating Properties of Synthetic Fluids," in *Proc. Conf. Synth. Lubr.*, Zakar, A. ed., 1989, pp. 213–19.

19. Perez, J.M., Hanson, R.C. and Klaus, E.E. "Comparative Evaluation of Several Hydraulic Fluids in Operational Equipment, A Full-Scale Pump Stand Test and the Four-Ball Wear Tester: Part II—Phosphate Esters, Glycols and Mineral Oils," *Lubr. Eng.*, 1990, 46, pp. 249–55.

20. Klaus, E.E. and Perez, J.M. "Comparative Evaluation of Several Hydraulic Fluids in Operational Equipment, a Full-Scale Pump Test Stand and the Four-Ball Wear Tester," *SAE Technical Paper Series*, Paper 831680, 1983.

21. Perez, J.M. "A Review of Four-Ball Methods for the Evaluation of Lubricants," in *Tribology of Hydraulic Pump Testing ASTM STP 1310*, Totten, G.E., Kling, G.H. and Smolenski, D.J. eds., American Society of Testing and Materials, Conshohocken, PA, 1996, pp. 361–71.

22. "Pump Test Procedure for Evaluation of Antiwear Hydraulic Fluids for Mobile Systems," Form M-2952-S, Vickers Incorporated, Troy, Michigan.

23. Perez, J.M., Klaus, E.E. and Hansen, R.C. "Comparative Evaluation of Several Hydraulic Fluids in Operational Equipment, a Full-Scale Pump Stand Test and the Four-Ball Wear Tester: Part III—New and Used Hydraulic Fluids," *Lubr. Eng.*, 1996, 52, pp. 416–22.

24. ASTM D2670-81, "Standard Method for Measuring Properties of Fluid Lubricants (Falex Pin and V-Block Method)," *American Society for Testing and Materials*, Conshohocken, PA.

25. Inoue, R. "Antiwear Characteristics of Fire Resistant Fluids—Results of the Gamma Falex Tests," *FRH J.*, 1982, 3, pp. 45–49.

26. Chu, J. and Tessman, R.K. "Antiwear Properties of Fire-Resistant Fluids," *FRH J.*, 1980, 1, pp. 15–20.

27. Eleftherakis, J.G. and Webb, R.P. "Correlation of Lubrication Characteristics to Pump Wear Using a Bench Top Surface Contact Test Method," in *Tribology of Hydraulic Pump Testing ASTM STP 1310*, Totten, G.E., Kling, G.H. and Smolenski, D.J. eds., *American Society for Testing and Materials*, Conshohocken, PA, 1996, pp. 338–48.

28. Inoue, R. "Surface Contact Wear—Part 2: Repeatability of the Gamma Falex Test," *BFPR J.*, 1983, 16, pp. 445–453.

29. Tessman, R.K. and Hong, T. *SAE Technical Paper Series*, Paper 932438, 1993.

30. ASTM D4992-95, "Standard Method for Evaluating Wear Characteristics of Tractor Hydraulic Fluids," *American Society for Testing and Materials*, Conshohocken, PA.

31. ASTM D2782-77, "Standard Method for Measurement of Extreme-Pressure Properties of Lubricating Fluids (Timken Method)," *American Society for Testing and Materials*, Conshohocken, PA.

32. Jung, S.-H., Bak, U.-S., Oh, S.-H., Chae, H.-C. and Jung, J.-Y. "An Experimental Study on the Friction Characteristics of Oil Hydraulic Vane Pump," in *Proc. of the Int. Tribology Conf.*, 1995, pp. 1621–25.

33. ASTM 5707-97, "Standard Method for Measuring Friction and Wear Properties of Lubricating Grease Using a High-Frequency Linear-Oscillator (SRV) Test Machine," *American Society for Testing and Materials*, Conshohocken, PA.

34. Reichel, J. "Mechanical Testing of Hydraulic Fluids," in *Industrial and Automotive Lubrication*, 11th Int. Colloquium at Technische Akademie Esslingen, 1998, Vol. III, pp. 1825–36.

35. DIN 51,354-2, "Prüfung von Schmierstoffen FZG–Zahnrad–Verspannungs-Prüfmaschine," *Deutches Institute für Normunge*. V., Berlin, 1992.

36. ASTM D2596-87, "Standard Test Method for Measurement of Extreme-Pressure Properties of Lubricating Grease (Four-Ball Method)," *American Society for Testing and Materials*, Conshohocken, PA.

37. ASTM D2783-03, "Standard Method for Measurement of Extreme-Pressure Properties of Lubricating Fluids (Four-Ball Method)," *American Society for Testing and Materials*, Conshohocken, PA.

38. Urata, E.Y., Iwaizumi, Y. and Iwamoto, K. "Assessment of the Antiwear Properties of High-Water-Content Fluids Using a Vickers V-104C Vane Pump," *Yuatsu to Kukiatsu*, 1986, 17(7), pp. 543–53.

39. Jacobs, G., Backe, W., Busch, C. and Kett, R. "A Survey on Actual Research Work in the Field of Fluid Power," in *1995 STLE National Meeting*, 1995.

40. Priest, M., March, C.N. and Cox, P.V. "A New Test Method for Determining the Antiwear Properties of Hydraulic Fluids," in *Tribology of Hydraulic Pump Testing, ASTM STP 1310*, Totten, G.E., Kling, G.H. and Smolenski, D.M. eds., American Society for Testing and Materials, Conshohocken, PA, 1995.

41. Lacey, P.I., Naegeli, D.W. and Wright, B.R. "Tribological Properties of Fire-Resistant, Non-Flammable and Petroleum-Based Hydraulic Fluids," in *Tribology of Hydraulic Pump Testing, ASTM STP 1310*, Totten, G.E., Kling, G.H. and Smolenski, D.M. eds., American Society for Testing and Materials, Philadelphia, PA 1995.

42. Kenny, P. and Yardley, E.D. "The Use of Unisteel Rolling Fatigue Machines to Compare the Lubricating Properties of Fire-Resistant Fluids," *Wear*, 1972, 20, pp. 105–21.

43. Young, K.J. "Hydraulic Fluid Wear Test Design and Development," in *Tribology of Hydraulic Pump Testing ASTM STP 1310*, Totten, G.E., Kling, G.H. and Smolenski, D.M. eds., American Society for Testing and Materials, Conshohocken, PA, 1995, pp. 156–64.

44. Wedeven, L.D., Totten, G.E. and Bishop, R.J. Jr. *SAE Technical Paper Series*, Paper 941752, 1994.

45. Wedeven, L.D., Totten, G.E. and Bishop, R.J. Jr. "Performance Map and Film Thickness Characterization of Hydraulic Fluids," *SAE Technical Paper Series*, Paper 952091, 1995.

46. Maamouri, M., Masson, J.F. and Marchand, N.J. "A Novel System to Study Wear, Friction and Lubricants," *J. Mater. Eng. Perf.*, 1994, 3, pp. 527–39.

47. Hogmark, S. and Jacobson, S. "Hints and Guidelines for Tribotesting and Evaluation," *Lubr. Eng.*, 1992, 48, pp. 569–79.

48. Robinson, G.H., Thomson, R.F. and Webbere, F.J. "The Use of Bench Wear Tests in Materials Development," *SAE Trans.*, 1959, 67, pp. 569–79.

49. Voitik, R.M. "Realizing Bench Test Solutions to Field Tribology Problems Utilizing Tribological Aspect Numbers," in *Tribology—Wear Test Selection for Design and Application*, Ruff, A.W. and Bayer, R.G. eds., American Society for Testing and Materials; Conshohocken, PA, 1993, ASTM STP 1199, pp. 45–59.

50. ASTM D2714-68, "Standard Method for Calibration and Operation of the Alpha Model LFW-1 Friction and Wear Testing Machine," *American Society for Testing and Materials*, Conshohocken, PA.

51. Ludema, K.C. "Cultural Impediments for Practical Wear Modeling of Wear Rates," in *Tribological Modeling for Mechanical Designers*, Ludema, K.C. and Mayer, R.G. eds., American Society for Testing and Materials; Conshohocken, PA, 1991, ASTM STP 1105, pp. 180–85.

52. Turski, A.B. "Studies of Engineering Properties of Fire Resistant Hydraulic Fluids for Underground Use," *Mining Miner. Eng.*, 1969, February, pp. 50–59.

53. Knight, G.C. "Experience with the Testing and Application of Fire-Resistant Fluids in the National Coal Board," *SAE Technical Paper Series*, Paper 810962, 1981.

54. Horiuchi, T. "Hydraulic Fluids and Trends in Oil Pumps and Motors," *Nisseki Rev.*, 1979, 21(3), pp. 151–58.

55. Feicht, F. "Factors Influencing Service Life and Failure of Hydraulic Components," *Oilhydraulik Pneumatic*, 1975, 20(12), pp. 804–6.

56. Ueno, H., Tanaka, K. and Okajima, A. "Wear in the Vanes and Cam Ring of a Vane Pump," *Nippon Kikai Gakkai Ronbunsho Bhen*, 1985, 52(480), pp. 2990–97.

57. Hemeon, J.R. "How to Evaluate Performance of a Hydraulic Fluid," *Appl. Hydraulics*, August, 1995, pp. 43–44.

58. Paul Schacht verbal communication regarding this procedure as the standard cycled pressure vane pump test recommended by (Robert Bosch) Racine Fluid Power, Racine, WI.

59. Arsenov, V.V., Sedova, L.O., Lapotko, O.P., Zaretskaya, L.V., Kel'bas, V.I. and Ryaboshapka, V.M. "Method for Investigating the Antiwear Properties of the Water -Glycol Liquids," *Vestnik Mashinostroeniya*, 1988, 68, pp. 32–33.

60. Thoenes, H.W., Bauer, K. and Herman, P. "Testing the Antiwear Characteristics of Hydraulic Fluids: Experience with Test Rigs Using a Vickers Pump," in *Performance Testing of Hydraulic Fluids*, Tourret, R. and Wright, E.P. eds., Heyden and Son Ltd; London, 1978.

61. Gent, G.M. "Review of ASTM D 2882 and Current Possibilities," in *Tribology of Hydraulic Pump Testing*, Totten, G.E., Kling, G.H. and Smolenski, D.J. eds., American Society for Testing and Materials, Conshohocken, PA, 1996, pp. 96–105.

62. Totten, G.E., Bishop, R.J. Jr. and Webster, G.M. "Water–Glycol Hydraulic Fluid Evaluation by ASTM D2882: Significant Contributors to Erroneous and Non-Reproducible Results," *SAE Technical Paper Series*, Paper 961740, 1996.

63. Glancey, J.L., Benson, E.R. and Knowlton, S. "A Low Volume Fluid Power Test for the Evaluation of Genetically Modified Vegetable Oils as Industrial Fluids," in *SAE Off-Highway Conference*, 1998.

64. Vickers Inc., "Vane Pump & Motor Design Guide for Mobile Equipment," Bulletin No. 353, revised 11-1-92, pp. 10.

65. Runhua, T. and Caiyun, Y. "The Vane Profile Improvement for a Variable Displacement Vane Pump," in *Proceedings of the International Fluid Power Applications Conference*, 1992, National Fluid Power Association, Milwaukee, WI.

66. Bishop, R.J. Jr. and Totten, G.E. "Comparison of Water–Glycol Hydraulic Fluids Using Vickers V-104 and 20VQ Vane Pumps," *Tribology of Hydraulic Pump Testing ASTM STP 1310*, Totten, G.E., Kling, G.H. and Smolenski, D.M. eds., American Society for Testing and Materials Philadelphia, PA, 1995.

67. Gengcheng, L. and Huanmou, M. "Test Stand Testing of Water–Glycol-Based, Non-Flammable Hydraulic Fluids," *Runhua Yu Mifeng*, 1985, 3, pp. 23–28.

68. Johnson, H.T. and Lewis, T.I. "Vickers' 35VQ25 Pump Test," in *Tribology of Hydraulic Pump Testing*, Totten, G.E., Kling, G.H. and Smolenski, D.J. eds., American Society for Testing and Materials Conshohocken, PA, 1996, pp. 129–39.

69. Broszeit, E.H., Steindorf, H. and Kunz, A. "Testing of Hydraulic Fluids with Cell Vane Pumps," *Tribol. Schmierungstechnik*, 1990, 37(4), pp. 202–09.

70. Kunz, A.J. and Broszeit, E. "Comparison of Vane Pump Tests Using Different Vane Pumps," in *Tribology of Hydraulic Pump Testing*, Totten, G.E., Kling, G.H. and Smolenski, D.J. eds., American Society for Testing and Materials, Conshohocken, PA, 1996, pp. 140–55.

71. Kunz, A., Gellrich, R., Beckmann, G. and Broszeit, E. "Theoretical and Practical Aspects of the Wear of Vane Pumps, Part B: Analysis of Wear Behavior in the Vickers Vane Pump Test," *Wear*, 1995, 181–83, pp. 868–75.

72. Maxwell, J.F., Schwartz, S.E. and Viel, D.J. "Flow Characteristics of Hydraulic Fluids of Different Viscosities II. Flow in Pumps: Internal Leakage and Loss of Efficiency," ASLE Prepr. No. 80-AM-713-2, 1980.

73. "Sundstrand Water Stability Test," Sundstrand Bulletin 9658. (The test Protocol described was conducted by Southwest Research Institute in San Antonio, TX.)

74. Totten, G.E. and Webster, G.M. "High Performance Water–Glycol Hydraulic Fluids," in *Proc. of the 46th Natl. Conf. on Fluid Power*, 1994, pp. 185–94.

75. Lefebvre, S. "Evaluation of High Performance Water–Glycol Hydraulic Fluid in High Pressure Test Stand and Field Trial," in *STLE Annual Conference*, 1993.

76. Melief, H.M. "Proposed Hydraulic Pump Testing for Hydraulic Fluid Qualification," in *Tribology of Hydraulic Pump Testing ASTM STP 1310*, Totten, G.E., Kling, G.H. and Smolenski, D.J. eds., American Society for Testing and Materials, Conshohocken, PA, 1995, pp. 200–07.

77. "Conduct Test-to-Failure on Hydraulic Pumps," Vickers AA-65560-ISC-4), NTIS No. AD 60224, 1963.

78. Janko, K. "A Practical Investigation of Wear in Piston Pumps Operated with HFA Fluids with Different Additive," *J. Synth. Lubr.*, 1987, 4, pp. 99–114.

79. Edghill, C.M. and Rubbery, A.M. "Hydraulic Pumps and Motors—Development Testing: Its Relationship With Field Failures," in First European Fluid Power Conference, 1973, Paper No. 31.

80. Ohkawa, S., Konishi, A., Hatano, H., Ishihama, K., Tanaka, K. and Iwamura, M. "Oxidation and Corrosion Characteristics of Vegetable-Base Biodegradable Hydraulic Oils," *SAE Technical Paper Series*, Paper 951038, 1995.

81. Hopkins, V. and Benzing, R.J. "Dynamic Evaluation of High Temperature Hydraulic Fluids," *Ind. Eng. Chem. Proad. Res. Dev.*, 1963, 2, pp. 77–78.

82. Gschwender, L.J. Snyder, C.E. Jr. and Sharma, S.K. "Pump Evaluation of Hydrogenated Polyalphaolefin Candidates for a –54°C to 135°C Fire-Resistant Air Force Aircraft Hydraulic Fluid," *Lubr. Eng.*, 1987, 44, pp. 324–29.

83. Paton, C.G., Maciejewski, W.B. and Melley, R.E. "Test Methods for Open Gear Lubricants," *Lubr. Eng.*, 1990, 46, pp. 318–26.

84. Frith, R.H. and Scott, W. "Wear in External Gear Pumps: A Simplified Model," *Wear*, 1994, 172, pp. 121–26.

85. Knight, G.C. "Experience with the Testing and Application of Fire-Resistant Fluids in the National Coal Board," *Trans. SAE*, 1981, 90, pp. 2958–69.

86. "Pump Test Procedure for Evaluation of Antiwear Fluids for Mobil Systems," Vicker's From No. M-2952-S.

87. Toogood, G.J. "The Testing of Hydraulic Pumps and Motors," in *Proc. Natl. Conf. Fluid Power*, 37th, National Fluid Power Association, Milwaukee, WI, 1981, Vol. 35, pp. 245–52.

88. Wanke, T. "A Comparative Study of Accelerated Life Tests Methods on Hydraulic Fluid Power Gear Pumps," in *Proc. Natl. Conf. Fluid Power*, 37th, National Fluid Power Association, Milwaukee, WI, 1985, Vol. 35, pp. 231–43.

89. American National Standard, "Hydraulic Fluid Power—Positive Displacement Pumps—Method of Testing and Presenting Basic Performance Data," ANSI/B93.27-1973.

90. Johnson, K.L. "Testing Methods for Hydraulic Pumps and Motors," in *Proc. Natl. Conf. Fluid Power*, 30th, 1974, Vol. 28, pp. 331–70.

91. Hunt, T.M. "Diagnostics in Fluid Power Systems—A Review," *Tech. Diagnost.*, November, 1981, pp. 89–99.

92. Bashta, T.M. and Babynin, I.M. "Determining the Operating Characteristics of Pumps," *Sov. Eng. Res.*, 1983, 3(5), pp. 3–5.

93. Avrunin, G.A. and Bakakin, G.N. "On Choosing the Conditions for Diagnosis of the Technical State of Hydraulic Motors," *Sov. Eng. Res.*, 1989, 9(10), pp. 37–39.

94. Maroney, G.E. and Fitch, E. in *3rd International Fluid Power Symposium*, 1973, pp. C5-81–C5-96.

95. Dowdican, M., Silva, G. and Lowery, R.L. Oklahoma State Univ.—Fluid Power Research Center, Report No. OSU-FPRC-A5/84, 1984. (Reports currently available from FES, Inc., Stillwater, OK.)

96. Feldmann, D.G. Hinrichs, J., Kessler, M. and Nottrodt, J. "Ermittlung der Anwendungseigen-schaften von Biologisch Schnell Abbaubaren Hydrailikflüssigkeiten durch Labortests," in *Industrial and Automotive Lubrication*, 11th Int. Colloquium at Technische Akademie Esslingen, 1998, Vol. I, pp. 271–80.

97. Gellrich, R., Kunz, A., Beckmann, G. and Broszeit, E. "Theoretical and Practical Aspects of the Wear of Vane Pumps, Part A: Adaptation of a Model for Predictive Wear Calculation," *Wear*, 1995, 181–83, pp. 862–67.

98. ASTM D4172-94, "Standard Test Method for Wear Preventive Characteristics of Lubricating Fluid (Four-Ball Method)," *American Society for Testing and Materials*, Conshohocken, PA.

99. AFNOR NF E 48-690 "Hydraulic Fluid Measurement of Filterability Without Water," Association Francaise De Normalisation. Plaine Saint-Denis Cedex France.

100. AFNOR NF E 48-691 "Hydraulic Fluid Measurement of Filterability in Presence of Water," Association Francaise De Normalisation. La Plaine Saint-Denis Cedex France.

101. A-TP-02100, "Procedure for Determining Filterability of Hydraulic Fluids," 2007, Parker Denison, Verizon, France.

102. ISO 13357-1, "Petroleum products—Determination of the Filterability of Lubricating Oils, Part 1: Procedure for Oils in the Presence of Water," 2002.

103. ISO 13357-2, "Petroleum products—Determination of the Filterability of Lubricating Oils, Part 2: Procedure for Dry Oils," 2005

104. ASTM D7043 – 04a, "Standard Test Method for Indicating Wear Characteristics of Non-Petroleum and Petroleum Hydraulic Fluids in a Constant Volume Vane" *American Society for Testing and Materials*, Conshohocken, PA.

105. ASTM D6973-08, "Standard Test Method for Indicating Wear Characteristics of Petroleum Hydraulic Fluids in a High Pressure Constant Volume Vane Pump," *American Society for Testing and Materials*, Conshohocken, PA.

106. A – TP – 30533 "Test Equipment and Instructions for Hydraulic Fluids Performance Evaluation on parker Pumps (Vane and Piston)," 2007, Parker Denison, Verizon, France.

11 Noise and Vibration of Fluid Power Systems

*Samir N.Y. Gerges, D. Nigel Johnston, and
Leonardo Zanetti Rocha*

CONTENTS

11.1 INTRODUCTION

A fluid power system can transmit high levels of power with high-power density and can deliver high force and torque with a good dynamic response, while providing more flexibility than a mechanical drive. However, the operation of the fluid power system components generates noise and vibration which may exceed the legislation requirements. Health, safety, and human comfort requirements with respect to noise and vibration levels have become increasingly demanding.

This chapter summarizes the fundamental concepts of acoustics and vibrations, describes measurement equipment and describes the sources of noise and methods of control.

11.2 FUNDAMENTALS OF NOISE AND VIBRATION CONTROL

Noise is a pressure wave caused by the vibration of fluid particles, and propagates in the medium at the speed of sound. Humans can generally perceive noise in the range of 20 Hz to 20 kHz, with acoustic pressure from the threshold of hearing, which is about 2×10^{-5} Pa, to the pain threshold, which is around 20 Pa [1]. The human ear's sensitivity to noise is not linear in frequency, but is often considered to follow the A weighting scale, and measurements are commonly expressed either in an unweighted dB scale or a weighted dBA scale.

The three main sound parameters are: (1) sound pressure level (SPL), (2) sound intensity level (SIL), and (3) sound power level (SWL). SPL and SIL depend on the distance between the sound source and the measurement point, and on the environment; whereas SWL is independent of any parameter except the source characteristics. Therefore SWL is a good parameter for source noise specification.

Noise and vibration control can be carried out at the source (which is the most effective means achieving a solution), at the transmission path, or at the receiver. Different mechanisms can be used for noise and vibration control such as:

- Acoustic energy absorption or vibration damping
- Noise isolation and attenuation
- Vibration transmission isolation by flexible mounting
- Noise and vibration resonance reduction

11.3 INSTRUMENTATION AND SENSORS FOR NOISE AND VIBRATION

Systems for measurement of noise and vibration consist of a sensor, signal conditioning, and signal analyzer.

The most robust and reliable sensor for the measurement of vibration is the piezoelectric accelerometer. Accelerometers have a frequency response function from a fraction of 1 Hz (depending on the signal conditioning) to about 15 kHz, and weigh from a fraction of a gram up to a few hundred grams. Figure 11.1 shows a wide range of accelerometers.

The most common noise sensor is an instrumentation microphone which can vary in diameter (1", 1/2", 1/4" or 1/8") and have a frequency range up to 100 kHz. These microphones are designed to be omni-directional (sensitivity independent of direction) especially if the microphone diameter is much less than the wavelength.

Sound level meters are classified by their precision as Type I (high precision), and Type II and Type III (less precision). Usually Type I is used for good quality measurements.

SPL can be measured by a sound level meter (Figure 11.2—left) whereas SIL can be measured by a sound intensity probe (Figure 11.2—right) which consists of two closely spaced, matched

FIGURE 11.1 Range of accelerometers.

FIGURE 11.2   A typical sound level meter (left) and intensity probe (right).

microphones with a two-channel analyzer. SWL can be measured using a sound level meter in a special acoustic room such as reverberation room or anechoic room, or by using a sound intensity meter without the need for a special acoustic room.

For the measurement of pressure ripple or fluid-borne noise, a dynamic pressure sensor needs to be immersed in the fluid in order to measure the dynamic variation of the pressure inside the fluid. Generally speaking, miniature piezoelectric transducers are most effective because they have a small sensor area and a very high-frequency response (typically 10 kHz or more). Typical piezoelectric pressure transducers are shown in Figure 11.3. Piezoelectric pressure transducers only detect pressure changes; they cannot normally be used for steady-state pressure measurement, and the transducer and charge amplifier normally have a lower-frequency limit of about 1 Hz. This is often an advantage as the steady-state pressure would result in a very large offset on the pressure ripple measurement. Ideally, the pressure transducer should be mounted with its sensor face flush with the inside of the fluid passageway, as closed tubes and chambers can result in unwanted resonances and trapped air.

The signal from an accelerometer, microphone, or pressure transducer passes through signal conditioning, which is often integrated into the sensor. It is then fed to the frequency analyzer to capture and analyze the measured signal. A calibrator is needed to calibrate the measurement system.

FIGURE 11.3   Dynamic pressure transducers. (Courtesy of PCB Piezotronics Inc.)

TABLE 11.1
1/1 and 1/3 Octave Band Central Frequencies and Bandwidth

Central Frequencies	1/3 Octave Band	1/1 Octave Band
25	–	22.4–28
31	22.4–45	28–35.5
40	–	35.5–45
50	–	45–56
63	45–90	56–71
80	–	71–90
100	–	90–112
125	90–180	112–140
160	–	140–180
200	–	180–224
250	180–355	224–280
315	–	280–355
400	–	355–450
500	355–710	450–560
630	–	560–710
800	–	710–900
1000	710 – 1400	900–1120
1250	–	1120–1400
1600	–	1400–1800
2000	1400 – 2800	1800–2240
2500	–	2240–2800
3150	–	2800–3550
4000	2800 – 5600	3550–4500
5000	–	4500–5600
6300	–	5600–7100
8000	5600 – 112000	7100–9000
10000	–	9000–11200
12500	–	11200–14000
16000	11200 – 22400	14000–18000
20000	–	18000–22400

The signal is captured in the time domain and can be analyzed in the time domain or frequency domain by Fourier transform or by narrow-band analysis, which is done by the frequency analyzer. Results can be displayed in linear or log scale (dB or dBA) in the frequency domain. Figure 11.4 show a typical frequency analyzer and a wide range of accelerometers.

Noise analysis can be carried out in 1/1, 1/3, 1/12 or 1/24 octave bands, which have a constant ratio bandwidth. Each band is 100%, 26%, 6% and 3% higher than the previous band for 1/1, 1/3, 1/12 or 1/24 octave bands, respectively. Table 11.1 shows the central frequencies and bandwidth of the 1/1 and 1/3 octave bands.

11.4 HYDRAULIC SYSTEM NOISE AND VIBRATION

Hydraulic system noise and vibration can be characterized into three different types: fluid-borne, structure-borne, and air-borne. The fluid-borne noise relates to the dynamics of the fluid and takes the form of pressure ripple and flow ripple, whereas the structure-borne noise considers the behavior of the structural vibration of the system. Pressure ripple can propagate in the fluid and excite

FIGURE 11.4 Typical frequency analyzers.

the system which radiates noise to the air. Pressure ripple is often considered to be independent of structural vibration but it is often a two-way process and the vibration can influence the pressure ripple through "fluid-structure interaction".

Excitation can be at resonance frequencies of the system, which only need a matching between the excitation frequency of ripples and the natural frequencies of the system to produce high noise levels even with very low excitation force amplitude. The system parameters that control the response to excitation are the geometry, mass distribution, stiffness, damping, and so forth. Therefore, these parameters can control the noise and vibration generation both in frequency and amplitude.

11.5 MECHANICAL AND ACOUSTIC RESONANCES

A fluid, mechanical, and acoustic system has dynamic characteristics. In the case of a mechanical system the dynamic characteristics are defined by the resonance frequencies, mode shapes, and damping, all of which depend on the mass distribution and stiffness distribution. Any excitation with frequency close to any of the resonance frequencies of the system will result in a high vibration response level. The same is true of an enclosed fluid system and an acoustic system; any space filled with fluid can exhibit resonant characteristics if the dimensions of the fluid space are coincident with multiples of half of the acoustic wavelength. The hydraulic fluid can usually be considered as a one-dimensional system with waves travelling along the length of the passageways, whereas an acoustic space is a three-dimensional system.

11.6 FLUID POWER SYSTEM COMPONENTS

Fluid power systems range in size from vehicle steering systems, which are very compact and relatively low power with a pressure typically up to 80 bar, to systems for industrial and off-road machinery with much larger size, power up to hundreds of kW, and pressures up to 300 bar. The primary components in the system are a pump, control valves, fluid reservoir, pipes, and the motor or actuator driving the load. Some of the main noise and vibration excitation components are described below.

11.6.1 PUMPS

Pumps are, in most cases, the main source of noise and vibration, due to the non-steady, cyclic nature of producing high-pressure fluid at the outlet [2–5]. Pumps generally produce periodic, repeatable noise. Hydraulic motors may also produce noise in a similar way.

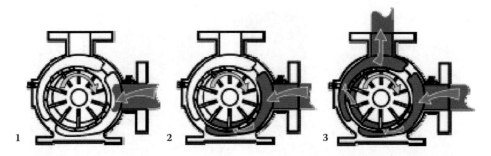

FIGURE 11.5   Typical pumps for vehicle steering system. (1) suction, (2) compression, and (3) discharge.

Different types of pump are used in hydraulic systems, such as piston pump, vane pump, external gear pump, internal gear pump, and lobe pump, among others [6]. Piston pumps are usually the noisiest, while internal gear pumps are relatively quiet. Figure 11.5 shows a typical vane pump, such as might be used in a vehicle steering system.

The noise generation can be fluid-borne noise (which is known as "pressure ripple") or structure-borne noise. Positive displacement pumps tend not to produce an absolutely steady flow rate. Instead, the flow consists of a mean value on which is superimposed a flow ripple. The magnitude of the flow ripple is dependent upon the pump type and operating conditions, but usually has a peak-to-peak amplitude of between 1% and 10% of the mean flow rate. Different classes of pump have different characteristic flow ripple waveforms. This flow ripple interacts with the characteristics of the connected circuit to produce a pressure ripple.

Figure 11.6 shows measured pressure ripple from a vane pump with 10 vanes at 900 RPM. The main component occurs at a frequency $f =$ RPM × No. of blades/(60) = 150 Hz, and harmonics are produced at multiples of this. The noise generated at this low rotational speed is often called "moan" noise, and at a higher rotational speed of about 3000 RPM the noise is referred as "whine" noise.

There are methods available for measuring pump and motor flow ripple and fluid-borne noise characteristics. For example, see references [7–9]. In these methods the flow ripple is determined indirectly from pressure ripple measurements. Some examples of measured flow ripple for an axial piston pump are shown Figure 11.7.

The pump's flow ripple produces a pressure ripple which is then transmitted throughout the circuit in a complex manner at the speed of sound in the fluid. The pump and its driving motor are also mechanical sources of noise and vibration due to force and moment generation. The

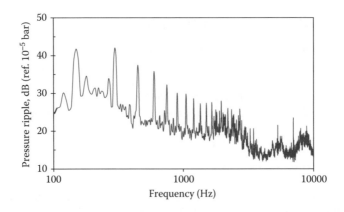

FIGURE 11.6   Typical pressure ripple spectrum at low rotational speed of 900 RPM (MOAN noise).

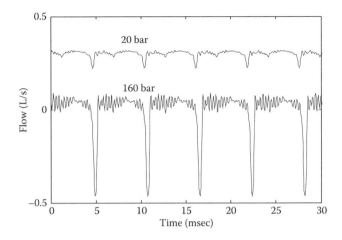

FIGURE 11.7 Flow ripple from an axial piston pump.

excitation energies in hydraulic systems are mainly due to pressure ripple which is generated by the pump.

11.6.2 VALVES

Valves produce broad band, high-frequency noise. A sudden pressure drop can produce cavitation, where vapor bubbles form and collapse violently and randomly, generating high-frequency, broad band fluid-borne noise. These pressure fluctuations propagate inside the system and can cause structural vibration. Valves can also suffer from self-excited oscillation or "chatter" due to dynamic instability under certain conditions. Also, if a valve is closed or opened rapidly, water hammer can occur, which can cause severe transient vibration, cavitation, and a knocking noise.

11.6.3 PIPES AND HOSES

These can amplify or reduce noise and vibration depending on the materials, length relative to wavelength, and mounting. Mechanical resonance of the pipe walls and also acoustic resonances in the fluid can generates high sound pressure levels. The pressure ripple characteristics of metal pipe are usually quite predictable, but the behavior of flexible hoses is more complex and there can be two distinct speeds of sound in the hose wall and the fluid.

11.7 NOISE AND VIBRATION CONTROL

Noise and vibration control at the source is the most effective solution but this is not always possible.

11.7.1 REDUCTION OF VALVE NOISE

Valve noise can be generated by cavitation, air release, turbulence, valve instability or "chatter", or water hammer. Cavitation relates to a sudden pressure drop condition that results in the formation of bubbles which collapse violently, releasing high-frequency noise in the fluid. This causes a structural vibration of all components and produces air-borne noise with a characteristic "hissing" sound. Cavitation noise occurs in a broad band typically from 400 Hz up to 8 kHz.

Cavitation noise can be reduced by increasing back pressure so that cavitation does not occur. This could be done by using two valves in series. Sometimes reducing the back pressure can also help [10] as this can generate an air pocket which damps out the pulsations from the collapsing bubbles. Flexible mounting of the valve and hose can be an effective way of noise and vibration reduction.

Some valves are more prone to instability than others, but it is often the interactions between the valve and system that lead to instability. Valve instability can be difficult to prevent but sometimes changing the pressure and flow, or small changes in the system layout, can help.

Water hammer occurs when a valve is opened or closed rapidly [11]. It can be prevented by using a slower opening or closing movement, or by using an accumulator or anti-surge device or pressure relief valve. However, these measures may slow down the response of the system.

11.7.2 REDUCING RESONANCE EFFECT

A high level of vibration and noise can occur at resonance frequencies. Mechanical and acoustic resonance can occur if the excitation frequency is close to the resonance frequency of the system.

Mechanical resonance frequencies depend on the system, specifically the mass and stiffness distribution and also materials damping. For the hydraulic fluid system and the external acoustics the resonance frequencies depend on the dimensions in relation to the wavelength. It is always recommended to avoid excitation frequencies near the resonance frequencies of the system.

Resonance frequencies for the hydraulic fluid system can be calculated from the wave propagation characteristics of the pipes and hoses, and the impedance characteristics of the pumps, valves and other components, but these characteristics are not always known with sufficient accuracy. Standard test methods are available for the measurement of flow ripple and the impedance characteristics of components. [7,9,12–14], and it is then possible to use specialized computer software to predict the resonant behavior [15,16]. However, these are involved techniques and often the "tuning" of a system to avoid resonance is done simply by trial-and-error.

Mechanical resonance frequencies can be calculated directly for a simple system or can use finite element modeling techniques for complex systems. Changing dimensions, materials, mass distribution, and stiffness distribution can alter the position of the resonance frequencies and therefore reduce high levels of noise and vibration.

11.7.3 SILENCERS

In principle, silencers can be active, dissipative, or reactive. Although active cancellation techniques are now commonly used for vibration and air-borne noise, there are currently no active silencers commercially available for fluid-borne noise in fluid power systems.

Reactive silencers (like car exhaust silencers or mufflers) are often used in hydraulic systems. A reactive silencer reflects noise back to the source by applying a mismatched impedance to the flow, effectively reflecting most of the noise back. This mismatching impedance is usually a geometric discontinuity. This silencer usually consists of a number of chambers with interconnecting pipes to give an impedance mismatch at several points. Orifices are often included to increase the impedance mismatch and also to provide some energy dissipation. Silencer efficiency can be measured by transmission loss (TL) in dB, which is the ratio of the reflected energy in relation to incident energy on a logarithmic scale. It should be noted that TL defines the performance under ideal conditions; the actual performance may be less efficient than this. Some typical reactive silencers are shown in Figure 11.8.

11.7.3.1 Simple Expansion Chamber

Figure 11.9 show a simple expansion chamber (middle and above). As can be seen, there is a sudden area change at the inlet and outlet to give impedance mismatch and reflect the wave. The area ratio

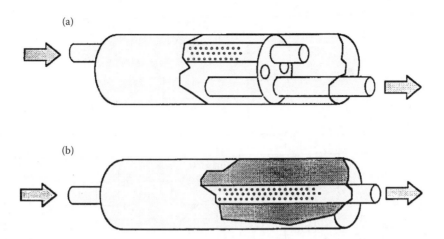

FIGURE 11.8   Silencer types: (a) twin chamber, (b) single chamber with damping orifices.

governs the maximum transmission loss, and the frequency at which this maximum occurs depends on the length. For good low-frequency attenuation a large volume chamber is needed.

The transmission loss of a simple expansion chamber is given by Equation 11.1 and is shown in Figure 11.9.

$$TL = 10\log_{10}\left[\cos^2\left(\frac{\pi.f}{2.f_n}\right) + \frac{1}{4}\left(\frac{S_2}{S_1} + \frac{S_1}{S_2}\right)^2 .\sin^2\left(\frac{\pi.f}{2.f_n}\right)\right] \qquad (11.1)$$

where $f_n = \dfrac{c}{4L}$,

L is the chamber length,
c is the speed of sound in the fluid.

The lowest cut-off frequency, below which the transmission loss is less than 3 dB, is given approximately by $f_0 = \dfrac{S_1 c}{\pi S_2 L}$ Hz.

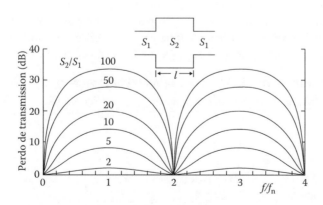

FIGURE 11.9   Simple expansion chamber silencer characteristics.

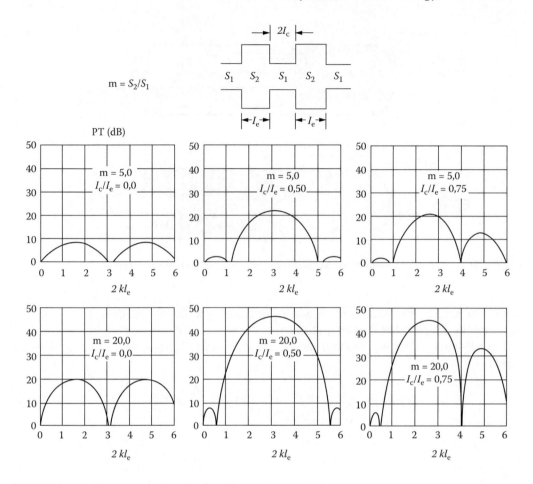

FIGURE 11.10 Double expansion chamber silencer characteristics.

11.7.3.2 Double Expansion Chamber

In this case, the behavior is more complex but a wider frequency band and better attenuation is possible than with a simple expansion chamber. By careful choice of the design parameters, transmission loss can be up to 40 dB. The transmission loss characteristics of some double expansion chamber silencers are shown in Figure 11.10.

Commercial silencers tend to be combinations of expansion chambers, side branch resonators, Helmholtz dampers, and orifices. An example is shown in Figure 11.11. They can be very effective over a broad frequency range, typically providing a broadband attenuation of 30 dB or more. However, they can be bulky and expensive, particularly if they are to be effective at low frequencies. They can also impair the dynamic response of the system due to the increased fluid volume.

11.7.3.3 Side branch Volume Resonator (Helmholtz Resonator)

The transmission loss of a Helmholtz resonator is given by Equation 11.2 and is shown in Figure 11.12.

$$TL = 10 \log \left[1 + \frac{\alpha}{4 \left(\dfrac{f}{f_0} - \dfrac{f_0}{f} \right)^2} \right] \qquad (11.2)$$

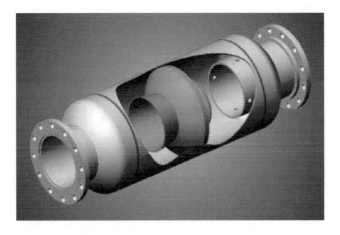

FIGURE 11.11 Example of a double expansion chamber silencer.

where:

$$f_0 = \frac{c}{2\pi}\sqrt{\frac{S_b}{VL}}$$ in Hz, $\alpha = \frac{S_b V}{LS^2}$, S_b is the neck cross-sectional area, L is the neck length, c is the speed of sound, and S is the cross-sectional area of the main flow passageway.

A Helmholtz resonator is a tuned device and only provides good attenuation over a narrow band near to the natural frequency f_0, which limits its range of applications.

11.7.3.4 Side Tube

The transmission loss of a side branch is given by Equation 11.3

$$TL = 10\log\left[1+\left[\frac{S_b}{2S}\tan\left(\frac{\pi f}{2f_0}\right)\right]^2\right]$$ (11.3)

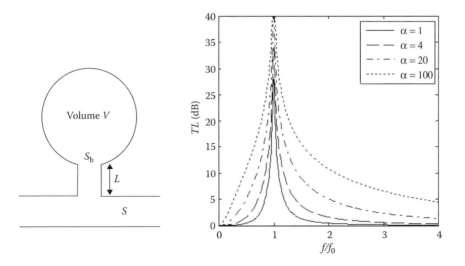

FIGURE 11.12 Helmholtz resonator characteristics.

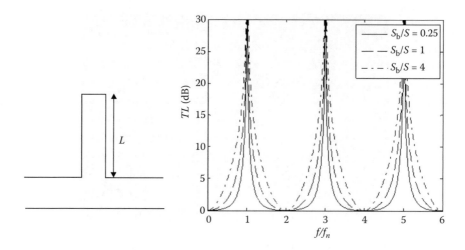

FIGURE 11.13 Side branch characteristics.

where: S_b is the side branch cross-sectional area, S is the cross-sectional area of the main flow passageway, $f_0 = \dfrac{c}{4L}$, L is the side branch length, and c is the speed of sound. This is shown in Figure 11.13.

Side branch resonators are sometimes known as "quarter wavelength" resonators, as they provide good attenuation when the length is equal to a quarter of the wavelength; that is, $f = f_0$.

Attenuation bands also exist at odd integer multiples of this frequency (i.e., $3f_0$, $5f_0$, etc.). These can be used to attenuate several harmonics produced by a pump or motor, but it is not possible to attenuate a complete harmonic series; it can be tuned to the first, third, and other odd harmonics, but not simultaneously to the even harmonics. It can be difficult to tune a side branch resonator accurately and the tuning will be affected by changes in c, due to temperature or pressure variations, or trapped air.

11.7.4 Flexible Mounting Using Vibration Isolators

Components such as the pump, motor, valves, hoses, and so on should be mounted on flexible isolators in order to reduce the vibration that is transmitted to the base. The isolator selection depends on the excitation frequencies and usually the lowest excitation frequency should be at least four times the mechanical mounting resonance frequency. Figure 11.14 shows transmissibility curves for a simple system of one degree of freedom (typical one point mounting) and it can be shown that an excitation frequency of about four to five times the mounting resonance frequency gives a transmissibility of about 5% transmitted vibration energy for low damping ratio. Based on this calculated resonance frequency and mass at this single point, one can determine the stiffness of the isolator.

11.7.5 Flexible Hoses

Pipes and hoses distribute fluid in the circuit and can carry fluid-borne noise from one part to another. Sound energy in the pipes and hoses can also be transmitted to the supports such as panels and walls, thereby extending the noise-radiating area. Metal and rigid pipes can be a very good transmission path from the source to the receiver. Isolating transmitted energy is a good way to reduce excitation and noise radiation. Isolation can be achieved by using flexible hoses which are inserted between the pump and valve, and at both inlet and outlet. Flexible hoses are very effective at isolating high-frequency structural vibration; however, flexible hoses will also transmit low-frequencies and may even amplify them. Flexible hoses are, however, also effective at absorbing high-frequency fluid-borne noise. Figure 11.15 shows typical flexible hoses which are used for a high-pressure system.

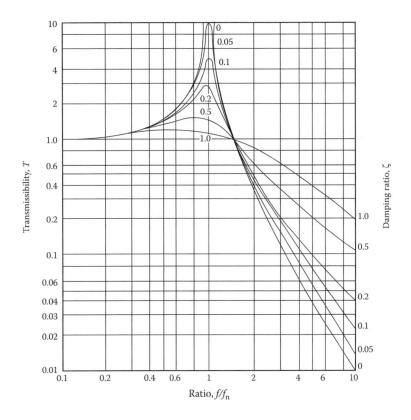

FIGURE 11.14   Design curves for the transmissibility vs. the frequency ratio, f/f_n, as a function of the damping ratio, ζ, for a linear single-degree-of-freedom system of a single point mounting.

In hydraulic power-assisted steering systems, flexible hoses are used as the main noise attenuation device. Highly compliant textile-braided hoses are generally used, as these provide good vibration isolation and noise reduction and the pressures are usually relatively low (<100 bar). To improve the fluid-borne noise attenuation further, tubular "tuner" inserts are often fitted inside the

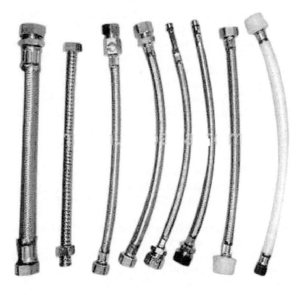

FIGURE 11.15   Flexible hoses.

hose. These usually consist of a flexible metal or plastic tube, which is attached inside one end of the hose, and provide a form of tuned attenuator. The plastic tubes may be perforated to improve the performance. Sometimes restrictors and sleeves are used in conjunction with the tubular insert. Tuner inserts can provide excellent attenuation at low cost but need to be "tuned" carefully to the system [15,16]. To be effective, these devices rely on compliant hoses, and are likely to be less effective in higher-pressure systems because stiffer hoses are needed.

11.8 CONCLUSION

The noise and vibration of fluid power systems are not only unpleasant and a possible cause of hearing damage, but may also be an indication of the well-being of the system; high noise levels or harsh noise may indicate faulty operation or damaged equipment. Therefore it is important to design a system with low noise and vibration, and to consider screening and external insulation as a last step in noise control.

REFERENCES

1. Gerges, S.N.Y. *Ruido: Fundamentos e Controle*, NR editora, 2000.
2. Skaistis, S. *Noise Control of Hydraulic Machinery*, Dekker, 1988, ISBN 0-824779-34-7.
3. BHRA (British Hydro-Mechanical Research Association), *Quieter Fluid Power Handbook*, Cranfield, UK, 1980, ISBN 0-906085-49-7.
4. British Fluid Power Association, *Guidelines to the Design of Quieter Hydraulic Fluid Power Systems*, 1986.
5. French Association of Pumps Manufacturers, *Guide Acoustique des Installations de Pompage*, CETIM, 1997, p. 239
6. Norvelle, F.D. *Fluid Power Technology*, West, 1995, ISBN 0-314-01218-4.
7. ISO 10767-1: "Hydraulic Fluid Power: Determination of Pressure Ripples Levels Generated in Systems and Components. Part 1: Precision Method for Pumps," 1996.
8. ISO 10767-2: "Hydraulic Fluid Power: Determination of Pressure Ripples Levels Generated in Systems and Components. Part 2: Simplified Method for Pumps," 1999.
9. ISO 10767-3: "Hydraulic Fluid Power: Determination of Pressure Ripples Levels Generated in Systems and Components. Part 3: Method for Motors," 1999.
10. Heron, R.A. "The Control of Cavitation in Valves," *7th Int. Fluid Power Symposium*, Bath, England, 1986. p. 275–283.
11. Wylie, E.B. and Streeter, V.L. *Fluid Transients*, Prentice Hall, 1993, ISBN 0-133221-73-3. New Jersey, USA.
12. ISO 15086-1:2001: "Hydraulic Fluid Power: Determination of the Fluid-Borne Noise Characteristics of Components and Systems. Part 1: Introduction,"
13. ISO 15086-2:2000: "Hydraulic Fluid Power: Determination of the Fluid-Borne Noise Characteristics of Components and Systems. Part 2: Measurement of the Speed of Sound in a Fluid in a Pipe,"
14. ISO 15086-3:2008: "Hydraulic Fluid Power: Determination of the Fluid-Borne Noise Characteristics of Components and Systems. Part 3: Measurement of Hydraulic Impedance,"
15. Drew, J.E., Longmore, D.K. and Johnston, D.N. "Theoretical Analysis of Pressure and Flow Ripple in Flexible Hoses Containing Tuners," *Proc IMechE*, Vol. 212, Pt. I, 1998, pp. 405–22.
16. Hastings, M.C. and Chen, C. "Analysis of Tuning Cables for Reduction of Fluid-Borne Noise in Automotive Power Steering Hydraulic Lines," *Trans. SAE*, 1995, Paper 931295, 1995.

BIBLIOGRAPHY

1. Vergara, E.F. Gerges S.N.Y. "Order Analysis of Noise and Pulsation Pressure on Vehicle Power Steering Pump," Paper presented in *SAE NVH Meeting*, 2006, Brazil.
2. British Hydro-Mechanical Research Association (BHRA). *Quieter Fluid Power Handbook*, Cranfield, UK, 1980.
3. Johnston, D.N. "Hydraulic System Noise Prediction and Control," In *Handbook of Noise and Vibration Control*, John Wiley & Sons, 2007, Chapter 76.
4. Johnston, D.N. and Edge, K.A. "Simulation of the Pressure Ripple Characteristics of Hydraulic Circuits," *Proc IMechE*, part C, Vol. 203, 1989, pp. 119–127.
5. Drew, J.E., Longmore, D.K. and Johnston, D.N. "The Systematic Design of Low Noise Power Steering Systems," Presented at the *4th Scandinavian Fluid Power Conference*, Tampere, Finland, 1995.

12 Failure Analysis

*Acires Dias**

CONTENTS

12.1 INTRODUCTION

Failure analysis aims to identify the events which inhibit a technical system from performing its function appropriately. It also indicates the pathways through which the cause of the failure can be eliminated or the effects on the function it carries out can be mitigated. In this context, failure analysis is an important requirement used by all who develop and operate technical systems, including designers, manufacturers, and users. It is a process that is constantly evolving and the more that the technology involved in the system under analysis develops, the more important and complex the analysis becomes.

The process of failure analysis in technical systems, in the form in which it is seen today, is relatively recent. It originated from scientific research focused on the development of technologies which characterized the industrial era starting around the sixteenth and seventeenth centuries. With the advent of the steam engine and with the use of fossil fuels as an energy source, a considerable development in the areas of production and transportation occurred. Coupled to this was the widespread advent of electricity in the nineteenth century. Certainly, the associated problems grew exponentially and it was necessary to learn from the mistakes and disseminate the knowledge acquired so that they did not reoccur.

However, it should be noted that the systemization of failure analysis occurred even more recently with the advent of the aeronautics industry, the nuclear industry, and the space race.

---

* This chapter is a revision of the chapter titled "Failure Analysis" by Steven Lemberger and George E. Totten from the *Handbook of Hydraulic Fluid Technology*, 1st Edition.

Based on these industries, confronted with the technology involved, some of the failure analysis techniques became popular in the machinery industry, in particular, hydraulic systems. In this regard, it is noteworthy that the strong relation between the electronics area and hydraulics and pneumatics formed the basis for automated systems and robotics. The automobile industry was also affected by these processes of analysis and subsequently, due to tertiarization, the affect spread to industry in general. In this way, failure analysis reached the different fields of work and technical processes, from the planning to the design and management processes, through to production and discontinuation.

Thus, failure analysis is a process which is currently integrated into the different industrial activities. This process refers to the systemization which industrial organizations use to carry out analysis of planning, design, and production systems, based on techniques such as Event Tree Analysis (ETA), Reliability Block Diagram (RBD), Bow-Tie Analysis (BTA), Bayesian Belief Networks (BBN), Fault Tree Analysis (FTA), Causal Network Event Analysis (CNEA), Failure Mode and Effect Analysis (FMEA), among others.

The analysis processes using any of the above-mentioned techniques can be developed at different levels of depth. In failure analysis it is common to identify the mode of failure, along with the causes and the effects on the systems, in order to increase the reliability and the safety of the systems. Thus, the analysis may be superficial or may develop at more fundamental levels in order to identify the route cause where the origin of the failures lie, as will be presented in this chapter. It is not always easy to correctly identify the causes. This is because they may be associated with the material, the physical or chemical characteristics of the lubricant, the roughness level of the components, or the organizational management of the work process; for instance, the design, production, or maintenance. It is for this reason that failure analysis is considered to be a process rather than a specific activity.

In hydraulic systems, the failure process is even more complex given that modern machines have become more powerful, faster, and more versatile through the use of hydraulic drives and controls. Heavier loads and faster cycles require that hydraulic systems have pumps with larger capacities and higher pressure ratings. Higher-pressure pumps create greater stresses on the system components. Therefore, hydraulic pump maintenance is becoming increasingly critical as a means of reducing failure rates and extending service life.

The most common cause of hydraulic pump failure is improper pump use. A root-cause analysis of pump failure has shown that 80% of failures were caused by improper operation and maintenance. The most common sources of failures are shown in Table 12.1.

12.2   STEPS IN FAILURE ANALYSIS

Failure is an event defined by the cessation of a function that an item should carry out. Therefore, failures can be characterized at different levels and different degrees of importance according to the intensity with which they are presented to the system, and the risks in terms of the product and the process.

If failure is defined by the cessation of a function, one can ask: what is failure analysis? In general, it can be stated that failure analysis is a process comprising a set of procedures to identify the mode of failure, to eliminate the cause, and to mitigate the effects on the function. The approaches to organizing the procedures reflect analysis methodologies. These methodologies, in turn, are permeated by the different analysis techniques, the choice of which is dependent on the context and the type of equipment or process to be analyzed. The choice of one technique or another also depends on the depth, context, and importance in terms of safety, the environment, the economy, and the time available for the analysis process. For this reason, in order to study the failure of a complex system, such as hydraulic systems, a plan of the actions to be carried out needs to be obtained. This applies both to complete systems and to subsystems such as a hydraulic pump. In either case, the following steps are recommended:

TABLE 12.1
Most Common Sources of Hydraulic Pump Failures

Source	Failure Frequency (%)
Design	2
Manufacture	6
Installation	12
Operation and maintenance	80

1. Know the equipment and where it has been used. It is important to carry out a survey on the use of the equipment in other applications, whether similar or not, and to obtain a history of the failures.
2. Know what a new part looks like so that changes can be readily identified. For this, consult the manufacturer's catalogs, contact the manufacturer, and consult specialists.
3. Carefully observe each component of the hydraulic system before disassembling it. If possible, take photographs of the system. It is recommended to develop a cause and effect analysis through brainstorming sessions. To aid the discussion and systemization of the results the technique of Ishikawa [1] can be used to register and organize the possible causes of each of the effects resulting from the main modes of failure. The main modes of failure can be identified by using the Paretto analysis [2].
4. Carefully disassemble each piece of equipment and make written notes during this process. If possible, photograph these steps. During the disassembling process it is important to re-input the information included in the Ishikawa diagram. This applies to any systemized process which has been adopted. If the reliability is to be calculated, the Reliability Block Diagram (RBD) technique should be used, which allows the fundamental relationships between components of the system under analysis to be obtained.
5. Thoroughly examine the internal parts making written notes of any observations. These observations must be supported by technical analysis such as FMEA, FTA, ETA, or BTA which allowed the structured relationship between the failure mode, causes, effects, criticality, and action to be obtained. The decision to analyze a component or subsystem in greater depth can be made by plotting the results in a Paretto diagram.
6. Identify the probable causes. When step 5 is carried out in a systemized form using failure analysis techniques, the causes of the failures can be organized as follows: proximate causes, intermediate causes, and root causes.
7. Plan the actions to eliminate the root causes or at least to mitigate the effects on the system functions. With the results of step 6, the actions to be carried out in the design, manufacturing, or equipment maintenance can be organized. The prioritization will depend on the importance of the failure in relation to safety, the costs of the failure, and the reliability required by the user.

The steps for the failure analysis described above will guide the actions of the technicians and planners in relation to two dimensions. The first, with a more managerial focus, is dependent on the systemized use of techniques such as those which will be described in Section 12.3, which lead to an integrated knowledge of the problems of the system under analysis. Theses techniques, such as FMEA, FTA, ETA, or BTA, will guide action planning at the levels of design, manufacturing, and use, and will also take into account the costs involved in eliminating the causes of the failures. The second dimension focuses on the study of the physics of the failure (described in Section 12.4) and provides guidance on the failure mechanisms with the aim of eliminating the causes which generate the failure modes. In this case, the objective is to contribute to the systemization of technical knowledge associated with actions related to capacitation and the training of personnel in order to improve the quality of the design, manufacturing, maintenance, and operation of hydraulic systems. The focus of Section 12.4 is on the hydraulic pump; however, the analysis process can be applied to any other hydraulic system.

12.3   QUALITATIVE FAILURE ANALYSIS TECHNIQUES

Before studying the physics of the failure it is important to plan the actions to be studied and to define the results to be obtained. Thus, in this section some techniques are described which are used to characterize the failure modes, to identify the causes and effects, and to organize the necessary actions. A summary of the techniques—Event Tree Analysis (ETA), Reliability Block Diagram (RBD), Fault Tree Analysis (FTA), Bow-Tie Analysis (BTA), and Bayesian Belief Networks (BBN)—is given in Table 12.2. The technique of Failure Mode and Effect Analysis (FMEA), since it is frequently used in analysis processes in all industrial sectors, is described in greater detail in order to highlight some definitions which are important in failure analysis.

Failure Mode and Effect Analysis (FMEA). FMEA is a technique used in the failure analysis of systems, and is widely used in quality, maintainability, and reliability engineering. This is because it is necessary in these areas to arrange the processes which systematize the knowledge of all who participate in the analysis of failure into systems, subsystems, and components. FMEA allows, in a dynamic way, the updating of information on the function of each item in a system in order to plan the actions required to ensure the continuity of the functions.

FMEA is a technique that aims to determine all of the possible failure modes and the effects resulting from them, and to identify the most probable causes for systems, subsystems, and

---

TABLE 12.2
Failure Analysis Techniques

Technique	Description	Age	Recommended literature
Event Tree Analysis (ETA)	The WASH-1400 Report (also known as Rasmussen Report) (USA/NRC, 1975, p. 1–2) presents ETA as "a modified form of the decision trees used in decision analysis" and explains that the application of the event tree is more appropriate than the decision tree in studies on reactor accidents, "since reactor safety systems are largely automated so that the sequence of initiating events plus safety system responses involves few decisions". ETA is an inductive technique (bottom-up approach) that supports an analysis of the possible consequences triggered by an initiating event. ETA presents graphically how a combination of some events can lead to a specific scenario, based on the occurrence (or not) of these events.	Possibly from the 1970s, during WASH-1400 studies.	[3–7]

TABLE 12.2   (Continued)
Failure Analysis Techniques

Technique	Description	Age	Recommended literature
Fault Tree Analysis (FTA)	Fault Tree Analysis (FTA) is a top-down analysis technique used to investigate cause-effect relationships in systems. For each significant effect that adversely affects the system, a function analysis of the causes, failure modes and possible combinations which lead to the final effect is carried out. This has the advantage of interacting with several factors including software and hardware failures and human error. It could become a widely used technique and already has well-defined rules and symbolism.	H.A. Watson of the Bell Telephone Company in 1961–62. System Safety Symposium of Washington University and Boeing in 1965.	[5]
Reliability Block Diagram (RBD)	The Reliability Block Diagram (RBD) is a technique that supports the analysis of which elements of a system are necessary to perform the required function, and which ones can fail without affecting it. A block in an RBD represents the physical component operating, and when it fails, interrupts the connection between the input and output points of this block. Therefore, there must be an alternative path connecting the input and output points of the system (using redundancy, for example) or the failure of this component will make the whole system stop working. The analysis of an RBD will show the cut-sets (combinations of failures that result in the interruption of the system) and path-sets (sets of components necessary to keep the system working).	Unknown.	[8–10]
Bow-Tie Analysis (BTA)	Bow-Tie Diagrams may be an evolution of the cause-consequences diagrams from the 1970s, and barrier diagrams from the 1980s. The central knot of the bow-tie represents the event that is being studied. The left-hand side shows the threats that could lead to the event, and the right-hand side shows the consequences. Barriers are then placed between the threats and the event, to reduce the probability of occurrence (called threat control measures), and between the event and its consequences, to mitigate the consequences (called recovery/mitigation measures).	Possibly in 1979, at the ICI Hazan Course Notes, presented by The University of Queensland, Australia.	[11–15]
Bayesian Belief Networks	Bayesian Belief Networks (BBN), or just Bayesian networks, are Directed Acyclic Graphs (DAG), in which nodes represent variables and arcs signify the existence of direct causal dependencies. The probability of occurrence of the basic nodes and the conditional probability that defines the nature of the causal dependencies between nodes, are presented in tables. BBN can be used for both inductive and deductive approaches (such as diagnosis).	The age of BBN is unknown, but the term "Bayesian networks" was possibly first used by Judea.	[16–19]

components. The technique has to be reapplied throughout the life cycle, in a systematic form, whenever some action is required to improve the quality, analyze the costs and production, or when a risk is imminent. This technique was developed in the 1960s as part of the Apollo project at NASA (National Aeronautics and Space Administration). The FMEA was initially standardized by MIL STD 1629A [20]. It was later applied in the aeronautical, military, and nuclear engineering sectors. In 1988, the ISO 9000 series was launched, encouraging organizations to develop systems for quality management where FMEA is very important, according to Bertsche [21].

SAEJ1739 [22] is one of the most important norms for the systemization of FMEA, establishing the terminology, and providing guidance for the application and systemization of results. According to this norm, FMEA can be described as a set of systematic activities which aim to:

1. Recognize and evaluate the potential failure modes of a product/process and the effects of this failure.
2. Identify actions which could eliminate or reduce the chances of a potential failure occurring.
3. Document the process.

When classifying the modes, causes, and effects according to the resulting risks it is recommended that one adopt the RPN (Risk Priority Numbers) criteria. This number defines a criticality criterion. When this occurs the FMEA is known as FMECA (Failure Mode, Effect and Criticality Analysis). The criticality criterion defined by the RPN takes into consideration the severity of the failure mode, the occurrence (that is, whether frequent or infrequent) and the ease or difficulty of detection. Thus, the RPN is referred to as SOD (Severity, Occurrence, and Detection).

When the concepts involved in the technique are well established by the FMEA team, application of the technique is very simple. Thus, one of the most important activities is to standardize the terms to be used, such as: failure mode, effect, causes, system, and component.

It should be noted that the clarity of the concepts must occur on two levels. The first refers to a knowledge and acceptance of the nomenclature related to the technique. The second deals with the domain of the application of the FMEA, which is directly related to the type of FMEA. Therefore, the nomenclature gains specific significance according to the domain of application. In this regard, some concepts related to the nomenclature and the domain of application are presented below.

Failure mode. The failure mode is defined in terms of the form of the failure and the manner in which the failure is presented; that is, the manner in which the component under study fails to carry out its function or lies outside its specifications.

The failure mode is a property inherent to each item, since each has its own particular characteristics such as function, work environment, materials, manufacturing, and quality. For example, the modes of failure for a shaft could be: rupture, warping, or wear, and so on; whereas those for a filter of a hydraulic power unit could be rupture or clogging.

There are two approaches to revealing the modes of failure: functional and structural. The first is centered on the functioning of the item. Thus, this approach is normally more qualitative. The use of this approach depends on the specificity of the item and the type of FMEA being applied. The functional approach is most recommended for the FMEA of the design, particularly when carried out in the initial phases of the design. For instance, for the analysis of a process in which a total

TABLE 12.3
Mode of Failure with Functional Approach [23]

Component	Function	Mode of Failure
Shaft	Transmit movement	No transmission of movement
	Transmit torque	No transmission of torque

TABLE 12.4
Mode of Failure with the Structural Approach [23]

Component	Function	Mode of Failure
Shaft	Transmit movement	Warping, shaft wear, bearing wear, abrupt rupture, fatigue rupture, overload rupture

productivity maintenance (TPM) system is to be implemented in a manufacturing environment, the failure mode can be simply nonfunctioning, as shown in Table 12.3. This idea indicates a functional approach.

The structural approach frequently requires information that allows some characterization of the failure mode to be established, whether related to the probability for this to occur or to the form in which it occurs. In this case, the mode of failure can be associated with the requirements such as loading, material, mechanical resistance, presence of surface treatment, hardness measurement, and so forth. Table 12.4 shows some failure modes selected in the structural approach for the same shaft considered in Table 12.3. For both the functional and the structural analysis it is very important to have a good description of the component function, since this is the starting point for the analysis of the failure mode.

The mode of failure cannot be viewed in isolation within a system and thus a relationship needs to be established between the effects presented in Figures 12.1 and 12.2.

Effects. The effect is the form or manner in which the failure mode is manifested or is perceived at the level of the system. While the failure mode occurs internally within an item, the effect is manifested externally, indicating that degradation of the system has occurred.

Figure 12.1 shows a diagram of a system comprising n items which need to act in order for a set of functions to be accomplished. Thus, the occurrence of a failure in item k has an effect on the system. The characteristic of the effect on the system that manifests is dependent on the dominant mode of failure of item k. In the case of a hydraulic system, if item k is a filter and the failure mode is a filter rupture, then the most probable effect will be leakage. However, depending on the configuration, a noise in the pump could occur due to cavitation, slow cylinder displacement or problems related to positioning precision. If the mode of failure is clogging, the effect on the system may be an increase in temperature, slow cylinder displacement, and so forth.

Figure 12.2 shows another approach to the relationship between cause and effect. It can be observed in Figures 12.1 and 12.2 that failure analysis using the FMEA technique is of the bottom-up type; that is, it moves from specific to general. It should be noted that the objective of Figure 12.2 is to call the reader's attention to the fact that there is a dynamic relationship between the mode of

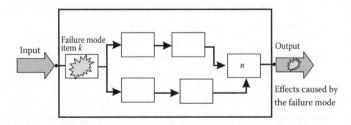

FIGURE 12.1 Relationship between the failure mode and its effect in a complex system. (From Sakurada, E.Y. *As Técnicas de Análise dos Modos de Falhas e seus Efeitos e Análise de Árvore de falhas no Desenvolvimento e na Avaliação do Produto*. Dissertação de Mestrado, Florianópolis: Universidade Federal de Santa Catarina, 2001.)

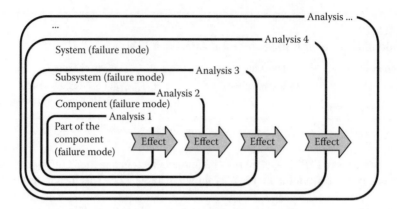

FIGURE 12.2 Successive analyses using the bottom-up approach. (From Dias, A. *Notes on Hazard Analysis Methods Review*. Center for Technology Risk Studies, Internal Report, University of Maryland, US, 2003.)

failure and its effect. The definition of one or another relationship will depend on the position of the analyst in relation to the part of the system under analysis.

The specification of an effect is also dependent on the type of analysis, with a function or structural focus, and the granulometry of the analysis. At the deepest level, when treating the item as a component or as part of a component, a structural focus is most appropriate. On the other hand, in systemic analysis a functional approach is recommended.

Cause. The study of the cause is important because it is the agent, the reason for the occurrence of the failure. It is generally considered that the origin of the occurrence of the failure mode or modes lies in the cause. In other words, each failure mode has one or more causes which may lead to its occurrence. The damages may be associated with environmental and human factors, or technical factors originating from the design, manufacturing process and use, influenced by neighboring items or those intrinsic to the functioning of the component. There are various terms used for the causes and different ways in which to characterize their dimension or importance. Regardless of the approach, the study of the causes is fundamental in order to understand the nature of the failure. It is from the study of the cause that the systems can be improved in order to be resistant in a wide variety of applications.

The study of causes allows for a deeper understanding of the relationship between the item and the function. In this way it is possible to generate more consistent procedures, eliminate the causes, or mitigate their influence on the generation of the failure mode. In other cases, if the effect is well characterized, its first manifestations can be exploited and the required measures can be taken in anticipation of the loss of function due to the occurrence of the failure mode.

It is important to note that the revealing of the causes is not carried out for all of the failure modes, as this demands considerable time and effort. Thus, the causes are investigated for the most critical failure modes, these being identified by defining the level of criticality of the failure modes in relation to safety issues and, in some cases, to economic importance.

Once the level of criticality has been defined, the causes are classified according to the different levels or stages of importance of the failure mode, depending on the level of detail or the granulometry of the analysis. For example, in Figure 12.3, there are three levels of causes of failure in relation to the functional mode of failure loss of efficiency of the hydraulic system. The first is called the "proximate causes", given that it is perceived by a person in the vicinity of the technical system, generally at the production level (the actual operator or a maintenance technician). There are cases when the failure mechanism itself is called the proximate causes, since it is more apparent. This is more visible, for example, in the case of an oil leak. The intermediate causes provides a greater level of detail and its definition requires more specific knowledge. It

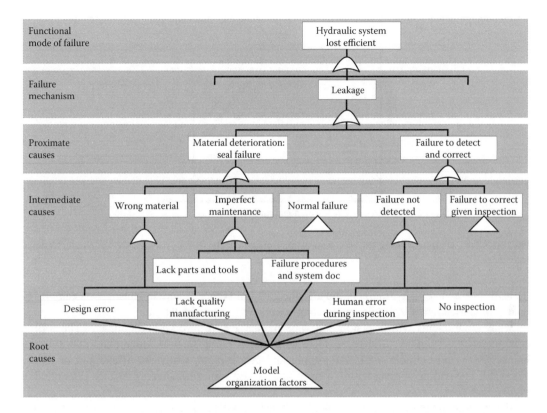

FIGURE 12.3 Fault tree of a hypothetical system to exemplify the relationship between the failure mode and its probable causes. (From Dias, A. *Notes on Hazard Analysis Methods Review.* Center for Technology Risk Studies, Internal Report, University of Maryland, US, 2003.)

is the result of the relationship between the failure mechanism and human actions within the system—at the level of design, manufacturing, production or maintenance. In conclusion, the root causes are the principal of all subsequent causes. It permeates the structure of the corporation and is generally called the "organizational causes", since all failures originate from the organization or the organizational management, whether from a lack of capacitation, training, investment, organizational culture, and so on.

The root causes are so important that the actions taken to eliminate them, when these are carried out with success, suppress definitively the failure mode under analysis.

The study of the failure mode, causes, and effects can be performed by using different approaches, depending on the life cycle phase of the product for which the analysis is being carried out and the required solution. Thus, there are different types of FMEA: system, design, service, and process. For instance, Section 12.4 presents an analysis in terms of the physics of hydraulic pump failure. The results of this analysis are organized into a design FMEA, since this focuses on the component. The objective of this FMEA is to eliminate the root causes at the component level, redefining the form, material, thermal treatment, surface finish, type of lubricant, and so on.

12.4 STRUCTURAL ANALYSIS AND DISCUSSION

12.4.1 HYDRAULIC PUMP LUBRICATION

One source of hydraulic pump failure is improper lubrication. The effect of fluid-film lubrication, material strength, heat balance of fluid film, rolling element bearing lifetimes, and cavitation on the

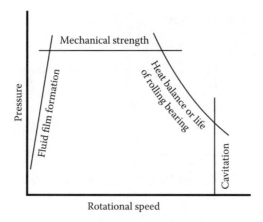

FIGURE 12.4 Hydraulic pump and motor operation limits. (Courtesy of A. Yamaguchi, Yokohama National University.)

limits of operation of a hydraulic pump are illustrated in Figure 12.4 [25]. Illustrative examples of fluid-film lubrication of hydraulic pump components will be provided here.

Many components within piston and vane pumps are in pure sliding contact. (The rolling component is negligible.) Sliding surfaces in a piston pump include valve plate and cylinder block, piston and cylinder wall, and slipper and swash plate, which are illustrated in Figure 12.5 [25]. The sliding contact within a vane pump is the vane-on-cam ring surface illustrated in Figure 12.6 [26].

Sliding wear at the valve plate-cylinder block surface will affect the performance of both swash-plate and bent-axis piston pumps. Wear at this interface may facilitate case leakage, which will increase with decreasing fluid viscosity [25].

The slipper-swash plate wear contact will significantly influence pump performance. Figure 12.5 shows that the slipper-piston contact may be treated as spherical bearing. Due to the small size, this contact operates in the mixed-film lubrication regime.

The shaft supports the force acting on the cylinder block and the fluid-film pressure acting on the valve plate with a hydrodynamic pad, as illustrated in Figure 12.7 [25]. A journal bearing may also be used to support the force, as shown in Figure 12.8 [25]. The total fluid-film pressure acting on the sliding valve plate-cylinder block surface illustrated in Figure 12.9 is composed of hydrostatic effect, wedge-film effect, and a squeeze-film effect [25]. The loading of the valve plate and cylinder block is primarily determined by the pressure of the high-pressure side of the pump. It has been shown that fluid-film thickness is relatively stable at rotational speeds of up to 105 rpm [25].

The piston and cylinder wall are also in sliding contact. For many swash-plate piston pumps, metal-metal contact is difficult to avoid because of the high lateral forces acting on the pistons [25].

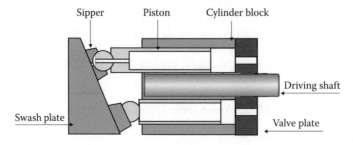

FIGURE 12.5 Sliding contacts of a swash-plate piston pump. (Courtesy of A. Yamaguchi, Yokohama National University.)

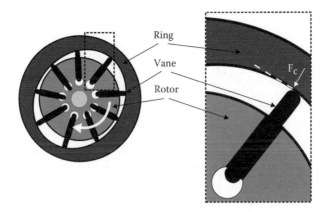

FIGURE 12.6   Vane-on-ring sliding line contact.

Lubrication at these surfaces can be improved by designing for enhanced hydrostatic lubrication and by varying the piston shape and size.

An example of a bushing that may be used in high-pressure gear pumps is illustrated in Figure 12.10 [27]. It has been shown that these bearings are lubricated hydrodynamically due to hydrostatic pressure generation arising from gear tilt and no flatness of the ends of the gear teeth and bushing surface [27].

The limits of lubrication of a representative bearing surface have been modeled and the results are illustrated in Figure 12.11 [26]. Metal-metal contact during the pump delivery stroke is represented by the shaded area No. 1. The suction to delivery stroke, which is the fluid-trapping period, is represented by the shaded area No. 2.

These examples show that hydraulic pump lubrication and performance are dependent on preventing pump and component damage and maintaining design clearances. Anything that produces a change in these clearances, such as lubrication failure and wear, may dramatically affect hydraulic pump performance.

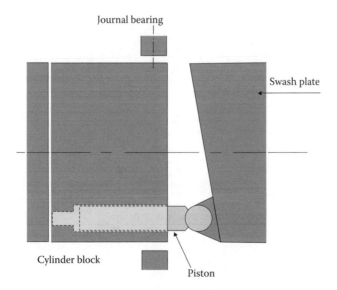

FIGURE 12.7   Force acting on a cylinder block. (Courtesy of A. Yamaguchi, Yokohama National University.)

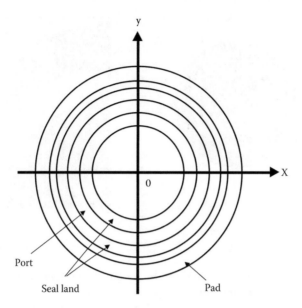

FIGURE 12.8   Valve plate with hydrodynamic pad. (Courtesy of A. Yamaguchi, Yokohama National University.)

12.4.2   HYDRAULIC PUMP FAILURE

12.4.2.1   Causes of Hydraulic Pump Failure

The pump is the "heart" of any hydraulic system. It is unusual for pump failures to be caused by manufacturing defects. Pump failures are usually a symptom of another problem in the system.

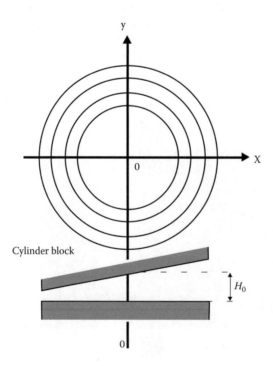

FIGURE 12.9   Fluid-film formation between valve plate and cylinder block. (Courtesy of A. Yamaguchi, Yokohama National University.)

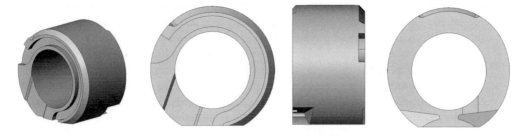

FIGURE 12.10   Schematic of a bushing-type bearing.

Preventing recurring failures requires a determination of the failure mechanism and the performance of preventive maintenance procedures. Typically, 85%–95% of pump failures can be attributed to one or more of the following causes [28]: foaming and aeration, cavitation, contamination, fluid oxidation, overpressurization, and improper viscosity.

Foaming and aeration. Although some fluid foaming is to be expected, excessive foaming may lead to the introduction of foam into the pump inlet, resulting in air-entrainment problems such as poor fluid compressibility and sluggish hydraulic response. Cavitation may also result. Foaming is caused by the following [29,30]:

1. Improper location of the return discharge line: Hydraulic systems should discharge the returning fluid below the oil surface in the reservoir in order to minimize foaming. Alternatively, the return line may discharge fluid onto an inclined plane or over a screen. Time is needed for the fluid to release entrained air to the surface before reentry into the system. Therefore, returning fluid should not be discharged directly to the reservoir outlet. If this occurs, baffles are needed to prevent direct discharge to the outlet.
2. Improperly sized reservoir: If the reservoir is too small or deep relative to the surface area, there may be insufficient time to permit the release of the entrained air before the fluid reenters the system.

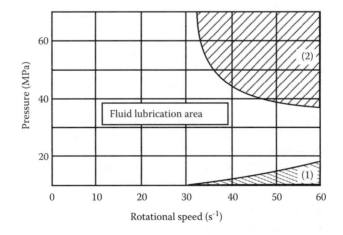

FIGURE 12.11   Fluid lubrication in a slipper bearing: (1) metal-metal contact during the pump delivery stroke, and (2) suction to delivery stroke. (Courtesy of A. Yamaguchi, Yokohama National University.)

3. Rapid pressure release: Rapid pressure release may result in the insolubilization of air that was soluble under the higher-pressure conditions, resulting in an entrained-air condition and increased foaming. Rapid pressure releases may be caused by sharp changes in pipe sizes or excessive vacuum on the discharge side of the pump inlet, discharge of metering valves and orifices to the reservoir, and a large pressure drop in the inlet line from the pressurized reservoir. In some cases, a negative suction head must be eliminated. The use of a supercharging (booster) pump may be required to pressurize the inlet. To minimize the potential for rapid pressure release and other inlet flow problems, abrupt changes in pipe sizes should be eliminated, if possible. Piping recommendations include the following [29,30]: (1) suction lift should be 610 mm (approximately 24 in.) maximum; (2) no more than two elbows in addition to a filter should be used in the suction line whose maximum length should not exceed 915 mm (approximately 36 in.); and (3) match the line size to the pump port size, if possible.
4. Insufficient fluid in the reservoir: If the oil level in the reservoir is too low, there may not be enough time for air release before the reentry of the fluid into the system. In addition, this condition may lead to vortexing at the suction inlet, causing increased air entrainment of the oil.
5. Hydraulic system leaks: Air may be drawn into the hydraulic fluid at packings or suction-line connections and from gas leaks from the accumulator. As the fluid is returned to the reservoir, excessive foaming will result. The suction-line strainer may also act as an air trap.
6. Moving parts: Crankshafts, gear teeth, and couplings, either too deep or shallow, may beat air into the hydraulic fluid that must be released upon return to the reservoir.
7. Fluid viscosity: As shown in Chapter 4, higher-viscosity fluids may exhibit greater foam stability, although lower foaming tendency. The use of higher temperatures, or lower fluid viscosity if it is still within the manufacturer's operating requirement, may minimize foam stability and facilitate air release.
8. Oil oxidation: In addition to increasing viscosity, some oxidation by-products may stabilize foam.
9. Contamination: A common cause of foaming is fluid contamination (Figures 12.61 and 12.63). Finely divided rust and scale may increase the foaming tendency. Detergent additives in mineral-oil hydraulic fluids may stabilize foaming with water contamination. Contamination with other fluids may also cause foaming.

Aeration and air entrainment. In addition to the sources of foaming discussed above, air ingression into the system may cause air entrainment. These include the following [29]:

1. Soluble air: Air is soluble in hydraulic fluid and air solubility increases with increasing temperature and pressure.
2. Mechanical introduction: Air ingression into the system may occur where negative pressure exists.
3. Improper system bleeding: Increased air entrainment may occur if the hydraulic system is not bled properly to release trapped air. Trapped air may occur during system filling, and vacuum filling may be necessary.
4. Improper fluid addition: Makeup fluid poured into the reservoir causing splashing or increased agitation may entrap air in the fluid.

In some cases, the only solution to air-entrainment problems is to use a de-aerator, where air is removed from a bypass stream under vacuum [33].

Cavitation. This may occur if the hydraulic fluid is too cold or the fluid viscosity is excessively high, which will lead to the formation of voids within the fluid caused by oil starvation on the suction side of the pump. In addition to excessive fluid voids, cavitation may also be caused by insufficient pumping capacity and restricted passages in the piping and strainers. Overall flow resistance should not create a net inlet vacuum in excess of 127–254 mm (5–10 in.) Hg maximum [34]. Additional potential sources of cavitation include excessive pump-shaft speeds, fluid-specific gravity higher than the system design accommodation, high altitude, boost or makeup pressure too low, and no baffle plate in the reservoir.

Fluid viscosity at the onset of cavitation may be calculated by the following [34]:

1. Adding or subtracting the head or lift to the pump suction capacity.
2. Express the bends, elbows, strainers, and values of the suction system as equivalent length of straight intake pipe and add this length to the suction pipe in feet.
3. From the following equation; Calculate the oil viscosity causing the maximum depression at the pump inlet (onset of cavitation):

$$\upsilon(cS) = \frac{843 \cdot a \cdot d^2}{b \cdot V}\left(\frac{0.87}{SPG}\right), \tag{12.1}$$

where
  a = the pump suction capacity (in. Hg)
  b = the equivalent length of suction system (ft)
  d = the diameter of the suction pipe (in.)
  V = the fluid velocity (in ft/s)
  SPG = the specific gravity of the hydraulic fluid

Some calculation examples are provided in Table 12.5 [34].

The above equation is derived from Fanning's equation for fluids with a viscosity in centistokes greater than $3.6 \cdot d \cdot V$ (d in inches and V in ft/s) and for fluids with a Reynolds number less than 2100.[*]

The equivalent length of pipe (in ft) for elbows, bends, gate valves, and strainers may be calculated from Table 12.6 [34].

Procedures to identify the sources of cavitation include the following:

1. Check the manufacturer's recommendation for start-up and operating viscosities of the fluids. These recommendations are specific of each pump manufacturer.
2. Check the inlet inside diameter and inlet flow velocity.
3. Check the size of the inlet strainer. The capacity of the inlet strainer should be at least two times the pump volume. (See subsection on Fluid Oxidation.)

---

[*] Fanning's equation, also known as the Darcy-Weisbach equation, is used to calculate friction loss in fluid flow through a pipe and is written as:

$$h_f = f \cdot \left(\frac{L}{d}\right)\left(\frac{v^2}{2 \cdot g}\right)$$

where h_f is the friction loss (in ft) of fluid, L is the length of pipe (ft), d is the average inside diameter of the pipe (ft), V is the average velocity (ft/s), g is the gravitational constant (32.174 ft/s²), and f is the friction factor [35].

TABLE 12.5
Illustrative Maximum Fluid Calculations

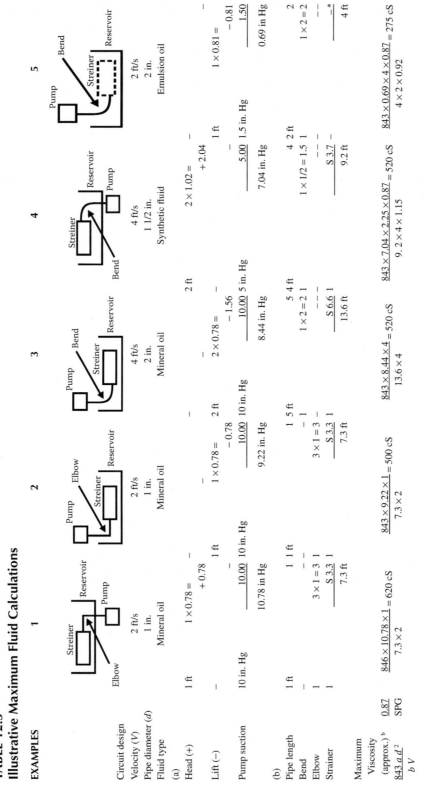

EXAMPLES	1	2	3	4	5
Velocity (V)	2 ft/s	2 ft/s	4 ft/s	4 ft/s	2 ft/s
Pipe diameter (d)	1 in.	1 in.	2 in.	1 1/2 in.	2 in.
Fluid type	Mineral oil	Mineral oil	Mineral oil	Synthetic fluid	Emulsion oil
(a)					
Head (+)	1 ft 1×0.78 = — +0.78	—	— 2 ft	2 ft 2×1.02 = +2.04	—
Lift (−)	— 1 ft	1×0.78 = 2 ft −0.78	2×0.78 = 2 ft −1.56	—	1 ft 1×0.81 = −0.81
Pump suction	10 in. Hg 10.00 10 in. Hg 10.78 in Hg	10.00 10 in. Hg 9.22 in. Hg	10.00 5 in. Hg 10.00 10 in. Hg 8.44 in. Hg	5.00 1.5 in. Hg 7.04 in. Hg	1.50 0.69 in Hg
(b)					
Pipe length	1 ft 1 1 ft	1 ft 1 1 ft	1 5 ft	4 2 ft	4 2 ft
Bend	—	—	1×2 = 2 1	1×1/2 = 1.5 1	1×2 = 2
Elbow	3×1 = 3 1	3×1 = 3 1	—	—	— —
Strainer	S 3.3 1	S 3.3 1	S 6.6 1	S 3.7 —	—[a]
	7.3 ft	7.3 ft	13.6 ft	9.2 ft	4 ft
Maximum Viscosity (approx.)[b]	846×10.78×1=620 cS / 7.3×2	843×9.22×1=500 cS / 7.3×2	843×8.44×4=520 cS / 13.6×4	843×7.04×2.25×0.87=520 cS / 9.2×4×1.15	843×0.69×4×0.87=275 cS / 4×2×0.92
SPG	0.87				

$843 \dfrac{a\,d^2}{b\,V}$

[a] Strainers not usually fitted owing to the low pump suction value permitted. Where fitted, a strainer with four times the normal capacity would be employed. In this case, the result = 150 cS (approx).

[b] $843\,a\,d^2/b\,V$ is derived form Fanning's equation for fluids with a centistokes viscosity greater than 3.7 dV, where d is the diameter in inches and V is the velocity in feet, and where flow is streamline having Reynolds numbers less than 2100.

TABLE 12.6
Equivalent Linear Length Factors

Component	Equivalent Length of Pipe (ft)
Elbow	$3 \times$ diameter (in.)
Bend	$1 \times$ diameter (in.)
Gate valve (open)	$0.5 \times$ diameter (in.)
Wire strainer (gap of 0.005–0.008 and aperture area of 2 in.^{2}/gal flow/min)[a]	$\dfrac{6.6d^2}{V}$ (ft),

[a] Adjust for other aperture areas: area X value = 13.2.

4. Be sure that the inlet strainer has been cleaned properly. (See subsection on Fluid Oxidation.)

One of the most common causes of hydraulic pump cavitation is improper pump inlet designs which do not provide the necessary NPSH (Net Positive Suction Head) for the fluid being used. The available NPSH at the pump inlet for a simple system may be calculated from the following equation (also see Figure 12.12) [36]:

$$NPSH_A = H_a + H_s - H_{vp} - H_f, \tag{12.2}$$

where

H_a = head pressure acting on the fluid surface in the tank. If the system is not pressurized, this will be atmospheric pressure.
H_s = head pressure between the surface of the fluid and centerline of the pump inlet. It is very important to note that this value will be negative if the fluid surface is below the pump inlet.
H_{vp} = vapor pressure of the fluid at the fluid temperature.
H_f = friction loss in the suction piping.

Alternatively, the following equation can also be used [29]. All units in m of head:

$$NPSH_A = H_a + H_g + \frac{V^2}{2g} - H_{vp}, \tag{12.3}$$

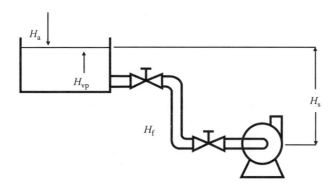

FIGURE 12.12 Illustration of contributing head pressures acting on a hydraulic pump inlet in a simple hydraulic circuit.

FIGURE 12.13 Surface damage pitting, and cratering due to cavitation. (Courtesy of IFAS, Aachen, Germany.)

where
  H_a = atmospheric pressure
  H_g = gauge pressure at the suction flange
  $V^2/2g$ = velocity head at the point of measurement. This adjustment is required because gauge
       pressures do not typically include velocity head
  H_{vp} = vapor pressure of the fluid at the fluid temperature

Cavitation damage is exemplified by the formation of pitting, holes, and craters on the wear surface. An example is shown in Figure 12.13.

Contamination. Any foreign material in a hydraulic fluid that exhibits a harmful effect is a fluid contaminant. Contaminants may be solid particles, liquids, or gases. Most solid contaminants cause an abrasive action, increasing the dimensions between the components. Solid contamination is discussed in detail in Chapter 3. Table 12.7 provides a summary of major contaminant types and the possible damage that they may cause [37].

TABLE 12.7
Common Fluid Contaminants and Possible Damage

Contaminant	Damage
Ingested dirt	Solids interfere with oil-film formation
	Hard particle abrasion
	Fine particles cause polishing wear
	Fatigue failure in rolling element bearings due to dents from particles
Water	Produces non-uniform fluid-film formation
	Causes rust
	Catalyzes oxidative oil degradation
Manufacturing debris	Metal chips penetrate oil film and initiate scuffing
Chemicals	Causes corrosion and oil degradation
Wear debris	Wear debris accumulation promotes oil degradation
Wrong oil	If viscosity is too low, oil film is too thin
	If viscosity is too high, pump efficiency decreases[a]
	If additives are too chemically active, corrosion and polishing will result
	If additives arc not sufficiently chemically active, risk of scuffing increases

[a] Efficiency increases with increasing viscosity due to reduced internal leakage. Excessive fluid viscosities
    produce cavitation.

FIGURE 12.14   Pressure plate is black due to excessive heat caused by high 150°C (>300°F) oil temperature. This plate cannot be reused. (Courtesy of Dam Corporation. Osaka, JAPAN.)

Fluid oxidation. It is normal for fluids to form acids and sludges due to oxidation over their lifetime. This is accelerated with extended operation at high temperature and thermal cycling [38]. Although the operating temperatures of mineral-oil hydraulic fluids will vary with the application, there are some guidelines: Typically, the maximum operating temperature at atmospheric pressure is 65°C (150°F), operating temperatures of 82–93°C (180–200°F) are possible, but the fluid must be changed two to three times as often, and operation at temperatures at 121°C (250°F) causes relatively rapid fluid decomposition. Fluid lifetimes, perhaps as short as 24 h, will result [38].

In addition to heat-exchanger failure, excessive heating of hydraulic fluids may be due to a sticking valve or relief valve set too low. If the sticking valve does not return to the neutral position, the system energy converts to heat instead of work, thus overheating the fluid. If the relief valve is set too low, a portion of the fluid will be dumped across the relief valve with each cycle, causing both overheating and slow operation [39]. The effect of high-temperature excursions on hydraulic equipment is illustrated in Figure 12.14 [40], Figure 12.15 [40], Figure 12.16 [41], and Figure 12.17.

One of the most common sources of hydraulic system problems is the deposition of oil oxidation by-products [42,43]. It has been shown that suction strainer clogging by resinous oil oxidation

FIGURE 12.15   Black coloration of gear and shaft due to excessive overheating of the oil 150°C (>300°F). This part is not usable. (Courtesy of Dana Corporation, Columbia, Mo.)

FIGURE 12.16 Flex plate discoloration from excessive heat. Some erosion damage is also observable. (Courtesy of Vickers Inc., Troy Inc., Troy, MI.)

by-products leads to cavitation and at least a 50% reduction in hydraulic pump lifetime [43]. The resinous by-products form an insulation coating on the strainer mesh as shown in Figure 12.18 [43]. This failure will lead to a buildup of static electricity and a spark discharge on full-flow filters (strainers), causing further oil breakdown [31,37]. This problem subsequently leads to pump failure by cavitation due to reduced flow rate. As a result of this study, it is recommended that suction filters and strainers be avoided in hydraulic systems [43].

FIGURE 12.17 Rotor, vanes, and bushing covered with fluid oxidation deposits. (Courtesy of DMT-Gesellschaft für Forschung und Prüfung mbH, Essen, Germany.)

(a)

(b)

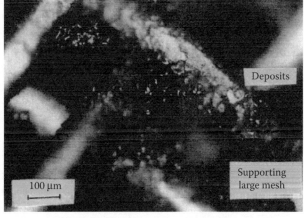

FIGURE 12.18 (a) Clogged inlet strainer. (Courtesy of A. Sasaki, Kleentek Inc., Cincinnati, OH.) (b) Illustration of sludge buildup on a strainer. (Courtesy of A. Sasaki, Kleentek Inc. Cincinnati, OH.)

Electrical discharge in mineral-oil hydraulic fluids has been reported previously and it was proposed that this is due to high fluid frictional effects and the length of the spark increased with fluid flow rate [44].

Overpressurization. A hydraulic pump should not be subjected to operating pressures greater than those for which the pump was designed. Overpressurization creates forces against various internal components and may cause premature failure. Overpressurization may also be caused by component failure. There are two possible causes for overpressurization [39]: (1) relief-valve failure, causing an extreme pressure surge and component failure, as illustrated in Figure 12.19 [39]; and (2) if the relief-valve setting is too high, repeated pressure breaks may result.

Viscosity. This is a measure of fluid internal friction or its resistance to flow. Fluid viscosity that is higher than recommended (or cold oil) may cause pump cavitation, increased pressure drop throughout the circuit, higher oil temperatures, sluggish hydraulic operation, or lower mechanical efficiency

FIGURE 12.19   Shaft breakage due to relief valve failure or by repeated excessive pressure surges. (Courtesy of Danfoss Fluid Power Inc., Wisconsin, WI.)

[45]. If fluid viscosity is too low, increased oil leakage, inability to hold pressure, lack of positive hydraulic control, or lower volumetric pump efficiency due to increased internal leakage and an accompanying increase in heat may result [45]. Therefore, the use of appropriate fluid with the recommended viscosity at the operating temperature is necessary. A summary of common hydraulic operation problems and solutions is provided in Table 12.8 [46].

TABLE 12.8
Troubleshooting Chart

Cause	What to do
Noisy Pump	
Air Leaking into system	Ensure that the oil reservoir is filled to normal level and that oil intake is below surface of oil. Check pump packing, pipe and tubing connections, and all other points where air might leak into system.
	One good way to check a point on the intake side suspected of leakage is to pour oil over it; if the pump noise stops, the leak has been found..
Air bubble in intake oil	If oil level is low or return line to reservoir is installed above oil level, air bubbles will form in oil in reservoir.
	Check oil level and return-line position.
Cavitation (the formation of vacuum in a pump when it does not get enough oil)	Check for clogged or restricted intake line, plugged air vent in reservoir. Check strainers in intake line. Oil viscosity may be too high. Check recommendations.
Loose or worn pump parts	Check manufacturer's maintenance instructions first. Tightening every nut in sight may not be the way to stop leakage. Look for worn gaskets and packings; replace if necessary. There is usually no way to compensate for wear in a part; it is always better to replace it. Oil may be of improper grade or quality;
	Check recommendations. Parts may be stuck by metallic chips, bits of lint, etc. If so, disassemble and clean thoroughly. Avoid the use of files, emery cloth, steel hammers, etc., on machined surfaces.
Stuck pump vanes, valves, pistons, etc.	Products of oil deterioration such as gums, sludges, varnishes, and lacquers may be the cause of sticking. Use solvent to clean parts and dry thoroughly before reassembling. If parts are stuck by corrosion or rust, they will probably have to be replaced. Ensure that oil has sufficient resistance to deterioration and provides adequate protection against rusting and corrosion.

TABLE 12.8   (Continued)
Troubleshooting Chart

Cause	What to do
Filter or strainer too dirty; filter too small	Filter and strainers must be kept clean enough to permit adequate flow. Check the filter capacity. Ensure that the original filter has not been replaced by one of smaller capacity. Use oil of high enough quality to prevent rapid sludge formation.
Pump running too fast	Determine the recommended speed. Check pulley and gear sizes. Ensure that no-one has installed replacement motor with other than recommended speed.
Pump out of line with driving motor	Check alignment. Misalignment may be caused by temperature variation.
Leakage Around Pump	
Worn packing	Tighten packing gland or replace packing. Trouble may be caused by abrasives in oil. If you suspect this sort of trouble, make a thorough check of points where abrasives may enter the system.
Head of oil on suction line	It is usually better to have the pressure on the suction side of the pump, although it may be necessary. With more than slight head, leakage may result. If head is not required and components can be rearranged, do so. Otherwise, do not worry about the leakage. Just wipe off the pump periodically and, if feasible, install a drip pan. Do not let oil leak onto floor!
Overheating	
Oil viscosity too high	Check oil recommendations. If you are not sure of the viscosity of oil in system, it may be worthwhile to drain the system and install oil of proper viscosity. Unusual temperature conditions may cause oil of proper viscosity for "working temperature" to thicken too much on way to pump. In this case, the use of oil with a higher viscosity index may solve problem.
Internal leakage too high	Check for wear and loose packings. Oil viscosity may be too low. Check recommendations. Under unusual working conditions, temperature may go high enough to reduce viscosity of recommended oil too much. Proceed with caution if tempted to try a higher viscosity oil.
Excessive discharge pressure	If oil viscosity is found to be OK, trouble may be caused by a high setting of relief valve. If so, reset.
Poorly fitted pump parts	Poorly fitted parts may cause undue friction. Look for signs of excessive friction; ensure that all parts are in alignment. On any machine equipped with an oil cooler, high temperatures are probably expected. If temperatures run high normally, they will go even higher if oil cooler passages are clogged.
Oil cooler clogged	If you find a clogged cooler, try blowing it out with compressed air. If ineffective, try solvents.
Low oil	If the oil supply is low, less oil will be available to carry away just as much heat. This will cause a rise in oil temperature, especially in machines without oil coolers. Ensure that the oil is up to level.
Pump Not Pumping	
Pump shaft turning in wrong direction	Shut down immediately. Some types of pumps can turn in either direction without causing damage; others are designed to turn in one direction only. Check belts, pulleys, gears, motor connections. Reversed leads on three-phase motor are most common cause of wrong rotation.
Intake clogged	Check line from reservoir to pump. Be sure filters and strainers are not clogged.
Low oil level	Ensure that the oil is up to recommended level in reservoir. Intake line must be below level of oil.
Air leak in intake	If any air at all is going through pump, it will probably be quite noisy. Pour oil over points suspected of leakage; if noise stops, the leak has been located.

(continued)

TABLE 12.8 (Continued)
Troubleshooting Chart

Cause	What to do
Pump shaft speed too low	Some pumps will deliver oil over a wide range of speeds; others must turn at a recommended speed to give appreciable flow. First determine the speed recommended by the manufacturer; then, with a speed counter, if possible, check the speed of the pump. If the speed is too low, look for trouble in the driving motor.
Oil too heavy	If the oil is too heavy, some types of pumps cannot pick up prime. A very rough check of viscosity can be made by first getting some oil known to have the right viscosity. Then, with both oils at the same temperature, pour a quart of each oil through a small funnel. The heavier oil will take a noticeably longer time to run through. Oil that is too heavy can do great harm to hydraulic systems. Drain and refill with oil of the right viscosity.
Mechanical trouble (broken shaft, loose coupling, etc.)	Mechanical trouble is often accompanied by a noise that can be located very easily. If you find it necessary to disassemble, follow the manufacturer's recommendations to the letter.
Low Pressure in System	
Relief valve setting too low	If relief-valve setting is too low, oil may flow from pump through relief valve and back to the oil reservoir without reaching point of use. To check relief setting, block discharge line beyond relief valve and check line pressure with pressure gauge.
Relief valve stuck open	Look for dirt or sludge in valve. If the valve is dirty, disassemble and clean. A stuck valve may indicate that the system contains dirty or deteriorated oil. Ensure that the oil has a high enough resistance to deterioration.
Leak in system	Check whole system for leaks. Serious leaks in the open are easy to detect, but leaks often occur in concealed piping. One routine in leak testing is to install a pressure gauge in the discharge line near the pump and then lock off the circuit progressively. When the gauge pressure drops with gauge installed at a given point, leak is between this point and the check point just before it.
Broken, worn, or stuck pump parts	Install pressure gauge and block system just beyond relief valve. If no appreciable pressure is developed and relief valve is satisfactory, look for mechanical trouble in pump. Replace worn and broken parts.
Incorrect control valve setting; oil "short-circuited" to reservoir	If open-center directional control valves are unintentionally set at neutral position, oil will return to the reservoir without meeting any appreciable resistance and very little pressure will be developed. Scored control-valve pistons and cylinders can cause this trouble. Replace worn parts.
Erratic Action	
Valves, pistons, etc., sticking or binding	First, check suspected part for mechanical deficiencies such as misalignment of a shaft, worn bearings, etc.; then look for signs of dirt, oil sludge, varnishes, and lacquers caused by oil deterioration. You can make up for mechanical deficiencies by replacing worn parts, but do not forget that these deficiencies are often caused by the use of the wrong oil.
Sluggishness when a machine is first started	Sluggishness is often cause by oil that is too thick at starting temperatures. If you can put up with this for a few minutes, oil may thin out enough to give satisfactory operation; but if oil thins out or if surrounding temperature remains relatively low, you may have to switch to oil with lower pour point, lighter viscosity, or, perhaps, higher viscosity index. Under severe conditions, immersion heaters are sometimes used.

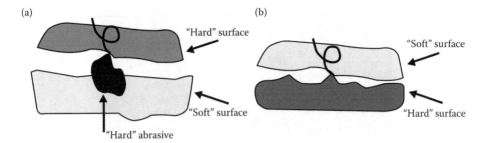

FIGURE 12.20 Abrasive wear: (a) three-body wear; (b) two-body wear.

12.4.3 WEAR MECHANISMS

To conduct successful failure analysis, it is first necessary to understand and recognize common wear mechanisms upon inspection. The objective of this section is to discuss commonly encountered wear processes, including: (1) abrasive wear, (2) adhesive wear, (3) erosive wear, (4) cavitation wear, (5) corrosive wear, (6) contact stress fatigue wear, and (7) other forms of wear.

12.4.3.1 Abrasive Wear

Abrasive wear refers to the cutting of a metal by hard particles or a rough surface by a ploughing or microcutting mechanism [37,47]. When abrasion wear is caused by a hard particle between two surfaces, it is termed "three-body wear", which is depicted in Figure 12.20 [48]. Hard particles causing three-body wear may be introduced into the system from the component manufacturing process, generated internally as wear debris, ingested through a breather or seals, or it may be added as a contaminant in the fluid upon addition to the system [49].

Abrasive wear is dependent on particle size distribution, shape, toughness, and hardness [48]. The mechanism illustrated by Figure 12.21 [48] shows that hard particles penetrate and become embedded into one of the wear surfaces. The embedded particle, tilted at an angle (Φ), then cuts

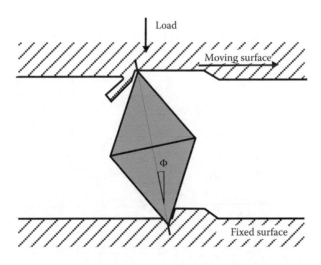

FIGURE 12.21 Three-body wear model.

FIGURE 12.22   Illustration of abrasion surface wear. (Courtesy of IFAS, Aachen, Germany.)

the opposite surface. The energy from the microcutting action causes plastic deformation and the formation of parallel furrows and ridges in the direction of surface movement [37]. A surface that has undergone abrasive wear is illustrated in Figure 12.22. Severe gear-tooth abrasion is illustrated in Figure 12.23 [50]. Additional examples of three-body wear are presented in Figure 12.24 through 12.28.

Two-body wear, shown in Figure 12.20b, is caused by a harder surface with asperity dimensions sufficiently large to penetrate the lubricating oil film causing a ploughing or microcutting action on the other softer surface, which is in relative motion. This is depicted in Figure 12.29 [48]. This form of wear is also known as cutting, ploughing, gouging, lapping, grinding, or broaching wear [48]. Abrasive wear may be minimized by the following [49]:

1. Removing potential residual manufacturing debris by proper draining and flushing procedures before starting.
2. Using wear-resistant construction materials.
3. Minimizing ingested particles by the use of proper breather filters and by keeping the system tight.
4. Using fine filtration to minimize particulate contamination of the hydraulic fluid.
5. Performing proper fluid maintenance and periodic fluid analysis.

FIGURE 12.23   Extreme abrasion wear. (Courtesy of American Gear Manufacturers Association.)

FIGURE 12.24 Vane-tip wear due to particle ingression through breather pipe. No air filter was used. (Courtesy of Denison Hydraulics Inc., Marysville. OH.)

12.4.3.2 Adhesive Wear

Adhesive wear, as depicted in Figure 12.30, occurs when surface asperities come into sliding contact under a load. If sufficient heat is generated, microwelding of the asperity with subsequent shearing and material transfer of the contact will be observed, as illustrated in Figure 12.31. This process will continue until larger surfaces are in contact and macrowelding or seizure occurs. The generation of adhesive wear debris may then cause abrasion wear. Therefore, it is possible for both wear processes to occur together. If the temperature is sufficiently high, metal flow and "smearing" will be observed. If the part is constructed from steel, tempering colors may be observed, and if the temperature of the total part is high enough, plastic deformation and fracture may result, as illustrated in Figure 12.32 [47]. Adhesive wear is dependent on the following [51]:

1. Surface characteristics, including hardness, shear strength, roughness, surface "waviness", geometry, elastic modulus, tensile strength, ductility, melting temperature, and interfacial metal compatibility.
2. Normal loading and tangential sliding force.
3. Lubrication conditions, including presence of oxide films, extreme pressure lubricants, oil films, and contaminants.
4. Environmental properties such as dust, salt, fog, water, temperature, In addition to the chemical substances, such as carbon dioxides, carbon monoxides, corrosive materials and other chemicals.

FIGURE 12.25 Scratch marks on the face of a piston shoe due to fluid particle contamination. (Courtesy of Denison Hydraulics Inc., Marysville, OH.)

FIGURE 12.26 Vane-rotor assembly damage due to large-particle contamination. Only one half of fluid flow was being filtered. The filter had a rating of β20 = 5, which was plugged due to water contamination. (Courtesy of Denison Hydraulics Inc., Marysville, OH.)

Fitch [51] has provided the following "Six Laws of Adhesive Wear":

1. The area of contact between the two surfaces is directly proportional to the applied normal load and inversely proportional to the indentation (Brinell) hardness of the softer material.
2. The volume of material worn away when one of the surfaces slides over the other is proportional to the true area of contact, the total sliding distance, and the wear coefficient of the material pair. Wear coefficients are provided in Table 12.9 in Fitcher et al. [51].
3. The average wear depth is obtained by dividing the volume of material wear by the nominal area of contact.
4. For surfaces of different materials, the ratio of wear depth of the harder material equals the square of the ratio of the yield strengths of the softer material to the harder material.
5. The wear coefficient remains constant as long as the pressure does not exceed one-third the material hardness. As long as the wear coefficient is constant, wear will vary linearly with

FIGURE 12.27 Internal wear contact surface with abrasive damage due to hard-particle contamination. (Courtesy of Kayaba Industry Co., Tokyo, Japan.)

FIGURE 12.28 Fine-particle abrasive wear on the vane tips is due to surface varnish formation from the oil. (Courtesy of Quaker Chemical Corporation, Conshohocken, PA.)

the distance traveled and the wear rate will be constant. At pressures in excess of one-third the hardness, the wear coefficient begins to increase and the wear increases nonlinearly and rapidly.

6. The material pair used determines the wear coefficient that is established experimentally. Oxide layers, surface contamination, surface finish, and experimental conditions affect this coefficient.

Adhesive wear may be reduced by increasing material hardness, surface hardening, avoiding metallurgically similar material pairs, avoiding highly soluble materials, using welding-resistant materials, and providing a soft layer containing compounds of sulfur and phosphorus [51].

Adhesive wear may occur in areas where the friction contacts are greatest due to tight fits, misalignment, high loading, poor lubrication, and high temperature [52]. Examples of adhesive wear are provided in Figure 12.33 (cam ring) [53], Figure 12.34 (bearing area of gear shaft) [52], Figure 12.35 (piston) [53], Figure 12.36 (bearing surfaces) [54], Figure 12.37 (piston-shoe metal transfer), and Figure 12.38 (piston-shoe/tin-brass layer adhesive failure).

Polishing. Polishing wear may occur by either of two mechanisms. One mechanism is chemical and occurs when the EP additives which are present in the hydraulic oil are too aggressive, leaving

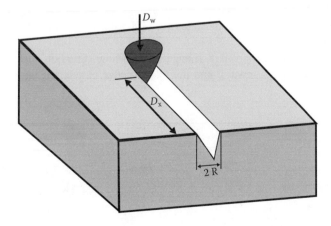

FIGURE 12.29 Model of two-body wear process.

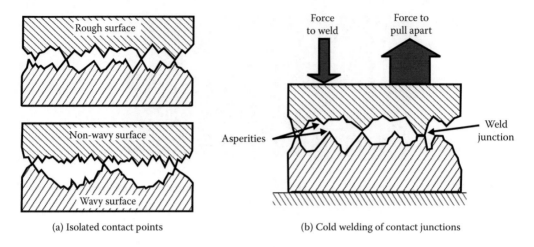

(a) Isolated contact points   (b) Cold welding of contact junctions

FIGURE 12.30   Illustration of (a) surface asperity contacts, and (b) adhesive wear.

a bright, mirrorlike surface [49]. This effect is undesirable because the polished surface is less resistant to wear.

Alternatively, polishing wear may occur during component break-in when surface asperity adhesive wear occurs resulting in a very fine, smooth surface [50,55]. This is caused by insufficient electro-hydrodynamic (EHD) lubrication to prevent such asperity contact. Polishing wear can be minimized or avoided with the use of a high-viscosity lubricant or reduced operating temperature. Polishing wear for the drive and gear teeth of a properly operating pump is illustrated in Figure 12.39 [40].

12.4.3.3   Erosion Wear

Moving particle surface impingement erosion wear (erosion wear), as illustrated in Figure 12.40 [56], may result in scratching, surface indentation, chipping, and gouging, as shown in Figure 12.41. From a mechanical point of view, erosion wear is a combination of two processes: surface fatigue and abrasive cutting [56]. Figure 12.42 illustrates that erosion wear is an exponential process including an initiation time, acceleration, and a period of maximum rate loss [56]. The parameters that affect erosion wear include fluid-flow velocity, particle and fluid density, drag coefficient, Reynolds number, and particle size. In general, if particles are ≤100 μm, erosion wear rate increases with particle size. If the particles are >100 μm, wear volume by erosion is independent of particle size [56].

Erosion wear occurs where there is a change in fluid-flow direction such as in orifices, line restrictions, turns in fluid passageways, and the leading edge of rotating parts [52]. Figure 12.43

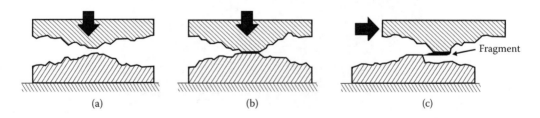

(a)   (b)   (c)

FIGURE 12.31   Model of adhesive wear process: a) surfaces being pressed together, b) cold weld achieved, and c) material transferred.

FIGURE 12.32 Damaged roller bearing from an axial piston pump with plastically deformed roller due to poor lubrication and adhesive wear. (Courtesy of DMT-Gesellschaft für Forschung und Prüfung mbH, Essen, Germany.)

illustrates this process, in which erosion wear above and below a piston bore is shown [54]. (Note that the erosion is greater at the top of the bore.)

12.4.3.4 Cavitation Wear

Cavitation arises when there is a collapse (implosion) of a gas bubble entrained in the hydraulic fluid. The bubble implosion process forms a high-velocity microjet capable of high-impact energy. The implosion process and subsequent microjet surface action is often referred to as the "water hammer" effect. Cavitation can occur when there are excessive gas bubbles, high loading, or abrupt surface contour changes. The mechanism of the cavitation process was explained in detail in Chapter 2.

After repeated bubble implosions, material fatigue damage results in surface damage with the formation of pitting and larger holes, as shown in Figure 12.44. The propensity for cavitation

TABLE 12.9
Approximate Wear Coefficients [51]

Material A	Material B	Wear coefficient $\times 10^4$
Cu	Pb	0.03
Ni	Pb	0.07
Fe	Ag	0.23
Ni	Ag	0.23
Al	Pb	0.47
Ag	Pb	0.84
Mg	Pb	0.87
Zn	Pb	0.87
Ag	Ag	1.14
Al	Zn	1.31
Al	Ni	1.58
Al	Cu	1.61
Al	Ag	1.78
Al	Fe	2.02
Fe	Zn	2.82

(continued)

TABLE 12.9 (Continued)
Approximate Wear Coefficients

Material A	Material B	Wear Coefficient $\times 10^4$
Ag	Zn	2.82
Ni	Zn	3.70
Zn	Zn	3.90
Mg	Al	5.24
Zn	Cu	6.21
Fe	Cu	6.41
Ag	Cu	6.65
Pb	Pb	8.00
Ni	Mg	9.60
Zn	Mg	9.76
Al	Al	10.00
Cu	Mg	10.25
Ag	Mg	10.90
Mg	Mg	12.25
Fe	Mg	12.92
Fe	Ni	20.00
Fe	Fe	26.00
Cu	Ni	27.20
Cu	Cu	42.30
Ni	Ni	95.90

is dependent on material properties. A general ranking of various properties with respect to the potential for cavitation damage is shown in Figure 12.45 [57]. Hobbs has reported a correlation of cavitation damage with "ultimate resilience", which was defined as [58]:

$$UR = \frac{0.5 \cdot \sigma_u^2}{E},$$

(12.4)

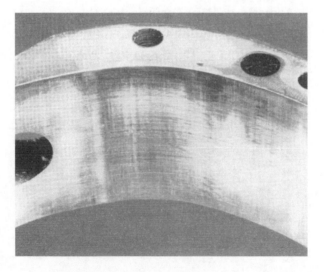

FIGURE 12.33 Cam ring smearing damage and discoloration. (Courtesy of Caterpillar Inc., Peoria. IL.)

FIGURE 12.34   Gear shaft with subsequent cracking due to heat from adhesive wear process. (Courtesy of Kayaba Industry Co., Tokyo, Japan.)

where

   UR = Ultimate resilience
   σ_u = Tensile strength
   E = Modulus of elasticity

Various examples of cavitation damage are provided in Figure 12.46 through 12.60.

12.4.3.5   Corrosive Wear

Corrosion wear is surface damage related to the electrochemical attack of the metal component. Corrosion may be classified as follows:

1. Crevice or cell corrosion occurs when the tendency for corrosion is greater in cracks or crevices.

FIGURE 12.35   Hydraulic pump piston which has undergone polishing due to smearing. (Courtesy of Caterpillar Inc., Peoria, IL.)

FIGURE 12.36 Polishing of lead-tin bearing surfaces due to smearing caused by poor lubrication. (Courtesy of Caterpillar Inc., Peoria. IL.)

2. Dezincification occurs when brass is used. Zinc, an alloying element in brass, dissolves preferentially as a soluble salt. Brasses with greater than 15% zinc are most susceptible to dezincification, particularly in a saltwater environment.
3. Stray electrical currents from motors or generators may come into contact with fabricated metal sections or hydraulic components, accelerating electrochemical attack.
4. Pitting is caused by contact with extreme temperature or acids. This will result in localized oxidative breakdown of the material. The corroded area is anodic and loses metal locally to the cathodic area, causing pitting. Severe abrasive wear damage due to oxidative pitting is illustrated in Figures 12.61 and 12.63.

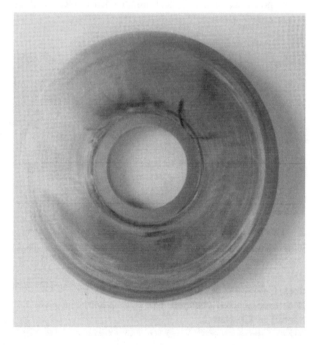

FIGURE 12.37 Yellow metal transfer from piston shoes to swash plate. (Courtesy of Southwest Research Institute, San Antonio, TX.)

FIGURE 12.38 Illustration of adhesive failure of carbon-steel piston shoes that have a sliding surface with a layer of sintered tin-brass. The pockets on the shoe surface are present to facilitate hydrostatic lifting. The pockets have disappeared due to excessive wear and the sintered tin-brass has transferred to the surface of the swash plate. Failure was due to inadequate adhesive strength of the tin-brass coating on the side of the steel pistons. (Courtesy of Kayaba Industry Co., Tokyo. Japan.)

5. Fatigue stresses caused by internal or external pressures, such as a nut on a bolt, may occur in a metal. This effect will create a pathway within the material that will corrode more rapidly. If the stresses are cyclic, accelerated corrosion or corrosion fatigue may result.
6. Intergranular corrosion may occur within grain boundaries of the metal.

FIGURE 12.39 Illustration of polishing of the loaded side of gear teeth. Some pitting will occur in the root of the gear teeth after extended operation. High gloss of the journals indicates a properly operating pump. (Courtesy of Caterpillar Inc., Peoria, IL.)

One result of corrosion is the formation of metal oxides, which have characteristic colors [52]. Ferrous alloys or steel often form reddish-brown oxides. Copper or brass forms bluish-green oxides. Oxides of aluminum are gray. Figures 12.56 and 12.62 illustrate the characteristic porous nature of an oxide coating on steel. Figure 12.57 illustrates a combination of bronze surface corrosion due to water contamination and additive hydrolysis and cavitation. Figure 12.58 illustrates abrasive wear failure when the root cause is corrosion.

Common causes of metal corrosion include the following [59]:

1. Water from humid air ingression and subsequent condensation, water from spray or splash contamination, water ingression from defective enclosure and cooling coil leakage, and water-contaminated lubricant, Figure 12.64.
2. Poor lubricant formulation using aggressive EP additives, presence of acids from decomposition or exposure to active metals, and lubricant and system incompatibility such as yellow metal attack.
3. Corrosive chemical vapors in the atmosphere.
4. Corrosive processing chemicals such as coolants and cleaning fluids.
5. Galvanic couple formation between metals with different electrochemical potential, such as yellow metal separators in martensitic steel roller bearings, stainless-steel surfaces contaminated with martensitic steel, and aluminum alloy components with unprotected surfaces in contact with steel and in the presence of an electrolyte which is any ion-conductive fluid.

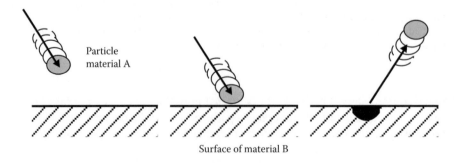

FIGURE 12.40 Model of erosive wear.

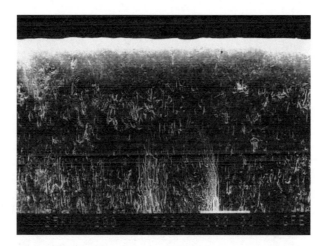

FIGURE 12.41 Illustration of erosion surface wear. (Courtesy of IFAS, Aachen, Germany.)

Corrosion may be avoided by using appropriate design methodology. Some guidelines for proper design include the following [60]:

1. Avoid galvanic couples by insulating dissimilar metals or increasing electrical resistance. This may be accomplished by using more noble metals or metals close to the metal of interest in the galvanic series shown in Figure 12.65 [61].
2. Avoid small anodic areas (the more active metal is the anode) relative to the cathode (the less active metal), such as fasteners, by using alloys that are cathodic with respect to the main structure.
3. Avoid concentration cells by welding to eliminate crevices and by using fluorocarbon gaskets between contact surfaces.
4. Prevent corrosive environments by providing a drop skirt or design to eliminate the collection of contaminants by geometry or capillarity; allow adequate drainage to prevent collection of corrosive solutions.
5. Avoid sections with high tensile stress.
6. Use protective coatings where possible.

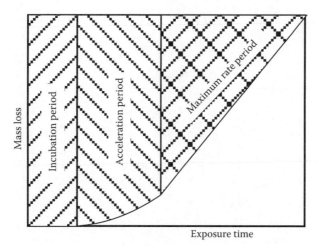

FIGURE 12.42 Mass removal kinetics of erosive wear process.

FIGURE 12.43   Illustration of erosive wear on a cylinder piston. (Courtesy of Caterpillar Inc., Peoria, IL.)

Galvanic corrosion. Galvanic corrosion is recognizable by the appearance of increased corrosion near the junction of dissimilar metals [62]. This is caused by the electrochemical transfer of one metal to another. The propensity for galvanic corrosion is dependent on the position of the two metals in the galvanic series, shown in Figures 12.59 and 12.65 [61]. Any metal will have a greater tendency to corrode when it is in contact with another metal in a lower position in the series and is in the presence of an electrolyte. The farther apart the two metals in the series are, the greater the potential for a galvanic attack will be. The example provided in Figures 12.59 and 12.65 were developed for steel rivets in copper plates in seawater. Electrons flow from steel to copper, resulting in a deposit of iron oxide on copper. Galvanic corrosion may be minimized by the following:

1. Use the same or similar metals, especially if an electrolyte is present.
2. Avoid combining dissimilar metals where the area of the less noble metal is relatively small.

FIGURE 12.44   Illustration of cavitation surface damage. (Courtesy of IFAS, Aachen, Germany.)

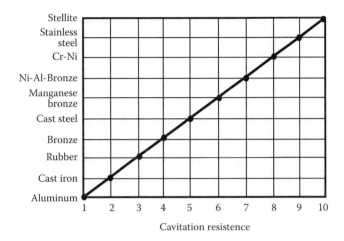

FIGURE 12.45 Cavitation resistance of various materials. (Courtesy of IFAS, Aachen, Germany.)

3. Use a dielectric material, paint, or coating to separate dissimilar materials where possible.
4. If a pair of dissimilar metals must be used, couple them to a piece of less noble metal, often zinc, which may be used for sacrificial corrosion [52].

Electrochemical erosion (pitting corrosion). This may be encountered with the use of fire-resistant phosphate ester hydraulic fluids with components with small orifices [63], such as spool valves [64].

FIGURE 12.46 Hydraulic pump cavitation damage. (Courtesy of Kayaba Industry Co., Tokyo, Japan.)

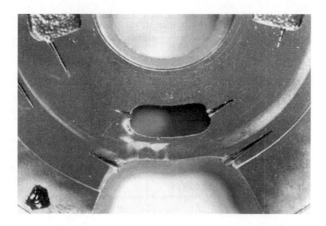

FIGURE 12.47 Cavitation damage of the port plate of a vane pump due to the insufficient release of entrained air in the fluid. This pump was operated at very fast pressure cycles. The pump was equipped with a flooded inlet and an inlet strainer. (Courtesy of Denison Hydraulics Inc., Marysville, OH.)

Electrochemical erosion occurs most rapidly when the valves are in the closed position, with the formation of a large pressure drop across the orifice by the edge of the slide and port in the sleeve [65]. This erosion is primarily a function of chloride ion contamination and electrical resistance [64]. Beck et al. [66,67] have reported that the mechanism of electrochemical erosion was driven by electrical currents which arose from the electrical double layer formed at the fluid interface, as shown in Figure 12.66 [68].

A streaming current (I) is formed when fluid flow which is parallel to the double-layer metal interface causes movement of the free charge, typically from chloride anions. The sum of the total charges at the fluid metal interface must be zero. Therefore, as additional current is formed from the streaming potential, additional current for electrical neutrality is usually provided by the metal, because the fluid exhibits low conductivity and is designated as wall current (I_W). This process leads to the corrosion observed below:

$$Fe^0 \rightarrow Fe^{2+} + 2e. \tag{12.5}$$

FIGURE 12.48 Two vanes exhibiting cavitation damage taken from the same pump as the port plate shown in Figure 12.47. The vanes exhibited cavitation in an area matching the port plate damage. Thus, cavitation not only occurs on the surface of one port but on different ports from the same area. (Courtesy of Denison Hydraulics Inc., Marysville, OH.)

FIGURE 12.49   Localized cavitation damage of the end plates of a piston pump.

The wall current, assuming that no current is provided by the fluid, may be calculated from [68]:

$$I_w = \frac{1.17 \cdot \varepsilon \cdot \xi \cdot v}{x^3} \cdot \left(\frac{Q}{V}\right)^{3/2},$$ (12.6)

where
  ε = fluid permittivity [F/m]
  ξ = the potential difference (V) between the electrical double-layer metal interface and the fluid
      outside of the double layer (zeta potential)
  v = the fluid viscosity [m²/s]
  Q = the laminar flow rate [m³/s]
  x = the distance from the metering edge [m] [69]

The total current normal to the metal surface (I_N) per unit length of the orifice is [68]:

$$I_N = \frac{0.585 \cdot \varepsilon \cdot \xi \cdot v}{g^2} \cdot \left(\frac{Q}{V}\right)^{3/2},$$ (12.7)

where g is the gap size.

FIGURE 12.50   Cavitation damage of the control plate of a piston pump.

FIGURE 12.51 Cavitation damage of a piston-pump port plate. (Courtesy of Kayaba Industry Co., Tokyo, Japan.)

Inspection of these equations shows that the current is greatest with the following [64]:

1. Small gaps or orifices.
2. High fluid velocity. The flow rate across the metering edge for an ISO VG 46 phosphate ester at 70 bar pressure is estimated to be 53.64 m/s (approximately 120 mph). Wall current is proportional to the pressure drop (Δp):

$$I_w \propto \frac{\Delta p^3}{4}, \tag{12.8}$$

3. High fluid permittivity. Fluids with a permittivity of $(2-5) \times 10^{-11}$ F/m exhibit no damage. Permittivities of $(12-15) \times 10^{-11}$ F/m produce significant erosion damage [64]. New phosphate ester fluids typically exhibit permittivities of $(7-8) \times 10^{-11}$ F/m [64].
4. High zeta potential. Although commercially available phosphate ester fluids exhibit zeta potentials of 25–150 mV [65], treatment with Fuller's earth may reduce the zeta potential to 0.5 mV. Fuller's earth treatment increases the electrical resistivity from 4×10^9 to 14×10^9 Ω.cm [64].

FIGURE 12.52 Cavitation damage on the inside of a cam ring with accompanying discoloration due to overheating. (Courtesy of Quaker Chemical Company, Conshohocken, PA.)

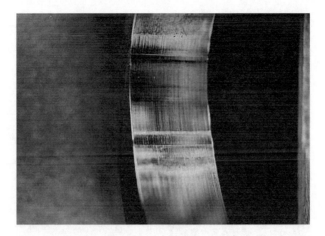

FIGURE 12.53   Vanes may exhibit a "chattering" noise on contact with the cam ring with fluid aeration and poor air release. The rippling effect observed here is due to vane chatter. Some discoloration due to overheating is also visible. (Courtesy of Quaker Chemical Company, Conshohocken, PA.)

Electrochemical erosion may be reduced by reducing the conductivity of the fluid. Nelson and Waterman [69] have reported that fluids exhibiting conductivities of 3×10^{-7} to 5×10^{-8} mho/cm will produce electrochemical erosion. In this case, the neutralizing current is supplied by the metal. However, fluids exhibiting higher conductivities ($>3 \times 10^{-7}$) do not exhibit electrochemical erosion. In this case, the neutralizing current is supplied by the fluid [69]. Table 12.10 provides a summary of the reported methods for reducing electrochemical erosion [64].

12.4.3.6   Contact-Stress Fatigue Wear

Contact-surface fatigue is favored by small contact areas, high loadings, and repeated flexing action under cyclic rolling or reciprocal sliding loads, although each occurs under different conditions, as illustrated in Figure 12.67 [70]. The applied stresses are less than the material yield stress and the process is often accompanied by frictional heat and plastic flow. Subsurface structural changes are also observed metallographically. Contact-stress fatigue wear processes are distinguishable from abrasion, adhesion, and corrosion, as shown in Table 12.11 [71].

FIGURE 12.54   Cavitation damage of an end plate from a Vickers V-104 vane pump.

FIGURE 12.55   Cavitation damage of a vane from same pump side plate as shown in Figure 12.51.

Reciprocal sliding action produces a cyclic shear load. Cracking initiates at the surface and progresses into the subsurface zones [54] and is characterized by pitting, as illustrated in Figure 12.68 [50].

Rolling action compresses the surface and cracks initiate below the surface at points of maximum Hertzian stresses, as shown in Figure 12.69 [72] (or deformation). For many materials, maximum shear stress occurs at a depth of 0.01 mm below the surface [72]. Traditionally, it was believed that crack initiation almost invariably occurs at subsurface nonmetallic inclusion sites or grain boundaries, or at surface stress risers such as a keyway, as illustrated in Figure 12.70 [70]. However, it has also been shown that cracking may initiate from surface that work hardened to about 20 μm, with the accompanying formation of tempered martensite [70]. Fatigue failure modes are summarized in Table 12.12 [47,73].

Surface fatigue is typically manifested by the following [74]:

1. Pitting: Pit sizes may vary from small (0.4–0.5 mm) to much larger destructive holes that often occur in the negative sliding condition. Surface pitting due to surface-contact fatigue is shown in Figure 12.68 [50].

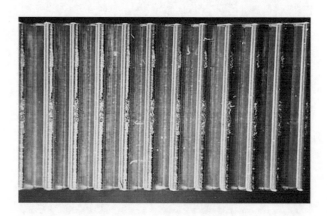

FIGURE 12.56   Vane-tip cavitation with some discoloration due to overheating. (Courtesy of Quaker Chemical Company, Conshohocken, PA.)

FIGURE 12.57 Although cavitation damage is observed on the flex plate of a vane pump, the major problem is inadequate lubrication due to severely aerated oil. Adhesive smearing and heat discoloration are also visible.

2. Spalling: Surface "spalls" may be much larger, shallower, and more irregular shapes than pits, as illustrated in Figure 12.71 [50].
3. Case crushing: This is a subsurface fatigue failure that occurs just below the hardened case [50] with longitudinal cracks ending on the surface, as shown in Figure 12.72 [28].

(a)
(b)

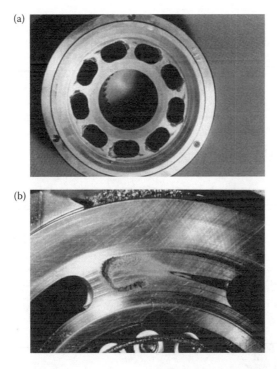

FIGURE 12.58 Cavitation erosion damage with subsequent fatigue for (a) a bronze control plate and (b) a bronze cylinder from axial piston pump (brass). Note that the highest pressure is on the suction side. As the pressure increases, bubbles of entrained air implode, eroding the material. The erosion is concentrated on the outer edge of the slots because the circumferential speed is slightly greater at this point. As the radius increases, the circumferential speed will also increase. (Courtesy of DMT-Gesellschaft für Forschung und Prüfung mbH, Essen, Germany.)

(a)

(b)

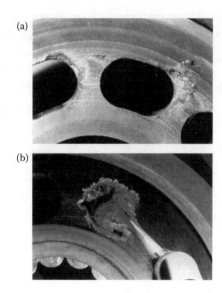

FIGURE 12.59 Cavitation erosion damage with subsequent fatigue for (a) a steel control plate and (b) a bronze cylinder plate. Note that the steel control plate exhibits much greater cratering than the bronze control plate shown in Figure 12.51. The reason for the greater damage was that the steel control plate was used in a pump under higher pressure. (Courtesy of DMT-Gesellschaft für Forschung and Prüfung mbH, Essen, Germany.)

Design factors that affect surface-contact fatigue include EHD film thickness, surface roughness, surface loading, sliding velocity, lubricant additives [73,74], and material properties, Figure 12.73 [72].

The fatigue life of roller bearings has been related to the lambda ratio (specific film thickness ratio), as shown in Table 12.13. The value lambda is calculated from [77]:

$$\Lambda = \frac{h_0}{\sigma}, \tag{12.9}$$

where h_0 is central EHD film thickness and σ is defined as:

$$\sigma = \left(\sigma_1^2 + \sigma_2^2\right)^{1/2}, \tag{12.10}$$

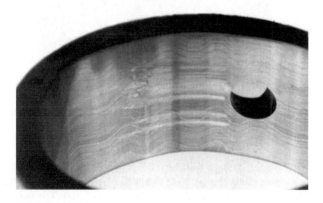

FIGURE 12.60 This cam ring exhibits two modes of wear: cavitation and adhesive smearing. The wavy marks are the original grinding marks. The root cause of failure was loss of antiwear additives in the hydraulic oil. (Courtesy of Denison Hydraulics Inc., Marysville, OH.)

FIGURE 12.61 Fluid contamination leads to the generation of surface pitting of seal ring and the formation of debris, which cause severe wear damage. (Courtesy of Kayaba Industry Co., Tokyo, Japan.)

where σ_1 and σ_2 are the root-mean-square (rms) surface finishes of contact surfaces 1 and 2, respectively. For many bearings, this value can be assumed to be 0.12 μm. For very large industrial bearings, σ can be assumed to be 0.65 μm [77].

An alternative to lambda, often designated as Λ_{exp}, is to use a corrected Λ_{adj} value, which is calculated from:

FIGURE 12.62 Illustration of porous nature of corroded surfaces. (Courtesy of IFAS, Aachen, Germany.)

FIGURE 12.63 Bronze port plate discoloration due to fluid contamination by water with subsequent additive hydrolysis. Cavitation damage is also observed. (Courtesy of Denison Hydraulics Co., Marysville, OH.)

$$\Lambda = \left(\frac{h_0}{\sigma}\right) \cdot k_{\mathrm{h}}, \tag{12.11}$$

where k_{h} is the EHD film correction factor, which may be determined from Figure 12.67 [77]. For a roller bearing, the value of the lubricant flow number GU in Figure 12.67 is calculated from [77]:

FIGURE 12.64 Vanes taken from a vane pump using a water-contaminated phosphate ester fluid. Water contamination caused corrosion and the corresponding formation of metal oxides that produced the observed abrasive wear. The course abrasive debris plugged the system tilter. Fine debris caused surface abrasion of the vanes. The two vanes on the left show the back side of the vanes and the two vanes on the right show coloration due to varnish deposits from the fluid. (Courtesy of Denison Hydraulics Co., Marysville, OH.)

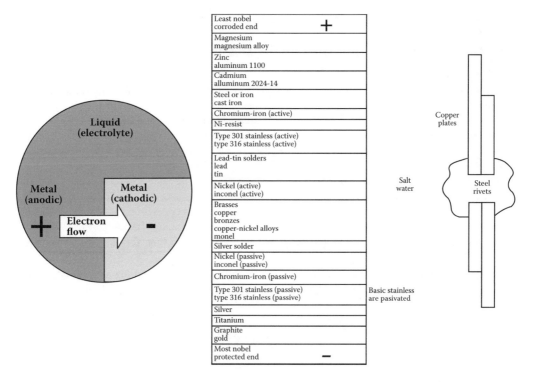

FIGURE 12.65 Illustration of galvanic corrosion and the galvanic series.

$$GU = k_{G} \cdot \alpha \cdot Z_{0} \cdot N \cdot \left(\frac{OD + ID}{OD - ID} \right), \qquad (12.12)$$

where α is in^2-lb. force and k_{G} is 1.52×10^{-10} min/cP (P is equivalent bearing load [in N] lbf/100).

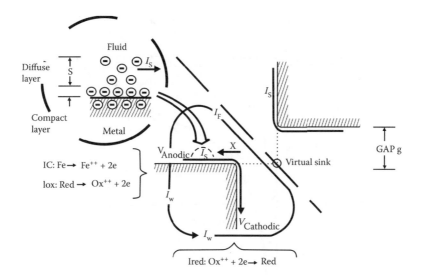

FIGURE 12.66 Generation of electrokinetic streaming current: streaming currents, wall currents, and fluid current flows.

TABLE 12.10
Methods to Minimize Electrochemical Valve Erosion

System Design	Fluid Type/Maintenance
Reduce flow rates or pressure drop across the valves	Use Fuller's earth treatment to reduce fluid acidity and increase resistivity
Minimize thermal stress on the fluid	Avoid contamination by halogenated solvents
Use corrosion-resistant steel	Use additives to reduce the zeta potential
Use zero or overlap designs for the servo valve	Use fluids of good thermal, oxidative, and hydrolytic stability
Minimize water contamination and maintain slight vacuum over the fluid in the storage tank	

Townsend and Shimski used correlated lambda (Λ) with gear surface fatigue life as shown in Figure 12.74 [75].

Moyer has reported that the use of a modified lambda ratio (λ_m) provides a better correlation for contact fatigue wear that may occur for both gears and roller bearings, which is calculated from [47]:

$$\lambda_m = h \cdot \left(\frac{L}{2 \cdot b} \right)^{1/2},$$

(12.13)

where L is the large end wavelength used to measure surface roughness and is usually equal to 0.8 mm (0.030 in.), and $2 \cdot b$ is the width of the contact in the direction of motion. A correlation of λ_m and contact fatigue wear for both gears and bearings is provided in Table 12.14 [47].

Hardened and unhardened steels behave differently when exposed to contact-surface fatigue. Plastic flow may arise in the surface layers of unhardened steels which will not occur in hardened steels. For example, unhardened spur gears may exhibit pitting along the pitch line at the point of

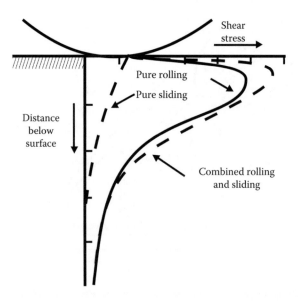

FIGURE 12.67   Model of shear stresses formed by rolling, sliding, and a combination of both processes.

TABLE 12.11
Differentiation of Surface-Contact Fatigue from Other Wear Processes

Mechanism	Surface-Contact Fatigue	Other Wear Mechanisms
Nature of contact	Counter formal	Conformal
Stress system	Hertzian and alternating	Dispersed and continuous
Lubrication and oil film	Partial to full EHD	Partial hydrodynamic to starvation
Relative motion to mating surface	Usually rolling	Usually sliding
Wear process	Crack initiation and propagation or material subsidence	Material adhesion, ploughing, corrosion, etc.
Wear particle characteristics	Lamina, sphere, and spall particles	Normal/severe sliding wear particles, cutting, wear particles, etc.

highest contact stress. Hardened gears, on the other hand, undergo pitting along the pitch line where traction forces are highest [72]. When traction occurs along the pitch line for hardened steels, the following statements are true [72]:

1. Surface fatigue is less at low slide-to-roll ratios than when pure rolling occurs.
2. More cracks propagate on the lower peripheral speed surface than at the edge.
3. Performance is sensitive to steel composition, heat treatment, and lubricant composition. There have been reports that surface-contact fatigue is significantly reduced by the addition of certain additives [75,76].
4. Surface-nucleated cracks produce earlier failures than subsurface cracks.
5. Surface-nucleated cracks tend to be located in asperity peaks.
6. Surface-nucleated failures are sensitive to surface temperature and therefore to EP reactions in the lubricant.

In summary, (1) fatigue cracks are caused by repetitive stress fluctuations, (2) failure times decrease with increasing stresses, (3) fatigue cracks occur at stresses less than the material yield stress, and

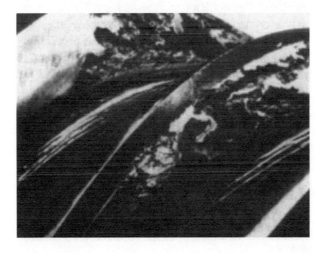

FIGURE 12.68   Destructive pitting due to surface-contact fatigue. (Courtesy of American Gear Manufacturers Association.)

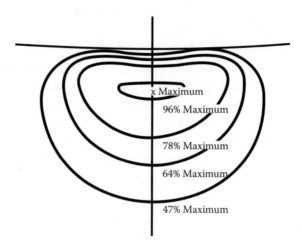

FIGURE 12.69   Location of maximum shear stress locations in a Hertzian contact.

(4) there is a finite time between the origination of the crack and final failure [70]. Various examples of fatigue failure are illustrated in Figures 12.75 through 12.79.

12.4.3.7   Other Forms of Wear

Brinelling. True brinelling: If the applied load exceeds the elastic limit of the material, elliptical "brinell marks", or brinelling, will be observed, as illustrated in Figure 12.80 [78]. Brinelling may occur when a hammer is used to install or remove a rolling element-bearing assembly or when it is dropped [79]. Brinelling is due to impact deformation rather than wear.

False brinelling: Elliptical wear patterns in the axial direction at each roller or ball position in the rolling element assembly, as illustrated in Figure 12.81 [79], is called "false brinelling". The elliptical pattern is often surrounded by debris. False brinelling is caused by vibration-induced wear.

Fretting corrosion. Fretting corrosion occurs when the tips of the asperities of adjacent moving surfaces come into contact and microwelding occurs. Continued surface movement causes the tips of the microwelded asperities to pull off, thus producing a pitting effect. The heat of the microwelding

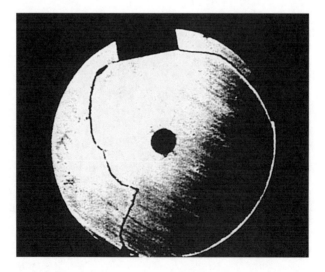

FIGURE 12.70   Fatigue cracking initiated by the keyway that acted as a stress riser. (Courtesy of Society of Tribologists and Lubrication Engineers [STLE].)

TABLE 12.12
Surface Fatigue Failure Modes

Mode	Characteristics
Wear-type failures	Surface removal
	Removal of loose wear particles
	Chemical or electrical surface removal
	Cumulative surface material transfer "smearing"
Plastic flow	Loss of contact geometry due to cold flow
	Material softening due to overheating
Contact fatigue	Pitting or spalling
	Surface distress (cracks)
Bulk failures	Overload cracking
	Overheat cracking
	Bulk fatigue
	Permanent dimensional changes
	Fretting of fit surfaces

process enhances oxidation of the fresh metal surfaces. If water is present, corrosion will result. Fretting corrosion, as illustrated in Figures 12.82 and 12.83 [80], may occur on the surfaces of moving tightly fitted assemblies such as roller bearings [52].

Electrical currents. A burning or arcing effect may be observed when electrical current contacts a component surface such as a bearing race. This may occur during a process such as welding or during long-term contact with lower currents. Either a "fluting" effect or "pitting", as illustrated in Figures 12.84 and 12.85, respectively, may be observed [78]. Therefore, it is essential that equipment be properly grounded.

Miscellaneous failures. In addition to those failures previously shown, there are a number of other possible causes. Some of the more commonly encountered sources include a poor or loose fitting, misalignment, and heat treatment. Examples of these failure modes are provided in Figures 12.86 through 12.88. A tabulated summary of the various failure mechanisms (and their characteristics) which may be encountered in hydraulic pump failure analysis is presented in Table 12.15 [81].

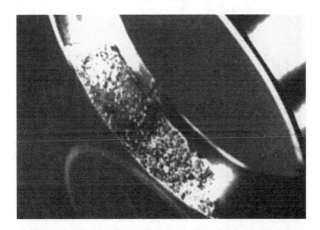

FIGURE 12.71   Bearing race damage by spalling. (Courtesy of Caterpillar Inc., Peoria, IL.)

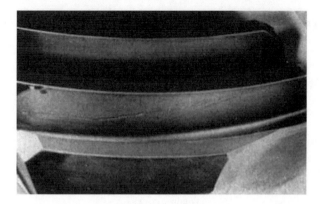

FIGURE 12.72 Illustration of case crushing of a carburized bevel gear. (Courtesy of American Gear Manufacturers Association.)

TABLE 12.13
Correlation of Lambda (Λ) and Fatigue Cracking of Roller Bearings

Λ value	Wear Observation
<1	Surface smearing or deformation
1–1.5	Surface distress accompanied by surface pitting
1.5–3	Surface glazing accompanied by subsurface failure
≥3	Minimal wear, long life, eventual subsurface-contact fatigue failure

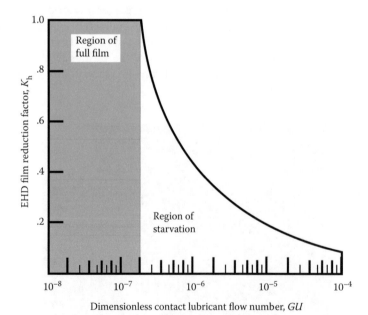

FIGURE 12.73 Elastohydrodynamic film reduction factor as a function of contact lubricant flow number. (Courtesy of Society of Tribologists and Lubrication Engineers [STLE].)

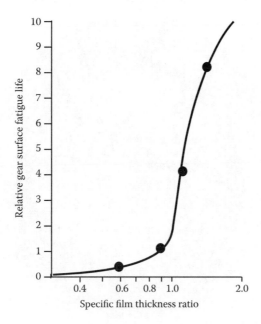

FIGURE 12.74 Relative gear fatigue life as a function of specific film thickness ratio.

TABLE 12.14
Correlation of Modified Lambda and Contact Fatigue Failure

λ_m Ratio	Initiation of Contact Fatigue	Material Influence	Surface Roughness Influence	Geometry Influence
>3.0	Subsurface fatigue inclusion[a]	Important	Minor	Important
3–1.0	Subsurface/near surface mostly inclusion	Important	"Sharp"/High Asperities Important	Important
1.0–1.3	Some inclusion; some surface related	Somewhat important	Important for surface and near surface fatigue	Somewhat important
<0.3	Surface related[b]	Minor	Important	Less important
Any ratio	Localized stress risers	Mixed	Mixed	Mixed
	PSO[c] – dents, grooves, and surface inclusions	Minor	Somewhat important	Minor
	GSG[d] – edge fatigue misalignment	Somewhat important	Minor	Somewhat important

[a] Fatigue originates at nonmetallic inclusion in the maximum shear zone below the surface for both bearings and gears.
[b] Called peeling or micropitting for bearings; spalling for gears.
[c] PSO (point surface origin): fan-shaped spall propagation starting in the surface.
[d] GSC (geometric stress concentration) starting at end of line contact.

FIGURE 12.75 Fatigue fracture at the weakest point was due to excessive fluid aeration, which was caused by flow turbulence occurring during fast pressure changes. The resulting aerated fluid possessed poor compressibility, creating pressure spikes that exceeded the design limits of the cam ring material. (Courtesy of Denison Hydraulics, Marysville, OH.)

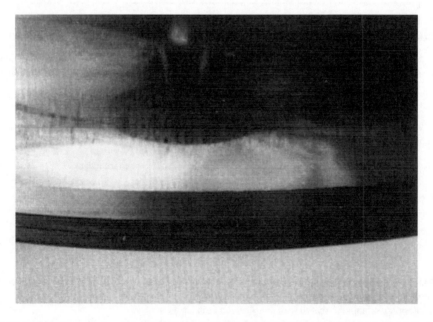

FIGURE 12.76 Fatigue crack on hydraulic pump flange caused by hammering impact during installation. (Courtesy of Kayaba Industry Co., Tokyo, Japan.)

(a)

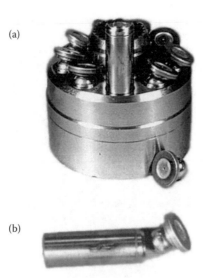

(b)

FIGURE 12.77 Surface fatigue of a piston from a swash-plate piston motor due to poor lubrication. In this case, the motor was tested without loading by increasing the rotational speed. Seizure occurred at 8000 rpm between the piston and the cylinder block. The piston was locked in the bore and the shoe was pulled off by the retainer plate. The failure was due to insufficient lubrication due to the high sliding speeds and increasing centrifugal force. (Courtesy of Kayaba Industry Co., Tokyo, Japan.)

(a)

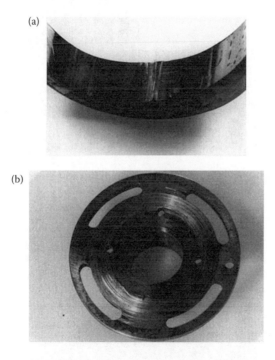

(b)

FIGURE 12.78 The (a) cam ring and (b) bushing failure were due to several wear mechanisms, including abrasion, cavitation, erosion. corrosion, and fatigue. Rippling was due to uneven pressure on the vanes (see Figure 53), producing a chattering effect from aerated fluid. Severe cavitation erosion produced debris and three-body abrasion wear. The abrasion process overheated the parts causing fatigue failure. (Courtesy of DMT-Gesellschaft für Forschung und Prüfung mbH, Essen, Germany.)

(a)

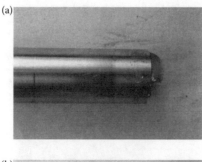

(b)

FIGURE 12.79   Fatigue failure due to non-uniform hardening of the hydraulic pump shaft. (Courtesy of Kayaba Industry Co., Tokyo, Japan.)

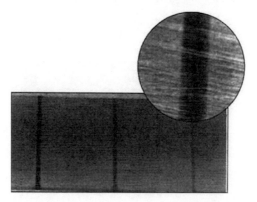

FIGURE 12.80   Brinelling impact damage to a bearing race. (Courtesy of The Timken Company, Canton, OH.)

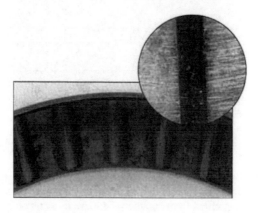

FIGURE 12.81   False brinelling of a bearing due to wear. Note the oxide deposits surrounding the wear damage. (Courtesy of The Timken Company, Canton, OH.)

FIGURE 12.82 Fretting corrosion on the outer surface of a rolling element bearing. (Courtesy of Caterpillar Inc., Peoria, IL.)

(a)

(b)

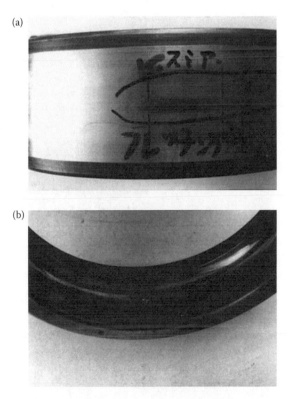

FIGURE 12.83 Axial movement during stopping produced temperatures in excess of 700°C causing non-uniform distribution of the lubricating grease, leading to fretting damage and localized smearing. (Courtesy of Kayaba Industry Co., Tokyo, Japan.)

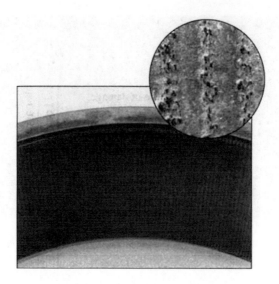

FIGURE 12.84   Fluting damage of a bearing race caused by exposure to electric current. (Courtesy of The Timken Company, Canton, OH.)

FIGURE 12.85   Pitting damage of a rolling element caused by exposure to electric current. (Courtesy of The Timken Company, Canton, OH.)

(a)

(b)

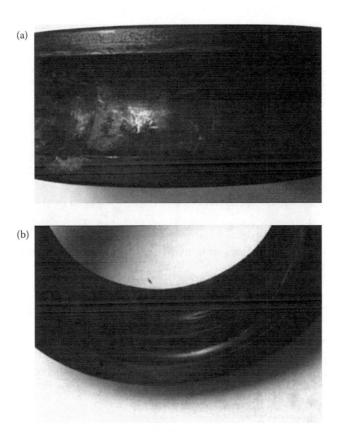

FIGURE 12.86   Bearing track damage caused by loose-fitting shaft resulting in a temperature rise to >700°C. (Courtesy of Kayaba Industry Co., Tokyo, Japan.)

FIGURE 12.87   Bearing misalignment resulting in very short lifetimes. (Courtesy of Kayaba Industry Co., Tokyo, Japan.)

FIGURE 12.88 Gear shaft for reducing gear box with a quench crack after induction hardening. (Courtesy of Kayaba Industry Co., Tokyo, Japan.)

TABLE 12.15
Common Wear Problems Related to Lubricants and Hydraulic Fluids

Names of Wear			Susceptible	Conditions
Preferred	Other	Definition	Machine Parts	Promoting Wear
Mild adhesion	Normal[a]	Generally, transference material from one surface to another due to adhesion and subsequent loosening during relative motion. Mild adhesion involves transfer and loosening of surface films only	All	Moderate loads, speeds and temperatures. Good, clean, dry lubricants Proper surface finish
Severe adhesion	Scuffing Galling Scoring[b]	Cold welding of metal surfaces due to intimate metal-to-metal contact	Piston rings and cylinder barrels Valve train Rolling & sliding bearings Gears Cutting tools Metal seals Chains	High loads, speeds and/or temperatures Use of stainless steels or aluminum Insufficient lubricant Lack of antiscuff additives No break
Abrasion[k]	Cutting Scratching "Wire wool" damage Gouging Scoring	Cutting and deformation of material by gard particles (three-body) or hard protuberances (two-body)	All surfaces in relative motion	Hard particles contaminating oil Insufficient metal hardness Hard metal with rough surface against soft metal

TABLE 12.15   (Continued)
Common Wear Problems Related to Lubricants and Hydraulic Fluids

Names of Wear		Definition	Susceptible Machine Parts	Conditions Promoting Wear
Preferred	**Other**			
Erosion	Solid particles Impact Erosion	Cutting of materials by hard particles in a high-velocity fluid impinging on a surface	Journal bearings near oil holes Valves Nozzles	High-velocity gas or liquid-containing solids impinging on a surface[f]
Polishing	Bore polishing	Continuous removal of surface films by very fine abrasives	Cylinder bores of diesel engines Gear teeth Valve lifters	Combination of corrosive liquid and tine abrasive on oil
Contact fatigue	Fatigue wear Frosting Surface fatigue Spalling	Metal removal by cracking and pitting due to cyclic elastic stress during rolling and sliding	Rolling and sliding bearings Valve train parts Gears	Cyclic stress over long periods Water or dirt in the oil Inclusions in steel
Corrosion[k]	Chemical wear Oxidative wear Corrosive film wear	Rubbing off of corrosion products on a surface	All bearings Cylinder walls Valve train Gears Seals and chains	Corrosive environment Corrodible metals Rust-promoting conditions[h] High temperatures
Fretting corrosion	False brinelling Fretting Friction oxidation	Wear between two solid surfaces experiencing oscillatory relative motion of low amplitude	Vibrating machines Bearing housing contacts Splines, keys, couplings Fasteners	Vibration causing relative motion
Electro-corrosion	"Erosion" Electrical erosion Electrochemical wear Electrical attack	Dissolution of a metal in a electrically conductive liquid by low-amperage currents	Aircraft hydraulic valves Hydraulic pumps and motors	High-velocity liquid flow causing streaming potentials Stray currents Galvanic metal combinations
Electrical discharge	Electrical pitting Sparking	Removal metal by high-amperage electrical discharge or spark between two surfaces	Bearing in high-speed rotation machinery such as compressors, atomizers Static charge producers	High-speed rotation High velocity two-phase fluid mixtures High-potential contacts Sparks
Cavitation damage	Cavitation erosion Fluid erosion	Removal of metal by bubble implosion in a cavitating liquid	Hydraulics parts, pump valves, gear teeth Cylinder lines, piston rings Sliding bearings	Sudden charges in liquid pressure due to changes in liquid velocity or to shape or motion of parts

(continued)

TABLE 12.15   (Continued)
Common Wear Problems Related to Lubricants and Hydraulic Fluids

| Name of wear | Symptoms | | | Prevention | |
	Unaided Eye	Microscopically	Oil analysis	Mechanical Changes	Lubricant Changes
Mild adhesion	Low rates of wear No damage Deeper original grinding marks still visible	Smooth microplateaus among original grinding marks Slight coloration due to films	1-5 ppm wear metals by emission spectroscopy Low % solids by filtration Metal salts (oxides, sulfides, phosphates, etc.) in wear fragments by x-ray diffraction	None	None
Severe adhesion	Rough, torn, melted or plastically deformed metal hand or streaks High-temperature oxidation High friction, high ratesof wear Possible seizure	Rough irregular surface Metal from one surface adhering to other surface by spot test or microprobe analysis	Large metallic wear fragments of irregular shape[c]	Reduce load, speed, and temperature Improve oil cooling Use compatible metals Apply surface coatings such as phosphating Modify surface such as ion implantation[d]	Use more viscous oil to separate surfaces Use "extreme pressure" additives such as a sulfur-phosphorus or borate compound[g]
Abrasion[k]	Scratches or parallel furrows in the direction of motion similar to "sanding" High rates of wear	Clean furrows, burrs, chips Embedded abrasive particles In sliding bearings with soft overlay embedded particles cause polished rings	High-metal contents in oil and high silicon (>10 ppm) by emission spectroscopy High % solids by filtration Chips and burrs by ferrography	Remove abrasive by improved air & oil filtering, clean oil handling practices, improved seals, flushing & frequent oil changes[e] Increase hardness of metal surfaces	Oil free abrasive particles Use more viscous oil
Erosion	Smooth broad grooves in direction of fluid flow Matte texture clean metal similar to sandblasting	Short V-shaped furrows by scanning electron microscopy Embedded hard particles	Element of hard particles by emission spectrograph Chips & burrs by ferrography	Same as above Reduce impact angle to less than 15°	Same as above
Polishing	High wear but a bright mirror finish Wavy profile	Featureless surface excepts scratches at high magnification by electron microscopy	Combination of fine-metal corrosion products & fine abrasive by x-ray diffraction	None	Choose less chemically active additive[g] Remove corrosive contaminant Remove abrasive

TABLE 12.15 (Continued)
Common Wear Problems Related to Lubricants and Hydraulic Fluids

Name of wear	Symptoms			Prevention	
	Unaided Eye	Microscopically	Oil Analysis	Mechanical Changes	Lubricant Changes
Contact fatigue	Cracks, pits & spalls	Combination of cracks & pits with sharp edges Subsurface cracks by metallographic cross section Numerous metal inclusions	Particles of metal with sharp edges Metal spheres by electron microscopy	Reduce contact pressure & frequency of cyclic stress Use high-quality vacuum-melted steels Use less abusive surface finish	Use clean, dry oil Use more viscous oil Use oil with higher-pressure viscosity coefficient
Corrosion[k]	Corroded metal surface	Scale, films, pits containing corrosion products Dissolution of one phase in two-phase alloy	Detection of corrosion products of worn metal Detection of anion such as chloride by x-ray diffraction	Use more corrosion-resistant metal (not stainless) Reduce operating temperature Eliminate corrosive material	Remove corrosive material such as too chemically active additive & contaminants Use improved corrosion inhibitor Use flesh oil
Fretting corrosion	Corroded stained surfaces[i] Loose colored debris around real contact areas Rouge (Fe_2O^2) colored films, debris, grease or oil for steel	Thick films of oxide of metal Red & black for steel	Identify metal oxide (α-Fe_2O_3 for steel) by x-ray diffraction	Reduce or stop vibration by tighter fit or higher load Improve lubrication between surfaces by rougher (than honed) surface finish	Use oil of lower viscosity Relubricate frequently[j] Use oxidation inhibitors in oil
Electro-corrosion	Local corroded areas Black spots such as made by a small drop of acid Corroded, worn metering edges	Corrosion pits, films, dissolution of metals	Detection of corrosion products Electrically conductive liquids[j]	Decrease liquid velocity & velocity gradients Use corrosion-resistant metals Eliminate stray currents Use nongalvanic couples	Decrease or increase electrical conductivity of lubricants or hydraulic fluids[j]
Electrical discharge	Metal surface appears etched In thrust bearing, sparks make tracks like an electrical engraver	Pits near edge or damage showing once molten state, such as smooth bottoms, rounded particles, gas holes Rounded particles near pits, welded to surface	Detection of large rounded particles by microscopy examination of filtrate or in ferrograph	Improve electrical insulation of bearings Degauss magnetic rotating parts Install brushes on shaft Improve machine grounding	Use of oil of higher electrical conductivity

(continued)

TABLE 12.15   (Continued)
Common Wear Problems Related to Lubricants and Hydraulic Fluids

Name of wear	Symptoms			Prevention	
	Unaided Eye	**Microscopically**	**Oil Analysis**	**Mechanical Changes**	**Lubricant Changes**
Cavitation damage	Clean frosted or rough appearing metal	Clean, metallic, bright, rough metal, pits	Observation of large chunks or spheres of metals in oils	Use hard, tough metals such as tool steel	Avoid low-vapor-pressure, aerated, wet oils
	Deep rough pits or grooves	Removal of softer phase from two-phase metal*		Reduce vibration, flow velocities, & pressures	Use noncorrosive oils
				Avoid restriction & obstructions to liquid flow	

[a] Mild adhesion is a desirable wear condition.

[b] Scoring is not recommended because it implies a scratch or furrow cut.

[c] Emission spectroscopy usually misses large (>5 μm) wear fragments.

[d] Increasing metal hardness does not reduce scuffing.

[e] Do not shot peen, bead or sandblast any surface in a lubricated machine because abrasive cannot readily be removed property

[f] Sandblasting embeds sand in surfaces

[g] A new additive reduces promotion of contact fatigue by water some extreme pressure additives are suspected of promoting contact Fatigue

[h] Rust (hydrated iron oxide $Fe_2O_3H_2O$) is common corrosion product of ferrous metal

[i] Damage on one surface is mirror image of damage on other.

[j] Highly compounded oils can be electrically conductive or electrolytes: phosphate ester hydraulic fluids are conductive

[k] Corrosion and abrasive in oil increase cavitation damage

* Graphite phase in cast-iron susceptible to removal by cavitation

12.5   FINAL CONSIDERATIONS

Each of the failure mechanisms described in the previous sections exhibits its own distinctive type of adverse effect. In order to properly troubleshoot, correct, and maintain hydraulic equipment, it is important to recognize and understand these different failure modes. The analysis process applied herein is called "failure analysis" and the objective is to determine the "root cause" of the problem through the inspection of hardware.

As mentioned in this chapter, several approaches can be taken to carry out failure analysis. For this reason a general overview of the main approaches is provided, with the focus on obtaining the root cause of the main failure modes. In order to also contribute with a practical view of the analysis of the physics of the failure, a hydraulic pump is considered as a reference and in order to demonstrate the analysis process, the definitions, the mechanisms, the failure modes, and the resulting effects of the failure modes.

ACKNOWLEDGMENTS

Sincere appreciation to Mr. Larans Kambol (handbook first edition) and Luis Alberto Galaz Mamani for his assistance with the preparation of the figures shown herein. We are grateful to professors Victor Juliano De Negri and Jonny Carlos da Silva for revising and editing the technical contents of the subjects presented in this chapter.

REFERENCES

1. Nancy R. *The Quality Toolbox*. ASQ Quality Press, 2004, Second Edition. RP 247–49.
2. Vincent J. Tarascio. Pareto's methodological approach to economics; a study in the history of some scientific aspects of economic thought. Studies in Economics and Business Administration, v. 6. University of North Carolina Press. 1968.
3. NRC (Nuclear Regulatory Commission)-USA (United States of America). *WASH-1400 (NUREG 75/014): Reactor Safety Study: An Assessment of Accident Risks in U.S. Commercial Nuclear Power Plants*, 1975.
4. Papazoglou, I.A. "Mathematical Foundations of Event Tree," *Reliability Engineering and System Safety*. Elsevier, Northern Ireland, vol. 61, no. 3, 1998, pp. 169–83.
5. Kumamoto, H. and Henley, E.J. *Probabilistic Risk Assessment and Management for Engineers and Scientists*, 2nd ed., New York: IEEE Press Marketing, 1996, ISBN 0780310047.
6. Ericson II, C.A. *Hazard Analysis Techniques for System Safety*. New Jersey: John Wiley & Sons Inc., 2005.
7. Leveson, N. *Safeware: System Safety and Computers*. New York: Addison-Wesley, 1995, p. 704, ISBN 0201119722.
8. IEC (Internationale Elektrotechnische Komission). *IEC 61078: Analysis Techniques for Dependability – Reliability Block Diagram Method*. IEC, 2006. p. 57
9. Rausand, M. and HØyland, A. *System Reliability Theory: Models, Statistical Methods, and Applications*. Wiley-IEEE, 2004. ISBN 9780471471332. p. 636.
10. Biroolini, A. *Reliability Engineering: Theory and Practice*, 5th ed. Springer, 2007. ISBN: 8181284518.
11. Lewis, S. and Hurst, S. *Bow-Tie: An Elegant Solution? Strategic Risk*, p. 8, November 2005.
12. Ramzan, A. "The Application of Thesis Bow-Ties in Nuclear Risk Management," *The Journal of the Safety & Reliability Society*, UK Safety and Reliability Society, vol. 26, no. 1, 2006.
13. Trbojevic, V. "Linking Risk Analysis to Safety Management," in International Conference on Probabilistic Safety Assessment and Management (PSAM), 7. Berlin, Germany, 2004.
14. Calil, L. et.al., "CNEA (Causal Network Event Analysis): Proposta de Técnica de Análise de Risco." In: *7o Congresso Brasileiro de Gestão de Desenvolvimento de Produto*. São José dos Campos, SP: Instituto de Gestão de Desenvolvimento de Produto (IGDP), 2009.
15. Ale, B.J.M. "Accidents in the Construction Industry in the Netherlands: An Analysis of Accident Reports Using Storybuilder," *Reliability Engineering and System Safety*, 93, pp. 1523–33. Elsevier 2008. doi:10.1016/j.ress.2007.09.004.
16. Pearl, J. "Bayesian Networks: A Model of Self-Activated Memory for Evidential Reasoning (UCLA Technical Report CSD-850017)," *Proceedings of the 7th Conference of the Cognitive Science Society*, University of California, Irvine, CA. 1985, pp. 329–34.
17. Pearl, J. *Probabilistic Reasoning in Intelligent Systems: Networks of Plausible Inference*. Revised second printing. San Mateo, USA: MorganKaufmann Publishers Inc., 1988.
18. Jensen, F.V. *Reliability in Engineering Design*. New York: Springer-Verlag, 2001.
19. Boerlage, B. "Link Strength in Bayesian Networks," Dissertation (Master of Science), University of British Columbia Vancouver, Vancouver, Canada, 1994, p.104.
20. DOD (Department of Defense) - USA (United States of America), *MILSTD-1629A: Procedures for Performing a Failure Mode, Effects and Criticality Analysis*, Washington, 1980.
21. Bertsche, B. *Reliability in Automotive and Mechanical Engineering*. Springer, 2008. ISBN 978-3-540-33969-4.
22. SAE (Society of Automotive Engineers). *J1739: Potential Failure Mode and Effects Analysis in Design (Design FMEA), Potential Failure Mode and Effects Analysis in Manufacturing and Assembly Processes (Process FMEA), and Potential Failure Mode and Effects Analysis for Machinery (Machinery FMEA)*. [S.l.], 2002.
23. Sakurada, E.Y. *As Técnicas de Análise dos Modos de Falhas e seus Efeitos e Análise de Árvore de falhas no Desenvolvimento e na Avaliação do Produto*. Dissertação de Mestrado, Florianópolis: Universidade Federal de Santa Catarina, 2001.
24. Dias, A. *Notes on Hazard Analysis Methods Review*. Center for Technology Risk Studies, Internal Report, University of Maryland, US, 2003.
25. Yamaguchi, A. "Tribology of Hydraulic Pumps," in *Tribology of Hydraulic Pump Testing*, Totten, G.E., Kling, G.H. and Smolenski, D.J., eds., 1996, American Society for Testing and Materials; Conshohocken, PA, 1996, pp. 49–61.

26. Totten, G.E. and Bishop, R.J. "Evaluation of Vickers V-104 and 20VQ5 Vane Pumps for ASTM D-2882 Wear Tests Using Water–Glycol Hydraulic Fluids," in *Tribology of Hydraulic Pump Testing*, Totten, G.E., Kling, G.H. and Smolenski, D.J., eds., American Society for Testing and Materials; Conshohocken, PA, 1996, pp. 118–28.

27. Koc, E., Kurban, A.0. and Hooke, C.J. "An Analysis of the Lubrication Mechanisms of the Bush-Type Bearings in High-pressure Pumps", *Tribology International,* 1997, 30(8), pp. 553–60.

28. Vickers Inc., "Pump Failure Analysis", Vickers Incorporated; Troy, MI, 1991.

29. "Mobile Oil, Foaming and Air Entrainment in Lubrication and Hydraulic Systems," Mobil Oil Co., Ltd.; London, 1971.

30. Fowle, T.I. "Problems in the Lubrication Systems of Turbomachinery," *Proceedings of the Institution of Mechanical Enginners,* 1972, 186, pp. 705–16.

31. Rosean, B. *A Study to Assure Proper Pump Environment When Using Fire Resistant Fluids and Effect of Piping and Filters in Suction Lines,* The Rosean Filter Company; Hazel Park, MI.

32. Rosean Filter Co., *Piping Recommendations*, The Rosean Filter Company; Hazel Park. MI, 1990.

33. Ingvast, H. "Deaeration of Hydraulic Oil Offers Many Effects," in *Third Scandinavian International Conference on Fluid Power,* 1993, vol. 2, pp. 535–46.

34. Jackson, T.L. "An Introduction to Industrial Hydraulic Oils," in *Fluid Power International*; 1965, pp. 17–23.

35. Heald, C.C. *Cameron Hydraulic Data*, 18th ed., Ingersoll-Dresser Pumps: Liberty Corner, NJ, 1994, 67p.

36. Mackay, R.C. "Pump Suction Conditions," *Pumps and Systems Magazine,* May,1993, pp. 20–24.

37. Godfrey, D. "Gear Wear Caused by Contaminated Oils," *Gear Tech.*, September/October, 1996, pp. 45–49.

38. Newingham, T.D. "Selecting the Best Hydraulic Fluid," *Power Transmission Design*, October, 1986, pp. 27–31.

39. Danfoss Fluid Power, *Failure Analysis-Hydraulic Gear Pumps*, Danfoss Fluid Power; Racine, WI, 1990.

40. Caterpillar Inc., *Diagnosing Tyrone Gear Pump Failures*, Caterpillar, Tory, MI, 1990.

41. Vickers Inc., *Pump Failure Analysis,* Vickers Incorporated, Troy, MI, 1991.

42. Sasaki, A. "A Study of Hydraulic Valve Problems," *Lubrication Engineering*, 1989, 433. pp. 140–46.

43. Sasaki, A. "A Review of Contamination Related Hydraulic Pump Problems in Japanese Injection Molding, Extrusion and Rubber Molding Industries," in *Tribology of Hydraulic Pump Testing*, Totten, G.E., Kling, G.H. and Smolenski, D.J., eds., American Society for Testing and Materials; Conshohocken, PA, 1996, pp. 277–87.

44. Green, W.L. "Lightning in Hydraulic Oil," *Fluid Power Int.*, December, 1968, pp. 51–52.

45. Zino, A.J. "What to Look for in Hydraulic Fluids," *Iron Steel Engineering*, 1951, September, 1951, pp. 119–23.

46. Tiffany, H.E. "Trouble-Shooting Hydraulic Systems", in *Hydraulics*, Wambach, W.E. Jr., ed., American Society of Lubrication Engineers; Park Ridge, IL, 1983, pp. 45–49.

47. Moyer, C.A. "Comparing Surface Failure Modes in Bearings and Gears: Appearances versus Mechanisms," *AGMA-American Gear Manufacturers Association*, Technical Paper 91 FTM 6, 1991.

48. Fitch, E.C., Hong, I.T. and. Xuan, J.L. "Abrasion Wear," *BFPR J*, 1988, 21, pp. 9–29.

49. Erichello, R. "Lubrication of Gears-Part 2," *Lubrication Engineering*, 1990, 46(2), pp. 117–21.

50. National Standard "Nomenclature of Gear Tooth Failures," *ANSI/AGMA-American National Standards Institute/ American Gear Manufacturers Association 1 10.04-1980*, AGMA-*American Gear Manufacturers Association*; Alexandria, VA, 1980.

51. Fitch, E.C., Hong, I.T. and Xuan, J.L. Adhesion Wear, *BFPR J.*, 1988, 21, pp. 31–45.

52. Caterpillar Inc., *Fundamentals of Applied Failure Analysis, Module 4: Analyzing Wear*, Caterpillar Inc.; Peoria, IL, 1990.

53. Caterpillar Inc., *Hydraulic Pumps and Motors: Applied Failure Analysis*, Caterpillar Inc.; Peoria, IL, 1990.

54. Caterpillar Inc., *Principles of Wear: Applied Failure Analysis*, Caterpillar Inc.; Peoria, IL, 1990.

55. Faure, L. "Different Types of Wear-How to Classify?" *American Gear Manufacturers Association*, Technical Paper 90 FTM 4 (1990).

56. Fitch, E.C., Hong, I.T. and Xuan, J. L. "Adhesion Wear," *BFPR J.*, 1988, 21, pp.93–106.

57. Fitch, E.C., Hong, I T. and Xuan, J.L. "Cavitation Wear," *BFPR J.*, 1988, 21, pp.107–18.

58. Hobbs, J.M. "Experience with a 20-KC Cavitation Erosion Test," in *Erosion by Cavitation or Impingement*, American Society for Materials Testing; Philadelphia, PA, 1967, pp. 159.

59. Tallian, T.E. *Failure Atlas for Hertz Contact Machine Elements*, American Society of Mechanical Engineers, New York, 1992.

60. Fitch, E.C., Hong, I.T. and Xuan, J.L. "Corrosion Wear," *BFPR J.*, 1988, 21, pp.119–28.

61. Scully, J.R. "Electrochemical," in *Corrosion Tests and Standards: Application and Interpretation,* Baboian, R., ed. American Society for Testing and Materials; Conshohocken, PA, 1995, pp. 75–90.
62. Henthorn, M. "Fundamentals of Corrosion: Part I," *Chemical Engineering*, May, 1971, pp. 127–32.
63. Snyder, C.E., Morris, G.J., Gschwender, L.J. and Campbell, W.B. "Investigation of Airforce Mil-H-5606 Hydraulic System Malfunctions Induced by Chlorinated Solvent Contamination," *Lubrication Engineering*, 1981, 37(8), pp. 457–61.
64. Phillips, W.D. "The Electrochemical Erosion of Servo Valves by Phosphate Ester Fire-Resistant Hydraulic Fluids," *Lubrication Engineering*, 1988, 44(9), pp. 758–67.
65. Beck, T.R. "Wear by Generation of Electrokinetic Streaming Currents," *ASLE Transactions,* 1983, 26(2), pp. 144–50.
66. Beck, T.R., Mahaffey, D.W. and Olsen, J.H. "Wear of Small Orifices by Streaming Current Driven Corrosion," *ASME Transactions, J. Basic. Eng.*, 1970, 92, pp. 782–91.
67. Beck, T.R., Mahaffey, D.W. and Olsen, J.H. "Pitting and Deposits with an Organic Fluid by Electrolysis and Fluid Flow," *Journal of The Electrochemical Society*, 1972, 119(2), pp. 155–60.
68. Beck, T.R., Curulla, J.F., Hainline, B.C., Lauba, A. and Sullivan, D.C. "Effect of Mixed Phosphate Ester Fluids on Aircraft Hydraulic Servo Valve Erosion," *SAE Technical Paper Series*, Paper 801 100, 1980.
69. Nelson, W.G. and Waterman, A.W. "Advances in Commercial Airplane Hydraulic Fluids," in *Meeting No. 76, SAE Committee A-6*, 1974.
70. Sachs, N.W. "Metal Fatigue," *Lubrication Engineering,* 1991, 47(12), pp. 977–81.
71. Jin, X.Z. and Kang, N.Z. "A Study on Rolling Bearing Contact Fatigue Failure by Macro-Observation and Micro-Analysis," *Proceedings Of the International Conference on Wear of Materials*, 1989, Denver, CO, Ludema, K., ed., American Society of Mechanical Engineers; New York, 1989, vol. 1, pp. 205–13.
72. Fitch, E.C., Hong, I.T. and Xuan, J.L. "Surface Fatigue Wear," *BFPR J.,* 1988, 21, pp. 47–62.
73. Talian, T.E. "On Competing Failure Modes in Rolling Contact," *ASLE Transactions*, 1967, 10, pp. 418–39.
74. Caterpillar Inc., *Diagnosing Hydraulic Pump Failures*, Caterpillar Inc.; Peoria, IL.
75. Townsend, D.P. and Shimski, J. "EHL Film Thickness, Additives and Gear Surface Fatigue," *Gear Technology,* May/June, 1995, pp. 26–31.
76. MacKenzie, K.D. "Why Bearings Fail," *Lubrication Engineering*, January, 1978, pp. 15–17.
77. Zaretsky, E.V. "STLE Life Factors for Roller Bearings," *Society of Tribologists and Lubrication Engineers*; Park Ridge, IL, 1992, pp. 199–201.
78. The Timken Co., *Bearing Maintenance Manual for Transportation Applications*, The Timken Company; Canton, OH.
79. The Barden Corp., *Bearing Failure Causes and Cures*, The Barden Corporation; Danbury, CT.
80. Caterpillar Inc., *Anti-friction Bearings: Applied Failure Analysis*, Caterpillar Inc.; Peoria, IL.
81. Godfrey, D. "Recognition and Solution of Some Common Wear Problems Related to Lubricants and Hydraulic Fluids," in *Starting from Scratch -Tribology Basics*, Litt, F., ed., Society of Lubrication Engineers, Park Ridge, IL, pp. 11–14. (Access in 2010 November) (www.stle.org/assests/doccument/starting_from_scratch.pdf).

13 Petroleum Oil Hydraulic Fluids

*John R. Sander**

CONTENTS

---

* This chapter is a revision of the chapter titled "Mineral-Oil Hydraulic Fluids" by Paul McHugh, William D. Stofey and George E. Totten from the *Handbook of Hydraulic Fluid Technology*, 1st Edition.

13.1 INTRODUCTION

There are several ways in which hydraulic oils are classified or described for marketing purposes. The two major class descriptions for hydraulic fluids are based upon the majority ingredient, the base fluid. These classes are mineral oils and synthetics. As a general class, mineral oil hydraulic fluids are the most common hydraulic fluids used in the fluid power industry. The main reasons for this include: tradition, comparative low cost, availability, good viscosity and viscosity–pressure properties, excellent lubricity, suitability for high temperature use, metal and seal compatibility, and resistance to corrosion. Disadvantages of mineral oil hydraulic fluids include: flammability characteristics, low biodegradability and ecotoxicity problems when leaked into open water or underground aquifers, poor compatibility with metalworking formulations, and—in some cases—poor viscosity-temperature properties when used over a broad temperature range [1].

The goal of this chapter is to take an in-depth look at mineral hydraulic oil properties, formulation, and production. The application of mineral oil hydraulic fluids will also be discussed. Specific subjects that will be addressed are: mineral oil composition and refining, hydraulic oil classification, physical properties, chemical properties, an overview of additive chemistry, performance testing, biodegradability, compatibility, oxidative stability, and recycling.

13.2 DISCUSSION

When mineral oil-based hydraulic fluid is formulated, it is composed of various ingredients. These ingredients can be broken down into two very general categories: base oil and additives. It was already noted that mineral oil base hydraulic oils are classified by their base fluid type. Yet, there are also common classifications based upon additive content. These classification descriptions include rust and oxidation (R&O), and antiwear (AW). Antiwear hydraulic fluids are then broken down into categories of zinc-containing and ashless. Finally, there are straight-grade and multigrade classes of mineral hydraulic oils. Various tests are employed both to evaluate the base fluids and the fully-formulated hydraulic fluids. Subsequent sections of this chapter will go into greater depth in regard to the comparisons between these types and how to discriminate them.

13.3 TERMS COMMONLY USED WHEN DISCUSSING MINERAL HYDRAULIC OILS

Ashless: Contains no metallic additives that contribute to ash in certain tests.

Antiwear: Additives or formulated fluid that provides reduced frictional properties enabling it to reduce metallic wear.

Base stock or base oil: Oil produced in the refining process, before blending and without additives.

Bright stock: Fully refined and dewaxed, high-viscosity lubricating oil produced from the residue (bottoms) of the vacuum distillation column.

Cracking: Breaking down of large oil molecules into smaller, lower-viscosity, lower boiling point molecules. At the same time, some of the more reactive molecules combine (polymerize) to give even larger molecules, forming tar and coke.

Cylinder stock: The residue remaining after the lighter portions of the crude oil have been removed by evaporation.

Distillate: Product of oil distillation collected by vapor condensation.

Finishing: Chemical-, solvent- and hydrogen-refining processes for removing undesirable components and improving quality, stability and appearance of the base oil.

Fraction or cut: The portion of a crude oil with a certain boiling temperature range separated out in the distillation process. Also applied to the part separated by precipitation, crystallization, and so forth.

Lubricating base oil: A range of hydrocarbons derived from atmospheric distillation of petroleum. They have high boiling points between 350° and 500°C, typically 25–35 carbon atoms per molecule and molecular weights between 400 and 550 (bright stocks 750). Corresponding viscosities would be 30 and 95 (bright stock 400) CST at 40°C. Typical base oil would be composed of 70% paraffinic, 25% naphthenic, and less than 5% aromatic hydrocarbons. There would be traces of organosulfur, organonitrogen, and organooxygen compounds.

Middle distillates: Boiling fractions from vegetable, animal or synthetic oils.

Multigrade: Oil that contains polymeric viscosity-modifying additives called viscosity index improvers that expand the viscometric operating temperature characteristics of the oil.

Petrolatum (petroleum jelly): A mixture of oil and microcrystalline wax obtained from petroleum residue by propane precipitation.

Rust and oxidation (R&O): A fully-formulated hydraulic oil that contains additives to to defend the oil and metal surfaces from oxidation (rust is metal oxidation).

Solvent neutral oil (SNO): Vacuum-distilled paraffinic base oil refined by solvent extraction containing no free acidity, and which has been dewaxed and finished ready for use in blending. Referred to as neutrals or SNO 40, 100, 150, 320, 850, and so on, where the number refers to SUS (Saybolt Universal Seconds at 37.8°C). For example, SNO 150 oils have viscosities of approximately 33 cS, and SNO 850 approximately 190 cS at 40°C.

Straight-grade: Formulated hydraulic fluid that contains no viscosity index improver additive.

Wax: High molecular weight hydrocarbons that separate from oil when the temperature is lowered below the pour point. Paraffinic waxes are crystalline, long, straight-chain, normal paraffins with 20–35 carbon atoms per molecule, molecular weights of 300–550 and melting points between 40°C and 70°C. Microcrystalline waxes with 35–75 carbon atoms per molecule and average molecular weights of 600–1000 are primarily branched-chain iso-paraffins with some naphthenic hydrocarbons. Melting points are between 55°C and 95°C. Wax is the material in asphaltene-resin-free oil, which is insoluble in the solvent methyl ethyl ketone (MEK).

13.4 MINERAL HYDRAULIC OIL BASE FLUID

13.4.1 Petroleum Mineral Oil Composition

Petroleum crude oil is pumped from below the earth's surface. It is a complex mixture of hundreds of compounds including solids, liquids, and gases which are separated through a process called "refining." Solid components include petroleum coke, asphalt bitumen, and inorganic materials. Liquids of increasing viscosity vary from gasoline, kerosene (paraffin oil), diesel oil, engine crankcase oil, light and heavy machine oil, and cylinder oils. Also included are materials that are gases at room temperature, often called "natural gas": methane, ethane, propane, and butane [2].

TABLE 13.1
Crude Oil Content and Suitability

Crude Type	Solvent Neutral Base Oil	Base Oil	Specialty Oil	Wax Content	S and N Content[a]	Asphalt	API Gravity[b]
Paraffinic base	Yes	Yes	No	Yes	Low	No	>40
Naphthenic base	No	Yes	Yes	No	Low	No	<33
Mixed base	No	Yes	No	Yes	Low	Yes	33 – 40
Asphaltic base	No	Yes	No	No	High	Yes	–

[a] S is sulfur and N is nitrogen.
[b] API gravity is scale in degrees adopted by the American Petroleum Institute where ° API = [141.5/specific gravity]-131.5.

Crude oils are often classified according to the amount of sulfur content as either sweet or sour crude (low sulfur—sweet, high sulfur—sour). When the crude is to be refined into mineral oil it is classified as: paraffinic, napthenic, intermediate (mixed), or asphaltic, as shown in Table 13.1 [2]. Paraffinic fractions are saturated linear or branched alkanes. Naphthenic fractions contain cyclic alkane (alicyclic) structures. Generic examples of typical chemical structures of these classes are illustrated in Figure 13.1 [3,4]. Also included in Figure 13.1 are aromatic (polyunsaturated cyclic) chemical structures which are derived from benzene:

In lubricating oil, aromatic components exhibit a greater tendency to form sludge and varnish than do paraffinic or naphthenic derivatives [5].

Paraffinic base oils contain 45%–60% paraffinic compounds. Naphthenic base oils contain 65%–75% of naphthenic compounds, and aromatic base oils contain 20%–25% aromatic compounds [3].

The relationships between density and molecular carbon content for paraffinic, naphthenic, and aromatic constituents of mineral oil are summarized in Table 13.2 [2].

Aromatic asphaltic materials are classified as resins, asphaltenes, or carbenes, depending on their solubility [2]:

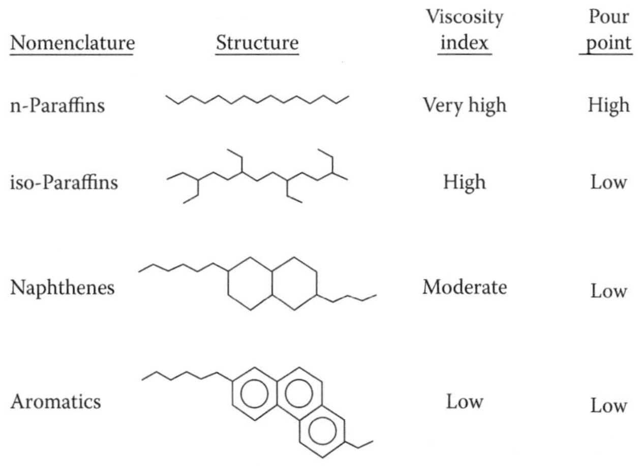

Nomenclature	Structure	Viscosity index	Pour point
n-Paraffins		Very high	High
iso-Paraffins		High	Low
Naphthenes		Moderate	Low
Aromatics		Low	Low

FIGURE 13.1 Typical organic structures present in a mineral oil.

TABLE 13.2
Density and Molecular Carbon Content

Base Oil Type	Paraffinic Carbon	Naphthenic Carbon	Aromatic Carbon	Density (kg/l)
Paraffinic	65–70	25–30	3–8	0.800–0.812
Naphthenic	50–60	30–40	8–13	0.834–0.844
Mixed	48–57	24–33	17–22	0.850–0.872
Aromatic	21–35	20–30	40–50	0.943–1.005

- Resins—pentane or heptane soluble
- Asphaltenes—pentane or heptane insoluble, benzene soluble.
- Carbenes—benzene insoluble, carbon disulfide soluble

In addition to paraffinic, naphthenic, and aromatic derivatives, petroleum oils also contain cyclic and heterocyclic derivatives of nitrogen, sulfur, and oxygen compounds such as carboxylic acids, amides, and aldehydes. Selected examples of some of the polar derivatives that may be found in petroleum oils are illustrated in Figure 13.2 [2–4]. Traces of phenolic and furan derivatives may also be present. A summary of the overall effect of these different components on the physical and chemical properties on the mineral oil are provided in Table 13.3 [3].

Yoshida et al. have shown that heterocylic nitrogen compounds provide a pro-oxidant effect. This effect is a function of the basicity of the nitrogen in the compound. Generally, only the more basic nitrogen heterocyclic compounds, pyridine, and quinoline derivatives, exhibit a pro-oxidant effect. This effect is accelerated in the presence of copper. The proposed mechanism involved a "ligand effect" through complexation with alkyl hydroperoxides that form during the oxidative break down of mineral oils [82].

The chemical composition of a petroleum base oil is often determined from physical properties; refractive index (n_D^{20}) at 20°C, density at 20°C (d_4^{20}), percentage sulfur concentration (S), and

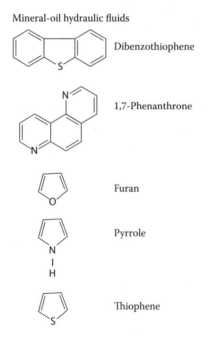

FIGURE 13.2 Illustration of heterocyclic derivatives typically present in a mineral oil.

TABLE 13.3
Effect of Composition on Base Stock Properties

Chemical Component	Viscosity Index	Pour Point (High/Low)	Oxidative Stability	Response to Antioxidants	Volatility (High = Poor Low = Good)
n-Alkane	Very High	High	Good	Good	Good
iso-Alkane	High	Low	Good	Good	Good/Average
Naphthene	Low	Low	Average	Good	Average
Aromatic	Low	Low	Average/Poor	Some Poor	Poor
Polar Compounds	Low	Low	S is Antioxidant; N and O are Pro-oxidant	Poor	Poor

average molecular weight (M), or by the "n-d-M Method" described in ASTM D 3238 [37]. Once these properties are measured the following computations can be performed:

Step 1: Calculate the factors v and w.

$$v = 2.51\left(n_D^{20} - 1.4750\right) - \left(d_4^{20} - 0.8510\right),$$

$$w = \left(d_4^{20} - 0.8510\right) - 1.1\left(n_D^{20} - 1.4750\right).$$

Step 2: Calculate % aromatic carbon (C_A).
If v is positive:

$$\%C_A = 430_v + \frac{3660}{M}.$$

If v is negative:

$$\%C_A = 670_v + \frac{3660}{M}.$$

Step 3: Calculate % total aromatic + naphthenic carbon (C_R).
If w is positive:

$$\%C_R = 820W - 3S + \frac{10,000}{M}.$$

If w is negative:

$$\%C_R = 1440W - 3S + \frac{10,600}{M}.$$

Step 4: Calculate % naphthenic carbon (C_N) and paraffinic carbon (C_P)

$$\%C_N = \%C_R - \%C_A,$$

$$\%C_P = 100 - \%C_R.$$

Figures 13.3a–c illustrate the effect of C_A, C_R, and C_N on the viscosity index, aniline point and volatility (K_{UOP}) [6,38]. Research has shown that base oil properties may be estimated from aromaticity since aromaticity is highly dependent on the degree of refining [6,39].

Two important parameters that relate to polar compound composition and reflect the base oil quality, are sulfur content and aniline point. Some sulfur-containing polar compounds have been found to provide significant antioxidant benefits. Therefore, base oils containing low sulfur contents may require additional antioxidants.

The aniline point is determined by mixing the base oil with the polar solvent aniline. The mixture is then heated until a homogenous solution is obtained. Next, the solution is allowed to cool. The temperature where the base oil separates from the solution is the "aniline point." As the paraffinic/naphthenic ratio increases the aniline point also increases [3].

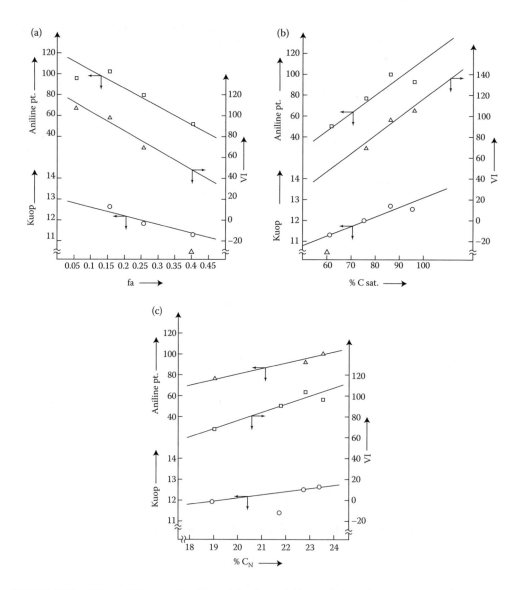

FIGURE 13.3 Effect of (a) aromatic, (b) naphthenic, and (c) parafinic carbon content on viscosity index, aniline point, and volatility of a mineral oil.

TABLE 13.4
Process Affects on Base Oil Composition and Properties

Constituent	De-asphalting	Distilling	Dewaxing	Finishing
Asphaltenes	*	*	+	*
Resins	*	*	+	*
Aromatics	*	*	+	V
Naphthenes	+	V	+	V
Paraffins	+	V	*	+
Wax Content	+	V	*	N
Nitrogen	*	*	+	*
Sulfur	*	*	+	*
Properties				
Specific Gravity	*	*	+	V
Flash Point	*	*	+	V
Viscosity	*	*	+	V
Viscosity Index	+	+	+	V
Pour Point	+	*	*	V
Color	*	*	*	*
Stability	+	+	V	+
Additive Response	+	+	V	+

Legend: * decreases + increases V Variable N no change

13.4.2 Petroleum-Refining Processes

Crude petroleum must be refined before it is used for lubricant formulation. In this section, distillation, de-asphalting, solvent extraction, hydrocracking, dewaxing, and hydrofinishing processes will be introduced. Table 13.4 provides a summary of the effects that de-asphalting, distilling, dewaxing, and finishing processes have on the composition and physical properties of mineral oil [41].

Distillation. The distillation of crude oil is conducted in different steps (see Figure 13.4) [4]. The first step is to remove inorganic salts. Then, after heating to approximately 350°C under pressure,

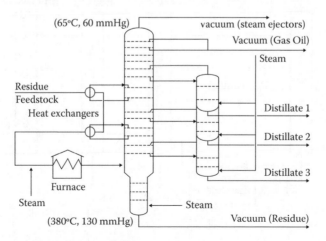

FIGURE 13.4 Vacuum distillation unit for lubrication oils.

an atmospheric distillation is conducted at which point the most volatile components are fractionated in the distillation column. Typically, fractions including C1 to C4 hydrocarbons, gasoline, naphtha, and kerosene are removed at this point [2]. The next step is a vacuum distillation process which is conducted at 10–40 mm Hg or 20–100 mm Hg in combination with steam up to 450°C. The fractions removed by vacuum distillation include light, medium and heavy neutral lubricating oil fractions.

De-asphalting. High-viscosity base oil may be removed from the residue remaining after distillation by a de-asphalting process. A volatile, low-molecular weight hydrocarbon, such as liquefied propane, is used to dissolve the bright stock, leaving an asphaltic residue (bitumen) behind.

Solvent extraction. To improve physical properties such as viscosity index and oxidative stability, solvent extraction is used to reduce aromatic content of the base stock obtained after distillation as shown in Figure 13.5 [2]. Some naphthenic derivatives and other undesirable materials are also removed. Solvents used for the extraction process are miscible with aromatic derivatives and include: phenol, furfural and N-methylpyrolidone (NMP). Of these solvents, furfural has traditionally been the most commonly used solvent [7].

Hydrocracking. Due to the high cost of solvents and improvement in refining technology, many refiners are now using a process called "hydrocracking" in lieu of solvent extraction. Hydrocracking increases paraffinic oil content, removes aromatics and napthenics, removes sulfur and nitrogen functional groups, and increases oxidation stability.

Hydrocracking is the most severe part of the base oil refining process. Hydrocracking is conducted by heating a base oil/hydrogen mixture to 370°C at 20 MPa pressure and passing the mixture over a catalyst bed [2]. The exothermic hydrogenation reactions are cooled by hydrogen. Examples of hydrocracking reactions are provided in Figure 13.6. During hydrocracking, large condensed aromatics are converted into more desirable hydroaromatics, the occurrence of cracking/dealkylation during hydrogenation leads to the formation of various carbonyl-containing compounds, and significant denitrogenation and desulfurization results [8].

As noted above, due to increased availability, there is increasing interest in the use of hydrocracked base stocks in hydraulic fluids. Their use has provided improved viscosity, thermal stability, and biodegradability properties [40].

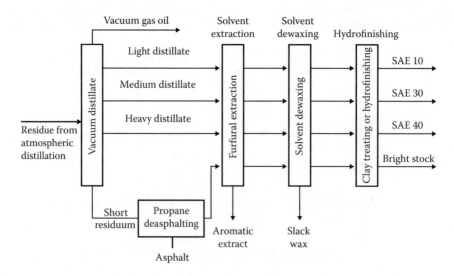

FIGURE 13.5 Simplified flow chart for solvent refining process.

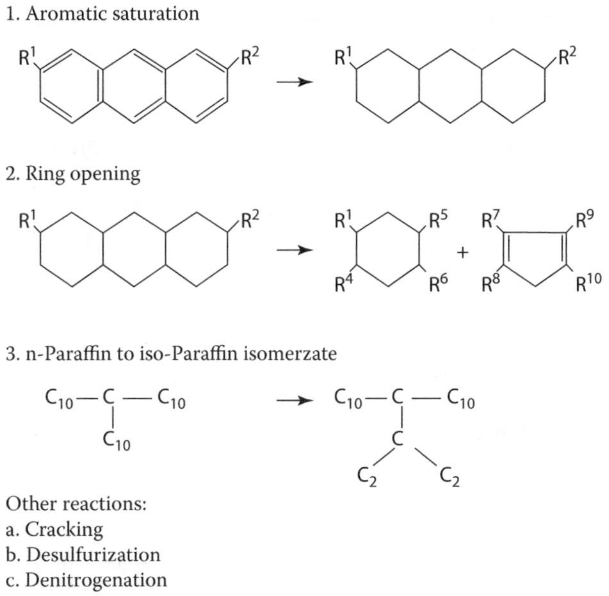

1. Aromatic saturation

2. Ring opening

3. n-Paraffin to iso-Paraffin isomerzate

$$C_{10}-\underset{\underset{C_{10}}{|}}{C}-C_{10} \longrightarrow C_{10}-\underset{\underset{C_2}{|}}{C}-C_{10}$$

Other reactions:
a. Cracking
b. Desulfurization
c. Denitrogenation

FIGURE 13.6 Illustrative hydrocracking reactions.

Clay treatment. Base oil color and thermal stability can be improved by a process called "clay treatment." In this process, the oil is percolated through clay (or Fuller's earth), which removes polar compounds by adsorption [2].

Dewaxing. Cold-flow properties of paraffinic base oils may be improved by the removal of waxy fractions in the oil. This is accomplished through several techniques.

The traditional technique is known as solvent dewaxing. This process involves dissolving the oil in a mixture of methyl ethyl ketone (or methyl isobutyl ketone) and toluene solvent, chilling the mixture, filtering the insoluble waxes out of the chilled mixture, and finally removing of the solvent mixture from the dewaxed base oil. [7]

Due to the increasing prices of solvents and improvements in catalyst and hydrotreating technologies, many modern refiners are now using processes called catalytic dewaxing. Essentially, catalytic dewaxing involves changing the molecular arrangement of the wax molecules which makes them more like average oil molecules. This is done by passing the hydrocracked oil through a catalyst material that chemically shortens the molecular length of waxy materials, then alkylating it to add side chains, and finally subjecting them to a hydrofinishing step to stabilize the molecules.

Hydrofinishing. The remaining polar compounds and further reduction in the aromatic content of base oils may be accomplished by catalytic hydrogenation at an elevated temperature. This will produce a base stock with improved color and oxidative stability. Hydrofinishing has largely replaced clay treatment in most modern refineries [2], although in some refining processes both hydrofinishing and clay treatment are performed [7].

13.4.3 BASE OIL ANALYSIS AND CHARACTERIZATION PROCEDURES

Numerous standardized testing procedures exist to characterize petroleum products, including hydraulic oils [10,11]. There are various organizations that produce and maintain these standards, but the following are some of the more well-known: ASTM International, the International Organization for Standards (ISO), Deutsches Institut fur Normung (DIN), and the Society of

Automotive Engineers (SAE). A detailed description of these procedures is beyond the scope of this text. There are also numerous analytical techniques that are being employed, or could be used, to both characterize specific chemical and physical properties of hydraulic oils and their components.

Nuclear magnetic resonance (NMR) spectroscopy. Both [1]PMR (Proton Magnetic Resonance) and [13]CMR (Carbon magnetic resonance) spectroscopy have been used to examine structure-performance relationships of petroleum oil base stocks [6,8,12]. NMR techniques are often employed to generically classify the paraffinic, naphthenic and aromatic content of oil for subsequent physical property studies. Additional tests are also available to elucidate various chemical moieties in these very complex mixtures.

Fourier transform infra-red spectroscopy (FTIR). FTIR spectroscopy can be used along with [13]CMR spectroscopy to conduct a relatively rapid characterization of the aromatic content of petroleum base oils [6]. Singh and Swaroop used FTIR to analyze the relative impact of aromatic carbon content on base stock oxidative stability [13].

Gas chromatography (GC). GC is a separation technique in which a sample is injected into a heated port attached to a separation column. The oil sample is heated until it boils, separating the oil into fractions that are qualified and quantified as the oil passes through the column and through a detector. One of the most common detectors are flame ionization detectors (FID) but some methods also used thermal conductivity detectors as well.

Thin layer chromatography (TLC). Thin layer chromatography using a flame-ionization detector (TLC-FID) has been used to quantitatively determine the ratios of paraffinic, naphthenic, aromatic and polar components in base oil [14]. It was proposed that this procedure was superior to the open-column gel chromatographic procedure of ASTM D 2007 [15].

Gel permeation chromatography (GPC). GPC analysis uses a diode-array detector and a RI (differential refractance) detector to identify and quantify fractionated oil components and characterize the aromatic functionality of the petroleum oil [16].

Combination of chromatography and spectral methods. Some have found advantages in employing some of the chromatographic techniques noted above in tandem with spectroscopic techniques. NMR and FTIR units have successfully been used to structurally identify the eluents obtained by chromatographic analysis such as high pressure liquid chromatography (HPLC), NMR [17], and mass spectrometrometry [18]. These analytical procedures provide much finer discrimination of the chemical structures present in a petroleum base oil. That information could be used to consider performance interdependence with respect to additive interactions, and so on.

13.4.4 HYDRAULIC OIL CLASSIFICATION

Sometimes hydraulic fluids are classified by application as mobile or industrial. These names are used to describe whether the fluid is either for use in moving equipment or stationary equipment. Since the same products can often be used in both applications, not much emphasis will be placed upon this type of classification. It has been noted that hydraulic oils are classified first as oil-based or synthetic-based, or simply as mineral oil or synthetic. Here, the focus is mineral oil hydraulic fluids, which are then classified by their additive makeup. Broadly considered, they are recognized as either R&O or antiwear. Sub-classifications of the antiwear type oils are recognized as ashless or zinc-containing. There are industrial detergent oils, which contain dispersant and detergent additives for some equipment that uses the oil in dirty environments or in combined hydraulic and engine applications. Finally, there are certain formulas that now contain viscosity index improvers to give them broader operating temperature properties. Each of these classes will be described subsequently.

13.4.5 CLASSIFICATION BY BASE OIL TYPE

It was previously described that mineral oil hydraulic fluid is a major classification of hydraulic oil. It turns out that there are classifications for mineral oil base stocks which are used for hydraulic fluid formulation. The main classifications are naphthenic or paraffinic [19]. However, mixed-base fluids are also available [20].

Naphthenic base stocks contain a high percentage of ring-containing cycloaliphatic compounds. Typically, these fluids exhibit relatively low viscosity indices and therefore are not useful for applications where wide use-temperatures will be encountered. These oils exhibit relatively low pour points. Generally, their maximum use-temperature is 150°F (121°C). These fluids are available in a full range of viscosities, exhibit good demulsibility characteristics, and resist foaming [19]. Hydrofinished naphthenic oils have found some application in incidental food contact hydraulic fluids.

Paraffinic oils contain a high percentage of straight-chain and branched aliphatic compounds. These fluids exhibit higher viscosity indices than naphthenic oils and therefore are more suitable for use at higher temperature. These oils exhibit somewhat higher pour points than naphthenic oils, although the gap is less with the increasing use of catalytic dewaxing. Properly stabilized, paraffinic hydraulic fluids may be used continuously at temperatures up to 285°F (285°C) [19].

Mixed-base oils, also referred to as "intermediate viscosity index oils," are blends of naphthenic and paraffinic base stocks. They may be used at either high or low temperatures, although they do not exhibit the best features of either of their component oils [20].

API classification has been used in recent years to describe base oils that are used in many industrial fluids, including mineral hydraulic fluids. There has been evolution in the way that base oils are refined from crude oil, driven primarily by increased demands by the engine crankcase oil market—the highest volume segment of the lubricant industry. New governmental requirements to reduce tailpipe emissions have forced refiners to refine base oils with improved oil volatility properties. To do this, the refiners have increased the use of severe hydrocracking and hydrofinishing processes. In 1993, the American Petroleum Institute (API), produced Publication 1509 "Engine Oil Licensing and Certification System," which provided a framework upon which engine oils could be classified. Table 13.5 is a reproduction of API base stock categories [154]. Today, many lubricant formulators will even reference these oil categories when formulating industrial oils, such as hydraulic oils.

TABLE 13.5
API Paraffinic Base Stock Categories

Group	Sulfur, Mass %	Saturates, Mass %	Viscosity Index
I	>0.03	<90	≥80 to <120
II	≤0.03	≥90	≥80 to <120
III	≤0.03	≥90	≥120
IV	All Polyalphaolefins		
V	All stocks not included in groups I–IV		

TABLE 13.6
Selection Chart Hydraulic Oils and Fluids

System Requirements	Straight Naphthenic	Straight Parafinnic	R&O Quality & Antiwear, Paraffinic	R&O Less-Than-Turbine Quality, Paraffinic Base	R&O Less-Than-Turbine Quality, Paraffinic Base	R&O & Antiwear Detergent, Paraffinic Base	Antileak, Naphthenic Base	Antileak & Antiwear, Naphthenic Base	R&O Quality & Antiwear Naphthenic Base	Combination Way-lube & Hydraulic Oil
Normal service life			X	X		X	X	X		
Long service life			X	X	X	X	X	X	X	X
Antirust and corrosion			X	X	X	X	X	X	X	
Exceptional antiwear			X			X		X		
Antifoam properties	X	X	X	X	X	X	X	X	X	X
High-temperature operation			X	X		X				
Low-temperature startup	X									
Continuous contamination, extraneous materials	X					X				
Continuous contamination, cutting fluids, with frequent changes	X									
Moderate leakage		X					X	X	X	
Excessive leakage	X						X	X		
Tight system, little or no makeup			X	X	X	X			X	
Active filter use	X	X								X
Inactive filter use	X	X	X	X	X	X	X	X	X	X
Way lubrication by hydraulic oil			1			1		1	X	
High degree of water separation	X	X	X	X						
Hydraulic system cleaner						X				

1 Will handle non-critical way lubricant

13.4.6 CLASSIFICATION BASED ON ADDITIVES

Hydraulic oils may also be classified based on the additive protection that they offer [20]. These include: rust-and oxidation (R&O), antiwear (AW), extreme-pressure (EP), ashless, and industrial detergent oils. A comparison of the advantages and disadvantages of these different oils is provided in Table 13.6 [20].

R&O hydraulic oils may be formulated with naphthenic, paraffinic, or mixed-base oils. Additives are used to provide rust and antioxidant protection.

Antiwear oils may be formulated with naphthenic, paraffinic, or mixed-base oils. They may also contain R&O inhibitors. They will contain an antiwear additive such as ZDDP (zinc dialkyldi-thiophosphate), which provides additional antiwear protection toward metal parts in the hydraulic system.

Extreme-pressure (EP) fluids are formulated to provide lubrication to both the hydraulic system components and the machine-tool ways to prevent stick-slip operation. Typically, these fluids only offer antirust protection without antileak, antioxidant or viscosity index (VI) improvers [20].

Ashless hydraulic oils are typically formulated without a zinc-containing additive and do contain phosphorus, nitrogen, and sulfur-containing additives for antiwear performance.

Industrial detergent oils are used in mobile equipment and robotic applications that are subjected to dirty environments. Some are even used in combination engine and hydraulic service. These fluids are typically formulated to provide R&O protection, clean and disperse contaminates, exhibit a high viscosity index, and are suitable for use over a broad range of temperatures.

13.4.7 CLASSIFICATION BASED ON VISCOSITY

Hydraulic fluids may also be classified as straight-grade or multi-grade oils. As has been noted previously, sometimes viscosity index improvers are added to increase the viscosity indices, which results in expanded operating temperature ranges.

Straight-grade oils are often classified according to ASTM D 2422 "Viscosity System for Industrial Fluid Lubricants," which is also often called the "ISO viscosity grade scale" [2,21]. The viscosity classification system is provided in Table 13.7. It should be noted that not all of the viscosity grades included in this table are commonly used in hydraulic fluids. This classification system only provides information on the viscosity at 40°C. Viscosities at other temperatures are dependent on the viscosity index properties of the fluid.

Multigrade oils are usually used in mobile hydraulic applications where dramatic temperature variations could be experienced. In these cases, the fluids are sometimes recognized by SAE viscosity grades described in the SAE J300 "Viscosity Classification" and shown in Table 13.8. As this table indicates, these categories require specific performance at high and low temperatures, thus the name multigrade. The important thing to point out is that the kinematic viscosity measurement

TABLE 13.7
ASTM D 2422 Viscosity Classification System

ISO Viscosity Grade	Kinematic Viscosity Limits cSt(mm²/s) at 40°C		
	Midpoint	Minimum	Maximum
ISO VG 2	2.2	1.98	2.42
ISO VG 3	3.2	2.88	3.52
ISO VG 5	4.6	4.14	5.06
ISO VG 7	6.8	6.12	7.48
ISO VG 10	10	9	11
ISO VG 15	15	13.2	16.5
ISO VG 22	22	19.8	24.2
ISO VG 32	32	28.8	35.2
ISO VG 46	46	41.4	50.6
ISO VG 68	68	61.2	74.8
ISO VG 100	100	90	110
ISO VG 150	150	135	165
ISO VG 220	220	198	242
ISO VG 320	320	288	352
ISO VG 460	460	414	506
ISO VG 680	680	612	748
ISO VG 1000	1,000	900	1,100
ISO VG 1500	1,500	1,350	1,650

TABLE 13.8
SAE J300 Viscosity Classification

SAE	Low-Temperature Viscosities		High-Temperature Viscosities		
Viscosity Grade	Cranking[a] (mPa.s) max at temp°C	Pumping[b] (mPa.s) max at temp°C	Kinematic[c] (mm²/s) at 100°C		High Shear[d] Rate (mPa.s) at 150°C
			min	max	min
0W	6200 at −35	60 000 at −40	3.8	−	−
5W	6600 at −30	60 000 at −35	3.8	−	−
10W	7000 at −25	60 000 at −30	4.1	−	−
15W	7000 at −20	60 000 at −25	5.6	−	−
20W	9500 at −15	60 000 at −20	5.6	−	−
25W	13 000 at −10	60 000 at −15	9.3	−	−
20	−	−	5.6	<9.3	2.6
30	−	−	9.3	<12.5	2.9
40	−	−	12.5	<16.3	2.9
40	−	−	12.5	<16.3	3.7
50	−	−	16.3	<21.9	3.7
60	−	−	21.9	<26.1	3.7

[a] ASTM D 5293
[b] ASTM D 4684
[c] ASTM D 445
[d] ASTM D 44683 or ASTM D 4741

is conducted at 100°C in contrast to 40°C for straight grade industrial oils. It is also important to note that not all of these grades are commonly used to classify hydraulic oils. Multigrade mineral hydraulic fluids will almost always have to contain viscosity index improvers for the oil to meet both high and low temperature requirements.

13.4.8 CLASSIFICATION SYSTEMS BASED UPON MAKEUP AND PERFORMANCE

Mineral oils may also be classified according to an ISO system which is based on the properties exhibited by the fluid [22,23]. The ISO classification categories are provided in Table 13.9. ASTM D 6158

TABLE 13.9
ISO 6743/4 Mineral Oil Classification System

Category	Composition Typical Properties	Applications Operating Temperatures
HH	Non-inhibited refined mineral oils	−10 to 90°C
HL	Refined mineral oils with improved antirust and antioxidant properties	−10 to 90°C
HM	Oils of HL type with improved antiwear properties	General hydraulic systems which include highly-loaded components −20 to 90°C
HR	Oils of HL type with improved viscosity/ temperature properties	−35 to 120°C
HV	Oils of HM type with improved viscosity/ temperature properties	Mobile applications −35 to 120°C
HS	Synthetic fluids with no specific fire-resistant properties	−35 to 120°C

TABLE 13.10
History of Hydraulic Fluids

Years	Type of Hydraulic Fluid	Remarks
1920	Water-based	Restricted working temperature, rust/corrosion problems
1920	Plain mineral oils	Gum formation, acidity buildup resulting in corrosion
1940	R&O oils	Wear problems with vanes pumps of sophisticated design, output of 1000 psi
1950	Motor oil-based	Excellent wear protection, poor demulsibility
1960	Conventional ZDDP-based oils	High wear of bronze piston shoes
1970	Thermally stable ZDDP-based oils	Meeting all international specifications
1980	Ashless (S-P-N)-based oils	Commercialized mostly for use in environmentally sensitive areas
1990	Multi-Grade	Increase operating temperatures
2000	Severely hydro-treated oils	Engine oils push large share of refinery production to Hydro-cracking for volatility

"Standard Specification for Mineral Hydraulic Oils" builds upon ISO 6743 by using the same classification convention, though it includes its own set of testing parameters for fluid performance as well [153].

13.5 BRIEF HISTORICAL OVERVIEW AND SYSTEM REQUIREMENTS

Table 13.10 provides a timeline for critical developments in hydraulic fluid technology since 1920 [24]. Although past problems have been successfully addressed, from this table, it is evident that fluid formulation and performance continue to provide significant challenges for the hydraulic fluid industry.

13.6 HYDRAULIC OIL PHYSICAL PROPERTIES

Knowledge of the physical properties of a hydraulic oil formulation is important in the design and operation of a hydraulic system. These properties include viscosity, viscosity index, pour point, compressibility, gas solubility, and foaming, in addition to various thermal and electrical properties [2,25]. In this section, the specific importance of these physical properties will be described.

13.6.1 VISCOSITY

Viscosity is the single most important physical property exhibited by a hydraulic fluid. If the hydraulic fluid viscosity is too high:

1. Flow resistance will be increased as the fluid passes through the clearances in the pump and valves.
2. System temperature will increase due to lack of lubrication.
3. System operation will then become sluggish.
4. There will be an increased pressure drop in the system.
5. In the end, power consumption will increase.

If the fluid viscosity is too low:

1. Internal and external leakage will increase.
2. Pump slippage will increase, thus reducing efficiency and increasing temperature.
3. The wear rate will increase due to thin hydrodynamic film strength of the fluid.
4. Pressure loss will occur.
5. Pressure loss results in loss of system control [29].

TABLE 13.11
Recommended Maximum Inlet Viscosities for Different Types of Hydraulic Pumps[a]

Pump Type	Maximum Inlet Viscosity, (mm²/s, cS)
Gear Pumps	2,000
Piston Pumps	1,000
Vane Pumps	500–700

[a] These are only general recommendations. They will vary with specific pump design and the manufacturer of the pump.

The maximum fluid viscosity for proper pump operation is dictated by the hydraulic pump design and is therefore manufacturer-specific. Some general guidelines are provided in Table 13.11. Pump manufacturers publish minimum and maximum specifications for their pumps [26–28]. These specifications typically include: minimum, optimum, and maximum startup viscosity. [26] An illustration of a recommended viscosity-temperature chart for hydraulic oil is provided in Figure 13.7 [26].

13.6.2 VISCOSITY INDEX

The viscosity of oils will vary with temperature, and that variation is dependent on oil composition. The magnitude of viscosity–temperature variation is quantified by the use of a single number called the "viscosity index" (VI). The calculation of VI from viscosity–temperature data is illustrated in Chapter 4. As indicated in Figure 13.7, the lower the VI the greater the viscosity varies with temperature. Most pump manufacturers recommend that the VI of hydraulic oil be at least 78 [27,28]. The VI value of an oil is especially important since it is used to determine the temperature range over which the oil may be used.

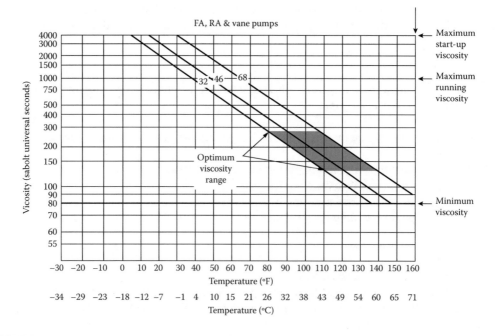

FIGURE 13.7   Vane pump viscosity recommendation chart for Racine FA, RA, and K vane pumps.

13.6.3 Viscosity-pressure Behavior

A general property of fluids is that they exhibit an increase in viscosity with increasing pressure. Interestingly, this viscosity increase is not linear. The viscosity becomes even greater at higher pressures than at lower pressures [29]. In addition, the viscosity increase with pressure is greater for naphthenic oils than it is for paraffinic. The effect of higher pressure becomes more exaggerated as the fluid viscosity increases. Thus, the result of the increasing fluid pressure is an increase of the VI of the oil [29]. Although a fluid viscosity increase can also cause even more of an increase in temperature, this effect is usually not significant up to system pressures of 2000 psi [19].

13.6.4 Compressibility and Bulk Modulus

Fluid compressibility (bar⁻¹) is defined by the following equation:

$$b = -\frac{1}{V}\left[\frac{dV}{dP}\right],$$

where b is the fluid compressibility, V is the initial volume, and P is fluid pressure. Another term, compressibility module (M), or bulk modulus, is occasionally used [25]:

$$M = \frac{1}{P},$$

where M is reported in pressure units of bar. The effect of pressure and temperature on the relative volume of a mineral oil (D.T.D. 585) is illustrated in Figure 13.8 [30]. Comparisons of isoentropic and isothermal bulk moduli as a function of temperature are illustrated in Figure 13.9.

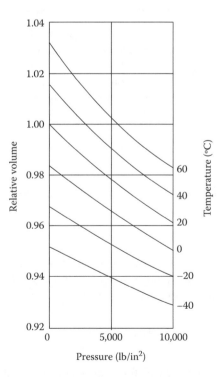

FIGURE 13.8 Effect of temperature and pressure on relative volume.

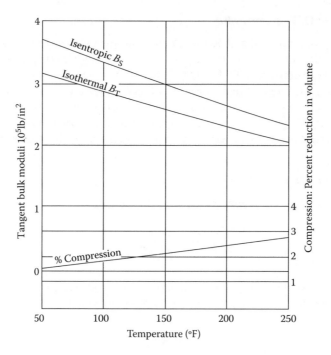

FIGURE 13.9　Isoentropic and isothermal tangent bulk modulus for a typical Shell Tellus oil at 5000 psi.

13.6.5　Estimation of Relative Cavitation Potential from Rho-C Values

Cavitation in a hydraulic system is a process by which gas bubbles in hydraulic fluid rapidly collapse and cause a shockwave. As many of these shockwaves occur over time, cavitation results in fatigue wear which can cause premature failure of hydraulic pump parts. One method that has been used to estimate the potential for cavitation is the "rho-c value" or "acoustical impedance." The isoentropic bulk modulus (M_I) is related to the velocity of sound (C) through the fluid and density (ρ) by:

$$M_r = \rho C^2.$$

Generally, the potential for a fluid to cavitate increases with the rho-c value [25,31]. Figure 13.10

13.6.6　Gas Solubility

Because of the agitation produced by a hydraulic pump during hydraulic system operation, it is possible for gases to find their way into the oil. The amount of dissolved gas in a fluid is related to the fluid pressure (P) and the solubility (Bunsen) coefficient (V_{gas}) of the gas by:

$$V_{gas} = V_{liquid}\alpha P.$$

Table 13.12 provides Bunsen coefficients for a mineral oil at 0, 50 and 100°C [25].

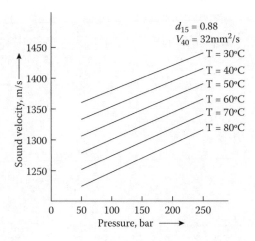

FIGURE 13.10   Sound velocity as a function of temperature and pressure (ISO 32 mineral oil).

13.6.7   AIR ENTRAINMENT

When the concentration of air in a hydraulic system exceeds the saturation level, the gaseous bubbles are suspended in the hydraulic fluid. Air entrainment must be minimized for proper fluid performance in the hydraulic pump. Air entrainment properties of a hydraulic fluid may be determined according to ASTM D 3427. In this test, air is blown into the fluid through a gas inlet at the bottom of a glass vessel at a constant flow rate and fluid temperature for seven minutes. The change in fluid density relative to the starting fluid condition is recorded until the total volume of entrained air is equal to 0.20% of the original volume. The time required to obtain this volume is known as the "gas separation time."

13.6.8   FOAMING

Once entrained air accumulates within a hydraulic fluid it eventually reaches a point where it tries to dissipate out the top of the fluid surface. Depending on the surface tension properties at the air-oil interface, sometimes oil-encased gas collects on the oil surface and forms a frothy look on the fluid surface. This froth is called foam. Foaming is illustrated in Figure 13.11. Excess foam can arise

TABLE 13.12
Bunsen Coefficients for Different Gasses in Mineral Oil as a Function of Temperature

Gas	Bunsen Coefficient $(V_{gas})^a$		
	0°C	50°C	100°C
Air	0.092	0.091	0.091
Oxygen	0.150	0.137	0.130
Nitrogen	0.081	0.088	0.090
Hydrogen	0.047	0.053	0.067
Carbon Dioxide	1.000	-----	-----

[a] Higher values of V_{gas} are obtained with low-viscosity fluids.

FIGURE 13.11 Hydraulic oil foam formation. (Courtesy of CITGO Petroleum Oil Company, Houston, TX.)

from various sources including: fluid contamination, poor reservoir design, and the formation of metallic soaps such as might be formed from the reaction of additives in the fluid or with lead coatings [30]. The propensity for foaming is measured according to ASTM D 892 [47]. In this test, air is injected into the hydraulic fluid at a constant temperature through a gas diffusion stone at a constant rate, as illustrated in Figure 13.12. The foam height is measured immediately at the conclusion of the fluid aeration period and timed to dissipation or once again measured after standing unagitated for ten minutes. The foam retention characteristics, measured in millimeters, are proportional to the propensity for foaming.

13.6.9 DENSITY

Fluid density at a given pressure (ρ_p) is determined from: [25]

$$\rho_P = \frac{\rho}{100 - \Delta V_P},$$

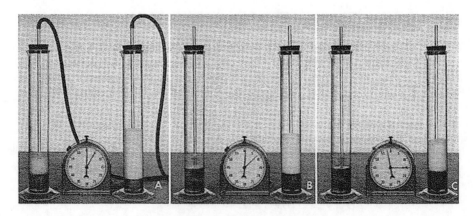

FIGURE 13.12 Illustration of ASTM D 892 foam stability test to measure foam formation, stability, and effectiveness of antifoam addition. (Courtesy of Shell Lubricants, Houston, TX.)

where: ρ is the density at atmospheric conditions and Δ, V_P is the volumetric change with increasing pressure (P)

The density of mineral oils may also be represented as $^\circ$ API, which is defined as [32]:

$$\text{Degrees API} = \frac{141.5}{\text{Sp.Gr. @ 15.6/15.6}^\circ\text{C (60/60}^\circ\text{F)}} - 131.5,$$

where Sp.Gr. is the specific gravity of the fluid.

13.6.10 POUR POINT

The pour point is the lowest temperature at which the fluid will flow [2]. The pour point is important to consider for a fluid to be used in mobile equipment operations where the equipment must be started after setting outside in cold weather. The oil used in such operations must have a pour point below the lowest startup temperature anticipated. However, pour points must be considered with the fluid viscosity within the pump manufacturer's highest recommended startup viscosity [33].

Pour points may be determined by various standard procedures, both manually and automatically, including: oil fluidity [42], automatic tilting method [43], auto-pulsing method [44], and the rotational method [45].

13.6.11 THERMAL EXPANSION

Since oil may expand as it heats up during use, it can create a potential overflow condition. Thus, it is necessary to account for the thermal expansion when designing a hydraulic system. Thermal expansion is expressed as the coefficient of thermal expansion, which is the ratio of volume change after heating to the initial volume after heating 1°C. It is expressed as °C^{-1}. The coefficient of thermal expansion for a mineral oil hydraulic fluid is approximately 6.4×10^{-4} °C^{-1} [32]. A procedure for obtaining more accurate values for thermal expansion is provided by ASTM D 1250 [46].

Figure 13.8 illustrates the change in relative volume of a mineral oil hydraulic fluid as a function of temperature and pressure. Figure 13.13 provides an estimate of the effect of pressure on the coefficient of thermal expansion [30].

13.6.12 THERMAL CONDUCTIVITY

Thermal conductivity is a measure of the rate of heat transfer through hydraulic oil. Heat transfer rates increase with thermal conductivity. The thermal conductivity for most mineral oils is approximately 0.1 Wm^{-1} °C^{-1} [32].

13.6.13 HEAT CAPACITY AND SPECIFIC HEAT

Heat capacity is the amount of heat which is required to raise the temperature of a given mass by one degree (°C or °F). The units of heat capacity are: Joules per kilogram degree Celsius. The specific heat of oil is the ratio of heat capacity of oil to that of water at 15°C (1 J kg°C^{-1}). The values of heat capacity and specific heat are numerically equal. Values for mineral oil hydraulic fluids have been reported to be approximately 0.444–0.456 [30].

13.6.14 ELECTRICAL CONDUCTIVITY/DIELECTRIC STRENGTH

It is not uncommon for mineral oil hydraulic fluids to be employed in mobile service equipment used by the electrical utility industry. As such, it is important to know the electrical conductivity or dielectric strength of the hydraulic fluid. Electrical conductivity is the amount of electricity per

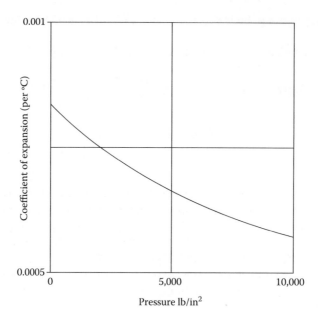

FIGURE 13.13   Effect of pressure on the coefficient of thermal expansion of D.T.D. 585.

unit area transferred through the oil at a given voltage [34]. The unit of electrical conductivity is mho.cm−1. The unit mho is the reciprocal of resistance (1/ohm = mho). A clean and dry mineral oil typically exhibits an electrical conductivity of 10^{-14} mho.cm. Dielectric constants for a mineral oil have been reported to be: 2.25 at 50°C, 2.21 at 75°C, and 2.17 at 100°C [30].

13.6.15 Flash Points and Fire Point

The flash point is defined as: "the temperature to which a fluid must be heated to give off sufficient vapor to momentarily form a flammable mixture with air when a small flame is supplied under specified conditions" [36]. An open-cup flash point test is conducted in the open air [50]. A closed-cup flash point test is conducted in a closed vessel [51]. The fire point is defined as the lowest temperature at which the fluid will sustain burning for five seconds in the ASTM D 92 test [50].

Although flash points and fire points do not affect the operation of in-service hydraulic fluid, these values should be considered in system design safety. In some applications there is a possibility of hot spots in the system where the fluid temperature may exceed the flash point or fire point of the fluid [36,49]. For example, an air pocket may develop when a source of ignition is present, such as air accumulation in the housing of a pump seal where the resulting air lock may cause oil starvation and a localized temperature rise which may create a condition sufficient for an explosion to occur [31]. Alternatively, a piece of hydraulic equipment may be performing its function in an area near a flame source or an extremely hot process function where misted lubricant spraying from a leaky hose produces an aerosol of oil that could result in a fire.

13.6.16 Seal Compatibility

If optimal seal strength is to be maintained and system leakage is to be minimized, the potential for a mineral oil to cause swelling or shrinkage of a seal is important. One of the oldest methods of estimating hydraulic seal compatibility is to measure the aniline point of the fluid. Generally, fluids with low-aniline points cause the highest degree of swelling to neoprene and BUNA-N nitrile rubber seals, as shown in Figure 13.14 [31,117]. Unfortunately, a direct correlation has not been made between the aniline point and actual swelling of various elastomeric seal composition with respect to different base oil and additive combinations [52]. Seal materials that are most prone to encounter problems with mineral oil hydraulic fluids are often made from natural rubber. Neoprene and nitrile seals, however, are generally acceptable for use with mineral oil hydraulic fluids [53].

Another method that was used was the seal compatibility index [52]. This method involves using a standard nitrile rubber (AHEM—American Hydraulic Equipment Manufacturers, Standard Nitrile Rubber No. 1) and measuring the volume uptake of the hydraulic fluid at equilibrium. The test specimens may be O-rings (1± 0.01 in., 25 mm ± 0.25 mm) or squares (0.060 ± 0.006 in., 1.5 ± 0.15 mm) [52]. The specimens are immersed into the test oil in a jar and placed in an oven at 100°C/24h ± 2°C for 24 hours after the initial ID is measured (D_1). After cooling in cold oil, excess oil is removed and the ID of the ring is measured (D_2) using a tapered gauge. The AHEM Seal Compatibility Index (SCI) is the percentage of volume swelling, to the nearest whole number, obtained from linear swelling (S_D) and the standard conversion values in Table 13.11 [52].

$$S_D = \frac{D_2 - D_1}{D_1} \times 100.$$

The SCI values range from 0–50 as shown in Table 13.13. While an interesting concept, the use of SCI values has not gained acceptance in the industry. Currently, the testing methodology recommended in ASTM D 6546 "Standard Test Methods for and Suggested Limits for Determining

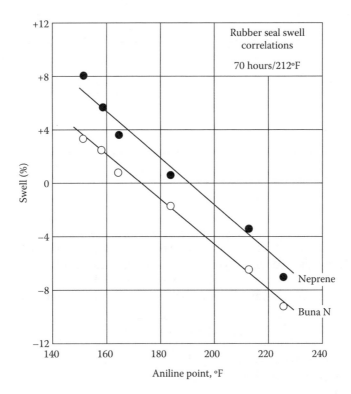

FIGURE 13.14 Correlation of volume swell and aniline point for neoprene and BUNA-N rubber.

TABLE 13.13
Conversion to Linear Swell to Seal Compatibility Index

SD %	D_2–D_1 Inch	SCI	SD %	D_2–D_1 Inch	SCI	SD %	D_2–D_1 Inch	SCI
0.2	0.002	0.6	5.2	0.05	16.4	10.2	0.1	33.8
0.4	0.004	1.2	5.4	0.05	17.1	10.4	0.1	34.6
0.6	0.006	1.8	5.6	0.05	17.8	10.6	0.11	35.3
0.8	0.008	2.4	5.8	0.05	18.4	10.8	0.11	36.1
1	0.01	3	6	0.06	19.1	11	0.11	36.8
1.2	0.012	3.6	6.2	0.06	19.8	11.2	0.11	37.5
1.4	0.014	4.3	6.4	0.06	20.5	11.4	0.11	38.3
1.6	0.016	4.9	6.6	0.06	21.2	11.6	0.12	39
1.8	0.018	5.5	6.8	0.06	21.8	11.8	0.12	39.8
2	0.02	6.1	7	0.07	22.5	12	0.12	40.5
2.2	0.022	6.8	7.2	0.07	23.2	12.2	0.12	41.3
2.4	0.024	7.4	7.4	0.07	23.9	12.4	0.12	42.1
2.6	0.026	8	7.6	0.07	24.6	12.6	0.13	42.8
2.8	0.028	8.6	7.8	0.07	25.3	12.8	0.13	43.5
3	0.03	9.3	8	0.08	26	13	0.13	44.3
3.2	0.032	9.9	8.2	0.08	26.7	13.2	0.13	45.1
3.4	0.034	10.6	8.4	0.08	27.4	13.4	0.13	45.8
3.6	0.036	11.2	8.6	0.08	28.1	13.6	0.14	46.6
3.8	0.038	11.8	8.8	0.08	28.8	13.8	0.14	47.4
4	0.04	12.5	9	0.09	29.5	14	0.14	48.2
4.2	0.042	13.2	9.2	0.09	30.3	14.2	0.14	49
4.4	0.044	13.8	9.4	0.09	31	14.4	0.14	49.7
4.6	0.046	14.4	9.6	0.09	31.7	14.6	0.15	50.5
4.8	0.048	15.1	9.8	0.09	32.4	14.8	0.15	51.3
5	0.05	15.8	10	0.1	33.1	15	0.15	52.1

Compatibility of Elastomer Seals for Industrial Hydraulic Fluid Applications" is much more commonly employed [155].

13.6.17 DEMULSIBILITY

Water contamination of hydraulic fluids can occur when there is a humid atmosphere above the hydraulic fluid in the reservoir, constant air leakage into the hydraulic circuit, or seepage of the heat exchanger coolant into the hydraulic fluid [56]. The contaminated oil may then form an emulsion via the high fluid shear motion produced in a hydraulic pump, which churns together the fluid and water [1]. Water-contaminated hydraulic systems may lead to numerous problems, such as:

- Formation of "thin and watery" or "thick and pasty" fluids
- Dirt and dust contamination
- Increase in the occurrence of valve malfunction
- Increase in wear and corrosion
- Oil oxidation
- Additive depletion
- Foaming and filter plugging [1,56]

Therefore, all possible attempts should be made to keep water contamination out of a hydraulic pump. However, if water contamination does occur, it must be removed as soon as possible.

Highly refined mineral oils are hydrophobic, but additives, or oil oxidation byproducts such as acids, may promote the emulsification of water in the fluid. An early test for water contamination enhanced by oil oxidation involved the measurement of the surface tension of the fluid. It was proven that as the fluid aged and oxidation occurred, surface tension of the fluid decreased, as is illustrated in Figure 13.15 [56]. Oil replacement is recommended when the surface tension is ≤15 dyne/cm.

Currently, many evaluate hydraulic oil's water separation properties by performing ASTM D 1401 [54,55]. Water separation ability is determined by mixing 40 mL of distilled and 40 mL of the oil in a 100 mL graduated cylinder at 54°C as illustrated in Figure 13.16. The time for complete separation is measured in 5-min intervals. Typically, oils that exhibit separation times of ≤30 min are suitable for continued use [1]. Factors that affect the accuracy of the test include: cleanliness of the test equipment, speed and position of the stirrer, bath temperature, test duration, and stray vibrations [55].

13.6.18 Filterability

Some hydraulic oils are formulated to contain additives, such as viscosity modifiers [58], which exist as very small particles in the fluid. Although these particles may not be visible to the naked eye, they may be removed by high efficiency fine (3μ) filtration [57]. These fine particles will actually cause plugging of the filters. Therefore, it is desirable to evaluate the "filterability" of a hydraulic fluid as both a measure of potential filter plugging and additive removal.

An evaluation of fluid filterability may be performed using the "Pall Filterability Test" [58,59,75]. This involves using a pump stand to pass the test fluid through a filter under a partial vacuum. In this test, the time it takes to plug the filter is determined. An alternative test currently under evaluation involves passing the fluid through a membrane instead of a filter on a pump [58].

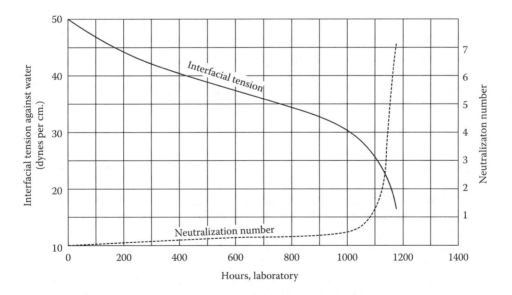

FIGURE 13.15 Effect of oil oxidation on surface (interfacial) tension.

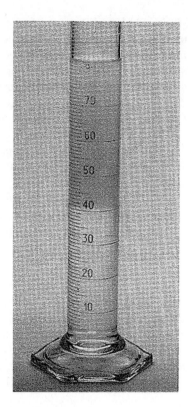

FIGURE 13.16 Illustration of ASTM D 1401 oil demulsibility test. (Courtesy of The Lubrizol Corporation, Wickliffe, OH.)

13.7 HYDRAULIC OIL PERFORMANCE PROPERTIES

13.7.1 Discussion Corrosion Properties

Although the terms "rusting" and "corrosion" are often used interchangeably, they are fundamentally different processes with different fluid formulation challenges [63]. Rusting is caused by the reaction of water and air with a ferrous surface in the hydraulic system [62]. Corrosion can be caused by the reaction of organic acids in the presence of peroxides with metals. Both originate from oxidation of the hydraulic oil [49]. Corrosion can also be caused by overly active additives. For example, some sulfur-containing wear-reduction additives can produce a harmful, dark, sulfide layer on copper parts in hydraulic systems.

13.7.2 Rust

Rust results from the buildup of an iron oxide film, typically accompanied by an increase in weight; while corrosion results in a weight loss and surface pitting [49,62]. Rusting is encountered when systems are contaminated with water and air, which then results in foaming [62]. This may lead to flaking of the rust into the fluid. In addition to causing abrasive wear, the presence of rust particles in the oil may plug passages or damage valves [33,68]. Rust can also catalyze oil oxidation [68]. Typically, protection against both corrosion and rusting requires the use of two inhibitors—oxidation inhibitors and rust inhibitors [62].

Although refined mineral oils exhibit poor rust preventative characteristics, rust inhibition additives are blended into mineral hydraulic fluids to provide preventative enhancement. There are two types of rust prevention additives: the type that protect a metal surface when water is introduced directly into the oil, and the type that provide a thin rust inhibitor coating in the airspace near the oil. The first type is called "rust-inhibited" oil. The second type are often referred to as "rust-preventatives." They may not necessarily be for in-service use, but rather as a preservative after equipment manufacture [57].

13.7.3 CORROSION

Corrosion occurs in systems that have undergone significant oil oxidation. It can also result from the use of an improperly formulated lubricant. Highly refined mineral oils that contain oxidation inhibitor additives typically exhibit improved corrosion prevention properties. These additives protect the oil from the formation of the carboxylic acids and oxidation byproducts that cause corrosion [49,68]. Metal deactivator additives can also be used to minimize the affects of overly aggressive oil ingredients. Chemical corrosion may also occur when the hydraulic oil is contaminated with halocarbons such as trichloroethane, which are sometimes used in plants as degreasing agents for parts. When this occurs, components with small orifices such as hydraulic actuators may fail due to corrosion and deposit formation [64]. When subjected to water contamination, heat and pressure, the presence of as little as 0.02% of the halocarbon may hydrolyze in the hydraulic oil to yield hydrochloric acid, which will then attack and etch the metal surface, forming iron chloride ($FeCl_3$). Iron chloride, along with the water contaminant in the oil, will catalyze the oxidative degradation of the oil and will result in the formation of weak organic acids, sludge, and varnish, which will form a brown, sticky residue on the metal surface.

13.7.4 TESTING FOR CORROSION AND RUSTING

A common test which is used to evaluate rust inhibition properties of hydraulic fluids is ASTM D 665, although other tests such as the "Fog (or Humidity) Cabinet," solubility, and colloidal stability, and the static water drop corrosion test have been reported [65]. One of the common corrosion tests is ASTM D 130, which measures the corrosion properties of mineral oil hydraulic fluids on copper-containing metals [66]. Only ASTM D 130 and ASTM D 665 will be discussed here since they are the most commonly utilized tests for mineral oil qualification [36].

13.7.5 RUST-PREVENTING CHARACTERISTICS TEST

Assuming that there would be cases where hydraulic oil would become contaminated with water, ASTM D 665 is a test method that is used to evaluate the ability of a mineral oil hydraulic fluid to inhibit rusting of ferrous surfaces. There are two variations of this procedure, A and B. Procedure A involves stirring 300 mL of the hydraulic oil with 30 mL of distilled water at 60°C (140°F) for four hours with a cylindrical steel rod immersed in the oil as shown in Figures 13.17 and 13.18. After the test is completed, the steel rod is inspected for the presence of rust. Procedure B is identical to procedure A except that synthetic seawater is used instead of distilled water. Some of the test variables to be controlled include: pH of the water, type and age in addition to grinding and drying of the steel rod (spindle), fitting of the rod in the plastic stem, stirring quality and magnitude, bath temperature, and, possibly, the laboratory temperature and humidity [55].

13.7.6 COPPER STRIP TARNISH TEST

ASTM D 130 has been recommended for the evaluation of the potential of oil to corrode copper [36]. This test is conducted by immersing a polished copper strip into 30 mL of the hydraulic oil in a 25×150 mm test tube, which is then placed in a stainless steel test bomb. After sealing, the test tube bomb assembly is heated in a constant temperature bath at 100°C for three hours. The test tube is then removed from the bomb after cooling in cold water and the cooper strip is removed. The degree

FIGURE 13.17   Illustration of the ASTM Oil Rust Testing Apparatus. (Courtesy of Falex Corporation, Aurora, IL.)

of corrosion is determined by a comparison to ASTM Copper Strip Corrosion Standards, which are available from ASTM and illustrated in Figure 13.19.* Provisions are made in the test procedure for higher test temperatures if more rigorous conditions are desired.

13.7.7 OXIDATION STABILITY

Oxidation stability refers to the "ability of hydraulic fluid to resist polymerization and thermal decomposition in the presence of air, water, heat and dissimilar materials" [49]. Additional variables that affect oil oxidation rates include: ambient temperatures, atmospheric conditions, contamination, oil viscosity, type of pump, pump pressures, and pressure cycling [63].

Oil decomposition products are either soluble or insoluble. Soluble oil degradation products generally thicken the oil [71]. Insoluble products form sludge deposits in the oil, which can lead to filter

---

* The ASTM Copper Strip Corrosion Standards may be obtained from ASTM Headquarters, 100 Barr Harbor Drive, West Conshohocken, PA 19428.

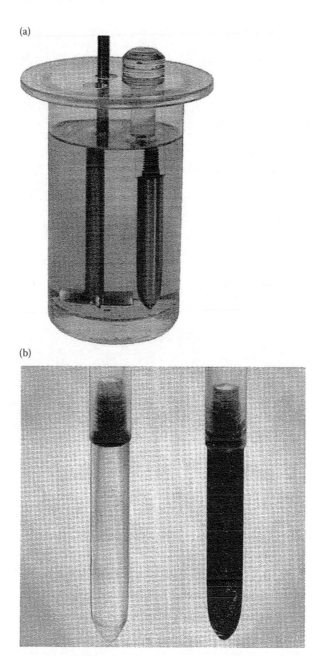

(a)

(b)

FIGURE 13.18 (a) Illustration of the oil rust testing cell showing agitator and steel rod, (b) Comparison of a good and bad rust testing result. (Courtesy of Shell Lubricants Houston, TX.)

clogging, valve and piston sticking, hydraulic lock [60,61,84], increased wear, and corrosion [71]. Oil degradation byproducts, water, wear debris, and corrosion debris often catalyze the overall oxidative degradation process. Table 13.14 shows that improperly inhibited hydraulic oil exposed to either water *or* a metal catalyst (copper or iron) accelerated degradation. However, when both water *and* the metal are present, the oil degradation rate was ten times greater with iron and thirty times greater with copper [70].

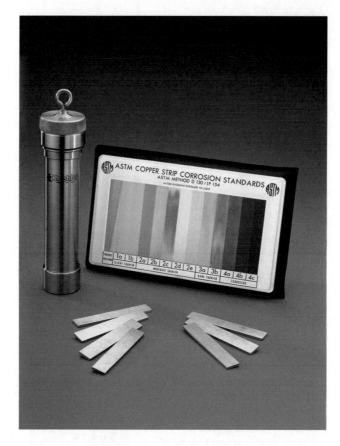

FIGURE 13.19 Illustration of the ASTM Copper Strip Corrosion Standards. (Courtesy of Koehler Instruments Inc., Bohemia, NY.)

The oxidative stability of hydraulic oil is largely a function of the base stock selected. For example, paraffinic oils exhibit better oxidative stability than naphthenic oils. Unfortunately, they also exhibit poorer "solvency," meaning less ability to dissolve additives [90]. The presence of alkylaromatic derivatives and some polar components further increase the susceptibility of petroleum oil to oxidation while others exhibit a stabilizing effect [73].

TABLE 13.14
Effect of Water and Catalysts on Oil Oxidation[a]

Catalyst	Water	Hours	Final TAN
None	No	3500	0.17
None	Yes	3500	0.9
Iron	No	3500	0.65
Iron	Yes	400	8.1
Copper	No	3000	0.89
Copper	Yes	100	11.2

[a] Tests were run at 200°F on a 150 SUS (at 100°F) turbine grade oil according to ASTM D 943 test procedure.

Temperature exhibits the greatest impact on oxidative stability. On the basis of temperature-viscosity performance, optimum fluid temperature ranges are often between 90°F and 160°F (32–71°C) [49]. Generally, below oil temperatures of 135°F (57°C), oil oxidation rates are relatively slow. However, above 145°C the oxidation rate doubles for each 10°C rise in temperature [71,76]. Some of the greatest thermal stress on the fluid may occur in valves where the fluid is forced through small orifices at high velocities. This will produce localized thermal excursions due to fluid friction [63]. It has been shown that even at oil temperatures as low as 100°F (38°C) the presence of varnish deposits on valves have indicated the presence of localized temperatures in excess of 200°F (93°C) [63].

Foaming and air entrainment increase oxygen contact with the fluid, and thus accelerate oxidative degradation. This is even further exacerbated at high temperatures. Oxidative degradation rates are dependent on the amount of dissolved and entrained air that is present in the fluid. Dissolved gas is released from the fluid as it flows through orifices within the system. Downstream pressure re-dissolves the air which is accompanied by high localized temperatures, since the time interval is too short for the bubbles to re-dissolve before compression. This facilitates the oxidation process [57]. For example, Table 13.15 illustrates that bubble compression temperatures, starting from an inlet temperature of 100°F (38°C), may be as high as 2100°F (1150°C) for a pump pressure of 3000 psi. In fact, in addition to oxidative degradation, thermal cracking and nitration of the oil are also possible [69].

Backé and Lipphardt studied the effect of entrained air on mineral oil oxidation. One of the variables examined was the effect of entrained air on the rate of oxidation by measuring the change in neutralization number over a certain amount of running time. Figure 13.20 shows that there is an initial oxidation rate that is controlled by the amount of air dissolved in the oil [151]. As the concentration of entrained air increases, the number of un-dissolved air bubbles increases. Un-dissolved air bubbles mean more oxygen is available to react with the oil, resulting in an increase in the rate of neutralization number increase [151]. In a hydraulic system, the high temperatures produced by compression of the air bubbles facilitate oil oxidation at the air-bubble-oil interface, thus accelerating the overall oil oxidation process. This process is retarded when a nitrogen atmosphere is provided or if antioxidants are used.

Backé and Lipphardt also showed that conditions such as dramatic pressure increases and bubble size can cause ignition to occur at the air-bubble/oil interface. This is called the "micro-dieseling effect" [151]. The conditions required to cause the micro-dieseling effect are illustrated by Figure 13.21 [151]. Based upon calculations, it was found that ignition would occur when the temperature on the bubble surface was typically 340–360°C. According to DIN 51,794, the oil had to be compressed when starting with oil at 25°C at compressed to 150,000 bar/sec in order to reach the "ignition temperature."

Roberton and Allen performed a detailed analysis of the potential for hydraulic system variables such as reservoir size, pump type, and so forth, to promote oil oxidation, thermal cracking, and nitration, and their conclusions were as follows [69]:

TABLE 13.15
Gas Bubble Compression Temperature
Variation with Pump Pressure

Pump Pressure (PSI)	Air Bubble Temperature (°F)[a]
1000	1410
2000	1820
3000	2100

[a] Calculations based on 100°F inlet temperature.

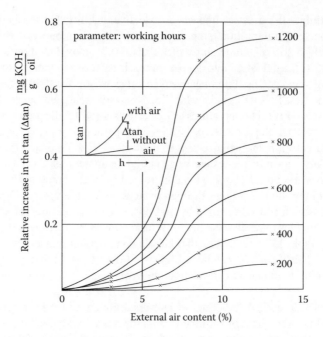

FIGURE 13.20 Relative increase in neutralization number with increasing air content.

1. Fluid aeration was the major factor contributing to accelerated oil degradation processes. Constant volume hydraulic pump systems with short residence times were the most prone to exhibit oil degradation and servo-valve problems.
2. It was shown that these problems could be modeled in the laboratory and were correlated to fluid air entrainment.
3. It is unlikely that laboratory bench tests for oxidation or thermal stability can completely assess the capabilities and limitations of a hydraulic fluid in the pump.
4. To minimize potential oxidative degradation, thermal cracking, and nitration problems with respect to hydraulic pump and servo-valve performance, it is recommended that you:

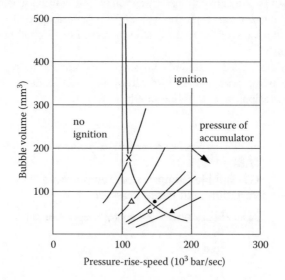

FIGURE 13.21 Backe and Lipphardt limiting curve for ignition during the compression of single air bubbles in petroleum oil.

a. Use variable volume pumps if possible;
b. Ensure that the reservoir volume is at least 2 1/2 times the pump capacity (This is especially important for constant volume pumps.);
c. Design reservoir and return lines to keep turbulence at a minimum and allow for oil deaeration;
d. Ensure that the bulk oil temperature is maintained below 120°F; and
e. Avoid the use of copper tubing.

Clearly, it is important that optimal separation of the entrained air be promoted, usually by the addition of an air release additive [62].

13.8 OIL OXIDATIVE DEGRADATION PROCESSES

Petroleum mineral oils degrade according to a reaction mechanism. This mechanism is driven by temperature, metal catalysts, and byproducts of oxidative degradation. The hydrocarbon degradation mechanism occurs in four steps:

1. Chain initiation
2. Chain propagation
3. Chain branching
4. Chain termination

Chain initiation is the first step, and it occurs when the oxygen partial pressure is greater than 50 torr. It is the slow, or rate determining, step that is catalyzed by trace concentrations of transition metals:

$$
\underset{\substack{|\\H}}{\overset{\substack{CH_3\\|}}{R-C-H}} \xrightarrow[k_1]{M^{N+},O_2} \underset{\substack{|\\H}}{\overset{\substack{CH_3\\|}}{R-C\bullet}} + HOO\bullet
$$

where R is an alkyl substituent, M^{N+} is a transition metal such as: Fe, Cr, and Cu. The rate of hydrogen radical abstraction follows the order:

$$
RCH_2-H < R_2CH-H < R_3C-H < RCH=\overset{\substack{H\\|}}{C}-\underset{\substack{|\\H}}{\overset{\substack{H\\|}}{C}}-R < \text{⬡}-\underset{\substack{|\\H}}{\overset{\substack{H\\|}}{C}}-R
$$

Chain propagation is the second step of the reaction mechanism. It is an irreversible reaction where the alkyl radical formed in Step 1 reacts with oxygen to form a peroxy radical:

$$
\underset{\substack{|\\H}}{\overset{\substack{CH_3\\|}}{R-C\bullet}} + O_2 \longrightarrow \underset{\substack{|\\H}}{\overset{\substack{CH_3\\|}}{R=C=OO\bullet}}
$$

The rate of reaction of the alkyl radical with oxygen follows the order:

$$
H_3C\bullet < \text{⬡}-\underset{\substack{|\\H}}{\overset{\substack{R\\|}}{C\bullet}} < RCH=\underset{\substack{|\\\bullet}}{\overset{\substack{H\\|}}{C}}-\overset{\substack{H\\|}}{C}-R < R_2\overset{\substack{H\\|}}{C}\bullet < R_3C\bullet
$$

Alkyl radical reactivity increases with increasing alkyl substitution on the carbon where hydrogen abstraction occurred. This is why branched hydrocarbons are more susceptible to oxidation than unbranched, n-paraffinic hydrocarbons. Chain propagation continues with the abstraction of a hydrogen radical by the peroxy radical to form a hydroperoxide:

$$\underset{\underset{H}{|}}{\overset{\overset{CH_3}{|}}{R-C-OO\bullet}} + \underset{\underset{H}{|}}{\overset{\overset{CH_3}{|}}{R-C-H}} \longrightarrow \underset{\underset{H}{|}}{\overset{\overset{CH_3}{|}}{R-C-OOH}} + \underset{\underset{H}{|}}{\overset{\overset{CH_3}{|}}{R-C\bullet}}$$

The relative rate of hydrogen abstraction by a peroxy radical is [80]:

	primary	secondary	tertiary
	$\underset{\underset{H}{|}}{\overset{\overset{H}{|}}{R-C-H}}$	$\underset{\underset{H}{|}}{\overset{\overset{R}{|}}{R-C-H}}$	$\underset{\underset{R}{|}}{\overset{\overset{R}{|}}{R-C-H}}$
Rel. Rate % H•Abstraction	1	30	300

There are numerous possibilities of intramolecular hydrogen abstraction which will not be discussed here. One possible intermediate from multiple intramolecular abstraction processes is [80]:

$$\underset{\underset{OOH}{|}}{\overset{\overset{H}{|}}{R-C}}-(CH_2)_x-\underset{\underset{R^2}{|}}{\overset{\overset{\overset{\bullet O}{\diagdown}O}{\|}}{C}}-R^1 \longrightarrow \underset{\underset{OOH}{|}}{\overset{\overset{\bullet}{}}{R-C}}-(CH_2)_x-\underset{\underset{R^2}{|}}{\overset{\overset{OOH}{|}}{C}}-R^1$$

Where X = 1,2

Hydroperoxides may undergo homolytic cleavage to yield an alkoxy and a hydroxy radical:

$$ROOH \longrightarrow RO\bullet + \bullet OH$$

Chain branching is the third step in the oxidation reaction mechanism:

$$HO\bullet + H_3C-R \longrightarrow H_2O + \bullet H_2CR$$

$$HO\bullet + R-CH_2R^1 \longrightarrow H_2O + R-\underset{\bullet}{C}H-R_1$$

$$RCH_2O\bullet + H_3C-R \longrightarrow RCH_2OH + R\underset{\bullet}{C}H_2$$

Secondary and tertiary alkoxy radicals preferentially form aldehydes and ketones:

$$\underset{\underset{R^1}{|}}{\overset{\overset{H}{|}}{R-C-O\bullet}} \longrightarrow \overset{\overset{O}{\|}}{R-C-H} + R^1\bullet$$

$$\underset{\underset{R^3}{|}}{\overset{\overset{R^1}{|}}{R^2-C-O\bullet}} \longrightarrow \overset{\overset{O}{\|}}{R-C-R^2} + R^3\bullet$$

Chain termination is the final step of the oxidation reaction mechanism. There are numerous possible radical combinations which will lead termination of the oxidation process. The following is one example:

$$2\,R-\overset{R^1}{\underset{H}{\underset{|}{\overset{|}{C}}}}-OO\bullet \quad\longleftrightarrow\quad R-\overset{R^1}{\underset{H}{\underset{|}{\overset{|}{C}}}}-O-O-O-O-\overset{R^1}{\underset{H}{\underset{|}{\overset{|}{C}}}}-R$$

$$\longrightarrow\quad R-\overset{O}{\overset{\|}{C}}-R^1 + O_2 + HO-\overset{R^1}{\underset{H}{\underset{|}{\overset{|}{C}}}}-R$$

$$R\bullet + ROO\bullet \longrightarrow ROOR$$

$$R\bullet + R\bullet \longrightarrow R-R$$

13.8.1 Metal Catalyzed Reactions

Transition metals (0.1–50 ppm) capable of a one-electron transfer process will act as a catalyst for hydrocarbon oxidation [80]. However, to be catalytically active, they must be in the form of a metal soap which results from the formation iron oxide with organic acids. Homolytic peroxide decomposition is catalyzed by metal soaps:

$$ROOH + M^{n+} \longrightarrow RO\bullet + M^{(n+1)^+} + \text{-}OH$$

$$ROOH + M^{(n+1)^+} \longrightarrow ROO\bullet + M^{n+} + H^+$$

$$2\,ROOH \xrightarrow{M^{n+}/M^{(n+1)^+}} ROO\bullet + RO + H_2O$$

In summary, these free radical oil degradation processes, and others not shown, are strongly affected by temperature and are catalyzed by metal soaps. They explain the formation of volatile byproducts, increasing oil viscosity with use, and the formation of sludge and other condensation byproducts. Figure 13.22 summarizes these processes [80].

13.8.2 Byproducts of Oxidative Degradation

Oil degradation byproducts are not only specific to the composition of the base oil and its source, but will also vary depending on the degradation conditions. However, for illustrative purposes, Table 13.16 provides a summary of those byproducts identified from an oxidative degradation study of a Middle Eastern oil [77].

13.8.3 Fluid Oxidation Testing Procedures

The amount of fluid oxidation, or remaining useful life of hydraulic oil, may be estimated or determined through laboratory analysis. The most commonly encountered oxidation tests used for petroleum hydraulic fluid characterization are neutralization number, TOST (Test for Oxidative Stability of Steam Turbine Oils), and RPVOT (Rotary Pressure Vessel Oxidation Test), and more recently, linear sweep voltammetry. The results of these tests are expressed as either a titer (mg KOH/g oil), amount of time required to achieve a predetermined amount of oxygen consumed, or amount of antioxidant depletion [90]. Although these tests will be discussed here, additional tests that have been utilized for either hydraulic fluid characterization or base oil oxidation studies will also be presented.

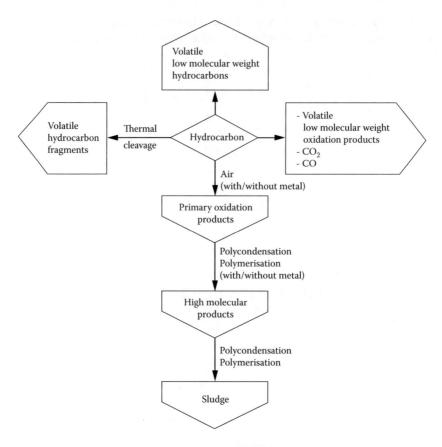

FIGURE 13.22 Rasberger model of lubricant degradation under high-temperature conditions.

13.8.4 NEUTRALIZATION NUMBER

The neutralization number, or total acid number (TAN), may be determined potentiometrically (ASTM D 664) [85] or colorimetrically (ASTM D 974) [86] and is expressed as "mg of potassium hydroxide required to neutralize the acid component in 1 gram of oil" [91].

An illustration of the variation of TAN with time in use for hydraulic oil is provided in Figure 13.23 [56]. Typically, the TAN number will increase slowly with time until it reaches a

---

TABLE 13.16
Oxidative Degradation Byproducts Obtained from a Middle Eastern Oil

Byproduct Type	Identity
Water	Water
Acids	Formic, Acetic
Alcohols	Ethanol, Isopropanol, 1- propanol, 2-butanol,1-pentanol, hexanol, 2-methyl propanol
Aldehydes	Acetaldehyde, butanal, 2-butanal, 2-pentanal, hexanal, 2-butanal, 2-pentenal
Ketones	Acetone, 2-butanone, 2-pentanone, 3-penten-2-one, 4-methyl-2-pentanone, cyclopentanone, 2-hexanone, 3-methylcyclohexanone, hexane-2,5-dione
Esters	Methyl acetate, ethyl acetate
Lactones	Dihydro-2-furanone, 2-cyclohexene-2-one, 5-methyl hihydro-2-furanone
Ethers	Tetrahydofuran, 2-methyl-dihydrofuran, dimethyldihydrofuran, 2-formylfuran
Aromatics	C-9 aromatics

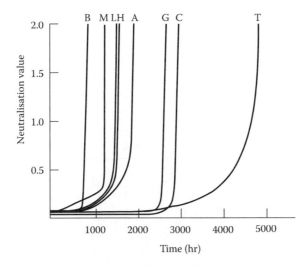

FIGURE 13.23 Comparison of TOST (Turbine Oil Stability Test) results for various petroleum oil hydraulic fluids. Note the varying induction times for oxidation to occur.

critical point where there is an exponential increase. This point is dependent on the oil, hydraulic system, and use conditions. This characteristic exponential increase in TAN is the onset of a dangerous use condition. Typically, it is recommended that the maximum allowable TAN in hydraulic systems is 2.0 [62].

Mang and Jünnemann have reported that the use of TAN values to monitor fluid oxidation is subject to misinterpretation and have recommended that this test only be used with oils of similar structure [79]. The reasoning provided was that TAN's are affected by both oxidation of the base oil and by the reaction mechanism of the additives.

13.8.5 Turbine Oil Stability Test (TOST)

Oxidative stability has been defined as "the ability of a fluid formulation to resist reaction with oxygen under a given set of environmental conditions including temperature and material of construction" [74]. ASTM D 943 [87], TOST, is an accelerated oxidative stability test used to compare the ability of a fresh oil to resist oxidation. The oxidation process is catalyzed using a mixture of copper, iron and water. TOST is conducted by bubbling oxygen (3 L/h) through 300 mL oil, 60 mL of water at 95°C in the presence of the iron-copper catalyst until a TAN value of 2 mg KOH/g oil is obtained. The ASTM D 943 TOST apparatus is illustrated in Figure 13.24. The test results are reported in hours. Although ASTM D 943 may be too severe for hydraulic oils operating at lower reservoir temperatures, it does provide a useful comparison between the relative inhibitory properties offered for hydraulic fluids containing different inhibitor additive packages, as illustrated in Figure 13.25 [63].

With the most recent version of DIN 51,587, the German standard for TOST, TAN characterization up to 2 mg KOH/g oil is not allowed due to potential additive interference. Mang and Jünnemann have suggested an alternative infrared analysis technique which can be used as an illustrative comparison of TOST results for a series of hydraulic oils. An example is provided in Figure 13.26 [76].

The TOST has been used to evaluate base oil components that aggravate color generation during use [81]. After chromatographic identification of the byproducts of the base oils evaluated in the TOST, it was found that the greatest contributors to color generation were byproducts formed by acid-catalyzed oiligomerization of aromatic sulfur compounds and other base stock components high in aromatic carbon. Color may be reduced by elimination strong-acids, which may act as catalysts for these oligomerization reactions. Results of this work are summarized in Table 13.17 [81].

FIGURE 13.24 Illustration of the ASTM TOST Apparatus. (Courtesy of Koehler Instruments Inc., Bohemia, NY.)

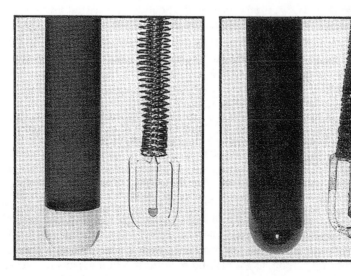

FIGURE 13.25 Comparison of oils possessing excellent and poor oxidative stability after testing according to ASTM D 943. (Courtesy of The Lubrizol Corporation, Wickliffe, OH.)

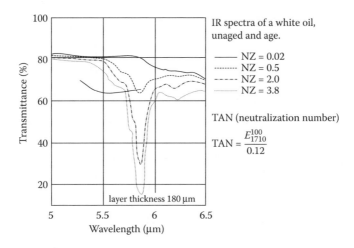

FIGURE 13.26 Aging of mineral oils according to DIN 51,587 (TOST) using IR-spectroscopic comparison.

13.8.6 ROTATING PRESSURE VESSEL OXIDATION TEST (RPVOT)

Oxygen tolerance is another very important property for hydraulic fluid. Oxygen tolerance has been defined as "the amount of oxygen that can be reacted with a test fluid under a given set of environmental conditions including temperature and materials of construction without causing dirtiness or property changes in the fluid" [74]. Oxygen tolerance is determined using the RPVOT (ASTM 2272) [88].

The RPVOT is conducted by heating 50 g of fresh, or used, oil, 5 mL of water and a copper catalyst coil in a steel bomb which is pressurized with 620 kPa (90 psi) of oxygen. The bomb is then heated in a constant temperature bath at 150°C while rotating at 100 rpm at a 30° angle from horizontal. The test is completed after a pressure drop of more than 175 kPa (25.4 psi) below the maximum pressure and the time to obtain this pressure drop, indicating the onset of oxidation, is recorded. Pressure-time plots of two test runs are illustrated in Figure 13.27 [88].

13.8.7 HOT-PANEL COKING TEST

One of the earliest tests conducted to determine the tendency of oil to form lacquer deposits at elevated temperatures was the hot-panel coking test [78]. Although this test has been used primarily to evaluate engine oils, it may be useful for evaluating hydraulic fluids since maintenance of fine tolerances are critical to the proper operation of a hydraulic system [55].

In the hot-panel coking test, oil is splashed from a reservoir onto a heated aluminum test panel that covers the reservoir. The test apparatus is illustrated in Figure 13.28. The data obtained include: weight of deposits on the panel, change in TAN and oil viscosity, and the weight of insolubles in the oil in the reservoir.

TABLE 13.17
Correlation of Base Stock Aromatic Composition and Color Generation during Oxidation

Lube	PPM – N	PPM – S	Wt% Aromatics	ASTM Color After Oven Treatment
A	27	6,200	26[a]	>8
B	1	92	4	5.5
C	<0.2	1	1	2.5

[a] This was an 84/12.5/3.5 mixture of: mono-/di-/tri-nuclear aromatics structures as determined by mass spectral analysis.

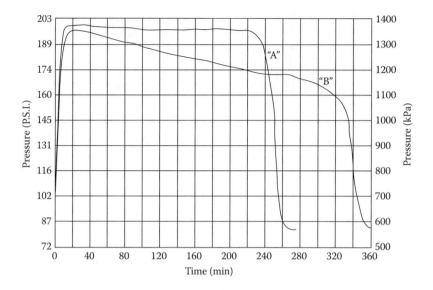

FIGURE 13.27 Pressure versus time plots for two RPVOT runs.

13.8.8 INFRARED (IR) SPECTROSCOPY

Mang and Jünnemann described the use of IR spectroscopy as a preferred alternative to TAN to analyze the oxidation resistance of HL and HL-P hydraulic oils. (See Table 13.9 for ISO hydraulic oil nomenclature.) Hydraulic oils were aged using TOST (DIN 51,387) procedures and a comparative IR absorbance of the C=O stretching vibrations at 1710 cm^{-1}, as shown in Figure 13.26, was used to estimate the degree of oxidation of the oil, even in the presence of additives [79].

IR analysis has also been used to identify and quantify [92]:

- Metal carboxylate salts—1600, 1400 cm^{-1}
- Carboxylic acids—1710 cm^{-1}
- Metal sulfates—1100, 1600 cm^{-1}
- Esters—1270, 1735 cm^{-1}

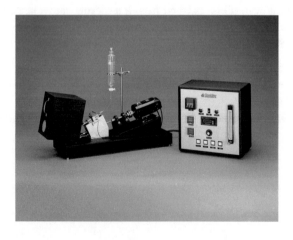

FIGURE 13.28 Hot-panel coking test apparatus. (Courtesy of Koehler Instruments Inc., Bohemia, NY.)

13.8.9 Cincinnati Milicron Thermal Stability Test

The Cincinnati Milicron Thermal Stability Test procedure is conducted by placing 200 mL of anti-wear hydraulic oil, a copper rod, and a steel rod in a 250 mL beaker, then placing the beaker into an aluminum block heated to 135°C [24]. Upon completion of the test, the used oils are observed for color change, filtered, and then submitted for elemental analysis. Figure 13.29 provides a comparison of good and poor multi-metal compatibility. The test is suitable for distinguishing oil additive performance. *Note*: There are three specifications of hydraulic fluids for this test: P-68, P-69, and P-70. Tests are all conducted in the same manner; the different specifications refer to different fluid viscosities at 40°C (104°F). The P-68 test is for 32 cSt fluids, P-69 is for 68 cSt fluids, and P-70 for 46 cSt fluids.

13.8.10 Differential Scanning Calorimetry (DSC)

DSC analysis has been used to study the relative oxidative stability of different classes of hydro-carbon structures [73]. In one study, this test was performed by oxidizing the base oils according to IP-48 over 48 hours at 120°C in a glass reactor, using an activated copper catalyst and varying levels of ZDDP as an antioxidant. Dry oxygen (1 L/h) was passed into the oil. DSC studies were performed on 5 mg samples using preprogrammed temperature ramping rates of 5, 10, and 20°C/min at a constant pressure of 100 psi. The characteristic temperature at the onset of degradation and the temperature at the maximum rate of degradation were determined [152]. A plot of the variation of onset times as a function of heating rate is provided in Figure 13.30 and with oil structure is provided

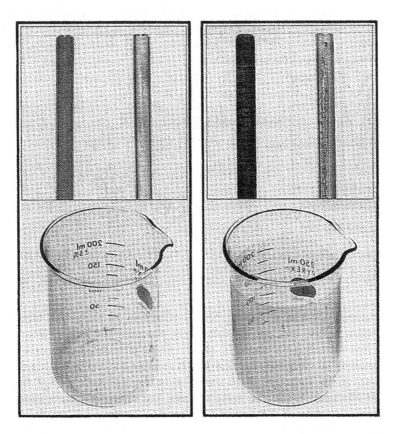

FIGURE 13.29 Illustration of good (left) and bad (right) multi-metal compatibility according to the Cincinnati Milacron Thermal Stability Test. (Courtesy of The Lubrizol Corporation, Wickliffe, OH.)

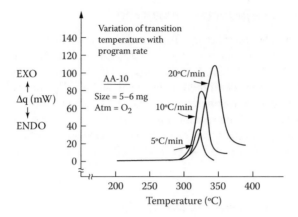

FIGURE 13.30 Illustration of DSC characterization of the oxidative stability of a petroleum oil.

in Figure 13.31 [73]. Pressure differential scanning calorimetry is commonly used to evaluate the oxidation resistance of lubricating oils according to ASTM D 6186 "Standard Test Method for the Oxidation Onset of Lubricating Oils by Pressure Differential Scanning Calorimeter (PDSC)" [156].

13.8.11 PENN STATE MICRO-OXIDATION (PSMO) TEST

The PMSO test is a thin-film oxidation test which can be calibrated to model various application conditions by varying testing conditions such as temperature [72]. The PSMO test apparatus, illustrated in Figures 13.32a and 13.32b, utilizes a glass reactor, 13 cm high × 2.5 cm inside diameter with a 24/40 ground-glass joint, and a low-carbon steel sample container. A 40 μL sample, which forms approximately a 160 μm thick film, is placed on the coupon and weighed. The "reactor" is placed in an aluminum block heater at 225°C. (see Figure 16.32b) The system is flushed with nitrogen (20 cc/min) for 30 min, then air (20 cc/min) for 10 min. The test coupon is inserted into the preheated and purged "reactor" and the test conducted with 20 cc/min of air purge. At the conclusion of the test, the weight loss due to evaporation is determined gravimetrically and the residue is characterized chromatographically. A "typical" result is shown in Figure 13.33.

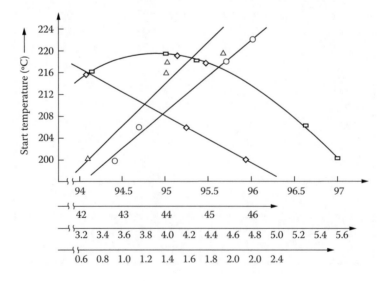

FIGURE 13.31 Comparison of the start of oxidative degradation by DSC measurement as a function of base oil composition.

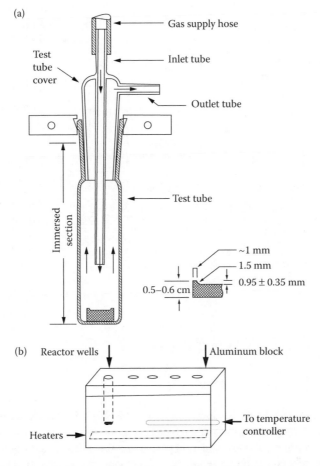

FIGURE 13.32 Illustration of the Penn State Micro-Oxidation (PSMO) test apparatus; a. Test cell, and b. Heating block.

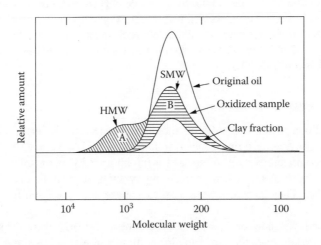

FIGURE 13.33 Typical PSMO test result.

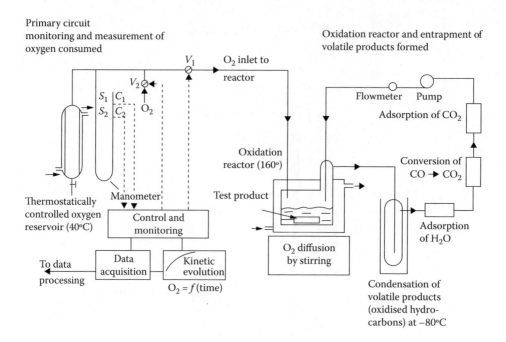

FIGURE 13.34 Schematic of IFP OXYTEST apparatus.

13.8.12 IFP OXYTEST

A schematic of the IFP OXYTEST is shown in Figure 13.34 [77]. This test measures oxygen uptake throughout the oxidation process to provide oxygen consumption data as a function of time. Other data that can be determined includes induction period and the point of maximum uptake. An additional feature of this test is that the gaseous byproducts can be collected in a cold-trap for subsequent identification.

13.8.13 THIN-FILM OXYGEN UPTAKE TEST (TFOUT)

The TFOUT, ASTM D 4742 [89], measures the oxidative stability of 1–2 g of fluid spread over a glass coupon in a rotating bomb under pure oxygen. The oxygen pressure is measured continuously and the induction period is the time elapsed before the oxygen pressure starts to drop. The ASTM TFOUT is conducted in the presence of oil-soluble fuel, metal catalysts, and water. As this test employs engine oil catalysts, it would be predominantly used to evaluate the oxidation resistance of Super Tractor Oil Universal (STOU).

13.8.14 LINEAR SWEEP VOLTAMMETRY

A more recent addition to the hydraulic oil monitoring repertoire is linear sweep voltammetry. There are currently two approved linear sweep voltammetry ASTM methods that are used to evaluate the amount of certain antioxidants often used as inhibitors in hydraulic oils—ASTM D 6810 and D6971, which are used to determine the remaining amounts of phenolic and aminic antioxidant, respectively [157,158]. They have also been used to analyze zinc dialkyldithiophosphate antioxidancy as well [159].

To perform a test, the new or in-service oil samples (400 μL) were diluted with an alcohol-formulated analytical test solution, containing dissolved electrolytes and a suspended solid substrate.

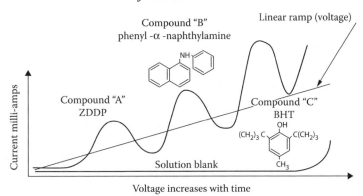

FIGURE 13.35 Voltage/current graph for voltammetric test method, with time/voltage in X-axis and current in Y-axis for three types of antioxidants, ZDDP, Amines, and Phenols in a neutral-type of voltammetric test solution.

When the hydrocarbon-based oil/solvent/solid substrate mixture was shaken, the suspended solids quickly precipitate the oil insolubles, leaving a clear solution for analysis, and free from interference from fuel, soot, or contaminants.

As the voltage potential increases the antioxidants and antiwear additives (such as ZDDP) become chemically excited (at their respective oxidation/reduction potential), causing the current to increase until it reaches a peak, and then decrease as the potential continues to increase (see Figure 13.35). In the voltage/current relationship, the voltage potential range is related to the identity of the type of the antioxidant/antiwear additive (such as ZDDP, amines, phenols, phenates, salicylates, etc.), and the current peak indicates the concentration of the additive [159].

13.9   ADDITIVES

In the simplest case, it can be described that most modern mineral oil hydraulic fluids are composed of base oil and additives. Generally, the base oil makes up the majority share of hydraulic oil composition. Yet, generalizing the additives into a single category in a hydraulic oil formulation is a dramatic oversimplification, because the additives are responsible for providing most of the performance properties desired to provide reliable hydraulic pump performance. There are various additives that provide specific performance functions. Examples of different additives include pour point depressants to enhance low temperature properties, viscosity index (VI) improvers to modify viscosity-temperature properties, antioxidants to increase oxidative stability of the oil, antiwear additives to protect moving pump parts, and rust and corrosion inhibitors to provide metal protection against oxygen and overly active additives. Hydraulic formulators have the option of purchasing pre-formulated additive packages from various suppliers or concocting their own formulation by purchasing additive components from many of the same suppliers. Often the decision for which formulation route to take is made based upon the formulator's resources of time, money, and desire for propriety.

Currently the emphasis in hydraulic system design for mobile equipment is for smaller reservoirs and higher pressures. Within any hydraulic system, mechanical energy is converted to thermal energy, meaning that the oil heats up. Thus, another function of the oil is to dissipate heat. The operating conditions in the newer systems significantly increase the operating severity to the hydraulic oil. In some cases, these conditions may result in temperature rises approaching the decomposition limits of the oil components and additives. If this occurs, oil and additive degradation byproducts may then attack component surfaces and lead to failure [93]. Therefore, hydraulic fluid additive technology is of critical importance when considering the proper fluid for selection and use.

Hydraulic fluid additive chemistry, including typical structures and performance mechanisms, will be discussed in another chapter and so only an overview will be provided here. The overview will include: base oil effects on additive performance, additive classification and function, history and use of ZDDP additives in antiwear hydraulic fluid formulation, ashless fluid technology, and finally, a brief discussion on additive analysis.

13.9.1 Additive Classification and Function

Although this entire chapter is dedicated to mineral oil hydraulic fluids, these fluids are also classified by additive type. The first type of additive classification description is rust and oxidation (R&O), or antiwear. Antiwear oils were originally developed for vane pumps which operate at high speed and high pressure. R&O-inhibited oils were generally used with piston pumps that typically utilized a yellow metal for the shoes which rode against a steel or ferrous plate [107]. Unfortunately, the antiwear additive ZDDP, which was commonly used for formulating AW oils, was susceptible to hydrolysis and the hydrolysis byproducts would attack the yellow metals in piston pumps precluding the use of a single oil for the two pumps. The antiwear hydraulic classification can be sub-classified into ashless or zinc-containing hydraulic fluids. The final additive-based classification for mineral oil hydraulic fluids is straight-grade and multi-grade hydraulic fluids, depending on whether the fluid contains a VI improver or not. Often, more than one of these classifications can be employed simultaneously for a single hydraulic fluid product, depending on the desirable functions of the hydraulic fluid. The primary function of hydraulic fluid is to act as an energy transfer medium, but it is important to realize that the fluid must also provide adequate lubrication of the moving surfaces in the hydraulic system, coat and protect the metal parts of the system from rust and corrosion, and must cool the system by acting as a heat transfer medium. Additives are used to provide and enhance the performance properties of the base oil used, including: viscosity-temperature properties, antiwear properties, rust inhibition, antioxidant performance, gum solvency (to assist in maintaining a clean system by solvating "sticky" degradation byproducts which cause sticking and hesitation of hydraulic system components), and antifoaming/air-release properties [103,104]. Table 13.18 provides a summary of the various types of additives used in mineral oil hydraulic fluids, with respect to function and chemical type [104,105].

13.9.2 Base Oil Effects on Additives

It has been previously mentioned that base oil composition exhibits some of the greatest contributions to finished fluid properties such as viscosity–temperature, pour point, and oxidative stability, as shown in Table 13.18 [94–96]. For example, oxidation characteristics of a base oil are dependent on the base oil chemistry itself, but they can be enhanced when antioxidant additives such as hindered phenols, alkylated aromatic amines, or various synergistic combinations of both interact with the base oil [94].

Paraffinic components of the base oil have been shown to improve synergism between antioxidants, thus improving oxidative stability; however, they also aggravate the antiwear and detergency properties of the oil [96]. Naphthenic and aromatic base oil components improve antiwear and detergency properties by improving the solubility of additives in the oil. Multiple linear regression equations have been developed to predict these effects based on base oil chemical composition [96].

Base oil composition is dependent on the refining process that is used. Earlier in this chapter, it was shown that processes; such as solvent refining, hydrocracking, and hydrotreating, will reduce the concentration of sulfur, nitrogen, and aromatic containing components present in the oil. These can exhibit substantial improvements in the base oil [97]. The amount of reduction of these components is dependent on the refining process. The additive solubility issues in base oils containing increased paraffinic content are typically addressed through the introduction of additives [97].

Additive response may vary within a certain fluid formulation, depending on the type and amount of oil degradation byproducts present during use. This has been shown to occur particularly when aldehyde degradation byproducts formed substantially and reduced antioxidant response properties

TABLE 13.18
Hydraulic Fluid Additives–Function and Chemical Type

Additive	Purpose	Function	Typical Compounds
Oxidation Inhibitors	Prevent varnish and sludge formation, extend fluid life by 10 to 100 times.	Terminates oil oxidation by the formation of inactive compounds or by oxygen scavenging	Organic compounds containing sulfur, phosphorous, or nitrogen such as: zinc dithiophosphates, organic sulfides and thiophosphates, hindered phenols and alkylated aromatic amines.
Viscosity Improver	Lower the rate of change of viscosity with temperature. Index (VI) Improver.	Because of the solubility differences, viscosity is increased more at high temperatures than at low temperatures.	Polymerized olefins or iso-olefins, butylene polymers, alkylated styrene polymers, polymethacrylate.
Pour Point Depressant	Lower the pour point temperature.	Modification of wax crystals to prevent growth with accompanying solidification at low temperatures.	Alkylated naphthylene or phenols and their polymers, methacrylate polymers.
Antifoamant	Prevent formation of stable foam and entrained air which may increase compressibility.	Change interfacial tension to permit bubble coalescence into larger bubbles which separate faster.	Silicone polymers, organic polymers (acrylate esters).
Antiwear Agent	Reduce wear.	Forms a film on metallic contacting surfaces.	Organic phosphates and phosphites, zinc dithiophosphate.
Extreme Pressure (EP) Additive	Prevent galling, scoring, and seizure.	Formation of low shear films on metal surfaces at point of contact.	Sulfur, chlorine, phosphorous containing materials.
Rust Inhibitor	Prevent or reduce rusting.	Preferential adsorption of polar, surface-active materials, neutralize corrosive acids.	Sulfonates, amines, fatty oils, oxidized wax, and halogenated ferivatives of some fatty acids.
Corrosion Inhibitor	Prevent corrosive attack on alloys or other metallic surfaces.	Inhibits formation of acidic bodies or forms a protective film over metallic parts.	Organic compounds containing sulfur, phosphorous or nitrogen such as phosphites, metal salts of thiophosphoric acid, and terpenes.
Detergent	Keep surfaces free of deposits.	Chemical reaction with sludge and varnish precursors to neutralize them and keep them soluble.	Metallo-organic compounds of barium, calcium and magnesium phenolates, phosphates, and sulfonates.
Dispersant	Keep insoluble contaminants dispersed.	Contaminants are bonded by polar attraction to dispersant molecules, preventing agglomeration and improving suspension.	Polymeric alkylthiophosphonates and alkylsuccinamides.
Oiliness Agent (Boundary Lubrication Additive)	Reduce friction under near-boundary conditions.	Adherence of polar materials to metal surfaces.	High-molecular weight compounds such as fatty oils, oxides, waxes or lead soaps.

of ZDDP, 2,6 - di(t-butyl)-4-methylphenol and N-phenyl-1-naphthylamine [98]. Carboxylic acids, ethers, alcohols, and ketones can also exhibit a negative effect, but to a lesser extent. Interestingly, ester-containing byproducts exhibited almost no effect [98].

13.9.3 ANTIWEAR ADDITIVES

One of the additive classification terms from Table 13.18, "oiliness," is seldom used in the industry today. Oiliness has been defined as: "a chemical feature of oil which permits it to wet metallic surfaces and establish a minimum coefficient of friction between two rubbing surfaces under conditions of boundary friction" [110]. Still used today though is film strength which is a property of a lubricant that "enables it to maintain protective films on the rubbing surfaces under conditions of boundary lubrication ..." [111]. Neither of these terms is rigorously correct since they do not properly define the required lubrication requirements for different wear surfaces in a hydraulic pump.

Papay has addressed the problem by providing the following terms and their proper definitions: "friction modification," "antiwear," and "extreme pressure" [108]. Boundary lubrication refers to the use of an additive, "friction modifier," which is chemisorbed to prevent two converging surfaces from asperity contact. An example of a friction-modifying additive is a fatty acid. Figure 13.36 illustrates a closely packed array of stearic acid molecules that is formed on a surface by chemisorption [108]. Friction-modified films typically provide coefficients of friction approximately 0.01–0.02 relative to an un-lubricated surface of 0.5. For reference, the coefficients of friction of fluid films being used for hydrodynamic lubrication are typically 0.001–0.006. Antiwear (AW)/extreme

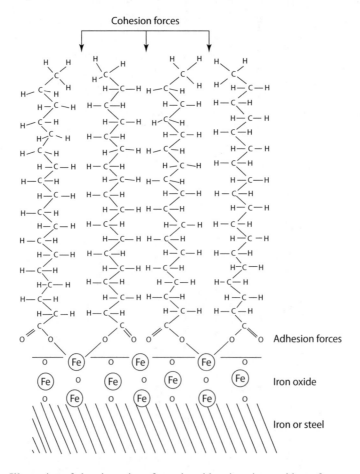

FIGURE 13.36 Illustration of chemisorption of stearic acid on iron-iron oxide surfaces.

pressure (EP) additives, such as ZDDP, function by a different mechanism. AW/EP additives form semi-plastic films which are difficult to remove by the shearing forces present with very close surface asperity contact. The coefficient of friction for an AW/EP film is approximately 0.1–0.2 [108].

Although, they are designed to be operated hydrodynamically, it has been shown that hydraulic pumps operate under conditions of mixed-film, sometimes called "elastohydrodynamic" (or "boundary lubrication") during certain modes of operation [114,112]. The moving surfaces within a hydraulic pump that are most susceptible to boundary or mixed-film lubrication include:

1. The interface between the leading edge of the vanes and the cam ring of a vane pump,
2. The line contact between the mating gear teeth of a gear pump,
3. The interface between the connecting rod and piston of certain types of piston pumps,
4. The interface between the piston and swash plate of other types of axial piston pumps, and
5. The interface between rotary valve plates and housings [110].

The lubrication of these controlled clearances demands the use of a lubricant that provides effective mixed-film and boundary lubrication along with proper surface finish. In addition, the selection of an additive package to provide the required lubrication effect is dependent on the surface being lubricated (material pair) and the base oil [113].

13.9.4 ZINC DIALKYLDITHIOPHOSPHATE (ZDDP)

One of the most popular antiwear additives used in antiwear mineral oil hydraulic fluid formulation is ZDDP. It has been found to provide the necessary lubrication demanded by highly loaded wear contacts. Antiwear (AW) additives, such as ZDDP, react with the metal asperities to form solid films which reduce wear at the contacting surface [101]. As the moving surfaces come into even closer contact, extreme pressure (EP) additives react with the metal surface by a mechanism whereby the heteroatom (S or P) present in the EP additive is actually extruded into the metal surface, thus forming a softer metal-heteroatom surface coating which inhibits adhesive failure (micro-welding of the surface aspirates as they come into contact with one another). The primary difference between AW and EP lubrication is the interfacial temperatures which are present within the wear contact [101].

ZDDP
(Zine dialkyldithiophosphate)

Some illustrative examples of ZDDP that are used as AW additives include [101]:

Zinc di-4-methylpentyl-dithiophosphate

Zinc di-iso-propyl-dithiophosphate

Zinc di-nonylphenyl-propyl-dithiophosphate

It has been shown that ZDDP compositions decomposed and reacted with metal surfaces to form films containing zinc, phosphorous, and sulfur. The rate of formation and thickness of the films was a function of the fluid temperature, additive concentration and interaction, base oil composition, atmosphere, and time [101]. In addition, decreasing thermal stability of the ZDDP resulted in increased load-bearing capacity.

Lubrizol Corporation evaluated the effect of varying the alcohol substituent (RO) on the ZDDP molecule on various performance properties, including hydrolytic stability, as shown in Table 13.19 [100]. Additives such as detergents and dispersants can also affect ZDDP performance [106].

13.9.5 Ashless Fluid Technology

As noted previously, some mineral oil hydraulic fluids are categorized as ashless antiwear hydraulic fluids. In the 1980s there was a great deal of interest in the development and use of non-zinc-containing additives for use in hydraulic fluids [105,108]. These additives are now called "ashless additives," and typically employ sulfur, phosphorous, and fatty ester chemistries in order to provide some of the same properties of ZDDP in a finished hydraulic fluid formulation. Recent work has shown that some hydraulic fluids formulated with ashless additives exhibit performance properties that are superior to ZDDP-formulated fluids, especially with respect to oxidative stability, protection of yellow metals, antiwear, and longer fluid lifetimes. Because it contains no heavy metals, ashless additive technology offers a significant advantage over conventional ZDDP additive technology respective to decreasing potential environmental and toxicological concerns due to fluid leakage and disposal [102,120].

13.9.6 Additive Analysis

There are various tests employed by mineral oil hydraulic fluid manufacturers and contract test labs to evaluate the amount of additive present in a new hydraulic fluid as well as in an in-service hydraulic fluid for quality control purposes or condition monitoring. Most of these tests involve the use of a scientific technique known as "spectroscopy." Spectroscopy is the study of the interactions between matter and radiation (or light). A spectrometer is a device that measures either absorbed or emitted light when samples of material are either bombarded with a light source or destroyed by high energy, such as a flame or plasma. Some of the common spectroscopic techniques used to

TABLE 13.19
Effect of ZDDP Structure on Thermal Performance

Zinc Type	Antiwear	Extreme Pressure	Antioxidant	Hydrolytic Stability	Corrosion of Soft Metals	Thermal Stability
Primary	Good	Good	Good	Best	Low to Moderate	Good
Secondary	Best	Best	Good to Excellent	Average	Moderate	Poor to Average
Aryl	Poor	Poor	Poor	Poor	Moderate	Best

evaluate hydraulic fluids include Fourier transform infrared (FTIR), atomic absorption (AA), arc spark, X-ray fluorescence, (XRF), and inductively coupled plasma (ICP).

One of the most common tests used to evaluate changes in additive concentration during use is FTIR spectroscopy [109]. This method is based on the infrared absorbency properties of the additives being utilized. Infrared absorbance follows Beer's law, which states that:

$$A = abc,$$

where A is the absorbency; a is absorptivity, which is dependent on the molecular structure; b is the path length of light through the sample in the spectrometer; and c is the concentration. A specific hydraulic oil formulation will have its own infrared-specific "fingerprint." Beer's law allows for this technique to be semi-quantitative for additive concentration when comparing new sample versus use sample. The FTIR mainly sees only the organic constituents in the hydraulic fluid, which is usually greater than 99% of the formulation.

All of the other spectroscopic techniques listed are used to evaluate the organic content of a hydraulic fluid samples, including: the evaluation of organometallic compounds used in antiwear additives; such as phosphorous, sulfur, molybdenum, zinc or other metals used as corrosion inhibitors or detergents; such as calcium. These techniques can also be used to evaluate wear metal levels when analyzing in-service oils.

13.10 DETERMINATION OF LUBRICATION PERFORMANCE

As hydraulic pump manufacturers change the designs of their systems, they want assurance that there will be hydraulic fluids available which will not cause the system harm. The lubricant formulators and additive suppliers must have ways to evaluate hydraulic fluid performance. Over the years, various tests have been employed to evaluate lubricant performance. When employing the scientific method it is always desirable to discriminate fluid performance. Some of the hydraulic fluid tests are laboratory bench tests, while others require the use of an actual pump stand. In any case, most tests are made more severe than most field service applications in order to accelerate the rate at which results are obtained.

One of the most common bench tests traditionally employed to evaluate the antiwear properties of hydraulic fluids is the 4-ball wear test. The 4-ball wear test is illustrated in Figure 13.37. This test can be affected by dissolved oxygen, as is illustrated in Figure 13.38 [118]. Although the 4-ball test (ASTM D 2783) has been proposed as a part of various performance specifications, no strong

FIGURE 13.37 4-ball wear test apparatus. (Courtesy Lubrication Engineers Inc., Wichita, KS.)

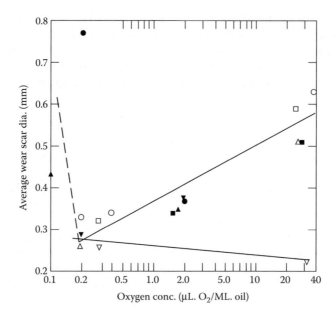

FIGURE 13.38   Effect of dissolved oxygen on wear exhibited by a mineral oil in a 4-ball wear scar test.

correlation has been found between the standard 4-ball wear test versus pump wear obtained while running ASTM D 2882, which utilizes a Vickers V-104 vane pump [35,121,122]. In fact, the only correlation data that has been published was performed at various test loads [115]. The best correlation was obtained with a 30 kg load, which is shown in Figure 13.39, but overall there was little to no correlation [115].

Tessmann et al. developed a modified Falex pin-and-V-block test, which was designated as the "gamma wear test." This method reportedly correlated with ASTM D 2882; however, it has never gained industry acceptance [195].

In general, bench performance tests have failed to model pump wear; therefore various pump tests have been developed [122,123]. In the past, the most commonly encountered test was based on the Vickers V-104 vane pump. A few years ago, the production of V-104 vane pump parts was discontinued. A new vane pump test, ASTM D 6973, has now been approved for the evaluation of mineral oil hydraulic fluids as a replacement to the V-104. For illustrative purposes, Figure 13.40 shows the difference between R&O oil and AW oil in the ASTM D 2882 wear test [117]. Figure 13.41 illustrates the typical differences in wear results obtained by different antiwear oils [117]. Langosch reported that V-104 vane pump wear (ΔG) is exponentially dependent on the rotation speed (v), as shown in Figure 13.42 [116]:

$$\Delta G = fv^2.$$

Interestingly, various specifications have been produced over the years that coupled bench wear and V-104 vane pump requirements, such as those shown in Table 13.20 [116]. Efforts are currently underway to replace it with the D6973 pump.

Table 13.21 provides the requirements of four common pump and chemical performance tests and the required "pass" criteria [102]. It also provides recent recommendations for overall hydraulic fluid testing requirements, including physical properties, rust and corrosion testing, oxidative stability, and pump lubrication performance [102].

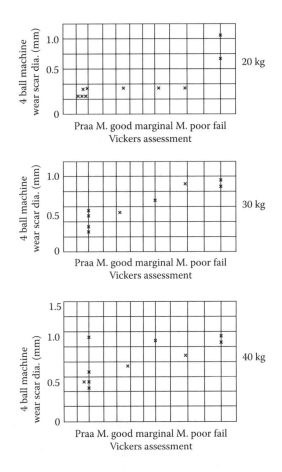

FIGURE 13.39 Attempted correlation of 4-ball machine wear and Vickers V-104 vane pump wear.

The Denison T6C pump wear test referenced in Table 13.21 has been developed in recent years to evaluate the effect of potential additive hydrolysis on yellow metal wear. [107, 123]. The T6C testing is conducted for 300 hours "dry" and then 300 hours "wet," which is accomplished by adding 1% water to the hydraulic oil. The test is conducted under 250 bar variable pressure for both dry and wet stages and a test temperature is 85°C. A 6-micron filter is employed on the system. If hydrolysis byproducts are present, they will cause filter plugging.

13.11 FLUID CONDITION MONITORING

Several things must be done to ensure reliable hydraulic fluid performance. First, it is important to keep the fluid clean. This can be done in several ways. One way is to remove contamination or keep it out of the system in the first place. One of the most popular ways to remove contamination is to use filtration. However, there are also techniques that can be used to keep contamination out of the system, such as air breather filters and desiccators. Either way, it is important to perform a periodic analysis of in-service hydraulic oil as part of any plant reliability maintenance program in order to optimize performance and minimize downtime. There are two primary reasons for performing oil analysis. The first is to determine if the oil quality is acceptable for continued use. The second is to assist in determining the root cause of problems within the hydraulic system or one of its components [124]. Both of these are often called "condition monitoring," because

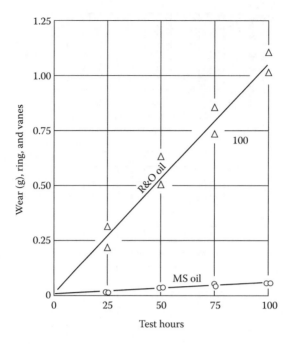

FIGURE 13.40   Comparison of R&O and AW oil performance using a Vickers V-104 vane pump test (ASTM D-2882).

the oil is used to determine the condition of both the oil and the hydraulic pump system while in-service.

   Various factors may affect the useful life of hydraulic oil. Some of the most common factors are summarized in Table 13.22 [124]. Of the factors cited, the most common factor is excessive operating temperatures, or system hot spots [124–127]. Optimal hydraulic oil temperature is 130°F–140°F (55°C–60°C). Excessive operating temperatures can significantly reduce the operating life of the

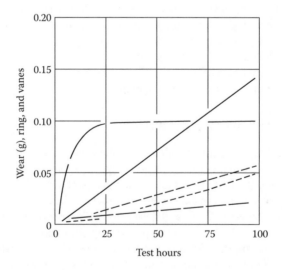

FIGURE 13.41   Comparison of antiwear performance of various mineral oil hydraulic fluids using a Vickers V-104 vane pump test (ASTM D-2882).

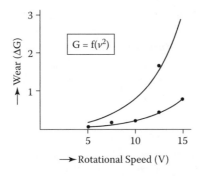

FIGURE 13.42 Dependence of Vickers V-104 vane pump test results on pump rotational speed.

oil from years to hours. A well-maintained hydraulic system is essential to minimize downtime and equipment loss and malfunction. Periodic fluid analysis should be an integral part of any maintenance program.

13.11.1 Fluid Sampling

Hydraulic Fluid samples should always be collected from a system while it is running and at operating temperature. A sample should be taken using a hand-pump or large syringe with a sufficient length of tubing to collect the sample from the middle of the reservoir. Routine sampling should not occur when oil is drained, from low spots in the system, or after the system has been shut down for more than two hours [130]. If a sample is collected from the pressure line, at least one quart of fluid should be flushed from the system prior to actually collecting the sample. In order to ensure sample accuracy, it is critical that all sampling equipment and bottles be scrupulously cleaned prior to sample collection [130]. The appropriate sampling frequency should be determined based upon the operating environment and system conditions. Initially, it could be weekly and then decreased to an appropriate level based on the results from the testing. Certain routine testing, such as fluid cleanliness, water content and viscosity, could be evaluated more frequently and detailed analyses performed at less frequent intervals, or as needed [130].

13.11.2 Basic Fluid Analysis Tests

A basic list of the tests that should be conducted on used hydraulic oil includes various physical and chemical tests for the following: clarity, color, odor, water, viscosity, specific gravity, neutralization number or differential infrared analysis, sediment, ash, elemental analysis, and foaming/air entrainment [124].

TABLE 13.20
Hydraulic Fluid Wear Classification Tests

Hydraulic Oil Classification	V-104 Vane Pump Total Wear (mg)	4-Ball Test 40 kg Load Wear Scar (mm)	Timken Test 15 lbs., 1 hour, wear scar (mm)	FZG Test A/8.3/90 Number of Stages Passed
HLP I	<50	<0.6	<0.8	m10
HLP II	<250	<1.0	<1.2	m10
HL	<1000	<1.3	<2.0	m6
H	>1000	>1.3	>2.0	<6

TABLE 13.21
Criteria for Common Pump Performance Tests

Test	Denison HF-0	Denison HF-2	ASTM D6158	DIN 51524 Part 2
Viscosity Index, min	90	90	90	
Viscosity $\leq$ 750 cP,°C, max			(1)	
Viscosity, cSt at 40°C	(2)	(2)	(2)	(2)
Rust Test (IP L35/ASTM D 665)				
Method A	Pass	Pass	Pass	Pass
Method B	Pass	Pass	Pass	
Steel Corrosion, (ISO 7120) max				Class 0-method A
Foam Test (IP/ASTM D 892)				
Seq I (stability/tendency)	0 in 10 mins	0 in 10 mins	150/0	150/0
Seq II (stability/tendency)	0 in 10 mins	0 in 10 mins	75/0	75/0
Seq III (stability/tendency)	0 in 10 mins	0 in 10 mins	150/0	150/0
Deaeration (ASTM D 3427 mins, max	7	7	5	5
Demulsibility (ISO 6614) mins, max				20
Demulsibility (ASTM D 1401), max	40/37/3 (30 mins)	40/37/3 (30 mins)	40/37/3 (30 mins)	
Oxidation at 1000 h (ASTM D 4310)				
Neutralization Value, mg KOH/g: max	1.0	1.0		
Copper oil/water/sludge, mg			Report	
Insoluble Sludge, mg: max	200	200	200	
TOST (ASTM D 943) TAN increase after 1000 hrs mg KOH/g, max			2.0	2.0
Thermal Stability Test (ASTM D 2070)				
Sludge, mg/100 mL max	100	100	25	
Steel Rod Rating			1	
Copper Wt. Loss: max	10	10		
Copper Rod Rating	Report	Report	5	
Hydrolytic Stability (ASTM D 2619)				
Cu Specimen Weight Loss, mg/cm^2: max	0.20	0.20		
Acidity of Water Layer, mg	4	4		
Filterability Test (TP 02100)				
(A) Max Filtration Time without Water, sec	600	600		
(B) Max Filtration Time with 2% Water, sec	2 × A	2 × A		
Filtration Test (ISO 13357-2) F1, min %/F2, min %				80/60
Filtration Test (ISO 13357-1) F1, min %/F2, min %				70/50
FZG A/8.3/90 (DIN 51354-2)				
Test Load Stage Fail, min	Report	Report	10	
Pump Tests				
Vane Pump Wear DIN ENISO 2160				

TABLE 13.21   (Continued)
Criteria for Common Pump Performance Tests

Test	Denison HF-0	Denison HF-2	ASTM D6158	DIN 51524 Part 2
Ring Wt. Loss, mg, max			120	
Vane Wt. Loss, mg, max			30	
Vickers 35VQ25			Pass	
Vane Pump Test (Denison T6H2OC)				
Wear (Vanes + Pins) mg, max	15	15	ASTM D7043 Report	
For 9 Pistons, mg, max	300	300		
Shear Test (high VI only)				
KRL (20 hrs) %-loss, max	15	15		
After 307 hrs in T6H2OC, % loss max		15		
After 608 hrs in T6H2OC, % loss max	15	15		
Seal compatibility SRE-NBR 1 (DIN53538 part 1)				
Relative volume, % change			(5)	(3)
Change Shore A Hardness			(6)	(4)
Pump Piston Test (Denison P46 assessment)	Pass			

(1) Varies from −33 to 16 depending on viscosity grade
(2) Viscosity sufficient for application
(3) Ranges vary from 0 to 18 depending on viscosity grade
(4) Ranges from 0 to −10 depending on viscosity grade
(5) Ranges from 0 to 15 depending on viscosity grade
(6) Ranges from 0 to −8 depending on viscosity grade

13.11.3   CLARITY, COLOR AND ODOR

The first few tests in the list are actually physical observations. They can be performed in a laboratory but might even be conducted while the oil is in operation by the operator or sample collector.

Fluid clarity is the first observation that is made in hydraulic fluid sampling. Uncontaminated or thermally abused fluid should be clear. Haziness or cloudiness is indicative of contamination, often by water or a water-containing source.

Color is perhaps the second observation. Sudden changes in color could be indicative of fluid contamination or degradation. For example, an increase in fluid darkness could be indicative of oil oxidation, as illustrated in Figure 13.43. The formation of sediment or sludge in the oil, either due to contamination or degradation, could result in a color change caused by the formation of a temporary suspension of small particulates within the oil.

The third physical observation is odor. Most hydraulic oils will have a certain distinctive odor when new, usually based upon the comprising mixture of ingredients. Certain chemicals that could be used near the hydraulic system may be drawn in as contaminants and might change the odor of the oil while in service. Oil degradation, such as oxidation, is typically accompanied by a rancid or pungent odor (burned fluid odor) [124,127]. Other sources of odor may be contaminants such as flushing fluids or strong sulfur odor, which may arise from contamination by other fluids [124].

TABLE 13.22
Major Factors Influencing Useful Hydraulic Oil Lifetime

Problem	Fluid Response
Excessive operating temperature and system hot spots.	Viscosity increase due to fluid degradation accompanied by presence of acidic byproducts that promote corrosion. Sludge and varnish is also formed, which reduces heat transfer and causes actuator malfunction.
Contaminants	
Water	Catalyzes fluid oxidation, promotes rusting, and may lead to emulsification of the fluid, which will lead to a reduction of lubrication effectiveness.
Airborne dirt and debris	Increase wear due to abrasion.
Metalworking fluids	May promote sludging or corrosion.
Foaming and aeration	Cause poor lubrication and sluggish hydraulic system response.
Wear metals	Promote fluid oxidation, cause abrasive wear and may stabilize the potential formation of oil-water emulsions when water contaminants are present.

13.11.4 WATER

Hydraulic systems have the opportunity to draw significant quantities of water into the reservoir through the breather and other sources. Once drawn in, the hydraulic oil may absorb some of the water. Because oil and water do not mix, at a certain point oil contaminated with water will take on a hazy appearance. In this case, the visual inspection noted above will provide a strong indication of water contamination. Figure 13.44 illustrates one of the determining factors for the amount of water absorbed in the base oil composition. Water contamination of hydraulic oil may result in metal

FIGURE 13.43 Comparison of color and TAN of new and oxidized oils from ASTM TOST. From the left: new inhibited hydraulic oil; same oil after 1000 hour TOST test (TAN is approximately 0.1); new straight mineral oil; same oil after 100 hour TOST (TAN is > 2.0, the maximum allowable value). (Courtesy of Shell Lubricants Houston, TX.)

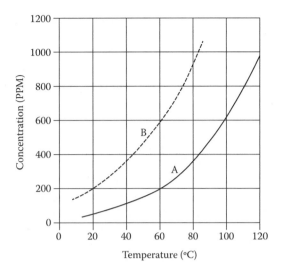

FIGURE 13.44 Effect of base oil composition and temperature on maximum water solubility concentration.

corrosion, enhanced bearing wear, and catalyzed fluid oxidation. Fluid oxidation results in reduced lifetime by as much as 70%, and adversely affects component operation [131]. One common source of water contamination is ingression through the reservoir's breather pipe. Generally, water concentrations should not exceed 200–300 ppm [131]. The effect of water concentration on fluid clarity is illustrated in Figure 13.45.

Several tests are used to measure the presence of water in hydraulic oil, such as:

* **Crackle Test**: This test provides only a qualitative measurement for the presence of water in hydraulic oil. It is possible for it to be performed tank-side. The crackle test is conducted by placing the fluid on a hot surface, such as a hot plate or even in a small piece of

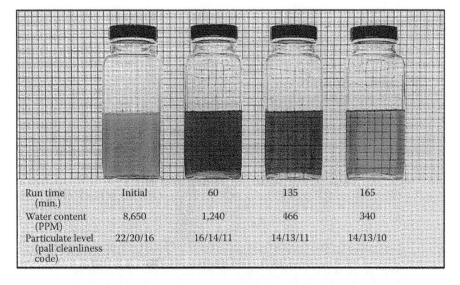

Run time (min.)	Initial	60	135	165
Water content (PPM)	8,650	1,240	466	340
Particulate level (pall cleanliness code)	22/20/16	16/14/11	14/13/11	14/13/10

FIGURE 13.45 Illustration of hydraulic fluid clarity and water concentration. Also illustrated is the effectiveness of the use of an oil purifier in reducing water and solids contamination. (Courtesy of Pall Industrial Hydraulics Company, Port Washington, NY.)

aluminum foil heated with a flame. It should be noted that anyone running this test should wear appropriate eye and face protection as the fluid will make a "crackling" sound and spatter if it contains water [124].

- **Distillation Method:** In this test, a sample of hydraulic fluid is heated with a solvent such as toluene or xylene in a round-bottomed flask equipped with a Dean-Stark trap and condenser as described in ASTM D 95 [134]. The mixture is heated until condensation of water into the trap ceases. The volume percentage of water in the oil is then calculated.
- **Karl Fischer Method:** This test method is a quantitative titration technique conducted according to ASTM D 6304. It is suitable for water concentrations in the range of 25–1000 ppm [135]. This method is much more precise than the crackle or distillation techniques and has become the test of choice for most in-service oil testing laboratories.
- **Centrifugation**: This test may be performed according to ASTM D 1796 if the hydraulic oil has a hazy or cloudy appearance [136]. In this test the sample is placed in a centrifuge and spun at a rate of 6000 rpm to clarify the oil. In the centrifuge, the ingredients in the sample will be separated according to specific gravity. The materials with the most density, such as metallic contaminants, will be on the bottom, followed by the carbonaceous debris, then the water, and finally the oil at the top. Sometimes, some of the water and oil stay mixed together in an emulsion layer which will reside above the water but below the oil phase [124].

13.11.5 Viscosity

A change in the fluid viscosity may be an immediate indication of a system problem. Fluid viscosity may be determined by ASTM D 445 (kinematic viscosity in centistokes (cS)) [137] or in Saybolt Universal Seconds (SUS) according to ASTM D 2161 [138]. Viscosity may also be readily determined in the field using an instrument such as that shown in Figure 13.46.

Low-viscosity values usually indicate contamination with a lower-viscosity oil or solvent but could also indicate additive deterioration [130]. Higher-viscosity values indicate contamination by a higher-viscosity oil or water, but could also result from oil oxidation [130]. If oxidative degradation

FIGURE 13.46 Illustration of a viscosity measurement gauge suitable for tank-side use. (Courtesy of Visgauge Inc., Glenwood Springs, CO.)

has occurred to a sufficient extent to lead to an increased viscosity, then a corresponding increase in TAN (total acid number) should also be observed [124]. It is essential that the fluid viscosity be maintained within the hydraulic pump manufacturer's limits.

13.11.6 SPECIFIC GRAVITY

The test method ASTM D 4052 is used to measure the specific gravity of in-service hydraulic oil. Specific gravity can be used as an important indicator of variations in fluid composition. It can also be used to identify contamination or oxidative degradation that can occur while hydraulic oil is used [127,135].

13.11.7 OIL DEGRADATION

Previously, it was discussed how hydraulic oils are formulated using additive and base fluid combinations which provide certain oxidation and degradation performance properties. Even well-formulated hydraulic oil will eventually degrade when used over long periods of time. Oil degradation byproducts can cause metal corrosion and/or the "gummy" sludge deposits which may result in sticking and malfunctioning hydraulic actuators [125,126]. Oil oxidation is evaluated using tests such as: TOST, RPVOT, TAN, FTIR, and linear sweep voltammetry, all of which were described earlier. When using these tests, a new oil sample is used as a reference and it is compared against used oil collected from an in-service hydraulic system. TOST, RPVOT, and linear sweep voltammetry have all been used as indicators of remaining useful life. FTIR and TAN are used to evaluate the amount of acidic oxidation material which has built up in used hydraulic oil. FTIR can also be used to evaluate certain additive depletion [124,128,133].

The presence of oil oxidation products may also be detected by various techniques such as the patch test illustrated in Figure 13.47 [126,130]. In this case, a small oil sample is diluted then filtered through a 0.8-micron patch filter. A predominantly yellow color indicates varnish-like byproducts. Gray color residues indicate dust and other debris [126]. Whirlpool Corporation has reported the use of a colorimetric test based on spectrophotometric analytical procedures which provide a numerical rating (1–10) of the amount of oxidation byproducts present in the sample [126].

13.11.8 ASH

During in-service hydraulic oil service it is important to determine additive depletion or the presence of contamination, such as wear metals, dirt, coolants, and other contaminants [124]. ASTM D 482 [140] is used for determining the ash content of R&O hydraulic oils that do not contain ash-forming additives [124]. ASTM D 874 [141] is used to determine the ash content of antiwear hydraulic oils formulated with ash-containing additives. In either case, these ash tests can be used to evaluate either additive depletion reduces ash, or contamination increases ash.

  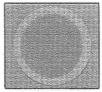 

| 0 hours | After 21 hours | After 45 hours | After 130 hours |
| Colorimetric value = 10 | Colorimetric value = 8 | Colorimetric value = 4 | Colorimetric value = 0 |

FIGURE 13.47 Illustration of the Kleentek patch test to quantify the presence of oil oxidation byproducts. (Courtesy of Kleentek - A CLARCOR Company, Glenwood Springs, CO.)

TABLE 13.23
Sources of Elemental Residue in Hydraulic Fluids

Element	Sources
Aluminum, silicon	Dust and airborne dirt
Boron, potassium, sodium	Coolant-inhibitor residues
Calcium, magnesium	Hard-water
Calcium, sodium	Saltwater (brine)
Chromium, copper, iron, zinc, lead, tin	Wear, corrosion, or assembly debris
Barium, calcium, magnesium, phosphorous, zinc	Engine oil additives

13.11.9 ELEMENTAL ANALYSIS

Either used hydraulic fluid or the ash residue obtained from ASTM D 482 or ASTM D 874 may be analyzed for elemental content using emission, atomic absorption (AA), or induction-coupled plasma (ICP) spectroscopy. The elemental content may provide some insight into the source of residue, as shown in Table 13.23 [124].

X-ray diffraction may be used to assist in identifying the source of crystalline, solid residues present in a hydraulic fluid. A select listing of crystalline solids which may be identifiable by x-ray diffraction, are presented by Young, and are listed in Table 13.24.

13.11.10 FOAMING/AIR RELEASE

Used hydraulic oils that have either been contaminated or have undergone oxidation are more susceptible to foaming and air entrainment. Foaming and air entrainment first and foremost lead to "spongy" hydraulic control due to increased fluid compressibility. They also result in accelerated oxidation rates which lead to increased susceptibility to sludge and varnish.

TABLE 13.24
X-Ray Diffraction Spectroscopy of Hydraulic Fluid Crystalline Residues

Element	Chemical Formula	Examples
Chromium	Cr	Wear metal
Copper	Cu	Wear metal or corrosion
Copper oxides	CuO and Cu_2O	Wear metal or corrosion
Iron	Fe	Wear metal
Iron oxide	FeO (wustite)	Fretting corrosion at high temperature
Iron oxide	αFe_2O_3 (hematite)	Fretting corrosion
Iron oxide	Fe_3O_4 (magnetite)	Corrosion of iron with limited supply of oxygen (under oil layer)
Iron oxide hydrate	$\gamma Fe_2O_3\,H_2O$ (lepidocrocite)	Light surface rusting
Iron oxide hydrate	$\alpha Fe_2O_3\,H_2O$ (goethite)	Bulk or heavy rusting
Iron oxide hydrate	$\beta Fe_2O_3\,H_2O$ (Beta iron oxide hydrate)	Halide corrosion or paint pigment residue
Silicon dioxide	SiO_2	
Sodium chloride	NaCl (salt)	
Tin	Sn	
Tin oxides	SnO and SnO_2	

13.11.11 Fluid Cleanliness

It is generally accepted that controlling fluid cleanliness is vitally important to the reliable operation and reduced maintenance of hydraulic systems [125,129]. One form of analysis of fluid cleanliness is particle counting. One of the most common particle tests is the ISO 4406 procedure, which uses a light blockage technique to measure and quantify the particles in a sample of hydraulic oil. In this procedure, it is common to present a three digit code. The code is based upon the particle counts collected at 2, 4, and 14-micron average diameters. Another particle analysis technique is ferrographic analysis. There are two types of ferrographic analysis—direct reading ferrography and analytical ferrography. Direct reading ferrography is used to produce a trendline analysis of large and small ferrous (magnetic) particles present in the sample. Analytical ferrography is a more qualitative technique. In this technique, debris is removed from the sample and deposited on a test slide. The debris on the slide is then analyzed under microscopic magnification to evaluate the actual condition of the pump (in contrast to most of the other tests, which were used to evaluate the condition of the oil). Figure 13.48 provides a comparison of the differences in the sizes of particles analyzed by these different procedures [133].

13.11.12 System Fluid Replacement

When hydraulic oil is in service it is possible for wear metals, environmental contaminants, and oil-oxidation films and deposits to build up in the hydraulic system. It is therefore necessary to

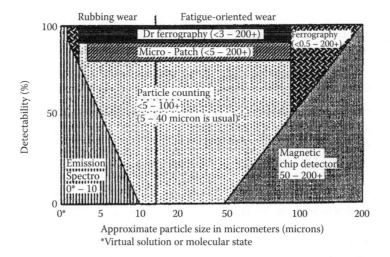

	Identification					Interference	
	Wear metals	Size	Shape	Abrasives	Other	Water	Opacity
Emission spectro	Excellent	Poor	No	V.Good	Additives coolant	Minimal	Minimal
Particle counting	Poor	Excell	No	Poor	----	Bad	Bad
Ferrography	Fair	V.Good	Excell	V.Good	Fibers heat	Minimal	Minimal
Direct reading ferrography	Iron	Fair	No	No	----	Minimal	Minimal
Micro - patch	Fair	Fair	Good	Good	Fibers rust	Minimal	Minor

FIGURE 13.48 Graphic illustration of estimated particle size operating range for various particulate characterization methods. (Courtesy of Lubricant Consultants Inc.)

periodically clean the system and replace the hydraulic oil. If deposits are films or light emulsion coatings on the metal surfaces, then the system can usually be cleaned by flushing [132]. However, if substantial buildup and deposits form, then the system will have to be dismantled and cleaned manually.

13.11.13 FLUSHING FLUID

A hydraulic system flush involves draining the used oil from the system, filling it with flushing oil, running it unloaded in order to splash the cleaner oil around in the system, then draining and refilling the system once again. The flushing fluid should [132]:

- Possess a viscosity high enough to lubricate moving parts and to keep solid contaminants suspended, typically 15–25 cSt at 40°C. The flushing fluid may be brand-new system operating fluid, or a special flushing fluid.
- Exhibit sufficient solvency to remove oily deposits. Usually naphthenic oils are better in this regard. The use of "gum or varnish removal solvents" is not recommended.
- Contain rust and oxidation inhibition so as not to negatively affect R&O properties when blended with the hydraulic oil when the system is refilled.
- Not be solvents or chemical cleaners, such as halogenated petroleum distillate solvents, caustic compounds, or water-based detergent blends [142].
- Compatible with all components of the hydraulic system including: seals, hoses, components, rust preventative paints, and preservatives in pipes or tubes which are normally not removed.

Steam cleaning is not recommended. It is impossible to remove all traces of water, so system rusting and oil contamination will result.

13.11.14 FLUSHING PROCEDURE

Although it may seem simple to fill the system with a flushing fluid and just turn it on, many have experienced problems in performing system flushes. As a result, a standardized procedure for performing a hydraulic system flush has been published by ASTM. The following is a summary of the procedure outlined in ASTM D 4174:

- Prior to flushing, all accessible areas of the system should be inspected.
- Do not flush through valves, but rather through bypasses.
- Close off all leads to cylinders and motors by using jumpers.
- The use of external pumps is recommended to assure a turbulent flow (Reynolds number > 20,000) through the system. Fluid flow should be at least three times the normal system flow velocity.
- All filtration and temporary humidity control devices that had been placed on the system before flushing should be removed.
- Solid contamination in the lines can be loosened by hammering and vibration.
- The flushing fluid should be preheated to between 140 and 180°F (60–80°C) prior to the flushing operations.
- Flushing times may vary from twelve hours to several days. The flow capacity should be 10%–20% of the total volume of the fluid during flushing.
- System pump discharge pressures should be reduced during flushing by lowering relief valve or compensator pressures.
- When flushing is completed, the flushing fluid should be completely drained from the system, coolers, and strainers.

- After flushing, the system is charged with operating fluid. If significant quantities of the flushing fluid remain, the system should be flushed again with the operating fluid. All fluids entering the hydraulic system should be pre-filtered to at least 3–10 microns absolute.

13.12 HYDRAULIC FLUID RECYCLING AND RECLAMATION

There has been increasing interest in extending the lifetime of hydraulic oils. The reasons for this include: conservation of natural resources (petroleum crude oil), decreased environmental liability for waste oil, reduced disposal costs, desire for fewer oil purchases, relatively low cost of reclamation, and the availability of excellent quality control procedures [149]. The four primary processes for reclaiming used hydraulic oil include: (1) recovery for use as fuel, (2) re- refining, (3) oil laundering, and (4) on-site reconditioning [143]. A brief overview of each of these processes will follow.

13.12.1 Recovery as Fuel

Recovery of used hydraulic oil as a fuel source is a very common reclamation procedure [143]. The first step in oil reclamation is collection. There are various companies who will go to sites and collect used oil. When oil is to be used as fuel, the next step in the process is filtration to remove emulsification and solids. The oil is then blended to the desired viscosity. Finally, it is pumped into equipment such as kilns and furnaces where it is burned as fuel. Although this procedure is still widely practiced, it is coming under increasing scrutiny because of worldwide goals to reduce emissions, reduce crude oil consumption, and increase oil recycling efforts [143].

13.12.2 Re-refining

One way to conserve natural resources is to recycle. Today, it is possible to recycle used hydraulic oil through a process called "re-refining." Re-refining of used hydraulic oil involves the removal of contaminants such as dust, moisture, combustion byproducts, corrosion products, and metal wear debris, followed by a separation and refining process similar to that originally used for the crude oil from which the original mineral hydraulic oil base stocks were processed [144]. One common re-refining process involves vacuum dehydration of the used oil at 650 mm Hg and 150°C. The oil is cooled to approximately 40°C and then treated with sulfuric acid to facilitate sludge separation and removal. After the sludge is removed, approximately 6% of clay is added to the oil to form slurry. Next, the slurry is heated to 300°C and held there for two to three hours. The vapors that cook off are condensed and the hot slurry is filtered. Once the re-refined oil cools down between 40–60°C it is treated with additives before reuse [144]. This process, however, is only applicable for large volume reclamation and, due to environmental concerns, it is not normally used. Re-refining is more commonly done by vacuum distillation followed by severe hydrotreating. Large volumes of oil are processed by this method and are used as base stock for many hydraulic oils.

13.12.3 Oil Laundering

Oil laundering refers to the recovery and cleaning of bulk loads of hydraulic oil and returning them to the same company for reuse [143,150]. In this process, water vapor, solids, and oil oxidation byproducts must be removed. The various processes that may be used in oil reclamation are typically classified as: mechanical separation, chemical separation, and electrostatic filtration [145]. The mechanical separation methods of dirt and debris removal most commonly include: filtration and centrifugation. Water vapor is removed by various processes, including distillation or vacuum dehydration [146,147]. A combination process illustrated in Figure 13.49 may also be used. Chemical separation processes include the use of adsorption and filtration of acidic oxidation byproducts, or chemical conversion to less harmful components by chemical reduction. Electrostatic filtration is sometimes used to remove charged oil-soluble oxidation byproducts.

(a)

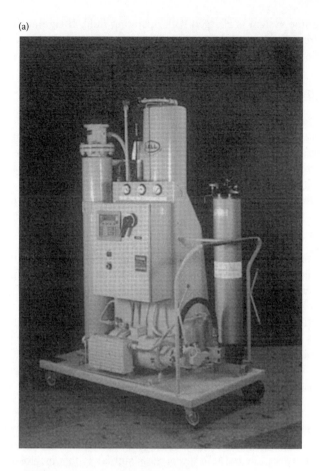

(b)

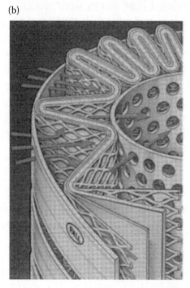

FIGURE 13.49 Illustration of a stationary (a) and portable (b) pall oil purifier for removal of water, dirt, and air contaminants from a hydraulic oil. The system operates under low (24 in. Hg) vacuum and employs a spinning disk (c) which facilitates contaminant separation by increasing surface area of the oil by formation of oil droplets during the separation process. (Courtesy of Pall Industrial Hydraulics Inc., Port Washington, NY.)

TABLE 13.25
Typical Results from Recycling Trials in a Cold Mill vs. New Oil Specs

Test	Reference	Specification	Waste Oil	Reclaimed Oil
Particle Count	ISO 4406	16/13 (Report)	Too Contaminated	16/13
Appearance	Visual	Clear and Bright	Black and Hazy	Clear and Bright
Water (ppm)	Karl Fischer	100 max	214.000	62.000
Viscosity (cSt @ 40°C)	ASTM D 445	41.4–50.6	46.700	44.000
Flash Point (°C)	ASTM D 92	204 min	not tested*	209.000
Total Acid No.	ASTM D 974	1.0 max	0.290	0.500
Demulsibility @ 54°C	ASTM D 1401	37 mL in 30 minutes	not tested*	37 mL in 30 minutes
Zinc (%)	ICP	0.04–0.06	0.029	0.050
Phosphorus (%)	ICP	0.04 min	0.029	0.046
Sulfur (%)	ICP	0.15 max	0.15	0.15
Copper Corrosion 3 hr.@ 100°C	ASTM D 130	1B max	not tested*	1a
Foam Sequence 2	ASTM D 892	100/0 max	not tested*	75/0
Foam Sequence 3	ASTM D 892	150/0 max	not tested*	100/0

*These tests are not normally conducted on waste oil, but should the sulfur content be greatly off specification, the oil recycler may, at his discretion, choose to do so.

13.12.4 ON-SITE RECLAMATION

Oil may also be reconditioned on-site, which usually involves a third party service organization or the purchase of permanent equipment to perform processes such as vacuum dehydration, filtration, and adsorption methods [143,148,149]. Table 13.25 illustrates the successful recondition of used hydraulic oil following the process illustrated in Figure 13.50 [149]. Although the process shown in

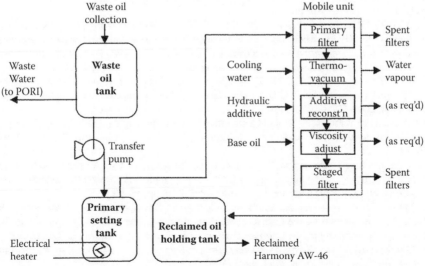

FIGURE 13.50 Schematic of a waste hydraulic oil reclamation process.

this figure is a relatively large on-site recondition procedure, it does provide an example of a successful process. Today, many are beginning to purchase their own equipment to clean oil for re-use.

REFERENCES

1. Villforth, F.J., ed., "Hydraulics," *Lubrication*, 1996, Vol. 82, No. 1, pp. 18–24.
2. Anon., "Product Review: Oil Refining and Lubricant Base Stocks," *Industrial Lubrication and Tribology*, 1997, Vol. 49, No. 4, pp. 181–88.
3. Hoo, G.H., and Lewis, E. "Base Oil Effects on Additives Used to Formulate Lubricants," Advanced Production and Application of Lubricant Base Stocks, *Proceedings from International Symposium*, eds. Singh, H., Rao, P., and Tata, T.S.R., *McGraw-Hill*, New Delhi, 1994, pp. 326–33.
4. Prince, R.J. "Base Oils From Petroleum," *The Chemistry and Technology Lubricants*, 1992, pp. 1–31.
5. Singh, H. "Characterization of Lube Oil Base Stock - Approach and Significance," Advanced Production and Application of Lubricant Base Stocks, Proceedings from International Symposium, eds. Singh, H., Rao, P., and Tata, T.S.R., McGraw-Hill, New Delhi, 1994, pp. 303–10.
6. Adhvaryu A., and Singh, I.D. "FT-NMR and FTIR Applications in Lubricant Distillates and Base Stocks Characterization," *Tribotest Journal*, 1996, 3–1, pp. 89–95.
7. Al-Banwan, M. "Base Stocks Properties/Characteristics, Additive Response and Their Interrelationship," Advanced Production and Application of Lubricant Base Stocks, Proceedings from International Symposium, Eds. Singh, H., Rao, P., and Tata, T.S.R., McGraw-Hill, New Delhi, India, 1994, pp. 303–10.
8. Singh, H., Adhvaryu, A., Singh D., and Chaudhary, G.S. "Influence of Refining on Base Oil Composition," *Symposium on Worldwide Perspectives on the Manufacture, Characterization, and Application of Lubricant Base Oils, Presented at the 213th National Meeting of the American Chemical Society*, San Francisco, Apr. 13–17, 1997, pp. 259–61.
9. Gazauskas, J.F., Abbott F.P., and Baumgartner, N.R. "Characteristics of Wax Extracted from Lubricant Base Stocks," *Lubrication Engineering.*, 1994, Vol. 50, No. 4, pp. 326–36.
10. *ASTM Manual on Significance of Tests for Petroleum Products*, 6th Ed., ed. by Dyroff, G.V., 1993, ASTM International, Conshocken, PA.
11. *ASTM Manual on Hydrocarbon Analysis*, 5th Ed., ed. Drews, A.W., 1993, ASTM International, Conshocken, PA.
12. Sarpal, A.S., Kapur, G.S., Mukherjee S., and Jain, S.K. "Characterization by ^{13}C NMR Spectroscopy of Base Oils Produced by Different Processes," *Fuel*, 1997, Vol. 76, No. 10, pp. 931–37.
13. Singh H., and Swaroop, S. "Oxidation Behavior of Base Oils and their Constituting Hydrocarbon Types," *Symposium on Worldwide Perspectives on the Manufacture, Characterization, and Application of Lubricant Base Oils, Presented at the 213th National Meeting of the American Chemical Society*, San Francisco, Apr. 13–17, 1997, pp. 218–20.
14. Barman, B.N. "Bias in the IP 346 Method for Polycyclic Aromatics in Base Oils and in ASTM D 2007 Method for Hydrocarbon Type Determination," *Symposium on Worldwide Perspectives on the Manufacture, Characterization, and Application of Lubricant Base Oils, Presented at the 213th National Meeting of the American Chemical Society*, San Francisco, Apr. 13–17, 1997, p. 263.
15. ASTM D 2007 "Standard Test Method for Characteristic Groups in Rubber Extender and Processing Oils and Other Petroleum Derived Oils by the Clay-Gel Absorption Chromatographic Method," ASTM International, 2009. Conshocken, PA.
16. Varotsis, N., and Pasadakis, N. "Rapid Quantitative Determination of Aromatic Groups in Lubricant Oils Using Gel Permeation Chromatography," *Industrial & Engineering Chemical Research*, 1997, 36, pp. 5516–19.
17. Daucik, P., Jakubik, T., Pronayova, N., and Zuki, B. "Structure of Oils According to Type and Group Analysis of Oils by the Combination of Chromatographic and Spectral Methods," *Mechanical Engineering*. 80 (Eng. Oils and Auto Lub), 1993, MCLEEF, pp. 48–58.
18. Ramakumar, S.S.V., Aggarwal, N., Madhusudhana Rao, A., Srivastava S.P., and Bhatnagar, A.K. "Effect of Base Oil Composition on the Course of Additive-Additive Interactions," *Symposium on Worldwide Perspectives on the Manufacture, Characterization, and Application of Lubricant Base Oils, Presented at the 213th National Meeting of the American Chemical Society*, eds. Singh, H., Rao, P., and Tata, T.S.R., McGraw-Hill, New Delhi, India, 1994, pp. 334–40.
19. Maas, P.A. "Selecting Hydraulic Fluids," *Plant Engineering*, 1973, October, pp. 122–25.
20. Tiffany, H.E. "How Additives Improve Hydraulic Fluids," *Machine Design*, Dec. 27, 1973, pp. 48–51.

21. ASTM DD 2422, "Standard Classification of Industrial Fluids Lubricants by Viscosity System," ASTM International, 2009. Conshocken, PA, 1986.
22. International Standard ISO 6743/4, "Lubricants, Industrial Oils and Related Products (Class L) - Classification Part 4: Family H (hydraulic systems)," First Edition - 1982–11–15.
23. Reichel, J. "Pump Testing Strategies and Associated Tribological Considerations - Vane Pump Testing Methods ASTM D 2882, IP281, and DIN 51389," in *Tribology of Hydraulic Pump Testing, 1310 STP*, eds. Totten, G.E., Kling, G.H., and Smolenski, D.J., ASTM International, Conshohocken, PA, 1996, pp. 85–95.
24. Saxena, D., Mooken, R.T., Srivastava, S.P., and Bhatnagar, A.K. "An Accelerated Agining Test for Anti wear Hydraulic Oils," *Lubrication Engineering*, Vol. 49, No. 10, pp. 801–9.
25. Klamann, D., in Lubricants and Related Products - Synthesis, Properties, Applications, International Standards, Verlag Chemie, Weinheim, Germany, pp. 327–323, 1984.
26. "Petroleum Hydraulic Fluids Recommendations," Brochure No. S-106, Available from *Racine Hydraulics & Machinery Inc.*, 2000 Albert Street, Racine, WI 53404, December 1966.
27. "Hydraulic Fluid Recommendations for Industrial Machinery Hydraulic Systems," Brochure No. 03-401-2010, Available from Eaton Hydraulics Group, Eden Prarie, MN, 2010, Rochester Hills, MI.
28. "Oil Recommendations," Bulletin 9000K, Available from *The Oilgear Company*, Milwaukee, WI, May 1969.
29. Anon., "Industrial Hydraulic Oils," *Lubrication*, 1956, Vol. 42, No. 7, pp. 89–100.
30. Bingham, A.E. "Some Problems of Fluids for Hydraulic Power Transmission," *Institute of Mechanical Engineers – War Emergency Proccdures*, 1951, Vol. 165, No. 69, pp. 254–77.
31. Smith, A.C. "Some Notes on the Data on Hydraulic Oil Properties Required by Systems Designers," *Scientific Lubrication*, February, 1965, pp. 63–69.
32. Godfrey, D.G., and Herguth, W.R. "Physical and Chemical Properties of Industrial Mineral Oils Affecting Lubrication - Part 2," *Lubrication Engineering*, 1995, Vol. 51, No. 6, pp. 493–96.
33. Wege, M.E. "Hydraulic Fluid Requirements - Construction Equipment," *SAE Technical Paper Series*, Paper Number 710723, 1971.
34. Godfrey, D.G., and Herguth, W.R. "Physical and Chemical Properties of Industrial Mineral Oils Affecting Lubrication - Part 3," *Lubrication Engineering*, 1995, Vol. 51, No. 10, pp. 825–28.
35. Hughs, C.W. "Designing to Accommodate High Temperature in Hydraulic Systems," *Machine Design*, Dec. 19, 1968, pp. 134–38.
36. Mently, A.A. "Basic Properties of Fluids for Hydrostatic Transmissions," September, 1979, pp. 93–96.
37. ASTM D 3238 2009, "Standard Test Method for Calculation of Carbon Distribution and Structural Group Analysis of Petroleum Oils by the n-d-M Method," *ASTM International*, Conshohocken, PA.
38. Singh, H., Adhvaryu, A., Singh, I.D., and Chaudhary, G.S. "NMR Based Characterization of Lubricant Base Oils," *Symposium on Worldwide Perspectives on the Manufacture, Characterization, and Application of Lubricant Base Oils, Presented at the 213th National Meeting of the American Chemical Society*, San Francisco, Apr. 13-17, 1997, pp. 255–258.
39. Singh, H., and Singh, I.D. "Use of Aromaticity to Estimate Base Oil Properties," Advanced Production and Application of Lubricant Base Stocks, *Proceedings From International Symposium*, eds. Singh, H., Rao, P., and Tata, T.S.R., New Delhi, India, 1994, pp. 288–294.
40. Metro, E. "Fuchs Study Sees trends Toward Hydrocracked Base Oil Manufacturing Within 5 Years," *Fuels & Lubes International*, Vol. 4, No. 3, p. 7.
41. Bhatnagar, A.K. "Base Oil Composition and Lubricant Performance," Advanced Production and Application of Lubricant Base Stocks, *Proceedings From International Symp*osium, eds. Singh, H., Rao, P., and Tata, T.S.R., New Delhi, India, 1994, pp. 402–417.
42. ASTM D 97 2009, "Standard Test Method for Pour Point Determination of Petroleum Oils," *ASTM International*, Conshohocken, PA.
43. ASTM D 5949 2009, "Standard Test Method for Pour Point of Petroleum Products (Automatic Tilt Method)," *ASTM International*, Conshohocken, PA.
44. ASTM D 5950 2009, "Standard Test Method for Pour Point of Petroleum Products (Automatic Pressure Pulsing Method)," *ASTM International*, Conshohocken, PA.
45. ASTM D 5985 2009, "Standard Test Method for Pour Point of Petroleum Products (Rotational Method)," *ASTM International*, Conshohocken, PA.
46. ASTM D 1250 2009, "Standard Guide for Petroleum Measurement Tables," *ASTM International*, Conshohocken, PA.
47. ASTM D 892 2009, "Standard Test Method for Foaming Characteristics of Lubricating Oils," *ASTM International*, Conshohocken, PA.

48. ASTM D 3427 2009, "Standard Test Method for Gas Bubble Separation Time of Petroleum Oils," *ASTM International*, Conshohocken, PA.

49. Zino, A.J. "What to Look For in Hydraulic Fluids - I.," *American Machinist*, November 6, 1947, pp. 93–96.

50. ASTM D 92 2009, "Standard Test Method for Flash and Fire Points by Cleveland Open Cup," *ASTM International*, Conshohocken, PA.

51. ASTM D 93 2009, "Standard Test Methods for Flash Point by Pensky-Martens Closed Tester," *American Society for Testing and Materials*, Conshohocken, PA.

52. Assn. of Hydraulic Equipment Manufacturers (AHEM) Working Party Report, "Proposed method of Classifying Mineral Oils by Seal Compatibility Index," *Journal of the Institute of Petroleum*, 1968, Vol. 54, No. 530, pp. 36–43.

53. "Industrial Hydraulic Fluids," Lubetext DG-2C, Brochure available from EXXON Company, U.S.A., Marketing Technical Services, P.O. Box 2180, Houston, TX 77252–2180

54. ASTM D 1401 2009, "Standard Test Method for Water Separability of Petroleum Oils and Synthetic Fluids," *American Society for Testing and Materials*, Conshohocken, PA.

55. Papay, A.G. and Harstick, C.S. "Petroleum-Based Industrial Oils - Present and Future Developments," *Lubrication Engineering*, January, 1975, pp. 6–15.

56. Zino, A.J. "What to Look for in Hydraulic Oils - IV Demulsibility," *American Machinist*, December 18, 1947, pp. 94–95.

57. Leslie, R.L. "Hydraulic Fluids for Extreme Service," *Machine Design*, January 13, 1972, pp. 114–117.

58. Hydrick, H. "Hydraulic Systems Benefit from Filtering Standard," *Lubricants World*, 1995, Vol. 5, No. 6, pp. 19–20.

59. Pall Corporation, "Pall Filterability Index Test for Paper Machine Oils," Pall FIT-PMO Revision 4, Glen Cove, New York, April 20, 1995.

60. Sasaki, A., Tobisu, T., Uchiyama, S., and Kawasaki, M. "GPC Analysis of Oil Insoluble Oxidation Products of Mineral Oil," *Lubrication Engineering*, 1991, Vol. 47, No. 7, pp. 525–27.

61. Sasaki, A., Tobisu, T., Uchiyama, S., and Kawasaki, M. "Evaluation of Molecular Weight and Solubility in Oil of Two Different types of Oils," *Lubrication Engineering*, 1991, Vol. 47, No. 10, pp. 809–13.

62. Dipple, R.H. "Hydraulic Oils - Their Physical Properties and Maintenance," *Hydraulic Power Transmission*, April, 1962, pp. 240–42.

63. Ross, F. "Why Treated Hydraulic Oils?" *Applied Hydraulics*, April, 1950, pp. 21–23, 42, 52.

64. Cappell, R.J. "Failure Analysis of Hydraulic Actuators - Break-down of Trichloroethane Used as Cleaning Agent Releases Chlorides," *Materials Protection*, December, 1962, pp. 30–31,33–34,36.

65. Baker, H.R., Jones, D.T., and Zisman, W.A. "Polar-Type Rust Inhibitors: Methods of Testing the Rust-Inhibition Properties of Polar Compounds in Oils," *Industrial & Engineering Chemistry Research*, 1949, 41, pp. 137–140.

66. ASTM D 130 2009, "Standard Method for Detection of Copper Corrosion from Petroleum Products by the Copper Strip Tarnish Test," *ASTM International*, Conshohocken, PA.

67. ASTM D 665 2009, "Standard Test Method for Rust-Preventing Characteristics of Inhibited Mineral Oil in the Presence of Water," *ASTM International*, Conshohocken, PA.

68. Smith, A.C. "Selection of Oils for Industrial Hydraulic Systems," *Scientific Lubrication*, July, 1951, pp. 20–26.

69. Roberton R.S., and Allen, J.M. "Study of Oil Performance in Numerically Controlled Hydraulic Systems," *Proceedings for National Conference for Fluid Power*, 30th Annual Meeting, Philadelphia, PA, Nov. 12–14, 1974, Vol. 28, pp. 435–54.

70. Farris, J.A. "Extending Hydraulic Fluid Life by Water and Silt Removal," Field Service Report 52, Available from Industrial Hydraulics Division, *Pall Corporation*, Glen Cove, New York, 11542.

71. Jones, C. "Properties of Hydraulic Fluids," *Mechanical World & Engineering Record*, January, 1964, pp. 3–5.

72. Gabilondo, P.A., Perez, J.M., and Lloyd, W.A. "Development of a Microreactor Bench Test for Lubricant Evaluation," *Symposium on Worldwide Perspectives on the Manufacture, Characterization, and Application of Lubricant Base Oils, Presented at the 213th National Meeting of the American Chemical Society*, San Francisco, Apr. 13–17, 1997, pp. 278–80.

73. Adhvaryu, A., Pandey, D.C., and Singh, L.D. "Effect of Composition on the Degradation Behavior of Base Oil," *Symposium on Worldwide Perspectives on the Manufacture, Characterization, and Application of Lubricant Base Oils, Presented at the 213th National Meeting of the American Chemical Society*, San Francisco, Apr. 13–17, 1997, pp. 225–228.

74. Klaus, E.E. "To Make Improved Hydraulic Fluids," *Hydrocarbon Processing*, 1966, Vol. 45, No. 6, pp. 167–170.

75. Antika, S., Wang, W.Y., and Dietz, T.G. "Development of Advanced Paper Machine Lubricant," *TAPPI Journal*, 1998, Vol. 81, No.4, pp. 62–74.

76. Chisholm, S.F. "Petroleum Hydraulic Fluids," *Scientific Lubrication*, May, 1960, pp. 35–38.

77. Malleville, X., Faure, D., Legros, A., and Hipeaux, J.C. "Oxidation of Mineral Base Oils of Petroleum Origin: The Relationship Between Chemical Composition, Thickening, and Composition of Degradation By-Products," *Lubrication Science,* 1996, Vol. 9, No. 1, pp. 3–60.

78. Rounds, F.G. "Coking Tendencies of Lubricating Oils," Paper presented at: *SAE National Fuels and Lubricants Meeting*, Cleveland, OH, Nov. 7–8, 1957.

79. Mang, T., and Jünemann, H., "Evaluation of the Performance Characteristics of Mineral Oil-Based Hydraulic Fluids," *Erdöl und kohle-Erdgas-Petrochemie verneigt mit Brennstoff-Chemie*, 1972, Vol. 25, No. 8, pp. 459–64.

80. Rasberger, M. "Oxidative Degradation and Stabilization of Mineral Oil Based Lubricants,"in *Chemistry and Technology of Lubricants*, eds. Mortier, R.M., and Orszulik, S.T., Blakie Academic & Professional, 1992, pp. 83–123.

81. Landis, M.E., and Murphy, W.R. "Analysis of Lubricant Components Associated with Oxidative Color Degradation," *Lubrication Engineering*, 1991, Vol. 47, No. 7, pp. 595–98.

82. Yoshida, T., Watanabe, H., and Igarashi, J. "Pro-oxidant Properties of Basic Nitrogen Components in Base Oil," *Proc. of 11th Int. Colloquium Industrial and Automotive Lubrication - Vol. 1*, ed. Bartz, W.J., Technische Academie Esslingen, Esslingen, Germany, Jan. 13-15, 1998, pp. 433–44.

83. Minami, I. "Influence of Aldehydes in Make-Up Oils on Antioxidant Properties," *Lubrication Science*, 1995, Vol. 7, No. 4, pp. 319–31.

84. Sasaki, A. and Yamamoto, T. "A Review of Studies of Hydraulic Lock," *Lubrication Engineering*, 1993, Vol. 49, No. 8, pp. 585–93.

85. ASTM D 664 2009, "Standard Test Method for Acid Number of Petroleum Products by Potentiometric Titration," *ASTM International*, Conshohocken, PA.

86. ASTM D 974 2009, "Standard Method for Acid and Base Number by Color-Indicator Titration," *ASTM International*, PA.

87. ASTM D 943 2009, "Standard Test Method for Oxidation Characteristics of Inhibited Mineral Oils," *ASTM International*, Conshohocken, PA.

88. ASTM D 2272 2009, "Standard Test Method for Oxidative Stability of Steam Turbine Oils by Rotating Bomb," *ASTM International*, Conshohocken, PA.

89. ASTM D 4742 2009, "Standard Test Method for Oxidation Stability of Gasoline Automotive Engine Oils by Thin-Film Oxygen Uptake (TFOUT)," *ASTM International*, Conshohocken, PA.

90. Godfrey, D., and Herguth, W.R. "Physical and Chemical Properties of Industrial Mineral Oils Affecting Lubrication - Part 4," *Lubrication Engineering*, 1995, Vol. 51. No. 12, pp. 977–79.

91. Godfrey, D., and Herguth, W.R. "Physical and Chemical Properties of Industrial Mineral Oils Affecting Lubrication - Part 5," *Lubrication Engineering*, 1995, Vol. 52. No. 2, pp. 145–48.

92. Watanabe, H. and Kobayashi, C. "Degradation of Turbine Oils - Japanese Turbine Lubrication Practices and Problems," *Lubrication Engineering* , 1978, Vol. 38, No. 8, pp. 421–28.

93. Anon., "Hydraulic Fluid: Making the Choice," *Engineering Materials and Design*, July-August, 1988, pp. 37–38.

94. Roell, B.C., and Cerda De Groote, C.L. "Turbine and Hydraulic Fluids," *Proc. of 11th Int. Colloquium Industrial and Automotive Lubrication - Vol. 3*, ed. Bartz, W.J., Technische Academie Esslingen, Esslingen, Germany, Jan. 13–15, 1998, pp. 1811–16.

95. Zhou, H., Li, K., Wang, X., Xu, Y., and Shen, F. "Pattern Recognition Studies on the Influence of Chemical Composition on Some Properties of Lubricating Base Oils," *Proceedings for International Conference on Petroleum Refining and Petrochemical Processing*, 1991, pp. 387–92.

96. Ramakumar, S.S.V., Affarwal, N., Madhusudhana Rao, A., Srivastava, S.P., and Bhatnagar, A.K. "Effect of Base Oil Composition on the Course of Additive-Additive Interactions," Advanced Production and Application of Lubricant Base Stocks, *Proceedings From International Symposium*, eds. Singh, H., Rao, P., and Tata, T.S.R., McGraw-Hill, New Delhi, India, 1994, pp. 334–41.

97. Galiano-Roth, A.S., and Page, N.M. "Effect of Hydroprocessing on Lubricant Base Stock Composition and Product Performance," *Lubrication Engineering*, 1994, Vol. 50, No. 8, pp. 659–64.

98. Minami, I. "Influence of Aldehydes in Make-Up Oils on Antioxidation Properties," *Lubrication Science*, 1995, Vol. 7, No. 4, pp. 319–31.

99. Papay, A.G. "Oil-Soluble Friction Reducers - Theory and Application," *Lubrication Engineering*, 1983, Vol. 39, No. 7, pp. 419–26.

100. "Industrial Lubricants At-a-Glance," 1989, Vol.2, No.2. Brochure available from *Lubrizol Corporation, Wickliffe*, Ohio.

101. Forbes, E.S. "Antiwear and Extreme Pressure Additives for Lubricants," Tribology, August, 1970, pp. 145–52.

102. Grover, K.B., and Perez, R.J. "The Evolution of Petroleum Based Hydraulic Fluids," *Lubrication Engineering*, 1990, Vol. 46, No. 1., pp. 15-20.

103. Flick, B.B. "Which Characteristics for Hydraulic Fluid?," *Product Engrg.*, August, 1953, pp. 149–53.

104. Newingham, T.D. "What You Should Know About Hydraulic Fluid Additives," *Hydraulics & Pneumatics*, November, 1986, pp. 62.

105. Chu, J., and Tessmann, R.K. "Additives Packages for Hydraulic Fluids," *The BFPR Journal*, 1979, Vol. 12, No. 2, pp. 111–17.

106. Jin, J., and Zhao, C-Z. "Applied Research of Detergents and Dispersants in Antiwear Hydraulic Oil," *Proc. of 11th Int. Colloquium Industrial and Automotive Lubrication - Vol. 3*, ed. Bartz, W.J., Technische Academie Esslingen, Esslingen, Germany, January 13–15, 1998, pp. 1837–45.

107. Lloyd, B.J. "Water, Water Everywhere," *Lubes 'n' Greases*, 1997, Vol. 3, No. 11, pp. 44–53.

108. Papay, A.G. "Advances in Hydraulic Oil Additives Technology," SAE Technical Paper Series, Paper No. 760573, 1976.

109. McHendry, W.D., and Littig, O.J. "Application of Infrared Spectrometric Techniques on the Quantitative Analysis of Hydraulic Fluids," *ASLE Trans*. 1969, Vol. 13, pp. 99–104.

110. Zino, A.J. "What to Look for in Hydraulic Oils. VI - Lubricating Value," *American Machinist*, January,, 1948, 15, pp. 97–100.

111. Bergstrom, A.G., and Sharp, R.Q. "Give Life to Hydraulic Systems," *Machine Design*, January, 1950, pp. 80–86,154,156.

112. Leslie, R.L. "Views on Oils for Hydraulic Service," Technical Paper No. 65-34L, Presented at National Fuels and Lubricants Session, *National Petroleum Refiners Assoc.*, New York, NY, September 15–16, 1965.

113. Anon., "The Graphoid Surface - An Aid to Oiliness," *Lubrication Science*, July, 1951, pp. 25–26.

114. Jung, S-H., Bak, U-S., Oh, S-H., Chae, H-C., and Jung, J-Y. "An Experimental Study on the Friction Characteristics of an Oil Hydraulic Vane Pump," *Proceedings of International Tribology Conference - Vol. III*, October 29-Novvember 2, 1995, Yokohama, Japan, pp. 1621–25.

115. Maccleod, A. "Developments in Hydraulic Oils," *Industrial Lubrication*, January, 1968, pp. 11–19.

116. Langossch, O. "Berechnen und Testen des Konstruktionselementes Hydrauliköl," Ölhydraulik und *Pneumatik*, 1972, Vol. 16, No. 12, pp. 498–501.

117. Griffith, J.Q., Reiland, W.H., and Williams, E.S. "Laboratory and Field Performance of Wear Resistant Antileak Hydraulic Oils," Paper No. F&L-69-64, Presented at the *National Fuels and Lubricants Meeting*, New York, NY, September 17–18, 1969.

118. Klaus, E.E., Tewksbury, E.J., and Fenske, M.R. "High-Temperature Hydraulic Fluids from Petroleum," *Industrial & Engineering Chemistry Research,* Product Research and Development, 1963, Vol. 2, No. 4, pp. 332–38.

119. Tessmann, R.K., Hong, I.T., and Fitch, E.C. "The Selection of Hydraulic Fluids," *Lubrication Engineering*, 1993, Vol. 49, No. 9, pp. 666–70.

120. Anon., "Hydraulic Fluids - Performance Criteria," *Fluid & Air Technology*, February/March, 1995, pp. 28–30.

121. ASTM D 2882 2009, "Standard Method for Indicating the Wear Characteristics of Petroleum and Non-Petroleum Hydraulic Fluids in a Constant Volume Vane Pump," *ASTM International*, Conshohocken, PA.

122. Totten, G.E., Bishop, R.J. Jr., McDaniels, R.L., and Kling, G.H. "Evaluation of Hydraulic Fluid Performance Correlation of Water-Glycol Hydraulic Fluid Performance by ASTM D-2882 Vane Pump and Various Bench Tests," SAE Technical Paper Series, Paper Number 952156, 1995.

123. Bishop, R.J., and Totten, G.E. "Tribological Testing with Hydraulic Pumps: A Review and Critique," *Tribology of Hydraulic Pump Testing - ASTM STP 1310*, eds. Totten, G.E., Kling, G.H., and Smolenski, D.M., ASTM International, Conshohocken, PA, 1995, pp. 65–84.

124. Young, C.H. "Used Hydraulic Oil Analysis," *Lubrication*, 1977, Vol. 63, No. 4, pp. 37–48.

125. Anon., "Proactive Hydraulic Oil Maintenance," *Maintenance Technology*, March, 1994, pp. 39–40.

126. Ogando, J. "A New Way to Look at hydraulic Oil Cleanliness," *Plastics Technology*, December, 1993, p. 42.

127. Anon., "Oil Maintenance in Today's Environment," *Fluid & Air Technology*, January/February, 1994, pp. 32–37.

128. Dong, J., Van De Voort, F.R., Ismail, A.A., and Pinchuk, D. "A Continuous Oil Analysis and Treatment (COAT) System for the Monitoring of Oil Quality and Performance of Additives," *Lubrication Engineering*, 1997, Vol. 53, No. 10, pp. 13–18.

129. Srimongkokul, V. "Why a Proactive Maintenance Program for Hydraulic Oil is Part of Statistical Quality Control," *Lubrication Engineering*, 1997, Vol. 53, No. 4, pp. 10–14.

130. Duncan, J.P. "How to Field-Evaluate Used Hydraulic Fluids," *Hydraulics & Pneumatics*, November, 1973, pp. 99–101.

131. Anon., "Is There Water in Your Oil?," *Fluid & Air Technology*, April/May, 1994, pp. 50–51.

132. Anon., "Keep Hydraulic Oils Clean," *Lubrication*, 1956, Vol. 42, No. 8, pp. 101–8.

133. Poley, J. "Oil Analysis for Monitoring Hydraulic Systems, A Step-Stage Approach," *STLE Preprints*, Preprint No. 89-AM-1E-1, 1989.

134. ASTM D 95, 2009, "Standard Test Method for Water in Petroleum Products and Bituminous Materials by Distillation," *ASTM International*, Conshohocken, PA.

135. ASTM D 6304, 2009, "Standard Test Method for Determination of Water in Petroleum Products, Lubricating Oils, and Additives by Coulometric Karl Fischer Titration," *ASTM International*, Conshohocken, PA.

136. ASTM D 1796 2009, "Standard Test Method for Water and Sediment in Fuel Oils by the Centriguge Method," *ASTM International*, Conshohocken, PA.

137. ASTM D 445 2009, "Standard Test Method for Kinematic Viscosity of Transparent and Opaque Liquids (and the Calculation of Dynamic Viscosity)," *ASTM Internatioanal*, Conshohocken, PA.

138. ASTM D 2161 2009, "Standard Test Method for Conversion of Kinematic Viscosity to Saybolt Universal Viscosity or to Saybolt Furol Viscosity," *American Society for Testing and Materials*, Conshohocken, PA.

139. ASTM D 4052, 2009, "Standard Test Method for Density and Relative Density of Liquids by Digital Density Meter," *ASTM International*, Conshohocken, PA.

140. ASTM D 482 2009, "Standard Test Method for Ash from Petroleum Products," *ASTM International*, Conshohocken, PA.

141. ASTM D 874 2009, "Standard Test Method for Sulfated Ash from Lubricating Oils and Additives," *ASTM International*, Conshohocken, PA.

142. ASTM D 4174 2009, "Standard Practice for Cleaning, Flushing, and Purification of Petroleum Fluid Hydraulic Systems," *ASTM International*, Conshohocken, PA.

143. Wilson, B. "Used Oil Reclamation Processes," *Industrial Lubrication and Tribology*, 1997, Vol. 49, No. 4, pp. 178–80.

144. Dang, G.S. "Rerefining of Used Oils - A Review of Commercial Processes," *Tribotest Journal*, 1997, Vol. 3, No. 4, pp. 445–457.

145. Stofey, W. and Horgan, M. "Reclaiming Hydraulic Oil Eliminates Oil Disposal Problems," *Hydraulics & Pneumatics*, December, 1993, pp. 36–37,76.

146. Topol, G.J. "Vacuum Dehydration of Oils," *Lubrication Engineering*, 1998.

147. Baranowski, L.B. "Degasification and Dehydration of Hydraulic Systems," Presented at the 23rd National Conference on Fluid Power, October 19–20, 1967.

148. Siegel, R. and Skidd, C. "Case Studies Utilizing Mobile On-Site Recycling of Industrial Oils for Immediate Reapplication," *Lubrication Engineering*, 1995, Vol. 51, No. 9, pp. 767–70.

149. Bowering, R.E., Davis, C.T. and Braniff, D.P. "Managing and Recycling Hydraulic and Process Oils at a Cold Mill," *Lubrication Engineering*, 1997, Vol. 53, No. 7, pp. 12–17.

150. Neadle, D.J. "Lubricants Recycling," *Industrial Lubrication and Tribology*, 1994, Vol. 46, No. 4, pp. 5–7.

151. Backé, W., and Lipphardt, P. "Influence of Dispersed Air on the Pressure Medium," *Proceedings from Contamination in Fluid Systems Conference*, Inst. Mech. Engr., Bath, UK, April 13-15, 1976/1977, pp. 77–84.

152. Gimzewski, E. "The Relationship Between Oxidation Induction Temperatures and Times for Petroleum Products," *Thermochinica Acta*, 1992, Vol. 198, pp. 133–40.

153. ASTM D 6158 2009, "Standard Specification for Mineral Oil Hydraulic Fluids," *ASTM International*, Conshohocken, PA, 2009.

154. API 1509 "Engine Oil Licensing and Certification System," 16th Edition, American Petroleum Institute, Washington, D.C., 2007.

155. ASTM D 6546 2009, "Standard Test Methods for and Suggested Limits for Determining Compatibility of Elastomer Seals for Industrial Hydraulic Fluid Applications," *ASTM International*, Conshohocken, PA.

156. ASTM D 6186 2009, "Standard Test Method for Oxidation Induction Time of Lubricating Oils by Differential Scanning Calorimetry (PDSC)," *ASTM International*, Conshohocken, PA.

157. ASTM D 6810 2009, "Standard Test Method for Measurement of Hindered Phenolic Antioxidant Content in Non-Zinc Turbine Oils by Linear Sweep Voltammetry," *ASTM International*, Conshohocken, PA.
158. ASTM D 6971 2009, "Standard Test Method for Measurement of Hindered Phenolic and Aromatic Amine Antioxidant Content in Non-zinc Turbine Oils by Linear Sweep Voltammetry," *ASTM International*, Conshohocken, PA.
159. Fentress, A., Sander, J., Ameye, J. "Using Linear Sweep Voltammetry for Engine Oil Condition Monitoring," *Society of Tribologists and Lubrication Engineers* 2009 Annual Meeting, Orlando, FL.
160. ASTM D 6973 2009, "Standard Test Method for Indicating Wear Characteristics of Petroleum Hydraulic Fluids in a High Pressure Constant Volume Vane Pump," *ASTM International*, Conshohocken, PA.

14 Emulsions

Brian B. Filippini, John J. Mullay, Yinghua Sun, and George E. Totten*

CONTENTS

---

* This chapter is a revision of the chapter titled "Emulsions" by George E. Totten and Yinghua Sun from the *Handbook of Hydraulic Fluid Technology*, 1st Edition.

The present chapter is an update of the corresponding chapter from the previous edition of this handbook [1]. In completing this work, the current authors left certain portions unchanged, but also rewrote several sections, reflecting the differing background and interests of the two sets of authors, and thus rounding out and updating the overall content.

14.1 INTRODUCTION

Although petroleum oil hydraulic fluids provide outstanding performance, there are a number of application areas where their high fire risk potential with respect to equipment damage and loss of lives precludes their use. For example, in a 1956 underground mine fire in Marcinelle, 262 Belgian miners were killed [2]. This fire was caused by damage to a hydraulic line containing mineral oil in the mine shaft. When the mineral oil spray contacted a nearby electrical cable, an electric arc ignited the oil. As a result of this and other disastrous fires, government regulations in Europe, United States, and in other countries, mandated that fire-resistant fluids be used.

Fire-resistance is imparted to hydraulic fluids in one of two ways. One method is in fluid molecular structure, which would include fluids such as phosphate esters. The second common method of imparting fire resistance is by the use of water in the hydraulic fluid formulation. In this case, water has been called a "snuffer" [3] and water-containing fluids have been called "snuffer fluids" [4]. Water may be emulsified in oil (water-in-oil or w/o emulsions) or oil may be emulsified in water (oil-in-water or o/w emulsions) to provide fire-resistant emulsion oil hydraulic fluids. In addition, water may be freely soluble when water-soluble components are used or it may be part of a microemulsion.

Since the 1956 Marcinelle fire, various water-containing emulsions have been among the most common types of fire-resistant fluids used in the coal mining industry [5–8]. Other industries that utilize emulsion hydraulic fluids include: zinc and aluminum die-casting [6], plastic injection molding [6], steel rolling [9], and continuous steel casting [10].

In this chapter, the formulation, properties, and use of emulsion hydraulic fluids will be discussed. This discussion will include:

- Fluid classification
- Fluid chemistry/formulation
- Fluid properties
- Fluid maintenance
- Fluid conversion

14.2 DISCUSSION

14.2.1 Fluid Classification

Emulsion hydraulic fluids are part of a larger ISO classification of fire-resistant fluids summarized in Table 7.2 (see Chapter 7) which include: HFA, HFB, HFC, and HFD fluids [11,12]. The fluids designated as HFC (water-glycol hydraulic fluids) were discussed elsewhere in this handbook, as are HFD fluids, polyol esters and phosphate esters. Fluids classified as HFA and HFB will be discussed in detail in this chapter.

14.2.1.1 HFA Fire-Resistant Hydraulic Fluids

There are two subclassifications of HFA fluids. One is HFA-E, which is an emulsion (E) that is typically formed on-site by the user by mixing 1%–5% oil and additives into water (95%) [12,13]. These fluids are often called soluble oils where water is the continuous phase. The oil is solubilized into water by a process which can be referred to as spontaneous emulsification [14,15]. This will be described in greater detail below. However, it can be noted that the term solubilized is a misnomer in this case since the oil actually forms an emulsion with the water. The amount and type of the soluble oil used will vary with the type of oil, fluid power application, and water hardness (in terms of calcium carbonate and chloride) [5]. These fluids may exhibit viscosities as low as 1 cP. Fluid viscosity can be increased by substantially decreasing water content [16].

The second subclassification of HFA-S is an emulsion or microemulsion of up to 20% of a synthetic polymer (S), additives, and water (>80%). These fluids are also known as high water-based fluids (HWBF) [16]. Thickened or viscous versions of these fluids (VHWBF) are also available which contain approximately 6% of a synthetic polymer thickener, 4% additives, and up to 90% water [10,16]. The polymeric thickeners that are used typically employ an associative thickening mechanism, which is dependent on surfactant-like hydrophobic interactions between either terminal or pendant hydrocarbon functional groups on the water-soluble polymer itself or with an emulsifier. The polymer-polymer or polymer-surfactant complex behaves as a much larger molecule which exhibits a significant increase in thickening efficiency. These associative complexes may be large and opaque (emulsions) or small and clear (microemulsions). Some polymers used for HFA-S formulations are water-soluble and do not employ associative thickening mechanisms. Solutions formed with these polymers are clear.

Therefore, HFA hydraulic fluids may be formulated as true solutions, microemulsions (transparent appearance), and macroemulsions (opaque or white appearance) [12]. (Spikes reports macroemulsion particle sizes of 0.5–10 μm for these fluids [17].) Table 14.1 provides a summary of emulsion particle size and appearance [18].

In general, HFA macroemulsions may not be thermally stable and may exhibit relatively poor antiwear properties in the hydraulic pump [12,19]. However, HFA microemulsions may exhibit improved thermal and viscosity stability relative to the macroemulsions in use [12].

14.2.1.2 HFB Fire-Resistant Hydraulic Fluids

HFB fluids are formulated by emulsifying 45%–50% water into a continuous oil phase [13] and therefore, as opposed to soluble oils (HFA-E), HFB emulsions are designated as invert emulsions. The water level for HFB recently increased to this range from 35–40% to meet updated industry standards. These fluids are supplied in ready-to-use form and are not mixed by the user on-site [17]. The colloidal particle size has been reported to be 3–8 μm [5].

The oil phase of a HFB fluid is formulated similarly to a typical hydraulic fluid, including antiwear agents, corrosion inhibitors, and defoamants. Unlike a hydraulic fluid, they also contain a low-HLB emulsifier which will support the emulsification of 45%–50% water and give the resulting emulsion stability such that it is a viable commercial hydraulic fluid. These emulsifiers are

TABLE 14.1
Effect of Dispersed Phase Size on Emulsion Appearance

Particle Size (μm)	Appearance
Greater than 1 μm	Milky white
1.0 to 0.1 μm	Blue-white emulsion
0.1 to 0.05 μm	Grey semi-transparent
Less than 0.05 μm	Transparent

typically derived from polyisobutenyl succinic anhydride (PIBSA) and have HLBs in the 2–6 range. Co-emulsifiers are also used to either assist in ease of emulsion formation or for emulsion stability.

An early disclosure of an invert emulsion used as a fire-resistant hydraulic fluid was made by Coleman in 1966. Although he disclosed many compositions, a typical oil phase consisted of a PIBSA-derived emulsifier, lecithin, an alkyl amine, a zinc dithiodialkylphosphate, barium sulfonate, a silicone antifoam agent, and a mineral oil. Into that mixture was emulsified 40 parts water [20]. Durr claimed in 1982 to have improved upon the Coleman compositions by adding ethoxylated alcohols and acids [21].

14.2.2 FLUID FORMULATION

An HFA-E or HFB emulsion hydraulic fluid is typically composed of mineral oil, surfactant, water, and additives. The surfactant that is selected is dependent on whether the final fluid will be an HFA-E or HFB. HFA-S fluids typically contain components similar to an HFA-E fluid with the exception that the mineral oil is replaced by a synthetic polymer. The additive package is selected to provide the desired emulsification, corrosion protection antifoam, dyes for leak detection, and antiwear protection. A list of additive performance requirements is provided in Table 14.2 [10,11].

Because the lubrication mechanisms are different for aqueous fluids, including emulsions such as HFA and HFB hydraulic fluids, some of the additives utilized are different to those used for mineral oils [10,22]. Additives that may be used in the formulations of HFA and HFB fluids may include alkyl sulfonates, fatty esters and carboxylates, alkylated phenolic derivatives, and non-ionic surface-active polymers for use as emulsifiers. Silicone polymers have been used as antifoam additives [23,24]. Antiwear properties have been provided by sulfurized sperm oil, metal dithiocarbamates, metal dithiophosphates, polychloronaphtha alkyl xanthates, dibenzyl sulfides, or other derivatives [24]. Possible antioxidants include a variety of substituted alkyl phenols (including di-t-butylcresol, also known as butylated hydroxytoluene, or BHT) and various aromatic amines (such as alkyl-substituted diphenylamines and alkylated phenyl-α-naphthylamine). Corrosion inhibitors, typically amine salts of acids and yellow metal passivators such as 2-mercaptobenzothiazole, are also used [23,24]. An illustrative additive formulation and recommended concentration ranges are shown in Table 14.3

14.2.3 EMULSIFICATION TECHNOLOGY

14.2.3.1 Surfactants

An emulsion is simply a dispersion of one immiscible liquid in another. In the present case it can be either oil-in-water (o/w) or water-in-oil (w/o). A w/o system is often referred to as an invert

---

TABLE 14.2
Development of an Advanced Emulsion HFA Hydraulic Fluid

Performance Requirement	Solution to Problem
Emulsion Stability	**Emulsifier Chemistry**
High volumetric	Viscosity > 1 mm^{2}s^{-1}
Shear stability	Emulsifier and polymer chemistry
Corrosion protection	Compatible corrosion inhibitor
Wear protection	Chemistry of: base oil, emulsifier, antipitter, antiwear, and polymer
Seal compatibility/wear	Emulsion in preference to solution
Low foaming	Compatible antifoam
Bacterial resistance	High pH, stable emulsion*

*Availability of compatible biocides for use if required.

TABLE 14.3

Selected Additives for Improving Physical, Chemical Rheological and Tribological Properties of Hydraulic Fluids

	Additive Name	Functional Application	Recommended Concentration Ranges (wt.%)
1	Alkylbenzotriazole	Copper passivator	0.1–1.0
2	Ethoxylated ethanolamide	Corrosion inhibitor	0.1–1.0
3	Polyalkylmethacrylate	Pour point depressant and viscosity index improver	1.0–2.0
4	Tetraethylene glycol	Antifoam and pour point depressant	0.1–1.0
5	ZDDP[a]	Antiwear agent	0.1–1.0
6	Nitroalkylmorpholine derivative[b]	Biocide	0.001–2.0
7	1-Phenyldodecane	Antiswelling agent	0.001–0.2

[a] For chemical formula.

[b] This additive consists of a mixture of 4-(2-nitrobutyl)morpholine and 4,4′-(2-ethyl-2-nitrotrimethylene)dimorpholine.

emulsion since o/w emulsions (e.g., milk, mayonnaise) are by far the most common in everyday use. Typically, there are three main material components present in every emulsion. These are: oil, water, and a surfactant. With hydraulic fluids these three ingredients all serve a specific purpose. The oil provides lubrication as well as hydraulic effects; the water provides the fire resistance. The surfactant is required to form the emulsion and to keep it stable.

In the case of an o/w emulsion, the oil is dispersed in the water. By comparison, with a w/o emulsion, the water is dispersed in the oil. It is readily apparent that the w/o type would be the more natural choice for hydraulic fluid applications as the oil is in direct contact with the metal surfaces in the hydraulic equipment. Thus it is more effective as a hydraulic fluid. In fact, in this case, the water can be considered as an oil extender as well as a fire suppressant.

An important fact about emulsions is that they are inherently unstable systems. This means that if the emulsion is left to stand for long enough it will revert to a state in which the oil and water phases are completely separated. For hydraulic fluids, this type of emulsion breakdown may lead to increased pump and component wear and corrosion as well as a decrease in fire resistance. These facts have two important implications: first, it is necessary for the emulsion formulator to choose the optimal surfactant system so that the separation is delayed for as long as possible. Second, the method of manufacture of the emulsion also has to be optimized in order to obtain maximal stability. These two aspects will be discussed below. However, a general introduction to emulsion stability will first be given.

Figure 14.1 shows the degradation of a normal w/o emulsion over time. The purpose of the surfactant and the mixing is to keep the emulsion in the configuration shown in the far left part of the figure. The four stages shown occur at a microscopic level but lead to some observable effects which are important for performance. As is shown, the original emulsion (leftmost picture) is a mixture of various sized droplets. For hydraulic fluids, the droplets typically have an average size of about five microns. Since this size particle scatters white light, this emulsion would look like milk.

The second and third pictures from the left show that the larger droplets will, over time, preferentially settle at the bottom. As this phenomenon occurs, the emulsion will look like it is separating into indistinct layers. In the case of the usual hydraulic fluid, at this stage the emulsion will appear darker on top and whiter near the bottom. Over time, even the smaller particles will settle. When this happens, a clear oil layer may be observed at the top, and a white layer or band may appear at the bottom. The same process occurs in milk or in oil-in-water hydraulic fluids, though in an opposite sense. In the case of milk, the oil separates to the top to form a cream layer. Thus this phenomenon is known as creaming. Since in the present case the water separates to the bottom, it is more appropriate to refer to this phenomenon as reverse creaming or sedimentation.

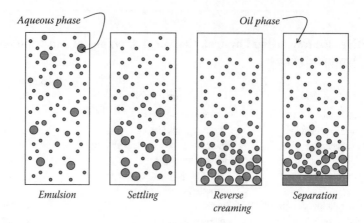

FIGURE 14.1   Depiction of normal aging processes in a w/o emulsion.

Both oil separation and sedimentation are readily reversible with agitation but can cause performance issues since there is a greater concentration of water at the bottom and oil at the top. In addition to this, the emulsion at the bottom is thicker than it was originally and will flow less easily. As will be discussed in more detail below, this phenomenon is minimized by emulsifier and process choices.

The most harmful process shown in Figure 14.1 is separation of the water from the oil (rightmost picture). This is a result of massive amounts of coalescence (merging of the water droplets). Its presence in an emulsion is visually apparent since a layer of water appears on the bottom of the container. This process can lead to significant product malfunction and is normally one of the chief concerns addressed in the formulation. This can be reversed only by applying high shear to the emulsion.

The underlying mechanisms that lead to all of the phenomena illustrated in Figure 14.1 are shown in Figure 14.2. These mechanisms will be briefly described. From the moment that the droplets are formed they undergo changes which can ultimately lead to emulsion instability and lack of product performance. The most important of these changes are those that are shown in Figure 14.2. Coalescence and settling are the most significant, as they could lead to the greatest performance difficulties.

1. Settling (creaming in o/w emulsions): Apart from coalescence, this is the most significant process that could lead to performance problems. It stems from the fact that water is denser than oil. This causes the water droplets to fall to the bottom of the container or oil droplets to rise to the surface. Increased viscosity of the emulsion and decreased size of the droplets can help offset settling. A thicker emulsion will cause the droplets to fall more slowly (or

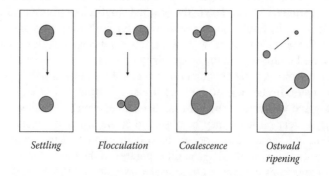

FIGURE 14.2   Droplet mechanisms that lead to normal aging processes in a w/o emulsion.

rise more slowly in the case of an o/w emulsion). The normal viscosities encountered in hydraulic fluids are high enough to inhibit settling as long as the droplets sizes are small enough. Smaller droplets also settle more slowly. The choice of emulsifier and process will greatly influence the droplet size.

2. Flocculation: This process leads to the joining of two or more droplets without the droplets merging (i.e., without coalescence). It effectively creates larger units out of smaller ones. Simple mixing or normal pumping of the emulsion is sufficient to reverse this process.

3. Coalescence: This occurs when two smaller droplets merge into one bigger droplet. It happens to flocculated droplets. As will be detailed below, one of the emulsifier's roles is to stop this from happening. If this process is not inhibited, it will ultimately lead to complete separation of the water and the oil, causing product failure. The product formulator should choose an emulsifier system that contains a mixture of chemistries, including larger molecules to provide a protective structure around the droplet.

4. Ostwald ripening: This process is of secondary importance but can lead to greater settling. Energy balances within the emulsion dictate that some of the water will travel from the smaller droplets to the bigger ones. This transfer causes a greater size imbalance, increasing the size of the bigger droplets while decreasing the size of the smaller ones. The overall effect is greater settling. Formulators can include proprietary ingredients which inhibit this process from occurring. In addition, as will be discussed in the processing section, the type of work input can also be optimized to minimize this effect.

14.2.3.2 Testing Emulsion Stability

It is critically important that hydraulic fluid emulsions remain stable during use. This means that they must be able to withstand various temperature conditions and remain essentially as is shown in the leftmost picture in Figure 14.1. There are various tests to evaluate this type of emulsion thermal stability. One common method is to store the emulsion at an elevated temperature (140°F) which serves as an accelerated test. The fluid being tested is observed for oil and/or water separation. This test was used to differentiate emulsion thermal stability of two o/w emulsion fluid formulations after use (see Figure 14.3). The same accelerated test was performed on the two fluids illustrated in Figure 14.3 without prior use. The results, shown in Figure 14.4, illustrate the amount of improvement that can be achieved through fluid formulation [25].

Typically, the HFB hydraulic fluid emulsion will encounter various cold-hot temperature cycles during shipment and storage, which can affect emulsion stability. This type of thermal stability is simulated by storing the fluid for a prescribed period of time at a cold (0°F) and hot (150°F) temperatures. The results of using this test with three different emulsion formulations are illustrated in

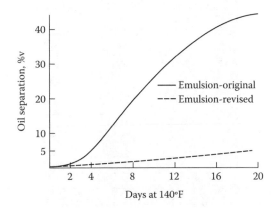

FIGURE 14.3 Effect of emulsion formulation on aging stability in use.

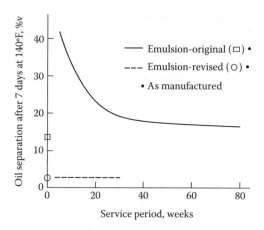

FIGURE 14.4 Effect of emulsion stability of fresh, as-manufactured emulsion fluids.

Figure 14.5 [26]. Note the water separation in Sample A, the settling in Sample B, and the emulsion homogeneity in Sample C.

Other tests that have been used to evaluate emulsion thermal stability have included more drastic temperature extremes in temperature cycling and centrifugation [10,26,27].

14.2.3.3 Surfactants

Surfactants are chemicals that are referred to as being "surface active." In the present chapter they will also be referred to as "emulsifiers." They are molecules or polymers which are made up of a water-loving (hydrophilic) portion and a oil-loving (lipophilic) portion. These hydrophilic groups are sometimes called lipophobic (oil-hating) and the lipophilic groups can be called hydrophobic (water-hating). Because of this dual nature, they seek to be in both the water and the oil phase. This makes them migrate to, and remain at, the oil-water interface and this stabilizes the emulsion. Figure 14.6 depicts a typical surfactant.

Illustrative surfactant molecules are depicted in Table 14.4. These include stearic acid, which has a heptadecyl, $CH_3(CH_2)_{16}-$, lipophilic end (lipophile) and a polar carboxyl, COOH, hydrophilic end. Another example is sodium oleate, where the lipophile is: $CH_3(CH_2)_7CH = CH(CH_2)_7-$, and the

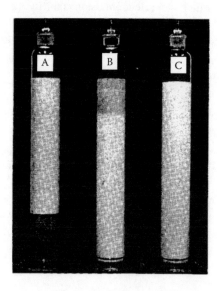

FIGURE 14.5 Effect of freeze-thaw cycling on three different emulsions.

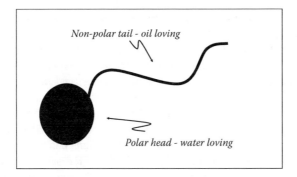

FIGURE 14.6   Depiction of a surfactant molecule.

hydrophile is the sodium salt of the carboxylic acid, $-COO^-Na^+$. Sodium oleate is an example of an ionic surfactant.

Examples of non-ionic surfactants in Table 14.4 are the ethoxylated oleyl alcohols, where the oleyl group is the lipophile and the poly(ethoxy) groups, $-(CH_2CH_2O)_n-$, are the hydrophiles. The larger the value of n, the more hydrophilic the poly(ethoxy) functionality. The other example shown is polyisobutenyl succinic acid. In this example, the polyisobutenyl tail is a very long, bulky lipophile and the succinic acid group is the lipophile.

A very useful measure of emulsifier (or surfactant) effectiveness is referred to as hydrophilic-lipophilic balance (HLB). As suggested by its name, HLB is a measure of the relative amounts and/or strengths of the two important parts of the surfactant molecule. It has been experimentally verified that this value is directly related to emulsion performance.

Typical HLB values range form 0 to about 40 with those between 0 and 20 being of the most importance in this chapter. Lower HLB values refer to more lipophilic surfactants. Specifically, values below about 9.0 indicate that the surfactant is more lipophilic in character, more oil-soluble, and tends to produce w/o emulsions. Hydrophilic surfactants (i.e., more water-soluble and tending to produce o/w emulsions) exhibit HLB values >11.0. HLB values in the range of 9–11 are intermediate.

Because of the usefulness of this concept, there is a relatively large literature associated with it. There are many ways to evaluate HLB values, both theoretically as well as experimentally. Some of these are given in Appendix A. The interested reader is referred to several excellent reviews on this topic for further information [28,29].

One of the first steps an emulsion technologist takes in formulating a new product is to determine the optimal HLB range for the application. This can be done by testing a standard set of surfactants

TABLE 14.4
Illustration of the Hydrophobic and Hydrophilic Portions of Commonly Encountered Surfactants

Molecule	Hydrophobe	Hydrophile
Stearic Acid	Heptadecyl $[CH_3(CH_2)_{16}-]$	-COOH
Sodium Oleate	Oleyl $CH_3(CH_2)_7CH=CH(CH_2)_7-$	$-COO^-Na^+$
Oleyl Alcohol, 2-ethoxylate	Oleyl $CH_3(CH_2)_7CH=CH(CH_2)_7-$	$-O(CH_2CH_2O)_2H$
Oleyl Alcohol, 20-ethoxylate	Oleyl $CH_3(CH_2)_7CH=CH(CH_2)_7-$	$-O(CH_2CH_2O)_{20}H$
Polyisobutenyl succinic acid	Polyisobutenyl $(CH_3)_3C-(CH_2C(CH_3)_2)_{20-40}-CH_2C(CH_3)_2-$	$-CH(CO_2H)CH_2CO_2H$

that represent a wide range of HLB values in a formulation that is similar to the one of interest. An excellent treatment of this topic can be found in reference 28. Once this is done, the next step is to determine the optimal chemistry within that range. As can be imagined, this study very often involves looking for the chemistry that provides the most stable emulsion at the lowest cost. This can translate into choosing the chemistry that can be used at a low level.

One surfactant will rarely be sufficient to provide both stability and ease/robustness of process for emulsion manufacturing. Formulators will generally choose a combination of surfactants to do the job and each surfactant will have a different HLB. The collective HLB value of the surfactants will be in the experimental HLB range discussed in the previous paragraph.

Inventors continue to develop compositions and emulsifier systems for these fire-resistant hydraulic fluids. Some of the more recent contributions in this area include compositions that contain dimerized/unsaturated fatty acids and salts [30], neutral calcium sulfonates [31], succinimide derivatives [32], as well as a variety of surfactant mixtures [33–37].

14.2.3.4 The Role of the Surfactant in Emulsion Stabilization

In order to form an emulsion it is necessary to break up one of the liquid phases into smaller droplets and disperse these in the other liquid. This obviously requires work input. Once the droplets are formed they will tend to coalesce again. The job of the surfactant is twofold—that is, to reduce the work input required and also to inhibit coalescence. Both functions are accomplished by the surfactant diffusing to, and remaining at, the interface between the water and the oil. Work input reduction will be discussed below. Stabilization versus coalescence then will be discussed.

Figure 14.7 depicts the two different emulsion types with an emphasis on the surfactant configuration. As indicated in Figure 14.7, for both emulsion types it is necessary for the surfactant to attach itself to the interface between the oil and the water. In the o/w case the polar head sticks out toward the water while in the w/o case the polar head points inward. The function of this layer of surfactant is to keep the droplets apart. There are two stabilization mechanisms that can do this: Electrostatic [Figure 14.7 (top)] and steric [Figure 14.7 (bottom)]. Both of these can be effective in o/w emulsions but steric stabilization is the primary mechanism with w/o systems.

With regard to HFA emulsions, it can be seen from Figure 14.7 (top) that the emulsion droplet contains an overall charge which is due to the charge on the surfactant (negative in the figure, e.g., a sulfonate ion). This then attracts positive ions which surround the droplet (these would be sodium ions in the case of sodium sulfonate). These ions form a positively charged shell around the droplet. The positive shells then repel one another. With this electrostatic effect it is important to have a charge on the surfactant so that one droplet will repel another more effectively. In aqueous systems, non-ionics work primarily by steric effects. There are probably some electrostatic effects (like dipole interactions) but they are of lesser importance in this discussion.

As shown in Figure 14.7 (bottom), steric stabilization requires the presence of a big, bulky group on each surfactant to force the droplets to remain at a certain distance from one another and prevent them from coalescing. For the case of an o/w emulsion, the bulky group would be the hydrophilic part of the molecule (i.e., it would be polar; e.g., a long EO group). For a w/o emulsion it would be the hydrophobe. For example, sorbitan monooleate is a w/o emulsifier with a oleyl tail of 18 carbons. Derivatives of polyisobutenyl succinic anhydrides (PIBSA) are also frequently used as w/o emulsifiers. They have hydrocarbon chains of 50 to 150 carbons. Experimentally, one finds that emulsion stability is enhanced by the inclusion of PIBSA-derived emulsifier. A reasonable explanation of this observation is that the long, bulky hydrocarbon tail of PIBSA provides steric hindrance between the emulsifier-coated droplets, thus improving emulsion stability.

It should also be apparent from the figures that not only the surfactant molecular size and charge are important, but also the surfactant concentration at the interface. The greater the number of surfactant molecules that surround the droplet, the greater will be the stabilizing effect. Thus it is important that the surfactant sticks or is anchored well to the interface. The proper choice of HLB and chemistry assures that the surfactant sticks and anchors to the interface.

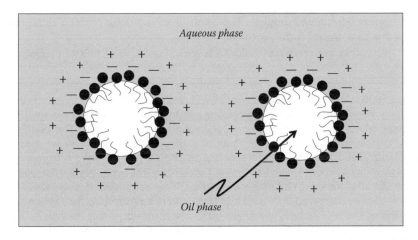

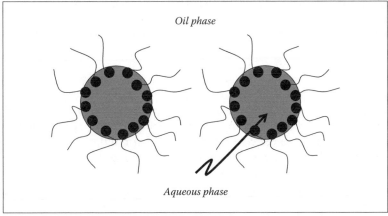

FIGURE 14.7 Depictions of o/w (top) and w/o (bottom) droplets.

Effect of hard water: The presence of salt (e.g., from hard water) decreases the stability of systems that rely on the electrostatic repulsion effect [38–40]. This is true primarily of HFA-type products. The choice of surfactant system can help to overcome this hard water effect [41].

14.2.3.5 The Role of the Surfactant on Ease of Processing and Droplet Size

As already discussed, the final emulsion should have small droplets. The surfactant plays a role in determining droplet size. As mentioned before, the surfactant seeks to be at the water-oil interface. The surfactant molecule makes the water appear more oil-like and the oil more water-like. This, in turn, will make it easier to form water-oil interfaces. If the surfactant can get to the interface relatively quickly, this will further ease processing by requiring less work to form the emulsion. Likewise, with the same reasoning, the surfactant will also allow the formation of smaller droplets.

14.2.3.6 The Role of Processing in Emulsion Stability

Since the emulsion is an inherently unstable system, the method of manufacture becomes quite important in determining final properties for the hydraulic fluid. Because this is an old technology area, numerous methods have been developed to form the best emulsion. The most common approach is to simply add the dispersed phase to the continuous phase while mixing. This is also the usual approach in the case of hydraulic emulsions. The process for HFA fluids (oil-in-water) involves adding the oil to the water while stirring the oil/water mixture sufficiently fast to break up the larger oil droplets into smaller droplets. The emulsification process for HFB fluids (water-in-oil) is reversed. In this case, the oil is the continuous phase.

By way of illustration, the procedure below is a typical example of how to prepare a HFB fluid in a laboratory.

> Combine in a 1000-mL beaker both the mineral oil and chemical additive (containing emulsifier, anti-wear agent, corrosion inhibitor, etc.) in the recommended ratio. In this specific example, 144 g of a group I mineral oil was combined with 15.3 g of commercially available additive. The resulting mixture was swirled to give coarse mixing, and for about 30 s using a Turrax mixer on low speed to ensure thorough mixing. Turrax mixer speed is increased to 3000–5000 rpm and water (141 g total) is added by means of an addition funnel. The water is initially added drop-wise, ensuring that the forming emulsion incorporates all of the water very rapidly once added. If the water addition gets ahead of the rate of incorporation, slow or stop the addition until all of the water is emulsified. As the addition progresses, the water addition can be increased from drop-wise to a slow stream, mixing continually throughout. The addition of the water takes 3–5 minutes. Once all of the water has been added, the mixing speed is increased to ~6000 rpm and mixed for 20–30 minutes. The emulsion will likely become warm to the touch. If the emulsion becomes hot to the touch, or if steam is evolved, mixing should be stopped. The final emulsion has droplets in the range of 1–10 microns, with the vast majority being in the range of 2–5 microns.

Both the type and the amount of mixing that is provided are critical to producing a good emulsion. The normal procedure employs a high shear mixing device to achieve small enough droplet sizes. Figure 14.8 is an example of how the water droplets in a w/o emulsion are formed under shear stress. In essence, the shear stretches out the droplet into a cylindrical form. This form then becomes unstable as its diameter becomes smaller and it ultimately breaks up into smaller droplets. Figure 14.8 depicts the result of one possible shear scenario. As is shown in the figure, a variety of droplet sizes are produced under these conditions.

The surfactant molecules require time to diffuse to the interface of the droplets which are formed. The surfactant molecule adsorbs at the water-oil interface such that the polar end penetrates into the water droplet surface and the hydrophobic end extends into the oil phase. In this way the surface of the water droplet is stabilized against the degradation processes described above.

Both droplet breakup and surfactant adsorption happen with all of the equipment commonly used in the production of these emulsions. It is apparent that this will lead to the presence of a distribution of droplet sizes in the final product and also to a time lag for all of the surfactant to get to the interface. These two effects are important in determining product performance. As was discussed above in both the emulsion stability and the surfactant stabilization sections, both small droplet size and good coverage of the droplet surface with surfactant are required for good performance.

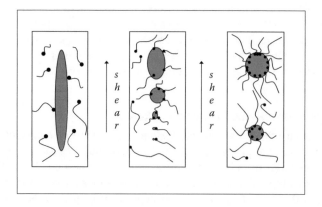

FIGURE 14.8   Example of the breakup of a water droplet under shear in oil.

Droplet size is determined by both the process and the chemistry (primarily the surfactant choice). The main parameters for processing are the work input type and the amount. For a given emulsion formulation, greater work input will translate to smaller droplets. Emulsion manufacturers will choose the equipment that delivers the greatest amount of work in the shortest time, thus costing the least amount of money. The most common piece of equipment is known generically as a colloid mill [42]. This usually includes a static element (stator) and a rotating element (rotor) in the emulsification part of the equipment. It utilizes relatively high speed measured by revolutions per minute (rpm) and a small gap between stator and rotor to achieve good droplet breakup. Other processes can utilize droplet impingement, rapid pressure drops, static mixers, or even lower-speed blenders in larger vessels.

There is one situation in which the chemistry is of more importance than processing—the case of soluble oils that are used to make o/w emulsions (HFA). In this situation the surfactants and other ingredients are chosen by the formulator such that the work required to form an emulsion is essentially zero. The oil spontaneously emulsifies in the water. Minimal agitation is required to make sure that the oil has contact with the water and to disperse the oil droplets that are formed.

One problem that is often encountered in mixing an emulsion is that insufficient mixing times are used [23]. Coleman showed that in one scenario an emulsion began to separate after mixing with a mechanical mixer at room temperature. This was due to inadequate mixing as indicated microscopically by non-uniform droplet formation, as shown in Figure 14.9a. This fluid was then mixed

(a)

(b)

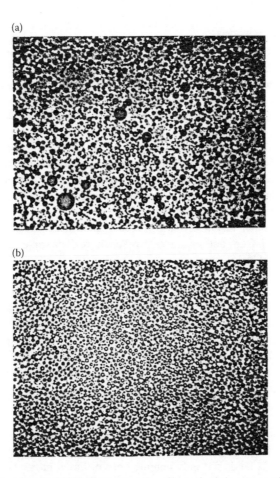

FIGURE 14.9 Photomicrographs (75X) illustrating the effect of mixing on particle size and uniformity. (a) Initial emulsion, (b) Emulsion after additional mixing.

further in a homogenizer, which provided a stable emulsion with uniform droplet size, as indicated in Figure 14.9b [23].

Figure 14.9 represents a readily observable case of the importance of mixing time. There are many instances of the same phenomena occurring without the accompanying emulsion separation. Figure 14.10 presents one of these. This is an instance of a w/o hydraulic fluid emulsion being made using the normal blend time. This emulsion was reprocessed two additional times. Figure 14.10 shows the results of particle size analysis after each processing step. The particle sizes were determined using a Coulter LS-230 instrument which determines droplet size by light-scattering techniques. The graphs plot droplet size diameter versus volume percent.

There are two aspects of Figure 14.10 that should be noted. First, the original distribution was trimodal, but was improved to mono-modal by additional work. The single reprocessing removed all of the larger-sized droplets and some of the smaller sizes (top plot). The double reprocessing eliminated both high and low sizes (bottom plot), and so the particle size distribution narrowed. As noted in the stability section, this translates to less Ostwald ripening and thus greater stability. Second, it can also be seen that the average size became smaller with further mixing. As noted in the stability section, this leads to less settling and, ultimately, to less coalescence.

It should also be noted that there can be too much mixing for a given formulation. In this case the droplets will be made too small, the surfactant concentration will be insufficient for the interfacial area created and the interface will become starved of surfactant. Thus, it is important to optimize processing conditions in order to achieve maximum performance.

14.2.3.7 W/O Microemulsions

Although HFB hydraulic fluids are typically macroemulsions (a common alternative name for a normal emulsion), there has been some interest in developing microemulsion HFB fluids since they would be less susceptible to the very large viscosity variations which may be encountered with macroemulsions. This technology needs relatively large amounts of surfactant to achieve a commercially viable product, usually in the range of 3%–10% by weight of the total composition. In addition, these microemulsions normally require a co-surfactant.

Garti et al. performed a systematic investigation of the effect of emulsifier type on the phase diagrams of water-in-oil mixtures of emulsifier/paraffinic mineral oil and distilled water [12]. The main surfactant in this case is a polyethoxylated alcohol with 12–18 carbon atoms and between 2 and 25 ethoxy groups. The co-surfactant is an alcohol such as pentanol. This work ultimately led to the development of viable commercial candidates [43].

There has also been other patent activity in this area which utilizes amidoalkyl betaines to achieve the same goal [44]. Garti and several co-workers have done some further fundamental work on these types of systems [45,46].

14.2.3.8 Viscosity

Because of the specific application in which the hydraulic fluid is used, its rheology is a critical property. It is important to have some understanding of the variables that affect this property in the case of hydraulic fluid emulsions. It is necessary to consider at least the following variables: amount of internal phase, viscosity of the external phase, droplet size, amount of shear acting on the fluid, temperature of the fluid and—as in the present case—the pressure on the fluid. These will be treated in this section.

14.2.3.8.1 Amount of Internal Phase

HFA and HFB fluids have the opposite internal phase materials: oil in the case of HFA and water with HFB. In addition, HFA fluids normally contain a much higher level of water. Figure 14.11 shows that HFA and HFB fluids not only contain very different amounts of water, but also that increasing water concentration exhibits the opposite effect on fluid viscosity. HFA fluids exhibit the expected decrease in viscosity with increasing water concentration. However, viscosity increases

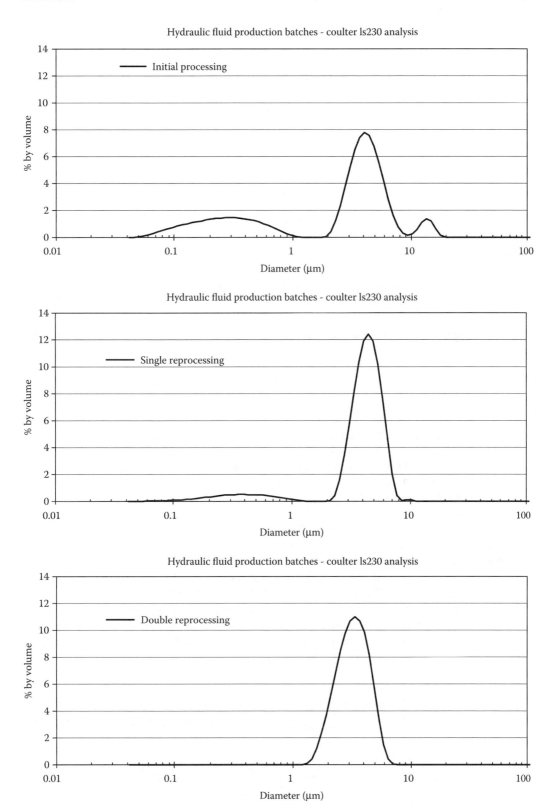

FIGURE 14.10 Effect of mix time on droplet size distribution.

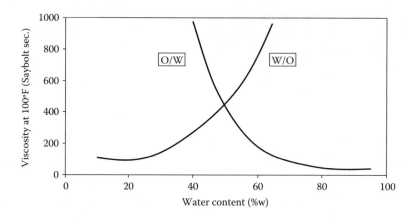

FIGURE 14.11   Effect of water content on O/W and W/O emulsion viscosity.

dramatically with increasing water concentration for HFB fluids. This is because the internal phase decreases with increasing water for HFA, and increases with increasing water amount with HFB fluids.

A useful equation for representing the effect of internal phase has been developed over the years and is given in Equation 14.1 [47].

$$\eta = \eta_0 \left(1 - \Phi/\Phi_{max}\right)^{-2}, \tag{14.1}$$

where η is viscosity of the fluid, η_0 is the viscosity of the external phase, Φ is the volume % of internal phase, and Φ_{max} is the maximum possible internal phase at close packing of the droplets (usually taken as ~0.64).

If this equation were plotted as in Figure 14.11 it would look very similar to the w/o line. This is because water is the internal phase. Likewise, in the case of the o/w emulsion, oil is the internal phase and thus its behavior would go in the opposite direction.

14.2.3.8.2   Effect of External Phase Viscosity

As noted in previous sections, HFA fluids contain >80% water as the external phase and more typically contain approximately 95% water. Without the addition of a thickener, the fluids exhibit viscosities as low as 1–2 cSt [48,49]. As is implied in Equation 14.1, this low viscosity is due to the fact that the viscosity of water is low and there is very little internal phase.

Figure 14.12 shows that for a w/o emulsion the viscosity increases with base oil viscosity and water content [50]. While base oil viscosity has some effect on the viscosity of w/o emulsions, the effect is minimal in view of the very high water concentration and the relatively low final formulated viscosities.

14.2.3.8.3   Droplet Size

Emulsion viscosity (η) is dependent on the diameter of the particle. Generally, emulsion viscosity increases with decreasing particle size, as shown Equation 14.2 (known as the Sherman equation) [51]:

$$\eta = X \frac{1}{d_m} + C, \tag{14.2}$$

where d_m is the average particle diameter and X and C are constants. This effect is illustrated by the data in Table 14.5 [51].

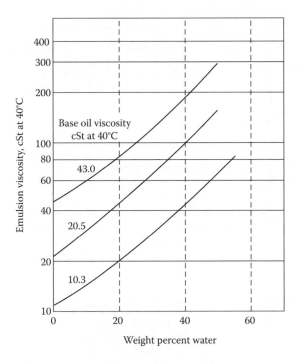

FIGURE 14.12 Effect of water content on w/o emulsion viscosity.

14.2.3.8.4 Effect of Imposed Shear Rate

Coleman showed that the viscosity of w/o emulsions (HFB fluids) decreased with increasing shear rate, as illustrated in Figure 14.13. The amount of viscosity loss increased with increasing initial fluid viscosity measured at shear rates $< 10^{-1}s^{-1}$, as illustrated in Figures 14.13(a) and (b) [23]. This viscosity loss was temporary and the fluids returned to their initial viscosity (non-Newtonian) when the high fluid shear rates were removed. Interestingly, the non-Newtonian behavior increased with increasing water content for emulsions with the same initial viscosity as illustrated in Figure 14.13(c) [23]. Due to these non-Newtonian effects, w/o emulsions were initially developed to provide a viscosity of 250 SUS (approximately 54 cSt) at 100°F when measured at a shear rate of 10,000 s^{-1}, a typical shear rate encountered in hydraulic pump applications. Therefore, the starting kinematic

TABLE 14.5
Effect of Internal-Phase Ratio (Φ) on
Particle Diameter and Emulsion Viscosity

Φ	$d_m\mu$	η (Poises)
0.1809	1.59	2.45
	1.15	3.77
0.3707	2.74	3.09
	1.41	3.91
0.5704	3.61	4.83
	1.38	5.82
0.6739	3.98	6.38
	1.56	7.12
0.7280	2.71	7.58
	1.46	8.18

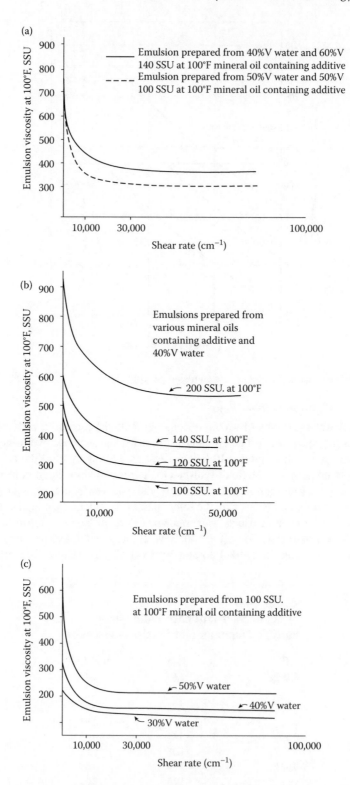

FIGURE 14.13 (a) Non-Newtonian (temporary) viscosity loss as a function of water content of a 100 SUS mineral oil containing an additive at 100°F. (b) Non-Newtonian behavior as a function of base-oil viscosity at 100°F. (c) Non-Newtonian viscosity properties as a function of base-oil viscosity at a constant initial viscosity.

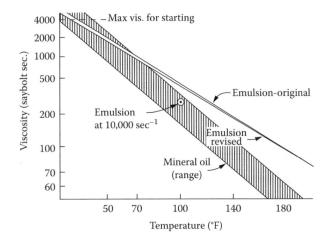

FIGURE 14.14   Effect of temperature on mineral oil and different emulsions at a low shear rate.

viscosity was increased, to 400 SUS (approximately 86 cSt) at 100°F which was higher than the 150–315 SUS (approximately 32–68 cSt) at 100°F typically used for petroleum oil formulations to provide proper pump operation and lubrication [25]. This is illustrated in Figure 14.14. Because of this higher viscosity, it was recommended that hydraulic pump suction be maintained at the minimum in order to avoid cavitation during cold startup [25]. Although the specific hydraulic pump requirement recommended by the manufacturer should be confirmed, the inlet vacuum values shown in Table 14.6 are typical of those often recommended [26].

The o/w emulsions (HFA-E) fluids typically do not exhibit shear thinning behavior, as was shown by Isaksson in Figure 14.15 [52]. This is not surprising in view of the relatively low initial viscosities exhibited by these fluids. However, thickened HFA fluids, HFA-S, may also be non-Newtonian as illustrated in Figure 14.16 [52]. The degree of non-Newtonian behavior is dependent on the molecular weight of the polymeric thickener, thickener concentration, initial fluid viscosity, shear rate, and degree of associative thickening involved.

14.2.3.8.5   Viscosity-Temperature Behavior

Viscosity-temperature is expressed numerically as the Viscosity Index (VI). The higher the value, the less the viscosity will vary with respect to temperature. For example, a representative VI for a group I petroleum oil is 95. A VI of 140 has been reported for an HFB fluid [53]. This means that the variation of viscosity is significantly less for an HFB fluid than for a petroleum oil as the temperature is increased. This is illustrated in Figure 14.14 [25].

Viscosity-temperature is seldom reported for HFA fluids because of the already low initial viscosity. The variation in viscosity with temperature would not seem great as is illustrated in Figure 14.17 [54] but the viscosity does, in fact, vary substantially. For example, a viscosity variation from 1.2 cSt at 20°C to 0.65 cSt at 50°C has been reported [55].

TABLE 14.6
Typical Inlet Vacuum Values

Type of Pump	Max. Intake Vacuum (in Hg)
Vane	5
Gear	7
Piston	11

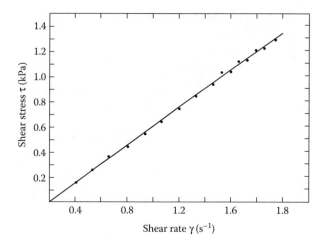

FIGURE 14.15 Shear stress as a function of shear rate for a water-glycol at $n = 0.974$, $h_m = 22.8°C$, $n = $ power law index, and h_m is the measured temperature.

Figure 14.18 illustrates the discharge fluid flow rate for an HFB fluid as a function of temperature [26]. This figure shows that the optimal fluid flow rate occurs between 125–150°F. The fluid flow rate decreases at lower temperatures due to increasing viscosity with decreasing temperature. Decreasing fluid flow rates at higher temperatures occur because of increased leakage due to very low viscosity. Most sources report 150°F as the maximum operating temperature for either HFA or HFB fluids due to the greatly increasing vapor pressure and high evaporation rate [26,50,56,57]. The minimum recommended operating temperature for these fluids is 40°F [56]. The optimal fluid temperature is 125°F [57].

14.2.3.8.6 Viscosity-Pressure Coefficients

The importance, calculation, and use of viscosity-pressure coefficients are described in detail in Chapter 5. The pressure-viscosity coefficients for a number of HFA fluids and an HFB fluid are provided in Table 14.7 [52,58]. Values for water and oil are provided for comparison. In general,

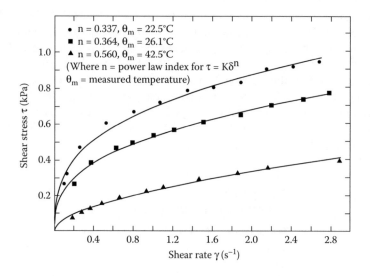

FIGURE 14.16 Shear stress as a function of shear rate for a HWBHF (High Water-Based Hydraulic Fluid) based on a water-soluble synthetic polymer.

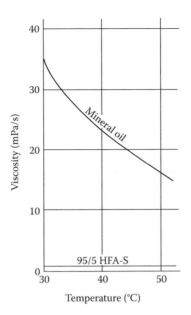

FIGURE 14.17   Viscosity-temperature relationship comparison of a mineral oil and an HFA hydraulic fluid.

the pressure-viscosity coefficients are dependent on the water concentration and the values for HFA fluids are little different from water itself. The pressure-viscosity coefficient for the water-in-oil emulsion shown is nearly equal to the value determined for oil itself. This data suggests that the lubrication film-forming properties for the HFA fluids are very poor. However, at least within the recommended temperature range, the lubrication properties for the HFB fluid are similar to the oil from which it is formulated.

Viscosity-pressure coefficients are most important for the high contact pressures encountered with EHD lubrication (see Chapter 6). Although petroleum oils do exhibit some increase in viscosity at typical hydraulic system pressures (see Chapter 16), there is minimal, if any, fluid viscosity increase observed at most hydraulic system pressures with water-containing fluids [59,60]. This is illustrated for an HFA fluid in Figure 14.19 [59].

14.2.3.9 Foaming and Air Entrainment
Two important properties for any hydraulic fluid are low foaming and minimal air entrainment. This subject is discussed extensively in Chapter 4, including mechanisms of foaming and air

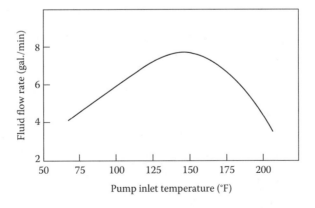

FIGURE 14.18   Effect of emulsion fluid temperature on the discharge flow rate for a hydraulic vane pump.

TABLE 14.7
Viscosity-Pressure Coefficients for HFA and HFB Hydraulic Fluids[a]

Fluid Classification	Note	Viscosity-Pressure Coefficient (U) mm²N⁻¹[b]
Water	–	7.5×10^{-4}
HFA-E	Emulsion, 95% H_2O	1.39×10^{-3}
HFA-E	Microemulsion, 95% H_2O	1.87×10^{-3}
HFA-S	Synthetic polymer solution, 80% H_2O	3.23×10^{-3}
HFA-S	Synthetic polymer solution, 88% H_2O	2.62×10^{-3}
HFA-S	Synthetic polymer solution, 90% H_2O	7.5×10^{-4}
HFB	Water-in-oil, 40% H_2O	20.7×10^{-9}[c,d]
	Mineral oil	$28,26 \times 10^{-9}$[c,e]

[a] All values taken from Reference 36 unless otherwise noted.
[b] All values determined at ~21–23°C unless otherwise noted.
[c] Values were taken from Reference 44.
[d] The ∪ valves was determined at 25°C.
[e] The ∪ valves was determined at 50°C.

entrainment, impact on fluid compressibility, and various testing procedures. Therefore, foaming and air entrainment will not be discussed in detail here other than to discuss two generally applicable test procedures.

Although there are various tests that can be used to determine the propensity of a fluid to exhibit foaming problems in use, usually ASTM D 892 [61], or an equivalent, is the test most often cited [62]. This test, which is described in detail in Chapter 4, involves bubbling air into the fluid through an air-stone at a prescribed pressure and for a specific time. The amount of foam and the stability are then determined.

Equally important is the evaluation of air entrainment of a hydraulic fluid. The relative amount of formation of entrained air and the failure to dissipate it leads to a number of very significant deleterious processes such as accelerated fluid oxidation and cavitation (see Chapter 4). Currently, the most

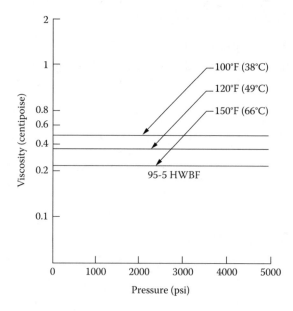

FIGURE 14.19 Viscosity-pressure relationship for an HWBHF at different temperatures.

commonly accepted test to measure air entrainment is ASTM D 3427 [63] or its equivalent [62]. This test, which is described in detail in Chapter 2, involves blowing compressed gas through the hydraulic fluid at a specified temperature, typically 25, 50 or 75°C. After the gas flow is stopped, to decrease in volume to 0.2% is determined as the air release time.

14.2.4 CHEMICAL PROPERTIES

14.2.4.1 Oxidative Stability

Various tests may be used to evaluate the oxidative stability of emulsion hydraulic fluids. One test that is used is a modified ASTM D 943 Turbine Oil Stability Test (TOST) [64]. (This test is modified by eliminating the water addition step [25].) Test results comparing two emulsion formulations shown in Table 14.8 illustrate that a fluid lifetime of greater than 1000 hours would be expected for a well-formulated HFB fluid [25].

Another test that has been reported is the "Work Factor Test", which is Method 345 of the Federal Specification VV-L-791, where the fluid is used to lubricate a six-inch babbit bearing loaded at 150 psi and rotating at 3000 rpm for 1175 hours. A four gallon reservoir is used with the fluid maintained at 140–150°F. Water, lost by evaporation, was replaced every four hours. Well-formulated HFB fluids should exhibit no apparent increase in acidity by this test.

14.2.4.2 Corrosion

There are various tests that have been used to evaluate the corrosion-inhibiting properties of HFA and HFB fluids toward ferrous and non-ferrous metals. The more commonly encountered tests will be discussed in this section.

14.2.4.2.1 ASTM D 665 (no added water)—Ferrous Metals

ASTM D 665 (see Chapter 14—Fluid Testing) is modified for use with water-containing fluids by eliminating the water addition step of the testing protocol [23,27,39,65–67]. This test involves immersing a cylindrical steel test specimen into 300 mL of stirred fluid (a dynamic test) at 60°C (140°F) for 24 hours. The steel test specimen is inspected for the presence of rust. Modifications of this test include: no stirring (a static test [23]), using a slightly higher temperature (65°C, 150°F) [23,65], longer immersion times (48 hours [27], seven days [65], and one month [23]) and partial immersion of the test specimen in order to obtain some measure of vapor phase corrosion inhibition properties [27,65].

14.2.4.2.2 Non-Ferrous Metals

Inhibition toward non-ferrous metals including: aluminum, brass, copper, and silver may be determined by partially immersing standard test coupons of the desired metal into the fluid contained in

TABLE 14.8
Oxidative Stability of Emulsion Hydraulic Fluids[a]

Fluid	Time to Develop Acidity of 2.0 mg KOH/g, Hr.		
	1	2	Average
No. 1	450	380	415
No. 2	1010	1050	1030

[a] This test was conducted by a modified ASTM D 943 (TOST), omitting the addition of water.

a bottle, beaker, or a sealed flask [62], without stirring, at 65°C for a period of time, typically seven days [23,62,65]. Changes in weight of the test specimen, changes in surface condition and the color of the fluid are recorded and classified as follows [62]:

0 = No Effect
1 = Slight color change or oxidation of less than 20% of the surface
2 = Strong color change
3 = Deposits or oxidation of more than 20% of the surface
4 = Corrosion or pitting
5 = Other effects yet to be specified

14.2.4.2.3   Multi-metal Testing

Another variation of the non-ferrous metals test (see Section 2.6.2.2 above) is to immerse metal pairs as shown in Figure 14.20a. The metal pairs that are tested include [62,67]:

1. Steel and zinc
2. Copper and zinc
3. Aluminum and zinc
4. Steel and aluminum

The metals are mounted as illustrated in Figure 14.20b with 1 mm plastic spacers between them. In this test, the metals are immersed into the fluid without stirring for 28 days at 35°C. At the conclusion of the test, the weight loss of the test specimens is determined; changes in the surface condition and color of the fluid are also recorded and classified.

14.2.5   Material Compatibility

Proper material choice is critical if optimal equipment failures are to be avoided and if optimal lifetimes are to be achieved. In this section, material compatibility—including paint, metal, elastomer, and filter compatibility—will be discussed base on the information provided in Table 14.9 [68].

14.2.5.1   Paint Compatibility

Epoxy and phenolic paints are generally suitable for emulsion hydraulic fluids [68]. However, it is prudent to verify compatibility either by testing or by contacting the paint manufacturer before painting hydraulic equipment and especially before painting the inside of a fluid reservoir.

14.2.5.2   Metal Compatibility

Emulsion hydraulic fluids are generally compatible with all metals except for lead and magnesium, as shown in Table 14.9.

14.2.5.3   Plastic and Elastomer Compatibility

Table 14.9 shows that emulsion hydraulic fluids are generally compatible with most common plastics and with the Buna-N seals that are typically used with mineral oil [68]. However, the designer/user is cautioned to confirm elastomer compatibility by testing or with the seal supplier since results may vary with the elastomer supplier. For example, one source found that Buna-N seals swelled approximately 9.6% while Viton only swelled 0.7% [65]. However, another report stated that Buna-N swelled less than 1% and the Viton seal material which were evaluated were not generally compatible with emulsion hydraulic fluids [69]. Unfortunately, such compatibility variations among seal suppliers for the same nominal fluid composition are not uncommon.

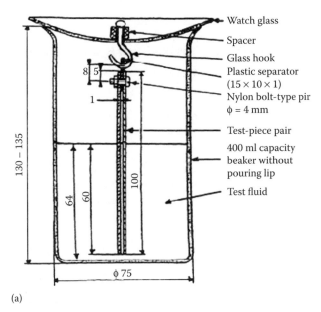

(a)

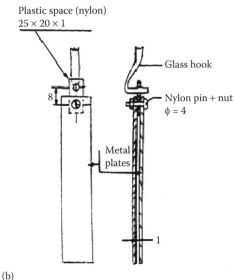

(b)

FIGURE 14.20 Illustrations of metal coupons in fluid testing.

As with other water-containing fluids, cork and leather should not be used as seal materials for either HFA or HFB fluids [69].

14.2.5.3.1 Elastomer Compatibility Testing
Seal compatibility testing strategies are provided in Chapter 10. In this chapter, some commonly encountered testing procedures will be briefly discussed.

14.2.5.3.1.1 Static Tests Static seal testing for hydraulic fluid compatibility typically involves immersion of the material of interest in a flask or bottle at 65°C [65] or 70°C [39] over extended periods of time (one to six weeks). Dimensional change such as volume swell and changes in length are typically measured relative to the initial size of the test specimen. Test specimens vary widely from the actual seal, O-ring, or small (50 mm × 25 mm) rectangular pieces cut from a sheet of the elastomer [39].

TABLE 14.9
Material Compatibility of Fire-Resistant Hydraulic Fluids

Material	Water-Glycol	Phosphate Ester	Emulsion
Paint:			
general industry	not suitable	not suitable	not suitable
epoxy and phenol	suitable	suitable	suitable
Metal:			
steel	suitable	suitable	suitable
bronze	low suitability	suitable	suitable
zinc	not suitable	suitable	suitable
cadmium	not suitable	suitable	suitable
lead	not suitable	suitable	not suitable
copper	suitable	suitable	suitable
aluminum	not suitable	suitable	suitable
anodized aluminum	suitable	suitable	suitable
magnesium	not suitable	suitable	not suitable
Plastic:			
poly(methyl methacrylate)	suitable	not suitable	suitable
poly(propylene)	suitable	not suitable	suitable
poly(styrene)	suitable	not suitable	suitable
epoxy	suitable	suitable	suitable
nylon	suitable	suitable	suitable
Rubber:			
polychloroprene	suitable	not suitable	suitable
butylene nitrile rubber	suitable	not suitable	suitable
poly(butylene)	suitable	suitable	not suitable
ethylene propylene rubber	suitable	suitable	not suitable
polyamine ester	not suitable	low suitability	not suitable
silicon rubber	suitable	suitable	suitable
ethylene polytetrafluoride	suitable	suitable	suitable
fluorine rubber	suitable	suitable	suitable
Filtering medium:			
celluloid (treated by phenol)	suitable	suitable	suitable

The European standard for the determination of seal compatibility with water-containing fire-resistant hydraulic fluids utilizes test specimens of uniform 2 mm thickness that may be rectangular (50 × 235 mm) or circular (36 mm diameter) [62]. The test materials vary with the fluid of interest and are summarized in Table 14.10 [62]. The test specimens are immersed in a sealed glass jar for 168 hours at a specified test temperature provided in Table 14.10. The percentage change in volume is measured at the conclusion of the test from [62]:

$$\Delta V = \frac{(m_3 - m_4) - (m_1 - m_2)}{(m_1 - m_2)} \times 100, \tag{14.3}$$

where:

m_1 is the initial mass of the elastomer test specimen in air,
m_2 is the apparent mass of the elastomer test specimen in water,

TABLE 14.10
Standard Elastomers and Test Conditions

Type of Fluid	Suitable Test Elastomers	Test Temperature (°C ± 2°C)	Test Duration (± 2h)
HFA	NBR 1[c] EPDM FPM 1	60	168[a]
HFB	NBR 1[c] FPM 1	60	168[a]
HFC	NBR 1 EPDM FPM 1	60	168[a]
HFDR	NBR 1[c] FPM 1[b, c]	100	168
HFDS	FPM 1	100	168

[a] The test duration is 168 hours, but equilibrium may be achieved in these fluids.
[b] FPM 1 is not suitable for use with alkyl phosphate esters.
[c] High swelling rates (> 25%) have been observed for these combinations of materials and fluids.

m_3 is the mass of the elastomer test specimen in air after the test is completed,
m_4 is the apparent mass of the elastomer test specimen in water at test completion.

The change in hardness is also determined according to ISO 48 [70] on the test specimens and calculated from [62]:

$$\Delta H = H_2 - H_1, \tag{14.4}$$

where H_1 is the measured hardness of the test specimen before testing, and H_2 is the measured hardness of the test specimen after testing.

A seal material is compatible if:

- The volume increase is less than 7% and if there is no volume contraction.
- The hardness does not increase by more than 2 IRH degrees or fall by more than 7 IRH degrees with respect to the initial hardness.

14.2.5.3.1.2 Compression Tests   Compression tests are also often used to determine the ability of the elastomer to recover to its initial dimensions before the compression test. Compression set values of 50 to 120% are considered satisfactory [65].

14.2.5.3.1.3 Lip Seal Tests   A significant problem with water-containing fluids such as hydraulic emulsions is lip seal failure [65]. Schmiede et al. devised the test rig illustrated in Figure 14.21 to evaluate lip seal fluid compatibility in use. This device is a hollow vessel with a hardened, ground steel shaft supported and sealed at each end by neoprene lip seals. The device rotates at 1450 rpm and the test fluid is circulated, as shown, under 103 kPa (15 psi) pressure. Internal friction typically causes the fluid temperature to rise to 35°C.

14.2.5.4   Strainers, Filters and Screens

Untreated cellulosic or paper filters should not be used with water-containing fluids [69,71]. However, treated cellulosic filters are reported to be more suitable [68]. Filter sizes as low as 5–10 microns have been reported as being suitable for w/o emulsions [69,72]. Filter sizes as low as 3 microns have been reported as being suitable for microemulsions [72].

Fine filtration should only be placed at the pressure or return side of the pump. On the suction side of the pump, only screens or filters large enough not to restrict flow should be used. Filters or strainers constructed with zinc or cadmium should only be used with the fluid manufacturer's approval [71].

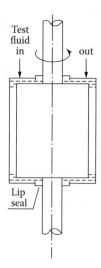

FIGURE 14.21 Test rig for lip seals with hydraulic emulsion fluids.

The use of Fuller's Earth (activated clay) or similar adsorbent filters should not be used because of the potential for additive removal by preferential adsorption [71].

Filtration may also be performed as a side stream. For example, a separate filter loop was installed on top of the reservoir which was independent of the main circulation system of the hydraulic circuit [72]. This system was used when the system was down for significant periods of time.

14.2.6 FLUID MAINTENANCE

It is recommended that HFA fluids be checked weekly for acidity and viscosity. Preventative measures should be taken to avoid contamination by salt. It is preferable for distilled or deionized water to be used for makeup [68]. The use of tap water will lead to hard metal contamination and reduced emulsion stability. HFA fluids are susceptible to biological attack and should be checked periodically.

HFB fluids, as with their HFA counterparts, may lose water by evaporation. Therefore water content must be monitored and maintained in order to sustain fire-resistance. Invert emulsion (HFB) fluids should be monitored for water content (see following section), viscosity, pH, particle count, and emulsion stability [68]. It is recommended that these analyses be performed biweekly [72].

Tramp oil, a common contaminant, will simply dissolve in the oil phase of the invert emulsion, generally not causing any significant effect on pump performance [73].

14.2.6.1 Water Analysis

Water content may be determined by distillation according to ASTM D 95 [74] after correction for the concentration of glycol, if present, which will co-distill with water. The correction procedure is described in Reference 62. Alternatively, water content may be determined by Karl Fischer analysis according to ASTM D 1744 [62,75].

Water content may also be readily determined in the plant or mine by mixing the fluid 1/1 by volume with water. A reagent is then added to assist in breaking the emulsion. The mixture is centrifuged until the water and oil separation is complete. The water content is determined by measuring the relative amount of that has been water separated.

Another recently reported test is the use of the water analysis apparatus illustrated in Figure 14.22 [50]. The fluid is accurately injected into the apparatus, which is instrumented to measure the small, but quantitative, temperature rise that occurs when water is adsorbed by the zeolite. An added

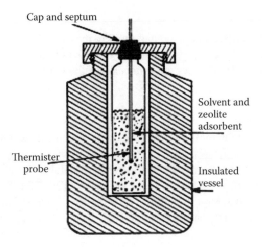

FIGURE 14.22 Water analysis apparatus.

advantage of this method is that it is not affected by the presence of glycols, which may be added to the fluid as a freeze point protectant. (This instrument is also reported to be applicable for use with HFC fluids [50].)

After the water level is determined, a chart (such as that shown in Figure 14.23) is used to determine the amount of water that must be added to the system to maintain fire resistance. Note: The water reading chart shown in Figure 14.23 is for illustrative purposes only and will vary with the manufacturer. Consult the fluid manufacturer for the correct chart for the fluid being used.

14.2.6.2 Viscosity

Kinematic fluid viscosity is determined by ASTM D 445 [76]. See Chapter 5 for details.

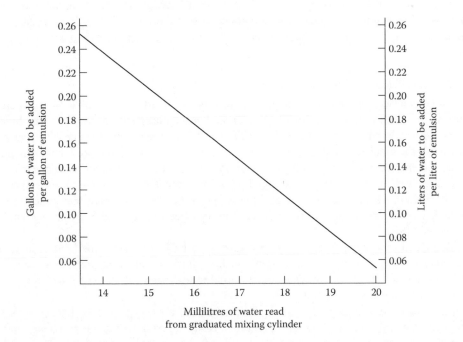

FIGURE 14.23 Example of a water makeup chart for a water-in-oil emulsion.

14.2.6.3 pH

The pH of a fluid can be easily, accurately, and quickly measured with inexpensive commercially available meters by simply dipping the probe into the fluid and reading the pH value from the analog or digital output. However, these devices must be periodically calibrated according to the manufacturer's instructions.

14.2.6.4 Fluid Cleanliness

There are numerous excellent methods of measuring the particle count of hydraulic fluids. Particle count analysis is discussed in Chapter 3. However, in many cases, a simple "patch test" is used to quantify the presence of particulates in the fluid. For HFA and HFB fluids, this analysis may be done according to ASTM F 311 [77] or ASTM 312 [78]. For both methods, 100 mL of the fluid is drawn through a membrane filter under vacuum. The collected particles are then counted and measured under a microscope [72]. Alternatively, a Millipore patch test may be used [50].

14.2.6.5 Emulsion Stability

Emulsion stability is discussed in detail in Section 14.2.3. There is a quick test which may be readily used in the plants to evaluate emulsion stability for water-in-oil emulsions. This test involves pouring drops of an emulsion into a beaker of water. A stable emulsion will form "beads" as it drops into the water and then rises to the surface. These beads will exhibit a strong tendency to resist breakage. However, if the emulsion is unstable, the drops will spread throughout the water forming a milky white appearance. This is a pass/fail test [72].

14.2.6.6 Biological Stability: Biodegradation Versus Biodeterioration

Biodegradation has been defined as: "the process by which a potential pollutant, such as a petroleum product, is converted by biological (usually microbiological) agents into simple, environmentally acceptable, derivatives" [79]. *Biodeterioration* has been defined as: "loss of product quality or performance and could be regarded as the initial stages of biodegradation, but in the wrong place at the wrong time; that is when the petroleum product is stored or in use" [79]. Biodegradation processes are discussed in Chapter 8. Biodeterioration processes and their prevention will be discussed in this section.

14.2.6.6.1 Fluid Biodeterioration Processes

Emulsion biodeterioration is a water-dependent process, as illustrated in Table 14.11, where the relationship between relative humidity and microbial susceptibility is shown [79]. The biodeterioration process, illustrated in Figure 14.24 [80], involves the reaction of water with a substrate such as oil or the emulsifier in the presence of bacteria or fungi to yield a product designated as "biomass." If this degradation process is not inhibited, enormous quantities of biomass may be present in the system in the form of sludge [81] or "microbial scums", which are composed of dead cells, gelatinous slimes, and fungal threads [38]. It has been reported that a bacterial cell may double in size and divide into two new cells every fifteen minutes until a limiting condition is encountered [81].

In addition to solid byproducts, obnoxious gases may be formed from the biodeterioration of certain additives acting as microbial nutrients such as: nitrites and nitrates which are converted to ammonia, and sulfur or sulfate which is converted to hydrogen sulfide (H_2S), which exhibits a characteristic "rotten egg" odor [82].

Biodeterioration processes that occur in the presence of air (oxygen) are enhanced by system agitation and are designated as "aerobic" processes. However, biodeterioration processes may also occur without air (oxygen) being present. These are called "anaerobic" processes and are inhibited by system agitation [82].

TABLE 14.11
Approximate Ranges of Water Activity Tolerated (Relative Humidity) and Microbial Growth

	Range of water activity of tolerated	
Microorganisms	**Lower**	**Upper**
Spoilage bacteria	0.92	0.99
Yeasts	0.86	0.98
Osmophilic yeasts	0.61	0.87
Moulds	0.61	0.098
Metalworking fluid examples tested		
Concentrates	0.69	0.97
Diluted fluid	0.98	0.99

Note: Reducing water activity within the range causes a decrease in growth rate, delay in growth initiation, and decrease in the size of the final population.

In summary, the biodeterioration process leads to [82]:

- Progressive step-wise oxidation of the hydrocarbon and continuous reduction of pH as shown in Figure 14.25 [81].
- Reaction with unsaturated bonds.
- Degradation of lubrication-enhancing additives and corrosion inhibitors causing greater wear and surface rusting of hydraulic system components.
- Emulsifier degradation leading to an increase in the size of the dispersed oil droplets as shown in Figures 14.25 and 14.26 and ultimately to emulsion breakage with free oil separation [81]. Interestingly, even a doubling of droplet diameter through the degradation process will cause an eightfold decrease in the number of droplets [81]. Therefore, it is possible to monitor the average diameter of the largest droplets as a means of predicting emulsion failure. However, this is dependent on the particular emulsion being used [81].
- Accumulation of slimes, sludges, overloading of filters and centrifuges.
- Unpleasant odors and fluid and surface discoloration [83].

14.2.6.6.2 Monitoring Procedures

Any effort to control biological attack, whether bacterial or fungal, on an emulsion hydraulic fluid is necessarily dependent on adequate procedures for detecting, identifying, and quantifying the microbial species present. It is not sufficient to simply wait until slime or sludge formation is visible or noxious vapors are present as, by this time, it may be too late to apply preventative or corrective procedures [80].

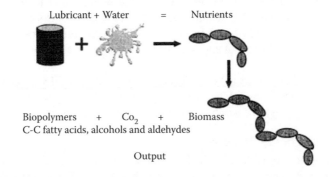

FIGURE 14.24 The biodeterioration process.

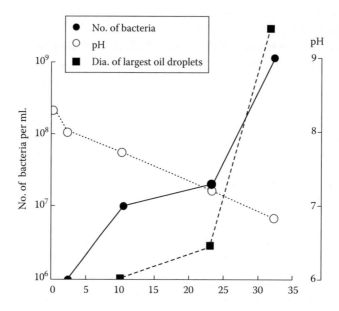

FIGURE 14.25 Bacterial degradation and related changes in pH and emulsions distribution.

There are four strategies for monitoring microbial contamination: (1) gross, (2) physical, (3) chemical, and (4) microbiological [80]. Gross detection procedures include the visual observation of slimes or the detection of foul odors. Physical detection procedures include the observation of haze and visible, non-metallic particulate matter in the fluid. Chemical tests that are often used include: pH, alkalinity and corrosivity. For example, if the alkalinity of a used fluid has dropped >25%, there is a strong potential that the cause is microbial contamination.

The fourth procedure is to conduct a microbiological test. One test is to directly observe the microbial species on a glass slide under a microscope [80,81]. However, this is not a practical test, especially for industrial use.

An older test which has been successfully applied for monitoring microbial contamination of hydraulic fluids is the "Red Spot Test" [82,84,85]. This test is conducted by taking a drop of fluid using a standard wire loop placed in a nutrient medium containing tri-phenyl tetrazolium chloride. After incubation at 37°C overnight, the initially colorless droplet is reduced by the bacteria to provide a red color. The intensity of the red color is proportional to the number of organisms present. However, this test is only applicable for aerobic organisms [85].

Currently, there are three bench test procedures that may be used for monitoring resistance to microbial growth: ASTM D 3946 [86], ASTM E 686 [87], and ASTM E 979 [88]. Alternatively, a commercial dip-slide test that is coated with a microbial growth media can be used. This is called a "viable titer method" where the population densities of the microbial species are estimated after incubation for 24–72 hours, as illustrated in Figure 14.27. Viable titer procedures may not detect microbial species that do not form colonies and therefore may not correlate with biodeterioration processes.

There are a number of simple tests being developed to provide rapid and accurate bioassays of industrial fluids. These include the catalase test [80,89] and the lipopolysaccahride assay [80,90]. It is beyond the scope of this text to describe the details of these test procedures.

14.2.6.6.3 Microbial Control Strategies

There are four strategies to provide microbial control in emulsion hydraulic fluids. These include: (1) use of fluid components resistant to biological growth, (2) microbial removal or kill procedures, (3) use of biocides/biostats, and (4) shock treatments.

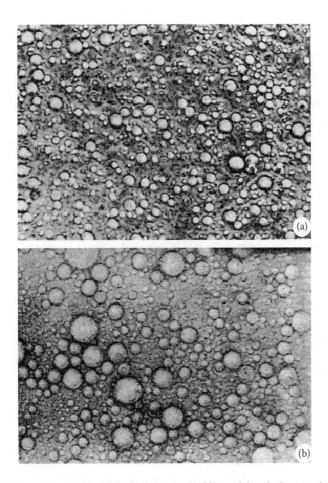

FIGURE 14.26   (a) Oil emulsion before biological attack, (b) Oil emulsion during attack but before complete failure (both images take at same scale).

14.2.6.6.3.1   Microbial Growth-Resistant Fluid Components   Chemicals that undergo microbial attack easily are referred to as biologically soft chemicals. Many emulsifiers are biologically soft, as are many paraffinic components of petroleum oil in the presence of sufficient amounts of water. Chemicals that are resistant to biological attach are known as biologically hard [79]. Some components in an emulsion, such as a glycol—which may be present to provide freeze point protection—may provide resistance to biological attack [79]. In some cases, if structurally similar biologically hard and biologically soft components are mixed, the mixture is biodegradable by a co-oxidation process.

One emulsion formulation strategy is to use components that are initially biologically hard, especially emulsifiers [81], but become biologically soft during use, or if mixed with the appropriate component to render them biologically soft after use. In addition, optimal quantities of nitrogen-, sulfur-, and phosphorous-containing additives should be used to provide satisfactory lubricity properties and corrosion inhibition while also minimizing the potential for biological attack [81].

14.2.6.6.3.2   Microbial Removal or Kill Procedures   There are a number of physical practices (e.g., filtration, heat, centrifugation) that may be used which will significantly aid in the control of biological growth, including [81]:

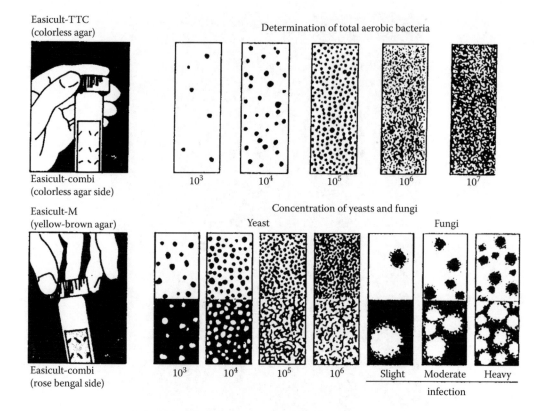

FIGURE 14.27 Dipstick test for bacteria and fungi detection.

1. Elimination of contaminated fluid from the system before recharging. The amount of contaminated residual fluid and the level of contamination will determine the lifetime of the new system. Ideally, the system will be sterilized before the new charge of fluid is added [79,81].
2. Maintenance of as high a fluid temperature as possible while still maintaining emulsion compatibility.
3. Avoid pockets of stagnation in pipes and tanks, especially for systems that are susceptible to anaerobic degradation.
4. Frequent cleaning and sterilization of cooling towers.
5. Use of microbe-free fluids.

There is a theoretical relationship between kill or removal procedures and biological growth rate. For net reduction in microbial species [68]:

$$\mu < A \times B, \tag{14.5}$$

where μ = the specific microbial growth rate (0.693/doubling time), A = fractional flow through the device (flow rate/volume), B = fractional kill (removal) in the device (microbial numbers - inflow - outflow/inflow)

14.2.6.6.3.3 Use of Biocide   A *biocide* is a chemical that is used to kill biological organisms. A *biostat* may also be used which is characteristically less chemically reactive than a biocide and functions to prevent the reproduction of biological organisms [83].

There are currently approximately 30 EPA-registered active ingredients for biocide use in lubricant applications [80]. These may be classified as: (1) target microorganisms(s), or (2) active ingredient chemistry. Targeted microorganisms include [80]:

- Bactericides—target bacteria and are generally ineffective against fungi.
- Fungicides—target fungi (yeasts and molds) and are not very effective toward bacteria.
- Microbiocides—broad spectrum, which kill both bacteria and fungi.

Chemical classification is based on chemical structure (e.g., alkanes, heterocyclic, phenolic, etc.) and is much more complicated. Since chemical classification may be confusing, it will not be discussed further here.

14.2.6.6.3.4 Shock Treatments   One effective method of delaying the onset of biodeterioration problems is to clean the system with very high concentrations (much higher concentrations than typically used for preservation) of a fast acting, broad-spectrum biocide [79]. In view of the high biocide concentrations used, health and safety and disposal problems must be carefully considered.

A second method that may be used is to treat the system periodically with very high biocide concentrations, also known as a shock treatment. Although this procedure may allow some degradation to occur, shock treatment would be used within a sufficiently short period of time to minimize any such occurrences [79].

14.2.7 FIRE RESISTANCE

HFB fluids are among the most commonly used fire resistance fluids for underground mining use [91]. HFB fluids are classified as fire-resistant because of their flammability properties relative to oil. Although HFB fluids burn, they burn with smaller flames than mineral oils [50]. Spray flammability tests have shown that the spray flammability of HFB fluids occurs with long intermittent bursts [92]. Flammability in the hot channel flammability test (see Chapter 7) exhibits a delayed ignition then burns continuously [92].

The results of these tests showed that there is a measure of improvement in the fire resistance of HFB fluids relative to mineral oil; however, they are significantly poorer than other fluids of this class. This is illustrated by comparing the escape time of different fire-resistant fluids. Escape time is defined as "the elapsed time between initial fluid-heat contact and ignition" [92]. The escape time data for various fire-resistant fluids illustrated in Table 14.12 shows that although the HFB fluid evaluated was better than mineral oil, it was substantially poorer than the water-glycol and phosphate ester. It should be noted that the polyol ester evaluated was little different than petroleum oil in this test. Also, although the phosphate ester did not burn, white smoke was formed [92].

Based on similar test results, Loudon has ranked the fire resistance offered by different fluids as: water-glycol > phosphate ester > invert emulsion > polyol esters > petroleum oil [93].

TABLE 14.12
Escape Time at 1400°F Comparison of
Various Fire-Resistant Fluids[a]

Fluid type	Typical escape times (s)
Petroleum oil	None—instant ignition
Polyol ester	2
Invert emulsion	16
Water-glycol	45
Phosphate ester	No ignition

[a] Test conducted by pouring 100 mL of fluid onto molten metal up to 1400°F.

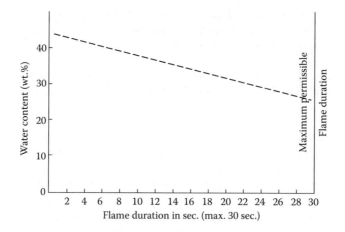

FIGURE 14.28   Effect of emulsion water content on fire resistance.

The fire resistance of HFB fluids (invert emulsions) is proportional to the water content of the fluid as shown in Figure 14.28 [5]. The water content must be maintained above 40% to maintain fire-resistant properties. If the water content is less than 35%, the fire resistance properties are lost [5]. Figure 14.29 [52] graphically illustrates the optimization of an HFB fluid formulation with respect to base oil viscosity, water content for fire resistance and optimal physical properties.

14.2.8   Lubrication Properties

14.2.8.1   Fluid Lubrication

One of the most critical functions of hydraulic fluids, in addition to power transmission, is fluid lubrication. Some of the contacts that require lubrication include: vane and ring for vane pumps,

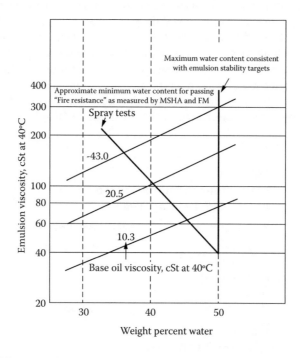

FIGURE 14.29   Establishment of minimum acceptable water content of an HFB emulsion hydraulic fluid.

piston slipper pads and brush plate for piston pumps, the gears of a gear pump, and bearing lubrication in all pumps. Invert emulsions, for example, provide lubrication challenges relative to mineral oils for which most pumps were designed. For example, the shear thinning viscosity behavior of HFB fluids has required design changes where hydrostatic lubrication is utilized. Since both HFA and HFB fluids may be subject to metallic contact, severe wear may result. Shear thinning or insufficient fluid viscosity may also affect internal leakage and hydraulic efficiency [51]. Therefore, it is important to assess fluid lubrication properties in order to determine if it is appropriate to use a particular fluid with a particular pump or hydraulic component.

Fluid lubrication fundamentals are discussed in detail in Chapter 6. In this section, the application of the fundamental properties of HFA and HFB fluids, particularly with respect to hydrodynamic and elastohydrodynamic (EHD) lubrication, will be discussed. Wherever possible, distinctions between the inherently different lubrication properties of HFA and HFB fluids will be made.

14.2.8.1.1 Film Thickness Studies

Hamaguchi et al. studied the EHD film thickness properties of a paraffinic oil (w/o) emulsion similar to an HFB hydraulic fluid [94]. The results of this study showed the film thickness was nearly independent of water concentration up to 80% water (see Figure 14.30), even though the fluid viscosity increased by ten-fold over this water concentration range as shown in Figure 14.31. A similar emulsion was prepared from Shell Vitrea 69 mineral oil. The film thickness versus water concentration relationship was similar to that obtained for the paraffinic oil as shown in Figure 14.32. The effect of base oil viscosity w/o emulsions prepared from these oils is shown in Figure 14.29 [50]. The EHD lubrication properties on emulsion are primarily determined by the properties of the base oil [94].

Hamaguchi reported that negligible film-forming properties were observed for oil-in-water emulsions such as HFA fluids [94].

14.2.8.1.2 Model of Film Formation

Since the fluid films formed by both HFA and HFB fluids are heterogeneous, a two-phase model of the hydrodynamic films formed by these fluids is necessary. The model proposed by Liu et al. for a heterogeneous film formed during rolling in the "x direction" is illustrated in Figure 14.33 [95]. For an HFB fluid, this model shows that a portion of the film consists of dispersed large water droplets which form patches of water on the contact surface. The remaining portion of the surface is covered by the continuous oil phase of the emulsion. Localized viscosity is either that of the water or that of the oil. The critical diameter defining large and small droplets is dependent on the local film thickness, water concentration in the continuous phase and the viscosity of the continuous phase [95].

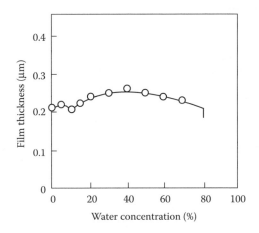

FIGURE 14.30 Effect of water concentration of lubricant (paraffin oil) film thickness.

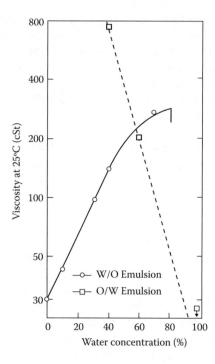

FIGURE 14.31 Effect of water content on o/w and w/o emulsion used for film-thickness studies.

A small element ($\Delta x \Delta y$) of the film shown in Figure 14.34 is large enough to contain a number of water patches but still small with respect to the EHD conjunction [95]. If the film thickness (h) is assumed to be constant over this element, the Reynolds equation is:

$$\frac{\partial}{\partial x}\left(\frac{h^3}{12\eta}\frac{\partial P}{\partial x}\right) + \frac{\partial}{\partial y}\left(\frac{h^3}{12\eta}\frac{\partial P}{\partial y}\right) = 0, \tag{14.6}$$

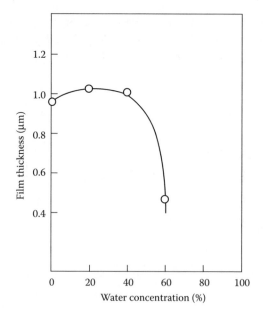

FIGURE 14.32 Effect of water content on lubricant (mineral oil) film thickness.

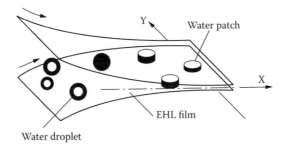

FIGURE 14.33   Two-phase hydrodynamic film.

where h is the film thickness, η is the viscosity, and P is the contact pressure [95]. After solving this equation, the equivalent viscosity of w/o emulsions relative to the viscosity of the base oil may be plotted against the total water concentration of the w/o emulsion (ϕ) and the fraction of water concentration that forms the patches (ξ) [95]. This relationship is shown in Figure 14.35 where the viscosity of the base oil is assumed to be 1.0.

Figure 14.35 shows that w/o films may be either thicker or thinner than films formed with the base oil, depending on the diameter of the water droplets. This means that water droplets may be trapped within the wear contact and are restrained from escaping by the high viscosity of the oil [95].

14.2.8.1.3   Pressure-Viscosity Effects

Although hydrodynamic fluid lubrication is primarily dependent on the bulk viscosity at the temperature of the wear contact, elastohydrodynamic lubrication is also dependent on the contact pressure. Therefore, the fluid film thickness at the wear contact is dependent on the viscosity-pressure coefficient. This relationship is evident from another form of the Reynolds equation [24]:

$$h_o = K \left(\eta_o\right)^{0.7} \left(\alpha\right)^{0.5},\tag{14.7}$$

where h_o is the minimum film thickness, η_o is the viscosity of the lubricant at atmospheric pressure, α is the pressure-viscosity coefficient, and K is a constant. Physically, this means that fluids which exhibit higher pressure viscosity coefficients will increase in viscosity faster with increasing contact pressure (loading). If the fluid viscosity increase with pressure is sufficient, the fluid film thickness will decrease less with increasing load, thus better inhibiting asperity contact and forming a higher wear boundary lubrication condition. Note that pressure–viscosity coefficients are described in detail in Chapter 4 and their effect on lubrication and wear is described in Chapter 6.

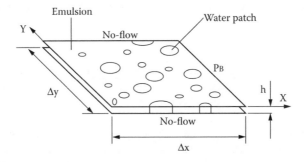

FIGURE 14.34   An element of the EHL film.

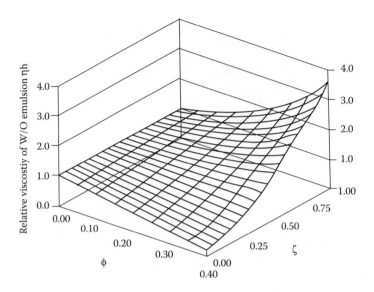

FIGURE 14.35   Relative viscosity (ηh) versus water concentration (f) and volume fraction (ξ) of small water particles (viscosity of base oil = 1.0).

In Section 14.2.3.8.6, pressure-viscosity coefficients for a number of w/o and o/w emulsions were provided. In general, it was observed that the pressure-viscosity coefficient was dependent on the water content of the fluid. High water content fluids (o/w emulsions) exhibited pressure-viscosity coefficients approaching those of water and w/o emulsions exhibited pressure-viscosity coefficients similar to those of the base oil. (A typical pressure-viscosity coefficient for a mineral oil is 14.7 GPa^{-1} [95].)

The effect of pressure-viscosity coefficient on bearing wear is modeled by the lubricant film parameter or lambda ratio (Λ) through the lubricant central film thickness (h_C') and surface roughness (σ). (See Chapter 12, "Failure Analysis.")

$$\Lambda = \frac{h_c'}{\sigma}. \tag{14.8}$$

This equation shows that wear is a function of both film thickness and surface roughness.

14.2.8.1.4   Fatigue Failure

Although there is a great deal of variation of fatigue wear among different fluids of the same class, there is a general trend for increasing wear with increasing water content. This is especially true for HFA and HFB fluids where the wear rate of HFA fluids is typically much greater than that for HFB fluids [24]. Spikes has suggested the following causes for greater wear of water-based hydraulic fluids relative to mineral oil [24]:

- Low EHD film thickness and lambda ratio.
- Chemical promotion of crack formation and growth leading to fatigue.
- Removal or prevention of protective chemical films due to low temperature or dissolution.
- Increased corrosion leading to corrosive/abrasive wear.

In many applications, such as rolling element bearing wear, the predominant failure mechanism is fatigue wear. Fatigue wear may be enhanced by water-based fluids because of: low EHD film thickness leading to asperity contact, chemical promotion of crack growth, and pressure pulsation of water entrapped with cracks [24]. At this time, the primary mechanism of fatigue crack growth promotion is not clear but is thought to be due to hydrogen embrittlement, stress corrosion cracking, and corrosion fatigue [24].

Yardley et al. studied fatigue failures of rolling element bearings with various fire-resistant fluids. Based on their studies they proposed the following load-life relationship [96]:

$$L_{10} = \left(\frac{C}{PD} \right)^B,$$ (14.9)

where L_{10} is the average lifetime in revolutions when 10% of the bearings will fail; C is the dynamic capacity of the bearing or "load that a bearing can carry for a life of one million inner-rave revolutions with a 90% probability of survival" [99]; P is the bearing load; and B and D are constants taken from Table 14.13. Tables such as this are used in the de-rating of hydraulic pumps for use with fluids other than mineral oil. However, it should be noted that the C/P ratio will vary with the bearing type and material. Nevertheless, this data shows that the relative order of bearing de-rating for use with fire-resistant fluids with respect to fatigue failures is: HFA (worst)>>HFC>HFB> phosphate ester ≈ mineral oil with a 90% probability of survival [97].

14.2.8.1.5 Fluid properties and pump design

As discussed previously, fluid viscosity is one of the most important variables to consider when converting a hydraulic system to a fire-resistant fluid from a mineral oil. Proper fluid viscosity is vitally important if optimal lubrication is to be maintained. However, this is a particularly critical design consideration for HFA fluids which may exhibits viscosities as low as 1 cSt compared to viscosities of up to 90 cSt (and higher for HFB fluids). In this section, a selected overview of some of the important rheological properties of a hydraulic fluid on hydraulic pump operation will be provided. Most of this review is based on a series of papers providing a more complete analysis by R. Taylor, et al. [98–101].

14.2.8.1.5.1 Importance of fluid viscosity HFA fluids, since they contain mostly water, provide many operational problems in a hydraulic system, particularly with respect to leakage flow and bearing lubrication. Leakage flow across the seal lands of a pump is directly proportional to the viscosity of the fluid. This is significant since an HFA fluid may exhibit a viscosity of 1/20, or less, of a typical mineral oil. HFB fluids provide an interesting contrast since they increase in viscosity with increasing water content. (This is why it is often recommended that HDFB fluids be replaced when the water content is greater than ± 5% of their correct operational range [103].)

Note that it is important to account for fluid density since absolute viscosity is an important design parameter for hydraulic systems. For example, compare mineral oil with a kinematic viscosity of 32 cSt and a density of 0.88. The absolute viscosity is (32 × 0.88 = 28.2 cP). However, an emulsion

TABLE 14.13
Constants in the Load-Life Relationship of Four Fire-Resistant Fluids

$$L_{10} = \left(\frac{C}{PD} \right)^B.$$

Fluid	B	Standard Error of B	D
Phosphate ester	3.19	0.54	1.05
O/W emulsion	2.64	0.37	1.71
W/O emulsion	2.25	0.21	1.25
Water-glycol	1.69	0.28	1.41

with the same kinematic viscosity but with a density of 1.27 will exhibit an absolute viscosity of $(32 \times 1.27 = 40.6 \text{ cP})$. Failure to provide this adjustment may lead to failure to accurately predict the transition from laminar to turbulent flow [99,103].

The effect of viscosity on flow rate is important when considering hydrostatic lubrication. For example, the flow rate through a slipper is an important contributor toward volumetric efficiency. The flow rate is a function of bearing geometry, port pressure, film thickness, and fluid viscosity, as shown by:

$$Q = q \cdot \frac{h^3}{\eta} \cdot P_P, \tag{14.10}$$

$$q = \frac{\pi}{6} \cdot \frac{1}{\ln(D/a)}, \tag{14.11}$$

where Q is the flow rate, h is film thickness, P_P is partial pressure, η is absolute oil viscosity, D is the port diameter, and α is the pad diameter. Figure 14.36 [99] illustrates the variation of flow rate versus viscosity at different film thicknesses and Figure 14.37 shows the variation of flow rate versus film thickness with varying viscosity [99]. These figures show that flow rate increases with increasing film thickness and decreasing viscosity. This illustrates one of the major problems with the use of HFA fluids.

Excessive flow rates may be reduced by pump design. For example, the film thickness of a hydrostatic bearing may be reduced by decreasing the film thickness as long as it is sufficient to prevent asperity contact.

The flow rate of a hydrostatic bearing with capillary compensation can be formulated thus [99]:

$$Q = \frac{K_c}{\eta}\left(P_s - P_p\right), \tag{14.12}$$

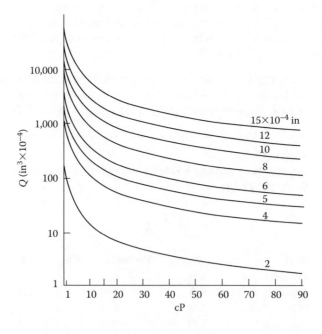

FIGURE 14.36 Flow rate versus fluid viscosity at different film thicknesses.

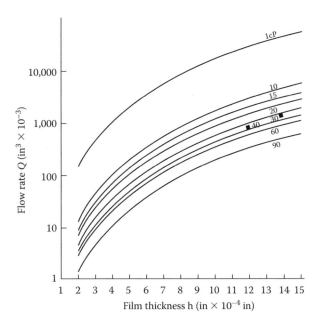

FIGURE 14.37 Fluid flow rate versus film thickness with ranging fluid viscosity.

where P_S is the supply pressure and K_C is the capillary coefficient which is a function of capillary diameter (d_C) and capillary length (l_C):

$$K_c = \frac{\pi d_c^4}{128 l_c}. \tag{14.13}$$

Since supply pressure and port pressure are constant, the capillary diameter can be reduced to compensate for leakage due to low viscosity. However, this is often difficult to practice because of the very small capillary diameters required [99].

The pumping power of a hydrostatic bearing is proportional to film thickness and inversely proportional to fluid viscosity as shown by:

$$H_f = \frac{\pi}{4} \cdot \frac{\eta - u_0^2}{h} \cdot \left(D^2 - d^2\right) + \frac{\left(1 - \dfrac{d}{D}\right)^4 \eta}{2} \cdot \frac{u_m^2 A}{h} \tag{14.14}$$

$$u_0 = \frac{\pi D_0 \eta}{60} \tag{14.15}$$

$$u_0 = \frac{\pi D \eta}{60} \tag{14.16}$$

where D is the port diameter, d is the pad diameter, D_0 is the pitch circle diameter, U_0 is the sliding velocity relative to the pitch circle diameter, U_m is the sliding velocity of the runner relative to the pad, and A is the pad area [99]. This is illustrated in Figure 14.38 [99]. When the fluid viscosity is 1 cP, the power loss increases dramatically with increasing film thickness. Interestingly, the optimum film thickness increases with increasing viscosity.

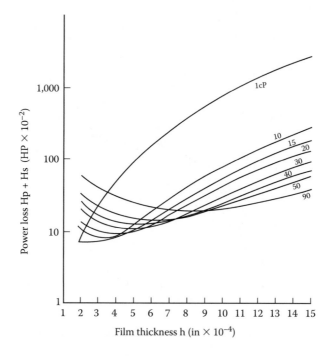

FIGURE 14.38 Variation of power loss with film thickness.

Figure 14.39 [99] illustrates total power loss versus viscosity. The solid line is minimum power loss and the dashed line is maximum power loss. This data shows that the difference between minimum and maximum power loss is approximately 0.5–1.0 Hp unless the viscosity is approximately 1 cP. This means that film thickness is critically important for HFA fluids. In general, film thickness should be kept as thin as possible for HFA fluids.

14.2.8.1.5.2 Hydrodynamic Bearings Hydrodynamic bearings are lubricated without external pumping equipment. This is accomplished by fluid viscosity, elliptical contact geometry (see Figure 14.40 [98]), and the relative motion of the moving surfaces. Figure 14.41 [98] illustrates the characteristic converging geometry of a contact for hydrodynamic lubrication. As the surfaces move in relative motion, the fluid, owing to its viscosity, adheres to the surfaces and is dragged into the contact. The critical parameters for hydrodynamic bearings are film thickness, pressure distribution, maximum load capacity (maximum load corresponding to the minimum film thickness), flow rate, friction loss, and temperature rise [98]. The equations governing film thickness, pressure distribution, load-carrying capacity and frictional force for hydrodynamic journal and thrust bearings are shown in Table 14.14 [98].

In Table 14.14, A is pad area; A_0 is h/h_0; B is breadth of bearing in direction of motion; c is radial clearance; F is friction force; h is film thickness; h_0 is minimum film thickness; K (h_i-h_0)/h_0; L is bearing length; R is journal radius; U is velocity; ε is eccentricity ratio; and η is oil viscosity.

One of the most important parameters for a journal bearing is the load-bearing capacity since it will determine the performance and lifetime of the bearing. The Sommerfield equation is typically used to analyze the performance of a journal bearing which is written as [100]:

$$\frac{W/L}{u \cdot \eta} \cdot \frac{C^2}{R^2} = \frac{6\pi\varepsilon}{\left(1-\varepsilon^2\right)^{3/2}\left(1+\dfrac{\varepsilon^2}{2}\right)} \tag{14.17}$$

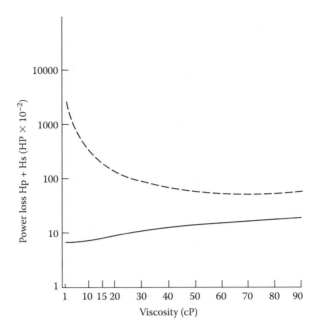

FIGURE 14.39   Total power loss with fluid viscosity.

where W is the load, L is the bearing length, U is the surface velocity, C is the radial clearance, R is bearing radius, ε is journal eccentricity, and η is the absolute viscosity.

The value on the right-hand side of this equation is called the Sommerfield Number. The inverse of this relationship is called the "duty parameter." As is evident from this equation, these values provide a measure of the effect of fluid viscosity, surface speed, and load on hydro-dynamic lubrication. Figure 14.42 provides a plot of the Sommerfield Number versus various L/D ratios.

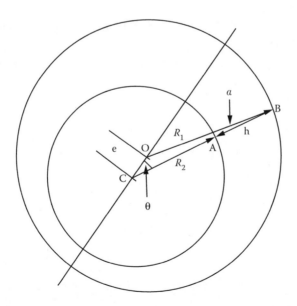

FIGURE 14.40   Geometry of a hydrodynamic bearing.

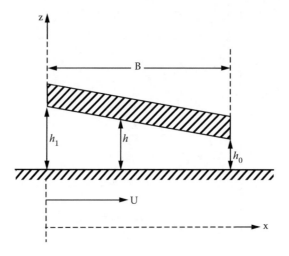

FIGURE 14.41 Geometry of a hydrodynamic bearing.

The load capacity of the bearing is dependent on the fluid viscosity from [100]:

$$\frac{W}{L} = 6W * u\eta \frac{R^2}{C^2},$$ (14.18)

where W^* is the non-dimensional load factor which is equal to

$$W* = \frac{1}{K}\left[\frac{\log\varepsilon(1+K)}{K} - \frac{2}{2+K}\right],$$ (14.19)

where K is the film thickness ratio. These equations show that the load capacity of a journal bearing is dependent on fluid viscosity. Thus, if a pump is designed to operate with a mineral oil, a low-viscosity HFA fluid will only provide a fraction of the load-bearing capacity. Therefore, the pump must be de-rated.

TABLE 14.14
Governing Equations for Hydrodynamic Lubrication of Hydrodynamic Journal and Thrust Bearings

Equations	Hydrodynamic Journal Bearing	Hydrodynamic Thrust Bearing
Film thickness	$h = C(1 + \varepsilon \cos \theta)$	$\dfrac{h}{h_0} = \left(1 + K - K\dfrac{x}{B}\right)$
Pressure distribution	$p = \dfrac{6u\eta R}{c^2}\left\{\dfrac{\varepsilon\sin\theta(2+\varepsilon\cos\theta)}{(2+\varepsilon^2)(i+\varepsilon\cos\theta)}\right\}$	$\dfrac{dp}{dx} = \dfrac{6u\eta h_0\left(1 + K - K\dfrac{x}{B}\right) - h_0 A_0}{\left(1 + K - K\dfrac{x}{B}\right)^3 h_0^3}$
Load-bearing capacity	$W = \omega A P_p$	$W = W \cdot \dfrac{6u\eta B^2 L}{h_0^2}$
Flow rate	–	$Q = Luh_0\left[\dfrac{(1+K)}{(2+K)}\right]$
Friction force	$F = \dfrac{2\pi\eta URL}{C}$	$F = \dfrac{\eta URB}{h_0}\left\{\dfrac{4\ln(1+K)}{K} - \dfrac{6}{2+K}\right\}$

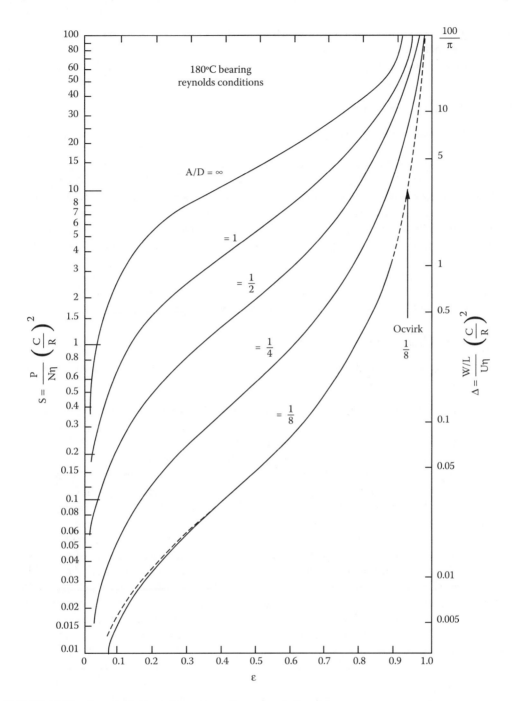

FIGURE 14.42 Sommerfield number versus ε for various L/D ratios.

14.3 HYDRAULIC PUMP TESTING

Over the years, it has been found that in order to obtain a reasonably reliable assessment of the anti-wear properties of a hydraulic fluid, it was necessary to perform pump tests [67,102–105]. These tests may be conducted with specialized pump tests including gear [67,102] and piston [67,104]. One of the most common tests is ASTM D 2882 [106] and its international equivalents such as DIN 51389 [107] which are based on the Vickers V 104 vane pump. Although HFB fluids may generally be tested under ASTM D 2882 conditions, HFA fluids cannot. Janko reported the maximum test pressure that could be utilized was 54 bar [104].

In addition to pump testing, it is often necessary to conduct valve tests with different fluids. Such work has been described by various authors, including Kelly [55,108] and will not be detailed here.

14.3.1 HYDRAULIC PUMP DE-RATING

For various reasons discussed throughout this chapter, it would not be expected that HFA and HFB fluids would be direct replacements for mineral oil as a hydraulic fluid. Therefore, these fluids, like most other fire resistant fluids, are de-rated with respect to maximum pressure and speed. An example of hydraulic pump de-rating is provided in Table 14.15 [109]. This table illustrates that Sundstrand routinely recommends the greatest de-rating for use of their pumps with HFA fluids. Although not de-rated to as large an extent as HFA fluids, HFB fluids are de-rated relative to polyol esters.

In addition to fluid derating, additional recommendations are also made such as the use of elevated and pressurized reservoirs, increased inlet size and reducing inlet vacuum below 2 in Hg and inlet pressurization [109,110].

The specific de-rating procedures vary with manufacturer and specific products and are influenced by fluid specific gravity and inlet suction head [111]. Therefore the equipment manufacturer should be consulted when the conversion from mineral oil is contemplated.

14.4 CONVERSION PROCEDURES

It has been suggested that the following steps be taken when converting from mineral oil to emulsion hydraulic fluids [112,113]:

1. Completely drain the system of the previous fluid. This includes reservoir, accumulator, coolers, lines, and cylinders. Low-pressure air may be used to assist in oil removal, particularly from remote parts of the system [114].
2. Disconnect, remove, and clean or replace filters and intake pipe strainers.
3. Replace seals with materials that are compatible with emulsion hydraulic fluids. Most of the materials used for petroleum oil are acceptable. In addition, silicone, Teflon, Viton, Fluorel and Kel-f are usually suitable [111]. However, all seal and gaskets materials of leather, cork, and paper must be replaced [115].
4. Wipe reservoir dry with lint-free rags. If there is paint on the inside surfaces, remove using commercial paint remover or hot caustic. Paint removal is very important since the paints that are typically used for oil hydraulic systems are readily attacked by water-containing fluids [114]. Paint residues will contaminate the fluid and interfere with the operation of hydraulic actuators.
5. Reconnect the system; clean and reinstall intake strainers. Inlet strainers are typically limited to 60 mesh or less. They may be constructed of brass or plated with copper or nickel. However, precautions should be taken to insure that they are compatible with the emulsion fluid being used [114].

TABLE 14.15
Illustration of Hydraulic Pump De-Rating for Fire-Resistant Fluids

Fluid type	Speed (% cataloy)	Pressure (% cataloy)	Maximum Temperature (°F)
Phosphate ester	100	100	180
Polyol ester	85	85	150
Invert emulsion (60/40 oil/water)	65	70	140
Water-glycol (60/40 oil/water)	65	60	140
HWCF (95/5 water/oil)	65	40	122

6. If a petroleum oil was used previously:
   a. Fill machine with light hydraulic oil.
   b. Operate machine for twenty minutes.
   c. Pump out sump; drain all accumulators, cylinders and coolers.
   d. Wipe reservoirs clean.
7. It is recommended that the system be flushed with the emulsion hydraulic fluid itself prior to final filling [10,114,116]. The use of solvents or chemical cleaners may affect fluid emulsion stability. Although residual petroleum oil, either hydraulic oil or flushing fluid is typically compatible in an invert emulsion [10,117], the presence of residual oils may exhibit an unfavorable effect on flammability [114,116,118]. One report recommends that one drain the mineral oil, flush with emulsion hydraulic fluid by running for thirty minutes, drain flushing charge and recharge with new emulsion fluid and run for one week, drain and recharge with new fluid and run normally [119].
8. Fill with emulsion hydraulic fluid.
9. Install water-compatible filter and do not place filter too close to pressure pulsation in order to avoid filter fatigue [120].
10. Cycle machine for ten minutes.
11. Check reservoir level.
12. Check all piping for leaks and tighten if necessary. The following guidelines with respect to piping have been recommended [121]:
    a. Loose, vibrating pipes are dangerous and prone to breakage. Only copper pipe should only be used, if at all, for low pressures <300 psi.
    b. Flexible hoses should be inspected very carefully and replaced if necessary. The use of a flexible hose can be avoided by the use of telescoping joints or unions which may be loosened for necessary adjustments. If a flexible hose is used, it should be enclosed in a shroud and vented to a waste pit.
    c. Fragile piping should not be placed in an area where it can be walked on, run over or used as a ladder. If it cannot be moved, it should be protected by adequate guards.
    d. Avoid temporary repairs. Malleable or wrought iron fittings in high-pressure lines should be replaced immediately. Strains introduced by bending or straining pipes to make them fit may lead to subsequent rupture.
    e. Long overhead runs of pipe should be avoided since rupture may spray fluid over danger areas.
    f. Pipe passageways through floors or between machines should be packed with a flexible, fire-resistant filling (ceramic) in order to prevent flames from spreading from one floor to another.
    g. Ensure that machines are operating properly. Rough hydraulic flow indicates improper circuit design, which may lead to subsequent leaks and breakdowns.

14.4.1 System Preparation

In addition to the changeover guidelines listed above, a number of additional precautions have been recommended [120]:

1. Use a reservoir design (preferably using stainless steel) that provides sufficiently quiet conditions so that particulate contaminants will settle into a sump for subsequent removal.
2. Owing to their reduced compressibility relative to mineral oil, emulsion hydraulic fluids are "stiffer." These stiffer systems may produce higher pressure ripples which may affect hydraulic components such as valves and seals. For example, if adequate design precautions are not taken, valve housings may undergo premature fatigue failures due to the high pressure ripple effect.

3. The pump inlet should have a positive inlet pressure with minimal bends to eliminate potential cavitation. The pump inlet pressure may be increased with the use of a feed pump or by increasing fluid head at the inlet. If the inlet vacuum is too high, precautionary measures must be taken. Small improvements can be made by increasing the size of the pipe into the inlet and, if necessary, reducing the pump speed to provide lower displacement [113]. "It is absolutely necessary that suction line restrictions be reduced to allow adequate flow of the fluid to the pump under all operating conditions" [116].

4. To prevent erosion, avoid continual fluid relief over a relief valve at high pressure.

5. Avoid using aluminum components.

6. Heat exchangers may be required to prevent reservoir temperatures from exceeding the recommended maximum temperature of 150°F [114].

14.4.2 Pumps: Selection and Preparation

Before any conversions are made, the hydraulic pump manufacturer should be consulted with respect to recommended de-rating practices and potential material incompatibilities (such as pump seals) with the emulsion hydraulic fluid to be used [113]. In some cases, the particular pump being used may not perform satisfactorily with a fire-resistant fluid. In such cases, a recommendation may be made to replace the pump with a model compatible for use with fire-resistant fluids while exhibiting higher performance and efficiency than the older model being used [111].

14.4.2.1 Cavitation

High inlet vacuum during normal operation may lead to cavitation. This condition may be detected by installing a vacuum gauge in the suction line immediately ahead of the pump inlet. Recommended maximum inlet vacuum conditions are provided in Table 14.6 [26].

If the inlet vacuum is higher than recommended, the suction line strainer or filter may be clogged. Other causes for pump cavitation include [26]:

1. Screen size is too small. It should not be smaller than 60 mesh.

2. If the pump suction line is too long or too small in diameter, cavitation may result. A collapsed suction hose will also lead to cavitation. The solution is to follow the manufacturer's recommendations.

3. Elevation of the pump above the reservoir may lead to cavitation. The pump or reservoir should be moved to provide a positive fluid head at the inlet.

4. An air leak on the suction side of the pump will cause fluid aeration and increase the potential for cavitation [122].

5. Fluid foaming in the reservoir may lead to cavitation. This may be caused by improper formulation chemistry or contamination.

APPENDIX A: HLB CALCULATION METHODS

There are various methods for the calculation of HLB values. One method is to calculate the HLB value from the chemical structure by summing a group contributing number from: [123,124] HLB = Σ (hydrophilic numbers) - m (group number per — CH_2 — group) + 7. Table 14.16 provides a summary of illustrative HLB group numbers. A comparison of the HLB values calculated by this method versus experimentally determined values is provided by Table 14.17.

Many emulsifiers are identified by a chemical name. This can be misleading as they are often mixtures of compounds. For example, naturally derived sorbitan monooleate or oleic acid will also frequently contain palmitic, stearic, erucic, and linoleic derivatives as well. In such cases, it is preferable to determine the HLB value experientially.

TABLE 14.16
HLB Group Number

Hydrophilic Groups	Group Number
sulfate (-SO$_4^-$Na$^+$)	38.7
potassium carboxylate (-CO$_2^-$K$^+$)	21.1
sodium carboxylate (-CO$_2^-$Na$^+$)	19.1
sulfonate (-SO$_3^-$X$^+$)	about 11
ester (sorbitan derived)	6.8
ester (non-sorbitan)	2.4
carboylic acid (-COOH)	2.1
hydroxyl (non-sorbitan)	1.9
ether linkage (-O-)	1.3
hydroxyl (sorbitan ring)	0.5

Lipophilic Groups	Group Number
- CH -	0.475
- CH$_2$ -	0.475
- CH$_3$	0.475

Derived Group	Group Number
ethoxy (-OCH$_2$CH$_2$-)	0.33

CALCULATING HLB VALUES

Non-ionic surfactants

I. For polyol fatty acid esters (like sorbitan monooleate), the HLB value can be calculated from [124]:

$$HLB = 20\left[1 - \frac{S}{A}\right], \tag{14.20}$$

where S is the saponification number of the ester and A is the acid number of the recovered ester.

II. If the emulsifier is a fatty ester of an ethoxylated polyol, the following equation may be used to calculate HLB [124]:

$$HLB = \frac{E+P}{5}, \tag{14.21}$$

where E is the weight percent of the poly(ethoxy) content and P is the weight percent of the polyol. The total value of the hydrophilic component is $(E + P)$.

III. In the case of non-ionic emulsifiers containing a poly(ethoxy) group as the lone hydrophilic moeity such as polyoxyethylene stearate, the HLB value can be calculated from [124]:

$$HLB = \frac{E}{5} \tag{14.22}$$

A wide range of HLB values may be obtained by blending emulsifiers. The HLB value of a blend may be calculated by multiplying the fraction of the emulsifier in the blend times by its HLB value.

TABLE 14.17
Comparison of the HLB Values Calculated by this Method Versus Experimentally Determined Values

Surface Active Agent	HLB From Expt.	HLB From Group Numbers
Sodium lauryl sulfate	40	(40)
Potassium oleate	20	(20)
Sodium oleate	18	(18)
Tween 80 (sorbitan monooleate, 20-ethoxylate)	15	16.5
Alkyl aryl sulfonate	11.7	–
Tween 81 (Sorbitan monooleate, 6-ethoxylate)	10	11.9
Sorbitan monolaurate	8.6	8.5
Methanol	—	8.3
Ethanol	7.9	7.9
n-Propanol	—	7.4
n-Butanol	7.0	7.0
Sorbitan monopalmitate	6.7	6.6
Sorbitan monostearate	5.9	5.7
Span 80 (Sorbitan monooleate)	4.3	5.7
Propyleneglycol monolaurate	4.5	4.6
Glycerol monostearate	3.8	3.7
Propylene glycol monostearate	3.4	1.8
Sorbitan tristearate	2.1	2.1
Cetyl alcohol	1	1.3
Oleic acid	1	(1)
Sorbitan tetrastearate	~0.5	0.3

For example, if an emulsifier blend contains 70% of Emulsifier A with an HLB of 15.0 and 30% of Emulsifier B with an HLB of 4.3, the HLB of the blend will be:

Emulsifier A	$0.7 \times 15.0 = 10.5$
Emulsifier B	$0.3 \times 4.3 = \underline{1.3}$

$$\text{HLB of the blend} = 11.8$$

REFERENCES

1. Sun, Y and Totten, G.E. *Handbook of Hydraulic Technology*, editor Totten, G.E., Chapter 17, pp. 847–916, Marcel Dekker, Inc., 2000.
2. Reichel, J. "Standardization Activities for Testing of Fire Resistance", in *Fire Resistance of Industrial Fluids ASTM STP 1284*, Eds. Totten, G.E. and Reichel, J., American Society for Testing and Materials, Philadelphia, PA, 1996, pp. 61–71.
3. Pollack, S.P. "Researching New Hydraulic Fluids", *Coal Age*, January, 1957, pp. 82–84.
4. Goodman, K.C. "Living with Fire-Resistant Hydraulic Fluids", June 1, 1965, Report Available from *Denison Engineering Division*, Columbus, OH. 43216.
5. Townshend, F. and Baker, P. "Factors Relating to the Selection and Use of Fire-Resistant Fluids in Hydraulic Systems", *Hydraulic Pneumatic Power*, April, 1974, pp. 134–40
6. Staley, C. "Fire-Resistant Hydraulic Fluids - Comparisons and Applications", *Petroleum Times*, November 24, 1967, pp. 1709–12.
7. Morris, A.E. and Lefer, H. "Changing Requirements in Hydraulic Fluids", *Hydraulics & Pneumatics*, February, 1965, pp. 69–74.

8. Anon., "Water-Based Fluids in Hydraulic Applications", *Industrial Lubrication and Tribology*, 1996, Vol. 48, No. 4, pp. 39–41.

9. Brooke, B. "Development of a High-Water-Based Fluid System for a Universal Beam Rolling Mill", in *Conference Proceedings: Hydraulics, Electrics and Electronics in Steel Works and Rolling Mills*, Brochure No. RE 00 252/09.94, Available from Mannesmann Rexroth GmbH, Jahnstrasse 3–5, D-97816 Lohr-am-Main, Germany.

10. Young, K.J. and Kennedy, A. "Development of an Advanced Oil-in-Water Emulsion Hydraulic Fluid, and its Application as an Alternative Mineral Hydraulic Oil in a High Fire Risk Environment", *Lubrication Engineering*, 1993, Vol. 49, No. 11, pp. 873–79.

11. Lubricants, Industrial Oils and Related Products (Class L) - Classification - Part 4: Family H (Hydraulic Systems), *International Standard ISO 6743/4*, Zurich Switzerland (1982).

12. Garti, N., Felkenkrietz, R., Aserin, A., Ezrahi, S. and Shapira, D. "Hydraulic Fluids Based on Water-in-Oil Microemulsions", *Lubrication Engineering*, 1993, Vol. 49, No. 5, pp. 404–11.

13. Millett, W.H. "Nonpetroleum Hydraulic Fluids - A Projection", *Iron and Steel Engineer*, May, 1977, Vol 54, No. 5, pp. 36–39.

14. Lopez-Montilla, J.C., Herrera-Morales, P.E. and Shah, D.O. "New Method to Quantitatively Determine the Spontaneity of the Emulsification Process", *Langmuir*, 2002, 18, pp. 4258– 62

15. Walstra, P. and Smulders, P.E.A. "Emulsion formation", in *Modern Aspects of Emulsion Science*, 1998, edited by Binks, B.P., The Royal Society of Chemistry, pp. 129–286.

16. Rasp, R.C. "Water-Based Hydraulic Fluids Containing Synthetic Components", *Journal of Synthetic Lubrication*, 1989, Vol. 6, pp. 233–52.

17. Spikes, H.A. "Wear and Fatigue Problems in Connection with Water-Based Hydraulic Fluids", *Journal of Synthetic Lubrication*, 1987, Vol. 4, No. 2, pp. 115–35.

18. Anon., "Types of Fire Resistant Hydraulic Fluid", *Industrial Lubrication and Tribology*, 1992, Vol. 44, No. 1, pp. 13–15

19. Janko, K. "A practical Investigation of Wear in Piston Pumps Operated with HFA Fluids with Different Additives", *Journal of Synthetic Lubrication*, 1987, Vol. 4, pp. 99–114.

20. U.S. Patent 3,281,356, Lester E. Coleman, Patented 10/25/1966.

21. U.S. Patent 4,360,443, Albert M. Durr Jr., patented 11/23/1982.

22. Ye, R.Z. "Water-Based Hydraulic Fluid Compositions and Processes", *The FRH Journal*, 1986, Vol. 6, pp. 137–44.

23. Coleman, L.E. "Development of Fire-Resistant Emulsion Hydraulic Fluid", *Journal of the Institute of Petroleum*, Vol. 50, No. 492, pp. 334–44.

24. Spikes, H.A. "Wear and Fatigue Problems in Connection with Water-Based Hydraulic Fluids", *Journal of Synthetic Lubrication*, 1987, Vol. 4, No. 2, pp. 115–35.

25. Francis, C.E. and Holmes, R.T. "New Developments Reflect Improved Performance", *Lubrication Engineering*, 1958, Vol. 14, No. 9, pp. 385–90.

26. Anon., "Fire Resistant Hydraulic Fluids", *Lubrication*, 1962, Vol. 48, No. 11, pp. 161–80.

27. Lapshina, L.N. and Chesnokov, A.A. "Fire-Resistant Emulsion Fluids for Hydraulic Systems (review)", *Chemistry & Technology of Fuels and Oils*, 1975, Vol. 11, pp. 902–07.

28. Shinoda, K. and Kunieda, H. "How to Formulate Microemulsions with Less Surfactant", in *Microemulsions-Theory and Practice*, Ed. Prince, M.L., Academic Press, 1977, pp. 64–86.

29. Becher, P. "HLB: Update III", in *Encyclopedia of Emulsion Technology*, Vol 4., Ed. Becher, P., Marcel Dekker, NY, 2001, pp. 337.

30. Steinmec, F. et al. "Emulsifiable Oil for Preparatioin of Noncombustible Oil-Water Hydraulic Emulsions", Patent # PL 183174, May 31, 2002.

31. Shitara, Y. and Yasutomi, S. "Production of W/O Emulsion Type Fire-Resistant Hydraulic Fluid", Patent App. # JP 2004217702, August 5, 2005.

32. Yoshizawa, H. "Undiluted and Diluted Emulsion as Cutting Oil, Hydraulic Fluid, and Agent for Preventing Scattering of Fiber Glass", Patent App. #JP 11071594, March 16, 1999.

33. Shitara, Y. "Water-in-Oil Emulsion Type Fire Retarding Hydraulic Oil", Patent App. #JP 2008127427, June 5, 2008.

34. Martin, D. W. "Compositions and Method for their Manufacture", Patent App. # US 20030134755, July 17, 2003.

35. Shitari, Y. "W/O Emulsion Type Fire-Resistant Hydraulic Fluid", Patent # JP 3919449, May 23, 2007.

36. Bekierz, G. et al. "Emulsifier for Mineral Oils", Patent # PL 164755, September 31, 1994.

37. Bekierz, G. et al. "Emulsifier for Hydraulic Fluids", Patent # PL 163467, March 31, 1994.

38. Deakin, P. "Fire Resistant Hydraulic Fluids", *Mining Technology*, November/ December, 1990, pp. 300–3.

39. "Emulsifying Oils for Dilute Emulsions for Hydraulic Purposes, N.B.C. Specification 463/1981", British Coal Corporation, London, 1981.
40. Greeley, M. and Rajagopalan, N. "Impact of Environmental Contaminants on Machining of Metal-working Fluids", *Tribology International*, 2004, 37(4), pp. 327–32.
41. Bekierz, G. et al. "Oil Concentrate for Production of Difficultly Flammable Hydraulic Fluids by Using Waters with High Salinity and Hardness", Patent # PL182003, October 31, 2001.
42. Becher, P. *Emulsions : Theory and Practice*, 3rd Edition, American Chemical Society, Oxford, Oxford University Press, 2001.
43. Garti, N., Ezrahi, S. and Aaserin, A. "Water-in-Oil Microemulsion", Patent App.# WO 95/33807, December 14, 1995.
44. Martin, D.W. "Amidoalkyl Betaines in Surfactant and Emulsifier Blends for Preparation of Fuel and Lubricant Water-in-Oil Microemulsions" Patent App GB2434372, July 25, 2007.
45. Garti, N. et al. "Solubilization of Active Molecules in Microemulsions for Improved Environmental Protection", *Colloids and Surfactants A: Physiochemical and Engineering Aspect*, 230, 2004, 99183–190.
46. Regev, et al. "A Study of the Microstructure of a Four-Component Nonionic Microemulsion by Cryo-TEM, NMR, SAXS and SANS", 1996, 12, pp. 668–74.
47. Quemada, D.E., in *Advances in Rheology: Vol. 2, Fluids*, edited by Mena, B. et al., Universidad Nacional Autonama de Mexico, Mexico City, 1984.
48. Berg, G.F. "The Oil in Hydraulic Drives", *Die Technik*, 1949, Vol.4, No. 12, pp. 545–48.
49. Trostmann, E. *Water Hydraulics Control Technology*, Marcel Dekker, New York, NY, 1996, pp. 57.
50. Law, D.A. "The Development and Testing of an Advanced Water-in-Oil Emulsion for Underground Mine Service", *ASLE Preprint*, Preprint No. 80-AM-88-1, 1980.
51. Pell, E.D. and Holtzmann, R.T. "Hydraulic Fluid Emulsions", in *Emulsions and Emulsion Technology - Part II*, Ed. Lissant, K.J., Marcel Dekker, New York, NY, 1974, pp. 639–99.
52. Isaksson, "Rheology for Water-Based Hydraulic Fluids", *Wear*, 1987, Vol. 115, No. 1–2, pp. 3– 17.
53. Schäfer, V.H."Schwer Brennbare Flüssigkeiten für hydraulische Systeme", *Geisseri*, 1964, Vol. 51, No. 26, pp. 817–19.
54. Stumpmeier, F. "Progressive Wasserhydraulik: Teil 1: Algemein und Pumpen/Motoren", *Ölhydraulik und Pneumatik*, 1979, Vol. 23, No. 3, pp. 185–88.
55. Kelly, E.S. "Fire Resistant Fluids: Factors Affecting Equipment and Circuit Design", *3rd International Fluid Power Symposium*, May 9–11, 1973, Sponsored by BHRA Fluid Engineering Bedford, England, held in Turin, Italy, Paper Number F1.
56. Guse, W. "Schwerentflammbare Druckflüssigkeiten - Eigenschaften und Verwendung", *Ölhydraulik und Pneumatik*, 1980, Vol. 24, No. 6, pp. 449–54.
57. Schmitt, C.R. "Fire Resistant Hydraulic Fluids for Die Casting - Part 3", Plant Maintenance Magazine (PMM), February, 1955, pp. 81–85.
58. Dalmaz, G. "Traction and Film Thickness Measurements of a Water Glycol and a Water-in-Oil Emulsion in Rolling-Sliding Point Contacts", in *Proceed. 7th Leeds-Lyon Symposium on Tribology, Friction and Traction*, September 1980, Paper IX, pp. 231–43.
59. Taylor, R. and Wang, Y.Y. "Lubrication Regimes and Tribological Properties of Fire-Resistant Hydraulic Fluids", *Lubrication Engineering*, 1984, Vol. 40, No. 1, pp. 44–50.
60. Law, P.J. "High Water Content Fluids - Products for the Future", in *Conference Proceedings: Hydraulics, Electrics and Electronics in Steel Works and Rolling Mills*, Brochure No. RE 00 252/09.94, Available from Mannesmann Rexroth GmbH, Jahnstrasse 3–5, D-97816 Lohr-am-Main, Germany.
61. Law, D.A. "The Development and Testing of an Advanced Water-in-Oil Emulsion for Underground Mine Service", *ASLE Preprint*, Preprint No. 80-AM-88-1, 1980.
62. "Requirements and Tests Applicable to Fire-Resistant Hydraulic Fluids Used for Power Transmission and Control (Hydrostatic and Hydrokinetic)", *European Safety and Heath Commission for the Mining and Other Extractive Industries*, Doc. No. 4746/10/91 EN, Luxembourg, April 1994.
63. ASTM D 3427, Approved 1986, "Standard Method for Gas Bubble Separation Time of Petroleum Oils", *American Society for Testing and Materials*, Conshohocken, PA.
64. ASTM D 943, Approved 1981, "Standard Test Method for Oxidation Characteristics of Inhibited Mineral Oils", *American Society for Testing and Materials*, Conshohocken, PA.
65. Schmiede, J.T., Simandiri, S. and Clark, A.J. "The Development and Applications of an Invert Emulsion, Fire-Resistant Hydraulic Fluid", *Lubrication Engineering*, 1985, Vol. 41, pp. 463–69.
66. ASTM D 665–92, "Standard Test Method for Rust-Preventing Characteristics of Inhibited Mineral Oil in the Presence of Water", *American Society for Testing and Materials*, Conshohocken, PA.

67. Young, K.J. "Development and Application of Advanced Emulsion Hydraulic Fluids", *SAE Technical Paper Series*, Paper Number 951196, 1995.

68. Wang, Z. "Use and Maintenance of Fire-Resistant Hydraulic Oil", Runhua Yu Mifeng, 1987, Vol. 5, 9. 60–64.

69. Henrikson, K.G. "Fire-Resistant Fluids and Mobile Equipment", *SAE Technical Paper Series*, Paper Number 650671, 1965.

70. ISO 48 - 1994. "Rubber Vulcanized or Thermoplastic Determination of Hardness between 10 IRHD and 100 IRHD."

71. "Design, Operation and Maintenance of Hydraulic Equipment for Use with Fire Resistant Fluids", Brochure available from *National Fluid Power Association*, Milwaukee, WI.

72. Borowski, J.L. "The Use of Invert Emulsion Hydraulic Fluid in a Steel Slab Caster", *ASLE Preprint*, Preprint No. 80-AM-88-2, 1980.

73. Foitl, R.J. and Kucera, W.J. "Formation and Evaluation of Fire-Resistant Fluids", *Iron and Steel Engineer*, July, 1964, pp. 117–20.

74. ASTM D 95 - 83, "Standard Method for Water in Petroleum Products and Bituminous Materials by Distillation", *American Society for Testing and Materials*, Conshohocken, PA.

75. ASTM D 1744 - 92, "Standard Test Method for Determination of Water in Liquid Petroleum Products by Karl Fischer Reagent", *American Society for Testing and Materials*, Conshohocken, PA.

76. ASTM D 446 - 89a, "Standard Specifications and Operating Instructions for Glass Capillary Kinematic Viscometers", *American Society for Testing and Materials*, Conshohocken, PA.

77. ASTM F 311, "Standard Practice for Processing Aerospace Liquid Samples for Particulate Contamination Analysis using Membrane Filters", *American Society for Testing and Materials*, Conshohocken, PA.

78. ASTM F 312, "Standard Test Methods for Microscopical Sizing and Counting Particles from Aerospace Fluids on Membrane Filters", *American Society for Testing and Materials*, Conshohocken, PA.

79. Hill, E.C. and Hill, G.C. "Biodegradable After Use But Not In Use", *Industrial Lubrication and Tribology*, 1994, Vol. 46, No. 3, pp. 7–9.

80. Passman, F.J. "Biocide Strategies for Lubricant Rancidity and Biofouling Prevention", *Proceed. AISE 1996 Annual Convention, Association of Iron and Steel Engineers*, Chicago, IL, Vol. 1, pp. 413–28.

81. Hill, E.C. "The Significance and Control of Microorganisms in Rolling Mill Oils and Emulsions", *Metals and Materials*, 1967, No.9, pp. 294–97.

82. Anon., "Microbiology of Lubricating Oils", *Process Biochemistry*, May, 1967, pp. 54–56.

83. Hill, E.C. "Degradation of Oil Emulsions", *Engineering*, June, 1967, pp. 983–84.

84. Hill, E.C., Davies, I., Pritchard, J.A.V. and Byron, D. "The Estimation of Micro-Organisms in Petroleum Products", *Journal of the Institute of Petroleum*, 1967, Vol. 53, No. 524, pp.275–79.

85. Hill, E.C., Graham Jones, J. and Sinclair, A. "Microbial Failure of a Hydraulic Oil Emulsion in a Steel Rolling Mill", *Metals and Materials*, 1967, Vol. 1, No. 12, pp. 407–9.

86. ASTM D 3946 - 92, "Standard Test Method for Evaluating the Bacteria Resistance of Water-Dilutable Metalworking Fluids", *American Society for Testing and Materials*, Conshohocken, PA.

87. ASTM E 686, "Standard Test Method for Evaluation of Antimicrobial Agents in Metalworking Fluids", *American Society for Testing and Materials*, Conshohocken, PA.

88. ASTM E 979, "Standard Test Method for Evaluation of Antimicrobial Agents as Preservatives for Invert Emulsion and Other Water Containing Hydraulic Fluids", *American Society for Testing and Materials*, Conshohocken, PA.

89. Gannon, J. and Bennett, E.O. "A Rapid Technique for Determining Microbial Loads in Metalworking Fluids", *Tribology*, 1981, Vol. 14, pp. 3–6.

90. Sloyer, J.D. "Rapid Determination (60 Seconds) of Bacterial Contamination in Industrial Fluids", In *Proceed. of the AAMA Metalworking Fluids Symposium: The Industrial Metalworking Environment Assessment & Control*, November 13–16, 1995, *American Automobile Manufacturers Association*, Detroit, 1996, pp. 362–63.

91. Blanpain, G.M.G. "Fire-Resistant Hydraulic Fluids in the French Mines", *Fluid Power Equipment in Mining, Quarrying and Tunneling*, February, 1974, 12–13, pp. 145–55.

92. "Houghton-Safe® Fire-Resistant Fluids Handbook", Booklet available from *Houghton International*, Valley Forge, PA.

93. Loudon, B.J. "Fire-Resistant Hydraulic Fluids", *Surface Coatings* Australia, July, 1989, pp. 23–29.

94. Hamaguchi, H., Spikes, H.A. and Cameron, A. "Elastohydrodynamic Properties of Water in Oil Emulsions", *Wear*, 1977, Vol. 43, pp. 17–24.

95. Liu, W., Dong, D., Kimura, Y. and Okada, K. "Elastohydrodynamic Lubrication with Water-in-Oil Emulsions", *Wear*, 1994, Vol. 179, pp. 17–21.

96. Yardley, E.D., Kenny, P. and Sutcliffe, D.A. "The Use of Rolling Fatigue Test Methods Over a Range of Loading Conditions to Assess the Performance of Fire-Resistant Fluids", *Wear*, 1974, Vol. 28, pp. 29–47.

97. Zaretsky, E.V. "Chapter 1 - Current Practice", in *STLE Life Factors for Roller Bearings*, Society of Tribologists and Lubrication Engineers, Park Ridge, IL, 1996, p. 7.

98. Taylor, R. and Wang, Y.Y. "The Effect of Fluid Properties on Bearing Parameters: Part 1", *The FRH Journal*, 1984, Vol. 4, No. 2, pp. 161–68.

99. Taylor, R. and Wang, Y.Y. "The Effect of Fluid Properties on Bearing Parameters: Part 2", *The FRH Journal*, 1984, Vol. 5, No. 1, pp. 31–37.

100. Taylor, R. and Wang, Y.Y. "The Effect of Fluid Properties on Bearing Parameters: Part 3", *The FRH Journal*, 1984, Vol. 5, No. 1, pp. 39–44.

101. Taylor, R. and Lin, Z.L. "The Application of Tribological Principals to the Design of the Valve Plate of an Axial Piston Pump: Part 3 - Fire Resistant Fluid Considerations", *The FRH Journal*, 1984, Vol. 5, No. 1, pp. 99–101.

102. Castleton, V.W. "Practical Considerations for Fire-Resistant Fluids", *Lubrication Engineering.*, 1998, Vol. 54, No. 2, pp. 11–17.

103. Anon., "Lubrication Engineers Take a Second Look at Fire-Resistant Fluids", *Coal Age*, July, 1971, pp. 118–19.

104. Janko, K. "A Practical Investigation of Water in Piston Pumps Operated with HFA Fluids with Different Additives", *Journal of Synthetic Lubrication*, 1987, Vol. 4, pp. 99–114.

105. Shrey, W.M. "Evaluation of Fluids by Hydraulic Pump Tests", *Lubrication Engineering*, February, 1959, pp. 64–67.

106. ASTM D 2882, "Standard Test Method for Indicating the Wear Characteristics of a Petroleum and Non-Petroleum Hydraulic Fluids in a Constant Volume Vane Pump", *American Society for Testing and Materials*, Conshohocken, PA.

107. DIN 51389 - 1981. "Testing of Cooling lubricants: Determination of the pH Value of Water-Mixed Cooling Lubricants."

108. Kelly, E.S. "Erosive Wear of Hydraulic Valves Operating with Fire-Resistant Emulsions", *Proceed. of the 2nd Fluid Power Symposium*, January, 1971, The British Hydromechanics Research Association., Paper F4, P. F4-45-F4-73.

109. "Sauer Sundstrand - All Series: Fluid Quality Requirements", Brochure BLN-9887, Rev. A, Available from Sauer Sundstrand Inc., Ames, IA.

110. Rynders, R.D. "Fire Resistant Fluids for a Component Builders Point of View" *National Conference of Industrial Hydraulics*, 1962, 16, p. 39–46.

111. Bonnell, G.C. "Fire-Resistant Fluids for Die-Casting", *Foundry*, 1967, Vol. 95, No. 5, pp. 224–29.

112. Schmitt, C.R. "Fire-Resistant Hydraulic Fluids for the Die-Casting Industry", Plant Maintenance Magazine (PMM), 1957, May, pp. 111–13.

113. Schmitt, C.R. "Fire-Resistant Hydraulic Fluids: Part 5 - Change-Over Practices", *Applied Hydraulics*, October, 1957, pp. 160–62.

114. Egan, E.J. "How to Install Fire-Resistant Hydraulic Fluids", *The Iron Age*, October 11, 1956, pp. 95–97.

115. Brink, E.C. "Fire-Resistant Hydraulic Fluids", *Lubrication*, Vol. 58, October/December, 1972, pp. 77–96.

116. Morrow, A.S., Sipple, H.E. and Holmes, R.T. "Fire-Resistant Hydraulic Fluids for the Die-Casting Industry: Part 1 - Emulsion Types", Plant Maintenance Magazine (PMM), January, 1957, pp. 133–44.

117. Foitl, R.J. "Formation and Evaluation of Fire-Resistant Hydraulic Fluids", *Iron and Steel Engineer*, July, 1964, pp. 117–20.

118. Klamann, D. "Chapter 11.9 Hydraulic Fluids", in *Lubricants and Related Products - Synthesis, Properties, Applications, International Standards*, Verlag Chemie, Basel, 1984, pp. 306–31.

119. Jackson, L. "Fire-Resistant Hydraulic Fluids for the Die Casting Industry: Part 6 - Safety & Use Factors", Plant Maintenance Magazine (PMM), 1957, June, pp. 41–42.

120. Pollock, S.P. "The User's Experience of Hydraulic Systems Incorporating High Water Based Fluids", in *Conference Proceedings: Hydraulics, Electrics, and Electronics in Steel Works and Rolling Mills*, Brochure No. RE 00 252/09.94, Available from Mannesmann Rexroth GmbH, Jahnstrasse 3–5, D-97816 Lohr-am-Main, Germany.

121. Anon. "Are Your Hydraulic Oil Lines Fireproof?", *Mill & Factory*, May, 1952, pp. 139–40.

122. Anon. The Texaco Company, Inc., *Operation and Care of Hydraulic Machinery*, New York, NY, 1962, pp. 95.

123. Davies, J.T. and Rideal, E.K. *Interfacial Phenomenon*, Academic Press, New York, NY, 1961, pp. 359–87.

124. "The ATLAS HLB System - A Time Saving Guide to Emulsifier Selection", Brochure available from *Atlas Chemical Industries Inc.*, Wilmington, DE, 1989 USA.

15 Water–Glycol Hydraulic Fluids

John V. Sherman

CONTENTS

15.1 INTRODUCTION

Fire-resistant hydraulic fluids are an essential component for the safe operation of hydraulic systems in many industries where there is a risk of fire due to the presence of necessary ignition sources. Water–glycol (W/G) hydraulic fluids are the most widely used fire-resistant type in the world today [1]. This is because they provide an optimum combination of fire resistance, lubrication, corrosion protection, and economy of use for medium to high pressures and most other conditions required in industrial hydraulic systems.

Prior to World War II, the most common hydraulic fluids used for all industrial and military applications were mineral oil-based and therefore flammable. By 1943, the U.S. Navy had experienced numerous on-board ship and aircraft fire disasters caused by hydraulic line rupture and subsequent ignition of the resulting mineral oil sprays. Prompted by these disasters, the U.S. Navy approved research at the U.S. Naval Research Laboratory to identify liquids that would be sufficiently non-flammable and yet could function as the base component for a hydraulic fluid. This and subsequent work led to the laboratory developing a new class of fire-resistant hydraulic fluids known as "hydrolubes" [2–6]. Hydrolubes were defined as polymer-thickened, corrosion-inhibited, aqueous fluids having one or more glycols as major constituents [4]. Hydrolubes, also known as water–glycol hydraulic fluids, were patented in 1947 by Roberts and Fife [7]. Although the naval research and initial trials were successful, the use of HFC fluids in naval operations did not gain acceptance at that time [8].

From 1940 to 1960, a number of fires originating with mineral-oil hydraulic fluids occurred, resulting in the loss of lives and destruction of equipment and buildings [9,10]. The most publicized was the Bois du Cazier mine disaster in 1956 at Marcinelle, Belgium where 267 miners died [11]. In this time period an increasing level of concern over these disasters among nations led to industries and companies pushing for the development of fire-resistant hydraulic fluids and protocols for their evaluation and classification. One of the first of these protocols implemented was called the "Luxembourg Report," which was first published in 1961 to set minimum fire-resistance standards for hydraulic fluids to be used in European coal mining [12].

Water–glycol hydraulic fluids were ideally suited for use as fire-resistant hydraulic fluids in industries where high-temperature processing was performed. A list of illustrative underground mining applications using different types of fire-resistant hydraulic fluids is provided in Table 15.1.

Industries utilizing high-temperature processing include metal heat treatment [9], plastics [10], zinc and aluminum die-casting [13], coal mining [13,14], iron and steelmaking [15,16], hydraulic presses [16], and machine tools for the automotive industry [16]. Early water–glycol compositions were only applicable for relatively low-pressure systems [17]. Currently, commercial formulations providing outstanding performance in high-pressure, >5000 psi, hydraulic pumps are well documented [18,19]. Over the years, various pumps have been developed for use with water–glycol hydraulic fluids [20,21], proper operating conditions have been established [22] and fluid performance has significantly improved due to formulation improvement [23,24].

The protocol used globally to designate different types of fire-resistant hydraulic fluids since first published in 1982 has been the ISO 6743/4 standard [14,25]. Water–glycol (W/G) hydraulic fluids, under this standard, are designated as "HFC" fluids. Other fire-resistant hydraulic fluid types identified in the standard include phosphate ester, oil-in-water (invert) emulsions, water-in-oil emulsions, high-water-base hydraulic fluids (HWBHF), polyol ester (POE), and poly(alkylene glycol) (PAG) fluids. The ISO 6743/4 classification designations for fire-resistant hydraulic fluids are provided in Table 15.2.

This chapter will provide an overview of water–glycol fluid composition, physical and chemical properties, lubrication evaluation, pump performance and practical considerations related to their usage, including maintenance and system conversions.

15.2 DISCUSSION

15.2.1 WATER–GLYCOL HYDRAULIC FLUID COMPOSITION

Water–glycol (W/G) hydraulic fluids contain water, glycol, thickener, and an additive package [16,26–29]. The general functions of W/G hydraulic fluid formulation components are summarized in Table 15.3. The additive package typically consists of amines for ferrous metal corrosion inhibition, aromatic azoles for yellow metal passivation, antiwear additives (typically aliphatic carboxylic acids), dyes to facilitate system leak detection, and antifoams to reduce foaming and

TABLE 15.1

Hydraulic Equipment and Machinery using Fire-Resistant Fluids in Underground Mining

Use	Hydraulically Driven Machinery and Equipment	Hydraulic Fluids Used					Drive	
		Water	HFA	HFC	HFD	Mineral Oil	Hydrostatic	Hydrodynamic
Off-highway	Drill carriages and drills			X			X	
	Percussion drills			X			X	
	Side-tipping loaders			X			X	
	Drilling/milling-type roadheaders			X			X	
	Impac rippers			X			X	
	TBMS			X			X	
	Fluid couplings	X						X
Coal mining	Drum shearer-loaders (winches)			X	X		X	
	Pumps for powered support		X				X	
	High-pressure water inclusions system	X					X	
	Fluid couplings	X						X
Infrastructure	Rope-driven transport systems		X	X		X		
	Diesel monorail locos		X			X		
	Hydraulic monorail transport units		X			X		
	Diesel locos (floorbound)					X	X	
	Chain creepers		X	X		X		X
	Fluid couplings	X						
Other	Trackless vehicles (drive line)			X			X	
	Small equipment (V.10L)		X				X	
	Electrohydraulic control units			X				X

TABLE 15.2
Fire-Resistant Hydraulic Fluid Classification According to ISO 6743/4

Category	Composition	Operational Temperature
Water-based fluids		
HFAE	Oil-in-water emulsions, >80% water in final diluted fluid. (concentrate portion ≤20%)	5–≤55°C
HFAS	Mineral-oil-free synthetic solutions, >80% water in final diluted solutions (concentrate portion ≤20%)	5–≤55°C
HFB	Water-in-oil emulsions, 35–40% water in final diluted solution (mineral oil portion ≤60%)	5–≤60°C
HFC	Aqueous polymer solutions (hydrolules), water portion ≤35% in final diluted solution	–20–≤60°C
Water-free fluids		
HFDR	Phosphate ester fluids	–20–70°C
HFDU	Synthetic fluids, other compositions	–35–≤90°C

ᵃ For hydrodynamic couplings, the maximum temperature is 150°C.

air entrainment . Recently, Singh has reported water–glycol hydraulic fluids containing additive packages which include sulfur and/or phosphorous-based antiwear compounds having improved lubricant properties [30].

The most common glycols used in water–glycol hydraulic fluid formulations are ethylene glycol (I), diethylene glycol (II), and propylene glycol (III):

$$HOCH_2CH_2OH \tag{I}$$

$$HOCH_2CH_2OCH_2CH_2OH \tag{II}$$

$$HOCH_2CH(CH_3)OH \tag{III}$$

The most widely used glycol in the world today for standard industrial water–glycol formulations is diethylene glycol [26]. For food processing applications propylene glycol is used as it is approved for incidental food contact [31] and diethylene glycol is not. It is important to note that some of the additives referenced, including amines, may not used in W/G hydraulic fluids used in food processing applications because they are not approved for use by the United States Food and Drug Administration (FDA) and the United States Department of Agriculture(USDA) where there is the risk of incidental food contact.

TABLE 15.3
Water–Glycol Hydraulic Fluid Composition

Component	Purpose
Water	Fire protection
Glycol	Freeze point reduction and some thickening
Thickener	Thicken the formulation and provide adequate film viscosity at the wear contact
Antiwear additives	Provide mixed-film and some boundary lubrication
Corrosion inhibitors	Vapor and liquid corrosion protection

The ability of glycol to reduce the freezing point of mixtures with water is illustrated in Figure 15.1. Figure 15.1 shows that the minimum freezing point of the ethylene glycol (also known as "monoglycol") is produced at −65°F when a eutectic mixture (67%/33% by volume of ethylene glycol/water) is formed. Although this composition produces the minimum freezing point, very small changes in water content produce relatively large increases in the freezing point.

There are two major problems with using a water–glycol solution as the basis for a hydraulic fluid. One is poor film thickness and the other is a low viscosity index (VI). This problem is addressed by the use of a thickener. The thickener most often used is a poly(alkylene glycol)—PAG (IV), which is a "random" copolymer of ethylene oxide and propylene oxide. Poly(alkylene glycol) polymer configurations may be branched [29] or linear [26], such as the linear polymer shown by (IV) and branched polymer shown by (V):

$$HO(CH_2CH_2O)_m (CH_2CH(CH_3)O)_n H, \tag{IV}$$

$$CH_3CH_2C(CH_2O[CH_2CH_2O]_m [CH_2CH(CH_3)O]_n H)_3. \tag{V}$$

One of the first poly(alkylene glycol) thickeners reported for use in hydrolube formulation was UCON® Lube 75H 90,000, which is a copolymer containing 75% by weight oxyethylene units designated as (CH_2CH_2O) and 25% by weight of oxypropylene units, which is designated as $(CH_2CH(CH_3)O)$ in (IV). The polymer exhibits a viscosity of 90,000 SUS at 100°F. The thickening and viscosity-temperature behavior of a 55 (ethylene glycol)/45 (water) mixture with varying quantities of 75H 90,000 added is shown in Table 15.4. Based on recommendations made by the Bureau of Aeronautics, the fluid-thickener combination was selected to provide a minimum viscosity of 10 cS at 130°F, maximum viscosity of 2000 cSt at −40°F, and a freezing point of <−50°F.

The correct ratio of thickener, glycol and water is extremely important to obtain the desired viscosity, viscosity index, pour point and fire-resistant properties of the W/G hydraulic fluid. The type and amount of glycol and thickener are determined which when blended with the required concentration of water result in the desired base stock properties. The antiwear, corrosion inhibition, defoaming, and air release additives are typically less than 4% of the final formulation.

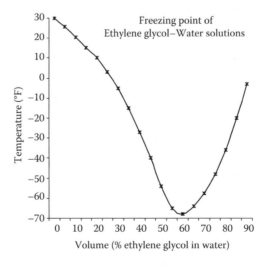

FIGURE 15.1 Freezing points of ethylene glycol–water mixtures.

TABLE 15.4

Viscosities of a Poly(Alkylene Glycol)-Thickened Ethylene Glycol–Water Mixture

% UCON® Lube 75H 90,000 in a 55%/45% by Weight Ethylene Glycol–Water Mixture	Temperature				
	–40°F	0°F	100°F	130°F	210°F
0	134.0	25.0	2.6	1.74	0.79
5	552.1	86.9	7.0	4.53	1.93
10	1500	223.0	16.5	10.06	3.66
20	–	1007.0	60.0	35.3	13.9
30	–	3730	170	98.9	34.9
40	–	–	–	236.2	79.1
50	–	–	–	498.7	158.8

In addition to the antiwear additives, it is important to recognize that both the water content and corrosion inhibitors that are present in a W/G hydraulic fluid may dramatically affect antiwear performance. In Figure 15.2, it is shown that water concentrations in excess of approximately 40% will result in corresponding decreases in pump wear performance [33]. Spikes (see Figure 15.3) has shown that fluid-film thickness decreases rapidly as water content increases, a phenomenon that will increase wear [34]. It is believed that water concentrations of greater than approximately 35% are necessary in order to provide industry-acknowledged levels of fire safety [35,36]. For these reasons, water–glycol hydraulic fluids typically contain 35–45% water.

Because W/G fluids contain water, corrosion inhibition is necessary. One corrosion-inhibitor system that may be used for water–glycol hydraulic fluids is based on an amine additive system. As originally reported by Brophy et al. [3], the concentration of the corrosion inhibitor, which may be quantitatively measured by "reserve alkalinity" determination, may also affect wear rates a shown in Figure 15.4 [33] .

15.2.2 CHEMICAL AND PHYSICAL PROPERTIES—GENERAL COMPARISON

To perform acceptably in a hydraulic system, any fire-resistant hydraulic fluid, including water–glycol fluids, must exhibit the following properties [37]:

- Good fire resistance
- Proper viscosity and viscosity index
- Good resistance to oxidation and deposit formation
- Good wear resistance
- Good resistance to rust and corrosion

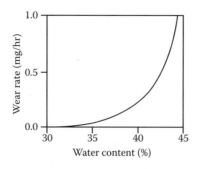

FIGURE 15.2   Effect of water content on ASTM D 2882 V-104 Vickers vane-pump wear for a water–glycol hydraulic fluid.

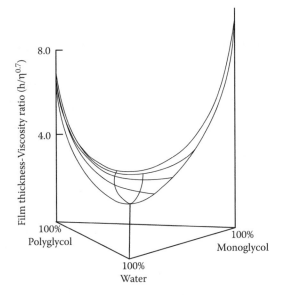

FIGURE 15.3 Effect of water, ethylene glycol, and poly(alkylene glycol) thickener composition on film thickness and viscosity.

- Low foaming
- Seal compatibility
- Paint compatibility

A general qualitative comparison of water–glycol hydraulic fluids to water-in-oil emulsions, phosphate esters, and an antiwear mineral oil is provided in Table 15.5 [20,38]. A more specific comparison of a water–glycol fluid to a mineral oil is provided in Table 15.6 [23]. The dependence of these properties on temperature and pressure for water–glycol hydraulic fluids is shown in Tables 15.7 and 15.8, respectively. *Note:* Physical property data are always dependent on the composition of the formulation

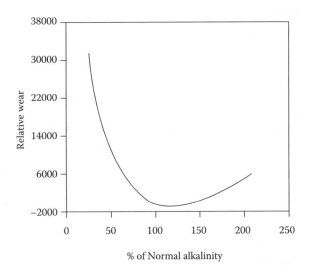

FIGURE 15.4 Effect of amine concentration (reserve alkalinity) on ASTM D 2882 V-104 Vickers vane-pump wear for a water–glycol hydraulic fluid.

TABLE 15.5
Generic Physical Property Comparison of Different Fire-Resistant Hydraulic Fluids with an Antiwear Petroleum Oil

	Antiwear Oil	Water–Glycols	Water-in-Oil Emulsions	Phosphate Esters
Properties				
Specific gravity	0.85	1.09	0.91	1.15
Flash point (°F)	420	None	None	500
Autogenous ignition temperature (°F)	–650	830	830	1100+
Bulk fluid temperature limits (°F)	–20–80	0–150	15–150	20–200
Low-temperature performance	Excellent	Excellent	Good	Fair
Lubricity	Excellent	Good	Good	Excellent
Oxidation stability	Excellent	Good	Good	Excellent
Rust Protection	Good	Good	Good	Fair to good
Bulk Modulus	Medium	Medium	Medium	High
Foaming resistance	Excellent	Good	Excellent	Good
Air release	Good to excellent	Fair to good	Fair to good	Fair
Compatibility				
Oil-resistant paint	OK	No	OK	No
Nonferrous metals	OK	Check[a]	OK	OK
BUNA N(nitrile)	OK	OK	OK	No
Butyl	No	OK	No	OK
Neoprene	OK	OK	OK	No
Silicone	No	OK	No	OK
Viton	OK	OK	OK	OK
Paper	OK	Check	Check	Check

[a] Not compatible with zinc, cadmium, lead, magnesium and unanodized aluminum.

being compared. Therefore, the specific data shown here should only be taken as illustrative examples. The reader should obtain the specific data of interest from the supplier of the fluid being considered.

15.2.2.1 Vapor Pressure

Above approximately 65°C (150°F), the vapor pressure of a water–glycol hydraulic fluid, like other water-containing fluids, rapidly increases with increasing temperature [23]. The pressure

TABLE 15.6
Quantitative Comparison of Physical Properties between a Petroleum Oil and a Water–Glycol Hydraulic Fluid

Property	Water–Glycol (41 cS at 50°C)	Petroleum Oil (36 cS at 50°C)
Density (g/cm³ at 15°C)	1.060	0.885
Vapor pressure (mbar at 50°C)	80	0.001
Vapor pressure (mbar at 70°C)	182	0.02
Air solubility% at 20°C/1030 mbar	27–30	9
Compressibility at 20°C/bar	51×10^{-6}	65×10^{-6}
Specific heat at 20°C (J/g°C)	3.3	1.8
Thermal conductivity (mW/cm°C)	3.0	1.3

TABLE 15.7
Dependence of Water–Glycol Fluid Physical Property on Temperature

Temperature (°F)	Viscosity (lb/ft min)	Density (lbs/ft³)	Specific Heat (BTU/lb.)	Thermal Conductivity (BTU/ft°F)	Vapor Pressure (psia)
160	0.546	65.0	0.737	0.26	3.7
130	0.842	65.8	0.729	0.26	1.7
100	1.39	66.5	0.719	0.26	0.74
0	20.7	68.5	–	–	–
–30	83.5	69.1	–	–	–
–35	113	69.1	–	–	–

dependence on temperature is shown in Figure 15.5. To minimize the potential for evaporative water loss in hydraulic systems using W/G fluids, pressurized reservoirs should be used if fluid temperatures are to exceed 65°C (150°F).

15.2.2.2 Viscosity

Fluid viscosity is not only an important parameter with respect to hydraulic system operation, but it is also important with respect to lubrication. Therefore, it is desirable that the hydraulic fluid viscosity exhibit as small a change in viscosity with varying temperature as possible. Figure 15.6 shows that water–glycol hydraulic fluids, as a class, exhibit significantly less variation in viscosity with respect to temperature than mineral oils, petroleum oil, and, especially, phosphate esters [39]. Water–glycol hydraulic fluids typically have viscosity indices greater than 180. Figure 15.7 shows that the variation of viscosity response of water–glycols with varying temperature is similar for different viscosity grades.

The viscosity of water–glycol hydraulic fluids is dependent on water concentration, as illustrated in Figure 15.8 [35]. If excessive water additions are made to a system containing a water–glycol fluid, low fluid viscosities will result, which may lead not only to reduced volumetric efficiency and excessive fluid leakage, but also to poor lubrication because of insufficient film thickness at the wear contact. (See Figures 15.2 and 15.3). Conversely, the loss of sufficient amounts of water by evaporation will cause excessively high fluid viscosities, resulting in low mechanical efficiency, sluggish operation, and, in extreme cases, loss of fire-resistance and cavitation.

Generally, W/G fluids are not used at temperatures above 60°C [40] however, if pressurized reservoirs are used, fluid temperatures of 70°C are possible [35]. Low-temperature limits vary from approximately 10°C to about –25°C, depending on the specific fluid [40].

TABLE 15.8
Dependence of Water–Glycol Fluid Physical Property on Pressure

Pressure (psi)	Compressibility (psi⁻¹ at 78°F)	Viscosity (cps) 100°F	150°F
1,000	3.58×10^{-6}	35	14.1
2,000	3.24×10^{-6}	36	14.3
3,000	3.14×10^{-6}	37	14.8
10,000	2.96×10^{-6}	44	18
20,000	–	56	23
30,000	–	70	28

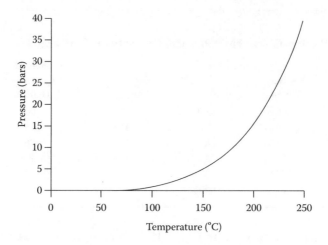

FIGURE 15.5 Effect of water–glycol hydraulic fluid temperature on vapor pressure.

15.2.2.3 Viscosity Stability

Newtonian versus non-Newtonian behavior. Although numerous water-soluble polymers have been used as thickeners for W/G fluids, the most common thickener in use from the time of initial development (1947) [3] until the present time is poly(alkylene glycol) (PAG). PAG thickeners provide Newtonian thickening behavior, which means that the fluid does not undergo significant viscosity loss due to shear thinning (temporary or permanent) during use; therefore, fluid viscosity is independent of shear rate.

The Newtonian behavior of a developed W/G fluid in shear fields which are typically found in hydraulic pumps was demonstrated using high-shear capillary viscometry [41]. The results shown in Figure 15.9 illustrate that a W/G hydraulic fluid containing a "typical" PAG polymer thickener does not undergo any significant shear-thinning behavior that would lead to hydraulic leakage and corresponding loss of pump efficiency. Similar high-shear viscometry results have been reported by Isaksson [42].

Mechanical shear degradation. Shear rate may be defined as the ratio of the velocity of the fluid divided by the clearance [43]. High shear rates may be encountered in numerous areas within the

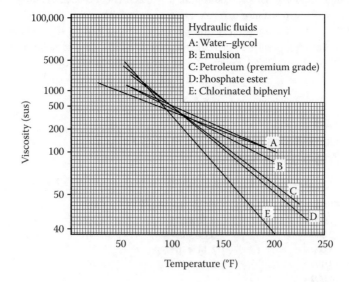

FIGURE 15.6 Viscosity-temperature relationships for a number of hydraulic fluids, including a water–glycol fluid formulation.

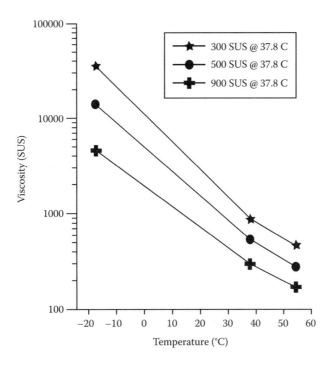

FIGURE 15.7 Effect of water–glycol hydraulic fluid viscosity on the viscosity-temperature relationship.

hydraulic pump, such as the relief valve seat and roller bearing assemblies. Very high-shear fields, approximately 10^6 s^{-1}, have been estimated to be present in hydraulic pumps [44].

A hydraulic fluid may not only undergo viscosity losses when subjected to high shear rates due to non-Newtonian behavior, but the polymer used to thicken some non-PAG-containing W/G hydraulic fluids may undergo molecular degradation (mechanodegradation), resulting in *permanent* viscosity loss. Shear stability is related to the molecular weight of the thickener as illustrated in Figure 15.10 [16].

The shear stability of PAG-thickened water–glycol hydraulic fluids is illustrated in Figure 15.11 for a series of fluids subjected to an ultrasonic shear test [45]. In this example, it is observed that typical PAG-thickened W/G hydraulic fluids were shear stable, as evidenced by no viscosity loss in the ultrasonic shear test, which means that the polymeric thickener is not expected to produce a permanent viscosity loss in the high-shear conditions encountered during operation in the hydraulic pump.

Shear stability testing procedures are reviewed in Chapter 3. In addition to the ultrasonic shear stability test shown above, Blanpain [35] had reported that water–glycol hydraulic fluids will easily pass a Fuel Injector Shear Stability Test (FISST), ASTM D 5275 [46], or the Kurt Orbahn method [47].

15.2.2.4 Foaming and Air Release

Although some water–glycols may exhibit longer air-release times than petroleum-oil hydraulic fluids [35,48], properly formulated W/G fluids should exhibit comparable air-release properties to a premium petroleum oil [37]. Foaming and air-release properties and testing procedures were discussed in Chapter 3 and will not be discussed in detail here. To reduce air entrainment, reservoir capacity should be increased and a baffle placed between the suction and return lines. The fluid return line should be placed below the air-liquid surface.

15.2.2.5 Nitrosamine Formation

Some of the earliest corrosion inhibitors used in W/G hydraulic fluid formulation included sodium nitrite and amine nitrites [49]. The technology available at the time required the use of these particular additives in order to achieve the necessary solution, vapor, and dry-film corrosion-inhibition

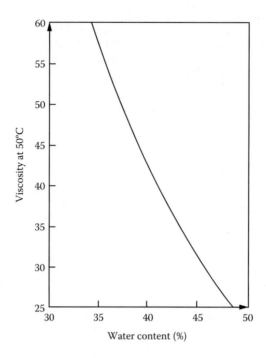

FIGURE 15.8 Effect of water content on water–glycol hydraulic fluid viscosity. Although slightly different charts will be obtained for different water–glycol fluids, the overall responses will be the same.

properties. However, concerns over potential nitrosamine formation [50] during the use of these additives, particularly in metalworking and cosmetic formulation, led to the development of non-nitrite-containing W/G hydraulic fluid formulations [18,19]. Nitrite-containing additives are no longer used in W/G hydraulic fluid formulations in the United States.

Currently, various amine-containing corrosion-inhibitor additives are used which provide outstanding corrosion protection in use. The vapor- and liquid-phase corrosion-inhibition properties of a typical W/G hydraulic fluid are illustrated in Figure 15.12. These inhibitory properties are maintained during use, as long as the concentration of the corrosion inhibitor is maintained [51] above the minimum levels required for satisfactory performance.

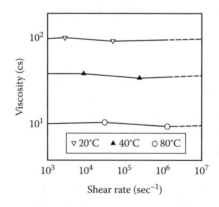

FIGURE 15.9 High-shear-rate viscosity profile of a water–glycol hydraulic fluid at different fluid temperatures.

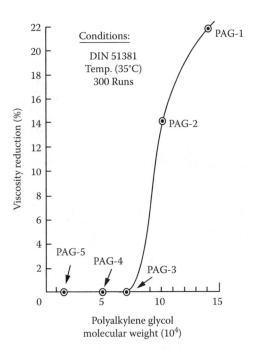

FIGURE 15.10 Stability of different poly(alkylene glycol) polymers to mechanodegradation.

15.2.2.6 Phenol Content

Increasingly stringent regulation of allowable wastes into water emitted from aluminum die-casting plants had caused the aluminum die-casting industry, a major user of fire-resistant hydraulic fluids, to limit the incoming "phenol content" of their hydraulic fluids. Hydraulic fluids are discharged from the plant as a result of fluid leakage from the hydraulic systems during use. The concept behind the regulation is that if there are no "phenolics" contained in the incoming fluid, then there should no phenolics present in the waste water used in the plant.

One of the specified analytical procedures for the determination of phenols and phenolic derivatives, according to the *Federal Register* (40 CFR 403 and 40 CFR 464), is the 4-aminoantipyrene (4-AAP) colorimetric titration procedure [52], which is illustrated in Figure 15.13.

In this procedure, 4-AAP is reacted with a phenolic substrate to yield a highly colored dye which is quantitatively analyzed by visible spectroscopy at 460 nm. Because this is a colorimetric

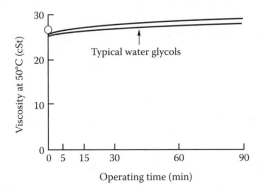

FIGURE 15.11 High-shear-rate stability of different water–glycol hydraulic fluids according to an ultrasonic high-shear stability test. The "typical" water–glycols were thickened with a poly(alkylene glycol) polymer.

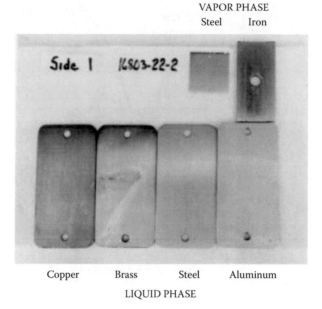

FIGURE 15.12 Illustration of vapor- and liquid-phase corrosion inhibition for a water–glycol hydraulic fluid obtained with an amine-containing additive.

procedure, any substrate capable of reacting with 4-AAP and producing a colored product will cause an interference and produce erroneous phenol-content values. Therefore, this test has been called the "phenol-response" test in the industry, because the analysis is based on the formation of a colored dye and is not necessarily indicative of the actual phenolic content of the fluid. Although some fluid formulations do contain very low levels of phenol-response, in many cases there were no actual phenolic derivatives and false interference phenol-response results have been obtained. Nevertheless, to assist the die-casting industry to comply with the federal wastewater requirement, W/G fluids were developed that contain no phenolics and do not produce any unacceptable phenol-response to the 4-AAP analysis.

15.2.2.7 Oxidative Stability

It is well known that a hydraulic fluid may undergo oxidation during prolonged use. Mineral oils, polyol esters, and other fluids may form sludge, resulting in poor lubrication if a sufficient amount of these oxidation by-products are present. Water–glycol hydraulic fluids, when used at temperatures less than 150°F, do not result in any sludge formation [36]. Higher temperatures (>150°F) may result in slight sludge formation, as shown in Figure 15.14 [36,53]. However, any sludging or "gumming" that may be observed is often attributable to contamination by petroleum oils which may be present after initial system conversions or by adding the wrong fluid to the water–glycol-containing

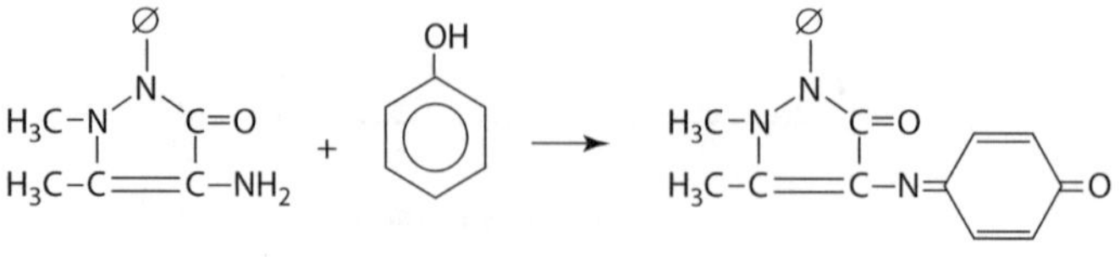

FIGURE 15.13 Phenol derivitization by the 4-aminoantipyrene (4-AAP) reaction.

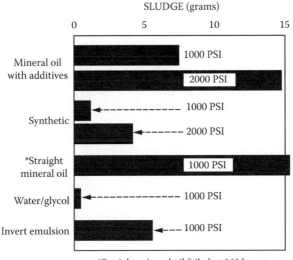

FIGURE 15.14 Relative propensity for sludge formation for various classes of hydraulic fluid.

reservoir [53]. Oxidation at higher temperatures may be simply due to co-evaporation of corrosion inhibitors, which may also inhibit oxidation, as will be discussed subsequently.

Iwamiya studied the thermal stability of a PAG-thickened water–glycol hydraulic fluid at elevated temperatures in a seal tube where evaporation could not occur. The results of this study are shown in Figure 15.15 [16]. From these results, it is apparent that fluid viscosity will decrease at points of localized high temperature. This should be contrasted to mineral oils, for which under the same conditions, fluid viscosity increases and sludging occurs.

Oxidation mechanism. Igarashi described some of the mechanisms encountered in the oxidation of mineral oils [54]. Oxidation of hydraulic fluids is accelerated when the fluid is exposed to extreme conditions of overheating or aeration through the use of poor defoaming or reservoir maintenance. Water–glycol hydraulic fluids are also potentially susceptible to various oxidative-degradation processes if they are operated under extreme, atypical conditions, or in a manner contrary to recommended supplier practice. In some cases, as with other classes of hydraulic fluids, by-products from degradation reactions may interfere with the wear-inhibiting mechanism of the additives used for fluid formulation.

Fuid degradation mechanism. The oxidation and thermal degradation processes for anhydrous poly(alkylene oxides) are well known and have been published previously [55]. Figure 15.16

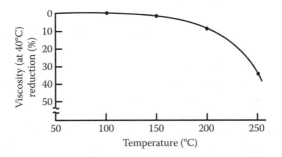

FIGURE 15.15 Thermal-oxidative stability of a water–glycol hydraulic fluid.

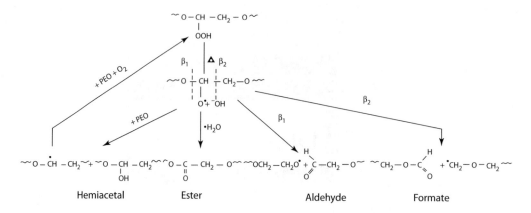

FIGURE 15.16 Mechanistic pathway for uncatalyzed thermal oxidation of an anhydrous poly(ethylene oxide).

illustrates the potential degradation processes for an uninhibited anhydrous poly(ethylene oxide) homopolymer [51]. This degradation pathway is also pertinent for diethylene glycol, which is present in most water–glycol hydraulic fluids today and is representative, although considerably less complex, to that of a poly(alkylene oxide) copolymer of ethylene and propylene oxide, which is present in most water–glycol hydraulic fluids as a thickener.

It is known that the oxidative-degradation processes illustrated in Figure 15.16 are slower in the presence of water than in their non-aqueous state [56]. During normal use, water–glycol hydraulic fluids do not form low-molecular-weight carboxylic acid byproducts, or at least not at a significant rate during the lifetime of the fluid. The corrosion literature provides at least two examples of degradation pathways that elucidate the potential oxidative-degradation mechanism for water–glycol hydraulic fluids [57,58].

The kinetics and severity of degradation of aqueous ethylene glycol coolant solutions were studied at 100°C and 180°C in solutions containing metals (presumably in a sealed tube) [58]. The conclusions from this study were as follows:

- Although degradation of aqueous ethylene glycol did not occur in the absence of metals, it was accelerated by their presence, although very slowly, even at these elevated temperatures.
- Degradation of ethylene glycol was accompanied by the generation of low-molecular-weight carboxylic acids and decreasing pH.
- Aeration only slightly accelerated the degradation process.
- The rate of degradation at 100°C and 180°C was essentially the same, which is unusual considering the temperature difference.

Subsequent work by Brown et al. showed that formic acid was one of the by-products in the oxidative degradation of aqueous ethylene glycol [57]. The rate of degradation was accelerated by the presence of copper(II) salts of these acids which arise from the corrosion of copper in the system [57]. Earlier work by Beavers showed that iron(III) salts similarly catalyzed the degradation reaction [58]. The presence and concentration of organic acids, including formic acid, was determined by ion chromatography [59].

Taken together, the results reported by Brown et al. [57] and Beavers [58] suggest that water–glycol hydraulic fluids degrade by a metal-catalyzed oxidative process that is almost independent of solution temperature but is slowly accelerated by aeration. These observations would suggest that the degradation process would be inhibited when the system is properly protected against corrosion and that degradation would then be accelerated by any process which would cause substantial depletion of the corrosion inhibitor below a "critical" level. This is the concentration below which there is insufficient inhibitor remaining to inhibit the formation of soluble corrosion metal by-products.

These results can be explained by a previously published metal-catalyzed peroxide degradation mechanism [60,61]. It is known that metal ions such as iron(III) and copper(II) will catalyze the degradation of peroxide species (ROOH), such as those shown in Figure 15.16, which are formed in the oxidative degradation of glycols, such as ethylene glycol, diethylene glycol, propylene glycol, and poly(alkylene glycol) copolymers [56]. In fact, this process can be catalyzed by various metal ions, including both iron(III) and copper(II):

$$Fe^{2+} + ROOH \rightarrow Fe^{3+} + RO\cdot + HO^- \tag{15.1}$$

$$Fe^{3+} + ROOH \rightarrow Fe^{2+} + H^+\cdot + RO_2\cdot \tag{15.2}$$

In Equations 15.1 and 15.2, both $RO_2\cdot$ and $RO\cdot$ are capable of undergoing further free-radical-generating reactions such as:

$$RO_2\cdot + RH \rightarrow RO_2H + R\cdot \tag{15.3}$$

$$RO\cdot + RH \rightarrow ROH + R\cdot \tag{15.4}$$

It is important to note that the presence of metal ions does not catalyze the formation of peroxide during the oxidation reaction. Instead, the metal ions catalyze the decomposition of the peroxide species after their formation by the process shown in Figure 15.16 [51]. Of course, the peroxide intermediate may also decompose without the presence of metal ions, but this process would be much slower.

The relative kinetic reactions of these two possible degradation routes, uncatalyzed and catalyzed, can be represented as [51]:

$$ROOH \xrightarrow{k_1} RO\cdot + HO\cdot \tag{15.5}$$

$$Fe^{3+} + RO_2H \xrightarrow{k_2} RO_2\cdot + H^+\cdot + Fe^{2+} \tag{15.6}$$

$$\frac{d[ROOH]}{dt} \, k'[ROOH] \tag{15.7}$$

$$= k_1[ROOH] + k_2\left[Fe^{3+}\right][ROOH] \tag{15.8}$$

where k' represents the overall hydroperoxide decomposition rate constant and k_1 and k_2 are uncatalyzed and catalyzed hydroperoxide decomposition rate constants, respectively. Lloyd had reported that the relative rate of k_2/k_1 is approximately 100,000/1 [56].

This mechanism explains why the oxidation rate of glycols in the absence of metal ions would be expected to be relatively slow, as observed by Brown et al. and Beavers [57,58] (and other water–glycol hydraulic fluid users). This was observed in practice when various amines and amine blends were used as corrosion inhibitors for water–glycol hydraulic fluids [62]. Fluid suppliers generally recommend that the level of the corrosion inhibitor in the system be maintained by periodic monitoring and replenishment, if necessary. The only systems where fluid degradation produced substantial quantities of formic acid were those that exhibited a corresponding loss of reserve alkalinity (amine corrosion inhibitors) [62]. Thus, the reserve alkalinity must be maintained if optimal fluid stability is to be achieved. Appropriate maintenance procedures are recommended by the fluid supplier, which will provide adequate control in operating systems.

Effect of formic and acetic acid on pump wear. The effect of formic acid and reserve alkalinity on the wear rate exhibited by a typical water–glycol hydraulic fluid in a hydraulic pump was studied by simply adding the desired quantity of the low-molecular-weight carboxylic acid (formic and acetic acid were used for this work) to the water–glycol fluid used in the study, using a statistically designed experimental matrix.

The reserve alkalinity, which is directly related to the concentration of the amine corrosion inhibitor present in the fluid, was determined for each fluid tested. Upon the addition of the formic or acetic acid, it was found that the reserve alkalinity decreased, which was expected. For the "high" amine-containing fluids, additional amine (in this case, morpholine) was added to bring the total reserve alkalinity of the formic-acid-spiked fluid into the recommended maintenance range. The "low" reserve alkalinity data points represent the fluids where formic acid was formed and reserve alkalinity was not maintained. The "high" reserve alkalinity data points represent fluids where reserve alkalinity was maintained, even though formic acid may be present. The experimental matrix and the data obtained from this study are summarized in Figure 15.17 [51].

The results shown in Figure 15.17 indicate that formic acid concentrations above 0.15% do in fact produce an accelerated wear rate, especially at concentrations above 0.30%, which is consistent with the results reported earlier [60]. (A typical wear rate of approximately 0.15–0.20 mg/h would be expected for this particular fluid when low-molecular-weight carboxylic acids are absent.) However, the data do show, as the results of Brown et al and Beavers suggest they should, that reserve alkalinity *must be maintained* to at least minimize the wear increase caused by the presence of formic acid.

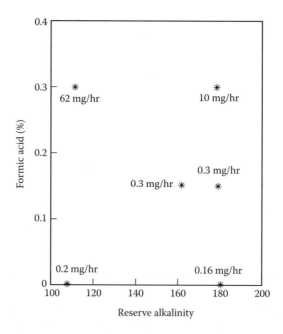

FIGURE 15.17 Two-dimensional experimental design matrix to illustrate the effect of formic acid and reserve alkalinity on pump wear.

Another low-molecular-weight acid that may be formed by oxidative degradation of the glycol and poly(alkylene glycol) is acetic acid. To evaluate the effect of acetic acid on wear rate, a pump test was performed on a fluid spiked with 0.3% acetic acid without adjusting for the corresponding decrease in reserve alkalinity arising from the acid addition. The wear rate obtained was 0.31 mg/h, which is substantially lower than the corresponding value for the same concentration of formic acid addition (61.5 mg/h, Figure 15.17).

Effect of the presence of formic acid on corrosion. Formic acid may act to increase hydraulic pump wear by selective adsorption at the active site on steel or other metallic surface, thus blocking the necessary chemisorption of the antiwear additive. This same mechanism would be expected to produce greater corrosion for formic-acid-containing fluids. To study this effect, the relative corrosion performance of three fluids (a typical water–glycol with no formic acid added, which had a reserve alkalinity [RA = 179], a similar fluid with 0.3% formic acid added but without reserve alkalinity adjustment [RA = 108], and the same fluid with 0.3% formic acid addition but with reserve alkalinity adjustment by morpholine addition [RA = 176]). The "200-Hour Corrosion Test" used for this work is described in Section 15.2.2.8. Figure 15.18 shows that the corrosion obtained with the formic-acid-spiked fluid was not significantly higher than that obtained for the unspiked fluid [51].

These observations are consistent with the corrosion results reported previously by Beavers [58]. They are also different from the relative order of pump wear rates discussed earlier. Even though formic acid may be selectively adsorbed relative to a less acidic carboxylic, also in the amine salt form, there is still sufficient protection to effectively inhibit corrosion, even though higher wear rates were obtained for this same level of formic acid. However, these results are limited and *do not* imply that sufficient corrosion protection would result in all possible formic acid concentrations.

15.2.2.8 Corrosion

There are numerous tests than have been used to determine the corrosion resistance offered by water–glycol hydraulic fluids. In this section, selected test procedures used to evaluate corrosion properties of water–glycol hydraulic fluids will be discussed.

Pump testing. One of the first tests was to run a pump test using the fluid of interest and inspect the different components both visually and by weight loss to determine material compatibility (corrosion) [63–65]. Although this test procedure does provide the necessary corrosion-inhibition information, there are at least three problems. One problem is the relatively high cost of the pump, components, and energy consumed to perform the test. The second problem is that different pumps utilize different materials of construction; therefore, more than one pump test will be required to obtain data on the range of materials of interest. This problem, however, has been addressed by mounting the test coupons of the materials of interest on a glass rack and immersing them in the 5-gal reservoir of the Vickers V-104 vane pump used to conduct the ASTM D 2882 pump test and monitoring the appearance and weight loss at the conclusion of the test. It should be noted that the ASTM D 2882 method stipulated the use of Eaton-Vickers ring and vanes, which are no longer available. The test method was replaced by the ASTM D 7043 method which stipulates use of Conestoga brand ring and vanes [66]. The third problem is that long running times, >100–200 h, are often required to obtain the necessary long-term compatibility information. This is not an accelerated test.

Static tests. In the static test, five metal coupons (aluminum, bronze, steel, copper, zinc, and brass) are mounted on a glass rack and placed in a 400-mL beaker as described in Section 4.4.4.1 of Procedure B of Mil-H-19457. After static (no fluid flow) immersion using 300 mL of fluid at 130°F (or 158°F) for a testing period of 90 h or one week, the corrosion resistance of the coupons is rated visually and gravimetrically [63].

Blanpain has reported on the use of a similar method in which the test metals, aluminum, copper, brass, zinc, cadmium, and copper/zinc are immersed in a flask containing the water–glycol

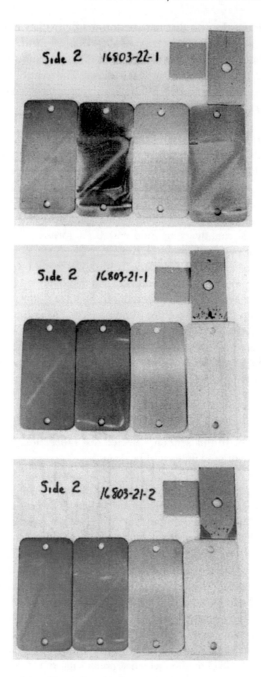

FIGURE 15.18　Illustration of the effect of the presence of formic acid on metal corrosion. Fluid 1 contains no additional formic acid, Fluid 2 contains 0.3% formic acid without adjusting reserve alkalinity, and Fluid 3 contains 0.3% formic acid with readjustment of reserve alkalinity to initial condition.

fluid of interest and the sample/fluid-containing flask is then stoppered and immersed in a liquid constant-temperature bath at 35°C. The method may also be modified to evaluate vapor-phase corrosion protection [48].

Modified turbine oil rust test. The metal coupons are cleaned, mounted on a glass rack, and then placed into the 400-mL Berzelius beaker used for the ASTM Turbine Oil Rust Test (ASTM D 665)

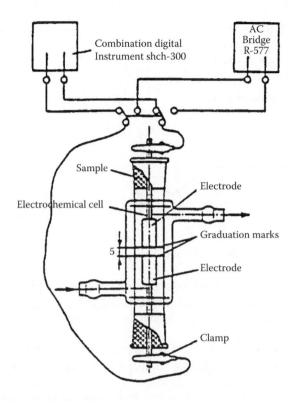

FIGURE 15.19 Electrolytic cell for determination of the rate of contact corrosion of metals.

[67]. The test coupons are then covered with 300 mL of the fluid. The fluid is agitated by using a shortened paddle attached to the drive spindle of the test apparatus. The paddle stem is shortened to provide clearance over the glass rack, which is immersed in the beaker. The plastic cover for the beaker is used and the fluid is held at 140°F while agitating at 1000 rpm. At the conclusion of the test, the coupons are inspected visually for rusting and weighed. In general, dynamic tests (those with agitation) provide much better correlation of corrosion results to a pump test than do static tests (no agitation) [63].

Electrochemical methodology. There are a number of electrochemical tests that have been reported to determine corrosion-resistance potential [68]. One test that has been specifically reported for use to determine the corrosion-protection potential of water–glycol hydraulic fluids was reported by Katorgin and Ramanova [69]. In this test, the rate of contact corrosion is calculated from the intensity of the electrical current between electrodes of different metals which are immersed into the water–glycol fluid. A schematic of the test cell is provided in Figure 15.19 [69].

The working electrodes, which are located parallel to one another, model the materials used for the hydraulic pump of interest. The remaining surfaces of the electrode are coated with an epoxy coating. To eliminate electrical resistance, the value of the current intensity reduced to zero resistance was used for the calculations. The rate of contact corrosion is calculated from:

$$V = \frac{3.627\,Ai}{zQ},$$

where:
  V = the corrosion rate of the anode metal (mm/yr)
  A = the atomic mass of the metal (g)

i = the current density (mA/cm^2)
z = the ionic charge on the dissolved metal
Q = the density of the anode metal (g/cm^3)
3.627 = the conversion factor (yr^{-1} × mA^{-1})

The current density is calculated from:

$$i = \frac{I_0}{S},$$

$$I_0 = \frac{UI}{U - IR},$$

where:
I_0 = the current reduced to zero resistance of the fluid (mA)
S = the electrode surface area (cm^2)
U = the equilibrium value of the potential difference between the electrodes (mV)
I = the magnitude of the current between the electrodes (mA)
R = the electrical resistance of the fluid between the electrodes

"200-Hour Corrosion Test." One of the most common tests in current use for the evaluation of liquid and vapor-phase water–glycol hydraulic fluid corrosion inhibition is the "200-Hour Corrosion Test." (Though variations of this test are known by other acronyms, the term "200-Hour Corrosion Test" will be used here.) The 200-Hour Corrosion Test consists of refluxing and aerating the solutions to be tested continuously for 200 h and measuring the solution corrosion by the weight loss of the immersed metal coupons of steel (SAE-1010, low carbon), cast aluminum (SAE-329), copper (CA-110), and brass (SAE-70C). Vapor-phase corrosion effects are also monitored by coupons of cast iron (G-3500) and steel (SAE-1010) suspended above the solution. Figure 15.20 illustrates a typical corrosion test cell [51,61].

The corrosion test procedure follows. To the test cell with the metal specimen placed in the order shown in Figure 15.20 was added 300 g of the water–glycol hydraulic fluid being tested. A water condenser was attached to the center ground glass connector and an air line was connected to the gas sparge tube. The total apparatus was placed into a constant-temperature bath and heated to 70 ± 2°C and held at this temperature for the duration of the test. A constant airflow rate of 793 cm^3/min was maintained through the sparge tube throughout the duration of the test. Periodically, de-ionized water was added to maintain a constant fluid volume in the test cell. (The periodic water addition was required because of losses through the condenser due to evaporation.) After the 200-h duration of the test, the metal specimens were removed, dried, weighed, and photographed.

ISO 4404-1. Another corrosion method titled: "Determination of the Corrosion Resistance of Fire-Resistant Hydraulic Fluids Part1: Water-Containing Fluids" [70], was established by ISO in 2001. Similar to the "200-Hour Corrosion Test" in some respects, a total of ten beakers containing the sample fluid are required. In five of these beakers, suspended by glass hooks, are single strips of steel, copper, zinc, aluminum and brass, which are totally immersed in the fluid. In four additional beakers, metal test strips are immersed in pairs (steel and zinc, copper and zinc, zinc and aluminum, aluminum and steel) whose surfaces are maintained 1 mm apart by means of a spacer. The tenth fluid-filled beaker sample is tested without test strips to evaluate the changes in the fluid over the test. Beakers are covered with watch-glasses to minimize fluid evaporation and placed in constant temperature bath or oven at 35°C ± 1°C for 672 hours ± 2 hours (equivalent to 28 days). Upon completion of the test, both the fluid and test strips are visually inspected and

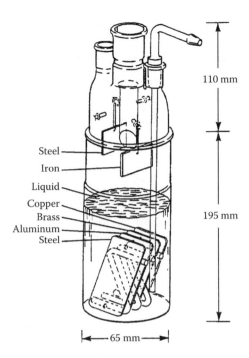

FIGURE 15.20 Corrosion test apparatus. This assembly is convenient for corrosion-inhibitor studies of water–gylcol hydraulic fluids under laboratory conditions. Immersed metal specimens are separated by a glass "Z-bar" in the specific order shown. Vapor-space test specimens are hung from the top of the glass test cell. Fluid temperature is monitored with an immersion thermometer, air is blown into the mixture using an aeration tube, and a cold-water condenser is used to reduce fluid loss by evaporation.

graded. Test strips are also weighted before and after the test in order to document any change in weight.

15.2.2.9 Compatibility

An extensive comparison of fluid compatibility for various materials including paints, metal, plastics, and elastomers is provided in Table 15.9 for petroleum oils, polyol esters, water–glycols, phosphate esters, and inverse emulsions [71]. However, the reader is cautioned that these compatibility values are affected by numerous variables, including fluid and material chemistries. Therefore, validation of these compatibility references with the appropriate fluid and seal manufacturer is recommended.

Seal compatibility. It is desirable that seals used in hydraulic equipment do not exhibit excessive volume swell (> +4 and/or −2%) in use. Similarly, Shore hardness should not exceed the ± −4 Shore hardness units. These values were established by testing the seal material at 60°C for 21 days [43]. Although natural rubber, BUNA N, neoprene, butyl, and silicone are, designated as "compatible" in Table 15.9 for water–glycols. Table 15.10 illustrates that each swells to a different degree [71,72].

More detailed information on seal materials, fluid compatibility, and compatibility testing is available in Chapter 2.

Metal compatibility. Water–glycol hydraulic fluids are compatible with steel, anodized aluminum, copper, and brass as long as the corrosion inhibitor is properly maintained. The use of

TABLE 15.9
Comparison of Material Compatibility of Fire-Resistant Fluids

Material	Oil/Polyol Ester	Water–Glycols	Phosphate Esters	Inverse Emulsions
Paint:				
common industrial	NC	NC	NC	NC
epoxy and phenolic	C	C	C	C
Metal:				
ferrous	C	C	C	C
bronze	C	LC[a]	C	C
zinc	C	NC	C	C
cadmium	C	NC	C	C
lead	C	NC	C	NC
brass/copper	C	C	C	C
aluminum, unanodized	C	NC	C	C
aluminum, anodized	C	N	C	C
magnesium	C	NC	C	NC
Plastic:				
"Lucite" type	NC	C	NC	C
acrylic	NC	C	NC	C
styrene	NC	C	NC	C
nylon	NC	C	C	C
PVC	NC	C	NC	C
Leather	_[b]	_[b]	_[b]	_[b]
Rubber:				
neoprene	NC	C	NC	C
BUNA N	NC	C	C	NC
butyl	NC	C	C	NC
ethylene propylene	NC	C	C	NC
polyurethane	NC	NC	LC[c]	NC
silicone	C	C	C	C
teflon	C	C	C	C
titon	C	C	C	C
Filter Media:				
cellulosic, phenolic treated	C	C	C	C
adsorbant (earth type)	NC	NC	NC[d]	NC

Note: C = compatible. LC = low compatibility. NC = not compatible.

[a] Bronze with lead content over 20% limited to 120°F.

[b] Leather compatibility depends on the type of impregnation and conditions. Consult leather packing manufacturer for specific recommendations.

[c] Compatibility marginally good, some sources better than others.

[d] Active earth media are unsatisfactory for routine use but acceptable for reclaiming purposes.

unanodized aluminum is not recommended. Also, the use of aluminum as a bearing material is not recommended [73].

Zinc and cadmium are dissolved by the relatively alkaline water–glycol solution forming insoluble soap complexes with the antiwear additive [72]. Therefore, the use of galvanized-metal-, zinc- or, cadmium-coated components with water–glycol hydraulic fluids is not recommended [16,

TABLE 15.10
"Typical" Rubber Swell Data for a Water–Glycol Hydraulic Fluid

Type of Rubber	Linear Swell (%) 5 Days at 158°F
Natural	1.0
BUNA S	1.6
BUNA N	3.5
Neoprene	1.1–2.0
Butyl	0.0–0.8
Silicone	0.0–2.0

72]. However, suitable alternatives include tin, nickel, or stainless steel [74]. Recently, it has been reported that a water–glycol fluid has been formulated that is compatible with zinc [75].

Paint compatibility. A problem often encountered in the fluid power industry is the attempted use of painted reservoirs which were previously in use with mineral oils. Water–glycol hydraulic fluids exhibit a solvent effect on these paints, thus acting as a paint remover [72]. Therefore, paint on surfaces which come into contact with the water–glycol should be removed [76]. Paints that *are not* compatible include phenolic and phthalic resin-based paints [16] and alkyd- and vinyl-based coatings. Compatible coatings are (from best to worst): catalyzed phenolic > catalyzed epoxy-phenolic > catalyzed epoxy coatings. The user is advised to test the compatibility of any paint that they plan to use. This is especially true if fluid temperatures in excess of 60°C (140°F) are to be encountered. It is now common practice to specify that surfaces which will come into contact with water glycols be unpainted [74].

15.2.3 PERFORMANCE TESTING

Perhaps the most striking feature of W/G hydraulic fluids is the progressive performance improvements that have been made since their introduction in 1947. Table 15.11 illustrates the progressive reduction in wear rates, approaching those obtained with mineral oil, achieved with W/G fluids containing >35% water from 1955 to 1974 when tested according to DIN 51,389 E (Vickers V-104C vane pump for 250 h at 10 MPa [1500 psi] and 1500 rpm) [23]. Historical results cannot be reproduced as the fluids were formulated with obsolete additives which are no longer available. In addition, the reported test conditions are significantly different to those reported today. For example, lower pressures and longer times and, in many cases, different pumps were used.

In another study comparing the ASTM D-2882 Vickers V-104 vane-pump test, 100 h at 13.7 MPa (2000 psi) and 1200 rpm, showed that it was possible to formulate high-performance W/G hydraulic

TABLE 15.11
Historical Wear Rates of Water–Glycol Hydraulic Fluids (1955–1974)

Year	Total Wear (mg)
1955–1965	4700–15100
1965–1970	1300–2800
1970–1974	300–1100
After 1974	Approx 100 (mineral oil) <100

Note: All tests conducted according to DIN 51,389 E using a Vickers V-104C vane pump for 250 h at 10 MPa (1500 psi) and 1500 rpm.

TABLE 15.12
Comparison of ASTM D-2882 Vane-Pump Test Results for Various Fire-Resistant Hydraulic Fluids and Mineral Oil

Fluid	Wear Rate (mg/h)
Typical W/G hydraulic fluid	0.65
Phosphate ester	0.05
Poly(alkylene glycol)	0.05 [a]
Polyol ester	0.10
High-Performance W/G hydraulic fluid	0.10
Rust and oxidation oil	7.0
Antiwear Oil	0.24
Low-wear invert emulsion	3.5
High-wear invert emulsion	10+[b]

Note: These tests were conducted over 100 h at 13.7 Mpa (2000 psi) and 1200 rpm using a Sperry-Vickers V-104 vane pump according to ASTM D-2882. The pump was equipped with a 30-L/min (8 ppm) ring. The "pass" criterion is 1.0 mg/h.

[a] Test run at different laboratory than other test results.
[b] Test incomplete because of rotor breakage.

fluids [25] that exhibit even lower wear rates than those achievable with mineral oils. The reported wear-rate data are provided in Table 15.12 [17].

In addition to exhibiting low wear rates, these high-performance W/G fluids could be also used at high pressures (≥34 MPa, 5000 psi) previously unattainable with other water-containing hydraulic fluids [17].

Cole had shown that W/G hydraulic fluids may be used in hydraulic equipment equipped with servo valves [77]. Although this study showed that the dynamic response was slightly reduced with W/Gs, the performance could be restored by using slightly larger piping, controls, and servo valves.

One historical area that has caused many problems in expanding the use of W/G hydraulic fluids was reduced roller bearing lifetimes obtained with these fluids due to fatigue failure [78,79]. However, studies have identified improved bearing materials and designs which are more suitable for use with W/G hydraulic fluids, thus permitting their use under more stringent lubrication conditions (e.g., higher pressure and greater rpm speeds) [14].

Water–glycol fluids may encounter mixed elastohydrodynamic (EHD) and boundary lubrication regimes at some moving surfaces during the operation of a hydraulic pump, including: (1) the interface between the leading edge of the vanes and the cam ring of a vane pump, (2) the line contact between the mating gear teeth of a gear pump, (3) the interface between the connecting rod and piston of certain types of piston pumps, (4) the interface between the piston and swash plate of other types of axial piston pumps, and (5) the interface between rotary valve plates and housings [80].

This section will discuss the effect of interfacial film formation and film-thickness properties of water–glycol hydraulic fluids on EHD, mixed-EHD, and boundary lubrication. Also included in this discussion are factors affecting the fatigue strength of surfaces lubricated by water–glycol hydraulic fluids.

15.2.3.1 Film-Forming and Lubrication Fundamentals of Water–Glycol Hydraulic Fluids

Various workers have studied the effect of the viscosity and pressure–viscosity effects on the film-thickness properties of water–glycol hydraulic fluids [81–90]. A summary of various pressure–viscosity coefficients for a range of fluids is provided in Table 15.13.

TABLE 15.13
Summary of Published Viscosity Coefficients for Various Hydraulic Fluids

Fluid	Water Conc. (%)	Pressure–Viscosity Coefficient (GPa⁻¹)	Ref
Water	100	0.75	42
Water–glycol	35	4.38	42
Water–glycol	36	2.04	86
Water–glycol	41	4.5	87
Water–glycol	48	1.96	86
Water–glycol	50	4.22	42
Polyol Ester	0	11.5	89
Mineral Oil	0	20.4	87

This study shows that the pressure–viscosity coefficient is lower for a water–glycol than for mineral oil and also lower than that exhibited by a water-in-oil emulsion (which is comparable to a mineral oil). This results in lower film thicknesses relative to mineral oil, as shown in Figure 15.21 [87]. The pressure–viscosity coefficient (and, thus, film thickness as shown in Figure 15.3) decreases with increasing water content, as shown in Figure 15.22 [81,86]. Taken together, both pressure–viscosity coefficient and film thickness are dependent on the water, glycol, and thickener content of the water–glycol hydraulic fluid.

Water–glycol fluids exhibit lower pressure–viscosity coefficients and film thickness than mineral oil. Vernesnyak et al. have studied the effect of load, temperature, and speed and found that the film thickness was much greater for mineral oil than a water–glycol at room temperature. However, at higher temperatures (≥4–50°C)—which are more representative of the actual fluid temperatures in use—the film thicknesses for water–glycol fluids may even exceed mineral-oil fluids. This was attributed to the much steeper viscosity-temperature relationship (lower VI, sometimes as low as 85–100) exhibited by mineral oils. It was concluded that the films formed for water–glycols were sufficient to provide adequate lubrication in EHD and mixed-EHD lubrication [82–85]. This is illustrated by Figure 15.23, which shows that the

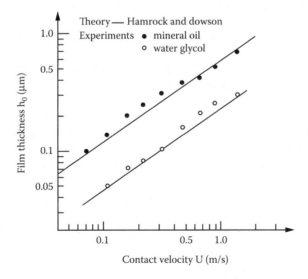

FIGURE 15.21 Central film thickness versus speed at a constant load of 2.6 N for a mineral oil and a water–glycol hydraulic fluid.

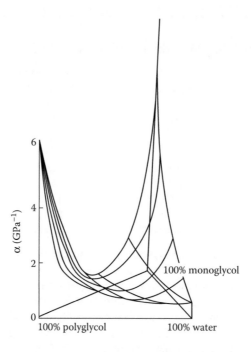

FIGURE 15.22 Pressure–viscosity coefficient as a function of ethylene glycol, poly(alkylene glycol), and water concentration.

effect of temperature on both fluid viscosity and the pressure–viscosity coefficient for mineral oil, phosphate ester, and a water–glycol are dependent on the use condition [91].

In another study, it was concluded that although thickened (or VI improved) fluids may exhibit higher viscosity–pressure coefficients and therefore greater film coefficients, this advantage may be lost for non-Newtonian fluids if sufficient shear rates are encountered to cause shear thinning [42].

Wedeven et al. have shown that although a water–glycol hydraulic fluid may exhibit lower film thicknesses than a polyol ester (consistent with pressure–viscosity coefficients for these fluids), it was still possible for the water–glycol to exhibit significantly lower friction coefficients, as shown in Figure 15.24 [89]. The difference observed in film-thickness values for different fluids is most critical as the film thickness approaches the asperity depth in the mixed friction to boundary lubrication transition.

One other caution is the source of pressure–viscosity coefficients. Many reported values are "effective pressure–viscosity" coefficients derived from lubrication tests using the Reynolds equation. Other data are obtained from viscometry results. However, it is common for significant differences to exist between the two sources of data, as shown in Table 15.14. Therefore, pressure–viscosity coefficient data obtained using the same methodology should be compared [89].

There are other factors that affect lubrication at the tribocontact in addition to film-thickness measurements. One especially important factor in sliding contact lubrication is additive chemistry as it relates to surface film formation. To some extent, this factor may counterbalance a less favorable pressure–viscosity coefficient and lower film thickness.

As indicated in the above discussion, there are a number of ways to address potentially unfavorable film-thickness problems. Fein and Villforth provided a number of guidelines in this regard that should be followed. These are summarized in Table 15.15 [85].

15.2.3.2 Rolling Contact Fatigue

In the previous subsection, it was shown that water-containing lubricants such as water–glycol hydraulic fluids exhibit lower pressure–viscosity coefficients, and therefore lower film thicknesses,

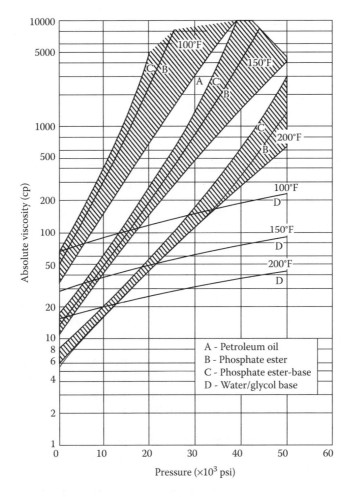

FIGURE 15.23 Pressure–viscosity coefficient isotherms for (A) petroleum oil, (B) phosphate ester, (C) phosphate ester blend, and (D) a water–glycol hydraulic fluid.

than mineral oil and some other synthetic fluids. One lubrication application area where this has been found to be a problem is in the lubrication of rolling element bearings. It has been shown that the bearing life is shorter using water–glycol hydraulic fluids than mineral oil [92–105]. The relative L_{10} fatigue lives of water–glycol fluids has been estimated as 0.17–0.40 of the value reported for mineral oils [108]. (The L_{10} life is the number of revolutions during which 10% of all nominally identical bearings would be expected to fail.) An example of this problem is illustrated in Figure 15.25 [92].

It has long been known that fluid viscosity will affect bearing lifetimes [98]. The effect of water–glycol viscosity on bearing wear was studied by Sullivan and Middleton [99]. The first step in this analysis was to determine the lubrication regimes for the steel twin-disk wear contact of the Amsler test machine shown in Figure 15.34 [99]. This was done by plotting the coefficient of friction versus the "bearing number," which is defined as:

$$\text{Bearing number} = \frac{\text{Viscosity (n)} \times \text{speed (s)}}{\text{Contact pressure (p)}}.$$

This relationship, illustrated by Figure 15.26, is the classical Stribeck curve [99]. (Interestingly, the formation of this curve required three viscosity grades of the water–glycol fluid.)

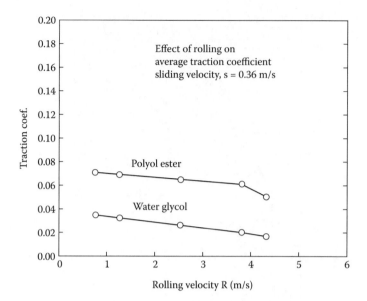

FIGURE 15.24 Effect of rolling velocity on average traction coefficient at a constant sliding speed of 35 cm/s for a polyol ester and a water–glycol hydraulic fluid.

The effect of bearing number on wear is illustrated in Figure 15.27 [99]. Similarly, the number of cycles for the first surface pit to appear was plotted versus bearing number, as shown in Figure 15.28 [99]. These data show that wear decreases with increasing bearing number (viscosity).

Surface roughness effects may be modeled using the lambda function (Λ), which is defined as:

$$\Lambda = \frac{\text{EHD film thickness } (h_o)}{\text{Composite relative roughness } (\sigma)}.$$

Figure 15.29 is a plot of Λ versus the number of cycles for the first surface pit to appear [99]. These data illustrate that the surface pitting rate increases with increasing surface roughness and correlates well with the bearing number (fluid viscosity) effects shown in Figure 15.29. In another study, it was found that bearing lifetimes was a more reliable ranking than wear measurement [109].

There have been many studies conducted in an attempt to quantify bearing lifetime for fire-resistant fluids from established handbook values for petroleum-oil lubricants. One approach utilized a lubricant factor (F) and the equation:

$$F = \left(\frac{\text{Modal life for petroleum oil}}{\text{Modal life for fire-resistant fluid}} \right)^{1/3}$$

TABLE 15.14
Comparison of Pressure–Viscosity Coefficients Obtained by Viscometry and Lubrication Experiments Using the Reynolds Equation

Fluid	Capillary Viscometry (GPa^{-1})	Optically Measured from EHD Contact (GPa^{-1})
Water–glycol	2.97	1.8
Polyol ester	11.5	7.8

TABLE 15.15
Guidelines for Increasing Film Thickness in Practice

Situation	Low Speed	High Speed
Frictional heat source	Increase viscosity; increase lubricant flow	Improve external cooling; optimize lubricant application, viscosity, and flow for minimum bearing temperature
External heat source	Increase lubricant flow; optimize viscosity for minimum bearing temperature; reduce heat from source	Increase lubricant flow; optimize viscosity for minimum bearing temperature; reduce heat from source
Low temperature	Decrease viscosity; increase temperature; increase lubricant flow to the loaded surfaces	Decrease viscosity; increase temperature; increase lubricant flow to the loaded surfaces
High-bearing load	Increase viscosity	Improve external cooling; optimize lubricant flow and viscosity for minimum bearing temperature
Contamination (dirt, wear, debris, etc.)	Increase viscosity; remove and/or inhibit formation	Remove and/or inhibit formation

where F values of 2.6 for water–glycol fluids have been reported.

Another relationship, developed by Kenny and co-workers, is:

$$L_{10} = \left(\frac{C}{KP} \right)^n \times 10^6 \text{ revolutions}$$

where C is the dynamic load rating, P is the dynamic bearing load, and n and K are constants. For a water–glycol, the values for n and K were reported to be 2.46 and 2.01, respectively [93,94,110].

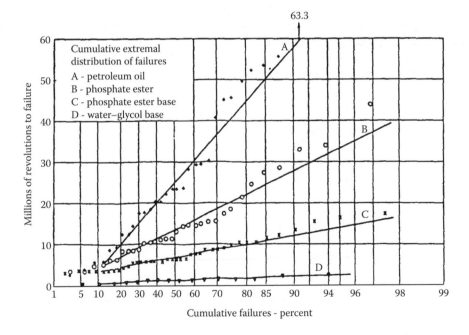

FIGURE 15.25 Comparison of the fatigue life of angular contact ball bearings lubricated with a petroleum oil and various fire-resistant hydraulic fluids.

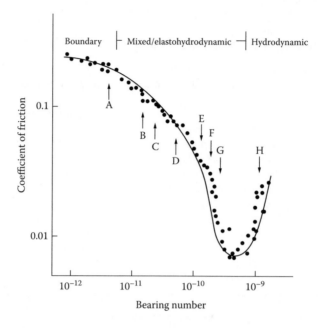

FIGURE 15.26 Stribeck curve generated from bearing number and coefficient of friction for a water–glycol hydraulic fluid. Three viscosity grades of the water–glycol hydraulic fluid were necessary to develop this curve.

There are various mechanisms that may possibly explain the observed lower fatigue strength exhibited by rolling element bearings lubricated by aqueous fluids such as a water–glycol. One possible explanation is hydrogen embrittlement [101,102]. Figure 15.30 illustrates that the fatigue life of steel balls pretreated cathodically in water to introduce hydrogen into the steel decreases with increasing hydrogen content in the steel [110]. These data were used to confirm a prior observation that the intergranular crack growth observed for beatings used with a water–glycol hydraulic fluid was indicative of hydrogen embrittlement (concentration of hydrogen at the grain boundaries). Interestingly, the simple exposure of steel to hydrogen gas did not induce this effect. Furthermore, in the absence of oxygen, water had little effect [110].

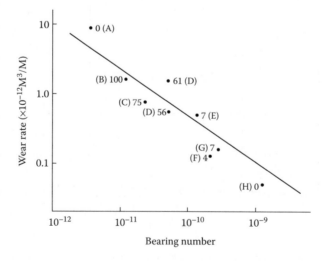

FIGURE 15.27 Correlation of wear rate versus bearing number for a water–glycol hydraulic fluid.

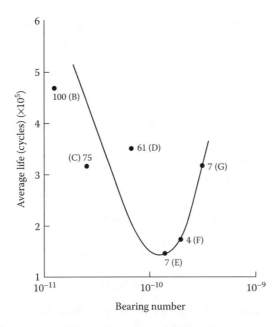

FIGURE 15.28 Correlation of average life cycles versus bearing number for a water–glycol hydraulic fluid.

Sullivan and Middleton showed that surface pitting rates increased with decreasing film thicknesses. However, the probability and rate of crack initiation can be significantly reduced with the appropriate antiwear additive selection. Therefore, it would appear that crack initiation is caused by a fatigue process related to asperity interaction. The crack propagation times decreased with an increase in film thickness and were additive independent. It was suggested that crack propagation was enhanced by a hydraulic pressure wedge during system pressurization. The lower compressibility of the water–glycol fluid relative to mineral oil may enhance pressure transfer into the crack [103].

Spikes had summarized four possible reasons for poorer fatigue strength provided by water-containing fluids [110]:

1. Low EHD film thickness illustrated by lambda ratio.
2. Chemical promotion of crack formation and growth.
3. Removal or prevention of protective films due to low temperature or dilution.
4. Increased corrosion and corrosion/abrasion wear.

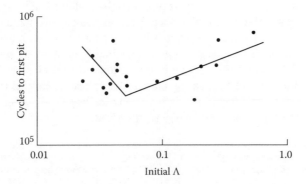

FIGURE 15.29 Correlation of cycles to obtain the first pit versus lambda (Λ) factor.

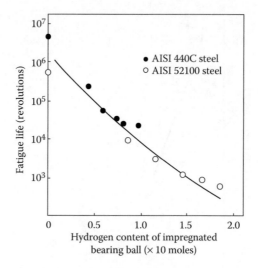

FIGURE 15.30 Effect of hydrogen content of impregnated ball bearings (× 10 moles) on fatigue life (revolutions).

Commentary on film-thickness/rolling contact fatigue data. Shitsukawa et al. have reported that improvements in fatigue life have been obtained because of improvements in steel quality and surface roughness of the bearing raceways [100]. In addition, the following were reported:

- Bearing life may depend on surface roughness rather than Λ, when Λ is very small and Λ alone does not fully define a lubrication condition.
- Previous studies have drawn different conclusions regarding fatigue life and the magnitude of Λ. For example, Skurka [111] stated that bearing life is not affected whether Λ is large or small, and Talian [112] reported that bearing life is always proportional to Λ.
- Test bearings have different geometry, manufacturing processes, and surface roughness. There has not been a complete series of tests reported using the same bearings from the same manufacturing process (same material chemistry and properties) in which only the film thickness was varied.

Yano et al. [107], using the Unisteel machine to evaluate fatigue life properties of water–glycol hydraulic fluids with subsequent pump test validation, reported that additives exhibited a "remarkable" effect on fatigue life and the effect as even more significant for fatty acid-amine pairs. One fluid exhibited a fatigue life of 48% that of a mineral oil. In addition, L_{10} lives increased with decreasing water content, with the most significant improvement occurring in the range of 35 to 40% water.

The effect of bearing material chemistry using water-glycol fluids had been reported by Reichel [106] where a piston-pump test rig, with two axial piston pumps was operated in single closed-loop system. One unit was operated as a pump, the other as a motor. Both were operated at 1500 rpm, a constant pressure of 280 bar, and at a fluid temperature of 40°C~. Figure 15.31 shows that bearings exhibiting L_{10} lives of 10%–90% with respect to a mineral oil were obtained. (The C/P values varied from 3.4 to 8.4.*) of the bearing material examined by Reichel [106], Cronidur 30, a high-nitrogen stainless steel (X 30 CrMoN~5), provided excellent L_{10} lifetimes obtained with eight different water–glycol formulations as shown in Figure 15.32.

---

* Fatigue life L_{10} is related to the ratio of C (the load that the bearing can carry for a life of 10^6 inner-race revolutions with a 90% probability of survival or dynamic load capacity) and P (equivalent bearing load or static load) by the Lundberg-Palmgren equation:

$L_{10} = (C)^p/(P)$

where p is the load-life exponent. See Reference 105.

Roller bearing design		c/p	L_{10} h	L_{HFC} h	L_{HFC} %	Axial piston unit/ swivel angle
Conical roller bearings 		8.42	2962	96	<10	bent shaft design 40°
		8.42	2200	222		
		3.0	58575			
		3.0	58575	2000	>4	swash plate design 14°
		2.9	52231			
		2.9	52231			
		3.2	817	233	<30	bent shaft design 28°
Cylindrical roller bearings (without cage)		8.38	10100	160		
		8.38	13000	122	1–2	swash plate design 14°
		8.90	16000	357		
Cylindrical roller bearings 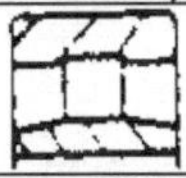		4.14	681	122	<20	bent shaft design 28°
Bevel ball bearings 		5.4	1732	850	<50	bent shaft design 25°
Double race ball bearings (plastic cage)		3.4	440	1180	>250	swash plate design 14°

FIGURE 15.31 Lifetime of different roller bearings in axial piston pumps using a water–glycol hydraulic fluid.

15.2.3.3 Fatigue Life Evaluation

In this section, a selected overview of different testing procedures that have been used to evaluate fatigue life of rolling element bearings will be discussed. In addition, actual pump tests such as the Reichel test [106], one of the oldest and most common testing methodologies used to evaluate the lifetime of an actual bearing assembly will be discussed. One such test was described by Bietkowski, who mounted two test bearings in sealed housings on a horizontal shaft supported by two large roller bearings [109]. The bearings were loaded with hanging weights, typically 3.34 kN. The shaft was rotated at the desired speed, typically 1450 rpm for 400 h, or failure, whichever came first. The lubrication rate was approximately 2 mL/min.

In some cases, a customized machine may be built to simulate the desired lubrication conditions: both contact stress and film thickness. The contact fatigue rig used by Danner is illustrated in Figure 15.33 [115]. A 1.25-in. test specimen illustrated in Figure 15.33a contacts a 3-in.-diameter driver, as illustrated in Figure 15.33b.

The test specimen was constructed from carburized AISI 4620 steel. The contact surfaces were ground to the approximate surface finish of the bearings being modeled. The load (0–4000 lbs) was

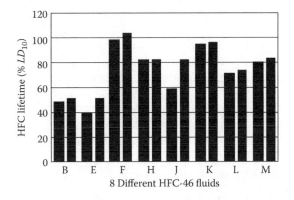

FIGURE 15.32 Lifetime of bevel ball bearings in an axial piston pump using a water–glycol hydraulic fluid. (Test conditions: pressure = 280 bar, rotational speed = 1500 rpm, and fluid temperature = 45°C [113°F].)

(a)

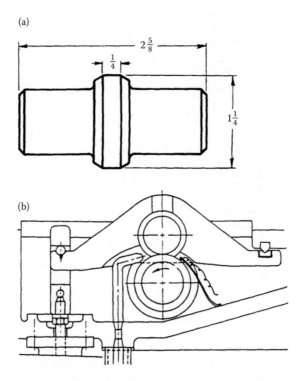

(b)

FIGURE 15.33   Rolling contact fatigue test rig: (a) fatigue test specimen; (b) cross section of test rig.

applied through a cantilever. The lubricant is supplied to the outlet side of the contact through a recirculating circuit with filtration. The contact temperature is measured by a thermocouple riding on an oil film on the inlet side. Rotational speeds of 100–3600 rpm were possible.

Sullivan and Middleton utilized an Amsler two-disk machine. The 40-mm diameter AISI 52100 bearing steel two-disk contact is illustrated in Figure 15.34 [102]. The Vickers hardness (HV) and the driven disk had a HV of 250, with surface roughnesses of 0.07 um and 0.25 um, respectively. The

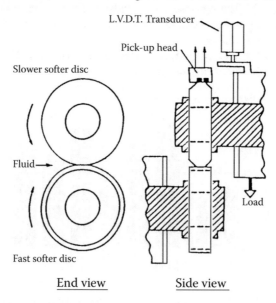

FIGURE 15.34   Amsler twin-disk test machine.

disks were loaded through a variable-compression spring to 900 N, which provided a Hertzian stress of 1.01 GPa. The disks were capable of being rotated independently at 20–4000 rpm with 10% slip.

Of the various bench tests that have been used, two of the most common are the 4-ball test rig [90,95] and the Unisteel test machine. In the 4-ball test, a hardened steel ball is held in a chuck at the end of a spindle which is loaded by a lever arm at 25–750 kg and rotated at a constant speed of 1500 rpm [98]. The upper ball is loaded against three balls held in a steel cavity and allowed to rotate freely, creating the desired rolling/sliding, as illustrated in Figure 15.35.

One of the most commonly used bench tests to model contact fatigue is the Unisteel Rolling Fatigue Machine illustrated in Figure 15.36. In this test, a flat ring test piece forms the upper half of

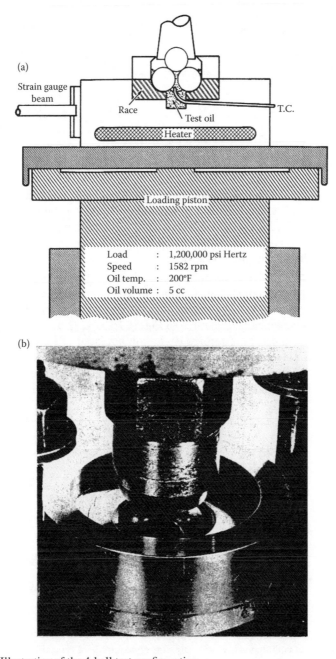

(a)

Strain gauge beam

Race

Test oil

T.C.

Heater

Loading piston

Load	:	1,200,000 psi Hertz
Speed	:	1582 rpm
Oil temp.	:	200°F
Oil volume	:	5 cc

(b)

FIGURE 15.35 Illustration of the 4-ball test configuration.

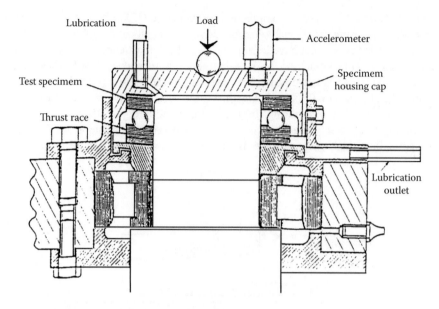

FIGURE 15.36 Schematic of Unisteel test machine.

a thrust bearing assembly. This is run with the cage, half of the balls, and one race from a standard production thrust bearing. The lubricant is drip-fed and the load is applied by a dead-weight lever arm. Typically, a 3.34-kN load is applied and rotated at 1500 rpm [105].

15.2.3.4 Hydraulic Pump De-rating Practice

Most hydraulic pump companies have traditionally de-rated the catalog pressure ratings for their pumps for use with fire-resistant fluids. Some pump products are not recommended for use with fire-resistant fluids at all. One illustration of pump de-rating is provided by Edgington for Lucas axial piston pumps, which is illustrated in Figure 15.37 [22]. Traditionally, boosted inlet suction pressure recommended to minimize cavitation potential has also been recommended for water–glycol fluids [114]. Another illustration of this is the general de-rating conditions published by Sundstrand and shown in Table 15.16 [115]. Sundstrand recommends that inlet vacuum not exceed 2 in. Hg and that an elevated reservoir be used with an increased inlet line size [115].

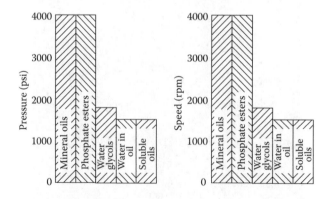

FIGURE 15.37 Maximum continuous operating pressures and speeds of typical Lucas axial piston pumps using mineral oils and various fire-resistant hydraulic fluids.

TABLE 15.16
Modified Operating Parameters for Fire-Resistant Fluids

Fluid Type	Speed (% catalog)	Pressure (% catalog)	Maximum Temp. (°F)
Phosphate ester	100	100	180
Polyol ester	85	85	150
Invert emulsion			
(60 oil/40 water)	65	70	140
Water–glycol (60/40)	65	60	140
HWCF (95 water/5 oil)	65	40	122

Similarly, Racine de-rated their pumps for use with water–glycol hydraulic fluids and recommended that a flooded inlet be used and that the inlet vacuum not exceed 3 in. Hg and that an elevated reservoir be used. Illustrative examples to assist in the calculation of inlet vacuum condition are provided in Table 15.17[116] and Figure 15.38. If the pump must be mounted on top of the reservoir, the liquid level should not be more than 12 in. below the pump. Strainers should be rated at least three times the pump outlet flow. Strainers should be no finer than 60 mesh (238 um) [117].

TABLE 15.17
Calculation of the Inlet Condition

A. The pressure loss at a pump suction inlet will be caused by the following:

1. Pressure loss due to length of piping = length (ft.) × graph value (Fig. 15.38) × flow (gal/min) × SPG
2. Pressure loss due to elbows = number × graph value (Fig. 15.38) × flow (gal/min) × SPG
3. Pressure loss due to bends = number × graph value (Fig. 15.38) × flow (gal/min) × SPG
4. Pressure loss due to suction strainers = number of elements × graph value (Fig. 15.38) × flow (gal/min) × SPG
5. Pressure loss due to head or lift = ft. × 0.88 × SPG

The pressure loss caused by the design of the suction system can be determined from the data given in Figure 15.38, for example:

Fluid viscosity	200 cS at 55°C
Flow rate	18 gal/min
Pipe size	1.5 in. in diameter
SPG	1.15
Head of fluid	2 ft.

3-element 0.005-in. suction strainer

1 bend and 4 ft. of pipe

From Figure 15.37:

$$\text{loss due to length of pipe} = 4 \times 0.026 \times 18 \text{ gal/min} \times 1.5 = 2.160$$

$$\text{loss due to bend} = 1 \times 0.04 \times 18 \text{ gal/min} \times 1.15 = 0.830$$

$$\text{loss due to suction strainer} = \frac{0.35 \times 18 \text{ gal/min} \times 1.15}{3} = 2.420$$

$$\text{positive head} = 2 \times 0.88 \times 1.15 = 2.02$$

Loss of Inlet 3.39 in. Hg

Notes: 1 ft height of a fluid with a SPG of 1 = 0.88 for other fluids in Hg = 0.88 × SPG

If the fluid is above the pump inlet, the pressure should be subtracted from the other losses and if the fluid is below the pump inlet, the pressure is to be added to the other losses.

SPG refers to specific gravity

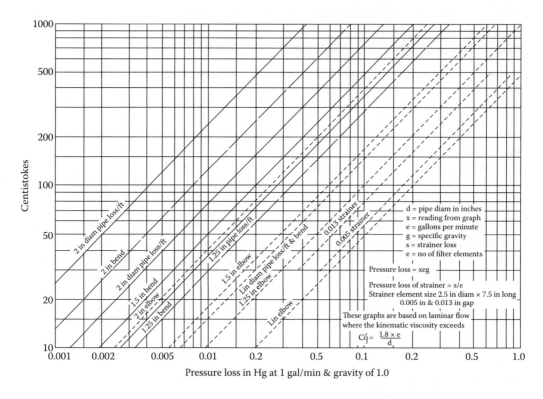

FIGURE 15.38   Correlation of hydraulic fluid viscosity and pressure loss in pipe elbows of different cross-section sizes.

In some cases, rolling element bearing lubrication problems with water–glycols can be successfully resolved by either changing the bearing design—for example, from uncaged to caged bearings for Rexroth Hydromafik and Breuninghaus pumps—or by the use of bronze sleeve bearings [118]. Recently Rexroth introduced an axial piston pump [119] specifically suited for operation with water–glycol (HFC) hydraulic fluids.

15.2.4   Fluid Maintenance Procedures

15.2.4.1   Fluid Contamination and Performance

Hydraulic pump lubrication is not only dependent on fluid chemistry, but it is also negatively affected by liquid and solid contamination [120–126]. In water–glycol hydraulic fluids, the most common liquid contamination is usually petroleum oils. Because water–glycols and petroleum oils are mutually insoluble, residual oil may be skimmed from the fluid reservoir, although often not before selectively extracting some additives from the fluid formulation. Although this is usually not a serious problem, every effort should be made to prevent this form of contamination from occuring. Solid contamination may have disastrous effects because solid debris, through abrasive action, will cause increased wear at the lubrication contact, as shown in Figure 15.39 [122–125].

Any wear in hydraulic systems is deleterious because optimal pump performance is dependent on the clearances within the system. Table 15.18 provides a summary of typical clearances encountered in a hydraulic system [120].

Fitch and Hong have illustrated the relative sensitivity of vane, gear, and piston pumps to solid contamination, which is shown in Figure 15.40 [122]. To address the problems of pump wear due to contamination, Sperry-Vickers developed filtration recommendations based on filter B-ratios developed earlier by Fitch and others [126,127]. These recommended values are provided in Table 15.19.

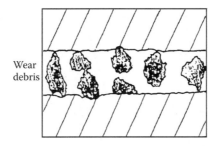

FIGURE 15.39 Mechanism of abrasive wear due to solid contamination.

These data show that no mater how good the lubrication properties, solid contamination can destroy a hydraulic system. Thus, fluid contamination must be removed using appropriate filtration if optimal performance is to be obtained.

15.2.4.2 W/G Fluid Analysis Procedures

The following discussion will focus on analytical procedures that may be used to monitor fluid chemistry and physical property variation. *Note:* All of the examples shown in the following discussions are for illustrative purposes only. Because these values are fluid specific, they will vary with the supplier. Therefore, the reader should consult with their fluid supplier to obtain the appropriate recommendations to use with these procedures.

Initial fluid observation. The first step of any analysis is to simply observe the sample. This is easily done by looking at the sample in a clear container such as a bottle. The sample should be clear without the presence of oil layers or solid debris. If solid debris is observed, a magnet should be used to determine if it is magnetic. Magnetic solids may be the result of either wear or corrosion. Nonmagnetic debris may be due to elastomeric seal erosion, carbonaceous deposits, or external nonmetallic contamination.

Water content and makeup. Antiwear properties and the fire resistance of a water–glycol hydraulic fluid is dependent on total water content. Therefore, water should not be added indiscriminately nor should wide variations in water content by evaporation during use be allowed. Water contained in a water–glycol fluid may be lost through evaporation during normal hydraulic operation. Water loss increases fluid viscosity. Therefore, water must be added back to the system to maintain fire resistance, assure proper viscosity, and, thus, system operation.

TABLE 15.18
Typical Hydraulic Component Clearances

Component	Clearance (μm)
Slide bearings	0.5
Vane pump (tip of vane)	0.5
Control valve	0.1–0.5
Rolling element bearing	0.1–1.0
Hydrostatic bearings	1–25
Gears	0.1–1.0
Gear pump (tooth to case)	0.5–5.0
Piston pump (piston to bore)	5–40
Servo valves	1–40

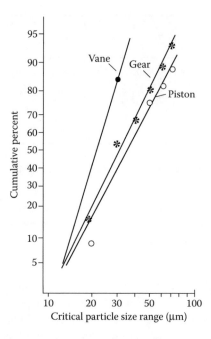

FIGURE 15.40　Relative sensitivity of different types of hydraulic pump to contamination.

The most common methods for the determination of water content of a water–glycol hydraulic fluid are the refractive index, viscosity, distillation, and Karl Fischer analysis. Of these, the most commonly used is the refractive index. The refractive index is readily determined using a portable temperature-compensated refractometer that provides readings in degrees Brix.

To perform this analysis using a digital refractometer, a drop of fluid is placed on the prism, as shown in Figure 15.41a (*Note:* the prism should be cleaned prior to use or incorrect readings may result.). The degrees Brix reading is then read directly from the digital readout as in Figure 15.41b. Digital handheld refractometers are generally more precise than traditional handheld refractometers. Water content is obtained from the refractometer reading and a calibration chart such as that shown in Figure 15.42. After the water concentration is determined,

TABLE 15.19
Suggested Acceptable Contamination Levels

Target Contamination Class to ISO Code			Sensitivity	Type of System	Suggested Filtration Rating
4 μm	6 μm	14 μm			βx > 100
14	11	9	Critical	Silt-sensitive control system with very high reliability; laboratory or aerospace	3
16	13	11	Semi-critical	High-performance servo and high-pressure long-life systems (i.e., aircraft, machine tool, etc.)	5
18	16	13	Important	High-quality reliable systems; general machine requirement	10
20	17	14	Average	General machinery and mobile systems	10
22	19	15	Crude	Low-pressure heavy industrial systems, or applications where long life is not critical	15–25

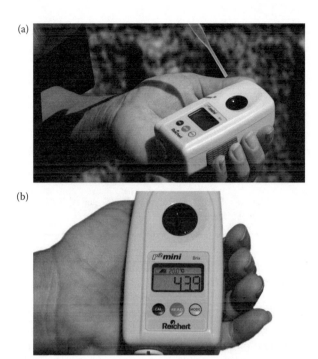

FIGURE 15.41 Determination of water content using a hand-held refractometer: (a) placing fluid on the prism, (b) reading the instrument. (Courtesy of Reichert Analytical Instruments, Depew, NY.)

additional water should be added if necessary. Some suppliers provide "water makeup" tables such as Table 15.20 or fluid-dependent plots such as Figure 15.43.

The principal limitation of water determination by refractive index is that the refractive index is affected by additive depletion, fluid degradation, or contaminants that may be present in the hydraulic fluid. Therefore, it is advisable to periodically cross-check water analyses obtained by refractive index by at least one other analytical method.

Water content may also be determined by viscosity measurement. One common method of viscosity measurement is to follow the ASTM D-445 procedure for kinematic viscosity [128]. The Cannon-Fenske tube used in this determination is shown in Figure 15.44. Because viscosity is temperature-dependent, it is essential that a constant-temperature bath be used for viscosity measurements. Plots of viscosity versus water content, such as Figure 15.45, are available from the water–glycol hydraulic fluid producer. If water additions are necessary, water makeup charts

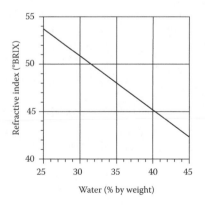

FIGURE 15.42 Water content versus refractive index in degrees Brix.

TABLE 15.20
Water Addition by Refractive Index

Water Makeup[a]	Refractive Index (degrees BRIX)
None	43.75–46.00
5	46.00–47.25
10	47.25–48.75
15	48.75–50.50
20	50.50–52.50
25	52.50–54.25

[a] Gallons of water added to each 100 gal of W/G hydraulic fluid.

such as that illustrated by Figure 15.46 are available. Alternatively, a water "makeup" table analogous to Table 15.21 may be obtained from the W/G hydraulic fluid supplier for the specific fluid being used.

The load-bearing capacity of a fluid film is dependent on fluid viscosity. Oxidative and thermal degradation processes will result in a decrease of fluid viscosity. Thus, routine viscosity measurement is one of the best methods of monitoring fluid stability. However, such comparative measurements must be made at the same total water content.

The water content may also be determined by azeotropic distillation from benzene or toluene [48]. This procedure, described in ASTM D95 [129], is relatively labor intensive and not commonly conducted on water–glycol hydraulic fluids at the present time.

The fourth method of water determination is by Karl Fischer analysis (ASTM D-1744) [130]. The advantage of Karl Fischer analysis is that it is a direct measure of water content, whereas viscosity and refractive index are both indirect measurements which are substantially affected by either contamination (refractive index) or fluid degradation (viscosity). Figure 15.47 illustrates the level of agreement that has been obtained with known water contents and the Karl Fischer procedure.

The quality of water used for system makeup is critically important because polyvalent hard metal ions (Ca^{2+}, Mg^{2+}, Mn^{2+}, etc.) present in tap water or city water will react with the antiwear

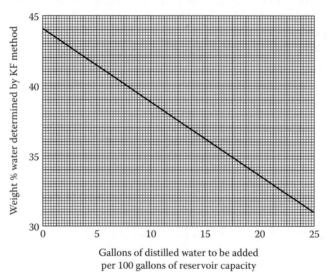

FIGURE 15.43 Water makeup graph.

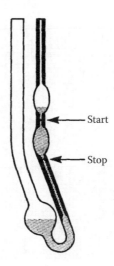

FIGURE 15.44 Cannon-Fenske viscometer.

additive, usually an organic carboxylic acid salt, to form a polyelectrolyte complex (V) [131], which appears as a "white, soapy" solid [132,133]. This process must be prevented for two reasons. The first is that it will lead to continuous depletion of the critically important antiwear additive. The second reason is that the presence of such precipitates, like any solid material, will increase wear and possibly plug filters:

$$2\,RCO_2H + Ca^{2+} \rightarrow Ca^{2+}\left(^-O_2CR\right)_2 \downarrow + 2H +$$

Only distilled or deionized water, with a conductance of less than 15 umho/cm (a maximum total water hardness of 5 ppm has also been recommended [35,133], should be added to a W/G hydraulic fluid system. It is recommended that total hardness of the system should not exceed 250 ppm [134]. The ionic content of water can be monitored by conductance. It is recommended that the maximum conductance of makeup water be less than 15 umho/cm.

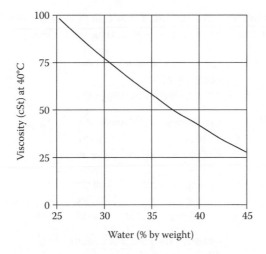

FIGURE 15.45 Viscosity versus water content.

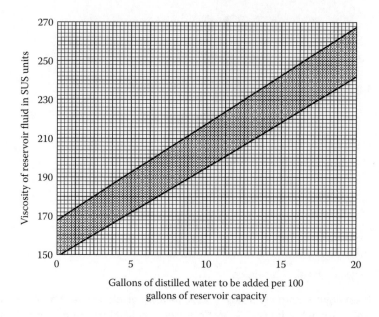

FIGURE 15.46 Water makeup graph based on fluid viscosity measurement.

Reserve alkalinity (corrosion inhibitor). Amine concentration in a W/G hydraulic fluid is designated as "reserve alkalinity" and is conventionally reported as the volume in milliliters of 0.1 N hydrochloric acid (HCl) required to titrate 100 mL of W/G fluid to pH 5.5. A typical titration plot is shown in Figure 15.48. Two breaks in the titration curve are observed because two chemical moieties are actually being analyzed—the amine carboxylate (VI) and the excess "free" amine (VII). Thus, acid titration also provides a method for quantifying the concentration of antiwear additive.

$$[RCO_2 \cdots +HNR_3]+HCl \rightarrow RCO_2H\,[Cl^- \cdots ^+ HNR_3] \tag{VI}$$

$$R_3N+HCl \rightarrow [Cl^- \cdots ^+ HNR_3] \tag{VII}$$

Changes in corrosion inhibitor concentration may also be monitored by pH [132–135]. It is recommended that the pH of the W/G system be greater than 8.0 [132,133].

TABLE 15.21
Water Addition by Viscosity

Water Makeup[a]	Viscosity (sSt at 40°C)
None	39–50
5	50–56
10	56–68
15	68–82
20	82–102
25	102–124

[a] Gallons of water added to each 100 gal of W/G hydraulic fluid.

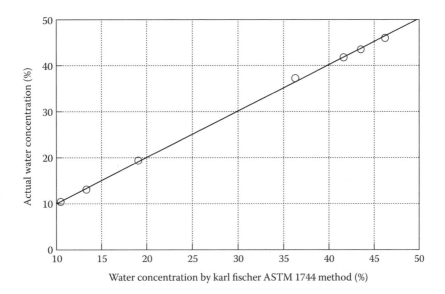

FIGURE 15.47 Correlation of water–glycol hydraulic fluid water content and Karl Fischer analysis (ASTM D01744), pyridine-cellosolve method).

Fluid degradation. The presence of formic acid is particularly deleterious because concentrations higher than 0.15% may lead to excessive wear, as discussed earlier in this chapter and shown in Figure 15.16. Because formic acid may result in increased wear, an analysis should be conducted to detect its presence in the used fluid. The analytical procedure most usually performed for acid detection in aqueous solution is ion chromatography [136].

Ferrography. It has thus far been shown that hydraulic fluid quality and performance is dependent on fluid cleanliness and chemistry variation. On occasion, it is necessary to troubleshoot fluid performance in improperly operating systems. In addition to the chemical and physical analyses described earlier, it is often of value to analyze any wear debris that may be formed. One of the principal methods of wear debris analysis is ferrography [137,138].

Ferrography may be used to determine the concentration and distribution of wear particles contained in a hydraulic fluid. This is illustrated in Figure 15.49, which shows the progression of wear as a function of total particle concentration [139]. Generally, "benign" wear occurs when the

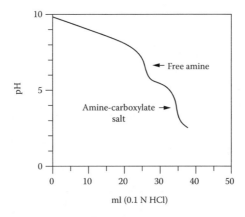

FIGURE 15.48 Determination of reserve alkalinity by acidic titration.

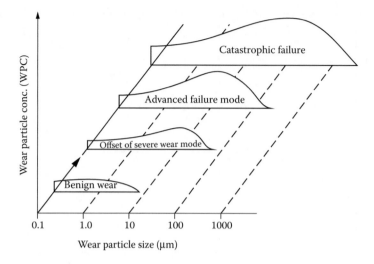

FIGURE 15.49   Typical progression of severe wear.

particles are less than 15 μm. Wear particles producing catastrophic failure are typically >200 μm. Ferrography can be used to detect and characterize wear particles in this size range.

Wear particles may also be profiled over time, as shown in Figure 15.50 [139]. Although all particles are susceptible to removal by filtration, large-particle filtration generally occurs more rapidly. Therefore, the larger particles indicated by ferrography are a good indicator of "current wear rate."

Ferrographs can be used to elucidate the identity of wear particles. This information is available from various collections published in handbooks [140]. For example, using ferrography, it is possible to identify carbonaceous material, rust, copper, soft metal (i.e., aluminum, zinc, etc.), and severe and normal wear. Some typical ferrographs are shown in Figure 15.51.

15.2.5   System Conversion Guidelines

Proper fluid conversion procedure is essential if optimal hydraulic system performance is to be achieved. In this section, conversion procedures from mineral oil and from other fire-resistant fluids will be discussed. The reader is encouraged to obtain a copy of an industry standard such as the NFPA T2.13.1 R4-2007 standard for more comprehensive compilations of recommended procedural guidelines [141].

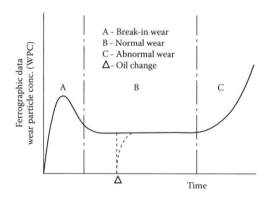

FIGURE 15.50   Ferrographic profile of large wear particles.

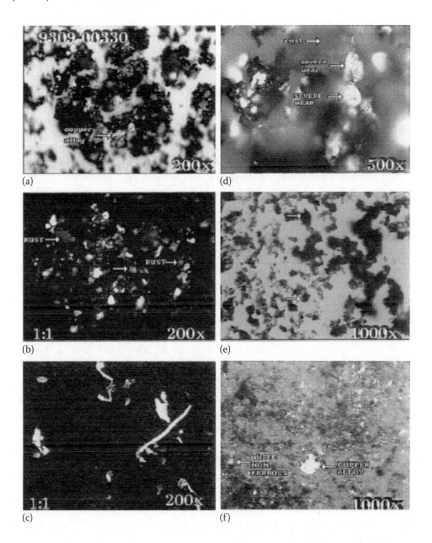

FIGURE 15.51 Typical ferrograms that may be obtained from ferrographic analysis: (a) copper-alloy wear debris, (b) rust contamination, (c) white non-ferrous and copper-alloy debris, (d) rust and severe wear particles, (e) amorphous carbon contamination, and (f) crystalline fiber contamination. (Courtesy of Engineered Lubricants Inc., St. Louis, MO.)

15.2.5.1 Piping Inspection and Repair

System conversion is an excellent time to perform a system inspection and repair hydraulic piping problems. The following list of inspection pointers, originally written by Oilgear for oil hydraulic systems, is appropriate for any hydraulic fluid system [146]:

1. Look for loose or vibrating pipes that may break due to work-hardening of the metal. Copper is especially vulnerable. Because copper is only recommended for low pressure (<300 psi), it should not be used for water–glycol systems, which are typically used at pressures higher than 300 psi.
2. Flexible hoses should be inspected very closely. A flexible hose is susceptible to sudden rupture or blowout at couplings and should not be used in potentially dangerous areas. Flexible hoses may often be avoided by telescoping joints, rotary joints, or unions, which may be loosened for repairs. If a flexible hose must be used, it should be enclosed in a manner that permits venting to a sump or waste pit.

3. Fragile piping should not be placed in an area where it can be run over, walked on, or used as a ladder. It should be moved or covered with adequate protection.
4. Avoid temporary repairs. Maleable or wrought-iron fittings should be replaced immediately. Avoid bending or straining pipes to "make them fit." The strain may lead to subsequent line breakage.
5. Long overhead runs of pipe should be avoided because rupture may lead to spraying of fluid over dangerous areas.
6. Pipe passageways through floors or between machines should be patched with nonflammable ceramics to prevent fluid from spreading from one floor or machine to another.
7. Be sure that machines are operating smoothly. Rough hydraulic flow may indicate improper circuit design and may result in future leaks and breakdowns in the system.

Pumps designed for oil hydraulic fluids may not have proper inlet flow conditions for the denser water–glycol fluid. Properly sized inlet piping and minimum distance between the reservoir and inlet and maximum distance of the reservoir placement above the inlet is essential to minimize pressure drops, which may lead to fluid starvation and cavitation [142]. If, after performing the calculations shown in Table 15.17, excessive inlet vacuum conditions exist, larger inlet pipe diameter (and possibly pump speed reduction to reduce flow) may be helpful. A chart such as that provided in Figure 15.52 can be used to determine proper pipe sizing when volume flow rate and velocity are known [142].

15.2.5.2 Additional Considerations

Water–glycol fluids as a class are incompatible with most paints used with hydraulic oil. Therefore, conventional paint inside of a fluid reservoir must be removed. Although it is desirable to use unpainted reservoir surfaces, there are some paints such as modified phenolic or epoxy-based paints which may be used. However, paint compatibility should be tested before use [148].

Water–glycol hydraulic fluids should be used at temperatures below 150°F; as a result, heat exchangers may be required in order to maintain optimum reservoir temperatures [143,144].

When switching to a water–glycol fluid, it is essential that as much residual oil as possible be removed. Low-pressure air may be used blowout residual oil from remote parts of the system [144]. The reservoir should be wiped down to remove all traces of the residual oil [71,73,143]. If it can be utilized, steam cleaning can be effective [71,73].

Flushing fluids are not recommended because residual amounts of the fluid are almost certain to remain and disrupt the overall chemistry of the fluid. (As shown previously, maintenance of fluid composition is critically important if optimal performance is to be guaranteed.) Instead, the water–glycol itself should be used as the flushing fluid [71,73,114,143].

The filters in the system must be compatible with the water–glycol hydraulic fluid being used. Do not use paper or treated-paper filters. Clean the filter housings and plug bypasses.

Additional cleaning guidelines are provided in Table 15.22 [145].

15.2.5.3 Converting from Petroleum Oil to Water–Glycol Fluids

In converting a hydraulic system from petroleum hydraulic oils to water–glycol hydraulic fluids, follow accepted engineering practices and be as thorough as possible [141]. A little extra attention at this point is necessary for safer and better performance and lower maintenance expense. The following procedure is recommended in changing over from petroleum oils to water–glycol:

1. Drain the oil from the system completely. Particular attention should be paid to the reservoir, pipe lines, cylinders, accumulators, filters, or other equipment in which oil might be trapped.

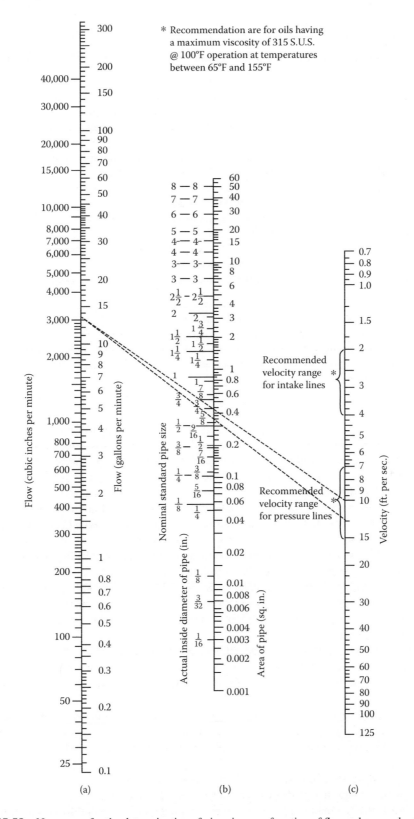

FIGURE 15.52 Nonogram for the determination of pipe size as a function of flow volume and velocity.

TABLE 15.22
Conversion Guidelines from Petroleum to Water–Glycol Hydraulic Fluids

Component	Procedure
Reservoir[a]	Drain and clean; remove internal gaskets; install a 10-μm breather
Inlet to pump	Check size of inlet pipe; supercharge or establish positive fluid head; install 60-mesh strainer four times pump capacity
Pump	De-rate all pumps except piston type by 125%; drain and clean
Valves	May require change in contro orifice size; drain and clean
Fluid motors and cylinders	May require size change; drain and clean
Accumulators	Drain and clean
Piping	Use Teflon thread sealant; clean by swabbing
Filters	Change to compatible elements clean housing
Coolers	Drain and clean
Start-up	Check with fluid supplier; follow his instructions
Fluid maintenance	Check water content and pH weekly at start
Component maintenance	Inspect pumps every 500 h, until a standard maintenance program is established; each inspection may vary

[a] Some designers use a closed reservoir with double-check valves to obtain supercharge pressure. Pressure is established by virtue of fluid and air expansion with temperature rise; 15–25 psi can be obtained by this method. One gravity-operated check valve in combination with a filter breather relieves the vacuum created by cooling after shutdown. A relief valve limits maximum pressure within the reservoir.

2. Clean the system of residual sludge and deposits and remove paint from the inside of the reservoir unless the paint has been tested and found to be resistant to the softening and lifting action of the hydrolube. Steam cleaning has been very effective in many instances. Carbon tetrachloride or other chlorinated metal cleaners should not be used.
3. Disconnect the filter.
4. Flush the system with a minimum amount of the water–glycol fluid being used. Flush initially by operating at no load or at minimum operating pressure, then bring the fluid up to normal temperature and operate all parts. Many users follow the practice of operating on the flush fill for several hours to provide complete circulation and take full advantage of the solvent cleaning characteristics of the hydrolube fluid.
5. Drain the flushing charge as completely as possible while it is still warm and without allowing it to settle. This fluid can be retained for further use in preparing other machines for service or for makeup purposes after suspended solid contaminants and residual petroleum oil have been removed.
6. Install a clean filter cartridge. Replace filter elements having zinc- or cadmium-plated parts with appropriate substitutes. Do not use a highly adsorptive filter medium, such as clay or Fuller's earth, because these filters may alter fluid composition by removing essential additives.
7. Examine pump parts, O-rings, and auxiliary equipment. Replace worn pump parts. Repair leaking pipe joints. Replace deteriorated gaskets, seals, and packings. Replace cork shaft seals and other water-susceptible packings and materials. Substitute waterproof materials.
8. Reconnect the system and tighten all joints and connections.
9. Fill system with proper grade of water–glycol hydraulic fluid.
10. Operate at reduced pressure to ensure proper lubrication of the hydraulic pump; then bring up to standard operating conditions.

During the first few weeks of operation, filters and inlet screens may become clogged by sludge and deposits which have been loosened by the solvent action of the hydrolube. The result may be pump starvation and cavitation, noisy operation, and high pump wear. Therefore, replace filter cartridges and clean inlet screens as needed.

Even with the most careful cleaning procedures, a small amount of petroleum oil may remain in the system. Small quantities will not interfere with the performance of the water–glycol fluid but will reduce the fire resistance, especially if residual oil collects in one part of the system such as the reservoir or accumulators. Hydraulic oil is lighter than the hydrolube fluids and will rise to the top of the reservoir. During the initial period of operation, periodically skim or siphon the residual oil from the surface of the fluid in the sump.

15.2.6   ENVIRONMENTAL AND TOXICOLOGICAL ASSESSMENT

There is increasing interest in the environmental consequences and toxicological issues relating to the use, incidental release, and disposal of all hydraulic fluids, including water–glycol hydraulic fluids [147]. Further information is available regarding these and other testing requirements in Chapter 8.

15.2.6.1   Toxicity

The health hazard of a fire-resistant hydraulic fluid in the conditions under which it is to be used is a key consideration in determining its applicability and safety in use. The Seventh Luxembourg Report [148] has as part of its requirements for certification of a fire-resistant hydraulic fluid, a comprehensive protocol for the evaluation of the fluid's hazard to health. This assessment of the hydraulic fluid's toxicity includes an evaluation based on the toxicological information available on the chemical compounds comprising the fluid and testing of the hydraulic fluid for acute toxicity, irritation of skin and mucous membranes, toxicity of the fluid as an aerosol, and the toxicity of the fluid's thermal decomposition products. The results of the tests are graded, weighted by coefficients, and combined in a cumulative score which must be below a stated value for that type of fluid tested to meet the health hazard requirements.

An evaluation method to determine the toxicological effects of the thermal decomposition products of a hydraulic fluid was described in the Sixth Edition of the Luxembourg Report [149]. The method used an apparatus by which a hydraulic fluid was sprayed onto a heated plate. The formed decomposition products were passed through a condensation system and into an inhalation chamber containing Wistar rats. Several classes of fire-resistant hydraulic fluids previously approved under the Luxembourg reports—chlorinated hydrocarbons and mixtures of phosphate esters and chlorinated hydrocarbons have been classified as dangerous substances in subsequent reports [150, 151] and their use has been made illegal due to the toxic dioxins and hydrofurans formed during their thermal decomposition. Van Dam reported the thermal degradation products of water–glycol (HFC) hydraulic fluids reduced respiration in Wistar rats [152]. Animals were exposed to the atmosphere resulting from spraying the HFC fluid on a plate heated to 500°C, 600°C, and 700°C and monitored for fourteen days. No mortalities occurred with the exposed animals [153]. The chemical compounds obtained in the thermal degradation of HFC fluids and their concentration levels were also determined by Van Dam [154].

15.2.6.2   Environmental Assessment

Water–glycol hydraulic fluids have been shown to be biodegradable [155] and environmentally friendly [156]. Today there are many standards and programs globally used to describe and register fluids as environmentally friendly. Those include the Seventh Luxembourg Report, Canadian Environmental Choice Program, German Blue Angel, Norwegian Nordic Swan, and the European Union Eco-label program. The following are two environmental standards whose requirements are met by water–glycol hydraulic fluids.

The Canadian Environmental Choice Program (ECP) [157,158] aims to reduce the stress on the environment by encouraging the demand for, and supply of, environmentally responsible products band services. Products and services meeting the EPC guidelines are given authority to label their product with the Environmental Choice EcoLogo. EPC has published a national guideline on synthetic industrial lubricants [159]. Several water–glycol hydraulic fluids have been certified to use the EPC EcoLogo.

The Seventh Luxembourg Report and subsequent Health and Safety Executive (HSE) report have, as part of their certification of a fire-resistant hydraulic fluid, a self-classification of the fluid into one of three existing water hazard classes based on the results obtained from testing of the fluid as to biodegradability and toxicity to higher and lower water-dwelling organisms including fish, algae, daphnia, and mammals. Many water–glycol hydraulic fluids meet these toxicological and biodegradability requirements as well as the other property, which is performance, and so fire-resistance tests and are approved according to the Seventh Luxembourg Report and HSE standards.

15.3 FIRE RESISTANCE

Even after 60 years of use [3,35,160], there is still no consensus on the best evaluation procedures to quantify the relative fire resistance provided by hydraulic fluids [161]. A review of fire resistance testing procedures and the results reported to date is provided in Chapter 9. The order of fire resistance offered by various major classes of fire-resistant fluids capable of high-pressure use is water–glycol > phosphate ester > polyol ester and poly(alkylene glycol) > mineral oil.

REFERENCES

1. Zink, M. "Selecting Hydraulic Fluids," *Hydraulics & Pneumatics*, May, 2000, pp. 31–35.
2. Millett, W.H. "Fire Resistant Hydraulic Fluids," *Appl. Hydr.*, June, 1957, pp. 124–28.
3. Brophy, J.E., Fitzsimmons, V.G., O'Rear, J.G., Price, T.R. and Zisman, W.A. "Aqueous Nonflammable Hydraulic Fluids," *Ind. Eng. Chem.*, 1951, 43(4), pp. 884–96.
4. O'Rear, J.G., Militz, R.O., Spessard, D.R. and Zisman, W.A. "The Development of the Hydrolube Non-Flammable Hydraulic Fluids," Naval Research Laboratory Report No. P-3020, April 1947.
5. Murphy, C.M. and Zisman, W.A. "Non-Flammable Hydraulic Fluids," *Lubr. Eng.*, October, 1949, pp. 231–35.
6. Zisman, W.A., Wolfe, J.K., Baker, H.R. and Spessar d, D.R. U.S. Patent 2,602,780 (1952).
7. Roberts, F.H. and Fife, H.R. U.K. Patent 2,425,755 (1947).
8. Anon., "Navy Announces Hydrolube Development," *Appl. Hydraul.*, September, 1948, p. 15.
9. Schmitt, C.R. "Fire Resistant Hydraulic Fluids for Near-Surface Systems Safeguard Life and Property at No Loss Efficiency," *Ind. Heating*, 1957, 24(9), pp. 1756–70.
10. Mathe, J. "Fire-Resistant Hydraulic Fluids for the Plastics Industry," SPE J., July, 1967, pp. 17–20.
11. Tutti Cadaveri, Le procès de la catastrophe du Bois du Cazier à Marcinelle (The Bois du Cazier Mine Disaster Trial) by Marie Louise De Roeck, Julie Urbain and Paul Lootens, Editions Aden, collection EPO, Brussels, 2006, p. 280.
12. Luxembourg Report, 20 December 1960 - Commission of the European Communities -Safety and Health Commission for the Mining and Other Extractive Industries. First Report on the Specifications and Testing Conditions Relating to "Fire Resistant Hydraulic Fluids Used for Power Transmission in Mines," Luxembourg.
13. Staley, C. "Fire-Resistant Hydraulic Fluids—Comparisons and Applications," *Petroleum Times*, 1967, 71(1834), pp. 1709–14.
14. Reichel, J. "Fluid Power Engineering with Fire Resistant Hydraulic Fluids—Experiences with Water-Containing Hydraulic Fluids," *Lubr. Eng.*, 1994, 50 952, pp. 947–52.
15. Rush, R.E. "Fire-Resistant Fluids in Basic Steelmaking—Which One?" *Iron and Steel Eng.*, 1980, 57(12), pp. 54–55.
16. Iwamiya, Y. "Water–Glycol Hydraulic Fluids," *Junkatsu*, 1987, 32(8), pp. 534–39.
17. Totten, G.E. and Webster, G.M. "High Performance Thickened Water–Glycol Hydraulic Fluids," in *Proc. of the 46th National Conference on Fluid Power,* March 23–24, 1994, National Fluid Power Association; Milwaukee, WI, pp. 185–94.

18. Lewis, W.E.F. "Water-Based Energy Transmitting Fluid Compositions," U.S. Patent 4,434,066 (1984).
19. Lewis, W.E.F. "Energy Transmitting Fluid," U.S. Patent 4,855,070 (1989).
20. Stewart, H.L. "Fire-Resistance Hydraulic Fluids," *Plant Eng.*, 1979, 33(4), pp. 157–60.
21. Bonnell, C.G. "Fire-Resistant Fluids for Diecasting Hydraulic Systems," *Foundry*, 1967, 95(5), pp. 224–29.
22. Edgington, R. "The use of Fire-Resistant Hydraulic Fluids in Axial Piston Pumps," *Eng. Digest*, 1965, 26(11), pp. 91–92.
23. Aengeneyndt, K.D. and Lehringer, P. "Schwerentflammbare Hydraulikiltissigkeiten auf Wasser–Glyko-Basis: Gestern-Heute-Morgen", *Giesserei*, 1978, 65(3), pp. 58–63.
24. Totten, G.E. and Bishop, R. Jr., "Historical Overview of the Development of Water–Glycol Hydraulic Fluids," *SAE Technical Paper Series*, Paper 952077, 1995.
25. International Standard ISO 6743/4, "Lubricants, Industrial Oils and Related Products (Class L)—Classification Part 4: Family H (Hydraulic Systems)," First Edition-1982-11–15.
26. Totten, G.E. "Thickened Water–glycol Hydraulic Fluids for Use at High Pressures," *SAE Technical Paper, Series*, Paper 921738, 1992.
27. Hosterrnan, F.O. "A Progress Report: Nonflammable Hydraulic Fluids," *Appl. Hydraul.*, September, 1951, pp. 66–71.
28. Millet, W.H. "Nonpetroleum Hydraulic Fluids—A Projection," *Iron Steel Eng.*, 1977, 54(5), pp. 36–39.
29. Rasp, R.C. "Water Based Hydraulic Fluids Containing Synthetic Components," *J. Synth. Lubr.*, 1989, 6(3), pp. 233–51.
30. Singh, T., Jain, M., Ganguli, D. and Ravi, K. "Evaluation of Water–Glycol Hydraulic Fluids: A Tribological Approach," *J. Synthetic Lubrication*, 2006, 23, pp.177–84.
31. Reference food Grade WGF paper.
32. Nassry, A, and Maxwell, J.F., 'Water-based hydraulic fluid and metalworking lubricant'. US Patent 4,151,009, 24 April 1979.
33. Totten, G.E., Bishop, R.J. Jr., McDaniels, R.L. and Wachter, D.A. "Water–Glycol Hydraulic Fluid Maintenance," *Iron Steel Eng.*, October, 1996, pp. 34–38.
34. Wan, G.T.Y., Kenney, P. and Spikes, H.A. "Elastohydrodynamic Properties of Water- Based ·:Fire-Resistant Hydraulic Fluids," *Tribol. Int.*, 1984, 17(6), pp. 309–15.
35. Blanpain, G. "The use of Polyglycols in French Coal Mines," in *IR Int. Symposium on Performance Testing of Hydraulic Fluids, Oct. 1978, London, England,* Tourret, R. and Wright, E.P., eds., Heyden and Son; London, 1978, pp. 389–403.
36. Papay, A.J. "Hydraulics," *Chem. Ind.*, 1993, 48, pp. 427–52.
37. Sharpe, R.Q. "Designing for Fire Resistant Hydraulic Fluids," *Product Eng.*, August, 1956, pp. 162–66.
38. Loudon, B.J. "Fire-Resistant Hydraulic Fluids," *Surf. Coatings Australia*, July, 1989, pp. 23–29.
39. Anon., "Fire Resistant Hydraulic Fluids," *Lubrication*, 1962, 48(11), pp. 161–80.
40. Staley, C. "Fire Resistant Hydraulic Fluids," *Chemicals for Lubricants and Functional Fluids Symposium*, 1979.
41. Totten, G.E. and Webster, G.M. "High-Performance Thickened Water–Glycol Hydraulic Fluids," in *Proc. of 46th National Conf. on Fluid Power,* National Fluid Power Association Milwaukee, WI, 1994, pp. 185–94.
42. Isaksson, O. "Rheology for Water-Based Hydraulic Fluids," *Wear*, 1987, 115(1–2), pp. 3–17.
43. Rynders, R.D. "Fire Resistant Fluids from a Component Builder's Viewpoint," *Natl. Conf. Ind. Hydroid,* 1962, 16, pp. 39–46.
44. Van Oene, H. "Discussion of Papers 73046 and 730487," *SAE Trans.*, 1973, pp. 1580.
45. Zaska'ko, P.P., Mel'nikova, A.V., Diment, O.N., Titurenko, S.G. and Stepanova, E.V. "Mechanical Stability of Nonflammable Hydraulic Fluids," *Chem. Technol. Fuels Oils*, 1974, 10(3–4), pp. 307–9.
46. ASTM D 3945-93 (with drawn 1998), "Standard Test Method for Shear Stability of PolynierlContaining Fluids Using Diesel Injector Nozzle," *American Society for Testing and Materials*, Conshohocken, PA.
47. IP 294/77 Kurt Orbahn Method, "Shear Stability of Polymer-Containing Oils Using a Diesel Injector Rig."
48. BIanpain, G.M.G. "Fire-Resistant Hydraulic Fluids in the French Mines," in *Conf. Proceed. Fluid Power Equipment in Mining, Quarrying and Tunneling*, 1974, pp. 145–55.
49. Zisman, W.A., Wolfe, J.K., Baker, H.R. and Spessard, D.R. U.S. Patent 2,602,780 (1952).
50. Anon., "Minutes of Nitrosamine Task Farce Meeting," Cosmetic, *Toiletry and Fragrance Association Inc.*; 1977.
51. Totten, G.E., Bishop, R.J., McDaniels, R.L., Braniff, D.P. and Irvine, D.J. "Effect of Low Molecular Weight Carboxylic Acids on Hydraulic Pump Wear," *SAE Technical Paper Series*, Paper 941751, 1994.

52. ASTM D1783-01 (2007), "Standard Test Method for Phenolic Compounds in Water," *American Society for Testing and Materials*, W. Conshohocken, PA.
53. Shrey, W.M. "Effect of Fire-Resistant Fluids on Design and Operation of Hydraulic Systems," *Iron Steel Eng.*, 1962, 39(9), pp. 191–97.
54. Igarashi, J. "Oxidative Degradation of Engine Oils," *Jpn. J. Tribol.*, 1990, 35, pp. 1095–105.
55. Costa, L., Gad, A.M., Camino, G., Cameron, G.C. and Qureshi, M.Y. "Thermal and Thermooxidative Degradation of Poly(ethylene oxide)—Metal Salt Complexes," *Macromolecules*, 1992, 25, pp. 5512–18.
56. Lloyd, W.G. "The Influence of Transition Metal Salts in Polyglycol Autoxidations," *J. Polymer Sci.*, Part A, 1963, 1, pp. 2551–63.
57. Brown, P.W., Galuk, K.G. and Rossiter, W.J. "Characterization of Potential Thermal Degradation Products from the Reactions of Aqueous Ethylene Glycol and Propylene Glycol Solutions with Copper Metal," *Solar Energy Mater.*, 1987, 16, pp. 309–13.
58. Beavers, J. A. and Diegle, R. B. "The Effect of Degradation of Glycols on Corrosion of Metals Used in Non-Concentrating Solar Collectors", *Battelle Columbus Labs*, 505 King Ave., Columbus, Ohio 43201. 1982, Corrosion 81/207, NACE, Houston, Tx.
59. Rossitter, W.J., Brown, P.W. and Godette, M. "The Determination of Acidic Degradation Products in Aqueous Ethylene Glycol and Propylene Glycol Solutions by Ion Chromatography," *Solar Energy Mater.*, 1983, 9, pp. 267–79.
60. Mannesmann Roth GmbH, "Hydraulic Power Units for Use with HFC Fluids," Mannesmann Rexroth GmbH.
61. West, C.W. "Additives for Corrosion Control," *Soap Chem. Specialties*, 1964, September, pp. 177, 178, 212, 213/October, pp. 192, 193, 226.
62. McGary, C.W. "Degradation of Poly(ethylene)Oxide," *J. Polym. Sci.*, 1960, 46, pp. 51–57.
63. Rakoff, P., Colucci, G.J. and Smith, R.K. "Development of Fire-Resistant Water-Based Hydraulic Fluids," NTIS Accession No. AD-608 564, Nov. 27, 1964.
64. Rakoff, P., Colucci, G.J. and Smith, R.K. "Development of Fire-Resistant Water-Based Hydraulic Fluids," NTIS Accession No. AD-605 910, Sept. 28, 1964.
65. Rakoff, P., Colucci, J. and Smith, R.K. "Development of Fire-Resistant Water-Based Hydraulic Fluids," Dept. of Navy, Contract No. 90269, January 27, 1965.
66. ASTM D 7043-10, "Standard Test Method for Indicating Wear Characteristics of Non-Petroleum and Petroleum Hydraulic Fluids in a Constant Volume Vane Pump," *American Society for Testing and Materials*, Conshohocken, PA.
67. ASTM D 665-06, "Standard Test Method for Rust-Preventing Characteristics of Inhibited Mineral Oil in the Presence of Water," *American Society for Testing and Materials*, Conshohocken, PA.
68. Baboian, R. (ed.) *Corrosion Tests and Standards*, American Society for Testing and Materials, Conshohocken, PA, 1995.
69. Katorgin, V.A. and Ramanova, T.V. "Estimation of the Rate of Contact Corrosion of Metals," *Chem. Technol. Fuels Oils*, 1989, 25(1–2), pp. 113–15.
70. ISO 4404-1:2001(E), "Petroleum and Related Products – Determination of the corrosion resistance of fire-resistant hydraulic fluids – Part1: Water containing fluids," International Organization for Standardization.
71 Millet, W.H. "Fire Resistant Hydraulic Fluids for Die Casting—Part 1, Aqueous Fluids," *Precision Metal Molding*, December, 1954, pp. 85–90.
72. Millet, W.H. "Fire-Resistant Hydraulic Fluids: Part 2—Aqueous Base Types," *Appl. Hydraul.*, June, 1957, pp. 124–28.
73. Kramer, G.F. and Natscher, J. personal communication, *Union Carbide Corporation*, Tarrytown, NY, 1988.
74. Snow, H.A. "Modern Fire Resistant Hydraulic Fluids for Industrial Use," *Sci. Lubr.*, May, 1960, pp. 39–42.
75 Pendergast, P. "Introduction of Next Generation UCON Advanta Water–Glycol Hydraulic Fluid," Commercial Marketing Forum, STLE Annual Meeting, May 21, 2008, Cleveland, OH.
76. Haden, V.G.I. "Fire-Resistant Hydraulic Fluids," *Metal Forming*, December, 1968, pp. 352–54.
77. Cole, G.V. "Investigation into the Use of 'Water–glycol as the Hydraulic Fluid in a Servo System," AERE R.11324, AERE Harwell—Engineering Projects Division, Harwell, UK, July 1984.
78. Kenny, P., Smith, J.D. and March, C.N. "The Fatigue Life of Ball Bearings When Used with Fire-Resistant Fluids," in *Inst. Petroleum Symp. on Performance Testing of Hydraulic Fluids*, 1978, paper 30.
79. Culp, D.V. and Widner, R.L. "The Effect of Fire Resistant Hydraulic Fluids on Tapered Roller Bearing Fatigue Life," *SAE Technical Paper Series*, Paper 770748, 1978.

80. Zino, J. "What to Look for in Hydraulic Oils. V1—Lubricating Value," *Am. Machinist*, January 15, 1948, pp. 97–100.
81. Spikes, H.A. "Wear and Fatigue Problems in Connection with Water-Based Hydraulic Fluids," *J. Synth. Lubr.*, 1987, 4(2), pp. 115–35.
82. Versnyak, V.P., Zaretskaya, L.V., Imerlishvili, T.V., Kel'bas, V.I., Lukashvili, N.V., Sedova, L.O., Ryanoshapka, V.M., Schvartsman, V.Sh. and Shoiket, V.Kh. "Behavior of Water Glycol Hydraulic Fluids in EHD Contacts," *Sov. J. Friction Wear*, 1989, 10(5), pp. 120–26.
83. Versnyak, V.E., Zaretskaya, L.V., Imerlishvili, T.V., Kerbas, V.L., Lukashvili, N.V., Sedova, L.O., Ryanoshapka, V.M., Schvartsman, V.Sh. and Shoiket, V.Kh. "Behavior of Aqueous Glycol Hydraulic Fluids in Elastohydrodynamic Contacts," *Trenie Iznos*, 1989, 10(5), pp. 919–27.
84. Versnyak, V.P., Zaretskaya, L.iV., Imerlishvili, T.V., Kel'bas, V.L., Lukashvili, N.V., Sedova, L.O., Ryanoshapka, V.M., Schvartsman, V.Sh. and Shoiket, V.Kh. "Thickness of a Lubricating Film of Aqueous Glycol Fluids Under Different Friction Regimes," *Trenie Iznos*, 1991, 12(1), pp. 144–53.
85. Fein, R.S. and Villforth, F.J. "Lubrication Fundamentals," *Lubrication*, October–December, 1973, 59, pp. 77–96.
86. Wan, G.T.Y. and Spikes, H.A. "The Elastohydrodynaniic Lubricating Properties of Water–Polyglycol Fire-Resistant Fluids," *ASLE Trans.*, 1994, 27(4), pp. 366–72.
87. Dalmaz, G. and Godet, M. "Film Thickness and Effective Viscosity of Some Fire Resistant Fluids in Sliding Point Contacts," *Trans. ASME*, 1978, 100, pp. 304–8.
88. Dalmaz, G. "Traction and Film Thickness Measurements of a Water–Glycol and a Water-in-Oil Emulsion in Rolling-Sliding Contacts," in *Proc. of 7th Leeds–Lyon Symposium on Tribology, Friction and Traction*, 1980, pp 231–42.
89. Wedeven, L.D., Totten, G.E. and Bishop, R.J. "Performance Map and Film Thickness Characterization of Hydraulic Fluids," *SAE Technical Paper Series*, Paper 952091, 1995.
90. Ratoi-Salagean, M. and Spikes, H.A. "The Lubricant Film-Forming Properties of Modern Fire Resistant Hydraulic Fluids," in *Tribology of Hydraulic Pump Testing—ASTM STP 1310*, Totten, G.E., Klhig, G.H. and Smolenski, D.J., eds., American Society for Testing and Materials; Conshohocken, PA, 1996, pp. 21–37.
91. Cordiano, H.V., Cochran, E.P. and Wolfe, R.J. "A Study of Combustion Resistant Hydraulic Fluids as 13a11 Bearing Lubricants," *Lubr. Eng.*, July/August, 1956, pp. 261–66.
92. Corcliano, H.V., Cochran, E.P. and Wolfe, R.J. "Effect of Combustion-Resistant Hydraulic Fluids on Ball-Bearing Fatigue Life," *Trans. ASME*, 1956, 78, pp. 989–96.
93. Kenny, P., Smith, J.D. and March, C.N. "The Fatigue Life of Bail Bearings When Used with Fire-Resistant Hydraulic Fluids," in *Int. Petroleum Symp. on Performance Testing of Hydraulic Fluids*, 1978, paper 30.
94. Yardley, E.D., Kenny, P. and Sutcliffe, D.A. "The Use of Rolling-Fatigue Test Methods over a Range of Loading Conditions to Assess the Performance of Fire-Resistant Fluids," *Wear*, 1974, 28, pp. 29–47.
95. March, C.N. "The Evaluation of Fire-Resistant Fluids Using the Unisteel Rolling Contact Fatigue Machine," in *Rolling Contact Fatigue: Performance Testing of Lubricants*, Tourret, R. and Wright, E.P., eds., Institute of Petroleum. London, 1976, pp. 217–29.
96. Hobbs, R.A. "Fatigue Lives of Ball Bearings Lubricated with Oils and Fire-Resistant Fluids," in *Elastohydrodynamic Lubrication*, Symposium Proc IME, 1972, Inst. of Mech. Eng., London, UK, pp. 1–4.
97. Culp, R.V. and Widner, R.L. "The Effect of Fire Resistant Hydraulic Fluids on Tapered Roller Bearing Fatigue Life," *SAE Technical Paper Series*, Paper 770748, 1977.
98. Burwell, F.T. and Scott, D. "Effect of Lubricant on Pitting Failure of Ball Bearings," *Engineering*, July 6, 1956, pp. 9–12.
99. Sullivan, J.L. and Middleton, M.R. "The Pitting and Cracking of SAE 52100 Steel in Rolling/Sliding Contact in the Presence of an Aqueous Lubricant," *ASLE Trans.*, 1985, 28, pp. 431–38.
100. Shitsukawa, S., Shibata, M. and Johns, T.M. "Influence of Lubrication on the Fatigue Life of Ball Bearings," *SAE Technical Paper Series*, Paper 972710, 1997.
101. Riddel, V., Pacor, P. and Appledorn, K.K. "Cavitation, Erosion and Rolling Contact Fatigue," *Wear*, 1974, 27, pp. 99–108.
102. Sullivan, J.L. and Middleton, M.R. "The Mechanisms Governing Crack and Pit Formation in Steel in Rolling Sliding Contact in Aqueous Lubricants," *J. Synth. Lubr.*, 1989, 6(1), pp. 17–29.
103. "The Development of Equipment and Techniques for Evaluating Effects of Oils on Bearing Fatigue Life," CRC. Project No. 413, Group on Gas Turbine Lubrication, of the Aviation Fuel, Lubricant and Equipment Research Committee of the Coordinating Research Council, Inc., May 1968.
104. Danner, C.H. "Relating Lubricant Film Thickness to Contact Fatigue," *SAE Technical Paper Series*, Paper 700560, 1970.

105. Kenny, P. and Yardley, E.D. "The Use of the Unisteel Rolling Fatigue Machines to Compare the Lubricating Properties of Fire-Resistant Fluids," *Wear*, 1972, 20, pp. 105–21.

106. Reichel, J. "Fluid Power Engineering with Fire Resistant Hydraulic Fluids," *Lubr. Eng.*, 1994, 50(12), pp. 947–51.

107. Yano, N., Ohnishi, T. and Saitoh, T. "Improvement in Rolling Contact Fatigue Performance of Water–Glycol Hydraulic Fluids," *STLE Annual Meeting in Kansas City*, MO, 1996.

108. Wan, G.T.Y., Kenny, P. and Spikes, H.A. "Elastohydrodynamic Properties of Water-Based Fire-Resistant Hydraulic Fluids," *Tribol. Int.*, 1984, 17(6), p. 309–16.

109. Bietowski, R. "Ball Bearing Lubricants—Use of Fire Resistant Hydraulic Fluids," *Colliery Guardian*, May, 1971, pp. 235–39.

110. Wakelin, R.J. "Life of Rolling Bearings in Contact with Fire Resistant Fluids," Final Report, M.O.D. (AS) Contract K78A1118/CB 78A, February 1975.

111. Skurka, J.C. "Elastohydrodynamic Lubrication of Roller Bearings," *J Lubr.*, 1970, 93, pp. 281–91.

112. Talian, T.E., Chiu, Y.P. and Van Amerogen, E., "Prediction of Traction and Microgeometry Effects on Rolling Contact Fatigue Life," *J Lubr. Tech.*, 1978, 100(1), pp. 156–66.

113. Boness, R.J., Crecelius, W.J., Ironside, W.R., Moyer, C.A., Pfaffenberger, E.E. and Poplawski, J.V. "Current Practice," in *STLE Life Factors for Rolling Bearings*, Enwin V.Z., ed., Society of Tribologists and Lubrication Engineers; Park Ridge, IL, 1992, pp. 1–45.

114. Townshend, F. and Baker, P. "Factors Relating to the Selection and Use of Fire-Resistant Fluids in Hydraulic Systems," Hydraulic Pneumatic Power, April, 1974, pp.134–40.

115. Sauer-Sundstrand, *Fluid Quality Requirements*, Rev. A, Sauer-Sundstrand Company; Ames, IA.

116. Jackson, T.L. and Alston, R.C. "The Selection and Application of Fire Resistant Hydraulic Fluids," *Plant Eng.*, 1968, 12(12), pp. 743–48.

117. Anon. *Fire Resistant Fluids*, Robert Bosch Group; Racine, WI.

118. Schmitt, C.R. "Fire resistant Hydraulic Fluids for Die Casting," *PMM*, February, 1995, pp. 81–83.

119. Bosch Rexroth WGF AG "Axial piston variable pump A4VSO for HFC Fluids," Technical Data sheet , RE 92 053/02.05 1/8.

120. Zingaro, A. "Walking the Fluid Cleanliness Tightrope," *Hydraulics Pneumatics*, April, 47(4), 1994, pp. 37–40.

121. Zingaro, A. "Walking the Fluid Cleanliness Tightrope [reprint]," Hydraulics Pneumatics, December, 47(12), 1994, pp. 25–26.

122. Fitch, E.C. and Hong, I.T., "Pump Contaminant Sensitivity: Part 1 - An Overview of the Omega Theory," *The Fire Resistant Hydraulics Journal*, 1986, 6, pp. 41–51.

123. Fitch, E.C. and Hong, I.T., "Pump Contaminant Sensitivity: Part 2 - Linear Omega Life Model." *The Fire Resistant Hydraulics Journal*, 1986, 6, pp. 53–61.

124. Xuan, J.L., "Pump Contaminant Sensitivity: Part 3 - An Updated Pump Contaminant Sensitivity Analysis Program," *The Fire Resistant Hydraulics Journal*, 1986, pp. 63–68.

125. Ou, R.F., "Pump Contaminant Sensitivity: Part 4 - The Effects of Errors in Flow Rate Measurements On The Omega Rating of a Pump," *The Fire Resistant Hydraulics Journal*, 1986, 6, pp. 69–75.

126. Fitch, E.C. *Fluid Contamination Control*, FES Inc.; Stillwater, OK, 1988.

127. Vickers Inc., "Contamination Control," in *Vickers Industrial Hydraulics Manual*, 3rd ed., Vickers Inc.; Maumee, OH, 1992, p. 6.22.

128. ASTM D 445-09, "Standard Test Method for Kinematic Viscosity of Transparent and Opaque Liquids (the Calculation of Dynamic Viscosity)," *American Society for Testing and Materials*, Conshohocken, PA.

129. ASTM D 1744-92 (withdrawn 2000), "Standard Test Method for Water in Liquid Petroleum Products by Karl Fischer Reagent," *American Society for Testing and Materials*, Conshohocken, PA.

130. ASTM D 4006-07, "Standard Test Method for Water in Crude Oil by Distillation," *American Society for Testing and Materials*, Conshohocken, PA.

131. Holiday, L. (ed.), *Ionic Polymers*, John Wiley & Sons, New York, 1975.

132. Skoog, P.N. "Care and Maintenance of Water–Glycol Hydraulic Fluids," *Foundry Manag. Technol.*, November, 1990, pp. 40–41.

133. Skoog, P.N. "The Care and Maintenance of Water–Glycol Hydraulic Fluids," *Hydraulics Pneumatics*, November, 1991, pp. 41–44.

134. Anon. *High Water Content Fluid*, The Oilgear Company; Milwaukee, WI.

135. Murata, T. "Fire Resistant Working Fluids: Monitoring of Water–Glycol System Working Fluids," *Nisseki Rehp*, 1989, 31(1), pp. 13–18.

136. Rossiter, W.J., Brown, P.W. and Godette, M. "The Determination of Acidic Degradation Products in Aqueous Ethylene Clycol and Propylene Glycol Solutions by Ion Chromatography," *Solar Energy Mater.*, 1983, 9, pp. 267–79.

137. Tessmann, R.K. "Ferrographic Measurement of Contaminant. Wear in Gear Pumps," in *Proc. Natl. Conf. Fluid Power*, 1978, 32, pp. 179–83.

138. Cickurs, P.V., S. A. and Kelley, V.T. "Prediction of Hydraulic Pump Failures Through Wear Debris Analysis," Naval Air Engineering Center Report, NAEC-92-171, July 19, 1983.

139. Anon., "Report on Ferrography and Its Application for Determining Wear Particle Equilibrium," Technical Report, Engineered Lubricants; Maryland Heights, MO, 1998.

140. Standard Oil, *Wear Particle Atlas—Revised*, SOI-HO Predictive Maintenance Series, BP America, Inc.; Cleveland, OH, 1976.

141. NFPA/T2.13.1 R4-2007, Recommended Practice—Hydraulic Fluid Power—Use of Fire Resistant Fluids in Industrial Systems, 5th edition.

142. Schmitt, C.R. "Changing Over to Fire-Safe Hydraulic Fluids," *Prod. Eng.*, November, 1956, pp.194–98.

143. Hemeon, J.R. "Changing to a Fire-Resistant Fluid," *Appl. Hydraul.*, July, 1955, pp. 66–68, 105.

144. Egan, E.J. "How to Install Fire-Resistant Hydraulic Fluids," *Iron Age*, October 11, 1956, pp. 95–7.

145. Goodman, K.C. *Living with Fire-Resistant Hydraulic Fluids*, Denison Hydraulics; Columbus, OH, 1965.

146. Anon., "Are Your Hydraulic Oil Lines Fireproof," *Mill Factory*, May, 1952, pp. 139–40.

147. Skoog, P.N. "Fire-Resistant Hydraulic Fluids in the 90s," *Die Casting Eng.*, 1989, 33(6), p. 30.

148. 7th Luxembourg Report, 3 March 1994 – Document Nr. 4746/10/91 EN – "Requirements and Tests Applicable to Fire-Resistant Hydraulic Fluids Used for Power Transmission (Hydrostatic and Hydrokinetic) in Mines," Editor: European Commission, Directorate General V Employment, Industrial Relations and Social Affairs Health and Safety Directorate GD V/E/4.

149. 6th Luxembourg Report, 3 November 1983 – Document Nr. 2786/8/81 EN, Commission of the European Communities-Safety and Health Commission for the Mining and Other Extractive Industries. Sixth Report on the Specifications and Testing Conditions Relating to Fire Resistant Hydraulic Fluids Used for Power Transmission in Mines, 2786/8SIE, Luxembourg.

150. 9l/3391EEC Council Directive of June 18, 1991 amending for the 11th time Directive 76/769/EEC on the approximation of the laws, regulations and administrative provisions of the Member States relating to restrictions on the marketing and use of certain dangerous substances and preparations.

151. Health and Safety Executive, "HSE Approved Specifications for Fire Resistance and Hygiene of Hydraulic Fluids for use in Machinery and Equipment in Mines" Reference HSE (M) File L11.6/3, October, 1999.

152. Van Dam, J. and Daenens, P. "The Effect of Thermal Degradation Products of Hydraulic Fluids Type C on the Carboxyhaemoglobin Level in Wistar SPF Rats," *Journal of Fire Sciences*, Vol. 11, September/October, 1993, pp 394–406.

153. Van Dam, J., Daenens, P. and Verbeken, E.K. "Acute Inhalation Toxicity in Rats Exposed to Thermal Degradation Products of Hydraulic Fluids Type C," *Journal of Fire Sciences*, Vol. 12, September/October, 1994, pp. 411–23.

154. Van Dam, J., Laruelle, L. and Daenens, P. "Qualitative and Quantitative Determination of Thermal Degradation Products of Hydraulic Fluids Type C," *Journal of Fire Sciences*, Vol. 12, July/August, 1994, pp. 376–87.

155. Totten, G.E., Cerf, J., Bishop, R.J. and Webster, G.M. "Recent Results of Biodegradability and Toxicology Studies of Water–Glycol Hydraulic Fluids," *SAE Technical Paper Series*, Paper 972744, 1997.

156. Union Carbide "UCON Hydrolube HP-5046 for Forestry Applications," Technical Bulletin, UC-1737 3M 4/98.

157. Canada's Environmental Choice Program, "Environmental Choice Program, Annual Report," 95/96.

158. Canada's Environmental Choice Program, "Environmental Technology Verification (ETV) Program," May, 1997.

159. Environmental Choice Program, Certification Criteria Document, CCD-069, published 1996/11.

160. Onions, R.A. "An Investigation into the Possibility of Using Fire-Resistant Hydraulic Fluids for the Royal Naval Systems," *Performance Testing of Hydraulic Fluids International Symposium*, Tourret, R. and Wright, E.P., eds., Institute of Petroleum; London, 1979, pp. 439–58.

161. Totten, G.E. and Webster, G.M. "Review of Testing Methods for Hydraulic Fluid Flammability," *SAE Technical Paper Series*, Paper 932436, 1993.

16 Water Hydraulics

Kari T. Koskinen and Matti J. Vilenius

CONTENTS

16.1 INTRODUCTION

Although the first water hydraulics applications are already known from 400–500 BC, the first real modern applications are only known from the end of eighteenth century, as we can see from Figure 16.1. After that, for over one hundred years all hydraulic systems used water as a pressure medium. In the early twentieth century hydraulics came into the picture when pressure medium lubricated pumps and motors and oil-resistant seal materials were developed.

In recent years, water hydraulic applications have increased and continue to increase. One reason for this growth is the growing importance of the environmental aspects of hydraulic fluids

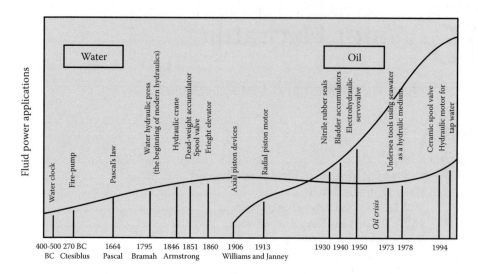

FIGURE 16.1   History of Fluid Power Technology.

throughout the world. Water, as a pressure medium, offers many other benefits compared to oil, including easy maintenance, good availability, low cost, non-flammability, and so forth. One other springboard for the development of water hydraulics has been recent developments in material technology. Many new materials have been developed which provide longer life expectancies with water without corrosion and erosion.

Applications in water hydraulics have traditionally been in the mining industry, steel industry, aluminium industry, automotive industry, and in water treatment plants. Examples of new application areas include rescue equipment, food industry, pulp and paper industries, and land and marine diesel engines [11]. Exotic new applications that are being developed include municipal engineering and fusion reactor remote handling equipment. Low-pressure water hydraulics has also been a very interesting area during last few years.

When thinking of the lifecycle of hydraulic oil, all of the investments related to drilling the oil, transportation, refining, marketing, delivery, use and disposal must be considered. When comparing the supply of oil and water to each other, the difference is quite essential. Water is easily available; for example, from sea water it is relatively easy to produce water for water hydraulic systems. One important global factor to consider is the packaging for oil products, which need extra energy.

The quality of water being used in hydraulic system needs careful consideration. Microbiological growth is a phenomenon that causes extra costs for maintenance and service. On the other hand, storage of water is much easier and normally, when tap water is used, no storage costs are generated. In addition to this, water does not wear out, so the fluid replacement will be less frequent than with oil.

The energy efficiency of machines and systems will also be a very important aspect in the future. The consumption of energy during system building, system use, and system disposal has to be minimized. So the life cycle of the whole system will have to be considered [6].

When thinking about water hydraulics, the situation is versatile. Due to the material requirements of components, the system building costs are higher than oil hydraulics at present. Some reasons for that include, for example, base materials and very small production quantities. In high-pressure water hydraulics the price difference can be significant. However, in low pressures the price difference is decreasing. The energy consumption per system can be evaluated as being higher, because the materials used require longer machining times and more complicated processing that in oil hydraulics.

16.2   WATER AS A PRESSURE MEDIUM IN FLUID POWER SYSTEMS

16.2.1   WATER HYDRAULIC FLUIDS [15]

When water hydraulic applications are discussed, they typically include tap water hydraulic replacements for current HFA-fluid (Hydraulic Fluid A) or HFC-fluid (Hydraulic Fluid C) applications. Seawater is not commonly used except in special offshore applications. HFA-fluids contain typically 95% water and 5% concentrate. Concentrate normally includes emulsification additives and wear inhibitors. HFA-fluids are also sometimes called HWCF- (High Water Content Fluid) or HWBF-fluids (High Water Based Fluid). HFC-fluids normally consist of 40% water and 60% glycol, additives, and thickener. The use of HFC-fluids has decreased in recent years. One reason might be the cost. HFB-fluids (Hydraulic Fluid B) have been used, but their use has also decreased.

Tap water is, of course, the genuine form of water hydraulic fluid. It is inexpensive, non-toxic, non-flammable and has very good availability. Water is clean, so storage and disposal are inexpensive and simple. The problems with water relate to component materials, which must be resistant to corrosion, erosion, cavitation, and so forth. Because the objective is that almost all water hydraulic systems are tap water systems, this chapter will concentrate on such systems.

Water is a very efficient pressure transmitter in a hydraulic system. Due to low compressibility and viscosity, a water hydraulic system has more energy inside, relative to an equivalent oil hydraulic system. Therefore, the dynamics are principally faster than in oil hydraulic systems. On the other hand, poorer lubrication and additional seals increase Coulomb friction, thus worsening the dynamic properties.

16.2.1.1   Physical Properties of Water [5]

The properties of water differ considerably from those of mineral oil. The viscosity, density, and bulk modulus of water and typical mineral oil have been determined. A typical mineral oil is ISO VG 32 (ESSO Univis 32). In Figure 16.2 the temperature dependency of the properties of water and typical mineral oil are shown. The change in density of mineral oil may be calculated from an empirical formula:

$$\rho_t = \rho_{15°C} \frac{1}{1 + \overline{\gamma}(t - 15)}, \tag{16.1}$$

where the thermal volume coefficient is

$$\overline{\gamma} = \frac{\Delta V_P}{V_0} \frac{1}{\Delta t_P}. \tag{16.2}$$

Value $\overline{\gamma} \approx 65 \cdot 10^{-5}$ 1/°C is generally accepted for mineral oils. The change of bulk modulus of mineral oil can be calculated from the following empirical formula:

$$\beta = 1000 \left(1.78 + 7 \left(\frac{\rho_0}{1000} - 0.86 \right) e^{\left(10 \left(0.0025(20 - T) \right) \right)} \right), \tag{16.3}$$

where ρ_0 is the density of mineral oil at $T = 20°C$. Equation 16.3 is valid when $20°C < T < 70°C$. The pressure dependence of water and typical mineral oil is shown in Figure 16.3. The viscosity change of mineral oil can be calculated from the following empirical formula:

$$\nu_p = \nu_{p-1} e^{0.0017(p_p - p_{p-1})}. \tag{16.4}$$

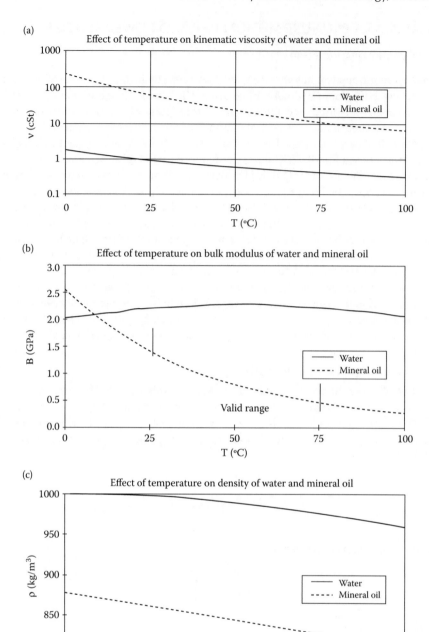

FIGURE 16.2 The effect of temperature on viscosity, density, and bulk modulus of water and typical mineral oil.

The bulk modulus of water can be calculated from the change of relative density:

$$\beta_p = \frac{\rho_p}{\rho_0} \beta_0. \tag{16.5}$$

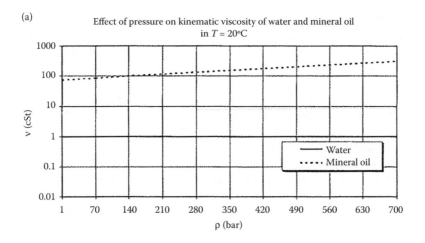

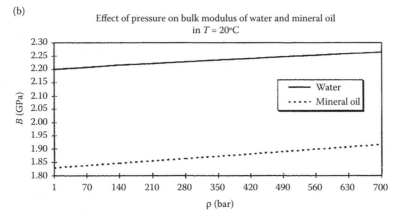

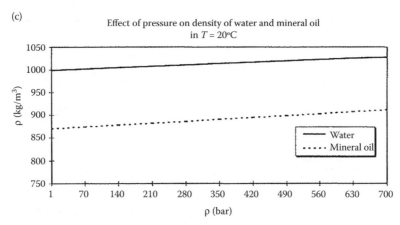

FIGURE 16.3 The effect of pressure on viscosity, density and bulk modulus of water and typical mineral oil.

The density change of mineral oil can be calculated from the following empirical formula:

$$\rho_p = \rho_{p0}\left(\frac{1}{1-65\cdot10^{-11}\,p\,10000}\right). \tag{16.6}$$

It is evident that the kinematic viscosity decreases with increasing temperature both with water and mineral oil. However, the decrease of the kinematic viscosity of mineral oil is much greater than water over the entire temperature range. For example, at 40°C the kinematic viscosity of water is approximately 0.7 cSt and the kinematic viscosity of mineral oil is approximately 32 cSt, so the ratio is approximately 1:46.

The variation of bulk modulus is also greater for mineral oil than for water. The bulk modulus of water is almost constant between 0°C and 100°C. On the other hand, the bulk modulus of mineral oil decreases quite strongly with temperature rise. Both bulk moduli are almost the same at 15°C, but at 50°C the difference is about 1.4 GPa.

The effect of temperature on the densities of water and mineral oil is almost the same. The density of water is about 150 kg/m³ greater than the density of mineral oil across the entire temperature range.

The effect of pressure on the kinematic viscosity, the bulk modulus and density of water, and of a typical mineral oil is presented in Figure 16.3. The pressure range selected is from 1 bar to 700 bar, because most normal water hydraulic systems use pressures in this range. The temperature is 20°C. Figure 16.3 shows that the kinematic viscosity of water is not affected by system pressure. However, the kinematic viscosity of mineral oil increases with the system pressure. This is a great benefit for water, especially in the high-pressure applications. The effect of pressure on the bulk moduli of water and mineral oil is about the same. They both increase with pressure, but the increase of water bulk modulus is slightly smaller. Also, the effect of pressure on the density of water is slightly less than with mineral oil.

The change of vapor pressure of water and mineral oil with temperature is shown in Figure 16.4. This data shows that the vapor pressure of water changes strongly with increasing temperature. At the same temperatures, the vapor pressure of mineral oil is nearly constant. This illustrates the cavitation sensitivity of water compared to mineral oil. The thermal conductivity of a fluid essentially

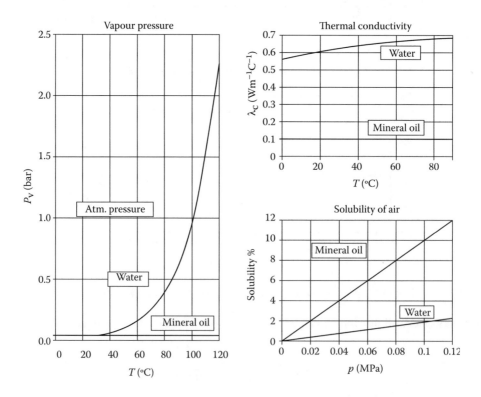

FIGURE 16.4 Vapor pressure, thermal conductivity, and solubility of water and typical mineral oil.

affects the heat balance of a hydraulic system. The advantage of water relative to mineral oil is illustrated in Figure 16.4. The thermal conductivity of water is about five to six times greater than the thermal conductivity of oil. In addition, the thermal conductivity of water increases slightly with increasing temperature.

The solubility of air in a fluid affects its compressibility, which may, in some cases, be significant for the operation of the system. The lower solubility of air in water relative to mineral oil as a function of pressure is illustrated in Figure 16.4.

From Figures 16.2 to 16.4 it is clear that water and mineral oil exhibit significantly different physical properties. These differences affect the system performance in different ways, depending on the application and operation conditions. In this context, it is also worth considering that water does not always exhibit the same properties because the hardness of water can vary in calcium and magnesium content. The acidity of water can also vary and the bacteria level in water also affects the performance of water in a system.

16.2.1.2 Chemical Properties of Water [14]

The chemical properties of water vary with the source and geographic location. The European Union has given the directive *Water Quality Standards in EU, 80/778/EEC*, where the requirements for drinking water are described. Some component manufacturers have based their component application limits on the chemical properties of water provided in this directive. In the following the most significant chemical parameters of water used as a pressure medium are listed:

- Hydrogen-ion concentration
- Chloride-content
- Hardness
- Bacteria and other microbiological organisms
- Solid-particle content

All of these values should be measured to ensure the quality of water. However, in practice, the measurement of all of these values may be expensive and time consuming. Monitoring the water quality should be continuous, which is difficult to arrange in many applications. Good and simple measurement and monitoring methods providing reliable information of the water, but which are still easy to apply to all systems, must be developed. In addition, general methods for cleaning the system must be developed. Because the requirements may be component dependent, this will increase the difficulty of identifying generally accepted purification methods.

16.2.2 STEADY STATE PIPE FLOW

The steady state pipe flow is the simplest parameter to identify differences between water and mineral oil as a hydraulic fluid. A smooth-surface pipe, 1 m long and 10 mm in diameter, is illustrated in Figure 16.5. When we use the values presented in Table 16.1 for the physical properties of water and mineral oil, the Reynolds number and pressure drop curves for water and typical mineral oil illustrated in Figures 16.6 through 16.8 are obtained.

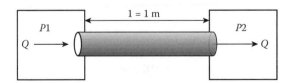

FIGURE 16.5 Smooth surface pipe.

TABLE 16.1
Densities and Viscosities

Water ($T = 40°C$)

   Density = 992 kg/m³

   Viscosity = 0.7 cSt

Mineral Oil ($T = 40°C$)

   Density = 870 kg/m³

   Viscosity = 32 cSt

Flow in a pipe changes gradually from laminar to turbulent across a transition region. This region contains the critical Reynolds number Re_{cr}. Normally Re_{cr} can be assumed to be in the range 1500–2500. The Reynolds number can be calculated from:

$$Re = \frac{v\, d_h}{v}, \tag{16.7}$$

where d_h is hydraulic diameter

$$d_h = \frac{4A}{P}, \tag{16.8}$$

For a pipe, the hydraulic diameter $d_h = d$. The flow velocity in a pipe can be calculated from:

$$v = \frac{Q}{A} = \frac{4Q}{\pi d^2}. \tag{16.9}$$

We can observe from Figure 16.6 that the Reynolds numbers of water and mineral oil have totally different magnitudes. The Reynolds number of water is about fifty times greater than the Reynolds number of mineral oil across the whole of the flow range presented. This means that the flow of water is, in practice, always fully turbulent, because the transition area from laminar to turbulent is somewhere between the values 0 l/min and 1 l/min. The Reynolds number of mineral oil is with small flows laminar and the transition area from laminar to turbulent flow is somewhere between the values 30 l/min and 60 l/min. The flow velocity curve is also presented in Figure 16.6 at the right hand side Y-axis. By studying the Reynolds numbers and flow velocity, it is evident that the water hydraulic

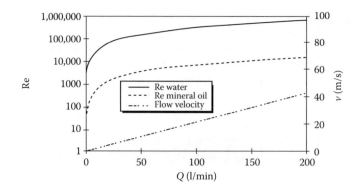

FIGURE 16.6   Reynolds numbers and flow velocity in smooth pipe.

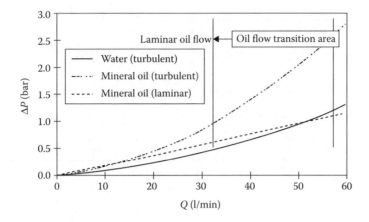

FIGURE 16.7 Mineral oil pipe flow transition area.

system has greater internal flow energy, even with the same flow velocity, compared to mineral oil flow. How that shows in pressure drop in the pipe is examined in Figure 16.7, where the pressure drop of laminar and turbulent mineral oil and turbulent water flow is presented for lower flows.

Figure 16.7 shows that pressure drop in pipe flow is always smaller with water than with mineral oil, even in the case when the mineral oil flow is laminar. The difference is not large with smaller flows, but as the flow increases, the two curves diverge even further. The transition area for mineral oil is also presented in Figure 16.7. This occurs between 30 l/min and 60 l/min, when the limit values are 2000 and 4000 as Reynolds numbers are used. Therefore, when modeling and simulating systems where the flows are in the right area and the small pressure drops in pipe flow are significant, the location of transition area must be taken into account. In Figure 16.8, the pressure drop in turbulent flow with water and mineral oil is provided for flows of up to 200 l/min. These data show that the benefits of using water are greater with larger flows.

The effect of the roughness of the pipe surface is presented in Figure 16.9. The relative roughness R_u is in X-axis and the loss head coefficient λ is presented in Y-axis. Figure 16.9 shows a completely smooth pipe with very small relative roughness; the difference in loss of head coefficients increases with increasing values of relative roughness. The difference decreases with increasing roughness. For small roughness values, the use of water instead of oil is more efficient than in very rough pipes.

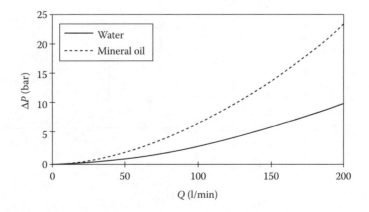

FIGURE 16.8 Pressure drop in turbulent pipe flow.

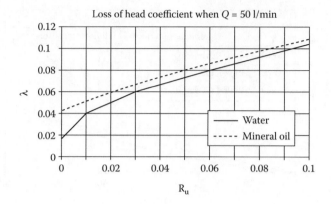

FIGURE 16.9   The effect of pipe surface roughness on loss of head coefficient.

16.2.3   Leakage Flow

Low-viscosity leakage is a greater problem in water hydraulic components than in oil hydraulic components. Leakage flow is significant; for example, in displacement pumps and control valves. In control valves, clearances are normally between 1 μm and 100 μm. Laminar leakage flow in an annular clearance is usually calculated from the equation:

$$Q = \frac{\left[1+1.5\left(\dfrac{\varepsilon}{h}\right)\right]\pi Dh^3}{12\rho v L}\,\Delta p. \tag{16.10}$$

The annular clearance between the piston and cylinder is illustrated in Figure 16.10. Also illustrated in Figure 16.10 is the leakage flow with water and mineral oil as a function of clearance width. Both flows have been assumed to be laminar. The leakage flow with water is much greater than with

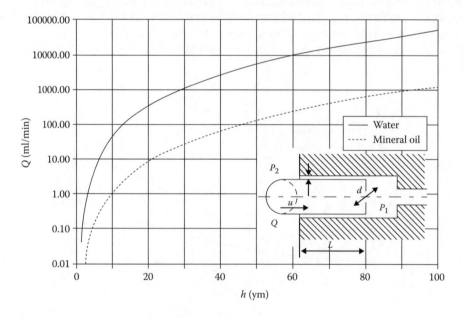

FIGURE 16.10   Laminar leakage flow in annular clearance.

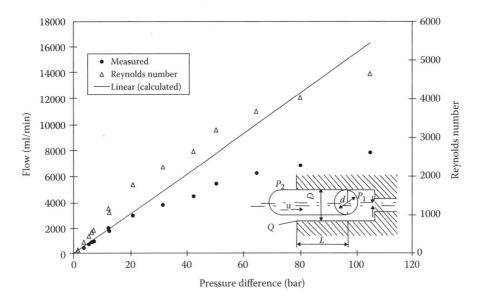

FIGURE 16.11 Leakage flow in totally eccentric annular clearance with sea water.

mineral oil. For the same leakage flow, (e.g., 10 ml/min), the clearance may be about 20 µm with oil but only 5 µm with water. The difference increases with increasing clearances.

On the other hand, some differences between laminar theory and practice may be observed. The flow-pressure curves and Reynolds number curve of a fully eccentric piston in an annular clearance is presented in Figure 16.11. The measured and calculated curves are similar for small pressure differences (small Reynolds numbers), but the difference between measured and calculated curves increases with increasing flows. (The measurements in this case were carried out with seawater.) The reason for this behavior is that the flow is turbulent with larger Reynolds numbers. The accuracy of the flow model can be improved by accounting for the annulus inflow and outflow effects.

Pressure losses within an annulus are composed of pressure losses on both ends and in the annular duct, as illustrated in Figure 16.12. The pressure loss within the annulus can be derived according to laminar or turbulent flow. For laminar flow:

$$\Delta p = \zeta \rho \frac{v^2}{2}. \tag{16.11}$$

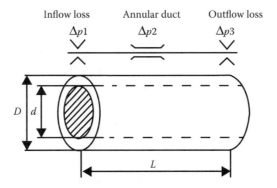

FIGURE 16.12 Pressure losses in an annular duct.

For calculating pressure loss due to a sudden change of the flow cross-section area, flow model 1 may be derived; it is presented in Equations 16.12, 16.13, and 16.14.

$$Q = \frac{k_2}{2}\left[\sqrt{\left(\frac{k_2}{k_1}\right)^2 + 4\Delta p} - \frac{k_2}{k_1}\right],$$

(16.12)

$$k_1 = \frac{\left[1 + 1.5\left(\frac{\varepsilon}{h}\right)\right]\pi D h^3}{12\rho v L},$$

(16.13)

$$k_2 = \frac{1}{\zeta}\pi D h\sqrt{\frac{2}{\rho}}.$$

(16.14)

Pressure-loss coefficients for different cases have been presented in [10]. The coefficient for sharp-edged inflow is 0.5 and the coefficient for outflow is 1.0. These can be combined to one pressure loss coefficient $\zeta = 1.5$.

For turbulent flow of a Newtonian fluid, the annular is considered as a pipe with diameter $2h$. Pressure loss in a pipe can be solved from equation

$$\Delta p = \xi\rho\frac{v^2}{2}\frac{L}{d_h}.$$

(16.15)

The friction factor ξ is a function of the Reynolds number and the relative surface roughness. For a smooth surface, empirical friction equations have been reported. [13] provides a method for the determination of friction factor for an annulus with fine clearance. The equation is valid in the range $2\times10^3 \le \text{Re} \le 2\times10^4$. The influence of piston eccentricity has also been studied and can be calculated from the equation:

$$\xi = \frac{0.316}{\text{Re}^{0.21}}\frac{1}{1 + 0.2\left(\frac{\varepsilon}{h}\right)}.$$

(16.16)

For turbulent flow, the flow model 2 may be derived by combining Equations 16.11, 16.15 and 16.16:

$$Q = A\sqrt{\frac{2\Delta p}{\rho}\left(\frac{d_h}{\xi L} + \frac{1}{\xi^2}\right)},$$

(16.17)

$$\xi = \frac{0.316}{\text{Re}^{0.21}}\frac{1}{1 + 0.2\left(\frac{\varepsilon}{h}\right)}.$$

(16.18)

The accuracy of flow models 1 and 2 are verifiable with laboratory measurements. The measurement and verification results are presented in Reference 2. The results show that the effect of inflow and outflow is very important in a short annulus. The radial elasticity of the housing has to be taken into account when considering small clearances. The conventional laminar pipe model for flow in an annulus is valid for short ducts when the Reynolds number is less than 1000. When the Reynolds number is >1000, flow is clearly turbulent.

Flow model 2 is a combination of a turbulent entrance loss and a turbulent pipe-loss model. It is valid for a high-Reynolds-number flow only because flow is laminar with a low Reynolds number. Modeling accuracy could be improved if the relative surface roughness was accurately known, which is not usually the case.

Flow model 1 is a combination of a turbulent entrance loss and a laminar pipe flow model. It provides the same result as the conventional pipe model for low-Reynolds-number flow and with satisfactory accuracy at high-Reynolds-number flow. Therefore, flow model 1 is suitable for modeling annular clearances with various Newtonian fluids.

From these relationships, it can be concluded that clearances must be smaller in order to achieve good efficiency in water hydraulic systems. The smaller clearances are more expensive to manufacture and increase the prices of water hydraulic components. However, the leakage flow in small clearances with water may also be turbulent with big pressure differences; in that case, the actual leakage flow is smaller than predicted by the laminar leakage flow equation. This must be considered when studying, for example, water hydraulic valves and pumps.

16.2.4 PRESSURE TRANSIENTS

Pressure transients can complicate the performance of hydraulic systems. Due to pressure transients, the system pressure rises momentarily far above the normal operation pressure, which may result in component damage. Pressure transients originate from sudden change in flow velocity. A sudden valve opening and closing and sudden actuator stops generally cause pressure transients. In water hydraulics, the importance of pressure transients is emphasized because the greater density and the greater bulk modulus compared to mineral oil means that the fluid is less compressive and the resulting velocity change results in a stronger pressure rise.

The results of computer simulation of hydraulic system pressure transients are illustrated in Figure 16.13. The symmetrical cylinder is controlled by a 4/3-valve. When the cylinder piston moves to the right in the central position, the valve spool is changed from position 1 to central

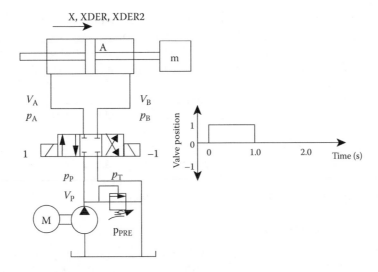

FIGURE 16.13 Simple hydraulic cylinder control system.

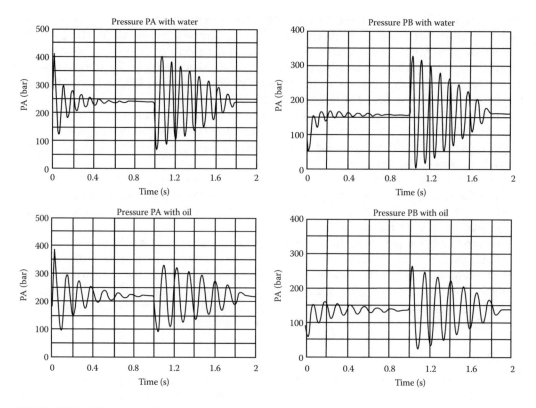

FIGURE 16.14 Simulated cylinder chamber pressures with water and oil.

position, at which point pressure transients occur in cylinder chambers. When the system is simulated using the same components and parameters with water and mineral oil, the effect of the fluid on the pressure transients can be approximated. The valve spool is controlled as presented in Figure 16.13. The pump output flow is modelled with a constant value. The pressure relief valve is modelled with first order model, but its behavior is not shown in the results because only cylinder pressures are examined.

In Figure 16.14 the pressures of the cylinder chambers PA and PB with water and oil are presented. When the piston moves to the right, the pressure PA is greater with water because of smaller flow resistance. In addition, pressure vibration in a water system exhibits a larger amplitude and frequency than in an oil system. Vibration is dampened slightly faster in the water system because of the larger internal leakage in the cylinder. For the water system, the pressure PB falls to zero almost instantly, whereas in the oil system, the minimum is at about 25 bar. This is one example of the cavitation sensitivity of a water hydraulic system, which may also result from excessive pressure transients.

It can be concluded that pressure transients and vibrations exhibit different behaviors in water hydraulic systems compared to oil hydraulic systems. However, the pressure transients are not an insuperable problem in water hydraulics. The generation of pressure transients can be avoided or minimized by the proper design of components and systems.

16.2.5 CAVITATION AND EROSION

One of the most difficult phenomena to research in hydraulic systems is cavitation. The role of cavitation in water hydraulic systems is much more significant than in oil systems. Cavitation may occur in pumps, valves and actuators.

Cavitation may be either gaseous cavitation or vaporous cavitation. Gaseous cavitation is caused by a rapid reduction of pressure, resulting in the release of dissolved air or gas. However, gaseous cavitation is less likely to occur in water hydraulics because the density and bulk modulus of water are greater than with oil. Also, the amount of dissolved air in water is less than that in oil at the same pressure. However, vaporous cavitation is far more likely to occur in water hydraulics than in oil hydraulics. Vaporous cavitation occurs when the pressure instantly drops to the fluid's vapor pressure. At this point, cavities will develop, and as they move downstream into a higher-pressure regime, they collapse, producing noise and vibration. Because the vapor pressure of water is higher than the vapor pressure of mineral oil, especially at higher temperatures, it is obvious that water is more susceptible to cavitation than is mineral oil.

When considering, for example, a simple orifice, cavitation may occur downstream of the restrictor due to the high velocity of the fluid just after the restrictor. The flow through the orifice with water is greater than with oil when the pressure difference over the orifice is the same. This means that the flow velocity is higher with water just after the orifice. The non-dimensional, cavitation index, K, for control valves gives the sensitivity of cavitation. K is defined as

$$K = \frac{P_1 - P_3}{P_1 - P_v}, \tag{16.19}$$

where:
P_1 = the supply pressure
P_3 = the system back pressure
P_v = the vapor pressure of the fluid

From this equation, it is evident that system back pressure dominates the cavitation index. When comparing water and mineral oil, it can be concluded that the term P_v is greater with water than with oil, especially at higher temperatures, and critical value K_c is greater with water at the same pressure difference. In Figure 16.15, the typical relationship between the cavitation index and fluid-borne noise is shown.

Erosion caused by cavitation is strongly dependent on the materials and on the design of components. One possibility for the reduction of cavitation is the use of cascading restrictors. With these, the pressure drop across one restrictor is reduced and the cavitation index stays low. This method is commonly used for valves for use in water hydraulic systems. One example is the poppet valve shown in Figure 16.16.

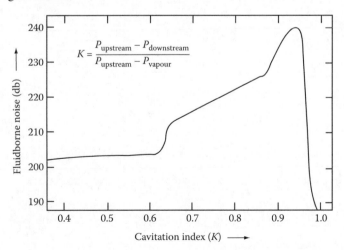

FIGURE 16.15 Typical relationship between cavitation index and fluid-borne noise.

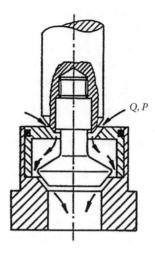

FIGURE 16.16   Cavitation reduction with double-seat construction in a poppet valve.

16.2.5.1   Cavitation in Seat Valves

A valve is a basic component in a hydraulic system. Its role is to control flow from one volume to another. A seat valve is a good construction for water hydraulics, because low viscosity makes it essential to avoid leakage gaps in the valve structure. Seat valves are easy and cost-effective to manufacture to meet this requirement.

The flow rate, pressure distribution, and cavitation occurrence in a seat valve can be studied by experimental measurements and Computational Fluid Dynamics (CFD) [9]. The experience shows that the appearance location and path of traveling vapor produced by the cavitation phenomenon can be located with both methods. The place where the cavitation has originated from may be very minor and the process of the bubble growth is unsteady. This makes the pressure measurement challenging and, in addition, increases the risk of cavitation errors. For instance an absolutely sharp corner of the seat in the model causes the results not to show local pressure drop around the corner, whereas any chamfer on the corner brings it out. Figure 16.17 represents the time dependency of

FIGURE 16.17   Unsteady situation of cavitating flow.

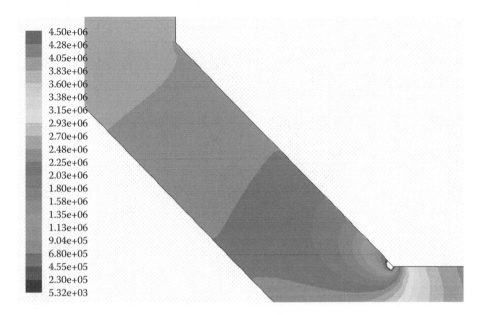

4.50e+06
4.28e+06
4.05e+06
3.83e+06
3.60e+06
3.38e+06
3.15e+06
2.93e+06
2.70e+06
2.48e+06
2.25e+06
2.03e+06
1.80e+06
1.58e+06
1.35e+06
1.13e+06
9.04e+05
6.80e+05
4.55e+05
2.30e+05
5.32e+03

FIGURE 16.18 Computational pressure distribution in chamfered orifice (Pa).

the vapor cloud behavior and the very local origin where the vapor is produced. At one moment the chamber can be full of vapor, while at another it can be full of liquid.

The differences in the flow characteristics are diverse. Cavitation occurs with the smallest pressure drop over the orifice in valve with chamfered seat. On the other hand, the discharge coefficient of chamfered valve structure is bigger than that of sharp valve structure, which leads to the conclusion that the cavitation sensitivity between these two valve types depends more on flow rate than just on pressure drop. The cavitation affects the chamfered seat valve by lowering its discharge coefficient, but the performance of the valve with the sharp seat endures despite of the vapor in the orifice. The shape of the poppet in a seat valve mainly affects the direction and speed behaviors of the vapor jet in the downstream side. This is significant when considering the risk of damage caused by cavitation if it cannot be fully eliminated by the shape optimization.

A computational investigation of pressure distribution in the valve orifice shows the benefits of a valve with multistage pressure drop. It reduces the risk of cavitation, but chamfer on a single corner of stage is still increasing it creating a local low-pressure area. Figure 16.18 shows the computational pressure distribution in a case of chamfered seat edge and Figure 16.19 shows pressure distribution with correspondingly defined pressure boundaries when there are two edges instead of continuous chamfer. Clearly, there is no such local small-pressure area if the continuous chamfer surface is replaced with the separate edges. Thus, cavitation appearance is less probable in the latter case.

16.2.4 Water Quality Aspects [12]

The growth and attachment of bacteria in water hydraulic systems is mainly a function of the nutrition concentration in the pressure medium. Hydraulic parameters and system materials have a slight effect on growth and attachment, but the best way to avoid problems with microbes is to maintain nutrition levels on a par with tap water (DOC 2-4 mg/l).

Lowering the nutrient concentration and increasing the system pressure have been reported as the most effective ways for controlling total bacterial numbers and counts of viable bacteria in the pressure medium. Low reductions have been obtained by adjusting the fluid flow velocity in the system. However, an increase in pressure resulted in enhanced microbial attachment on the reservoir, as shown in Figure 16.20.

FIGURE 16.19   Computational pressure distribution in two-step orifice (Pa).

Experiments showed that biological or physical contamination can be reduced to acceptable levels in water to secure machine operation when they are not simultaneously present in significant amounts. Based on the experiments, it is difficult to determine the precise combined effect of particles and biofilm on a filter. Parallel experiments might contribute to statistical analyses, but statistical analyses do not work well in this case due to the nature of microbial growth. For example, enhanced microbial growth causes problems with the determination of β (microbe)-value because of its bioreactor effect in the filter (Figure 16.21). That is, biologically active material, together with water and a suitable temperature, promotes an environment for microbes to reproduce in a filter cartridge.

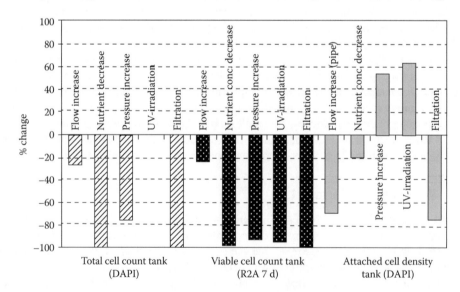

FIGURE 16.20   Changes in number of microbes in water hydraulic systems due to adjusted operational parameters and control treatments. (From Jay D., Rantanen O., and Koskinen K.T. "Diesel Engine - Combustion Control with High Pressure Water Injection," The Fourth Scandinavian International Conference on Fluid Power, Tampere, Finland, September 27–29, 1995.)

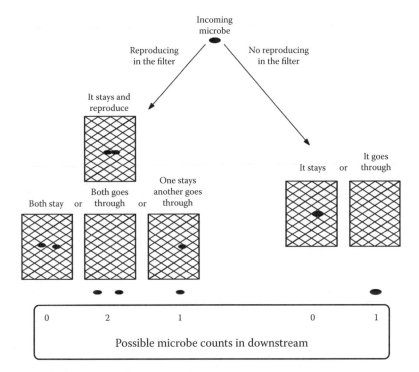

FIGURE 16.21 Bioreactor effect on a filter cartridge.

Because of the incomplete knowledge of both physical and biological system parameters, the working conditions of filters cannot be accurately predicted. Preventive maintenance requires more detailed data about hydraulic systems than can be achieved even using pilot-scale systems in laboratory conditions.

The best and only way to deal with filtering problems caused by both microbial growth and particles (Figure 16.22) is to use two-phased filtration and make sure that the system has minimum

FIGURE 16.22 Particle and microbe contamination on filter fibres.

interface with the surrounding environment. A large filtering area ensures a better long-term operation of filters, because abrupt malfunctions are more likely to happen with a filter system with a low dirt-holding capacity.

16.3 WATER HYDRAULIC COMPONENTS

16.3.1 PUMPS

New materials and design innovations have recently increased the types of water hydraulic pumps that are available. Axial and radial piston pumps, and more traditional plunger-, centrifugal- and piston-type pumps are now obtainable. A water hydraulic axial piston pump is shown in Figure 16.23. Axial piston pumps are the newest pumps on the market. There are many newly developed materials used in these pumps, including reinforced plastics, over-molded stainless steel, and special seals. These pumps are normally designed hydraulic systems with a maximum pressure of 140 bar. All parts of these pumps are water lubricated, which means that no oil is used inside the pump.

A water hydraulic triplex-piston pump is illustrated in Figure 16.24. These pumps have been used for many years in the mining and steel industries for high pressure generation in various applications. The three pistons are driven by an oil-lubricated crankshaft mechanism. The oil side and water side are totally separated from each other to prevent the fluids from mixing. In the water side, all the parts are made of corrosion-resistant materials such as stainless steel or industrial ceramics. The pumping unit is a normal piston with suction and pressure valves. The valves are normally seat valves. The typical flow rate for this kind of pump is up to 700 l/min and pressure up to 800 bar.

One possibility of producing high-pressure water is to use a pressure intensifier such as that illustrated in Figure 16.25. The pump unit is driven by oil pressure, which is supplied to the central part of the intensifier [4]. The water pistons are located at both ends and water is pumped in upon every stroke. When one end is pumping, the other end is priming. The change of direction is automatically controlled by the oil valve, which is integrated to the oil piston.

When using a pressure intensifier, oil hydraulic power is needed. In some applications, this might lead to problems and higher costs, but there are also many applications where oil hydraulic power is

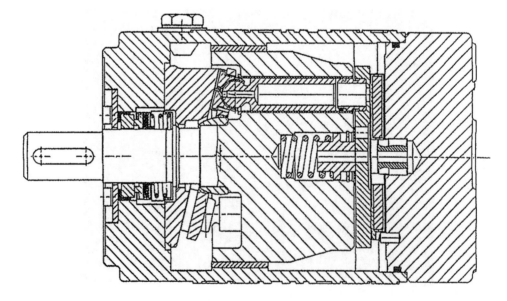

FIGURE 16.23   Water hydraulic axial piston pump. (Danfoss A/S Nordborgvej 81, Nordborg, Denmark.)

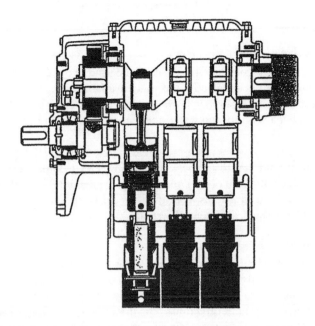

FIGURE 16.24 Water hydraulic triplex pump. (Hauhinco Machinenfabrik G. Hausherr, Jochums GmbH & Co. KG, Beisenbuchstratsse 10, Sprochövel, Germany.)

already present for other reasons. In such cases the pressure intensifier may be economically acceptable. Pressure intensifiers have been traditionally used in water-jet cutting systems, where the pressures may be as high as 4000 bar. The intensifier shown in Figure 16.25 is typically designed for 200–800 bar.

In low-pressure applications, where the pressure is between 10 and 50 bar, a centrifugal pump is a potential alternative. Centrifugal pumps are simple and reliable, and with multi-stage pumps, 30 bar pressure can be easily achieved. The centrifugal pump also simplifies the hydraulic circuit as no pressure relief valve is needed: The pump cannot increase the pressure beyond its capacity. However, low-pressure water hydraulics is still a relatively new idea and there are not many examples of the use of centrifugal pumps in water hydraulic systems.

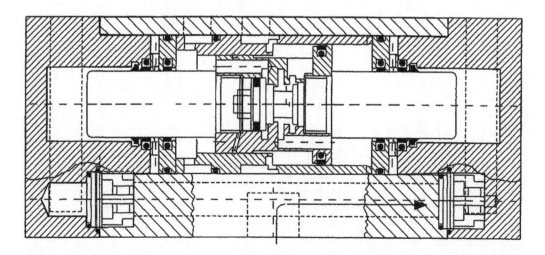

FIGURE 16.25 Water hydraulic pressure intensifier. (Dynaset Oy, Menotie 3, Ylöjärvi, Finland.)

16.3.2 Control Valves

16.3.2.1 On-Off Valves

On-off valves are the simplest control valves that can be used in cylinder position control. There are a few different on-off valves suitable for tap water available in the market. Traditionally, ball seat valves have been used with smaller flows and piloted seat valves in larger flow systems. Spool valves have not been used for tap water until recently, when the ceramic spool valve was introduced.

One new on-off valve for tap water use is presented in Figure 16.26 [1]. The valve is electrically activated, which includes built-in pressure peak dampening. The maximum pressure is 140 bar and the maximum flow is either 30 l/min or 60 l/min. The valve opening time with maximum pressure is 70 ms and closing time is about 350 ms. On-off valves can be used for the positioning of a water hydraulic cylinder with a sophisticated control system.

16.3.2.2 Flow-Control Valves

The easiest way to control cylinder speed is to use flow-control valves. The speed of a hydraulic cylinder can be controlled easily with a pressure-compensated flow-control valve. This is especially true in applications where the speed is the same in every working cycle—the correct flow can be adjusted beforehand for both directions. The direction control can be handled with separate arrangements, such as on-off valves. However, there are only a limited number of pressure-compensated flow-control valves suitable for use with tap water. One example is provided in Figure 16.27.

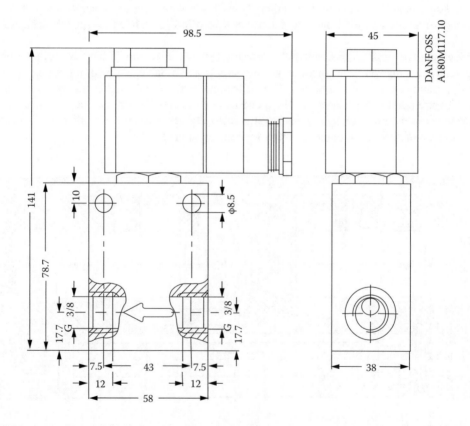

FIGURE 16.26 Electrically operated on-off valve. (Danfoss A/S Nordborgvej 81, Nordborg, Denmark.)

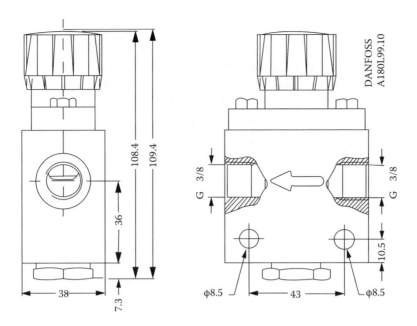

FIGURE 16.27 Manually operated flow-control valve. (Danfoss A/S Nordborgvej 81, Nordborg, Denmark.)

16.3.2.3 Proportional Valves

Ball seat valves, as shown in Figure 16.28 are quite often used as proportional tap water valves. Ball seat has the advantage of small leakage and reliable operation. The manufacturing is also relatively easy as parts can be designed to be very simple and advanced materials can be used. The balls, for example, may be constructed from stainless steel or industrial ceramics. Ball seat valves are available in both 2/2-way and 3/2-way versions. In a position servo system, the 4/3 function can be established with two pieces of 3/2-way valves or four pieces of 2/2-way valves.

The spool valve illustrated in Figure 16.29 is one of the newest developments in the range of water hydraulic proportional valves. There are few spool valves available that can be used with tap water, although spool valves possess many advantages over other valve types. When used with water, the main problems in spool valves are high leakage, high flow forces, cavitation, and erosion caused by the high flow velocity. The spool and the housing sleeve can be constructed

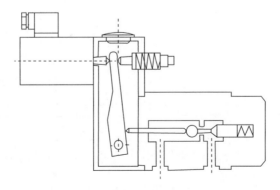

FIGURE 16.28 Proportional ball seat valve. (Hauhinco Machinenfabrik G. Hausherr, Jochums GmbH & Co. KG, Beisenbuchstrasse 10, Sprochövel, Germany.)

FIGURE 16.29   Proportional spool valve. (Sitek-Palvelu Oy, Laukaantie 4, Jyväskylä, Finland.)

from industrial ceramics or stainless steel. In order to achieve high pressure, it is necessary to have low leakage rates, long lifetimes, and very fine manufacturing tolerances. The valve presented in Figure 16.29 is designed for low-pressure use (under 40 bar) and it is based on originally pneumatic spool and sleeve.

16.3.2.4   Servo-Valves

Servo valves which are especially designed for tap water are (at the moment) difficult to obtain due to a lack of availability. Some manufacturers have stainless steel versions of their normal oil hydraulic servo valves. The characteristics of these valves are guaranteed for HFA-fluids but are not necessary for tap water. Although these valves operate properly with tap water, their lifetimes are unknown, because there is relatively little application experience. One of the sophisticated developments for water use is the valve based on a smart idea of using hydrostatic spool support on servo-valve spool (Figure 16.30) [15]. Using this arrangement, the friction forces have been minimized and good linearity has been achieved.

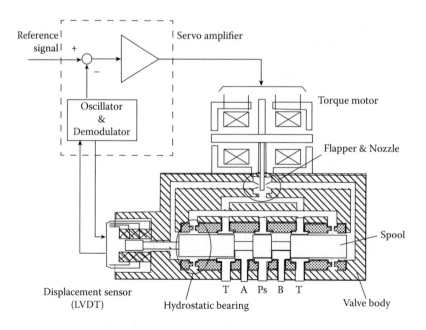

FIGURE 16.30   Water hydraulic servo-valve (Ebara).

16.3.3 ACTUATORS

16.3.3.1 Cylinders

The variety of reliable water hydraulic cylinders is still quite limited. There are mainly two critical factors in manufacturing a water hydraulic cylinder: materials and seals. The materials must be able to work with water, which is not the best possible lubrication fluid. Different types of stainless steel are most often used in piston, rod, and cylinder tubes. Ceramic materials and different surface coatings may also be used.

The selection of materials in piston and piston-rod seals is also important when designing a water hydraulic cylinder. The seals must exhibit low-friction characteristics or the seal friction of the cylinder will be too high (for example high-accuracy applications). Excessive friction generates heat and increased wear reduces the lifetime of the cylinder. Normal materials used for these seals include Perbunan, PTFE and PEEK.

The use-pressure significantly affects the construction of the cylinder. The present commercial cylinders are designed primarily for pressure levels of 70–160 bar. Higher-pressure cylinders, up to 310 bar, are custom constructed in most cases. When using lower pressures (10–50 bar) the cylinder construction is simplified and more plastic materials may be used. This will significantly affect the component cost. One example of a water hydraulic cylinder is illustrated in Figure 16.31.

16.3.3.2 Motors

The use of tap water for hydraulic motors is still quite rare. Although some axial-piston-type motors have recently become available, there are only a limited number of choices with respect to sizes and operation parameters. Some motors have been specifically designed to use tap water to lubricate

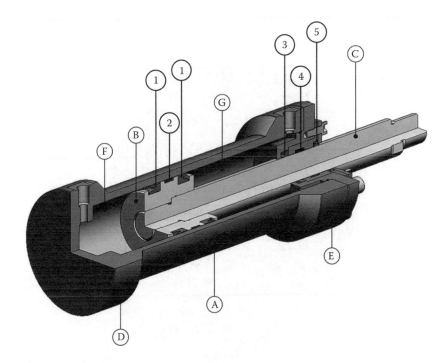

FIGURE 16.31 Water hydraulic cylinder. A, cylinder tube; B, piston; C, piston rod; D, Cylinder head; E, Cylinder head; F, piston chamber; G, piston rod chamber; 1, piston sealing; 2, piston bearing; 3, rod sealing; 4, rod bearing; 5, wiper ring.

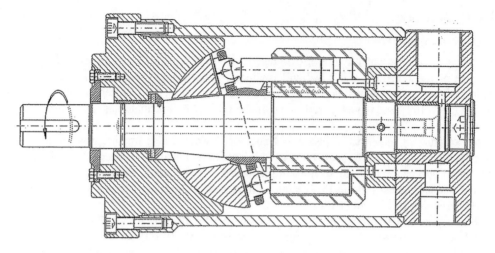

FIGURE 16.32   Water hydraulic axial piston motor. (Hytar Oy, Turjankatu 1, 33100 Tampere, Finland.)

moving parts. To minimize leakage against the inherent low viscosity of water, small clearances between moving parts and longer leakage paths are incorporated.

In Figure 16.32 a recent design of an axial-piston motor is illustrated. The motor has been developed for tap water and seawater use. This means that the materials used are very carefully selected, because seawater is a much more corrosive fluid than tap water. The maximum pressure of the motor is 210 bar.

Another development is the vane motor shown in Figure 16.33. This motor is designed for a maximum pressure of 50 bar and the speed is between 10–200 rpm. The motor is bi-directional and the total efficiency is about 80%.

FIGURE 16.33   Water hydraulic vane motor. (Danfoss A/S Nordborgvej 81, 6430 Nordborg, Denmark.)

16.4 WATER HYDRAULIC APPLICATIONS

16.4.1 ITER Fusion Reactor Maintenance Tools

International Thermonuclear Experimental Reactor (ITER) is an international project with the aim of developing fusion as a clean and sustainable energy source. The ITER machine will be the first fusion device to produce thermal energy at the level of an electricity-producing power station. Its design is based on the "Tokamak" principle, whereby the plasma (mixture of charged particles) is maintained at a very high temperature within a toroidal vacuum chamber surrounded by a magnetic field. A major issue for the successful operation of ITER is the maintenance and/or exchange of in-vessel components by remote handling methods.

The maintenance operations are very demanding due to the weight of the components and the constricted space around them. The operations require high position accuracy, high forces, and compact size from the actuators. Hydraulics meets these requirements. The use of oil hydraulics is not permitted in a fusion reactor because of the risk of reactor contamination due to possible oil leakage and also because of the danger of oil becoming active by the radiation. Water hydraulics is thus the only feasible solution.

Successful accomplishment of the operations requires the utilization of remote operation at the control of the water hydraulic maintenance robots. Several water hydraulic tools have been developed for maintenance operations, such as Cassette Multifunctional Mover (CMM), Second Cassette End Effector (SCEE), and water hydraulic prototype manipulator, which can be seen in Figure 16.34.

16.4.2 Water Injection in Large Diesel Engines

In large diesel engines, high-pressure water is directly injected into the combustion chamber in order to reduce nitrogen oxides (NO_x) [3]. This is necessary because NO_x is the main pollutant coming from the diesel engines. Both the international and local legislations have become more restrictive with respect to NO_x emissions.

The principle of the injection system is shown in Figure 16.35. High-pressure water is supplied by the power pack and injected via a separate needle to the combustion chamber. The pressure level is in the range 250–400 bar. The injection timing and duration is controlled by a separate rail valve, which controls the needle.

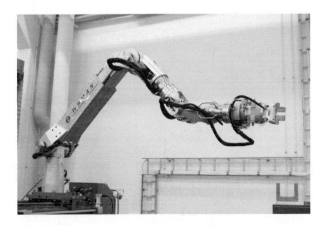

FIGURE 16.34 Water hydraulic prototype manipulator.

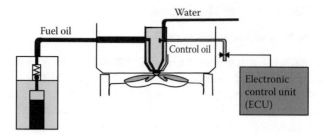

FIGURE 16.35   Water injection to the combustion chamber.

Under laboratory conditions, the water-injection system was tested and measured. The schematic of the hydraulic test system is presented in Figure 16.36. Water flow was generated by a water hydraulic power pack. The pressure level is about 400 bar. The seal oil pressure was generated by a small oil hydraulic power pack. Seal oil was used between the water part and oil part to prohibit the fluids from mixing. The control oil pressure was generated by a small oil hydraulic power pack. The orifice in the water line simulated the flow fuse in test measurements. The actual flow-fuse design parameters were obtained from the flow characteristics of the orifice. The electronic rail valve used was a fast-switching rail valve. The pressures from both sides of the orifice were measured and, in addition, the control-oil pressure, seal-oil pressure, and water-injection pressure were measured. The needle lift was also measured.

The system was measured with many different variations of the accumulator pipe. The effect of the accumulator pipe between the orifice and injector was discovered to be essential for injection characteristics. The pressure drop in the accumulator pipe during the injection depends on the volume, shape, and material of the pipe. Also, the pressure pulsation occurring during the injection can be strongly affected by the design of the accumulator pipe.

The test system operation was also computer simulated. The model takes into account the dynamic behavior of the accumulator pipe without acoustic characteristics. In Figure 16.37 the simulated and measured orifice flows and accumulator pipe pressures during the injection are shown. The flow through the orifice is about 20 l/min and the pressure drop during the injection is about

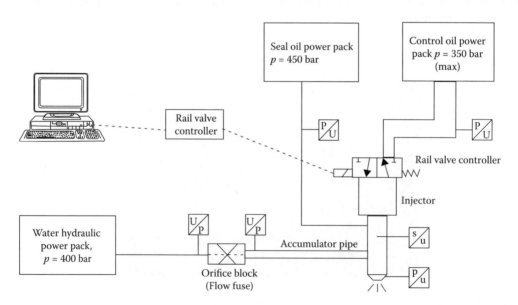

FIGURE 16.36   Schematic of the water-injection test system.

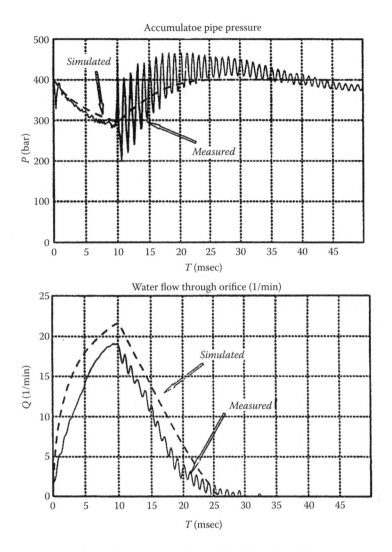

FIGURE 16.37   The accumulator pipe pressure and the orifice block flow of the test system.

100 bar. The simulated curves correlate quite well with the measured curves if the acoustic pressure ripples are included.

In the high-pressure line before the injector, there is a flow fuse, which is a safety component that prevents water overflow to the diesel-engine combustion cylinder. The overflow is possible only in some kind of malfunction; for example, in the case of a needle sticking in open position. In this situation, the flow fuse prevents the filling of the combustion chamber with water, which could cause failure. Normally, flow fuses are switch type, meaning that they receive the closing and opening signals from the pressure difference over the fuse. The pressure difference corresponds to a certain flow rate through the flow fuse. Switch-type flow fuses exhibit stability problems and are often very sensitive to pressure transients in the system.

An active-type flow fuse (AFF) operates by a different principle from the switch type. The prototype of the AFF is provided in Figure 16.38. AFF operates by a displacement principle. This means that in every injection, the piston moves and displaces a certain amount of water to the injector. In normal injection operation, the injected flow consists of the flow through the metering orifice and the flow displaced by the piston. The operation of the flow fuse is active because the displacement piston moves actively in every injection. When the injector needle stays open, the flow demand in

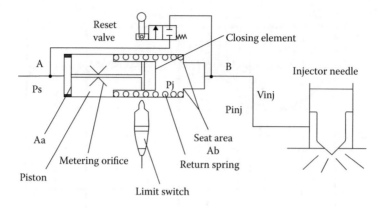

FIGURE 16.38 Schematic of the active type flow fuse.

the injector increases and the pressure P_{inj} decreases. In this situation, the piston moves the whole stroke and the closing element closes the fuse. The closing can be observed by the electrical switch. When the fuse closes, it remains closed because the pressure forces are keeping the piston in its close position until the reset valve is opened.

One problem with direct water injection to a diesel engine is the pressure pulsation that occurs in the injector and pipe during the injection. The active-type flow fuse reduces the pressure pulsation

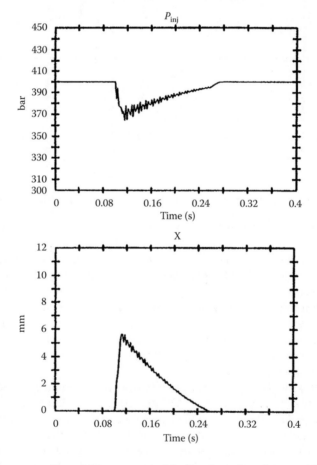

FIGURE 16.39 The pressure P_{inj} and the movement of the flow-fuse piston in normal injection.

because it displaces a specific amount of fluid in every injection. AFF efficiently reduces high-frequency vibrations and the supply pressure drop decreases. Therefore, the active-type flow fuse increases the injection volume and smaller piping dimensions can be used, which is an important factor in diesel-engine design.

From the simulation shown in Figure 16.39 it is apparent that the pressure P_{inj} in volume V_{inj} is decreasing by about 30 bar and the flow-fuse piston is moving about 5 mm. The maximum stroke of the flow fuse is 10 mm. The pressure drop without the flow fuse was about 100 bar. The effect of the flow fuse on pressure ripple damping can also be shown in Figure 16.39.

The situation when the needle is sticking in the open position is illustrated in Figure 16.40. The pressure decreases until the flow-fuse piston reaches the other end and the flow is closed and does not reopen until the reset valve is reopened.

16.4.3 WATER POWERED ROCK DRILLS [16]

Although the mining industry traditionally employs the use of water hydraulics, there are also new applications for modern water hydraulics in this area. In water hydraulic rock drills the fluid used is pure water, which is replacing traditionally used pneumatics. Water hydraulic down-to-hole drills

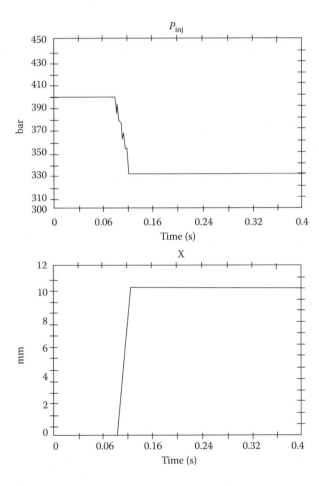

FIGURE 16.40 The pressure P_{inj} and the movement of the flow-fuse piston when the needle is stuck in the open position.

makes the drilling process more profitable and environmentally friendly compared to pneumatic drilling. This is based on the high rate of penetration and low consumption of diesel fuel. Also, the drilling depths are greater and more precise than with pneumatic drills. In addition, more benefit is achieved in dust prevention, because the same water used as a power medium can be also injected to drilling hole. Low noise level is another important factor; water hydraulic rock drills are designed so that they are easy to maintain on-site.

16.4.4 WATER-JET CUTTING IN PAPER INDUSTRY

Water hydraulics can also be seen in use as water-jet cutting systems in paper-making machines. The edge of the paper reel is usually trimmed by conventional mechanical knives. The problem is with wearing of the knives, which leads to frequent knife-changing operations. By using water-jet cutting, the edge can be trimmed quite cleanly and without wearing parts. The pressure normally used is 800–1600 bar. The system is presented in Figure 16.41.

The critical part in the system is the pump, which produces the water hydraulic power. The lifetime and reliability must be very good as the system operates 24 hours per day and service is allowed only once a year. This provides a high demands for the design and manufacture of materials for the pump. One possibility is to use a Dynaset pressure converter (Figure 16.42) as a pressure source. The converter has a very compact design which enables reliable operation.

Water-jet cutting is also used in many other industrial processes for cutting different materials including steel plates, leather, and plastic. In these applications the pressures may be higher (depending on the material); in the case of 4000 bar pressure, for example, a normal piston-type pressure converter would generally be used.

16.4.5 WATER MIST SYSTEMS

Water mist systems (Figure 16.43) are considered to be new technology in fire protection compared to conventional sprinkler systems. The idea is to use high-pressure water through small-hole nozzles to create small drop water mist. The pressure level is normally between 80 and 200 bar, which makes the technology special.

The major benefits of using water mist instead of low-pressure water include lower levels of water used, more environmentally friendly, greater fire cooling effect, and oxygen displacement effect. The technology is thus very well-suited to environments such as tunnels, ships, and hospitals.

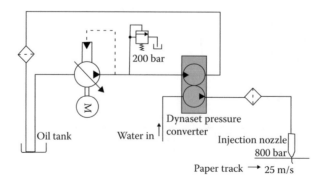

FIGURE 16.41   Operation of water-jet cutting system.

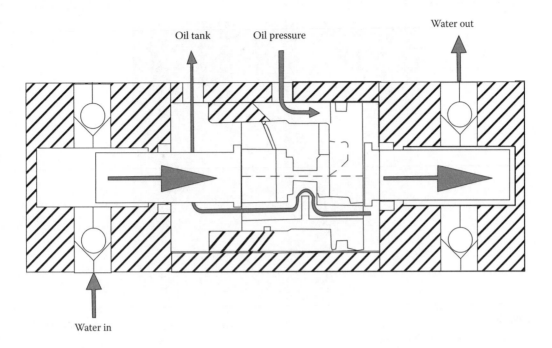

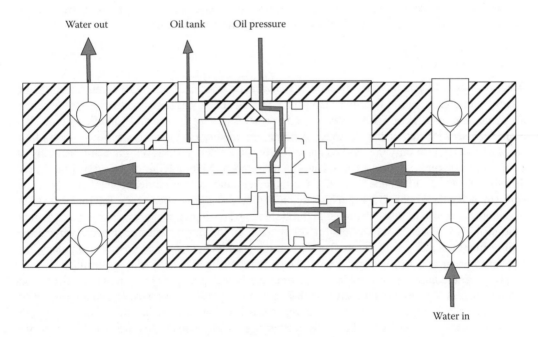

FIGURE 16.42   Operation principle of Dynaset pressure converter.

16.4.6 OTHER APPLICATIONS

Water hydraulics has also been successfully applied to various applications in the food industry. For example a rib-top saw system has been designed. This system is manually operated and used for sawing pork ribs and backs in meat-packing plants. The water hydraulic system has been developed to replace the conventional pneumatic saw-system, which has some challenges concerning noise, weight and cold air. Also conveyor and many washing applications are developed for food industry.

FIGURE 16.43 Water mist nozzle (Courtesy of Fogtec Brandschutz GmbH & Co. KG, Schanzenstraße 19A, 51063 Köln (Cologne), Germany.)

There are also many water hydraulic systems used in reverse osmosis applications, humifidication applications, wood processing applications, and hydropower applications.

16.5 SUMMARY

Water is a very efficient pressure transmitter in hydraulic systems. Due to low compressibility and viscosity, a water hydraulic system contains more intrinsic energy as the same size oil system. Thus, the dynamic properties of water systems are, in principle, faster than in oil hydraulic systems. For example, a servo system using tap water as a pressure medium is theoretically faster system than an oil servo system if only the fluid compressibility is considered. However, the compressibility of component structures are also a major factor in the efficient bulk modulus of the servo system, and therefore the actual difference is smaller. On the other hand, poorer lubrication and possible extra seals can make the Coulomb friction higher, making dynamic properties worse. Internal leakage in the components also increase damping in the system, and cavitation and erosion have more significant meaning in water hydraulics than in oil systems. In the future, when component variety and quality improves along with developing design, material technology, and control technology, the benefits of water as a pressure medium will be far better utilized in industrial applications.

The general component variety in water hydraulics is still quite poor compared to oil hydraulics and pneumatics. Due to the more difficult design, more expensive materials, and small production series, the price level of water hydraulic components is today quite high. This, of course, affects the willingness to apply water hydraulic technology in mechanical engineering. However, the market for water hydraulics is expanding slowly and the development of new, more intelligent but inexpensive components will probably accelerate this trend. Also, the introduction of low-pressure hydraulics in the market will open many new application areas for water hydraulics. Therefore, in the future, water hydraulics will offer an environmentally friendly, complementary alternative for the traditional power transmission methods: electrical drives, oil hydraulics, and pneumatics.

It can be stated that the main challenges for wider usability of water hydraulics are reliability, controllability, and price level.

Reliability means that the components must be able to work for longer periods without service and without a risk of breakage. It includes the control of water quality so that the components can operate with optimal pressure fluid quality. It also includes the development of components' characteristics so that they are not sensitive for fluid quality, temperature, pressure peaks, cavitation, erosion and wear. And it also means having successful references, where water hydraulics is in operation in demanding and long time tasks.

Controllability means that overall ability of water hydraulics to realize more accurate and dynamic control systems have to be improved [7]. This can be done by developing better control valves, control methods, and actuators. At the moment, fairly accurate position control systems can be achieved with low pressure by using servo-valves or digital hydraulic controls. However, more challenges are in higher pressures (over 20 MPa), where the component supply is very limited. Also, components and methods for realizing pressure controls will be in greater demand in the future.

The price level of water hydraulic components is certainly a big challenge. Low-pressure water hydraulics offers the possibility of cheaper costs by using cheaper materials like polymers. However, in general the biggest factors are more expensive materials and small production amounts. Even though the components and systems are technically perfect, the price level still strongly affects the machine builders' choices. Therefore, increasing production amounts, along with increasing practical applications, will slowly lead to decreasing price level of the components. It is important that the machine builders always bear in mind the total cost for the whole system life cycle including purchase, use, disposal, and the fluid itself. In addition, societal aspects such as laws, environmental taxes, insurance costs, and so forth, may also decrease water hydraulics' relative price level in the fluid power market.

REFERENCES

1. Danfoss, "NESSIE-Water Hydraulic Programme," Website www.danfoss.com, 2009.
2. Ellman A, Koskinen K.T. and Vilenius M.J. "Model for Steady State Flow in Circular Annulus," *Automotive and Agricultural Applications of Fluid Power*, 1995 ASME Winter Annual Meeting, San Francisco, California, USA, November 12–17, 1995.
3. Jay D., Rantanen O. and Koskinen K.T. "Diesel Engine - Combustion Control with High Pressure Water Injection," *The Fourth Scandinavian International Conference on Fluid Power*, Tampere, Finland, September 27–29, 1995.
4. Karppinen R. "Dynaset-HPW-Pressure Converter," Fluid Power Theme Days in IHA. 1997.
5. Koskinen K.T. "Proposals for Improving the Characteristics of Water Hydraulic Proportional Valves Using Simulation and Measurement," PhD Dissertation, Tampere University of Technology, Tampere University of Technology Publications 189. p. 131, 1996.
6. Koskinen K.T., Leino T. and Riipinen H. "Sustainable Development with Water Hydraulics – Possibilities and Challenges," *7th JHPS International Symposium on Fluid Power*, Toyama, Japan, September 15–18, 2008.
7. Koskinen K.T., Mäkinen E., Vilenius M.J. and Virvalo T. "Position Control of a Water Hydraulic Cylinder," *Third JHPS International Symposium on Fluid Power*, Yokohama, Japan, November 4–6, 1996.
8. Koskinen K.T. and Vilenius M.J. "Water as a Pressure Medium in Fluid Power Systems," *IFAC Workshop on Trends in Hydraulic and Pneumatic Components and Systems*, November 8–9, 1994, Chicago, IL, USA, p. 13.
9. Leino, T. "On the Flow and Cavitation Characteristics of Water Hydraulic Seat Valve Structures," PhD Dissertation, Tampere University of Technology, Tampere University of Technology Publications 722, p. 131, 2008.
10. Miller D.S. *Internal Flow Systems,* 1990 BHRA; Cranfield, Bedford, U.K, 1990.
11. Brochure, *Modern Water Hydraulics - Your Choice for the Future.* National Fluid Power Assosiation, USA, 1995.
12. Riipinen, H. "Life in the Water Hydraulics System," PhD Dissertation, Tampere University of Technology, Tampere University of Technology Publications 716, p. 54, 2008.
13. Tao N.L. and Donovan W.F. "Through-Flow in Concentric and Eccentric Annuli of Fine Clearance with and without Relative Motion of the Boundaries," *Trans. ASME*, Vol 77, USA, 1955.
14. Trostman, E. *Water Hydraulics Control Technology,* Marcel Dekker Inc., New York, 1996.
15. Urata, E. Miyakawa, S. and Yamashina, C. "Hydrostatic Support of Spool for Water Hydraulic Servo Valves - Its Influence on Flapper-Nozzle Characteristics," *The Fourth Scandinavian International Conference on Fluid Power*, Tampere, Finland, September 27–29, 1995.
16. Wassara rock drills, www.wassara.com, 2009.

17 Polyol Ester Fluids

*Paula R. Vettel**

CONTENTS

---

* This chapter is a revision of the chapter titled "Polyol Esters" by Richard A. Gere and Thomas V. Hazelton from the *Handbook of Hydraulic Fluid Technology*, 1st Edition.

17.1 INTRODUCTION

Fluid power engineers and users have traditionally selected hydraulic fluids based on mineral oil because they have high-performance characteristics and are economically favorable. External factors, however, often dictate an alternative selection. One major concern is the potential flammability of mineral-oil-based hydraulic fluids. Major capital and personnel losses have occurred because of hydraulic-fluid-related fires [1]. More recently, potential environmental problems associated with mineral-oil fluids have gained notoriety. These problems include low rates of biodegradability, aquatic toxicity, handling of contaminated material, and high waste-treatment costs.

A simple and attractive alternative to mineral-oil-based hydraulic fluids that addresses these problems are fluids based on polyol esters.

Esters are compounds produced by the reactions between organic or inorganic acids and alcohols. Naturally occurring esters (glycerides) are a major component of all fats and vegetable oils. Synthetic esters can be constructed from many organic and inorganic acids and alcohols, and the choices made determine the physical and chemical properties and the cost of the resultant ester.

This chapter describes hydraulic fluids based on polyol ester technology. The term "polyol ester" relates to the base stock used in the final product. Polyol ester hydraulic fluids, like most commercial hydraulic fluids, use performance-enhancement additives which can account for up to 10% of the total product. These fluids are typically prepared from long-chain carboxylic fatty acids (derived from natural fats and oils) and polyhydric alcohols that contain two or more hydroxyl groups.

Polyol-ester-based hydraulic fluids have been commercially available for many years and are considered a major category of fire-resistant or *less hazardous hydraulic fluids* [2,3]. They are classified as HFD-U under ISO standard 6743-4 for fire-resistant hydraulic fluids. Factory Mutual Research Corporation (FMRC) initially listed polyol ester fluids as *less hazardous hydraulic fluids* in 1973 [4]. With the growing need for fluids that are environmentally compatible, polyol ester hydraulic fluids have found a niche in this market because they are readily biodegradable and have low aquatic toxicity. The products also exhibit excellent lubrication characteristics and have found applicability in most general-purpose power units. In 1996, approximately three million gallons of polyol ester hydraulic fluids were produced worldwide.

The cost of a polyol ester hydraulic fluid is approximately two to three times that of mineral-oil fluids. However, if fire resistance is required or if the fluid is used in environmentally sensitive areas, the selection of a polyol ester fluid can be a cost-effective choice. Figure 17.1 shows the relative cost of a polyol ester hydraulic fluid compared with other popular hydraulic fluids.

In this chapter, polyol ester chemistry, the method of manufacture, and the role of performance-enhancement additives will be reviewed. Physical and chemical properties and operational characteristics such as fire resistance, environmental properties, and general compatibility with hydraulic power units will be discussed. Fluid limitations and maintenance requirements, including conversion procedures, will be described.

17.2 POLYOL ESTER TECHNOLOGY

17.2.1 BASIC CHEMISTRY

Polyol esters are members of the general ester family formed by the reaction between alcohols and inorganic or organic acids. The reaction process is called "esterification", with water formed as a

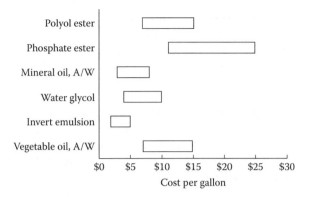

FIGURE 17.1 Hydraulic fluid costs.

by-product. Figure 17.2 is a representation of the esterification process and illustrates the large number of potential ester products possible and the route to polyol esters.

The alcohol portion of the ester can have a single functional group (—OH) or multiple ones (polyhydric). The structure of the compound may be an open-carbon-chain (aliphatic) or a closed-ring (aliphatic or aromatic) structure. Polyol esters typically employ open-chain alcohols. Phosphate esters are fire-resistant hydraulic fluids formed by a closed-ring-structure alcohol and an inorganic acid. Organic acids form the acid portion of a polyol ester.

Organic acids are generally called "fatty acids" because most organic acids are obtained from naturally occurring materials such as animal and marine fats and seed oils (e.g., rapeseed oil and sunflower oil). They are present in these sources as a glyceride and are, in fact, polyol esters. Here,

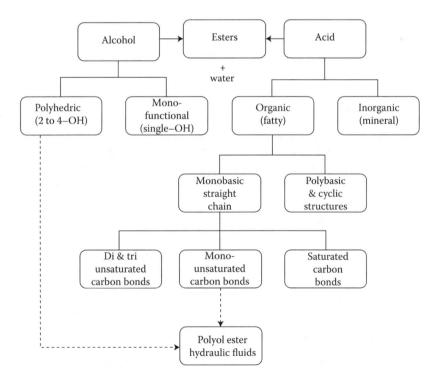

FIGURE 17.2 Formation of esters.

the alcohol portion is glycerol, $C_3H_5(OH)_3$. Many seed-oil-based hydraulic fluids currently being marketed as "environmentally compatible lubricants" are natural glyceride polyol ester fluids. A large variety of fatty acids are recovered from these fat and seed sources, varying in carbon chain length, structure, and number of carboxyl groups [5]. Most fatty acids are of a straight-carbon-chain structure with a single carboxyl group (—COOH). The number of carbon elements may vary from 10 to 20 + but generally contain 14, 16, and 18 carbon atoms. Polybasic (more than one carboxyl group) and cyclic structures are not as common. Monobasic straight-chain fatty acids also vary with respect to their level of saturation. Fatty acids that contain double bonds between carbons are said to be unsaturated. A monounsaturated acid contains one such bond. Di- and tri-unsaturated structures are also common, and although possible, more than three is unusual. Fatty acids with single carbon-to-carbon bond linkages are said to be saturated.

In a polyol ester, the fatty acid portion represents approximately 80% of the total ester because of its large size compared with most polyhydric alcohols, and two or more acid groups are used for each alcohol molecule. As a result, many properties characteristic of the acid, particularly oxidative resistance, are conferred upon the final ester.

In the commercial recovery of fatty acids, many different acids are obtained from a given glyceride source and separation of specific acids is difficult; mixtures of acids are typically realized. The level of the desired fatty acid will vary significantly, based on cost and purity of the glyceride source and on the efficiency of the separation processes.

The selection of a fatty acid is based on several factors, including the following:

- Higher carbon lengths produce esters with higher viscosity, flash/fire points, vapor pressure, and pour points.
- Saturated acids are more thermally stable and oxidation resistant than unsaturated acids.
- The higher the degree of unsaturation, the more oxidatively unstable the ester becomes.
- Saturated acids produce esters with higher viscosities and pour points than unsaturated acids.

Polyol esters are generally prepared from monounsaturated fatty acids containing 16 and 18 carbon-length chains because they offer the best combination of physical and chemical properties. Figure 17.3 illustrates a typical reaction to form a polyol ester from glycerol and a C_{18} monounsaturated fatty acid (oleic acid).

17.2.2 Polyol Ester Base Stocks

Although many aliphatic alcohols are multifunctional, the three most commonly used to produce polyol esters for hydraulic fluid service are shown in Figure 17.4. The major difference among them is the number of hydroxyl groups. The viscosity of the polyol ester formed from these polyols is directly related to the number of hydroxyl sites and the choice is generally based on the desired ISO class of the hydraulic fluid.

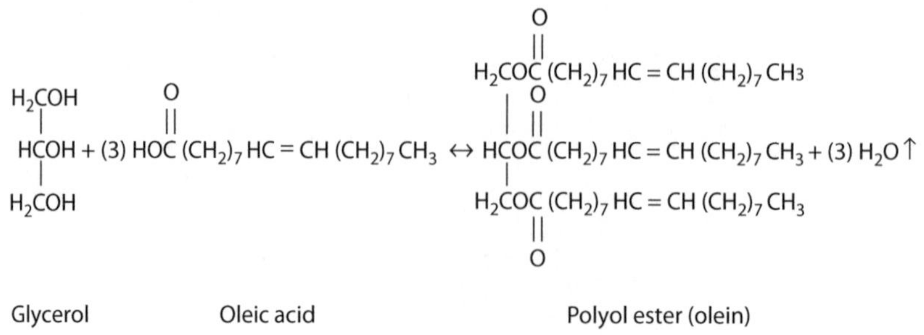

FIGURE 17.3   Polyol ester reaction.

To prepare polyol esters suitable for typical hydraulic fluid service, the fatty acid of choice is 18-carbon-length monobasic and monounsaturated. The chemical name for this compound is oleic acid. Selection of this fatty acid, found in abundant supply and until recently with a relatively stable price history, is based primarily on the physical and chemical properties of the esters formed with the polyols shown in Figure 17.4. Figure 17.4 also shows the chemical structure for oleic acid; its reaction equation with a polyol is shown in Figure 17.3. Oleic acid is present in virtually every natural fat and oil. Table 17.1 shows the major sources for oleic acid and the approximate distribution of the major fatty acids present. The presence of acids other than oleic acid is important because certain acids will not be readily separated in the separation processes to purify the acid. Esters formed from these acids may detract from the properties of the oleic acid ester. Saturated acids such as stearic and palmitic are readily separated from the unsaturated acids because of significant differences in their melting points. Separating unsaturated acids is more difficult because their melting points are similar and fractional distillation processes must be employed. This is not only a costly step but is also difficult because their vapor pressures are similar. All commercial grades of oleic acid will be approximately 75% pure, with the balance containing a variety of other fatty acids. The composition will depend on the original source.

Table 17.2 shows the nominal acid composition of oleic acid derived from three sources—beef tallow, tall oil, and a vegetable oil. The major difference in the fatty acids listed is the number of double bonds in the carbon–carbon chain. Oleic acid, with one double bond, forms a polyol ester with properties well suited to serve as a hydraulic fluid base for most fluid power applications. Esters formed from the saturated acids will have higher titer points than the oleic acid ester. This means that at low temperatures, these esters may partially desolubilize and cloud the fluid. In a severe case, these esters may coalesce and settle out of the fluid as deposits. Fatty acids with more than one double bond will form esters that are inherently more susceptible to oxidation and degradation [6]. There are also grades of products within each source category that vary in oleic acid content and undesirable acids. The compositions

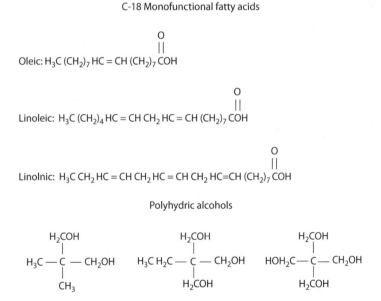

FIGURE 17.4 Polyol ester starting materials.

TABLE 17.1
Oleic Acid Sources

	Source (%)					
Acid	**Lard**	**Tallow**	**Palm Oil**	**Corn Oil**	**Soya Oil**	**Tall Oil**
Oleic (C-18$_1$)[a]	43	43	37.5	26.5	22.5	59.5
Palmitic (C-16$_0$)	26	26	47	11.5	–	–
Stearic (C-18$_0$)	13.5	22.5	4	2	3	2
Linoleic (C-18$_2$)	9	1.5	–	59	54.5	37
Palmitoleic (C-16$_1$)	4	2.5	–	–	–	–
Myristic (C-14$_0$)	1.5	3	1	–	–	–
Linolenic (C-18$_3$)	–	–	–	1	8.5	–

Source: Technical Bulletin 140A9, Henkel Corporation, Emery Group Cincinnati, OH. With permission.
[a] Denotes length of carbon chain (C-18) and number of double bonds, 0, 1, 2, or 3 (subscript).

shown in Table 17.2 are representative and may vary between fatty acid suppliers and the level of distillation efficiencies employed.

Tallow is the preferred source for oleic acid based on its cost and availability and generally contains a minimum level of other undesirable acids. Oleic acid from tallow is typically used in high-quality polyol ester hydraulic fluids. Commercial fluids are available that are based on other acid sources because of lower cost and with some sacrifice in fluid quality.

The three primary polyol ester base stocks described may be used singularly or in combination to optimize certain properties, particularly viscosity. Mixtures may also be formed in situ by combining different alcohols during the reaction. Table 17.3 reviews the structure of these three polyol esters and the nominal characteristics associated with each.

17.2.3 Manufacturing Process

17.2.3.1 Reactor Considerations

Polyol ester reactions are normally carried out in batch processing vessels. The following criteria are considered in the design of an appropriate reactor:

TABLE 17.2
Oleic Acid Composition

	Source (%)		
Acid	**Tallow**	**Tall Oil**	**Vegetable Oil**
Oleic (C-18$_1$)[a]	73	67	60
Palmitic (C-16$_0$)	5	10	3
Stearic (C-18$_0$)	–	3	2
Linoleic (C-18$_2$)	8	3	24
Palmitoleic (C-16$_1$)	6	2	–
Myristic (C-14$_0$)	4	–	–
Linolenic (C-18$_3$)	1	5	8

Source: Technical Bulletin 140A9, Henkel Corporation, Emery Group. With permission.
[a] Denotes length of carbon chain (C-18) and number of double bonds, 0, 1, 2, or 3 (subscript).

TABLE 17.3
Polyol Ester Base Stocks

Polyol Ester	Formula	Molecular Weight	Flash/Fire Point[a] (°C)	Viscosity @ 40°C[b] (cSt)	Pour Point[c] (°C)
TMPO	$C_{63}H_{110}O_6$	962	316/354	50	−25
NPGO	$C_{43}H_{76}O_4$	656	288/316	37	−20
PEO	$C_{81}H_{140}O_8$	1240	316/354	68	−30

Note: Trimethylol propane oleate (TMPO): CH_3—CH_2—C—$(CH_2$—O—R$)_3$; neopentyl glycol oleate (NPGO): $(CH_3)_2$—C—$(CH_2$—O—R$)_2$; pentaerythritol oleate (PEO): C—$(CH_2$—O—R$)_4$, where R = C_{18} oleic side chain.

[a] Based on ASTM D 92.
[b] Based on ASTM D 445.
[c] Based on ASTM D 97.

- **Capacity:** Most commercial manufacturing vessels are between 3000 and 10,000 gal. Dedicated equipment should be used whenever possible.
- **Material of construction:** The reactor should be fabricated from stainless steel or glass-lined steel because of the corrosive nature of the raw materials at the elevated temperatures employed.
- **Heating and cooling:** Polyol ester reactions are carried out at temperatures up to 288°C (550°F) and precise control of heating and cooling profiles is required to yield a high-quality product. Steam-heated reactors are inadequate for preparing quality esters.
- **Water removal:** Water, the major byproduct produced, must be removed continuously to force the reaction to completion. Removal of the water formed may also promote the removal of nonreacted starting materials. Provision for continuous reflux of these materials along with water removal should be provided. The reactor should have vacuum capability to help in water removal. Inert-gas sparging is also recommended in order to minimize the risk of oxidation to both the unreacted acid and the ester.

17.2.3.2 Reactants

The two reactants are oleic acid and the appropriate alcohol. Depending on the desired polyol ester, the alcohol will be either neopentyl glycol, pentaerythritol, or trimethylol propane, or a mixture of the three. The quantity of alcohol used is typically 5–15% above the stoichiometric requirement. The excess alcohol facilitates completion of the reaction; it is later stripped.

17.2.3.3 Reaction Catalysts

The basic polyol ester reaction can be achieved without the use of a catalyst. However, to maximize the degree of reaction completion, temperatures in excess of 260°C (500°F) and extended reaction times must be employed. Both factors contribute to oxidative and thermal degradation, higher product viscosity because of undesirable polymerization, and incomplete utilization of the available hydroxyl sites. These all contribute to the attenuation of many of the critical ester properties (e.g., flash/fire points, hydrolysis stability, and resistance to oxidation). The use of a catalyst reduces both the reaction time and reaction temperature.

There are a number of potential catalysts, though the most commonly used are sulfonic acid derivatives and certain organometallic compounds. Catalysts, such as the organometallics, are generally used at concentrations less than 0.15% and may or may not be recoverable from the final product. Each type of catalyst has advantages and disadvantages, and choice becomes a matter of trade-off. Benefits to be derived from the use of a catalyst are as follows:

- Reduced reaction temperature and reaction time.
- Higher reaction efficiencies (utilization of more hydroxyl sites).

- Enhancement of most performance characteristics.
- Improved product quality.

In addition, the beneficial interaction between additives and the base stock increases significantly with improved quality of the base stock.

17.2.3.4 Manufacturing Process Cycle

The overall process is best described by reviewing the time versus temperature profile shown in Figure 17.5. The profile is for illustrative purposes only and will vary significantly depending on equipment, batch size, and starting materials.

The major steps are as follows:

- Charge the appropriate quantities of starting materials. Because the final product will require certain physical and chemical characteristics to be within precise specifications, exact control of the starting materials is critical.
- The chemical reaction begins early in the "heat-up" cycle and, therefore, the time–temperature profile should be precise and conform to previous optimization studies. Automatic control is highly recommended.
- The length of time at the reaction temperature is critical because it must be sufficiently long to complete the reaction yet minimize oxidative and thermal degradation that may contribute to the formation of the unwanted side chemical reactions. The duration of this period is best determined by monitoring free acid and ending the reaction phase when the acid content has stabilized or stalled.
- Cooling periods should be as short as possible. Because some additives employed may have different optimum solubility temperatures and heat sensitivities, two additive addition levels are shown. Having low acid and hydroxyl numbers is important. Residual alcohol can be removed by vacuum stripping during the initial cooling period. Unreacted acids are more difficult to remove because of their low vapor pressures. Ideally, all acids should be used during the reaction phase but, if not, neutralizing with an agent such as calcium or sodium hydroxide to form an insoluble acid–salt can be effective.

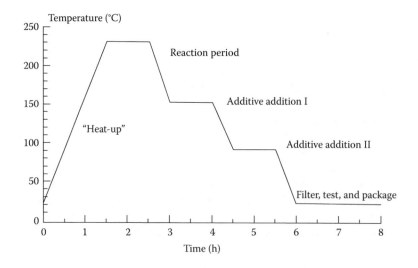

FIGURE 17.5 Reaction profile.

- After cooling, the final steps are filtration and packaging. The product is filtered to remove residual particulates formed during the reaction (e.g., neutralization salts, scale, and undissolved additives). The filtration process is typically multistage. Gross particles are removed by simple screening. Fine filtration is accomplished by means of plate and frame filters, bag filters, cartridge media, sand filters, and so forth. Ideally, the filtering process should be done with process control monitoring to ensure that the final product is within cleanliness specifications. When the product is within all specifications, it may be held in intermediate bulk storage or packaged directly for delivery to the end user. All shipping containers should satisfy stringent cleanliness standards.

17.2.3.5 Process Control and Product Specifications

The goal of the manufacturing process is to create a product that satisfies all fluid performance goals and conforms to a rigorous quality control standard. This can be accomplished through an in-process monitoring program that regulates the critical steps of the process and assures product compliance to final specifications. Industry-accepted statistical process control (SPC) principles should be incorporated.

Acid number history during the reaction is perhaps the most important in-process control parameter. Acid number reduction is a direct measure of reaction completion and is related to final ester quality. This parameter must be stabilized during the reaction phase of the process.

The product must satisfy all of the specifications established by the manufacturer to ensure that the product will meet all required performance characteristics. Typical parameters which might be incorporated into a final product specification are as follows:

- Total acid number: < 5 mg KOH/g
- Strong acid number: 0 mg KOH/g
- Hydroxyl number: < 1 mg KOH/g
- Viscosity at 40°C: ISO grade ± 10% (cSt)
- Cleanliness: NAS Class 7 (ISO 4406 16/13)
- Moisture content: < 0.05% (500 ppm)
- Color and odor: clear light amber, burnt almond
- Pour point: < −26°C (−15°F)

17.2.4 Fluid Additives

Commercial polyol ester hydraulic fluids incorporate performance-enhancement additives into the base ester to provide the requisite properties of a quality hydraulic fluid. The selection and level of these additives along with the quality of the base stock often account for the differences between fluids. The cost of additives is a major component of the product price structure.

17.2.4.1 Viscosity Index Improvers

Polyol esters exhibit a temperature–viscosity relationship similar to all hydraulic fluids—a tendency to have high viscosities at low temperatures and low viscosities at high temperatures. The viscosity index (VI) is a mathematical expression that describes the rate of change in viscosity as a function of temperature [7]. Viscosity index improvers are additives that reduce this rate of change to provide a more uniform viscosity over the operating temperature.

Viscosity index improvers are high-molecular-weight polymers such as polyisobutylenes, polymethacrylates, ethylene–propylene copolymers, styrene–isoprene copolymers, styrene–butadiene copolymers, and styrene–maleic ester copolymers. At low temperatures, these polymers are marginally soluble in the fluid, have a compact structure, and exert a small influence on fluid viscosity. As the fluid temperature increases, the solubility increases and the polymer uncoils or expands [8]. This decreases the mobility of the fluid and results in higher viscosity.

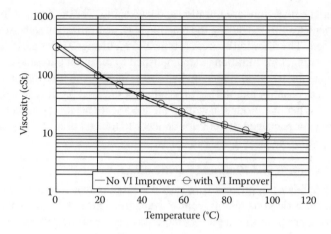

FIGURE 17.6   Viscosity index of polyol esters.

Figure 17.6 compares the temperature–viscosity properties of ISO Grade 46 polyol esters with and without a VI improver. The benefit derived from the additive appears small, as both fluids exhibit high viscosity indexes. The VI values with and without the additive are 200 and 170, respectively. Viscosity index values for mineral-oil hydraulic fluids are typically 75–150. Figure 17.7 compares the viscosity properties of a high-VI antiwear ISO 46 Grade mineral oil (VI ≈ 95) fluid with a polyol ester containing a VI improver (VI ≈ 230). Viscosity differences at low temperatures are significant.

Although the value of a VI improver may be marginal with respect to increasing the viscosity index of a polyol ester, it does play a major role with respect to fire resistance. A major feature of a fire-resistant hydraulic fluid is the reduction of damage caused by spray-mist fires. Ruptures in high-pressure lines or hoses in a hydraulic system operating on mineral-oil-based fluids often create a spray mist, which if ignited, may burn throughout the spray envelope. In such a spray mist, the oil particle sizes formed are very small, depending on the operating pressure and the geometry of the rupture. Small mist particles, because of a high ratio of surface area to mass, are susceptible to ignition [9]. Thermal properties of the fluid are also important factors. This subject is discussed more fully in Section 4.2.

Fire-resistant polyol ester hydraulic fluids incorporating VI improvers influence the spray characteristics of a fluid under these conditions. The polymer acts as a binder and causes the fluid under expulsion to form large spray particles (lower surface area to mass ratio) which are more difficult to ignite.

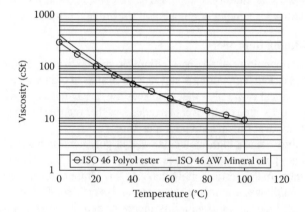

FIGURE 17.7   Viscosity index comparison.

Because VI improvers are high-molecular-weight polymers, shear stability is an important consideration in their selection. Table 17.4 illustrates the shear stability of three representative VI improvers with varying molecular weights. The polymers were incorporated into an NPG oleate at 3% concentration. The differences between the initial molecular weight and preshear value are attributable to the breakdown caused by incorporating the polymer into the base ester. This often requires extensive mixing at an elevated temperature. Polymer B was readily solubilized, whereas polymers A and C required substantial mixing energy, resulting in molecular-weight breakdowns. Based on molecular-weight changes, polymer B would not be suitable for use as an additive because of excessive shear loss. Polymers A and C show less shear loss in this test and would be viable candidates for use in polyol ester fluids. Field experience has shown that with molecular-weight loss of this scale, a compromise in fire resistance would not be expected.

17.2.4.2 Antioxidants

The operating environments for all hydraulic fluids—irrespective of type—are severe, considering the high mechanical energy imparted, exposure to high localized temperatures, and, sometimes, sustained high bulk temperatures. Additional factors, such as air entrainment, external contaminates, and internal contaminations brought on by wear processes and corrosion, exacerbate the situation and create an atmosphere conducive to fluid degradation.

The major fluid degradation process is oxidation—particularly in relation to fluids based on hydrocarbon structures—and the subsequent formation of oxygen-containing hydrocarbons. Fluids are often differentiated by their ability to resist oxidation. Fluid oxidation results primarily in fluid viscosity increases and the generation of particulate matter and varnish. Fluid viscosity increases are caused by the solubility of high-molecular-weight polymers formed in the oxidation process; insoluble oxidative products lead to varnish and sludge. Fluid degradation of this form leads to a sluggish hydraulic system response, plugged filters and screens, and sticking valves and actuators.

Polyol ester fluids are no exception and effects from oxidation can be extreme. The mechanism of oxidation is similar to that associated with mineral-oil hydraulic fluids. The generally accepted theory on hydrocarbon oxidation is that the process is initiated by the formation of a primary alkyl radical (R·) at the most vulnerable carbon–hydrogen bonds [10]. This radical is rapidly oxidized to its peroxide free radical (ROO·) and initiates a chain reaction. Figure 17.8 is

TABLE 17.4
Shear Stability of VI Improvers

	Polymer		
	A	B	C
Molecular weight	91,750	44,200	107,000
		Preshear	
Fluid viscosity (cSt)	45.6	49.2	47.1
Molecular weight	54,200	126,100	107,000
		Postshear	
Fluid viscosity (cSt)	43.4	41.9	44.5
Molecular weight	44,200	67,100	62,500
Molecular-weight loss (%)	18.5	41.3	18.7
Shear stability index (SSI)[a]	11.9	33.2	13.3

[a] $SSI = \dfrac{\text{Preshear viscosity} - \text{Postshear viscosity}}{\text{Preshear viscosity} - \text{Base viscosity}} \times 100.$

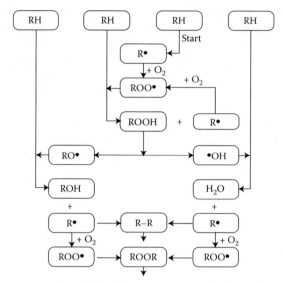

RH - alkyl group   R• & ROO• - alkyl & peroxide free radicals

FIGURE 17.8   Polyol ester oxidation steps. RH = alkyl group; R· = alkyl free radical; ROO· = peroxide free radical.

a flowchart outlining the steps involved in this chain reaction. The initial peroxide radicals subsequently react with additional energized hydrogens in the polyol ester molecule to form hydroperoxides (ROOH) and additional peroxide radicals. The process continues in a chain reaction. This phase of the oxidation process, sometimes called an "induction period", is relatively slow and innocuous to the hydraulic fluid. There is a point in which the hydroperoxide concentration (or concentration of oxygen into the fluid) reaches a level where other oxidative processes are initiated, ending the benign induction period. The major reaction now becomes one of molecular degradation (severing the O—O bond) of the peroxides to form new aggressive radicals (RO· and ·OH). They, in turn, react with the polyol ester, which may lead to the formation of complex branched chains, polymers, acidic oxidative products, insoluble condensates, and a variety of varnish and sludge-like materials. This phase of the process can be rapid, as the rate of oxidation accelerates.

All commercial polyol ester hydraulic fluids incorporate antioxidants which are designed to disrupt the chain reaction and extend the induction period. One type of antioxidant "caps" the initially formed peroxide radicals and prevents subsequent chain-reaction steps. These antioxidants are hydrogen donors that recombine with the alkyl and peroxide radicals to form stable low-energy radicals which do not subsequently react with the base ester. Commonly used materials are sterically hindered phenols and, occasionally, secondary aromatic amines. There are many compounds designed for this process and the formulator must optimize the specific product with the available commercial antioxidant products. The antioxidant will be gradually consumed, depending on the specific fluid and the operating conditions encountered.

Another type of antioxidant additive functions on a different mechanism and prevents decomposition of the hydroperoxides into RO· and ·OH radicals. They react stoichiometrically with the hydroperoxides to form stable compounds; thioethers and phosphites represent this type of antioxidant. These materials are not regenerative and oxidation rates increase rapidly after the additive has been depleted. Antioxidants are typically formulated into the fluid at levels from 0.5% to 2%.

Beyond the role of antioxidant additives, the oxidation process is sensitive to other factors, including the following:

- The major portion of the polyol ester molecule is derived primarily from oleic acid. Because of its original source, low levels of linoleic acid, linolenic acid, and other polyunsaturated acids are also present. The major difference between these acids is the number of double bonds in their structures. The number and position of double bonds drastically affect the rate of oxidation. Methylene groups positioned between double-bond carbon linkages are readily activated and subjected to oxidation. The monounsaturated version (oleic) provides the greatest resistance, while linolenic (three double bonds) is the most vulnerable. The relative rates of oxidation of oleic, linoleic, and linolenic acids are 6, 64, and 100, respectively [6]. The higher-quality polyol fluids use fatty acid sources that maximize oleic acid content (see Table 17.1).
- Dissolved metal ions, particularly copper, in the polyol ester can affect the oxidation process [8]. Divalent copper (Cu^{2+}) is the predominate form of dissolved copper and is not considered a pro-oxidant; however, if reduced to the cuprous state (Cu^{1+}), it becomes an active catalytic agent and accelerates the decomposition of the hydroperoxides into additional chain-promoting free radicals. Compounds employed as the primary antioxidant (sterically hindered phenols) are capable of acting as a reducing agent in converting Cu^{2+} to Cu^{1+}. The net effect is that the dissolved copper ions contribute to the degeneration of the antioxidant and adversely affect the life of the fluid. A chelating agent such as disalicylidene propylenediamine forms copper adducts which have no pro-oxidant characteristics. These additives, however, have no effect in slowing the rate of migration of additional copper ions into solution. Additives that promote the formation of a passivation film on nonferrous metal surfaces will reduce such migration. Examples of these materials include mercaptobenzothiazole, mercaptobenzimidazole, and aromatic triazole derivatives. These additives may also limit the corrosive effects of sulfur.

17.2.4.3 Corrosion Inhibitors

Corrosion inhibitors are used to reduce chemical attack on ferrous and nonferrous metal surfaces. Polyol esters are generally benign to the common materials of construction. However, once in service, several factors contribute to increases in their chemical aggressiveness. The major one is fluid oxidation and the resultant formation of potentially corrosive acids. Metal attack by these materials contributes to further fluid oxidation and continuation of the corrosion process. This process is accelerated by water contamination and high operating temperatures.

Ferrous-metal corrosion inhibitors prevent the formation of rust by forming a protective film barrier. Highly polar compounds such as sulfonic and succinic acid derivatives and certain amides are commonly used. The dosage level in the hydraulic fluid is typically 0.25%. It should be noted that these barrier films are susceptible to mechanical wear or abrasion and the polar nature of the additive may interfere with other polar-type additives in the fluid [10]. Polyol esters are highly polar and provide a natural level of corrosion protection. Most polyol ester hydraulic fluids do not incorporate rust inhibitors.

Nonferrous inhibitors also provide protection by the formation of film barriers. Chemical compounds such as heterocyclic sulfur–nitrogen compounds, mercaptobenzothiazole, and dimercaptothiadiazole derivatives are commonly used. Dosage levels range from 0.1% to 0.5%. All polyol ester products contain nonferrous corrosion inhibitors, with a range of effectiveness among fluids from different suppliers. This effect is related to the overall efficiency of the antioxidant system and the quality of the base ester.

It is particularly important that the moisture level and total acid number (TAN) in the hydraulic fluid be monitored and controlled, as they are major influences in nonferrous-metal attack. The presence of water is detrimental because it can solubilize acid products and increase their chemical activity. If possible, the moisture level should be maintained at 0.2% or less and the TAN less than 5 mg KOH/g.

17.2.4.4 Antiwear and Extreme-Pressure Additives

The basic function of a hydraulic fluid is to transmit power, a function that is readily achieved by virtually any liquid. However, the practical application of fluid power requires the fluid to operate under high pressures in high-speed equipment. This creates the additional fluid requirement of being a lubricant that satisfies the needs of modern high-speed hydraulic pumps and motors and other system components.

Like mineral-oil base stocks, a polyol ester base provides lubrication (hydrodynamic) between sliding and rotating surfaces through a film of lubricant that separates the surfaces. Fluid viscosity is the principal fluid attribute that maintains the film and prevents wear. However, under high pressure and reduced relative speeds, the lubricating film is forced out or becomes too thin to separate the surfaces, and metal-to-metal contact occurs. Without additional protection, this contact can lead to catastrophic failures of the surfaces involved. Under these conditions, the type of lubrication changes from hydrodynamic to boundary (see Chapter 6). Polyol ester-based hydraulic fluids are formulated with antiwear additives to address this problem. Briefly, these additives are (1) adsorbed onto the metal surfaces to provide a lubricating film and (2) decomposed under the high temperatures generated by frictional heat to form reactive products that combine chemically with the metal surfaces to form a separating film. Examples of antiwear additives used include the following:

- Metal dialkyl dithiophosphates
- Metal diaryl dithiophosphates
- Alkyl phosphates
- Phosphorized fats and olefins
- Phospho-sulfurized fats and olefins
- Sulfurized fats, fat derivatives, and carboxylic acids
- Chlorinated fats, fat derivatives, and carboxylic acids
- Fatty acids, other carboxylic acids, and their metal salts
- Esters of fatty acids
- Oxidized paraffins and oils

Besides reducing metal wear, some types of antiwear agents may also function as antioxidants and corrosion preventives [11]. The fluid supplier must therefore optimize the selection of an antiwear additive with other ingredients in the additive package. Additive interactions may also synergistically enhance or antagonistically diminish certain performance characteristics [12]. Also, depending on the application, ashless (no heavy metals) additives should be used [13].

17.2.4.5 Combustion Inhibitors

For applications requiring the use of fire-resistant hydraulic fluids, polyol esters are finding wide acceptance. This is based in part on flash/fire points and autoignition temperatures higher than mineral-oil fluids. VI improvers augment these properties and contributes to the fluid's overall fire resistance (see Section 2.4.1), particularly spray-mist ignition. An acceptable level of fire resistance was established in Factory Mutual Research Corporation (FMRC) testing, and polyol-ester-based fluids have been listed in the FM Approval Guide for years [4].

17.2.4.6 Others

A variety of other additives may be incorporated into polyol-ester-based fluids to address problem areas unique to specific applications:

- **Foam inhibitors:** Entrained and dissolved gases can affect fluid performance. Entrained air is usually introduced at the low-pressure portion of the system (e.g., pump-suction piping and the fluid reservoir) and will be partially soluble at operating pressures. Residual entrained air can create a "spongy" hydraulic system operation and lead to a variety of system problems. Depressurization of the fluid releases dissolved air and may generate foam in the reservoir. These problems are properly addressed by ensuring that suction piping does not ingest air and all return lines extend well below the liquid level in the reservoir to reduce surface turbulence. Other consequences of dissolved and entrained air are increased pump wear through cavitation and fluid oxidation.

Generally, polyol-ester-based fluids do not incorporate foam inhibitors or air-release agents. For applications prone to air ingestion, their use should be a consideration. Materials commonly used are silicones (polysiloxanes or dimethylsiloxanes) and polyacrylates. These materials are used at low concentrations (1–50 ppm) and form a fine dispersion. Overdosage can result in excessive foam and loading of system filters. Foam inhibitors function by changing the surface-tension properties of the fluid and the coalescing of the small air bubbles into large ones.

- **Demulsifiers:** Water contamination can be detrimental to polyol-ester-based hydraulic fluids and should be avoided if possible. Because of the high shear rates imposed by the hydraulic pumps, water is readily dispersed or even emulsified into the fluid. This water is not easily separated and may contribute to fluid and system degradation. One approach to reducing the effect of water is the use of additives that coalesce water droplets and ease their removal. Certain polyesters, phosphorous-containing chemicals, Ca and Mg sulfonates, and heavy-metal soaps are effective demulsifiers. Additives used for other functions in the lubricant formulation should be evaluated for potential demulsification response.
- **Dispersants:** Particulate matter contaminating the fluid often settles in dead areas of the hydraulic system and eventually builds up to a point where large aggregates may break loose and cause damage. Dispersant additives help in the suspension of insoluble products formed by fluid oxidation and corrosion. Polybutenylsuccinic acid derivatives are commonly used. The use of a dispersant will ease the removal of these particles by the hydraulic system filters.
- **Pour point depressants:** Polyol esters are primarily derived from oleic acid (Section 2.2). Small quantities of other fatty acids are also involved in the esterification process. These esters may have high titer and melting points and come out of solution at a higher temperature than the oleic acid ester. Pour point depressants will suppress this early precipitation and improve low-temperature flow properties. This additive technology is well established for mineral-oil fluids and has been extended to vegetable oils and ester fluids. Compounds such as polymethacrylates, wax alkylated naphthalene polymers, wax alkylated phenol polymers, and chlorinated polymers are candidates for this group of additives.

17.2.4.7 Fluid Cost Implications

The performance improvements realized by these additives are significant and directly influence the cost of the fluid. The use of performance-enhancement additives is a value-added feature that varies widely among suppliers. It is estimated that additives account for 10–30% of the cost of the fluid. Other factors affecting the cost are raw material source and manufacturing quality standards. Each of these factors influences the final product cost and partially explains why the price of a polyol-ester-based hydraulic fluid may range from $6.00 to $20.00/gal.

17.3 SUMMARY OF PHYSICAL AND CHEMICAL PROPERTIES

Three ISO grade fluids, –46, –68, and –100, satisfy most of the industrial applications. Engineering properties specific to each ISO class are summarized in Table 17.5. Figures 17.9 and 17.10 show viscosity and specific gravity as a function of temperature, respectively. Properties that are applicable

TABLE 17.5

Typical Properties of ISO 46, 68, and 100 Polyol Ester Fluids

Property	ASTM Method	ISO Viscosity Grade		
		46	68	100
Viscosity (cSt) at	D 445			
25°C (77°F)		82	100	168
40°C (104°F)		46	68	100
100°C (212°F)		9	10	13
Viscosity index	D 2270	230	235	>200
Specific gravity at	D 1298			
25°C (77°F)		0.91	0.91	0.92
50°C (122°F)		0.89	0.89	0.91
Coefficient of thermal expansion (per°F)	D 1903	4.06×10^{-4}	4.08×10^{-4}	4.17×10^{-4}
Flash point (COC) (°C)	D 92	260	260	257
(°F)		(500)	(500)	(495)
Fire point (COC) (°C)	D 92	288	293	354
(°F)		(550)	(560)	(670)
Autoignition temperature (°C)	E 659	404	404	404
(°F)		(760)	(760)	(760)

Source: Product Brochure: Qunitolubric 822 Fire Resistant Hydraulic Fluids, Quaker Chemical Corporation, Conshohocken, PA. With permission.

to the general class of polyol ester-based fluids are shown in Table 17.6. Besides the bulk modulus data shown in Table 17.6, Figure 17.11 shows the compressibility (percent volume reduction) of an ISO grade 68 fluid at 20°C. Figure 17.12 shows the effect of pressure (up to 10,000 psi) on fluid viscosity at various temperatures. It should be noted that these properties represent high-quality commercial products which are currently available. These properties are dependent on the specific chemistry involved with a given fluid and the level and quality of the various additives used. It is recommended that specific suppliers be contacted for the engineering properties applicable to their products.

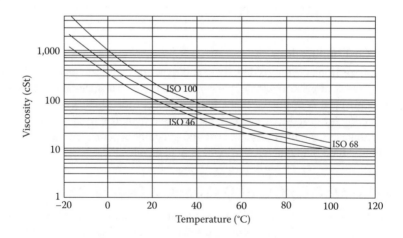

FIGURE 17.9 Viscosity of ISO 46, 68, and 100 polyol ester fluids.

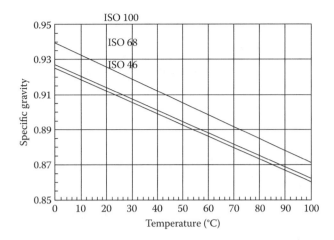

FIGURE 17.10 Specific gravity of ISO 46, 68, and 100 polyol ester fluids.

17.4 RATIONALE FOR SELECTING POLYOL ESTER-BASED HYDRAULIC FLUIDS

The vast majority of power units, either in service or in the planning phase, are designed assuming the use of mineral-oil hydraulic fluids. This is based on many years of experience, performance, and cost. Most hydraulic systems and components use the established properties of mineral-oil lubricants for design criteria. The decision to use an alternative hydraulic fluid, for whatever reason, requires an insight into the suitability of that fluid for all aspects of the hydraulic system. This requires analysis in areas such as the fluid's antiwear properties, metals and elastomer compatibility, temperature limits, environmental compatibility, presence of hazardous ingredients, special handling procedures, costs, and other considerations.

In existing hydraulic systems, the decision to change fluids is motivated by factors such as previously unrecognized fire hazards or new regulations regarding potential spills. These factors will dictate the use of fire-resistant fluids or ones that are environmentally compatible. Regarding new systems, the design process is simpler in that suitable components can be selected for any candidate fluid. Within these broad alternative fluid categories are additional choices that relate to the type of chemistry employed. A critical step, therefore, in the selection of a replacement fluid is an analysis of the potential impacts the fluid might have on the operation, life, maintenance, and operating cost of the system.

Polyol-ester-based hydraulic fluids offer certain advantages over other fluid types and often represent a simple and economic alternative to mineral-oil-based lubricants. The factors that should be considered in selecting an alternative fluid, along with the advantages and disadvantages associated with polyol-ester-based hydraulic fluids, will be discussed.

17.4.1 PUMP TESTING AND LUBRICATION PROPERTIES

The trend in fluid power technology has been continually driven toward more compact, higher-performance hydraulic systems. As a result, higher operating pressures, higher pump speeds, and more complex duty cycles have become standard with increasingly higher demands on all aspects of hydraulic fluid performance. The ability of a fluid to provide extended pump and component life is arguably its most important attribute.

Polyol-ester-based hydraulic fluids have shown a high level of antiwear characteristics and acceptable service in nearly all types of commercial hydraulic pumps at fully rated operating conditions. Performance history has varied because of product differences in additives and base stocks with varying specifications. A less-than-ideal set of operating conditions has also been a factor.

TABLE 17.6
Typical Properties—Polyol Ester Fluids

Property	ASTM Method	Value
Total acid number	D 974	<2 mg KOH/g
Strong acid number	D 974	0
Vapor pressure at		
20°C (68°F)		3.2×10^{-6} mm Hg
66°C (150°F)		7.5×10^{-6} mm Hg
Pour point	D 97	<−26°C (<−15°F)
Bulk modulus at 20°C (68°F) at		
207 bar (3,043 psi)		1.84×10^4 bar (27.05×10^4 psi)
345 bar (5,072 psi)		1.94×10^4 bar (28.52×10^4 psi)
89 bar (10,128 psi)		2.18×10^4 bar (32.05×10^4 psi)
Thermal conductivity at	D 2717	
19°C (66°F)		1.439 cal/h° cm²°C/cm
		(0.0971 Btu/h/ft.²°F/ft.)
71°C (160°F)		1.385 cal/h° cm²°C/cm
		(0.0934 Btu/hft.²°F/ft.)
Specific heat at 20°C (68°F)	D 2766	0.49 cal/g°C
Solubility in water		Negligible
Water separation	D 1401	40:35:5 (30) mL oil: water: emulsion (min.)
Foaming tendency	D 892	Pass
Dielectric breakdown voltage	D 877	30,000 V (30 kV)

Source: Product Brochure: Quintolubric 822 Fire Resistant Hydraulic Fluids, Quaker Chemical Corporation, Conshohocken, PA. With permission.

Fluid suppliers have conducted extensive laboratory and formulation studies to maximize the lubrication and pump antiwear properties of polyol esters. Unfortunately, traditional bench-scale lubrication test procedures do not satisfactorily reflect the wear processes in a hydraulic pump [15]. One exception to this generalization is the FZG gear test (DIN 51354). It has been suggested that this test and a standard vane-pump test be used to establish minimum requirements for hydraulic fluid specifications [16]. Polyol-ester-based hydraulic fluids do very well in the FZG test (+12 stages). The major product development tool, therefore, has been small-scale pump testing as defined in ASTM

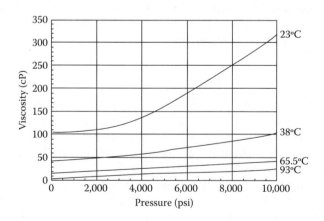

FIGURE 17.11   Compressibility of ISO 68 polyol ester fluids.

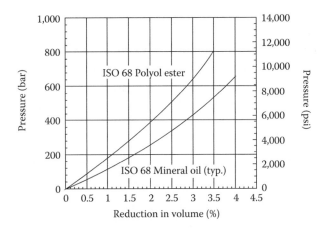

FIGURE 17.12 Pressure–viscosity relationship of ISO 68 polyol ester fluid.

D 2882 [15] and its successor D 7043. This method is based on the Vickers 104C/105C vane pump and the evaluation criterion is the loss in weight created by wear between the pump vanes and cam ring. It is based on the assumption that weight loss of the rotating parts is directly related to the antiwear properties of the hydraulic fluid. The wear process is accelerated by operating the pump at 2000 psi instead of the nominal design pressure of 1000 psi, thereby increasing the pressure loading of the vanes by 100%. These tests have been used for many years, and although there are acknowledged reservations about the extension of results to other pump types [17], it has served as a significant test for fluid development and qualification. This test is inexpensive to run and provides a rapid response. D 7043 test conditions are as follows:

- Duration: 100 hours
- Operating pressure: 2000 psi
- Operating temperature: 66°C (150°F) for fluid viscosity <50 cSt and 79°C (175°F) for fluid viscosity >50 cSt
- Fluid volume: 5 gal
- Pump: Vickers 104C or 105C (rated at 7.5 gal/min flow at 1200 rpm, 49°C, and 1000 psi) with internal parts from Conestoga

The test measurement is the combined loss in weight of the pump vanes and cam ring over the test period. Because of the differences in commercial polyol-ester-based hydraulic fluids, it is difficult to present precise information on the results obtained in this test. Table 17.7 illustrates representative weight losses for ISO 46, 68, and 100 fluids. Three values are given: the minimum that could be expected from a superior fluid, the maximum weight losses associated with marginal fluids that have sufficient antiwear properties to survive the test, and a value that represents the better polyol-ester-based fluids. There are no statistically significant differences in the wear characteristics among the ISO fluid classes.

Polyol-ester-based fluids have also been tested under the conditions specified in ASTM D 2882/D 7043 in which the running time was significantly extended. An ISO Grade 68 fluid was tested for 8000 h and the total ring and vane weight loss was 19.1 mg. An interim weight loss measurement after 167 hours was 5.1 mg.

Despite a fluid's performance as measured under ASTM D 2882/D 7043, many pump manufacturers and system designers remain skeptical of a fluid's ability to perform satisfactorily, particularly in pump designs other than the vane type, unless it satisfies certain test criteria specific to their equipment. Warranty considerations are a major factor for this position.

TABLE 17.7
Polyol Ester ASTM D 2882/D 7043 Pump Test Results

Fluid	Minimum Wear[a]	Maximum Wear[a]	Typical Wear[a]
ISO 46	<10 mg	<100 mg	<20 mg
ISO 68	<10 mg	<100 mg	<20 mg
ISO 100	<10 mg	<100 mg	<20 mg

[a] Weight loss of cam ring and vanes.

Several manufacturers conduct "in-house" tests on selected new fluids or provide specifications for an acceptable test protocol to be conducted by a fluid supplier or a third party.

Certain polyol-ester-based hydraulic fluids have been evaluated under various manufacturers' test programs and have been judged satisfactory. Individual fluid suppliers should be contacted about their status with the following established tests.

- Denison HF-O
- Eaton-Vickers 35VQ25A
- Rexroth Piston Pump Test
- Racine Wear Test

Refer to Chapter 10 for a more detailed description of these tests. Based on many factors, including results from the above tests and extensive field experience, many pump manufacturers have endorsed the use of polyol-ester-based hydraulic fluids. It should also be noted that some manufacturers have placed limitations on the operating conditions and certain applications when using a polyol-ester-based fluid. These limitations are often temporary, pending results over a trial period. These issues should be reviewed with the applicable original equipment manufacturer. The following pump manufacturers have endorsed the use of polyol-ester-based fluids based on the above:

- Eaton-Vickers
- Parker Hannifin Corp
- Sauer-Sundstrand
- Oilgear Co

Some equipment manufacturers may be reluctant to approve or recommend the general class of polyol-ester-based fluids and may restrict approvals to specific brands.

17.4.2 HYDRAULIC SYSTEM COMPATIBILITY

The possible conversion from one fluid type to another requires an analysis of the hydraulic system and an assessment of its compatibility with the new fluid. Historically, converting from a mineral-oil fluid to a fire-resistant type was viewed with skepticism. A common view was that a fluid change inevitably created hydraulic system compromises. Retrofitting seals, hoses, and elastomers with more expensive, chemical-resistant types was one concern. Derating pump speed and pressure and restricted operating temperatures were also associated with the use of fire-resistant fluids. These types of problems are not a major consequence when converting to polyol-ester-based hydraulic fluids. The major considerations are presented in the following subsections.

17.4.2.1 Components

The suitability of each component in the hydraulic system must be established. Experience with the fluid and recommendations by the fluid suppliers and the OEM become important factors. Because of the chemistry of polyol-ester-based fluids and their typical properties, few components are excluded for use. Based on experience, OEMs may express a preference for the type of fluid to be used and possibly, the supplier. These preferences may influence warranty conditions.

17.4.2.2 Metals

The basic metals used in hydraulic pumps, valves, and other components are generally compatible with polyol-ester-based hydraulic fluids. These metals include iron, steel, aluminum, nickel, tin, copper, magnesium, titanium, and silver. Three incompatible metals are lead, zinc, and cadmium [18]. Zinc and cadmium are occasionally used as cladding materials and should not be used in applications in which they are continuously exposed to the fluid. Reservoir coating is one example. Pipe and tube fittings are often plated with zinc or cadmium. Any corrosive attack would more than likely be cosmetic and not affect the functionalism or life expectancy of the fitting. These metals are also used as alloy constituents and could present potential problems. Lead is used in certain bronze alloys and, under continual exposure, may be leached out of the alloy. Surface pitting would be apparent. Generally, alloys with a lead content of less than 5% are acceptable. Sintered metal components may show the effects of lead dissolution more rapidly and should not be used.

These general compatibility comments pertain principally to the interaction between fresh fluid and the individual metals. Under service conditions, the fluid may undergo degradation through oxidation that can significantly alter its interaction with metals. Products of fluid oxidation are generally acidic and more corrosive than the original fluid. High operating temperatures and water contamination are major factors that may initiate metal attack. Dissolved metal ions and products of metal corrosion, acting as catalysts, accelerate the oxidation process and cause additional fluid breakdowns. Polyol-ester-based fluids generally contain additives to reduce these effects. Federal Test Method Standard 791B, 5308.7 is a useful test method for evaluating a metal's compatibility with polyol-ester-based fluids because it combines a simulated oxidation environment with metals commonly used in hydraulic systems arranged to form a galvanic cell. In this test, 1-in.-square coupons of steel, copper, aluminum, magnesium, and cadmium-plated steel are immersed in a 100-mL sample of fluid. Air, at 5 L/h, bubbles through the fluid, maintained at 250°F, for 168 hours (seven days). At the end of this period, the metal specimens are inspected for visual appearance and any changes in weight are determined. Additionally, the fluid is tested for changes in total acid number (TAN) and viscosity. Table 17.8 summarizes the weight losses associated with each level of attack.

This is a useful test for evaluating fluid additives, comparing competitive polyol-ester-based products, and evaluating the effects of service on fluids. There are commercial-grade polyol-ester-based fluids that rank in the "zero to light" category for all metals except cadmium. The cadmium specimens exhibit excessive weight losses and surface pitting.

TABLE 17.8
Metal Corrosion—Federal Test Method No. 791B, 5308.7

Metal	Weight Loss (mg/cm^2)		
	Mild	Moderate	Heavy
Steel	0 to ± 0.077	± 0.077 to ± 0.38	> ± 0.38
Copper	0 to ± 0.077	± 0.077 to ± 1.16	> ± 1.16
Aluminum	0 to ± 0.077	± 0.077 to ± 0.38	> ± 0.38
Magnesium	0 to ± 0.077	± 0.077 to ± 0.38	> ± 0.38
Cadmium	0 to 27	27 to 78	> 78

17.4.2.3 Seals, Packing, and Hose Elastomers

Seals, hoses, packing, accumulator bladders, and other nonmetal components used with mineral-oil-based hydraulic fluids are usually compatible with polyol-ester-based hydraulic fluids. This can be a major factor in the conversion from a mineral-oil-based fluid to one requiring fire resistance. The OEMs responsible for the various system components should be consulted and recommendations from the fluid supplier solicited before making a final decision.

Table 17.9 lists most of the elastomers employed in hydraulic systems. Those listed as "recommended" elastomers are suitable for service with polyol-ester-based fluids with rare exceptions. These materials have been tested for compatibility by most of the fluid suppliers and have had a successful history of usage under normal conditions. "Conditional" elastomers should be used only under certain conditions. The elastomer may undergo excessive swelling and softening because of fluid absorption and should not be used in dynamic applications. These materials may be suitable for static seal applications. However, replacement with a "recommended" elastomer is advised during the fluid changeover procedure or during the next regularly scheduled maintenance procedure. "Others" are elastomers that do not have an established history of service or have been proven unsatisfactory with polyol-ester-based hydraulic fluids. If there are compelling reasons to use these materials, the fluid and seal suppliers should be consulted. Compatibility testing should be conducted before use.

With all materials, the fluid operating temperature may be a major factor. Fluid and elastomer suppliers usually recommend the operating temperature limits.

17.4.2.4 Coatings

Reservoir surfaces generally do not need protective coatings. If required, epoxy coatings and thermally cured phenolic or polyurethane paints may be used. Again, compatibility tests should be conducted. Latex, acrylic, and alkyd paints are not suitable. These paints are often applied to exterior surfaces for cosmetic purposes and could be subjected to blistering and peeling. Stainless-steel reservoirs may be used under extreme conditions.

TABLE 17.9
Compatibility of Elastomers with Polyol Ester Fluids

Recommended for all Applications	
Fluorocarbon rubber (FPM)	BUNA N/nitrile[a]
Epichlorohydrin Rubber (CO, ECO)	Polyacrylate rubber (ACM)
Fluorosilicone (FSI)	Polyurethane rubber (AU, EU)
Polytetrafluoroethylene (TFE)	

Conditional Compatibility[b]	
Ethylene–propylene rubber (EPDM, EPM)	Butyl rubber (IIR)
Buna N/nitrile (NBR)[c]	Neoprene rubber (chloroprene, CR)

Others[d]	
BUNA S rubber (SBR)	Chlorosulfonated polyethylene (CSM)
Polychloroprene (CR)	Polysulfide (T)
Butadiene rubber (BR)	Chlorinated polyethylene (CM)
Ethylene acrylic	

[a] Compounded with >30% nitrile copolymer.
[b] Should not be used for dynamic applications; provides reduced life expectancy for static seals.
[c] Compounded with <30% nitrile.
[d] Check with seal and/or fluid supplier for recommendations.

17.4.2.5 Plastics

Most thermoplastics are compatible with polyol-ester-based fluids and are suitable for components such as molded reservoirs, tubing, and fittings. Polyvinyl chloride (PVC), polyethylene, polyurethane, polypropylene, polytetrafluoroethylene (Teflon) and polyamide have been used successfully. Specific recommendations and operating limits should be solicited from both the fluid and seal suppliers. If necessary, compatibility testing should be carried out.

17.4.2.6 Filter Media

Polyol-ester-based lubricants are compatible with all commercial filter media including sintered powder, wound and woven metal, natural and synthetic fiber, and impregnated fiber. There are no restrictions on micron ratings. Adsorbent cartridge filters that incorporate Fuller's earth, activated alumina, or other exchange resins are often used with phosphate-ester-based fluids to maintain a low level of acidity [19]. This type of medium is not necessary with polyol-ester-based fluids. Water-absorbent media, however, are useful and should be employed in applications subject to water contamination.

17.4.3 Oxidation and Thermal Stability

The major factor in the practical life of a hydraulic fluid is its ability to resist oxidative and thermal degradation. These changes can cause a fluid to undergo significant increases in viscosity and total acid number (TAN), generate varnish and sludge, and cause an accelerated corrosive attack on metals. Polyol-ester-based hydraulic fluids, by nature of their chemistry, are vulnerable to oxidation because of polyunsaturation in the fatty acid portion of the ester (see Section 17.2.1). Fortunately, two factors reduce this susceptibility. Additive technology has progressed significantly in recent years, and fluid oxidation can be effectively reduced. In addition, base ester quality has improved since the commercial introduction of polyol-ester-based fluids, making the ester more responsive to additives and less susceptible to oxidative attack.

Most of the commonly used test procedures for measuring hydraulic fluid oxidation resistance and thermal stability have been developed for mineral-oil fluids. These include Federal Test Method Standards No. 791B, 5308.7 and ASTM methods D 943, D 2070, D 2272, and D 4310. The two that are most applicable to polyol-ester-based fluids are the Federal Test Method 791B, 5308.7 and the Cincinnati Milacron Thermal Stability Test (ASTM D 2070). Compared with other tests, these procedures do not incorporate water into the test. Polyol esters, like phosphate esters, are acknowledged to be more water sensitive than oil-based fluids.

The test conditions of the two methods are summarized in Table 17.10. The Federal method is directed toward measuring oxidation and incorporates an air sparge as the oxygen source. This test was briefly described in Section 17.4.2.3. The Cincinnati Milacron procedure, on the other hand, is directed toward the thermal stability of the fluid and is conducted at a higher temperature, 135°C (275°F) versus 121°C (250°F). Test duration and sample sizes are the same in both tests. Metal specimens are used in both tests and serve a dual purpose. Metals act as catalysts in the oxidation process and the metals listed are commonly found in hydraulic systems. Also, oxidized or thermally degraded hydraulic fluids produce corrosive by-products and accelerate metal attack. These tests offer a measure of this effect.

The evaluation criteria are similar for each test. In the Federal method, metal corrosion is judged by weight change and surface inspection. Fluid oxidation is monitored by changes in TAN and viscosity over the test period.

Thermal breakdowns of hydraulic oil generally result in the formation of particulate matter and metal corrosion. In the Cincinnati Milacron test, the sludge formed is separated by filtration, weighed, and expressed as milligram of sludge per 100 mL of fluid. The metal specimens are weighed for weight loss and assessed for surface effects (e.g., pitting, discoloration, etc.).

TABLE 17.10
Oxidation and Thermal Stability Test Methods

Test Parameter	Federal Test Method Standard No. 791B	ASTM D 2070: Thermal Stability of Hydraulic Oils
Duration	168 h (7 days)	168 h (7 days)
Temperature	121°C (250°F)	135°C (275°F)
Airflow	3 L/h	None
Fluid sample size	200 mL	200 mL
Metals		
Steel	✓	✓
Aluminum	✓	
Magnesium	✓	
Cadmium	✓	
Copper	✓	✓
Metal configuration	2.54-cm × 2.54-cm × 0.813 mm thick	6.35-mm-diameter × 7.6-cm-long rod

Results from the metal corrosion phase of the Federal test method were shown in Table 17.8. In addition, the changes in fluid properties, because of oxidation, are shown in Table 17.11. Again, representative numbers are shown for three levels of fluid oxidation: mild, moderate, and severe. The results shown in the "mild" category represent high-quality polyol fluids in all ISO viscosity grades.

Table 17.12 shows the results from the Cincinnati Milacron Thermal Stability test using a similar format (i.e., three levels of fluid stability). As in the Federal test method, the results shown under "mild" represent the higher grades of polyol-ester-based fluids.

One shortcoming of standard oxidation and thermal stability testing is the inability to correlate the results with field experience. Monitoring fluid conditions during extended pump testing provides added insight into a fluid's resistance to the thermal and oxidative stresses induced by such a test. Again, these effects will vary widely based on the particular polyol-ester-base fluid being evaluated. The ISO 46 and 68 fluids have been tested under ASTM D 2882/D 7043 conditions continuously for

TABLE 17.11
Fluid Oxidation—Federal Test Method 791B, 5308.7

	Degree of Fluid Oxidation[a]		
	Mild	Moderate	Heavy
Viscosity[b] at 40°C (cSt)			
Initial	68	68	68
Final	60–75	41–60:75–85	<41: >85
Change[c] (%)	0–10	10–25	>25
Total acid number[d]			
Initial	2.0	2.0	2.0
Final	1.6–2.4	1.2–1.6:2.4–2.8	<1.2 to >2.8
Change[c] (%)	0–20	20–40	>40

[a] Fluid oxidation rates are representative of ISO 46, 68, and 100 fluids.
[b] ASTM D 445.
[c] Fluid property changes are generally positive (+).
[d] ASTM D 974.

TABLE 17.12
Thermal Stability of Polyol Ester Fluids (ASTM D 2070)

	Degree of Fluid Degradation[a]		
	Mild	Moderate	Heavy
Viscosity[b] at 40°C (cSt)			
Initial	68	68	68
Final	60–75	41–60:75–88	<41:>88
Change[c] (%)	0–10	>10:<30	>30
Sludge formation[d]	0–8	8–20	>20
Total acid number[e]			
Initial	2.0	2.0	2.0
Final	1.7–2.3	1.4–1.7:2.3–2.6	<1.4 >2.6
Change[c] (%)	0–15	>15:<30	>30

[a] Fluid degradation rates are representative of ISO 46, 68, and 100 fluids.
[b] ASTM D 445.
[c] Fluid property changes are generally positive (+).
[d] In units of mg of sludge formed per 100 mL of sample.
[e] ASTM D 974.

5000 h (approximately seven months) and monitored for TAN and viscosity change. Viscosity and TAN as a function of time are shown in Figure 17.13. Given the severity of the test conditions and duration, the results reflect a stable fluid. These results also show the influence of fluid temperature on the condition of the fluid under high-stress conditions. The ISO 68 fluid was tested at 79°C (175°F) and the ISO 46 at 65.5°C (150°F); the higher temperature caused higher increases in acid number and viscosity because of an increased rate of fluid oxidation.

In recent years, another oxidation test has been developed which shows good field correlation with good and bad hydraulic fluids. Pressure Differential Scanning Calorimetry (PDSC) measures the exotherm produced by thin-film uptake of oxygen by a fluid. ASTM D 6186 establishes standard conditions for testing a variety of fluids. Vegetable oil and polyol-ester-hydraulic fluids are tested at the lower end of the evaluated temperature range.

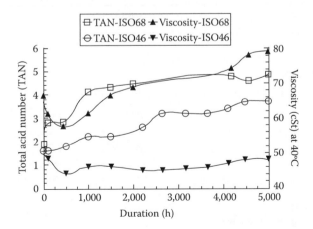

FIGURE 17.13 Extended pump testing of polyol ester hydraulic fluids.

ASTM is working to establish a test method for running the D 943 oxidation test on ester-based hydraulic fluids without added water. The "dry-TOST" test has been used for many years in Europe and the US to evaluate polyol-ester and vegetable oil-based hydraulic fluids. Good correlation with the field performance of these fluids has been found.

17.4.4 OPERATIONAL TEMPERATURE RANGE

The liquid range of polyol-ester-based hydraulic fluids is broad, less than 0°C (32°F) and greater than 93°C (200°F); however, the practical operating range is much narrower. The recommended range is between 10°C (50°F) and 66°C (150°F). The fluid is functional beyond these limits; however, certain characteristics become compromised. At the high end, continuous exposure at elevated temperatures will cause oxidation and thermal breakdowns of the fluid. Excursions into temperatures as high as 93°C (200°F) are possible, but some fluid degradation will take place, depending on the temperature reached and the exposure time. In these instances, fluid monitoring programs should be started. At the low-temperature end, the limitation is one of equipment compatibility with the fluid viscosity at that temperature.

17.4.5 FIRE RESISTANCE

The major growth in the use of polyol-ester-based hydraulic fluids has been in applications with the potential for fire. A fire hazard associated with the use of mineral-oil-based hydraulic fluids is the ignition potential of spray mists. Spray particles are often formed under high-pressure expulsion of the fluid from the hydraulic system and are readily ignited; the flame is then propagated throughout the spray envelope. This relative ease of ignition is attributed to the fluid's low fire point and the small particle size of the mist. Other hazards include pool fires, ignition of oily waste material, and secondary fire damage. Polyol-ester-based fluids derive their fire resistance from a combination of factors. The flash, fire, and autoignition temperatures are relatively high compared with mineral-oil fluids (Table 17.13), and the products may be formulated with modifiers that alter the spray-mist characteristics. A larger-particle-sized coarse spray is produced, which is more difficult to ignite, and the flame does not readily propagate throughout the spray [20].

Polyol-ester-based hydraulic fluids have been tested by Factory Mutual Research Corporation (FMRC) and have been listed in their Approval Guide [4] since 1973. FMRC has changed its approval standards to use total heat release from a spray fire and ease of ignition from the fire point to calculate a Spray Flammability Parameter (SFP) [14]. Under the July 2008 Approval Standard for Flammability Classification of Industrial Fluids Class 6930, fluids with a fire point and a normalized SFP of 5×10^4 or less are classified as Approved. Fluids with SFP greater than 5×10^4 but less than 10×10^4 are classified as Specification Tested [22,26]. Polyol ester hydraulic fluids on the market today could be either Factory Mutual Approved or Specification Tested, depending on their composition.

TABLE 17.13
Thermal Properties of Polyol Ester Fluids

	Flash Point[a, b]	Fire Point[a, b]	Autoignition[b, c]
Polyol ester	260–295°C	280–350°C	388–404°C
	(500–563°F)	(536–662°F)	(730–760°F)
Mineral oil	102–200°C	110–225°C	230–350°C
	(215–392°F)	(230–437°F)	(446–662°F)

[a] ASTM D 92.
[b] Representative values.
[c] ASTM D 2155/E 659.

The Mine Safety & Health Administration (MSHA) employs a spray-mist ignition as part of their test protocol. Polyol-ester-based lubricants have been approved for mining applications [21,27].

The advantages and disadvantages of polyol-ester-based fluids as fire-resistant hydraulic fluids can be summarized as follows:

Advantages

- Within manufacturers' specifications, the fluids exhibit excellent spray-mist fire protection.
- Polyol esters exhibit high fire point, >288°C (550°F), and very low vapor pressure.
- When burning under continuous ignition, excessive quantities of smoke or particulate matter are not generated and are not considered toxic. Other synthetic fluid types, notably phosphate esters, under similar conditions may significantly reduce visibility and habitability without assisted breathing equipment.
- Because of low smoke generation, radiant energy is low with reduced secondary fire damage.
- In "pool fires," the low vapor pressure of polyol esters contributes to a longer ignition delay.

Disadvantages

- As with all hydraulic fluids, fire resistance may be dependent on fluid condition and service history. A fluid monitoring program should be started with the fluid supplier.
- The autoignition temperature (404°C, 740°F) is low compared with other non-water-containing fire-resistant fluids and may present some risk in certain applications involving molten light metals and superheated manifolds. Water-containing fluids offer other problems in these applications.

17.4.6 FLUID LOSS AND ENVIRONMENTAL CONSIDERATIONS

All hydraulic systems are subject to fluid leakage caused by unpredictable component failures, routine system and fluid maintenance procedures, and other factors. In stationary systems, this leakage can be readily contained through dikes or drip pans. With mobile applications, contamination of the surroundings with hydraulic fluid is inevitable and the net effect on waste-treatment facilities, natural waterways, and the general environment must be considered. Excessive contamination could result in fines, the need to restore damaged property, increased waste-treatment costs, and higher insurance rates.

Except for vegetable-oil-based fluids, this is one area in which polyol-ester-based lubricants have clear advantages over other fluids, particularly the historic fire-resistant types. These advantages are as follows:

- **Biodegradability:** The chemical structure, similar to natural glycerides, lends itself to biodegradability. All of the test protocols under consideration by the global community for rating biodegradability rank polyol esters within proposed guidelines. For examples, CEC-L-33-T-82 is widely used to measure the biodegradability rates of lubricants in water. This test measures the biological depletion rate of a hydrocarbon chain over a 21-day period. Polyoleates exhibit greater than 80% degradation [23] in this test, generally superior to most other lubricant base stocks. OECD 301B (modified Sturm) test measures conversion to CO_2 over 28 days. This is a more rigorous test that is now becoming the industry standard for ready biodegradability. To pass the test, a lubricant must show > 60% biodegradation. Polyol-ester-based fluids exhibit 70–90% biodegradation.

  It should be noted that biodegradability studies generally relate to the base stock of the fluid under consideration. Additives may compromise the biodegradability of the base stock because they generally do not have comparable degradation rates. Usually, the low level of additives used reduces this concern. Nevertheless, before a specific fluid is selected or specified, the original manufacturer should be responsible for supplying biodegradability data specific to that fluid.

- **Aquatic toxicity:** The LC_{50} for a fully formulated polyol-ester-based fluid has been reported as greater than 2000 ppm under testing done according to OECD 203, "Fish Acute Toxicity Test" [24]. This characteristic, however, is also dependent on the additives used in the product and results specific to a given product should be obtained from the supplier.
- **Nonhazardous components:** Many commercial products are free of hazardous materials listed under EPA and OSHA (40 CFR and 29 CFR). State regulatory requirements should be consulted as necessary.
- **Water miscibility:** Polyol-ester-based fluids are not miscible with water and have a density less than water. This eases removal of the fluid from settling ponds and waste-treatment ponds using conventional skimming techniques. In contrast, many other fire-resistant fluids are either soluble in water (water–glycol) or heavier than water (phosphate ester) and add to waste-treatment chemical costs. The "floating" feature of a polyol-ester-based fluid also allows visual detection of leakage.
- **Incineration:** Reclaimed or recovered hydraulic fluids are often disposed of through incineration to recover their Btu content. Most polyol-ester-based fluids can also be treated in this manner. Many other fire-resistant fluids are not so easily accommodated because of water content or chemical toxicity.

17.4.7 Comparison of Polyol-Ester-Based Hydraulic Fluids with Other Fluids

Under certain circumstances, ester-based hydraulic fluids are a potential replacement for mineral-oil hydraulic fluids. Other fluid types can also be considered, each with certain advantages and disadvantages. Based on characteristics generally thought to be important, Table 17.14 summarizes how various fluids compare. Such a ranking can be subjective based on the particular bias of the scorer. Nevertheless, an objective rating was attempted based on a collection of experiences. It is acknowledged that one characteristic may be all-important, despite that fluid's ranking in other categories. It is simply intended as a general guide.

17.4.8 Conversion Considerations and Procedures

Converting from the different types of hydraulic fluids to a polyol-ester-based hydraulic fluid is a straightforward process. Specific recommendations and suggestions should be made by the manufacturer of the replacement fluid. These recommendations should include a detailed analysis of the

TABLE 17.14
Comparison of Polyol Esters with Other Fluids

Fluid Property	Relative Ranking[a]
Fire resistance	WaG > PhE > PoE > IvE > VgO > MnO ≈ Pao
Oxidation resistance	WaG ≈ Pao > MnO > PhE ≈ PoE > IvE > VgO
Lubrication	MnO > PhE ≈ PoE ≈ Pao > VgO > WaG > IvE
Biodegradability	VgO ≈ PoE > WaG > PhE > IvE > MnO ≈ Pao
Hydrolytic stability	WaG > IvE > MnO ≈ Pao > PoE > VgO > PhE
Unit cost ($/gal)	Pao ≈ PhE > PoE ≈ Vgo > WaG > MnO > IvE
Ease of disposal	VgO ≈ PoE > WaG > PhE ≈ IvE ≈ MnO ≈ Pao
Seal compatibility	MnO > IvE > WaG > VgO ≈ PoE ≈ Pao > PhE
Thermal stability	Pao ≈ PhE > MnO > PoE > VgO > IvE ≈ WaG

[a] WaG = water–glycol; PhE = phosphate ester; IvE = Invert emulsion; VgO = vegetable oil; MnO = mineral oil; Pao = polyalphaolefin; and PoE = polyol ester.

hydraulic system regarding its general compatibility and any changes that should be made to adapt to the new fluid. Often, the supplier will be actively involved in the conversion. The conversion procedures are summarized in Table 17.15 and general comments on various aspects of the procedures are as follows:

- **Seal and hose compatibility:** Fluid and elastomer suppliers should verify the compatibility of a given material with a polyol-ester-based fluid. Generally, seals and hoses used with mineral-oil, vegetable-oil, synthetic ester, or invert emulsion fluids will be satisfactory. Seals and hoses used with other fluid types should be checked for their compatibility with polyol-ester-based fluids and, if necessary, replaced with suitable ones.
- **Drain and purge requirements:** For all conversions, the recommended practice is to drain the original fluid from the reservoir and lines. A polyol-ester-based fluid is generally compatible with mineral-oil- and vegetable-oil-based fluids and can be used as makeup. However, all of the benefits of the new fluid will not be fully realized, particularly fire resistance. In these applications, at least 95% of the old fluid should be removed. Polyol-ester-based fluids are also compatible with most phosphate-ester-based fluids *if* the fluid is in excellent condition (moisture <0.05% and total acid number <0.05). A polyol-ester-based fluid will chemically react with strong mineral acids in phosphate esters to form a varnish-like material which will cause valve sticking and reduced pump life. Water-based fluids must be completely removed from the system with a target efficiency of at least 97%.
- **System flushing:** For systems containing water-based fluids, a flushing fluid must be used before adding a polyol-ester-based hydraulic fluid. This fluid replaces the quantity of water-based fluid initially drained and may be an inexpensive mineral-oil- or polyol-ester-based fluid. The system is operated at a reduced pressure and all functions actuated for approximately thirty cycles.

  The purpose is to displace any residual fluid trapped in lines and components with a non-water-containing fluid. The contaminated flushing medium is then drained and discarded. Some fluid suppliers provide a flushing compound that will absorb or emulsify the residual water-based fluid. This fluid can be saved and reused for subsequent conversions; it may have to be processed to remove excess water according to the supplier's recommendations.

TABLE 17.15
Hydraulic System Conversion Practices

	Mineral Oil	Water Glycol	Phosphate Ester	Invert Emulsion	Vegetable Oil	HWCF (95:5)
Compatibility						
• Seals	OK		Verify		OK	Verify
• Hoses	OK		Verify		OK	Verify
• Coatings	OK		Verify		OK	Verify
Drain fluid & purge	Not req'd	Yes	Not req'd	Yes	Not req'd	Yes
System flush	Not req'd	Yes	Yes	Yes	Not req'd	Yes
Clean components			Yes			
Replace filter elements			Yes			
System startup			Yes			
Monitor fluid			Yes			
Monitor filters			Yes			

Abbreviation: HWCF, high water content fluid.

For conversions involving mineral-oil- or vegetable-oil-based fluids, a flushing procedure is generally not required. In a phosphate ester conversion, the need for a flushing procedure is based on the fluid condition. If the fluid's acid level is high, flushing is necessary. For hydraulic systems that have had a history of varnish buildups, use additives that, when added to the flushing medium, will dissolve or loosen accumulated varnish. This is a recommended procedure and the material is generally available from the fluid supplier.

- **System cleanliness:** To the extent possible, residual fluid and particulate matter should be removed from system components. In particular, the reservoir should be wiped down and any deposits such as scale and other foreign matter removed from the internal surfaces. It is also recommended that heat exchangers and fluid coolers be inspected and serviced if necessary.
- **Filter elements:** After the final flushing, or just before adding the new fluid, all filter elements should be replaced. Screens should be cleaned and replaced if damaged.
- **System startup:** Startup procedures should follow the recommendations provided by the original equipment manufacturer. This generally involves low initial loads and operating pressures, cycling valves, and bleeding-off any entrained air.
- **Fluid monitoring:** This is the most critical phase of the conversion. Fluid samples should be obtained within 24 hours of the changeover and tested for the amount of the previous fluid retained and dirt content. If water content is more than 2%, the fluid should be treated to remove water or dumped and the system recharged with fresh fluid. If dirt content is beyond the fluid supplier's specifications, replace filter elements and process the fluid through an external filter unit. A fluid change generally results in loosening and entrainment of contamination which has built up over time and will be circulated with the fluid. If the amount of retained old fluid is excessive, the fluid should be drained and recharged with fresh fluid. The fluid condition should be monitored weekly for the first month and regularly after that. The fluid supplier should provide guidelines on sampling frequency and condition criteria.
- **Filter monitoring:** For the above-cited reasons, filter elements may load up initially at a higher rate than normal and should be serviced as required.

17.5 FLUID MAINTENANCE

A major factor in the operation of a trouble-free fluid power system is fluid cleanliness and the retention of its original properties. This section will identify problems often encountered with polyol-ester-based fluids and present recommended practices that will maximize fluid life expectancy and retention of basic properties.

17.5.1 Types of Contamination

The sources of fluid contaminants are either generated from within the hydraulic system or originate from outside it. Contaminations generated within the system are wear particles and products caused by chemical attack on seals and hoses, corrosive attack, and degradation of the fluid through oxidation. Examples of external contamination are moisture or water ingested into the system, dirt, sand, and airborne process contaminants. Contaminants may also be introduced during routine fluid additions and maintenance procedures.

17.5.2 Fluid Variables to Monitor

All of the above-described potential contaminants may be identified through fluid monitoring programs and, usually, the cause or source readily identified and appropriate corrective actions taken. A fluid analysis program includes both physical testing and spectrochemical analysis.

The following physical variables should be monitored and compared with self-imposed standards or those recommended by the fluid supplier:

- **Kinematic viscosity at 40°C (104°F):** Viscosity monitoring is perhaps the most important measure of the fluid condition. Changes in viscosity that deviate from the manufacturer's standard suggest possible contamination from other fluids, excessive fluid oxidation, or degradation of additives. Listed below are the normal viscosity ranges for the three common polyol-ester-based fluids. Any fluid with a viscosity outside of this range should be examined for potential problems.

    ISO 46 viscosity grade: 41–51 cSt (191–237 SUS)

    ISO 68 viscosity grade: 61–75 cSt (282–348 SUS)

    ISO 100 viscosity grade: 90–110 cSt (418–510 SUS)

- **Water (% or ppm):** Water can be detrimental to a polyol-ester-based fluid because of its role in fluid oxidation and metal corrosion (see Sections 17.2.4 and 17.4.3). The target for maximum water content should be 0.2% to ensure minimum influence. A polyol-ester-based fluid can hold up to 2% water without short-term detrimental effects; however, this condition should not be allowed to continue without corrective action. Fluids contaminated with water can be treated with water-absorbent filter media or short-term elevations in fluid operating temperature. The important response is the identification and correction of the conditions leading to the contamination.

- **Total acid number (ASTM-D 974):** Increases in TAN values show fluid degradation has taken place and suggests metal corrosive attack rates may be accelerated. The allowable TAN will vary with fluids from different manufacturers; however, as a rule, the TAN should be < 5mg KOH/g. A more realistic monitor should be the increase in TAN. Rate increases of 0.1 mg KOH/g per month or larger suggest potential fluid problems.

- **Particle count:** Automated optical particle counters are used for determining particle size distribution. These data can be used to establish a level of contamination as defined by an ISO 4406 Solid Contamination Code [25]. In this code, the particle size distribution is defined by three ISO range numbers (e.g., 18/15/12). The first number represents the number of particles greater than 4 μm in size, the second number represents the number of particles greater than 6 μm, and the third number represents the number of particles greater than 14 μm. ISO range numbers correspond to the maximum number of particles per milliliter of fluid, as shown in Table 17.16. Figure 17.14 shows suggested ISO cleanliness levels for different types of components and operating pressures. Fluid contamination

TABLE 17.16
Cleanliness Codes

ISO Contamination Level	Maximum Number of Particles	
	>6 μm	>14 μm
12/8	40	2.5
13/9	80	5
14/10	160	10
15/11	320	20
16/12	640	40
17/13	1,300	80
18/14	2,500	160
19/15	5,000	320
20/16	10,000	640

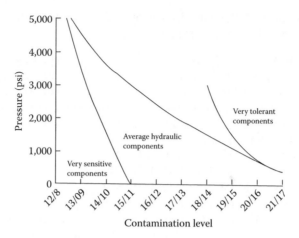

FIGURE 17.14 Suggested cleanliness levels. (Reprinted from *Effective Contamination Control in Fluid Power Systems,* Eaton-Vickers Inc., 1980. With permission.)

levels above the ranges shown may result in excessive component wear, and improved filtration practices should be employed.

- **Total solids:** Gravimetric analysis is a useful test for determining the total weight of particulate matter in the fluid. In this test (ASTM D 4898), a known volume of the fluid is passed through a filter patch and the amount of solids retained is weighed. This test is particularly useful when used with particle count data. Also, the filter patch lends itself to microscopic and photographic examination for identification of metallic and nonmetallic contaminants. Retained solids more than 1.5 mg/mL indicate a possible fluid problem.

Spectrochemical analysis is useful in that it provides evidence and identification of metals in the fluid caused by wear processes within the hydraulic system. Also, the analysis can be used to monitor the level of metal-based additives. Metals that might be present and can be identified through spectrochemical analysis include the following:

Iron	Chromium	Nickel
Aluminum	Lead	Copper
Tin	Silver	Titanium
Silicon	Boron	Sodium
Potassium	Molybdenum	Phosphorus
Zinc	Calcium	Barium
Magnesium	Antimony	Vanadium

In addition to assessing the condition of hydraulic fluids, the value of fluid analysis is to provide insight into the cause of system problems. Although certain tests can provide measurements of fluid condition, the major tool is in interpreting the overall test results and the possible interactions between different tests. Fluid analysis programs are often conducted in concert with the supplier, who is familiar with the allowable ranges of certain variables and is knowledgeable in the specific product chemistry. The fluid analysis program is also enhanced by knowledge of the fluid history based on previous analysis and trends in certain parameters.

17.5.3 Fluid Life Expectancy

The projected life of a polyol-ester-based hydraulic fluid, or any fluid, is impossible to quantify because of the many variables that might affect its potential life expectancy. The most reliable method for evaluating a fluid's potential for continued high performance is a rigorous fluid monitoring program. Fluid degradation is directly related to extreme operating temperatures and contamination with alien fluids and particulate matter.

Figure 17.15 is a general guide to the expected life expectancy of a polyol-ester-based hydraulic fluid as a function of its average operating temperature assuming no other negative variables. Because the operating temperature has such a major impact on fluid condition, continuous monitoring of heat exchangers, coolers, and fluid reservoir levels is important.

As noted in Figure 17.15, 57°C (135°F) should be regarded as a nominal, average fluid temperature. Above that temperature, more rapid fluid degradation can be expected and, conversely, below that temperature, extended fluid life is to be expected.

17.5.4 Fluid Testing Program

A formal fluid sample and analysis program is recommended. This can be carried out with the fluid supplier but is often done independently to ensure objectivity of the results and conclusions. Because the test methodology is sophisticated and requires specialized equipment, most users will not have the resources to do "in-house" testing. Many test laboratories do have the capability and will generally supply test results within one week. A supplier's response to fluid analysis should be compared.

A fluid sample analysis program is inexpensive and often has a positive return in reduced maintenance cost, reduced unscheduled downtime, and extended fluid life. Many fluid suppliers provide fluid sample analysis as a service item or charge a nominal fee. Third-party laboratories typically provide sample analysis at $25 to $100 per sample, depending on the scope of testing required.

The frequency of obtaining samples will vary depending on the application. Quarterly sampling is adequate unless a history of system problems suggests otherwise, in which case weekly or biweekly sampling would be appropriate. For recent conversions, weekly sampling is suggested until a history of trouble-free operation has been established.

The point of fluid sampling should be selected to ensure the acquisition of representative samples. If possible, the sample valve should be upstream of the return-line filter and the valve should be of an unrestricted orifice style (ball or gate). This reduces high velocities through the valve which can often separate or concentrate solids. Reservoir sampling should be a last resort, but if done, the power unit must be operating to reduce any settling effects and sampled at mid-depth of the fluid.

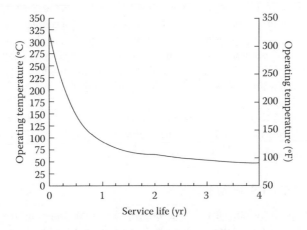

FIGURE 17.15 Polyol ester fluid life expectancy.

Maintaining a history of fluid sample reports is integral to a successful program. Trends in the various parameters are equally as important as the measurements themselves. Many fluid sample report forms incorporate a historical profile that may tabulate previous results or present them in a graph format.

17.5.5 FLUID DISPOSAL PROCEDURES

The important consideration with a polyol-ester-based hydraulic fluid spill is that the material is not regarded as hazardous by Federal Standards, although all spills to the environment are to be cleaned up. The fluid is often marketed as an environmentally safe product and is considered readily biodegradable and nontoxic to aquatic life. This does not, however, relieve the user from exercising responsible and meaningful treatment of any materials spilled or discharged from the hydraulic system. Specific fluid hazardous material information, treatment procedures, and other safety data should be obtained from the fluid supplier before its use.

Floor spills should be treated with standard industrial absorbents and disposed of according to local, state, or federal regulations. This material is not generally considered a hazardous material. Residual spilled materials can be washed away using mild detergent and copious amounts of water.

Any significant quantity of fluid that reaches waste-treatment ponds or settling basins can be skimmed using equipment normally employed for oily wastes. The fluid recovered can be processed by waste-treatment companies or incinerated at the plant site to recover its energy value. Used fluid drained from hydraulic reservoirs can also be treated in this way.

Spills from mobile equipment or at construction sites should be avoided if possible; large spills can be partially cleaned up by using absorbent materials. Minimum ground contamination will be consumed by biological processes. Ground contaminated with a polyol-ester-based fluid is not regarded as hazardous waste.

17.5.6 FLUID RECOVERY

A fluid analysis report often concludes that the fluid is no longer suitable for continued use; disposition of the contaminated fluid must be determined. Two alternatives are possible: discard the fluid or attempt to recover the fluid for continued service. Recovery is the preferred route, particularly with large fluid volume systems where the replacement cost could be significant.

A polyol-ester-based fluid can be reconditioned for additional use depending on the nature of fluid degradation and the type of application. If fluid properties such as viscosity and TAN are outside of the suppliers recommended range, oxidation and shear breakdown are the probable causes and the fluid cannot be salvaged, especially when fire resistance was the reason for fluid selection. If particle count and total solids were the reasons for fluid rejection, the fluid can be cleaned by using an external filter system or by upgrading the hydraulic system filters. The source of contamination should be determined and corrected. For cases where excess water has been found in the fluid, vacuum flashing techniques or filtering through water-absorbing media will correct the problem. The cause for water contamination should be thoroughly investigated. The advice of the supplier should always be solicited.

It is not recommended that polyol-ester-based hydraulic fluids be sent to a commercial fluid recycle facility, especially when the fluid was fire-resistant. Only the original supplier can attest to the fluid meeting fire-resistant fluid industry standards.

17.6 SUMMARY OF APPLICATIONS

Polyol-ester-based fluids have been used in a variety of applications and are usually associated with processes that impose a potential fire hazard. More recently, the fluid has found application in environmentally sensitive areas in which biodegradability characteristics are considered as highly important. A major factor in the use of polyol-ester-based fluids in diverse applications is the relative ease of replacing mineral-oil fluids with one that meets newly imposed safety and environmental requirements.

Listed below are the major Standard Industrial Classifications (SIC) in which polyol-ester-based hydraulic fluids have been used. This list is representative and does not include many applications where the fluid has been successfully employed. The intent is to show the diversity of applicability for this type of fluid.

	SIC No.
Blast furnace and basic steel products	331
Industrial machinery and equipment	35
Iron and steel foundries	332
Motor vehicles and equipment	371
Fabricated metal products	34
Nonferrous rolling and drawing	335
Metal forging and stampings	346
Nonferrous foundries and die casting	336
Aircraft and parts	372
Heavy construction	16
Oil and gas extraction	13
Rubber and miscellaneous plastic parts	30
Electric, gas, and sanitary services	49
Lumber and wood products	24
Petroleum and coal products	29
Coal	12
Railroad equipment	374
Metal cans and shipping containers	341
Chemicals and allied products	8
Food and kindred products	20
Metal mining	10
Space propulsion units and parts	376
Water transportation	44
Pipelines except natural gas	46
Motorcycles, bicycles, and parts	375
Ship- and boat-building and repair	373
Paper and allied products	26
Miscellaneous primary metal products	339
Stone, clay, and glass products	32
Forestry	08
Miscellaneous amusement and recreation services	799

REFERENCES

1. Totten, G.E. and Webster, G.M. "Fire Resistance Testing Procedures: A Review and Analysis," in *Fire Resistance of Industrial Fluids, ASTM 1284,* Totten, G.E., and Reichel, J. eds., American Society for Testing and Materials; Philadelphia, 1996.
2. Evanoff, T. "Selection of Hydraulic Fluids for a Rolling Mill," STLE Annual Meeting, 1996.
3. Gere, R.A. and Hazelton, T.V. "Rules for Choosing a Fire Resistant Hydraulic Fluid," *Hydraulic Pneumatics*, April, 1993.
4. Factory Mutual Research Corporation, *Factory Mutual System Approval Guide, 1973–96,* Factory Mutual Research Corporation; Norwood, MA, 1996.
5. Sonntog, N.O.V. "Structure and Composition of Fats and Oils," in *Bailey's Industrial Oil and Fat Products, Vol. I,* 4th ed., Swern, D. ed., John Wiley & Sons, New York, 1979.

6. Sonntog, N.O.V. "Reactions of Fats and Fatty Acids," in *Bailey's Industrial Oil and Fat Products Vol. 1*, 4th ed., Swern, D. ed., John Wiley & Sons, New York, 1979.

7. Klaus, E.E. and Tewksbury, E.J. "Liquid Lubricants," in *CRC Handbook of Lubrication (Theory and Practice of Tribology), Volume II: Theory and Design*, Booser, E.R. ed., CRC Press; Boca Raton, FL, 1984, pp. 229–54.

8. Rizvi, S.Q.A. "Lubricant Additives and Their Functions, Lubricants and Lubrication," in *Friction, Lubrication and Wear Technology, ASM Handbook*, Blau, P.J., volume chairman, ASM International; Metals Park, OH, 1992, pp. 98–112.

9. Khan, M.M. "Spray Flammability of Hydraulic Fluids," in *Fire Resistance of Industrial Fluids, ASTM STP284*, American Society for Testing Materials; Philadelphia, 1996.

10. Hamblin, P.C. and Kristen, U. "Ashless Antioxidants, Copper Deactivators and Corrosion Inhibitors: Their Use in Lubricating Oils," *Lubr. Sci.*, 1990, 2(4), pp. 287–318.

11. Smalheer, C.V. and Smith, R.K. *Lubricant Additives, Section I: Chemistry of Additives*, The Lezius-Hiles Co.; Cleveland, OH, 1967.

12. O'Brien, J.A. "Lubricating Oil Additives," in *CRC Handbook of Lubrication (Theory and Practice of Tribology), Volume II: Theory and Design*, Booser, E.R. ed., CRC Press, Inc. Boca Raton, FL, 1984, pp. 301–15.

13. Gehrmann, W.A. "Non-Zinc Hydraulic Oils: Technology, Applications, and Trends," in *International Off-Highway & Power Plant Congress & Exposition*, 1992.

14. Khan, M.M. "Spray Flammability of Hydraulic Fluids and Development of a Test Method," Technical Report FMRC J.1. OTOW3.RC, Factory Mutual Research Corp., Norwood, MA, 1991.

15. Totten, G.E. Bishop, R.J. and Kling, G.H. "Prediction of Hydraulic Fluid Performance: Bench Test Modeling," in *Proceedings of the 47th National Conference on Fluid Power, Volume 1*, NFPA; Milwaukee, WI, 1996.

16. Reichel, J. "Importance of Mechanical Tests with Hydraulic Fluids," in *Tribology of Hydraulic Pump Testing*, Totten, G.E., Kling, G.H., and Smolenski, D.J. eds., American Society for Testing and Materials: Philadelphia, 1996.

17. Melief, H.M. "Proposed Hydraulic Pump Testing for Hydraulic Fluid Qualification," in *Tribology of Hydraulic Pump Testing*, Totten, G.E., Kling, G.H., and Smolenski, D.J. eds., American Society for Testing and Materials; Philadelphia, 1996.

18. The Rexroth Corporation, Information Bulletin KD012, The Rexroth Corporation, Bethlehem, PA.

19. Marino, M.P. and Placek, D.G. "Synthetic Lubricants, and Applications," in *CRC Handbook of Lubrication and Tribology, Volume III: Monitoring Materials*, Booser, E.R. ed., CRC Press; Boca Raton, FL, 1984.

20. Khan, M.M. "Spray Flammability of Hydraulic Fluids," in *Fire Resistance of Industrial Fluids, ASTM STP 1284*, Totten, G.E., and Reichel, J. eds., American Society for Testing Materials; Philadelphia, 1996.

21. U.S. Mine Safety and Health Administration, CFR Part 35, "Fire Resistant Fluids," 40 CFR 796. 3260 Quintolubric 823-300, Report to Quaker Chemical Corporation, 1992.

22. Brandao, A.V. "Implementation of Revised Evaluations of Less Flammable Hydraulic Fluids," in *Fire Resistance of Industrial Fluids, ASTM STP 1284*, Totten, G.E., and Reichel, J. eds., American Society for Testing Materials; Philadelphia, 1996.

23. International Technology Corp. "Ready Biodegradability: Modified Sturm Test, 40 CFR 796.3260 Quintolubric 822-300, Report to Quaker Chemical Corp., 1992.

24. McNair, J. "Rainbow Trout Acute Toxicity Tests, Report to Quaker Chemical Corporation, Re. Quintolubric 822-300 & EAL-224 (Mobil Oil Corporation)," Report to Academy of Natural Sciences of Philadelphia, 1993.

25. Exxon. "The ISO Cleanliness Code for Lubricating and Hydraulic Oil Systems," Exxon Marketing Technical Bulletin, No. MTB 85-15, Series RIM-4, 1985.

26. Factory Mutual Approvals, "Approval Standard for Flammability Classification of Industrial Fluids, Class 6930", July, 2008.

27. U.S. Mine Safety and Health Administration, "Commercially Available Fire-Resistant Hydraulic Fluids Approved by MSHA under Code of Federal Regulations Title 30, Part 35", www.msha.gov.

18 Biobased and Biodegradable Hydraulic Oils

Lou A.T. Honary

CONTENTS

18.1 INTRODUCTION

In the United States, biobased products are a part of the country's national policy to help preserve petroleum resources. The term biobased refers to products that are made of a portion that is renewable. The U.S. Department of Agriculture has a labeling process for biobased products; biobased-labeled products receive purchase preference from the U.S. federal purchasers.

The use of vegetable oils as fuel, lubricating fluid, and energy transfer media has been known for many years. Their usage, however, has depended on the availability of petroleum. Both World Wars, for example, saw an increased usage of vegetable oils for fuel, lubrication, and energy transfer. Similarly, the oil embargo of 1973 resulted in an increased interest in the use of vegetable-based oils as well as other alternatives such as water-based hydraulic fluids.

Concern for the environment has created a new impetus for the use of vegetable oils in industrial applications. The environmental factor promises to be a more steady and persistent force in promoting the use of vegetable oils than the fluctuating availability of petroleum. Either because of imposed environmental regulations or society's desire for a morem lubricant, the interest and increased use of vegetable-based hydraulic fluids and industrial lubricants will continue. Therefore, the engineer dealing with industrial lubricants should become familiar with vegetable oils and their characteristics related to industrial use. Because this chapter deals with *hydraulic fluids,* the emphasis will be placed on the use of vegetable oils for hydraulic systems. Furthermore, the chapter provides information on the chemistry of vegetable oils, the general performance requirement of hydraulic fluids, standard test procedures required for the oil approval, and the performance reports on some of the vegetable-based hydraulic oils. Issues of biodegradability and toxicity will also be discussed.

18.2 PAST INDUSTRIAL USES OF VEGETABLE OILS

The use of vegetable-based lubricant or fuel is not new and since the invention of the wheel animal fats and vegetable oils have been used to lubricate the wheel and its axel. In more recent history, in 1900 during a exhibition in Paris, Rudolf Diesel used peanut oil to power one of his diesel engines. The use of biolubricants and biofuels was put on hold, however, due to the abundance and low cost of petroleum.

The industrialist Henry Ford I and Dr. George Washington Carver pioneered agricultural research for industrial purposes. Henry Ford I had a vision of using crop-based materials in making cars and tractors and creating a closed circle of cradle-to-grave renewable products. George Washington Carver's research resulted in over 500 products from peanuts, sweet potatoes, and pecans. He graduated in 1894 from Iowa State College of Agriculture and Mechanic Arts (now Iowa State University, in Ames, Iowa). Later, in 1896, he joined the faculty and assumed the directorship of the Department of Agricultural Research at Tuskegee Normal (now Tuskegee University). His research involved a series of experiments with peanuts, sweet potatoes, soybeans, and Georgia clay.

Henry Ford also worked closely with George Washington Carver, recognizing their mutual visions. They shared a vision of a future in which agricultural products would be put to new uses in order to create products and industries.

Ford believed that agriculture could supply the industry with renewable raw materials. He paid attention to the mechanization of farm machinery and was convinced that farmers could become self sufficient in creating their own food and fuels, and lubricants, from their farm-renewable products. Towards that end, Henry Ford sponsored the research activities of George Washington Carver, whom he considered as another visionary in the use of biobased products. He even manufactured biobased plastic car bodies to show the potential of renewable products.

During the 1980s, European researchers, encouraged by the agricultural community in Europe, began to explore the use of vegetable oils as hydraulic fluids and other industrial lubricants. The concern for the environment and lobbying by farmers' organizations led to mandates for the use of biodegradable products in certain parts of Europe. For example, in the 1980s, the German government required the use of biodegradable hydraulic oils in Black Forest regions. During this period, the European community created environmental seals and emblems to identify the "environmentally aware" lubricants.

In the 1990s, many of the companies in North America began to follow the Europeans' lead on creating biodegradable products. In the 1990s meetings of the American Society for Testing and Materials, over 40 North American companies had representatives present to discuss their efforts in creating biodegradable products and help in establishing standards.

In the U.S., for example, The Lubrizol Corporation invested significant amounts of resources in creating additive packages for vegetable oils (specifically sunflower oil) lubricants. The list of additive packages and products from The Lubrizol Corporation was comprehensive and included food grade products, two-stroke-cycle engine oils, and universal tractor hydraulic fluids.

Eventually, the relatively low price of petroleum, along with the lack of mandates for using biodegradable or renewable products, diminished the investments in research and development (R&D) for these products. By the late 1990s, only farmers' groups like the United Soybean Board (representing U.S. soybean farmers), or the U.S. Department of Agriculture were continuing to fund the research and developmental activities of industrial products and lubricants made from renewable biomass materials.

In 2000, the United States' Farm Bill, which was a five-year plan for the advancement of agriculture, included provisions for the promotion and use of biodegradable and renewable products. In an attempt to avoid mandates and allow the free market enterprise to bring about the success of renewable products, the U.S. government selected "the leadership by example" approach. Federal agencies were required to purchase and use biobased products so as to prove viability of performance and eventually lead to commercial success in competitive private sector markets. Biobased and renewable-based lubricants now have a significant presence in the world market and are anticipated to grow in technology and use.

In order to better understand the future of biobased lubricants vis-à-vis petroleum lubricants, it is important to understand the nature and types of plant oils and petroleum oils. Within that context, basic concepts relating to these two oils will be covered.

18.3   BASIC STRUCTURE OF HYDROCARBONS

Carbon can uniquely form chemical bonds with itself. This property of carbon allows it to form carbon atom chains of various lengths, which results in different physiochemical properties. This property of carbon is observed in petroleum where carbon is chemically bound to hydrogen and to itself to form hydrocarbon chains. It is not the purpose of this chapter to go into the complex system of hydrocarbons; however, general concepts are presented here for reference and use in the presentation of vegetable oils.

Below is a list of general points that are known to confuse students and chemistry knowledge seekers.

1. Carbon is unique in the fact that it can bond to itself, meaning two carbon atoms can each donate an electron to the bond to form a single bond; thus a single *bond* is made up of *two* electrons, one from each atom.
2. A *double bond* between carbon atoms requires that each atom contribute two electrons to the bond, resulting in four electrons being shared in total.
3. Fatty acids are made of carbon chains containing mostly single carbon-to-carbon bonds. Fatty acid chains vary in length from ten carbon atoms to as many as thirty-six carbon atoms.
4. Glycerol is a triahydric alcohol meaning that it contains three hydroxy groups.
5. Vegetable oils and fats can be in the form of monoglycerides, diglycerides, or more commonly, triglycerides, depending on how many of the three hydroxy groups on glycerol have been replaced by fatty acids.

6. If a fatty acid has *no* double bonds, then it is considered a *saturated* fatty acid. Saturation impacts the melting point of the fatty acid. If there is only *one* double bond in its carbon chain, then it is *monounsaturated*; and if more than one double bond exist in its chain, it is considered *polyunsaturated*. Examples are shown in Table 18.1.
   Table 18.2 shows mono- and polyunsaturated fatty acids in selected common vegetable oils [1].
7. The melting point of triglycerides is important in formulating industrial lubricants. The variation of *chain length* and the *degree of unsaturation* combine to form many triglycerides with different melting points.
8. To describe fatty acids, the *total number of carbon atoms* is stated; for example, "C 18" for stearic and "C 16" for palmitic.

TABLE 18.1
Saturated, Mono- and Polyunsaturated Fatty Acids and Melting Points

Saturated Fatty Acids

Number of Carbon Atoms	Common Name of Acid	Systematic Name	Melting Point °C
C 4	butyric	butanoic	−8
C 6	capoic	hexanoic	−3.4
C 8	caprylic	octanoic	16.5
C 10	capric	decanoic	31.5
C 12	lauric	dodecanoic	43.5
C 14	myristic	tetradecanoic	54.4
C 16	palmitic	hexadecanoic	62.9
C 18	stearic	octadecanoic	69.6
C 20	arachidic	eicosanoic	75.4
C 22	behenic	docosanoic	79.9
C 24	lignoceric	tetracosanoic	84.2
C 26	ceratinic	hexacosanoic	87.7

Polyunsaturated Fatty Acids

Number of Carbon Atoms	Common Name of Acid	Systematic Name	Melting Point °C
C 18	linolenic	cia-(9, 12)-octadecadienoic	−5
C 18	linolenic	cia-(9,15,12)-octadecatrienoic	−11.3
C 20	arachidonic	cia-(5,8,11,14)-eicosatetraenoic	−49.5

Monounsaturated Fatty Acids

Number of Carbon Atoms	Common Name of Acid	Systematic Name	Melting Point °C
C 10	caproleic	decenoic	–
C 12	lauroleic	dodecanoic	–
C 14	myristoleic	*cis*-9-tetradecenoic	−4
C 16	palmitoleic	*cis*-9-hexadecenoic	0–5
C 18	oleic	*cis*-9-octadecenoic	13
C 18	elaidic	*trans*-9-octadecenoic	44
C 18	vaccenic	*trans*-11-octadecenoic	39
C 22	erucic	*cis*-13-docosenoic	33

TABLE 18.2
Mono and Polyunsaturated Fatty Acids in Common Vegetable Oils

Oil	Monounsaturates	Polyunsaturates	Saturates
Olive	75%	11%	14%
Canola	58%	36%	6%
Peanut	48%	34%	18%
Palm	39%	10%	51%
Corn	25%	62%	13%
Soybean	24%	61%	15%
Sunflower	20%	69%	11%
Cottonseed	19%	54%	27%
Safflower	13%	78%	9%
Coconut	6%	2%	92%

Source: Gapinski, R.E., Joseph, I.E. and Layzel, B.D. "A Vegetable Oil Based Tractor Lubricant", SAE Technical Paper Series, Paper 941758, Warrendale, PA, SAE Publications, 1994.

18.4 AVAILABILITY OF VEGETABLE OILS

The world production of edible oils is made up of 71% vegetable oils, 26% animal fat, and 2% fats from marine species. It is estimated that soybean oil, for example, contributes to nearly one-third of the world's oilseeds, with almost one-third produced in the United States.

According to the United States Department of Agriculture, vegetable oils may be classified into three categories based on their production, use, and volume: (1) major oils, (2) minor oils, and (3) non-edible oils.

Oils that are known for human or animal feed consumption and which often play important economic roles in the regions producing them are considered major oils. These include: sunflower, soybean, palm, rapeseed, cottonseed, coconut, peanut, olive, palm kernel, corn, linseed, and sesame.

Oils that are known for their uses but do not match the large production magnitude of major oils are referred to as minor oils. These oils have fatty acid profiles that could make them effective for industrial uses. They include: poppy, cocoa bean, niger, mango kernel, shea, hempseed, grapeseed, perilla, Chinese tallow, Ethiopian mahogany, German sesame, watermelon seed, avocado, and apricot seed.

A number of non-edible oils such as linseed, castor, tung, and tall are commercially grown for their unique chemical makeup which is important for the industry. These non-edible oils are used in industrial applications such as soaps, varnishes, resins, and in many agrochemicals. However, their use is also being considered for industrial lubricants applications. These oils include: linseed, castor, neem, mahua, and karanja.

Due to controversy over food versus fuel, there has been a significant investment of public and private capital for the development of non-edible, non-food alternative crop oils. Although different in end-use, many of the industrial crops and special processes developed for biofuels have applications in biobased hydraulic fluids as well. Additionally, the attention given to the negative health effects of trans fats has reinvigorated the development of special varieties of oilseeds like the low linolenic and high oleic soybeans by major U.S. seed companies.

Most noteworthy are crops like cuphea, camelina, castor, lesquerella, and pennycress, which are being investigated by the oil crop division of the Association for Advancements of Industrial Crops (AAIC). Some of these crops (such as camelina) have reached the commercial production stage and reasonably large acreages are being produced in the western United States.

The future technologies will encompass the known vegetable oils, the newer genetically enhanced high oleic varieties, and the more sophisticated and economical chemically modified

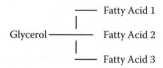

FIGURE 18.1 Structure of a triglyceride.

high-functioning esters which can be derived from a large variety of raw materials. These developments will not completely replace the use of petroleum for industrial and automotive lubricants, but they will capture a significant portion of those markets.

18.5 CHEMISTRY OF VEGETABLE OILS

Fats and oils are made up of an ester between glycerol and three fatty acids. Depending on the length of the carbon chain making up the fatty acid, the resulting fat can be either liquid (oil) or solid (fat or butter). Since glycerol is capable of bonding with three molecules of fatty acid the result is referred to as a triglyceride, as shown in Figure 18.1. A triglyceride, also called triacylglycerol (TAG), is a chemical compound formed from one molecule of glycerol (also called glycerin) and three fatty acids.

Glycerol, a trihydric alcohol (containing three —OH hydroxyl groups) can combine with up to three fatty acids. Fatty acids also combine with any of the three hydroxyl groups to create a wide diversity of compounds. Monoglycerides, diglycerides, and triglycerides are classified as *esters*, which are compounds created by the reaction between (fatty) acids and alcohols (glycerol) that release water (H_2O) as a byproduct.

Fatty acids contain a chain of carbon atoms combined with hydrogen (forming hydrocarbon). They terminate in a carboxyl group. If the three fatty acids are alike, the molecule is a *simple* triglyceride; if they are different it is a *mixed* triglyceride.

Table 18.3 presents the general composition of several oils and fats and the ratio of unsaturated to saturated fats [2].

Unsaturation affects the liquidity of the oil or its melting point; and is related to its soluability and chemical reactivity. With an increase in unsaturation, the melting point goes down (higher liquidity) while solubility in certain solvents and chemical reactivity increases. This usually results in oxidation and thermal polymerization. Saturated oils, in general, show more oxidative stability but have high melting points (lower liquidity), such as palm oil, which has high oxidation stability but is solid at room temperature, thus limiting its use for liquid lubricants applications unless modified.

For a normal single-bond atom there is freedom of rotation around the bond, but there is rigidity at the site of a double bond. Thus, two fixed positions of *cis* (meaning on same side) and *trans* (meaning across) are possible (see Figure 18.2). Trans forms of fatty acids "pack" much closer together than do the *cis* forms. Therefore the *trans* forms more closely resemble the saturated fatty acids making "trans fat" more undesirable as a food oil. Because the saturated fatty acids have no double bonds to distort the chain, they pack more easily into crystal forms and therefore have higher melting points than unsaturated fatty acids of the same length. They are also less vulnerable to oxidation. This property also allows for winterization of the vegetable oils, as will be described later.

18.6 PHYSIOCHEMCIAL PROPERTIES OF VEGETABLE OILS

When considered for use in industrial lubricants and hydraulic fluids, vegetable oils have many advantages and some shortcomings. Unless modified, they lack oxidation stability, which is dependent on the position and degree of unsaturation of the fatty acids that are attached to the glycerol molecule. The majority of soybean oil fatty acid composition, for example, is comprised of conjugated carbon-to-carbon double

TABLE 18.3

Fatty Acid Composition of Selected Fats and Oils Percent by Weight of Total Fatty Acids

Oil or Fat	Unsat./Sat. Fat Ratio	Saturated					Monounsaturated	Polyunsaturated	
		Capric Acid	Lauric Acid	Myristic Acid	Palmitic Acid	Stearic Acid	Oleic Acid	Linoleic Acid (ω6)	Alpha Linolenic Acid (ω3)
		C10:0	C12:0	C14:0	C16:0	C18:0	C18:1	C18:2	C18:3
Almond Oil	9.7	–	–	–	7	2	69	17	–
Beef Tallow	0.9	–	–	3	24	19	43	3	1
Butterfat (cow)	0.5	3	3	11	27	12	29	2	1
Butterfat (goat)	0.5	7	3	9	25	12	27	3	1
Butterfat (human)	1	2	5	8	25	8	35	9	1
Canola Oil	15.7	–	–	–	4	2	62	22	10
Cocoa Butter	0.6	–	–	–	25	38	32	3	–
Cod Liver Oil	2.9	–	–	8	17	–	22	5	–
Coconut Oil	0.1	6	47	18	9	3	6	2	–
Corn Oil (Maize Oil)	6.7	–	–	–	11	2	28	58	1
Cottonseed Oil	2.8	–	–	1	22	3	19	54	1
Flaxseed Oil	9	–	–	–	3	7	21	16	53
Grapeseed Oil	7.3	–	–	–	8	4	15	73	–
Illipe	0.6	–	–	–	17	45	35	1	–
Lard (Pork fat)	1.2	–	–	2	26	14	44	10	–
Olive Oil	4.6	–	–	–	13	3	71	10	1
Palm Oil	1	–	–	1	45	4	40	10	–
Palm Olein	1.3	–	–	1	37	4	46	11	–
Palm Kernel Oil	0.2	4	48	16	8	3	15	2	–
Peanut Oil	4	–	–	–	11	2	48	32	–

(continued)

TABLE 18.3 (Continued)
Fatty Acid Composition of Selected Fats and Oils Percent by Weight of Total Fatty Acids

Oil or Fat	Unsat./Sat. Fat Ratio	Saturated						Monounsaturated	Polyunsaturated	
		Capric Acid C10:0	Lauric Acid C12:0	Myristic Acid C14:0	Palmitic Acid C16:0	Stearic Acid C18:0		Oleic Acid C18:1	Linoleic Acid (ω6) C18:2	Alpha Linolenic Acid (ω3) C18:3
Safflower Oil[a]	10.1	–	–	–	7	2		13	78	–
Sesame Oil	6.6	–	–	–	9	4		41	45	–
Shea Nut	1.1	–	1	–	4	39		44	5	–
Soybean Oil	5.7	–	–	–	11	4		24	54	7
Sunflower Oil[a]	7.3	–	–	–	7	5		19	68	1
Walnut Oil	5.3	–	–	–	11	5		28	51	5

[a] Non-high oleic variety.

Percentages may not add to 100% due to rounding and other constituents not listed.
Where percentages vary, average values are used.

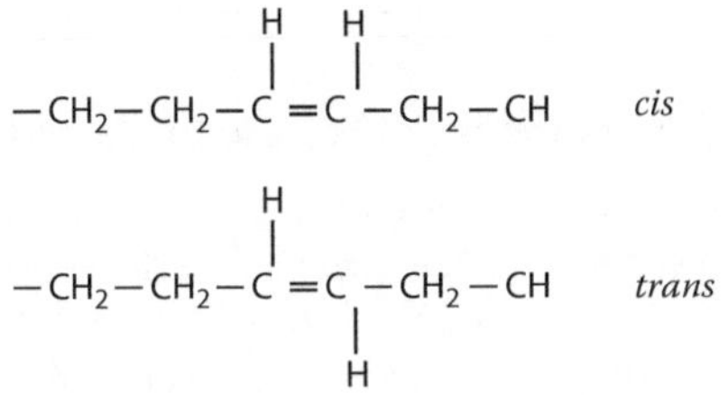

FIGURE 18.2 Hydrogen atoms on one side (*cis*) or on both sides of the chain (*trans*).

bonds which make it more susceptible to oxidation. "Conjugated" is a term used to describe a condition where two double bonds in a carbon chain are close to each other. Conventional soybean oil contains approximately 52% linoleic acid, which has two conjugated double bonds and 7%–8% linolenic acid which contains three conjugated double bonds. If left untreated, the use of these oils could lead to increased oxidation and consequently to increased viscosity. In extreme cases, if the oil continues to oxidize in use, it could lead to polymerization and the formation of polymer films in the oil.

To avoid oxidation in use, the vegetable oil is either chemically modified and/or antioxidants are used to increase oxidation stability. Hydrogenation—chemically adding hydrogen to the double bonds—is one method used to increase oxidative stability. Unfortunately, the melting point is also increased and can result in a product that is solid or semisolid at room temperature. Oilseeds that are genetically enhanced and have higher oxidation stability are more conducive for use in industrial lubricants and hydraulic oils.

Double bonds within the carbon chain lower the melting point significantly. The longer the fatty acid carbon chain, the higher the melting point. Vegetable oils, due to their fatty acid structure, tend to freeze at relatively higher temperatures than their mineral oil counterparts. A pour point comparison of hydraulic fluid using both mineral oil and soybean oil as base fluids is shown in Table 18.4. For applications where hydraulic oil or industrial lubricants are exposed to subzero temperatures, a mixture of vegetable oils and mineral or synthetic oils could be used. Mixing, however, impacts other properties of vegetable oils including viscosity index and flash/fire points as well as compatibility with elastomers and other components.

Vegetable oils, due to their polarity, adhere to metal surfaces for better metal-to-metal separation. Also, due to a higher viscosity index relative to petroleum oils, they are more stable as the temperature changes. For example, soybean oil has a viscosity index of about 220, with a viscosity of 30.69 at 40°C (104°F) and a viscosity of 7.589 at 100°C (212°F). Comparable naphthenic base oil with a viscosity of 37.95 at 40°C (104°F) and a viscosity of 5.295 at 100°C (212°F) would have a viscosity index of 53. Since the high viscosity index results in a more stable viscosity when

TABLE 18.4

Viscosity, Viscosity Index, and Pour Points of Selected Oils and Identical Hydraulic Fluids Utilizing Soybean Oil and Mineral-Oil-Based Fluids

Description of Base Oil or Hydraulic Fluid	Pour Point (°C) (ASTM D 6749)	Viscosity @ 40°C (ASTM D 445)	Viscosity @ 100°C (ASTM D 445)	Viscosity Index (ASTM D 2270)
Refined				
High Oleic Soybean Oil	−16	31.19	8.424	200
Crude Conventional Soybean Oil	−6	31.69	7.589	222
Mineral Oil Near Equal Viscosity	−49	37.95	5.295	53
Hydraulic Fluid with Crude Conventional Soy	−4	32.26	7.592	217
Hydraulic Fluid with High Oleic Soy	−4	39.14	8.412	199

TABLE 18.5

Tribological Performance Characteristics of Selected Oils and Finished Hydraulic Fluids

Description of Base Oil or Hydraulic Fluid	Loadwear Index (Weld Point Kg) ASTM D2783	4-Ball Wear Scar (mm) ASTM D4172	Pin and Vee Force (lb) - Torque (lb-in) ASTM D3233 A	Tapping Torque (N-m) ASTM D5619
High Oleic Soy Oil	21.87 (160)	0.626	Broke @ 1755.64 lb-f Torque = 31.8	8.198
Crude Conventional Soy	26.74 (160)	0.589	Broke @ 1656.94 lb-f Torque = 53.1	8.027
Mineral Oil Near Equal Viscosity	13.96 (126)	0.810	Broke @ 215.41 lb-f Torque = 51.1	10.99
Hydraulic Fluid with Crude Conventional Soy	33.98 (200)	0.510	N/A	N/A
Hydraulic Fluid with High Oleic Soy	26.82 (160)	0.529	N/A	N/A

temperatures change, lower-viscosity vegetable-oil-based hydraulic fluid could be used in applications where higher-viscosity petroleum oil is required. As an example, ISO Viscosity Grade (VG) 46 hydraulic fluid made from vegetable oil may be suitable for applications where an ISO Viscosity Grade (VG) 68 from petroleum oil is specified. The viscosity of base soybean oils, formulated soybean-oil-based tractor hydraulic fluid, and petroleum-oil-based tractor hydraulic fluids are presented in Table 18.4. This table shows the viscosity, viscosity index, and pour points of soybean oils, mineral oils, and identical hydraulic fluid packages utilizing both soybean oil and mineral oil as base fluids. The mixture of ISO VG 100 and ISO VG 500 was prepared to create a viscosity range closer to soybean oils. The difference in viscosity index is significant with the soybean oil showing almost four times higher viscosity index than the petroleum mineral oils.

In Table 18.5 the tribological performance characteristics of the selected base oils and finished hydraulic fluids are shown. Soybean oil shows better lubricating properties as indicated by the 4-ball wear (ASTM D 4172), 4-ball Extreme Pressure (ASTM D 2783), Pin and Vee, (ASTM D 3233A) and tapping torque (ASTM D 5619) results.

The flash and fire points of vegetable oils are consistently and considerably higher than equivalent viscosity mineral oils. Typically, the fire points of vegetable oils are greater than 300°C (572°F). This property is suitable for the creation of hydraulic fluids and industrial lubricants that could meet some fire retardancy standards including those of the Factory Mutual specification standards in the U.S. Metalworking fluids made from vegetable oils show less tendency to burn, and hydraulic applications like building elevators could benefit from the fire safety aspect of this property of vegetable oils. Table 18.6 presents viscosity representation for mineral oils [3].

TABLE 18.6

Viscosity Ranges for Mineral Oils

Viscosity Ranges for Industrial Fluid Lubricants

Viscosity System Grade Identification	Mid-point Viscosity cSt (mm²/s) at 40°C	Kinematic Viscosity Limits, cSt (mm²/s at 40°C)	
		Minimum	Maximum
ISO VG22	22	19.8	24.2
ISO VG32	32	28.8	35.2
ISO VG46	46	41.4	50.6
ISO VG68	68	61.2	74.8

There are many factors that affect the fatty acid makeup of vegetable oils. In addition to their natural structure, changes in the growing conditions and geographic location as well as factors such as exposure to daylight, light intensity, and quality can impact the properties of vegetable oils. Because the fatty acid composition of oils and fats is unique, their characteristics are different. One important process that can be used to affect the types of fatty acid present is *partial hydrogenation* in which only some of the double bonds present in the carbon chain are given their full complement of hydrogen. This allows for liquidity while improving stability.

18.7   OXIDATION

Most materials will have some degree of reaction with oxygen resulting in oxidation. Vegetable oils, for example, are used for frying applications and exposure to heat, moisture from food, light, and air would cause them to oxidize. For frying applications, attempts are made to stabilize the oil through hydrogenation or the use of more saturated oils like palm oil. One indication of oxidation is the onset of rancidity which can be recognized by smell. In industrial applications, as indicated earlier, increased viscosity and the rancidity odor could be used to note the onset of oxidation.

The change in viscosity mentioned above is irreversible. This is different to when the oil thickens up due to exposure to cold temperature. In the latter case, heating the oil reverses it back to its original viscosity. In the case of oxidation thickening, a change in molecular structure takes place resulting in the initiation of the polymerization process. Once initiated, larger molecules continue to form, thus propagating the polymerization. This will then continue until terminated by intervention or by the completion of the full polymerization.

Figure 18.3 shows crude soybean oil which was exposed to air and ambient temperatures. Over a period of a few months, the oil fully polymerized. The same crude oil when sealed in a bottle away from light would remain liquid for several years. Creating a barrier to oxygen is an effective way of reducing oxidation. The containers of some edible oils or beauty products that contain vegetable oils or animal fat may be topped off with nitrogen in a process called "nitrogen dosing." The nitrogen replaces the air in the container before capping it thus preventing exposure of the product to air. Figure 18.4 shows product poorly formulated polymerized in container.

18.8   ESTERIFIED VEGETABLE OILS

Esters, especially the complex esters, have higher viscosity and higher molecular weight than the common esters, thus offering advantages for some applications. The *Journal of Synthetic Lubrication* has published numerous papers on synthetic esters. Some of these complex esters, for example, may be made via the reaction of a polyol, dicarboxylic acid, and monoalcohol as an end-capping agent. Meng and Dresel explained the process for complex esters as first esterifying the diol with dicarboxylic acid, and then, depending on the desired product, reacting this ester with either carboxylic acid or a monoalcohol [4].

FIGURE 18.3   Soybean oil oxidized and polymerized naturally (left) or by using heat and air.

FIGURE 18.4   Vegetable oil oxidized in storage.

18.9   FORMATION OF ESTOLIDES

High oxidation stability and lower cold-temperature flowability are most desired for vegetable oils in lubricant uses. In an attempt to synthesize base oils that are suitable for use in formulating bio-based lubricants, efforts are made to create flexibility in the use of the base raw materials. Estolides are designed with the idea of being able to use various vegetable oils or animal fats as starting materials. A patent by Steve Cermak et al. [5], describes the formation of estolides for industrial and automotive lubricants.

The unsaturated oleic acid estolides showed pour points of −22°F for the unsaturated oleic estolides and the saturated estolides showed −40°F. In terms of oxidation stability, using the Rotary Bomb Oxidation Test (RBOT), the oxidative breakdown was 200 minutes for unsaturated oleic acid estolide and 400 minutes for comparable mineral engine oil."

The oxidation stability and viscosity-related data for three commercially supplied estolides are shown in Table 18.7. These are: distilled oleic estolide 2-ethylhexyl ester, oleic estolide with monomer, and distilled coco-oleic estolide 2-ethylhexyl ester.

Cermak and Isbell [6] provided comparative data on commercial petroleum-based hydraulic oils, regular soybean oil, and estolides. The estolides without property-enhancing additives showed comparable viscosity index and lower cloud and pour points than formulated products, Table 18.8.

Estolides are one example of chemical modification schemes that provide flexibility in the raw input materials while providing consistent quality base oils with high oxidation stability and low pour point.

TABLE 18.7
OSI Time and Viscosity Data of Three Estolides

ID	Estolide Type	OSI Time	Viscosity at 40°C	Viscosity at 100°C	Viscosity Index
1	Oleic Estolide	18.18	98.17 cSt	15.40 cSt	167
2	Coco-oleic Estolide	22.25	76.58 cSt	12.59 cSt	165
3	Oleic Estolide with Monomer	15.45	40.30 cSt	8.20 cSt	185

TABLE 18.8

Comparison of Low-Temperature Properties and Viscosity Index of Coconut-Oleic Estolide 2-ethylhexyl Esters to that of Commercial Lubricants

Lubricant	Pour Point (°C)	Cloud Point (°C)	Viscosity @ 40 (°C)/(cSt)	Viscosity Index
Commercial petroleum Oil[a]	−27	2	66	152
Commercial synthetic Oil[a]	−21	−10	60.5	174
Commercial soy-based Oil[a]	−18	1	49.6	220
Commercial hydraulic fluid[a]	−33	1	56.6	146
Coco-oleic estolide[b]	−33	−26	55.2	162
Oleic estolide[b]	−33	−33	92.8	170

[a] Commercial formulated products
[b] Unformulated estolides

18.10 OTHER CHEMICAL MODIFICATIONS

There are many other processes being explored to modify the properties of fats or fatty acids and add value for industrial application. An example of such a process is "metathesis," which essentially means to change places.

This was the focus of 2005's Nobel Prize in Chemistry when two compounds react with each other to form two new compounds. In olefin metathesis (olefin is another name for alkene, a carbon chain with double bonding) the double-bonding atom groups will change places with one another.

A company by the name of Elevance Renewable Sciences [7], which deals with metathesis, describes the process of metathesis as a "reaction in which one of the propene molecules exchanges its CH_2 group for the CH_3CH group in the other propene molecule. The result is butene and ethene. A catalyst, which is not consumed in the reaction, is required for the reaction to occur." Producing new substances through metathesis is not new. However, learning which catalyst to use and the role of the catalyst in metathesis is significant because it provides an enormous diversity for the chemicals when organic compounds could be transformed into chemicals that have for years been produced by inorganic compounds.

As described earlier, in organic chemistry, carbon has the ability to form bonds with other carbon atoms and with other elements such as hydrogen, oxygen, chlorine and sulphur. A carbon atom can bind other atoms with single, double or triple bonds and form chains, branched structures and rings of different forms and sizes. So far, only a small proportion of the different organic molecules have been investigated. Still, we have already gained new pharmaceuticals, materials, coatings, and so forth, which were undreamed-of a few years ago.

Pharmaceutical and biotechnical industries use organic synthesis which involves producing different substances by making substances react in certain ways. That is making new molecules with the help of other molecules. Using special catalysts, two new double bonds could be obtained to make the large ring from the two double bonds at the terminals of the chain. One of these has been used to join two carbon atoms and form the large ring. The other double bond turns up in the by-product, ethene. Synthesizing the large ring in any other way is very complicated and requires many more reaction steps.

According to the company information from Elevance Renewable Resources, metathesis was discovered in industry as early as the 1950s. A number of patents described the catalyzation of olefin polymerization. Among these documents was a report from 1957 by H. S. Eleuterio at DuPont, USA describing the formation of carbon chains with double bonds—that is unsaturated polymers (olefins). Earlier attempts to polymerize the olefin ethene to polyethene had produced saturated polymers (no double bonds).

Also in 1957, another patent showed that propene could be converted into butene and ethene when treated with a mixture of triisobutyl aluminium and molybdenum oxide on aluminum oxide. This reaction, described as the "Phillips triolefin process," is shown in Figure 18.5.

More detailed information on the catalytic metathesis is beyond the scope of this chapter. However, its importance in creating new base oils for industrial lubricants and hydraulic oils needs to be emphasized.

18.11 ENZYMATIC HYDROLYSIS OF FATTY ACIDS

The use of enzymes to mimic the digestive system's breakdown of fats and triglycerides has been reported. Triglycerides can be enzymatically hydrolyzed to fatty acids and glycerol by the use of lipases. Most industrial hydrolyses involve high-pressure steam stripping to hydrolyze the triglyceride esters. Unfortunately this process destroys some of the more useful fatty acids found in the more exotic plant oils. The use of lipases for the enzymatic hydrolysis of oils can provide a more efficient approach that is less energy intensive and does not alter the fatty acids that result.

A U.S. Patent #5089403 by Hammond and Lee [8] describes using moistened dehulled oat seeds or oat caryopses which contains lipases. When the oil is exposed to the moistened caryopsis, the fatty acids dissolve in the oil phase and the glycerol is absorbed into the moist dehulled oil seeds. According to Hammond and Lee, in a single cycle, about 20% of the oil can be catalyzed. Increasing the number of contact cycles will result in the hydrolysis of more oil.

Using lipases for splitting the triglycerides into free fatty acids and glycerin will increase the potential for the creation of better lubricants and greases. The removal of glycerin from the fatty acids could reduce the hydrophilic properties of the formulated product. Also, if the free fatty acids are further processed into individual free fatty acids, more uniform grease soaps could be manufactured with predictable performance.

18.12 GENETIC MODIFICATION AND INDUSTRIAL CROPS

In addition to the chemical modification of oils which will play an important role in the future of biobased lubricants, there are other promising technologies that are showing promise. The genetic modification of crops, for example, was initially used primarily as a means of creating healthier oils that do not require hydrogenation for stability in cooking. Today, the genetic modification of crops has taken a new direction with new and diverse goals. Genetic modification is considered for creating crops that are resistant to pests, herbicides, or aphids; or for including vitamins and pharmaceutical properties as well as transgenic properties for draught resistance or for season or climate independence. Interestingly, agronomists and food scientists, working on the genetics of oilseeds for the development of healthier food, may have contributed to the increased potential for oilseed use in industrial lubricant applications. It is now clear that through genetic modification of the seeds, the fatty acid profile of the oilseeds can be altered with great benefits in terms of stability, cold-temperature flowability, and reduced need for chemical additives. This concept eliminates the need for hydrogenation or other chemical modification and thus reduces the cost of the base oil and finished products for use in industrial applications.

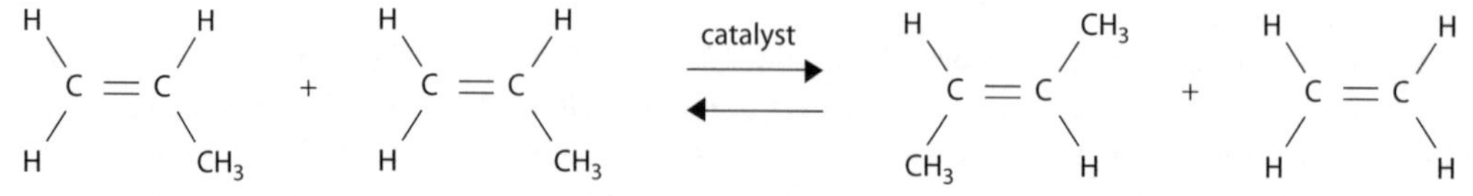

FIGURE 18.5 Two propene molecules undergo olefin metathesis with the help of a catalyst, producing two new alkenes, butene and ethene (ethylene).

Genetic modification has been accomplished for some of the vegetable oils, including rapeseed, sunflower, and soybean. Salunkhe et al. [9] reported on the changes in the genetic makeup of rapeseed oil starting in the late 1960s. Accordingly, prior to the 1970s, rapeseed oil contained 20%–50% erucic acid. The first Canadian low-erucic acid variety, containing about only 3% erucic acid, was licensed in 1968. In 1974, a lower-acid variety containing less than 0.3% erucic acid was introduced, following the Canadian government's encouragement to switch to low-acid varieties. Canada Oil Low Acids (CANOLA) is the Canadian version of rapeseed with distinctly low erucic acid and low glucosinolate. It has a linoleic to linolenic acid ratio of 2:1. In the United States, various genetically modified seeds for canola, sunflower, and soybean have been developed by major seed and chemical companies. Currently, the availability of these oils with high oleic acids of 80%–90% and excellent oxidation stability promise to create many new opportunities for vegetable-based industrial lubricants.

Perhaps the most promising new research advance to date is the development of new mutant lines of soybeans, which lead to new lines of high oleic soybeans. These mutant lines with *improved fatty acid profiles* of the oil can be cloned and then integrated into high-yielding elite lines by providing molecular markers. Clone genes are introduced into soybeans to create transgenetic lines with increased lysine, oleic, or stearic acid contents. These are oils with a very low content of polyunsaturated fatty acids and show high oxidative stability.

The interest in biofuels and bioproducts has created an opportunity for exploring the potential use of crops that are naturally grown in various parts of the world and are not necessarily edible. Investigating the use of non-edible secondary crops has further intensified due to the food versus fuel controversy. As petroleum prices increased to over $140 per barrel in 2008, the increase in food prices was attributed to the growth of biofuels during the first half of the decade. As a result, considerable efforts have been made in identifying and qualifying natural oil crops with oils suitable for industrial and fuel uses. When such crops are grown in arid lands, they offer a particularly attractive alternative to both edible oils and petroleum.

In the United States, the Association for the Advancement of Industrial Crops (AAIC) has been leading the exploration of the use of non-edible native crops which have oil and potential for use in biofuels and biobased products. AAIC has a list of several alternative industrial crops that its members are working on. Most noteworthy are crops like cuphea, camelina, canola, castor, lesquerella, and pennycress. Some of these crops (e.g., camelina) have reached the commercial production stage and reasonably large acreages are being produced in Montana. Table 18.9 presents the fatty acid content of industrial crops along with canola, soybean and sunflower oils.

Over the last 30 years, the commercial viability and economic and technical performance of biobased lubricants have been established. As the worldwide demand for petroleum increases—and with further attention given to the carbon footprint of industrial products—there will be continued development and enhancement of biobased products. The future technologies will include the old plant oils, the newer genetically enhanced high-oleic varieties, and the more sophisticated and economical chemically modified high-functioning esters derived from a large variety of raw materials. These developments will not completely replace the use of petroleum for industrial and automotive lubricants, but they will capture a significant portion of those markets.

18.13 BIOBASED HYDRAULIC OIL TECHNOLOGY

The initial technology for biobased lubricants, primarily referred to as biodegradable lubricants, was based on vegetable oils with minor chemical modification and performance-enhancing additives. Since vegetable oils generally face inherent challenges when it comes to industrial lubricant uses, their performance properties must be carefully studied. Soybean oil, for example, shows a significant lack of oxidation stability, with an OSI (oxidative stability index) value of about seven hours. In one case, this oil was partially hydrogenated to improve its oxidation stability and then winterized to improve its pour point performance. The addition of hydraulic oil additive packages,

TABLE 18.9

Fatty Acid Contents of Industrial Crops along with Canola, Soybean and Sunflower Oils

	Fatty Acid Content (% of Oil)							
Fatty acid	Canola	Soybean	Sunflower	Crambe	Flax	Camelina	Babasu	Jojoba
Caprylic (C 8)							4.5	
Capric (C 10)							7	
Lauric (C 12)							45	
Myristic (C 14)							16	
Palmitic (C 16:0)	6.19	10.44	6.05	2.41	5.12	7.8	7	
Stearic (C 18:0)	0	3.95	3.83	0.4	4.56	2.96	4	
Oleic (C 18:1)	61.33	27.17	17.36	18.36	24.27	16.77	14	10
Linoleic (C 18:2)	21.55	45.49	69.26	10.67	16.25	23.08	2.5	
Linolenic (C 18:3)	6.55	7.16	0.5	5.09	45.12	31.2		
Arachidic (C 20:0)	0	0.5	0.5	0.5	0	0		
Eicosenoic (C 20:1)	1.5	0.5	0	2.56	0	11.99		68.5
Erucic (C 22:1)	0.5	0	0	54	0.88	2.8		
Benenic (C:22)								17
Other FA	2.38	4.79	2.5	6.01	3.8	3.4	2.5	4.5

antioxidants, and point depressants resulted in hydraulic oils that were capable of working in high-performance hydraulic systems [10].

Perhaps the most important development for U.S. biobased lubricants was the introduction of high oleic acid soybeans by the DuPont Corporation in the early 1990s. This genetically enhanced soybean had oil with a fatty acid profile considerably superior to that of conventional soybean oils. Originally designed for frying applications, this oil showed an OSI value of 192 or about 27 times more stability than conventional soybean oils. This helped in the creation of a number of highly successful lubricants and greases. Table 18.4 showed the physical as well as rheological performance differences between conventional soybean oils and the high oleic soybean oils. In the 1990s, The Lubrizol Corporation had built its additive and lubricants technology based on high oleic and ultra-high oleic sunflower oils. Today, for many industrial lubricant applications, high oleic soybean oil, high oleic sunflower oil, and high oleic canola oils are still the base oils of choice.

18.14   OXIDATION STABILITY

Oxidation stability refers to the ability of the oil to maintain its properties when exposed to specific operating conditions. Since vegetable oils are a main ingredient in the foodstuffs, the majority of methods dealing with the oxidative stability of vegetable oils have been created through the efforts of the food industry, and by association, the chemical industry. Oxidation stability methods used in the lubricants and grease industry are based on petroleum and its derivative oils and are often not able to determine the stability of vegetable oils. Biobased lubricants researchers and developers have been relying on the use of standards created by the American Oil Chemist Society (AOCS) as well-modified versions of standards created by the American Society for Testing and Materials (ASTM). Others at the University of Northern Iowa's National Ag-Based Lubricants (NABL) Center have used hydraulic pump tests and field evaluations to create reference materials for possible use with vegetable-based lubricants. Below are examples of some of the test methods used to determine the oxidation stability of vegetable oils.

18.14.1   ACTIVE OXYGEN METHOD (AOM)

Based on the AOCS Cd 12-57 method in the AOM, oxygen is bubbled through an oil or fat which is held at 36.6°C (97.8°F) and samples are withdrawn at regular intervals and the peroxide value (PV) is determined. The AOM is expressed in hours and is the length of time needed for the PV to reach a certain level. AOM is used as a specification for fats and oils. AOM hours tend to increase with the degree of saturation or hardness. This method has been largely replaced by the oxidative stability index.

18.14.2   OIL STABILITY INDEX (OSI)

This is based on AOCS Method Cd 1 2b-92. As in the AOM, the OSI uses regular shop air instead of pure oxygen as this makes it simpler to operate. A conductivity probe monitors the conductivity of de-ionized water as evaporatives from test oil are emitted into the de-ionized water.

18.14.3   PEROXIDE VALUE (PV)

Based on (AOCS 8b-90) method, this is a test for measuring oxidation in fresh vegetable oils, and is highly sensitive to temperature. Peroxides are unstable radicals formed from triglycerides. A peroxide value (PV) over 2 is an indication that the product has a high rancidity potential and could fail on the shelf (Figure 18.6).

Like AOM, oil stability index (OSI) values are expressed in hours and the lower number of hours indicates a lower stability for the oil. For lubricants like hydraulic fluids that tend to reside in the system for hundreds or thousands of hours, a high OSI value for the base oil and a yet higher value for the formulated products will be needed. OSI values can be correlated with other oxidation tests. However, for vegetable-oil-based lubricants, it is best to establish a relationship between field test results and the OSI values. Table 18.10 shows the OSI values for selected vegetable oils.

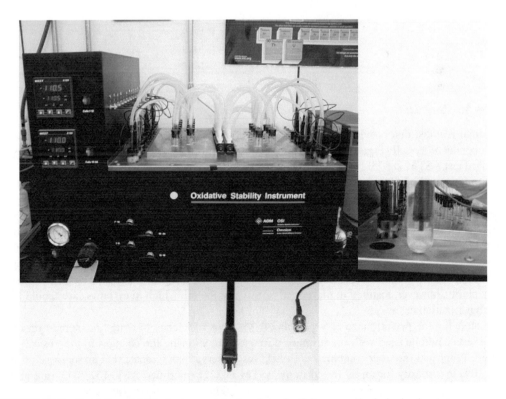

FIGURE 18.6   Oxidation stability instrument (top) and conductivity sensor in de-ionized water.

TABLE 18.10
OSI Values for Selected
Vegetable Oils

Oil	OSI (hours)
Apricot Kernel	23.42
Avocado	18.53
Babassu	57.8
Castor	105.13
Coconut	75.38
Corn	3.73
Cottonseed	4.35
Flaxseed	1.17
Grapeseed	2.83
Hempseed	0.10
Jojoba – Refined	42.15
Jojoba – Golden	38.3
Macadamia	6.87
Olive	5.08
Palm	21.52
Poppyseed	17.86
Ricebran	20.82
Ricinoleic Acid	117.1
Safflower	17.98
Sesame	5.8
Soy	17.67
Sunflower	10.23
Walnut	16.48

18.14.4 RANCIMAT

Rancimat is a test that provides similar information to the OSI test. It too uses de-ionized water and a conductivity cell (Figure 18.7). It was recently approved by ASTM as a part of the biodiesel standard test ASTM D 6751.

A unique method for an evaluation of the oxidative stability of vegetable oils was proposed by the UNI-NABL Center researchers. Using a hydraulic pump test to perform long-term tests on vegetable oils would expose the oil to industrial conditions that are difficult to duplicate in many standard test methods. For example, most untreated vegetable oils meet ASTM wear protection requirements for hydraulic pumps because of their naturally higher lubricity, but the oils also break down and oxidize. In ASTM D 2882, currently designated as modified ASTM D 7043, the untreated crude soybean oil showed 40 mg wear, which is under the 50 mg passing level. The lack of oxidation stability, however, resulted in *increased viscosity* of the oil, which in extreme cases could lead to polymerization of the oil.

In studying the performance of vegetable oil, vis-à-vis resistance to oxidation, some hydraulic pump tests could be used with the primary purpose of monitoring the changes in the viscosity of test oil. Empirical research combined with field test observations indicate that an increase of less than 10% in viscosity for an oil tested in an ASTM D 2271 (modified ASTM D 7043) or equivalent is desirable. It is also observed that although the addition of additive packages to vegetable oil improves many of the oil's characteristics, it could negatively impact its natural lubricity.

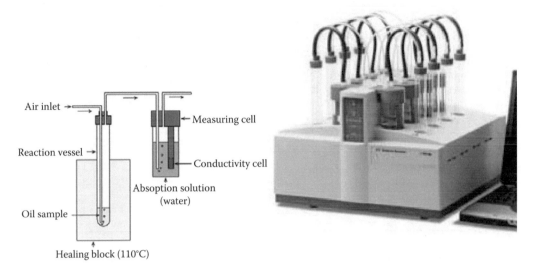

Air inlet →

Reaction vessel →

← Measuring cell

← Conductivity cell

Absoption solution
(water)

Oil sample

Healing block (110°C)

FIGURE 18.7 Components of Rancimat with oil sample and conductivity probe in de-ionized water.

Tests that are longer in duration, such as ASTM D 7043 (1000 hours), are more desirable than shorter-term tests such as modified ASTM D 7043 (100 hours). Increased oleic acid content combined with a reduced percentage of linolenic acid, as developed through new genes, could present improved oxidative stability as well as a reduced cost of the base oil. More stable base oils reduce the need for chemical modification such as hydrogenation.

Figure 18.8 shows the arrangement of the components of the hydraulic system setup according to the ASTM D2271 (modified ASTM D7043). The pump cartridge components are also shown separately comprising two brass side plates, a rotor, and cam ring and vanes. The cam ring and the vanes are weighted before and after the test and the weight loss is used as a determining factor for a pass or fail. To pass the test, a weight loss of less than 50 mg would be required.

The speed for running the pump in this test is 1200 rpm, and the reservoir is stainless steel placed two feet about the inlet of the pump. A pressure-relief valve allows for the adjustment of the pressure and a cooler is used to maintain the temperature of the oil at test conditions. The pump has about an eight gallon per minute flow rate.

FIGURE 18.8 System setup for the modified ASTM D 7043 (formerly ASTM D 2882 and ASTM D 2271).

TABLE 18.11
Changes in the Viscosity of Oils as Tested in ASTM D 2271, 1000-Hour Pump Test

Oil Type	% Oleic Acid	Δ Viscosity (cSt)
Crude Soybean Oil (Hexane Extracted)	23.4	43.86
Low Linolenic Soybean Oil	32.9	20.57
Partially Hydrogenated Soybean Oil	37.5	24.18
Crude Soybean Oil (Mechanically Extracted)	23.4	19.15
High Oleic Sunflower Oil	78.2	19.24
Ultrahigh Oleic Sunflower Oil	86.8	16.23
High Oleic Canola Oil (Supplier 1)	76	14.08
High Oleic Canola Oil (Supplier 2)	76.5	19.53
Palm Oil	38.8	12.97
Meadowfoam	62.5	32.55

Table 18.11 shows the change in viscosity for a number of vegetable oils when exposed to the hydraulic pump test conditions requiring 1000 hours of exposure to 1000 psi at 79°C (174°F). In order to show the relationships between fatty acid makeup and stability, as shown by change in the viscosity, the percent of oleic acid (18:1) was listed for each oil. From this table, it is shown that the higher oleic acid content improves oxidation stability, resulting in less of an increase in the viscosity of the oil. In cases where the oleic acid contents are the same, the percent of other fatty acids, especially the linoleic (18:2) and linolenic acid (18:3), was considered the reason for the higher change in viscosity.

Table 18.12 shows changes in the viscosity of five formulated hydraulic oils. Two were formulated as universal tractor hydraulic fluids and one was antiwear hydraulic fluid. For reference, an antiwear petroleum-based hydraulic oil was also tested. Petroleum oils do not show the same changes in viscosity as vegetable oils do. Often the viscosity of petroleum-based oil may actually go down due to a breakdown of the additives or shearing of viscosity modifiers.

Solubility. Vegetable oils are miscible with many organic solvents, particularly those that do not form hydrogen bonds (that are polar and aprotic). Several solvents have been experimented on for miscibility because of their use in the solvent extraction process of the seed oil. Pryde [11] indicates that in order to have miscibility the oil and the solvent should have properties in the same order of magnitude. Solubility characteristics can be estimated from dielectric constants or solubility parameters as a measure of polarity. Examples of the dielectric constants of some of the common liquids are as follows: water (dielectric constant = 8.5), ethanol (dielectric constant = 24.3), hexane (dielectric constant = 1.89), and vegetable oils (dielectric constant = 3.0–3.2) (p.43).

TABLE 18.12
Changes in the Viscosity of Formulated Oils as Tested in ASTM D 2271, 1000-Hour Pump Test

Oil Type	Δ Viscosity (cSt)
Soybean-Oil-Based Tractor Hydraulic Fluid	4.60%
Canola-Oil-Based Tractor Hydraulic Fluid	2.93
Rapeseed-Oil-Based Hydraulic Fluid	37.2
Petroleum-Based Hydraulic Fluid	1.5
50-50 Blend of Soybean-Oil-Based Tractor Hydraulic Fluid and Petroleum-Based Hydraulic Fluid	1.6

18.15   COMPATIBILITY WITH RUBBER HOSES

In general, vegetable oils are shown to be compatible with common hydraulic hoses. Several tests of compatibility were conducted to determine the interaction of various vegetable oils with rubber hoses. The test procedure included exposing a segment of a hydraulic hose to a specific volume of the test oil for a period of 72 hours at 100°C and measuring the increase in the volume of the hose. This test was described by Hedges [12] for use when no information is available on the compatibility of the elastomeric material for a given fluid. Accordingly, a clean 1000 mL glass flask is filled with 700 mL of the fluid to be tested and the rubber hosing to be used is cut into a convenient segment. The chunk of rubber hosing should be measured, obtaining values for mass, width, length, and height of the sample. The volume of the sample before testing is determined by immersing it in water to establish the amount of water displaced by the sample. Observations should be made and recorded for the appearance and feel of the sample. A hot plate is set up with a thermal control device to heat the fluid to a constant temperature of 100°C (212°F). A segment of the rubber hosing is placed into the flask with the fluid, and the flask is placed on the hot plate to be heated at 100°C (212°F) for 72 hours, magnetically stirring the hose. The flask may be heated for 72 continuous hours or the testing time may be broken up into convenient time periods equal to 72 hours. After the test is completed, the rubber sample should be removed from the fluid and washed in soap and water. The sample should be inspected for any stickiness or visual swelling. Any change in appearance or consistency, such as wearing away of the rubber or gumminess due to chemical reaction, should be recorded. Volumetric expansion of each sample can be determined by immersing it in water and determining the amount of water displaced. Length, width, height, and mass of the sample should be measured. All measurements should be recorded and compared to the original values and any significant differences should be noted. Honary [13] reported on the results of three tests of hose compatibility as follows:

1. Rubber hosing tested at 104°C (219°F) for 94 hours in crude soybean oil. The sample showed an increase of less than 10% for all the measurements with 0% water volume displacement (as measured). No significant signs of wear were evident on the sample. No corrosion of the rubber or burning was noted. Some swelling of the ends of the hosing was noted, but the amount was small. Compatibility with the rubber hosing does not appear to be a significant problem for the crude soybean oil.
2. Rubber hosing tested at 100°C for 100.5 hours. The expansion for all measurements was less than 10%, with 0% water volume displacement (as measured). When the rubber hosing was wiped with a paper towel, some rubber wiped off on the towel. However, it was discovered that rubber hosing which had undergone no testing performed in the same manner.
3. Rubber hosing tested in a commercially available rapeseed-based hydraulic oil containing full additive package at 103°C (217.4°F) for 78.25 hours. Hose volume increased from 7.5 cm³ to 8.0 cm³, an increase of 6.25%. This expansion, while within the acceptable limits of under vegetable oils 10% could be attributed to the presence of additives in the oil.

Volume expansions of 40%–50% were noted when the same test was repeated for methyl esters derived from various vegetable oils. This is mainly due to the solvent properties of the esters, which when used as biodiesel could interact with the rubber fuel lines.

18.16   OTHER PROPERTIES OF VEGETABLE OILS

18.16.1   FILM THICKNESS

Film thickness tests of three soybean oils and a fully formulated soybean-based mobile hydraulic fluid were conducted using an optical interferometry elastohydrodynamic (EHD) rig [14]. Furthermore, a conventional petroleum-based mobile hydraulic fluid was tested for comparison

TABLE 18.13
Viscosity and Density of the Test Oils

Test Oil	Viscosity (cSt) 40°C	at 100°C	Density (g/cm³)
Crude Soybean Oil #1	30.7	7.8	0.89
Crude Soybean Oil #2	31.1	7.4	0.885
Partially Hydrogenated Soybean Oil	35.6	7.9	0.83
Formulated Soybean Hydraulic Oil	48.1	10.5	0.897
Formulated Petroleum Hydraulic Oil	56.9	9.4	0.874

purposes. The rig incorporated a 3/4-in. steel ball rolling against a 6-in glass disk. The film thickness interference fringes were obtained under the following conditions:

Load W = 40 lbf (max. hertzian pressure Ph = 153 psi)
Temperature (T) = 40°C (104°F) and 75°C (167°F)
Rolling Speed Um = 10, 20, and 40 in./s.

The viscosity and density of the test oils are presented in Table 18.13.

Table 18.13 presents the film thickness of the test oils at two temperatures and three rolling speeds. In lubricants, the measured film thickness increases with the rolling speed and decreases with the temperature. The lower thermal reduction factors (defined as $\phi = - dH/dT$) the higher the resistance to film thickness variation as temperature changes. In this case soybean oil showed more favorable film thickness-temperature characteristics than the formulated petroleum sample. Both crude soybean oils (both were expelled oils) showed high resistance to temperature-induced film thickness variations. Under elastohydrodynamic conditions, the viscosities of lubricants tend to increase drastically with pressure. To describe such behavior, an experimental function may be used to model the *viscosity–pressure* relationship:

$$\eta = \eta_{oe}{}^{ap},$$

where η is the viscosity at pressure p, η_o denotes the viscosity at ambient pressure, and α is the *viscosity-pressure* coefficient. Direct measurement of α is a difficult task since high pressures of I GPa are required. This is not often available from the lubricants' suppliers. Table 18.14 shows film thickness at selected temperatures and rolling speeds. This is not often available from product suppliers and therefore, a safe estimated value for mineral oils was used ($\alpha = 2 \times 10^{-2}$ MPa⁻¹). This is a rather safe estimate since the viscosity–pressure coefficients of most lubricants are around this value. Using regression from the measured film thickness, values which were shown in Table 18.14, the estimated viscosity–pressure coefficients of the test oils were determined and are presented in Table 18.15.

Finally, another property of vegetable oils reported [15] is the ultrasonic velocities in some vegetable oils. Using a simple Hartley circuit with a piezoquartz in the oscillating circuit, ultrasonics of about 7 mcycles/s were generated. To determine the ultrasonic velocities in the oils, the following procedure was used: (1) a parallel beam of monochromatic light was passed through the oil; (2) the soundwaves were passed through the light, focusing onto a photographic plate; and (3) the diffraction rate of effect produced was used to calculate the sound velocities. Using this method, Table 18.16 shows the sound velocities determined for several vegetable oils.

The data presented in the Table 18.16 indicate that despite their higher viscosity than other organic compounds, the sound velocity in vegetable oils is below 1.00 m/s [16]. However, due to the fact that: (a) lowering of velocity occurs in esters, and (b) glycerin having three OH groups takes three acid radicals during esterification, then the observed velocities conform to previous works.

TABLE 18.14
Film Thickness of the Test Oils at Selected Temperatures and Rolling Speed

Test Oil	Temperature °C	Rolling Speed in/s	Film Thickness (μ.in)
Crude Soybean Oil #1	40	10	3
		20	3.6
		40	6.7
	75	10	2.1
		20	3.1
		40	3.8
Crude Soybean Oil #2	40	10	3.1
		20	5.4
		40	6.5
	75	10	2.2
		20	3
		40	5.8
Partially Hydrogenated Soybean Oil	40	10	2.3
		20	4.9
		40	7.5
	75	10	2.2
		20	2.7
		40	4.3
Formulated Soybean Hydraulic Oil	40	10	2.5
		20	6.5
		40	10.3
	75	10	2.4
		20	2.7
		40	3.3
Formulated Petroleum Hydraulic Oil	40	10	2.7
		20	6.9
		40	11.8
	75	10	2.2
		20	2.3
		40	3.1

Source: Ai, X. and Nixon, H.P. "Soybean-Based Lubricants Evaluation", unpublished report, University of Northern Iowa's Ag-Based Industrial Lubricants Research Program, 1996.

The lower velocity in such high-viscosity oils gives absorption coefficients very near those of the theoretical values.

Using the determined sound velocities and the densities of these oils, Pancholy et al. calculated the adiabatic compressibilities of the vegetable oils. Table 18.17 presents the results of their calculations.

Because the chemical compositions of the oils are not definite, the values of velocities and compressibilities of these oils can be taken as representative for each case.

18.17 OTHER FACTORS TO BE CONSIDERED FOR BIOBASED OILS

Viscosity-pressure properties. To determine the viscosity-pressure relationship, various tools could be used to pressurize the oil while studying changes in its viscosity. A falling-piston or falling-ball

TABLE 18.15
Estimated Viscosity–Pressure Coefficients (MPa⁻¹)

Test Oil	Estimated Viscosity–Pressure Coefficient (MPa⁻¹)
Crude Soybean Oil #1	1.91×10^{-2}
Crude Soybean Oil #2	1.17×10^{-2}
Partially Hydrogenated Soybean Oil	2.01×10^{-2}
Formulated Soybean Hydraulic Oil	2.71×10^{-2}
Formulated Petroleum Hydraulic Oil	2.64×10^{-2}

viscometer is one method which uses a tube with a measured liquid, piston, or ball inside the tube, an electrical magnet, and a magnetic switch. The piston or ball is made of a ferromagnetic material. It is first lifted to the top of the tube by the magnet and is then permitted to drop to the bottom of the tube under the force of gravity. The moment that the piston or ball reaches the bottom is sensed by a magnetic switch. The time required to pass the length of the tube is proportional to the viscosity, as illustrated in Figure 18.9. Other methods, such as the laser Doppler technique, have also been used. The tube can be pressurized to determine the impact on the viscosity as observed by the change in time for the ball to drop. To measure the viscosity-pressure relationship the ASTM D 6278 standard test method is used.

Table 18.18 shows the pressure–viscosity coefficients for different oils.

18.18   BULK MODULUS

Bulk modulus is a property that indicates the compressibility of a fluid which impacts the response time of hydraulic systems operating at high pressures. It can be described as a measure of resistance to compressibility of a fluid. Bulk modulus is the index of stiffness among the physical properties of fluid. While liquids are considered incompressible, as pressure increases all fluids compress proportional to the amount of pressure applied. Also, the presence of air and evaporation of chemicals within the oil can increase compressibility. Compression of a fluid causes its temperature to increase which also causes its volume to expand and, in a confined space, results in a further increase in pressure. If compression is slow the generated heat can have time to dissipate (approaching isothermal process). However, if compression takes place rapidly so that heat cannot dissipate (approaching adiabatic) then the pressure increases correspondingly, which is common to most hydraulic systems [17].

The term "bulk modulus" refers to the reciprocal of compressibility and is defined by the slope of the curve when plotted against the specific volume (at various pressures). Since specific volume is dimensionless, units of bulk modulus are the same as pressure, like pounds per square inch (psi), bar, pascal,

TABLE 18.16
Sound Velocities in Selected Vegetable Oils

Oil	Temperature (°C)	Frequency of Oscillation (mcycles/s)	Sound Velocity (m/s)
Olive Oil	32.5	7.03	1381
Sesame Oil	32.5	7.189	1432
Rapeseed Oil	30.75	6.977	1450
Coconut Oil	31	6.97	1490
Peanut Oil	31.5	7.054	1562
Linseed Oil	31.5	7.105	1772
Mustard Oil	31.5	6.934	1825

TABLE 18.17
Calculated Adiabatic Compressibility of Selected Vegetable Oils

Oil	Temperature (°C)	Density	Adiabatic Compressibility ($\times$ 10/atm)
Olive Oil	32.5	0.904	58.73
Sesame Oil	32.5	0.916	53.94
Rapeseed Oil	30.75	0.912	52.71
Coconut Oil	31.5	0.916	49.79
Peanut Oil	31.5	0.914	44.35
Linseed Oil	31.5	0.922	34.97
Mustard Oil	31.5	0.904	33.61

or N/m^2. A flat slope indicates a fairly compressible fluid having a low bulk modulus. Bulk modulus is determined either under isothermal (constant temperature) or adiabatic (isentropic) conditions.

There are two ways to define the slope of pressure-specific volume curve, or the bulk modulus: secant bulk modulus and tangent bulk modulus.

Secant bulk modulus can be calculated as the product of the original volume and the slope of the line drawn from the origin to any specified point on the curve of pressure-specific volume (or the slope of the scant line to the point). It can be written as:

Secant bulk modulus = [(Volume at zero pressure) × Pressure] ÷ (Volume at zero pressure)

− (Volume at desired pressure)]

Tangent bulk modulus can be calculated as the product of the original volume at any specified pressure and the derivative of fluid pressure with respect to volume at the point (the slope of the tangent line to the point). Thus:

Tangent bulk modulus = $V_0 \times (dP / dV)$

Sonic bulk modulus is derived from the sonic velocity in the fluid and its density. Assuming the constant density of the fluid, it is easily obtained from the propagation velocity of pressure wave, so

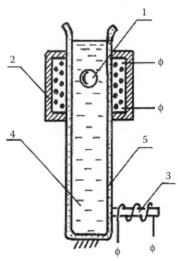

FIGURE 18.9 Falling ball pressure–Viscosity measuring unit. Falling-piston or falling-ball viscometer. 1 = ferromagnetic ball, 2 = electrical magnet, 3 = magnetic switch, 4 = liquid, 5 = container.

TABLE 18.18
Reference Pressure–Viscosity Coefficient for Selected Oils

	Pressure–Viscosity Coefficient, α (GPa^{-1})			
Lubricant	20°C	40°C	60°C	80°C
Naphthenic mineral Oil	26.5	23.4	20.0	16.4
Mix of paraffinic/ naphthenic Oil, 50–50	23.0	20.8	18.5	16.1
Paraffinic mineral Oil	19.8	18.2	16.6	15.0
Rapeseed Oil	18.9	17.5	16.3	14.6
Polyglycol	18.7	16.0	13.2	10.5
Pine tree Oil	17.2	15.6	13.7	12.5
TMP-ester	15.5	14.4	13.1	12.2
Diester	14.6	13.6	12.8	11.6
Polyalphaolefin	15.5	13.8	12.2	10.5

it is widely used to measure the dynamic bulk modulus. The sonic bulk modulus has the same value as the adiabatic bulk modulus.

Research is underway to study the differences between the bulk modulus of vegetable oils and conventional fluids. Many of the hydraulic fluid specifications for military fluids have bulk modulus requirements of 200,000 to 250,000 psi for conventional petroleum or synthetic oils.

The viscosity and pressure–viscosity coefficient of a lubricant has a major influence on film thickness formation and shear stresses in an elastohydrodynamic lubrication contact. It is therefore important to determine the influence of pressure and temperature on these parameters.

18.19   THERMAL CONDUCTIVITY OF VEGETABLE OILS

Thermal conductivity of hydraulic oil is an important property as it describes the ability of a fluid to transport thermal energy via molecular motion. Werner et al. [18] investigated the effects of pressures up to 400 MPa on the thermal conductivity of plant oils using a high-pressure adapted transient hot wire method. Table 18.19 shows the thermal conductivity of selected vegetable oils using a temperature range of about 20 degrees K (273–293 °K) or (0–20 °C) and showing the narrow range values for thermal conductivity (λ).

TABLE 18.19
Thermal Conductivity of Selected Vegetable Oils

Oil	T/K	λ/W·m^{-1}·K^{-1}	$(d\lambda/dT)$/10^{-4} W·m^{-1}·K^{-2})
Olive Oil	273.2 to 473.2	0.1696 to 0.1574	−0.61
Olive Oil	311.7	0.1679	
Olive Oil	296.13	0.164	
Castor Oil	273.2-433.2	0.183 to 0.1683	−0.91
Canola Oil	293.15	0.16	
Canola Oil	298.15	0.22 to 0.23	
Linseed Oil	303.15	0.17	2.1
Almond Oil	283.15	0.17	
Peanut Oil	297.5	0.1679	
Poppy seed Oil	283.15	0.16	
Sesame seed Oil	283.15	0.16	
Butter Oil	293.13	0.17	

Over the years, the technology for the preparation of petroleum base oils has been developed and perfected whereby a uniform quality can be expected from differing suppliers. However, in vegetable oils, factors such as climate, geography, plant variety, processing, purity, age, and storage conditions influence their composition and, as a result, their physiochemical properties.

Thermal conductivity of different pressurized liquids shows a strong sensitivity to specific volume. Werner et al. [18] studied the pressure effect up to 400 MPa on the thermal conductivity of olive oil, sunflower oil, linseed oil, and castor oil. Table 18.20 shows the physiochemical properties of these oils at 293.2K (20.05°C), where ρ is thermal conductivity.

For some ambient temperatures the values of the vegetable oils (olive, safflower, and linseed) are around 7% lower in comparison to those of castor oil. This is an indication that the significant difference in the molecular chain and the subsequent geometrical arrangement of the fatty acids can be used to determine behavior under pressure or at varied temperatures. For example, the density of vegetable oils increases with the increase in the unsaturated fatty acids content. Castor oil appears to be different due to a high percent (nearly 77%) of content of the monoacid triglyceride triricinolin and a polar hydroxyl group near one double bond in the ricinoleic acid. Castor oil showed the highest density, thermal conductivity, and viscosity among the tested plant oils as well as good solubility with ethanol. Correspondingly, the iodine value (an indicator of the number of double bonds and the amount of unsaturated fatty acids), correlates with the compressibility—with compressibility decreasing at higher iodine values. High pressures tend to pack the molecules into a denser formation thus increasing the melting point of triglycerides. It could be assumed that high pressure also promotes a change in *cis*-to-*trans* configuration in the unsaturated fatty acids of triglycerides. Finally, since increased pressure results in "volume reduction," the melting point increases with the reduced volume as well.

Werner et al. [18] conclude that:

1. The relative thermal conductivity of tested vegetable oils increases with pressure.
2. The slope of the isothermal flattens with rising pressures and decreasing temperatures.

Additionally, the temperature coefficient of the thermal conductivity of vegetable oils depends on its thermal expansion coefficient and is linearly proportional to it. The pressure coefficients of the thermal conductivity of a vegetable oil are a function of its compressibility and are near linearly proportional almost independent of temperature.

18.20 THERMAL CONDUCTIVITY AT HIGH TEMPERATURES

Cecil, Koerner, and Munch [19] described a refined modification of the hot-wire technique that provides absolute values for the thermal conductivity of liquids. Using this technique Cecil et al. determined the thermal conductivity of a series of liquids over temperatures ranging up to 217°C (102.8°F).

TABLE 18.20
Physiochemical Properties of the Investigated Plant Oils at 293K (10.85°C)

Oil	ρ	$d\rho/dT^{30}$	Fatty Acid Concentrations (g per 100 g of total fatty acids)						
	kg·m^{-3}	10^{-3} kg·m^{-3}· K^{-1}	C16	C16:1	C18	C18:1	C18:2	C18:3	Other
Olive Oil	912.7	−554.1	11.6	0.8	3.7	75.3	6.7	0.7	0.1
Safflower	922.8	−629.7	5.6	0.1	2.3	10.9	79.1	0.2	0.0
Linseed	927.7	−641.9	5.0	0.0	3.5	16.3	15.2	58.7	0.0
Castor	959.2	−648.4	1.2	0.0	1.1	3.3	5.7	0.6	87.6 Ricinoleic

TABLE 18.21
Thermal Conductivity of Selected Materials

Material	Estimated Thermal Conductivity W/m °C	Standard Error	Literature Value W/m °C
Olive Oil	0.1702	0.0016	0.17
	0.1736	0.005	
Vacuum Pump Oil	0.1586	0.0016	0.145 (engine Oil)
De-ionized Water	0.5426	0.0006	0.602
Ethanol	0.1871	0.0036	0.18
Ethylene Glycol	0.2569	0.0065	0.249
Mineral Oil	0.1307	0.0021	0.14
Silicone Oil	0.1439	0.0023	0.134
Fuji Apple	0.3989	0.0023	0.422-0.513
Wheat Flour	0.1357	0.0011	0.117
Beef (rounds)	0.4074	0.0051	0.475
Ground Beef (93% lean)	0.3777	0.0049	0.453
Chicken Breast Meat	0.4683	0.0083	0.47

To measure thermal conductivity of hydraulic fluids the ASTM D 2717 standard test method for thermal conductivity of liquids is used. This test method covers the determination of the thermal conductivity of nonmetallic liquids. According to ASTM, it is applicable to liquids that are: (1) chemically compatible with borosilicate glass and platinum; (2) moderately transparent or absorbent to infrared radiation; and (3) have a vapor pressure less than 200 torr at the temperature of test. The thermal conductivity of a substance is a measure of the ability of that substance to transfer energy as heat in the absence of mass transport phenomena. Materials that have vapor pressures of up to 345 kPa (50 psi) can be tested provided that adequate measures are taken to repress volatilization of the sample by pressurizing the thermal conductivity cell. The usual safety precautions for pressure vessels should be followed under these circumstances.

18.21 THERMAL CONDUCTIVITY AND THERMAL DIFFUSIVITY

Huang and Liu [20] developed a new transient plane-source method to measure thermal conductivity K and thermal diffusivity α. The thermal diffusivity is defined as the ratio of the thermal conductivity to the heat capacity of a material. It is considered the rate at which thermal energy diffuses by conduction through a material. Using a calibrated probe, constructed Resistive Temperature Detector (RTD), Huang and Liu were able to simultaneously measure K and α of olive oil, food, and other materials. Table 18.21 and Table 18.22 show the thermal conductivity of selected vegetable oils and materials.

Bernal-Alvarado et al. [21] experimented with a thermal lens technique to measure thermal diffusivity. Using this method Bernal-Alvarado et al. showed the thermal diffusivity of five test oils as shown in Table 18.22. (p. 698)

Facina and Colley [22] reported on the viscosity and specific heat of vegetable oils as a function of temperature, using a temperature range of 35°C (95°F) to 180°C (356°F). Table 18.24 presents viscosities of twelve vegetable oils at various temperatures. The viscosities at 35°C (95°F) were about ten to fifteen-fold larger than viscosities at 180°C (356°F), while the percent increase in specific heat from 35–180°C (95°F, 356°F) was about 17% (p. 741).

TABLE 18.22
Thermal Diffusivity of Tested Materials

Material	Estimated Thermal Diffusivity ($\times 10^{-7}$ m²/S)	Standard Error	Literature Value ($\times 10^{-7}$ m²/S)
Olive Oil	0.715	0.016	
	0.733	0.05	
Vacuum Pump Oil	0.7156	0.019	0.85 (engine Oil)
Water	1.553	0.048	1.44
Ethanol	0.806	0.039	0.8519
Ethylene Glycol	0.938	0.058	0.937
Mineral Oil	0.525	0.017	0.56
Silicone Oil	0.701	0.046	
Apple	1.299	0.02	0.77–1.5
Book Papers	1.449	0.04	1.44
Wheat Flour	0.988	0.023	0.99–1.104
Beef (rounds)	1.138	0.036	1.23 = 1.33
Ground Beef (93% lean)	1.085	0.036	
Chicken Breast Meat	1.524	0.072	10.3 (0°C)

18.22 DIELECTRIC CONSTANT AND DIELECTRIC STRENGTH OF VEGETABLE OILS

Dielectric constant is an important property of hydraulic oils because changes in this property of the oil are used to indicate the presence of contamination. The dielectric constant of a material is a measure of its ability to transmit electrical energy. Dielectric constant can also be considered a measure of the relative ratio of the speed of an electric field in a material compared to the speed of the electric field in a vacuum. The dielectric constant of vacuum is exactly 1.0, while metals as conductors have infinite dielectric constant. The presence of water, gasses, wear metals, and other by-products in a hydraulic oil during use continue to change the dielectric constant of the oil.

Dielectric strength of a product is measured by applying voltage to two standard plates submerged in the test product. At some voltage the test material cannot act as an insulator anymore and begins to conduct electricity. In effect, dielectric strength is the maximum insulating strength the test product shows in the face of applied voltage. For some vegetable oils, for example, voltages as high as nearly 50 kV may be needed to overcome insulating strength, Table 18.23. The higher the dielectric strength, the better insulating property the product has. Hydraulic oils with higher dielectric strength

TABLE 18.23
Dielectric Strength of Selected Vegetable Oil

Oil*	Dielectric Strength (kV)
Soybean Oil	35.4
Canola Oil	38
Olive Oil	42.2
Castor Oil	44.42
Corn Oil	47.44
Sunflower Oil	48.48

* *Note*: These are representative sample oils and do not cover all varieties of different vegetable oils.

TABLE 18.24
Viscosities of Twelve Vegetable Oils at Various Temperatures

Oil Source	Sample Temperature (°C)									
	35	50	65	80	95	110	120	140	160	180
Almond	43.98	26.89	17.62	12.42	9.15	7.51	6.54	5.01	4.02	3.62
Canola	42.49	25.79	17.21	12.14	9.01	7.77	6.62	5.01	4.29	4.65
Corn	37.92	23.26	15.61	10.98	8.56	6.83	6.21	4.95	3.96	3.33
Grapeseed	41.46	25.27	16.87	11.98	9	10.37	9.18	7.5	6.1	4.78
Hazelnut	45.55	27.4	17.83	12.49	9.23	7.56	6.69	5.25	4.12	3.48
Olive	46.29	27.18	18.07	12.57	9.45	7.43	6.49	5.29	4.13	3.44
Peanut	45.59	27.45	17.93	12.66	9.4	7.47	6.47	5.14	3.75	3.26
Safflower	35.27	22.32	14.87	11.17	8.44	6.73	6.22	4.77	4.11	3.44
Sesame	41.44	24.83	16.8	11.91	8.91	7.19	6.25	4.95	4.16	3.43
Soybean	38.63	23.58	15.73	11.53	8.68	7.17	6.12	4.58	3.86	3.31
Sunflower	41.55	25.02	16.9	11.99	8.79	7.38	6.57	4.99	4.01	3.52
Walnut	33.72	21.2	14.59	10.51	8.21	6.71	5.76	4.8	3.99	3.36

are suitable for applications that come in proximity of voltage systems; for instance, boom trucks used in maintaining high voltage electric lines (Figure 18.10).

To measure the dielectric property of hydraulic oils, the ASTM D 877 "Standard Test Method for Dielectric Breakdown Voltage of Insulating Liquids Using Disk Electrodes" is used.

The dielectric breakdown voltage is a measure of the ability of an insulating liquid to withstand electrical stress. The breakdown test uses AC voltage in the power-frequency range from 45 to 65 Hz. The suitability for this test method has not been determined for a liquid's viscosity higher than 900 cSt at 40°C (104°F).

18.23 ENVIRONMENTAL CONSIDERATIONS: BIODEGRADABLE AND BIOBASED

Vegetable oils are, in general, considered to be biodegradable and meet the biodegradability standards of the American Standard for Testing and Materials (ASTM) and the Organization for Economic Cooperation and Development (OECD).

FIGURE 18.10 Dielectric tester showing disk electrodes.

TABLE 18.25
Typical Biodegradability Results for Base Oils

Mineral Oil Type	Amount Biodegraded
White Mineral-Based Oil	25%–45%
Natural and Vegetable Oil	70%–100%
PAO	5%–30%
Polyether	0%–25%
Polyisobutylene (PIB)	0%–25%
Phthalate and Trimellitate Esters	5%–80%
Polyols and Diesters	55%–100%

Biobased: In the United States, vegetable-oil-based lubricants fall under the definition of biobased as defined by the United States Department of Agriculture (USDA). According to Honary (2007), discussions of biobased materials must first draw the distinction between "biodegradable" and "biobased." Two widely used designations are "readily" and "inherently" biodegradable:

- According to the OECD pass levels for a readily biodegradable product, there must be 70% removal of dissolved organic carbon (DOC) and 60% removal of theoretical oxygen demand and theoretical carbon dioxide within 28 days (OECD 301 Guideline for Testing of Chemicals).
- Inherently biodegradable defines no timing or degree of biodegradability.

Harold [23] reported on the biodegradability of mineral oil and mineral-oil-derived base oils, as shown in Table 18.25.

The United States Department of Agriculture (USDA) has introduced the term "biobased," initially intended as a label for products comprised of 51% (or more) renewable materials. Later, USDA revised the approach to identify products based on their percentage of renewable materials. A broader definition of both "biobased" and "biodegradable" follows:

- **Biobased**: Products containing natural renewable bio content like those made with agricultural materials. (These products could be biobased and biodegradable.)
- **Biodegradable:** Products that meet U.S. (ASTM) or European (OECD) biodegradability requirements and could be made of biodegradable and or biobased materials.

These definitions make clearer the distinction between the terms biobased and biodegradable. It is essential to understand these terms, as they are not necessarily interchangeable. For example, there are some synthetic esters which are petroleum-derived and are biodegradable (meets standards of biodegradability) but are not biobased. In the case of lubricants and greases, they are tested for their carbon 14 content to determine what percentage of their content comes from renewable carbon (bio) or what percentage of carbon comes from fossilized materials; this determines whether they are biobased. USDA now relies on ASTM D 6866-04 standards using radiocarbon and isotope ratio mass spectrometry analysis in order to determine the biobased content of these items. These test methods look at content alone and do not address environmental impact, functionality, or product performance.

18.24   BIODEGRADABILITY TEST

In this test a sample of the oil is incubated at 77°F (25°C) for 28 days. The biodegradation of the oil is induced by an inoculum of a mixed population of micro-organisms. As an inoculum, the test may use bacteria from the sewage-sludge of a domestic sewage-treatment plant, from natural water, soil bacteria or a combination of these.

Common oil biodegradability tests are performed in the presence of oxygen and water (aerobic aquatic biodegradation).

Primary biodegradation is measured by using infrared spectrometer. Ultimate biodegradation is determined according to the evolution of carbon dioxide (CO_2) from the tested sample over that produced in a blank sample, which contains inoculum only.

18.25   STANDARD BIODEGRADABILITY TESTS

- ASTM D-5864 Standard Test Method for Determining Aerobic Aquatic Biodegradation of Lubricants. This method is used for testing nonvolatile oils, which are not inhibitory to the inoculum micro-organisms.
- CEC-L-33-A-94 of the Coordinating European Council (CEC). This method is applicable for the determination of primary biodegradability. It is widely used for testing engine oils.
- OECD 301B, or Modified Sturm Test of the Organization for Economic Cooperation and Development (OECD). This method determines only ultimate biodegradability by measuring evolving carbon dioxide.
- OECD 301D, or Closed Bottle Test of the Organization for Economic Cooperation and Development (OECD). This method is used when the oxygen concentration in the test oil is not the limiting factor for degradation.
- EPA 560/6-82-003, or Shake Flask Test of the U.S. Environmental Protection Agency (EPA). This method is used to determine the ultimate biodegradability of test materials in 28 days and includes soil inoculums.

18.26   TOXICITY TYPES AND TESTING METHODS

Toxicity: Biobased fluids can be tested to determine if they present a health hazard when used as intended. Toxicity is measured according to the following:

(A) Aquatic toxicity: The product's acute toxicity level of LC-50 is tested using ASTM D 6081 "Toxicity Studies – Daphnia Magna "for aquatic toxicity." Figure 18.11 shows Daphnia Magna, which is a sensitive vertebrae used a one test for aquatic toxicity.
(B) Toxicity Studies – Fish: For the biobased products to be considered nontoxic it should have acute fish toxicity levels of LC-50 or greater than 1000 ppm for amounts measured in forty-eight hours.

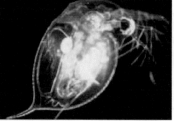

FIGURE 18.11   Daphnia Magna used as a test of aquatic toxicity.

TABLE 18.26
Biodegradability Requirements

Title	Test Parameter	Definition or Pass Criteria
Ready Biodegradability		
301A: DOC die-away	% DOC removal	≥ 70% DOC removed within 28 days and within 10 days window after 10% CO_2 production is reached
301B: CO_2 evolution (Modified Sturm Test)	% CO_2 production	≥ 60% of theoretical CO_2 production within 28 days and within 10-days window after 10% CO_2 production is reached
301C: MITI (I)	% BOD removal	≥ 60% theoretical BOD removal within 28 days
301D: Closed bottle	% BOD removal	≥ 60% theoretical BOD removal within 28 days and within 10- or 14-days window after 10% BOD removal is reached
301E: Modified OECD screening	% DOC removal	≥70% DOC removal within 10% 10-days window after DOC removal is reached
301F: Manometric respirometry	% BOD removal	≥ 60% theoretical BOD removal within 28 days and within 10-days window after 10% BOD removal is reached
Inherent Biodegradability		
302A: Modified SCAS Test	% DOC removal in daily cycles	20%–70% daily DOC removal during 12-week testing
302B: Zahn-Wellens/EMPA	% DOC removal	20%–70% DOC removal within 28 days
302C: Modified MITI Test (I)	% BOD removal and/or % loss of parent compound	20%–70% BOD or parent compound removal within 28 days
304A: Inherent biodegradability in soil	Production of 14CO_2 from radiolabeled substrate	Not specified
Simulation (confirmation)		
303A: Aerobic sewage treatment: coupled units test	% DOC removal	Degradation rate is calculated

18.27   BIODEGRADABILITY REQUIREMENTS

Table 18.26 presents the requirements for biodegradability, including test parameters and pass criteria.

REFERENCES

1. Gapinski, R.E., Joseph, I.E. and Layzel, B.D. "A Vegetable Oil Based Tractor Lubricant", *SAE Technical Paper Series*, Paper 941758, Warrendale, PA, SAE Publications, 1994.
2. http://www.scientificpsychic.com/fitness/fattyacids.html.
3. http://www.katangroup.com.my/Know_Lubricants.htm.
4. Meng, T. and Dresel, W. *Lubricants and Lubrication,* Wiley-VCH GmbH, Weinheim, Germany, 2001.
5. Cermak, S.C. and Isbel, T.A. "Synthesis and Physical Properties of Estolide-Based Functional Fluids", Elsevier, *Journal of Industrial Crops and Products*, No. 18, 2003, pp. 183–196.
6. Cermak, S.C. and T.A. Isbell, "Estolides—The Next Bio-Based Functional Fluid", *Inform*, 2004, 15:515–517.
7. Elevance Renewable Sciences: http://www.elevance.com/
8. US Patent #5089403 by Hammond and Lee, "Process for Enzymatic Hydrolysis of Fatty Acid Triglycerides with Oat Caryopses", February 18, 1992.
9. Salunkhe, J.K., Chavan, J.K., Adsule, R.N. and Kadem, S.S. *World Oilseeds: Chemistry, Technology and Utilization*, Van Nostrand Reinhold, New York, 1992.
10. US Patent #5972855 by Honary, L.A., "Soybean-Based Hydraulic Fluid", October 16, 1999.

11. In Erickson, P.E.H., D.R., Pryde, E.H., Brekke, O.L., Mounts, T.L. and Falb, R.A., *Handbook of Soybean Oil Processing and Utilization.* American Soybean Association, St. Louis, MO/AOCS Press, Champaign, IL, 1985.

12. Hedges, C.S. *Fluid Power in Plant and Field.* Womack Educational Publications, Dallas, Texas, 1981, p.136

13. Honary, L.A.T. (1995), "Performance of Selected Vegetable Oils in ASTM Hydraulic Tests", *SAE Technical Papers*, Paper 952075, 1995

14. Ai, X. and Nixon, H.P. "Soybean-Based Lubricants Evaluation", unpublished report, University of Northern Iowa's Ag-Based Industrial Lubricants Research Program, 1996.

15. Pancholy, M., Pande, A. and Parthasarathy, M. "Ultrasonic Velocities in Some Vegetable Oils", *Journal Science Industrial Research*, 1944, 3, p. 111.

16. Zachooal, I.J. Phys. Res., 1939, 10, P. 350, in Partasarathy, M. et al., "Ultrasonic Velocities in Some Vegetable Oils", *Journal Science Industrial Research*, 1944, 3, p. 112.

17. George, H.F. and Barber, A. In "Hydraulic and Pneumatic Magazine", *What is bulk modulus, and when is it Important?* July 2007.

18. Werner, M., Baars, A., Eder, C. and Delgado, A. "Thermal Conductivity and Density of Plant Oils under High Pressure", *Journal of Chemical and Engineering Data* 2008, 53, 1444–52.

19. Cecil, O.B., Koerner, W.E. and Munch, R.H. "Thermal Conductivity of Some Organic Liquids High Temperature Measurements", *Industrial and Engineering Chemistry 2002*, Vol. 2, No.1, pp. 54–56.

20. Huang, L. and Liu, L. "Simultaneous Determination of Thermal Conductivity and Thermal Diffusivity of Food and Agricultural Materials Using a Transient Plane-Source Method", *Journal of Food Engineering*, 95, 2009, pp. 179–85.

21. Bernal-Alvarado, J., Mansanares, A.M., da Ailva, E.C. and Moreira S.G.C. "Thermal Diffusivity Measurements in Vegetable Oils with Thermal Lens Technique", *Review of Scientific Instruments 2003*, Vol. 74, No. 1, pp. 697–99.

22. Facina, O.O. and Colley, Z. "Viscosity and Specific Heat of Vegetable Oils as a Function of Temperature: 35°C to 180°C", *International Journal of Food Properties*, 11, 2008, 738–746.

23. Harold, S. *Biodegradability: Review of the Current Situation*, The Lubrizol Corporation, company publication, 1993.

GLOSSARIES

Acid Number: A measure of the amount of KOH needed to neutralize all or part of the acidity of a petroleum product.

Additive: Any material added to base stock to change its properties, characteristics, or performance.

Anhydrous: A lubricating grease without water (as determined by ASTM D 128).

Aniline Point: The lowest temperature at which equal volumes of aniline and hydrocarbon fuel or lubricant base stock are completely miscible. A measure of the aromatic content of a hydrocarbon blend, used to predict the solvency of a base stock or the cetane number of a distillate fuel.

Apparent Viscosity: A measure of the viscosity of a non-Newtonian fluid under specified temperature and shear rate conditions.

Bactericide: Additive to inhibit bacterial growth in the aqueous component of fluids, preventing foul odors.

Bases: Compounds that react with acids to form salts plus water. Alkalis are water-soluble bases, used in petroleum refining to remove acidic impurities. Oil-soluble bases are included in lubricating oil additives to neutralize acids formed during the combustion of fuel or oxidation of the lubricant.

Base Number: The amount of acid (perchloric or hydrochloric) needed to neutralize all or part of a lubricant's basicity, expressed as KOH equivalents.

Base Stock: The base fluid, usually a refined petroleum fraction or a selected synthetic material, into which additives are blended to produce finished lubricants.

Bleeding: Separation of liquid lubricant from grease.

Blending: Blending is the process of mixing fluid lubricant components for the purpose of obtaining desired physical properties.

Boundary Lubrication: Lubrication between two rubbing surfaces without the development of a full fluid lubricating film. It occurs under high loads and requires the use of antiwear or extreme-pressure (EP) additives to prevent metal-to-metal contact.

Bright Stock: A heavy residual lubricant stock with low pour point, used in finished blends to provide good bearing film strength, prevent scuffing, and reduce oil consumption. Usually identified by its viscosity, SUS at 210°F, or cSt at 100°C.

Brookfield Viscosity: Measure of apparent viscosity of a non-Newtonian fluid as determined by the Brookfield viscometer at a controlled temperature and shear rate.

Bulk Appearance: Appearance of an undisturbed grease surface. Bulk appearance is described by:

- **Bleeding:** Free oil on the surface (or in the cracks of a cracked grease.)
- **Cracked:** Surface cracks.
- **Grainy:** Composed of small granules or lumps of constituent thickener.
- **Rough:** Composed of small irregularities.
- **Smooth:** Relatively free of irregularities.

Cetane Number: A measure of the ignition quality of a diesel fuel, as determined in a standard single cylinder test engine, which measures ignition delay compared to primary reference fuels. The higher the cetane number, the easier a high-speed, direct-injection engine will start, and the less "white smoking" and "diesel knock" after startup.

Cloud Point: The temperature at which a cloud of wax crystals appears when a lubricant or distillate fuel is cooled under standard conditions. Indicates the tendency of the material to plug filters or small orifices under cold weather conditions.

Coefficient of Friction: Coefficient of static friction is the ratio of the tangential force initiating sliding motion to the load perpendicular to that motion. Coefficient of kinetic friction (usually called "coefficient of friction") is the ratio of the tangential force sustaining sliding motion at constant velocity to the load perpendicular to that motion.

Cohesion: Molecular attraction between grease particles contributing to its resistance to flow.

Complex Soap: A soap crystal or fiber formed usually by co-crystallization of two or more compounds. Complex soaps can be a normal soap (such as metallic stearate or oleate), or incorporate a complexing agent which causes a change in grease characteristics—usually recognized by an increase in dropping point.

Consistency: The resistance of a lubricating grease to deformation under load. Usually indicated by ASTM Cone Penetration, ASTM D 217 (IP 50), or ASTM D 1403.

Copper Strip Corrosion: A qualitative measure of the tendency of a petroleum product to corrode pure copper.

Corrosion: The wearing away and/or pitting of a metal surface due to chemical attack.

Corrosion Inhibitor: An additive that protects lubricated metal surfaces from chemical attack by water or other contaminants.

Demulsibility: A measure of the fluid's ability to separate from water.

Density: Mass per unit volume.

Dispersant: An additive that helps keep solid contaminants in a crankcase oil in colloidal suspension, preventing sludge and varnish deposits on engine parts. Usually nonmetallic ("ashless"), and used in combination with detergents.

Dropping Point: The temperature at which grease becomes soft enough to form a drop and fall from the orifice of the test apparatus of ASTM D 566 (IP 132) and ASTM D 2265.

Dry Film Lubricant: A low shear-strength lubricant that shears in one particular plane within its crystal structure (such as graphite, molybdenum disulfide and certain soaps).

Elastohydrodynamic Lubrication (EHD): A lubricant regime characterized by high unit loads and high speeds in rolling elements where the mating parts deform elastically due to the incompressibility of the lubricant film under very high pressure.

Emulsifier: Additive that promotes the formation of a stable mixture, or emulsion, of oil and water.

Evaporation Loss: The loss of a portion of a lubricant due to volatization (evaporation). Test methods include ASTM D 972 and ASTM D 2595.

Extreme Pressure Property: That property of a grease that, under high applied loads, reduces scuffing, scoring, and seizure of contacting surfaces. Common laboratory tests are Timken OK Load (ASTM D 2509 and ASTM D 2782) and 4-Ball Load Wear Index (ASTM D 2596 and ASTM D 2783).

Flash Point: Minimum temperature at which a fluid will support instantaneous combustion (a flash) but before it will burn continuously (fire point). Flash point is an important indicator of the fire and explosion hazards associated with a petroleum product.

Friction: Resistance to motion of one object over another. Friction depends on the smoothness of the contacting surfaces, as well as the force with which they are pressed together.

Fretting: Wear characterized by the removal of fine particles from mating surfaces. Fretting is caused by vibratory or oscillatory motion of limited amplitude between contacting surfaces.

Fuel Ethanol: Ethanol (ethyl alcohol, C_2H_5OH) with impurities, including water but excluding denaturants.

Homogenization: The intimate mixing of grease to produce a uniform dispersion of components.

Hydrolytic Stability: Ability of additives and certain synthetic lubricants to resist chemical decomposition (hydrolysis) in the presence of water.

Kinematic Viscosity: Measure of a fluid's resistance to flow under gravity at a specific temperature (usually 40°C or 100°C).

Lubricating Grease: A solid to semifluid dispersion of a thickening agent in liquid lubricant containing additives (if used) to impart special properties.

Naphthenic: A type of petroleum fluid derived from naphthenic crude oil, containing a high proportion of closed-ring methylene groups.

Neutralization Number: A measure of the acidity or alkalinity of an oil. The number is the mass in milligrams of the amount of acid (HCl) or base (KOH) required to neutralize one gram of oil.

Neutral Oil: The basis of most commonly used automotive and diesel lubricants; they are light overhead cuts from vacuum distillation.

Newtonian Behavior: A lubricant exhibits Newtonian behavior if its shear rate is directly proportional to the shear stress. This constant proportion is the viscosity of the liquid.

Newtonian Flow: Occurs in a liquid system where the rate of shear is directly proportional to the shearing force. When shear rate is not directly proportional to the shearing force, flow is non-Newtonian.

NLGI Number: A scale for comparing the consistency (hardness) range of greases (numbers are in order of increasing consistency). Based on the ASTM D 217 worked penetration at 25°C (77°F).

Non-Newtonian Behavior: The property of some fluids and many plastic solids (including grease), of exhibiting a variable relationship between shear stress and shear rate.

Non-Soap Thickener: Specially treated or synthetic materials (not including metallic soaps) dispersed in liquid lubricants to form greases. Sometimes called "synthetic thickener," "inorganic thickener," or "organic thickener."

Oxidation: Occurs when oxygen attacks petroleum fluids. The process is accelerated by heat, light, metal catalysts and the presence of water, acids, or solid contaminants. It leads to increased viscosity and deposit formation.

Oxidation Inhibitor: Substance added in small quantities to a petroleum product to increase its oxidation resistance, thereby lengthening its service or storage life; also called "antioxidant."

Oxidation Stability: Resistance of a petroleum product to oxidation and, therefore, a measure.

Paraffinic: A type of petroleum fluid derived from paraffinic crude oil and containing a high proportion of straight chain saturated by hydrocarbons; often susceptible to cold-flow problems.

Poise: Measurement unit of a fluid's resistance to flow (i.e., viscosity), defined by the shear stress (in dynes per square centimeter) required to move one layer of fluid along another over a total layer thickness of one centimeter at a velocity of one centimeter per second. This viscosity is independent of fluid density and directly related to flow resistance.

$$\text{Viscosity} = \frac{\text{shear stress}}{\text{shear rate}}$$

$$= \frac{\text{dynes/cm}^2}{\text{cm/s/cm}}$$

$$= \frac{\text{dynes/cm}^2}{\text{s}} = 1 \text{ poise}$$

Pour Point: An indicator of the ability of an oil or distillate fuel to flow at cold operating temperatures. It is the lowest temperature at which the fluid will flow when cooled under prescribed conditions.

Pour Point Depressant: Additive used to lower the pour point or low-temperature fluidity of a petroleum product.

Pumpability: The low temperature, low shear stress-shear rate viscosity characteristics of an oil that permit satisfactory flow to and from the engine oil pump and subsequent lubrication of moving components.

Rheology: The deformation and/or flow characteristics of grease in terms of stress, strain, temperature, and time (commonly measured by penetration and apparent viscosity).

Rust Preventative: Compound for coating metal surfaces with a film that protects against rust. Commonly used to preserve equipment in storage.

Saponification: The formation of a metallic salt (soap) due to the interaction of fatty acids, fats, or esters generally with an alkali.

Sludge: A thick, dark residue, normally of mayonnaise consistency, that accumulates on nonmoving engine interior surfaces. Generally removable by wiping unless baked to a carbonaceous consistency. Its formation is associated with insolubles overloading of the lubricant.

Stoke (St): Kinematic measurement of a fluid's resistance to flow defined by the ratio of the fluid's dynamic viscosity to its density.

Synthetic Lubricant: Lubricating fluid made by chemically reacting materials of a specific chemical composition to produce a compound with planned and predictable properties.

Texture: The texture of a grease is observed when a small portion of it is pressed together and then slowly drawn apart. Texture can be described as:

- **Brittle:** ruptures or crumbles when compressed
- **Buttery:** separates in short peaks with no visible fibers
- **Long fibers:** stretches or strings out into a single bundle of fibers
- **Resilient:** withstands a moderate compression without permanent deformation or rupture
- **Short fiber:** short break-off with evidence of fibers
- **Stringy:** stretches or strings out into long fine threads, but with no evidence of fiber structure

Thickener: The structure within a grease of extremely small, uniformly dispersed particles in which the liquid is held by surface tension and/or other internal forces.

Tribology: Science of the interactions between surfaces moving relative to each other, including the study of lubrication, friction, and wear.

Viscosity: A measure of a fluid's resistance to flow.

Viscosity Index: Relationship of viscosity to temperature of a fluid. High-viscosity-index fluids tend to display less change in viscosity with temperature than low-viscosity-index fluids.

Viscosity Modifier: Lubricant additive, usually a high-molecular-weight polymer, that reduces the tendency of an oil's viscosity to change with temperature.

Water Resistance: The resistance of a lubricating grease to adverse effects due to the addition of water to the lubricant system. Water resistance is described in terms of resistance to washout due to submersion (see ASTM D1264) or spray (see ASTM D4049), absorption characteristics and corrosion resistance (see ASTM D1743).

White Oil: Highly refined lubricant stock used for specialty applications such as cosmetics and medicines.

Yield: The amount of grease (of a given consistency) that can be produced from a specific amount of thickening agent; as yield increases, percent thickener decreases.

W.D. Phillips

CONTENTS

19.1 INTRODUCTION

The term "phosphate ester" is used to describe an enormous range of chemicals, from gases to solids, which vary considerably in their properties and applications. Depending on their composition, they may be used in many different applications—for example, in foodstuffs, toothpaste, fertilizers, as fuel or oil additives, or as insecticides. However, the purpose of this chapter is to examine their use as fire-resistant hydraulic fluids.

When it is considered that phosphorus, either in elemental form or as sulphide, has long been used in the manufacture of matches, it is perhaps surprising that the phosphorus content of a molecule can make it less combustible. However, phosphate esters have been used as fire-resistant hydraulic fluids for over 50 years in a wide variety of industrial processes where a fire hazard exists and they have made a significant contribution to a safe working environment for operating personnel.

During this period, performance requirements have steadily become more severe while, concurrently, regulatory pressures on toxicity, environmental behavior, and waste disposal have also increased. The fluid suppliers have reacted to the changes by developing improved products and techniques for substantially extending fluid life. These developments have had a very positive effect on life-cycle costs and a noticeable change in the perception that these products are noxious and difficult to handle.

In this chapter, the chemistry, manufacturing processes, and performance of the different types of phosphate ester in commercial use will be reviewed together with comments on their different applications. Additionally, information on in situ purification techniques, fluid maintenance, the current position on fluid toxicity, ecotoxicity, and disposal aspects will be discussed.

19.2 HISTORICAL BACKGROUND

In the field of fire-resistant hydraulic fluids, the main products in commercial use are neutral trialkyl phosphates, triaryl phosphates, and mixed alkyl aryl phosphates. The chemical composition of these types of phosphate esters has been known for about 150 years [1,2], but their actual industrial use is much more recent. Tricresyl phosphate (TCP) was initially sold as a plasticizer for cellulose nitrate shortly after World War I [3] as a result of a shortage of camphor. Subsequent investigations revealed that not only was the product a good plasticizer, but it also rendered the product "nonflammable" [3]. Commercial production of TCP commenced in 1919, and although its use in camphor was eventually discontinued, it later found use as a "nonflammable" plasticizer in pyroxalin lacquers. In the 1930s, it was introduced into poly(vinyl chloride) formulations, which, today is still the largest market for triaryl phosphates.

In the oil industry, triaryl phosphates were first reported as antiwear additives for mineral oils in 1940 [4] and have since become firmly established in this application, providing enhanced lubrication for a range of circulatory oils and also for selected synthetic lubricants. In the 1950s and 1960s the products were also used as ignition control additives for gasoline in order to avoid pre-ignition arising from the deposition of lead salts formed by the interaction of the antiknock additive and alkyl halide scavengers. As a result of concern over catalyst poisoning by phosphorus and the move away from leaded fuel, the application in motor gasoline has since disappeared. However, there is still some interest in their use as antiwear additives in other fuel types due to the removal of sulfur-containing materials because of stricter emission regulations.

The first recorded consideration of "phosphate esters" as potential hydraulic fluids arose toward the end of World War II, following a series of fires and explosions in military hardware—particularly aircraft. The U.S. Navy instituted a study of the flammability of different liquids [5], and phosphate esters and water-glycol fluids were eventually selected for further investigation [6]. A research program sponsored by the U.S. Navy and Air Force and carried out by the Shell Development Company subsequently focused on formulations based on blends of trialkyl phosphates and TCP [7], the trialkyl derivative being chosen chiefly to reduce the density (and hence weight) of the fluid and to improve the low-temperature performance for operation at high altitude. In 1946, and independent of these studies, the Douglas Aircraft Company began an intensive search for a fire-resistant hydraulic fluid for use in commercial air transports [8]. Although blends of triaryl and trialkyl phosphates were considered [9], the fluid eventually marketed was based on an alkyl aryl phosphate [9–12] supplied by the Monsanto Chemical Company. The Lockheed Aircraft Corporation also used trialkyl phosphates in the development of low-density hydraulic fluids [13,14] but apparently did not pursue their studies to commercialization. By the end of the 1950s, most of the passenger jet aircraft being built were operating on phosphate esters and this is still the situation almost sixty years later. However, despite the appearance of specifications for fire-resistant hydraulic fluids for use on aircraft [15,16], the main military application for phosphate ester fluids has been on aircraft carriers (e.g., for hydraulic lifts) [17] and, for a short time, in submarines.

In the field of general industrial hydraulic fluids, density and low-temperature viscosity are normally less important than fire resistance and, although triaryl phosphates were promoted as fire-resistant fluids in the mid-1940s, they were initially more often used as blending components for the even more stable (and fire-resistant) polychlorinated biphenyls (PCBs) [18]. In these formulations, phosphates were used to reduce the very high fluid density of the PCBs and therefore ease the pumping requirements.

Other applications investigated included the development of hydraulic fluids based on trialkyl phosphates for gun recoil systems and shock absorbers [19].

By the early 1960s, several fluid suppliers were offering a range of phosphate esters covering different viscosity levels [20–22] in competition with the PCBs and PCB/phosphate blends [23]. Environmental concerns associated with the chlorinated products eventually resulted in their withdrawal in the 1970s. Although an attempt was made to replace them in these blends by other chlorinated aromatics, such as chlorinated benzyltoluene, use of the latter was restricted. Only in the German mining industry were such blends sold in any quantity. They were eventually banned by the European Commission in 1991 [24] because of environmental concerns similar to those associated with the PCBs.

Over the years, other attempts have been made to reduce the cost of phosphate esters—particularly the triaryl phosphates—and to improve certain physical characteristics by blending with other materials (e.g., carboxylate esters and even mineral oil). Such blends are, of course, also less fire resistant. Early developments of this type were carried out by the Cities Service Oil Co. [25]. Other fluid suppliers subsequently introduced similar products [26–29] and these are attractive for those applications where the leakage rate is so high that the user's prime consideration becomes cost rather than safety. As the properties of such blends are largely predictable from knowledge of the performance of the individual components and their relative amounts, no detailed examination of these products will be made in this chapter. Instead, the focus will be on 100% phosphate-based materials, which are used in applications where a high level of fire resistance is essential.

19.3 STRUCTURE AND ITS INFLUENCE ON FLUID PROPERTIES

Phosphate esters are also known as esters of phosphoric acid and can be considered as having the following general structure:

$$R_1O \longrightarrow \overset{\displaystyle\overset{O}{\|}}{\underset{\displaystyle OR_3}{P}} \longrightarrow OR_2$$

For *trialkyl phosphates*, R_1, R_2, and R_3 are normally the same and are alkyl groups with a chain length of C_4–C_{10}, but principally C_4, C_8, and C_{10}. The chain may be straight as in tri-*n*-butyl phosphate (TBP), or branched, as in tri-isobutyl phosphate (TiBP). Also used occasionally in hydraulic fluid formulations is a trialkoxyalkyl phosphate, specifically tributoxyethyl phosphate, in which case R_1, R_2, and R_3 in the above structure are $C_4H_9OC_2H_5$—.

For *triaryl phosphates*, R_1–R_3 are phenyl or substituted phenyl (e.g., methylphenyl, also known as cresyl or tolyl), dimethylphenyl (better known as xylyl), isopropylphenyl, or tertiarylbutylphenyl groups. Other substituted phenyl groups, such as nonylphenyl or cumylphenyl, have also been used in the past but are not currently available.

In addition to those triaryl phosphates where $R_1 = R_2 = R_3$, (e.g., trixylyl phosphate) there are also products where the aryl groups are different (e.g., cresyldiphenyl phosphate or xylyldiphenyl phosphate). As will be shown later, mixed aryl phosphates are not widely used as general industrial hydraulic fluids.

For mixed alkyl aryl phosphates, the groups attached to the phosphorus are also different. The product may be a monoalkyldiaryl phosphate (e.g., octyldiphenyl phosphate) or a dialkylmonoaryl phosphate such as dibutylphenyl phosphate. Early interest in these products has since declined and the properties sought are achieved today by blending trialkyl- and triaryl phosphates.

Although the phosphorus-containing nucleus of the molecule largely determines the fire resistance of the product, it is the "organic" or the hydrocarbon part of the molecule which mainly determines the variation in the physical and chemical properties, in addition to having some effect on the flammability characteristics. For example, the alkyl groups used in trialkyl or alkyl aryl phosphates are either straight or branched chain. Use of the former favors viscosity–temperature properties, thermal stability, volatility, and so forth, whereas branched chains normally improve hydrolytic stability. Similarly, the properties of aryl phosphates are heavily influenced by four factors:

1. *Whether or not the aryl group is substituted.* The unsubstituted phenyl group gives the best thermal and oxidative stability but confers the worst hydrolytic stability and viscosity–temperature properties on the molecule. Triphenyl phosphate, however, is a solid at ambient temperature and therefore cannot be used alone in hydraulic applications. The general effect on phosphate performance of a single substituent on the aromatic ring is indicated in Figure 19.1 [30]. As can be seen, if an attempt is made to improve one fluid property, this can have a detrimental effect on another aspect of behavior.
2. *The structure of the substituent.* This impacts on physical properties as a result of its size (e.g., the greater the number of carbon atoms, the higher the viscosity of the fluid, provided the substituent is attached to the same place on the ring) and also on chemical properties as a consequence of its stability. Thus, the isopropylphenyl group is less oxidatively stable than the tertiarybutylphenyl group.
3. *The position of the substituent on the aromatic ring.* Alkylated phenol feedstocks used in the manufacture of triaryl phosphates are available in different "isomeric" forms, where the material has the same chemical formula and molecular weight, but the substituent is attached to a different place on the aryl ring. Figure 19.2 shows the isomerism of monoisopropylphenol.

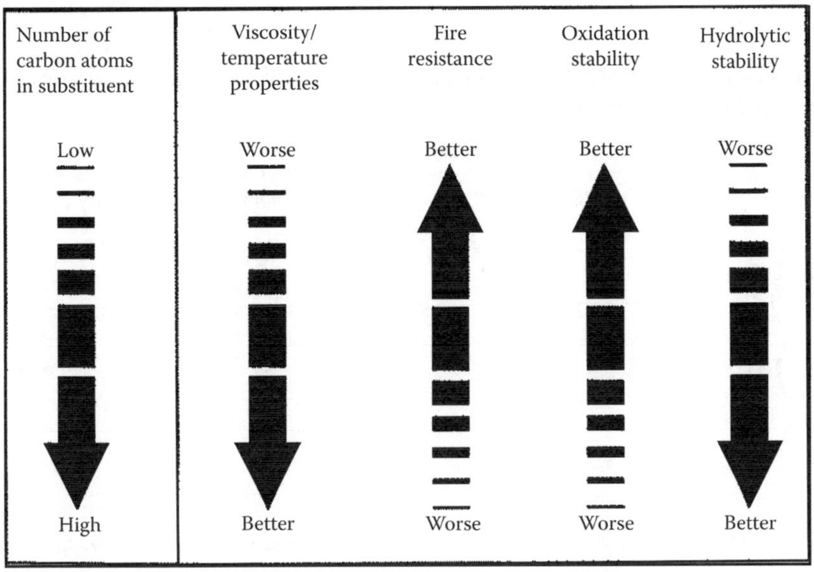

FIGURE 19.1 Effect of aryl group substituents on physical/chemical properties.

The pure phosphate esters produced from the individual isopropylphenyl isomers vary considerably in performance. Table 19.1 [30] shows the effect on viscosity and oxidation stability. The *ortho*- isomer, which has the highest rate of change of viscosity with temperature (referred to as a "low viscosity index"), has the best oxidation stability. By contrast, the *para*- derivative has better viscosity–temperature properties (a higher viscosity index) but poor oxidation stability. These differences in behavior are due to: (a) the effect of steric hindrance of the —P—O bond when in the *ortho* position, and (b) its ability when in the *para* position to destabilize the molecule so that less energy is required for its breakdown.

4. *The number of substituents on the aromatic ring of the alkylated phenol feedstock.* Where more than one substituent is present, the number of isomeric phosphates that can, theoretically, be components of the finished product is significantly increased. With xylenol, for example, which has two methyl groups and six different isomeric forms, it is possible to produce over 200 isomeric phosphates. Again, the physical and chemical properties are affected, and trixylyl phosphates display higher viscosities and also better hydrolytic stability than tricresyl phosphates.

It will be appreciated from this brief explanation of the impact of ring substituents on fluid behavior that although it is possible to produce a product with, for example, excellent oxidative stability or better

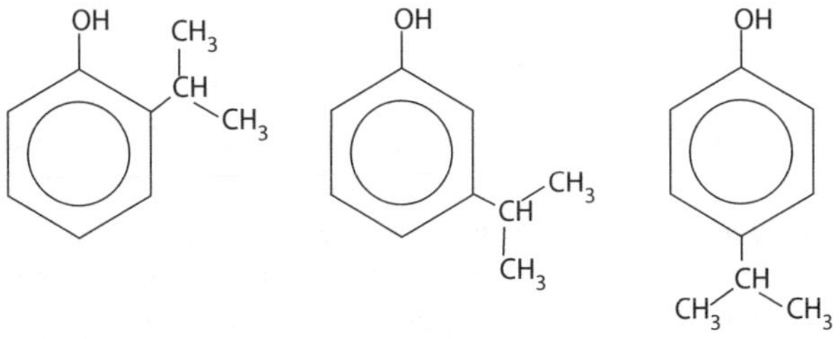

FIGURE 19.2 Isomeric forms of monoisopropylphenol.

TABLE 19.1
Variation in Viscosity/Temperature Properties and Oxidation Stability for Isomeric Forms of Tris-isopropylphenyl Phosphate

Phosphate Isomer	Viscosity (cSt)			Viscosity Index	Oxidation Stability[a] (% Viscosity Change at 37.8°C)
	99.8°C	37.8°C	0°C		
(isopropylphenyl-O–)₃PO	8.2	120	14,540	<0	4
(isopropylphenyl-O–)₃PO	5.0	34.8	508	63	800
(isopropylphenyl-O–)₃PO	6.3	56.7	1,450	52	2,000

[a] After oxidation at 150°C for seven days in the presence of metals (modified FTMS VV-L-791C, method 5308.6).

viscosity–temperature properties, it is difficult to achieve this without some adverse impact on other fluid characteristics. A degree of compromise is, therefore, almost inevitable when selecting a triaryl phosphate base stock to meet the technical requirements of a specific application. For example, it may be appropriate to select the fluid which is either the most hydrolytically or the most oxidatively stable, depending on whether water contamination or high temperature stability is the most important requirement.

19.4   MANUFACTURE

Although phosphate esters can be regarded as "salts" of orthophosphoric acid, they are not currently produced from this raw material because the yields are low (~70% for triaryl phosphates) [31]. Instead phosphorus oxychloride ($POCl_3$) is reacted with either an alcohol (ROH), a phenol (ArOH) or an alkoxide (RONa), as indicated in the following reaction schemes:

$$3ROH + POCl_3 \rightarrow (RO)_3 PO + 3HCl \tag{19.1}$$

$$3ArOH + POCl_3 \rightarrow (ArO)_3 PO + 3HCl \tag{19.2}$$

$$3RONa + POCl_3 \rightarrow (RO)_3 PO + 3HCl \tag{19.3}$$

where R represents an alkyl group and Ar, an aryl group.

The above reactions pass through intermediate steps in the production of the phosphate, as shown in scheme (19.4):

$$ROH + POCl_3 \xrightarrow{} ROPOCl_2 + HCl \xrightarrow{ROH} (RO)_2 POCl + HCl \xrightarrow{ROH} (RO)_3 PO + HCl \tag{19.4}$$

The intermediate products are called phosphochloridates and it is possible to stop the reaction at each step by changing the ratio of reactants. If it is desired to produce mixed products this can be achieved by reacting an intermediate with a different alcohol or a phenoxide (scheme 19.5):

$$(RO)_2 POCl \xrightarrow{ArONa} (RO)_2 PO.OAr + NaCl \tag{19.5}$$

19.4.1 Trialkyl Phosphates

Trialkyl phosphates can be prepared by either of routes shown in reactions 19.1 and 19.3, although in reaction 19.1, a considerable excess of alcohol is required to drive the reaction to completion and the hydrogen chloride (HCl) by-product must be removed as rapidly as possible—usually by vacuum and/or water washing. The reaction temperature is not allowed to rise too high in order to minimize the thermal degradation of the phosphates. When the alkoxide route (19.3) is chosen, the chlorine precipitates as sodium chloride, somewhat simplifying the purification treatment. Normally, this consists of a distillation step to remove excess alcohol, an alkaline wash, and a final distillation to remove water [32]. With the alkoxide method, however, any residual sodium chloride can be removed by water washing followed by a final distillation under vacuum. If mixed alkyl phosphates are required (i.e., using more than one alcohol), they can be produced either in one step using an alcohol mix or by the reaction of the phosphochloridate intermediate with an alkoxide.

19.4.2 Triaryl Phosphates

Triaryl phosphates, which form the basis of the majority of hydraulic fluid formulations, are manufactured almost exclusively by reaction 19.2. Phosphorus oxychloride is added slowly to the reaction mass containing an excess of phenol in the presence of a small amount of catalyst—typically aluminum chloride or magnesium chloride—before heating slowly [32]. The hydrogen chloride is removed as it is formed, by heating under vacuum followed by absorption in water. On completion of the reaction, the product is distilled to remove most of the excess phenol(s), the catalyst residue, and traces of polyphosphates. Finally, the product is steam-stripped to remove volatiles, including residual phenol(s). Depending on the application, it may be necessary to reduce the level of acidity by subsequent alkaline washing or by a solids treatment.

As with trialkyl phosphates, it is possible to manufacture mixed aryl phosphates as long as the reactivity of the different phenolic materials is similar. Such mixed phosphates can have improved low-temperature properties (e.g., pour point) as a result of the asymmetrical nature of the molecule, and the technique is particularly important when trying to manufacture products to meet a particular viscosity level.

The phenolic feedstocks used for triaryl phosphate manufacture were originally obtained from the distillation of coal. The resulting "coal tar" contained a mixture of phenol, cresols, and xylenols which was fractionated to provide the raw materials for the phosphate process. However, these fractions or "distillation cuts" were still complex mixtures—often containing up to twenty different cresol, xylenol, and other alkylphenol isomers—and were frequently referred to as "cresylic acids."

In the 1960s, it became increasingly difficult to obtain suitable quality feedstocks from this source because of a decline in the coal distillation industry. The wider use of natural gas for domestic heating purposes and the concurrent contraction in the steel industry led to a fall in demand for "coke", the solid residue from the distillation of coal. As a result, many coal-tar distillers closed and the quality of available phenolic feedstocks deteriorated.

In order to ensure supplies of suitable raw materials, the phosphate ester manufacturers subsequently introduced feedstocks based on phenol (from petroleum sources) which was reacted with propylene or iso-butylene to form mixtures of isopropylphenols or tertiarybutyl phenols [33,34] (see reactions 19.6 and 19.7).

o-isopropyl
phenol

m-isopropyl
phenol

p-isopropyl
phenol

(19.6)

m-tert-butylphenol   p-tert-butylphenol   (19.7)

No *ortho-* derivative is normally produced when reacting phenol with isobutylene.

Depending on the reaction conditions, it is possible to control the relative amounts of the isomeric alkylated phenols to optimize the properties of the resulting phosphate. However, if the mixtures in reactions 19.6 and 19.7 are converted into phosphates, the viscosity level of the resulting liquid would be too high for most hydraulic applications. To reduce the viscosity, the alkylated phenol is mixed with a quantity of pure phenol before phosphorylation, which enables the manufacture of fluids meeting a wide range of viscosity grades. Unfortunately, the phenol content results in the production of triphenyl phosphate, particularly in the lower-viscosity fluids (see Table 19.2) [35]. This is not very hydrolytically stable and although the stability of the mixed product is adequate for most applications where significant contact with moisture is likely, other phosphates may be preferred for more critical uses.

In order to minimize the TPP content of products based on tertiarybutylphenol, an alternative process has been suggested [36] which involves a two-stage reaction. In the first stage phosphorus oxychloride is reacted with only enough phenol or alkylated phenol to replace one or two of the chlorine atoms; and the reaction is finished in the second stage by further reacting with either alkylated phenol or phenol itself (see reactions 19.8 and 19.9).

$$ArOH + POCl_3 \rightarrow ArOPOCl_2 + HCl \tag{19.8}$$

$$AlkArOH + ArOPOCl_2 \rightarrow ArOPO(OArAlk)_2 + 2HCl \tag{19.9}$$

The TCP and TXP produced from cresylic acid or "natural" raw materials have been described as "natural" products, whereas feedstocks based on isopropylphenols and tertiarybutylphenols are known as "synthetic" feedstocks (because of their chemical synthesis) and the resulting phosphates are known as "synthetic" phosphates. Unfortunately, these terms still persist, even though the cresol and xylenol feedstocks used today are obtained by the refining of petroleum (a natural product) and synthetically from phenol (cresols and xylenols), toluene (cresols), or isophorone (xylenols) [37].

TABLE 19.2
The Triphenyl Phosphate Content of Triaryl Phosphates Manufactured by Reaction 19.2

	TPP Content for Different ISO Viscosity Grades (%)			
Phosphate Ester	22	32	46	68
TCP	–	0.1	–	–
TXP	–	–	None	–
IPPP	26–34	14–18	8–10	5–8
TBPP	40–48	31–40	14–21	8–12

Note: TCP = tricresyl phosphate; TXP = trixylyl phosphate; IPPP = isopropyl-phenyl phosphate; TBPP = tertiarybutylphenyl phosphate.

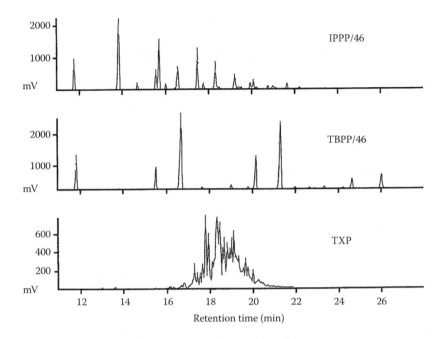

FIGURE 19.3 Gas–liquid chromatograms for different triaryl phosphates.

It was indicated earlier that the phenolic feedstocks are frequently complex mixtures of different isomers. This, in turn, results in phosphates based on many different combinations of phenolic groups. The gas–liquid chromatograms in Figure 19.3 show the number of major components of the three principal triaryl phosphate types and the significant differences between the products. The simplest product is a tertiarybutylphenyl phosphate (TBPP) containing only about seven significant components. By comparison, the isopropylphenyl phosphate (IPPP) contains about ten major components and trixylyl phosphate, over twenty [35]. Differences in retention times for the IPPP and TBPP products indicate a broader spread of molecular weights of the component phosphates, whereas for TXP, the molecular weight range is relatively small.

19.4.3 Mixed Alkyl Aryl Phosphates

Small quantities of both alkyldiaryl- and dialkylmonoaryl phosphates are used in the production of specialty hydraulic fluids, particularly where some improvement in low-temperature properties is required with a measure of stability at high temperatures. These products can be most satisfactorily produced from the appropriate phosphochloridate intermediate by two-stage processes similar to 19.8 and 19.9 above.

$$ArOH + POCl_3 \rightarrow ArOPOCl_2 + HCl \tag{19.10}$$

$$2ROH + ArOPOCl_2 \rightarrow \underset{\text{a dialkylaryl phosphate}}{ArOPO(OR)_2} + 2HCl \tag{19.11}$$

$$2ArOH + POCl_3 \rightarrow (ArO)_2 POCl + 2HCl \tag{19.12}$$

$$ROH + (ArO)_2 POCl \rightarrow \underset{\text{an alkyldiaryl phosphate}}{(ArO)_2 PO.OR} + HCl \tag{19.13}$$

TABLE 19.3
Phosphate Esters Used in Hydraulic Applications

Common name	Abbreviation
Trialkyl Phosphates	
Tributyl phosphate[a]	TBP
Tri-isobutyl phosphate	TiBP
Tributoxyethyl phosphate	TBEP
Trioctyl phosphate[a]	TOP
Alkyl aryl Phosphates	
Dibutylphenyl phosphate[a]	DBPP
Butyldiphenyl phosphate[a]	BDPP
Octyldiphenyl phosphate[a]	ODPP
Isodecyldiphenyl phosphate[a]	DDPP
Triaryl Phosphates	
Triphenyl phosphate	TPP
Tricresyl phosphate	TCP
Trixylylphosphate	TXP
Cresyldiphenyl phosphate	CDP
Isopropylphenyl phosphate	IPPP[b]
Tertiarybutylphenyl phosphate	TBPP[b]

[a] Unless otherwise stated, the alkyl groups used in the manufacture of trialkyl phosphates and alkylaryl phosphates are of the following types: butyl is n-butyl, octyl is 2-ethyl-hexyl, and decyl is iso-decyl.

[b] In the subsequent tables and figures, the IPPP and TBPP abbreviations are usually followed by the ISO viscosity grade. Thus, IPPP/46 is an isopropylphenyl phosphate meeting ISO viscosity grade 46.

An alkoxide may also be used as an alternative to the use of alcohol, primarily for the preparation of alkyldiaryl phosphates:

$$2ArONa + ROPOCl_2 \rightarrow ROPO(OAr)_2 + 2NaCl \qquad (19.14)$$

The purification of alkyl aryl phosphates can be carried out using techniques already identified for the other two classes of phosphate esters. See Gamrath [10] for more details on the preparation of these materials.

To summarize, the phosphate esters which are in commercial production and used in hydraulic fluid applications are listed in Table 19.3 together with their common abbreviations.

19.5   PROPERTIES OF PHOSPHATE ESTERS

In this section, the performance of phosphate esters will be examined against the principal technical requirements for hydraulic fluids (see Table 19.4). These requirements impact on both the physical and chemical properties of the fluids and will, therefore, be discussed under those headings, with the exception of fire resistance and lubrication performance, which are termed "performance properties."

19.5.1   PERFORMANCE PROPERTIES

19.5.1.1   Fire Resistance

Because the fire resistance of these fluids is the principal reason for their use, it is appropriate to discuss this aspect of their behavior first. Before clarifying why phosphate esters are difficult to ignite,

TABLE 19.4
Main Technical Requirements for Fire-Resistant Hydraulic Fluids

Fire resistance	Good resistance to shear breakdown
Low compressibility	Low foaming properties
Appropriate viscosity	Rapid air release
Adequate low-temperature behavior	Good water separation
Good lubrication performance	Non-corrosive
Adequate stability (thermal, oxidative, and hydrolytic)	Good compatibility with system materials

it is useful to recall that fire, or combustion, is essentially a vapor-phase oxidation process that requires the presence of a fuel, sufficient energy to volatilize the fuel, and oxygen. Fluid characteristics which therefore determine ignitability include vapor pressure (and, hence, viscosity/temperature properties, molecular weight, and heat of vaporization) and thermal/oxidation stability [38], all of which are very structure dependent. The measurement of ignition delay time (i.e., the time it takes for the fluid to ignite on a hot surface at a specific temperature) is affected by other factors, such as specific heat, thermal conductivity, and surface tension. A variety of physical and chemical properties can therefore affect the flammability characteristics. However, the term "fire resistance" should not be equated with "non-flammable". If sufficient energy is injected into almost any organic molecule, it will ignite. Even water has been known to explode in contact with very hot surfaces!

At first glance, it might seem surprising that phosphate esters are much less flammable than mineral oils. Apart from a relatively high molecular weight, they only differ in terms of composition by several oxygen atoms and a phosphorus atom, the latter being present in a phosphate to the extent of only a few percent (~7–12% for commercially available materials). The reason why these materials display such favorable characteristics lies, as suggested above, in a combination of their structure (and its effect on stability) and physical properties. However, whereas a low vapor pressure and generally high thermal/oxidative stability will favorably affect fire resistance, in the case of phosphate esters a further factor, the phosphorus content, plays an important role.

Irrespective of the mode of decomposition, lower-molecular-weight materials, such as alcohols, alkylphenols, and ketones, are initially produced during degradation. These are highly volatile and much easier to oxidize. When present in sufficient concentration (i.e., above the lower limit of flammability), they will combust and release heat.

The remainder of the molecule will contain one or more —P—O· radicals, which will polymerize in a reaction that is highly endothermic (or energy absorbing) to form polyphosphates or, in an extreme case, phosphorus oxides. This reduces the total heat released and is the principal reason why phosphate esters do not propagate flame.

High phosphorus content alone, however, will not significantly increase the fire resistance if the breakdown products are themselves readily oxidizable. The effect of phosphorus content and other physical properties on some flammability characteristics are given in Table 19.5 [32,38–40].

TABLE 19.5
Relationship between Different Physical Properties and Flammability Behavior

Phosphate Ester	Molecular Weight	Phosphorus Content (%)	Viscosity at 40°C (cSt)	Vapor Pressure at 150°C (mm Hg)	Flash/Fire points, ISO 2592 (°C)	Autoignition ASTM D 2155 (°C)
TBP	266	12.0	2.5	7.3	165/182	388
TOP	434	7.37	7.5	0.9	195/238	370
TPP	327	9.8	Solid	0.15	224/310	635
TXP	410	7.8	46	0.009	250/360	570

Examination of the data reveals that none of the common physical properties by themselves correlate satisfactorily with the listed flammability data.

The measurement of fire resistance in the laboratory is difficult because no single test can satisfactorily reproduce all the variables that constitute the specific industrial hazard [38]. The best that can normally be achieved is a comparative assessment of different fluids under the same test conditions. However, as a result, a large number of tests have been devised to simulate particular hazards, many of which rank products in the same order and do not add to an understanding of the behavior of the material in use.

The different tests currently available for determining fire resistance can be classified according to the mode of fluid ignition (i.e., bulk fluid ignition, hot surface ignition, spray ignition, and ignition when adsorbed on a substrate) [38]. These tests essentially measure ignitability (temperature of ignition or time to ignition at a given temperature) and the tendency to propagate flame. An exception is a test which attempts to measure some intrinsic property of combustion, such as heat of combustion. Unfortunately, the conditions used for measuring this property are not suitable for products that contain elements other than carbon, hydrogen, oxygen, nitrogen and sulfur [41] and can give misleading information on phosphate esters [38]. In spite of the limitations of the tests used [38], most flammability data on fluids are reported in terms of flash and fire points and autoignition temperatures. Table 19.6 [32,42–45] provides this information on commercially available fluids. More recently, ISO 12922, the specification for industrial fire-resistant hydraulic fluids, has indicated the minimum performance requirements of different fluid types under spray ignition, hot surface ignition, and wick flame persistence conditions, tests which are regarded as more relevant to the hazards found in use.

TABLE 19.6
Flash Point, Fire Point, and Autoignition Temperatures of Phosphate Esters

Phosphate Ester	Flash Point (°C) ISO 2592	Fire Point (°C) ISO 2592	AutoIgnition Temperature (°C) ASTM E 659
TBP	165	182	400[a]
TiBP	155	170	476
TOP	190	238	370[a]
TBEP	190	230	260[a]
BDPP	205	246	–
DBPP	171	–	>430
ODPP	224	238	–
DDPP	240	260	–
TPP	224	310	635
CDP	255	340	620
TCP	240	338	600
TXP	250	360	535
IPPP/22	255	335	570
IPPP/32	245	330	545
IPPP/46	250	335	500
IPPP/100	260	335	515
TBPP/22	240	340	590
TBPP/32	255	350	545
TBPP/46	250	350	535
TBPP/68	255	350	525
TBPP/100	255	360	520

[a] ASTM D 2155. This method usually gives slightly higher values than ASTM Method E 659.

Many of the applications where phosphates are used require the better fire resistance and high-temperature stability provided by the triaryl phosphates. As a consequence, the vast majority of fire-resistance data generated under tests more closely representing specific hazards is limited to an evaluation of these fluids.

Table 19.7 [30,32,38,42–44,46] indicates the performance of TXP and IPPP/46 fluids in the most important industry-related tests.

As will be seen from Table 19.7, the majority of tests do not discriminate between the different triaryl phosphate esters. This is mainly due to the pass/fail requirements of specifications in which these tests appear. These tests not only fail to differentiate between triaryl phosphate esters (unless increased in severity) but may also fail to discriminate between phosphate esters and other types of fire-resistant fluids.

In order to be able to rate the performance of *all* the different types of fluid under identical conditions, two spray ignition tests have been developed: in the United States by the Factory Mutual Research Corporation [47] and detailed in a new "Flammability Classification of Industrial Fluids" [48], and the other in the United Kingdom by the Health and Safety Executive, which has been published as ISO/TS 15029-2 [49]. Both tests measure the heat released by a stabilized flame of the ignited fluid. Treatment of the results is different in that Factory Mutual (FM) quote data in terms of a "Spray Flammability Parameter" (SFP) which takes into consideration the efficiency of combustion based on heat of combustion measurements and the critical heat flux for ignition as measured by the fire point. By comparison, the ISO test [49,50] reports the "Ignitability Factor," which is related directly to the heat released. Unfortunately, the FM test now has different procedures for water-based and non-aqueous fluids, thus making comparisons impossible; is

TABLE 19.7
Performance of Triaryl Phosphates in the Most Widely Used Fire Tests for Hydraulic Fluids

Test method	Source	Fluid type		
		TXP	**IPPP/46**	**TBPP/46**
Bulk fluid ignition				
Fire point (°C)	ISO 2592	360	335	350
Hot-surface ignition				
Manifold ignition test	ISO 20823 (at 704°C)	Pass	Pass	Pass
Autoignition temperature	ASTM D 2155	580	540	585
	ASTM E 659	535	500	535
Molten metal ignition	Rheinisch Westfälischer TÜV	16 s	16 s	–
Spray Ignition				
Spray flame persistence – hollow cone nozzle method	ISO 15029-1	Pass	Pass	Pass
Spray test – stabilized flame heat release method	ISO 15029-2	Class D	Class E	Class D
Factory Mutual Spray Test	FM Standard 6930	Approved	Approved	Approved
Wick Ignition				
Wick flame persistence	ISO 14935	Pass	Pass	Pass
Linear flame propagation rate of lubricating oils and hydraulic fluids	ASTM D 5306	No ignition	No ignition	No ignition
Compression–ignition test	Section 4.5.1 MIL-PRF-19457D	>42:1	41:1	47:1

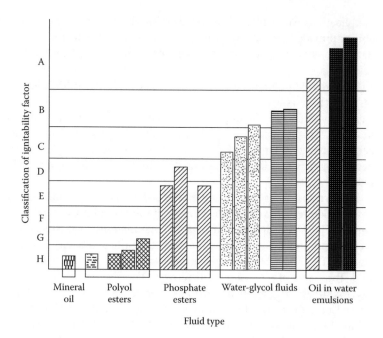

FIGURE 19.4　Results of ISO 15029-2 stabilized-flame heat-release spray test on different fluid types.

not suitable for phosphate ester fluids and does not take the persistence of burning behavior into consideration. Currently, therefore, the ISO test method, which can rank all the different fluid types, is preferred.

Figures 19.4 [51] and 19.5 [48] indicate some of the results obtained by these new procedures and how the different fluid types perform. However, the calculation of SFP values and

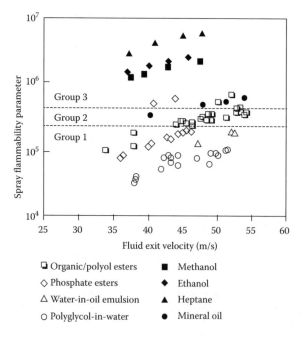

FIGURE 19.5　Early Factory Mutual spray flammability test data. (Anon., "Unveiling a New Protocol for Less Flammable Hydraulic Fluids," Factory Mutual Approved Product News, 1996, pp. 6–8.) (Reprinted with permission, copyright FMRC., 1996.)

TABLE 19.8
Fire-Resistance Test Data on a Trialkyl Phosphate-Based Aircraft Hydraulic Fluid

Property	Result	Test Method
Flash point (°C)	159	ASTM D 92
Fire point (°C)	183	ASTM D 92
Autoignition temperature (°C)	466	ASTM D 2155
Manifold ignition at 700°C	Burns on tube but does not propagate to tray	BMS 3-11
Spray ignition	No increase	DMS 2014
Wick cycling ignition	>26 cycles	BMS 3-11
High-temperature/high-pressure spray test	Will not ignite	BMS 3-11

Note: BMS = Boeing Material Specification; DMS = Douglas Material Standard.

the limits for the different categories have changed in recent years. Today, Factory Mutual only fully approve products with an SFP of $\leq 5 \times 10^4$ (previously these were referred to as "Group 1 fluids"); products with values between 5×10^4 and 10×10^4 (previously "Group 2 fluids") are classified as "Specification Tested" and FM reserves the right to insist on additional mechanical fire protection when these fluids are used. All the aryl phosphates tested to date are FM Approved products.

Although trialkyl phosphates are more flammable and are not FM approved, they have been the basis for fire-resistant fluids used in commercial aircraft for many years. Obviously in this latter application, as well as requiring good fire resistance, they also need to have good low-temperature properties in view of the ambient conditions found at high altitudes. Table 19.8 [52] lists typical fire-resistant test data on a hydraulic fluid that is widely used in commercial aircrafts.

One advantage of the latest spray tests is that they can eliminate the effects of additives (e.g., polymeric materials) which may be used to increase droplet size in a spray (and therefore provide some temporary improvement to the spray ignition performance) but which have no significant effect on the flammability under other tests conditions.

Owing to the lack of a suitable definition for a fire-resistant fluid, a number of non-aqueous fluids with high fire points have been introduced as "fire-resistant". Such fluid types have included synthetic organic esters, vegetable oils (triglycerides), and polyalkyleneglycols. That these fluids do not possess the same level of fire resistance as the phosphate esters can be seen in Table 19.9, which shows the behavior in current standard fire-resistance tests.

19.5.1.2 Lubricating Properties

Hydraulic fluids are predominantly used in dynamic applications that involve continuous circulation around the system at pressures up to 400 bar. The ability to lubricate metal surfaces moving relative to one another (particularly steel on steel) as found in pumps, valves, and other components is therefore of paramount importance. As the lubrication performance significantly influences system operating pressures and capacity, it can also impact the capital cost of the system. Although fluid viscosity and its change with temperature have an obvious impact on pump and valve efficiency, the ability of a fluid to lubricate satisfactorily depends mainly on its potential for forming a film between the metal surfaces which is thick enough to separate them, and the ability of that film to resist displacement under load and through relative motion of the surfaces. As in the case of fire resistance, the lubricating performance of the fluid depends on a variety of physical and chemical properties; for example, compressibility, viscosity, specific heat and thermal conductivity (which

TABLE 19.9

A Comparison of the Fire-Resistance Properties of Different Non–Aqueous Fluid Types

Fire Resistance Test	Min Oil ISO VG 46	Triaryl Phosphate ISO VG 46	Polyol Ester ISO VG 46	Vegetable Oil ISO VG 46	Polyalkylene-Glycol ISO VG 46
Flash Point (°C) - ASTM D 92	230–250	240–270	260–320	>290	270
Fire Point (°C) - ASTM D 92	260–285	335–370	335–378	>345	310
Autoignition (°C) - ASTM D 2155	380–400	545 >625	400–435	390	395
Manifold Ignition (°C) - ISO 20823	370–395	~750	440–480	420–465	400–420
Spray Ignition - Ignitability Class - ISO 15029-2 Rating (H is worst)	H	D–E	G–H	H	F
Spray Ignition - Spray Flamm. Factor - FM 6930 Std.	Not classfd.	Approved	Approved/ Spec. Tested	Approved/ Spec. Tested	Approved/ Spec. Tested
Wick Flame Persistence (secs) - ISO 14935	Cont. burning	4	◄———————— Continuous burning ————————►		

affects the temperature of the lubricating film and hence its viscosity), the polarity and shape of the molecule, and its thermal and hydrolytic stability.

The good antiwear properties of phosphates were referred to earlier. These are attributed to the polarity of the molecule and its acidic decomposition products, which initially form a physically adsorbed film on the metal surface, and also to the reactivity of the decomposition products with the metal surface to form iron phosphides or phosphates [4,53–55]. It was originally thought that these lower-melting products flowed and filled the gaps between the asperities or peaks on the metal surface. Consequently, the area carrying the load was increased and the pressure/unit area ratio was reduced. This enabled a higher load to be supported before breakdown of the lubricating film. However, more recent studies have suggested that the antiwear behavior may be due to the formation of a mixture of polyphosphate and carbon [56]. The antiwear performance of the phosphates also varies with stability—particularly hydrolytic stability. The more readily the phosphate breaks down, the better the antiwear behavior. This fact may account for the variable performance of phosphate esters in some reports [57].

Although phosphates are recognized as effective antiwear agents, they have limited ability to prevent scuffing at high loads without enhancement by additives. Aryl phosphates are reported as more effective extreme-pressure agents than alkyl phosphates [58], the difference being attributed to the formation of iron phosphides by aryl phosphates and to the less effective iron phosphate by the alkyl derivatives. Table 19.10 [42,43] illustrates vane-pump wear test data on different triaryl phosphates, which confirms their good antiwear behavior; Table 19.11 [59] compares the different types of fire-resistant hydraulic fluids in terms of their pump-performance capabilities relative to mineral oil. Although the data have been provided by equipment manufacturers, they are probably conservative estimates and, in reality, somewhat higher performance limits for the fire-resistant products may be achieved.

In addition to the effect on the conventional wear process, the composition of fluids can significantly impact the pitting fatigue life of ball and roller bearings. With non-aqueous fluids, the life of bearing surfaces depends mainly on the fluid-film thickness and, hence, viscosity [60,61]. Phosphate esters have been found to have lives comparable with those of mineral oils of similar viscosity [60].

With water-based fluids, the water accelerates the propagation of surface cracks as well as considerably reducing the pitting life, probably by an embrittlement process [60,61].

TABLE 19.10
Vickers V104C Vane-Pump Wear Test Data for Some Triaryl Phosphates at 140 Bar; Test Methods IP281 and ASTM D 7043

Test Data	Phosphate Ester								
	IPPP/22		IPPP/32		IPPP/46		IPPP/68		TBPP/46
Fluid temperature (°C)	53	53	60	60	66	66	72	72	NA
Test duration (h)	100	250	100	250	250	1000	100	250	100
Ring weight loss (mg)	3.6	6.5	2.5	2.6	3.7	8.2	3.7	5.4	14.0
Vane weight loss (mg)	14.5	15.6	34.8	36.7	7.4	8.5	2.0	3.3	55.3
Total weight loss (mg)	18.1	22.1	37.3	39.3	11.1	16.7	5.7	8.7	69.3
Pump speed (rpm)	◄─────────────────── 1440 ───────────────────►								1200

The fatigue life of fluids is normally expressed as the L_{10} life, which is the time after which 10% of bearings would have failed. Bearing L_{10} lives characteristic of the different ISO classes of fire-resistant fluids are given in Figure 19.6 for an axial piston pump at 1500 rpm. Again, depending on the fluid composition, pump type, and operating conditions, these values will show some variation [62,63].

TABLE 19.11
Effect of Hydraulic Fluid Type on Pump Performance

	Pump Type				
		Piston			
Fluid Category	Gear	Hydrostatic Bearing	Hydrodynamic Bearing	Screw	Vane
1. HFA (E and S)[a]					
Max. continuous pressure rating (bar)	200	200	150	No known experience	Not suitable[b]
Max. continuous speed[c] (%)	100	50	50		
Bearing life[c] (%)	100	10	30		
2. HFB[a]					
Max. continuous pressure rating (bar)	200	250	200	250	70
Max. continuous speed[c] (%)	100	100	100	100	6
Bearing life[c] (%)	100	30–70	20–40	100	100
3. HFC[a]					
Max. continuous pressure rating (bar)	200	200	300	250	105
Max. continuous speed[c] (%)	100	100	100	100	100
Bearing life[c] (%)	100	20–60	20	100	100
4. HFD (R and U)[a]					
Max. continuous pressure rating (bar)	250	350	350	250	140
Max. continuous speed[c] (%)	100	100	100	100	100
Bearing life[c] (%)	100	80–100	70–100	100	100
Bearing type	Plain	Roller	Roller	Roller (but no fluid contact)	Plain (sleeve)

Note: All data taken from pump manufacturers' catalogs.

[a] The classification of fluid types is as given in ISO standard 6743-4

[b] The problems with water-based fluids in vane pumps relate to vane tip wear rather than bearing lubrication.

[c] Values indicated are relative to mineral-oil performance in the same pump.

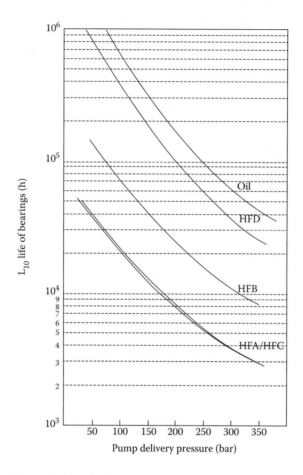

FIGURE 19.6 The variation in bearing L_{10} lives with pump delivery pressure for an axial piston pump operating on fire-resistant hydraulic fluids at 1500 rpm.

The behavior of triaryl phosphate esters in gears has been studied in some detail [64,65]. Tests comparing the behavior of IPPP/46 and TXP on an FZG gear rig with mineral oil show the following:

- **A similar wear behavior for TXP and mineral oil (see Figure 19.7).** The inferior viscosity–temperature and viscosity–pressure characteristics of the phosphates are largely compensated for by polar effects and chemical reactivity at the metal surface. For example, the fact that an IPPP/46 fluid displays better wear characteristics than TXP [65] is probably related to its poorer hydrolytic stability. This more readily allows the production of acidic species, which subsequently react with the surface.
- **Inferior frictional characteristics and, as a result, slightly higher power losses**. These properties are also influenced by film thickness. Although load-dependent losses are about 20% higher for triaryl phosphates than for mineral oils, about 80% of the total losses in high-speed gears are related to no-load losses. This means, for example, that in turbine gears, the total gear losses for a phosphate would be about 4% higher than the losses for mineral oil of comparable viscosity [66].
- **Inferior running-in properties**. This is not surprising in view of the lack of extreme-pressure performance of this type of fluid. Mineral oil, of course, usually contains sulfur-containing impurities which assist this process. As a result, it is important to clarify the optimum conditions for running-in new gears from the manufacturer when phosphates are used.

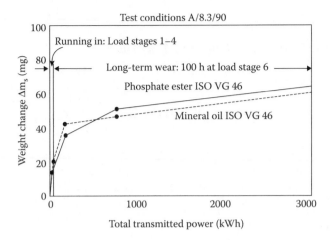

FIGURE 19.7 Long term wear tests on an FZG gear machine for TXP and an ISO VG46 mineral oil.

- **A better scuffing performance**. In FZG tests, ISO VG 46 mineral oils (without additives) typically fail at load stages three to five. Triaryl phosphates of similar viscosity typically fail at stages seven and eight [64,65]. Improving the scuffing behavior of mineral oils is, however, fairly easy; with phosphate esters, this is more difficult in view of competition between the phosphate (or its decomposition products) and the additive for the metal surface. Careful selection of the latter is necessary when upgrading the extreme-pressure performance of the phosphate.

Aluminum and its alloys are not recommended as suitable bearing materials for plain bearings in components using these fluids owing to the inability of phosphates to "wet" aluminum surfaces. Copper/lead bearings can be used, but the preferred lining material is white metal. Extended tests on TXP at bearing temperatures of 110°C and fluid inlet temperatures of 95–98°C showed no significant attack on an 80/12/5 tin/antimony/copper bearing [67].

The scuffing resistance of alkyl or alkyl aryl phosphates in gears has not been investigated in such detail. One report [68] investigated the mechanism of wear associated with this product and identified scoring as the mode of failure.

19.5.2 Physical Properties

19.5.2.1 Viscometric Properties and Low-Temperature Behavior

The viscosity of a fluid and its change with temperature and pressure significantly affect a number of aspects of fluid use and performance, such as pump efficiency. If a fluid has a very low viscosity, internal leakage and pump slippage can occur. This results in a power loss and an increase in fluid temperature. On the other hand, a high-viscosity fluid may show better internal sealing but require more power for circulation; it may display a higher pressure drop in lines and, under extreme conditions, even reduce the volume of flow to the pump, causing erratic system operation. Selection of a fluid with an appropriate viscosity for the pump design is therefore essential.

Other properties and performance parameters directly affected by viscosity include heat removal, wear properties, and air release. Obviously, secondary effects can be found on oxidation stability (and, hence, fluid life), frictional losses in bearings, cavitation, and so forth.

Depending on their chemical structure, phosphate esters are available in a wide range of viscosities, with trialkyl phosphates falling generally into ISO viscosity grades 2–7, alkyl aryl phosphates in grades 10–15, and substituted aryl phosphates in grades 22–100 (Table 19.12)

TABLE 19.12
Typical Viscometric Properties of Phosphate Ester Fluids

Phosphate Ester	Viscosity (cSt) - ISO 3104						Viscosity index ISO 2909	Critical pumping temperature (°C)	Pour point (°C) ISO 3016
	0°C	20°C	37.8°C	40°C	99°C	100°C			
Trialkyl phosphates									
TBP	7	3.7	2.7	2.6	1.10	1.05	118		<−90
TiBP	10	4.9	3.1	3.0	1.12	1.10			<−90
TOP	96	40	8.0	7.9	2.20	2.20	145		−70
TBEP	80	26	6.8	6.7	2.15	2.10	90		<−70
Alkyl aryl phosphates									
ODPP	66		9.95		2.43				−57
DDPP	89		12.80		2.88				−54
DBPP			4.66			1.34			<−70
Triaryl phosphates									
TPP	← Solid →								Solid, melts at 49°C
CDP	220	44	17.5		3.20				−34
TCP	1,000	80	28.3		4.00				−28
TXP	1,700	170	46.0	43	5.2	5.0			−21
IPPP/22	380	70	23.5	22	4.0	3.8	30	−8	−35
IPPP/32	990	90	36	32		4.7	30	0	−27
IPPP/46	1,600	155	47	43		5.3	15	0.6	+6
IPPP/68	7,600	245	78	66		5.7	<0	0.13	+13
IPPP/100	14,400	550	115	103		6.2	<0	0.22	+22
TBPP/22	445	75		24		4.0	35		−30
TBPP/32	1,500	104	35	33	4.8	4.5	25		−26
TBPP/46	2,500	160	47	44	5.8	5.2	15		−18
TBPP/68	9,000		75	65	6.8	6.3	5		−15
TBPP/100		448	104	95	8.3	7.8	<0		−7

[10,32,39,40,42–44,69–71]. The variation in viscosity with temperature is shown in Figure 19.8 [32] for most of the alkyl and aryl phosphates.

The rate of change of viscosity with temperature, otherwise known as viscosity index, varies considerably; trialkyl phosphates exhibit values comparable to those of paraffinic mineral oils, whereas commercial triaryl phosphate mixtures show very low values—usually between 0 and 30.

Symmetry within the molecule, steric effects, and the number of individual components all assist in determining whether the product is solid or liquid and, in the latter case, its viscosity index. Triphenyl phosphate, for example, is a solid at ambient temperatures, as is tris-*ortho*-isopropylphenyl phosphate. Pure tris-*para*-isopropylphenyl phosphate, by comparison, has a viscosity index of 50.

Depending on the fluid type and the pump selected for the hydraulic system, the viscosity may need to be reduced (by heating) before the pump can be operated. In the past, a critical pumping viscosity of 850 cSt maximum has sometimes been quoted and the temperature at which this is reached has been termed the "critical pumping temperature" [42]. Table 19.10 also indicates the critical pumping temperatures for some of the substituted aryl phosphate fluids where this aspect is more important.

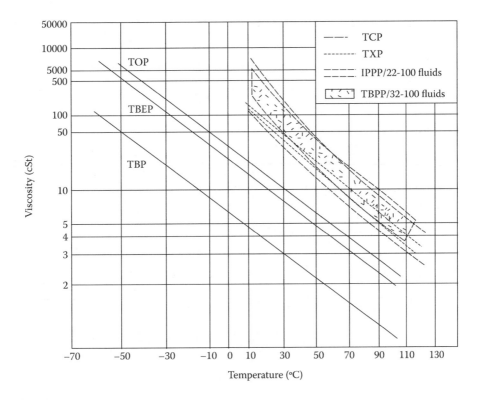

FIGURE 19.8   Viscosity–temperature relationships for commercial phosphate esters.

The effect of pressure on viscosity is to increase frictional losses in gears and bearings, but this can help resist the displacement of the lubricant film under load. Figure 19.9 shows the variation in absolute viscosity of an ISO VG 46 phosphate with pressure up to ~800 bar and temperature up to ~100°C [42]. The data are intermediate between that of a paraffinic and a naphthenic mineral oil [71]. In the future, the viscosity at high pressures is likely to assume greater importance, as operating pressures are continually increasing.

In selecting a fluid for an application requiring a wide operating temperature range, the user has the option of taking, for example an ISO VG 46 fluid and then providing tank heating in order to minimize the effect of the poor low-temperature properties, or of selecting a product based on a thinner fluid but containing a polymeric thickener to provide an adequate working viscosity. In reality, apart from applications involving very low temperatures (where the alkyl or alkylaryl phosphates are preferred), the latter approach is not favored because: (1) there are concerns over the shear stability and, hence, pump performance of the thickened fluid; (2) the polymer can be removed by some types of filtration media; and (3) some thickeners can hydrolyze to give degradation products that can catalyze the hydrolysis of the phosphate ester base.

The pour points of phosphates, like their viscosities, vary considerably, depending on their structure (Table 19.12) and are generally superior to mineral oils of similar viscosity. However, unlike mineral-oil-based products, pour point depressants are not used in commercial phosphate ester fluids as there is normally no need to inhibit the crystallization of a fluid component.

19.5.2.2   Air Release, Foaming, and Demulsibility Characteristics

Air is a common "contaminant" of hydraulic fluids that is often overlooked or is thought of only in terms of its effect on compressibility. Its presence, however, can have major implications for both the performance of the fluid and the system [72–75]. Air can, of course, be present in a fluid in either the dissolved or dispersed (entrained) form. The solubility of air depends on the pressure and

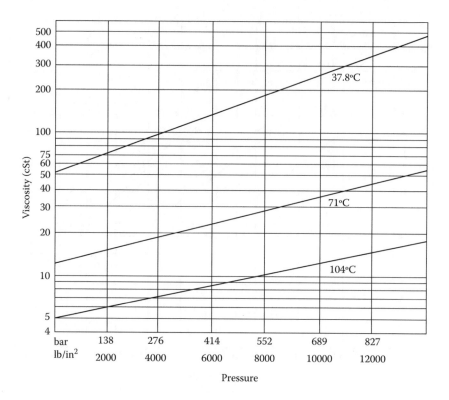

FIGURE 19.9 The viscosity–pressure relationship of an IPPP/46 phosphate ester.

temperature of the fluid and is quoted in terms of an absorption coefficient (known as the "Bunsen coefficient"). The higher the value, the greater the solubility of the gas.

Dissolved air normally has no effect on a fluid's physical properties, but it is possible for dissolved air to come out of solution because of local pressure changes in the system and promote cavitation. Whereas some air bubbles may re-dissolve relatively quickly, others may combine to form larger bubbles, which are more difficult to solubilize and therefore remain as entrained air.

In contrast, dispersed or entrained air, in addition to the effect on compressibility noted earlier (and, therefore, on actuator response) will reduce the energy efficiency of the system. It will also adversely affect lubrication performance, including the possible loss of pump suction, and will increase fluid oxidation. Therefore, it is vital that entrained air be allowed to escape from the fluid or be encouraged to do so by the application of a slight vacuum. Air should preferably be kept out of the system, but as this normally cannot be avoided, the air content should be minimized by a combination of good tank and pipework design and by selecting a fluid with low air entrainment or solubility.

The air-release properties of a fluid are measured in the laboratory by first saturating the fluid with air and then measuring the time (at a constant temperature) for the air to be released to a level of 0.2% volume. However, actual measurements of the air content in the system at the pump inlet are more relevant [75].

A comparison of air-release values and Bunsen coefficients for a limited range of phosphates is given in Table 19.13. The data indicate that Trixylyl phosphate has a very low air-release value and also a low Bunsen coefficient in comparison with other phosphates and mineral oils. The figures quoted are, of course, for fresh fluids. Air entrainment is very dependent on fluid temperature and fluid viscosity; according to Stoke's Law, a thinner fluid releases air more quickly. Figure 19.10 [35] shows the typical changes in air-release values with temperature for unused natural and synthetic phosphates.

TABLE 19.13

Typical Air-Release and Foaming Test Values for Unused ISO VG 46 Phosphate Esters

Phosphate Ester	Air-release Value at 50°C (min) ISO 9120	Foaming Tendency/ Stability at 24°C (mL) ISO 6247	Bunsen Coefficient for Air at 40°C	Surface Tension at 20°C (dyn/cm)
TXP	1	25/0	0.014	40
IPPP/46	5–8	40/0	0.020	33
TBPP/46	6–10	40/0	–	43
Mineral hydraulic oil (ISO VG 46)	3–5	30/0	0.085	23

In use, air-release values can increase as a result of fluid degradation and the generation of small amounts of surface-active materials (e.g., metal soaps). They may also be adversely affected by the presence of particulate matter—for example, dirt and fluid degradation products.

Whereas entrained air consists of bubbles separated by a thick layer of oil or fluid, foam consists of bubbles separated by a relatively thin film [73], the breakdown of which is determined by temperature and surface tension. To control the generation of foam and reduce its stability, it has been common practice for many years to incorporate small amounts of an antifoam. Most frequently, this is an organopoly-siloxane (silicone), which is effective in parts per million. However, the effectiveness of an antifoam normally depends on its insolubility in the fluid. Thus, it is necessary to produce a homogenous dispersion of the antifoam in the fluid, and with such small quantities this is not easy to achieve. If homogeneity is not achieved, then it is possible for the antifoam to "plate out", usually on the walls of the tank. Depletion also occurs in use when the antifoam accumulates at the air/oil interface on the bubble as it rises to the surface. Here, after assisting the bubble to burst, it migrates to the reservoir wall. A further problem is that the addition of too much antifoam can adversely affect air entrainment [76]; therefore, a careful balance has to be maintained.

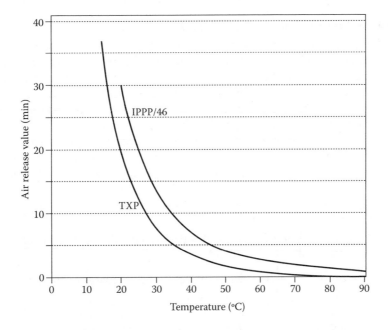

FIGURE 19.10 Air-release values and their variation with temperature for a natural and a synthetic triaryl phosphate.

Typical foam test values for fresh phosphates are also found in Table 19.13 together with surface-tension data. The former are recorded in terms of foaming tendency and its stability at different temperatures after generating small air bubbles in the sample for a fixed period of time.

In service, the foaming properties can deteriorate as a result of antifoam depletion, or as a result of the production of surface-active species in the degradation products. The laboratory measurements are not very precise and are to be regarded more as evidence of a trend in performance, especially as the tank design heavily influences foaming behavior. In service, the presence of a stable foam needs investigation.

Demulsification or water separability characteristics is the third property in the group broadly categorized as "surface-active properties" and which are sensitive to the presence of small amounts of polar impurities or degradation products.

Mixtures of new phosphate ester and water normally separate quickly. This is advantageous for avoiding the formation of emulsions, which could cause a rapid increase in the rate of hydrolysis. As the fluid degrades, the tendency for the water to be emulsified may increase, but this is a rare occurrence because of the small amounts of water normally present in the system.

19.5.2.3 Bulk Modulus and Compressibility

The bulk modulus of a fluid is defined as the ratio of pressure on the fluid to the resulting decrease in volume, and is the reciprocal of compressibility. The ability of a fluid to resist compression is of prime importance when it is used as a hydraulic medium. Highly compressible oils result in sluggish operation, a build up of heat, and significant energy losses [77,78] while fluids that are difficult to compress usually display a "harder" or "noisier" operation and are more susceptible to pressure pulsations. Table 19.14 details the isothermal secant bulk moduli and compressibility values of an IPPP/46 triaryl phosphate ester over a wide range of pressures [42], whereas Figure 19.11 [79] shows that phosphate esters are less compressible or "elastic" than a mineral oil but more compressible than a water-glycol fluid. This property can sometimes influence the choice of fluid, as a high elasticity can adversely affect the accuracy and precise operation of equipment.

19.5.2.4 Vapor Pressure, Volatility, and Boiling Point

In addition to its impact on volatility—and therefore the operating temperature range of a hydraulic fluid—vapor pressure is also an important factor in determining the cavitation tendency of the fluid. Vaporous cavitation is caused by the evaporation of a liquid as a result of a sudden drop in pressure and is followed by bubble collapse when the vapour pressure falls below the value at its boiling point. This can arise from changes in flow rate (e.g., in restrictors, pumps, and valves). Liquids that have high vapor pressures are, therefore, more likely to promote cavitation, cause damage to the surfaces where the bubbles collapse, and cause pressure variations in the circuit. The vapor pressure of phosphates, as expected, varies with chemical structure and molecular weight; the trialkyl phosphates are the most volatile, whereas the triaryl derivatives have such a low vapor pressure

TABLE 19.14
Bulk Modulus and Compressibility of an ISO VG 46 Phosphate Ester Hydraulic Fluid at 37.8°C

Pressure (bar)	Bulk Modulus ($\times 10^4$) bar	Compressibility ($\times 10^{-5}$)/bar
138	1.99	5.02
344	2.12	4.71
689	2.34	4.27
1034	2.53	3.95

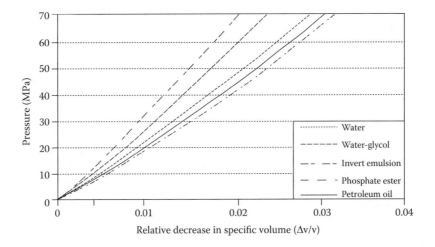

FIGURE 19.11   Generalized isentropic compressibility curves for various types of hydraulic fluids at 20°C. (Reprinted with permission. Elsevier.)

that accurate measurement is difficult below about 200°C. Under normal use conditions, therefore, aryl phosphate ester fluids would be unlikely to produce a significant concentration of vapor in the operating environment.

The vapor pressure characteristics of a range of isopropylphenyl phosphates are given in Figure 19.12 [42]. Additional data are listed in Table 19.15 [45].

Allied to vapor pressure is boiling point and Table 19.16 gives available data on a variety of products at different pressures [41,42,45,71].

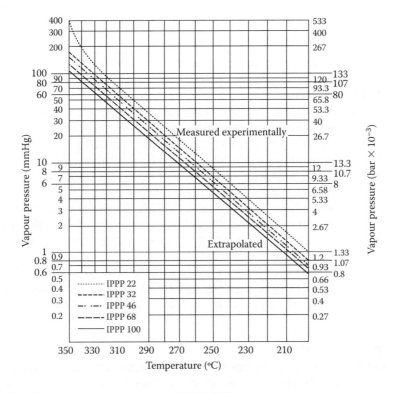

FIGURE 19.12   Vapor pressure characteristics of IPPP fluids.

TABLE 19.15
Vapor Pressure Data for Phosphate Ester Fluids

	Vapor Pressure (mm Hg)				
	10°C	**70°C**	**110°C**	**210°C**	**310°C**
Trialkyl Phosphates					
TBP	1.1×10^{-3}	0.1325	1.398	91.42	
TiBP	2.4×10^{-3}	0.256	2.648	157.6	
TBEP	1.7×10^{-8}	6.2×10^{-5}	3.4×10^{-3}	4.35	472.8
TOP	7.8×10^{-5}	1.1×10^{-2}	0.1251	9.41	160.8
Alkyl aryl Phosphates					
DBPP	1.1×10^{-4}	0.0205	0.2698	26.17	528.2
Triaryl Phosphates					
TPP	1.1×10^{-6}	4.68×10^{-4}	9.2×10^{-3}	1.8	59.9
CDP	4.2×10^{-7}	2.4×10^{-4}	5.4×10^{-3}	1.39	52.85
TCP	1.1×10^{-7}	8.62×10^{-5}	2.0×10^{-3}	0.768	35.12
TXP	2.4×10^{-9}	7.05×10^{-6}	3.6×10^{-4}	0.38	37.4
IPPP/22	4.2×10^{-10}	3.02×10^{-6}	2.4×10^{-4}	0.562	92.14
IPPP/32	1.2×10^{-8}	3.05×10^{-5}	1.0×10^{-3}	0.921	70.34
IPPP/46	2.6×10^{-9}	7.77×10^{-6}	4.0×10^{-4}	0.441	43.84
IPPP/68	3.4×10^{-8}	4.25×10^{-5}	1.0×10^{-3}	0.743	45.09
IPPP/100					33
TBPP/32	1.1×10^{-7}	1.0×10^{-4}	3.0×10^{-3}	1.12	56.15
TBPP/46	8.6×10^{-8}	7.69×10^{-5}	2.0×10^{-3}	0.831	41.25
TBPP/68				0.06	45.1
TBPP/100				0.02	31.2

19.5.2.5　Thermal Properties

As the fluid is circulated around the hydraulic or lubrication system, heat is generated by the compression of the fluid in the pump, at restrictions in the circuit, by frictional losses in bearings, and so forth. To avoid excess thermal stress on sensitive components (e.g., bearing surfaces), it is important that the lubricant assists with the removal of heat. The efficiency of this process depends on the flow rate, the thermal conductivity, density and specific heat of the fluid or lubricant.

A comparison of the specific heat and thermal conductivity of a triaryl phosphate ester with mineral oil and other fire-resistant fluids is given in Table 19.17 [59]. The lower values for the phosphates indicate a slightly inferior cooling behavior, although this is somewhat offset by the higher density.

Specific heat (C_p) data on a range of phosphates is given in Table 19.18 [42–44,71] and shows quite a wide variation in results but no clear trend.

The coefficient of thermal expansion for a number of phosphates between 25°C and 50°C is detailed in Table 19.19 [42,44], together with data on mineral oil and a polyol ester. Although there is some variation in the reported data, it can be seen that values are very low and similar to those of mineral oil.

19.5.2.6　Electrical Properties

In order to avoid the electrochemical erosion of servo-valves in steam turbine control systems, it has become common practice to specify minimum values for the volume resistivity of both fresh and used fluids. Many turbine builders now require at least 50 MΩm on fresh fluid (at 20°C) and

TABLE 19.16
Boiling Point Data for Phosphate Ester Fluids

	Boiling Point (°C)		
Phosphate Ester	**at 760 mmHg**	**at 10 mmHg**	**at 4 mmHg**
Trialkyl Phosphates			
TBP	284	155	139
TiBP	264	137	
Tri-*n*-propyl	252		
TOP	384	211	215
TBEP	320	220	222
Alkyl aryl Phosphates			
ODPP		239 (decomp.)	
DDPP		245 (decomp.)	
DBPP	325	185	
Triaryl Phosphates			
TPP	413	254	
CDP	414	265	
TCP	427	271	
TXP	402	276	
IPPP/22	365	255	
IPPP/32	385	258	
IPPP/46	396	262	
IPPP/68	407	267	
IPPP/100	415	272	
TPBB/32	402	260	
TPBB/46	416	269	
TBPP/68	424	270	
TBPP/100	435	271	

40 MΩm on product in service. These values are normally achieved by removing polar species such as water or acidic products from the fluid. In use, this is effectively achieved by in situ purification of the fluid with an adsorbent medium. Traditionally, Fuller's earth or activated alumina have been used, but ion-exchange resins are now widely used. However, a low value of resistivity does not necessarily result in erosion. Other factors, such as the hardness of the metal surface and type of surface finish, can also influence the process.

Volume resistivity is very temperature dependent and Figure 19.13 [35] shows the variation of $\log_{10}$ resistivity with the reciprocal of temperature for TXP.

TABLE 19.17
Specific Heat and Thermal Conductivity Data on Different Fire-Resistant Hydraulic Fluids

	Fluid type			
Property	**Triaryl Phosphate**	**Water-Glycol**	**Polyol Ester**	**Mineral Oil**
Specific heat (C_p) at 25°C (J/g °C)	1.6	3.2	2.0	1.89
Thermal Conductivity at 20°C (W/mK)	0.132	0.444	0.160	0.134

TABLE 19.18
Specific Heat Data for Phosphate Ester Fluids (ASTM D 3947-80, Differential Scanning Calorimetry)

Phosphate Ester	Specific Heat (C_p) J/g °C		
	at 25°C	at 60°C	at 150°C
TBP	1.40	1.58	2.04
TiBP	–	2.23	–
TBEP	2.00	2.17	2.20
TOP	–	1.97	2.20
CDP	1.45	1.89	2.60
TCP	1.45	1.85	2.13
TXP	1.50	1.66	1.88
IPPP/22	1.50	1.93	2.33
IPPP/32	1.50	1.75	1.99
IPPP/46	1.60	1.70	1.97
IPPP/68	1.60	1.84	2.09
IPPP/100	1.60	1.85	2.10
TBPP/22	1.50	1.78	2.10
TBPP/32	1.50	1.78	1.85
TBPP/46	1.60	1.87	2.11
TBPP/68	1.06	1.69	1.86
TBPP/100	1.70	1.94	2.09

The reciprocal of resistivity is conductivity, and this parameter is sometimes specified as an alternative. The turbine governor application, however, is the only hydraulic or lubricant application to date requiring the measurement of electrical characteristics. As trialkyl or alkyl aryl phosphates are not used in this application, data on this aspect of their performance are not available.

The permittivity (or dielectric constant) of a triaryl phosphate is high (for IPPP/46 at 20°C, the value is 6.75 [35]). Because of their polar nature and sensitivity to moisture, the dissipation factors of triaryl phosphates are also high (e.g., 7% at 20°C [35] for IPPP/46). Breakdown-voltage values can be ≥60 kV (at 20°C), depending on product purity [35].

TABLE 19.19
Coefficients of Thermal Expansion for Different Phosphate Esters

Phosphate Ester	Coefficient of Thermal Expansion ($\times 10^{-4}$°C)
TBP	8.6
TiBP	8.7
TBEP	8.1
TOP	8.0
TCP	5.8–6.7
TXP	6.3
IPPP 22-100	6.9
TBPP 22-100	7.0
Polyol ester (ISO VG 32)	7.5
Mineral oil (ISO VG 32)	7.5

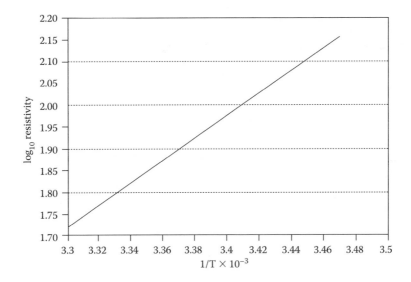

FIGURE 19.13 The variation in volume resistivity with temperature for TXP.

19.5.2.7 Shear Stability

The importance of retaining the viscosity of the hydraulic fluid in service has previously been mentioned. Any significant change can, of course, affect the pump operating efficiency and should be avoided.

One of the principal reasons for viscosity changes in used hydraulic fluids is their breakdown on exposure to the shearing forces present in pumps, valves, bearings, and so forth, and in view of the trend toward higher system operating pressures and smaller system capacities, the use of shear-stable fluids is increasingly required.

Although the triaryl phosphates have low viscosity indices, they are seldom formulated with polymeric thickeners—largely because of the existence of products meeting all the widely used viscosity grades and because minor changes to system design (e.g., the incorporation of tank heating) can enable the disadvantage of their high viscosity at low temperatures to be overcome. Polymeric thickeners can also be hydrolytically unstable and some types are removed from solution when the fluid is treated by an adsorbent solid. Therefore, except for when it is necessary to operate in very cold conditions, phosphate esters do not contain viscosity index improvers and, as a result, retain their viscosity stability over long periods of time at high pressures. In contrast, the trialkyl phosphate-based aircraft hydraulic fluids are polymer-thickened in view of the need to provide a lubricating film over a very wide temperature range.

19.5.2.8 Miscellaneous Physical Properties

Specific gravity

The specific gravity (and density) of phosphate esters also depends on chemical structure: The trialkyl phosphates exhibit values slightly less than 1 at ambient temperature, whereas the triaryl derivatives are some 10–18% higher. The effect of the high specific gravity of the latter, which is up to 30% greater than mineral oil, is to require more energy for fluid circulation. The dirt suspension capacity is also greater, resulting in cleaner systems but necessitating adequate sizing of the filters or more frequent changes, at least during commissioning and outages. Table 19.20 [38,39,42–44] lists typical specific-gravity data on a variety of phosphates at ambient temperatures. As many of the triaryl phosphates are complex mixtures, there is some variation in the data and average values are quoted. Figure 19.14 [35] indicates the variation of specific gravity with temperature of TXP.

TABLE 19.20
Specific Gravity and Refractive Index Data on Phosphate Ester Fluids

Phosphate Ester	Specific Gravity at 20/20°C (ASTM D 1298)	Refractive Index at 25°C (ASTM D 1218)
Trialkyl Phosphates		
TBP	0.980	1.423
TiBP	0.965	1.417
TOP	0.920	1.444
TBEP	1.020	1.434
Alkyl aryl Phosphates		
ODPP	1.091	1.508
DBPP	1.069	
IDPP	1.070	1.510
DDPP	1.087	1.506
Triaryl Phosphates		
TPP	1.767	–
CDP	1.195	1.561
TCP	1.140	1.552
TXP	1.132	1.553
IPPP/22	1.175	1.553
IPPP/32	1.153	1.552
IPPP/46	1.130	1.545
IPPP/68	1.121	1.547
IPPP/100	1.140	1.555
TBPP/22	1.180	1.556
TBPP/32	1.170	1.555
TBPP/46	1.155	1.551
TBPP/68	1.145	1.554
TBPP/100	1.135	1.555

Refractive index

Refractive index can sometimes help in identifying a pure phosphate ester. It has also been used to provide an approximate value for the degree of contamination of a phosphate by a mineral oil, but to do this, the refractive index of both components must be accurately known. Table 19.20 [38,39,42–44] lists the refractive indices at ambient temperatures of most phosphates in commercial use.

Solubility data

The solubility of phosphate esters in water and mineral oil and vice versa is given in Table 19.21 [35]. The oil used in this study was a solvent-refined paraffinic-type product. Greater compatibility would be expected from oils containing a higher concentration of naphthenic or aromatic components.

19.5.3 CHEMICAL PROPERTIES

19.5.3.1 Thermal Stability

The thermal stability of a fluid can provide an approximate guide to its upper operating temperature, but then only in terms of its ability to withstand breakdown in the absence of air or oxygen—a situation which rarely, if ever, exists in practice. In reality, the presence of a small amount of dissolved oxygen can result in degradation at lower temperatures—particularly in the presence of

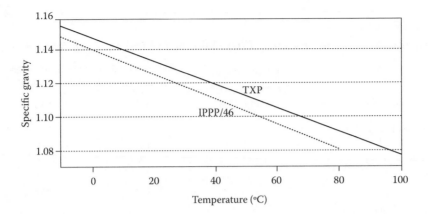

FIGURE 19.14   Typical specific gravity/temperature variation of TXP and IPPP/46.

metals—and apparent changes in the physical/chemical properties of the fluid may be due to oxidation rather than pure thermal breakdown. Where changes in fluid vapor pressure are measured, results are influenced by the presence of volatile impurities or base-stock components, particularly for products which are complex mixtures. Most of the tests will, at best, provide a ranking of different fluids rather than absolute values.

An early assessment of the thermal stability of an alkyl and an aryl phosphate was provided by Blake et al. [80] using an isoteniscope for measuring the change of vapor pressure with temperature. When a sudden change in rate occurred (or when a value of 0.014 mmHg/s was reached), it was assumed that this was the decomposition temperature. With this technique, values of 195°C and 423°C were recorded for tri-*n*-octyl phosphate and TPP, respectively. By comparison, the decomposition temperature of a refined mineral oil was given as ~355°C. These figures compare with 485°C (for TPP) and 375–379°C for TCP reported by Raley [81], using the same technique.

More recently, Lhomme et al. [82] reported that TBP starts to decompose below its boiling point of 284°C under helium, whereas trimethyl and triethyl phosphates exhibit initial degradation at 300°C and 200°C, respectively. In contrast, TPP was very stable and only began to decompose at ~600°C and was not completely thermally degraded even at 1000°C [83].

TABLE 19.21
Solubility Data for Phosphate Esters in Mineral Oil and Water

Phosphate Ester	Phosphate in Water at 25°C (mg/L)	Phosphate in Oil at 25°C (%) v/v	Water in Phosphate at 25°C (%) v/v	Oil in Phosphate at 15°C (%) v/v
TBP	280	Miscible	7	
TiBP	<1000			
TBEP	1100	7	7.3	
TOP	<1000	Miscible		
ODPP	1.9			
DBPP	96			
DDPP	0.75			
TPP	2.1		Nil (TPP is solid)	Nil (TPP is solid)
CDP	0.24			
TCP	0.36	1–2	0.2–0.3	2–4
TXP	0.11	1–2	0.2–0.3	2–4
IPPP/46	0.7	1–2	0.2–0.3	2–4
TBPP/46	2.1	1–2	0.2–0.3	2–4

TABLE 19.22
Thermal Stability of Phosphate Ester Fluids in an Inert Nitrogen Atmosphere

Phosphate Ester	DSC-Initiation of Decomposition (°C) (ASTM E 537)	TGA-Temperature Required for given Weight Loss (°C) (ASTM D 3850)	
		5%	10%
Trialkyl Phosphates			
TBP	283	169	197
TiBP	–	126	141
TOP	281	205	231
TBEP	276	157	181
Alkyl aryl Phosphates			
ODPP	252	226	231
DDPP	264	233	246
Triaryl Phosphates			
CDP	306	236	254
TCP	333	272	306
TXP	311	276	302
IPPP/22	314	264	293
IPPP/32	311	282	307
IPPP/46	311	281	307
TBPP/22	347	274	305
TBPP/32	345	277	305
TBPP/46	338	278	306

Marino [32] reported on two studies using differential scanning calorimetry (DSC) and thermogravimetric analysis (TGA) for comparing the thermal stability of a range of phosphate esters. Table 19.22 lists these results and other data produced by these methods [45,84]. Both methods give a similar ranking of the relative stability of these materials. The tertiarybutylatedphenyl phosphates are the most stable, followed by other triaryl phosphates, and, finally, by the alkyl and alkyl aryl phosphates. It is to be expected that there would be no significant difference between the latter two groups: the stability of the alkyl group in the mixed alkyl aryl phosphate being the weakest link in the chain. Therefore, the low TGA value for ODPP is a surprise and may be linked to product purity.

A number of studies of the pyrolysis and combustion of phosphate esters have been made, mainly to clarify if any highly toxic decomposition products were being produced. Lhomme et al. [83] examined the degradation products under helium of trimethyl, triethyl, and triphenyl phosphate and, as a result, proposed a general pyrolysis scheme (Figure 19.15). Depending on the phosphate structure and temperature, different degradation pathways were followed. At "low" temperatures, the alkyl phosphates followed reaction (a) as a result of the cleavage of the —C—O bond and the production of olefins, monohydrogen phosphates, dihydrogen phosphates, and so forth. With increasing temperature, path (b) was followed, involving the breakage of the —P—O bond, whereas at very high temperatures, the phosphate residue forms phosphorus pentoxide (probably after passing through an intermediate phase involving the formation of polyphosphates and pyrophosphates).

With triphenyl phosphate (and, presumably, other triaryl phosphates), both reactions (a) and (b) take place simultaneously but at much higher temperatures (>600°C).

Under oxidative conditions in excess air, the differences in stability between alkyl and aryl phosphates persist, although degradation commences at lower temperatures and appears complete (for both types) by about 900°C based on the production of carbon dioxide (Figure 19.16) [83].

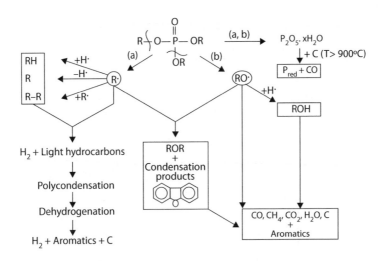

FIGURE 19.15 General pyrolysis scheme of phosphates. (Reprinted with permission from Lhomme, V., Bruneau, C., Soyer, N. and Brault, A. "Thermal Behavior of Some Organic Phosphates," *Ind. Eng. Chem. Prod. Res. Dev.*, 1984, 23(1), pp. 98–102. Copyright 1984, American Chemical Society.)

Pyrolysis studies of IPPP/46 under a helium atmosphere have also been carried out at temperatures between 500°C and 1000°C [35]. Measurements were made of: (1) the amounts of carbon monoxide and dioxide formed using non-dispersive infrared analysis; (2) the organic volatiles, which were identified using gas chromatography/mass spectrometry; and (3) the amount of phosphorus pentoxide generated, which was collected in aqueous potassium hydroxide and determined as orthophosphate by ion chromatography. For comparison, the same product was also examined for the production of carbon dioxide and phosphorus pentoxide under combustion conditions by passing air over the sample. Comparative results on the generation of oxides of carbon and phosphorus are given in Table 19.23; the main changes in the production of organic volatile degradation products are shown in Figure 19.17. It can be seen that virtually no degradation takes place up to 1000°C under pyrolytic conditions. Under combustion or oxidative decomposition, the product shows significant degradation at the latter temperature, but the values for phosphorus pentoxide content are lower than might be expected.

Unfortunately, neither of the above studies included an assessment of the stability of phosphates in comparison with that of other fire-resistant hydraulic fluids—or even with mineral oil. A highly detailed comparison of the former group was, however, made as part of an investigation into "Coal mine combustion products, their identification and analysis" by Paciorek et al. [85]. In this study, three water-glycol fluids, a water-in-oil emulsion, a synthetic organic ester fluid, and several triaryl phosphates—including a blend with mineral oil—were heated in the presence of a limited amount of air at 370°C and 420°C. This was followed by separation of all products by vacuum line fractionation and analyses of the fractions by gas chromatography/mass spectroscopy. The report concluded that at the temperatures investigated, "pure" phosphate esters "represent probably the least dangerous of all the compositions tested insofar as flammability and the overall toxicity of the degradation productions are concerned." The authors, however, did point out that blends of phosphates and mineral oil could evolve the toxic product phosphine.

The conclusion was based on the fact that the phosphate esters showed very little decomposition under these conditions, and although phenolic species were evolved, the concentration and toxicity of these were judged to be less important than the amounts of carbon monoxide, acrolein, and so forth, which were produced by the other fluids.

19.5.3.2 Oxidation Stability

Oxidation stability is an important property for fluids that are exposed to high temperatures, both from within the hydraulic system—in pumps, valves, bearings, and restrictors—and occasionally

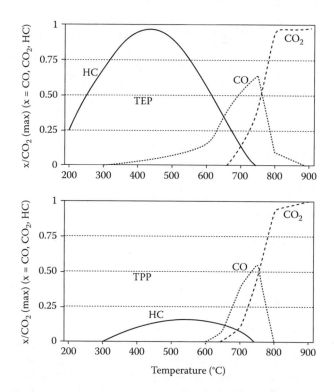

FIGURE 19.16 CO/CO_2 (max.), CO_2/CO_2 (max.), and hydrocarbon/CO_2 (max.) ratios as a function of temperature for triethyl phosphate (TEP) and triphenyl phosphate (TPP). (Reprinted with permission from Lhomme, V., Bruneau, C., Soyer, N. and Brault, A. "Thermal Behavior of Some Organic Phosphates," *Ind. Eng. Chem. Prod. Res. Dev.*, 1984, 23(1), pp. 98–102. Copyright 1984, American Chemical Society.)

also from external heat sources (e.g., steam lines). In addition, compression of free air in the pump can cause bubble-wall temperatures to rise by several hundred degrees [83] exceeding the thermal stability of the fluid and causing rapid oxidation. Although the temperature increase is highly localized, continuous operation under these conditions can, within months, result in considerable degradation of the fluids. Such a condition is often the result of a combination of poor system design and high air entrainment in the fluid, and confirms the need for stable products in view of the trend to smaller systems and higher operating pressures.

The initial step in the oxidation process is normally the reaction of oxygen with the alkyl groups (of alkyl phosphates) or the alkyl part of substituted aryl groups (e.g., the methyl or isopropyl groups). Where the aryl group is unsubstituted, the initial degradation arises through thermal breakdown

TABLE 19.23

Development of Carbon Oxides and Phosphorus Pentoxide under Pyrolysis and Combustion Conditions for an IPPP/46 Phosphate

Yield of Degradation Products (%)	Temperature (°C)					
	500		700		1000	
	Pyrolysis	Combustion	Pyrolysis	Combustion	Pyrolysis	Combustion
Carbon monoxide	0.63	–	0.56	0.1	1.1	11.23
Carbon dioxide	<0.001	–	<0.001	0	<0.001	49.63
Phosphorus pentoxide	<0.01	–	<0.01	0.57	<0.01	1.42

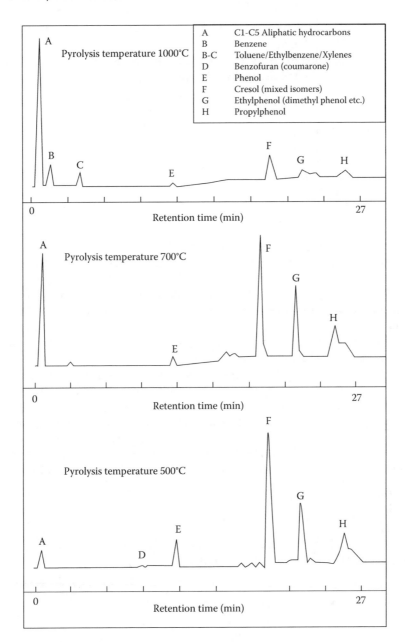

FIGURE 19.17 Chromatograms of pyrolysis products from IPPP/46.

(fission) of the P—O bond and the evidence suggests that this is also the main degradation process for the tertiarybutylphenyl phosphates because of the high stability of the substituted *t-butyl* group. The ease of oxidation of the phosphate molecule therefore varies approximately with the structure as follows:

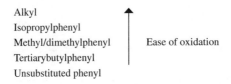

TABLE 19.24

Rotating Pressure Vessel Oxidation Test Results on Triaryl Phosphates; Test Method: ASTM D 2272

Phosphate Ester	Time to a Pressure Drop of 175 kPa (25.4 psi) - min
TXP type B (uninhibited)	600
TXP type A (inhibited)	455
TXP type A (uninhibited)	450
TBPP/46 (uninhibited)	300
TBPP/32 (uninhibited)	300
IPPP/46 (inhibited)	250

Many tests exist to evaluate and compare the oxidation stability of phosphate esters. Some of the procedures used with mineral oils (e.g., ASTM D 943 [TOST] test and the ASTM D 2272 [Rotating Pressure Vessel] test) are unsuitable in their existing form because they involve significant quantities of water. This causes hydrolytic breakdown to dominate over oxidative breakdown, and the test results do not necessarily reflect the ranking obtained under dry test conditions. Table 19.24 [35] shows rotating pressure vessel stability test data on a range of phosphates, with the most hydrolytically stable product exhibiting the longest test life. Furthermore, an inhibited xylyl phosphate shows no improvement over the unstabilized base.

One of the first comparisons of oxidation stability of the modern range of phosphate esters was carried out in 1979 [87] on a micro-oxidation tester using uninhibited fluids. Very small samples (40 μL) were placed in a mild-steel cup and heated to temperatures of 225–270°C while a current of air was circulated over the sample. At the end of the prescribed period, the fluid was dissolved in a solvent and analyzed using gas-phase chromatography. The results are shown in Table 19.25. All of the phosphates showed good stability under these conditions, with the TBPP ester exhibiting essentially no oxidation. As expected, TBP was the least stable and little difference was seen between TCP and TXP.

TABLE 19.25

Oxidative Stability of Phosphate Esters in a Micro-Oxidation Tester: Test Conditions: Air Blown in Presence of Steel at 225°C–270°C

Phosphate ester	Time (min)	Temperature (°C)	Percent of Original Product		
			Unoxidized	Oxidized	Evaporated
TBP	5	225	43	6	51
	10	225	19	8	73
TCP	30	225	85	1	14
	60	225	67	2	31
	30	250	57	3	40
	15	270	60	5	35
TXP	15	250	84	2	14
	30	250	65	4	31
	15	270	55	6	39
t-butylphenyl-	360	250	77	<1	22
diphenyl phosphate	180	270	81	<1	15

Source: Cho, L. and Klaus, E.E. "Oxidative Degradation of Phosphate Esters," *ASLE Trans.*, 1979, 24(1), pp. 119–124. With permission.

TABLE 19.26

Oxidative and Thermal Stability of Commercial Phosphate Esters by DSC (ASTM E 537) and TGA (ASTM D 3850)

Phosphate Ester	DSC - Onset of Oxidation (°C) (ASTM E 537)	TGA - Temperature Required for Given Weight Loss in Oxygen (°C) (ASTM D 3850)		
		1 wt%	5 wt%	10 wt%
TBP	175	76	113	127
TiBP	192	84	116	130
TBEP	155	61	169	187
TOP	160	122	189	203
TPP	a	188	236	252
CDP	265	198	230	246
TCP	215	184	255	278
TXP	210	224	268	286
IPPP/22	215	200	239	263
IPPP/32	215	201	252	272
IPPP/46	210	202	265	287
IPPP/68	210	218	265	288
IPPP/100	180	224	243	258
TBPP/22	295	213	262	280
TBPP/32	295	222	268	286
TBPP/46	300	227	272	292
TBPP/68	300	230	275	293
TBPP/100	305	234	277	295
ISO VG 22 mineral oil	167	155	205	225

a Does not oxidize under these conditions.

A similar picture is seen in the results of DSC/TGA data (Table 19.26) [44,88] and also in DIN 51373 oxidation tests (Table 19.27) [35].

The effect of the position of the substituent on the aryl ring was shown, with respect to isopropylphenyl phosphates, in Table 19.1. The advantages of the *ortho-* isomer from the point of view of stability are, of course, offset by its poor viscosity index. Because commercial products contain a mixture of different isomeric products, the resultant behavior of the finished product will depend on the proportion of each isomer present.

The isopropylphenyl phosphates and TCP/TXP are also responsive to the addition of stabilizers, which can improve performance and significantly extend life. In contrast, the tertiarybutylphenyl

TABLE 19.27

Comparison of the Oxidation Stability of Different Types of Uninhibited ISO VG 46 Triaryl Phosphate Esters: Test Method: DIN 51373 - Air Blown at 120°C for 164 hour in the Presence of Metals

Phosphate Ester	Viscosity Change at 40°C (%)	Total Acidity Increase (mg KOH/g)	Metal Weight Changes (mg)	
			Copper	Iron
TBPP/46	−0.6	0.01	0.1	0.2
TXP	3.0	0.47	0.2	0.2
IPPP/46	46.5	8.45	−0.3	0

phosphates are not as responsive to classical antioxidants but have such good oxidation stability that this is rarely a disadvantage. A comparison of the high-temperature stability of uninhibited TBPP/32 and TBPP/46 products in comparison with various formulated conventional and highly refined mineral gas turbine oils is given in Table 19.28 [89].

Although the mineral oil products show substantial increases in viscosity and acidity as the test duration increases, there is no significant change in the viscosity of the TBPP products and only moderate increases in acidity.

One additional aspect of oxidation stability that is of concern in some applications is the deposit-forming tendency at high temperatures. This property, also known as "coking", occurs when a thin film of fluid is heated on a metal surface while exposed to air. Depending on the stability of the product, the deposit can vary from soft and carbonaceous to a hard, brittle layer, or to a lacquer. The formation of such deposits can occur on heater and valve surfaces and can reduce component efficiency. However, in situ conditioning has been shown to control the deposit formation [90]. Another form of deposit, commonly known as "varnish", occurs when fluid degradation products precipitate onto the surface of metal components and then harden under the influence of heat. It can be difficult to differentiate between the two forms of deposit without complex analytical investigations. Table 19.29 [35] shows the variation in coking tendencies for a range of triaryl phosphates and commercial mineral gas turbine oils.

It was mentioned earlier in Section 19.5.1.1 that the flammability characteristics were more likely to correlate with oxidation stability than other physical properties. An examination of the ranking of ease of oxidation in this section will be seen to broadly correlate with the fire-resistance data.

19.5.3.3 Hydrolytic Stability

In the presence of small amounts of water, phosphate esters tend to hydrolyze (i.e., break down into their constituent acids and alcohols [or phenols]). It is the most common form of degradation of phosphates and the one which frequently dictates their service life.

The hydrolysis reaction is not rapid at ambient temperatures but accelerates with increasing temperatures and is catalyzed by the presence of strong acids. As these are normally produced by phosphate hydrolysis, the reaction is said to be "autocatalytic". Strong bases can also catalyze the reaction.

The hydrolysis of triaryl phosphates (e.g., triphenyl phosphate) takes place in steps, as indicated in Figure 19.18. The replacement of successive aryl groups with —OH groups becomes increasingly difficult and the last step is not normally achieved without using very severe conditions, such as boiling with dilute acid. In view of the relatively low acidities found with these products in practice, it is extremely unlikely that, under most service conditions, phosphoric acid would be generated.

The reaction scheme shown in Figure 19.18 shows the strong acids (pH < 7)—also known collectively as partial phosphates—that are produced. Although these are of greatest importance in view of their reactivity, weak acids (pH > 7) such as phenols are also formed in this process. Although they are not normally thought to have a major effect on the performance of the fluid, there is some evidence to suggest that they may adversely affect foaming when present in significant quantities.

The presence of strong acids can have different effects on fluid and system performance. Some of these are beneficial, whereas others can cause operational problems. The acids may, for example, inhibit the corrosion of ferrous metal surfaces but promote corrosion of nonferrous components; they may adversely affect the electrical properties of the fluid but assist in reducing wear. In general, the uncontrolled generation of acidic products is harmful to the life and performance of the fluid, and in certain critical applications, such as the power generation industry, it is customary to remove them as they are produced by circulating the fluid through an adsorbent solid.

Because of the effect of hydrolytic stability on the performance of both the fluid and the system, a number of studies of the stability of phosphate esters in the presence of water have been carried out [10,32,91,92]. Hydrolytic stability tests are not the most precise because they normally require an

TABLE 19.28

High-Temperature Oxidation Tests on Tertiarybutylphenyl Phosphates and Commercial Industrial Gas Turbine Lubricants: Test Method: Federal Test Method Standard VV-L-791C, Method 5308.6 at 175°C

Phosphate Fluid	Test Duration (days)	Viscosity Increase (%)	Acidity Increase (mg KOH/g)	Metal Weight Changes (mg/cm²)				
				Fe	Cu	Cd/Fe	Al	Mg
TBPP/32	3	3	0.11	-0.11	0.17	-0.03	-0.07	-0.03
	6	4.4	0.37	0.02	0.04	0.02	Nil	0.02
	9	3.8	0.85	-0.01	0.1	0.03	Nil	-0.02
TBPP/46	3	6	0.27	-0.01	0.03	-0.01	-0.11	-0.02
	6	6.4	0.23	0.11	0.29	0.17	0.08	0.13
	9	7.3	1	-0.04	0.02	0.18	-0.11	0.07
Gas turbine oil A	3	3.6	0.44	Nil	0.02	-0.04	Nil	0.01
(ISO VG 32)	6	4.2	0.8	0.03	0.04	0.02	Nil	0.03
	9	25	5.3	0.14	0.01	-0.15	0.09	0.06
Gas turbine oil B	3	18.3	2.2	0.03	0.08	-0.1	0.05	0.06
(ISO VG 46)	6	24.3	3.1	0.21	0.13	0.07	0.02	0.07
	9	49.5	5	0.56	0.04	0.97	0.32	0.33
Gas turbine oil C	3	23.4	3	0.02	0.03	0.05	0.03	0.02
(ISO VG 46)	6	28.6	4.5	0.04	-0.1	-0.05	0.03	0.01
	9	80.6	7	0.18	0.47	-11.7	0.05	0.04
Gas turbine oil D	3	13.4	2	0.06	0.02	-0.04	Nil	0.01
(ISO VG 46)	6	21.2	5.1	0.24	0.13	0.02	0.52	0.35
Turbine industry limits	3	-5 to 15	2.5 max.	-	-	-	-	-
ASTM D 4293 limits	3	-5 to 15	3.0 max.	-	-	-	-	-

TABLE 19.29
Coking Tendency Data for Triaryl Phosphates:
Test Method: Federal Test Method Standard
VV-L-791C, Method 3462

Phosphate Ester	Deposit Formation (mg) at		
	300°C	316°C	325°C
TXP (uninhibited)	460	1820	1750
TXP (inhibited)	43	1006	1420
TBPP/32 (uninhibited)	3	16	6
TBPP/46 (uninhibited)	3	4	2
IPPP/46 (inhibited)	4		
Mineral gas turbine lubricants			
–ISO VG 32	25	170	130
–ISO VG 46	49	185	227

excess of water present in order to accelerate the degradation, and the degree of contact between the two fluids—which determines the rate of hydrolysis—is difficult to control. A test in which the two fluids are stirred together will normally result in the generation of more acidity than one in which the two layers are not in motion. Another important variable is the initial acidity of the fluid; if this is not the same as that of other fluids being tested (within the precision of the test) then a comparison can be meaningless. Consequently, it is sometimes difficult to compare data from different sources.

The available data indicates that hydrolytic stability is very structure dependent. Triaryl phosphates are generally more stable than trialkyl phosphates, whereas increasing the length of the alkyl chain (whether or not attached to an aromatic ring) and increased branching on the alkyl chain also results in increased stability (Table 19.30) [71].

In comparing the different types of unstabilized triaryl phosphates (Table 19.31) [32], it is apparent that TXP is by far the most resistant to hydrolysis and this is also evident in data on formulated fluids (Table 19.32) [32]. The inferior stability displayed by some of the IPPP- and TBPP-based flu-

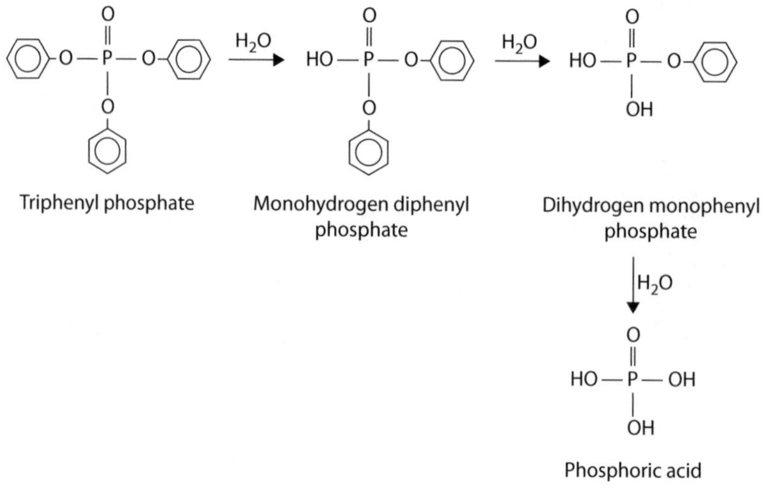

FIGURE 19.18 The hydrolysis of triphenyl phosphate.

TABLE 19.30
Hydrolytic Stability of Selected Phosphate Esters: Test Conditions: 24-Hour Reflux Period with Water

Phosphate Ester	mL of Normal Sodium Hydroxide/mol
TBP	4
TOP	0.2
TCP	1.2
n-Butyldiphenyl	27.8
t-Butyldiphenyl	7.3
n-Octyldiphenyl	14.6
6-Methylheptyldiphenyl	7.6
ODPP	3.6
n-Octyldicresyl	4.4
2-Ethylhexyldicresyl	1.7

Source: Gamrath, H.R., Horton, R.E. and Weesmer, W.E.
Ind. Eng. Chem., 1954, 46, p. 208.

ids is due to the presence of triphenyl phosphate (see Table 19.2), which is unstable in the presence of moisture (see Table 19.33).

In addition to the effects of the different substituents on the aryl ring, the position of these substituents (or the isomer distribution) also impacts on the stability of the fluid. Table 19.33 [35] shows the effect of hydrolysis on various pure cresyl and xylyl phosphate isomers. In this test, the fluid (5g) was maintained at 90°C for two weeks in the presence of 5 mL of water, after which the total increase in acidity generated was determined. As can be seen, the xylyl phosphates are generally significantly more stable than the cresyl derivatives, with the exception of the *p*-cresyl phosphate isomer. Single-point determinations, however, do not necessarily reveal the whole picture and additional data would ideally be required to confirm the stability ranking.

The sensitivity of phosphates to moisture has, in the past, limited their use, as it significantly reduced their service life and increased fluid costs. In the power-generation industry, an "on-line" purification technique has been used for many years to remove acid as it is formed, thereby prolonging fluid life. However, by selecting a particular xylenol isomer distribution, products with superior

TABLE 19.31
Hydrolytic Stability of Unstabilized Phosphate Esters: Test method: ASTM D 2619

Phosphate Ester	Increase in Fluid Acidity (mg KOH/g)	Increase in Water Acidity (mg KOH)	Copper Weight Loss (mg/cm$_2$)
TCP	0.10	9.1	0.3
TXP	Nil	Nil	0.03
IPPP/46	0.01	3.63	0.15
TBPP/46	0.05	6.17	0.18
TBPP/68	0.09	8.98	0.34
TBPP/100	0.2	11.2	0.45
Standard criteria	0.2 max.	5 max.	0.3 max.

TABLE 19.32

Hydrolytic Stability of Formulated Phosphate Ester Hydraulic Fluids: Test Method: ASTM D 2619

Phosphate Ester	Increase in Fluid Acidity (mg KOH/g)	Increase in Water Acidity (mg KOH)	Copper Weight Loss (mg/cm²)
TXP	0.02	Nil	−0.008
IPPP/46	0.05	0.4	0.05
TBPP/46	0.03	2.7	0.05
Standard criteria	0.2	5	0.3

hydrolytic stability can be produced [93]. Figure 19.19 [35] shows, for example, the hydrolytic stability of two synthetic phosphates, TBPP/46 and IPPP/46, in comparison with two xylyl phosphates, including one fluid which has been specifically developed with improved stability.

Fluids based on such feedstocks have now been used successfully for extended periods as main bearing lubricants in steam turbines without the need for in situ purification.

19.5.3.4 Radiation Stability

Lubricants in nuclear applications must be able to withstand radiation without significant deterioration and be non-reactive toward the materials used in reactor construction. Past studies of resistance to radiation for both alkyl and aryl phosphates indicated that they were not very stable in such an environment [91–93] and degrade to form significant amounts of acid.

Recently, as a result of an interest in the use of triaryl phosphates as reactor coolant pump lubricants, further studies were carried out in which the fluids were subjected to a gamma radiation dose of 0.6 Mrad. This level is thought to be representative of about eighteen months' operation—the minimum acceptable life—during which time no fluid maintenance would be possible because of its location in the reactor. After this length of operation, the radiation level of the fluid should still be low enough not to require disposal as special waste. Table 19.34 shows the fluid properties before and after radiation for three uninhibited triaryl phosphate esters and one containing a stabilizer package. As can

TABLE 19.33

Hydrolytic Stability of Different Cresyl and Xylyl Phosphate Isomers: Test Conditions: 90°C for Two Weeks with 5 g each of Fluid and Water

Phosphate Isomer	Melting Point (°C)	Total Acid Value Increase (mg KOH/g)
Tri-*o*-cresyl	Liquid[a]	27.2
Tri-*m*-cresyl	Liquid[a]	34.8
Tri-*p*-cresyl	78	0.21
Tri-2,3-xylyl	60–61	0.13
Tri-2,4-xylyl	Liquid[a]	0.74
Tri-2,5-xylyl	79–80	0.18
Tri-3,4-xylyl	71–72	0.11
Tri-3,5-xylyl	43–45	0.33
Triphenyl	49	171.3

[a] product is liquid at ambient temperature

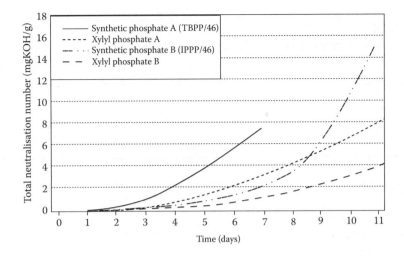

FIGURE 19.19   Hydrolytic stability of different phosphates under extended ASTM D 2619 test conditions.

be seen, the main effect is on hydrolytic stability. Oxidation stability and physical properties are not significantly changed. The stability of the TXP sample was thought adequate for service operation.

19.5.3.5   Rusting and Corrosion

The rusting (of ferrous metals) and corrosion (of non-ferrous components) can both take place in hydraulic systems. The former phenomenon is possible in either the liquid or vapor phase, whereas the latter normally occurs only in the liquid phase.

With non-aqueous fluids like phosphate esters, rusting occurs when free water is present. Water has a much higher solubility in phosphate esters than in mineral oil; for example, up to ~0.25% at ambient temperatures or ~0.5% at temperatures of 50–60°C and these levels would need to be exceeded before rusting commences. If the temperature of the fluid cycles up and down, then free water could separate out at the lower temperature. Therefore, it is important to monitor the water content and check that it remains below the solubility level at the lowest temperature likely to be experienced (unless inhibitors are incorporated). High water levels, of course, reflect mechanical problems in the system and require investigation as soon as possible.

Phosphate esters, and particularly their acidic degradation products, are highly polar and form an adsorbent layer on the metal surface which helps to reduce contact with water. As a result of these properties, phosphate fluids seldom require the use of rust inhibitors, and systems containing them can be made from mild steel. As rust inhibitors can also interfere with the foaming and surface-active properties of the fluid, they should not be used unless the fluid is likely to operate very wet. Rusting can occur above the liquid layer in the tank if condensation is a regular feature and if fluid has not wetted the surface. This can be easily overcome by ensuring a flow of dry air through the tank. The same technique can also be used to remove water from the bulk fluid should the content increase as a result of, for example, cooler leaks. If gross contamination of phosphates leads to a layer of free water on the top of the phosphate ester, it is possible to remove this water by siphoning or skimming. Unlike other fluids, free water is less likely to collect in other parts of the system because of the higher density of the phosphates.

The corrosion of ferrous metals is not normally a concern with phosphates and, when fluid acidities are kept low (preferably <0.5 mg KOH/g), the corrosion of non-ferrous metals, particularly copper, zinc, lead, and their alloys is also minimized. Table 19.35 [35] indicates the performance of an IPPP/46 fluid in the ISO 4404-2 corrosion test and no significant weight changes are seen on any of the metals. Where the stress on the fluid is likely to lead to significant degradation, phosphate esters are generally used with either continuous or intermittent on-line conditioning involving passing the fluid through an adsorbent solid, which removes the acid.

TABLE 19.34
Effect of Gamma Irradiation on Different Triaryl Phosphates

Physical Property	TXP (uninhibited)	TXP (inhibited)	TBPP/46	IPPP/46
Pre-irradiation				
Neutralization No. (mg KOH/g)	0.03	0.03	0.01	0.01
Viscosity at 40°C (cSt)	49.68	53.58	42.01	44.56
Foam sequence 1 at 24°C (mL)	150-0	190-0	0-0	430-10
Air release (min.)	2.0	7.0	9.0	3.0
Flash point (°C)	250	250	253	238
Fire point (°C)	344	352	340	328
DIN 51373 - oxidation stability				
-Cu weight change (mg)	−0.1	0.0	0.1	−0.2
-Fe weight change (mg)	0.0	0.0	−0.1	0.2
-Fluid final neutralization No. (mg KOH/g)	0.55	0.11	0.02	2.64
-Volatiles neutralization No. (mg KOH/g0	0.07	Nil	0.03	0.90
-Total neutralization No. change (mg KOH/g)	0.59	0.08	0.04	3.53
ASTM D 2619 - hydrolytic stability				
-Cu weight change (mg/cm^2)	Nil	Nil	−0.008	−0.27
-Fluid final neutralization No. (mg KOH/g)	0.08	0.06	0.09	0.05
-Neutralization No. change (mg KOH/g)	0.05	0.03	0.08	0.04
-Acid content of aqueous layer (mg KOH)	3.44	2.89	5.62	7.29
Post-irradiation				
Neutralization No. (mg KOH/g)	0.19	0.17	0.15	0.13
Viscosity at 40°C (cSt)	51.31	53.52	41.86	45.08
Foam sequence 1 at 24°C (mL)	170-0	210-0	0-0	540-530
Air release (min)	2.0	4.5	6.0	7.5
Flash point (°C)	252	254	246	232
Fire point (°C)	350	356	338	328
DIN 51373 - oxidation stability				
-Cu weight change (mg)	−0.4	−0.3	0.1	−0.8
-Fe weight change (mg)	0.0	0.5	0.0	·0.3
-Fluid final neutralization No. (mg KOH/g)	0.92	0.23	0.18	1.76
-Volatiles neutralization No. (mg KOH/g)	0.14	Nil	Nil	0.58
-Total neutralization No. change (mg KOH/g)	0.87	0.06	0.03	2.21
ASTM D2619 - hydrolytic stability				
-Cu weight change (mg/cm^2)	−0.015	−0.105	−0.285	−0.89
-Fluid final neutralization No. (mg KOH/g)	0.39	0.32	1.76	0.36
-Neutralization No. change (mg KOH/g)	0.20	0.15	1.61	0.23
-Acid content of aqueous layer (mg KOH)	7.23	6.29	43.85	18.59

19.6 COMPATIBILITY

There are several different aspects of phosphate ester compatibility that need to be considered:

- With system constructional materials, both metallic and non-metallic;
- With mineral oil, in case of contamination;
- With preservative fluids used during system assembly and storage;
- Between different phosphate types in the event of topping-up a "synthetic" fluid with a "natural" fluid or vice versa;

TABLE 19.35
Determination of the Corrosivity of an IPPP/46 Fluid: Test Method: ISO 4404-2

Test no.	Test Metal(s)	Metal Weight Change (mg)	Metal Appearance		Test Fluid Appearance
			In Fluid	In Air	
1	Steel	0.5	Very slight tarnish	No change	No deposits or color change
2	Copper	0.1	No change	No change	No deposits or color change
3	Zinc	0.1	No change	No change	No deposits or color change
4	Aluminum	0.1	No change	No change	No deposits or color change
5	Cadmium	0.1	No change	No change	No deposits or color change
6	Brass	0.2	Slight tarnish	No change	No deposits or color change
7	Steel/Cadmium	0.3	No change	No change	No deposits or color change
		0.2	No change	No change	No deposits or color change
8	Copper/ Zinc	Nil	No change	No change	No deposits or color change
		0.2	No change	No change	No deposits or color change
9	Aluminum/Zinc	0.1	No change	No change	No deposits or color change
		0.2	No change	No change	No deposits or color change
10	Steel/Aluminum	0.2	Very slight tarnish	No change	No deposits or color change
		0.2	Very slight tarnish	No change	No deposits or color change

Note: Test conditions: (a) 250 cm^3 of test fluid with test specimen(s) 100 mm x 20 mm immersed in the fluid; (b) temperature 35°C; (c) duration 28 days.

- Between phosphate esters and other types of hydraulic fluids in the event that one fluid replaces the other in the system or is used inadvertently as a top-up.

Phosphate esters are compatible with all common constructional metals, but they do not "wet" the surface of aluminum. Therefore, the bearing manufacturers should be consulted prior to the use of aluminum alloys as plain bearing materials. When used for other purposes in equipment, the aluminum surface should be hard anodized. Phosphate esters can be used in conjunction with copper and its alloys, provided that the acidity levels in service do not exceed those recommended by the equipment manufacturers.

Bearing materials suitable for use with phosphates include white metal and copper/lead babbit, with the former type preferred.

The compatibility of phosphate esters with different types of paints, packings, seals, hoses, and filtration media is given in Table 19.36 [32,97,98]. Such a table can only provide general guidance. There are, for example, a variety of different types of fluorocarbon rubber that do not all perform equally in compatibility tests and the manufacturer should be contacted for more specific recommendations.

Mineral oil has variable solubility in phosphate esters, depending on the temperature, the type of oil (i.e., whether paraffinic, naphthenic or aromatic), and on the type of phosphate. Unsurprisingly, paraffinic oils have a greater solubility in trialkyl phosphate and naphthenic or aromatic hydrocarbons in triaryl phosphates. While the major problem arising from oil contamination is a loss of fire resistance, the fluid surface-active properties can also be adversely affected. Mineral oil is difficult to remove from phosphates, and contamination may require replacement of the fluid charge.

Mineral oil-based preservatives, which are used to prevent rust on system components during storage, can promote foaming and increase air entrainment if allowed to contaminate phosphates. If these are used, it will be advantageous to flush through a new system before filling with the operating charge. A phosphate ester-based preservative could be advantageous if suitable facilities for storage are available at the equipment manufacturer.

TABLE 19.36
Phosphate Ester Compatibility with Standard Constructional Materials

Material	Triaryl Phosphate Ester	Trialkyl Phosphate Ester	Alkyl aryl Phosphate Ester
Elastomers			
Butadiene acrylonitrile (BUNA N or nitrile)	Unsatisfactory	Unsatisfactory	Unsatisfactory
Chlorosulfonated polyethylene	Unsatisfactory	Unsatisfactory	Unsatisfactory
Ethylene propylene (EPR)	Unsatisfactory	Unsatisfactory	Unsatisfactory
Ethylene propylene diene (EPDM)	Recommended	Recommended	Recommended
Epichlorohydrin	Unsatisfactory	Unsatisfactory	Unsatisfactory
Fluorinated hydrocarbon	Recommended	Unsatisfactory	Recommended
Isobutylene isoprene (butyl rubber)	Recommended	Recommended	Recommended
Isoprene	Unsatisfactory	Unsatisfactory	Unsatisfactory
Polyacrylate	Unsatisfactory	Unsatisfactory	Unsatisfactory
Polyisochloroprene (Neoprene)	Unsatisfactory	Unsatisfactory	Unsatisfactory
Polyurethane	Unsatisfactory	Unsatisfactory	Unsatisfactory
Silicone rubber	Acceptable	Acceptable	Acceptable
Styrene butadiene (BUNA S)	Unsatisfactory	Unsatisfactory	Unsatisfactory
Plastics			
ABS	Unsatisfactory	Unsatisfactory	Unsatisfactory
Acrylic	Unsatisfactory	Unsatisfactory	Unsatisfactory
Polyamide (Nylon)	Recommended	Recommended	Recommended
Polycarbonate	Acceptable	Acceptable	Acceptable
Polyester	Acceptable	Acceptable	Acceptable
Polyethylene	Acceptable	Acceptable	Acceptable
Polypropylene	Acceptable	Acceptable	Acceptable
Polystyrene	Unsatisfactory	Unsatisfactory	Unsatisfactory
Polysulfone	Unsatisfactory	Unsatisfactory	Unsatisfactory
Polyvinyl chloride (PVC)	Unsatisfactory	Unsatisfactory	Unsatisfactory
Polytetrafluoroethylene (PTFE)	Recommended	Recommended	Recommended
Paints and finishes			
Acrylic	Unsatisfactory	Unsatisfactory	Unsatisfactory
Alkyd resin (stoved)	Acceptable	Acceptable	Acceptable
Epoxy resin	Recommended	Recommended	Recommended
Latex	Unsatisfactory	Unsatisfactory	Unsatisfactory
Phenolic resins	Unsatisfactory	Unsatisfactory	Unsatisfactory
Polyurethane paint	Unsatisfactory	Unsatisfactory	Unsatisfactory
Vinyl ester	Recommended	Recommended	Recommended
Metals			
Aluminum	Acceptable	Acceptable	Acceptable
Brass	Acceptable	Acceptable	Acceptable
Bronze	Acceptable	Acceptable	Acceptable
Cadmium	Acceptable	Acceptable	Acceptable
Cast iron	Recommended	Recommended	Recommended
Copper	Acceptable	Acceptable	Acceptable
Magnesium	Acceptable	Acceptable	Acceptable
Nickel	Recommended	Recommended	Recommended
Steel (all grades)	Recommended	Recommended	Recommended

(continued)

TABLE 19.36 (Continued)
Phosphate Ester Compatibility with Standard Constructional Materials

Material	Triaryl Phosphate ester	Trialkyl Phosphate ester	Alkyl aryl Phosphate ester
Silver	Recommended	Recommended	Recommended
Titanium	Recommended	Recommended	Recommended
Zinc	Acceptable	Acceptable	Acceptable
Filter media			
Activated clays, (e.g., Fuller's earth)	Acceptable	Acceptable	Acceptable
Activated alumina	Acceptable	Acceptable	Acceptable
Activated carbon	Acceptable	Acceptable	Acceptable
Cellulose	Acceptable	Acceptable	Acceptable
Ion-exchange resins	Recommended	Recommended	Recommended
Gasket materials			
Cork	Unsatisfactory	Unsatisfactory	Unsatisfactory
Leather, impregnated	Acceptable	Acceptable	Acceptable
Graphite laminates/foils	Recommended	Recommended	Recommended
Proprietary flexible gaskets	Acceptable	Acceptable	Acceptable

Note: Materials listed as acceptable may be used under certain conditions; the manufacturer or fluid supplier should be consulted prior to use.

Different types of aryl (or alkyl) phosphate esters from different sources can normally be mixed. Indeed, mixtures of synthetic and natural triaryl phosphate fluids are in wide commercial use, as well as mixtures of trialkyl and triaryl phosphates. If there is any doubt, advice should be sought from the fluid or manufacturing site.

If a new phosphate fluid charge is to replace one which has been significantly degraded, then it is normally advisable to flush the system before filling with the new fluid. This need not be carried out using the same quality of fluid required for normal use, but the flushing fluid should be clean, dry, of a relatively low acidity (a value of 0.15 mg KOH/g maximum is regarded as acceptable), and the same chemical type as the operating charge. Fluid suppliers should be consulted for the availability of a suitable product. Details of appropriate flushing procedures can be found in fluid suppliers' or manufacturers' literature [97,98] or in national/international use guides (Appendix 1).

With regard to the compatibility of phosphates with other hydraulic fluid types, the following general rules apply when topping-up a system or changing from one fluid to another:

- Phosphate esters should never be added to systems containing mineral oil or water-containing fluids, and vice versa. Quite apart from the immiscibility of the two fluids, it is probable that, in the former case, the phosphate will be incompatible with the seals, paints and gaskets used in the system.
- Although phosphate esters have greater compatibility with carboxylate (polyol) esters, no addition of phosphate should be made without consulting the manufacturer because of possible incompatibility with seals, gaskets, and paints. Adding carboxylate esters to phosphate fluids is unlikely to be a problem from the seal compatibility aspect, but it will reduce fire resistance and probably both oxidative and hydrolytic stability. In addition, small amounts of polyol esters in phosphates can adversely affect the foaming properties.
- When changing from one fluid type to another, particularly from aqueous to non-aqueous or vice versa, certain system modifications may be necessary and prior discussions with the equipment builders are essential.

After modification, a system flush is necessary and recommendations on suitable procedures are available in relevant national/international use guides (e.g., ISO 7745 [Appendix 1]) or in fluid manufacturers' literature [97,98].

19.7 HEALTH, SAFETY, AND ENVIRONMENTAL BEHAVIOR

19.7.1 TOXICITY

The toxicity profile of the commercially available phosphate esters that are used in hydraulic fluids is essentially benign. However, some differences in behavior between fluids do exist and these are outlined below.

Despite the positive picture available from the existing data, misconceptions regarding the health and environmental hazards of phosphate esters persist. Some of the concern is due to confusion over nomenclature as the description "phosphate esters", in addition to generically describing the products used as hydraulic fluids, is also used in connection with pesticides, insecticides, and so forth. The similarity between the terms "phosphate esters" and "phosgene" (the nerve gas) has also led to completely incorrect conclusions. It must be emphasized, therefore, that the phosphate esters used in lubricant applications are not biologically active materials. However, many years ago, TCP was responsible for causing neurotoxicity when ingested [99]. This problem was initially thought to arise as a result of the presence of a specific isomer, tri-orthocresyl phosphate (TOCP). Subsequent investigations of other "natural" phosphates indicated a similar but less severe behavior and, for a period, it was assumed that all aryl phosphate isomers were neurotoxic. However, because of the lack of sophistication of the analytical techniques then available, most (if not all) of the aryl phosphates tested were impure and contained small amounts of either TOCP or other components containing *ortho*-cresyl substituents as contaminants. Some of these other *ortho*-substituted products, although present at low levels, were subsequently found to be even more neurotoxic than TOCP.

Structure–activity studies for neurotoxicity have since been completed on over seventy substituted triaryl phosphates [100]. Based on these studies, it has been found that the primary structural requirement for neurotoxicity is a substituent in the ortho position on the phenyl ring with at least one hydrogen atom on the alpha carbon atom, and the neurotoxicity increases with the number of hydrogen atoms on the ortho substituent. As a result, industry has since taken considerable care to reduce or eliminate *o*-cresol from commercially available feedstocks. Today, TCP is derived from a feedstock with extremely low levels (only a few parts per million) of *ortho*-cresol, whereas the synthetic triaryl phosphates are based on phenol obtained from petrochemical sources. This ensures that they have a very low neurotoxic potential [101]—significantly below levels which would require labeling the materials as hazardous.

In order to assess toxicity and ecotoxicity behavior, it is necessary to have an agreed set of test methods and limits for determining acceptable performance. Unfortunately, there is still no international agreement in the area and national requirements can differ. Health and safety data sheets must, of course, include any data which might suggest the product is toxic.

In Europe, the principal health hazards associated with the use of hydraulic fluids are identified in CEN Technical Report 14489 "Fire-resistant hydraulic fluids—Classification and specification—Guidelines on selection for the protection of safety, health, and the environment", and are as follows:

- Acute oral toxicity (LD_{50})
- Skin and eye irritation
- Toxicity of hot and cold aerosols
- Toxicity of thermal decomposition products
- Allergic reactions or other results of repeated or long-term exposure

The standard states that the health hazards which are potentially present in the situation under examination shall be determined.

Triaryl phosphates readily meet these requirements. However, although trialkyl phosphates have not been fully evaluated, some products are known to be skin irritants. The test methods to be used for assessing the above properties are included in European Council Directive 67/548/EEC.

In addition, European legislation on product labeling (Directive 93/21/EEC) [102] requires that its toxicity be assessed and the product labeled accordingly. Included in the acute tests accepted for classification purposes are the following:

- Acute oral toxicity (LD_{50})
- Skin and eye irritation
- Skin sensitization
- Mutagenicity data (Ames test)

Most of the data on triaryl phosphates given in Table 19.35 [103] have been obtained against the CEN/TR 14489 guidelines. Data on the alkyl and alkyl aryl products may have been produced under dissimilar test conditions and, therefore, not be strictly comparable [104–109].

It should also not be assumed that the highest LD_{50} data automatically equate with the lowest acute toxicity. In order to reduce animal tests, products today, are evaluated at much lower dose levels than in the past. A product with a test pass at a single dose of 2 g/kg is now accepted as having a very low acute toxicity.

In general, the results tabulated on acute toxicity are low. The skin irritant behavior of TBP and TiBP, however, means that they require additional care in their handling and use. As a result, TBP and TiBP currently require a hazard warning label under EU guidelines.

19.7.2 ECOTOXICITY

CEN/TR 14489 also indicates procedures for assessing the environmental behavior of hydraulic fluids. As a minimum, it advises the production of acute fish toxicity and information on biodegradability. Methods for assessing acute Daphnia toxicity, bacterial toxicity, and bioaccumulation are detailed for possible use by competent authorities when determining their specific requirements.

Included in the recommendations is the determination of the Water Hazard Classification. This rating scheme originated from the chemical industry in Germany (where it is known as the Wassergefährdungsklasse, or WGK [110]) and uses acute toxicity and ecotoxicity data to rank products in terms of their pollution potential. The categories range from 0 (no pollution potential) to 3 (severe pollution potential) and most phosphates fall into category 1 (slight potential) and are, therefore, similar to or better than mineral oil in this respect (which falls into categories 1–2).

In Europe, under Directive 93/21/EEC [102], the environmental behavior of chemicals has to be assessed for labeling purposes. This takes into consideration biodegradability data on the fluid, together with acute fish, daphnia and algal toxicity.

The ecotoxicity data on phosphate esters are given in Table 19.38 [106–109,111,112]. Where adequate data exist, none of the products appears to have a serious pollution potential. It should be noted, however, that testing of some products which have a very low solubility in water (for example, aryl phosphates), can give misleading results if dispersants are used to "solubilize" the fluids. These additives can, themselves, influence the data—hence, the variation in fish toxicity data on IPPP/46. It has now also been found that the daphnia toxicity tests for insoluble products exhibit anomalous results due to the physical immobilization of the organisms rather than as a result of chemical "poisoning".

Biodegradability data on TIPP, TIBP and TXP under OECD 301F manometric respirometer test conditions are given in Table 19.39. In this test biodegradation is measured as the net oxygen uptake over that occurring in blank tests containing only inoculated medium. The extent of biodegradation is calculated from the mass of test material added to the test vessels and its theoretical oxygen demand for complete biodegradation. The test was carried out in triplicate on the ISO VG 46

TABLE 19.37
Toxicity Behavior of Phosphate Esters

Property					Fluid				
	TBP	TiBP	TBEP	TOP	ODPP	DDPP	TXP	IPP/46	TBPP/46
Acute toxicity LD_{50} (g/kg)	1.2	>5	>3	>5	>15	>15	>30	>20	>30
Irritancy effects									
-Skin	Irritant	Irritant	None	Irritant	Slight	Moderate	None	None	None
-Eye	None	None	None	None	Slight	Slight	None	None	None
Exposure to hot and cold aerosols.	–	–	–	No effect	–	Moderate	No effect	No effect	No effect
Toxicity of thermal decomposition products.	–	–	–	–	–	–	Moderate–reversible	Moderate–reversible	Moderate–reversible
Neurotoxicity	Negative	Negative	Negative	Negative	–	Negative	Negative >10 g/kg	Negative >25 g/kg	Negative >25 g/kg
Mutagenicity	Equivocal	–	Negative	Negative	–	Negative	Negative	Negative	Negative
Reprotoxicity	–	–	–	–	–	–	Positive-Reversible	Positive-Reversible	Negative

TABLE 19.38
Ecotoxicity Behavior of Phosphate Esters

Property	Fluid							
	TBP	TBEP	TOP	ODPP	DDPP	TXP	IPPP/46	TBPP/46[b]
Biodegradability	Moderate–high	Readily	Readily	Readily	–	Inherently	Inherently	Readily
Acute fish toxicity (mg/L)	(2.4–13)	Moderate (16–44)	Low–moderate (20–>100)	High	Moderate (7.6–18)	Low–high (0.16–>100[a])	Low (>1,000)[a]	(High) (1–10)[c]
Acute Daphnia toxicity (mg/L)	High (2.6)	Low (>75)	Moderate (6.5–36.5)	–	High (0.48)	Moderate 3.9[a]	Low >100[a]	(Moderate) (>10)[c]
Algal toxicity (mg/L)	–	–	–	–	Low (71–79)	–	–	–
Bacterial toxicity (mg/L)	–	–	>100	>10,000	–	–	>100[a]	–
WGK value: water pollution hazard	2	1	2	2	–	1	1	(1)[c]
EU environmental label required?	No	No	No	–	–	Yes	No	No

[a] Values are well above the solubility of product in water at ambient temperature.
[b] Values are based on product manufactured according to reaction 19.2.
[c] Values given are based on the evidence available to date.

TABLE 19.39
OECD 301F Biodegradability Data on Different Types of Aryl Phosphate

Product (ISO VG 46 base stocks)	% Biodegradability after:		
	10 Days	28 Days	68 Days
Trixylyl phosphate (TXP)	5	29	70
Isopropylatedphenyl phosphate (IPPP)	18	47	65
Tertiarybutylatedphenyl phosphate (TBPP)	25	62	72

grades of the different types of aryl phosphates manufactured according to reaction scheme (19.2) (see Section 4).

The results are initially in the order of their hydrolytic stability but it is interesting to note that TXP, after a slow start, eventually reaches the same level as the synthetic fluids and may possibly have exceeded them had the test been extended.

In view of this data, the tertiarybutylatedphenyl phosphate would be regarded as readily biodegradable (Pw1) while the TXP and isopropylatedphenyl phosphate would be classified as inherently biodegradable (Pw2).

The low water pollution potential of phosphates has been confirmed by a U.S. industry aquatic surveillance program, which sampled river water close to sites where phosphates were manufactured or used. The results showed an absence of a significant concentration of aryl phosphates in surface water [113]. A study carried out by the Japanese authorities produced similar findings [114].

In spite of relatively benign ecotoxicity of the "higher" viscosity grades of aryl phosphates, these products are classified as marine pollutants according to the UN Marine Pollutant Classification. However, because they are used at low concentrations they are unlikely to contribute significantly to the ecotoxicity of the finished product.

Testing continues on many of these products, particularly in view of the difficulty in evaluating products that have a very low solubility in water, and the user is advised to consult the fluid supplier for the latest health, safety, and environmental information.

19.7.3 DISPOSAL OF USED PHOSPHATE ESTERS

One aspect of the use of phosphate esters that has become increasingly important to end users is their disposal, normally as a result of excessive degradation but also due to contamination. Depending on the extent of degradation and/or contamination, reclamation by the fluid supplier or by a specialist reclaimer may be feasible, and several alternative procedures are available. The most widely used is high-temperature incineration, either in admixture with mineral oil in an industrial incinerator, or possibly in a cement kiln. An alternative technique has been to hydrogenate the phosphate and use the product as a component of fuel. Landfill, although no longer environmentally favored, is still occasionally used in secure and approved sites, and phosphates are said to be beneficial in assisting the biodegradation process by acting as a nutrient for the bacteria. In all cases, the local regulations regarding the disposal of chemicals must be met.

The disposal or recycling of used fluid normally requires transportation of the fluid to the disposer or reclaimer. However, strict regulations now control the labeling and transportation of waste—particularly if this involves movement across national borders and the waste is classified as hazardous. As the regulations are still evolving and vary from country to country, it is essential that the user consult an authorized waste disposal company, as well as the fluid supplier, when considering the most appropriate means of disposal.

Obviously, the quantity of phosphate fluid requiring disposal can be minimized by good maintenance and particularly by the use of in situ conditioning.

19.8   ADDITIVE SOLVENCY AND THE FORMULATION OF HYDRAULIC FLUIDS

The polar nature of phosphate esters makes them good solvents for most types of additives, with the possible exception of non-polar products based on aliphatic hydrocarbons (e.g., polymeric thickeners).

In the same way that additives are used to improve the performance of mineral oil in many applications, it is also possible to improve the behavior of phosphate esters [87]. Not all additive types have a beneficial effect and their use will depend on the application. For example, polymeric thickeners and other highly polar additives, such as rust or corrosion inhibitors, can promote foaming, emulsification, and a reduction in volume resistivity. They are also removed quite rapidly from solution by some adsorbent solids used in fluid conditioning. The benefits of these additives—in certain applications—are therefore outweighed by their disadvantages. On the other hand, the use of stabilizers (i.e., antioxidants, metal passivators, and acid scavengers) can be very beneficial, particularly in applications where no on-line conditioning is available. Small amounts of antifoams and dyes (for identification purposes or to detect leaks) may also be added to phosphate esters. Dyed fluids, however, can be more difficult to reclaim for future use because of the variety of colors produced. The dyes can also interfere with the measurement of acidity by colorimetric techniques.

19.9   FLUID OPERATING LIFE

The economics of using a hydraulic fluid should not only take purchase price into consideration but also other fluid-related variables, including operating life, maintenance, and disposal costs. However, predicting fluid life is difficult because of the use of different system designs and the varying quality of both system and fluid maintenance, but the important factors that affect fluid life in service can be identified, as indicated in Figure 19.20.

19.9.1   System Design

Many aspects of system design can affect fluid life. These include operating pressures and temperatures, tank design and fluid circulation rate, the type and design of the pump, the routing of the pipe work, and so forth [115]. If a problem arises with fluid performance, the actual origin of the problem is often design related.

19.9.2   Leakage/Top-up Rates

In heavy engineering processes, leakage rates can be high. Obviously, it may be less costly to suffer leaks than to shut down equipment and effect repairs. If a low-cost fluid is being used, there is also less incentive to ensure that adequate maintenance is carried out. In some cases, the rate of leakage is so high that the fluid is never physically replaced; instead, the amount of top-up compensates for any fluid degradation and depletion of additives.

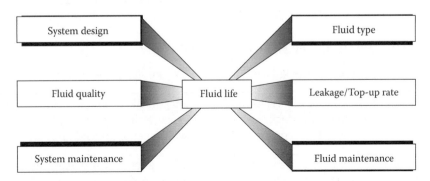

FIGURE 19.20   Major factors influencing fluid life.

19.9.3 FLUID TYPE

As indicated earlier in this chapter, the chemistry of the different fluid types available can play an important part in deciding their suitability for specific applications. Thus, in an environment where significant water contamination is likely, a hydrolytically stable product should be used.

19.9.4 FLUID QUALITY

To control the quality of fluids used in hydraulic systems, there are national and international specifications available which stipulate limits on the most important fluid parameters. A list of these standards is given in Appendix 1. In some cases there are also limits on the quality of used fluid.

19.9.5 SYSTEM MAINTENANCE

It is widely accepted that effective mechanical filtration significantly improves the life of system components and, in certain cases, of the fluid itself. Mention has already been made of the adverse effects of solid particles on resistivity, foaming, and demulsification. Depending on the size and hardness of the particles, the wear performance of the fluid can also be impaired. The use of filters appropriate to the degree of cleanliness required for the correct functioning of the system and to the tolerances found in system components is therefore essential.

19.9.6 FLUID MAINTENANCE

In service, phosphate esters can degrade either due to contamination or through interaction with their environment (e.g., through thermal stress). In practice, there is considerable overlap between the effect of contaminants and the degradation caused by system operation. Indeed, one can lead to the other. If, for example, the phosphate is contaminated by mineral oil, the air content of the fluid is increased. This, in turn, can lead to fluid oxidation by compression of the air bubbles in the pump. Although fluid maintenance is often less for phosphate esters than for other types of fire-resistant fluids, it is still necessary to monitor and control the contamination of the fluid—whether from internal sources such as wear particles or from the external environment (e.g., moisture, dust, etc.)—to ensure satisfactory operation of equipment and extend fluid life [116–118].

In the case of phosphate esters, considerable attention has also been given to controlling the generation of acidic degradation products, as they normally impair the performance and life of the fluid. This aspect is discussed in greater detail in Section 11.

The most common contaminants are, of course, water, dirt, and mineral oil. Even air (certainly when in excess) can be regarded as a contaminant. Less frequently found, but still detrimental, are chlorinated solvents, as they can promote servo-valve erosion [119] when they are converted to ionic chloride. When present in significant quantities they can also make disposal more difficult.

The principal effects of contaminants on phosphate ester performance are shown in Figure 19.21 [115].

19.10 MONITORING OF FLUID CONDITION

The following determinations are routinely carried out on phosphate ester fluids:

Viscosity change is often used as a measure of oxidative degradation. However, phosphate esters do not normally display significant viscosity changes in use. If such changes are noted in a short period, they are more likely to result from contamination (e.g., with water or mineral oil).

Acidity is an indicator of both oxidative and hydrolytic breakdown. It is perhaps the single most important parameter for monitoring phosphate ester condition. A decision to replace the fluid or the conditioning media is usually made on the basis of this determination.

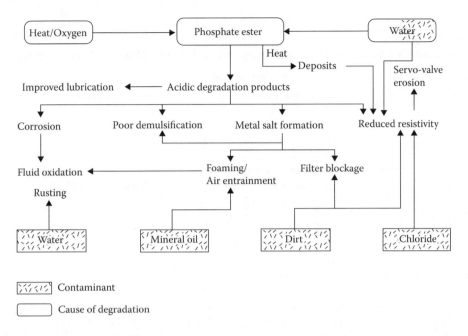

FIGURE 19.21 Impact of contaminants and fluid degradation products on phosphate ester performance.

Water content is monitored in view of the sensitivity of esters to moisture and the subsequent development of acidity.

Particulate levels are monitored for the reasons outlined in Section 19.5.

Appearance is a simple check to see if the color is changing quickly, which may indicate excessive degradation or, if the fluid is becoming turbid, of water contamination.

These checks are required by equipment builders for general fluid applications. Additional tests, such as foaming, air release, and volume resistivity, would be required for specialist uses. The frequency of testing will depend on the severity of the operating conditions and the sophistication of the equipment (e.g., whether fine tolerance valves are used). The testing will also be more frequent during commissioning of the system or after refurbishment. In normal operation, samples of phosphate esters might be evaluated on a monthly or quarterly basis, but not all the tests would necessarily have the same frequency. Guidance on the frequency of testing and proposed operating limits is given in the standards listed in Appendix 1. In addition to these national/international standards, individual equipment builders frequently have their own specifications and it is important for the end user to follow the manufacturer's requirements, particularly during the warranty period. A general list of methods used to assess fluid quality is given in Appendix 2.

19.11 CONTROL OF FLUID DEGRADATION AND PERFORMANCE

As contaminants are often responsible for initiating fluid degradation (oxidative and hydrolytic) or impairing the physical/chemical properties, which then lead to reduced performance, one method of maintaining fluid quality and extending fluid life is to remove the contaminants as soon as they enter the system. (It is, of course, preferable to avoid their entry, but this is not completely possible because of system design and, often, inadequate maintenance.) There are techniques for removing most of the common contaminants either on-line or off-line. Table 19.40 [120] lists the most widely used methods together with those recommended for removing acidic degradation

TABLE 19.40
Common Hydraulic Fluid Contaminants and Techniques Used for Their Removal

Contaminant	On-line removal	Off-line removal
Water	Vacuum dehydration	Vacuum dehydration
	Water-absorbing polymers	
	Adsorbent solids	
	Physical separation (e.g., centrifuging, siphoning)	
	Increased bulk fluid temperature	
Acid	Adsorbent solids (e.g., fuller's earth, ion-exchange resins)	Water or alkali washing
Particulates	Cellulose or glass-fibre filters	
	Adsorbent solids	
	Electrostatic precipitation/filtration	
	Magnetic filtration	
Chlorinated solvents	Vacuum techniques	Distillation
	Degradation products can be removed by adsorbent solids	
Mineral oil	No suitable procedure for removing small quantities	Precipitation by cooling
		Distillation
	Large amounts may be removed by centrifuging	

products. With the exception of mechanical filtration, most of the techniques are rarely used. Only in power-generation applications with phosphate esters has there been any attempt to introduce on-line conditioning to date. The reasons for this include cost and the fact that it makes little sense in systems with high leakage rates. In the power-generation industry, however, leakage rates are more tightly controlled and the treatment significantly improves the cost-effectiveness of the fluid. Steam turbine hydraulic systems are also more likely than most to suffer from water ingress, and with this treatment fluid life can be considerably extended—indeed, experience to date suggests that with the correct treatment, adequately sized for the volume of fluid, there may be no need to replace the fluid during the life of the turbine. Certainly, fluid is now lasting over twenty-five years and is still in excellent condition after this time. As a result of an increasing emphasis on reducing process costs and the expense of disposing of waste fluid, a greater use of on-line conditioning in other industries seems inevitable.

The use of adsorbent solids to remove acid degradation products has not been without its problems. For many years, fuller's earth and activated alumina have been used to remove the acid from phosphates by circulating a small quantity of fluid on a bypass loop from the main reservoir through the adsorbent solid. Unfortunately, the solids contain impurities which react with the acid degradation products to form metal soaps, which are initially soluble. These soaps, which can reach high molecular weights [120,121], adversely affect the surface-active properties of the fluid and, as a result of a deterioration in the air-release properties, fluid oxidation increases with the generation of more acid [121,122]. Thus, a cycle of degradation (Figure 19.22) [120] begins. Under certain conditions, the soaps precipitate and form gelatinous deposits in the tank, valves, filters, and so forth. Systems that seem clean when operational can, on cooling, show evidence of deposits. These problems have more frequently been associated with the "synthetic" phosphates, probably because of their inferior hydrolytic stability.

Fortunately, new techniques for acid removal, such as the use of ion-exchange resins, are now available which do not have the above-indicated side effects [115,123], and where these are in use, phosphate ester life has been extended to such an extent that it approaches that of the equipment.

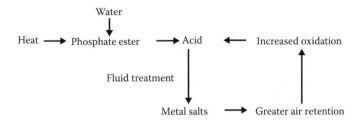

FIGURE 19.22 The fluid degradation cycle.

Quite apart from the savings on the cost of replacement fluid, the difficulty (and expense) of the disposal of used material should now be almost eliminated.

19.12 APPLICATIONS

Phosphate ester fire-resistant hydraulic fluids are ideally suited for most industrial processes involving high temperatures, or where system designs are unsuitable for the use of water-based fluids. These applications include the following:

Continuous casting machines	Hydraulic mining equipment
Die-casting machinery	Ship control systems
Hydraulic presses	Gate valve actuators
Centrifugal compressors	Steam turbine governor systems
Glass-working machinery	Aircraft hydraulic systems
Hydraulic jigs	Clay guns
Injection-molding machines	Ingot manipulators
Oven controls	Foundry equipment
Automatic welders	Forklift trucks
Billet loaders	Hydraulic lifts
Furnace mechanisms	Aerial bucket trucks
Gas turbine hydraulics	Fluid couplings
Hot-roll mills	

There are also some applications where the fluid serves as a combined hydraulic fluid and lubricant; for example, in steam and gas turbines [89].

REFERENCES

1. Williamson and Scrugham, Ann., 1854, 92, pp. 316.
2. Vogeli, F. Ann., 1849, 69, pp. 190.
3. Egan, E.G. "A Synthetic Lubricant for Hydraulic Fluid," *Lubr. Eng.*, 1947, 3, February–March, 1947, pp. 24–26.
4. Beeck, O. Givens, J.W. and Williams, E.C. *Proc. Roy. Soc.*, 1940, 177A, pp. 103–18.
5. Sullivan, M.V., Wolfe, J.K. and Zisman, W.A. "Flammability of the Higher Boiling Liquids and their Mists," *Ind. Eng. Chem.*, 1947, 39(12), pp. 1607–14.
6. Murphy, C.M. and Zisman, W.A. "Synthetic Hydraulic Fluids," *Product Eng.*, 1950, September, pp. 109–13.
7. Watson, F.J. U.S. Patent 2,549,270 (1951), U.S. Patent 2,636,861 (1953) (to Shell Development Co.).
8. Moreton, D.H. "*Development and Testing of Fire-Resistant Hydraulic Fluids*," SAE Technical Paper, Paper 490229, 1949.
9. Moreton, D.H. U.S. Patent 2,566,623 (1951), U.S. Patent 2,834,733 (1958), U.S. Patent 2,894,911 (1959) (to Douglas Aircraft Co.).
10. Gamrath, H.R. Hatton, R.E. and Weesner, W.E. "Chemical and Physical Properties of Alkyl Aryl Phosphates," *Ind. Eng. Chem.*, 1954, 46, pp. 208–12.

11. Gamrath, H.R. and Craver, J.K. U.S. Patent 2,596,140 (1952), U.S. Patent 2,596,141 (1952) (to Monsanto Chemical Co.).
12. Gamrath, H.R. and Hatton, R.E. U.S. Patent 2,678,329 (1954) (to Monsanto Chemical Co.).
13. Hamilton, W.F., George, M.F. and Weible, G.B. U.S. Patent 2,392,530 (1946) (to Lockheed Aircraft Co.).
14. George, M.F. and Reedy, P. U.S. Patent 2,659,699 (1953) (to Lockheed Aircraft Co.).
15. U.S. Military Specifications MIL-F-7100 (1950) and MIL-H-83306 (1971).
16. U.K. Aircraft Material Specifications DTD 5507 (1956) and DTD 5526 (1962).
17. U.S. Military Specification MIL-H-19457 (Ships); originally published 1961.
18. Gamrath, H.R. and Hatton, R.E. U.S. Patent 2,698,537 (1955), U.S. Patent 2,707,176 (1955) (to Monsanto Chemical Co.).
19. Morgan, J.D. U.S. Patent 2,410,608 (1946) (to Cities Service Oil Co.).
20. Stauffer Chemical Co., "*Cellulube® Fire-Resistant Fluids and Lubricants*," *Technical Bulletin*, Stauffer (now Akzo) Chemical Co., 1966.
21. Houghton, E.F. & Co., "Fire-Resistant Hydraulic Fluids," *Technical Bulletin*, E. F. Houghton & Company, Philadelphia, PA, 1960.
22. The Geigy Co., "Fire-Resistant Hydraulic Fluids," *Technical Service Bulletin*, The Geigy Co., Ltd., Manchester, UK, 1962.
23. Monsanto Chemical Co., "Pydraul®150," Technical Bulletin, 1959; "Pydraul®625," Technical Bulletin, 1959; "Pydraul®AC," *Technical Bulletin*, 1957; "Pydraul®F-9," *Technical Bulletin*, 1959; Monsanto Chemical Co., St. Louis, MO.
24. European Council Directive 76/769/EEC, Eleventh Amendment, L186, 12/7/91, p. 64.
25. Morgan, J.D. and Lowe, R.E. U.S. Patent 2,395,380 (1946), U.S. Patent 2,396,161 (1946); U.S. Patent 2,396,192 (1946), U.S. Patent 2,409,443 (1946); U.S. Patent 2,409,444 (1946); U.S. Patent 2,423,844 (1947) (to Cities Service Oil Co.)
26. Stauffer Chemical Co., "Fyrlube® Fire-Resistant Hydraulic Fluids," *Technical Bulletin, Stauffer Chemical Co.*, Westport, CT.
27. Stauffer Chemical Co., "Fyrtek® Fire-Resistant Hydraulic Fluid," *Product Data Sheet, Stauffer Chemical Co.*, Westport, CT.
28. Houghton, E.F. & Co., "Vital® Hydraulic Fluid 29," *Technical Data Sheet*, E. F. Houghton & Co., Philadelphia, PA, 1971.
29. Stuart, D.A. Co., "DASCO FR 300 Fire-Resistant Hydraulic Fluids," *Product Data Bulletin*, D. A. Stuart Co., Chicago, IL, 1970.
30. Phillips, W.D. and Schade, R. "Nicht-wässrige Schwerbrennbare Hydraulikflüssigkeiten," in *Hydraulikflüssigkeiten*, W. J. Bartz, ed., Expert Verlag, Renningen-Malmsheim, Germany, 1995.
31. Prahl, W.H. "Triaryl esters of orthophosphoric acid," British Patent No. 763,311 (1956).
32. Marino, M.P. "Phosphate Esters," in *Synthetic Lubricants and High-Performance Functional Fluids*, Shubkin, R.L. ed., Marcel Dekker; New York, 1992.
33. Garrett, K.M. British Patent 1,165,700 (1965) (to Bush Boake Allen Ltd.).
34. Randell, D.R. and Pickles, W. British Patent 1,146,173 (1966) (to J. R. Geigy A. G.).
35. FMC Corporation (UK) Ltd., unpublished data.
36. Hombek, R. and Marolewski, T.A. US Patent 6,242,631 (Akzo Nobel NV)
37. *Chemical Economics Handbook,* SRI International, Menlo Park, CA, 1993.
38. Phillips, W.D. "Fire-Resistance Tests for Fluids and Lubricants—Their Limitations and Misapplication", ASTM STP1284 (1996).
39. FMC Corp., "Trioctyl Phosphate," *Technical Data Sheet*, FMC Corporation, Philadelphia, PA, 1991.
40. FMC Corp., "Tributyl Phosphate," *Technical Data Sheet*, FMC Corporation, Philadelphia, PA, 1987.
41. ASTM Test Method D240–92—Standard Test Method for Heat of Combustion of Liquid Hydrocarbon Fuels by Bomb Calorimeter 1992.
42. Chemtura Corp., "Reolube® HYD Fluids," *Technical Bulletin*, Manchester, UK, 2005.
43. Supresta LLC, "Fyrquel® Fire-Resistant Hydraulic Fluids," Product Bulletin, Supresta LLC, Ardsley, NY, USA, 2008.
44. Houghton, E.F. & Co., "Houghto-Safe® 1120," *Technical Bulletin*, E. F. Houghton & Co., Philadelphia, PA, 1990.
45. Placek, D.G. and Marino, M.P. "Phosphate Esters," in *STLE/CRC Tribology Data Handbook, Synthetic Oil Properties,* CRC Press; Boca Raton, FL, 1996.
46. Beyer, R. and Bauer, K. "Schwerentflammbarkeit von Drückflüssigkeiten beim Kontakt mit Metallschmelzen," *Sonderdruck Ölhydraulik and Pneumatik*, 32, 6, 1988.

47. Khan, M.M. and Brandao, A.V. "Method for Testing the Spray Flammability of Hydraulic Fluids," *SAE Technical Paper*, Paper 921737, 1992.
48. Anon., "Unveiling a New Protocol for Less Flammable Hydraulic Fluids," Factory Mutual Approved Product News, 1996, pp. 6–8.
49. ISO/TS 15029-2:2011 Petroleum and Related Products – Determination of Spray Ignition Characteristics of Fire-Resistant Fluids.
50. Yule, A.J. and Moodie, K. "A Method for Testing the Flammability of Sprays of Hydraulic Fluid," *Fire Safety J.*, 1992, 18, pp. 273–302.
51. Data provided by Health and Safety Executive, Buxton, U.K. and by the Laboratorie Centrale de Houilleres du Bassin de Lorraine, Marienau, France.
52. *Solutia* Inc., "Skydrol 5 High Temperature Hydraulic Fluid," *Solutia*, 2006.
53. Barcroft, F.T. and Daniel, S.G. "The Action of Neutral Organic Phosphates as E. P. Additives," *ASME Trans.*, 1964, 64-Lub 22.
54. Godfrey, D. "The Lubrication Mechanism of Tricresyl Phosphate on Steel," *ASLE Preprint* 64 LC-1, 1964.
55. Bieber, H.E., Klaus, E.E. and Tewkesbury, E.J. "A Study of Tricresyl Phosphate as an Additive for Boundary Lubrication," *ASLE Trans.*, 1968, 11, pp. 155–161.
56. Saba, C.S. and Forster, N.H. "Reactions of Aromatic Phosphate Esters with Metals and Their Oxides', *Tribol. Lett.*, 2002, 12(2), pp. 135–146.
57. Perez, J.M., Hansen, R.C. and Klaus, E.E. "Comparative Evaluation of Several Hydraulic Fluids in Operational Equipment, A Full-Scale Pump Stand Test and the Four-Ball Wear Tester. Part II. Phosphate Esters, Glycols and Mineral Oils," *Lubr. Eng.*, 1990, 46(4), pp. 249–255.
58. Yamamoto, Y. and Hirano, F. "Scuffing Resistance of Phosphate Esters," *Wear*, 1978, 50, pp. 343–348.
59. Phillips, W.D. "A Comparison of Fire-Resistant Hydraulic Fluids for Hazardous Industrial Applications," *Journal of Synthetic Lubrication*, 1998, 14(3), pp. 211–235.
60. Hobbs, R.A. "Fatigue Lives of Ball Bearings Lubricated with Oils and Fire-Resistant Fluids," in *Elastohydrodynamic Lubrication Symposium*, Institute of Mechanical Engineers, 1972, paper C1.
61. Hobbs, R.A. and Mullett, G.W. "Effects of Some Hydraulic Fluid Lubricants on the Fatigue Lives of Roller Bearings," *Proc. Inst. Mech. Eng.* (1968–69), 183 (Pt. 3P), pp. 23–29.
62. Culp, D.V. and Widner, R.L. "The Effect of Fire-Resistant Hydraulic Fluids on Tapered Roller Bearing Fatigue Life," SAE Technical Paper, Paper 770748, 1977.
63. Yardley, E.D., Kenny, P. and Sutcliffe, D.A. "The Use of Rolling Fatigue Test Methods over a Range of Loading Conditions to Assess the Performance of Fire-Resistant Fluids," *Wear*, 1974, 28, pp. 29–47.
64. Michaelis, K. "Reibungs, Verschleiss und Fresshalten natürlicher Phosphatester in Zahnradgetrieben," *Antriebstechnik*, 1988, 27(8), pp. 43–46.
65. Winter, H. and Michaelis, K. "Gutachtliche Beurteilung der Fresstragfähigkeit von Reolube® Turbofluid 46," Internal Report No. 1313, *FZG Institute*, München, Germany, 1983.
66. Michaelis, K. "Efficiency of Gears Lubricated with Phosphate Esters," private communication, 1995.
67. Peeken, H. and Niester, Th. "Verträglichkeit des Synthetischen Schmierstoffes, Reolube® 46T mit dem Gleitlagerwerkstoff Tego V738," Internal Report 9/89 Institute für Maschinenelemente und Maschinengestaltung, Aachen, 1989.
68. Borsoff, V.N. "Wear Studies with Radioactive Gears," *Lubr. Eng.*, 1956, 12, pp. 24–28.
69. Ferro corp., "Santicizer 141," Technical Data Sheet, Polymer Additives Div., Ohio, USA, 2010.
70. Ferro Corp., "Santicizer 148," Technical Data Sheet, Polymer Additives Div., Ohio, USA, 2010.
71. Gamrath, H.R., Horton, R.E. and Weesmer, W.E. "Chemical and Physical Properties of Alkyl Aryl Phosphates," *Ind. Eng. Chem.*, 1954, 46(1), pp. 208–212.
72. Fowle, T.I. "Aeration in Lubricating Oils," in *Tribology International*, 1981, 14(3), pp. 151–157.
73. Mobil Oil Co., "Foaming and Air Entrainment in Lubrication and Hydraulic Systems," Mobil Oil Co., Mobil Oil Co. Ltd., London, UK, 1971.
74. Döllinger, L. and Vogg, H."Untersuchungen zum Luftaufnahme-und-Abgabeverhalten (LAAV) von Schmierölen und Hydraulikflüssigkeiten in Maschinenanlagen," DGMK Forschungsber., 1981, 221.
75. Schöner, W. "Betriebliche Luftgehaltsmessungen in Schmier-und-Steuerflüssigkeit-skreisläufen," *Elektrizitätswirtschaft*, 1982, 81 (17/18), pp. 564–567.
76. Hatton, D.R. "Some Practical Aspects of Turbine Lubrication," *Can. Lubr. J.*, 1984, 4(1), pp. 3–8.
77. Hayward, A.T.J. "The Compressibility of Hydraulic Fluids," *J. Inst. Petrol.*, 1965, 51, pp. 35–47.
78. Smith, L.H., Peeler, R.L. and Bernd, L.H. "Hydraulic System Bulk Modulus—Its Effect on System Performance and Techniques for Physical Measurement," in *16th National Conference on Industrial Hydraulics*, 1960, Vol. 14, pp. 179–197.

79. Scott, W. "The Influence of Fire-Resistant Hydraulic Fluids on System and Component Design," *Wear*, 1979, 56, pp. 105–121.

80. Blake, E.S., Hammann, W.C., Edwards, J.W., Reichard, T.E. and Ort, M.R., "Thermal Stability as a Function of Chemical Structure," *J. Chem. Eng. Data*, 1961, 6, 87–98.

81. Raley, C.F. Jr., WADC Technical Report 53-337, Wright Air Development Center, Wright-Patterson Air Force Base, OH, 1955.

82. Lhomme, V., Bruneau, C., Brault, A., Chevalier, G. and Soyer, N. "Degradation Thermique du Tributylphosphate et de quelques Homologues," Report C. E. A., R-5095, Service de documentation du C. E. N. Saclay, Gif-sur-Yvette, France, 1981.

83. Lhomme, V., Bruneau, C., Soyer, N. and Brault, A. "Thermal Behavior of Some Organic Phosphates," *Ind. Eng. Chem. Prod. Res. Dev.*, 1984, 23(1), pp. 98–102.

84. Shankwalkar, S.G. and Cruz, C. "Thermal Degradation and Weight Loss Characteristics of Commercial Phosphate Esters," *Ind. Eng. Chem. Res.*, 1994, 33(3), pp. 740–743.

85. Paciorek, K.L., Kratzer, R.H., Kaufman, J. and Nakahara, J.H. "Coal Mine Combustion Products, Identification and Analysis," U.S. Bureau of Mines Open File Report 104-77, 1976.

86. Lohrentz, H.-J. "Die Entwicklung extrem hoher Temperaturen in Hydrauliksystemen und die Einflüsse dieser Temperaturen auf die Bauteile und ihre Funktionen," Mineralöltechnik, 13, 14/15, 1968; for Mahoney, Barnum et al it is Proc. 5th World Petroleum Congress, May 30–June 5, 1959, pp147–161, New York, USA.

87. Cho, L. and Klaus, E.E. "Oxidative Degradation of Phosphate Esters," *ASLE Trans.*, 1979, 24(1), pp. 119–124.

88. Shankwalkar, S.G. and Placek, D.G. "Oxidation and Weight Loss of Commercial Phosphate Esters," *Ind. Eng. Chem. Res.*, 1992, 31, pp. 1810–1813.

89. Phillips, W.D. "Triaryl Phosphates—The Next Generation of Lubricants for Steam and Gas Turbines," ASME Paper 94-JPGC-PWR-64, 1994.

90. Anzenberger, J.F. Sr. "Evaluation of Phosphate Ester Fluids to Determine Stability and Suitability for Continued Service in Gas Turbines," ASLE Paper 86-AM-IE-2, 1986.

91. Westheimer, F.H. "The Hydrolysis of Phosphate Esters," *Pure Appl. Chem.*, 1977, 49, pp. 1059–1067.

92. (a) Mhala, M.M. and Patwardhan, M.D. "Hydrolysis of Organic Phosphates, I. Hydrolysis of p-chloro-m-tolyl Phosphate," *Indian J. Chem.*, 1968, 6(12), pp. 704–707. (b) Mhala, M.M., Patwardhan, M.D. and Kasturi, T.R. "Hydrolysis of Organic Phosphates. II. Hydrolysis of p-chloro- and p-bromophenyl Orthophosphates," *Indian J. Chem.*, 1969, 7(2), pp. 145–148. (c) Mhala, M.M., Holla, C., Kasturi, G. and Gupta, K. "Hydrolysis of Organic Phosphates. III. Hydrolysis of o-methoxy-, p-methoxy- and p-ethoxy-phenyl Dihydrogenphosphates," *Indian J. Chem.*, 1970, 8(1), pp. 51–56. (d) Mhala, M.M., Holla, C., Kasturi, G. and Gupta, K. "Hydrolysis of Organic Phosphates. IV. Hydrolysis of di-o-methoxy-, di-p-methoxy-, and di-p-ethoxyphenyl Hydrogen Phosphates," *Indian J. Chem.*, 1970, 8(4), pp. 333–336. (e) Mhala, M.M. and Prabha, S. "Hydrolysis of Organic Phosphates. V. Hydrolysis of 2, 3-dimethoxyphe-nyl Dihydrogen Phosphate," *Indian J. Chem.*, 1970, 8(11), pp. 972–976. (f) Mhala, M.M. and Saxena, S.B. "Hydrolysis of Organic Phosphates. VI. Hydrolysis of Monoallyl Orthophosphate (Disodium Salt)," *Indian J. Chem.*, 1971, 9(2), pp. 127–130. (g) Mhala, M.M. and Saxena, S.B. "Hydrolysis of Organic Phosphates. VII. Hydrolysis of Diallyl Orthophosphate (Sodium Salt)," *Indian J. Chem.*, 1972, 10(7), pp. 703–705. (h) Mhala, M.M. and Prabha, S. "Hydrolysis of Organic Phosphates. VIII. Hydrolysis of 2,6-dimethoxyphenyl Hydrogen Phosphate," *Indian J. Chem.*, 1972, 10(10), pp. 1002–1005. (i) Mhala, M.M. and Prabha, S. "Hydrolysis of Organic Phosphates. IX. Hydrolysis of tris(2,6-dimethoxyphenyl) Phosphate," *Indian J. Chem.*, 1972, 10(11), pp. 1073–1076. (j) Mhala, M.M. and Nand, P. "Hydrolysis of Organic Phosphates: Part IX—Hydrolysis of 1-nitro-2-naphthyl- and 4-nitro-1-naphthyl-phosphate Monoesters," *Indian J. Chem.*, 1976, 14A(5), pp. 344–346.

93. Vilyanskaya, G.D., Lysko, V.V., Fragin, M.S., Kazanskii, V.N. and Vainstein, A.G. "Improving Fire Protection in Turbine Plants by Using Fire-Resistant Oils," *Therm. Eng.*, 1988, 35(4), pp. 193–195.

94. Mahoney, C.L., Barnum, E.R., Kerlin, W.W., Sax, K.J. and Saari, W.S. "Effect of Radiation on the Stability of Synthetic Lubricants," in *Proceedings Fifth World Petroleum Congress*, 1959, pp. 147–161.

95. Vaile, P.E.B. "Lubricants for Nuclear Reactors," *Proc. Inst. Mech. Eng.*, 1962, 176(2), pp. 27–59.

96. Wagner, R.M., Kinderman, E.M. and Towle, L.H. "Radiation Stability of Organophosphorus Compounds," *Ind. Eng. Chem.*, 1959, 51(1), pp. 45–46.

97. Chemtura Corp., "Reolube® Turbofluids—A Guide to Their Maintenance and Use," Technical Bulletin, Chemtura Corporation (U.K.) Ltd., 2005.

98. Supresta Inc., "Fyrquel® Compatibility Guide," Technical Bulletin, Supresta, Inc., Ardsley, NY.

99. Morgan, J.P. and Tulloss, T.C. "The Jake Walk Blues," *Ann Intern. Med.*, 1976, 85, pp. 804–808.

100. Johnson, M.K. "Organophosphorus Esters Causing Delayed Neurotoxic Effects: Mechanism of Action and Structure/Activity Studies," *Arch. Toxicol.*, 1975, 34, p. 259.
101. Johnson, M.K. "Organophosphates and Delayed Neuropathy—Is NTE Alive and Well?" *Toxicol. Appl. Pharmacol.*, 1990, 102, p. 385.
102. European Standard EN/TR 14489: Fire-Resistant Hydraulic Fluids—Classification and Specification—Guidelines on Selection for the Protection of Safety, Health and the Environment, 2005.
103. Directive 93/21/EEC, Official Journal of the European Communities, 36, No. L110A, 1993.
103. Benthe, H.F. Pharmacological/Toxicological Reports on Reolube® HYD 46 (1975), Turbofluid 46XC (1982), Reolube® MF46 (1988), University of Hamburg.
104. Society of Chemical Manufacturers, Washington, DC—Task Force on Tributyl Phosphate.
105. Deetman, G. U.S. Patent 5,464,551 (1995) (to Monsanto Co.).
106. *Joint Assessment of Commodity Chemicals, No. 21, Tri-(2-butoxethyl)-phosphate,* European Chemical Industry—Ecology and Toxicology Centre, Brussels, 1992.
107. *Joint Assessment of Commodity Chemicals, No. 20, Tris-(2-ethylhexyl) phosphate,* European Chemical Industry—Ecology and Toxicology Centre, Brussels, 1992.
108. Bayer A.G. "Disflamoll® DPO," Safety Data Sheet, Bayer A. G., Leverkusen, Germany, 1993.
109. Monsanto Chemical Co., Santiciser® 148, Material Safety Data Sheet, Monsanto Chemical Co., St. Louis, MO, 1982.
110. "Bewertung wassergefährdender Stoffe", Beirat "Lagerung und Transport wassergefährdender Stoffe" beim Bundesminister für Umwelt, Naturschutz und Reaktorsicherheit LTwS-Schrift Nr. 10, September 1979.
111. "Determination of the Acute Toxicity of Reolube® HYD 46 to Zebra Fish," Test Report, Institut National de Recherche Chimique Appliqueé, vert-le-Petit, France, 1991.
112. "Determination of the Acute Toxicity of REOLUBE® HYD 46 to Daphnia Magna," Test Report, Institut National de Recherche Chimique Appliqueé, vert-le-Petit, France, 1991.
113. Michael, P.R. and Adams, W.J. "Final Report of the 1982 Industry—EPA Phosphate Ester Aquatic Surveillance Program," Monsanto Chemical Co., 1983.
114. Japanese Government Agency, "Chemicals in the Environment," 1985, Office of Health Studies, Japanese Government Agency.
115. Staniewski, J.W.G. "The Influence of Mechanical Design of Electro-Hydraulic Steam Turbine Control Systems on Fire-Resistant Fluid Condition," *Lubr. Eng.*, 1996, 52(3), pp. 255–258.
116. Brown, K.J. "Condition Monitoring and Maintenance of Steam Turbine Generator Fire-Resistant Triaryl Phosphate Control Fluids," STLE Special Publication SP-27, Park Ridge, IL, 1989, pp. 91–96.
117. Staniewski, J.W.G. "Maintenance Practices for Steam Turbine Control Fire-Resistant Fluids: Part 1", *Jnl. Syn. Lub.*, 2006, 23, 109–121
118. Staniewski, J.W.G. "Maintenance Practices for Steam Turbine Control Fire-Resistant Fluids: Part 2", *Jnl. Syn. Lub.*, 2006, 23, 121–135.
119. Wolfe, G.E., Cohen, M. and Dimitroff, V.T. "Ten Years' Experience with Fire-Resistant Fluids in Steam Turbine Electrohydraulic Controls," *Lubr. Eng.*, 1970, 26(1), pp. 6–14.
120. Phillips, W.D. and Sutton, D.I. "Improved Maintenance and Life Extension of Phosphate Esters Using Ion Exchange Treatment," in *10th International Tribology Colloquium*, Technische Akademie Esslingen, 1996.
121. Grupp, H. "Aufbau von schwer entflammbaren Hydraulikflüssigkeiten auf Phosphorsäureesterbasis, Erfahrungen aus dem praktischen Einsatz im Kraftwerk," *Der Maschinen Schaden*, 1979, 52(3), pp. 73–77.
122. Tersiguel-Alcover, C. "Problems Encountered With Phosphate Esters on Hydraulic Systems of EDF Power Plants," *Proc. Int. Tribol. Congress*, 1981, 3, pp. 296–307.
123. Collins, K.G. and Duchowski, J.K. "Effectiveness of the Ion Exchange/Vacuum Dehydration Treatment of Phosphate Ester Fluids", *Jnl. Syn. Lub.*, 2002, 19, 31.

APPENDIX 1

International Specifications and Use Guides for Fire-Resistant Hydraulic Fluids Including Phosphate Esters

Organization	Standard Number	Title
ISO	6743–4	Lubricants, industrial oils and related products (class L). Classification—Part 4: Family H (Hydraulic systems)
ISO	7745	Hydraulic Fluid Power—Fire-resistant (FR) fluids—Guidelines for use
ISO	10050	Lubricants, industrial oils and related products (class L)— Family T (Turbines) — Specifications of triaryl phosphate ester turbine control fluids (category ISO-L-TCD)
ISO	12922	Lubricants, industrial oils and related products (class L)— Family H (hydraulic systems)—Specifications for categories HFAE, HFAS, HFB, HFC, HFDR and HFDU
CEN	TR14489	Fire-resistant hydraulic fluids—Classification and specification—Guidelines on selection for the protection of safety, health and the environment
ISO	11365	Maintenance and use guide for triaryl phosphate ester turbine control fluids

Additional National Specifications and Use Guides for Fire-Resistant Fluids Including Phosphate Esters

Country	Organization	Standard number	Title
Canada	Canadian Standards	CSA M423–M87	Fire-resistant hydraulic fluids
China	Chinese National Standards	DL/T 571–95	Guide for acceptance, in-service supervision, and maintenance of fire-resistant fluid used in power plant
Germany	DIN	24320	Schwerentflammbare Flüssigkeiten—Flüssigkeiten der Kategorien HFAE and HFAS—Eigenschaften und Anforderungen
India	Indian Bureau of Standards	IS: 10531	Code of practice for the selection and use of fire-resistant fluids
USA	ANSI/(NFPA)	T2.13.8	Hydraulic fluid power—Fire-resistant fluids— Definitions, classifications and testing
		T2.13.1	Practice for the use of fire-resistant hydraulic fluids for industrial fluid power systems
		T2.13.5	Hydraulic fluid power—Industrial systems—Practice for the use of high water content fluids

Key to Appendix 1

ISO	International Standards Organization
IEC	International Electrotechnical Commission
ANSI	American National Standards Institute
NFPA	National Fluid Power Association (USA)

APPENDIX 2

Suitable Test Methods for Monitoring Phosphate Ester Quality

Fluid property	Test method
Kinematic viscosity	ISO 3104
Neutralization no.	ISO 6618/6619
Pour point	ISO 3016
Density	ISO 3675
Foaming	ISO 6247
Air release	ISO 9120
Rust prevention	ISO 7120
Corrosion protection	ISO 4404-2
Water content	ISO 760
Flash/fire points	ISO 2592
Spray ignition	ISO 15029-2
Hot surface ignition	ISO 20823
Wick flame persistence	ISO 14935
Particulate levels	ISO 11500/4406
Emulsion stability	ISO 6614
Color	ISO 2049
Volume resistivity	IEC 60247
Chlorine content	IP 510
Mineral oil	Thin-layer chromatography
Metal content	ASTM D2788 (mod)

---

20 Polyalphaolefins and Other Synthetic Hydrocarbon Fluids

Ronald L. Shubkin, Lois J. Gschwender, and Carl E. Snyder, Jr.

CONTENTS

20.1 INTRODUCTION

Synthetic hydrocarbon fluids are fluid compounds or mixtures of compounds that are characterized as having specific molecular compositions containing only carbon and hydrogen atoms. Synthetic hydrocarbons may be distinguished from highly refined mineral oils because they are manufactured from specific raw materials by known chemical transformations. Mineral oils, on the other hand, are produced from petroleum base stocks by a variety of refining processes which may include cracking, extraction, dewaxing, distillation, isomerization, hydrocracking, and hydrorefining [1].

Although many synthetic hydrocarbon fluids have physical and chemical properties that are superior to those of equiviscous petroleum-based mineral oils, they also tend to be totally miscible in mineral oils and compatible with systems designed for mineral oils. This characteristic sets synthetic hydrocarbons apart from many other classes of synthetic fluids in that they may often be used to replace mineral oils (fully or partially) without the need to retrofit equipment.

There are five families of fluids classified as synthetic hydrocarbons:

1. **Polyalphaolefin (PAO) fluids**. First and foremost in commercial importance among the synthetic hydrocarbons are the polyalphaolefin (PAO) fluids [2]. These fluids are enjoying a rapidly growing market for a wide variety of applications, including hydraulics.
2. **Alkyl aromatic (AA) fluids**. Second in importance are a class of fluids manufactured by attaching hydrocarbon chains (alkyl groups) to aromatic rings [3]. Alkyl aromatic (AA) fluids gained a degree of prominence in the 1970s during the building of the Alyeska pipeline in Alaska. Although still an important class of commercial fluids, AAs have not undergone the rapid growth of the PAO fluids.
3. **Silahydrocarbon (SiHC) fluids**. Silahydrocarbon (SiHC) fluids are not hydrocarbons in the strict sense of the word, but are often classified as such because their chemical and physical properties are so similar to hydrocarbon fluids [4]. SiHCs have never been manufactured on a commercial scale, but their superior performance, especially in aeronautic hydraulic fluid applications, makes it worthwhile to include them in this chapter.
4. **Polybutene (PB or PIB) fluids**. Polybutene fluids are important commercially [5], but they are not used in hydraulic applications and will not be discussed in this chapter.
5. **Cycloaliphatic fluids**. Cycloaliphatic fluids are the fifth class of hydrocarbon fluids [6]. A wide variety of cycloaliphatic fluids have been synthesized and evaluated, and many have excellent properties. Only one, however, has been commercialized, and that low-volume application is as a bearing lubricant in spacecraft. Cycloaliphatic fluids will not be further considered in this chapter.

20.2 HISTORY

Synthetic hydrocarbon compounds were being developed and patented for use as lubricating fluids as early as 1928 in Germany [7,8] and in the United States [9,10]. The first attempt to commercialize synthetic hydrocarbon oil was made by Standard Oil Company of Indiana in 1929. The project was unsuccessful because of a lack of demand.

The onset of World War II and the subsequent shortages of petroleum feed-stocks in Germany, France, and Japan revitalized interest in synthetic lubricants. Moreover, the German disaster at the Battle of Stalingrad in 1942 demonstrated the inadequacy of then current petroleum products to perform satisfactorily in extremely cold weather. During the winter of 1942, the lubricants used in tanks, aircraft, and other military vehicles gelled, and the engines used in the vehicles could not be started. An intense German research effort to find alternative lubricants followed the Stalingrad debacle and led to the first manufacture of synthetic products derived by olefin polymerization.

Alkylaromatic fluids were also produced in Germany during World War II, and they later became important as low-temperature lubricants during the Alaskan oil explorations in the 1960s and the building of the Alyeska pipeline in the 1970s. A dialkylbenzene product was used as an engine crankcase oil and as a hydraulic fluid.

During the Southest Asia conflict in the 1960s, the need was recognized for a less flammable hydraulic fluid for military aircraft [11]. Although commercial aircraft had switched to hydraulic fluids based on phosphate esters, the military felt it was necessary to develop a fluid that was completely compatible with mineral-oil-based MIL-H-5606 [12] and which would require no retrofit of equipment. A PAO-based fluid was developed at the Wright Materials Laboratory, Wright-Patterson Air Force Base, which was eventually designated as MIL-H-83282 [13]. MIL-H-83282 replaced MIL-H-5606 in most Air Force applications except those requiring operation at very low temperatures. More recently, MIL-H-87257 [14], also PAO based, has been approved for low-temperature hydraulic applications.

As in the case of PAO fluids, the U.S. Air Force Materials Laboratory, starting in the 1950s, played a pivotal role in the development of silahydrocarbon fluids [15–17]. Again, the driving force for this program was the recognition by the Air Force that fluids operating in high-temperature/high-load environments would be required in the high-performance military weapons systems under development.

20.3 POLYALPHAOLEFIN FLUIDS

Polyalphaolefin fluids are gaining rapid acceptance as high-performance lubricants and functional fluids because they have certain inherent and highly desirable characteristics relative to mineral oils. Among these favorable properties are the following:

- A wide operational temperature range
- Good viscometrics (high viscosity index)
- Thermal stability
- Excellent response to conventional antioxidants
- Hydrolytic stability
- Shear stability
- Low corrosivity
- Compatibility with mineral oils
- Compatibility with various materials of construction
- Low toxicity
- Good to moderate relative biodegradability
- Low deposit formation
- Ability to be "tailored" to specific end-use application requirements

20.3.1 HISTORICAL DEVELOPMENT

In 1931, Standard Oil, in an article by Sullivan et al. [18], disclosed a process for the polymerization of olefins to form liquid products. These workers employed cationic polymerization catalysts such as $AlCl_3$ to polymerize olefin mixtures obtained from the thermal cracking of wax. At about the same time that the work at Standard Oil was being carried out, Zorn of I.G. Farben Industries independently discovered the same process [19].

The first use of a linear α-olefin to synthesize an oil was disclosed by Montgomery et al. in a patent issued to Gulf Oil Company in 1951 [20]. $AlCl_3$ was used in these experiments as it was in the earlier work with olefins from cracked wax.

The use of free-radical initiators as α-olefin oligomerization catalysts was first patented by Garwood of Socony Mobil in 1960 [21]. Coordination complex catalysts, such as the ethylaluminum sesquichloride–titanium tetrachloride system, were disclosed in a patent issued to Southern et al. at Shell Research in 1961 [22].

The fluids produced by the various catalyst systems described above contained oligomers with a wide range of molecular weights. The compositions and internal structures of these fluids resulted in viscosity–temperature characteristics that gave them no particular advantage over the readily available—and significantly less expensive—mineral oils of the day.

A patent describing the use of boron trifluoride (BF_3) as a catalyst for the oligomerization of linear α-olefins to make fluids suitable for use in lubricating oils was issued to Hamilton et al. of Socony Mobil in 1964 [23]. This patent describes a variety of oligomerization catalysts and gives only one example of the use of BF_3. There is no mention in the patent regarding the unique properties of the product prepared using the BF_3 catalyst. In 1968, Brennan, also at Socony Mobil, patented a process for the oligomerization of α-olefins using a BF_3 catalyst system [24]. Prior to that time, BF_3 catalysis had given irreproducible results. Brennan showed that the reaction could be controlled if two streams of olefins were mixed in the reactor. The first stream contained the olefin plus a $BF_3 \cdot ROH$ complex, where ROH is an alcohol. The second stream contained the olefin saturated with gaseous BF_3. Of particular interest was the fact that this catalyst system produced a product consisting of a mixture of oligomers that was markedly peaked at the trimer. There was no mention of a lubricating oil application in this patent.

Shubkin of the Chemicals Group of Ethyl Corporation (spun off as Albemarle Corporation on March 1, 1994) showed that H_2O [25], as well as other protic cocatalysts such as alcohols and carboxylic acids [26], could be used in conjunction with BF_3 to produce oligomers of uniform quality. The experimental technique employed a molar excess of BF_3 in relation to the cocatalyst. The excess was achieved by sparging the reaction medium with BF_3 gas throughout the course of the reaction or by conducting the reaction under a slight pressure of BF_3. These studies showed that the oligomerization products exhibited pour points that were well below those anticipated for such compounds, even when dimeric products were allowed to remain in the final mixture. Until then, the molecular structure of the dimer was believed to consist of a straight carbon chain containing a single methyl group near the middle. Such branched structures were known to exhibit relatively high pour points. These patents were the first to address the unique qualities, and thus the potential importance as synthetic lubricants, of PAOs derived from $BF_3 \cdot ROH$ catalyst systems. Shubkin et al. later showed that the unique low-temperature properties could be attributed to a high degree of branching in the molecular structure [27].

20.3.2 MANUFACTURE

Polyalphaolefins are manufactured by a two-step process from linear α-olefins, which are themselves manufactured from ethylene. The first synthesis step entails oligomerization, which simply means a polymerization to relatively low-molecular-weight products:

$$\alpha - \text{Olefin} \rightarrow \text{Dimer} + \text{Trimer} + \text{Tetramer} + \text{Pentamer, and so on.}$$

For the production of low viscosity (2–10 cSt) PAOs, the catalyst for the oligomerization reaction is usually boron trifluoride. (*Note*: PAOs are commonly classified by their approximate kinematic viscosity at 100°C. That convention will be used throughout this chapter. Thus, a fluid referred to as PAO 4 has a viscosity at 100°C of ~4 cSt.) The BF_3 catalyst is used in conjunction with a protic cocatalyst such as water, an alcohol, or a weak carboxylic acid. The $BF_3 \cdot ROH$ catalyst system is unique because of its ability to form highly branched products with the oligomer distribution peaking at the trimer. Figure 20.1 shows a gas chromatography (GC) trace indicating the oligomer distribution of a typical reaction product from 1-decene and $BF_3 \cdot C_4H_9OH$ as the catalyst system [26]. Higher-viscosity (40 and 100 cSt) PAO fluids are manufactured using alkylaluminum catalysts in conjunction with an organic halide [28] or by the use of $AlCl_3$.

The second step in the manufacturing process entails hydrogenation of the unsaturated oligomer. The reaction is carried out over a metal catalyst such as nickel or palladium. Hydrogenation gives the final product enhanced chemical inertness and added oxidative stability. Distillation, either before or after hydrogenation, is usually employed to separate the total product into different viscosity grades.

One of the distinct advantages in the manufacture of PAO fluids is that they can be "tailor-made" to fit the requirements of the end-use application [29]. This customizing is done by manipulation of the reaction variables, which include the following:

- Temperature
- Time
- Pressure
- Cocatalyst type
- Cocatalyst concentration

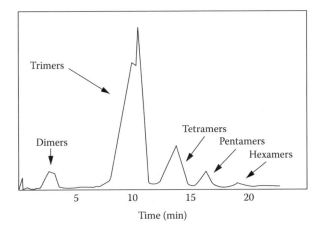

FIGURE 20.1 Gas chromatography trace of total product from the oligomerization of 1-decene using a BF$_3$ · ROH catalyst.

- Distillation of final product
- Chain length of olefin raw material

Today, the commercial PAO market is dominated by decene-derived material because these products have the broadest temperature range of desirable properties. Manipulation of the first six variables (see above) is the common practice for manufacturing the usual grades of PAO (PAO 2, PAO 4, PAO 6, PAO 8, and PAO 10).

Some PAO fluids manufactured from other olefin streams (mainly 1-octene and 1-dodecene) are being offered to fill the needs of certain niche markets.

Figure 20.2 compares the gas chromatography (GC) trace of a commercial 4.0-cSt PAO (PAO 4) with that of a hydrotreated VHVI (very high-viscosity index) mineral-oil base stock having the same approximate viscosity at 100°C. The trace from the mineral oil shows that it consists of a broad range of different kinds of molecules. Included in the mineral oil are low-molecular-weight materials which adversely affect volatility, and high-molecular-weight components which adversely affect low-temperature properties. By comparison, the PAO 4 is primarily decene trimer, with small amounts of decene tetramer and pentamer present.

The fine structure in the trace on Figure 20.2 is attributable to isomers of the different oligomers. (*Note*: Oligomers are low-molecular-weight polymers such as dimers, trimers, and so forth. Isomers are molecules with identical formulas and molecular weights, but with different skeletal structures). A knowledge of reaction variables can be used to control the relative abundance of the various

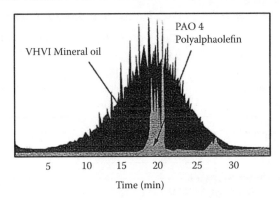

FIGURE 20.2 Gas chromatography comparison of PAO 4 and a VHVI mineral oil.

TABLE 20.1
Typical Physical Properties of Commercial PAO Fluids

	PAO 2	PAO 4	PAO 6	PAO 8	PAO 10	PAO 40	PAO 100
KV at 100°C (cSt)	1.80	3.90	5.90	7.80	9.60	40.0	100
KV at 40°C (cSt)	5.54	16.8	31.0	45.8	62.9	395	1,250
KV at −40°C (cSt)	310	2,460	7,890	18,160	32,650	–	–
Viscosity index[a]	–	129	138	140	134	151	168
Pour point (°C)	<−63	−70	−68	−63	−53	−34	−20
Flash point (°C)[b]	>155	215	235	252	264	272	288
NOACK (% loss)[c]	99	12	7.0	3.0	2.0	0.8	0.6

[a] VI is the average of samples. It is not calculated from viscosities given.
[b] Cleveland Open Cup, ASTM D92.
[c] Volatility at 250°C, 1.0 h, DIN 51581.

isomers and provides the producer with another method of influencing the physical properties of the final product [30].

20.3.3 PHYSICAL PROPERTIES

Table 20.1 lists the physical properties of various grades of PAO base fluids. These products are produced from decene, and the differences in properties illustrate what can be accomplished by the manipulation of the reaction parameters. Some of these products are coproduced and separated by distillation. The properties are typical of what is currently available and do not represent the specifications of any particular producer. It should be noted that the data represent typical values collected over time. Because of this, the listed viscosity index (VI), which is calculated from the kinematic viscosities (KVs) at 40°C and 100°C by a very complex relationship, is not the same as would be calculated from the viscosities shown.

Table 20.2 is a brief listing of the physical properties of PAO fluids prepared from different olefin raw materials. Each of these fluids was prepared in the laboratory using the same recipe, which included distilling of the dimer product and hydrogenating the final fluid [31]. None of the fluids in Table 20.2 is offered commercially.

TABLE 20.2
Physical Properties of PAO Fluids from Various Olefins

	C8[a]	C10[a]	C12[a]	C14[a]
KV at 100°C (cSt)	2.77	4.10	5.70	7.59
KV at 40°C (cSt)	11.2	18.7	27.8	41.3
KV at −180°C (cSt)	195	409	703	1150
Viscosity index	82	121	152	154
Pour point (°C)	<−65	<−65	−45	−18
Flash point (°C)[b]	190	228	256	272
NOACK volatility (%)[c]	55.7	11.5	3.5	2.3

[a] Not a commercial product.
[b] Cleveland Open Cup, ASTM D 92.
[c] Volatility at 250°C, 1.0 h, DIN 51581.

TABLE 20.3
PAO 4 Compared to Equiviscous Mineral Oils

	PAO	100N	100N	100NLP	VHVI
KV at 100°C (cSt)	3.90	3.81	4.06	4.02	3.75
KV at 40°C (cSt)	16.8	18.6	20.2	20.1	16.2
KV at –40°C (cSt)	2460	Solid	Solid	Solid	Solid
Viscosity index	137	89	98	94	121
Pour point (°C)	–70	–15	–12	–15	–27
Flash point (°C)[a]	215	200	212	197	206
NOACK (% loss)[b]	12.0	37.2	30.0	29.5	22.2

[a] Cleveland Open Cup, ASTM D 92.
[b] Volatility at 250°C, 1.0 h, DIN 51581.

Table 20.3 compares the physical properties of a commercial 4.0-cSt PAO with those of two 100N (neutral) mineral oils, a 100NLP (low pour) mineral oil, and a hydrotreated VHVI mineral oil. The PAO shows markedly better properties at both high and low temperatures. At high temperatures, the PAO has lower volatility and a correspondingly higher flash point. Low volatility is an important property in order for a fluid to "stay in grade" (i.e., retain original viscosity) during its working life. At the low end of the temperature scale, the differences are equally dramatic. The pour point of the PAO is less than −65°C, and pour points for the three 100N mineral oils and the VHVI oil are −15°C, −12°C, −15°C, and −27°C, respectively.

Table 20.4 compares a commercial 6.0-cSt PAO with a 160HT (hydrotreated) mineral oil, a 240N oil, a 200SN (solvent neutral) mineral oil, and a UHVI (ultra high-viscosity index) fluid. The broader temperature range of the PAO is again apparent. Table 20.5 makes similar comparisons for 8.0-cSt fluids.

20.3.4 ENVIRONMENTAL IMPACT

The use of environmentally friendly fluids has become an important issue in a number of hydraulic fluid applications. Particularly important are those applications where there is a potential for large-scale contamination of either soil or water by the accidental rupture of hydraulic lines. Off-road construction and mining are examples where very large hydraulic machinery is used. Contamination by oil spills can result in serious long-term damage to the environment.

TABLE 20.4
PAO 6 Compared to Equiviscous Mineral Oils

	PAO	160HT	240N	200SN	UHVI
KV at 100°C (cSt)	5.90	5.77	6.96	6.31	5.49
KV at 40°C (cSt)	31.0	33.1	47.4	40.8	25.9
KV at –40°C (cSt)	7890	Solid	Solid	Solid	Solid
Viscosity index	135	116	103	102	156
Pour point (°C)	–68	–15	–12	–6	–9
Flash point (°C)[a]	235	220	235	212	226
NOACK (% loss)[b]	7.0	16.6	10.3	18.8	14.3

[a] Cleveland Open Cup, ASTM D 92.
[b] Volatility at 250°C, 1.0 h, DIN 51581.

TABLE 20.5
PAO 8 Compared to Equiviscous Mineral Oils

	PAO	325SN	325N
KV at 100°C (cSt)	7.80	8.30	8.20
KV at 40°C (cSt)	45.8	63.7	58.0
KV at –40°C (cSt)	18,160	Solid	Solid
Viscosity index	136	99	110
Pour point (°C)	–63	–12	–12
Flash point (°C)[a]	252	236	250
NOACK (% loss)[b]	3.0	7.2	5.1

[a] Cleveland Open Cup, ASTM D 92.
[b] Volatility at 250°C, 1.0 h, DIN 51581.

Polyalphaolefin fluids are exceptional in their low toxicity. They have Food and Drug Administration (FDA) approval for both indirect and incidental food contact, and they are used in a number of personal-care products. In addition, PAO fluids do not bioaccumulate and, depending on viscosity, biodegrade readily [32]. These characteristics, combined with excellent viscometrics and lubricity, have made low-viscosity PAO fluids a significant player in off-shore drilling mud formulations. Figure 20.3 shows the biodegradability by the CEC-L-33-A-94 test. The range of results shown on the graph indicates the range of data obtained from three independent laboratories on multiple samples over a period of time. PAO 2 and PAO 4 show very good biodegradability under these test conditions, but the higher-viscosity fluids do not.

Demonstration formulations have indicated that high-performance, biodegradable hydraulic fluids can be prepared using PAO 2, a styrene/isoprene comonomer viscosity index improver (8.0–18.5%), an ashless antiwear agent (1.0%), and an ester (10%). Three grades were formulated: ISO 15, ISO 32, and ISO 46. Biodegradability by the CEC-L-33-A-94 test ranged from 76% to 79% for the fully formulated fluids. Excellent performance was found in the Turbine Oil Stability Test (TOST), the Kurt Orbahn Shear Stability Test, and the FZG Antiwear Gear Test. Finally, unlike many ester-based fluids, the PAO is hydrolytically stable [33].

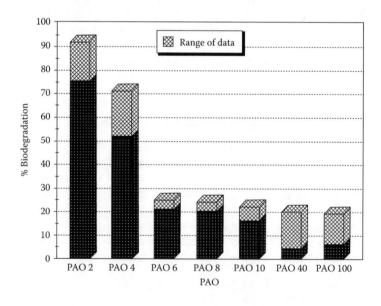

FIGURE 20.3   Biodegradability of PAO fluids by the CEC-L33-A94 test.

20.4 ALKYLATED AROMATIC FLUIDS

The class of fluids commonly referred to as "alkylated aromatics" (AAs) includes several different, but related, chemical types. The class may conveniently be broken down into three categories or subclasses, which are the alkylbenzenes, the alkyl-naphthalenes, and the alkylated multiring aromatics other than naphthalenes [3].

Alkylated aromatic fluid first gained prominence because of the following qualities:

- Exceptionally good low-temperature performance
- Excellent thermal stability
- Excellent oxidative stability
- Hydrolytic stability
- Miscibility with mineral oils
- Ability to solubilize additives
- Seal compatibility

20.4.1 HISTORICAL DEVELOPMENT

The early development of AA fluids for use as functional fluids dates back to 1928–1936 [7–10]. The first commercial production of AA fluids for use as lubricants appears to have been at the Chemische Werke Rheinpreussen, where alkynaphthalene oils were produced on a 7.2 MM lb./year scale from 1942 to 1946 [34]. These products were made by the alkylation of naphthalene with chlorinated aliphatic hydrocarbons in the presence of aluminum chloride catalysts.

In the period following World War II, relatively low-cost dialkylbenzenes became available as by-products of the manufacture of sodium alkylbenzenesulfonate detergents. In the 1950s, these were branched-chain alkylbenzenes made by the alkylation of benzene with propylene oligomers (trimers to pentamers). In the 1960s, the detergent industry turned to longer, linear alkyl chains, and the types of dialkylbenzene available for functional fluid use changed accordingly.

In the 1980s, AA fluids began to lose ground to PAO fluids, which were found to be more cost-effective, less toxic, and to operate over a wider temperature range in many applications. Nevertheless, a good deal of research has been published on improved products and processes in recent years, and AA fluids remain the product of choice for a variety of applications. Unimolecular 1, 3, 5-tri-n-alkyl substituted alkyl benzenes were found to uniquely possess exceptional thermal stability, acceptable oxidative stability, and adequate low-temperature viscosities to provide acceptable temperature range capability for military applications. However, commercially feasible synthetic routes have not yet been developed to produce the 1, 3,-tri-n-alkyl substituted alkyl benzenes [35].

20.4.2 MANUFACTURE

In general, the manufacture of alkyl aromatic fluids may be described as the alkylation of aromatic compounds with haloalkanes, alcohols, or olefins in the presence of a Lewis or Bronsted acid catalyst. The nature of the resulting fluid, however, is very dependent on the type of aromatic nucleus, the nature of the alkylating agent, the degree of alkylation, the position of the alkyl groups on the aromatic ring, the position of the aromatic ring on the alkyl chain and the mixture of molecules in the final product. The degree of alkylation and the molecular structure are, in turn, determined by the type of catalyst and the reaction conditions.

As mentioned earlier, branched-chain alkyl benzenes are the by-product of detergent manufacture. Excess benzene is alkylated with propylene tetramer at 20–50°C in the presence of hydrofluoric acid or aluminum chloride as the catalyst. The monoalkylate is obtained in about 75% yield and is used as a precursor for the sulfonated detergent [36]. The remaining 25% is available as lubricant base stock.

Dialkyl benzenes incorporating linear alkyl chains are also by-products of detergent manufacture [37]. A C_{10} to C_{14} paraffin feed is chlorinated and then used to alkylate benzene. A variety of products are formed in this process, including the monoalkyl benzenes, the dialkyl benzenes, diphenylalkanes, tetralin, and indane derivatives. The benzene ring is not attached to the linear alkyl chain at the chain's terminal carbon atom. Rather, it is attached randomly to carbon atoms located along the chain. The process can be tailored to give relatively less or more dialkylbenzenes for the lube base-stock market.

An alternative process for the manufacture of dialkylbenzenes is by alkylation of the benzene with linear olefins. A great deal of progress has been made in recent years in accomplishing this reaction over solid catalysts [38].

A great many patents and papers have been published on alternative methods for the production of AA fluids. A good review of the literature from the mid-1970s to 1990 is available and recommended for those interested in obtaining more detail [3].

20.4.3 PHYSICAL PROPERTIES

The physical properties of AA fluids vary widely, depending on structure and purity. In 1969, Conoco began production of a series of formulated products based on a dialkylbenzene base stock which was designated DN-600 Stock. From this material, Conoco produced CN-600 Synthetic Motor Oil, DN-600 Antiwear Hydraulic Oil, DN-600 Gear Oil, and Polar Start DN-600 SRI Grease.

Table 20.6 shows the specifications and composition for Conoco's DN-600 Base Stock and Table 20.7 gives the specifications for DN-600 Antiwear Hydraulic Fluid.

In 1981, Conoco was acquired by DuPont, and the alkylbenzene business was spun off to Vista Chemical. Vista was acquired by RWE-DWE Co. of Germany in 1991 [39]. Table 20.8 gives typical values for the Vista dialkylbenzene (DAB) base oil.

20.4.4 ENVIRONMENTAL IMPACT

The toxicity of aromatic compounds varies, depending on structure. In general, however, they cannot compete in applications where extremely low toxicity is important.

TABLE 20.6
Conoco DN-600 Base-Oil Properties and Specifications

	Specification	Test Method
Specific gravity	0.8628–0.8576	ASTM D-2422
Gravity (°API)	32.5–33.5	ASTM D-287
lbs/gal, Typical	7.16	
Flash point (COC) (°C)	≥224	ASTM D-92
Pour point (°C)	≤–65	ASTM D-97
KV at 100°C (cSt)	≥4.9	ASTM D-445
KV at 40°C (cSt)	26–27 typical	
KV at –40°C (cSt)	≤9500	ASTM D-2602
VI	100	ASTM D-2270
Composition (wt %)		
Monoalkylbenzenes	≤1.5	
Diphenylalkanes	≤12.5	
Dialkylbenzenes	≥80.0	
Polyalkylbenzenes	≤6.0	
Chlorine	≤0.01	

Source: Dressler, H. "Alkylated Aromatics," in *Synthetic Lubricants and High-Performance Functional Fluids*, Shubkin, R.L. ed., Marcel Dekker; New York, 1993, pp. 125–144.

TABLE 20.7
Conoco DN-600 Antiwear Hydraulic Oil

	Specification	Typical
Gravity (°API)	31	
Flash point (°F)	≥300	325
KV at −40°C (cSt)	≤3000	2600
KV at 100°C (cSt)		3.3
VI		160
Pour point (°F)	≤−70	−80
ASTM rust, A&B	Pass	
Foam test	Pass	
Vickers vane pump	Pass	

Source: Dressler, H. "Alkylated Aromatics," in *Synthetic Lubricants and High-Performance Functional Fluids,* Shubkin, R.L. ed., Marcel Dekker; New York, 1993, pp. 125–144.

The same may be said for biodegradability, where micro-organisms normally found in soil or water cannot digest the aromatic ring.

20.5 SILAHYDROCARBON FLUIDS

Silahydrocarbons (SiHCs), or tetraalkylsilanes, may be characterized as having a central silicon atom to which four hydrocarbon chains are attached. The silicon atom provides a thermally more stable center than would a smaller, isovalent carbon atom. The resultant SiHC molecule has better low temperature fluidity and higher thermal stability than the same molecule with a carbon atom in the middle of the molecule instead of the silicon atom. Although SiHC fluids have never been produced on a commercial scale, they have been extensively studied by the U.S. Air Force for aerospace hydraulic fluid applications.

20.5.1 HISTORICAL DEVELOPMENT

The first syntheses of tetraalkylsilane molecules were reported more than 130 years ago by the distinguished chemists Charles Friedel and James Mason Crafts [40,41]. There appears to have been little

TABLE 20.8
Comparison of Dialkylbenzene to Mineral Oils

	DAB	Paraffinic	Naphthenic
KV at 250°F	2.42	2.40	2.05
KV at 210°F	5.10	5.10	5.10
KV at 100°F	29.7	32.9	45.1
KV at 0°F	910	No flow	No flow
KV at −40°F	9540	No flow	No flow
Pour point (°F)	−70	0	−30

Source: Dressler, H. "Alkylated Aromatics," in *Synthetic Lubricants and High-Performance Functional Fluids,* Shubkin, R.L. ed., Marcel Dekker; New York, 1993, pp. 125–144.

practical interest in these chemically inert curiosities until the U.S. Air Force Materials Laboratories (Wright-Patterson Air Force Base, OH) began to take an interest in the 1950s [15–17]. Researchers at Wright-Patterson first coined the term "silahydrocarbons" to reflect the similarity in behavior between these compounds and branched, aliphatic hydrocarbons of similar molecular weight [42].

The interest of the Air Force in the development of SiHC fluids was largely the same as their interest in PAO fluids. There was a general recognition that new high-performance aircraft would require lubricants and fluids capable of operating in a high-temperature/wide-liquid-range environment. Specifically, the Air Force was interested in developing a hydraulic fluid for a –54°C to 315°C missile application. Later, the Air Force was interested in developing a hydraulic fluid incorporating the low-temperature requirements of MIL-H-5606 (mineral-oil based) and the high-temperature requirements of MIL-H-83282 (PAO based). By the late 1970s, the Air Force had developed SiHC-based formulated fluids which very nearly met the target properties [42,43]. The particular tetraalkylsilane that allowed the properties of both applications to be met was a random mixture of compounds containing a 50:50 composition of octyl ($C_8H_{17}-$) and decyl ($C_{10}H_{21}-$) groups on the silicon.

20.5.2 SYNTHESIS

Silahydrocarbons have been prepared by a variety of synthetic routes:

The earliest preparation of SiHCs by Friedel and Crafts employed the reaction of an alkyl magnesium halide (Grignard reagent) with silicon tetrachloride [40,41]:

$$4 \, RMgX + SiCl_4 \rightarrow R_4Si + 4 \, MgXCl.$$

In 1946, Gilman and Clark studied the reaction of Grignard reagents with alkyl trichlorosilanes [44]:

$$3 \, R'MgX + RSiCl_3 \rightarrow RSiR'_3 + 3 \, MgXCl.$$

The Air Force developed a route based on the reaction of mixed alkyl magnesium halides or mixed alkyl lithium reagents with tetrachlorosilane or methyltrichlorosilane [45]. The development of these SiHCs with mixed alkyl groups was important in obtaining low melting points.

Similar preparations based on the use of alkyl lithium, alkyl zinc, and alkyl aluminum compounds have been reported. A more detailed discussion with references may be found in Reference 4.

A different approach to the preparation of silahydrocarbon fluids is the catalyzed hydrosilation of alkenes with alkylsilanes. This was first demonstrated by Austin et al., who studied the photocatalyzed addition of alkenes to trialkylsilanes in the presence of trinuclear metal carbonyl catalyst precursors [46]. Onopchenko and Sabourin at Gulf Research and Development (now part of Chevron) later used platinum-based catalysts for hydrosilation [47,48]. Finally, researchers at Ethyl Corporation (now Albemarle Corporation) developed a route to ethyltrialkylsilanes starting from silane itself [49,50]:

$$SiH_4 + 1 - R^{2-} + NaAlR_4/LiCl(cat.) \rightarrow R_2SiH_2 + R_3SiH$$
$$R_2SiH_2 + R_3SiH + 1 - R^{2-} + Co_2(CO)_8(cat.) \rightarrow R_3SiH$$
$$R_3SiH + CH_2 = CH_2 + H_2PtCl_6(cat.) \rightarrow R_3SiCH_2CH_3.$$

20.5.3 PHYSICAL PROPERTIES

The physical properties of SiHC fluids show that they are suitable for applications over a wide temperature range. The properties depend, of course, on the four alkyl groups attached to the central silicon atom. Because there are no commercially available SiHC fluids, Table 20.9 is a listing of properties for laboratory-produced fluids [51]. Table 20.9 lists only methyltrialkylsilanes, $CH_3SiRR'R''$, where R, R', and R'' are linear alkyl chains that may be the same carbon number or a mixture of two

TABLE 20.9
Physical Properties of Methyltrialkylsilane Fluids

| | Carbon Nos. of Alkyl Groups | | | |
	C6/C8	C8/C10	C10	C12/C14
KV at −54°C (cSt)	906	2550	Solid	Solid
KV at −40°C (cSt)	256	627	1020	Solid
KV at 40°C (cSt)	5.84	9.84	12.7	25.1
KV at 100°C (cSt)	1.92	2.82	3.41	5.82
Viscosity index	Undefined	139	151	186
Pour point (°C)	<−65	<−65	−54	−9
Oxidation onset (°C)	197.6	197.7	198.0	197.8
Flash point (COC) (°C)	186	214	234	274
Fire point (COC) (°C)	188	232	254	>283

Note: Fluids are laboratory samples, not commercial fluids.

carbon numbers as indicated. It may be noted from the data in the table that SiHCs have, in general, high VIs, low pour points, high flash points, and good thermal and oxidative stability.

20.5.4 Environmental Impact

Silahydrocarbons, like PAO, have very low toxicity. Oral and dermal LD_{50} values for octyl/decyl silahydrocarbons have been reported to be >5.0 g/kg and > 2.0 g/kg, respectively, and the inhalation LC_{50} was >4.8 mg/L [52]. Little or no information is available on the biodegradability of SiHC compounds, but they would be expected to biodegrade in a fashion similar to PAO fluids—with better biodegradability for the lower-viscosity products.

20.6 APPLICATIONS

20.6.1 Background

The major outlet for synthetic hydrocarbon hydraulic oils as fire-resistant fluids has been in U.S. military hydraulic applications. Since the 1940s, aircraft used mineral oil for hydraulic fluid, but the industry was plagued from the beginning with fire hazards. The mineral oils were required to operate in extreme cold conditions, which require a relatively low-flash-point oil. In the 1950s, the commercial aircraft switched to fire-resistant phosphate ester hydraulic fluids but the U.S. military remained with mineral oils mainly because of high- and low-temperature operational requirements and the reluctance to switch seals, paints, and other materials to phosphate ester-compatible materials.

For many years, MIL-H-5606 hydraulic fluid was the military standard hydraulic fluid for aircraft and MIL-H-6083 [53], the rust-inhibited version, was the standard hydraulic fluid for use in ground vehicles. Table 20.10 shows comparative properties of military hydraulic fluids. MIL-PRF-5606 and MIL-PRF-6083 were comprised of a very low-molecular-weight mineral oil base fluid, in order to meet −54°C (−65°F) low-temperature operational requirements. These base fluids were thickened with approximately 17% viscosity index improver, typically a poly(methy methacrylate), allowing the fluid to meet 135°C (275°F) high-temperature viscosity requirements for a Type II MIL-H-5440 military hydraulic system. The inherent danger of this fluid is obvious in that the flash point of this oil is 100°C (212°F), by the Cleveland Open Cup method [54], well below the upper operational temperature for the aircraft of 135°C (275°F). For the Air Force alone, noncombat fire damage was costing over a $20 million per year.

TABLE 20.10
Selected Typical Physical Properties of Military Hydraulic Fluids

Property	MIL-PRF-5606	MIL-PRF-83282	MIL-PRF-6083	MIL-PRF-46170	MIL-PRF-87257	MIL-PRF-27601	Army SHF[a]
Kinematic viscosity (mm/s), ASTM D445							
At −54°C	2,450	20,000	3,480	25,000	2,480	20,000	3,450
−40°C	490	2,140	800	2,500	520	3,950	650
38°C	14.2	14.2	13.2	19.3	9.0	14.0	9.0
99°C	5.1	3.6	5.0	3.5	2.6	3.5	2.5
Pour point (°C) ASTM D97	<−59	<−59	<−59	<−54	<−59	<−40	<−60
Elastomer compatibility NBR-L[b] Rubber Swell (vol %), ASTM D4289	24	18	23	21	20	N/A[c]	22

[a] SHF = single hydraulic fluid.
[b] NBR-L is the designation for Buna N rubber.
[c] N/A indicates that there is no elastomer compatibility requirement for this hydraulic fluid.

Designations for military specification materials were changed in order to reflect that they are based on performance-based requirements. Therefore, instead of designating it as MIL-H-5606, for example, it was re-designated as MIL-PRF-5606. All of the military specification fluids and lubricants were changed accordingly.

20.6.2 POLYALPHAOLEFINS

In the 1960s, both the Air Force and the Navy embarked on research programs to develop fire-resistant hydraulic fluids as drop-in replacements for MIL-H-5606, the standard mineral-oil-based DoD hydraulic fluid. This was a very difficult task because the replacement fluid had to be completely compatible with MIL-H-5606 as well as with all of the other hydraulic system materials. The replacement fluid also had to provide equivalent hydraulic system performance. The Navy developed fluids called MS5 and MS6, which were chlorophenylsilicone based [55]. These fluids were exceptionally fire resistant but did not have sufficient bulk modulus for the flight controls of operational aircraft. In F-4 iron bird testing with MS6, the hydraulic controls experienced excessive flutter and the fluids were deemed unsafe for use in existing aircraft. No further development was performed with these fluids. The Air Force program led to the development of MIL-H-83282, based on the newly emerging commercialization of polyalphaolefins [56]. The U.S. Army developed a rust-inhibited version, MIL-H-46170, for ground vehicle applications [57]. Table 20.11 shows the comparative flammability properties of MIL-H-5606 and MIL-H-83282 and other DoD hydraulic fluids. MIL-H-83282 has a Cleveland Open Cup flash point of 224°C (435°F), well above the upper use temperature of 135°C (275°F). Considerable controversy over the flammability tests and their interpretation delayed the use of MIL-H-83282 in the Navy until 1976, in the Army until 1977, and in the Air Force until 1980. Even then, the Air Force did not convert all of their aircraft for many

TABLE 20.11
Flammability Properties of Military Hydraulic Fluids

Property	MIL-PRF-5606	MIL-PRF-83282	MIL-PRF-6083	MIL-PRF-46170	MIL-PRF-87257	MIL-PRF-27601	Army SHF[a]
Use temperature (°C)	−54 to 135	−40 to 200	−54 to 110	−40 to 135	−54 to 200	−40 to 290	−54 to 135
Flash point, D92 (°C)	102	224	126	220	166	224	182
AIT E659 (°C)	235	365	243	363	235	365	356
Linear flame propagation rate, D5306 (cm/s)	0.73	0.20	0.95	0.23	0.30	0.20	0.25

[a] SHF = single hydraulic fluid.

years based on the higher viscosity of MIL-H-83282 at low temperature, versus MIL-H-5606. (Even today, in extreme cold climates, MIL-H-5606 and MIL-H-6083 are used because they are −54°C [−65°F] lower-temperature-use fluids compared to MIL-H-83282 and MIL-H-46170, which are −40°C [−40°F] low-temperature-use fluids.) Significant decrease in Air Force fire damage, from an average of $20 million per year to an average of approximately $600,000 per year, even with some of the aircraft still using MIL-H-5606, has borne out the predictions for savings with MIL-H-83282.

MIL-H-83282 is composed of 4-cSt PAO hydrogenated decene trimer with approximately 30% diester as a rubber swell agent. Smaller amounts of other performance-improving additives are used as required.

As mentioned earlier, some aircraft and tanks did not convert to MIL-H-83282 or to the rust-inhibited counterpart, MIL-H-46170, used in most U.S. Army tank and other ground vehicle applications. In response to the need for a low-temperature version of a fire-resistant fluid, the Air Force developed MIL-PRF-87257 hydraulic fluid in the 1980s [43,58]. Nearly all of the aircraft that were not converted to MIL-PRF-83282 have been successfully converted to MIL-PRF-87257. MIL-PRF-87257 hydraulic fluid is composed of a blend of 2-cSt PAO and 4-cSt PAO along with the needed rubber swell ester and other additives. The low-temperature viscosity requirement in the military specification is less than or equal to 2500 cSt at −54°C (−65°F).

The U.S. Army embarked on the development of a low-temperature fire-resistant hydraulic fluid for tank use based on somewhat different flammability and low-temperature limits and also containing the required rust inhibition for tank and other ground vehicle applications. This fluid, like MIL-H-87257, is also based on a blend of PAO hydrogenated decene dimer and trimer blends, but formulated with a rust inhibitor in addition to the other performance-improving additives. The intent of the development of this new fluid is to replace both MIL-H-6083 and MIL-H-46170 with one hydraulic fluid for logistic reasons. It is therefore called the Army "single hydraulic fluid" (SHF). This new fluid has not yet been adopted. [59,60].

Another military hydraulic fluid based on PAOs is MIL-H-27601B [61]. This fluid specification in previous versions was based on a highly refined, deep dewaxed paraffinic mineral oil and the fluid was only used in the once classified SR-71 aircraft. A PAO was developed by the Air Force as a no-retrofit, direct drop-in replacement for the mineral oil but has yet to be fielded because the aircraft was decommissioned. MIL-H-27601B is based on the 4-cSt PAO with some performance-improving additives, but with no rubber swell ester. Because of the imposed mission of the SR-71, the MIL-H-27601 temperature range was −40°C (−40°F) to 288°C (550°F). At that high upper temperature, no elastomeric seals were used in the aircraft, so no rubber swell ester was needed nor could it be used at that upper temperature. With the military emphasis on higher temperatures and increased performance, MIL-H-27601 and other high-temperature hydraulic fluids continue to be considered for future applications [62].

20.6.3 Alkylaromatics and Silahydrocarbons

Both alkylaromatics [35] and silahydrocarbons [46,47,49], as previously described, have been investigated by the Air Force as high-temperature, fire-resistant hydraulic fluids in the past. The silahydrocarbon was extensively developed as a −54°C (−65°F) to 316°C (600°F) Advanced Strategic Air Launched Missile fluid, but the missile was never fielded because of demilitarization agreements between the United States and the Soviet Union in the 1970s.

Silahydrocarbon-based candidates were considered for both the MIL-H-87257 and the MIL-H-27601B applications and showed superior properties to the PAO-based replacements [61], but they could not compete with the PAO in cost, commercial availability, and use history. In addition to the excellent thermal stability exhibited by the n-alkyl-substituted silahydrocarbons, they also demonstrate excellent viscosity indices. If either alkylated aromatics or silahydrocarbons are used in the near future, it will likely be in a niche application where PAOs are unacceptable for some reason. However, as requirements become increasingly stringent, silahydrocarbons would be a natural candidate replacement base stock for the PAOs when the PAOs can no longer meet the necessary requirements.

REFERENCES

1. Sequeira, A. Jr. *Lubricant Base Oil and Wax Processing,* Marcel Dekker; New York, 1994.
2. Shubkin, R.L. "Polyalphaolefins," in *Synthetic Lubricants and High-Performance Functional Fluids,* Shubkin, R.L. ed., Marcel Dekker; New York, 1993, pp. 1–40.
3. Dressler, H. "Alkylated Aromatics," in *Synthetic Lubricants and High-Performance Functional Fluids,* Shubkin, R.L. ed., Marcel Dekker; New York, 1993, pp. 125–44.
4. Pettigrew, F.A. and Nelson, G.E. "Silahydrocarbons," in *Synthetic Lubricants and High-Performance Functional Fluids,* Shubkin, R.L. ed., Marcel Dekker; New York, 1993, pp. 205–14.
5. Fotheringham, J.D. "Polybutenes," in *Synthetic Lubricants and High-Performance Functional Fluids,* Shubkin, R.L. ed., Marcel Dekker; New York, 1993, pp. 271–318.
6. Venier, C.G. and Casserly, E.W. "Cycloaliphatics," in *Synthetic Lubricants and High-Performance Functional Fluids,* Shubkin, R.L. ed., Marcel Dekker; New York, 1993, pp. 241–69.
7. British Patent 323,100 (December 3, 1928), to IG Farbenindustrie A.G.
8. Zorn, H., Mueller-Cunradi, M. and Rosinski, W. German Patent 565,249 (March 26, 1930), to IG Farbenindustrie A.G.
9. Davis, G.H.B. U.S. Patent 1,815,072 (July 14, 1931), to Standard Oil Development Company.
10. McLaren, F.H. U.S. Patent 2,030,832 (February 11, 1936), to Standard Oil Company in Indiana. Synthetic Lubricating Oils.
11. Snyder, C.E. Jr. and Gschwender, L.J. "Aerospace," in *Synthetic Lubricants and High-Performance Functional Fluids,* Shubkin, R.L. ed., Marcel Dekker; New York, 1993, pp. 525–32.
12. MIL-PRF-5606H Military Specification (2002), Hydraulic Fluid, Petroleum Base, Aircraft Missile and Ordinance, NATO code number H-515.
13. MIL-PRF-83282D Military Specification (1997), Hydraulic Fluid, Fire Resistant, Synthetic Base, Aircraft Metric, NATO code number H-537.
14. MIL-PRF-87257B Military Specification (2003), Hydraulic Fluid, Fire Resistant, Low Temperature, Synthetic Hydrocarbon, Base, Aircraft and Missile, Metric, NATO code number H-538.
15. Rosenberg, H., Groves, J.D. and Tamborski, C. *J. Org. Chem.,* 1960, 25, p. 243.
16. Tamborski, C. and Rosenberg, H. *J. Org. Chem.,* 1960, 25, p. 246.
17. Baum, G. and Tamborski, C. *J. Chem Eng. Data,* 1961, 6, p. 142.
18. Sullivan, F.W. Jr., Vorhees, V., Neeley, A.W. and Shankland, R.V. *Ind. Eng. Chem.,* 1931, 23, p. 604.
19. Boylan, J.B. "Synthetic Base Stocks for Use in Greases," NLGI Spokesman, 1987, 51(5), pp. 188–95.
20. Montgomery, C.W., Gilbert, W.I. and Kline, R.E. U.S. Patent 2,559,984 (1951), to Gulf Oil Co.
21. Garwood, W.E. U.S. Patent 2,937,129 (1960), to Socony Mobil Oil Company.
22. Southern, D., Milne, C.B., Moseley, J.C., Beynon, K.I. and Evans, T.G. British Patent 873,064 (1961), to Shell Research.
23. Hamilton, L.A. Pittman and Seger, F.M. U.S. Patent 3,149,178 (1964), to Socony Mobil Oil Company.
24. Brennan, J.A. U.S. Patent 3,382,291 (1968), to Socony Mobil Oil Company.

25. Shubkin, R.L. U.S. Patent 3,763,244 (1973), to Ethyl Corp.
26. Shubkin, R.L. U.S. Patent 3,780,128 (1973), to Ethyl Corp.
27. Shubkinl, R.L., Baylerian, M.S. and Maler, A.R. "Olefin Oligomers: Structure and Mechanism of Formation," Symposium on Chemistry of Lubricants and Additives, Div. of Petroleum Chemistry, ACS, 1979; also published in *Ind. Eng. Chem.*, Product Res. Dev., 1980, 19, p. 15.
28. Loveless, F.C. U.S. Patent 4,469,910 (1984), to Uniroyal Inc.
29. Shubkin, R.L. and Kerkemeyer, M.E. "Tailor Making PAOs," 7th International Colloquium on Automotive Lubrication, 1990; also *J. Synth. Lubr.*, 1991, 8(2), pp. 115–34.
30. Theriot, K.J. and Shubkin, R.L. "A Polyalphaolefin with Exceptional Low Temperature Properties," 8th International Colloquium, TRIBOLOGY 2000, 1990; also *J. Synth. Lubr.*, 1993, 10(2), pp. 133–42.
31. Kumar, G. and Shubkin, R.L. "New Polyalphaolefin Fluids for Specialty Applications," 47th Annual Meeting of the Society of Tribologists and Lubrication Engineers, 1992; also *Lubr. Eng.*, 1993, 49(9), pp. 723–25.
32. Carpenter, J.F. "Assessment of Environmental Impact of PAOs," STLE Annual Meeting, 1993.
33. Unpublished data, Albermarle Corporation.
34. Koelbel, H. "Synthesis of Lubricants via the Alkylation of Naphthalene," *Erdoel Kohle*, 1948, 1, pp. 308–18.
35. Gschwender, L.J., Snyder, C.E. and Driscoll, G. "Alkyl Benzenes—Candidate High-Temperature Hydraulic Fluids," *Lubr. Eng.*, 1990, p. 377.
36. Kosswig, K. "Tenside," in *Ullmann's Encyclopedia of Technical Chemistry,* 4th ed., Verlag Chemie; Weinheim, Vol. 22, 1982, p. 455.
37. Tokoaka, S. SRI-PEP (Process Economics Program) Report No. 59A Supplement, Aliphatic Surfactants, Menlo Park, 1974.
38. *Oil Gas J.*, July 16, 1990, p. 48.
39. *Chem. Week*, January 16, 1991, p. 48.
40. Friedel C. and J. M. Crafts, *Ann. Chem.*, 1863, 127, p. 28.
41. Friedel, C. and Crafts, J.M. *Ann. Chem.*, 1870, 259, p. 334.
42. Snyder C.E.Jr., Gschwender, L.J., Tamborski, C., Chen, G.J. and Anderson, D.R. "Synthesis and Characterization of Silahydrocarbon—A Class of Thermally Stable Wide Liquid Range Functional Fluids," *ASLE Trans.*, 1982, 25(3), pp. 299–308.
43. Gschwender, L.J., Snyder, C.E. Jr. and Fultz, G.W. "Development of a –54° to 135°C Synthetic Hydrocarbon-based, Fire-Resistant Hydraulic Fluid," *Lubr. Eng.*, 1986, pp. 485–90.
44. Gilman, H. and Clark, R.N. J. *Am. Chem. Soc.*, 1946, 68, p. 1675.
45. Tamborski, C. and Snyder, C.E. Jr. U.S. Patent 4,367,343 (1983), to the United States of America.
46. Austin, R.G., Paonessa, R.S., Giordano, P.J. and Wrighton, M.S. U.S. NTIS, AD Report, 1977. Available form Gov. Rep. Announce. Index (U.S.), 1977, 77(25), 92 74–1. *Chem. Abstr.*, 88 (20), p. 144251x.
47. Onopchenko, A. and Sabourin, E.T. U.S. Patent 4,578,497 (1986), to Gulf Research and Development Co.
48. Sabourin, E.T. and Onopchenko, A. *Bull. Chem. Soc. Jpn.*, 1989, 62, p. 3691.
49. Pettigrew, F.A., Plonsker, L., Nelson, G.E., Malcolm, A.J. and Everly, C.R. Society of Tribologists and Lubrication Engineers, 1989.
50. Malcolm, A.J., Everly, C.R. and Nelson, G.E. U.S. Patent 4,670,574 (1987), to Ethyl Corporation.
51. Unpublished data, Albermarle Corporation.
52. Kinkead, E.R., Bunger, S.K., Wolfe, R.E. and Doarn, C.R. Harry G. Armstrong Medical Research Laboratory Report, AAMRL-TR-89-026, 1989. (Available from DODSSP, Subscription Services Desk, Building 4D, 700 Robbins Avenue, Philadelphia, PA.)
53. MIL-H-6083 Military Specification (1986), Hydraulic Fluid, Petroleum Base, for Preservation and Operation.
54. ASTM D 92, Flash and Fire Points by Cleveland Open Cup.
55. DeMarchi, J.N. and Haning, R.N. NADC-79120-60, 1981. (Available from DODSSP, Subscription Services Desk, Building 4D, 700 Robbins Avenue, Philadelphia PA.)
56. Loving, B.A., Adamczak, R.L. and Schwenker, H. AFML-TR-71-5, 1971. (Available from DODSSP, Subscription Services Desk, Building 4D, 700 Robbins Avenue, Philadelphia PA.)
57. MIL-PRF-46170D Military Specification (2004), Hydraulic Fluid, Rust Inhibited, Fire Resistant Synthetic Hydrocarbon Base.
58. Gschwender, L.J., Snyder, C.E. Jr. and Sharma, S.K. "Pump Evaluation of Hydrogenated Polyalphaolefin Candidates for a –54°C to 135°C Fire-Resistant Air Force Aircraft Hydraulic Fluid," *Lubr. Eng.*, 1988, 44, p. 324.

59. Purdy, E.M. Technical Report, USA-BRDEC-TR/2540, 1993. (Available from DODSSP, Subscription Services Desk, Building 4D, 700 Robbins Avenue, Philadelphia, PA 19111-5094.)
60. Purdy, E.M. and Rutkowski, D.M. Technical Report, USA-TARDEC-TR-13620, 1994. (Available from DODSSP, Subscription Services Desk, Building 4D, 700 Robbins Avenue, Philadelphia, PA.)
61. MIL-PRF-27601B Military Specification. (1993). Hydraulic Fluid, Fire Resistant, Hydrogenated Polyalphaolefin Base, High Temperature, Flight Vehicle, Metric.
62. Gschwender, L.J. and Snyder, C.E. Jr. "High Temperature Hydraulic Fluids," *J. Synth. Lubr.*, 1992, 9, p. 115–25.

21 Food-Grade Hydraulic Fluids

Sibtain Hamid

CONTENTS

21.1 A BRIEF HISTORY OF FOOD-GRADE LUBRICANTS

Food-grade lubricants are liquid, semi-fluid, or solid compounds that are acceptable for use in and around food-processing equipment. Such compounds are used on food-processing machinery as a protective antirust film and as a lubricant for machine parts.

Food-grade lubricants are made for meat and poultry and other food-processing equipment. These lubricants are generally registered with the National Sanitary Foundation (NSF).

In the mid-1970s the need for lubricants suitable for food-processing equipment led to the introduction of the first food-grade lubricants. These early food-grade lubricants satisfied the requirements for acceptability in the food-processing environment, but were generally considered less satisfactory as lubricants, in part because additives such as those that inhibit oxidation or reduce wear on metal surfaces were not very effective. Some food and beverage plant managers were reluctant to convert from the industrial lubricants which were then in use to food-grade lubricants. In the mid-1980s, the introduction of improved food-grade lubricants which met both acceptability and lubrication requirements meant that food-grade lubricants became more widely used in the food-processing industry.

The commercial performance of food-grade lubricants has now been proven in hydraulic systems, compressors, and gear lubrication systems. The evolution of food-grade lubricants from mostly white oils with very few additives to advanced high-performance lubricants which incorporate very effective additives took two decades or more. This dynamic product development and product improvement continues today as the Food and Drug Administration FDA (Food and Drug Administration) qualifies and approves more additives. The producers of additives have carried out extensive toxicity testing programs to verify that proposed new additives meet the FDA's non-toxicity requirements and thus should be considered for FDA approval.

At the same time, base stock producers have also contributed by producing and testing base stocks that were acceptable to the FDA. The formulators of lubricants have displayed their ingenuity by selecting from a wide range of additives from the FDA list of approved additives. They have also used as additives substances from the GRAS (Generally Recognized As Safe) list, which is a list of substances that have not been specifically approved as additives but which have long been used as ingredients in foods or beverages. An example of an additive from the GRAS list is ammonium phosphate. By using additives from both of these sources, formulators have been able to design the high-performing products needed today for food and beverage machinery.

Hydraulic fluids (most liquid lubricants) are composed of a base stock and additives. The base stock provides the basic properties of the lubricant:

- Viscosity and viscosity index (Index can be improved to 200 with VI improvers)
- Pour point and flash point
- Oil or water compatibility
- Seal and paint compatibility

The additives supplement or add new properties to the base stock, such as:

- Steel and copper corrosion protection
- De-foaming
- Antiwear protection
- Pour point and viscosity index improvers

21.2 ADVANTAGES OF FOOD-GRADE LUBRICANTS

The threat of lubricant contamination of foods and beverages from operating machinery such as hydraulic systems and gear boxes is a great concern. Any food or beverage plant that is still using non-food-grade (i.e., industrial-grade) lubricants is at risk of an expensive product recall which could damage the brand or the company's reputation. Minor leaks due to normal wear and tear could result in batches being contaminated with industrial-grade lubricants. The problem becomes even more severe when the product has already been packaged or distributed to the market. However, when food-grade lubricants are used, the FDA allows contamination of up to ten parts per million. Many food plants, slaughter houses, bakeries, and food packagers are required to undergo regular inspections by the USDA to ensure the quality of the food being prepared.

In order to reduce the risk of contamination or fines, many companies are converting to food-grade lubricants registered with the NSF (National Sanitary Foundation), which is an independent not-for-profit organization. NSF registers lubricants based on FDA pre-approved ingredients under 21 CFR 178.3570, entitled "Lubricants with Incidental Food Contact". In 1999, the NSF took over the approval process for non-food compounds including lubricants from the U.S. Department of Agriculture. NSF toxicologists determine product acceptability based on FDA's list of approved ingredients/raw materials against CFR 9, 21, and 40. This approval or listing assures lubricant users and inspection officials that formulations and food labels meet food safety requirements. The NSF lists all approved products complying with these requirements in its online "White Book" (www. nsf.org/usda/psnclistings.asp).

The NSF program for non-food compounds includes these three categories:

H-1: This category includes compounds and lubricants that are permitted on equipment where food may potentially be exposed to the lubricated parts of a machine. These lubricants are also referred to as "incidental food contact lubricants". These are the recommended lubricants for food and beverage plants.

H-2: This category includes compounds and lubricants that may be used in food-processing plants on equipment locations where there is no possibility of the lubricant or parts lubricated with such lubricant coming in contact with the food that is being processed. One example of this equipment is an isolated air compressor that feeds air to food-processing equipment.

H-3: This category refers to water-soluble lubricants that are used during the manufacture of food-processing equipment and which are typically present on gears and other surfaces when new food-processing equipment is delivered to a plant. These lubricants should not be used in food-processing plants and must be cleaned off of the equipment before it is put into use.

Some food plants require kosher and/or halal certification along with the NSF registration. Kosher means sanctioned by Jewish law or ritually fit for use according to Jewish law. Halal means permissible or sanctioned by Islamic law. Both of these certifications prohibit the use of pork and pork by-products. Halal additionally excludes the use of alcohol in its products. Lubricant manufacturers can obtain kosher certification from a number of organizations. Two of the largest are the Orthodox Union and the Star-K Kosher Certification organization. Halal certification is obtained from the Islamic Food and Nutrition Council of America. These certifications require documentations from the raw material suppliers in support of the origin of the ingredients.

In addition to the above, suppliers and manufactures must follow good manufacturing practice protocols. These include having dedicated blending kettles, piping, and storage and packing equipment to avoid any possible contamination of food from non-kosher or non-halal lubricants or additives.

21.3   SELECTING THE RIGHT FOOD-GRADE LUBRICANT

When a food or beverage processing plant has made the decision to use food-grade lubricants, one of the first tasks is to locate the most suitable lubricant for a specific piece of equipment.

Traditionally, very few original equipment manufacturers (OEM) have approved lubricants in category H-1 for the food and beverage equipment that they manufacture. This is mainly due to deficiencies exhibited by earlier generations of food-grade lubricants. The more recent development of synthetic food-grade lubricants has alleviated this problem for many applications.

Many companies have successfully developed both mineral oil and synthetic lubricants for food and beverage plants. These food-grade lubricants encompass major applications such as compressors, hydraulics, gears, bearings, and chains. A number of food-grade lubricants registered as H-1 can now perform the same technical functions as many non-food-grade conventional lubricants. Food-grade lubricants can now provide protection against wear and friction, corrosion and oxidation, heat dissipation, and elastomer compatibility. In addition, food-grade lubricants should be able to resist degradation caused by plant food ingredients, plant process chemicals, and water. For example, a food-grade lubricant might come into contact with an acid being used to control the pH of the food being processed; or it might come in contact with KOH, which is often used for cleaning. The presence of water and food ingredients also requires inhibition of micro-organisms such as bacteria and fungi. The final selection of a lubricant must be based on technical data and case studies relating to the equipment in question, the specific application, and the supplier's reputation.

21.4   APPLICATIONS

A number of the industries that make use of food-grade lubricants are listed below. Many of these plants use fluid power in their operations and employ food-grade hydraulic fluids.

- Meat- and poultry-processing plants
- Egg-processing plants
- Fish and seafood-processing plants
- Breweries and wineries
- Soft drink and bottling plants
- Vegetable and fruit processors
- Cheese and cheese product producers
- Bakeries

- Snack food manufacturers
- Pasta manufacturers
- Pet food and animal feed producers
- Oil mills and seed cake processors
- Pharmaceutical and drug manufacturers
- Cosmetic manufacturers
- Food and beverage container manufacturers
- Paper and paperboard manufacturers
- Water well drillers
- Drinking and potable water-treatment plants

21.5 FOOD-GRADE HYDRAULIC BASE FLUID TYPES

Lubricants are an integral part of food and beverage machinery. They reduce friction and wear and protect parts against corrosion caused by moisture and food chemicals. Food-grade lubricants are composed of a base fluid as the main body of the lubricant which provides the basic properties such as lubricity, viscosity, pour and flash points. The additives used with this base material enhance some of these properties and add a few more properties to the final product. Base fluids that have been approved for H-1 category incidental food contact lubricants are:

1. White mineral oil
2. Polyalphaolefins (PAO)
3. Polyalkylene glycols (PAG)
4. Natural oils

21.5.1 WHITE MINERAL OILS (USP OR TECHNICAL GRADE)

The United States Pharmacopoeia (USP) listed "white mineral oils" in 1926. In 1965, following the enactment of regulations governing food additives, the FDA issued regulations covering the application of food-grade lubricants in food and beverage plants.

White mineral oils are made from hydrocracking petroleum base oil and advanced chemical dewaxing. In their chemical structure, these oils are saturated paraffinic and naphthenic hydrocarbons and are essentially free of sulfur and nitrogen compounds. White oils are colorless, odorless and transparent liquids available in a broad range of viscosities. They exhibit moderate thermal and oxidative stability and lubricity. White mineral oils have low to medium viscosity indices and moderate to poor pour points. White mineral oils meet the requirements of the provisions of CFR 21 section 178.3570 for lubricants with incidental food contact (or category H-1 incidental food contact). This section limits their presence in food to a level not exceeding ten parts per million (Table 21.1).

Hydraulic fluids that were made several years ago from white mineral oil base fluids contained as an additive small amounts (up to 3%) of oleic acid to aid in lubricity. This fluid did not perform adequately in some systems and frequent oil changes were required. Advances in additive technology and the approval of key food-grade additives by the FDA have increased the life expectancy and performance of food-grade lubricants. Specifically, antioxidants far superior to oleic acid have been developed to stabilize the lubricant. Properly formulated white mineral-oil-based hydraulic fluids can now offer better performance and longer drain intervals (Table 21.2).

White mineral-oil-based hydraulic fluids are now formulated with FDA approved antioxidants such as butylated hydroxyl toluene and/or diphenylamine. Antiwear and extreme pressure additives such as phosphorothioc acid, triphenyl esters, and others listed in section 178.357 are used. For corrosion protection, phosphoric acid mono- and di-isooctyl esters reacted with C12-C14 primary amines are used. These additives are not allowed to constitute more than 0.5% by weight of the lubricant. To boost the viscosity of the base fluid, polybutene with a minimum average molecular

TABLE 21.1
Properties of White Mineral Oil Base Fluids

Typical Values			Typical Properties			
Viscosity		Specific Gravity	Flash Point		Pour Point	
ASTM D 445		ASTM D 4052	ASTM D-92		ASTM D-97	
SUS @ 100°F	CST @ 40°C	@ 77°T	°F	°C	°F	°C
21/25	67/80	430/439	505	263	10	−12
35/37	667/712	857/873	440	227	5	−15
68/73	329/355	864/881	445	229	0	−18
37/41	48/53	429/436	490	254	10	−12
52/55	200/211	214/219	435	224	5	−15
40/43	384/415	853/876	410	210	5	−15
18/19	349/373	71/73	395	202	5	

Source: Typical physical properties of White Oils.

weight of 80,000 is sometimes used. The lubricants can also be prepared by selecting additives from one or more of the following categories:

1. Substances generally recognized as safe (GRAS) for use in food.
2. Substances used in accordance with the provisions of a prior sanction or approval.
3. Substances identified in CFR 21 section 178.3570.

21.5.2 POLYALPHAOLEFIN (PAO) SYNTHETIC BASE FLUID

This base fluid is made from polymerization of decene/linear alpha-olefins. After polymerization, the process entails hydrogenation of the unsaturated oligomers to make them more thermally stable. PAOs have been approved by the FDA under subsection 178.3570 for incidental food contact or H-1 applications. Select grades of PAO having designated viscosities are used to make food-grade hydraulic fluids. PAOs are thermally stable and have good pour points. When these fluids are fortified with FDA-approved food-grade antioxidants, antiwear, and rust and corrosion inhibitors, the performance and life expectancy of the fluid can be increased two- to three-fold compared to white mineral-oil-based hydraulic fluids (Table 21.3 and 21.4).

TABLE 21.2
Properties of a Typical White-Oil-Based Hydraulic Fluid

Property	Typical Results
Viscosity @ 40°C, cSt	46.5
Viscosity @ 100°C	6.54
Viscosity Index	86
Pour Point °C	−20
Flash Point °C	204
Fire Point °C	216
RBOT Minutes	200
Vane Pump Test ASTM 2882	<50

TABLE 21.3
Typical Physical Properties of PAO Fluids

Property	PAO 2	PAO 4	PAO 6	PAO 10
KV 100°C cSt	1.86	3.95	5.94	9.8
KV 40°C cSt	5.6	16.88	32.4	64.0
KV −40°C cSt	290	2,400	7,900	32,500
Pour Point °C	<−64	<−69	<−65	<−52
Flash Point °C	157	218	240	260
Viscosity Index	-	127	135	134

TABLE 21.4
Properties of a Typical PAO-Based Hydraulic Fluid

Property	Typical Results
Viscosity @ 40°C, cSt	46.0
Viscosity @ 100°C	7.44
Viscosity Index	132
Pour Point °C	−50
Flash Point °C	252
Fire Point °C	262
RBOT Minutes	420
Vane Pump Test ASTM 2882	<25

Polyalphaolefin fluids offer certain inherent properties that are desirable relative to white mineral oils. Some of these properties are listed below:

- Thermal stability
- Hydrolytic stability
- Shear stability
- Wide operating temperature range
- Good low temperature properties
- Low toxicity
- Compatibility with mineral oils

Finished food-grade hydraulic fluids are made with select viscosity grades of PAOs. For ISO 46 formulation, PAO 4 and PAO 10 are combined in a ratio that will give a viscosity of around 46 cSt at 40°C. The additives required will be an amine-type antioxidant, an antiwear additive such as phosphorothioate, a rust inhibitor such as amine phosphate, a metal deactivator/passivator such as a benzotrizole derivative, and a polysiloxane-type defoamer with more than 300 cSt viscosity. These additives are blended with the base fluid. All of these additives must comply with the FDA 21 CFR section 178.3570. Performance evaluation tests include RBOT and Thermal and oxidative stability according to ASTM D-2070. A performance evaluation test could also include a vane pump test according to ASTM D-2882.

21.5.3 POLYALKYLENE GLYCOLS (PAGs)

These base fluids are polymers made from the reaction of alkylene oxide monomers. Commercial PAGs are composed of ethylene oxide and propylene oxide and may be polymerized as block or random copolymers. A butanol-initiated 50/50 ethylene oxide/propylene oxide random polymer is an example of a common PAG base stock used for lubricant formulation. Glycol-initiated 75/25 ethylene oxide/propylene oxide random polymers are used as thickeners for water–glycol hydraulic fluids.

TABLE 21.5
Physical Properties of Polyalkylene Glycols

Property	EO/PO 1500 MW	EO/PO 2500 MW	EO/PO 4000 MW
Viscosity @ 40°C, cSt	130	402	965
Viscosity @ 100°C, cSt	25.6	73.8	165
Viscosity Index	233	261	274
Pour Point °C	−45	−39	−33
Flash Point °C	495	500	500
Density @ 20°C	1.0462	1.0530	1.0579

PAGs are gaining greater acceptance as food-grade lubricants because their overall performance is better than the performance of mineral-oil-based fluids in the food plant environment. This may be attributed to their ability to tolerate some amount of water and still function as adequate lubricants. The water tolerance of PAGs comes from their makeup of alkylene oxide monomers. The greater the percentage of ethylene oxide in the polymer, the greater the water solubility; the greater the percentage of propylene oxide in the polymer, the lesser the water solubility. This allows product customization to meet specific needs for viscosity and other properties through blending different PAG stocks.

Unlike mineral oils and PAOs, when PAGs are heated to above-operating-temperatures they do not leave hard carbon residue or varnish. Equipment lubricated with PAG-based lubricants therefore remains clean even after accidental over-heating or equipment malfunction. Parts lubricated with PAG show no carbon buildup or gumming, even after prolonged use (Table 21.5 and 21.6).

The advantages of polyalkylene glycol base stock include:

- Excellent lubricity
- High viscosity index
- Low toxicity
- Biodegradability
- Good cold flow behavior
- Clean burning with minimal residue or sludge
- Good oxidative and thermal stability

21.6 POLYALKYLENE GLYCOL-BASED HYDRAULIC FLUIDS

1. Food-grade PAG-based fire-resistant hydraulic fluids
2. Food-grade anhydrous PAG (non-water containing) hydraulic fluids

TABLE 21.6
Physical Properties of EO/PO Thickeners

Property	EO/PO Thickener
Viscosity @ 40°C, cSt	27,582
Viscosity @ 100°C, cSt	4,137
Viscosity Index	468
Pour Point °C	+6
Flash Point °C	520
Density @ 20°C	1.0931

TABLE 21.7
Properties of a Typical PAG-Based Hydraulic Fluid

Property	Typical Results
Viscosity @ 40°C, cSt	46.0
Viscosity @ 100°C	9.2
Viscosity Index	190
Pour Point °C	−50
Flash Point °C	252
Fire Point °C	262
RBOT Minutes	520
Vane Pump Test ASTM 2882	<25

21.6.1 FOOD-GRADE PAG-BASED FIRE-RESISTANT HYDRAULIC FLUID

The development of food-grade fire-resistant hydraulic fluid is an example of a fully formulated product that satisfies all of the requirements generally met by an industrial-grade non-food-grade fire-resistant hydraulic fluid. Most industrial-grade glycol-based fire-resistant hydraulic fluids meet the FM (Factory Mutual) international protocol for fire resistance. They protect metal parts and provide boundary lubrication to the bearing and vanes. In addition, some of these products also provide vapor phase corrosion protection to the rust-prone sump overhead and other internal parts (Table 21.7).

Until recently, commercially available food-grade glycol-based fire-resistant hydraulic fluids offered boundary lubrication to hydraulic systems and had FM international approval but did not provide vapor phase corrosion and rust protection to the hydraulic system. The conversion from a mineral-oil-based hydraulic system to food-grade glycol-based hydraulic system presented a severe corrosion and rust potential to some parts—especially on sump overheads, which are generally made up of carbon steel.

FIGURE 21.1 Vapor phase corrosion test: In Figure 21.1, Photo 1 is a photograph showing the results of a vapor phase corrosion test conducted on a commercially available food-grade glycol-based fire-resistant hydraulic fluid. Photo 2 shows the food-grade hydraulic fluid made with the food-grade vapor phase corrosion inhibitor additive. The corrosion inhibitor has fully protected coupon #2.

TABLE 21.8
Vapor Phase Corrosion Test @ 145°F

	VPC Test 24 hrs	VPC Test 48 hrs	VPC Test 96 hrs
Commercial FG (Standard)	Fail	Fail	Fail
Commercial Non-Food Grade ISO 46 Hydraulic Fluid	Pass	Pass	Pass
Food-Grade Fire-Resistant Hydraulic Fluid	Pass	Pass	Pass

The lubricant industry has now developed food-grade glycol-based fire-resistant hydraulic fluids which provide boundary lubrication and have been approved by FM international. In addition, these products inhibit rust and vapor phase corrosion.

Commercial food-grade glycol-based hydraulic fluids are tested in the laboratory by the vapor phase corrosion test (Figure 21.1). Three freshly polished carbon steel coupons are hung in a manner that the bottom coupon is half submerged in a 100 mL sample placed in an 18-inch cylinder. The cylinder is closed on top with a condenser. The cylinder is placed in an oil bath heated to 145°F. Rust and corrosion are observed every twenty-four hours for up to ninety-six hours (Table 21.8 and 21.9).

The addition of these additives did not alter any other properties such as boundary lubrication. Several vane pump tests were carried out to confirm that the lubrication performance of this fluid is on par with the industrial product's:

- Protection against wear and friction.
- Protection against corrosion and oxidation.
- Heat dissipation.
- Compatibility with elastomers and seals.
- Protection against degradation from plant food ingredients, plant process chemicals, and water.
- Inhibition of the growth of micro-organisms such as bacteria and fungi.

21.6.2 Food-Grade Anhydrous PAG (Non-Water Containing) Hydraulic Fluids

These fully formulated hydraulic fluids are made from water-soluble FDA-sanctioned ingredients for use in the food, drug, and cosmetic industries. The benefit of such a lubricant is that all spills or leaks from hydraulic systems can be washed away with water. Since these products do not contain water, the pH control or water evaporation in the system is not of concern. Fluid can be used in these systems for higher pressure and temperature with little maintenance. The life of these fluids is usually greater than the life of fluids that contain water. Anhydrous PAG-based hydraulic fluids are thermally more stable and offer longer life.

TABLE 21.9
Seal Compatibility of Hydraulic Fluids (Polyalkylene Water-Containing Fluids)

	% Swell Volume	Hardness Shore A	Remarks
Neoprene	+10.5	66/70	Compatible
Viton A	+4.8	71/67	Compatible
EPDM	+8.3	68/70	Compatible
Buna N	+11.0	76/78	Acceptable
Fluorosilicone	0.6	64/65	Compatible
Silicone	+4.9	52/49	Acceptable
Polyurethane	12.5	66/75	Not Recommended

TABLE 21.10
Physical Properties of Food-Grade Anhydrous PAG Hydraulic Fluids

Property	Typical Results ISO 32 Fluid	Typical Results ISO 46 Fluid
Viscosity @ 40°C, cSt	30	46.0
Viscosity @ 100°C	5.3	6.8
Viscosity Index	114	100
Pour Point °C	<−8	<−15
Flash Point °C	176	183
Water Solubility	Soluble	Soluble
Specific Gravity 20C/20C	1.126	1.147
Vane Pump Test ASTM 2882	25	<25

Applications: These products are usually used on smaller hydraulic systems where fire resistance is not critical. Some large systems also use non-water-containing or anhydrous hydraulic fluids on regular basis (Table 21.10).

21.7   NATURAL-OIL-BASED HYDRAULIC FLUIDS

A number of natural oils have been considered and are approved under CFR section 178.3570 for use as hydraulic fluids. These oils will be discussed in detail in another chapter.

21.8   CONCLUSION AND FUTURE OF FOOD-GRADE HYDRAULIC FLUIDS

Depending on specific requirement of a hydraulic system, the food industry has many options from which to select hydraulic fluids—Standard mineral oils, PAO's or synthetic water glycols based NSF approved products for their needs. As the food industry sees the satisfactory performance of these products, more companies will be inclined to use them in the future.

REFERENCES

1. Hamid, S. "No-Compromise Synthetic Food-Grade Lubricants," *Food Manufacturing*, May 2008, pp. 12–14.
2. Morawek, R., Tietze, P.G. and Rhodes, R.K., "Food-Grade Lubricants and Their Applications," Presented to ASLE 33rd Annual Meeting 1078.
3. Stewart, H.L. "Fire-Resistant Hydraulic Fluids," *Plant Engineering*, 1979, 33 (4), pp. 157–60.

Appendix 1 Temperature Conversion Table

Equivalent Temperature Readings for Fahrenheit and Celsius Scales [°F = (9/5) °C + 32; °C = (5/9)(°F − 32)]

°Fahrenheit	°Celsius	°Fahrenheit	°Celsius	°Fahrenheit	°Celsius	°Fahrenheit	°Celsius
−459.4	−273	−39	−39.4	−18.4	−28	2	−16.7
−436	−260	−38.2	−39	−18	−27.8	3	−16.1
−418	−250	−38	−38.9	−17	−27.2	3.2	−16
−400	−240	−37	−38.3	−16.6	−27	4	−15.6
−382	−230	−36.4	−38	−16	−26.7	5	−15
−364	−220	−36	−37.8	−15	−26.1	6	−14.4
−346	−210	−35	−37.2	−14.8	−26	6.8	−14
−328	−200	−34.6	−37	−14	−25.6	7	−13.9
−310	−190	−34	−36.7	−13	−25	8	−13.3
−292	−180	−33	−36.1	−12.0	−24.4	8.6	−13
−274	−170	−32.8	−36	−11.2	−24	9	−12.8
−256	−160	−32	−35.6	−11	−23.9	10	−12.2
−238	−150	−31	−35	−10.0	−23.3	10.4	−12
−220	−140	−30	−34.4	−9.4	−23	11	−11.7
−202	−130	−29.2	−34	−9	−22.8	12	−11.1
−184	−120	−29	−33.9	−8	−22.2	12.2	−11
−166	−110	−28	−33.3	−7.6	−22	13	−10.6
−148	−100	−27.4	−33	−7	−21.7	14	−10
−139	−95	−27	−32.8	−6	−21.1	15	−9.4
−130	−90	−26	−32.2	−5.8	−21	15.8	−9
−121	−85	−25.6	−32	−5	−20.6	16	−8.9
−112	−80	−25	−32.9	−4	−20	17	−8.3
−103	−75	−24	−31.7	−3	−19.4	17.6	−8
−94	−70	−23.8	−31.1	−2.2	−19	18	−7.8
−85	−65	−23	−30.5	−2	−18.9	19	−7.2
−76	−60	−22	−30	−1	−18.3	19.4	−7
−67	−55	−21	−29.4	−0.4	−18	20	−6.7
−58	−50	−20.2	−29	0	−17.8	21	−6.1
−49	−45	−20	−28.9	+1	−17.2	21.2	−6
−40	−40	−19	−28.3	1.4	−17	22	−5.6
23	−5	54	12.2	85	29.4	116.6	47
24	−4.4	55	12.8	86	30	117	47.2
24.8	−4	55.4	13	87	30.6	118	47.8
25	−3.9	56	13.3	87.8	31	118.4	48
26	−3.3	57	31.9	88	31.1	119	48.3

(continued)

Equivalent Temperature Readings for Fahrenheit and Celsius Scales [°F = (9/5) °C + 32; °C = (5/9)(°F − 32)] (Continued)

°Fahrenheit	°Celsius	°Fahrenheit	°Celsius	°Fahrenheit	°Celsius	°Fahrenheit	°Celsius
26.6	−3	57.2	14	89	31.7	120	48.9
27	−2.8	58	14.4	89.6	32	120.2	49
28	−2.2	59	15	90	32.2	121	49.4
28.4	−2	60	15.6	91	32.8	122	50
29	−1.7	60.8	16	91.4	33	123	50.6
30	−1.1	61	16.1	92	33.3	123.8	51
30.2	−1	62	16.7	93	33.9	124	51.1
31	−0.6	62.6	17	93.2	34	125	51.7
32	0	63	17.2	94	34.4	125.6	52
33	+0.6	64	17.8	95	35	126	52.2
33.8	1	64.4	18	96	35.6	127	52.8
34	1.1	65	18.3	96.8	36	127.4	53
35	1.7	66	18.9	97	36.1	128	53.3
35.6	2	66.2	19	98	36.7	129	53.9
36	2.2	67	19.4	98.6	37	129.2	54
37	2.8	68	20	99	37.2	130	54.4
37.4	3	69	20.6	100	37.8	131	55
38	3.3	69.8	21	100.4	38	132	55.6
39	3.9	70	21.1	101	38.3	132.8	56
39.2	4	71	21.7	102	38.9	133	56.1
40	4.4	71.6	22	102.2	39	134	56.7
41	5	72	22.2	103	39.4	134.6	57
42	5.6	73	22.8	104	40	135	57.2
42.8	6	73.4	23	105	40.6	136	57.8
43	6.1	74	23.3	105.8	41	136.4	58
44	6.7	75	23.9	106	41.1	137	58.3
44.6	7	75.2	24	107	41.7	138	58.9
45	7.2	76	24.4	107.6	42	138.2	59
46	7.8	77	25	108	42.2	139	59.4
46.4	8	78	25.6	109	42.8	140	60
47	8.3	78.8	26	109.4	43	141	60.6
48	8.9	79	26.1	110	43.3	141.8	61
48.2	9	80	26.7	111.1	43.9	142	61.1
49	9.4	80.6	27	111.2	44	143	61.7
50	10.0	81	27.2	112	44.4	143.6	62
51	10.6	82	27.8	113	45	144	62.2
51.8	11	82.4	28	114	45.6	145	62.8
52	11.1	82.9	28.3	114.8	46	145.4	63
53	11.7	84	28.9	115	46.1	146	63.3
53.6	12	84.2	29	116	46.7	147	63.9
147.2	64	179	81.7	210	98.9	241	116.1
148	64.4	179.6	82	210.2	99	242	116.7
149	65	180	82.2	211	99.4	242.6	117
150	65.6	181	82.8	212	100	243	117.2
150.8	66	181.4	83	213	100.6	244	117.8
151	66.1	182	83.3	213.8	101	244.4	118
152	66.7	183	83.9	214	101.1	245	118.3

Equivalent Temperature Readings for Fahrenheit and Celsius Scales [°F = (9/5) °C + 32; °C = (5/9)(°F − 32)]

°Fahrenheit	°Celsius	°Fahrenheit	°Celsius	°Fahrenheit	°Celsius	°Fahrenheit	°Celsius
152.6	67.7	183.2	84	215	101.7	246	118.9
153	67.2	184	84.4	215.6	102	246.2	119
154	67.8	185	85	216	102.2	247	119.4
154.4	68	186	85.6	217	102.8	248	120
155	68.3	186.8	86	217.4	103	249	120.6
156	68.9	187	86.1	218	103.3	249.8	121
156.2	69	188	86.7	219	103.9	250	121.1
157	69.4	188.6	87	219.2	104	251	121.7
158	70	189	87.2	220	104.4	251.6	122
159	70.6	190	87.8	221	105	252	122.4
159.8	71	190.4	88	222	105.6	253	122.8
160	71.1	191	88.3	222.8	106	253.4	123
161	71.7	192	88.9	223	106.1	254	123.3
161.1	72	192.2	89	224	106.7	255	123.9
162	72.2	193	89.4	224.6	107	255.2	124
163	72.8	194	90	225	107.2	256	124.4
163.4	73	195	90.6	226	107.8	257	125
164	73.3	195.8	91	226.4	108	258	125.5
165	73.9	196	91.1	227	108.3	258.8	126
165.2	74	197	91.7	228	108.9	259	126.1
166	74.4	197.6	92	228.2	109	260	126.7
167	75	198	92.2	229	109.4	260.6	127
168	75.6	199	92.8	230	110	261	127.2
168.8	76	199.4	93	231	110.6	262	127.8
169	76.1	200	93.3	231.8	111	262.4	128
170	76.7	201	93.9	232	111.1	263	128.3
170.6	77	201.2	94	233	111.7	264	128.9
171	77.2	202	94.4	233.6	112	264.2	129
172	77.8	203	95	234	112.3	265	129.4
172.4	78	204	95.6	235	112.8	266	130
173	78.3	204.8	96	235.4	113	267	130.6
174	78.9	205	96.1	236	113.3	267.8	131
174.2	79	206	96.7	237	113.9	268	131.3
175	79.4	206.6	97.7	237.2	114	269	131.7
176	80	207	97.2	238	114.4	269.6	132
177	80.6	208	97.8	239	115	270	132.2
177.8	81	208.4	98	240	115.6	271	132.8
178	81.1	209	98.3	240.8	116	271.4	133
272	133.3	300	148.9	327.2	164	355	179.4
273	133.9	300.2	149	328	164.4	356	180
273.3	134	301	149.4	329	165	357	180.6
274	134.4	302	150	330	165.6	357.8	181
275	135	303	150.6	330.8	166	358	181.1
276	135.6	303.8	151	331	166.1	359	181.6
276.8	136	304	151.1	332	166.7	359.6	182
277	136.1	305	151.7	332.6	167	360	182.2
278	136.7	305.6	152	333	167.2	361	182.8

(continued)

Equivalent Temperature Readings for Fahrenheit and Celsius Scales [°F = (9/5) °C + 32; °C = (5/9)(°F − 32)] (Continued)

°Fahrenheit	°Celsius	°Fahrenheit	°Celsius	°Fahrenheit	°Celsius	°Fahrenheit	°Celsius
278.6	137	306	152.2	334	167.8	361.4	183
279	137.2	307	152.8	334.4	168	362	183.3
280	137.8	307.4	153.3	335	168.3	363	183.9
280.4	138	308	153.3	336	168.9	363.2	184
281	138.3	309	153.9	336.2	169	364	184.4
282	138.9	309.2	154	337	169.4	365.6	185
282.2	139	310	154.4	338	170	366	185.6
283	139.4	311	155	339	170.6	366.8	186
284	140	312	155.6	339.8	171	367	186.1
285	140.6	312.8	156	340	171.1	368	186.7
285.8	141	313	156.1	341	171.7	368.6	187
286	141.1	314	156.7	341.6	172	369	187.2
287	141.7	314.6	157	342	172.2	370	187.8
287.6	142	315	157.2	343	172.8	370.4	188
288	142.2	316	157.8	343.4	173	371	188.3
289	142.8	316.4	158	344	173.3	372	188.9
289.4	143	317	158.3	345	173.9	372.2	189
290	143.3	318	158.9	345.2	174	373	189.4
291	143.9	318.2	159	346	174.4	374	190
291.2	144	319	159.4	347	175		
292	144.4	320	160	348	175.6		
293	145	321	160.6	348.8	176		
294	145.6	321.8	161	349	176.1		
294.8	146	322	161.1	350	176.7		
295	146.1	323	161.7	350.6	177		
296	146.7	323.6	162	351	177.2		
296.6	147	324	162.2	352	177.8		
297	147.2	325	162.8	352.4	178		
298	147.8	325.4	163	353	178.3		
298.4	148	326	163.3	354	178.9		
299	148.3	327	163.9	354.2	179		

Appendix 2   SI Unit Conversions

Conversions of USCS to SI Units

Energy

To convert from	To	Multiply by
British thermal unit (ISO/TC 12)	Joule	$1.055\ 06 \times 10^3$
British thermal unit (International Steam Table)	Joule	$1.055\ 04 \times 10^3$
British thermal unit (mean)	Joule	$1.055\ 87 \times 10^3$
British thermal unit (thermochemical)	Joule	$1.054\ 350\ 264\ 488 \times 10^3$
British thermal unit (39°F)	Joule	$1.059\ 67 \times 10^3$
British thermal unit (60°F)	Joule	$1.054\ 68 \times 10^3$
Calorie (International Steam Table)		
Calorie (mean)	Joule	$4.190\ 02 \times 10^0$
Calorie (thermochemical)	Joule	4.184×10^0
Calorie (15°C)	Joule	$4.185\ 80 \times 10^0$
Calorie (20°C)	Joule	$4.185\ 90 \times 10^0$
Calorie (kilogram, International Steam Table)	Joule	4.1868×10^0
Calorie (kilogram, mean)	Joule	4.190×10^3
Calorie (kilogram, thermochemical)	Joule	4.184×10^3
Electronvolt	Joule	$1.602\ 10 \times 10^{-19}$
Erg	Joule	1.00×10^{-7}
Foot-pound force	Joule	$1.355\ 817\ 9 \times 10^0$
Foot poundal	Joule	$4.214\ 011\ 0 \times 10^{-2}$
Joule (International of 1948)	Joule	$1.000\ 165 \times 10^0$
Kilocalorie (International Steam Table)	Joule	4.1868×10^3
Kilocalorie (mean)	Joule	$4.190\ 02 \times 10^3$
Kilocalorie (thermochemical)	Joule	4.184×10^3
Kilowatt hour	Joule	$3.600\ 59 \times 10^6$
Kilowatt hour (International of 1948)	Joule	$3.600\ 59 \times 10^6$
Ton (nuclear equivalent of TNT)	Joule	4.20×10^9
Watt hour	Joule	3.60×10^3

Energy/Area Time

To convert from	To	Multiply by
Btu (thermochemical)/foot^2 second	Watt/meter^2	$1.134\ 8931 \times 10^4$
Btu (thermochemical)/foot^2 minute	Watt/meter^2	$1.891\ 488\ 5 \times 10^2$
Btu (thermochemical)/foot^2 hour		$3.152\ 480\ 8 \times 10^0$
Btu (thermochemical)/inch^2 second	Watt/meter^2	$1.634\ 246\ 2 \times 10^6$
Calorie (thermochemical)/cm^2 minute	Watt/meter^2	$6.973\ 333\ 3 \times 10^2$
Erg/centimeter^2 second	Watt/meter^2	1.00×10^{-3}
Watt/cm^2	Watt/meter^2	1.00×10^4

(continued)

Conversions of USCS to SI Units (Continued)

Force

To convert from	To	Multiply by
Dyne	Newton	1.00×10^{-5}
Kilogram force (kgf)	Newton	$9.806\ 65 \times 10^{0}$
Kilopound force	Newton	$9.806\ 65 \times 10^{0}$
Kip	Newton	$4.448\ 221\ 615\ 260\ 5 \times 10^{3}$
lbf (pound force, avoirdupois)	Newton	$4.448\ 221\ 615\ 260\ 5 \times 10^{0}$
Ounce force (avoirdupois)	Newton	$2.780\ 138\ 5 \times 10^{-1}$
Pound force lbf (avoirdupois)	Newton	$4.448\ 221\ 615\ 260\ 5 \times 10^{0}$
Pounds	Newton	$1.382\ 549\ 543\ 76 \times 10^{-1}$

Acceleration

To convert from	To	Multiply by
Foot/second^2	Meter/second^2	3048×10^{-1}
Free fall, standard	Meter/second^2	$9.806\ 65 \times 10^{0}$
Gal (galileo)	Meter/second^2	1.00×10^{-2}
Inch/second^2	Meter/second^2	2.54×10^{-2}

Area

To convert from	To	Multiply by
Acre	Meter^2	$4.046\ 856\ 422\ 4 \times 10^{3}$
Circular mil	Meter^2	$5.067\ 074\ 8 \times 10^{-10}$
Foot^2	Meter^2	$9.290\ 304 \times 10^{-2}$
Inch^2	Meter^2	6.4516×10^{-4}
Mile^2 (U.S. statute)	Meter^2	$2.589\ 988\ 110\ 336 \times 10^{6}$
Yard^2	Meter^2	$8.361\ 273\ 6 \times 10^{-1}$

Density

To convert from	To	Multiply by
Gram/centimeter^3	Kilogram/meter^3	1.00×10^{3}
lbm/inch^3	Kilogram/meter^3	$2.767\ 990\ 5 \times 10^{4}$
lbm/foot^3	Kilogram/meter^3	$1.601\ 846\ 3 \times 10^{1}$
Slug/foot^3	Kilogram/meter^3	$5.153\ 79 \times 10^{2}$

Appendix 3   Commonly Used Pressure Conversions; Fraction Notation

---

Commonly Used Pressure Conversions

1 bar = 100,000 Pa	1 MPa = 10 bar
1 bar = 100 kPa	1 MPa = 1,000 kPa
1 bar = 0.1 MPa	1 MPa = 1,000,000 Pa
1 kPa = 1000 Pa	1 bar = 14.5 psi
1 kPa = 0.01 bar	100 kPa = 14.5 psi
1 kPa = 0.001 MPa	1 MPa = 14.5 psi

Examples for Using Fractions and Multiples of Base Units

Fract./Mult	Symbol	Pa	m	L	N	W
10^{-3}	m		mm	mL		
10^{-2}	c		cm			
10^{-1}	d		dm	dL		
10	da				daN	
10^{2}	h			hL		
10^{3}	k	kPa	km			kW
10^{6}	M	MPa			MN	mW

Appendix 4   Volume and Weight Equivalents

Volume and Weight Equivalents (Example: 20 U.S. Gallons × 3.7854 = 75.708 Liters)

Convert from	Convert to	Volume and Weight Equivalents						Weight Equivalent Basis Water at 60°F (15.6°C)		
		U.S. Gallons	Imperial Gallons	Cubic Inches	Cubic Feet	Liters	Cubic Meters	Pounds	U.S. Tons	Kilograms
U.S. Gallons		1	0.8327	231	0.13368	3.7854	0.0037854	8.338	0.00417	3.782
Imperial Gallons		1.20094	1	277.39	0.16054	4.546	0.004546	10.0134	0.005	4.542
Cubic inches		0.004329	0.003605	1	0.0005787	0.016387	0.000016387	0.036095		0.016372
Cubic Feet		7.48052	6.229	1,728	1	28.317	0.02832	62.3714	0.03119	28.291
Liters		0.2642	0.22	61.024	0.035315	1	0.001	2.2029	0.0011	0.1
Cubic Meters		264.2	220	61,024	35.315	1,000	1	2,202.65	1.10133	1,000
Pounds		0.1199	0.09987	27.71	0.016033	0.4539	0.000454	1	0.0005	0.45359
U.S. Tons		239.87	199.7	55,409	32.066	907.9	0.908	2,000	1	907.2
Kilograms		0.2644	0.2202	61.08	0.03534	1	0.001	2.205	0.0011	1

Note: The capacity of a barrel varies in different industries. for instance:

1 bbl of beer = 31 U.S. gallons.
1 bbl of wine = 31.5 U.S. gallons.
1 bbl of oil = 42 U.S. gallons.
1 bbl of whiskey = 45 U.S. gallons.

Drums: The drum is not considered to be a unit of measure as is the barrel. Drums are usually built to specifications and are available in sizes from 2.5 gallons to 55 gallons; the most popular sizes are the 5 gallon, 30 gallon, and 55 gallon drums.

Appendix 5   Head and Pressure Equivalents

Equivalents of Head and Pressure (Example: 15 lb/ft² × 4.88241 = 73.236 kg/m²)

Convert from	Convert to lb/in.²	lb/ft²	Atmospheres	kg/cm²	kg/m²	Water (68°F)[a] in.	ft	Mercury (32°F)[b] in.	mm	bar[c]	Mega Pascals MPa[c]
lb/in.²	1	144	0.068046	0.070307	703.07	27.7276	2.3106	2.03602	51.715	0.06895	0.006895
lb/ft²	0.0069444	1	0.00473	0.000488	4.88241	0.1926	0.01605	0.014139	0.35913	0.000479	0.0000479
Atmospheres	14.696	2,166.2	1	1.0332	10,332.27	407.484	33.957	29.921	760	1.01325	0.101325
kg/cm²	14.2233	2,048.155	0.96784	1	10,000	394.38	32.865	28.959	735.559	0.98067	0.098067
kg/m²	0.001422	0.204768	0.0000968	0.0001	1	0.03944	0.003287	0.002896	0.073556	0.000098	0.0000098
in. Water[a]	0.036092	5.1972	0.002454	0.00253	25.375	1	0.08333	0.07343	1.8651	0.00249	0.000249
ft Water[a]	0.432781	62.3205	0.029449	0.03043	304.275	12	1	0.88115	22.3813	0.029839	0.0029839
in. Mercury[b]	0.491154	70.7262	0.033421	0.03453	345.316	13.6185	1.1349	1	25.40005	0.033864	0.0033864
mm Mercury[b]	0.0193368	2.7845	0.0013158	0.0013595	13.59509	0.53616	0.04468	0.03937	1	0.001333	0.0001333
Bars[c]	14.5038	2,088.55	0.98692	1.01972	10,197.2	402.156	33.513	29.53	750.062	1	0.1
MPa[c]	145.038	20,885.5	9.8692	10.1972	101,972	4,021.56	335.13	295.3	7,500.62	10	1

[a] Water at 68°F (20°C).

[b] Mercury at 32°F (0°C).

[c] 1 MPa (mega Pascal) = 10 bars = 1,000,000 N/m² (Newtons/meter²).

Appendix 6   Flow Equivalents

Flow Equivalents—for Any Liquid (Example: 100 U.S. gal/min × 0.0631 = 6.31 liters/sec)

Convert from \ Convert to	U.S. gal/min	Imp. gal/min	U.S. Million gal/day	Cu. ft per sec (sec-ft)	Cu meters per hour	Liter per sec	Barrels (42 gal) per min	Barrels (42 gal) per day
U.S. gal/min	1	0.8327	0.00144	0.00223	0.2271	0.0631	0.0238	34.286
Imp. gal/min	1.201	1	0.00173	0.002676	0.2727	0.0758	0.02859	41.179
U.S. Million gal/day	694.4	578.25	1	1.547	157.7	43.8	16.53	23,810
Cu ft/sec	448.83	373.7	0.646	1	101.9	28.32	10.686	15,388
Cu m/sec	15,852	13,200	22.83	35.35	3,600	1,000	377.4	543,462
Cu m/min	264.2	220	0.3804	0.5886	60	16.667	6.29	9,058
Cu m/hr	4.403	3.67	0.00634	0.00982	1	0.2778	0.1048	151
Liter/sec	15.85	13.2	0.0228	0.0353	3.6	1	0.3773	543.3
Liter/mm	0.2642	0.22	0.00038	0.000589	0.06	0.0167	0.00629	9.058
Barrels (42 gal)/min	42	34.97	0.0605	0.0937	9.538	2.65	1	1,440
Barrels (42 gal)/day	0.0292	0.0243	0.000042	0.000065	0.00662	0.00184	0.00069	1

Note: 1 miners inch of water = 8.977 gpm (in Idaho, Kansas, Nebraska, New Mexico, N. Dakota, and Utah)

        = 11.22 gpm (in Arizona, California, Montana, Nevada and Oregon)

        = 11.69 gpm (in Colorado).

Barrel per day = 31 gal × 0.02153 = gpm (beer)

               31.5 gal × 0.02188 = gpm (wine)

               42 gal × 0.02917 = gpm (oil)

               45 gal × 0.03125 = gpm (whiskey).

Appendix 7   Viscosity Conversion Charts

Approximate Viscosity Conversions

Seconds Saybolt Universal (SSU)	Kinematic Viscosity		Seconds Saybolt Furol (SSF)	Seconds Redwood 1 Standard	Seconds Redwood 2 Admiralty	Degrees Engler	Degrees Barbey
	Centistokes (cSt)	ft²/s					
31	1.0	0.00001076	—	29	—	1.00	6,200
31.5	1.13	0.00001216	—	29.4	—	1.01	5,486
32	1.81	0.00001948	—	29.8	—	1.08	3,425
32.6	2.00	0.00002153	—	30.2	—	1.10	3,100
33	2.11	0.00002271	—	30.6	—	1.11	2,938
34	2.40	0.00002583	—	31.3	—	1.14	2,583
35	2.71	0.00002917	—	32.1	—	1.17	2,287
36	3.00	0.00003229	—	32.9	—	1.20	2,066
38	3.64	0.00003918	—	33.7	—	1.26	1,703
39.2	4.00	0.00004306	—	35.5	—	1.30	1,550
40	4.25	0.00004575	—	36.2	5.10	1.32	1,459
42	4.88	0.00005253	—	38.2	5.25	1.36	1,270
42.4	5.00	0.00005382	—	38.6	5.28	1.37	1,240
44	5.50	0.00005920	—	40.6	5.39	1.40	1,127
45.6	6.00	0.00006458	—	41.8	5.51	1.43	1,033
46	6.13	0.00006598	—	42.3	5.54	1.44	1,011
46.8	7.00	0.00007535	—	43.1	5.60	1.48	885
50	7.36	0.00007922	—	44.3	5.83	1.58	842
52.1	8.00	0.00008611	—	46.0	6.03	1.64	775
55	8.88	0.00009558	—	48.3	6.30	1.73	698
55.4	9.00	0.00009688	—	48.6	6.34	1.74	689
58.8	10.00	0.0001076	—	51.3	6.66	1.83	620
60	10.32	0.0001111	—	52.3	6.77	1.87	601
65	11.72	0.0001262	—	56.7	7.19	2.01	529
70	13.08	0.0001408	—	60.9	7.60	2.16	474
75	14.38	0.0001548	—	65.1	8.02	2.37	431
80	15.66	0.0001686	—	69.2	8.44	2.45	396
85	16.90	0.0001819	—	73.4	8.87	2.59	367
90	18.12	0.0001950	—	77.6	9.30	2.73	342
95	19.32	0.0002080	—	81.6	9.71	2.88	321
100	20.52	0.0002209	—	85.6	10.12	3.02	302
120	25.15	0.0002707	—	102	11.88	3.57	246
140	29.65	0.0003191	—	119	13.63	4.11	209
160	34.10	0.0003670	—	136	15.39	4.64	182

(continued)

Approximate Viscosity Conversions (Continued)

Seconds Saybolt Universal (SSU)	Kinematic Viscosity		Seconds Saybolt Furol (SSF)	Seconds Redwood 1 Standard	Seconds Redwood 2 Admiralty	Degrees Engler	Degrees Barbey
	centistokes (cSt)	ft²/s					
180	38.52	0.0004146	—	153	17.14	5.12	161
200	42.95	0.0004623	—	170	18.90	5.92	144
300	64.6	0.0006953	32.7	253	28.0	8.79	96
400	86.2	0.0009278	42.4	338	37.1	11.70	71.9
500	108.0	0.001163	52.3	423	46.2	14.60	57.4
600	129.4	0.001393	62.0	507	55.3	17.50	47.9
700	151.0	0.001625	72.0	592	64.6	26.44	41.0
800	172.6	0.001858	82.0	677	73.8	23.36	35.9
900	194.2	0.002090	92.1	762	83.0	26.28	31.9
1,000	215.8	0.002323	102.1	846	92.3	29.20	28.7
1,200	259.0	0.002788	122	1,016	111	35.1	23.9
1,400	302.3	0.003254	143	1,185	129	40.9	20.5
1,600	345.3	0.003717	163	1,354	148	46.7	18.0
1,800	388.5	0.004182	183	1,524	166	52.6	15.6
2,000	431.7	0.004647	204	1,693	185	58.4	14.4
2,500	539.4	0.005806	254	2,115	231	73.0	11.5
3,000	647.3	0.006967	305	2,538	277	87.6	9.6
3,500	755.2	0.008129	356	2,961	323	102	8.21
4,000	863.1	0.009290	408	3,385	369	117	7.18
4,500	970.9	0.01045	458	3,807	415	131	6.39
5,000	1,078.8	0.01161	509	4,230	461	146	5.75
6,000	1,294.6	0.01393	610	5,077	553	175	4.78
7,000	1,510.3	0.01626	712	5,922	646	204	4.11
8,000	1,726.1	0.01858	814	6,769	738	234	3.59
9,000	1,941.9	0.02092	916	7,615	830	263	3.19
10,000	2,157.6	0.02322	1,018	8,461	922	292	2.87
15,000	3,236.5	0.03438	1,526	12,692	—	438	1.92
20,000	4,315.3	0.04645	2,035	16,923	—	584	1.44

Viscosity—Unit Conversions

Multiply	By	To Obtain
Kinematic Viscosity[a]		
ft²/s	92,903.04	cSt
ft²/s	0.092903	m²/s
m²/s	10.7639	ft²/s
m²/s	1,000,000	cSt
cSt	0.000001	m²/s
cSt	0.0000107639	ft²/s
Absolute or Dynamic Viscosity[b]		
lbf s/ft²	47,880.26	cP
lbf s/ft²	47.8803	Pa s

(continued)

Viscosity—Unit Conversions (Continued)

Multiply	By	To Obtain
cP	0.000102	kgf s/m^2
cP	0.0000208854	lbf s/ft^2
cP	0.001	Pa s
Pa s	0.0208854	lbf s/ft^2
Pa s	1,000	cP
Absolute to Kinematic Viscosity		
cP	1/density (g/cm^2)	cSt
cP	0.00067197/density (lb/ft^3)	ft^2/s
lbf s/ft^2	32.174/density (lb/ft^3)	ft^2/s
kgf s/m^2	9.80665/density (kg/m^3)	m^2/s
Pa s	1000/density (g/cm^3)	cSt
Kinematic to Absolute Viscosity		
cSt	Density (g/cm^3)	cP
m^2/s	0.10197 × density (kg/m^3)	kgf s/m^2
ft^2/s	0.03108 × density (lb/ft^3)	lbf s/ft^2
ft^2/s	1488.16 × density (lb/ft^3)	cP
cSt	0.001 × density (g/cm^3)	Pa s
m^2/s	1000 × density (g/cm^3)	Pa s

[a] See previous page for conversion in SSU, Redwood, and so forth.

[b] Sometimes absolute viscosity is given in terms of pounds mass; in this case: cP × 0.000672 = lbm/ft^2.

Appendix 8 Water Vapor Pressure Chart

Temp. (°C)	Vapor Pressure mm Hg	lb/in.	Head (ft)	Temp. (°C)	Vapor Pressure mm Hg	lb/in.	Head (ft)	Temp. (°C)	Vapor Pressure mm Hg	lb/in.	Head (ft)	Temp. (°C)	Vapor Pressure mm Hg	lb/in.	Head (ft)
0	4.579	0.09	0.21	26	25.209	0.49	1.15	52	102.09	2.01	4.6	78	327.3	6.42	14.8
1	4.926	0.10	0.23	27	26.739	0.53	1.22	53	107.20	2.11	4.9	79	341.0	6.68	15.5
2	5.294	0.10	0.24	28	28.349	0.56	1.3	54	112.51	2.22	5.1	80	355.1	6.92	16.1
3	5.655	0.11	0.26	29	30.043	0.59	1.35	55	118.04	2.33	5.4	81	369.7	7.25	16.8
4	6.101	0.12	0.28	30	31.284	0.62	1.4	56	123.80	2.44	5.6	82	384.9	7.55	17.5
5	6.543	0.13	0.30	31	33.695	0.66	1.5	57	129.82	2.55	5.9	83	400.6	7.85	18.2
6	7.013	0.14	0.32	32	35.663	0.70	1.6	58	136.08	2.67	6.2	84	416.8	8.16	18.9
7	7.513	0.15	0.35	33	37.729	0.74	1.7	59	142.60	2.80	6.5	85	433.6	8.50	19.6
8	8.045	0.16	0.37	34	39.898	0.78	1.8	60	149.38	2.94	6.8	86	450.9	8.83	20.4
9	8.609	0.17	0.39	35	42.175	0.83	1.9	61	156.43	3.07	7.1	87	468.7	9.15	21.2
10	9.209	0.18	0.41	36	44.563	0.88	2.0	62	163.77	3.21	7.4	88	487.1	9.55	22.1
11	9.844	0.19	0.44	37	47.067	0.93	2.1	63	171.38	3.36	7.8	89	506.1	9.92	22.9
12	10.518	0.21	0.48	38	49.692	0.98	2.3	64	179.31	3.51	8.1	90	525.8	10.3	23.8
13	11.231	0.22	0.51	39	52.442	1.03	2.4	65	187.54	3.68	8.5	91	546.1	10.7	24.8
14	11.987	0.24	0.55	40	55.324	1.09	2.5	66	196.09	3.84	8.9	92	567.0	11.1	25.6
15	12.788	0.25	0.58	41	58.34	1.15	2.7	67	204.96	4.02	9.3	93	588.6	11.6	26.8
16	13.643	0.27	0.62	42	61.50	1.21	2.8	68	214.17	4.20	9.7	94	610.9	12.0	27.7
17	14.530	0.29	0.66	43	64.80	1.27	2.9	69	223.73	4.38	10.2	95	633.9	12.4	28.6
18	15.477	0.30	0.69	44	68.26	1.34	3.1	70	233.7	4.57	10.7	96	657.6	12.9	29.8
19	16.477	0.32	0.74	45	71.88	1.41	3.3	71	243.9	4.78	11.1	97	682.1	13.4	31.0
20	17.535	0.34	0.78	46	75.65	1.49	3.5	72	254.6	5.00	11.6	98	707.3	13.9	32.1
21	18.650	0.37	0.84	47	79.60	1.57	3.7	73	265.7	5.21	12.1	99	733.2	14.3	33.0
22	19.827	0.39	0.90	48	83.71	1.65	3.8	74	277.2	5.43	12.5	100	760	14.7	34.0
23	21.068	0.41	0.95	49	88.02	1.73	4.0	75	289.1	5.67	13.1				
24	22.377	0.44	1.02	50	92.51	1.82	4.2	76	301.4	6.00	13.9				
25	23.756	0.47	1.08	51	97.20	1.91	4.4	77	314.1	6.15	14.2				

Index

A

Abrasive wear. *See also* Wear mechanism processes
  minimized by, 486
  on surface, 486
  three-body and two-body, 485–486
Accelerators, 56
Accelerometers, 448
Acceptable water content, 642
Accumulator, 47–48
Acrylonitrile (ACN), 59–60
Activators, 56
Active oxygen method (AOM), 811
Additives
  analysis
    atomic absorption (AA)/arc spark, 582–583
    FTIR spectroscopy, 583
    spectroscopy, 582–583
    XRF/ICP, 582–583
  antiwear
    boundary lubrication, 580
    elastohydrodynamic, 581
    extreme pressure, 580–581
    friction modifier, 580–581
    oiliness, 580
  ashless fluid technology, 582
  base oil effects
    function and chemical type, 578–579
    naphthenic and aromatic components, 578
    paraffinic components, 578
    refining process, 578
    response variation in, 578, 580
  classification and function
    energy transfer medium, 578
    performance properties enhancement, 578
    rust and oxidation (R&O)/antiwear, 578
    straight-grade and multi-grade, 578
  hydraulic system design, 577
  mineral oil hydraulic fluids, 577
  performance, 577
  stearic acid, chemisorption of, 580
  ZDDP
    alcohol substituent (RO), 582
    antiwear, 581
    AW and EP lubrication, difference, 581
    detergents and dispersants, 582
    films, formation and thickness, 582
    illustrative examples, 581–582
    thermal performance, effect on, 582
Adhesive wear. *See also* Wear mechanism processes
  laws of, 488–489, 491–492
  model, 490
  polishing, 489–490
  surface asperity contacts, 489–490
Aeronautical Material Standard, 379
AFNOR Test, OECD 301A, 333

Air-borne noise, 450–451
Air entrainment
  experimental procedures of
    Allgemeine Elek Tricitats—Gesellschaft (AEG)
      method, 135
    bubbly oil viscometer, 135
    dead weight apparatus, 137
    Deutsche Shell air release test, 135
    dynamic L.T.G, 137
    dynamic test methods, 134–135
    hydrostatic balance, 137–138
    in-line analysis of fluid aeration, 140–141
    Rowland's air entrainment, 136
    sample bottle method, 138–140
    Technischer Überwachungs/Verein (TÜV),
      135–136
    turbidity measurement, 140
    Zander method, 137–138
Air injection method, 134
Air release specifications, 140
  measurement apparatus, 141
Alkylated aromatic fluids, 905
  application, 912
  categories, 905
  environmental impact
    toxicity, 906–907
  historical development
    commercial production of, 905
    cost-effective, 905
    feasible synthetic routes, 905
    use, 905
  manufacture
    branched-chain, 905
    compounds, 905
    methods for, 906
    monoalkylate, 905
  physical properties
    Conoco DN-600 base, 906
    dialkylbenzene and mineral oils, 906–907
  qualities, 905
Allgemeine Elek Tricitats-Gesellschaft (AEG)
    method, 135
Alternate TOW (ALTOW), 210–211
American National Standards Institute (ANSI),
    303–304
American Standard for Testing and Materials (ASTM),
    65, 71–72, 303–304, 824
Analytical ferrography, 595
Antidegradants, 56
Antioxidants, 769–771
Antiwear and extreme-pressure additives, 772
Aquatic toxicity, 786
Argon plasma torch, 248–249
Asperity-scale phenomena, 261
ASTM Copper Strip Corrosion Standards,
    559–560, 562

Q

R